GEORGE WASHINGTON HIGH SCHOOL
3535 E. 114th STREET
CHICAGO, ILLINOIS 60617
BOOKROOM

GEOMETRY

Eugene D. Nichols
Mervine L. Edwards
E. Henry Garland
Sylvia A. Hoffman
Albert Mamary
William F. Palmer

Holt, Rinehart and Winston, Inc.

Harcourt Brace Jovanovich, Inc.

HBJ

Austin • Orlando • San Diego • Chicago • Dallas • Toronto

Acknowledgments

PHOTO CREDITS

Abbreviations used: (l)left; (t)top; (b)bottom; (tl)top left; (bl)bottom left; (ml)middle left.

Chapter 1: page 1, © California Institute of Technology, 1961; 5, HRW Photo by Bryan Tumlinson; 15, Lewis Portnoy/Spectra-Action Inc.; 31, NASA.

Chapter 2: page 48, Dave Stock; 83, HRW Photos by Bryan Tumlinson.

Chapter 3: page 106, Russ Kinne/Comstock; 119(t) Mark Antman/The Image Works; 119(b), Bill Varie/The Image Bank.

Chapter 4: page 147, Audrey Gibson; 152, Tony Freeman/PhotoEdit; 153, Phil Degginger/TSW-Click/Chicago.

Chapter 5: page 177, Cameramann International Ltd.; 182, Robert Brown/Allsport USA; 215, HRW Photos by Eric Beggs.

Chapter 6: page 229, Mark Antman/The Image Works; 230, IBM/PhotoEdit; 235, Tony Freeman/PhotoEdit; 239, HRW Photo by Eric Beggs; 247 HRW Photos by Eric Beggs.

Chapter 7: page 261, Frank Cezus/TSW-Click/Chicago; 283(ml)(b), HRW Photos by Eric Beggs; 283(t), Patrick Fischer Photography; 285, Patrick Fischer Photography.

Copyright © 1991, 1986 by Holt, Rinehart and Winston, Inc.

All rights reserved. No part of this publication may be reproduced or transmitted in any form or by any means, electronic or mechanical, including photocopy, recording, or any information storage and retrieval system, without permission in writing from the publisher.

Requests for permission to make copies of any part of the work should be mailed to: Permissions Department, Holt, Rinehart and Winston, Inc., 1627 Woodland Avenue, Austin, Texas 78741.

Printed in the United States of America

ISBN 0-03-005407-9

45678 032 98765

Chapter 8: page 307, HRW Photo by Russell Dian.

Chapter 9: page 348, Four By Five; 349, Robert Essel/The Stock Market; 364, Paul Gero/Journalism Services; 365, Comstock.

Chapter 10: page 388, Tom Grill/Comstock; 397, Jake Rajs/The Image Bank.

Chapter 11: page 415, John Patsch/Journalism Services; 450, HRW Photo by Eric Beggs; 461, HRW Photo by Eric Beggs.

Chapter 12: page 474, *Drawing Hands*, © 1989 M. C. Escher Heirs/Cordon Art—Baarn—Holland; 493, Photo Courtesy of the New York Life Insurance Company, New York; 502, Mike Schneps/The Image Bank; 520, 521, Landiscor Aerial Photo, Inc.

Chapter 13: page 529, Kirk Schlea/Allsport USA.

Chapter 14: page 556, Hal Clason Photography/Tom Stack and Associates; 574, Mike Mazzaschi/Stock Boston.

Chapter 15: page 599, HRW Photos by Eric Beggs; 612, Jim Lund/After Image Inc.

ILLUSTRATORS

Gary Eldridge, pages 414, 474, 598

Jimmy Longacre, pages xiv, 176, 224, 302, 338, 528

Joe Ruszkowski, pages 20, 25, 30, 62, 68, 74, 76, 96, 97, 98, 101, 118, 120, 133, 146, 157, 204, 214, 235, 246, 256, 283, 284, 297, 317, 339, 347, 354, 359, 360, 365, 391, 417, 538, 570, 579, 582, 606, 622

John A. Wilson, pages 42, 92, 138, 268, 378, 554

About the Authors

Eugene D. Nichols
Distinguished Professor
of Mathematics Education
Florida State University
Tallahassee, Florida

Mervine L. Edwards
Chairman, Department of Mathematics
Shore Regional High School
West Long Branch, New Jersey

E. Henry Garland
Head of Mathematics Department
Developmental Research School
DRS Professor
Florida State University
Tallahassee, Florida

Sylvia A. Hoffman
Resource Consultant in Mathematics
Illinois State Board of Education
State of Illinois

Albert Mamary
Superintendent of Schools for Instruction
Johnson City Central School District
Johnson City, New York

William F. Palmer
Professor of Education and Director
Center for Mathematics and Science Education
Catawba College
Salisbury, North Carolina

Chapter Contents

1 GEOMETRIC FIGURES xiv

2 PROOF 42

3 PARALLELISM 92

4 CONGRUENT TRIANGLES 138

5 CONGRUENCE 176

6 POLYGONS 224

11 CIRCLES 414

12 AREA 474

15 TRANSFORMATIONS 598

Symbol List

$\overleftrightarrow{MN}$	line MN
MN	distance between M and N
$\overline{MN}$	segment MN
$\overrightarrow{MN}$	ray MN
$\cong$	is congruent to
$\angle AOB$	angle AOB
$m\angle AOB$	degree measure of $\angle AOB$
$\therefore$	therefore
$\perp$	is perpendicular to
$\not\perp$	is not perpendicular to
$\sqrt{}$	square root
$\parallel$	is parallel to
$\nparallel$	not parallel to
$\ncong$	is not congruent to
$\triangle XYZ$	triangle XYZ
$\square ABCD$	parallelogram $ABCD$
$\sim$	is similar to
$\nsim$	is not similar to
$\odot$	circle
$\overparen{AB}$	arc AB
$m\overparen{AB}$	degree measure of arc AB
$\wedge$	logical "and"
$\vee$	logical "or"
$\sim$	logical "not"
$\approx$	approximately equal to
$\neq$	not equal to
$\stackrel{?}{=}$	possibly equal to
$\rightarrow$	logical "If . . ., then . . ."

1 GEOMETRIC FIGURES

Sonya Kovalevskaya (1850–1891)—As a child, this famous mathematician was fascinated by the mathematics on the temporary wallpaper in her room. Her father had covered the walls with the calculus notes he bought as a student.

1.1 Introduction to Geometric Figures

Objectives To identify and name points, lines, and planes
To name the intersection of two geometric figures

When you look at the sky on a clear night, you can see thousands of stars. They look so small that they appear to be dots. Each of these tiny dots suggests the simplest figure studied in geometry—a *point*. The three basic figures of geometry are the point, the line, and the plane.

A point is represented by a dot and named by a capital letter. A point has no size but is used only to indicate *position*. All geometric figures consist of points.

A B C

The points are named A, B, and C and are read as point A, point B, and point C.

A line is represented by a straight mark with an arrowhead at each end. A line consists of an endless, or infinite, number of points. It has no width or thickness and extends infinitely in two directions.

A B k

The line is named as $\overleftrightarrow{AB}$, $\overleftrightarrow{BA}$, or k and read as line AB, line BA, or line k.

A plane is represented by a four-sided figure and named by a capital letter in script. It is a flat surface, like the image of a picture projected onto a screen. It has length and width, but no thickness. A plane continues without end. There are an infinite number of points and lines in a plane.

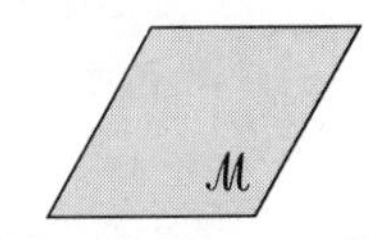

The plane is named $\mathcal{M}$ and is read as plane $\mathcal{M}$.

EXAMPLE 1 Give seven different names for the line represented at the right.

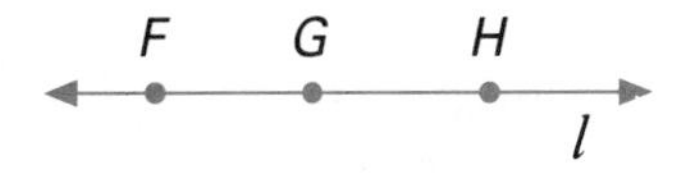

Solution A line may be named by any two points on the line or by a lowercase letter. Thus, the line may be named $\overleftrightarrow{FG}$, $\overleftrightarrow{GF}$, $\overleftrightarrow{FH}$, $\overleftrightarrow{HF}$, $\overleftrightarrow{GH}$, $\overleftrightarrow{HG}$, or l.

The undefined terms *point*, *line*, and *plane* can be used to define other terms.

Definition

Space is the set of all points.

Definition

Points are **collinear** if there exists a line that contains the points.

In the figure at the right, points A, B, and C are on the same line. These three points are collinear. However, there is no line that can contain points A, B, C, and D. These four points are **noncollinear**.

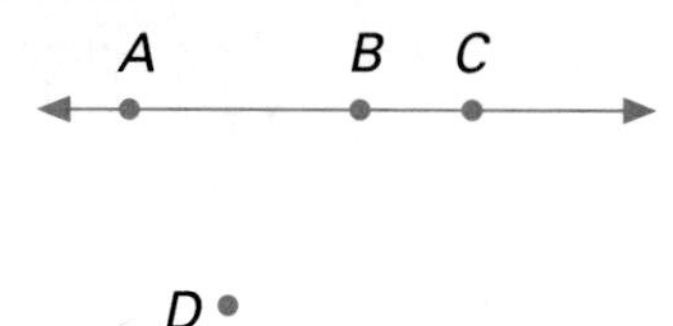

Definition

Points are **coplanar** if there exists a plane that contains the points.

The diagram at the right shows portions of three planes. (NOTE: A plane continues without end, so only *part* of a plane can be illustrated.) This figure might be interpreted as a double picture frame on a shelf. Points H, E, and F are in the same plane—the left frame. These points are coplanar. However, no plane can contain points H, E, F, and I. These four points are **noncoplanar**.

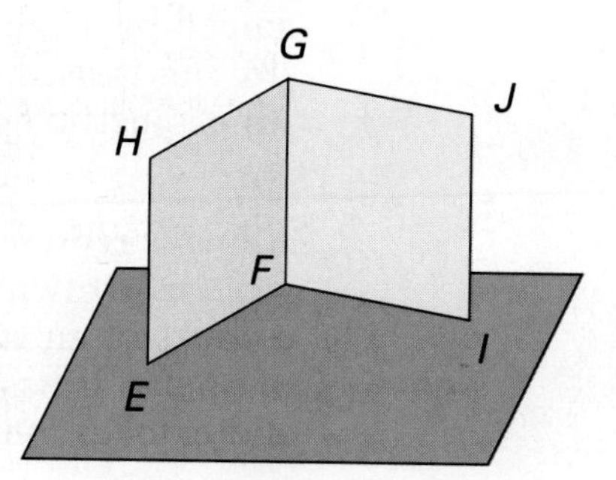

In all of the diagrams in this book, you should assume that points that *appear* to lie on a line are collinear. Similarly, assume that points that *appear* to lie on a plane are coplanar.

EXAMPLE 2 In the figure, find 3 collinear points, 3 noncollinear points, 4 coplanar points, and 4 noncoplanar points.

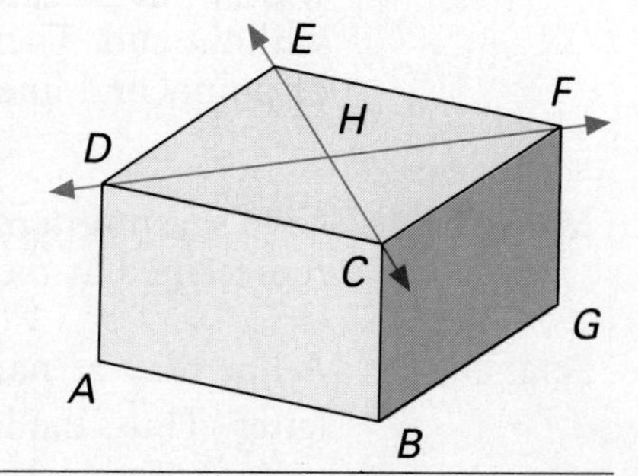

Solution D, H, and F are collinear.
D, E, and F are noncollinear.
C, B, F, and G are coplanar.
D, A, B, and H are noncoplanar.
(Other answers are possible.)

Following are some commonly used expressions describing various relationships among points, lines, and planes.

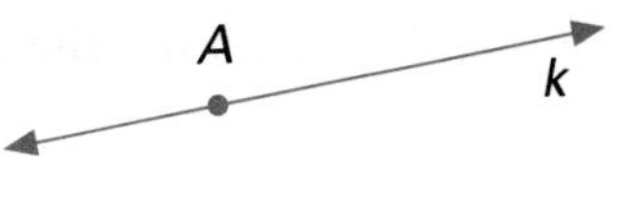

A is on k.
k passes through point A.
k contains point A.

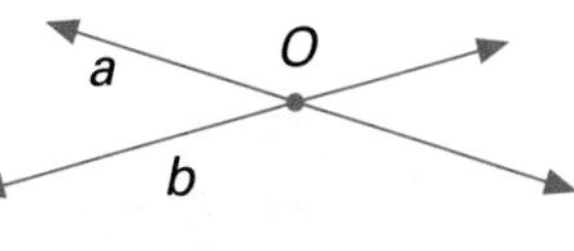

Intersect means "cut" or "meet."
a intersects b at O.
The intersection of a and b is {O}.

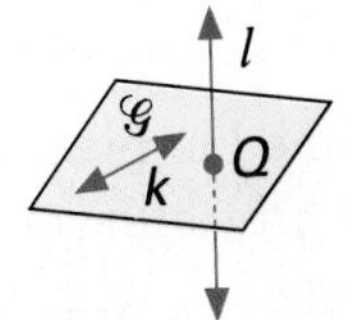

Plane $\mathcal{G}$ contains line k and point Q.
k and Q lie on plane $\mathcal{G}$.
l intersects plane $\mathcal{G}$ at $\{Q\}$.
Q is the intersection of l and plane $\mathcal{G}$.

Definition

The **intersection** of two geometric figures is the set of points that are contained in both figures.

Think of the drawing at the right as a diagram of a classroom. The front wall and ceiling of the room are portions of two intersecting planes. They meet at a line.

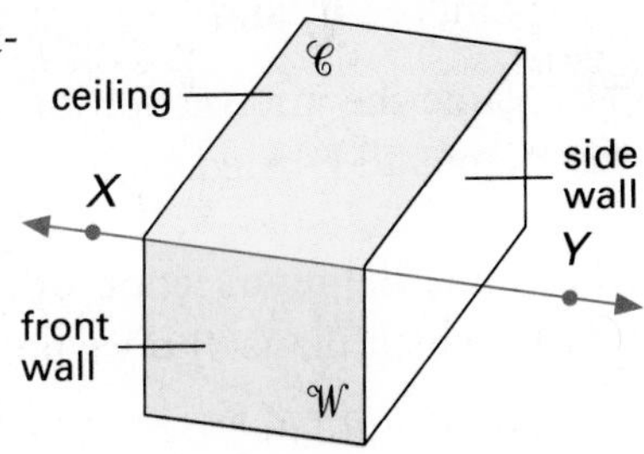

Planes $\mathcal{C}$ and $\mathcal{W}$ contain $\overleftrightarrow{XY}$.
Planes $\mathcal{C}$ and $\mathcal{W}$ intersect at $\overleftrightarrow{XY}$.

Focus on Reading

Match the term on the right with its correct description on the left.

1. a flat surface	**a.** line
2. indicates position in space	**b.** plane
3. points that are on the same line	**c.** coplanar
4. extends in exactly two directions	**d.** point
5. the set of all points	**e.** collinear
6. points contained in the same plane	**f.** space

Classroom Exercises

Give a physical illustration of the set of points.

1. plane **2.** point **3.** line **4.** noncollinear points
5. intersecting lines **6.** intersecting planes **7.** noncoplanar points

Written Exercises

Write all possible names for the figure.

1. 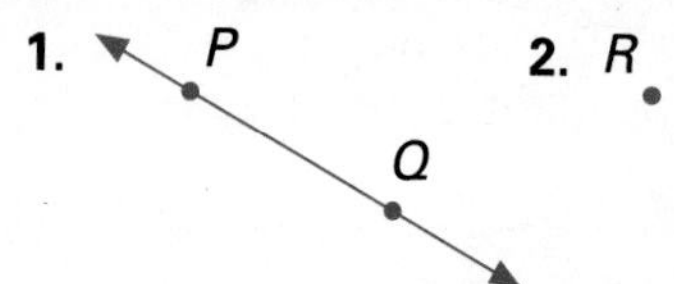

2. 3. 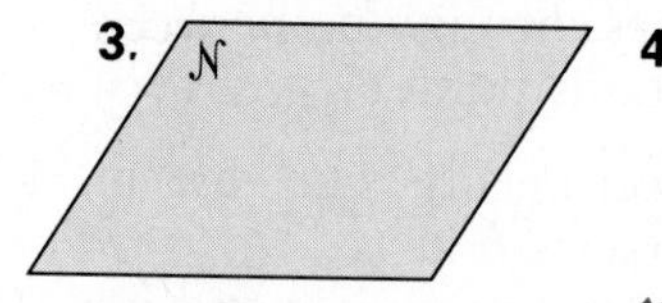

4. 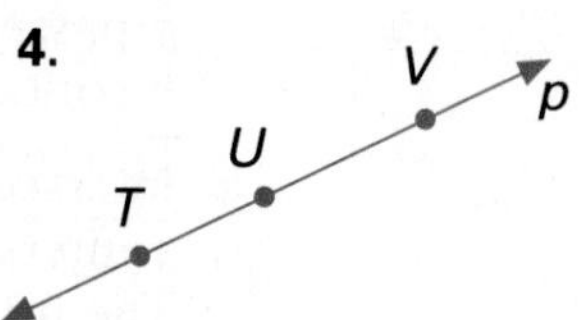

True or false? If false, indicate why.

5. $\overleftrightarrow{AB}$ is contained in plane $\mathcal{M}$.
6. F, B, and A are collinear.
7. Plane $\mathcal{M}$ contains point E.
8. Name a point that is collinear with points F and G.
9. What is the intersection of $\overleftrightarrow{DE}$ and plane $\mathcal{M}$?
10. Name two points that are coplanar with points F, B, and A.

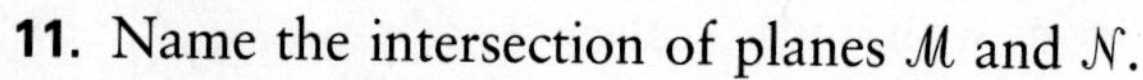

11. Name the intersection of planes $\mathcal{M}$ and $\mathcal{N}$.
12. Name all labeled points that are coplanar with S, Q, and R.
13. Name the intersection of $\overleftrightarrow{KL}$ and $\overleftrightarrow{GH}$.
14. In which plane(s) does point Q lie?

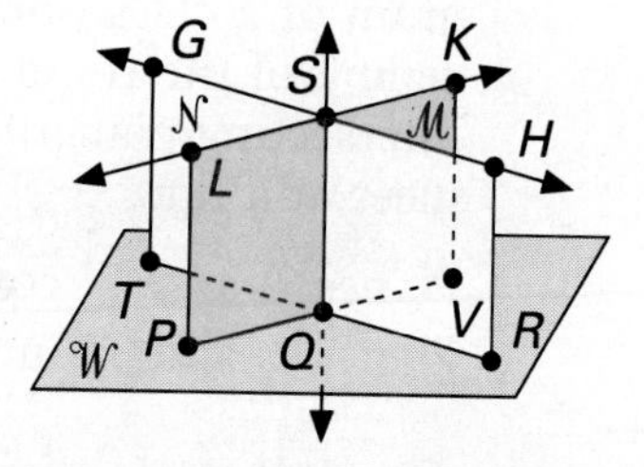

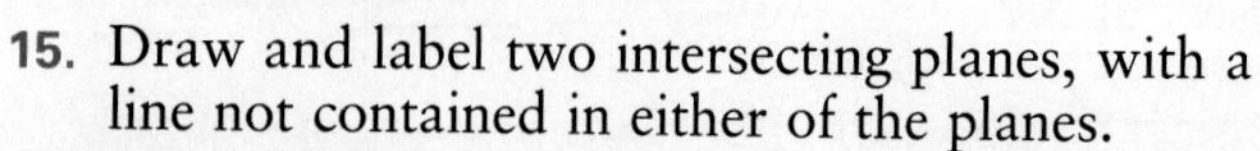

15. Draw and label two intersecting planes, with a line not contained in either of the planes.
16. Draw a diagram to show that three planes may intersect in a point.
17. Three lines may lie in the same plane in four ways such that there are 0, 1, 2, or 3 points of intersection. Draw all four cases.

Algebra Review

In geometry, absolute value can be of use in finding the distance between points. $|x|$ means "the **absolute value** of a number x." That is, $|x| = x$, if $x > 0$ and $|x| = -x$, if $x < 0$.

Example $|-8 + 3| = |-5| = -(-5) = 5$

Simplify.

1. $|-9 + 5|$
2. $|-7 - 8|$
3. $|-4 + 6 - 10|$
4. $|-3 - 7 + 12|$
5. $|5| - |-6|$
6. $|4| + |7 - 8|$
7. $-4 - |-8|$
8. $|-8 + 0| - |-7|$

1.2 Distance, Segments, and Rays

Objectives

To find the distance between two points on a number line
To apply the Ruler Postulate and the Segment Addition Postulate
To identify and name segments and rays

The length of the paper clip at the right can be given by different numbers of units depending upon the measurement system used.

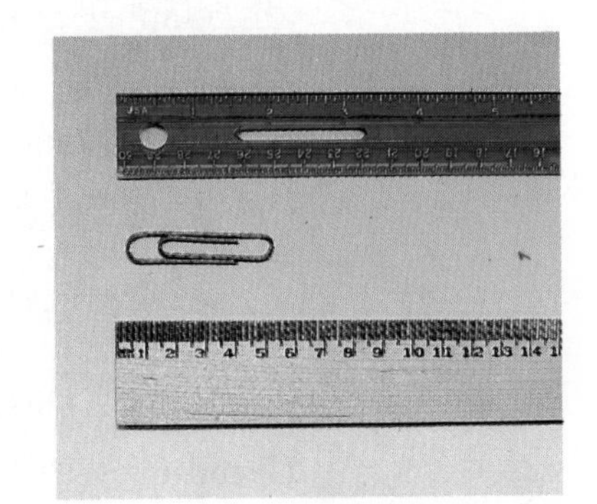

To find the distance between any two points on a line, it is necessary to agree upon a measuring device or ruler. Pick any two points P and Q on a line, with Q to the right of P. Assign the number 0 to P and the number 1 to Q.

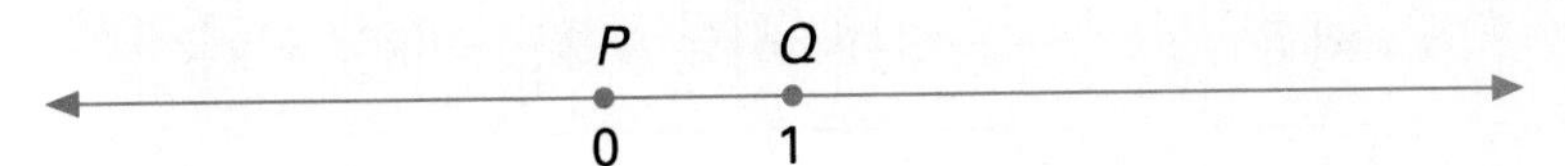

The distance from P to Q is 1. Write $PQ = 1$ or $QP = 1$, where the symbols PQ and QP mean "the distance between P and Q." Using PQ as a reference, the set of real numbers can now be associated with points on the line. When this is done, the line is called a **number line**.

Below, point T corresponds to 3, which is called the **coordinate** of T. The coordinate of a point on a number line is the number associated with that point.

X W P Q R S T U V

$-\sqrt{2}$ -1 0 1 2 $\frac{5}{2}$ 3 π 4

On the number line above, the distance between the two points W and R with coordinates -1 and 2, respectively, is defined to be equal to the absolute value of the difference of their coordinates.

$$WR = |2 - (-1)| = |3| = 3, \text{ or } WR = |-1 - 2| = |-3| = 3$$

Definition

The **distance** between any two points A and B with coordinates m and n is $|m - n|$ or $|n - m|$.

A B

m n

The ideas above are summarized in the Ruler Postulate. A **postulate** is a statement that is accepted without proof.

Postulate 1

Ruler Postulate

1. Any two distinct points on a line can be assigned coordinates 0 and 1.
2. There is a one-to-one correspondence between the real numbers and all points on the line.
3. To every pair of points, there corresponds exactly one positive number called the distance between the two points.

EXAMPLE 1 Find the distance between each pair of points listed below.

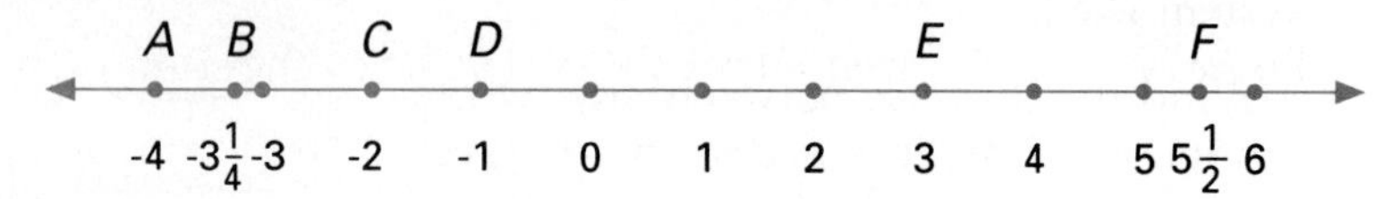

Points	**Distance**
D and C	$\lvert -2 - (-1) \rvert = \lvert -2 + 1 \rvert = \lvert -1 \rvert = 1$
B and F	$\left\lvert 5\frac{1}{2} - \left(-3\frac{1}{4}\right) \right\rvert = \left\lvert 5\frac{1}{2} + 3\frac{1}{4} \right\rvert = \left\lvert 8\frac{3}{4} \right\rvert = 8\frac{3}{4}$

In the figure, note that C is between A and B. There are an infinite number of points between A and B.

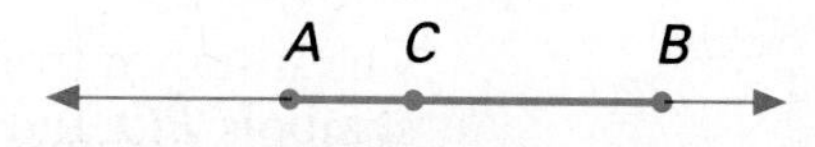

Definition

Segment AB ($\overline{AB}$) is the set of points consisting of points A, B, and all points between A and B. A and B are called the **endpoints** of the segment. Either $\overline{AB}$ or $\overline{BA}$ can be used to name the segment.

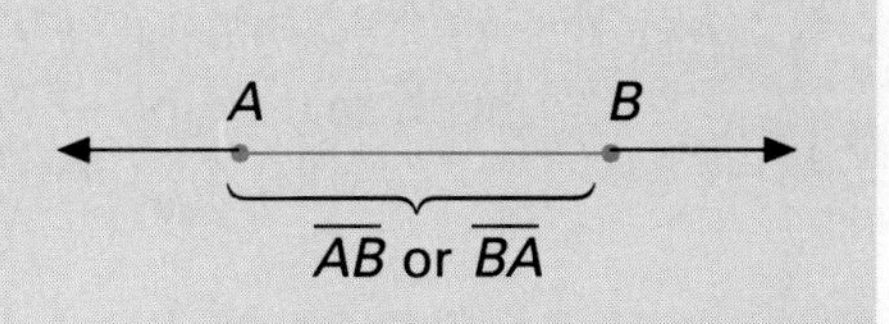

Keep in mind that:

(1) $\overline{AB}$ is a set of points.
(2) AB is the distance between points A and B, which is a number.
(3) The length of $\overline{AB}$ is AB—the distance between A and B.

In the figure, C is between A and B.

$AC = 3 \quad CB = 6 \quad AB = 9$

$3 + 6 = 9$

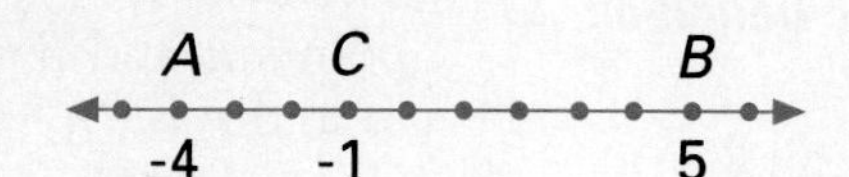

Therefore, $AC + CB = AB$

This suggests the Segment Addition Postulate stated on the next page.

Postulate 2

Segment Addition Postulate

If C is between A and B, then
$$AC + CB = AB.$$

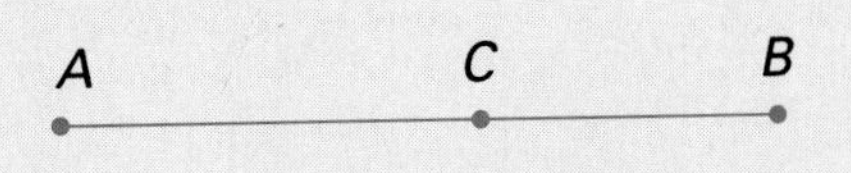

EXAMPLE 2 Points A, G, and R are collinear. Point R is between A and G. Draw a diagram. Use the Segment Addition Postulate to write an equation.

Solution Point A can be placed either to the left or to the right of point R.

A R G

$AR + RG = AG$

G R A

$GR + RA = GA$

Thus, the equation is $AR + RG = AG$, or $GR + RA = GA$.

EXAMPLE 3 G, R, and A are three collinear points such that A is between G and R. $GA = \frac{3}{5}AR$ and $GR = 24$. Find AR.

Plan Let $x = AR$. Then $GA = \frac{3}{5}x$.

G A R

Solution

Write an equation.	$GA + AR = GR$
Substitute.	$\frac{3}{5}x + x = 24$
Multiply both sides of the equation by 5.	$3x + 5x = 120$
	$8x = 120$
	$x = 15 \longrightarrow AR = 15$

Definition

Ray XY ($\overrightarrow{\mathbf{XY}}$) consists of $\overline{XY}$ and all points P such that Y is between X and P. X is called the **endpoint** of the ray. The symbol $\overrightarrow{XY}$ is used to name the ray.

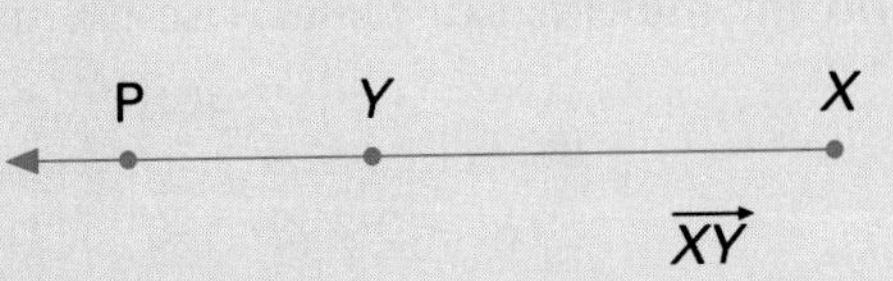

It is often possible to name a ray in more than one way. However, the first letter always names the endpoint of the ray, and the arrow above the two letters points to the right.

EXAMPLE 4 Write three names for the ray that has endpoint P and contains point T.

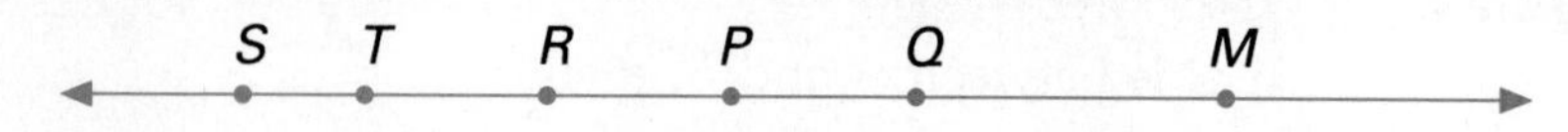

Solution The ray begins at P and passes through T. Shade the drawing to show the ray.

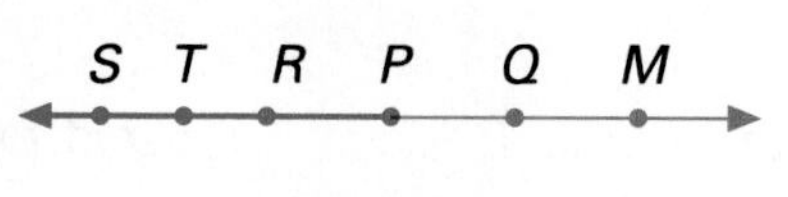

Three names for the ray are $\overrightarrow{PR}$, $\overrightarrow{PT}$, and $\overrightarrow{PS}$.

Summary

line	ray	segment
A B	A B	A B
$\overleftrightarrow{AB}$	$\overrightarrow{AB}$	$\overline{AB}$

The distance between points C and D corresponds to a unique number, CD. CD is the length of $\overline{CD}$.

$$CD = |4-(-2)| = |6| = 6$$

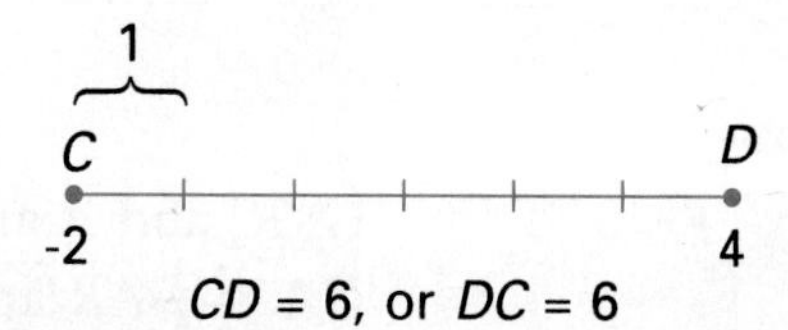

Classroom Exercises

Use the Segment Addition Postulate to write an equation for the segment.

1. A T Y

2. W U O

3. G L N

Written Exercises

Find the indicated distance. Use the number line.

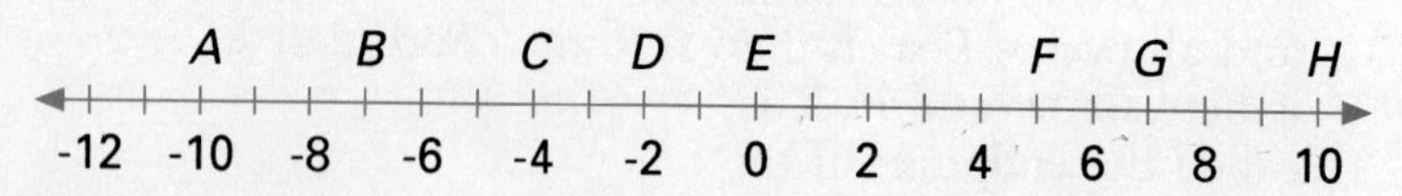

1. AC **2.** CF **3.** EB **4.** GF **5.** GB **6.** HD **7.** GA **8.** DA

For the given coordinates of G and H, find GH.

9. $G{:}-7;\ H{:}5$ **10.** $G{:}4\frac{5}{7};\ H{:}8$ **11.** $G{:}-6.2;\ H{:}-2.5$

For the given set of collinear points, draw a diagram. Write an equation using the Segment Addition Postulate.

12. G, K, and T such that T is between G and K

13. R, A, and V such that R is between A and V

Give three names for the indicated ray.

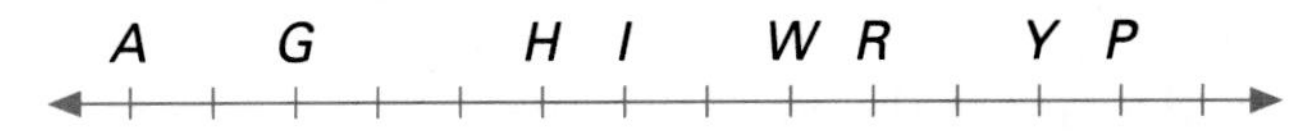

14. Endpoint I containing H

15. Endpoint W containing Y

16. Endpoint R containing I

Write a name for each of the following.

17. A B

18. C D

19. Distance between points H and W

20. R S

21. A, B, and C are three collinear points such that B is between A and C. $AB = \frac{3}{4}BC$ and $AC = 28$. Find AB.

22. P, Q, and R are three collinear points such that Q is between P and R. $PQ = \frac{4}{7}QR$ and $PR = 33$. Find QR.

23. For $\overline{AB}$, the coordinate of A is -6 and $AB = 7$. Find all possible coordinates of point B.

Find the distance ST for the following coordinates.

24. $S{:}6.38$; $T{:}-7.91$

25. $S{:}19.07$; $T{:}4.63$

26. $S{:}-0.45$; $T{:}-8.54$

27. A, B, and C are three collinear points such that B is between A and C. $AB = \frac{3}{7}AC$ and $AB = 9$. Find BC.

28. Point T is on $\overrightarrow{MG}$, but T is not on $\overline{MG}$. $MT = \frac{5}{4}GT$ and $MT = 18$. Find MG.

29. For $\overline{GH}$, the coordinate of G is $2x - 6$ and the coordinate of H is $x - 5$. Find the coordinate of G if $GH = 6$.

Mixed Review

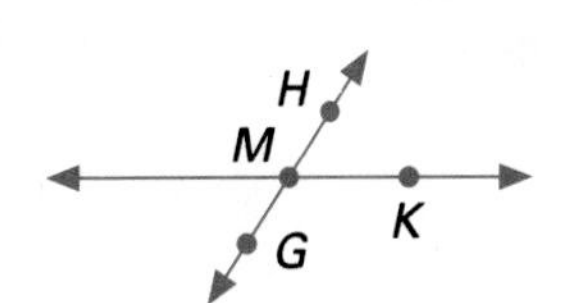

1. Name three collinear points. *1.1*
2. Name three noncollinear points. *1.1*
3. Name the intersection of $\overleftrightarrow{GH}$ and $\overleftrightarrow{MK}$ *1.1*

1.3 Congruent Segments and Constructions

Objectives

To identify and construct congruent segments
To locate the midpoint of a segment by construction
To apply the Midpoint Formula

In the diagram at the right, the two segments, $\overline{AB}$ and $\overline{CD}$, have the same length, 2 cm. The segments are said to be *congruent*. The symbol for congruence is $\cong$. The symbol $\ncong$ is read "is not congruent to."

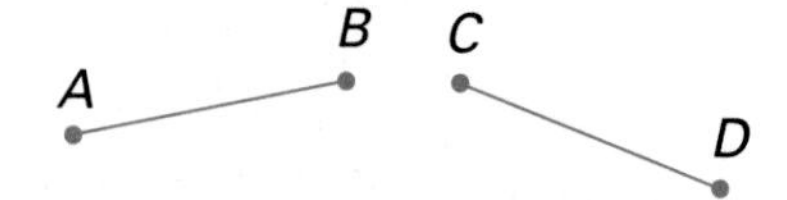

$AB = CD$: equal lengths
$\overline{AB} \cong \overline{CD}$: congruent segments

Definition

Congruent ($\cong$) **segments** are segments that have the same length. $\overline{AB} \cong \overline{CD}$ is read "$\overline{AB}$ is congruent to $\overline{CD}$."

In the figure, markings (tick marks) are used to indicate that $\overline{AB}$ and $\overline{CD}$ are congruent. You should not assume congruence of segments unless they are indicated as congruent. By definition, *congruent segments* are segments that have the *same length*. Therefore, the statements $\overline{AB} \cong \overline{CD}$ and $AB = CD$ are equivalent.

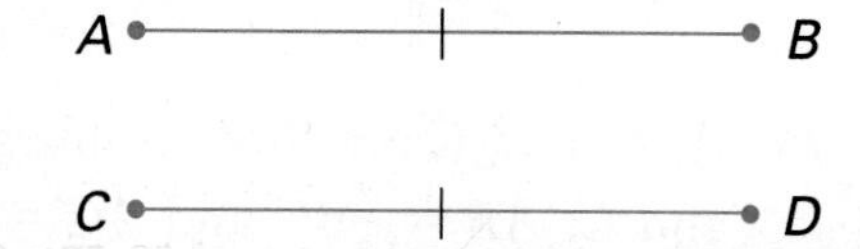

In both figures below, R is between P and S. In the figure on the right, R divides $\overline{PS}$ into two congruent segments.

P R S

$PR + RS = PS$
$\overline{PR} \ncong \overline{RS}$

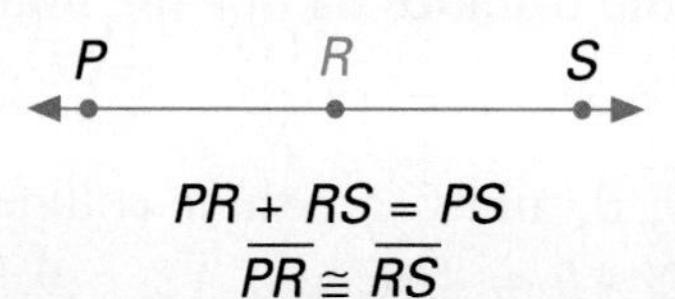

Definition

M is the **midpoint** of $\overline{AB}$ if M lies on $\overline{AB}$ and $\overline{AM} \cong \overline{MB}$ ($AM = MB$).

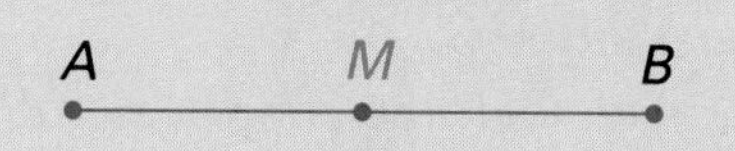

Definition

A **bisector of a segment** is a line, ray, segment, or plane that intersects the segment at its midpoint. A bisector divides the segment into two congruent segments.

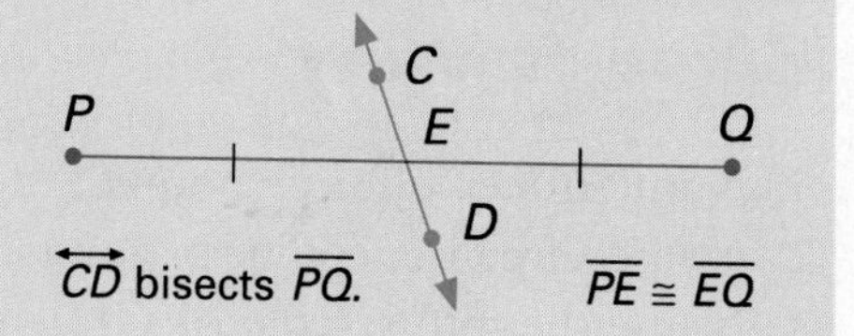

$\overleftrightarrow{CD}$ bisects $\overline{PQ}$.
$\overline{PE} \cong \overline{EQ}$

Postulate 3

Any segment has exactly one midpoint.

By agreement, a compass and straightedge are the only tools that may be used in the construction of a geometric figure. A straightedge is used for drawing segments. It need not have the coordinate markings of a ruler. A compass is used for drawing arcs, which are unbroken parts of a circle. A compass and straightedge can be used to construct a segment congruent to a given segment.

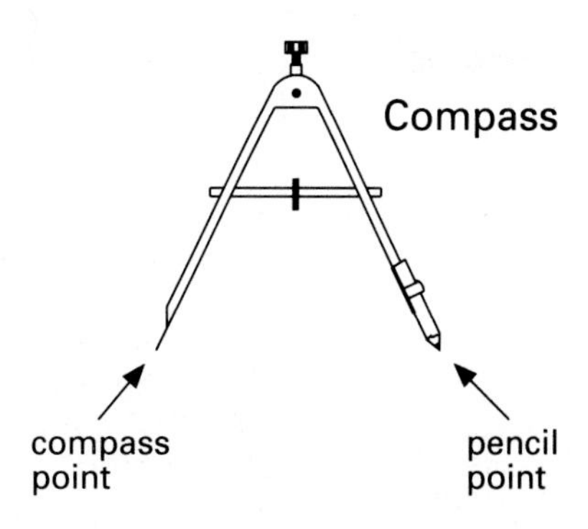

Construction Construct a segment congruent to a given segment.

Construct $\overline{RS}$, where $\overline{RS} \cong \overline{PQ}$, on line k.

Use a straightedge to draw line k. Adjust the compass opening to correspond to the distance from P to Q.

Choose any point on line k and label it R. Set the compass point at R and draw an arc intersecting k. Label the point of intersection as point S.

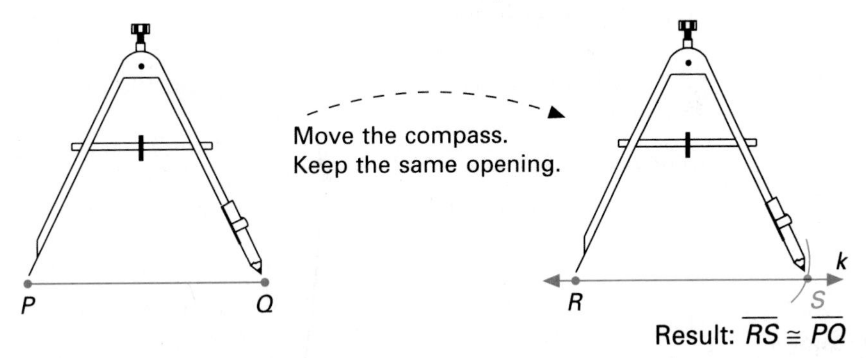

A compass and a straightedge can also be used to construct a point *equidistant* from two points. **Equidistant** means "the same distance."

Construction Locate a point equidistant from two given points.

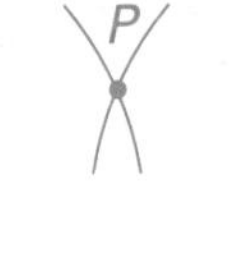

Let A and B be two given points. Open the compass wider than one-half of AB. Place the compass point on A. Draw an arc. Using the same compass opening, place the compass point on B. Draw a second arc to intersect the first arc at point P. Point P is equidistant from the two given points, A and B.

Construction Locate the midpoint of a given segment.
Locate M, the midpoint of $\overline{AB}$.

A B

Locate a point P equidistant from A and B. | Locate a second point Q equidistant from A and B. | Draw $\overleftrightarrow{PQ}$ intersecting $\overline{AB}$ at point M.

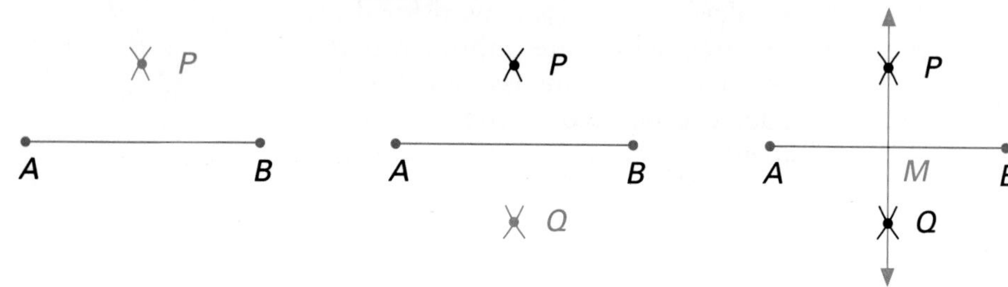

Result: M is the midpoint of $\overline{AB}$.

It is possible to find the coordinate of the midpoint of a segment when the coordinates of the endpoints are given. The procedure is the same as that used to find the arithmetic mean (average) of two numbers.

EXAMPLE 1 M is the midpoint of $\overline{AB}$. Find its coordinate, m.

A M B
-5 m 8

Solution Point M is halfway between A and B. Find the arithmetic mean of -5 and 8.

Thus, $m = \dfrac{-5 + 8}{2} = \dfrac{3}{2}$, or $1\frac{1}{2}$.

The general formula for finding the coordinate of the midpoint of a segment when the coordinates of the endpoints are given is stated below.

Midpoint Formula

If the coordinates of the endpoints of $\overline{AB}$ are a and b, and m is the coordinate of the midpoint M, then $m = \dfrac{a + b}{2}$.

A M B
a m b

EXAMPLE 2 M is the midpoint of $\overline{AB}$. Find the coordinates of points A and B.

A M B
(p - 5) 6 (p + 2)

Solution

$$6 = \frac{(p - 5) + (p + 2)}{2}$$

$$6 = \frac{2p - 3}{2}$$ ← Multiply each side of the equation by 2 to eliminate fractions.

$$12 = 2p - 3$$
$$15 = 2p$$
$$7\tfrac{1}{2} = p$$

The coordinate of A is $p - 5 = 7\frac{1}{2} - 5 = 2\frac{1}{2}$.

The coordinate of B is $p + 2 = 7\frac{1}{2} + 2 = 9\frac{1}{2}$.

Classroom Exercises

1. Tell how to construct a point equidistant from the endpoints of a segment.
2. Tell how to construct the midpoint of a segment.
3. What is a line called that intersects a segment at its midpoint?
4. According to our definition, what does it mean to say that two segments are congruent?
5. State the Midpoint Formula.

Written Exercises

Write a statement of congruence and an equation.

1. L is the midpoint of $\overline{TU}$.

T L U

2. $\overleftrightarrow{RW}$ bisects $\overline{FG}$.

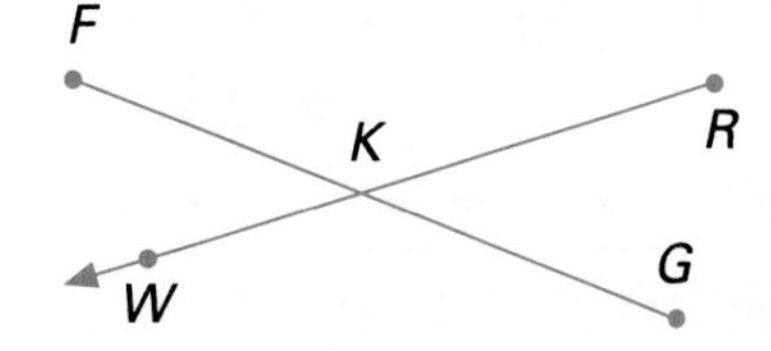

Draw a segment $\overline{PQ}$. (Exercises 3–5)

3. Construct a segment $\overline{GH}$ congruent to $\overline{PQ}$.
4. Locate a point T equidistant from the endpoints of $\overline{PQ}$.
5. Locate the midpoint of $\overline{PQ}$.

Find the coordinate of the midpoint M of $\overline{AB}$ whose endpoints have the given coordinates.

6. 4 and 7
7. −5 and 9
8. 0 and 13
9. 5 and 12
10. 3.2 and 7.6
11. −3.4 and −1.2
12. $3\frac{1}{2}$ and $7\frac{1}{2}$
13. $-4\frac{1}{5}$ and $8\frac{3}{5}$
14. $-2\frac{1}{2}$ and 7

M is the midpoint of $\overline{AB}$. Find the coordinates of A, B, or M for the given data.

A M B

15. $A{:}p - 4;\ M{:}8;\ B{:}p + 12$
16. $A{:}3p - 1;\ M{:}10;\ B{:}2p + 6$
17. $A{:}p - 8;\ M{:}10;\ B{:}p + 6$
18. $A{:}2p - 2;\ M{:}14;\ B{:}4p + 6$
19. $A{:}-4;\ M{:}6 - p;\ B{:}2p - 4$
20. $A{:}12 - 2p;\ M{:}p - 6;\ B{:}16$

Find the coordinate of the missing endpoint of $\overline{GH}$.

G M H

21. Midpoint $M{:}8;\ H{:}12$
22. Midpoint $M{:}-5\frac{1}{2};\ H{:}0$

23. For $\overline{GH}$ above, if the coordinates of M and H are positive, is it reasonable that G must be positive?
24. Let a and b be the coordinates of the endpoints of $\overline{AB}$. Let m be the coordinate of the midpoint M of $\overline{AB}$. Assume that $a < m < b$. Use the definition of midpoint to derive the Midpoint Formula: $m = \frac{a + b}{2}$.
25. The coordinates of the endpoints A and B of $\overline{AB}$ are 6 and 21, respectively. Find the coordinate of point C between A and B such that $AC = \frac{2}{3}(CB)$.
26. C, D, and E are the midpoints of $\overline{AB}$, $\overline{AC}$, and $\overline{CB}$, respectively. D and E have coordinates 6 and 13, respectively. Find the coordinates of A and B.

Mixed Review

Identify the meaning of each symbol. *1.2*

1. $\overleftrightarrow{XY}$
2. $\overline{XY}$
3. $\overrightarrow{XY}$
4. XY

Brainteaser

Draw some four-sided, closed figures. Construct the midpoint of each side. Then connect the midpoints consecutively.

Is there a pattern in the shape of the figure formed by connecting the midpoints?

1.4 Angle Measurement and Constructions

Objectives

To classify angles by their measures

To determine the measure of an angle by applying the Protractor Postulate

To construct an angle congruent to a given angle

The distance a football player can kick a football is determined by the force of the kick and the *angle* formed by the beginning path of the football and a horizontal ray on the ground.

Definition

An **angle** is a geometric figure consisting of two distinct rays with a common endpoint. The rays are the *sides* of the angle. The common endpoint is the *vertex* of the angle.

The symbol for angle is $\angle$.

In the diagram, the sides of the angle are $\overrightarrow{BA}$ and $\overrightarrow{BC}$.

The vertex is the common endpoint B.

The angle may be named in three ways: by three capital letters—$\angle CBA$ or $\angle ABC$—where the middle letter names the vertex; by the vertex alone, when there is no possibility of confusion—$\angle B$; or by a numeral placed between the rays—$\angle 1$.

EXAMPLE 1 Name each angle in the figure.

Solution Since there are three angles in the figure, it is unclear what is meant by $\angle Q$. None of these angles can be named by its vertex alone. Thus, the angles are $\angle PQR$, $\angle 1$ (or $\angle SQR$), and $\angle 2$ (or $\angle PQS$).

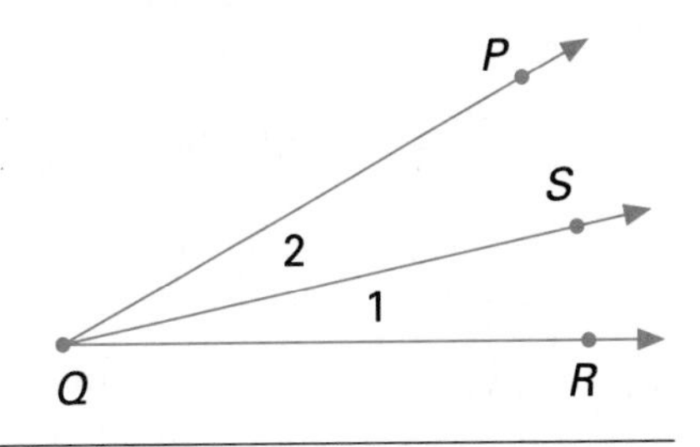

A **protractor** is used to measure an angle in **degrees** (symbol: °). In the figure at the right, the measure of angle BOC equals 70. This is written "m $\angle BOC = 70$."

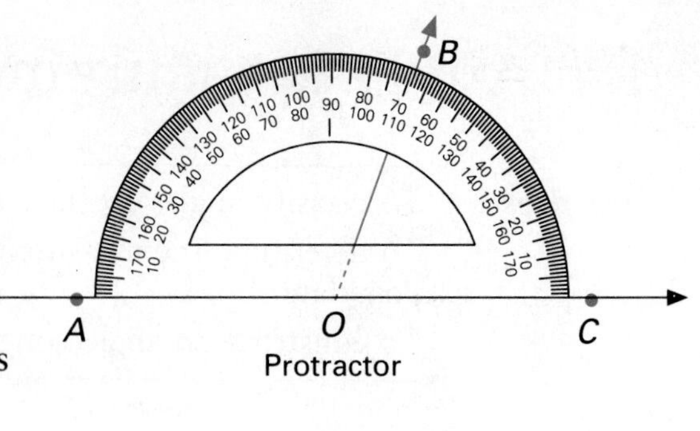

In this book, you may assume that all angle measures are in degrees.

Using the protractor to measure an angle suggests that every angle has a measure, a basic assumption in geometry.

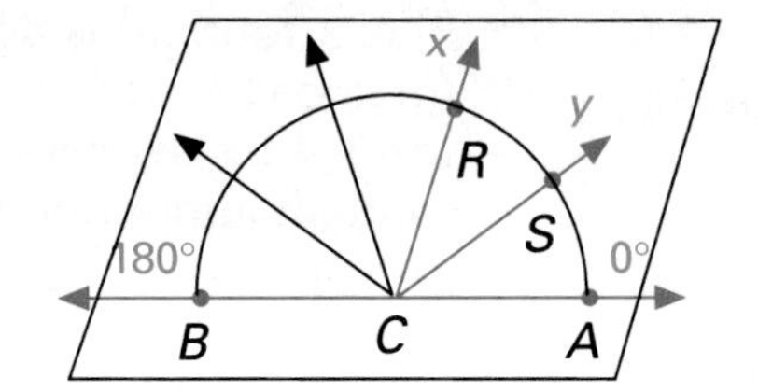

Postulate 4

Protractor Postulate

In a given plane, select any line $\overleftrightarrow{AB}$ and any point C between A and B. Also select any two points R and S on the same side of $\overleftrightarrow{AB}$ so that S is not on $\overrightarrow{CR}$. Then, there is a *pairing* of rays to real numbers from 0 to 180 in the following way:

1. $\overrightarrow{CA}$ is paired with 0 and $\overrightarrow{CB}$ is paired with 180.
2. If $\overrightarrow{CR}$ is paired with x, then $0 < x < 180$.
3. If $\overrightarrow{CR}$ is paired with x and $\overrightarrow{CS}$ is paired with y, then m $\angle RCS = |x - y|$.

In general, the Protractor Postulate says that to every angle there corresponds exactly one real number n such that $0 < n \leq 180$. The measure of the angle is n. We assume that the rays of an angle are distinct and cannot coincide. Therefore, an angle cannot have measure 0.

Angles may be classified into four categories by their measure.

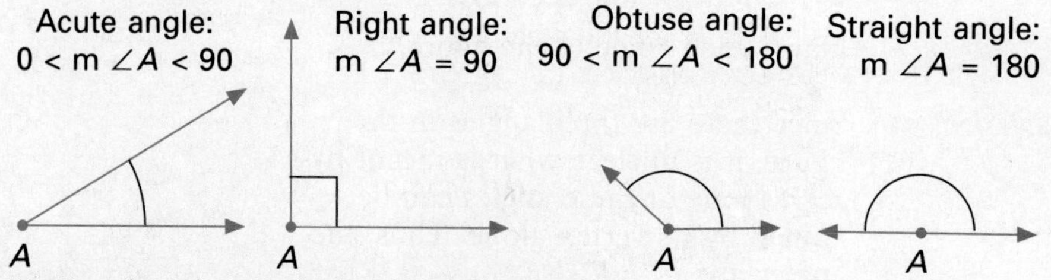

Notice that the symbol ⌉ is used to indicate a right angle, and that a straight angle is a line with a labeled point indicating the vertex.

EXAMPLE 2 Find the measure of $\angle POR$, $\angle FOH$, and $\angle GOH$. Classify each angle as acute, right, obtuse, or straight.

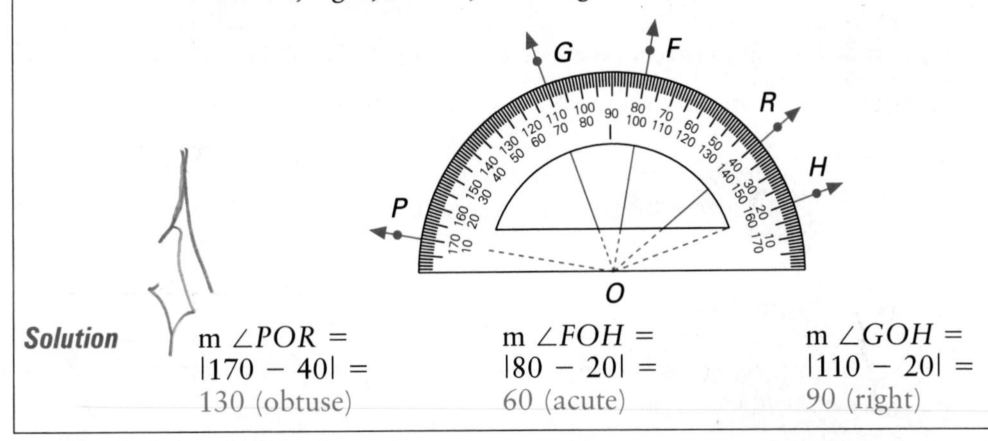

Solution

m $\angle POR$ = $|170 - 40|$ = 130 (obtuse)

m $\angle FOH$ = $|80 - 20|$ = 60 (acute)

m $\angle GOH$ = $|110 - 20|$ = 90 (right)

Definition

Congruent (≅) angles are angles that have the same measure. $\angle A \cong \angle B$ means "$\angle A$ is congruent to $\angle B$."

In the figure, similar markings (arcs) are used to indicate that $\angle A$ and $\angle B$ are congruent. Unless marked as such, angles should not be assumed to be congruent. By definition, congruent angles are angles that have the same measure. Therefore, the statements $\angle A \cong \angle B$ and m $\angle A$ = m $\angle B$ are equivalent.

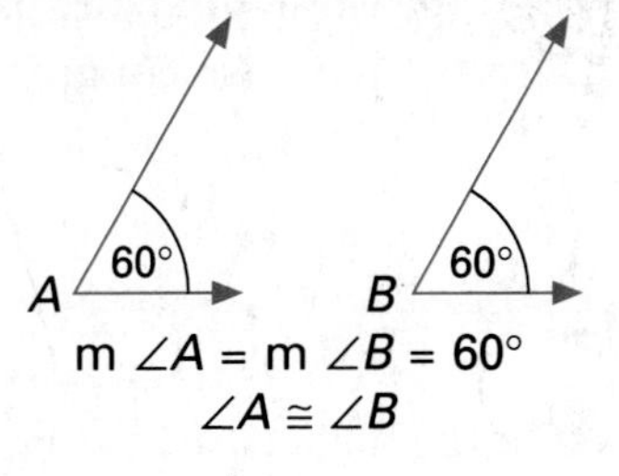

Construction Construct an angle congruent to a given angle.

Construct $\angle B$ congruent to the given $\angle A$.

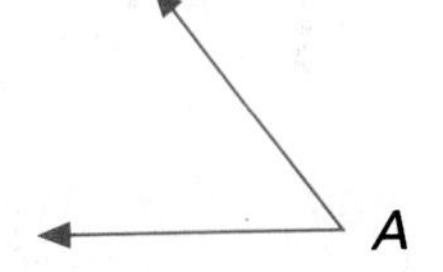

Using any compass opening, construct an arc with center A. Label points P and R on the line as shown. Draw line l; choose a point B. Using the same compass opening, construct an arc with center B, intersecting l at S.

Open the compass to the length PR. Construct an arc with center S, intersecting the other arc at point T. Draw $\overrightarrow{BT}$.

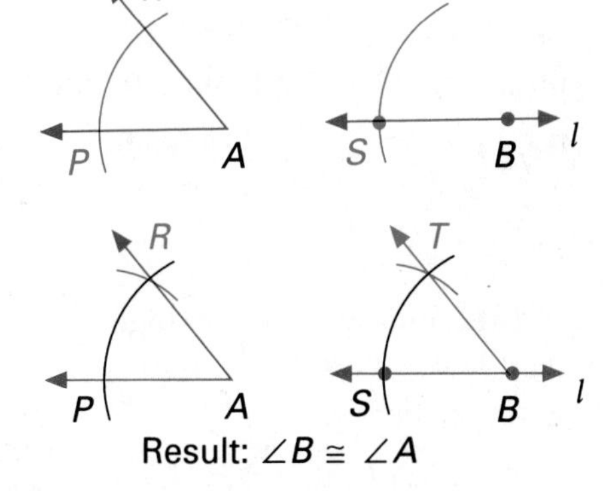

Result: $\angle B \cong \angle A$

Focus on Reading

Complete the sentence with a correct word or expression. (Exercises 1–3)

1. The equation m $\angle A$ = m $\angle B$ is equivalent to the statement ______.
2. An angle with a measure of 180 is called a ______ ______.
3. An ______ angle has a degree measure between 90 and 180.
4. What is a protractor?
5. What is a degree?
6. What is a postulate?

Classroom Exercises

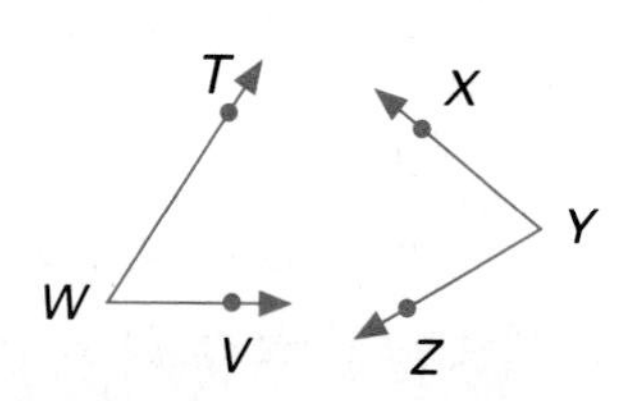

1. $\angle W$ has sides ______, ______.
2. The vertex of $\angle XYZ$ is ______.
3. $\angle W$ can also be called ______ or ______.
4. To measure an angle, use a ______.
5. Two angles are congruent if ______.
6. To construct an angle congruent to $\angle W$, use a ______.

Written Exercises

Name the angle(s) in the figure in as many ways as possible.

1.

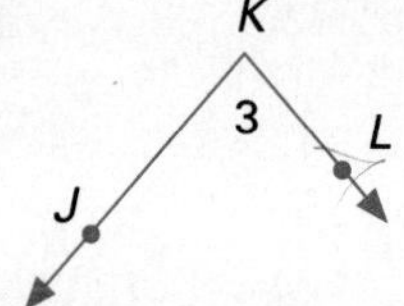

2.

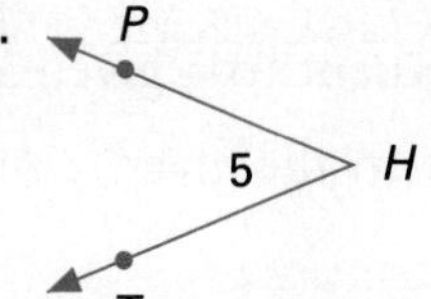

3.

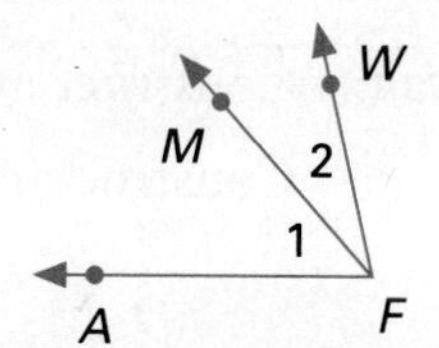

Use a protractor to draw an angle with the given measure. Then use a compass and straightedge to construct an angle congruent to that angle.

4. 25 5. 160 6. 95 7. 15

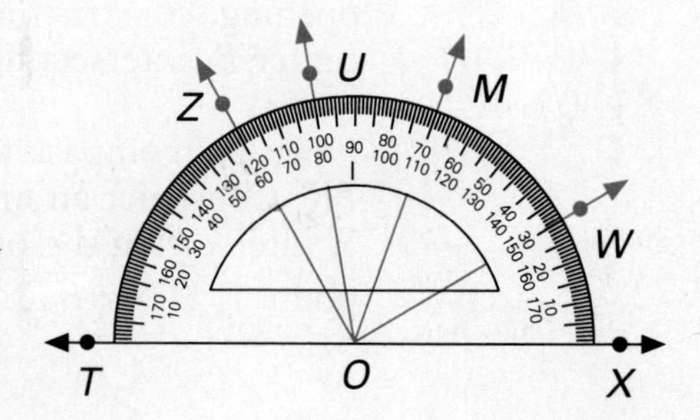

Find the measure of the angle. Tell whether the angle is acute, obtuse, right, or straight.

8. $\angle MOX$ 9. $\angle TOZ$ 10. $\angle UOX$ 11. $\angle TOM$
12. $\angle UOM$ 13. $\angle ZOU$ 14. $\angle ZOW$ 15. $\angle TOX$

16. Name a right angle.

17. Name an acute angle with measure given in degrees.

18. Name an obtuse angle with measure given in degrees.

19. Find the measure of $\angle GFE$.

20. Find the measure of $\angle AFB$.

21. Name four straight angles.

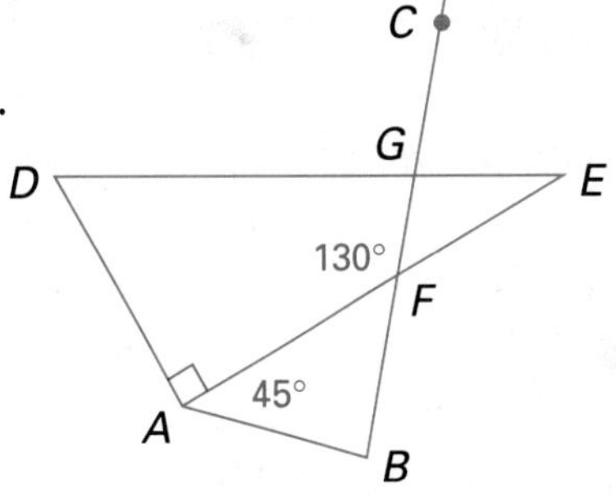

Refer to Figure 1 for Exercise 22. Use the Protractor Postulate to solve.

22. $m\angle TOA = 3x + 50$. $m\angle SOA = x + 40$. $m\angle TOS = 50$. Find $m\angle TOA$.

23. $\angle TUV \cong \angle XYZ$, $m\angle TUV = 3x - 20$, and $m\angle XYZ = \frac{5x + 20}{3}$. Determine whether $\angle TUV$ is a right angle.

24. A portion of $\angle BAC$ is shown below. Construct an angle congruent to $\angle BAC$.

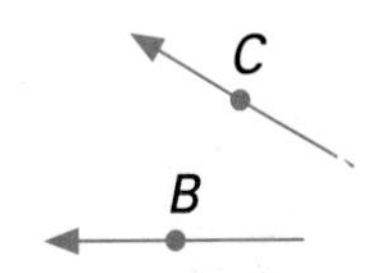

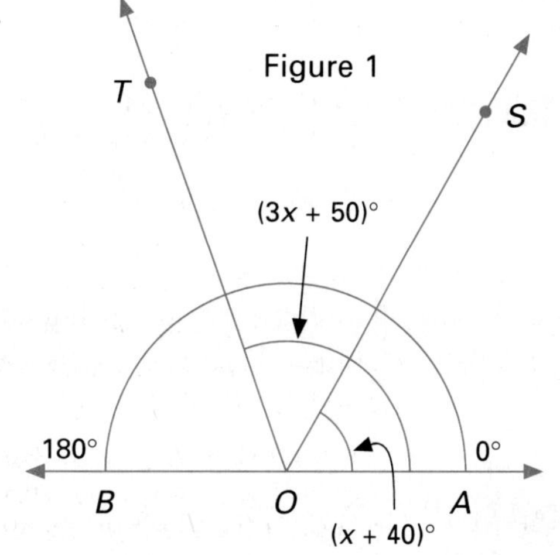

Midchapter Review

Points P, T, and Q are collinear points such that T is between P and Q. (Exercises 1–3)

1. Use the Segment Addition Postulate to write an equation. *1.2*
2. $PT = \frac{4}{5}TQ$, $PQ = 18$. Find PT. *1.2*
3. Find the coordinate of the midpoint of $\overline{PQ}$ for coordinates $P{:}-8$ and $Q{:}10$. *1.3*

For Exercises 4–6, use A and B to represent the following with a diagram and with symbols.

4. a line *1.1*
5. a ray with endpoint A *1.2*
6. a segment *1.2*
7. Draw a segment. Locate its midpoint. *1.3*

Computer Investigation

Supplementary and Complementary Angles

Use a computer software program that constructs angles of a given measure, labels and moves points, extends line segments, and measures line segments and angles.

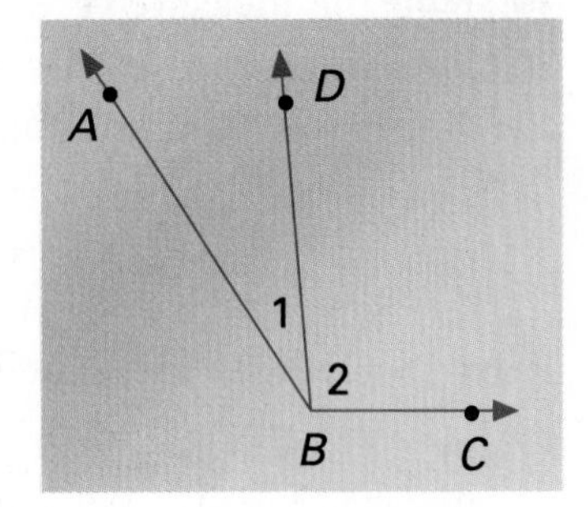

Notice in the figure at the right that there are three angles, $\angle 1$, $\angle 2$, and $\angle ABC$. $\angle 1$ and $\angle 2$ are called **adjacent angles.** Such angles will be formally defined in the next lesson. The activity below will help you discover an important property of adjacent angles.

Activity 1

Draw and label an angle of any measure as $\angle ABC$. Locate a point in the interior of $\angle ABC$. It will be labeled D. Draw $\overrightarrow{BD}$.

1. Find each of the following: m $\angle ABD$, m $\angle DBC$, and m $\angle ABC$.
2. Is there a numerical relationship between the measures of the three angles?
3. Construct a new angle. Then repeat Exercises 1–2 above for the new angle.
4. Repeat the process for a third angle.
5. Make a generalization about the relationship between two adjacent angles and the angle they form together.

Activity 2

In the lesson following this page, you will study two pairs of special types of adjacent angles. You will discover the property of one of these special types in the following activity.

Draw an angle with a measure of 150. The angle will be labeled $\angle CBA$. Extend side $\overline{CB}$ to a point D such that B is between C and D. Notice that $\angle CBA$ and $\angle DBA$ are adjacent angles. Although $\overrightarrow{BD}$ and $\overrightarrow{BC}$ may not actually appear as rays, they are called the outer rays of these adjacent angles.

1. Find m $\angle CBA$ and m $\angle DBA$.
2. Find m $\angle CBD$.
3. Repeat the activity above using the steps of Exercises 1 and 2 for a different angle, say with measure 70.
4. Repeat the process for a third angle of measure 140.
5. Can you generalize the relationship between the measures of two adjacent angles whose **outer rays** form a **line**?

1.5 Adjacent Angles and Angle Bisectors

Objectives

To identify adjacent angles
To apply the Angle Addition Postulate
To apply the definition of angle bisector

An angle in a plane separates the plane into three sets of points: the *angle* itself, the *interior* of the angle, and the *exterior* of the angle. (This does not apply to a straight angle.) A point is an **interior point of an angle** if it is not on either side and is between two points of the angle, one on each side.

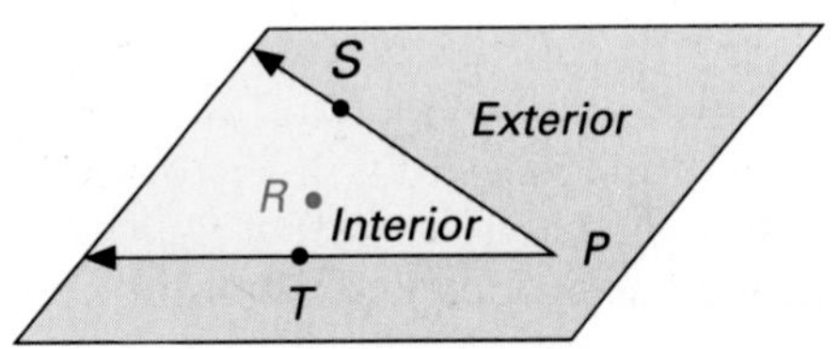

Interior point R lies *between* points S and T.

The figure at the right consists of three angles: $\angle 1$, $\angle 2$, and $\angle CAB$. $\angle 1$ and $\angle 2$ share vertex A and side $\overrightarrow{AD}$. No common interior points are shared by $\angle 1$ and $\angle 2$. $\angle 1$ and $\angle 2$ are called *adjacent angles*.

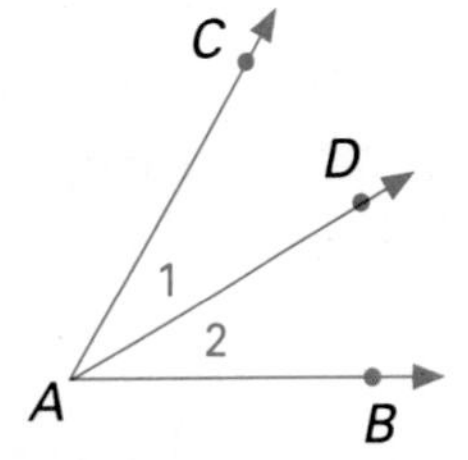

Definition

Adjacent angles are two coplanar angles that have one common side and a common vertex, but no common interior points.

The figure at the right shows adjacent angles, $\angle ABD$ and $\angle DBC$. The common side of the angles is $\overrightarrow{BD}$. The rays $\overrightarrow{BA}$ and $\overrightarrow{BC}$ are called the **outer rays** of the angles. Notice that m $\angle ABC = 50 + 30$, or 80. This suggests the following postulate.

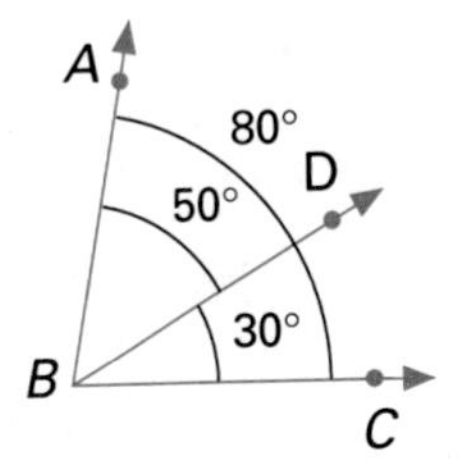

Postulate 5

Angle Addition Postulate

If D is in the interior of $\angle ABC$,
then m $\angle ABC$ = m $\angle ABD$ + m $\angle DBC$

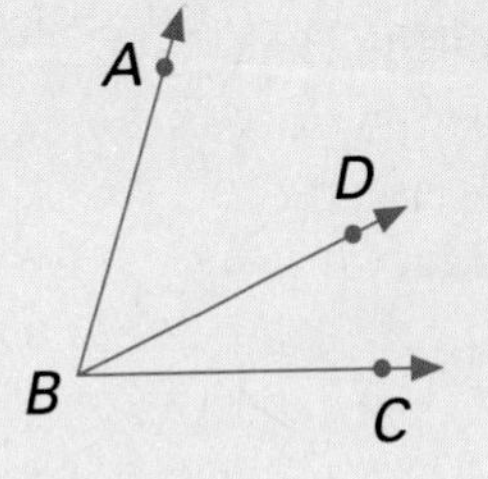

EXAMPLE 1 Given: m $\angle ABD = 130$, m $\angle 2 = 30$
Find m $\angle 1$.

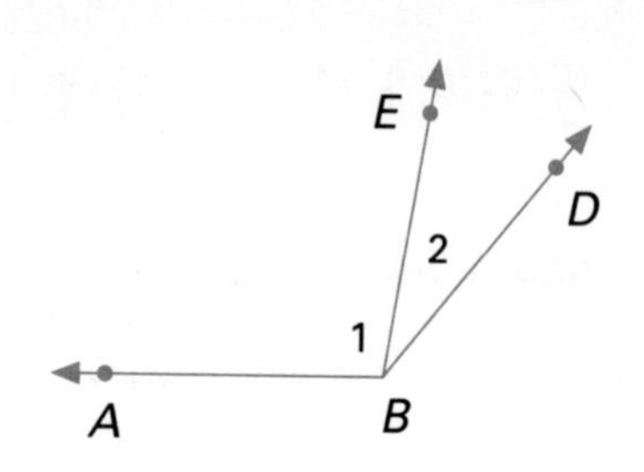

Plan Use the Angle Addition Postulate to write an equation.

Solution m $\angle 1$ + m $\angle 2$ = m $\angle ABD$

Solve for m $\angle 1$. $\quad$ m $\angle 1 + 30 = 130$
m $\angle 1 = 100$

EXAMPLE 2 Given: m $\angle 1 = \frac{2}{3}$(m $\angle 2$), m $\angle ABC = 140$
Find m $\angle 1$ and m $\angle 2$.

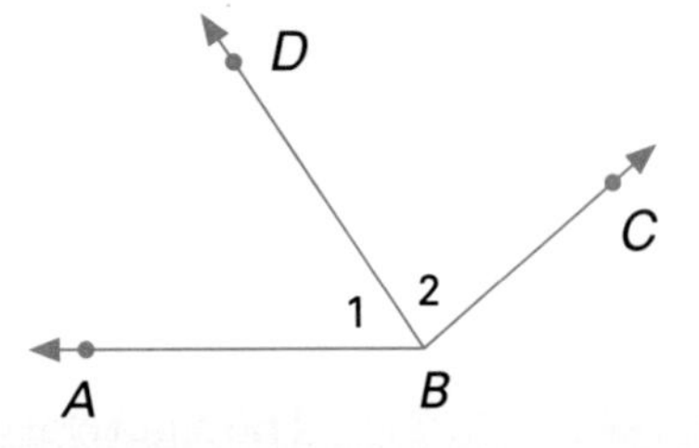

Plan Let x = m $\angle 2$. Then m $\angle 1 = \frac{2}{3}x$.

Solution

Write an equation.	m $\angle 1$ + m $\angle 2$ = m $\angle ABC$
Substitute.	$\frac{2}{3}x + x = 140$
Multiply both sides of the equation by 3.	$2x + 3x = 420$
	$5x = 420$
	$x = 84 \longrightarrow$ m $\angle 2 = 84$
	$\frac{2}{3}x = 56 \longrightarrow$ m $\angle 1 = 56$

In the figure at the right, $\angle 1$ and $\angle 2$ are adjacent angles. Suppose m $\angle 1$ = m $\angle 2$ = 30. Then $\overrightarrow{OC}$ divides $\angle AOB$ into two congruent angles. $\overrightarrow{OC}$ is called an **angle bisector**.

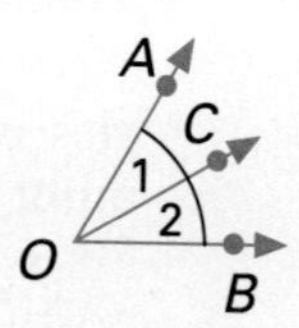

Definition

$\overrightarrow{OC}$ bisects $\angle AOB$ if C is in the interior of $\angle AOB$ and $\angle AOC \cong \angle COB$ (m $\angle AOC$ = m $\angle COB$). $\overrightarrow{OC}$ is the **bisector** of $\angle AOB$.

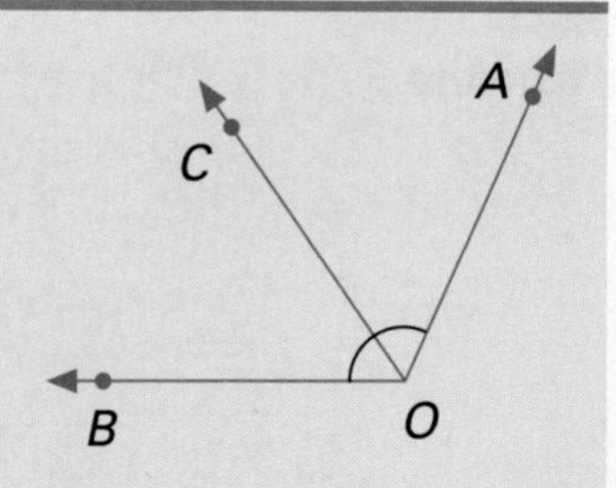

Construction Construct the bisector of a given angle.

Using O as the center, draw an arc, intersecting the sides of $\angle POQ$. Label the points of intersection G and H.

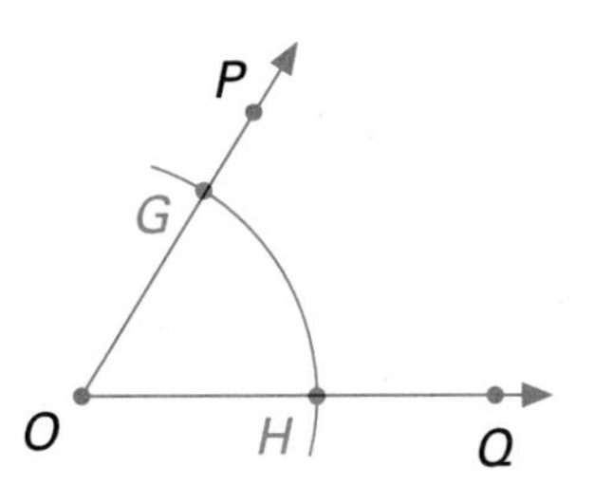

Next, use either the same or an enlarged compass opening to draw two arcs, one with center G, the other with center H.

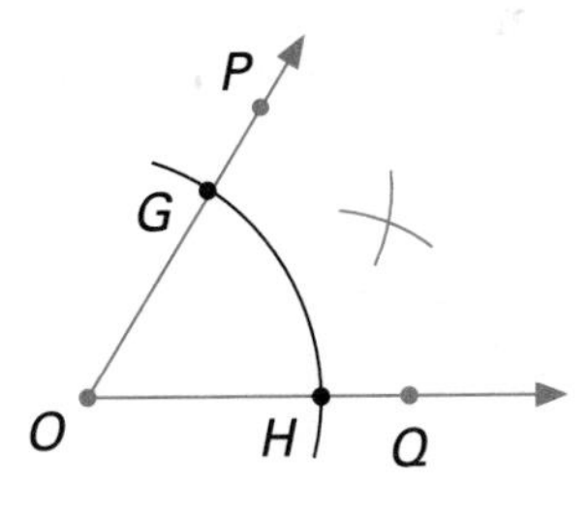

Label the point where the two arcs intersect T. Draw $\overrightarrow{OT}$.

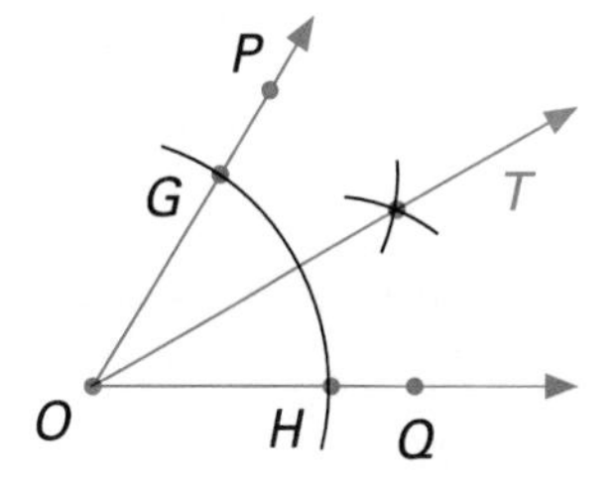

Result: $\overrightarrow{OT}$ is the bisector of $\angle POQ$.

Postulate 6 Every angle, except a straight angle, has exactly one bisector.

EXAMPLE 3 Assume that $\overrightarrow{PS}$ bisects $\angle RPQ$, $m\ \angle RPS = 5x + 4$, $m\ \angle SPQ = 7x - 10$. Find $m\ \angle RPS$.

Solution

$$m\ \angle RPS = m\ \angle SPQ \quad (\overrightarrow{PS} \text{ is a bisector.})$$
$$5x + 4 = 7x - 10$$
$$-2x + 4 = -10$$
$$-2x = -14$$
$$x = 7$$

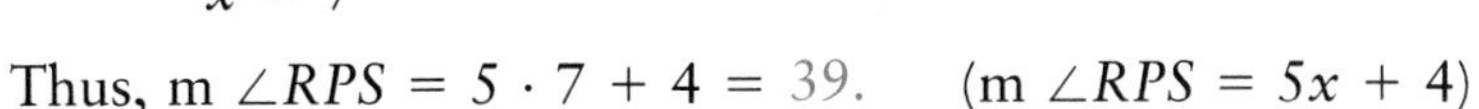

Thus, $m\ \angle RPS = 5 \cdot 7 + 4 = 39$. $\quad (m\ \angle RPS = 5x + 4)$

Classroom Exercises

Ex. 1–8

1. $m\ \angle 1 = 35$, $m\ \angle 2 = 40$. Find $m\ \angle STU$.
2. $m\ \angle 1 = 20$, $m\ \angle STU = 75$. Find $m\ \angle 2$.
3. $m\ \angle 2 = 32$, $m\ \angle STU = 68$. Find $m\ \angle 1$.
4. Name two adjacent angles.
5. Name two non-adjacent angles.

6. $\overrightarrow{TV}$ bisects $\angle STU$, m $\angle 1 = 37$. Find m $\angle STU$.
7. $\overrightarrow{TV}$ bisects $\angle STU$, m $\angle 2 = 37$. Find m $\angle STU$.
8. $\overrightarrow{TV}$ bisects $\angle STU$. Name two angles that are not congruent.

Written Exercises

State whether $\angle 1$ and $\angle 2$ are adjacent. If not, tell why not.

1.
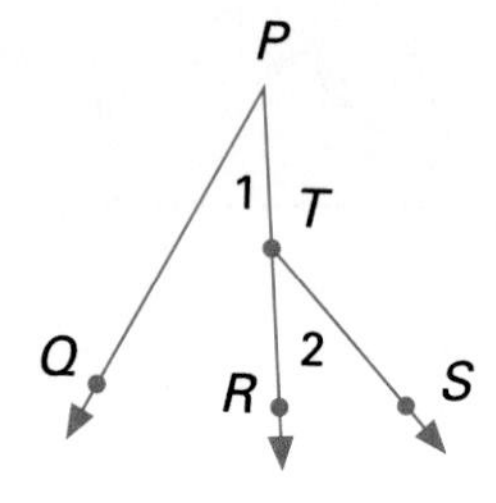

2.
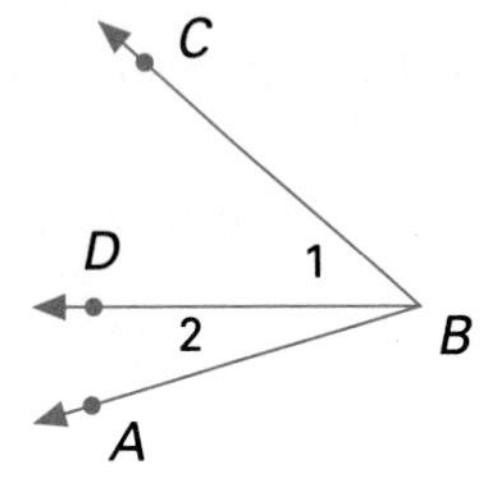

3.
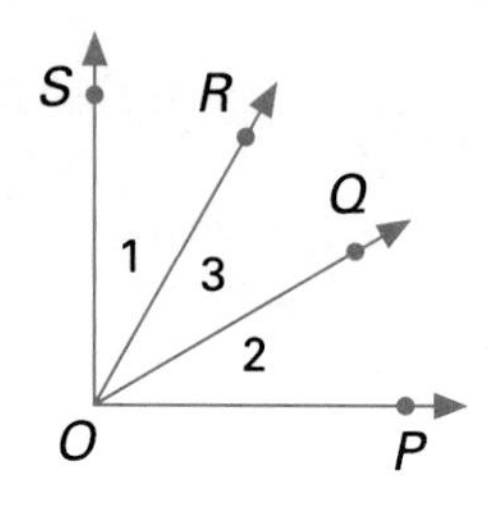

Find the indicated measure.

4. m $\angle POR = 80$, m $\angle 2 = 25$ Find m $\angle 1$.
5. m $\angle QOS = 100$, m $\angle 3 = 65$ Find m $\angle 2$.
6. m $\angle POR = 125$, m $\angle 3 = 35$ Find m $\angle POS$.
7. m $\angle QOS = 49$, m $\angle 1 = 23$ Find m $\angle POS$.

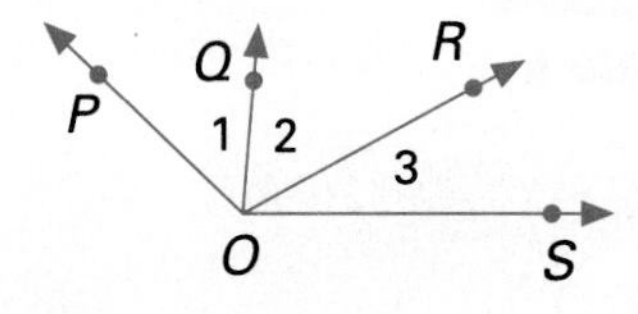

In the figure below, m $\angle FDB = 85$, m $\angle DBA = 100$, m $\angle 2 = 30$, m $\angle 3 = 50$, m $\angle 4 = 25$, and m $\angle 5 = 40$. Find each measure.

8. m $\angle EDC$
9. m $\angle 6$
10. m $\angle EBC$
11. m $\angle 1$
12. m $\angle FDC$

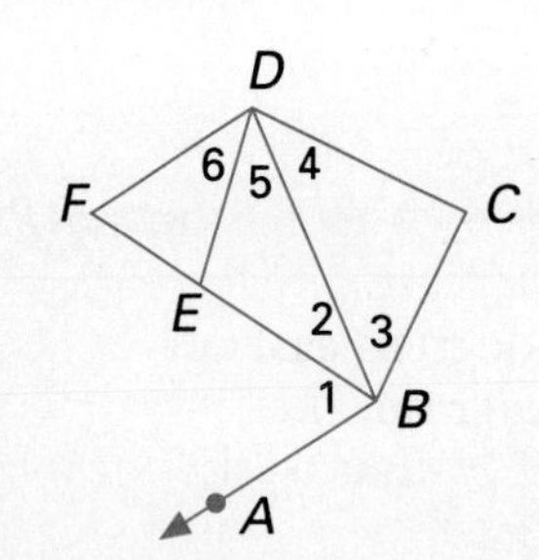

13. Draw an angle with a measure of 70. Construct an angle with $\frac{1}{2}$ of this measure.

14. Draw an angle with a measure of 130. Construct an angle with $\frac{1}{4}$ of this measure. (HINT: $\frac{1}{4} = \frac{1}{2} \cdot \frac{1}{2}$)

15. Given: m $\angle 1 = \frac{2}{5}$ (m $\angle 2$), m $\angle PQR = 49$. Find m $\angle 1$.
16. Given: m $\angle 2 = \frac{3}{5}$ (m $\angle 1$), m $\angle PQR = 64$. Find m $\angle 2$.
17. Given: m $\angle 1$ is twice m $\angle 2$, m $\angle PQR = 78$. Find m $\angle 1$.

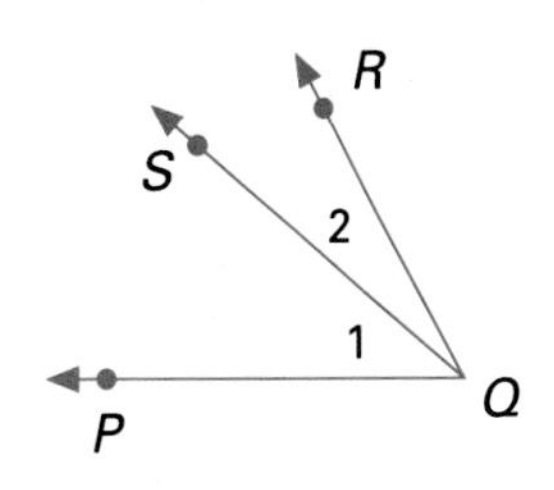

For Exercises 18–20, $\overrightarrow{QS}$ bisects $\angle PQR$.

18. Given: m $\angle 1 = 4x + 30$, m $\angle 2 = 2x + 40$. Find m $\angle 1$.
19. Given: m $\angle 1 = 42 - 2x$, m $\angle 2 = 30 + 4x$. Find m $\angle PQR$.
20. Given: m $\angle 1 = 6x + 18$, m $\angle 2 = 9x$. Find m $\angle 1$.

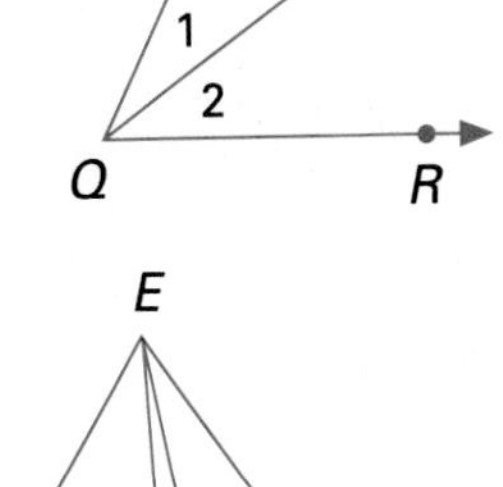

21. $\angle AOT$ and $\angle TOG$ are adjacent angles, m $\angle AOG = 100$, and m $\angle AOT = 3$(m $\angle TOG$). Draw the figure and find m $\angle TOG$.
22. In the figure at the right, $\overrightarrow{EB}$ bisects $\angle AED$, m $\angle AEB = 3x + 12$, m $\angle BED = x + 32$, and m $\angle BEC = \frac{1}{6}$m $\angle CED$. Find m $\angle AEC$.

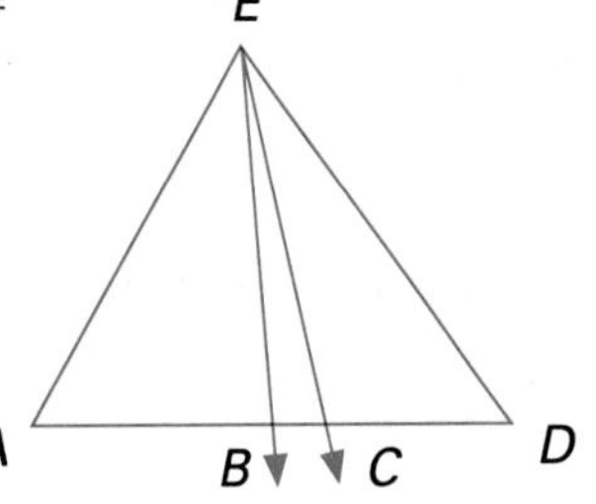

Mixed Review

Define the following. *1.4, 1.5*

1. acute angle
2. right angle
3. obtuse angle
4. straight angle
5. congruent angles
6. adjacent angles

Application: *Surveying*

Surveyors provide accurate measurements of both distance and direction. *Bearing* is used to indicate direction. Bearing is stated as the number of degrees east or west of the north or south line. In the diagram, the bearing from Town A to Town B is N49°E, that is, 49° east of due north. The bearing from Town B to Town C is S20°E, or 20° east of due south.

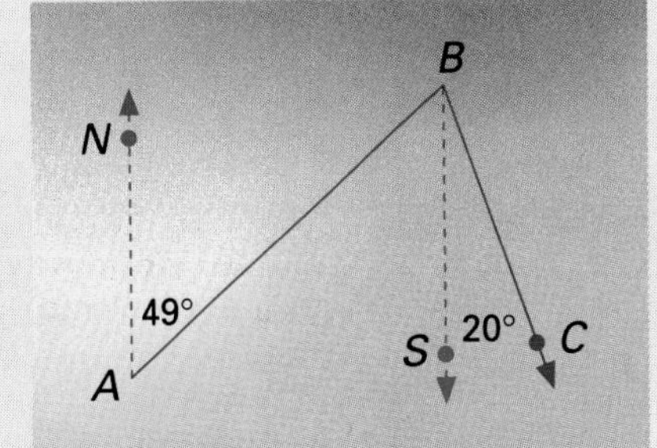

1. Using a protractor, find the bearing from Town B to Town A.
2. Explain how to find the bearing from Town A to Town C, and then find it. Use a straightedge and protractor. Find m $\angle NAC$.

1.6 Supplementary and Complementary Angles

Objectives

To find the measures of two supplementary angles
To find the measures of two complementary angles
To solve problems involving supplementary and complementary angles

A straight angle has a measure of 180. The sides are rays that together form a line. In the diagram, O is between A and B. Notice that the rays extend in *opposite* directions.

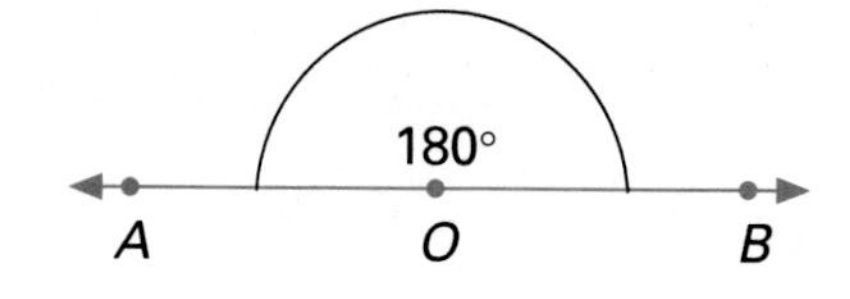

Definition

If point O is between points A and B on $\overleftrightarrow{AB}$, then $\overrightarrow{OA}$ and $\overrightarrow{OB}$ are called **opposite rays.** These rays are sides of $\angle AOB$, a straight angle.

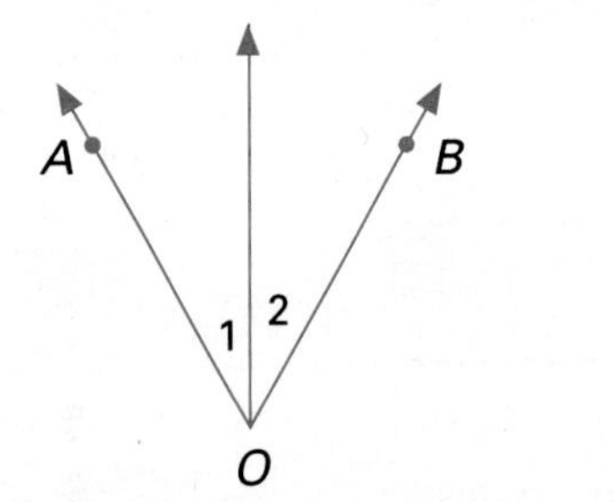

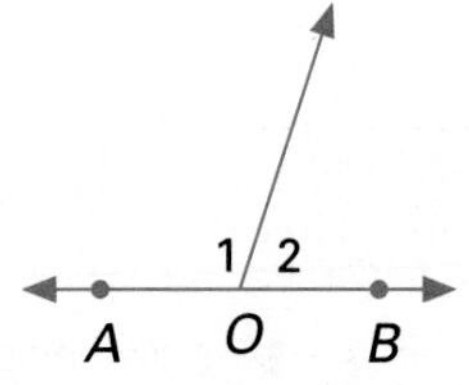

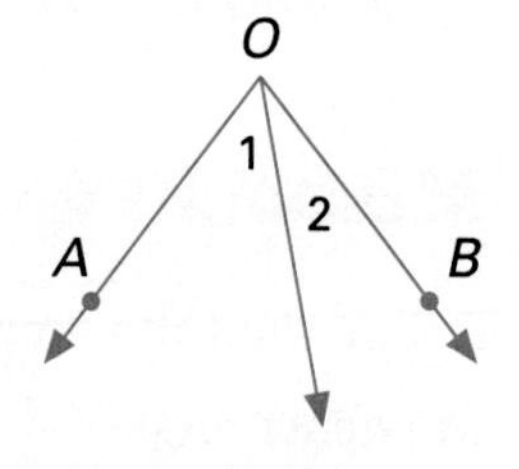

Each figure above shows a pair of adjacent angles, $\angle 1$ and $\angle 2$. In the second figure, the outer rays $\overrightarrow{OA}$ and $\overrightarrow{OB}$ form a line. In this case, $\angle AOB$ is a straight angle and the outer rays are opposite rays.

Definition

Two adjacent angles whose outer rays are opposite rays are called a **linear pair** of angles.

Postulate 7

If the outer rays of two adjacent angles form a straight angle, then the sum of the measures of the angles is 180.

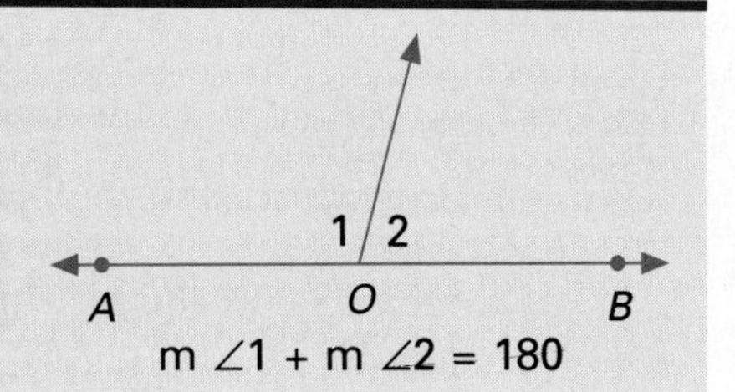

Definition

Two angles are **supplementary** if the sum of their measures is 180. Each angle is called a **supplement** of the other.

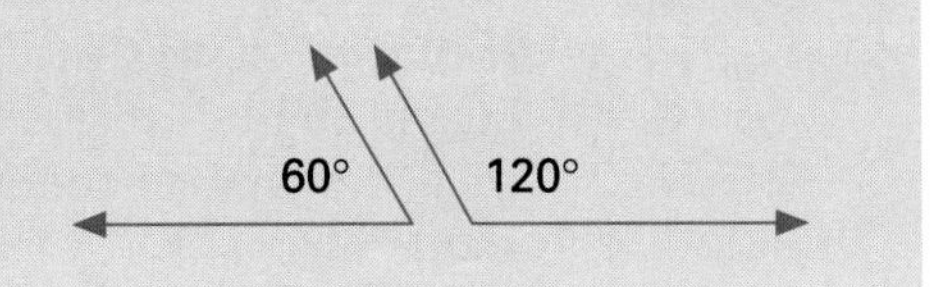

If supplementary angles are adjacent, then their outer rays form a straight angle. Postulate 7 may be restated as follows: If the outer rays of two adjacent angles form a straight angle, then the angles are supplementary. In this book, assume that angles that appear to be straight angles are straight angles.

EXAMPLE 1 Are the pairs of angles supplementary? Explain why or why not.

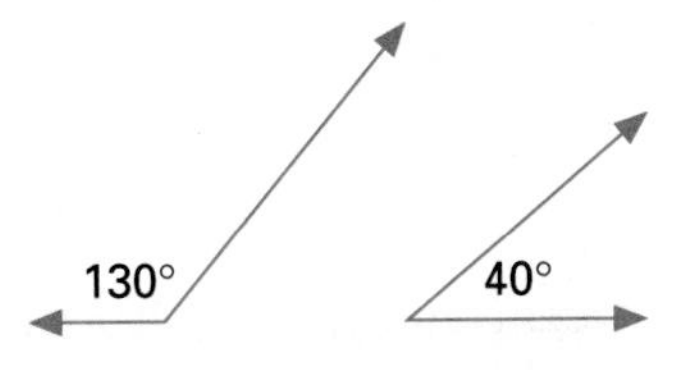

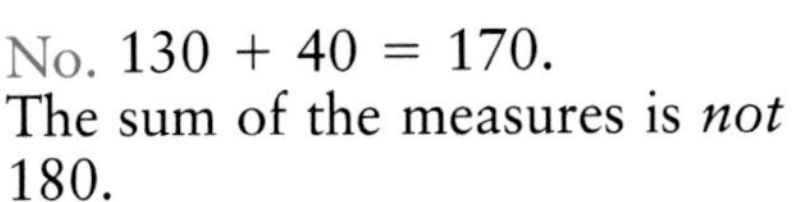

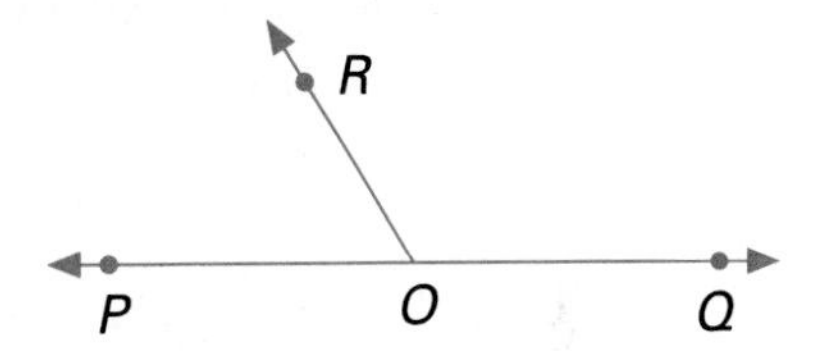

Solution No. 130 + 40 = 170. The sum of the measures is *not* 180.

Yes. The outer rays form a straight angle. The sum of the measures is 180 by Postulate 7.

Definition

Two lines are **perpendicular** if they intersect to form a right angle.

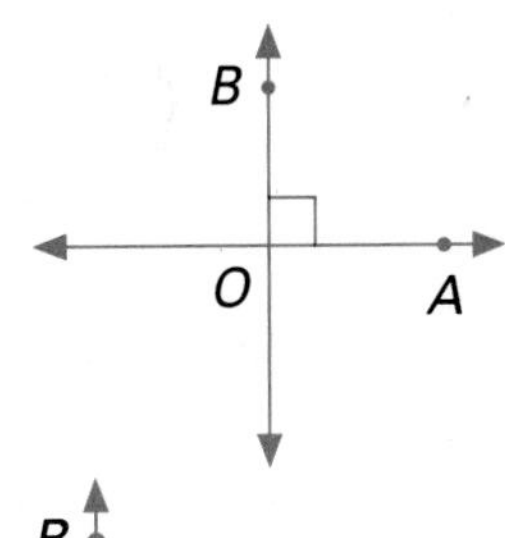

The symbol for perpendicular is $\perp$. Thus, $\overleftrightarrow{OB} \perp \overleftrightarrow{OA}$ means "$\overleftrightarrow{OB}$ is perpendicular to $\overleftrightarrow{OA}$." Since segments and rays are parts of lines, then intersecting segments and rays are perpendicular if the lines that contain them are perpendicular.

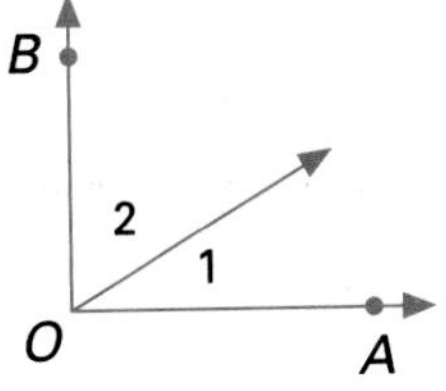

Sometimes, adjacent angles have their outer rays perpendicular. In the figure at the right, $\overrightarrow{OA} \perp \overrightarrow{OB}$ and $\angle 1$ and $\angle 2$ are acute angles. So $\angle AOB$ is a right angle with m $\angle AOB = 90$. By the Angle Addition Postulate, m $\angle 1$ + m $\angle 2$ = m $\angle AOB$. Therefore, m $\angle 1$ + m $\angle 2 = 90$, by substituting 90 for m $\angle AOB$.

Consider the process of reasoning used above. Such a logical sequence of statements leading to a conclusion is called a **proof.** A statement that has been proved true, not just taken for granted, is called a **theorem.**

Theorem 1.1 If the outer rays of two acute adjacent angles are *perpendicular*, then the sum of the measures of the angles is 90.

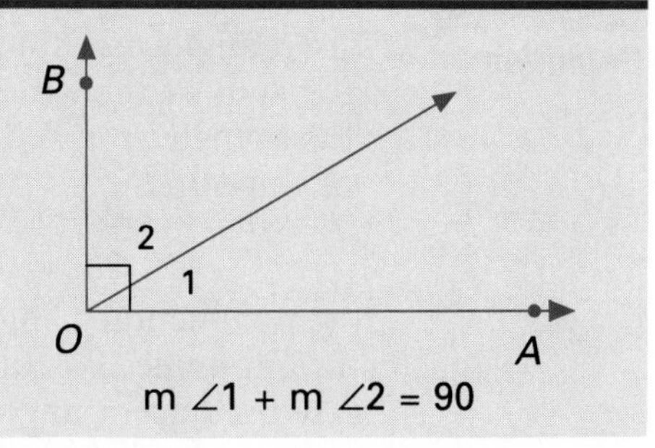

Definition Two angles are **complementary** if the sum of their measures is 90. Each angle is called a **complement** of the other.

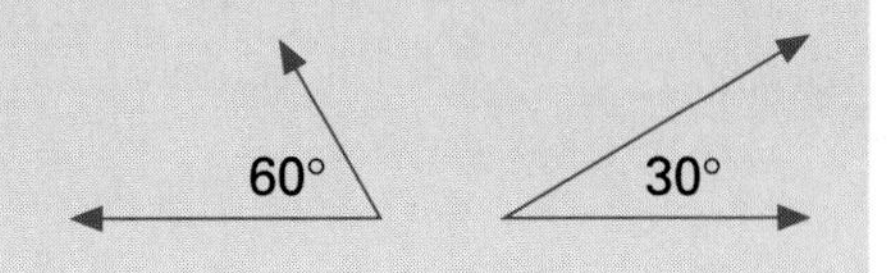

Theorem 1.1 may now be stated as follows:
If the outer rays of two acute adjacent angles are perpendicular, then the angles are *complementary.*

EXAMPLE 2 Is $\angle 1$ complementary to $\angle 2$? Explain why or why not.

Given: m $\angle 1 = 45$, m $\angle 2 = 55$ Given: $\overrightarrow{ED} \perp \overrightarrow{EF}$

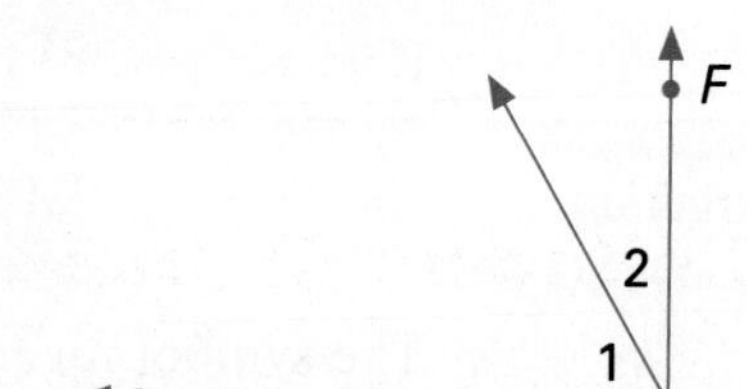

Solution No. $45 + 55 = 100$. The sum of the measures is *not* 90.

Yes. The outer rays are perpendicular. The sum of the measures is 90 by Theorem 1.1.

EXAMPLE 3 Find the measure of a complement and supplement, if possible.

Angle	Measure	Measure of Complement	Measure of Supplement
A	35	55	145
B	97	Not possible	83
C	x	$90 - x$ for $\angle C$ acute	$180 - x$ for $\angle C$ not straight
D	$3x - 20$	$90 - (3x - 20) =$ $110 - 3x$ for $\angle D$ acute	$180 - (3x - 20) =$ $200 - 3x$ for $\angle D$ not straight

EXAMPLE 4 The measure of an angle is 60 less than twice the measure of its complement. Find the measure of the angles.

As you go about solving the problem, proceed in a step-by-step fashion. Determine exactly what you are looking for—the measures of two complementary angles. Then:

(1) Represent the *data*. Let x = measure of complement.
$2x - 60$ = measure of the angle.
(60 less than twice x)

(2) Write the *equation*. $x + (2x - 60) = 90$
Sum of measures of complementary angles is 90.

(3) *Solve* the equation. $3x - 60 = 90$
$3x = 150$
$x = 50$ (measure of the complement)
Find the measure of the angle.
$2x - 60 = 2 \cdot 50 - 60$
$= 100 - 60 = 40$

(4) Check *solutions*. Are the angles complementary?
$40 + 50 = 90$ ✔
Is angle measure 60 less than twice measure of complement?
$40 = 2 \cdot 50 - 60$
$40 = 100 - 60$
$40 = 40$ ✔

(5) Label the *answer*. Thus, the measures of the angles are 40 and 50.

Classroom Exercises

Find the measure of a complement of an angle with the given measure.

1. 42 **2.** 65 **3.** 9 **4.** 25 **5.** 72 **6.** 16

Find the measure of a supplement of an angle whose measure is given.

7. 120 **8.** 80 **9.** 135 **10.** 70 **11.** 104 **12.** 139

Written Exercises

Are the indicated pairs of angles supplementary, complementary, or neither? Explain why or why not.

1.

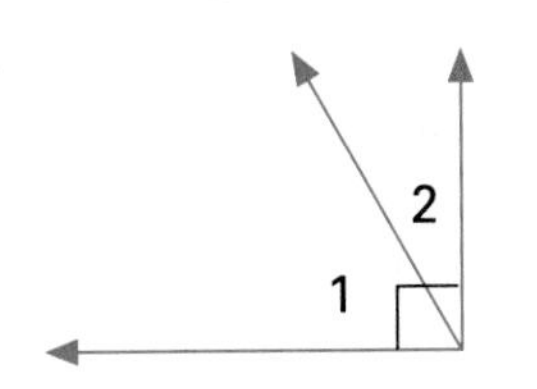

2.

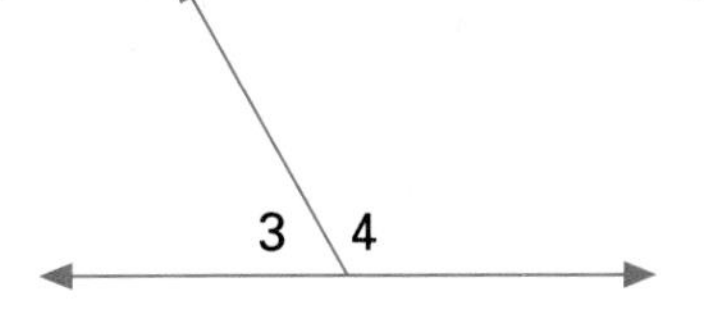

3.

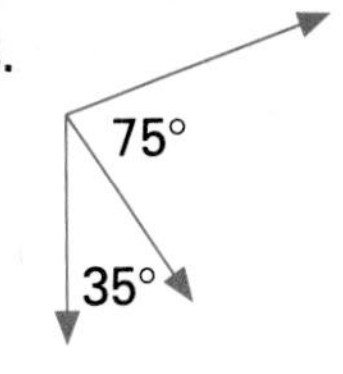

Find the measure of a complement and a supplement, if possible, of an angle with the given measure.

4. 32 **5.** 74 **6.** 19 **7.** 99 **8.** 136 **9.** 180

10. t **11.** 31.7 **12.** $66\frac{1}{2}$ **13.** $m - 3$ **14.** $8 - x$ **15.** $2x - 10$

16. The figure at the right shows pieces of a square-cornered picture frame. What must be the relationship between $\angle 1$ and $\angle 2$? Why?

17. Is it reasonable to expect that a supplement of an obtuse angle can be a right angle? Why?

18. Is it reasonable to expect to find a pair of complementary angles that are also supplementary? Why?

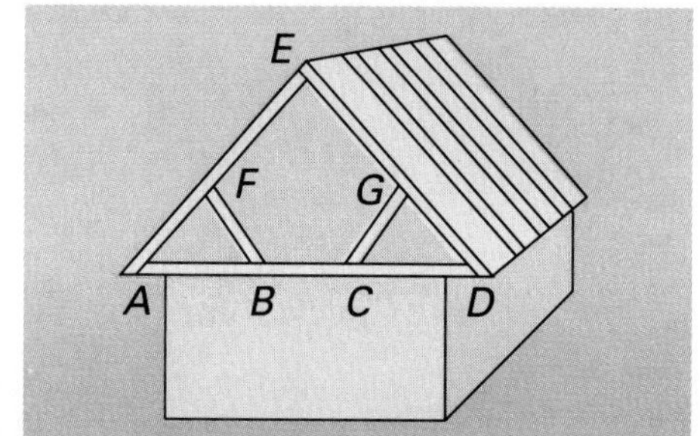

19. The figure at the right shows rafters $\overline{AE}$ and $\overline{ED}$ of a roof with support braces $\overline{BF}$ and $\overline{CG}$. What relationship must exist between $\angle BFA$ and $\angle BFE$?

20. The measure of an angle is 50 more than that of its complement. Find the measure of each angle.

21. The measure of an angle is 70 more than that of its supplement. Find the measure of each angle.

22. The measure of an angle is 3 times that of its supplement. Find the measure of each angle.

23. The measure of an angle is equal to that of its supplement. Find the measure of each angle.

24. The measure of an angle is 30 less than 5 times the measure of its complement. Find the measure of each angle.

25. The measure of an angle is 20 more than 3 times the measure of its supplement. Find the measure of each angle.

26. The measure of an angle is 26 less than 3 times the measure of its complement. Find the measure of each angle.

27. The measure of an angle is $\frac{1}{5}$ the measure of its complement. Find the measure of each angle.

28. The measure of an angle is $\frac{2}{3}$ the measure of its supplement. Find the measure of each angle.

29. Four times the measure of the complement of an angle is 12 more than twice the difference of the measures of its supplement and its complement. Find the measure of each angle.

30. Can the measure of the complement of an angle be $\frac{1}{2}$ the measure of the supplement of the angle? Why or why not?

31. Prove: If complementary angles are adjacent, then their outer rays are perpendicular.

Mixed Review

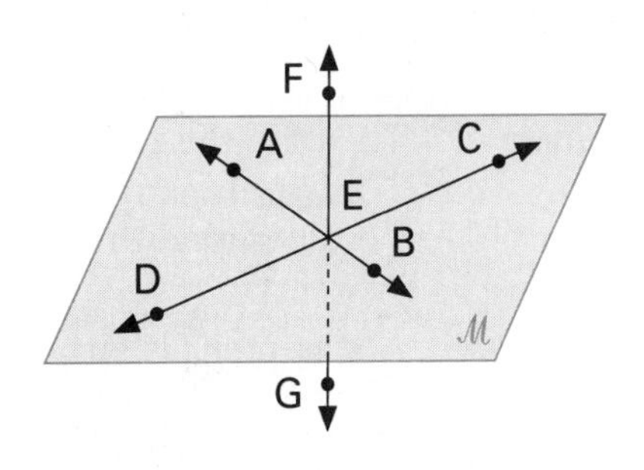

1. Name the intersection of $\overleftrightarrow{FG}$ and plane $\mathcal{M}$. *1.1*
2. Name a point that is collinear with points C and E. *1.1*
3. Name 4 pairs of adjacent angles not contained in $\mathcal{M}$. *1.5*
4. Name 4 noncoplanar points. *1.1*

For the given coordinates of A and B, find AB. *1.2*

5. $A{:}-3$; $B = 8$
6. $A{:}2\frac{1}{2}$; $B{:}-\frac{3}{4}$

Find the coordinate of the midpoint, M of $\overline{AB}$ whose endpoints have the given coordinates. *1.3*

7. 2 and 7
8. $-3\frac{1}{2}$ and 6

Application: *Astronomical Units*

Astronomers measure distance within the solar system in astronomical units (A.U.). By the Ruler Postulate, the point 0 is assigned to the Sun and the point 1 is assigned to the Earth.

One astronomical unit, therefore, is the mean distance from the Sun to Earth. The mean distances of the other planets in astronomical units are shown in the table below.

Sun	0
Mercury	0.39
Venus	0.72
Earth	1
Mars	1.5
Jupiter	5.2
Saturn	9.5
Uranus	19.2
Neptune	30.1
Pluto	39.4

1. If one A.U. equals 92,960,000 mi, what is the distance in miles from the Sun to Mars?
2. At its closest, Venus is about 26 million miles from Earth. But can it also be as far as 159 million miles from Earth? Why?
3. What is the distance from the Earth to Saturn?

1.7 Logic: Conjunction and Disjunction

Objectives

To write the conjunction (disjunction) of two statements
To determine if a conjunction (disjunction) is true
To determine the truth table of a conjunction (disjunction)

A statement is either true or false, but never both. The statement "$5 = 9 - 4$" is true. The statement "A line has an endpoint" is false. Two statements may be combined into a single statement connected by the word *and*.

Definition

If p and q are statements, the statement "p *and* q" is called the **conjunction** of p and q and written as $p \wedge q$. $p \wedge q$ is true when p and q are both true. $p \wedge q$ is false when at least one of the statements, p or q, is false.

Statement p: $5 = 9 - 4$ Statement q: A line has an endpoint.
Conjunction: $5 = 9 - 4$ *and* A line has an endpoint.

p $\wedge$ q

The conjunction "$5 = 9 - 4$ *and* A line has an endpoint" is false because one of the two statements is false ($p \wedge q$ is false).

The conjunction "Bread is a food *and* Blue is a color" is true because both parts are true. ($p \wedge q$ is true).

There are four different combinations of truth values for the two statements. The truth table below summarizes these combinations.

p	q	$p \wedge q$	
T	T	T	⟵ Both statements are true.
T	F	F	⟵ Only one statement is true.
F	T	F	
F	F	F	⟵ Both statements are false.

EXAMPLE 1 Determine whether each of the following conjunctions is true.

a. A segment is a set of points (T) *and* $9 > 2$ (T)

Solution The conjunction is true since both parts are true.

b. $7 = 2 \cdot 3 + 1$ (T) *and* $-4 > 2$ (F)

Solution The conjunction is false since one part is false.

Two statements may also be combined into a single statement by the word *or*. In everyday language, this word is often used in the *exclusive* sense. Consider a parent saying to a child, "Tonight, you may go to the movies or you may go bowling." The parent means *either . . . or*: only one of the two choices, not both, may take place. In mathematics, however, the *inclusive* meaning of "or" is used. This means that if one or the other or *both* statements are true, then the single statement combining them by the word "or" is true.

Definition If p and q are statements, the statement "p *or* q" is called the **disjunction** of p and q and written as $p \vee q$. $p \vee q$ is true when at least one of the statements, p or q, is true. $p \vee q$ is false when p and q are both false.

The truth table for a disjunction is given below.

p	q	$p \vee q$	
T	T	T	← Both statements are true.
T	F	T	← At least one statement is true.
F	T	T	
F	F	F	← Both statements are false.

EXAMPLE 2 Determine if each of the following disjunctions is true.

a. A ray has two endpoints *or* A plane contains a finite number of points.

Solution (F) (F)

Thus, the disjunction is false since both of the statements are false.

b. Cows can fly *or* $2 + 3 = 5$.

Solution  (T)

Thus, the disjunction is true since at least one of the statements is true.

Classroom Exercises

Indicate whether the statement is true or false.

1. $3 = 1 + 2$ *or* $3 = 1 \cdot 2$

2. $3 = 1 + 2$ *and* $3 = 1 \cdot 2$

3. $6 = 4 \cdot 2$ *or* $12 = 7 \cdot 5$

4. $6 = 4 + 2$ *and* $12 = 7 + 5$

5. A disjunction is true if exactly one of its two parts is false.

6. A conjunction is true if one of its two parts is false.

7. If one part of a disjunction is true, then the disjunction is true.

8. A conjunction is false if one of its two parts is false.

Written Exercises

Write the conjunction and disjunction of the pair of statements. Determine whether the conjunction (disjunction) is true or false.

1. A potato is a vegetable. West Berlin is a country.
2. A line consists of a finite number of points. $4 = 6 \cdot 3$.
3. A plane is a flat surface. $3 + 7 \leq 10$.
4. A ray contains an infinite number of points. $(-3)^2 = 6$.

Is the conjunction (disjunction) true? If it is false, explain why. (Exercises 5–8)

5. A line extends without ending in two directions *or* a line segment has two endpoints.
6. A plane is the set of all points *or* noncoplanar points are collinear.
7. $8 - 4 \cdot 2 < 0$ *or* $(-5)^2 = 25$
8. $5 \cdot 4 - 12 > 1$ *and* $-8 < -5$
9. When is $p \vee q$ false?
10. When is $p \wedge q$ true?

Write the conjunction and disjunction of the pair of statements. Determine whether the conjunction (disjunction) is true or false.

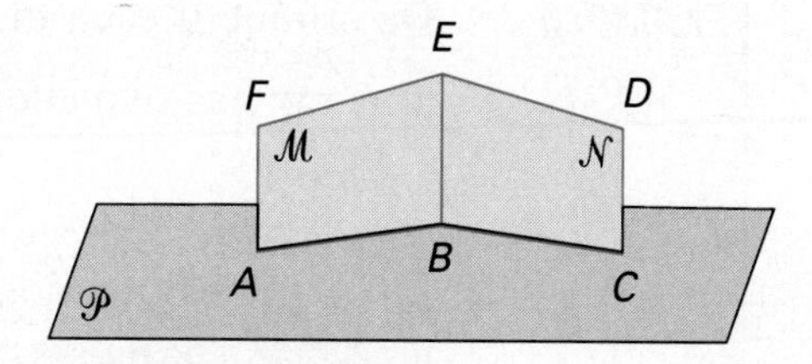

11. Plane $\mathcal{M}$ intersects plane $\mathcal{N}$ in $\overleftrightarrow{BE}$. Points F, A, and B are collinear.
12. $\overleftrightarrow{AB}$ is the intersection of planes $\mathcal{P}$ and $\mathcal{M}$. Points A, E, and B are coplanar.
13. $\overline{DE}$ lies in the plane $\mathcal{P}$. $\overleftrightarrow{BE}$ contains only two points, E and B.
14. Explain how to determine whether a given conjunction is true and whether a given disjunction is true. For which truth values of p and q are both the conjunction $(p \wedge q)$ and the disjunction $(p \vee q)$ true?
15. Construct a truth table for $(p \vee q) \vee r$. When is $(p \vee q) \vee r$ false?
16. Construct a truth table for $(p \wedge q) \wedge r$. When is $(p \wedge q) \wedge r$ true?
17. For two mathematical statements, there are four true-false combinations. How many true-false combinations are there for four mathematical statements?

Mixed Review

For the given coordinates of C and D, find CD. *1.2*

1. $C{:}-6$; $D{:}5$
2. $C{:}-8$; $D{:}-11$
3. $C{:}5\frac{1}{2}$; $D{:}9$
4. $C{:}-2.1$; $D{:}3.6$

Application: *Electric Circuits*

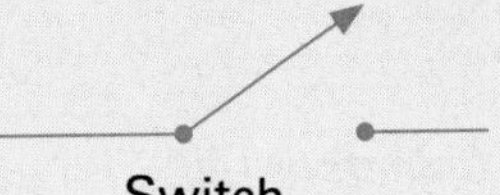

The electric circuits that make computers possible work on logic like that used in conjunction or disjunction.

Conjunctions can be used to represent two switches that are lined up in a row (series). The circuit will be closed only if both switches are closed. If either switch is open (or both are open), the current cannot flow across. If we use F for open and T for closed, we get the same truth table as for a conjunction of statements.

Switch *A*	Switch *B*	Circuit
T	T	T
T	F	F
F	T	F
F	F	F

If switches are set up side by side (parallel), then the current can flow if either one of the switches is closed. In this case, the truth table is the same as for a disjunction.

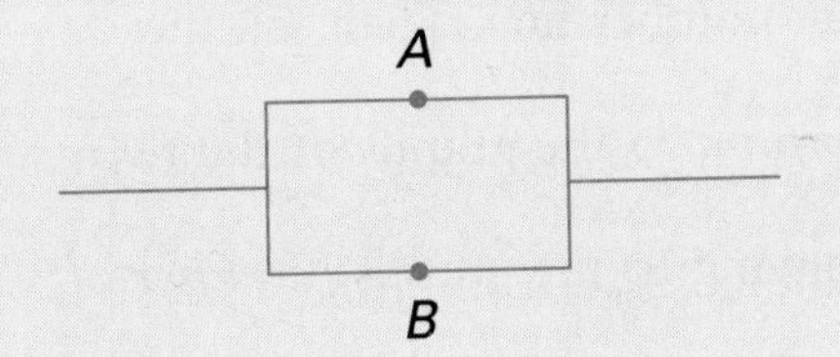

Switch *A*	Switch *B*	Circuit
T	T	T
T	F	T
F	T	T
F	F	F

Two switches in a row (series) can be represented as $A \wedge B$. Two switches side by side (parallel) can be represented as $A \vee B$.

The diagram to the right can be represented as $A \wedge (B \vee C)$.

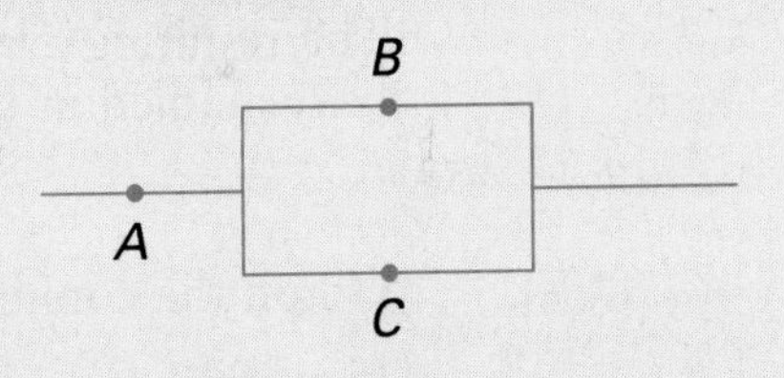

Represent each of the following in conjunction/disjunction form.

1.

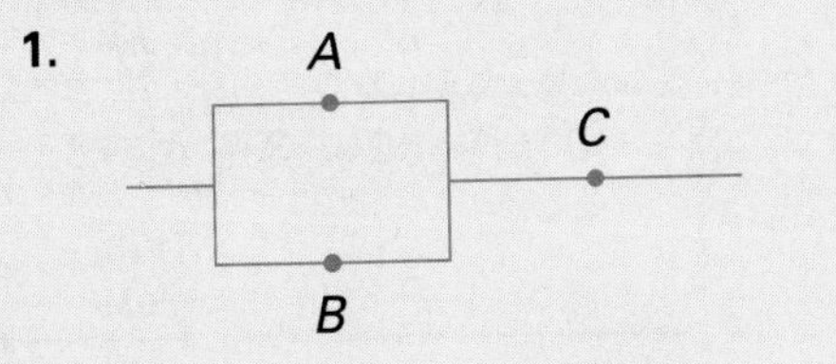

2.

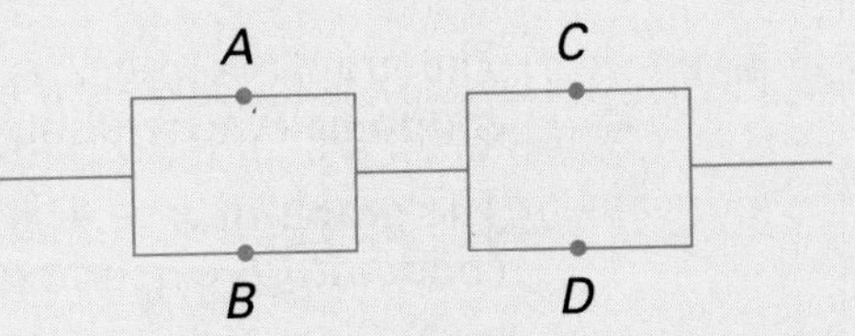

3.

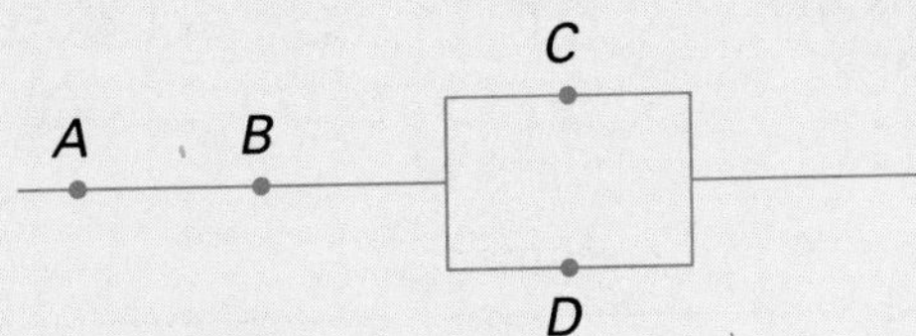

4.

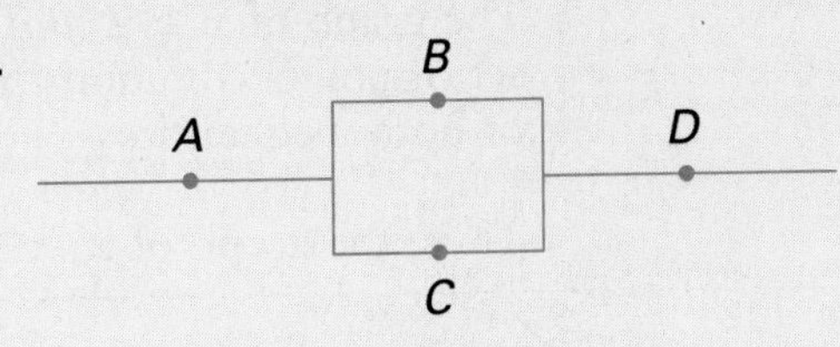

1.8 Graphing Conjunctions and Disjunctions

Objective To graph the conjunction or the disjunction of two inequalities

Inequalities can be graphed on a number line. The graph of $x \geq 5$ is shown below. A solid dot is placed at A. Then all points to the right of A are shaded.

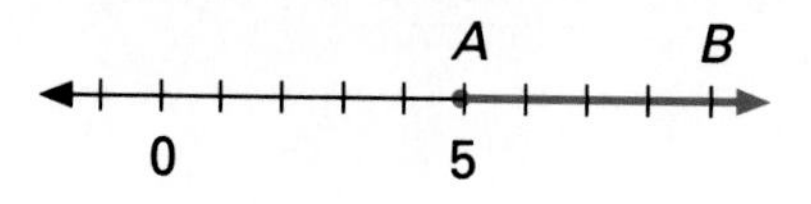

The graph is a ray: $\overrightarrow{AB}$

The graph of $x \leq 3$ is shown below. A solid dot is placed at C. Then all points to the left of C are shaded.

D C

0 3

The graph is a ray: $\overrightarrow{CD}$

The graph of the conjunction (disjunction) of two inequalities can be drawn as follows.

First, graph each inequality on a separate number line. Then, on the third number line, locate either:

(1) the conjunction—the points *common to* the graphs of the two inequalities; or
(2) the disjunction—the points belonging to the graphs of *either* of the two inequalities, or both.

The resulting graph may be one of the four geometric figures that you have studied in this chapter—point, line, segment, or ray.

EXAMPLE 1 Graph the conjunction $x \geq 2$ *and* $x \leq 4$.

Identify the resulting geometric figure, if possible.

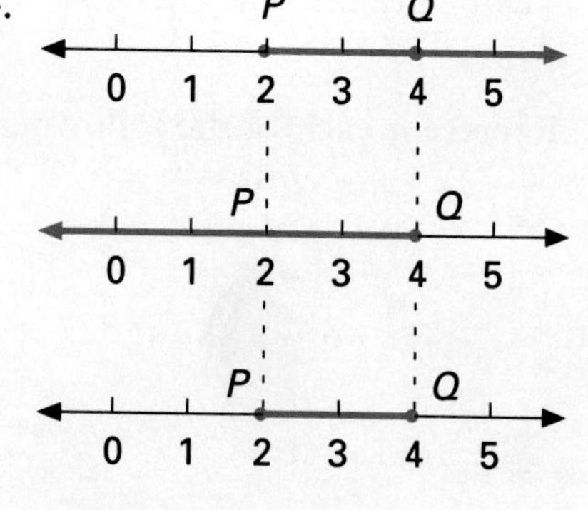

Solution The graph of $x \geq 2$ is the ray with endpoint P, extending to the right.

The graph of $x \leq 4$ is a ray with endpoint Q, extending to the left.

The graph of $x \geq 2$ *and* $x \leq 4$ is the set of points common to the graphs of both inequalities.

Thus, the graph of $x \geq 2$ *and* $x \leq 4$ is the segment $\overline{PQ}$.

EXAMPLE 2 Graph the disjunction $x \leq -1$ *or* $x \leq 3$.

Identify the resulting geometric figure, if possible.

Solution The graph of $x \leq -1$ is a ray with endpoint A, extending to the left.

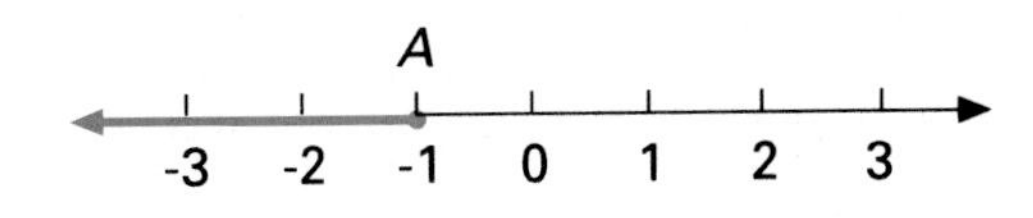

The graph of $x \leq 3$ is the ray with endpoint B, extending to the left.

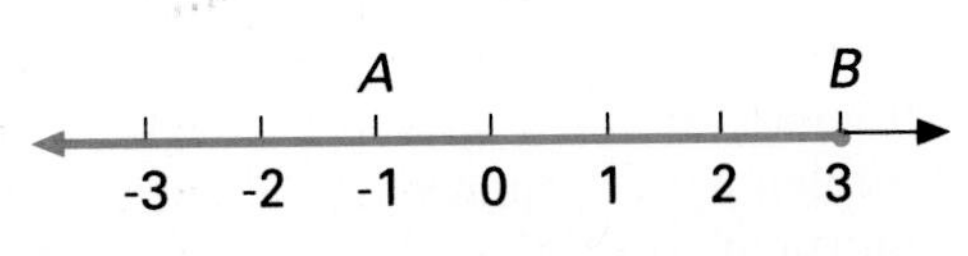

The graph of the disjunction is the set of points belonging to the graph of $x \leq -1$, or to the graph of $x \leq 3$, or to both.

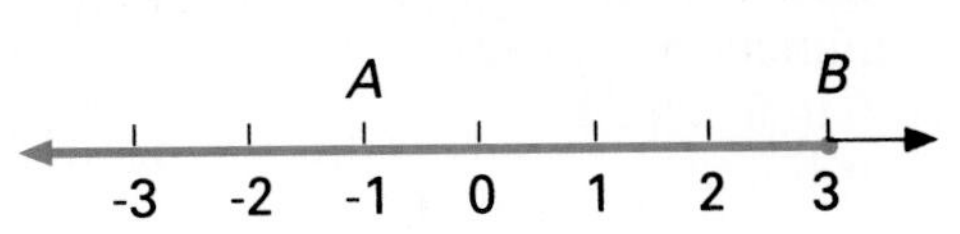

Thus, the graph of $x \leq -1$ *or* $x \leq 3$ is the ray $\overrightarrow{BA}$.

Classroom Exercises

Describe the graph of each of the following.

1. $x \leq 5$
2. $x \geq -6$
3. $x \leq -7$
4. $x \geq -2$ *and* $x \leq 2$

Written Exercises

Graph the inequality on a number line.

1. $x \geq 4$
2. $x \leq 2$
3. $x \geq -7$
4. $x \leq 0$
5. $x \geq -6$

Graph each of the following. Identify the resulting geometric figure(s).

6. $x \geq 1$ *and* $x \leq 7$
7. $x \geq 4$ *or* $x \geq 8$
8. $x \leq 5$ *or* $x \leq 12$
9. $x \leq -3$ *and* $x \geq -5$
10. $x \geq -6$ *or* $x \leq 8$
11. $x \geq -3$ *and* $x \geq -2$
12. $x \geq 3$ *and* $x \leq 3$
13. $x \geq 3$ *or* $x \leq 3$
14. $x \geq 1$ *or* $x \leq 6$
15. $x \leq 4$ *or* $x \geq 5$
16. $x \leq -3$ *or* $x \geq 5$
17. $x \geq -6$ *and* $x \leq -6$
18. $x \leq -1$ *or* $x \leq 3$
19. $x \leq -1$ *or* $x \geq -1$
20. $|x| \leq 7$
21. $[x \leq 3$ *or* $x \geq 5]$ *and* $[x \geq 3$ *or* $x \geq 5]$
22. $x \leq -2$ *and* $x \geq 4$

Mixed Review

Is the conjunction (disjunction) true? If false, explain why. 1.7

1. A ray has two endpoints *or* a plane is flat.
2. A ray has two endpoints *and* a plane is flat.

Chapter 1 Review

Key Terms

acute angle (p. 16)
adjacent angles (p. 21)
angle (p. 15)
angle bisector (p. 23)
bisector (p. 10)
collinear (p. 2)
complementary angles (p. 28)
congruent (p. 10)
conjunction (p. 32)
coplanar (p. 2)
coordinate (p. 5)
disjunction (p. 33)
distance (p. 5)
equidistant (p. 11)
intersection (p. 3)
line (p. 1)
linear pair (p. 26)
midpoint (p. 10)
noncollinear (p. 2)
noncoplanar (p. 2)
obtuse angle (p. 16)
perpendicular (p. 27)
plane (p. 1)
point (p. 1)
postulate (p. 5)
proof (p. 27)
ray (p. 7)
right angle (p. 16)
segment (p. 6)
space (p. 2)
straight angle (p. 16)
supplementary angles (p. 27)
theorem (p. 27)
vertex (p. 15)

Key Ideas and Review Exercises

1.1 A line can be named by any two points that lie on it.

1.2 To find the **distance** between two points on a number line, find the absolute value of the difference of their coordinates.

1. Give six possible names for the line shown.

2. Find AB for A:-7 and B:5.

3. Use the Segment Addition Postulate to write an equation.

4. $AB = \frac{2}{5}BC$ and $AC = 21$. Find BC.

1.3 The **midpoint** of a segment divides the segment into two congruent segments.

To find the **coordinate of the midpoint** of a segment, find the arithmetic mean of the coordinates of the endpoints.

A **bisector** of a segment is a line, ray, segment, or plane that intersects the segment at its midpoint.

5. Find the coordinate of the midpoint of $\overline{AB}$ for A:-8 and B:12.

6. C is the midpoint of $\overline{AB}$. Find the coordinates of A and B for A:$(p - 8)$, B:$(3p + 4)$, and C:6.

Write a statement of congruence and an equation for the figure, using the Segment Addition Postulate.

7. Y is the midpoint of $\overline{XZ}$.

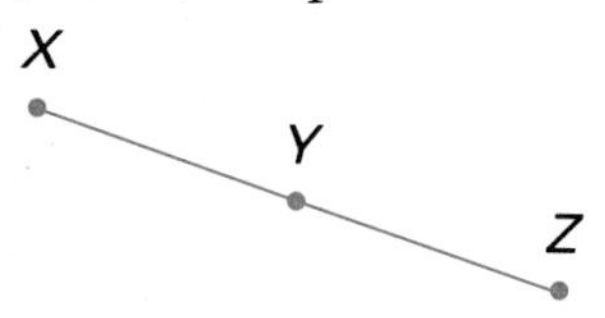

8. $\overleftrightarrow{SU}$ bisects $\overline{VW}$.

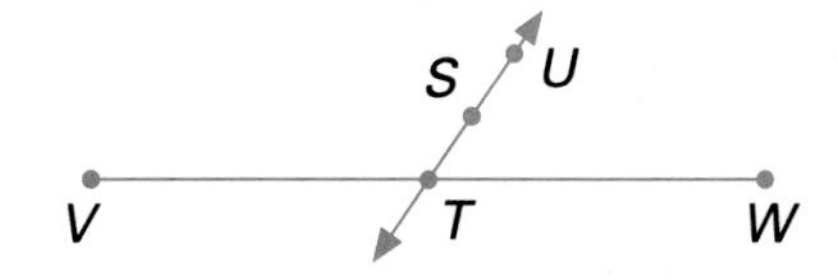

9. Draw a segment. Locate its midpoint.

1.4, 1.5 To **classify angles,** refer to the following:

acute	right	obtuse	straight
$0 < m < 90$	$m = 90$	$90 < m < 180$	$m = 180$

Given: m $\angle PRS = 130$, m $\angle 2 = 20$, m $\angle URS = 40$

10. Find m $\angle 3$. Classify the angle.
11. Find m $\angle PRT$. Classify the angle.
12. Name an angle bisector.
13. Name a pair of adjacent angles.
14. Draw an obtuse angle. Construct its bisector.

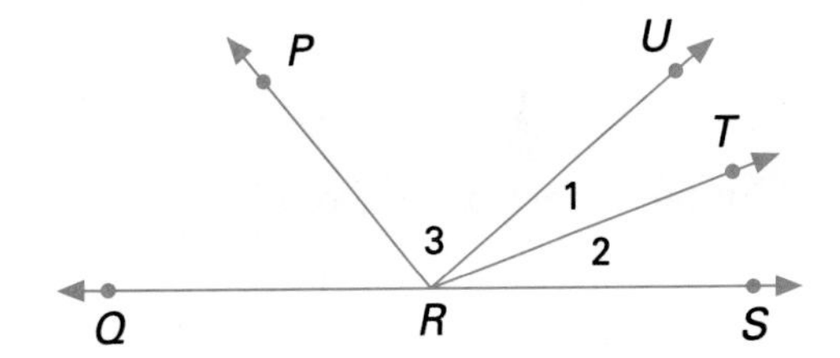

1.6 Two angles are **supplementary** if the sum of their measures is 180.
Two angles are **complementary** if the sum of their measures is 90.

Find the measure of a complement and a supplement, if possible, of the angle with the given measure.

15. 37 **16.** 117 **17.** 2 **18.** $3x - 30$

19. The measure of an angle is 10 more than 7 times the measure of its complement. Find the measure of the angle.

1.7 $A \wedge B$ (**conjunction**) is true when A and B are both true. It is false otherwise.
$A \vee B$ (**disjunction**) is false when A and B are both false. It is true otherwise.

20. Write the conjunction and disjunction. Determine if each is true.
Collinear points are noncoplanar. Adjacent angles have points in common.

1.8 The **graph of the conjunction** of two inequalities is drawn by locating the points common to the graphs of the two inequalities.
The **graph of the disjunction** of two inequalities is drawn by locating the points belonging to the graphs of either inequality or both.

21. **a.** Graph the disjunction $x \leq 4$ *or* $x \leq 0$.
b. Graph the conjunction $x \geq -2$ *and* $x \leq 6$.

Chapter 1 Test

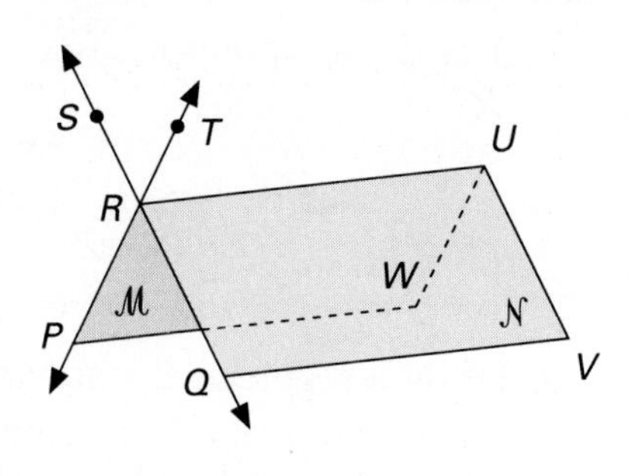

1. Identify the intersection of $\overleftrightarrow{PT}$ and $\overleftrightarrow{SQ}$.
2. Identify the intersection of planes $\mathcal{M}$ and $\mathcal{N}$.
3. Name three collinear points.
4. Name four coplanar points.
5. Find AB for $A{:}-8$ and $B{:}6$.
6. Find the coordinate of the midpoint of $\overline{AB}$ for $A{:}-4$ and $B{:}12$.
7. $AC = \frac{4}{5}CB$ and $AB = 27$. Find CB.
8. C is the midpoint of $\overline{AB}$. Find the coordinates of A and B for $A{:}(16 - 2p)$, $B{:}(p + 12)$, and $C{:}12$.

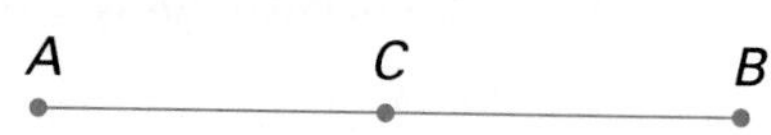

Draw an angle and label it $\angle ABC$. (Exercises 9–10)

9. Construct an angle congruent to $\angle ABC$.
10. Construct the bisector of $\angle ABC$.
11. Draw a segment. Locate the midpoint of the segment.

Use the diagram for Exercises 12–15.

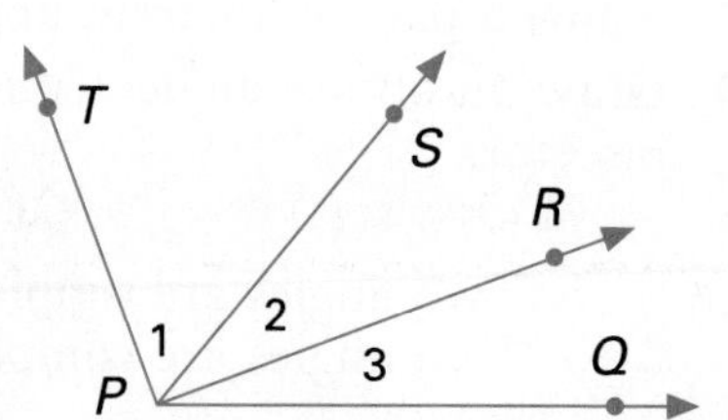

12. By the Angle Addition Postulate, $m\angle 2 + m\angle 3 =$ ______

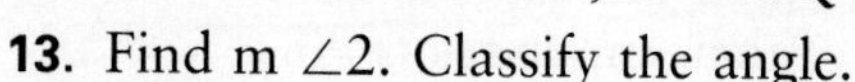

Given: $m\angle TPR = 110$, $m\angle SPQ = 50$, $m\angle 3 = 20$

13. Find $m\angle 2$. Classify the angle.
14. Find $m\angle 1$. Classify the angle.

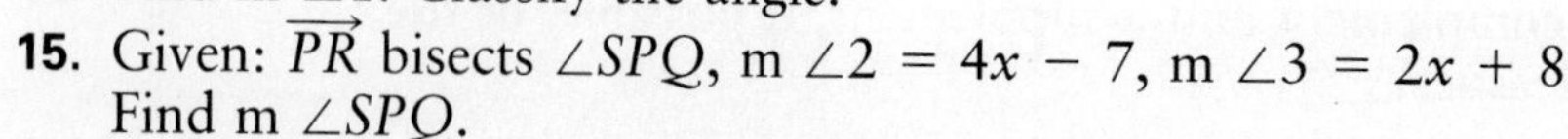

15. Given: $\overrightarrow{PR}$ bisects $\angle SPQ$, $m\angle 2 = 4x - 7$, $m\angle 3 = 2x + 8$. Find $m\angle SPQ$.

Find the measure of a complement and a supplement, if possible, of the angle with the given measure.

16. 42
17. 150
18. $2x - 10$
19. The measure of an angle is 10 less than 3 times the measure of its complement. Find the measure of the angles.
20. * $\angle PTK$ and $\angle KTG$ are adjacent angles. $m\angle PTK$ is 60 more than $\frac{2}{3}m\angle KTG$. $m\angle PTG = 110$. Draw the figure and find $m\angle PTK$.
21. Write the conjunction and the disjunction for statements (a) and (b). Determine if each is true or false.
    (a) A point has no length.
    (b) The sum of the measures of two complementary angles is 180.

Graph the conjunction or disjunction. Identify the resulting geometric figure.

22. $x \geq 4$ *and* $x \leq 9$
23. $x \geq 5$ *or* $x \geq 8$

College Prep Test

Strategy for Achievement in Testing

In each item, you are to compare a quantity in Column 1 with a quantity in Column 2. Write the letter of the correct answer from these choices:

A—The quantity in Column 1 is greater than the quantity in Column 2.
B—The quantity in Column 2 is greater than the quantity in Column 1.
C—The quantity in Column 1 is equal to the quantity in Column 2.
D—The relationship cannot be determined from the given information.

(NOTE: Information centered above both columns refers to one or both of the quantities to be compared.)

Sample Question		Answer
Column 1	**Column 2**	
$\angle 1$ is complementary to $\angle 2$		$m\angle 1 + m\angle 2 = 90$
$m\angle 1 + m\angle 2$	80	Since $90 > 80$, the answer is **A**.

Column 1	**Column 2**

1.

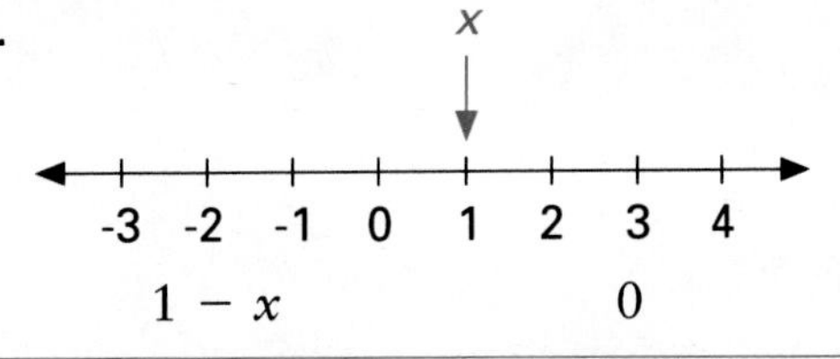

$1 - x$	0

2. Three lines intersect in a point.

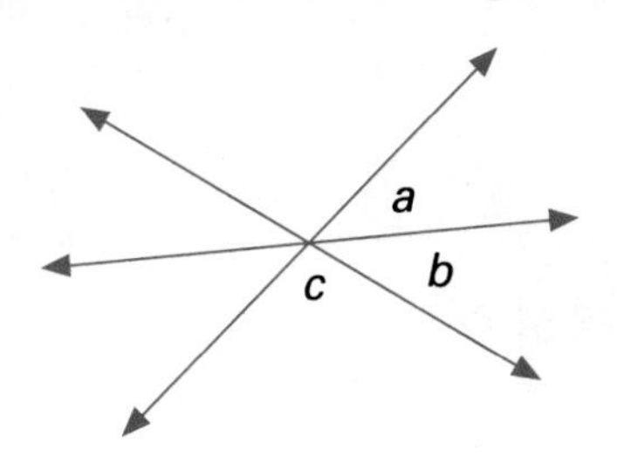

150	$c + a + b$

3. $\angle 1$ and $\angle 2$ are supplementary.

$m\angle 1$	$m\angle 2$

Column 1	**Column 2**

4.

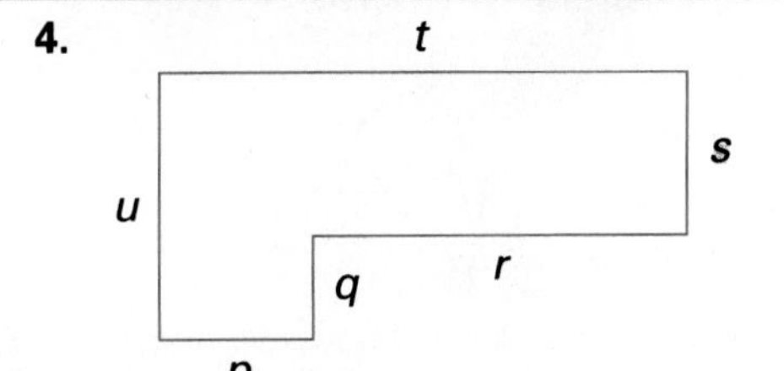

$p + q + r + s$	$t + u$

5. $x \neq 0$

$\frac{1}{x} \cdot x$	$x + (-x)$

6. A M B

M is the midpoint of $\overline{AB}$.

AB	$2 \cdot MB$

2 PROOF

Lewis Carroll (1832–1898)—He used his real name, Charles Dodgson, at Oxford University in England where he taught mathematics for over 20 years. As Lewis Carroll, he put complicated logical arguments into his fantastic stories and humor into his mathematics publications.

2.1 Drawing Conclusions

Objectives To draw conclusions from given statements
To state reasons for conclusions

Chapter 2 is an introduction to the methods used to prove geometric statements. When a lawyer presents his defense to a jury, he must be able to justify or give a convincing reason for every conclusion he draws. In geometry, you will be asked to draw conclusions and give reasons for them. A statement in geometry can be justified if you can provide a reason for the statement. Such reasons may include:

Definitions and geometric properties
Postulates
Theorems that have been proved
Algebraic properties
Arithmetic facts
Given or assumed information

The process of drawing conclusions and giving reasons is called deductive reasoning. Diagrams are provided for many examples and exercises in this text. The diagrams often supply the *given* information. In this lesson you will learn what may and what may not be assumed from a diagram.

EXAMPLE 1 What conclusion can be drawn from the diagram? Give a reason for the conclusion.

Plan C is between A and B on $\overline{AB}$. Betweenness may be assumed from a diagram.

A C B

Solution Conclusion: $AC + CB = AB$
Reason: Segment Addition Postulate

In Example 2, some information is given *in addition* to the diagram.

EXAMPLE 2 Given: F is the midpoint of $\overline{DE}$.
What conclusion can be drawn? Why?

D F E

Plan More than one conclusion is possible. First, F is between D and E. Second, more specifically, a midpoint divides a segment into two congruent segments.

Solution Conclusion: $DF + FE = DE$
Reason: Segment Addition Postulate
Conclusion: $\overline{DF} \cong \overline{FE}$ or $DF = FE$
Reason: Definition of midpoint

In both diagrams below, you may assume that $\angle 1$ and $\angle 2$ are a pair of adjacent angles. Other information may or may not be assumed for each diagram as indicated.

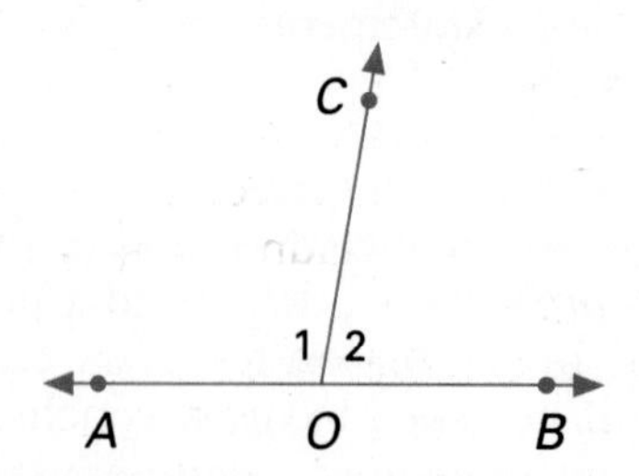

Assume: $\overrightarrow{OA}$ and $\overrightarrow{OB}$ are opposite rays forming a line.
$\angle AOB$ is a straight angle.
Do *not* assume: $\angle 1 \cong \angle 2$, $\overrightarrow{OC} \perp \overleftrightarrow{AB}$, or m $\angle 1 >$ m $\angle 2$.

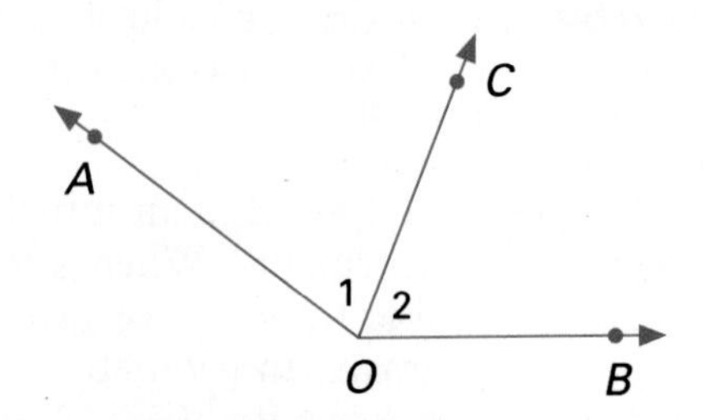

Assume: $\overrightarrow{OA}$ and $\overrightarrow{OB}$ are not opposite rays.
C is in the interior of $\angle AOB$.
Do *not* assume: $\overrightarrow{OC}$ is an angle bisector, or $\angle 1 \cong \angle 2$.

EXAMPLE 3 Given: $\overrightarrow{OC}$ bisects $\angle AOB$.
What conclusions can be drawn?
Give a reason for each conclusion.

Plan C is in the interior of $\angle AOB$, so m $\angle AOB$ = m $\angle 1$ + m $\angle 2$. More specifically, since $\overrightarrow{OC}$ is a bisector of $\angle AOB$, it divides the angle into two congruent angles.

Solution Conclusion: m $\angle AOB$ = m $\angle 1$ + m $\angle 2$
Reason: Angle Addition Postulate
Conclusion: $\angle 1 \cong \angle 2$
Reason: Definition of angle bisector

EXAMPLE 4 Given: The figure as shown, with $\overrightarrow{SR} \perp \overrightarrow{ST}$.
Draw a conclusion and give a reason.

Plan You can assume from the figure that $\angle 1$ and $\angle 2$ are acute adjacent angles. Use the fact that $\overrightarrow{SR} \perp \overrightarrow{ST}$ to draw a conclusion.

Solution Conclusion: m $\angle 1$ + m $\angle 2$ = 90

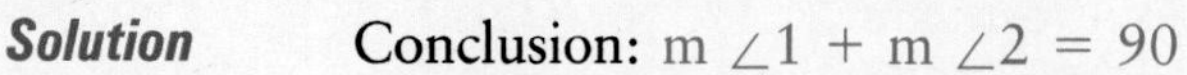

Reason: If the outer rays of two acute adjacent angles are perpendicular, then the sum of the measures of the angles is 90 (Theorem 1.1).

Summary

Information that may be assumed from a diagram:
- Angles are adjacent.
- Outer rays of two adjacent angles form a line or straight angle.
- A point is between two points.
- Points are collinear or coplanar.
- A point is in the interior of an angle.
- Geometric figures intersect.

Information that may *not* be assumed from a diagram:
- An angle is bisected.
- A segment is bisected.
- A point is the midpoint of a segment.
- Rays are perpendicular.
- Segments are congruent.
- Angles are congruent.
- An angle has a specific measure—for example, 60.

Classroom Exercises

In each exercise, indicate whether the stated conclusion may be drawn from the figure above it. Give a reason.

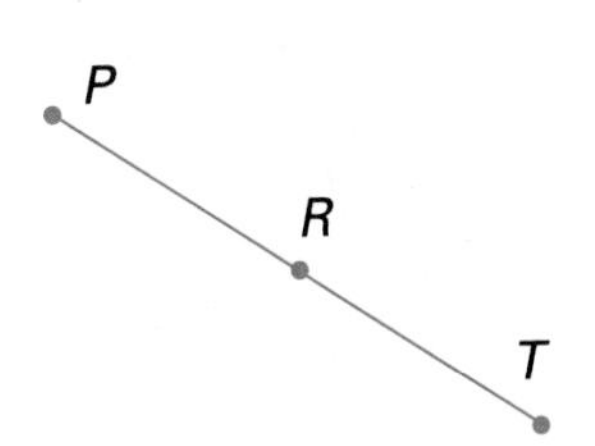

1. Conclusion: $PR + RT = PT$
2. Conclusion: R is between P and T

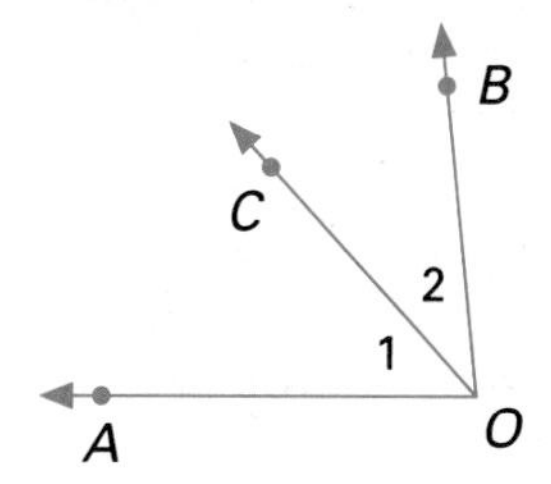

3. Conclusion: $\angle 1 \cong \angle 2$
4. Conclusion: $m\angle 1 + m\angle 2 = m\angle AOB$

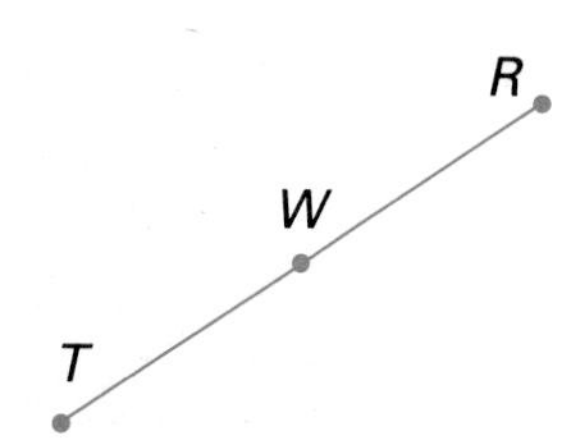

5. Conclusion: $TW + WR = TR$
6. Conclusion: $TW = WR$

Written Exercises

From the given information, draw a conclusion and express it as an equation. Give a reason for your answer.

1. Given: The figure below

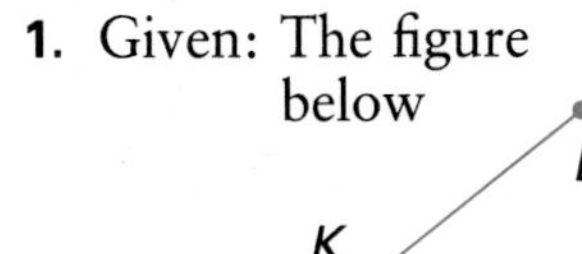
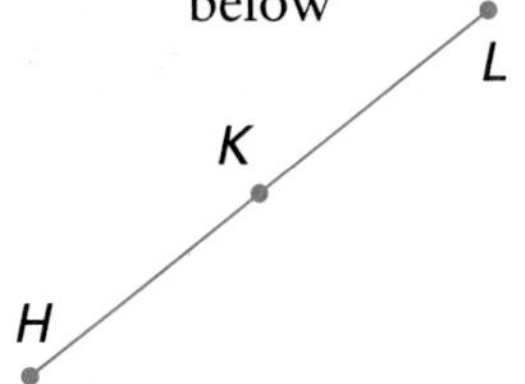

2. Given: W is the midpoint of $\overline{TG}$.

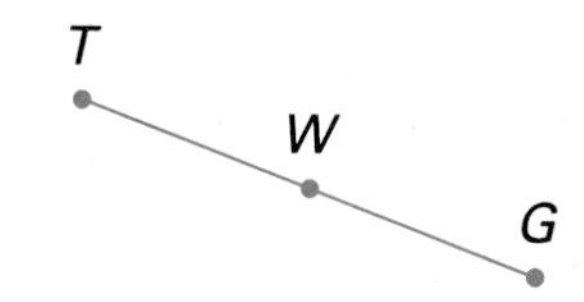

3. Given: $\overline{BD}$ bisects $\overline{AC}$.

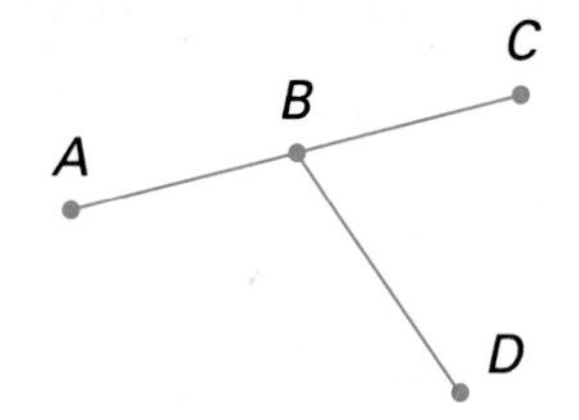

4. Given: The figure below

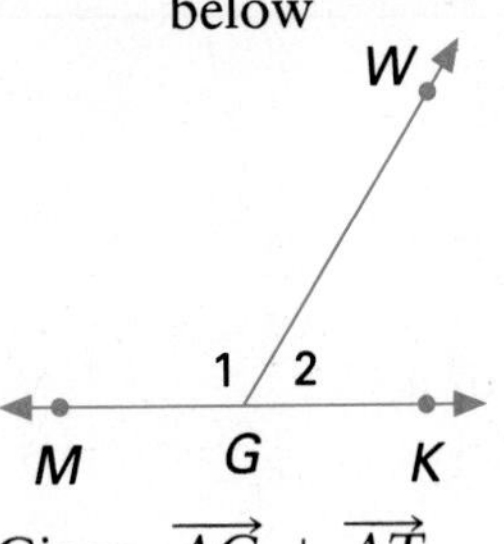
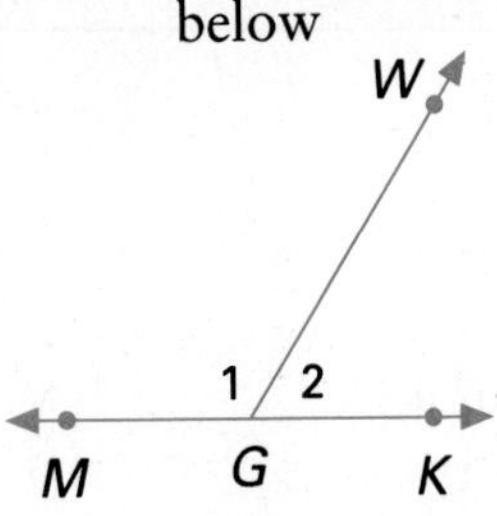

5. Given: The figure below

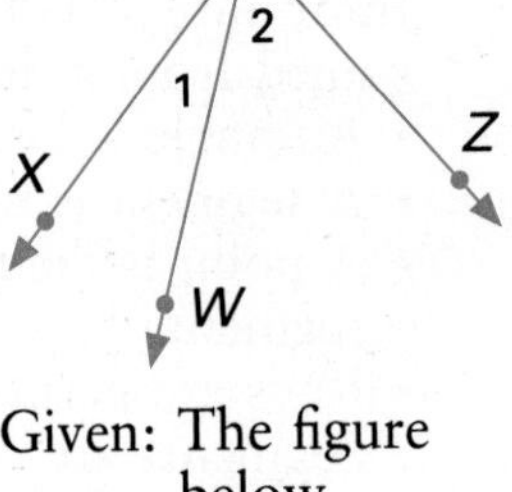

6. Given: $\overrightarrow{OR}$ bisects $\angle QOT$.

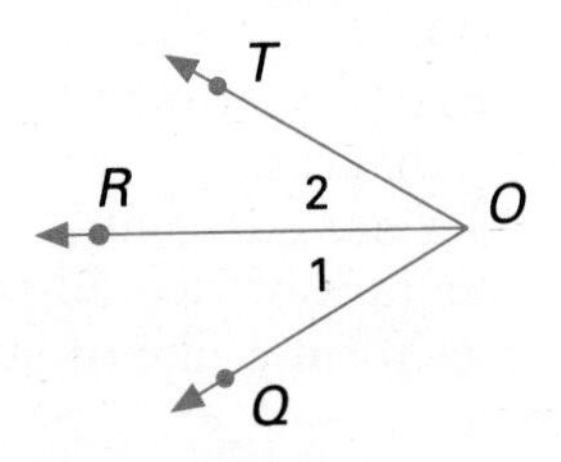

7. Given: $\overrightarrow{AG} \perp \overrightarrow{AT}$

A
G
2
1
H
T

8. Given: The figure below

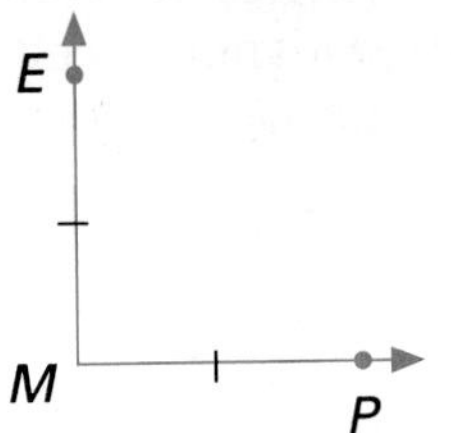

9. Given: The figure below

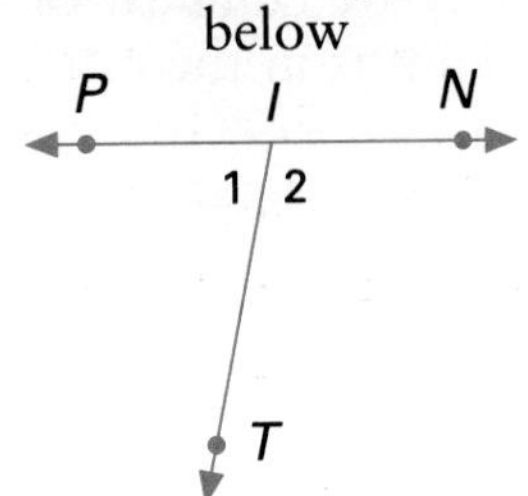

10. Given: $\overrightarrow{BD} \perp \overrightarrow{BE}$

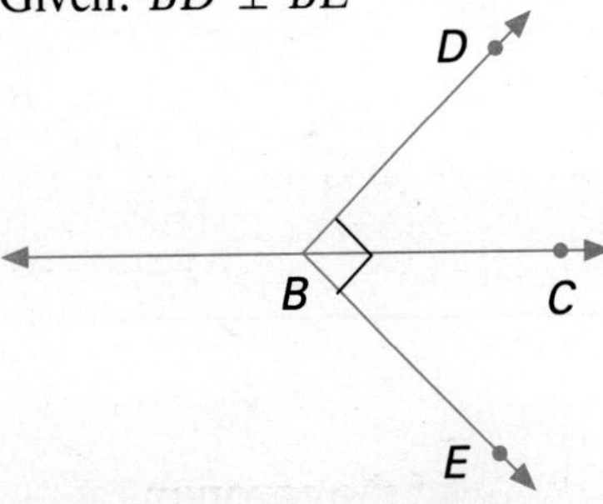

11. Given: $\overrightarrow{EC}$ bisects $\angle BED$.

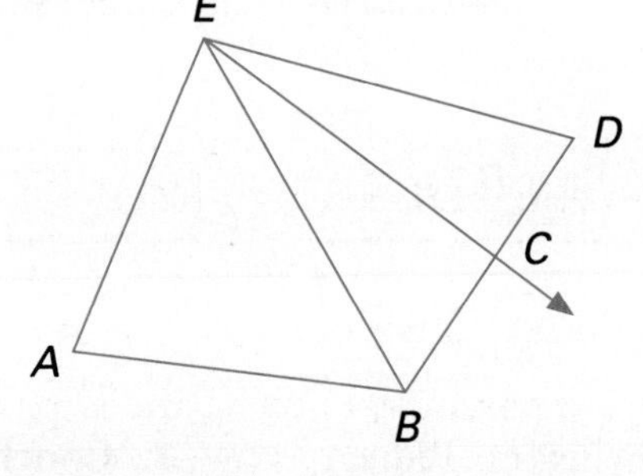

12. Given: $\overline{SQ}$ bisects $\overline{PR}$.

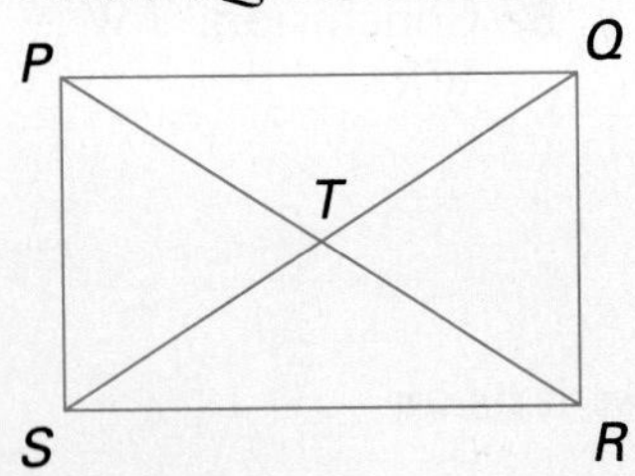

13. Given: $\overline{QP} \perp \overline{QR}$

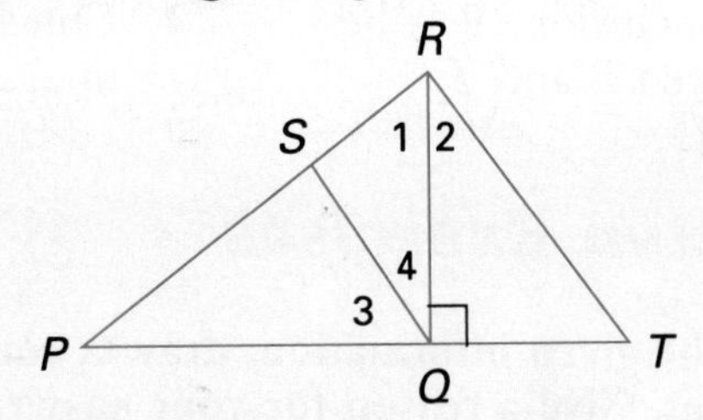

Sketch a figure from the given information. Then determine whether the conclusion is correct. If it is, give the reason.

14. Given: $\angle GHK$, $\overrightarrow{HL}$ with L in the interior of $\angle GHK$, $\overrightarrow{HG} \perp \overrightarrow{HK}$
Conclusion: $\angle GHL \cong \angle LHK$

15. Given: $\angle APT$ and $\angle TPQ$ are adjacent angles, and $\overrightarrow{PA}$ and $\overrightarrow{PQ}$ are opposite rays.
Conclusion: $m\,\angle APT + m\,\angle TPQ = 180$

16. Given: $\angle CAB$ with $\overline{BD}$ such that D is on $\overline{AC}$
Conclusion: $\overline{AC} \perp \overline{DB}$
17. Given: $\angle WXY$, Z on $\overline{XY}$, V on $\overline{XW}$, and $\overrightarrow{ZV}$ bisects $\angle XZW$.
Conclusion: m $\angle VZY$ = m $\angle WZY$ + m $\angle XZV$
18. Given: $\angle PQR$, T on $\overline{QR}$, S is a point in the interior of $\angle PQR$; and $\overline{PT}$ bisects $\overline{QS}$ at U.
Conclusion: $QS = QU + US$

Mixed Review

1. If m $\angle A = 113$, what is the measure of its supplement? *1.8*
2. The measure of an angle is $\frac{4}{5}$ the measure of its complement. Find the measure of each angle.

The coordinates of the endpoints of a segment are −8 and 5. (Exercises 3, 4)

3. Find the length of the segment described above. *1.2*
4. Find the coordinate of the midpoint of the segment. *1.5*

Algebra Review

To solve linear equations:

(1) remove parentheses using the Distributive Property.
(2) put only those terms containing the variable on one side of the equation by adding and/or subtracting.
(3) get the variable alone on one side by multiplying or dividing.

Example: Solve. $2a + 3(4 - 2a) = 2(a + 3)$

$$\begin{aligned} 2a + 12 - 6a &= 2a + 6 \\ -4a + 12 &= 2a + 6 \\ 12 - 6a &= 6 \\ -6a &= -6 \\ a &= 1 \end{aligned}$$

Solve.

1. $a + 9 = 15$
2. $s - 8 = 4$
3. $-7 + n = -5$
4. $6 + 2k = 10$
5. $8d - 14 = 26$
6. $17 - 3k = -10$
7. $6 + 10c = 8c + 12$
8. $7x - 11 = -10 + 8x$
9. $-9 + 8x = x - 30$
10. $a + 11 = 2(a + 3)$
11. $-5(y + 2) = 20$
12. $x + 2 = -3(x - 6)$
13. $5(r - 1) = 2r + 4(r - 1)$
14. $-2(3 - 4z) + 7z = 12z - (z + 4)$

2.2 Introduction to Proof

Objectives To identify properties of congruence and equality
To give missing reasons in proofs
To write proofs using properties of equality

Jane wants to prove that she qualifies to try out for J.V. volleyball. The following two-column format helps her to organize her thinking about her qualifications.

Qualifications for J.V.

(1) Team members can only be freshmen or sophomores.
(2) They must be physically fit.
(3) They must be passing all academic subjects.

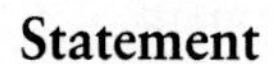

Statement	Reason
1. I am a freshman or sophomore.	1. I am in the 9th grade.
2. I am physically fit.	2. I have a fitness statement signed by a doctor.
3. I am passing all subjects.	3. My average grade in every subject is B.
4. Therefore, I qualify to try out.	4. I meet all requirements.

An argument such as the one above is called a *proof.* You will learn in this lesson how to write a *proof* of a conclusion in mathematics.

Lengths of segments and measures of angles are real numbers. Therefore, the properties of real numbers may be applied to geometric situations. Four of these properties are listed below.

Addition Property of Equality: If $a = b$, then $a + c = b + c$ for all real numbers a, b, and c.

Subtraction Property of Equality: If $a = b$, then $a - c = b - c$ for all real numbers a, b, and c.

Multiplication Property of Equality: If $a = b$, then $ac = bc$ for all real numbers a, b, and c.

Division Property of Equality: If $a = b$, then $\frac{a}{c} = \frac{b}{c}$ for all real numbers a, b, and c $(c \neq 0)$.

These properties can be used to justify steps in solving an equation.

EXAMPLE 1 Given: $3x - 4 = 11$
Prove: $x = 5$

Steps in the process are listed in the column labeled *Statement*. Reasons, or justifications, for each step are listed in the *Reason* column.

Proof

Statement	Reason
1. $3x - 4 = 11$	1. Given
2. $3x = 15$	2. Add Prop of Eq
3. $\therefore x = 5$	3. Div Prop of Eq

The table in Example 1 is a *two-column form* for writing a proof. The symbol $\therefore$ means "therefore." It indicates the conclusion.

EXAMPLE 2 Given: $\frac{1}{2}x - 4 = 11$
Prove: $x = 30$

Proof

Statement	Reason
1. $\frac{1}{2}x - 4 = 11$	1. Given
2. $\frac{1}{2}x = 15$	2. Add Prop of Eq
3. $\therefore x = 30$	3. Mult Prop of Eq

Other properties of equality used in geometry are listed below.

Substitution Property: If a and b are any two real numbers and $a = b$, then a may replace b in an equation or inequality.

Reflexive Property: For any real number a, $a = a$.

Symmetric Property: For all real numbers a and b, if $a = b$, then $b = a$.

Transitive Property: For all real numbers a, b, and c, if $a = b$ and $b = c$, then $a = c$.

EXAMPLE 3 Which properties do the statements illustrate?

Solution

Statement	Property Illustrated
If $t = y$, then $y = t$	Symmetric
If $AB = CD$ and $CD = GH$, then $AB = GH$	Transitive
$m\angle 1 = m\angle 1$	Reflexive
If $a = 3$ and $a + b = 5$, then $3 + b = 5$	Substitution

Recall that $\overline{AB} \cong \overline{CD}$ is equivalent to $AB = CD$. Similarly, $\angle A \cong \angle B$ is equivalent to m $\angle A$ = m $\angle B$. Thus, the following properties of congruence may be stated:

Reflexive Property of Congruence: For any segment $\overline{AB}$, $\overline{AB} \cong \overline{AB}$. For any angle $\angle A$, $\angle A \cong \angle A$.

Symmetric Property of Congruence: For any segments $\overline{AB}$ and $\overline{CD}$, if $\overline{AB} \cong \overline{CD}$, then $\overline{CD} \cong \overline{AB}$. For any angles $\angle A$ and $\angle B$, if $\angle A \cong \angle B$, then $\angle B \cong \angle A$.

Transitive Property of Congruence: For any segments $\overline{AB}$, $\overline{CD}$, and $\overline{PQ}$, if $\overline{AB} \cong \overline{CD}$ and $\overline{CD} \cong \overline{PQ}$, then $\overline{AB} \cong \overline{PQ}$. For any angles $\angle A$, $\angle B$, and $\angle C$, if $\angle A \cong \angle B$ and $\angle B \cong \angle C$, then $\angle A \cong \angle C$.

The properties of equality and congruence can be used in proofs. The two properties below follow from the Addition and Subtraction Properties of Equality, along with the Substitution Property.

For any real numbers a, b, x, and y: if $a = b$ and $x = y$, then
$a + x = b + y$
and $a - x = b - y$.

These two properties may be used as reasons in proofs. When they are used they may be stated as "equations may be added" and "equations may be subtracted."

In the following examples you will be asked to write either reasons for given statements, or statements corresponding to given reasons in sample proofs. These examples will pave the way for writing complete geometric proofs on your own in the next lesson. Notice in the following examples that the algebraic properties of this lesson are used in geometric settings.

EXAMPLE 4 Given: m $\angle 2$ = m $\angle 3$
Prove: m $\angle 1$ + m $\angle 3$ = m $\angle ABC$

Give the missing reason in the proof below.

Statement	Reason
1. m $\angle 2$ = m $\angle 3$	1. Given
2. m $\angle 1$ + m $\angle 2$ = m $\angle ABC$	2. $\angle$ Add Post
3. $\therefore$ m $\angle 1$ + m $\angle 3$ = m $\angle ABC$	3. ________

Solution According to Statement 1, m $\angle 2$ = m $\angle 3$. Therefore, m $\angle 3$ can be substituted for m $\angle 2$ in the equation, and m $\angle 1$ + m $\angle 2$ = m $\angle ABC$ of Step 2 above yields the following equation:

m $\angle 1$ + m $\angle 3$ = m $\angle ABC$ in Step 3.

Therefore, the missing reason (Step 3) is Substitution, abbreviated *Sub*.

EXAMPLE 5 Given: $\overline{AB} \perp \overline{BC}$, m $\angle 1$ = m $\angle 2$
Prove: m $\angle 3$ + m $\angle 2$ = 90

Give the reason for Step 2 in the proof below.

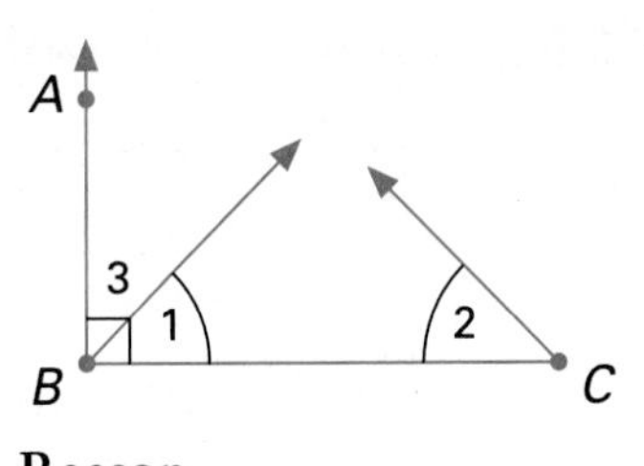

Statement	Reason
1. $\overline{AB} \perp \overline{BC}$	1. Given
2. m $\angle 1$ + m $\angle 3$ = 90	2. ______
3. m $\angle 1$ = m $\angle 2$	3. Given
4. $\therefore$ m $\angle 2$ + m $\angle 3$ = 90	4. Sub (Sub m $\angle 2$ for m $\angle 1$ in statement 2)

Solution Use $\overline{AB} \perp \overline{BC}$ given in Step 1 to conclude that m $\angle 1$ + m $\angle 3$ = 90. The reason is: "If the outer rays of two adjacent acute angles are perpendicular, then the sum of the measures of the angles is 90."

Classroom Exercises

Give the missing reason.

1.

Statement	Reason
1. $x - 8 = 14$	1. Given
2. $x = 22$	2. ______

2.

Statement	Reason
1. $x = y$	1. Given
2. $3x = 3y$	2. ______

3.

Statement	Reason
1. $\overline{AB} \cong \overline{ST}$	1. Given
2. $\overline{ST} \cong \overline{YZ}$	2. Given
3. $\overline{AB} \cong \overline{YZ}$	3. ______

4.

Statement	Reason
1. m $\angle 1$ = m $\angle 2$	1. Given
2. m $\angle 3$ = m $\angle 4$	2. Given
3. m $\angle 1$ + m $\angle 3$ = m $\angle 2$ + m $\angle 4$	3. ______

Written Exercises

Which property does each statement illustrate?

1. $AB = AB$
2. If m $\angle 1$ = m $\angle 2$, then m $\angle 2$ = m $\angle 1$
3. If $\overline{AB} \cong \overline{PQ}$ and $\overline{PQ} \cong \overline{TW}$, then $\overline{AB} \cong \overline{TW}$
4. If $x = 5$ and $x + y = c$, then $5 + y = c$
5. If $x = y$ and $a = b$, then $x + a = y + b$

Write a proof for each of the following.

6. Given: $x - 7 = 4$
Prove: $x = 11$

7. Given: $7x = 28$
Prove: $x = 4$

8. Given: $3x + 7 = 22$
Prove: $x = 5$

Give the missing reason(s) for each of the following proofs.

9. Given: $AT = BG$

Prove: $AT - BT = BG - BT$

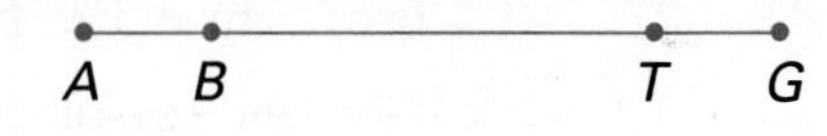

Statement	Reason
1. $AT = BG$	1. Given
2. $\therefore AT - BT = BG - BT$	2. ______

10. Given: $\overrightarrow{AB}$ bisects $\angle CAD$, $\angle 2 \cong \angle 1$.

Prove: $\angle 3 \cong \angle 1$

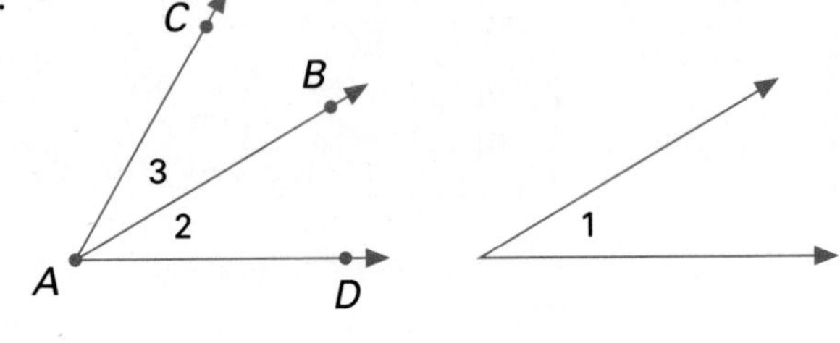

Statement	Reason
1. $\overrightarrow{AB}$ bisects $\angle CAD$.	1. Given
2. $\angle 3 \cong \angle 2$	2. ______
3. $\angle 2 \cong \angle 1$	3. Given
4. $\therefore \angle 3 \cong \angle 1$	4. ______

11. Given: $m\angle 1 = m\angle 3$

Prove: $m\angle 2 + m\angle 3 = 180$

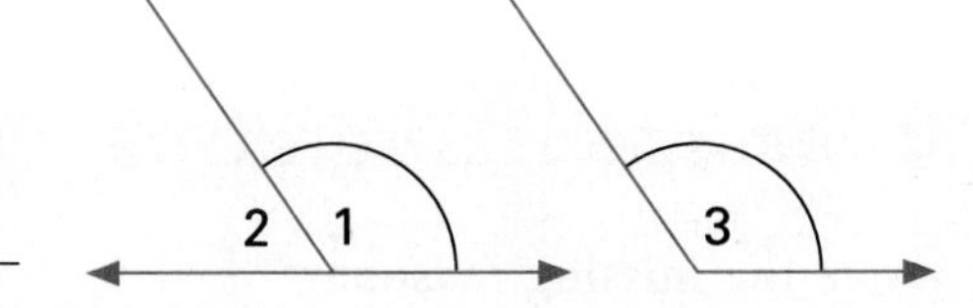

Statement	Reason
1. $m\angle 1 = m\angle 3$	1. Given
2. $m\angle 1 + m\angle 2 = 180$	2. ______
3. $\therefore m\angle 3 + m\angle 2 = 180$	3. ______

12. Given: M is the midpoint of $\overline{AB}$.

Prove: $AM = \frac{1}{2}AB$

Statement	Reason
1. M is the midpoint of $\overline{AB}$.	1. ______
2. $AM = MB$	2. ______
3. $AM + MB = AB$	3. ______
4. $AM + AM = AB$	4. ______
5. $2AM = AB$	5. ______
6. $\therefore AM = \frac{1}{2}AB$	6. ______

Equality (=) and congruence (≅) are examples of relations. A relation that is reflexive, symmetric, and transitive is an equivalence relation. Which of the following are equivalence relations? Why or why not?

13. $>$ **14.** is the brother of **15.** $\neq$ **16.** is a complement of

Mixed Review

1. Find AB. *1.2*

2. Find the coordinate of the midpoint of $\overline{AB}$. *1.3*

3. $AC = \frac{3}{4}CB$. Find C. *1.2*

2.3 Writing Proofs in Geometry

Objective To write proofs using algebraic and geometric properties

In the last lesson you wrote algebraic proofs and supplied missing reasons in geometric proofs. Now you will form a plan of reasoning and write complete geometric proofs.

(1) What is given in writing? What can be assumed from the diagram?
(2) What conclusions can be made from each given statement?
(3) What reason can you give to justify each conclusion?
(4) In what order can you prove these conclusions to arrive at what you are asked to prove?

EXAMPLE 1 Given: $\overrightarrow{QP} \perp \overrightarrow{QR}$, m $\angle 1$ = m $\angle 3$
Prove: m $\angle 3$ + m $\angle 2$ = 90

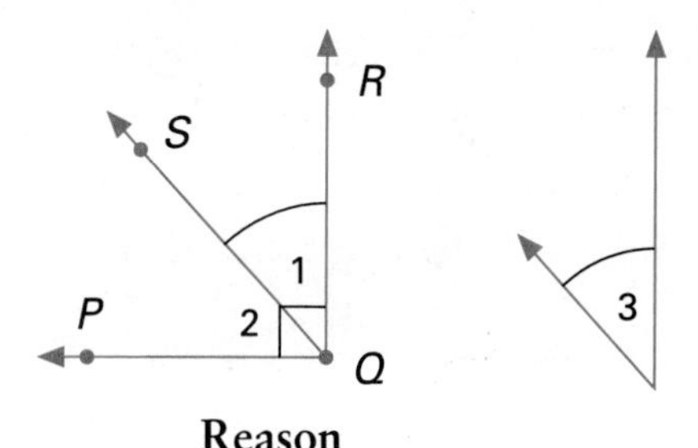

Proof

Statement	Reason
1. $\overrightarrow{QP} \perp \overrightarrow{QR}$	1. Given
2. m $\angle 1$ + m $\angle 2$ = 90	2. If outer rays of two adj acute $\angle$s are $\perp$, then the sum of the $\angle$ meas is 90.
3. m $\angle 1$ = m $\angle 3$	3. Given
4. m $\angle 3$ + m $\angle 2$ = 90	4. Sub

EXAMPLE 2 Given: $\overline{PQ}$ bisects $\overline{AB}$ at R; $\overline{AR} \cong \overline{RQ}$.
Prove: $\overline{RB} \cong \overline{RQ}$

Plan Use the key word *bisects* to write an equation: $\overline{RB} \cong \overline{AR}$. Knowing $\overline{RB} \cong \overline{AR}$ and given $\overline{AR} \cong \overline{RQ}$, you may conclude $\overline{RB} \cong \overline{RQ}$, by the Transitive Property (or Substitution).

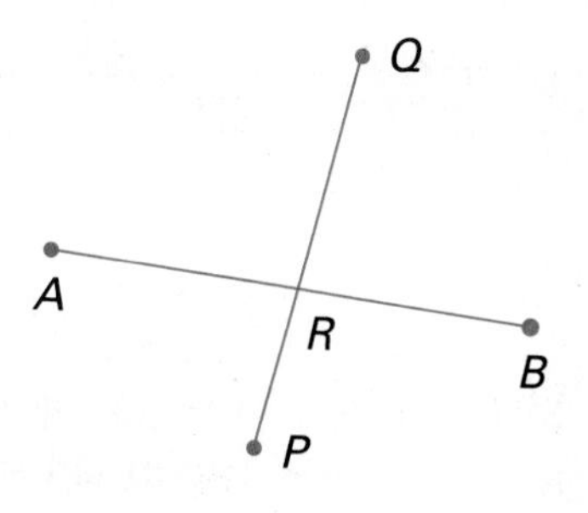

Proof

Statement	Reason
1. $\overline{PQ}$ bisects $\overline{AB}$ at R.	1. Given
2. $\overline{RB} \cong \overline{AR}$	2. A bis divides a seg into two $\cong$ segs.
3. $\overline{AR} \cong \overline{RQ}$	3. Given
4. $\therefore \overline{RB} \cong \overline{RQ}$	4. Trans Prop

Classroom Exercises

Draw a conclusion from the diagram.

1.

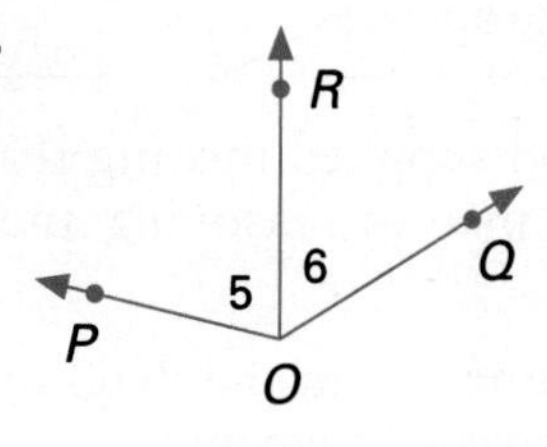

2.

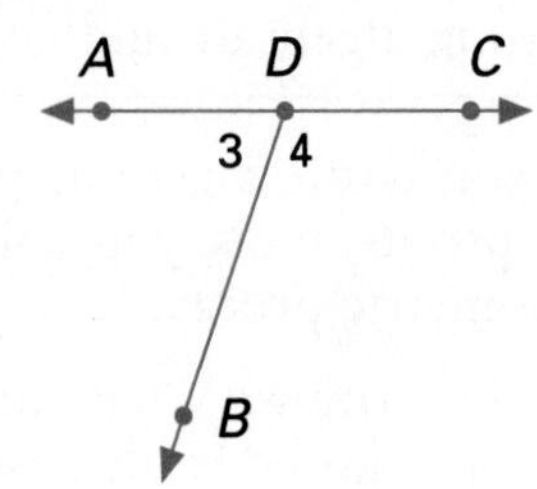

3.

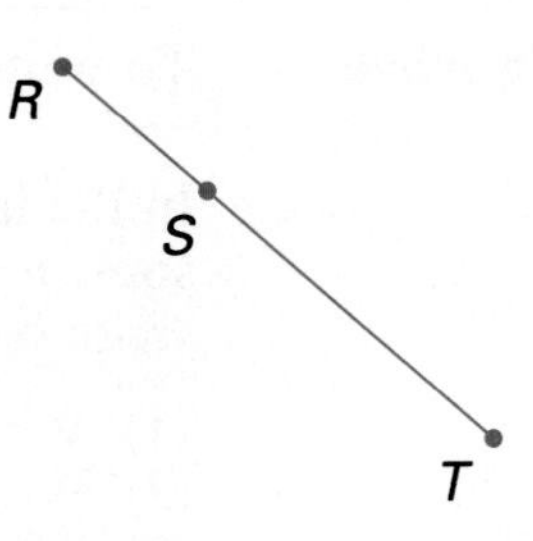

Written Exercises

1. Given: $\overrightarrow{OB}$ bisects $\angle TOY$.
Prove: m $\angle 1$ = m $\angle 2$

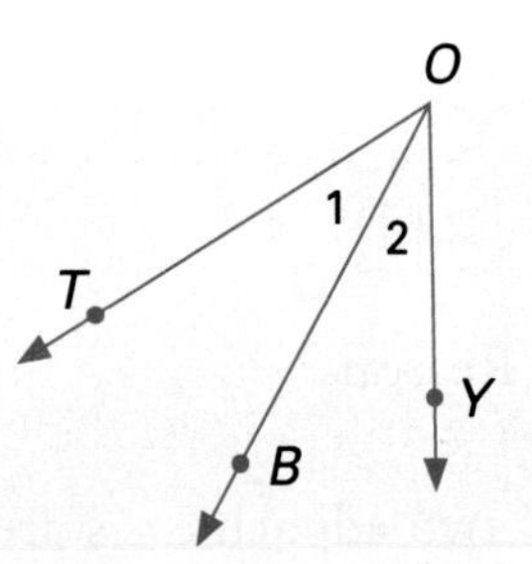

2. Given: $\overrightarrow{GK} \perp \overrightarrow{GL}$
Prove: m $\angle 3$ + m $\angle 4$ = 90

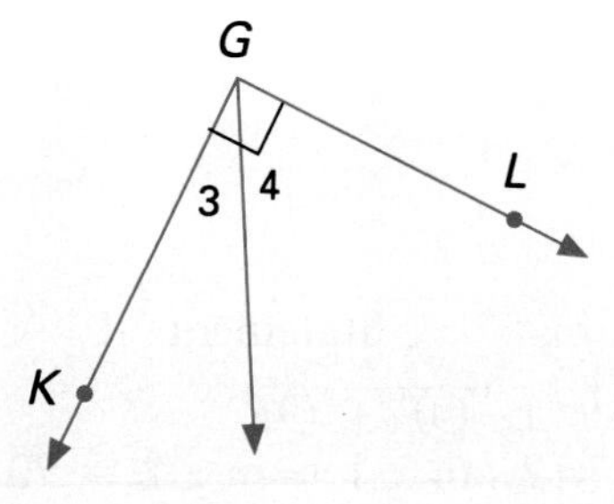

3. Given: $\overline{UR}$ bisects $\overline{YK}$.
Prove: $\overline{YU} \cong \overline{KU}$

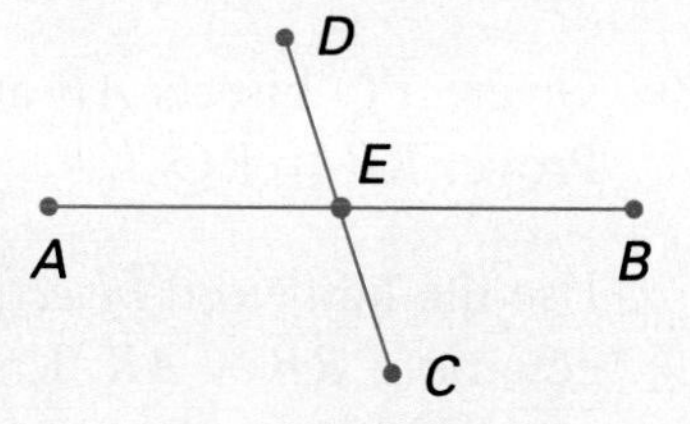

4. Given: $\overline{AB}$ and $\overline{CD}$ bisect each other.
Prove: $\overline{AE} \cong \overline{EB}$ and $\overline{EC} \cong \overline{ED}$

5. Given: m $\angle 3$ = m $\angle 4$
Prove: m $\angle 2$ + m $\angle 4$ = 180

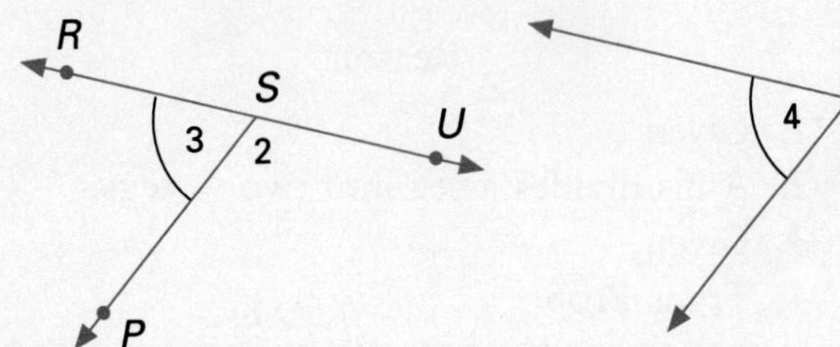

6. Given: $\overline{TY} \perp \overline{TW}$, m $\angle 5$ = m $\angle 6$
Prove: m $\angle 4$ + m $\angle 6$ = 90

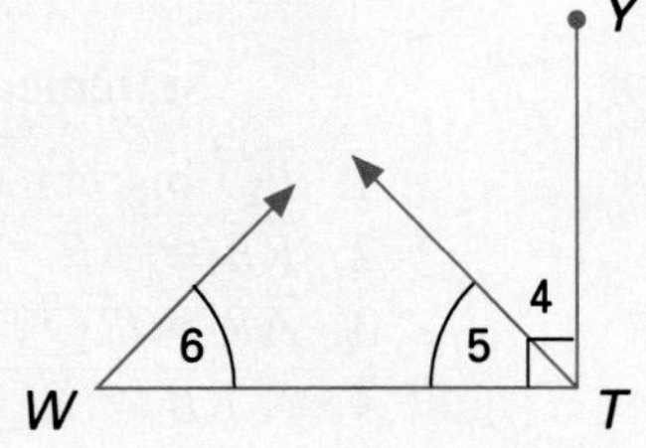

7. Given: $\overrightarrow{AB}$ bisects $\angle PAR$, and $\angle 3 \cong \angle 2$
Prove: $\angle 1 \cong \angle 2$

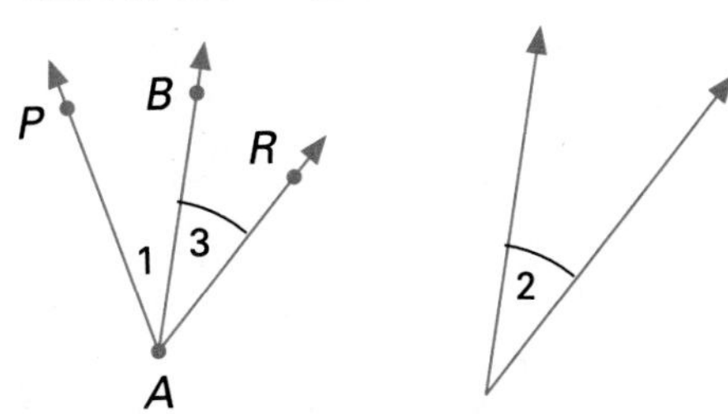

8. Given: $\angle 1 \cong \angle 2$, and $\angle 2 \cong \angle 3$
Prove: $\angle 1 \cong \angle 3$

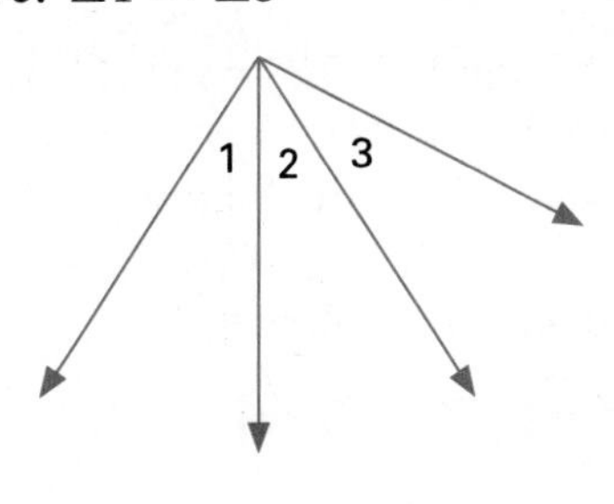

9. Given: $\overline{TL}$ bisects $\overline{BG}$; $\overline{TA} \cong \overline{BA}$.
Prove: $\overline{TA} \cong \overline{AG}$

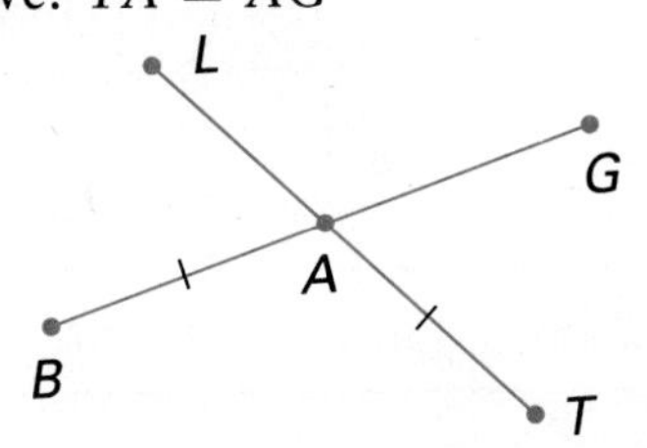

10. Given: $AD = AB$
Prove: $AD + BC = AC$

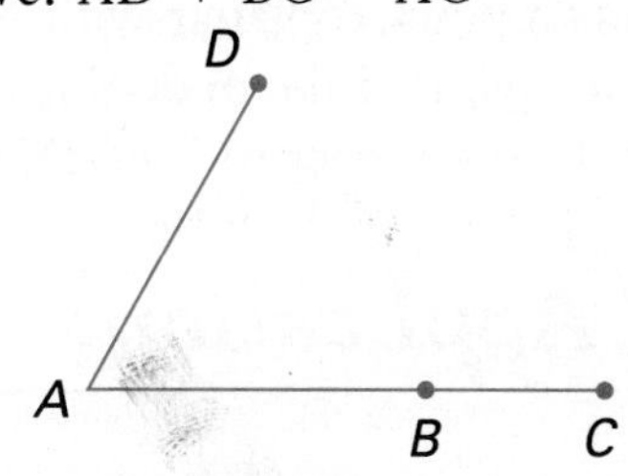

11. Given: B is the midpoint of $\overline{AC}$.
Prove: $AB + CD = BD$

12. Given: $\overrightarrow{OD}$ bisects $\angle AOC$.
Prove: $m\angle 4 + m\angle 5 = m\angle DOB$

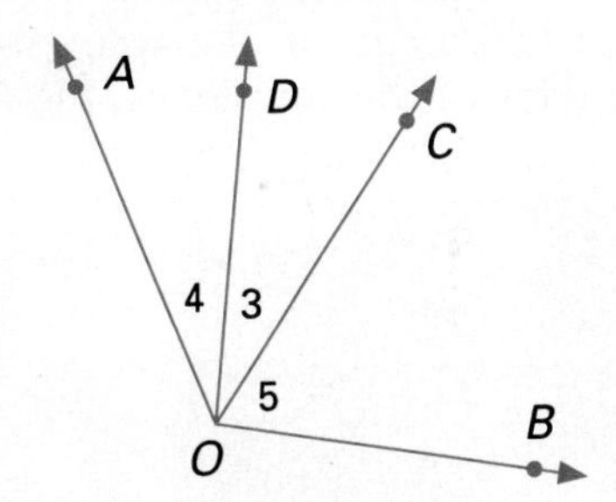

13. Given: $\overline{BA} \perp \overline{BD}$, and $\overline{BD}$ bisects $\angle EBC$.
Prove: $m\angle 1 + m\angle 3 = 90$

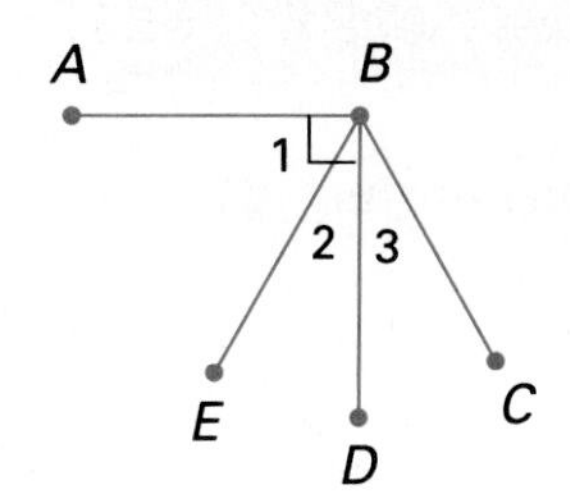

14. Given: $\overline{AB}$ bisects $\angle CAE$, $m\angle 2 = m\angle 4$.
Prove: $m\angle 1 + m\angle 3 = 180$

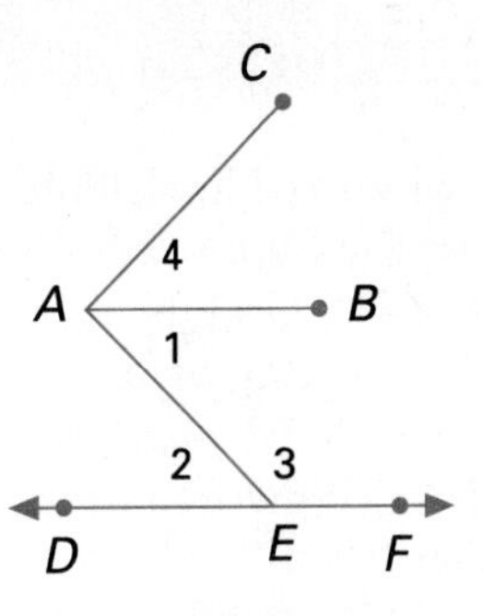

Draw a figure for each of the following. Prove what is asked.

15. Given: $\overline{AC} \perp \overline{AB}$, D is a point in the interior of $\angle CAB$ such that $m\ \angle ADB = m\ \angle CAD$.
    Prove: $m\ \angle ADB + m\ \angle DAB = 90$
16. Given: $\overline{GH}$ with P between G and H, and W between P and H, such that W is the midpoint of $\overline{PH}$.
    Prove: $GP + WH = GW$

Mixed Review

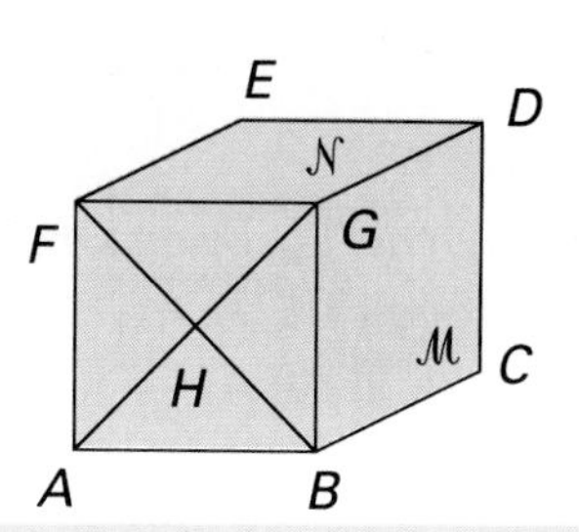

1. Name the intersection of planes $\mathcal{M}$ and $\mathcal{N}$. 1.1
2. Name a point collinear with F and B. 1.1
3. Name a point coplanar with F, G, and E. 1.1
4. $\overline{AB} \cong \overline{BC}$. Is B the midpoint of $\overline{AC}$? 1.3
5. Is $\angle AGF$ adjacent to $\angle FGD$? 1.5

Application: *Tilt of the Earth*

If we could watch the Earth from space as it revolves around the Sun, we would see that the planet's polar axis is tilted at a constant angle. The hemisphere (northern or southern) closer to the Sun at a given time receives sunlight longer during the 24-hour day. On the days of the solstices, the difference in the amount of light received is at a maximum. On the days of the equinoxes, the amount of light received is the same.

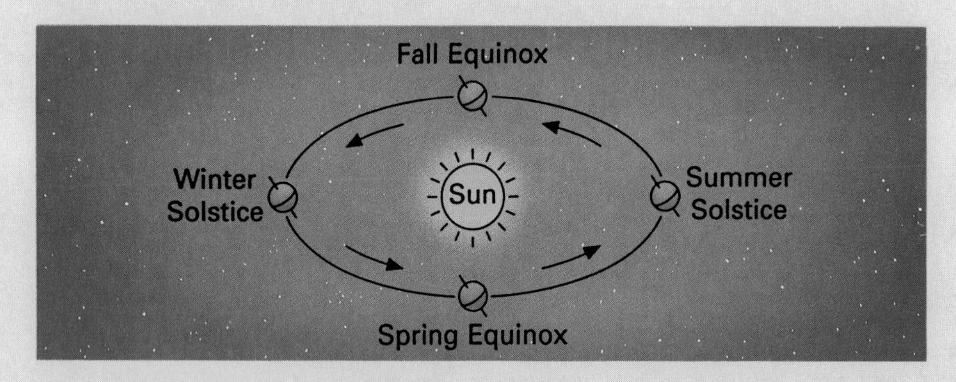

1. Explain why the North Pole has sunlight all day long on June 21, the day of the Summer Solstice.
2. Why are day and night the same length all over the world on March 21, the day of the Spring Equinox?
3. The diagram is labeled in relation to the seasons in the Northern Hemisphere. How would it need to be changed for the Southern Hemisphere?

2.4 Proofs and More Complex Figures

Objectives To apply alternate forms of the Segment Addition and Angle Addition Postulates

To use algebraic properties in multistep geometric proofs

The Subtraction Property of Equality can be used to obtain three equivalent statements of the Segment Addition Postulate. Similarly, the Angle Addition Postulate can be stated in three equivalent ways.

Segment Addition Postulate

Given: $\overline{AB}$, with C on $\overline{AB}$

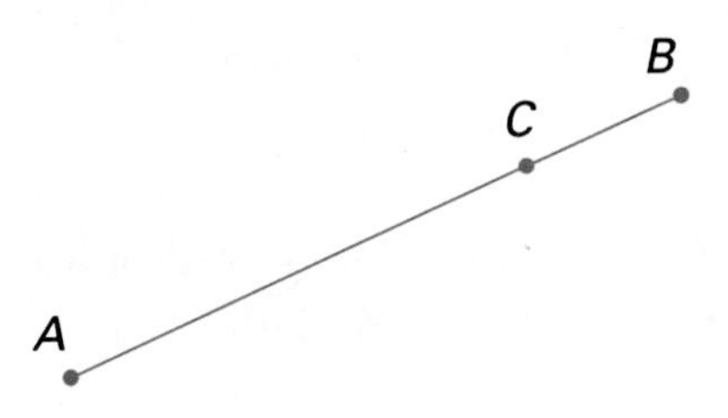

1. $AC + CB = AB$
2. $AC = AB - CB$
3. $CB = AB - AC$

Angle Addition Postulate

Given: $\angle ACB$, $\overrightarrow{CD}$ with D in the interior of $\angle ACB$

1. $m\angle 1 + m\angle 2 = m\angle ACB$
2. $m\angle 1 = m\angle ACB - m\angle 2$
3. $m\angle 2 = m\angle ACB - m\angle 1$

EXAMPLE 1 Given: $m\angle ABC = m\angle HGF$,
$m\angle 1 = m\angle 3$

Prove: $m\angle 2 = m\angle 4$

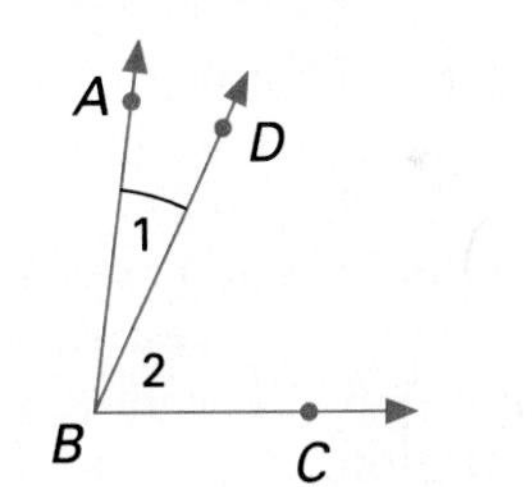

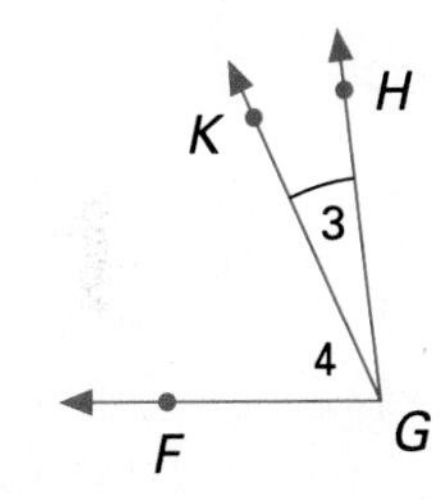

Plan Decide what to do with $m\angle ABC$ and $m\angle 1$ in the *given* to get $m\angle 2$ in the *prove* part. Notice that $m\angle ABC - m\angle 1 = m\angle 2$ and $m\angle HGF - m\angle 3 = m\angle 4$.

Proof

Statement	Reason
1. $m\angle ABC = m\angle HGF$	1. Given
2. $m\angle 1 = m\angle 3$	2. Given
3. $m\angle ABC - m\angle 1 = m\angle 2$; $m\angle HGF - m\angle 3 = m\angle 4$	3. $\angle$ Add Post
4. $m\angle ABC - m\angle 1 = m\angle HGF - m\angle 3$	4. Equations may be subtracted.
5. $\therefore m\angle 2 = m\angle 4$	5. Sub

EXAMPLE 2 Given: $\overline{AC} \cong \overline{BD}$
Prove: $\overline{AB} \cong \overline{CD}$

A B C D

Plan Look at the parts of the figure. Notice the relationships of the segments in the *given* and the *prove* statements.

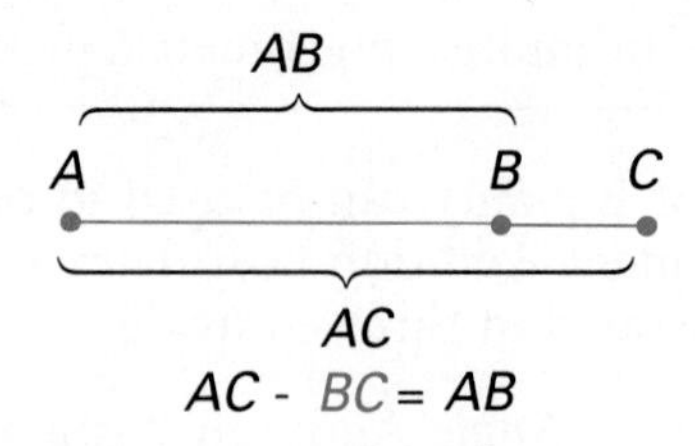

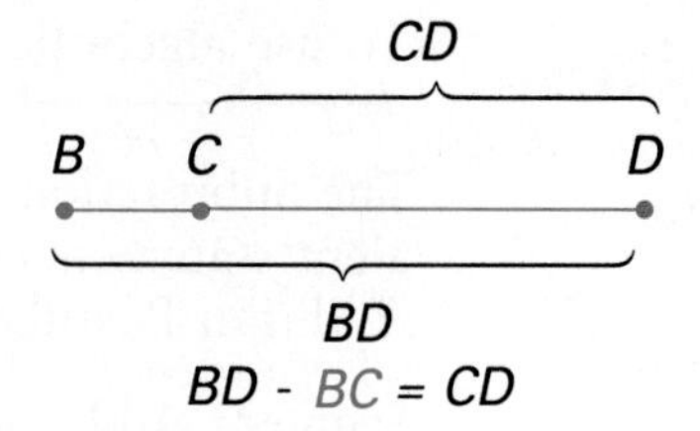

This tells you to subtract BC from each side of the equation $AC = BD$.

Proof

Statement	Reason
1. $\overline{AC} \cong \overline{BD}$ $(AC = BD)$	1. Given
2. $AC - BC = AB$; $BD - BC = CD$	2. Seg Add Post
3. $AC - BC = BD - BC$	3. Subt Prop of Eq
4. $\therefore AB = CD$ $(\overline{AB} \cong \overline{CD})$	4. Sub

EXAMPLE 3 Given: $\overrightarrow{BA} \perp \overrightarrow{BC}$, and $\overrightarrow{BC}$ bisects $\angle DBE$.
Prove: $m\angle 1 + m\angle 3 = 90$

Plan It is often helpful to *work backwards* in developing proofs. In this example, you can show that $m\angle 1 + m\angle 2 = 90$. Therefore, if you can also show that $m\angle 3 = m\angle 2$, you can prove the conclusion—$m\angle 1 + m\angle 3 = 90$—by substitution.

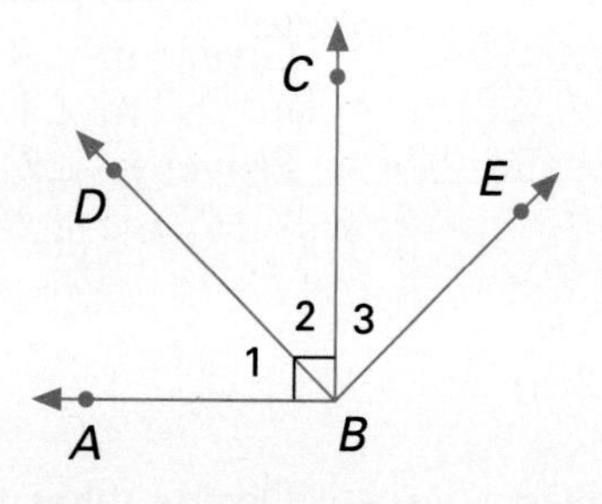

Notice that because $\overrightarrow{BC}$ bisects $\angle DBE$, $m\angle 2 = m\angle 3$.

Proof

Statement	Reason
1. $\overrightarrow{BA} \perp \overrightarrow{BC}$	1. Given
2. $m\angle 1 + m\angle 2 = 90$	2. If outer rays of acute adjacent angles are perpendicular, then the sum of the angle measures is 90.
3. $\overrightarrow{BC}$ bisects $\angle DBE$.	3. Given
4. $m\angle 3 = m\angle 2$	4. Def of $\angle$ bis
5. $\therefore m\angle 1 + m\angle 3 = 90$	5. Sub

Classroom Exercises

Complete each equation from the information provided by the diagram above it.

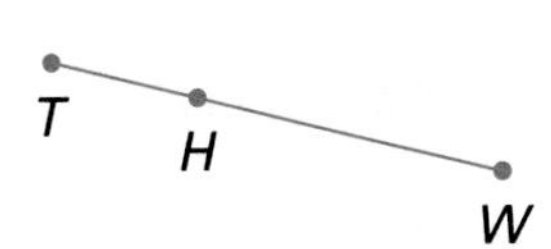

Q
P
U
T
R

1. $TW = TH +$ ______

2. $TH = TW -$ ______

3. ______ $= TW - TH$

4. $m\ \angle QUR = m\ \angle QUT +$ ______

5. $m\ \angle PUR - m\ \angle PUT =$ ______

6. $m\ \angle RUP -$ ______ $= m\ \angle TUR$

Written Exercises

Copy the proof. Write the missing statements and reasons.

1. Given: $AP = AQ$, $AM = AN$
Prove: $MP = NQ$

P Q M N A

Statement	Reason
1. $AP = AQ$	1. Given
2. $AM = AN$	2. ______
3. $AP - AM = AQ - AN$	3. ______
4. $AP - AM =$ ______, $AQ - AN =$ ______	4. ______
5. $\therefore MP = NQ$	5. ______

2. Given: $PR = QS$
Prove: $PQ = RS$

P Q R S

Statement	Reason
1. $PR = QS$	1. ______
2. $PR -$ ______ $= QS -$ ______	2. Subt Prop of Eq
3. $PR - QR = PQ$, $QS - QR = RS$	3. ______
4. $\therefore PQ = RS$	4. ______

3. A carpenter cuts off a 1-ft piece from each of two boards of the same length. Explain why the remaining board lengths must be equal.

Write the proof. Give statements and reasons.

4. Given: $SW = GY$, $TW = HY$
Prove: $ST = GH$

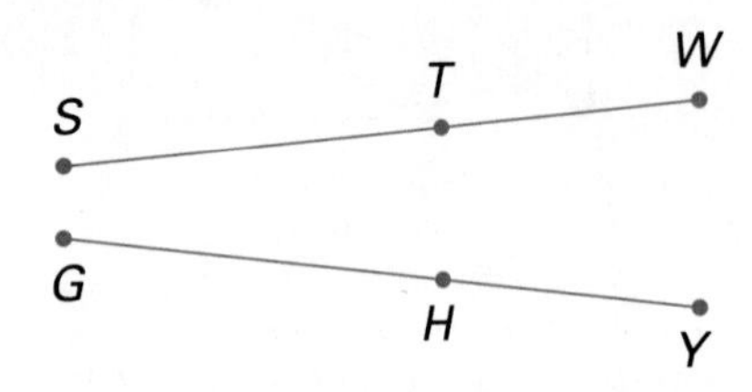

5. Given: m $\angle 5$ = m $\angle 8$, m $\angle 6$ = m $\angle 7$
Prove: m $\angle ABC$ = m $\angle EFG$

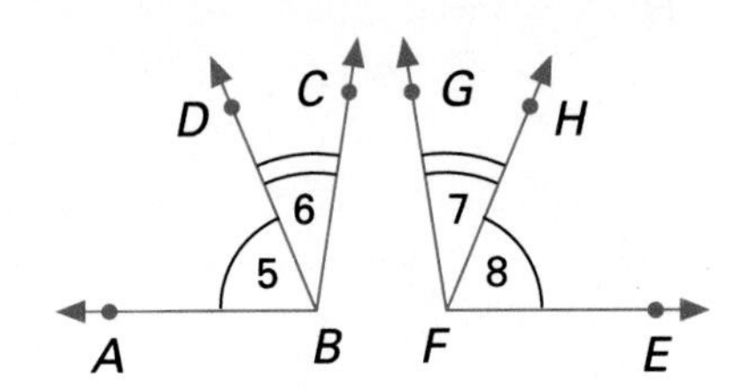

6. Given: $\overrightarrow{QS}$ bisects $\angle PQR$, and $\angle 1 \cong \angle 3$
Prove: $\angle 2 \cong \angle 3$

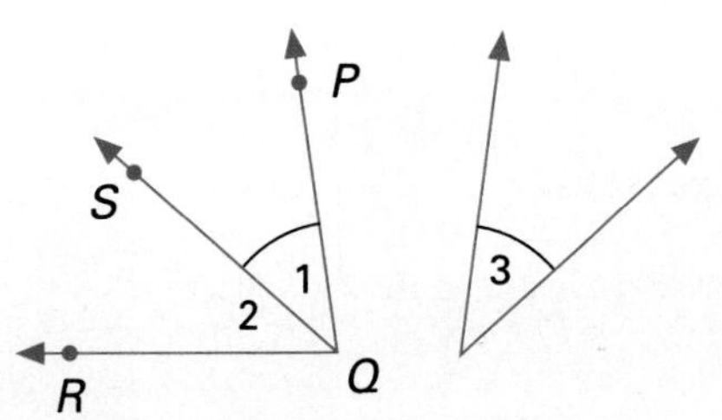

7. Given: m $\angle 3$ = m $\angle 5$
Prove: m $\angle 4$ + m $\angle 5$ = 180

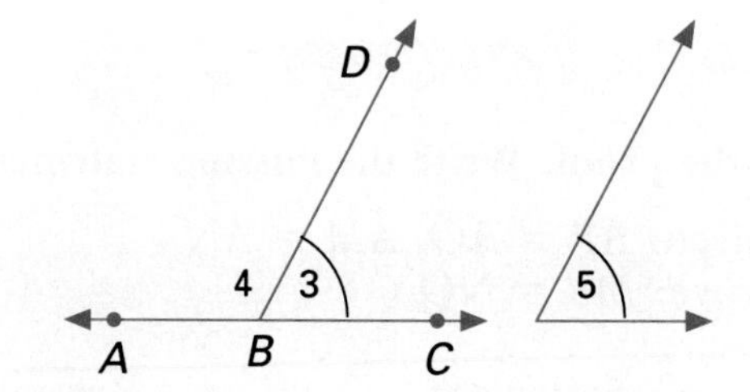

8. Given: $\angle 3 \cong \angle 5$
Prove: $\angle ABE \cong \angle DBC$

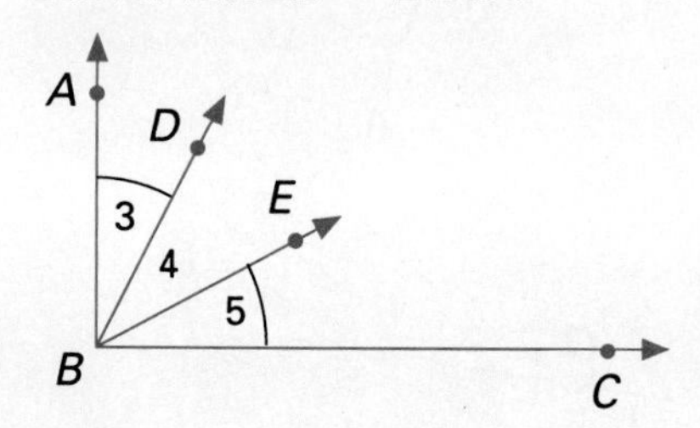

9. Given: $AB = RS$
Prove: $AR = BS$

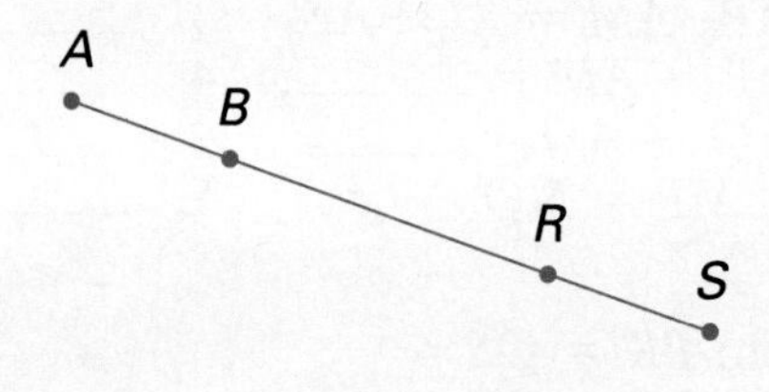

10. Given: $AB = BD$, $BC = BE$
Prove: $DE = AC$

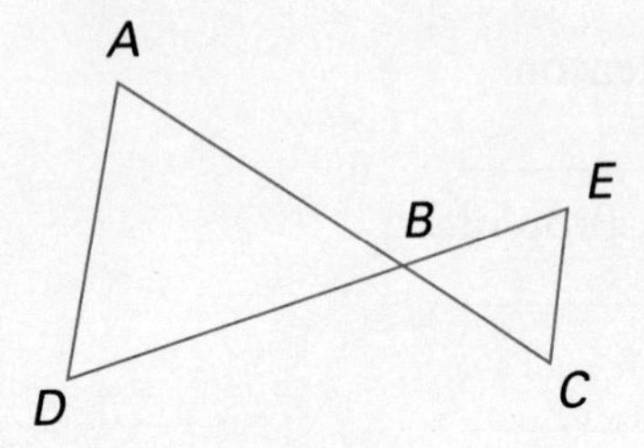

11. Given: $\overline{WX} \perp \overline{WY}$, m $\angle 5$ = m $\angle 7$
Prove: m $\angle 6$ + m $\angle 7$ = 90

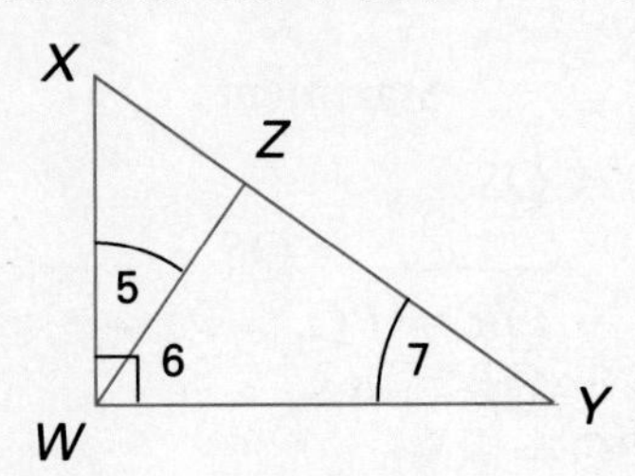

12. Given: $AB = AE$, $AC = AD$
Prove: $BC = ED$

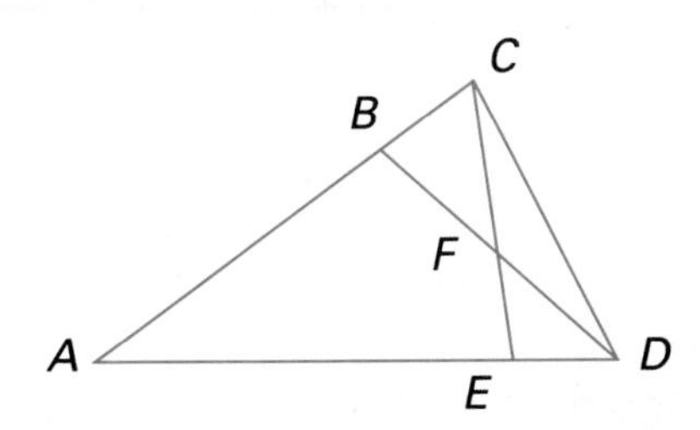

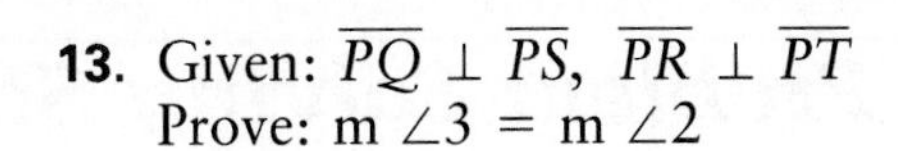

13. Given: $\overline{PQ} \perp \overline{PS}$, $\overline{PR} \perp \overline{PT}$
Prove: m $\angle 3$ = m $\angle 2$

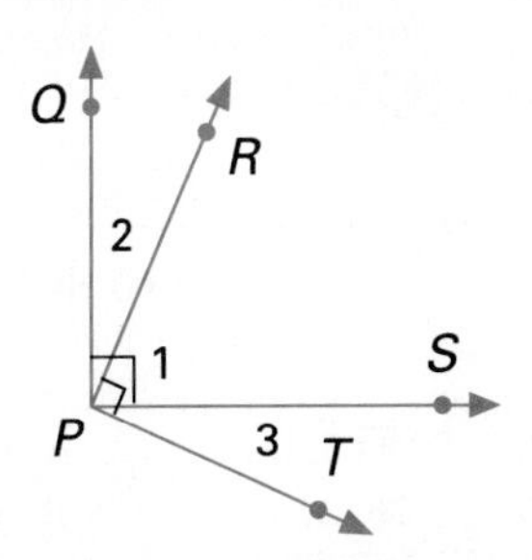

14. Given: $\overrightarrow{ST} \perp \overrightarrow{SY}$, m $\angle 4$ = m $\angle 5$
Prove: m $\angle 5$ + m $\angle 6$ = 90

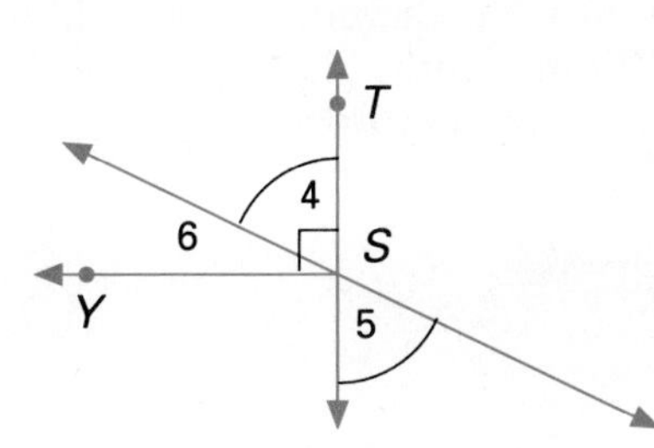

15. Given: $\overrightarrow{BE}$ bisects $\angle ABD$.
Prove: m $\angle 6$ + m $\angle 8$ = m $\angle EBC$

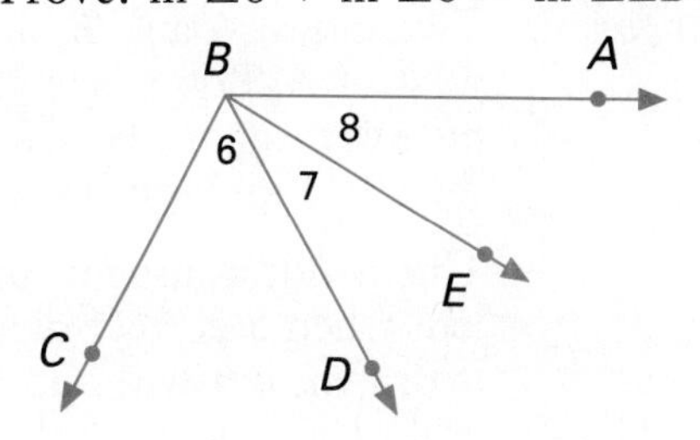

16. Given: $\overrightarrow{BF}$ bisects $\angle ABE$, and $\overline{BF} \perp \overline{BD}$.
Prove: m $\angle 1$ + m $\angle 3$ = 90

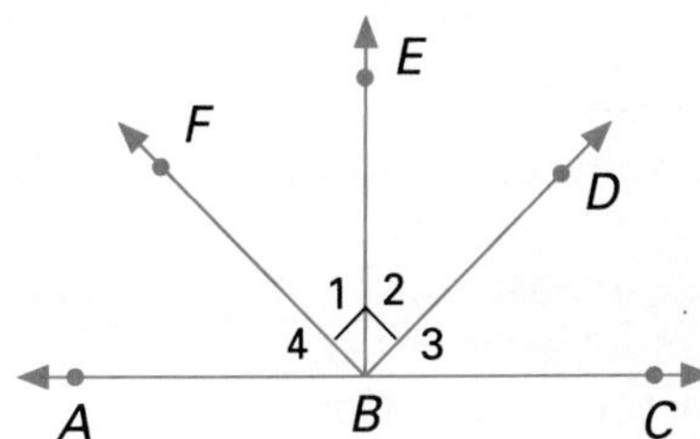

17. Given: B is the midpoint of $\overline{AC}$, $AC = AD$, $AB = BE$, and $BE = AE$.
Prove: $BE + ED = AC$

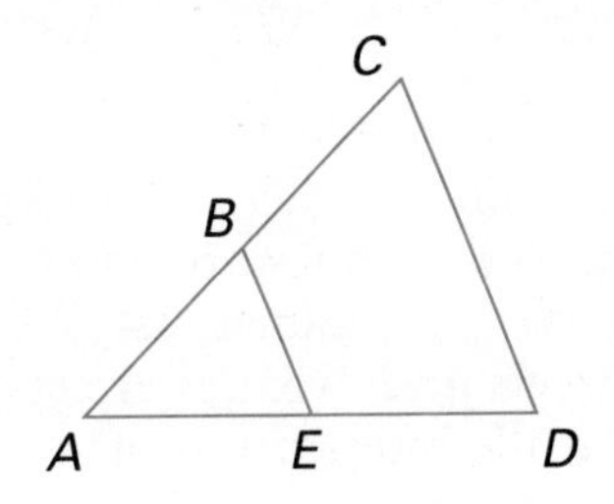

Midchapter Review

Which property is illustrated? (Exercises 1–2) *2.2*

1. $PQ = PQ$

2. If $x = y$, then $x + c = y + c$

3. Given: m $\angle 1$ = m $\angle 2$. Draw a conclusion about $\angle 2$ and $\angle 3$. Give a proof for it. *2.3, 2.4*

3 1 2

Problem Solving Strategies

Drawing a Diagram

Objective To solve problems by drawing a diagram

At times, you may need to draw a diagram according to given information. It is important that your diagram represent the given situation accurately. The problems that follow will give you practice in drawing diagrams.

Example Two cities, *A* and *B*, are located 1,200 mi from each other. *B* is directly west of *A*. Bob starts at *B* and drives 300 mi south. He then drives 200 mi east. Adam lives 200 mi south of *A*. He drives 100 mi south and then 100 mi west. How far apart are the two men?

This is not a difficult problem when a correct diagram is drawn, but without a diagram it is easy to make mistakes. The figure to the right shows the distance and the answer, 900 mi.

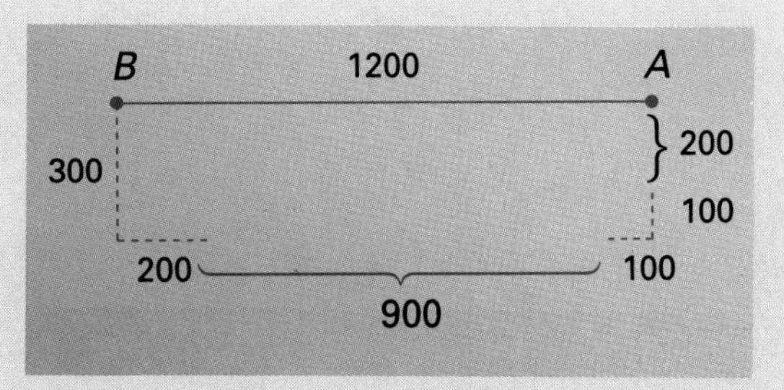

1. In a certain city the avenues run east/west and the streets run north/south. The avenues are named *A*, *B*, *C*, etc. from south to north. The streets are named 1st, 2nd, 3rd, etc. from west to east. Both the streets and avenues are spaced one-tenth of a mile apart. Alice lives at the intersection of *A* and 1st; Betty lives at the intersection of *F* and 5th. How far do they live from each other? No short cuts are possible because of buildings.
2. Towns *D*, *E*, and *F* are located along the same straight road. *E* is between *D* and *F*. *D* is 4 mi from *E* and 11 mi from *F*. The three towns are planning to build a movie theater. An equal number of people are expected to attend from each town. Where should they build the theater so that the total miles driven will be the least? (Consider one person traveling from each town.)
3. A radio station sends out signals in all directions (circular) for 60 mi. Another radio station 100 mi away also sends out signals for 60 mi. If you are driving along a road that directly connects these two stations, what is the distance in miles through which you can hear both stations?

2.5 Conditional Statements

Objectives To rewrite statements in "If . . . , then . . . " form
To determine whether conditional statements are true

Although theorems and problems in geometry are often written with separate *given* and *prove* (conclusion) statements, this format is not required. Consider the following example.

Given: M is the midpoint of $\overline{AB}$.
Conclusion: $AM = MB$

A M B

The given (*hypothesis*) and the conclusion can be written as a single conditional statement.

If M is the midpoint of $\overline{AB}$, then $AM = MB$.
given or hypothesis conclusion

Often letters such as p, q, r, s, and t are used to represent statements or parts of statements.

The statement "If p, then q" is written $p \rightarrow q$.

A statement that is in the form "If p, then q" is a **conditional** statement. The part of the statement following "if" is the **hypothesis.** The part of the statement following "then" is the **conclusion.**

Many statements that are not in "if . . . , then . . ." form can be written in this form.

Statement:	A right angle has a measure of 90.
In "If . . . , then . . ." form:	If an angle is a right angle, then its measure is 90.
Statement:	The midpoint of a segment divides the segment into two congruent segments.
In "If . . . , then . . ." form:	If a point is the midpoint of a segment, then it divides the segment into two congruent segments.

EXAMPLE 1 Write the statement in "If . . . , then . . . " form.
The angles formed when a ray bisects an angle are congruent.

Solution If a ray bisects an angle, then the angles formed are congruent.

The truth table at the right shows all possible truth values for p and q. A conditional is false only when the hypothesis p is true and the conclusion q is false.

p	q	$p \rightarrow q$
T	T	T
T	F	F
F	T	T
F	F	T

This can be illustrated by the following example: A salesman is told, "*If* you sell \$10,000 worth of furniture, *then* we will give you a \$500 bonus." The only situation that breaks this promise is the one in which the salesman sells \$10,000 worth of furniture (making the hypothesis, or p, true), and he is *not* given a \$500 bonus (making the conclusion, or q, false).

A case that proves the conditional false is a **counterexample**. Notice that if the salesman does *not* sell \$10,000 worth of furniture and is *not* given \$500, the promise is still kept. The conditional is true even though the hypothesis and conclusion are both false. This example can also be used to show that a conditional is true for the other two cases given in the truth table.

The conditional $p \rightarrow q$ can be read in any one of the following ways:

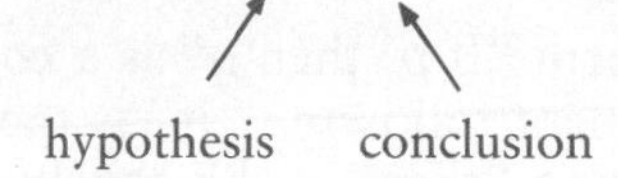

(1) If p, then q (3) q if p
(2) p implies q (4) p only if q

EXAMPLE 2 Copy the statement. Underline the hypothesis once and the conclusion twice. Then rewrite the statement in two other equivalent forms.

Two angles are supplementary if the sum of the angle measures is 180.

Solution The format as stated is "q if p."

Two angles are supplementary if the sum of the angle measures is 180.

If $\underline{p}$, then $\underline{q}$	If the sum of two angle measures is 180, then the angles are supplementary.
$\underline{p}$ implies $\underline{q}$	The fact that the sum of two angle measures is 180 implies that the angles are supplementary.

Recall that in any proof, the *given* is the hypothesis, and the *proof* is the conclusion.

EXAMPLE 3 Examine the proof below. State what is proved in "If . . . , then . . ." form.

Statement	Reason
1. m $\angle 2$ = m $\angle 3$	1. Given
2. $\overrightarrow{OR}$ bisects $\angle POQ$.	2. Given
3. m $\angle 1$ = m $\angle 2$	3. Def of $\angle$ bis
4. $\therefore$ m $\angle 1$ = m $\angle 3$	4. Sub

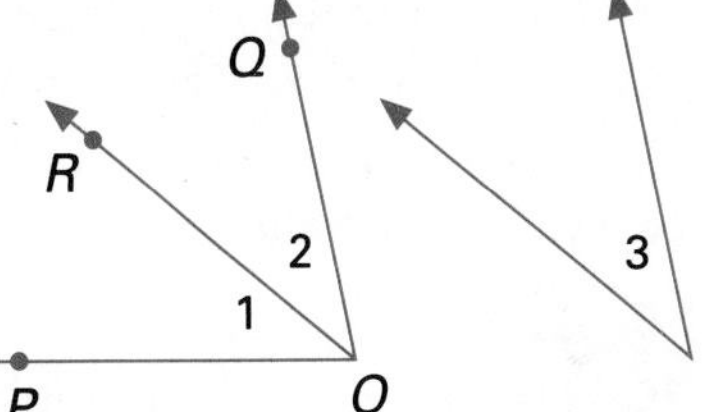

Plan There are two *given* statements. Hypothesis: m $\angle 2$ = m $\angle 3$ and $\overrightarrow{OR}$ bisects $\angle POQ$. The conclusion is the last statement: m $\angle 1$ = m $\angle 3$.

Solution Conditional: If m $\angle 2$ = m $\angle 3$ and $\overrightarrow{OR}$ bisects $\angle POQ$, then m $\angle 1$ = m $\angle 3$.

Focus on Reading

Copy and complete each statement.

1. A statement of the form "If . . . , then . . ." is called a __________.
2. In $p \rightarrow q$, p is called the __________.
3. In $p \rightarrow q$, q is called the __________.
4. $p \rightarrow q$ is false only if q is __________ and p is __________.

Classroom Exercises

Name the hypothesis and the conclusion of the statement.

1. $r \rightarrow s$
2. If $x = 4$, then $5 > x$
3. If $a < b$ and $b < c$, then $a < c$
4. m $\angle A < 90$ if $\angle A$ is acute
5. $AB = BC$ only if $BC = CD$

Written Exercises

Use the information to write a statement in "If . . ., then . . ." form.

1. Given: $\angle A$ is obtuse.
   Conclusion: $90 <$ m $\angle A < 180$
2. Given: Two segments are congruent.
   Conclusion: The segments have the same length.

Copy the conditional. Underline the hypothesis once and the conclusion twice. State whether the conditional is true.

3. If $x^3 = -27$, then $x = -3$
4. If $x^2 + 1 = 1$, then $x = 2$
5. If $x < 4$, then $x < 5$
6. If $x < -4$, then $x < -5$

Write the statement in "If . . . , then . . ." form.

7. Supplementary angles are two angles the sum of whose measures is 180.
8. A ray that bisects an angle divides the angle into two congruent angles.

Copy the statement. Underline the hypothesis once and the conclusion twice. Rewrite the statement in two other equivalent forms.

9. An angle is acute if its measure is less than 90.
10. The measure of an angle is 90 only if the sides of the angle are perpendicular.

Examine the proof. State what is proved in "If . . . , then . . ." form.

11.

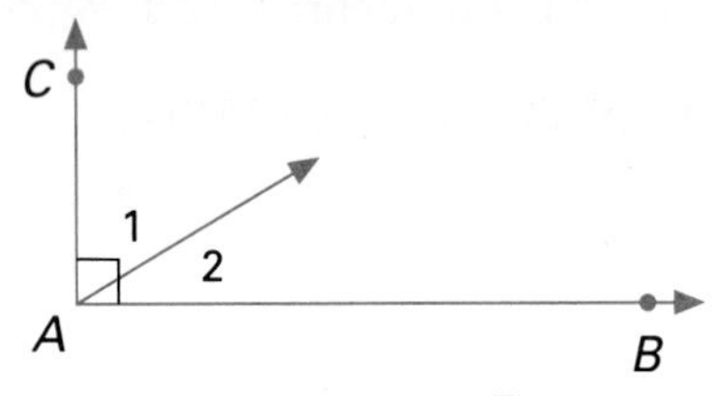

Statement	Reason
1. $\overrightarrow{AB} \perp \overrightarrow{AC}$	1. Given
2. $\therefore m\angle 1 + m\angle 2 = 90$	2. If outer rays of two acute adjacent $\angle$s are $\perp$, then the sum of measures is 90.

12.

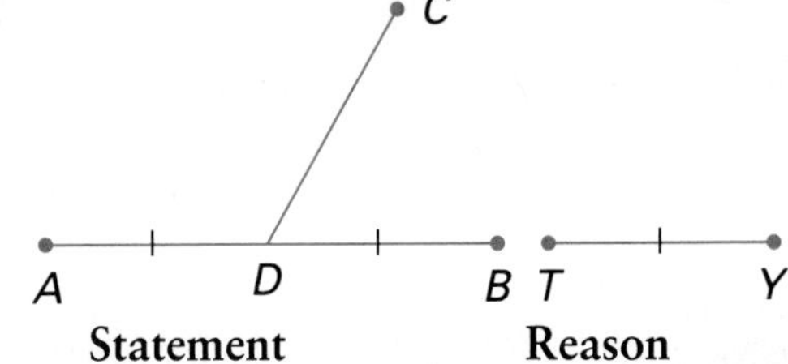

Statement	Reason
1. $\overline{CD}$ bisects $\overline{AB}$.	1. Given
2. $AD = BD$ ($\overline{AD} \cong \overline{BD}$)	2. Def of bis
3. $BD = TY$	3. Given
4. $\therefore AD = TY$	4. Trans Prop (or sub)

Copy and complete the truth table.

13.

p	q	$p \to q$	$p \wedge (p \to q)$
T	T	T	T
T	F		F
F	T		
F	F		

14.

p	q	$p \to q$	$p \vee (p \to q)$	$[p \vee (p \to q)] \to q$
T	T		T	T
T	F			F
F	T			
F	F			

Mixed Review 1.1, 1.4, 1.6, 1.7

Use the statements in Exercises 1–4 below.

p: A right angle has perpendicular sides.
q: The sum of the measures of two supplementary angles is 180.
r: A plane consists of a finite number of points.

1. Write the conjunction $p \wedge r$. Is the conjunction true? Give a reason.
2. Write the disjunction $p \vee r$. Is the disjunction true? Give a reason.
3. Write the conjunction $p \wedge q$. Is the conjunction true? Give a reason.
4. Write the disjunction $q \vee r$. Is the disjunction true? Give a reason.

2.6 Deductive and Inductive Reasoning

Objectives

To draw conclusions from given statements
To determine whether a conclusion is reached deductively or inductively
To use inductive reasoning to draw conclusions

The proofs you have been writing in this chapter are examples of *deductive reasoning*. Using deductive reasoning, you draw conclusions that follow from other statements.

EXAMPLE 1 Draw a conclusion that follows from the given statements.

Given: If a person studies geometry, then he or she develops an appreciation for logic. Maria studies geometry.

Solution Conclusion: Maria develops an appreciation for logic.

The given statements are often called **premises**. If the premises are true, then the conclusion drawn deductively is true, provided the argument used is valid. This example illustrates the importance of establishing a collection of statements in geometry that are accepted as true. Recall that such statements include undefined terms, definitions, and postulates.

EXAMPLE 2 Draw a conclusion from the given statement.

Given: $\overrightarrow{AB}$ bisects $\angle XAY$.

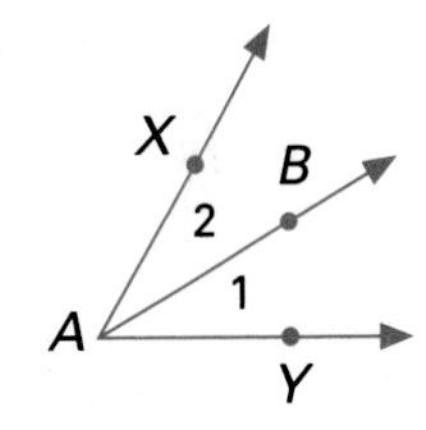

Solution Conclusion: $\angle 1 \cong \angle 2$. The definition of angle bisector, an accepted statement since Lesson 1.5, applies in this situation.

Another kind of deductive argument is illustrated by the following.

Given: If you live in Chicago, then you live in Illinois. $(p \rightarrow q)$

If you live in Illinois, then you live in the U.S.A. $(q \rightarrow r)$

If you live in Chicago, then you live in the U.S.A. $(p \rightarrow r)$

In this way, conditionals can be linked together.

Given: $p \rightarrow q$ *and* $q \rightarrow r$ Conclusion: $p \rightarrow r$

Another way of reaching conclusions is to use *inductive reasoning*. Scientists use this reasoning when, after repeating an observation or experiment, they make a generalization. For example, by repeated observations it is established that water freezes at 32°F or 0°C.

EXAMPLE 3 In each diagram of intersecting lines, use a protractor to find the measure of a pair of nonadjacent angles. Draw a conclusion inductively.

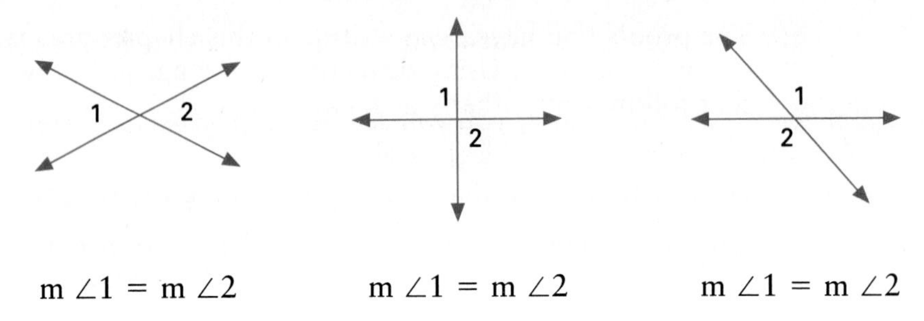

$m\angle 1 = m\angle 2$ $m\angle 1 = m\angle 2$ $m\angle 1 = m\angle 2$

Solution Conclusion: If two lines intersect, the pair of nonadjacent angles formed are equal in measure, or congruent.

Although the conclusions drawn inductively may *seem* true, they may not always *be* true. The ancient Egyptians farmed land in the shape of long narrow triangles. They observed that the formula

$$A = \frac{1}{2} \times \text{base length} \times \text{side length},$$

worked for finding the area of each of their triangular fields. They then concluded that this formula was true for all triangles. Actually, the correct formula is

$$A = \frac{1}{2} \times \text{base length} \times \text{height length}.$$

In the land the Egyptians measured, there was very little difference between the height length and true length of a side.

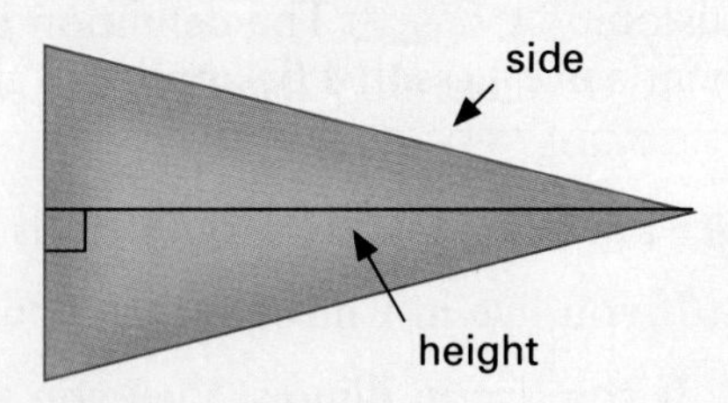

In geometry, inductive reasoning often serves to discover a new concept or theorem. However, to prove or to establish the truth of a statement, deductive reasoning is used.

Classroom Exercises

True or false. If false, give a reason.

1. Inductive reasoning is used to prove statements in geometry.
2. If $a \rightarrow b$ and $b \rightarrow c$, then it follows that $a \rightarrow c$.

Draw a valid conclusion from the given statements.

3. If M is the midpoint of $\overline{AB}$, then $AM = MB$. G is the midpoint of $\overline{WY}$.
4. If you go to the West Coast, you will see beautiful sunsets. If you see beautiful sunsets, you will feel exhilarated.

Written Exercises

Write a conclusion that follows from the given statements. (Exercises 1–5)

1. If a person brushes his or her teeth with No-Cav toothpaste, then he or she will not develop cavities. Mary brushes with No-Cav.
2. If today's date is December 31, then tomorrow is New Year's Day. Today's date is December 31.
3. If two angles are complementary, then they are congruent. $\angle A$ and $\angle B$ are complementary.
4. If you are under 16 years of age and live in New Jersey, then you cannot get a driver's license. If you cannot get a driver's license, then you cannot legally drive a car in New Jersey.
5. If you have a cold, then you will need rest. If you need rest, then you should not exercise strenuously.
6. Draw a triangle. Measure each of the angles. Find the sum of the measures of the three angles. Repeat this exercise with four more triangles. Inductively draw a conclusion.
7. In each of the following figures, $\overline{CD}$ bisects $\angle ACB$. Use a ruler to find AD and BD for the first three figures. Draw a conclusion. Measure $\overline{AD}$ and $\overline{BD}$ in figure **d.** What does the result illustrate about inductive reasoning?

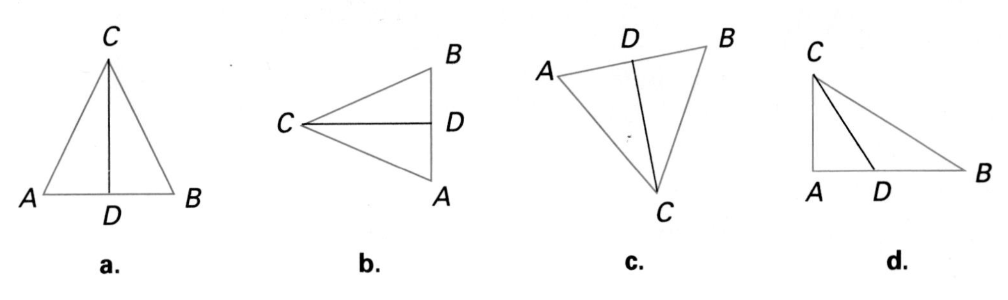

Tell whether the conclusion drawn is reached deductively or inductively. (Exercises 8–11)

8. A student finds the sum of the measures of the angles of 5 four-sided figures. In each figure, the sum is 360. The student concludes that in any four-sided figure, the sum of the measures of the angles is 360.

9. A person reading a map observes that the distance between two Towns *A* and *B* is equal to the sum of the distances from *A* to *C* and from *B* to *C*. The person concluded that Town *C* is between *A* and *B*.

10. A student has had good experiences with his high-school math teachers. He concludes that all high school math teachers are nice people.

11. A wall 10 ft wide consists of 20 bricks across. A mason concludes that a wall 15 ft wide will consist of 30 bricks.

There is a pattern for determining the number of segments that can be drawn between each of the points in a set of points, no three of which are collinear. Find the number of such segments for the given number of points, no three of which are collinear. To get started, draw figures like the ones shown. Then see if you can discover a pattern.

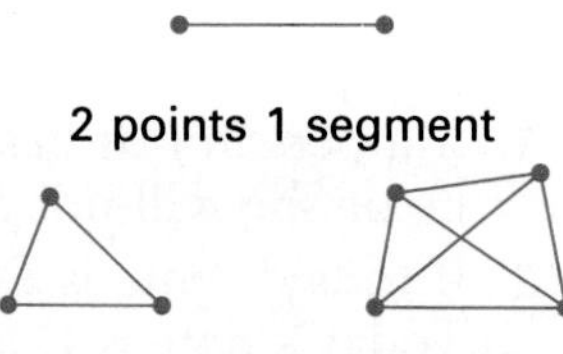

12. 5 points **13.** 6 points **14.** 7 points **15.** 20 points

Mixed Review

Find the measure of a complement and a supplement, if possible, of an angle with the given measure. *1.6*

1. $m \angle A = 110$ **2.** $m \angle B = 35$ **3.** $m \angle A = x$ **4.** $m \angle A = x + 30$

5. Graph the conjunction: $15 < x - 30 < 45$. *1.8*

Brainteaser

The "proof" below claims to show that 2 = 0. Find the error in reasoning.

1. Let $x = 1$ and $y = 1$.	
2. $x = y$	Substitute.
3. $x^2 = y^2$	Square both sides.
4. $x^2 - y^2 = 0$	Subtract y^2 from both sides.
5. $(x + y)(x - y) = 0$	Factor.
6. $x + y = 0$	Divide each side by $(x - y)$.
7. $2 = 0$	Substitute.

2.7 Proving Theorems About Angles

Objective To prove and apply theorems about supplementary angles and complementary angles

Two angles are supplementary if the sum of their measures is 180. Suppose that m $\angle A$ = 110 and m $\angle B$ = 110. The measure of a supplement of $\angle A$ is 70. Similarly, the measure of a supplement of $\angle B$ is 70. The supplements of the two angles are congruent.

Theorem 2.1 If two angles are supplements of congruent angles, then they are congruent. (Supplements of congruent angles are congruent.)

Given: $\angle B \cong \angle C$, $\angle A$ is a supplement of $\angle B$, and $\angle D$ is a supplement of $\angle C$.

Prove: $\angle A \cong \angle D$

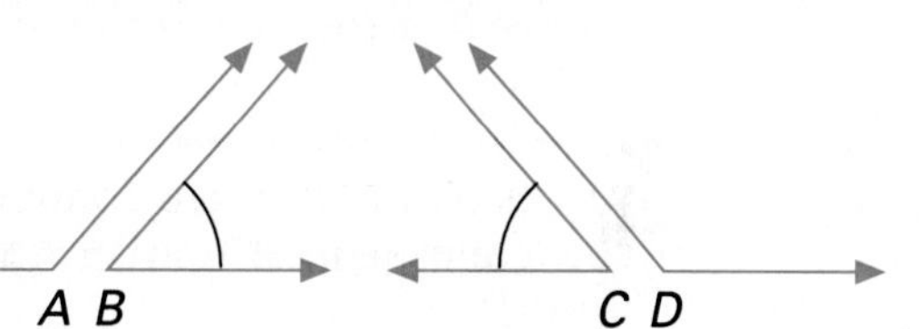

Plan m $\angle A$ + m $\angle B$ = 180 and m $\angle D$ + m $\angle C$ = 180, so m $\angle A$ + m $\angle B$ = m $\angle D$ + m $\angle C$. Use subtraction to get m $\angle A$ = m $\angle D$.

Proof

Statement	Reason
1. $\angle A$ is a supplement of $\angle B$; $\angle D$ is a supplement of $\angle C$.	1. Given
2. m $\angle A$ + m $\angle B$ = 180, m $\angle D$ + m $\angle C$ = 180	2. Def of supp $\angle$s
3. m $\angle A$ + m $\angle B$ = m $\angle D$ + m $\angle C$	3. Sub
4. m $\angle B$ = m $\angle C$	4. Given
5. $\therefore$ m $\angle A$ = m $\angle D$ ($\angle A \cong \angle D$)	5. Equations may be subtracted.

A **corollary** of a theorem is a theorem whose proof follows from the original theorem in a few steps. The statement below is a corollary of Theorem 2.1.

Corollary If two angles are supplements of the same angle, then they are congruent. (Supplements of the same angle are congruent.)

A theorem or a corollary may be used as a reason in a proof. In Example 1, notice how Theorem 2.1 is used as a reason in Step 3.

EXAMPLE 1 Given: $\angle 1 \cong \angle 4$
Prove: $\angle 2 \cong \angle 3$

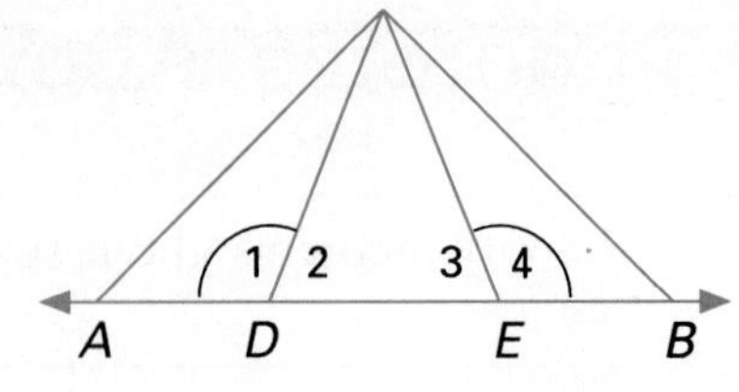

Proof

Statement	Reason
1. $\angle 1 \cong \angle 4$	1. Given
2. $\angle 2$ is a supplement of $\angle 1$; $\angle 3$ is a supplement of $\angle 4$.	2. If outer rays of two adj $\angle$s form a st $\angle$, then the $\angle$s are supp.
3. $\angle 2 \cong \angle 3$	3. Supp of $\cong$ $\angle$s are $\cong$.

You know that two angles are complementary if the sum of their measures is 90. A proof similar to that of Theorem 2.1 can be given to prove a theorem about complements of congruent angles.

Theorem 2.2 If two angles are complements of congruent angles, then they are congruent. (Complements of congruent angles are congruent.)

Given: $\angle B \cong \angle C$, $\angle A$ is a complement of $\angle B$, and $\angle D$ is a complement of $\angle C$.

Prove: $\angle A \cong \angle D$

You will be asked to prove this theorem in Exercise 15.

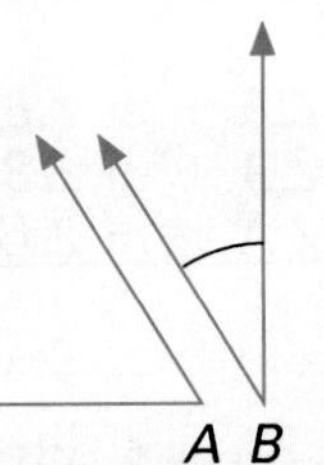
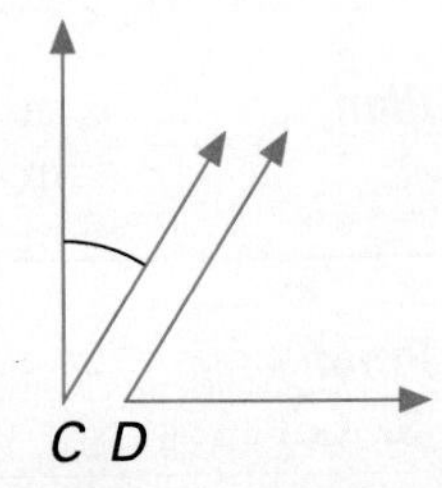

Corollary If two angles are complements of the same angle, then they are congruent. (Complements of the same angle are congruent.)

EXAMPLE 2 Given: m $\angle UQT = 25$, m $\angle SQT = 25$, $\overrightarrow{QT} \perp \overleftrightarrow{PR}$

What conclusion can be drawn about angle measures? Why?

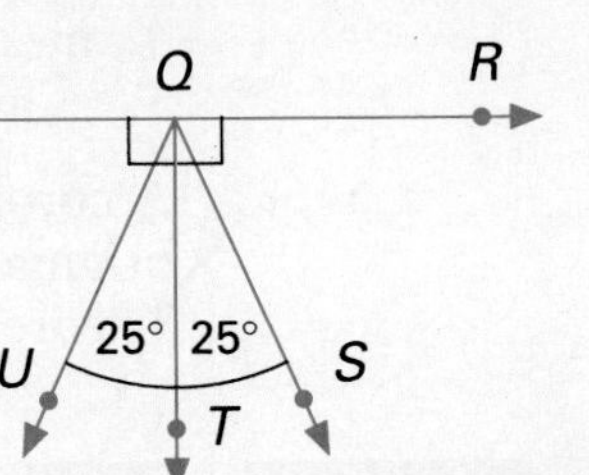

Plan Since $\overrightarrow{QT} \perp \overleftrightarrow{PR}$, $\angle PQU$ is a complement of $\angle UQT$. Similarly, $\angle RQS$ is a complement of $\angle SQT$. Apply Theorem 2.2.

Solution $\angle PQU \cong \angle RQS$ since they are complements of congruent angles. Thus, m $\angle PQU = 90 - 25 = 65$, and m $\angle RQS = 65$.

Theorem 2.3 If two angles are right angles, then they are congruent.

You will be asked to prove this theorem in Exercise 17.

Focus on Reading

Write the term for each abbreviation.

1. adj	**2.** bis	**3.** coll	**4.** comp	**5.** $\perp$
6. cor	**7.** def	**8.** $\angle$	**9.** midpt	**10.** Post
11. prop	**12.** rt	**13.** st	**14.** supp	**15.** Thm

Classroom Exercises

Complete the statement of each conclusion. Give a reason for each conclusion.

1. Given: $\angle 5 \cong \angle 6$
Conclusion: $\angle 3 \cong$ __________

2. Given: $\angle 6$ is a supplement of $\angle 8$.
Conclusion: $\angle 8 \cong$ __________

3. Given: $\angle 5$ is a supplement of $\angle 7$.
Conclusion: $\angle 3 \cong$ __________

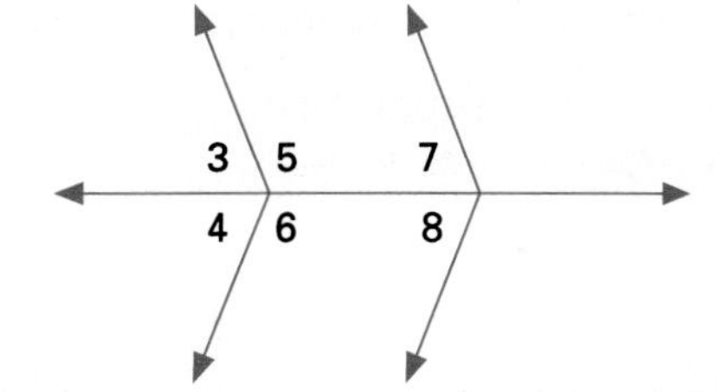

Written Exercises

State a conclusion that may be drawn. Give a reason.

1. Given: m $\angle 1 = 60$, m $\angle 2 = 60$

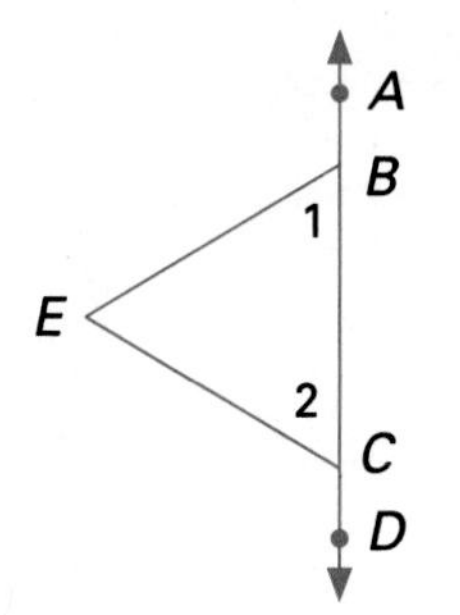

2. Given: m $\angle ABF = 42$, m $\angle ECD = 42$, $\overrightarrow{BA} \perp \overline{BC}$, $\overrightarrow{CD} \perp \overline{BC}$

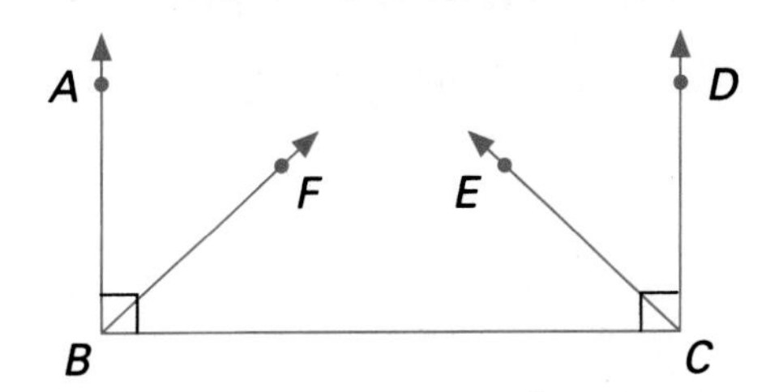

3. Given: $\overline{AB} \perp \overline{AD}$, $\overline{BC} \perp \overline{CD}$, and $\angle DAC \cong \angle ACB$

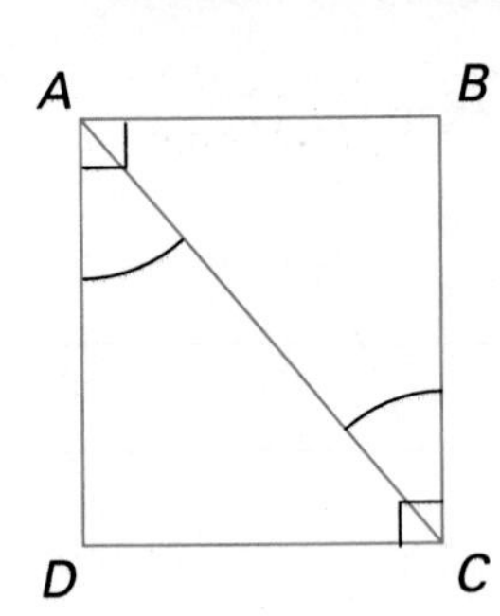

4. Given: $\angle FBC \cong \angle ECD$

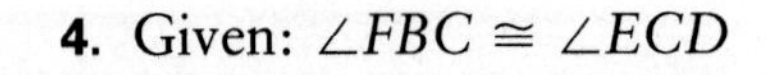

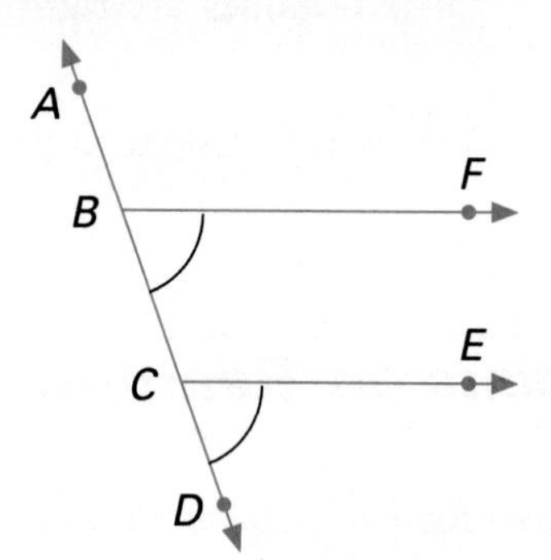

5–8. Write a proof of each conclusion in Exercises 1–4.

9. The figure at the right shows a roof-truss system. The beam from A to B is perpendicular to the crossbeam from C to D, and $\angle ABE \cong \angle ABF$. Draw a conclusion about m $\angle 1$. Prove it.

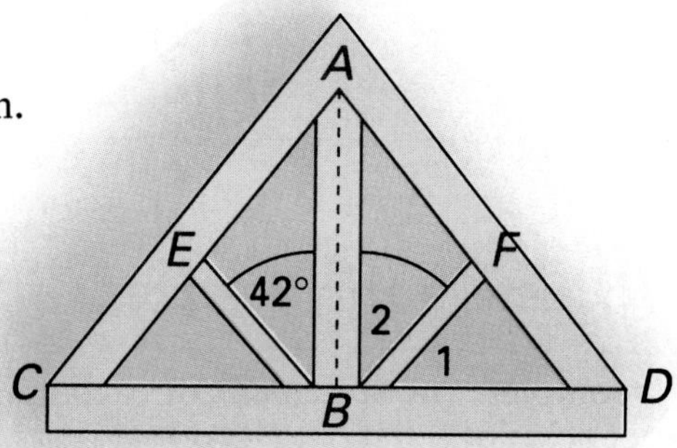

Write the proof. Give statements and reasons.

10. Given: $\angle 3$ is a supplement of $\angle 2$.
Prove: $\angle 2 \cong \angle 1$

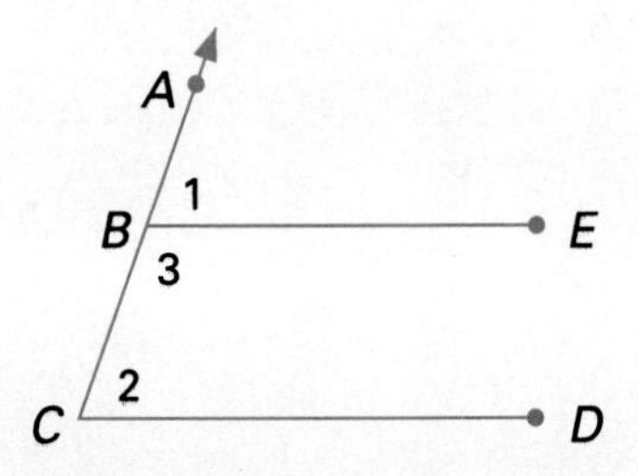

11. Given: $\overrightarrow{QP} \perp \overrightarrow{QR}$, and $\angle 3$ is a complement of $\angle 2$.
Prove: $\angle 1 \cong \angle 3$

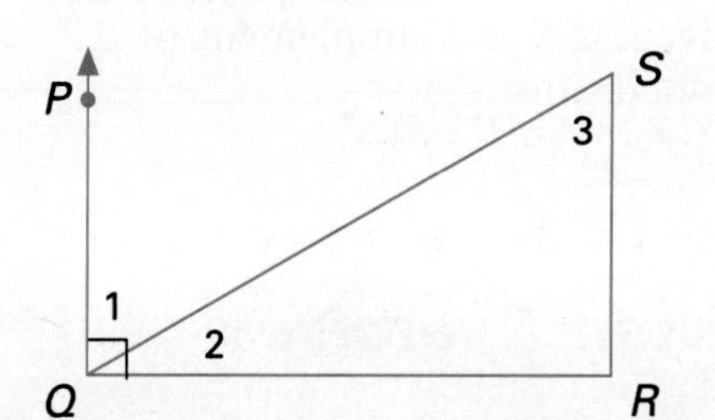

12. Given: $\overrightarrow{CA}$ bisects $\angle ECF$, and $\overrightarrow{CA} \perp \overleftrightarrow{BD}$.
Prove: m $\angle DCE$ = m $\angle BCF$

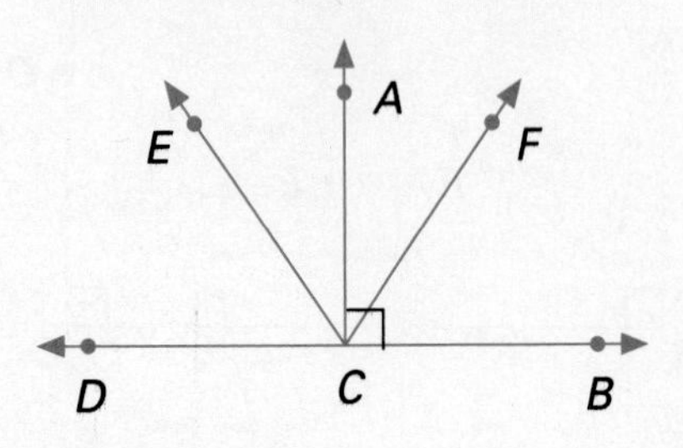

13. Given: m $\angle 2$ = m $\angle 5$, m $\angle 3$ = m $\angle 5$
Prove: m $\angle 4$ = m $\angle 1$

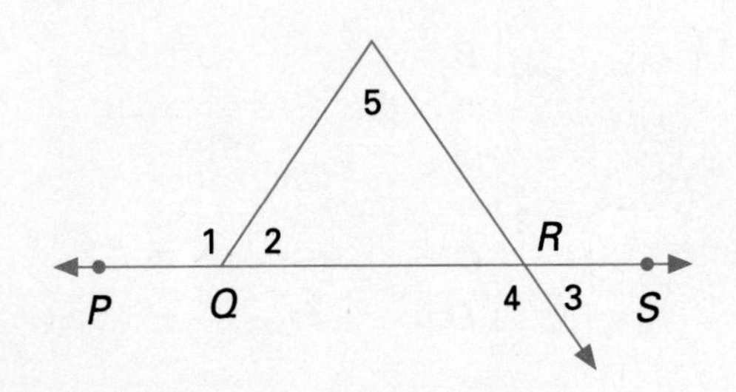

14. Prove the corollary to Theorem 2.1.

15. Prove Theorem 2.2.

16. Prove the corollary to Theorem 2.2.

17. Prove Theorem 2.3.

18. Given: $\overrightarrow{QP}$ bisects $\angle SQT$.
Prove: $\angle 2$ is a supplement of $\angle 3$.

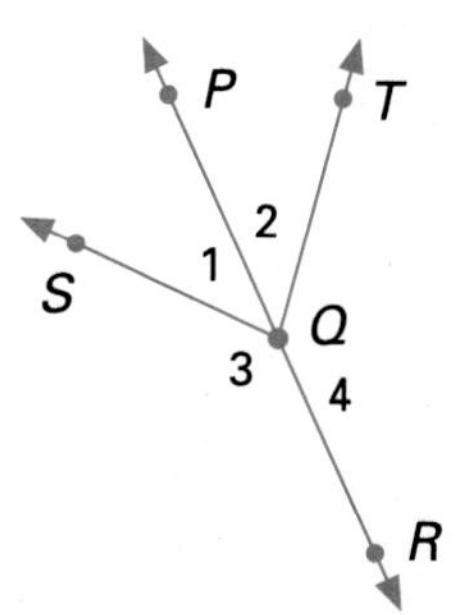

19. Given: $\angle 1$ is a supplement of $\angle 4$.
Prove: $\angle 2$ is a supplement of $\angle 3$.

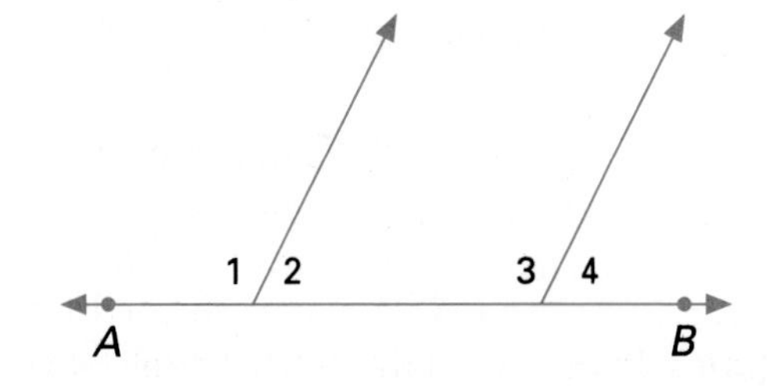

20. Given: $\overline{BA} \perp \overline{BC}$, $\overline{CD} \perp \overline{CB}$, $\angle 2 \cong \angle 3$
Prove: $\angle 1$ is a supplement of $\angle DCF$.

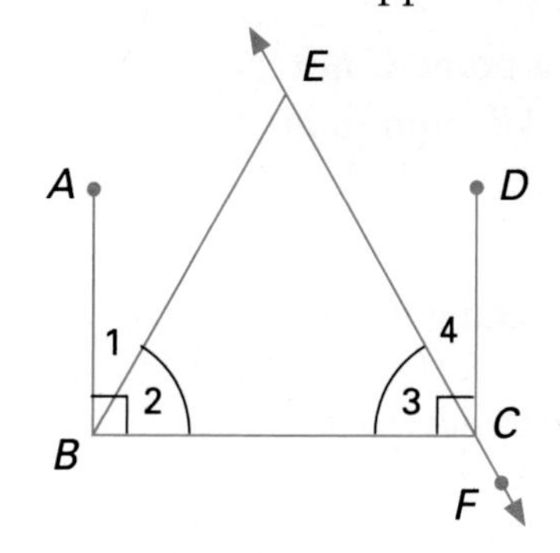

Mixed Review

Determine if the statement is true or false. *1.7, 2.5*

1. $2 \cdot 5 + 4 = 15$ *or* the measure of the complement of $\angle A$ is $180 - m\angle A$.
2. If $\overrightarrow{AB}$ is a ray, then a segment has one endpoint.
3. For what truth values of p and q is $p \wedge q$ true?
4. For what truth values of p and q is $p \vee q$ false?
5. For what truth values of p and q is $p \rightarrow q$ false?

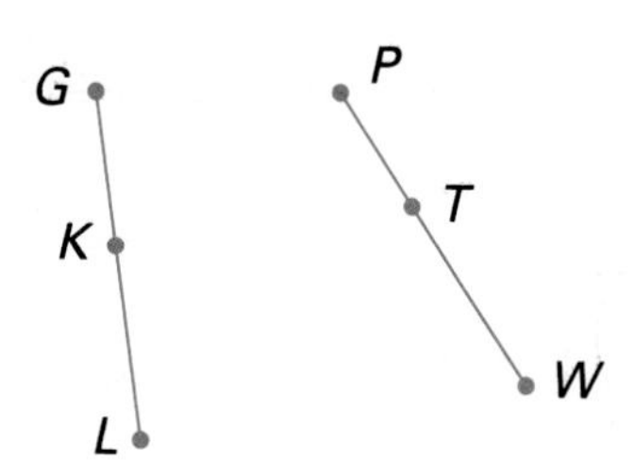

6. Can you conclude from the figure that $GK + KL = GL$? Why or why not? *2.1*
7. Can you conclude that $GK = PT$? Why or why not? *2.1*

Computer Investigation

Vertical Angles

Use a computer software program that constructs line segments, labels and moves points, extends line segments, and measures line segments and angles.

The figure to the right shows two segments, $\overline{AB}$ and $\overline{DC}$, intersecting at point E. The activities are designed to help you discover a relationship between certain pairs of angles formed by two intersecting lines. This relationship will be stated and proved in the next lesson.

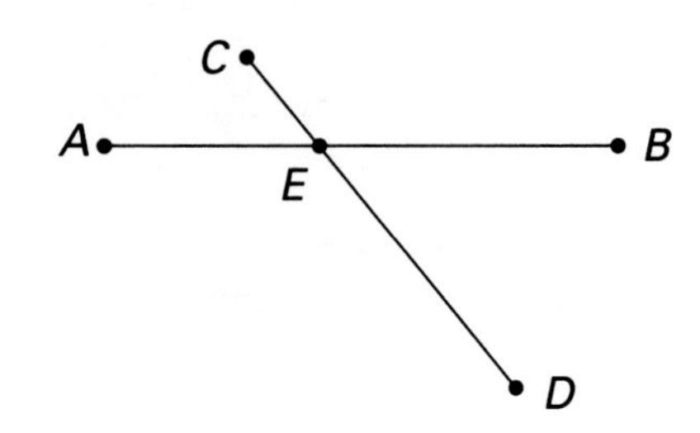

Activity 1

Draw a segment, $\overline{AB}$, of some length, say 8 units. Draw a point C not on $\overline{AB}$. Draw a second point, D, on the opposite side of $\overline{AB}$ from point C. Draw $\overline{CD}$. Label the point of intersection of the two segments E.

1. Find m $\angle DEA$, m $\angle BEC$, m $\angle DEB$, and m $\angle AEC$.
2. Clear the screen and draw a new pair of intersecting segments. Repeat the directions of Exercise 1 for this new drawing.
3. Repeat the process above for a third pair of intersecting segments. (Keep this pair of intersecting segments on the screen.)

Activity 2

Label a point F somewhere in the interior of $\angle AED$. Draw $\overline{EF}$.

4. Find m $\angle FED$ and m $\angle FEA$.
5. Does either angle measure of Exercise 4 equal m $\angle CEB$? Can you guess why these angle measures are not equal when those of the previous activity were equal?

Summary

Try to state the generalization of the discovery activities of this page. Use the following questions to help make the generalization.

1. What are $\angle AEC$ and $\angle CEB$ called?
2. Can you think of a description of $\angle CEB$ and $\angle AED$ to distinguish them from the type above?
3. Complete the following generalization:
   If two lines intersect, then __________ are congruent.

2.8 Vertical Angles

Objective To identify vertical angles
To apply the Vertical Angles Theorem

As shown at the right, two intersecting lines form two pairs of *nonadjacent* angles. $\angle 1$ and $\angle 2$ are nonadjacent. $\angle 3$ and $\angle 4$ are nonadjacent.

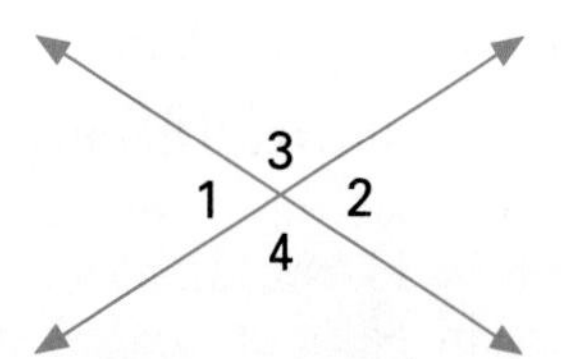

Definition **Vertical angles** are two nonadjacent angles formed by intersecting lines.

EXAMPLE 1 State whether the given angles are vertical. Give a reason.

$\angle XOW$ and $\angle YOZ$

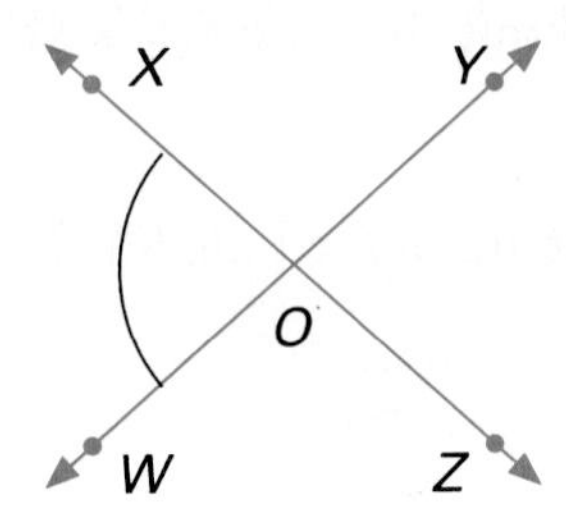

$\angle AOB$ and $\angle DOC$

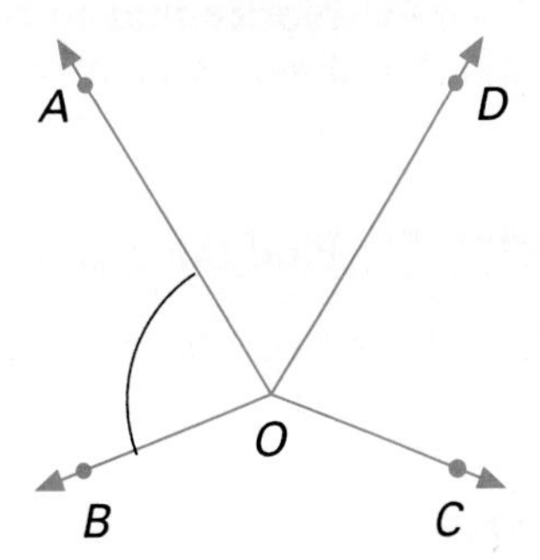

Solutions $\angle XOW$ and $\angle YOZ$ are vertical angles. They are nonadjacent angles formed by two intersecting lines.

$\angle AOB$ and $\angle DOC$ are not vertical angles. They are not formed by intersecting lines.

EXAMPLE 2 Name a second angle to form a pair of vertical angles: $\angle DOC$ and __________.

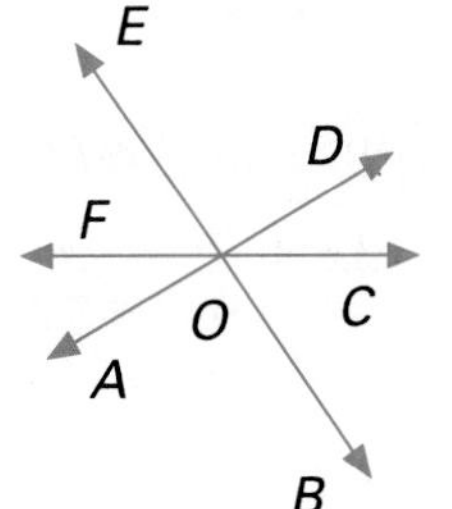

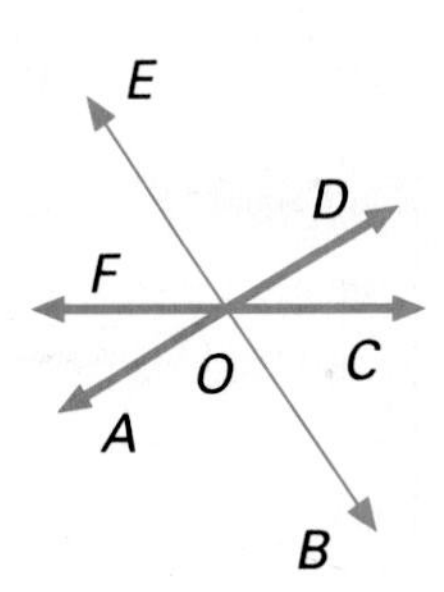

Plan Copy the diagram. Shade the lines that intersect to form $\angle DOC$.

Solution Thus, $\angle DOC$ and $\angle AOF$ are vertical angles.

You may have noticed that vertical angles always appear to have the same measure. The following theorem proves this is true.

Theorem 2.4 **Vertical Angles Theorem:** Vertical angles are congruent.

Given: A pair of vertical angles ($\angle 1$ and $\angle 2$)

Prove: The angles are congruent ($\angle 1 \cong \angle 2$).

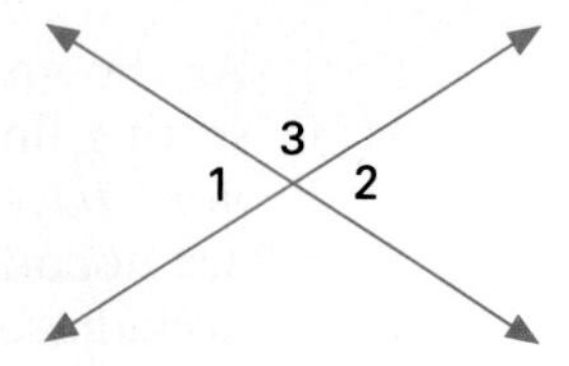

Proof

Statement	Reason
1. $\angle 1$ and $\angle 2$ are vert $\angle$s.	1. Given
2. $\angle 1$ is a supp of $\angle 3$; $\angle 2$ is a supp of $\angle 3$.	2. If the outer rays of two adj $\angle$s form a st $\angle$, then the $\angle$s are supp.
3. $\therefore \angle 1 \cong \angle 2$	3. Supp of same $\angle$ are $\cong$.

Notice that to prove the theorem, you only need to show that $\angle 1 \cong \angle 2$. Every statement in the proof about these angles is also true for any other pair of vertical angles.

EXAMPLE 3 Find the measure of each of the numbered angles.

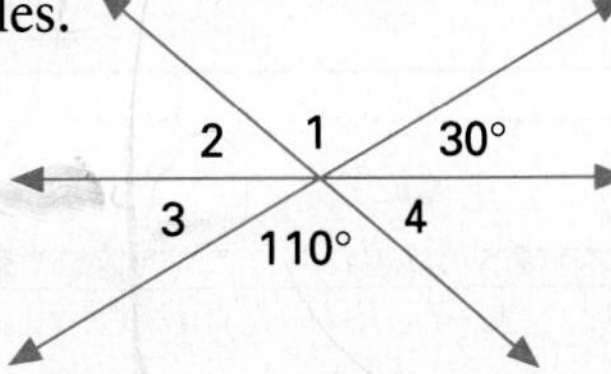

Solutions Look for vertical angles. m $\angle 1 = 110$, m $\angle 3 = 30$ by the Vertical Angles Theorem. m $\angle 2 = 180 - (110 + 30) = 40$. m $\angle 2$ and m $\angle 4$ are vertical angles. m $\angle 4 =$ m $\angle 2 = 40$.

EXAMPLE 4 Find m $\angle APB$.

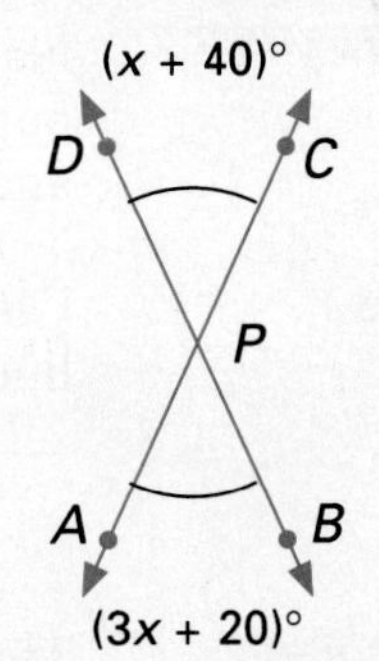

Solution m $\angle APB =$ m $\angle DPC$ by Vertical Angles Theorem

$$3x + 20 = x + 40$$
$$2x = 20$$
$$x = 10$$

Thus, m $\angle APB = 3x + 20$
$= 3 \cdot 10 + 20 = 50$.

EXAMPLE 5 Given: $\overrightarrow{PQ} \perp \overrightarrow{PR}$

Prove: $m\angle 2 + m\angle 3 = 90$

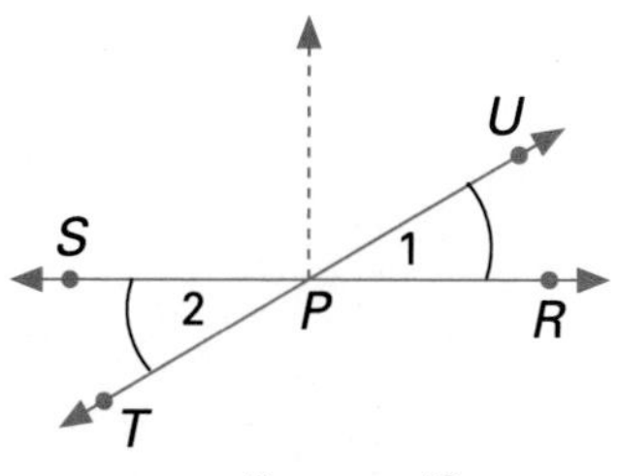

$m\angle 1 = m\angle 2$

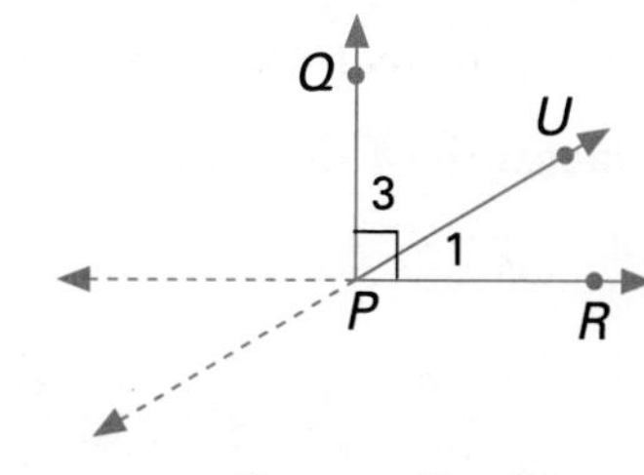

$m\angle 1 + m\angle 3 = 90$

Proof

Statement	Reason
1. $\overrightarrow{PQ} \perp \overrightarrow{PR}$	1. Given
2. $\angle 1$ and $\angle 2$ are vert$\angle$s.	2. Def of vert $\angle$s
3. $m\angle 1 = m\angle 2$ $(\angle 1 \cong \angle 2)$	3. Vert $\angle$s are $\cong$
4. $m\angle 1 + m\angle 3 = 90$	4. If outer rays of two acute adj $\angle$s are $\perp$, then the sum of the measures is 90.
5. $m\angle 2 + m\angle 3 = 90$	5. Sub

Corollary If two lines are perpendicular, then four right angles are formed.

Classroom Exercises

State whether the given pair of angles is vertical. Give a reason for the answer.

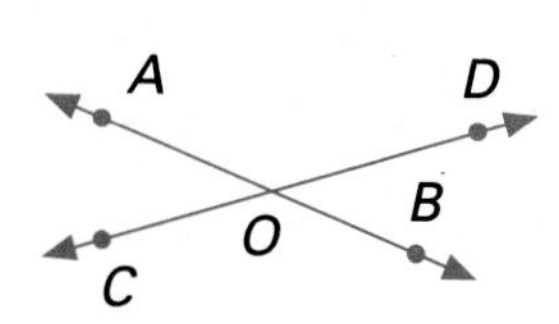

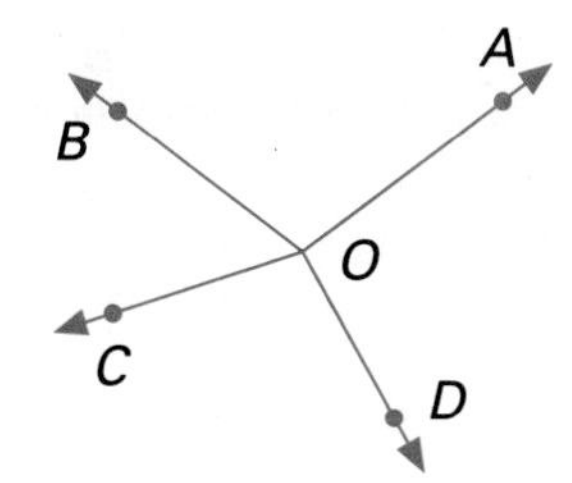

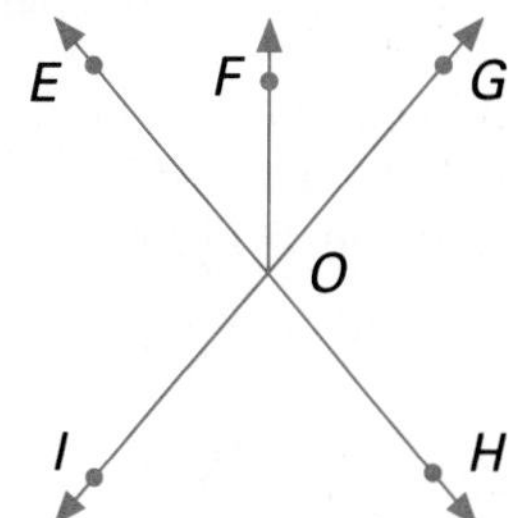

1. $\angle AOD$ and $\angle BOC$

2. $\angle AOC$ and $\angle COB$

3. $\angle AOB$ and $\angle DOC$

4. $\angle BOC$ and $\angle BOA$

5. $\angle GOH$ and $\angle HOI$

6. $\angle IOH$ and $\angle EOG$

Find the measure of the angle.

7. m $\angle 1$
8. m $\angle 2$

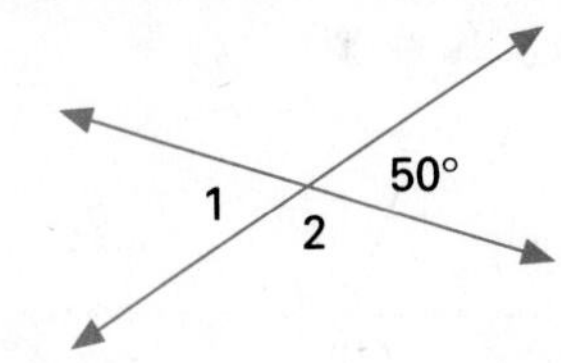

Written Exercises

Find the measure of the angle.

1. m $\angle 4$ 2. m $\angle 2$

3. m $\angle 1$ 4. m $\angle 3$

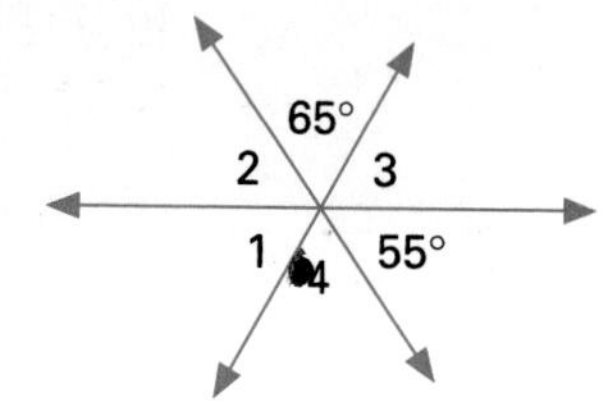

Find the indicated measure.

5. m $\angle AOB$

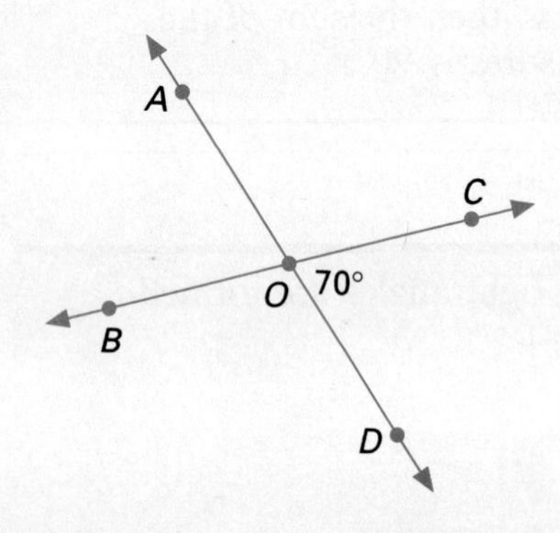

6. m $\angle RST$

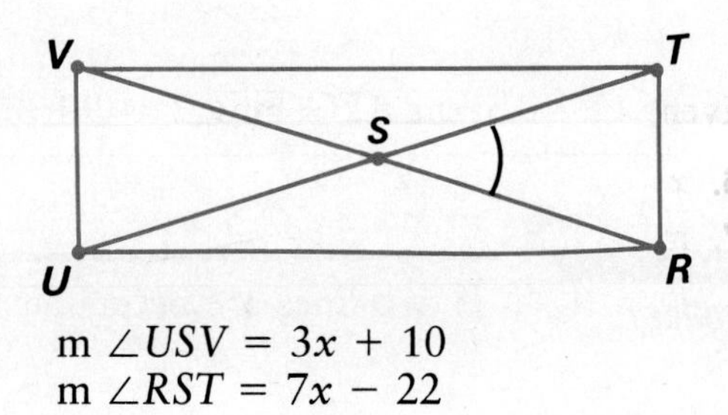

m $\angle USV = 3x + 10$
m $\angle RST = 7x - 22$

Use the figure at the right for Exercises 7–10.

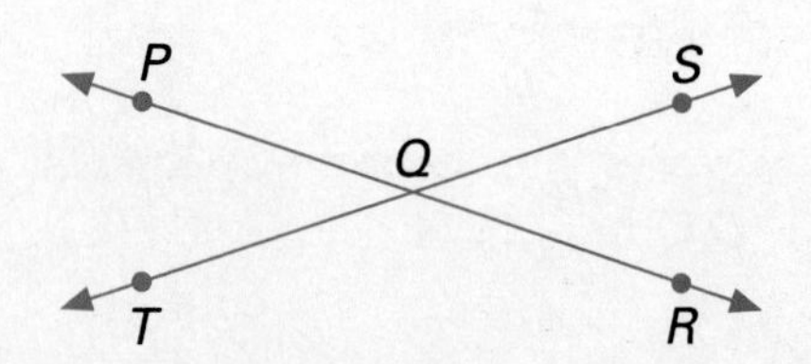

7. Given: m $\angle PQT = 2x - 20$,
   m $\angle SQR = x + 10$
   Find m $\angle TQR$.
8. Given: m $\angle PQT = 3y - 20$,
   m $\angle SQR = y + 20$
   Find m $\angle SQR$.
9. Given: m $\angle SQR = 5x - 10$,
   m $\angle TQP = x + 30$
   Find: m $\angle PQS$.
10. Given: m $\angle PQT = 8y + 8$,
    m $\angle RQS = 10y - 12$
    Find: m $\angle PQS$.

In the figure at the right, $\overleftrightarrow{US} \perp \overleftrightarrow{PR}$ and m $\angle VQU = 23$. Find the indicated measure.

11. m $\angle SQT$
12. m $\angle TQR$
13. m $\angle PQT$

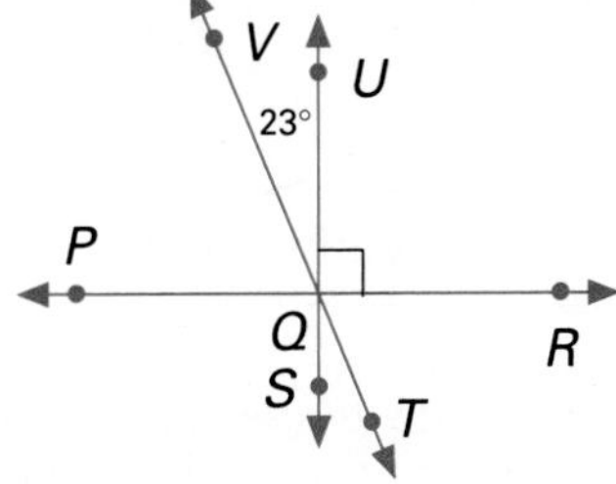

14. Given: m $\angle 2$ = m $\angle 3$
Prove: m $\angle 2$ = m $\angle 1$

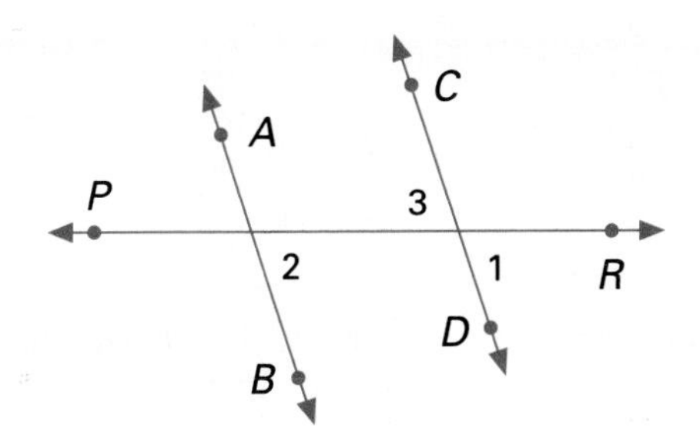

15. Given: The figure with $\overleftrightarrow{AC}$ intersecting $\overleftrightarrow{DE}$ at B
Prove: m $\angle 5$ + m $\angle 4$ = m $\angle DBF$

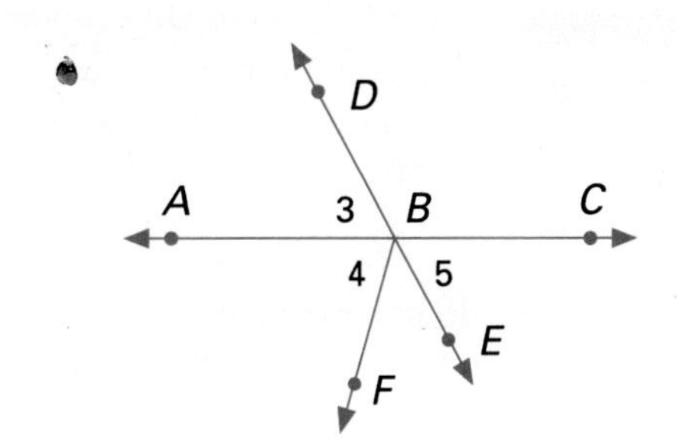

Given: $\overrightarrow{FB}$ bisects $\angle AFC$. Find each of the following values.

16. x
17. m $\angle DFE$
18. m $\angle DFA$

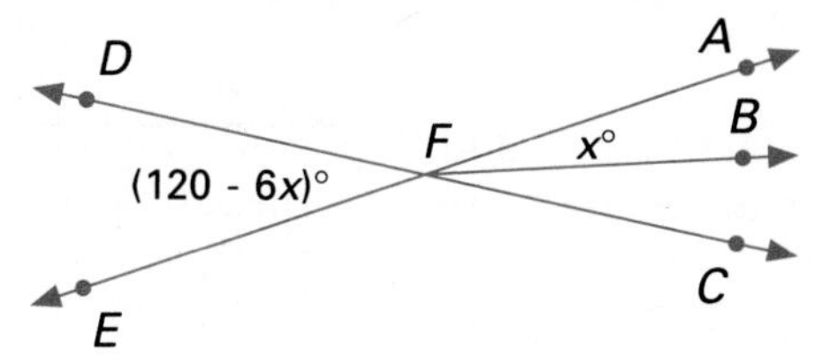

19. Write an explanation of how to identify vertical angles.
20. Prove the corollary to the Vertical Angles Theorem.

Mixed Review 1.1

1. What is the intersection of planes $\mathcal{M}$ and $\mathcal{P}$?
2. Points A, B, G, and ___ are coplanar.
3. Points C, D, and ___ are collinear.
4. Which labeled planes contain point C?
5. What is the intersection of planes $\mathcal{M}$ and $\mathcal{Q}$?
6. Is B between F and E? **1.2**
7. Name a pair of supplementary angles. **1.6**

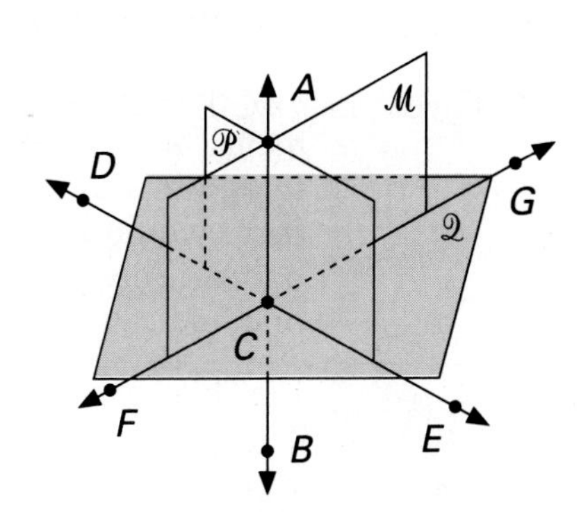

2.9 Postulates and Theorems: Points, Lines, Planes

Objective To apply postulates and theorems relating points, lines, and planes

In Chapter 1, *point*, *line*, and *plane* were accepted as undefined terms. *Space* was defined as the set of all points. Now that the idea of proof has been introduced, theorems that relate points, lines, and planes can be developed. First, it is necessary to state some postulates.

Postulate 8

A line contains at least two points.
A plane contains at least three noncollinear points.
Space contains at least four noncoplanar points.

If you walk in a straight line between any two points in the classroom, there is exactly one path. This suggests the following postulate.

Postulate 9

For any two points, there is exactly one line containing them.

The phrases "exactly one" and "one and only one" are used interchangeably in mathematics. Also, in this book phrases such as "two points" or "two lines" mean two distinct, or different, points or lines.

EXAMPLE Can two lines intersect in more than one point? Give an informal argument to support the answer.

Solution No, they cannot. The following *indirect* argument can be used. Suppose lines l and m had two points of intersection, P and Q. Then there would be two lines containing these two points. This contradicts Postulate 9, since only one line can contain any two points.

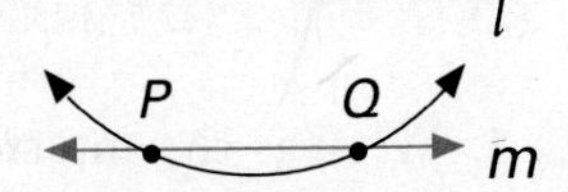

The argument in the Example is an **indirect proof**. This type of proof will be studied in greater detail in the next chapter.

Theorem 2.5 Two lines intersect at exactly one point.

A ruler (line) placed on a desk top (plane) illustrates the basis for another postulate about lines and planes. If any two points of the ruler touch the plane of the desk top, then the entire ruler (without bending) must lie on the flat surface.

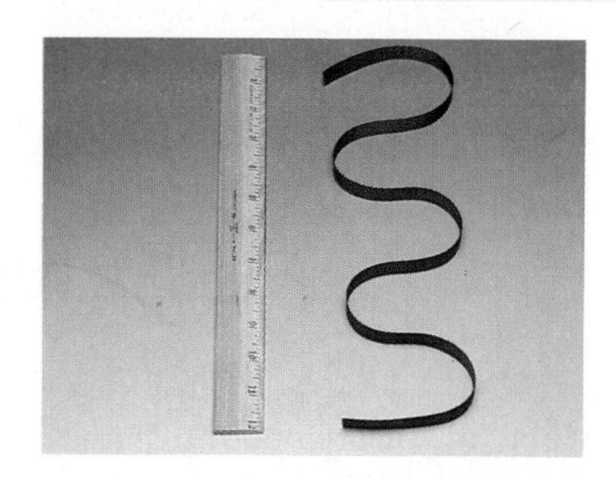

Postulate 10 If two points of a line are in a given plane, then the line itself is in the plane.

In the figure, $\overleftrightarrow{PQ}$ intersects plane $\mathcal{M}$ in point O. But the line is not contained in the plane. It has one and only one point in common with the plane. This result is stated below as a theorem. Proof is based on the kind of indirect argument used in Example 1.

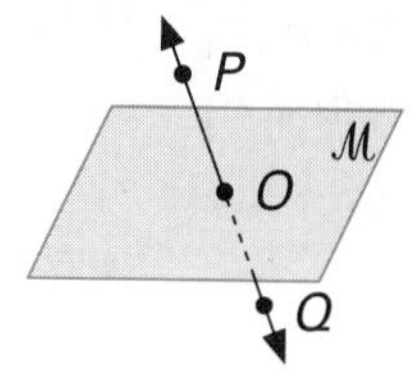

Theorem 2.6 If a line intersects a plane but is not contained in the plane, then the intersection is exactly one point.

Proof Suppose line l intersects plane $\mathcal{P}$ in two or more points. Then, according to Postulate 10, the line is contained in the plane. This contradicts the information given that the line is *not* contained in the plane. Therefore, the line intersects the plane in exactly one point.

Postulate 11 If two planes intersect, then they intersect in exactly one line.

You may have noticed that some four-legged chairs wobble a bit even when placed on a level floor. On the same floor, a three-legged stool does not wobble. This is because the ends of the legs of a three-legged stool always lie in exactly one plane, whereas the ends of the legs of a four-legged stool need not.

Postulate 12 Three noncollinear points are contained in exactly one plane.

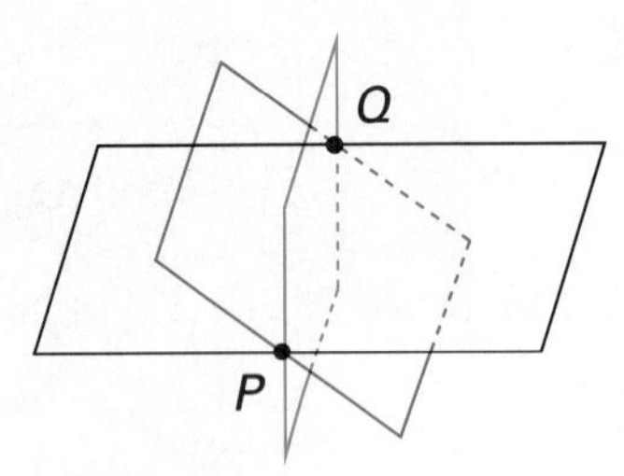

Two points: infinite number of planes

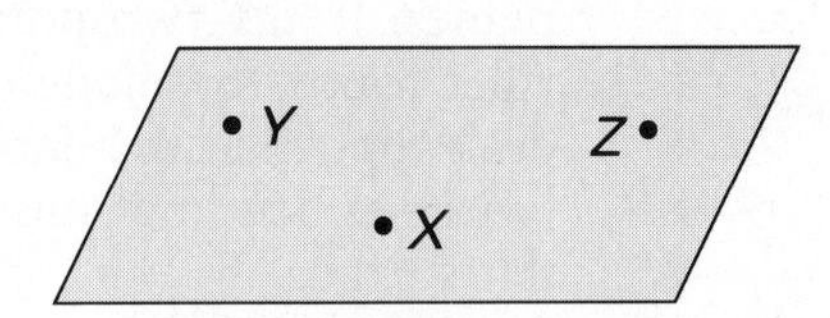

Three noncollinear points: exactly one plane

Postulates 8, 10, and 12 can be used to prove the following theorem.

Theorem 2.7 A line and a point not on the line are contained in exactly one plane.

Given: Line l and point P not on l

Prove: Exactly one plane contains P and l.

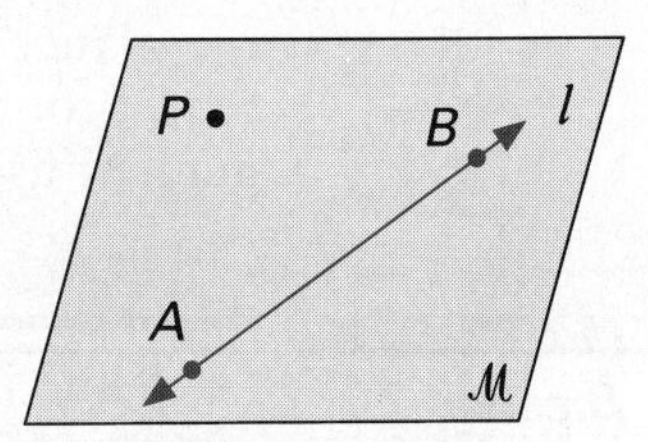

Plan A line contains at least two points, say, A and B. Exactly one plane contains the three noncollinear points P, A, and B.

Proof

Statement	Reason
1. Line l and point P not on l	1. Given
2. l contains at least two points. Name these points A and B.	2. A line contains at least two points.
3. P, A, and B are noncollinear.	3. Def of noncoll points
4. Exactly one plane contains P, A, and B (plane $\mathcal{M}$).	4. Three noncoll points are contained in exactly one plane.
5. ∴ Exactly one plane contains P and l.	5. If two points of a line lie in a plane, then the line lies in the plane.

The following theorem is a direct consequence of Theorem 2.7.

Theorem 2.8 Two intersecting lines are contained in exactly one plane.

Classroom Exercises

Which postulate or theorem guarantees the truth of the statement?

1. $\overleftrightarrow{AB}$ and $\overleftrightarrow{CD}$, two different intersecting lines, have only one point in common.
2. Given two points R and S, there is one and only one line containing them.
3. If r and s are two different intersecting lines, then one and only one plane can contain both of them.

Written Exercises

1. Can a line that does not lie in a plane intersect the plane in more than one point? Give a reason for your answer.
2. Can two planes intersect in exactly one point? Give a reason for your answer.
3. How many different planes can contain the same line?
4. How many different planes can contain a given obtuse angle?
5. Given any three points in space, is it possible that there is no plane that contains all three of the points? Give a reason for your answer.
6. Given any four points in space, is it possible that there is no plane that contains all four points? Give an example to support your answer, using points and planes of the classroom.

Indicate whether the statement is always true, sometimes true, or never true. (Exercises 7–9)

7. Three coplanar points are collinear.
8. Three collinear points are coplanar.
9. Two lines are contained in exactly one plane.
10. Draw a figure illustrating that it is possible for four planes to have only one point in common.
11. Prove Theorem 2.8.

Mixed Review

For each of the Exercises 1–3, draw a conclusion. Then prove it. *2.1, 2.2*

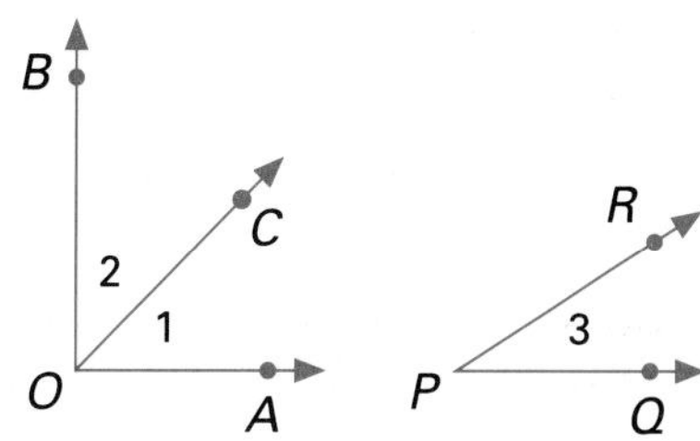

1. Given: $\angle AOB$
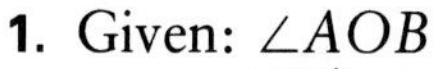
2. Given: $\overrightarrow{OC}$ bisects $\angle AOB$.
3. Given: $\overrightarrow{OB} \perp \overrightarrow{OA}$
4. Given: $m\angle 1 = m\angle 3$
   Prove: $m\angle 2 + m\angle 3 = m\angle AOB$

Chapter 2 Review

Key Terms

conditional (p. 63)
corollary (p. 71)
deductive reasoning (p. 67)
inductive reasoning (p. 68)
Reflexive Property (p. 49)
Substitution Property (p. 49)
Symmetric Property (p. 49)
Transitive Property (p. 49)
vertical angles (p. 77)

Key Ideas and Review Exercises

2.1, 2.2 To draw a conclusion from given data and to write a two-column proof

Information is *given* in two ways:
(1) in writing or (2) as implied in a geometric figure.

Draw a conclusion and write a proof.

1. Given: The figure at the right
2. Given: T is the midpoint of $\overline{PW}$.

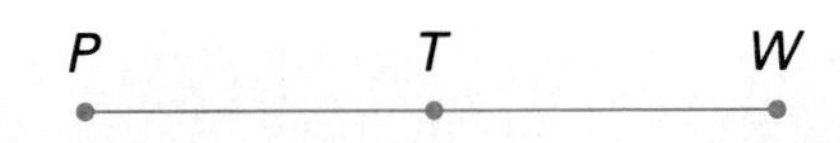

2.2, 2.3, 2.4 To use algebraic properties as reasons in proofs

- If $a = b$, then $a + c = b + c$, $a - c = b - c$, $ac = bc$, and $\frac{a}{c} = \frac{b}{c}$ $(c \neq 0)$
- Equations may be added to or subtracted from each other.
- An equation may be substituted for its equal.

To use any of the three equivalent statements of the Segment Addition Postulate or the Angle Addition Postulate in a proof

3. Given: $QR = BV$, $PR = AV$
Prove: $PQ = AB$

4. Given: $m\angle 1 = m\angle 3$
Prove: $m\angle GHL = m\angle KHM$

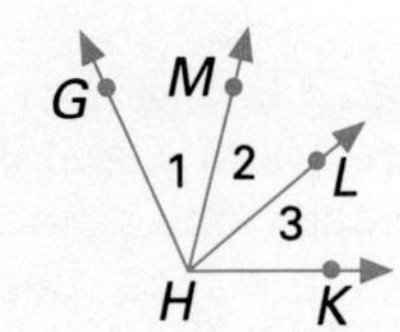

2.5 A conditional, "If p, then q," can be written as $p \rightarrow q$.
A conditional is false only when p is true and q is false.

Rewrite the given and the conclusion as a conditional. Is the conditional true?

5. Given: $\angle A$ and $\angle B$ are supplementary.
Conclusion: $\angle A$ and $\angle B$ are congruent.

2.6 Use deductive reasoning to draw conclusions.
Use inductive reasoning to make generalizations.

6. $p \rightarrow q$ *and* $q \rightarrow r$

7. 101, 88, 75, 62, ______

2.7 Complements of the same angle (or congruent angles) are congruent.
Supplements of the same angle (or congruent angles) are congruent.

8.

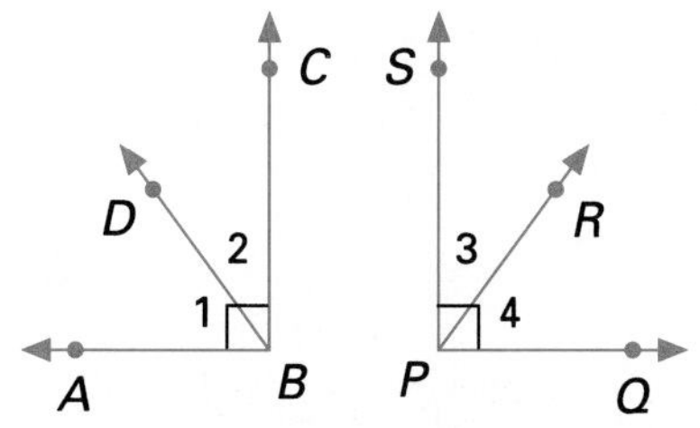

Given: $\overrightarrow{BA} \perp \overrightarrow{BC}$, $\overrightarrow{PS} \perp \overrightarrow{PQ}$,
$m\angle 1 = m\angle 4$
Prove: $m\angle 2 = m\angle 3$

9.

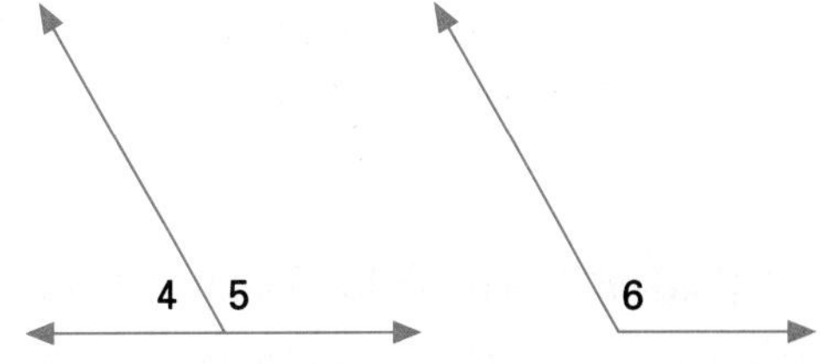

Given: $\angle 6$ is supplementary to $\angle 4$.
Prove: $\angle 5 \cong \angle 6$

2.8 Vertical angles are angles formed by two intersecting lines that are neither adjacent nor straight angles. Vertical angles are congruent.

10.

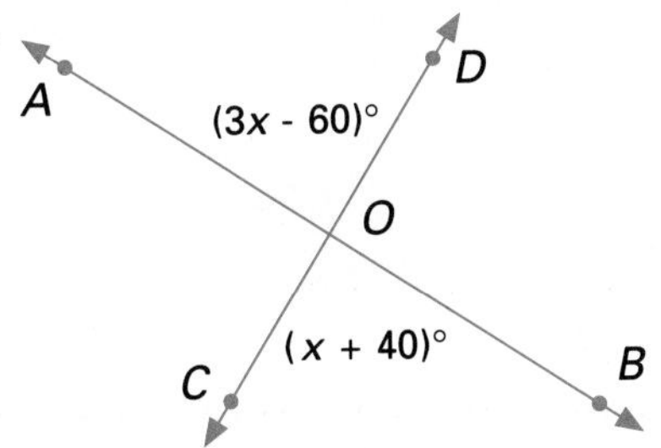

Find $m\angle BOD$.

11.

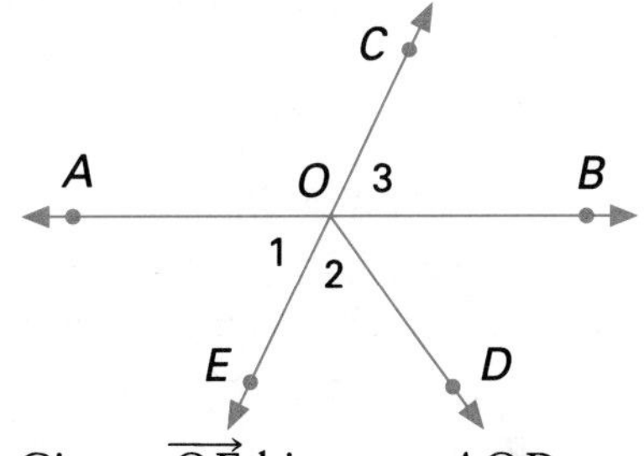

Given: $\overrightarrow{OE}$ bisects $\angle AOD$.
Prove: $m\angle 2 = m\angle 3$

2.9 The postulates and theorems that follow relate points, lines, and planes.

- For any two points, there is exactly one line containing them.
- Two lines intersect in at most one point.
- If a line intersects a plane but is not contained in the plane, then the intersection is exactly one point.
- If two planes intersect, then they intersect in exactly one line.
- Three noncollinear points are contained in exactly one plane.
- Two intersecting lines are contained in exactly one plane.
- A line and a point not on the line are contained in exactly one plane.

Indicate whether the statement is always true, sometimes true, or never true.

12. Two different lines can have several points in common.

13. Three points are contained in exactly one plane.

14. Two planes have exactly two points in common.

Chapter 2 Test

1. How many lines can be drawn that contain two points P and Q? Why?

2.

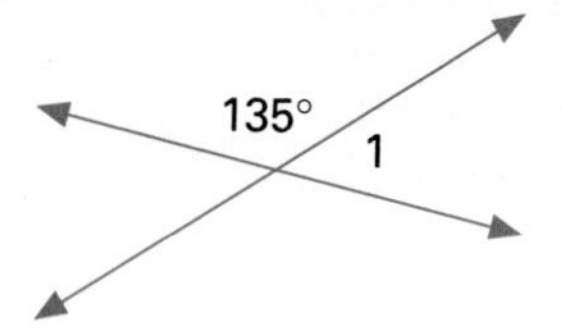

Find m $\angle$ 1.

3.

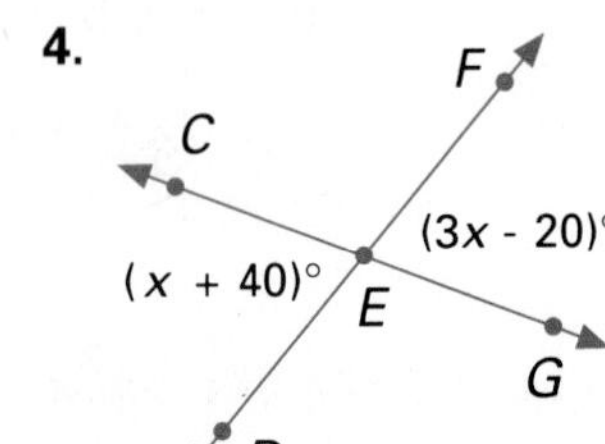

Given: $\overrightarrow{OA} \perp \overrightarrow{OB}$
Find m $\angle 1$.

4.

Find m $\angle CEF$.

5. Write a statement in "If . . . , then . . ." form. Is the conditional true?
Given: $\angle K$ and $\angle A$ are supplementary.
Conclusion: $\angle K$ and $\angle A$ are adjacent.

For Items 6 and 7, draw a conclusion. Then prove it.

6. R S T

7. Given: $\overrightarrow{GH}$ bisects $\angle MGR$.

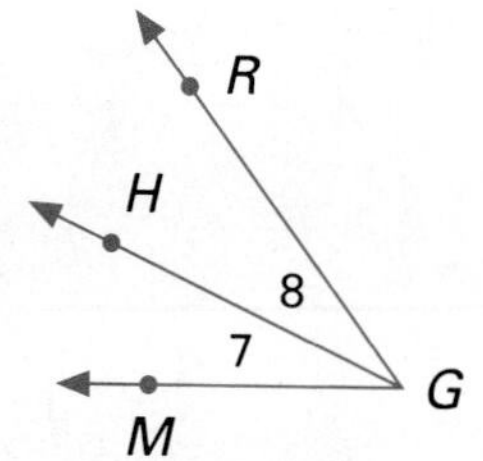

8. Given: $PW = SY$, $PT = YU$
Prove: $TW = US$

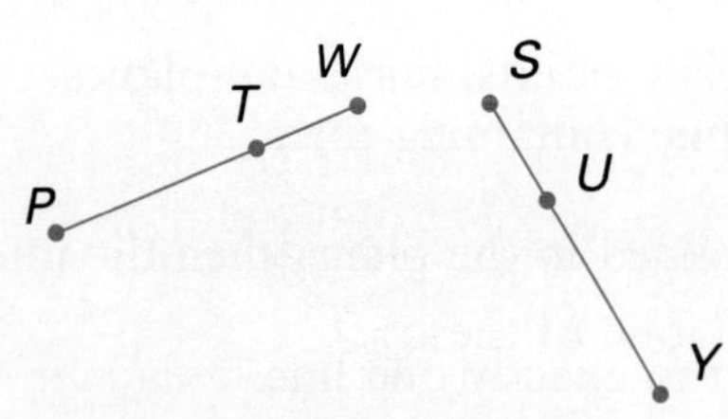

9. Given: $\angle 2$ is supplementary to $\angle 3$.
Prove: $\angle 1 \cong \angle 4$

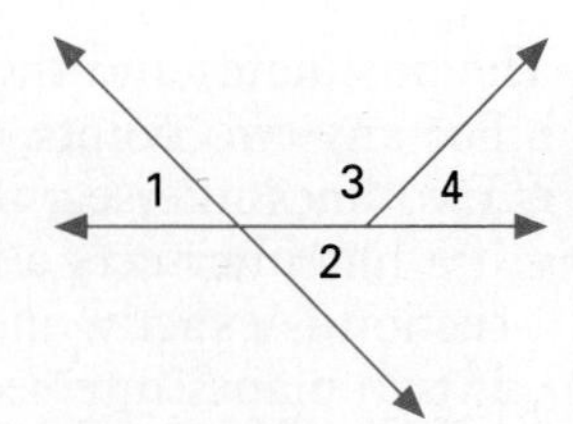

10. Rewrite the statement in two equivalent forms. Two acute adjacent angles are complementary only if their outer rays are perpendicular.

11. Give the format of a conditional statement and explain how to determine if it is true.

*** 12.** Given: $\angle BAC$ is an acute angle, $\overline{CB} \perp \overline{CA}$, D lies on $\overrightarrow{AB}$ such that B is between A and D, and $\angle BCD \cong \angle A$.
Prove: $\angle A$ is complementary to the supplement of $\angle ACD$.

College Prep Test

Strategy for Achievement in Testing

A helpful test-taking strategy is to discard any given information *that is not relevant*. In Item 1 below, the value $4x + 3y$ is not needed for the solution of the problem. Instead, use the fact that $\angle AQP$ and $\angle BQP$ are supplementary to find z. Then w is found using the congruence of vertical angles.

Choose the one best answer to each item.

1.

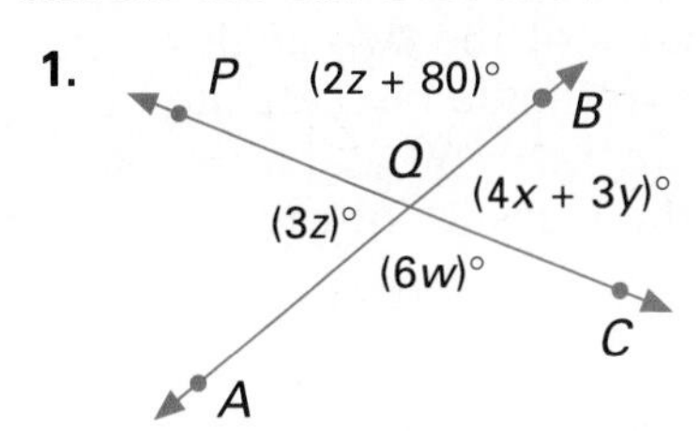

Find the value of w.

(A) 120 (B) 20 (C) 60 (D) 80
(E) none of these

2.

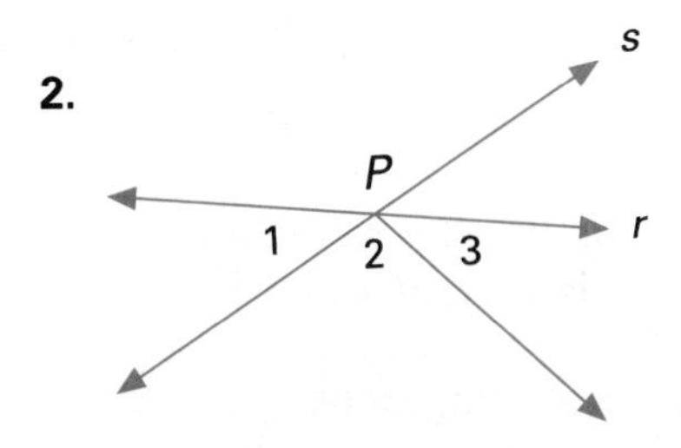

Lines r and s intersect at P; m $\angle 1 = 3x$, and $\angle 3 \cong \angle 1$.
Express m $\angle 2$ in terms of x.

(A) x (B) $180 - 3x$ (C) $6x$
(D) $180 - 6x$ (E) $3x$

3. Five bananas cost as much as three pears. If bananas cost 30 cents each, what is the cost of each pear?

(A) \$1.50 (B) 18¢ (C) 10¢
(D) 30¢ (E) 50¢

4.

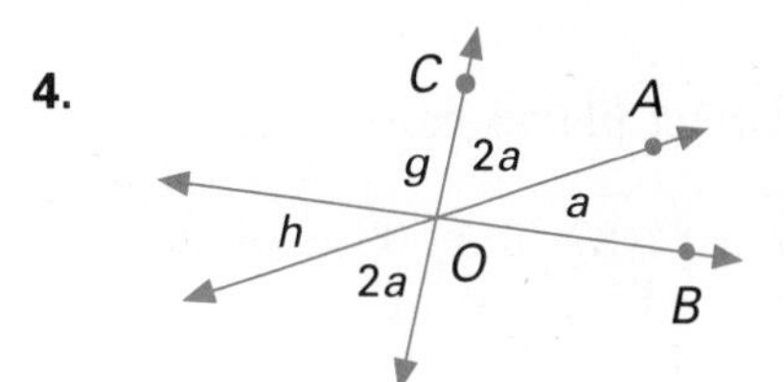

Find the value of $h + g$.

(A) a (B) $180 - \frac{1}{2}a$ (C) $2a$

(D) $180 - 2a$ (E) $\frac{1}{2}a$

5. The formula for the area of a rectangle is $A = lw$, where l is the length and w is the width. If the area of the rectangle below is 1, then $y =$ ______.

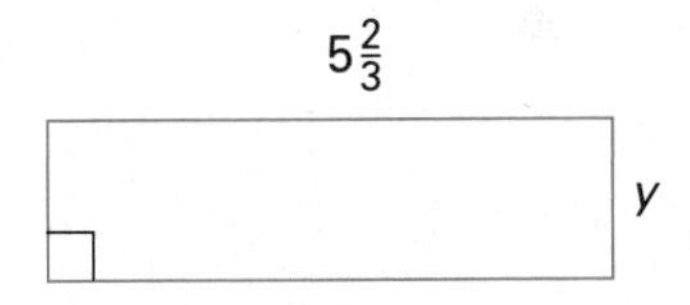

(A) $\frac{3}{17}$ (B) $\frac{17}{3}$ (C) 17 (D) 1
(E) None of these

6. In the figure below, all segments intersect at right angles. Then $x + y =$ ______.

(A) $w + r$
(B) $e + t$
(C) $t + r + \text{w} + \text{e}$
(D) $2t + 2w$
(E) None of these

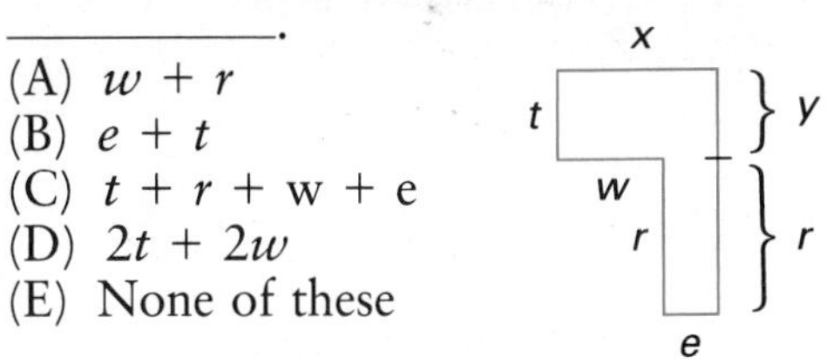

Cumulative Review *(Chapters 1–2)*

Choose the best possible answer for each exercise. (Exercises 1–5)

1. What does $\overline{AB}$ represent? *1.2*
(A) plane (B) line
(C) segment (D) distance
(E) ray

2. Two distinct planes can intersect in a ________. *1.1*
(A) point (B) line (C) plane
(D) ray (E) segment

3. If m $\angle$A = x, then the measure of a supplement of $\angle A$ is ________. *1.8*
(A) x (B) $180 + x$ (C) $90 - x$
(D) $x - 180$ (E) $180 - x$

4. The coordinates of the endpoints of a segment are -6 and 10. Find the coordinate of the midpoint of the segment. *1.5*
(A) 4 (B) 6 (C) 2 (D) 8
(E) none of these

5. A line that does not lie in a given plane can intersect the plane in ________. *2.8*
(A) exactly one point
(B) exactly two points
(C) a line
(D) two or more points
(E) a ray

6. Graph the conjunction. Identify the resulting geometric figure. *1.4*
$x \geq 4$ *and* $x \geq 9$

7. Draw an obtuse angle. Construct the bisector of the angle. *1.6*

8. The measures of the two angles formed by an angle bisector of an angle are $3x - 60$ and $x + 10$. Find the measure of each angle. *1.7*

9. The measure of an angle is twice the measure of its complement. Find the measure of each angle. *1.8*

10. Indicate whether the following conditional is true or false: If two angles are vertical, then the angles are not congruent. *2.5, 2.7*

Use the figure for Exercises 11 and 12.

11. Given: m $\angle 1 = 45$,
m $\angle 2 = 3x$
Find x.

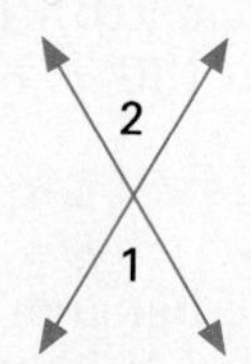

12. Given: m $\angle 1 = 4x - 30$,
m $\angle 2 = x + 30$
Find: m $\angle 2$.

13. Write a statement in "If . . ., then . . ." form. *2.5*
Given: A ray bisects an angle.
Prove: The two angles formed are equal in measure.

14. Given: $\overrightarrow{GH} \perp \overrightarrow{GL}$, m $\angle 1$ = m $\angle 2$
Prove: m $\angle 2$ + m $\angle 3$ = 90

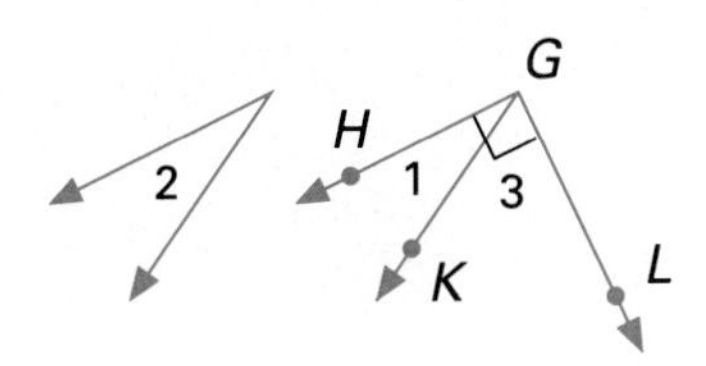

15. Given: $TW = TZ$, $RW = YZ$ *2.4*
Prove: $TR = TY$

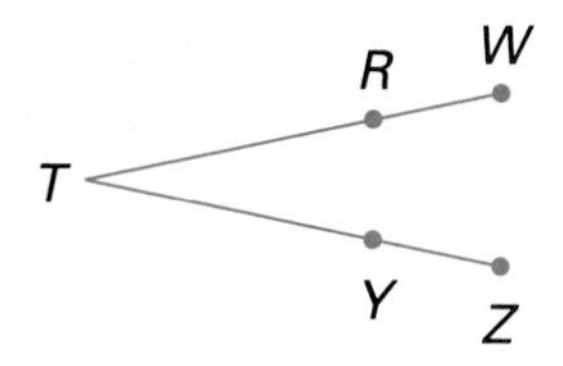

16. Given: $\angle 2$ is supplementary to $\angle 3$. *2.6*
Prove: $\angle 1 \cong \angle 3$

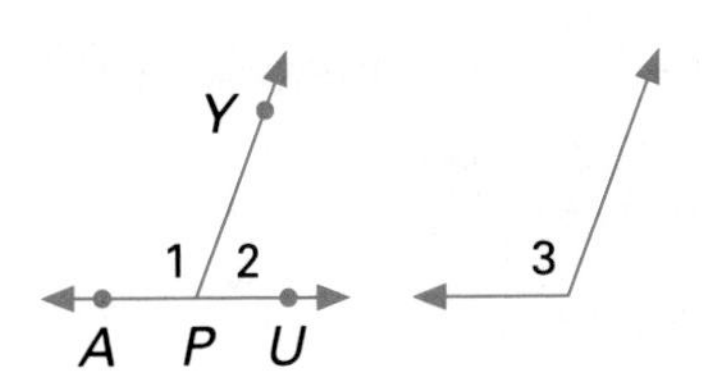

17. When is $p \vee q$ false? *1.7*
When is $p \wedge q$ true?

18. Draw a conclusion for each figure below and express it as an equation. Give a reason for your answer. *2.1, 2.8*

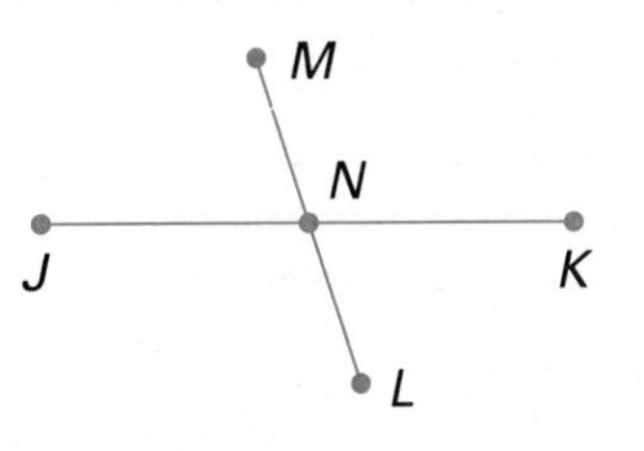

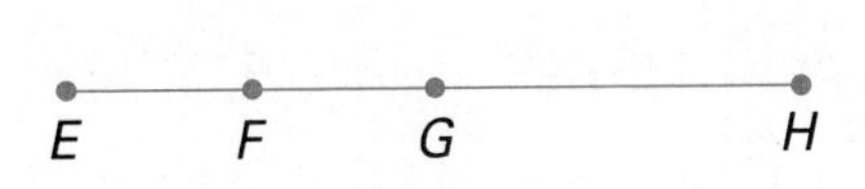

19. Given: $\overrightarrow{OU}$ bisects $\angle ROT$. *2.3*
Prove: m $\angle 4$ + m $\angle 5$ = m $\angle UOS$

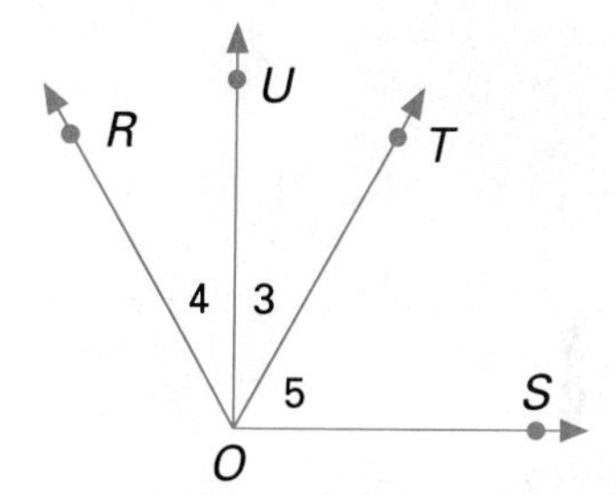

20. Given: m $\angle RSU$ = m $\angle VST$ *2.4*
Prove: m $\angle 5$ = m $\angle 6$

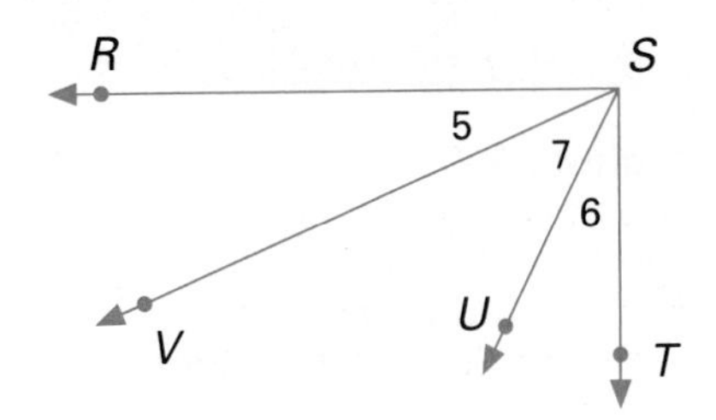

3 PARALLELISM

Euclid (about 300 B.C.)—Writing in Greek on rolls of Egyptian papyrus, Euclid produced the world's most enduring geometry text. This book, called *The Elements*, has been printed in more than 1,000 different editions worldwide. We know very little about Euclid's life.

3.1 Parallel and Skew Lines

Objective To identify relationships among parallel lines and skew lines

Definition **Coplanar** lines are lines that lie in the same plane. Coplanar lines either intersect or are parallel.

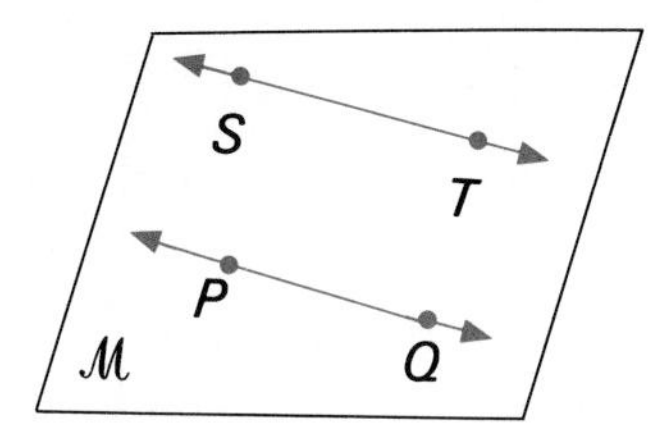

$\overleftrightarrow{PQ}$ and $\overleftrightarrow{ST}$ do not intersect. $\overleftrightarrow{PQ}$ is parallel to $\overleftrightarrow{ST}$ ($\overleftrightarrow{PQ} \parallel \overleftrightarrow{ST}$).

$\overleftrightarrow{AB}$ and $\overleftrightarrow{CD}$ intersect. $\overleftrightarrow{AB}$ is not parallel to $\overleftrightarrow{CD}$ ($\overleftrightarrow{AB} \nparallel \overleftrightarrow{CD}$).

Definition **Parallel lines** are coplanar lines that do not intersect.

In space there exists a third possible relationship between two lines. $\overleftrightarrow{AB}$ and $\overleftrightarrow{CD}$ do not intersect, and they are not parallel since they are not coplanar. $\overleftrightarrow{AB}$ and $\overleftrightarrow{CD}$ are *skew lines.*

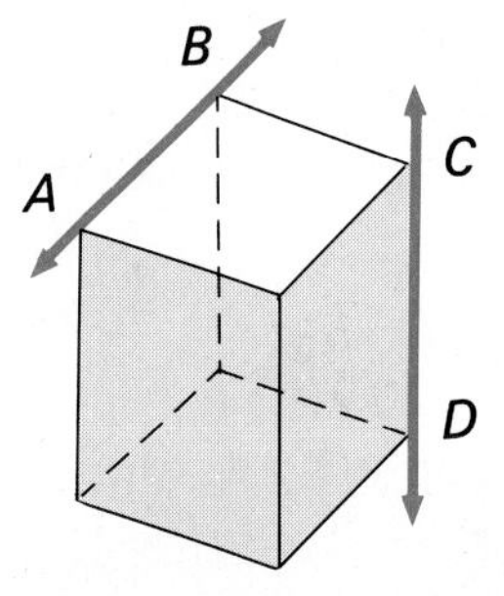

Definition **Skew lines** are noncoplanar lines.

EXAMPLE 1 Identify the indicated pair of lines as parallel or skew.

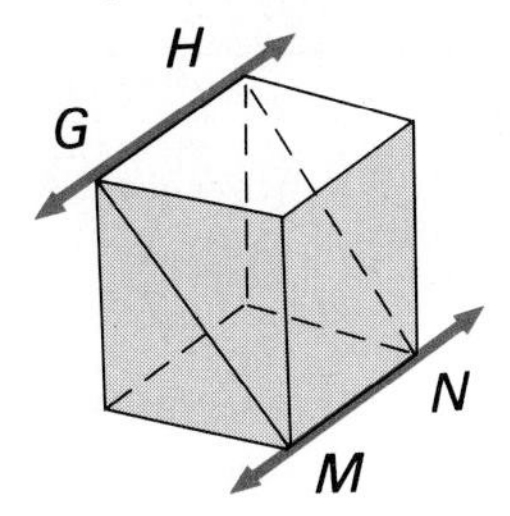

$\overleftrightarrow{GH}$ and $\overleftrightarrow{MN}$ are parallel.

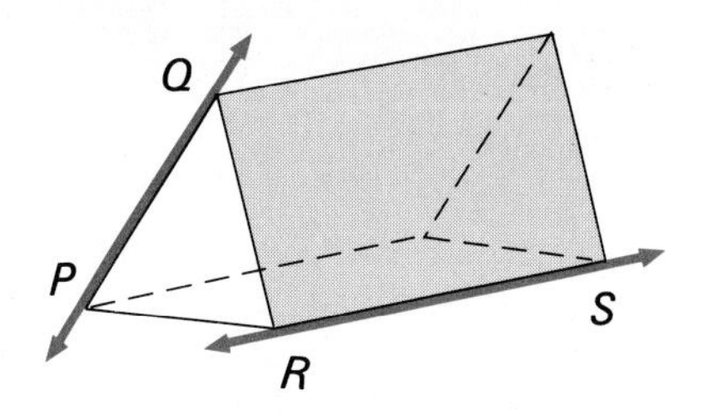

$\overleftrightarrow{PQ}$ and $\overleftrightarrow{RS}$ are skew.

Segments or rays can also be parallel. In the figure, $\overline{GH} \parallel \overrightarrow{VW}$ since $\overleftrightarrow{GH} \parallel \overleftrightarrow{VW}$

segment $\parallel$ ray

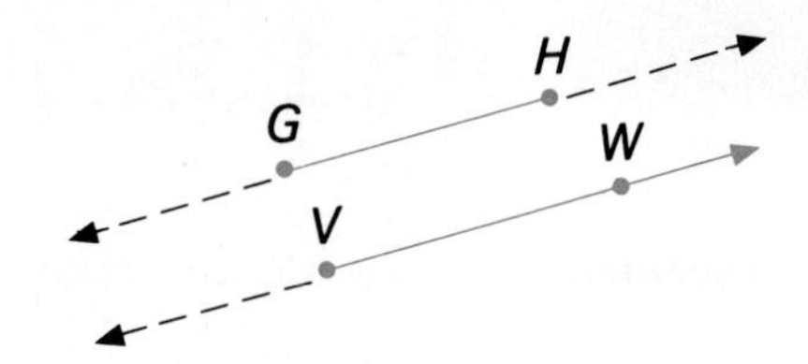

Definition

Segments or *rays* are *parallel* if the lines that contain them are parallel.

Similarly, *segments* or *rays* are *skew* if the lines that contain them are skew.

By definition, if two coplanar lines do not intersect, then they are parallel. This is not true for segments or rays, as seen below.

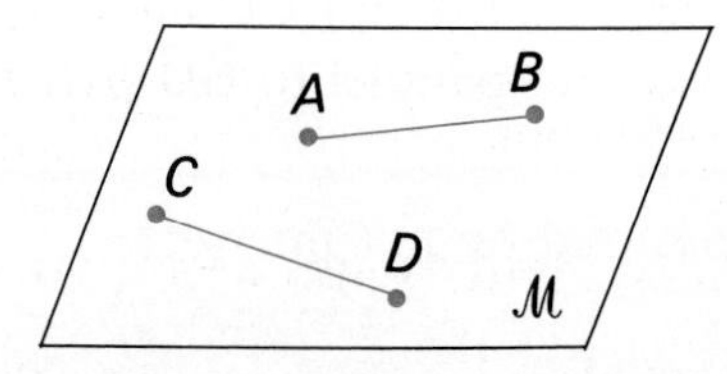

$\overline{AB}$ and $\overline{CD}$ are segments that do not intersect. But, $\overline{AB} \nparallel \overline{CD}$.

$\overrightarrow{PQ}$ and $\overrightarrow{RS}$ are rays that do not intersect. But, $\overrightarrow{PQ} \nparallel \overrightarrow{RS}$.

EXAMPLE 2 Which pairs of segments or rays appear to be parallel?

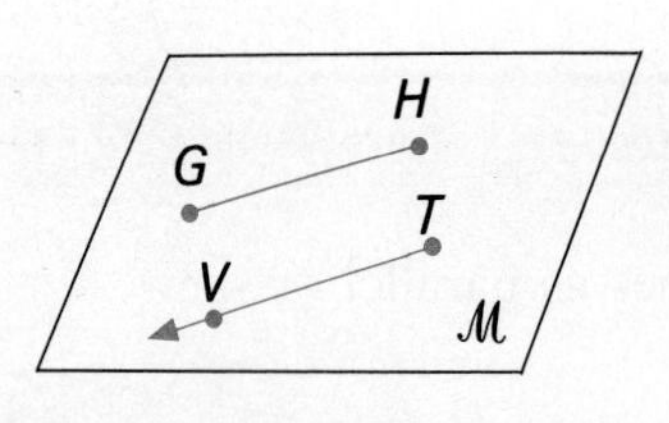

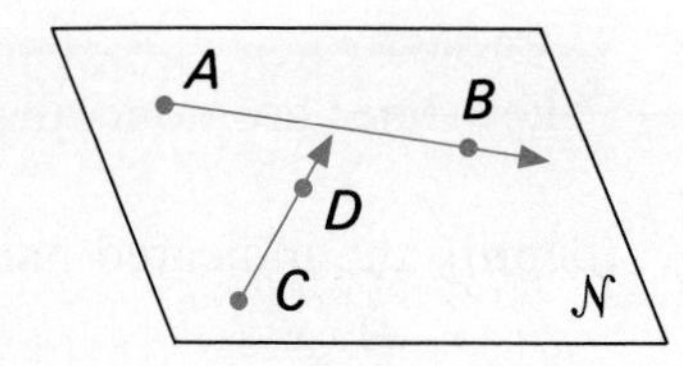

$\overline{GH} \parallel \overrightarrow{TV}$.

$\overrightarrow{CD}$ intersects $\overrightarrow{AB}$. $\overrightarrow{CD} \nparallel \overrightarrow{AB}$.

It is frequently helpful to draw a box to illustrate geometric relationships between lines in space. A diagram of a four-sided box with top and bottom can be drawn by following these steps.

Step 1
Draw the front of the box, as shown. Then draw the back slightly above and to the right of the front, using two dashed segments, as shown.

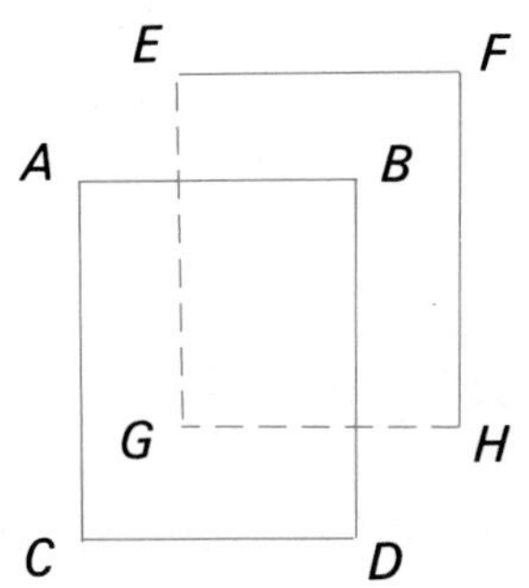

Step 2
Draw four segments to complete the box as shown. Use dashed segments for any segments that would be invisible.

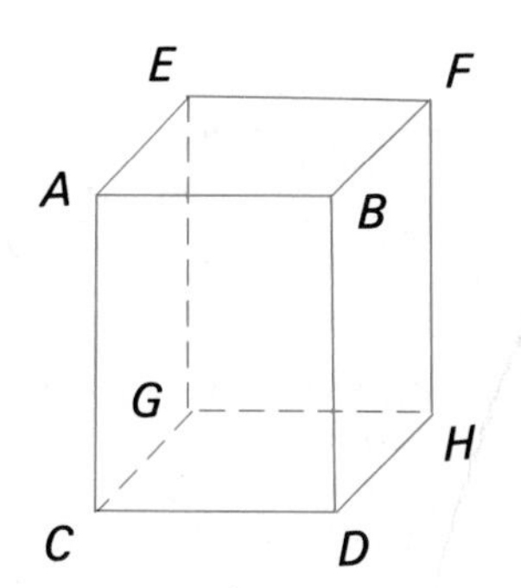

Practice drawing this figure. It is referred to in the next example. Exercises occasionally require that reasons be given why a statement is sometimes, always, or never true. In such cases, a geometric figure may be drawn to illustrate when the statement is true and when it is false.

EXAMPLE 3 Indicate whether the statement is always, sometimes, or never true. Give a reason for your answer.

	Parallel lines are skew.	Parallel segments do not intersect.	Nonparallel segments intersect.
Solution	Never true. Reason: by definition of skew lines.	Always true. Reason: by definition of parallel lines and segments.	Sometimes true. Reason: Refer to the diagram above. $\overline{AB} \nparallel \overline{BF}$. $\overline{AB}$ and $\overline{BF}$ intersect. $\overline{AB} \nparallel \overline{HD}$. $\overline{AB}$ and $\overline{HD}$ are skew.

Classroom Exercises

Tell whether the pair of lines, rays, or segments appears to be parallel, intersecting, skew, or none of these.

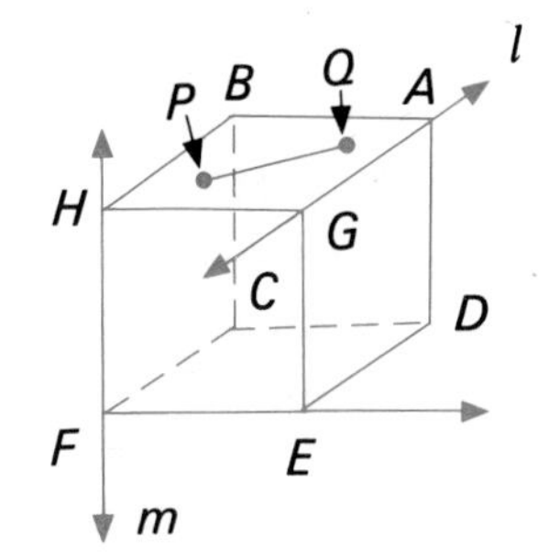

1. l and m
2. $\overline{BC}$ and m
3. $\overline{AB}$ and l
4. $\overline{QP}$ and $\overline{AB}$
5. $\overline{FE}$ and $\overline{AB}$
6. $\overline{BC}$ and l
7. $\overline{AD}$ and m
8. $\overrightarrow{FE}$ and $\overrightarrow{AG}$
9. $\overline{BH}$ and m

Written Exercises

For each of the following refer to labeled segments or lines in the figure of the highway and overpass shown at the right. Give an example.

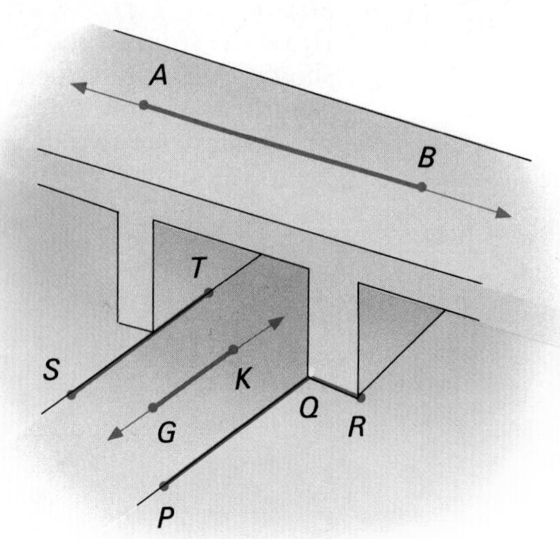

1. parallel segments
2. skew lines
3. intersecting segments
4. coplanar segments which do not intersect and are not parallel

Draw a diagram to illustrate each of the following descriptions.

5. two nonintersecting segments that are not parallel
6. two intersecting lines
7. two skew lines
8. a segment that is parallel to a ray

Indicate whether the statement is always true, sometimes true, or never true. Give a reason for your answer.

9. Two coplanar lines that do not intersect are parallel.
10. Two lines that are not skew are parallel.
11. A segment in the plane of the ceiling of your classroom intersects a segment in the plane of the floor.
12. Two lines in space that do not intersect are parallel.

Draw the figure described.

13. Lines p and q are skew. Lines q and t are skew. But $p \parallel t$.
14. $\overleftrightarrow{AB}$ and $\overleftrightarrow{GH}$ are skew. $\overleftrightarrow{GH}$ and $\overleftrightarrow{PQ}$ are skew. $\overleftrightarrow{AB} \perp \overleftrightarrow{PQ}$.

Mixed Review

1. The measure of one of two complementary angles is $\frac{2}{3}$ the measure of the other. Find the measure of each angle. *1.7*
2. The measure of one of a pair of vertical angles is $3x - 40$. The measure of the other is $x + 20$. Find the measure of each angle. *2.8*
3. The coordinates of the endpoints of $\overline{AB}$ are -8 and 12. Find the coordinate of the midpoint of $\overline{AB}$. *1.4*

3.2 Transversals and Special Angle Relationships

Objective To identify special pairs of angles formed by the intersection of two lines by a transversal

In the figure at the right, line t intersects line m at S and line n at W. Line t is called a *transversal* of m and n.

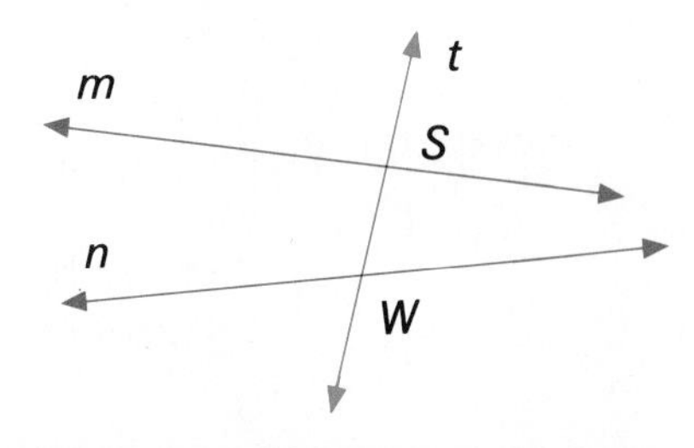

Definition A **transversal** is a line, ray, or segment that intersects two or more coplanar lines, rays, or segments, each at a different point.

EXAMPLE 1 Identify each transversal in the figure and the lines to which it is a transversal.

$\overleftrightarrow{PQ}$ is a transversal of $\overleftrightarrow{PR}$ and $\overleftrightarrow{QR}$.
$\overleftrightarrow{PR}$ is a transversal of $\overleftrightarrow{PQ}$ and $\overleftrightarrow{QR}$.
$\overleftrightarrow{QR}$ is a transversal of $\overleftrightarrow{PQ}$ and $\overleftrightarrow{PR}$.

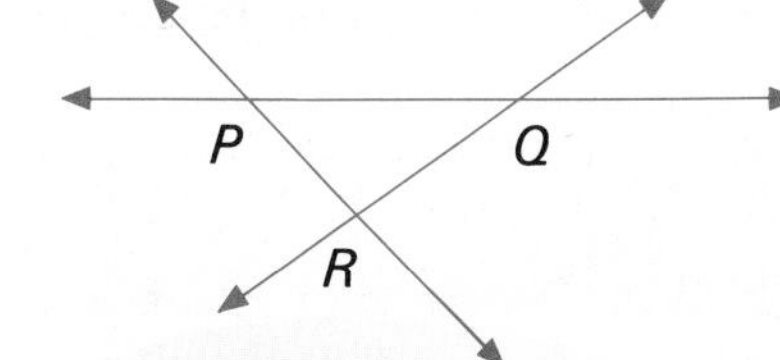

In the definitions of the special angles that follow, the *interior* and *exterior* of two lines cut by a transversal will be used. These regions are shown in the figure at the right. In each case, line t is a transversal of lines m and n.

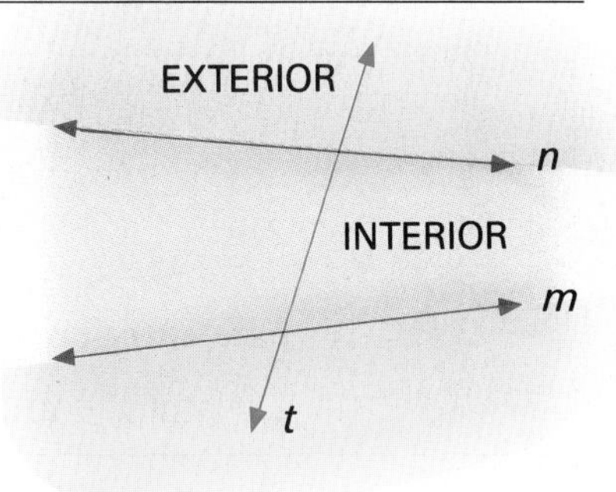

When a transversal intersects two lines, special pairs of angles are formed. Such pairs will be identified by arcs in illustrations that follow.

$\angle 3$ and $\angle 5$ are **alternate interior angles.**
Two angles are alternate interior angles if:
(1) they lie on opposite sides of a transversal;
(2) they are both interior angles; and
(3) they are nonadjacent.
Another pair of alternate interior angles is $\angle 4$ and $\angle 6$.

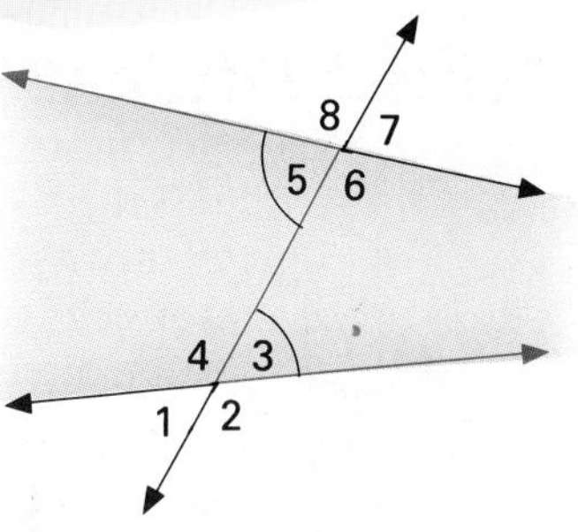

$\angle 4$ and $\angle 8$ are **corresponding angles.**
Two angles are corresponding angles if:
(1) they lie on the same side of a transversal;
(2) one angle is interior, one exterior; and
(3) they are nonadjacent.
Three other pairs of corresponding angles are: $\angle 1$ and $\angle 5$, $\angle 2$ and $\angle 6$, $\angle 3$ and $\angle 7$.

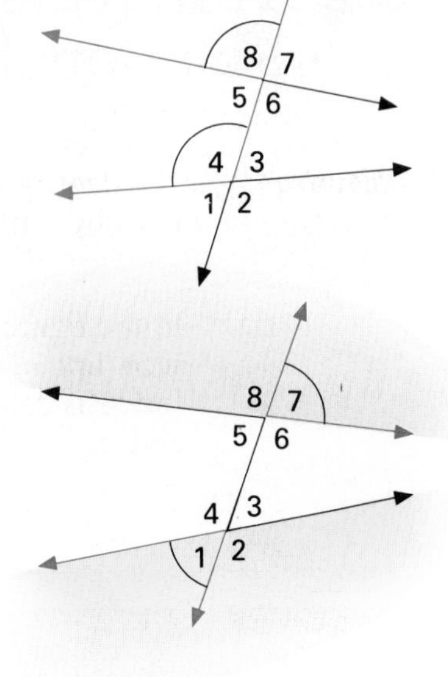

$\angle 1$ and $\angle 7$ are **alternate exterior angles.**
Two angles are alternate exterior angles if:
(1) they lie on opposite sides of a transversal;
(2) they are both exterior angles; and
(3) they are nonadjacent.

EXAMPLE 2 Identify the given pair of angles as alternate interior, alternate exterior, corresponding, or none of these:
$\angle 4$ and $\angle 6$, $\angle 2$ and $\angle 8$, $\angle 4$ and $\angle 8$

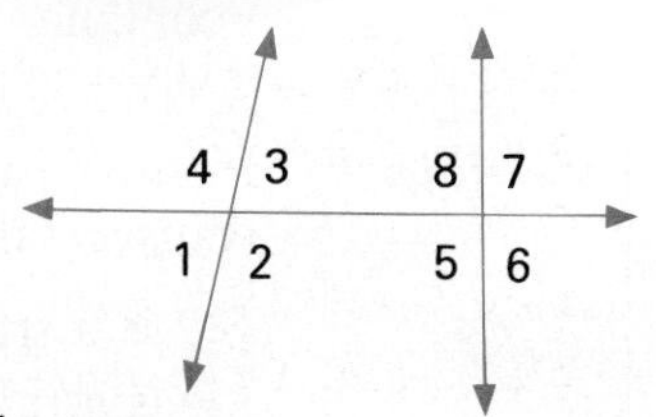

Plan Redraw the figure to show each pair of angles.

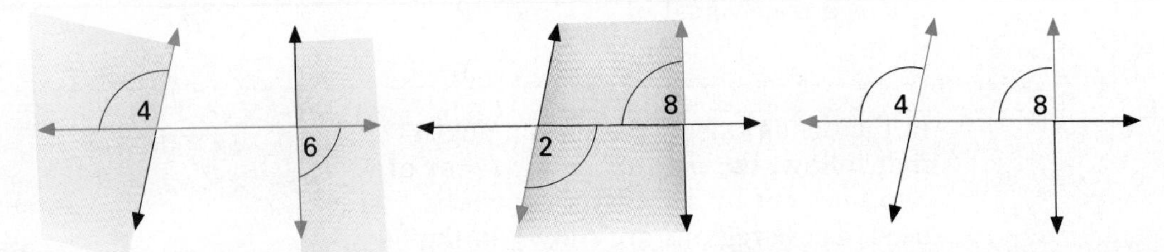

Solution $\angle 4$ and $\angle 6$ are alternate exterior angles. $\angle 2$ and $\angle 8$ are alternate interior angles. $\angle 4$ and $\angle 8$ are corresponding angles.

$\angle 1$ and $\angle 8$ are none of these special pairs of angles.

A special pair of angles can be identified more easily by shading and extending their sides. By extending the sides of $\angle 1$ and $\angle 3$, their angle measures are not changed, but they are clearly seen to be alternate interior angles.

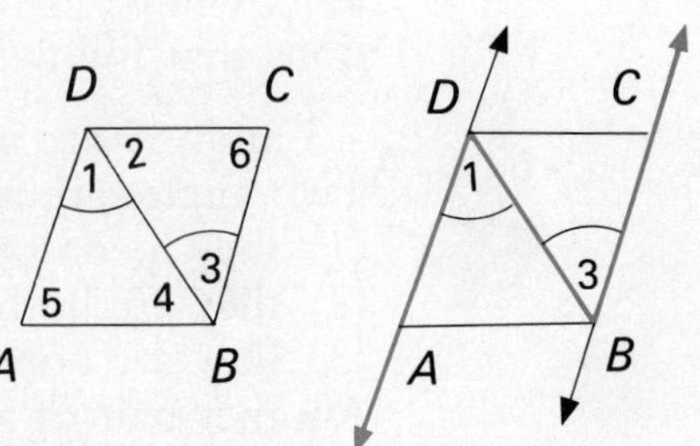

Classroom Exercises

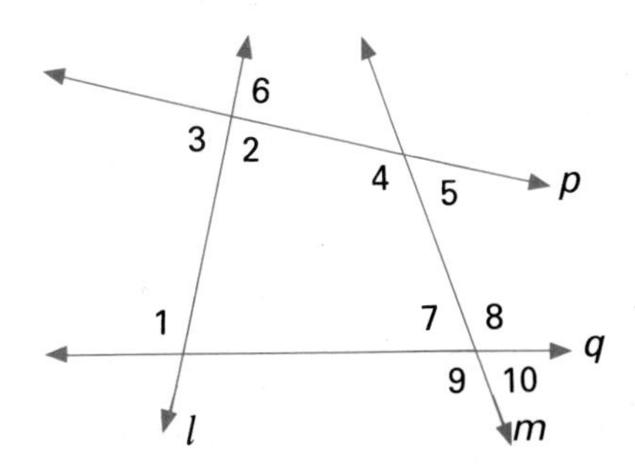

1. l is a transversal to ________.
2. m is a transversal to ________.

Identify the pair of angles as alternate interior, alternate exterior, corresponding, or none of these.

3. $\angle 1$ and $\angle 2$
4. $\angle 3$ and $\angle 4$
5. $\angle 4$ and $\angle 9$
6. $\angle 1$ and $\angle 10$
7. $\angle 5$ and $\angle 6$
8. $\angle 4$ and $\angle 8$

Written Exercises

Identify the pair of angles as alternate interior, alternate exterior, corresponding, or none of these.

1. $\angle 6$ and $\angle 10$
2. $\angle 7$ and $\angle 9$
3. $\angle 6$ and $\angle 12$
4. $\angle 5$ and $\angle 10$
5. $\angle 8$ and $\angle 12$
6. $\angle 5$ and $\angle 7$

7. $\angle 1$ and $\angle 2$
8. $\angle 2$ and $\angle 5$
9. $\angle 1$ and $\angle 3$
10. $\angle 5$ and $\angle 1$
11. $\angle 2$ and $\angle 4$
12. $\angle 6$ and $\angle 2$

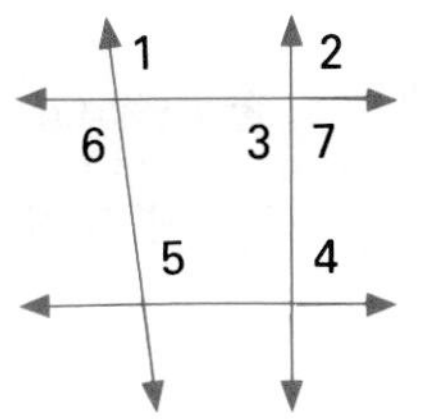

13. $\angle 1$ and $\angle 3$
14. $\angle ABD$ and $\angle 2$
15. $\angle E$ and $\angle 6$
16. $\angle 7$ and $\angle 5$
17. $\angle 4$ and $\angle FDB$
18. $\angle 7$ and $\angle 3$
19. $\angle DBA$ and $\angle 5$

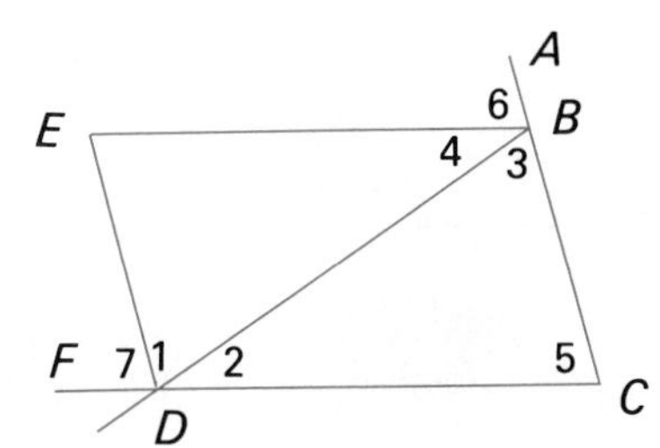

Mixed Review

1. Given: m $\angle 8$ = m $\angle 3$
   Prove: m $\angle 6$ = m $\angle 3$ ***2.8***
2. Given: m $\angle 3$ = m $\angle 8$
   Prove: m $\angle 1$ = m $\angle 6$ ***2.8***
3. Given: $\angle 3$ is supplementary to $\angle 5$.
   Prove: $\angle 6 \cong \angle 3$ ***2.7***
4. Given: $\angle 4$ is supplementary to $\angle 8$.
   Prove: $\angle 2$ is supplementary to $\angle 8$. ***2.8***

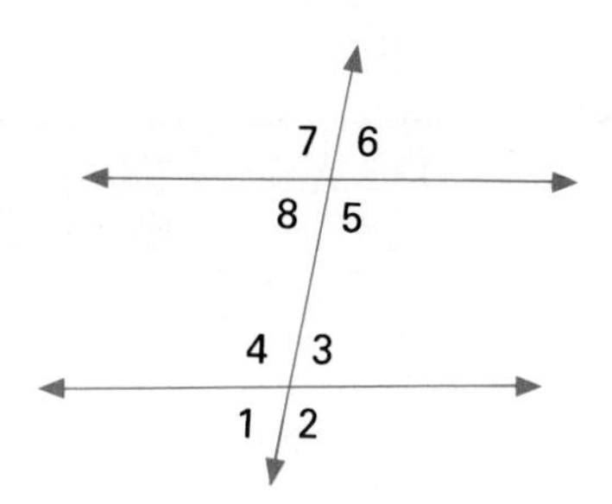

3.3 Proving Lines Parallel

Objectives To prove that under certain conditions lines are parallel
To apply the Alternate Interior Angle Postulate

You learned in Lesson 1.4 how to construct an angle that has the same measure as a given angle. A similar procedure can now be used to construct a congruent pair of alternate interior angles.

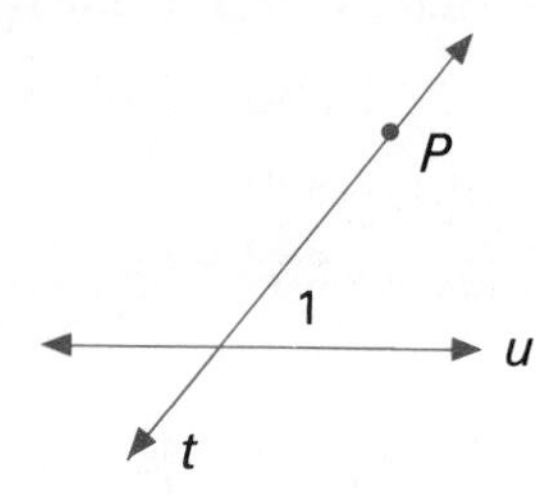

Construction Given: Point P on line t, line u forming $\angle 1$ with t

Construct: A line v through point P forming $\angle 2$ such that $\angle 1 \cong \angle 2$, and $\angle 1$ and $\angle 2$ are alternate interior angles.

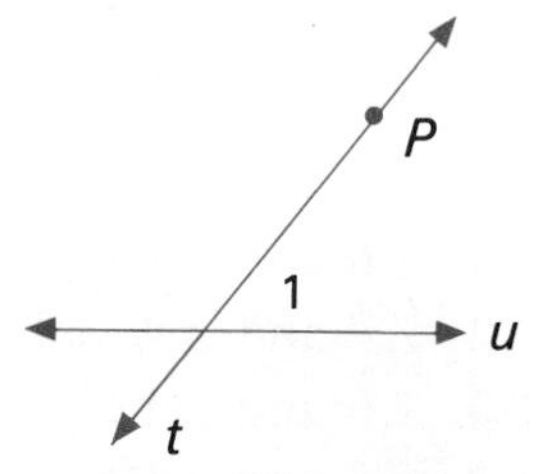

Construct $\angle 2$ congruent to $\angle 1$, with vertex P, on the left side of the transversal t. Extend the ray to form line v.

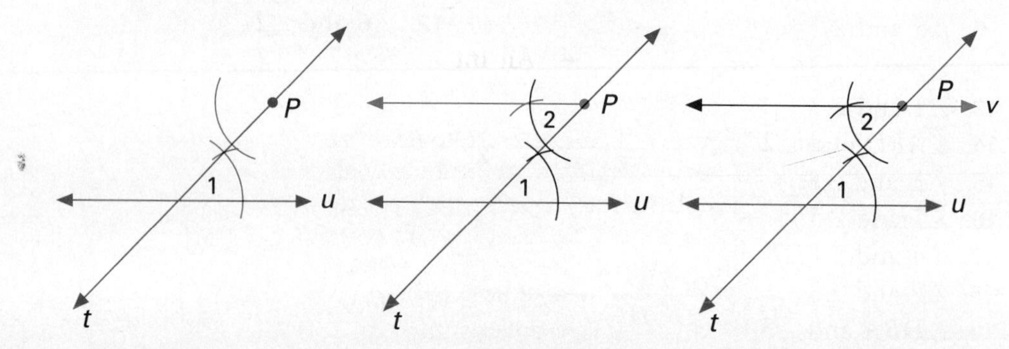

Result: $\angle 1$ and $\angle 2$ are congruent alternate interior angles.

Notice in the construction above that line v appears to be parallel to line u. This construction suggests the following postulate.

Postulate 13 **The Alternate Interior Angles Postulate:** If a transversal intersects two lines so that alternate interior angles are congruent (equal in measure), then the lines are parallel.

The Alternate Interior Angles Postulate can be used to show that lines are parallel if certain other pairs of angles are congruent.

In the figure, let m $\angle 1 = 130$ and m $\angle 3 = 130$. Now, $\angle 1$ and $\angle 3$ are corresponding angles, not alternate interior angles. But because $\angle 1$ and $\angle 2$ are vertical angles, m $\angle 2 = 130$. Therefore $\angle 2$ and $\angle 3$ are *congruent* alternate interior angles (m $\angle 2$ = m $\angle 3 = 130$). Therefore, $m \parallel l$. This suggests the following theorem.

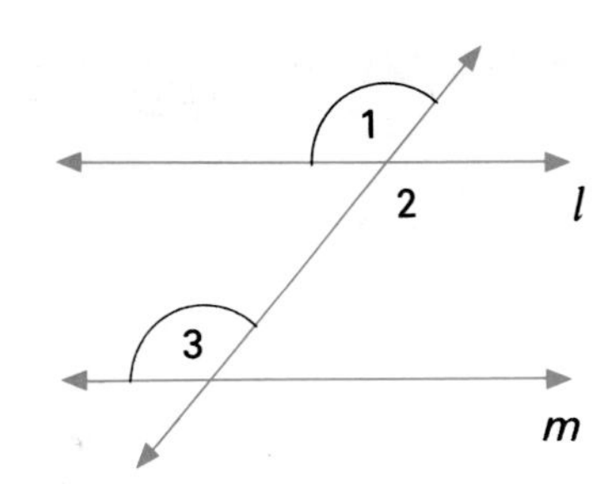

Theorem 3.1

If a transversal intersects two lines so that corresponding angles are congruent, then the lines are parallel.

Given: $\angle 1 \cong \angle 3$
Prove: $m \parallel n$

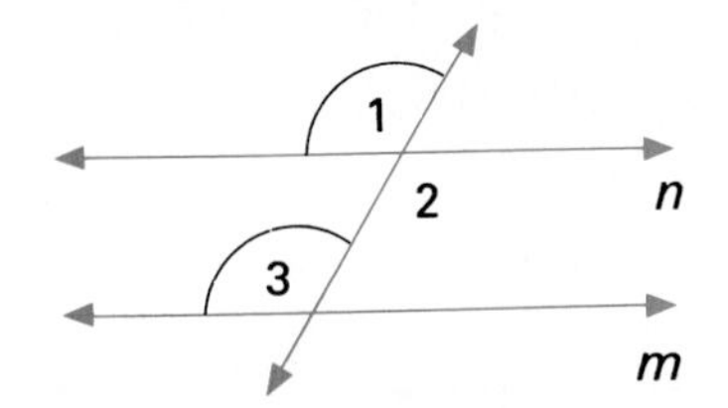

Proof

Statement	Reason
1. $\angle 1 \cong \angle 3$	1. Given
2. $\angle 2 \cong \angle 1$	2. Vert $\angle$s are $\cong$.
3. $\angle 2 \cong \angle 3$	3. Trans Prop of $\cong$ (or Sub)
4. $\therefore m \parallel n$	4. Alt Int $\angle$s Post

In the figure, $\angle 1$ and $\angle 2$ are **interior angles on the same side of the transversal.** If $\angle 2$ is supplementary to $\angle 1$, then it can be shown that $m \parallel n$. A numerical example of this is shown below. This relationship will then be proved as a theorem. Suppose that $\angle 2$ is supplementary to $\angle 1$, and m $\angle 1 = 120$. The three sequential diagrams below illustrate that $m \parallel n$.

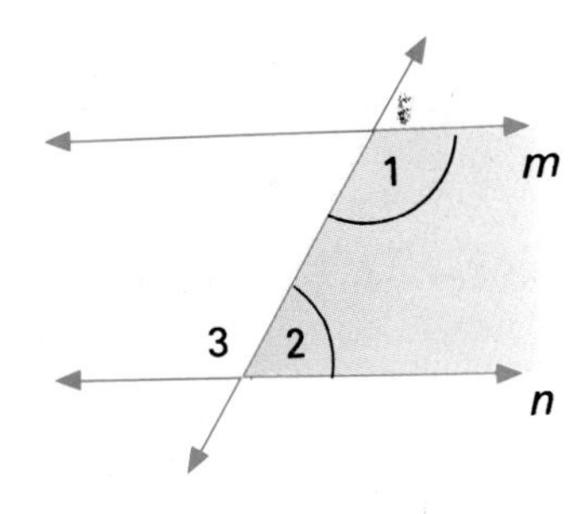

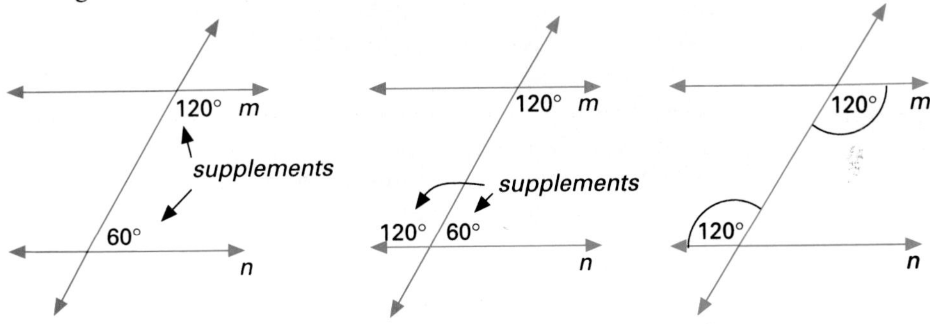

Alternate interior angles are $\cong$. Therefore $m \parallel n$.

Theorem 3.2 If two lines are intersected by a transversal so that interior angles on the same side of the transversal are supplementary, then the lines are parallel.

Given: $\angle 1$ and $\angle 2$ are supplementary.
Prove: $\overleftrightarrow{AB} \parallel \overleftrightarrow{CD}$

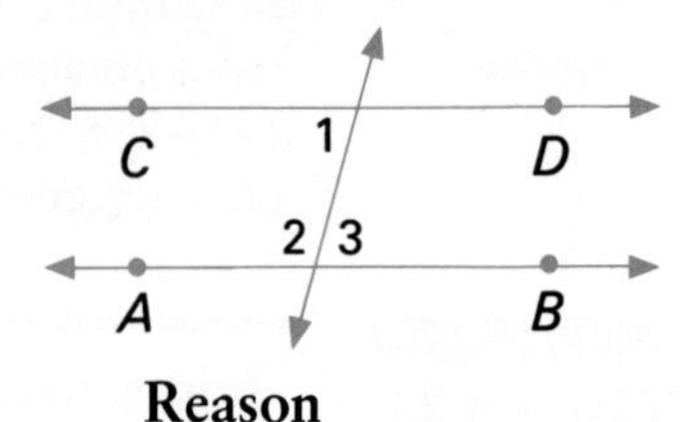

Proof

Statement	Reason
1. $\angle 1$ and $\angle 2$ are supplementary.	1. Given
2. $\angle 3$ and $\angle 2$ are supplementary.	2. If the outer rays of the two adj $\angle$s form a st $\angle$, then the $\angle$s are supp.
3. $\angle 1 \cong \angle 3$	3. Supp of the same $\angle$ are $\cong$.
4. $\therefore \overleftrightarrow{AB} \parallel \overleftrightarrow{CD}$	4. Alt Int $\angle$s Post.

In the Figure, $\overleftrightarrow{AB} \perp \overleftrightarrow{PQ}$ and $\overleftrightarrow{CD} \perp \overleftrightarrow{PQ}$. The corresponding angles, $\angle 1$ and $\angle 2$, therefore have the same measure, 90. So, the two lines are parallel. This is stated below as a theorem.

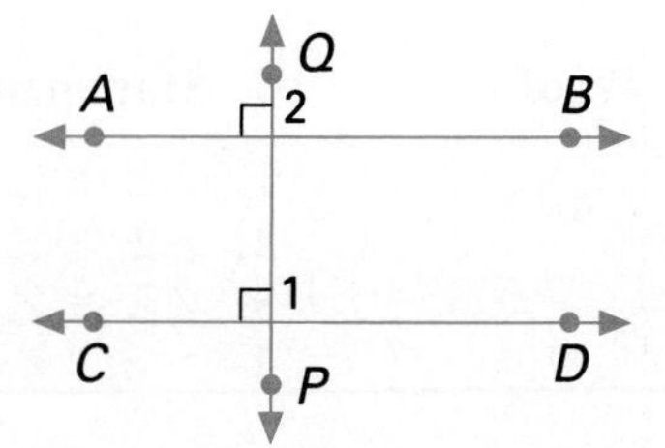

Theorem 3.3 In a plane, if two lines are perpendicular to the same line, then they are parallel.

EXAMPLE 1 Given: $m\angle 1 = m\angle 3$, $m\angle 2 = m\angle 3$
Prove: $l \parallel n$

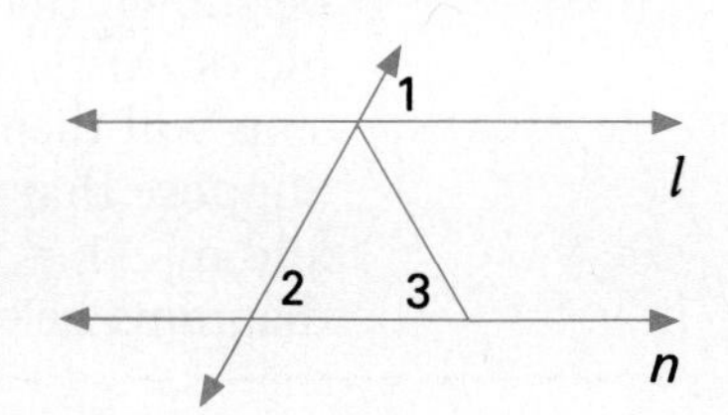

Plan Look for a special pair of angles.
$\angle 1$ and $\angle 2$ are *corresponding* angles.

Proof

Statement	Reason
1. $m\angle 1 = m\angle 3$	1. Given
2. $m\angle 2 = m\angle 3$	2. Given
3. $m\angle 1 = m\angle 2$ ($\angle 1 \cong \angle 2$)	3. Sub
4. $\therefore l \parallel n$	4. If corr $\angle$s $\cong$, then lines $\parallel$.

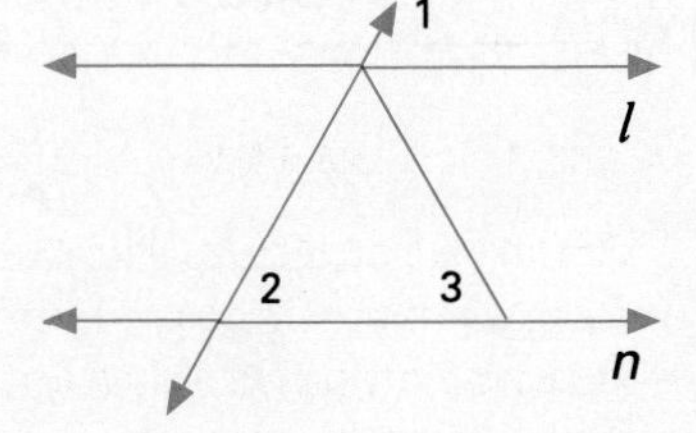

All the properties of parallel lines in this lesson apply to segments and rays as well.

EXAMPLE 2 Find the value of x so that $m \parallel n$.

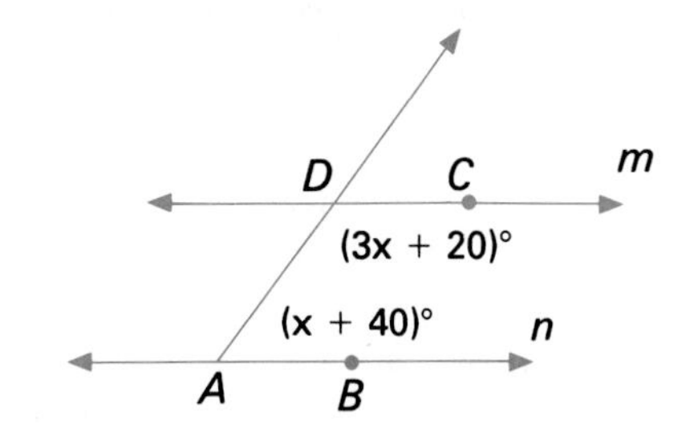

Solution The *sum* of measures of $\angle ADC$ and $\angle DAB$ must be 180.

$$(3x + 20) + (x + 40) = 180$$
$$4x + 60 = 180$$
$$4x = 120$$
$$x = 30$$

EXAMPLE 3 Given: $\overline{AB} \parallel \overline{ED}$. Find m $\angle BCF$.

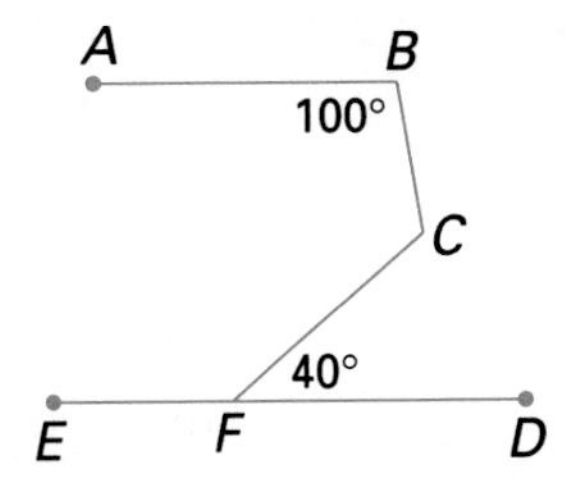

Plan Through C, draw an auxiliary line segment $\overline{CG}$ that is parallel to $\overline{AB}$ and $\overline{ED}$.

Solution

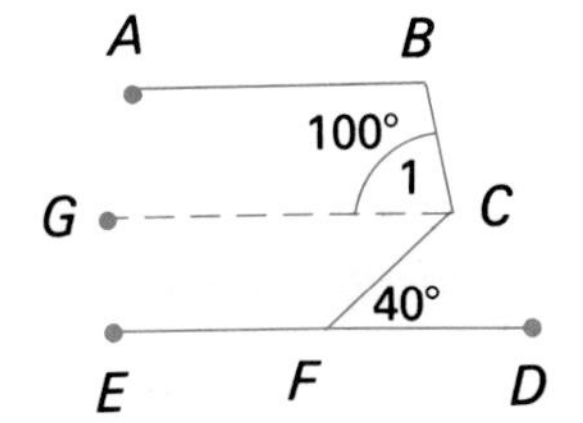

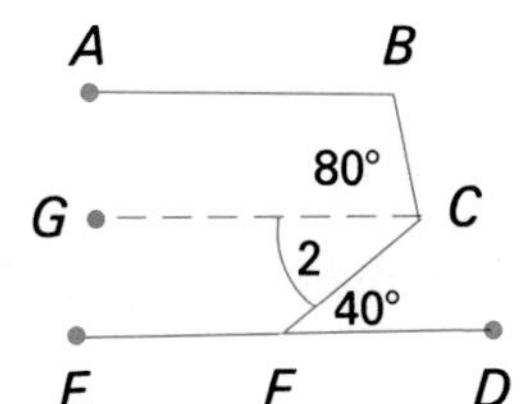

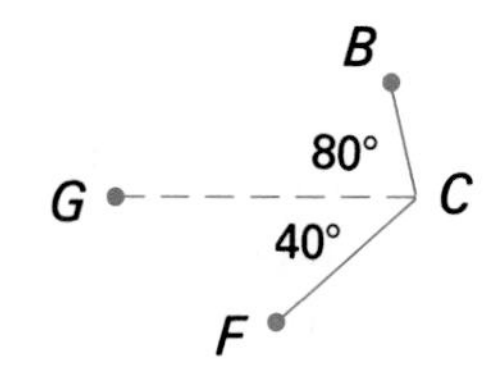

$\angle ABC$ and $\angle 1$ are supplementary if $\overline{AB} \parallel \overline{CG}$.

Therefore, m $\angle 1 = 180 - 100 = 80$.

m $\angle 2 = 40$ (alternate interior angles of parallels $\overline{CG}$ and $\overline{ED}$)

Then m $\angle BCF$ = m $\angle 1$ + m $\angle 2 = 80 + 40 = 120$.

Thus, m $\angle BCF = 120$.

Summary

Ways to prove lines, rays, or segments parallel:

- Alternate interior angles congruent
- Corresponding angles congruent
- Interior angles on same side of transversal supplementary
- Two lines, segments, or rays perpendicular to the same transversal

Classroom Exercises

State the reason why the lines, rays, or segments are parallel.

1.

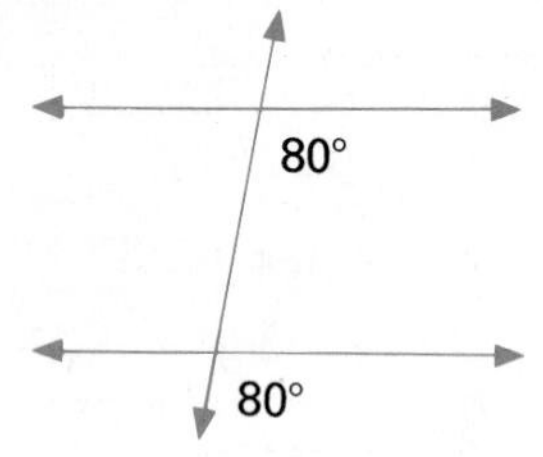

2.

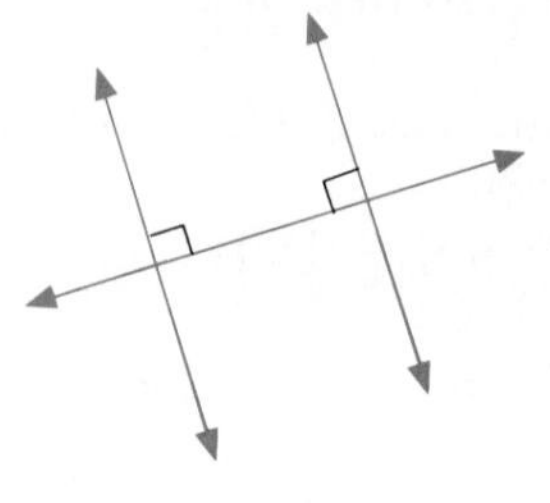

3.

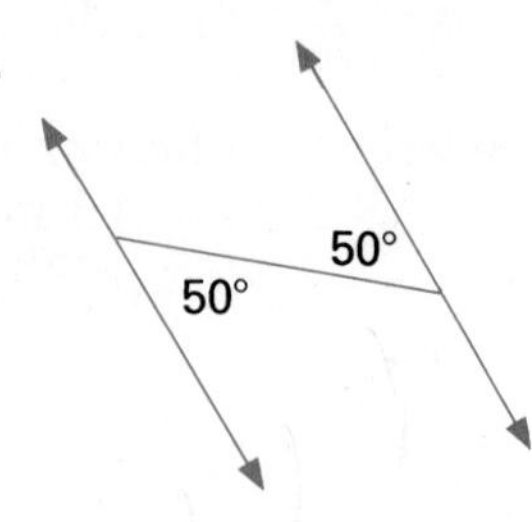

4.

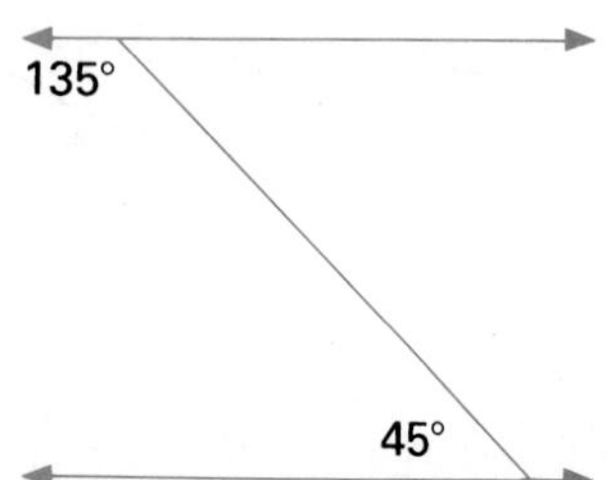

5.

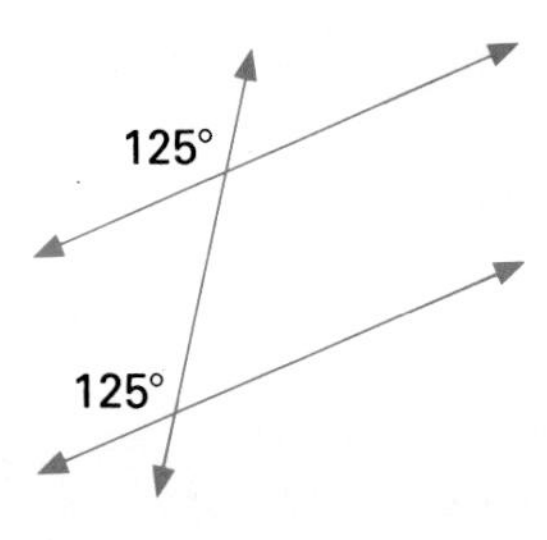

6.

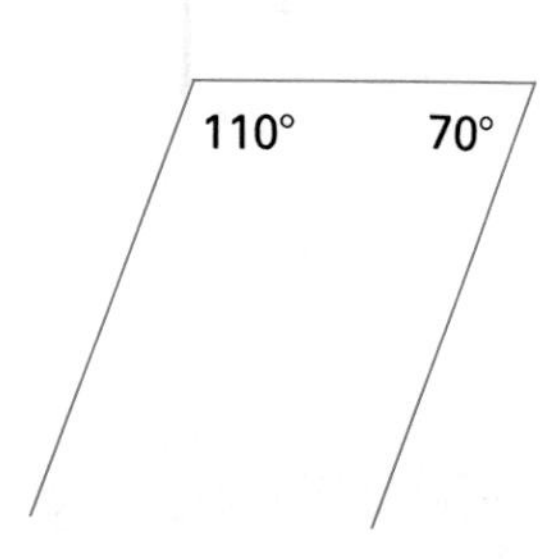

Written Exercises

Determine which lines or segments are parallel. Give a reason for your answer.

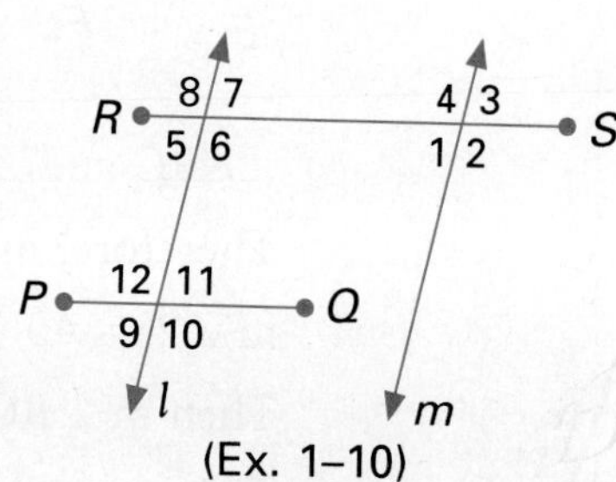

(Ex. 1–10)

1. m $\angle 5 = 65$, m $\angle 9 = 65$

2. m $\angle 1 = 65$, m $\angle 7 = 65$

3. m $\angle 5 = 65$, m $\angle 12 = 115$

4. m $\angle 2 = 115$, m $\angle 6 = 115$

5. m $\angle 6 = 115$, m $\angle 4 = 115$

6. m $\angle 6 = 115$, m $\angle 1 = 65$

7. m $\angle 10 = 115$, m $\angle 6 = 115$

8. m $\angle 8 = 115$, m $\angle 4 = 115$

9. m $\angle 5 = 65$, m $\angle 1 = 65$

10. m $\angle 11 = 115$, m $\angle 6 = 65$

11. In the drawing of the partial garage frame, the ceiling joist $\overline{PQ}$ and soleplate $\overline{RS}$ are parallel. How is the "let-in" corner brace $\overline{PT}$ related to $\overline{PQ}$ and $\overline{RS}$?

12. How are $\angle RTP$ and $\angle QPT$ related?

13. How is the corner brace $\overline{PT}$ related to the vertical studs it crosses? How are $\angle 1$ and $\angle 2$ related?

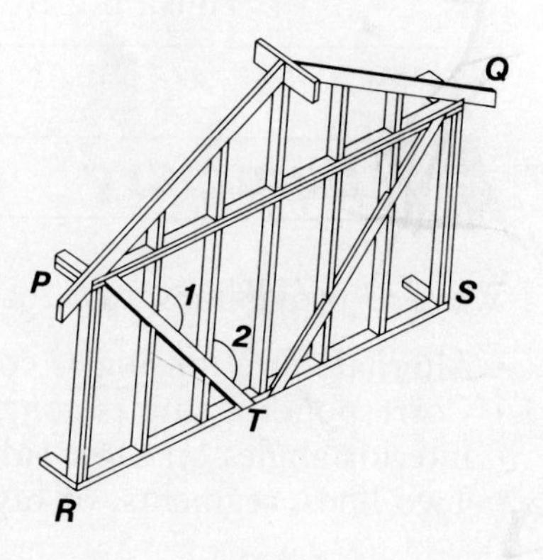

Find the value of x so that $m \parallel n$. (Exercises 14–16)

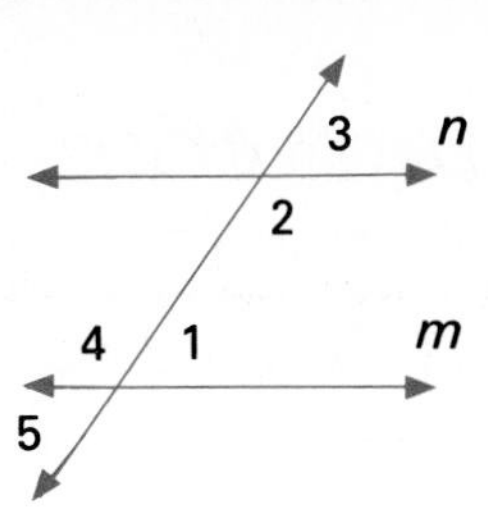

14. $m\angle 4 = 3x - 10$, $m\angle 2 = x + 80$

15. $m\angle 1 = 3x - 10$, $m\angle 2 = 2x + 40$

16. $m\angle 5 = 5x - 40$, $m\angle 3 = 3x$

17. $m\angle 1 = \frac{4}{5}\, m\angle 2$. Find $m\angle 1$, $m\angle 2$.

Find $m\angle ABC$ so that $\overline{AE} \parallel \overline{CD}$. (Exercises 18–20)

18.

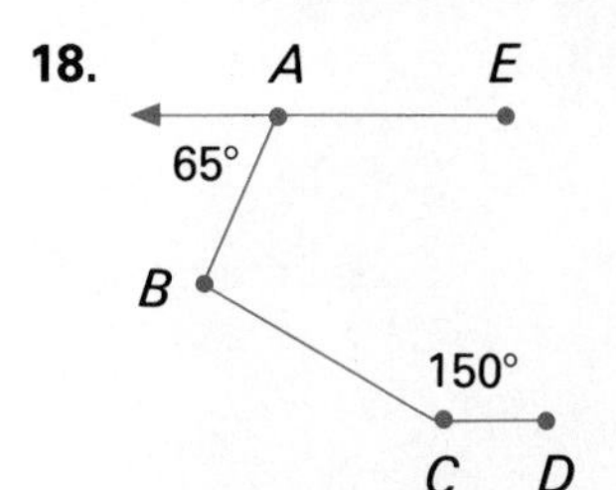

19.

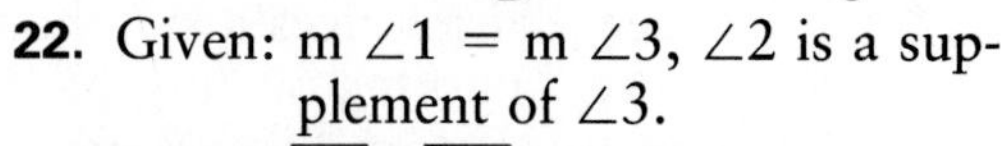

20.

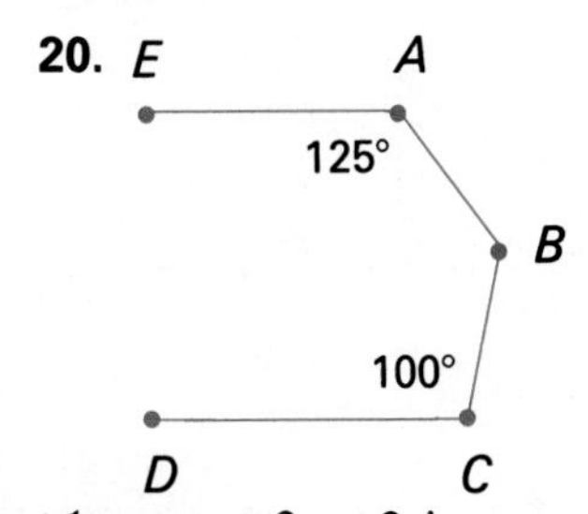

21. Given: $m\angle 3 = m\angle 2$, $\overline{BC}$ bisects $\angle DBE$.
Prove: $\overline{AD} \parallel \overline{BC}$

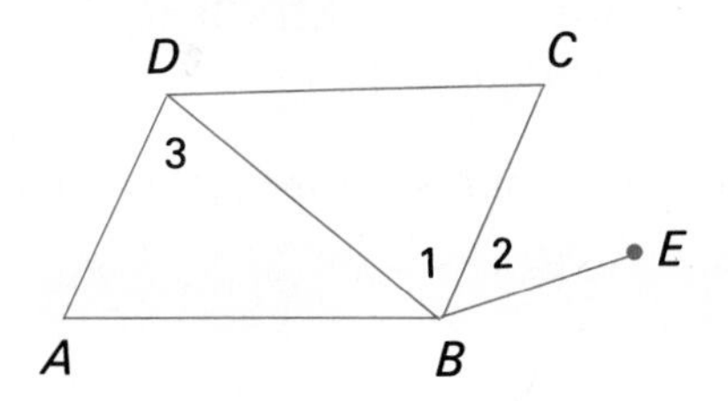

22. Given: $m\angle 1 = m\angle 3$, $\angle 2$ is a supplement of $\angle 3$.
Prove: $\overline{AB} \parallel \overline{CD}$

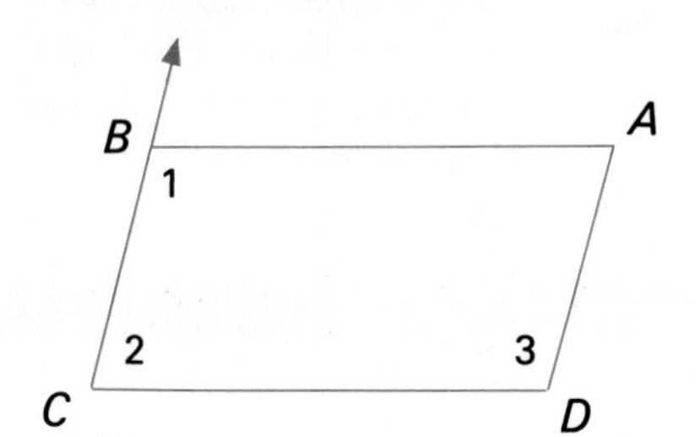

23. Prove Theorem 3.3.

24. Given: $m\angle VQT = m\angle WTQ$, $m\angle 1 = m\angle 2$
Prove: $\overline{PR} \parallel \overline{SU}$

25. $m\angle 3 = x^2$, $m\angle 5 = 12x + 72$
Find x so that $\overline{PR} \parallel \overline{SU}$.

26. $m\angle 3 = x$, $m\angle 5 = 20y + 60$, $m\angle 4 = 5y + 20$
Find x and y so that $\overline{PR} \parallel \overline{SU}$.

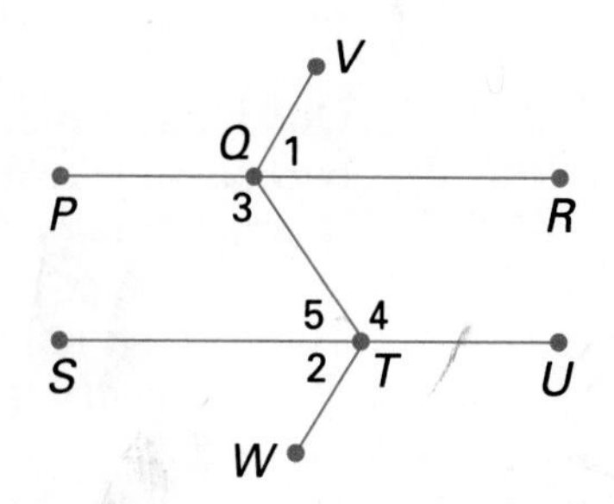

Mixed Review

Write the statement using logic symbols. (Exercises 1–3)

1. If p, then q *2.5* **2.** *p and q* *1.7* **3.** *p or q* *1.7*

4. Identify the hypothesis and the conclusion of the conditional.
If a transversal cuts two lines so that a pair of corresponding angles is congruent, then the lines are parallel. *2.5*

3.4 Introduction to Indirect Proof

Objective To write indirect proofs

Up until now, you have been writing direct proofs. Step by step, such proofs lead to a true conclusion. Another method of proving theorems is that of the *indirect* proof. The indirect proof shows that a conclusion cannot possibly be false. *Indirect* proof is frequently used by lawyers.

For example, suppose that Mr. Al Lee Bigh receives a summons for illegal parking on July 13. His lawyer argues *indirectly*. Mr. Bigh can prove that on July 13 his car was parked in another city 1,000 miles away. This leads to a *contradiction*. His car cannot be in two cities at the same time. The assumption of guilt must be false. Mr. Bigh, therefore, did not park illegally, and his alibi clears him.

Writing indirect proofs involves recognizing contradictory statements. For example:

(1) $\angle 1$ is complementary to $\angle 2$.
(2) $m\angle 1 + m\angle 2 = 120$

Statement 2 contradicts the definition of complementary angles. The sum of the measures of complementary angles is 90, not 120.

One kind of indirect proof makes use of the fact that there are only two possible truth values for any mathematical statement. A statement is either true or false, but not both.

To write an indirect proof, proceed as follows:

(1) Accept the given information as true. Assume the *opposite* of what is to be proved.
(2) Reason directly until there is a contradiction of the given or another known fact.
(3) State that the assumption of the opposite of what was to be proved must be false. So, the original conclusion must be true because it is the only other possibility.

Sometimes it is difficult or impossible to find a direct method of proof. In such cases, indirect proof may be used. An example of indirect proof is illustrated on the next page.

EXAMPLE Write an indirect proof.

Given: m $\angle 1 \neq$ m $\angle 2$
Prove: m $\angle 1 \neq$ m $\angle 3$

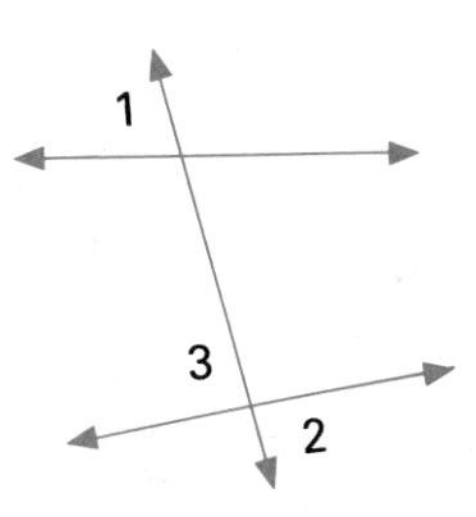

Proof (Indirect)

1. Assume the opposite of what is to be proved.
   Given: m $\angle 1 \neq$ m $\angle 2$
   Assume: m $\angle 1 =$ m $\angle 3$
2. Reason directly until a contradiction is reached.
   m $\angle 2 =$ m $\angle 3$ because vertical angles have the same measure. Then, by substitution, m $\angle 1 =$ m $\angle 2$.

Now, at this point, m $\angle 1 \neq$ m $\angle 2$ is given, and m $\angle 1 =$ m $\angle 2$ is concluded. This is a contradiction!

3. State that the assumption in the first step must be false. So, the original conclusion must be true.
   So, the assumption that m $\angle 1 =$ m $\angle 3$ must be *false*. Therefore, m $\angle 1 \neq$ m $\angle 3$.

Classroom Exercises

Explain why the two statements are contradictory.

1. m and n are skew. m intersects n at point B.

2. The sides of $\angle 1$ are $\perp$. m $\angle 1 = 60$

3. $\angle 1$ is a supplement of $\angle 2$. m $\angle 1 +$ m $\angle 2 = 165$

4. $\angle 1$ and $\angle 2$ are vertical angles. m $\angle 1 \neq$ m $\angle 2$

5. $m \parallel n$. The two lines m and n meet at point P.

6. $\angle A$ is obtuse. m $\angle A = 40$

Written Exercises

For Exercises 1–2, write an indirect proof. (NOTE: $\not\perp$ **means "is not perpendicular to."**)

1. Given: P is not the midpoint of $\overline{AB}$.
Prove: $AP \neq PB$

B
P
A

2. Given: m $\angle 1 \neq$ m $\angle 2$
Prove: $r \not\perp s$

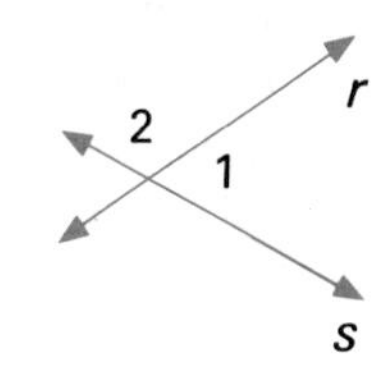

For Exercises 3–8, write an indirect proof. Use the diagram below.

3. Given: $l \not\perp m$
Prove: $m\angle 1 \neq m\angle 3$

4. Given: $m \nparallel n$
Prove: $\angle 3 \not\cong \angle 2$

5. Given: $l \perp m$, $m \nparallel n$
Prove: $l \not\perp n$

6. Given: $m \nparallel n$
Prove: $m\angle 4 + m\angle 2 \neq 180$

7. Given: $m \nparallel n$
Prove: $m\angle 1 + m\angle 2 \neq 180$

8. Given: $m\angle 3 \neq m\angle 2$
Prove: $m\angle 1 + m\angle 2 \neq 180$

9. Prove indirectly that if $m\angle 1 \neq m\angle 2$, then $\angle 1$ and $\angle 2$ are not vertical angles.

Essay.

10. Write an explanation of how an indirect proof is written. Indicate the three major steps.

11. Given: $\overleftrightarrow{AB}$ and $\overleftrightarrow{CD}$ are skew.
Prove: $\overleftrightarrow{AC}$ and $\overleftrightarrow{BD}$ are skew.
(HINT: If two lines are not skew, they either intersect or are parallel.)

12. The assertion that the given is true and the conclusion is false in an indirect proof may be stated as $p \wedge \sim q$. The contradiction of this assertion is $\sim(p \wedge \sim q)$. Use truth tables to show that $\sim(p \wedge \sim q)$ is logically equivalent to $p \rightarrow q$.

Mixed Review

Indicate whether each of the following is always true, sometimes true, or never true.

1. Three coplanar points are collinear. *2.9*
2. Exactly one plane contains two lines. *2.9*
3. Two distinct nonparallel planes intersect in one line. *2.9*
4. Two distinct lines intersect in several points. *2.9*
5. If $m\angle A = m\angle B$ and $m\angle B = m\angle C$, then $m\angle A = m\angle C$. *2.2*

Brainteaser

Ms. Brown, Ms. Green, and Ms. Blue live on the same street and wear coats of these three colors, but none wears a color matching her name. The one wearing the brown coat lives across the street from the other two. Ms. Brown lives next to the woman who drives to work. Ms. Blue takes the train to work. Match the women's names with the color of coat that each wears.

3.5 Converses and the Parallel Postulate

Objectives

To form the converse of a conditional and to determine whether it is true

To prove and apply theorems about angles formed by a transversal of parallel lines

Interchanging the hypothesis and the conclusion of a conditional produces another conditional, called the converse. The converse of a true conditional may or may not be true. Consider this example:

Conditional
If you live in California, then you live in the United States. True

Converse
If you live in the United States, then you live in California. False

If a conditional is represented symbolically as $p \rightarrow q$, then $q \rightarrow p$ is the converse. $q \rightarrow p$ is false in the example above.

Definition

The **converse** of a conditional is the statement formed by interchanging the hypothesis and the conclusion.

EXAMPLE 1 A compact form of the Alternate Interior Angle Postulate is this: If alternate interior angles are congruent, then two lines are parallel. Write the converse of this conditional.

Solution If two lines are parallel, then alternate interior angles are congruent.

In the figure, line m contains point P and $m \parallel k$. It would seem that any other lines containing P, such as l or n, would intersect k. This suggests that m is the only line through P that is parallel to k.

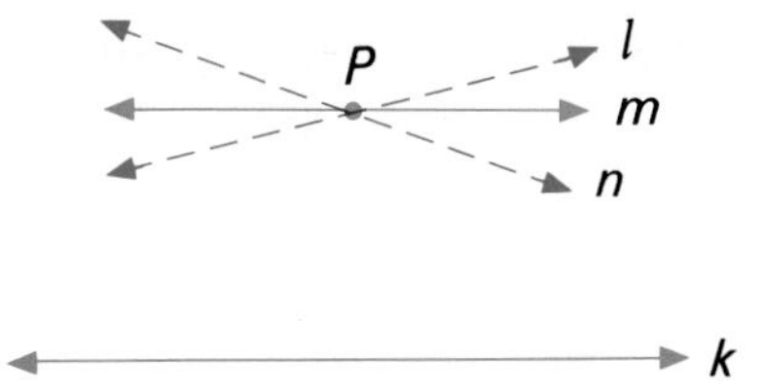

Postulate 14

The Parallel Postulate: Through a point not on a line, there is exactly one line parallel to the given line.

With the help of an *auxiliary line*, the converse of the Alternate Interior Angle Postulate can now be proved by using the Parallel Postulate. An auxiliary line is a line added in a diagram to help in a proof.

Theorem 3.4 If two parallel lines are intersected by a transversal, then alternate interior angles are congruent.

Given: $\overleftrightarrow{AB} \parallel \overleftrightarrow{CD}$
Prove: $\angle DEF \cong \angle EFB$

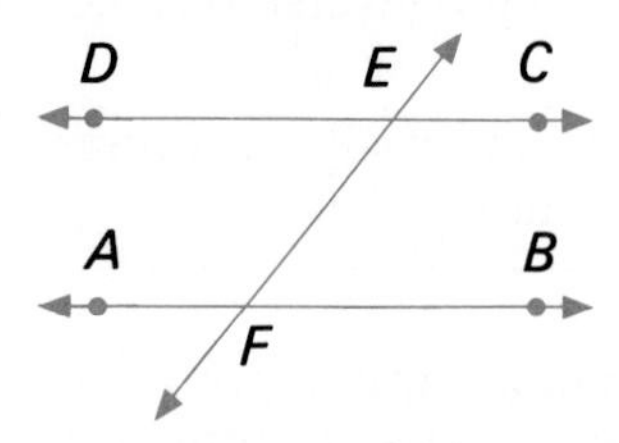

Proof (Indirect) Given: $\overleftrightarrow{AB} \parallel \overleftrightarrow{CD}$

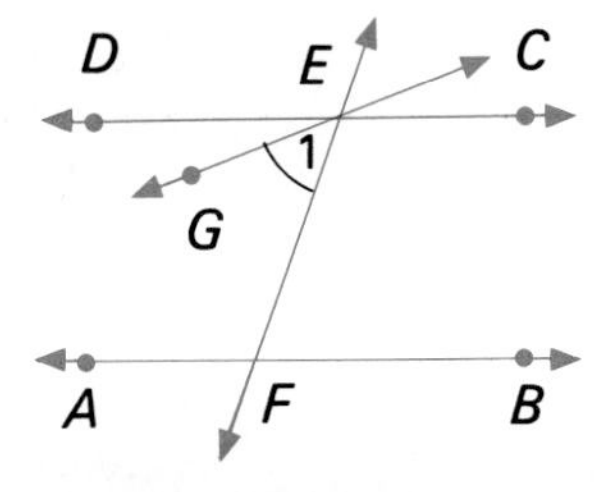

Assume that $\angle DEF \not\cong \angle EFB$. At E, construct $\angle 1$ such that $\angle 1 \cong \angle EFB$. Call the new line formed by this angle $\overleftrightarrow{GE}$. Then $\overleftrightarrow{GE} \parallel \overleftrightarrow{AB}$ by the Alternate Interior Angle Postulate. So, $\overleftrightarrow{DC}$ and $\overleftrightarrow{GE}$ both contain point E and are both parallel to $\overleftrightarrow{AB}$. This contradicts the Parallel Postulate. The assumption that $\angle DEF \not\cong EFB$ is false. Therefore, $\angle DEF \cong \angle EFB$.

Theorem 3.4 can be used to prove the converse of Theorem 3.1.

Theorem 3.5 If two parallel lines are intersected by a transversal, then corresponding angles are congruent.

Given: $\overleftrightarrow{PQ} \parallel \overleftrightarrow{RS}$
Prove: $\angle 1 \cong \angle 3$

Proof

Statement	Reason
1. $\overleftrightarrow{PQ} \parallel \overleftrightarrow{RS}$	1. Given
2. $\angle 2 \cong \angle 3$	2. Alt int $\angle$s of $\parallel$ lines are $\cong$.
3. $\angle 1 \cong \angle 2$	3. Vert $\angle$s are $\cong$.
4. $\therefore \angle 1 \cong \angle 3$	4. Trans Prop

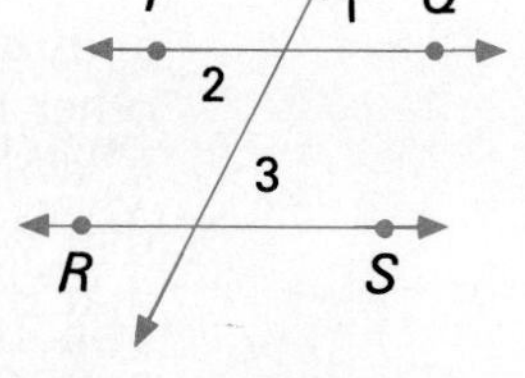

Recall that if interior angles on the same side of a transversal are supplementary, then the lines are parallel. The converse of this is stated as Theorem 3.6. You will be asked to prove this theorem in Exercise 24.

Theorem 3.6 If two parallel lines are intersected by a transversal, then interior angles on the same side of the transversal are supplementary.

EXAMPLE 2 Given: $\overleftrightarrow{AB} \parallel \overleftrightarrow{CD}$, m $\angle 1 = \frac{2}{3}$ m $\angle 2$.
Find m $\angle 1$.

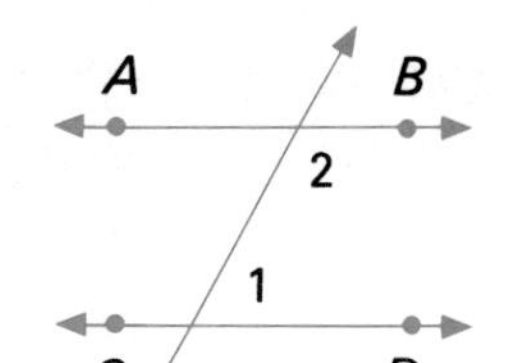

Plan By Theorem 3.6, $\angle 1$ and $\angle 2$ are supplementary:
m $\angle 1$ + m $\angle 2 = 180$.

Solution Let m $\angle 2 = x$ and m $\angle 1 = \frac{2}{3}x$. m $\angle 1 = \frac{2}{3}$ m $\angle 2$.

$$\frac{2}{3}x + x = 180$$
$$2x + 3x = 540$$
$$5x = 540$$
$$x = 108$$

Therefore, m $\angle 1 = \frac{2}{3} \cdot 108 = 72$

EXAMPLE 3 Given: $p \parallel q$, m $\angle 1$ = m $\angle 3$
Prove: $r \parallel s$

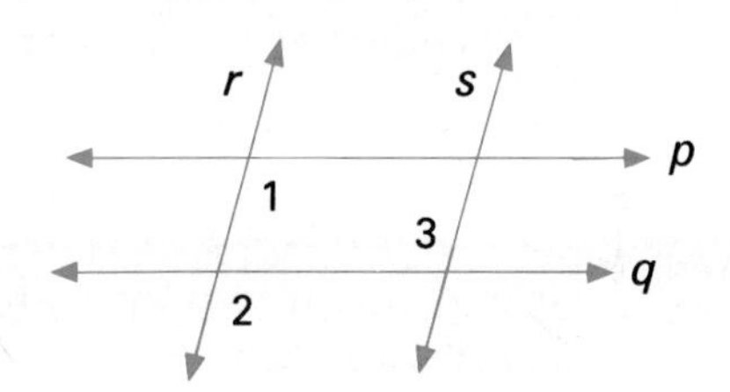

Plan r is parallel to s if m $\angle 2$ = m $\angle 3$.

Proof

Statement	Reason
1. $p \parallel q$	1. Given
2. m $\angle 1$ = m $\angle 2$	2. Corr $\angle$s of $\parallel$ lines are $\cong$ (have equal measure).
3. m $\angle 1$ = m $\angle 3$	3. Given
4. m $\angle 2$ = m $\angle 3$	4. Sub
5. $\therefore r \parallel s$	5. If alt int $\angle$s are $\cong$, then lines are $\parallel$.

Sometimes you have to decide whether you have enough information given to draw a conclusion. For example, before using a theorem about parallel lines to find an unknown measure, make sure that all of the conditions of that theorem are satisfied.

EXAMPLE 4 Based on the given information, find the measures of $\angle 1$ and $\angle 2$. If not enough information is given, state this fact.

Given: $\overline{AD} \parallel \overline{BC}$
Find m $\angle 1$ and m $\angle 2$.

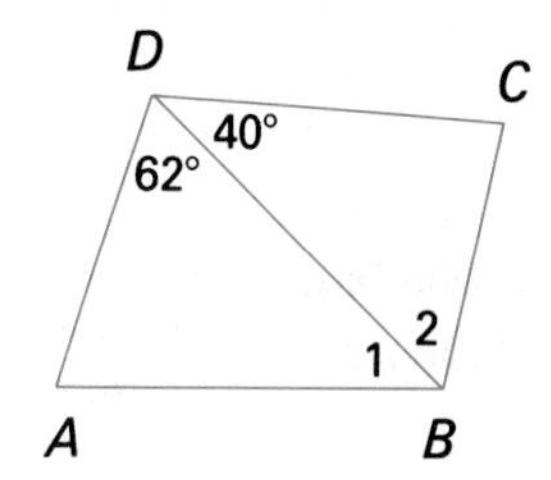

Solution Shade the parallel segments $\overline{AD}$, $\overline{BC}$ and transversal $\overline{BD}$. $\angle 2$ and $\angle ADB$ are alternate interior angles of $\overline{AD}$ and $\overline{BC}$. So, m $\angle 2 = 62$.

You do not know whether $\overline{AB}$ and $\overline{CD}$ are parallel. The measure of $\angle 1$ cannot be found.

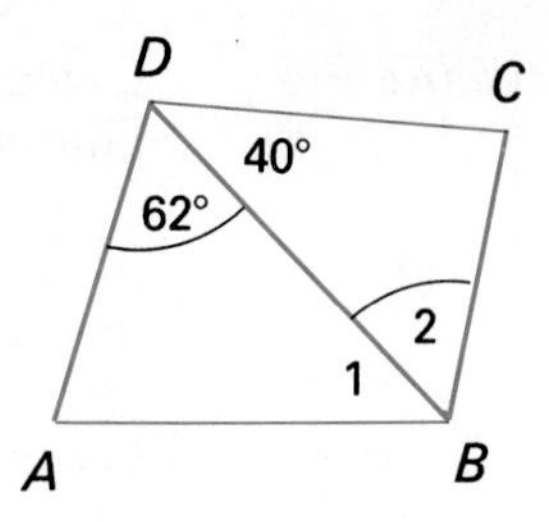

So, m $\angle 2 = 62$, but not enough information is provided to determine m $\angle 1$.

Theorem 3.7 If a transversal is perpendicular to one of two parallel lines, then it is perpendicular to the other.

Theorem 3.8 In a plane, if two lines are parallel to the same line, then they are parallel to each other.

Classroom Exercises

Given $l \parallel m$. Find the measures.

1. m $\angle 1$ = ______
2. m $\angle 2$ = ______
3. m $\angle 3$ = ______

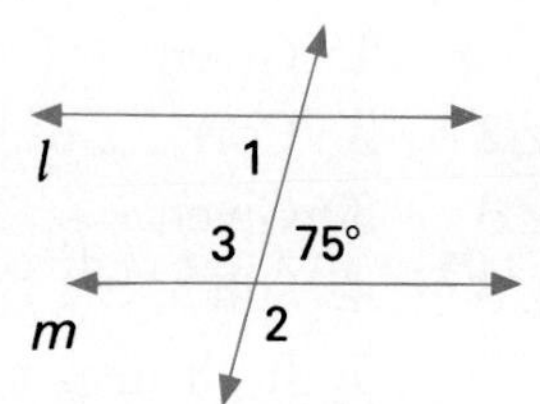

Written Exercises

If $\overleftrightarrow{AB} \parallel \overleftrightarrow{CD}$, find the measure of the indicated angle.

1. Given: m $\angle 1 = 37$
   Find m $\angle 8$.
2. Given: m $\angle 3 = 40$
   Find m $\angle 5$.
3. Given: m $\angle 7 = 100$
   Find m $\angle 4$.
4. Given: m $\angle 8 = 30$
   Find m $\angle 2$.
5. Given: m $\angle 8 = 3x + 30$,
   m $\angle 3 = x + 80$
   Find m $\angle 8$.
6. Given: m $\angle 4 = 4x + 20$,
   m $\angle 8 = 3x + 90$
   Find m $\angle 4$.
7. Given: m $\angle 3 = \frac{4}{5}$ m $\angle 5$
   Find m $\angle 5$.
8. Given: m $\angle 4 = \frac{2}{7}$ m $\angle 8$
   Find m $\angle 4$.

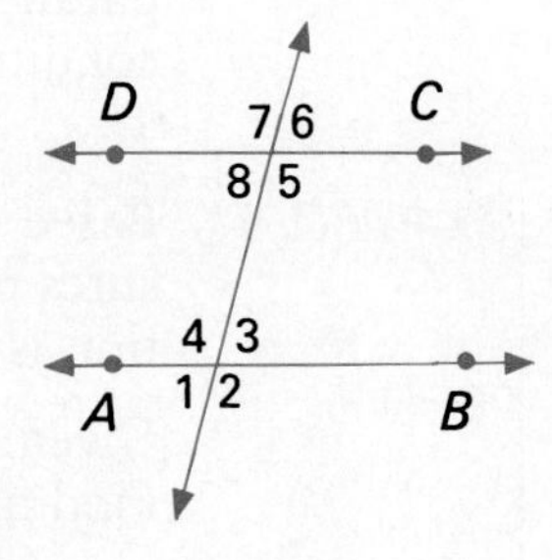

Write the converse of the conditional. Determine whether the converse is true. (Exercises 9–11)

9. If two angles are complementary, then the sum of the angle measures is 90.

10. If two lines are parallel, then the two lines do not intersect.

11. If two lines are skew, then they are not parallel.

Based on the given information, find the indicated angle measures. If not enough information is given, state this fact. (Exercises 12–15)

12. Given: $\overline{DC} \parallel \overline{AB}$, m $\angle 2 = 60$, m $\angle 1 = 40$
Find m $\angle 3$ and m $\angle 4$.

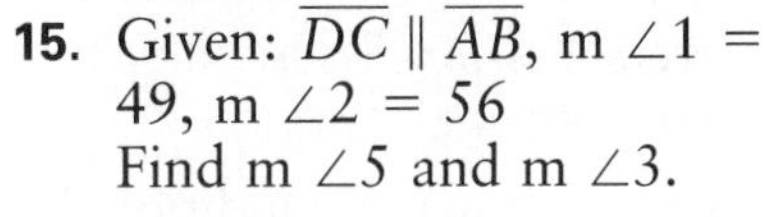

13. Given: $\overline{AD} \parallel \overline{BC}$, m $\angle 3 = 35$, m $\angle 4 = 25$
Find m $\angle 1$ and m $\angle 2$.

14. Given: $\overline{AD} \parallel \overline{BC}$, m $\angle 5 = 50$, m $\angle 3 = 20$
Find m $\angle ABC$ and m $\angle ADC$.

15. Given: $\overline{DC} \parallel \overline{AB}$, m $\angle 1 = 49$, m $\angle 2 = 56$
Find m $\angle 5$ and m $\angle 3$.

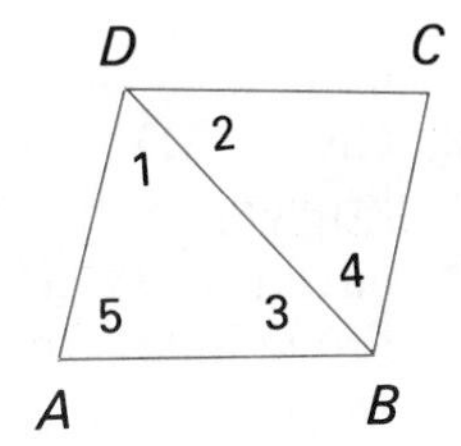

16. Given: $r \parallel s$
Prove: $\angle 1 \cong \angle 2$

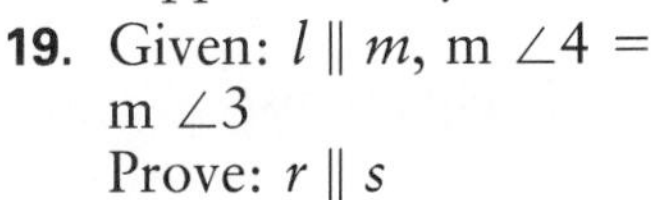

17. Given: $r \parallel s$
Prove: $\angle 1$ and $\angle 3$ are supplementary.

18. Given: $l \parallel m$
Prove: $\angle 1$ and $\angle 4$ are supplementary.

19. Given: $l \parallel m$, m $\angle 4 =$ m $\angle 3$
Prove: $r \parallel s$

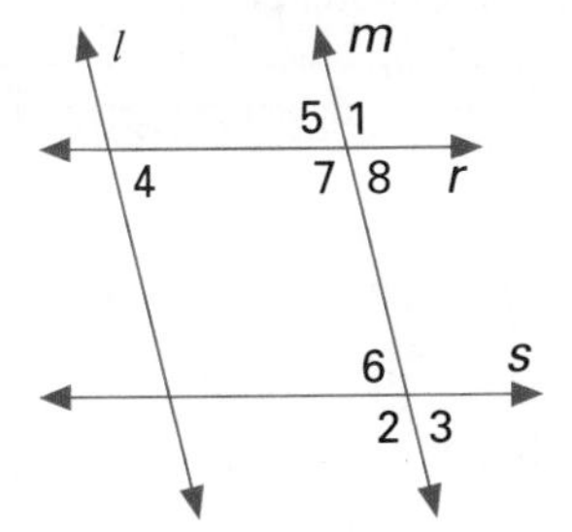

20. Given: $\overline{AB} \parallel \overline{CD}$
Prove: m $\angle 3$ + m $\angle 2 =$ m $\angle ABD$

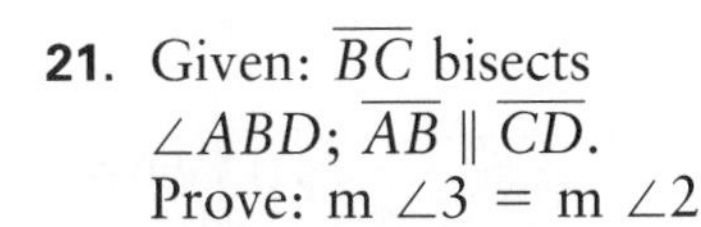

21. Given: $\overline{BC}$ bisects $\angle ABD$; $\overline{AB} \parallel \overline{CD}$.
Prove: m $\angle 3 =$ m $\angle 2$

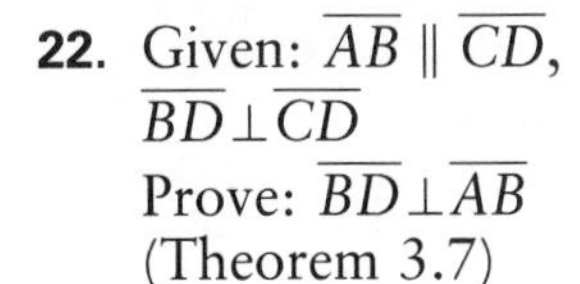

22. Given: $\overline{AB} \parallel \overline{CD}$, $\overline{BD} \perp \overline{CD}$
Prove: $\overline{BD} \perp \overline{AB}$ (Theorem 3.7)

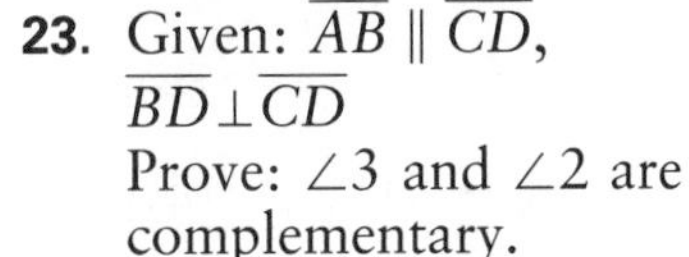

23. Given: $\overline{AB} \parallel \overline{CD}$, $\overline{BD} \perp \overline{CD}$
Prove: $\angle 3$ and $\angle 2$ are complementary.

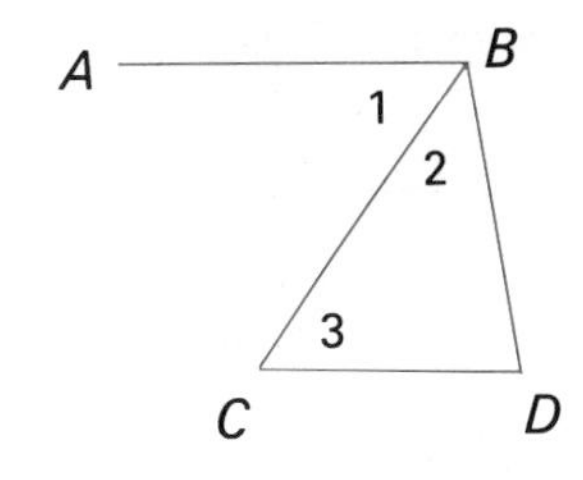

24. Prove Theorem 3.6.

25. Write a direct proof of Theorem 3.8. (HINT: Draw a transversal.)

26. Write an indirect proof of Theorem 3.8.

27. Prove: If two parallel lines are intersected by a transversal, then the bisectors of a pair of alternate interior angles are parallel.

Midchapter Review

True or false? If false, draw a diagram to illustrate why. *3.1*

1. If two lines are not parallel, then they must intersect.
2. Two lines parallel to the same plane are parallel.
3. A line parallel to a plane must be parallel to any line in that plane.
4. Given: $\angle 4 \cong \angle 5$ *3.2, 3.3*
   Which segments are parallel? Why?
5. Given: $\overline{ED} \parallel \overline{AC}$, m $\angle 2 = \frac{4}{5}$m $\angle 8$.
   m $\angle 3$ = ______ *3.5*
6. Given: $\overline{ED} \parallel \overline{AC}$, $\angle 6 \cong \angle 2$
   Prove: $\angle 6 \cong \angle 3$ *3.5*
7. Given: $\overline{ED} \nparallel \overline{AC}$
   Prove indirectly: $\angle 1 \ncong \angle 2$ *3.4*
8. Write the converse of the following conditional.
   If $\angle 6 \cong \angle 7$, then $\overline{EF} \parallel \overline{AB}$ *3.5*

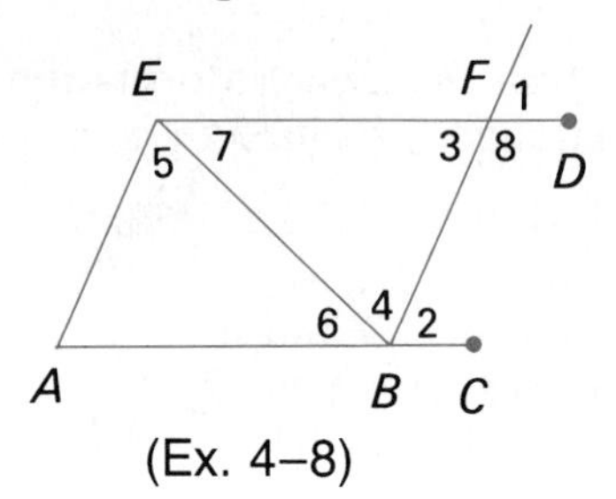

(Ex. 4–8)

Algebra Review

Objective: To solve fractional equations

To solve a fractional equation:
(1) Multiply each side by the LCD.
(2) Solve the resulting equation.

Example: Solve.

$$\frac{3x-5}{4} = \frac{2x+3}{3} \qquad \text{(LCD = 12)}$$

$$12 \cdot \frac{3x-5}{4} = 12 \cdot \frac{2x+3}{3} \qquad \text{Multiply each side by 12.}$$

$$3(3x-5) = 4(2x+3) \qquad \text{Solve.}$$

$$9x - 15 = 8x + 12$$

$$x = 27$$

Solve for x.

1. $\frac{4x-2}{3} = \frac{5x+1}{4}$
2. $\frac{2y-5}{3} = \frac{3y+2}{4}$
3. $\frac{z+2}{3} = \frac{4z-3}{6} - \frac{z}{2}$
4. $\frac{2w-3}{6} - \frac{w+5}{9} = \frac{w-1}{2}$
5. $\frac{4}{t} + \frac{3}{5} = 3$
6. $\frac{3}{4v} = \frac{5}{6} - \frac{2}{3v}$

3.6 The Angles of a Triangle

Objective To apply the theorem about the sum of the measures of the angles of a triangle

Triangle ABC ($\triangle ABC$) is formed by three segments joining three noncollinear points, A, B, and C. Each of these points is a **vertex** (plural: **vertices**) of the triangle. The segments are the **sides** of the triangle.

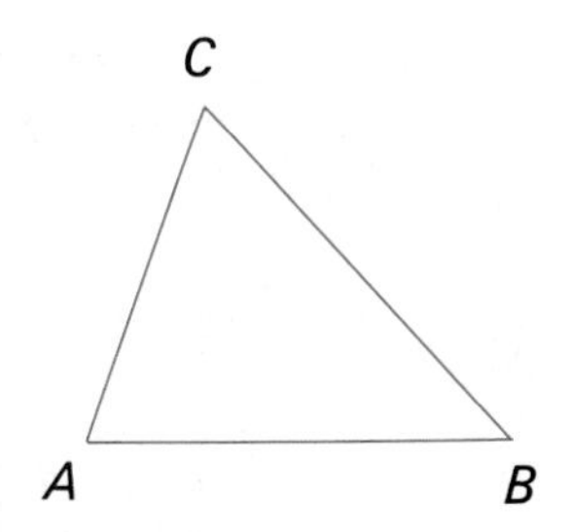

Sides of $\triangle ABC$: $\overline{AB}$, $\overline{BC}$, $\overline{CA}$
Vertices of $\triangle ABC$: A, B, C
Angles of $\triangle ABC$: $\angle A$, $\angle B$, $\angle C$

Definition A **triangle** is a figure formed by three segments joining three noncollinear points.

Recall that a straight angle has a measure of 180. In the figure, m $\angle AOC$ + m $\angle 3$ = 180, or by the Angle Addition Postulate, (m $\angle 1$ + m $\angle 2$) + m $\angle 3$ = 180.

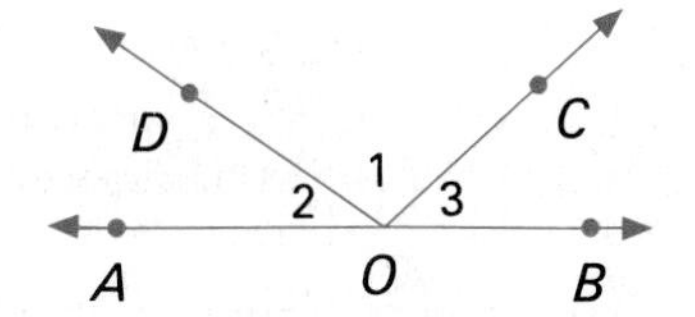

This concept of straight angle, together with properties of parallel lines, can be used to discover the sum of the measures of the angles of a triangle.

In the figure, $\overleftrightarrow{AB} \parallel \overline{CD}$, a side of $\triangle COD$.

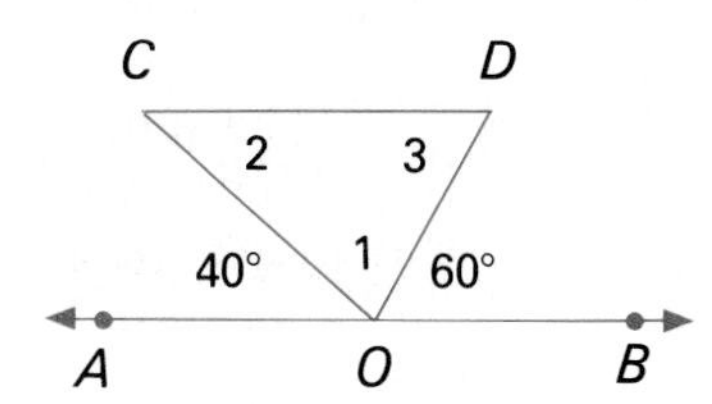

(1) Because $\angle AOB$ is a straight angle,

$$40 + m\,\angle 1 + 60 = 180$$
$$100 + m\,\angle 1 = 180$$
$$m\,\angle 1 = 80$$

(2) Since $\overleftrightarrow{AB} \parallel \overline{CD}$, m $\angle 2$ = 40 and m $\angle 3$ = 60 (alt int $\angle$s of $\parallel$ lines are $\cong$).

From (1) and (2) above, m $\angle 1$ + m $\angle 2$ + m $\angle 3$ becomes 80 + 40 + 60, or 180.

Therefore, m $\angle 1$ + m $\angle 2$ + m $\angle 3$ = 180.

If you were to perform this experiment with other pairs of sample values for m $\angle AOC$ and m $\angle BOD$, say 50 and 70, or 20 and 55, you would be able to *inductively* make a generalization about the sum of the measures of the angles of a triangle.

Theorem 3.9 The sum of the measures of the angles of a triangle is 180.

Given: $\triangle DOC$
Prove: $m\angle 1 + m\angle 2 + m\angle 3 = 180$

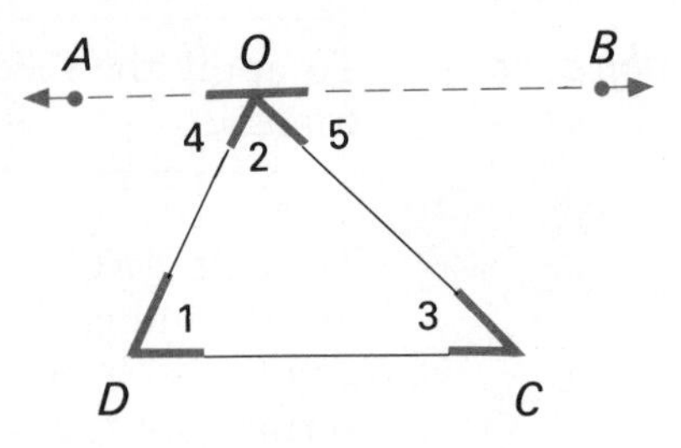

Proof

Statement	Reason
1. $\triangle DOC$	1. Given
2. Through O, draw $\overleftrightarrow{AB} \parallel \overline{DC}$.	2. Par Post
3. $m\angle 1 = m\angle 4$, $m\angle 3 = m\angle 5$	3. Alt int $\angle$s of $\parallel$ lines have equal measure.
4. $m\angle 4 + m\angle DOB = 180$	4. If outer rays of two adj $\angle$s form a st $\angle$, then the sum of their measures is 180.
5. $m\angle DOB = m\angle 2 + m\angle 5$	5. $\angle$ Add Post
6. $m\angle 4 + m\angle 2 + m\angle 5 = 180$	6. Sub
7. $\therefore m\angle 1 + m\angle 2 + m\angle 3 = 180$	7. Sub (Steps 3 and 6)

EXAMPLE 1 Given: $m\angle A = 2x + 15$, $m\angle B = 3x - 5$, $m\angle C = 4x + 35$

Find $m\angle A$, $m\angle B$, and $m\angle C$.

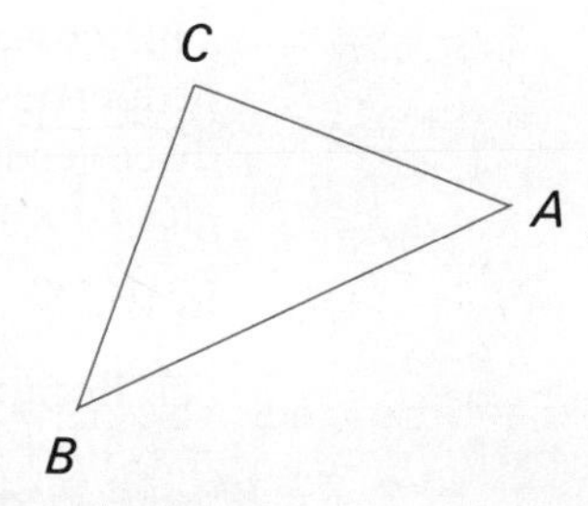

Solution The sum of measures of the angles of a triangle is 180.

$$(2x + 15) + (3x - 5) + (4x + 35) = 180$$
$$9x + 45 = 180$$
$$9x = 135$$
$$x = 15$$

Check

$m\angle A = 2x + 15 = 2 \cdot 15 + 15 = 45$
$m\angle B = 3x - 5 = 3 \cdot 15 - 5 = 40$
$m\angle C = 4x + 35 = 4 \cdot 15 + 35 = 95$

$45 + 40 + 95 = 180$

Recall that if the sides of an angle are perpendicular, the measure of the angle is 90. This idea is used in the next example.

EXAMPLE 2 Given: $\triangle PQR$, with $\overline{PQ} \perp \overline{PR}$, m $\angle Q = \frac{2}{3}$ m $\angle R$
Find m $\angle R$.

Plan $\overline{PQ} \perp \overline{PR}$. So, m $\angle P = 90$.
Since m $\angle Q = \frac{2}{3}$ m $\angle R$, let m $\angle R = x$ and m $\angle Q = \frac{2}{3}x$.

Solution

$$\text{m} \angle P + \text{m} \angle Q + \text{m} \angle R = 180$$
$$90 + \frac{2}{3}x + x = 180$$
$$3(90 + \frac{2}{3}x + x) = 3 \cdot 180$$
$$270 + 2x + 3x = 540$$
$$5x = 270$$
$$x = 54 \Rightarrow \text{m} \angle R = 54$$

Classroom Exercises

Find the measure of the third angle of $\triangle ABC$.

1. m $\angle A = 60$, m $\angle B = 40$, m $\angle C =$ ______
2. m $\angle A = 70$, m $\angle B = 70$, m $\angle C =$ ______
3. m $\angle A = 90$, m $\angle B = 30$, m $\angle C =$ ______
4. m $\angle A = 45$, m $\angle B = 45$, m $\angle C =$ ______

Find the measure of each angle of $\triangle ABC$.

5. m $\angle A = x$, m $\angle B = 2x$, m $\angle C = 3x$
6. m $\angle A = 40$, m $\angle B = x$, m $\angle C = x + 10$

Written Exercises

Find the measure of each angle of $\triangle ABC$ for the given data.

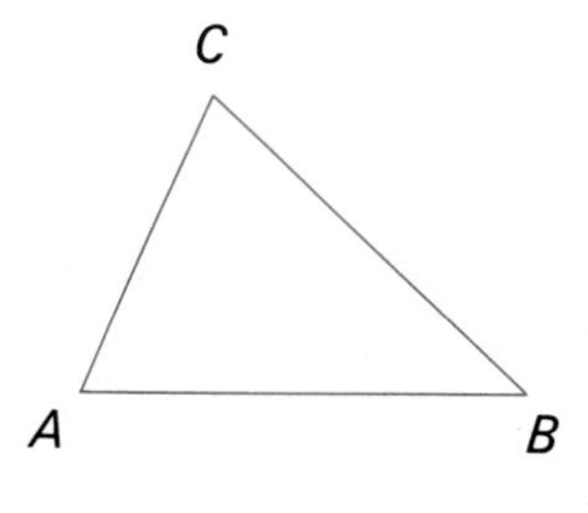

1. m $\angle A =$ m $\angle B =$ m $\angle C$
2. $\overline{AC} \perp \overline{BC}$, m $\angle A = 20$
3. m $\angle A = x$, m $\angle B = x$, m $\angle C = 3x$
4. m $\angle A = x + 2$, m $\angle B = 3x - 10$, m $\angle C = 4x - 4$
5. m $\angle A = x$, m $\angle B = x + 20$, m $\angle C = x + 40$
6. m $\angle A = 3x + 20$, m $\angle B = x - 10$, m $\angle C = 90 - 2x$
7. m $\angle A = \frac{3}{7}$ m $\angle B$, $\overline{AC} \perp \overline{BC}$
8. $\overline{AC} \perp \overline{BC}$, m $\angle A = \frac{5}{4}$ m $\angle B$
9. In $\triangle PQR$, m $\angle P = 70$. Angles Q and R have equal measures. Find m $\angle Q$.
10. In $\triangle ABC$, m $\angle C = 150$. Find m $\angle A$ if m $\angle A = 2(\text{m} \angle B)$.

In the figure of the frame of a house under construction, $\overline{EG} \perp \overline{AB}$, $\overline{FH} \perp \overline{AB}$, $\overline{CE} \perp \overline{AD}$, $\overline{CF} \perp \overline{DB}$, $\overline{CD} \perp \overline{AB}$. (Exercises 11–13)

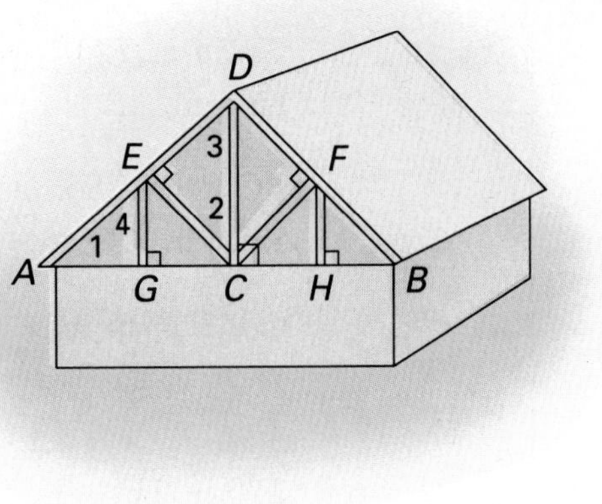

11. If the pitch of the roof ($\angle 1$) measures 44, find the measure of the angle between roof rafters, m $\angle ADB$. (Assume m $\angle DAB$ = m $\angle DBA$.)

12. If the pitch of the roof measures 30, find m $\angle 4$. ($\angle 4$ is the angle formed by brace $\overline{EG}$ and the roof rafter $\overline{AD}$.)

13. Find the measure of the angle between the roof rafters (m $\angle ADB$) if m $\angle 2$ = 20.

14. Given: $\overline{PS} \parallel \overline{QR}$, $\overline{PQ} \parallel \overline{RS}$
Prove: m $\angle 1$ + m $\angle 2$ + m $\angle 3$ = 180

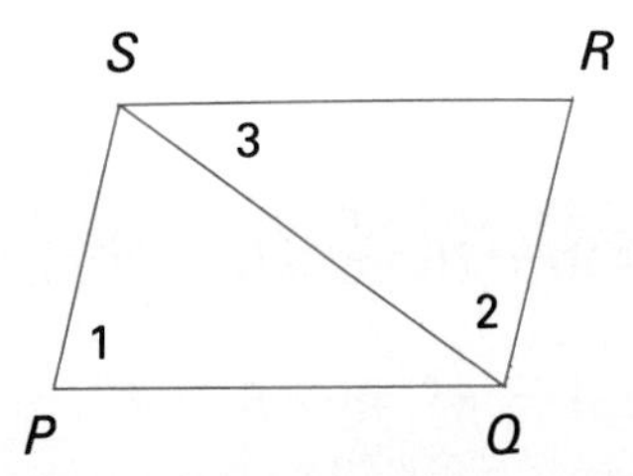

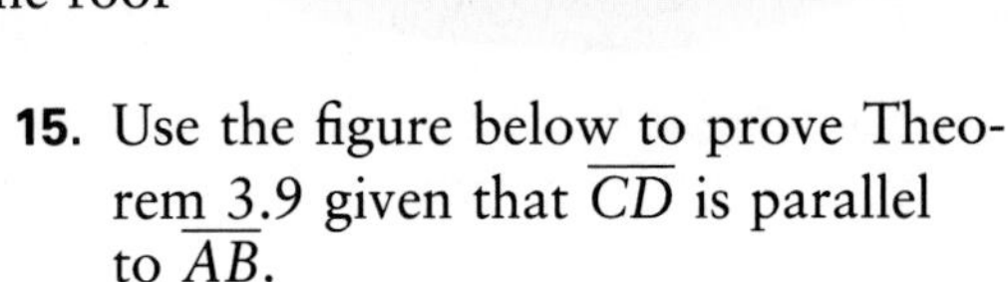
15. Use the figure below to prove Theorem 3.9 given that $\overline{CD}$ is parallel to $\overline{AB}$.

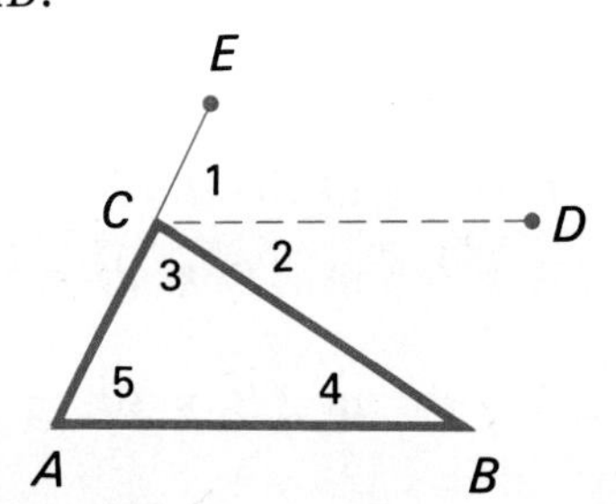

Find the measure of each numbered angle. (Exercises 16–18)

16. Given: $\overleftrightarrow{AB} \parallel \overleftrightarrow{CD}$

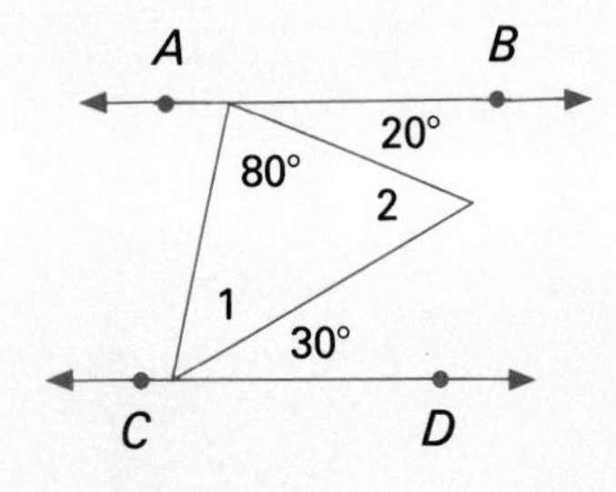

17. Given: $\overline{CD}$ bisects $\angle ACB$.

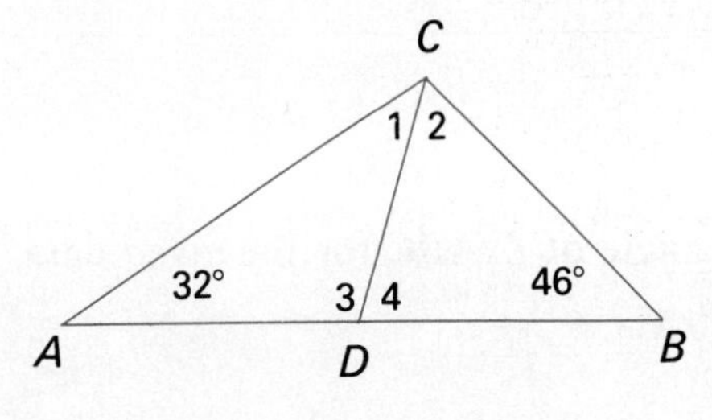

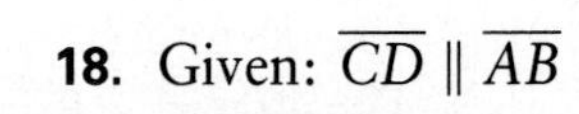
18. Given: $\overline{CD} \parallel \overline{AB}$

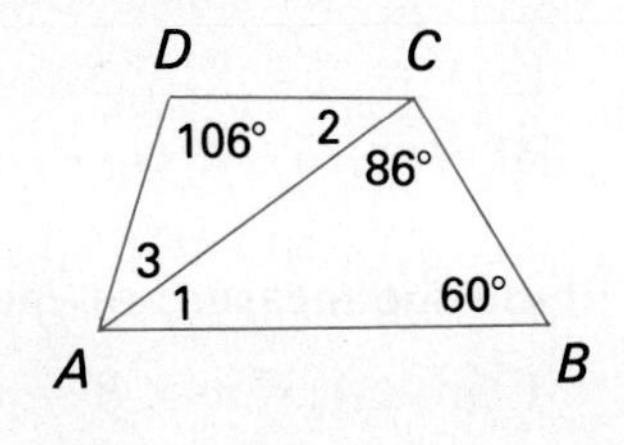

In the figure below, $\overline{AD}$ bisects $\angle CAB$, and $\overline{BD}$ bisects $\angle CBA$.

19. Given: m $\angle C$ = 70, m $\angle 1$ = 20
Find m $\angle D$.

20. Given: m $\angle C$ = 50, m $\angle 1$ = 30
Find m $\angle D$.

21. Given: m $\angle 1$ = m $\angle 3$, m $\angle D$ = 100
Find m $\angle C$.

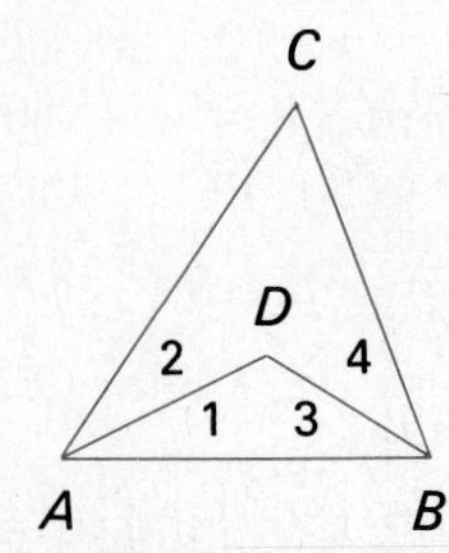

In the figure, $\overline{ED} \parallel \overline{AB}$, $\overline{GF}$ bisects $\angle DGC$; $\overline{CF}$ bisects $\angle BCG$.

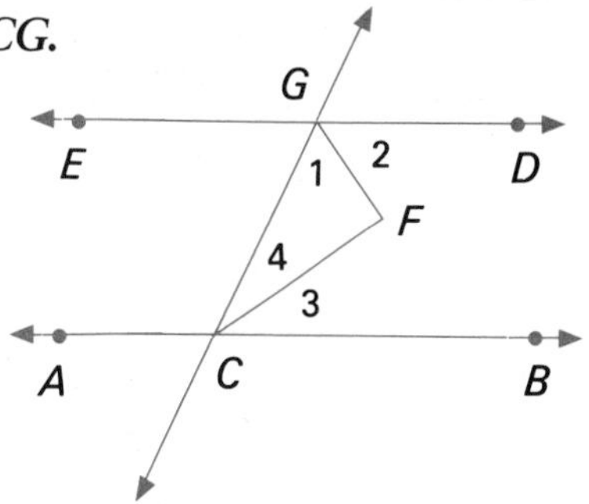

22. Given: m $\angle 2 = 40$. Find m $\angle F$.
23. Given: m $\angle GCB = 70$. Find m $\angle F$.
24. Given: m $\angle CGD = 100$. Find m $\angle F$.
25. The results of Exercises 22–24 inductively suggest the following property: If two lines are parallel, then the bisectors of interior angles on the same side of a transversal are ______.

Mixed Review

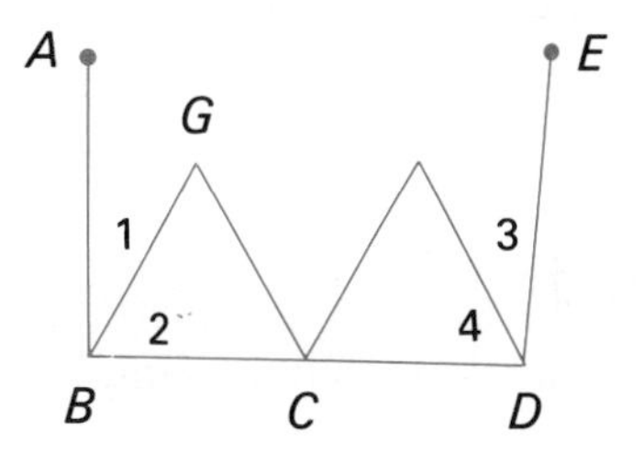

1. Given: C is the midpoint of $\overline{BD}$; $AB = BC$.
   Prove: $AB = CD$ **1.3, 2.2, 2.4**
2. Given: m $\angle 2$ = m $\angle 4$, m $\angle 1$ = m $\angle 3$
   Prove: m $\angle ABC$ = m $\angle EDC$ **2.2, 2.4**
3. Given: $\overline{AB} \perp \overline{BD}$, $\overline{ED} \perp \overline{BD}$, m $\angle 1$ = m $\angle 3$
   Prove: m $\angle 2$ = m $\angle 4$ **1.6, 2.3, 2.4**
4. Given: $\overline{BG}$ bisects $\angle ABC$; $\angle 2 \cong \angle 3$.
   Prove: $\angle 1 \cong \angle 3$ **1.5, 2.3, 2.4**

5. In the portion of the stairway shown, m $\angle PQR = 120$. Find m $\angle 1$. Explain why m $\angle 1$ can be found. **3.5**

6. This photo shows parking stripes and concrete dividers in a parking lot. How are the positions of the stripes related? What name can be applied to a line drawn through the center of the row of dividers as it crosses the parking stripes? How can this photo be used to illustrate postulates or theorems you have studied in this chapter? **3.5**

Computer Investigation

Angles of a Triangle

Use a computer software program that constructs random triangles by classification (acute, obtuse, etc.), labels points, extends segments, and measures angles.

The figure at the right shows a triangle with one side, $\overline{BC}$, extended to point D in the *exterior* of the triangle. $\angle ACD$ is called an *exterior angle* of $\triangle ABC$.

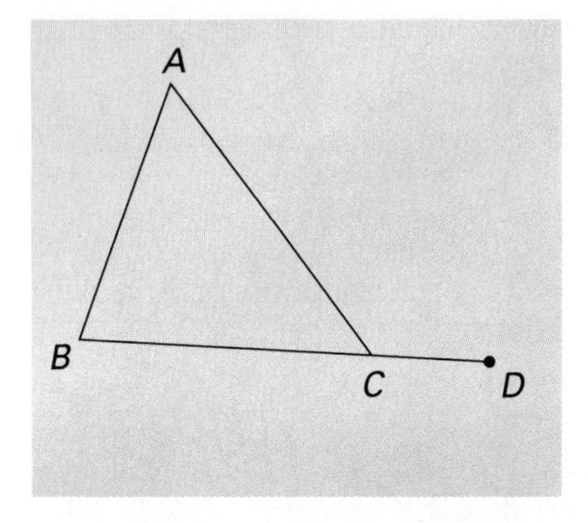

The activity below will help you discover some properties of the angles of a triangle and of an exterior angle of a triangle.

Activity 1

Draw and label an acute triangle *ABC*.

1. Find the measures of the three angles of the triangle.
2. Find m $\angle A$ + m $\angle B$ + m $\angle C$.
3. Draw a right triangle. Find the measures of the three angles. Find the sum of their measures.
4. Draw an obtuse triangle and find the sum of the measures of the three angles.
5. Draw an isosceles triangle. Find the sum of the measures of the three angles.
6. What general property of the angles of any triangle is suggested by the results of the five exercises above?

Activity 2

Draw and label an obtuse triangle *ABC*. Extend $\overline{BC}$ to a point *D* so that *C* is between *B* and *D*. As mentioned above, $\angle ACD$ is an exterior angle of $\triangle ABC$.

7. Find m $\angle ACD$ and the measure of each of the two angles of $\triangle ABC$ that are not adjacent to $\angle ACD$.
8. What numerical relationship exists between the three angle measures of Exercise 7?
9. Clear the screen and draw and label an acute triangle ABC, with an exterior angle. Then find the measure of the exterior angle and each of the two angles of $\triangle ABC$ that are not adjacent to the exterior angle. What relationship exists between the three angles?

3.7 Exterior and Remote Interior Angles of a Triangle

Objectives

To find the measures of exterior and remote interior angles of a triangle
To write proofs using the Exterior Angle Theorem

Side $\overline{AB}$ of $\triangle ABC$ is extended to point D. $\angle 1$ is adjacent and supplementary to $\angle 2$. $\angle 1$ is called an *exterior angle* of $\triangle ABC$. $\angle A$ and $\angle C$ of $\triangle ABC$ are called the *remote interior angles* of the exterior $\angle 1$.

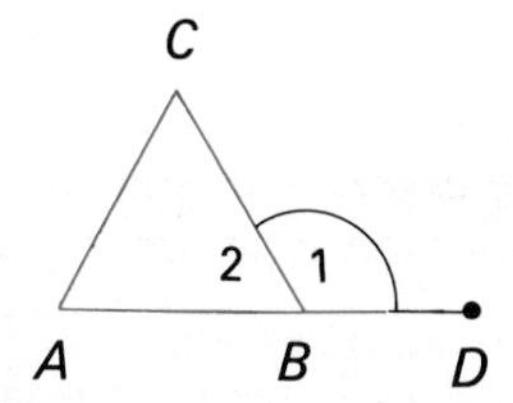

Definitions

An **exterior angle** of a triangle is an angle that is adjacent and supplementary to one of the angles of the triangle. The other two angles of the triangle are called the **remote interior angles** of that exterior angle.

Theorem 3.10

Exterior Angle Theorem: The measure of an exterior angle of a triangle is equal to the sum of the measures of its two remote interior angles.

Given: $\triangle ABC$ with exterior $\angle 1$
Prove: $m\angle 1 = m\angle A + m\angle C$

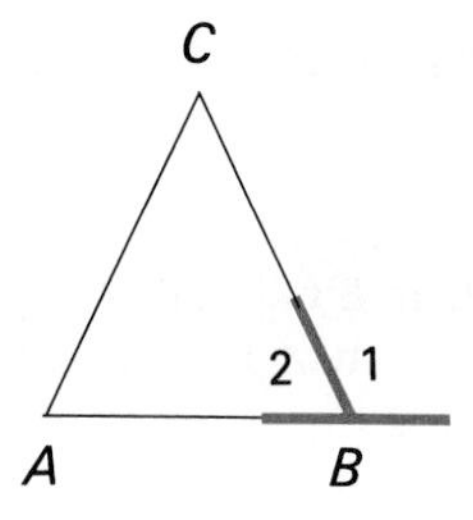

Proof

Statement	Reason
1. $\triangle ABC$ with exterior $\angle 1$	1. Given
2. $\angle 1$ is supplementary to $\angle 2$.	2. Def of ext $\angle$
3. $m\angle 1 + m\angle 2 = 180$	3. If outer rays of two adj $\angle$s form a st $\angle$, then sum of their meas is 180.
4. $m\angle A + m\angle C + m\angle 2 = 180$	4. Sum of measures of $\angle$s of a $\triangle$ = 180.
5. $m\angle 1 + m\angle 2 = m\angle A + m\angle C + m\angle 2$	5. Sub
6. $\therefore m\angle 1 = m\angle A + m\angle C$	6. Subt Prop of Eq

EXAMPLE 1 Find m $\angle B$.

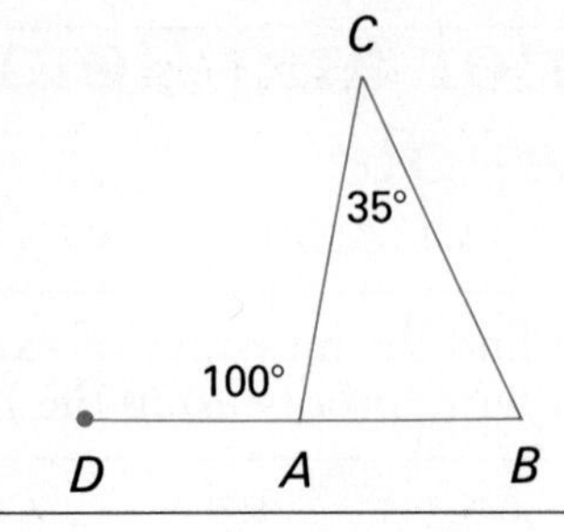

Solution $\angle DAC$ is an exterior angle.

$$m\ \angle DAC = m\ \angle C + m\ \angle B$$
$$100 = 35 + m\ \angle B$$
$$m\ \angle B = 65$$

EXAMPLE 2 Given: m $\angle 1 = 10x - 6$
m $\angle 3 = 3x + 4$
m $\angle 4 = 4x + 2$
Find m $\angle 2$.

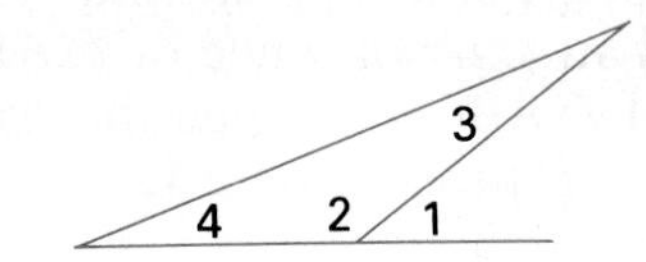

Plan $\angle 1$ is an exterior angle. Use Theorem 3.10 to write an equation. Solve for x to find m $\angle 1$. Then use the fact that $\angle 2$ is supplementary to $\angle 1$ to find m $\angle 2$.

Solution

$$m\ \angle 1 = m\ \angle 3 + m\ \angle 4$$
$$10x - 6 = (3x + 4) + (4x + 2)$$
$$10x - 6 = 7x + 6$$
$$3x = 12$$
$$x = 4$$
$$m\ \angle 1 = 10x - 6 = 10 \cdot 4 - 6 = 34$$
$$m\ \angle 1 = 34$$

Then, since $\angle 1$ and $\angle 2$ are supplementary, m $\angle 2 = 180 - 34 = 146$.

EXAMPLE 3 Given: $\overline{DC} \parallel \overline{AB}$
Prove: m $\angle 1$ = m $\angle 3$ + m $\angle 2$

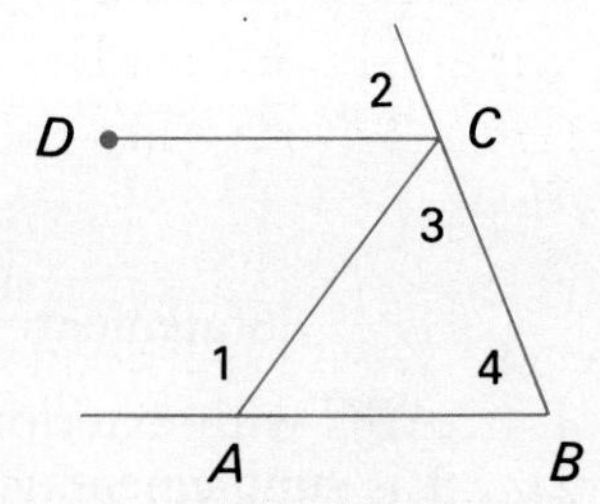

Proof

Statement	Reason
1. $\overline{DC} \parallel \overline{AB}$	1. Given
2. m $\angle 2$ = m $\angle 4$	2. Corr $\angle$s of $\parallel$ lines have equal measures.
3. m $\angle 1$ = m $\angle 3$ + m $\angle 4$	3. Ext $\angle$ Theorem
4. $\therefore$ m $\angle 1$ = m $\angle 3$ + m $\angle 2$	4. Sub

Focus on Reading

Indicate whether the statement is always true, sometimes true, or never true.

1. An exterior angle of a triangle is equal in measure to either of its remote interior angles.
2. An exterior angle of a triangle is supplementary to the angle of the triangle that is not one of its remote interior angles.
3. The angle that is adjacent to an exterior angle of a triangle is one of the remote interior angles.

Classroom Exercises

Find the missing angle measures.

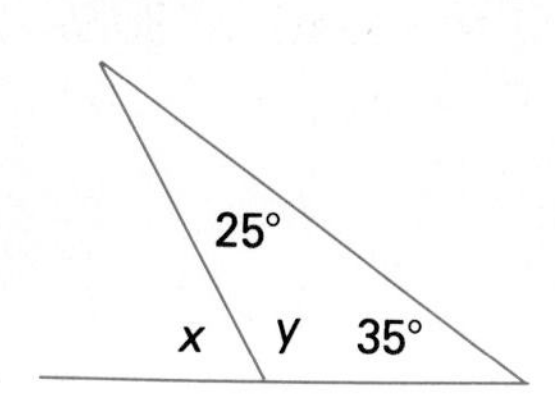

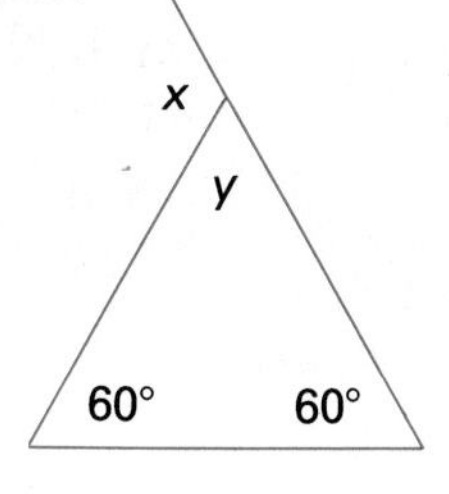

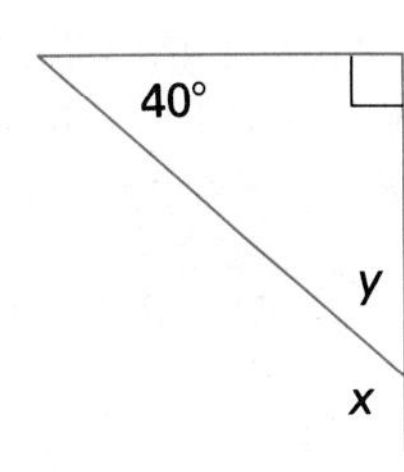

1. x
2. y
3. x
4. y
5. x
6. y

Written Exercises

Find the indicated angle measure. (Exercises 1–6)

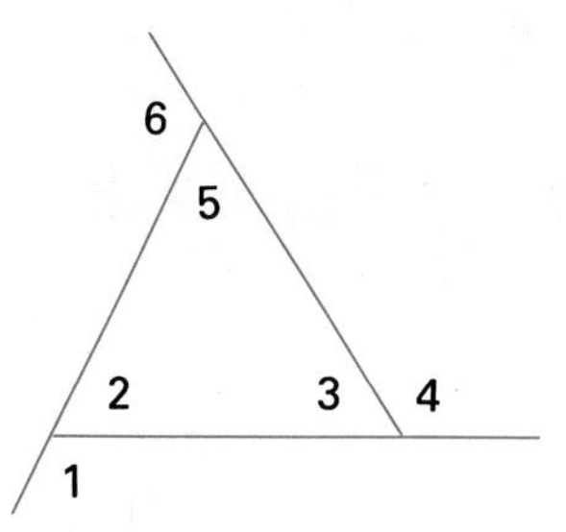

1. m $\angle 5 = 43$, m $\angle 3 = 37$. Find m $\angle 1$.
2. m $\angle 3 = 60$, m $\angle 2 = 40$. Find m $\angle 6$.
3. m $\angle 5 = 32$, m $\angle 2 = 70$. Find m $\angle 4$.
4. m $\angle 1 = 160$, m $\angle 3 = 100$. Find m $\angle 5$.
5. m $\angle 6 = 130$, m $\angle 3 = 75$. Find m $\angle 2$.
6. m $\angle 4 = 60$, m $\angle 2 = 39$. Find m $\angle 5$.

Find the indicated angle measure. (Exercises 7–16)

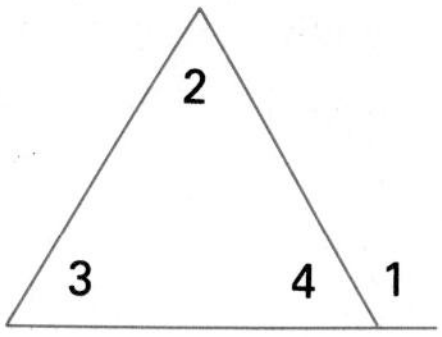

7. m $\angle 2 = 7x$, m $\angle 3 = 3x$, m $\angle 1 = 60$
   Find m $\angle 2$.
8. m $\angle 2 = 3x$, m $\angle 3 = 3x$, m $\angle 1 = 120$
   Find m $\angle 2$.
9. m $\angle 3 = 40$, m $\angle 2 = 5x$, m $\angle 1 = 7x$
   Find m $\angle 2$.
10. m $\angle 3 = x + 8$, m $\angle 2 = 2x + 3$, m $\angle 1 = 5x - 11$. Find m $\angle 1$.

11. m $\angle 1 = 10x - 4$, m $\angle 2 = 2x + 6$, m $\angle 3 = 4x + 2$
Find m $\angle 2$.
12. m $\angle 1 = 110$, m $\angle 2$ is 10 less than 3 times m $\angle 3$.
Find m $\angle 3$.
13. m $\angle 1 = 70$, m $\angle 3$ is 10 more than twice m $\angle 2$.
Find m $\angle 3$.
14. m $\angle 1 = 140$, m $\angle 2 = \frac{5}{9}$ m $\angle 3$. Find m $\angle 3$.
15. m $\angle 3 = 60$, m $\angle 1$ is twice as large as m $\angle 2$.
Find m $\angle 1$.
16. m $\angle 4 = 70$, m $\angle 2$ is three times as large as m $\angle 3$.
Find m $\angle 2$.

17. Given: $\overline{PQ} \parallel \overleftrightarrow{ST}$
Prove: m $\angle 1$ = m $\angle 2$ + m $\angle 4$

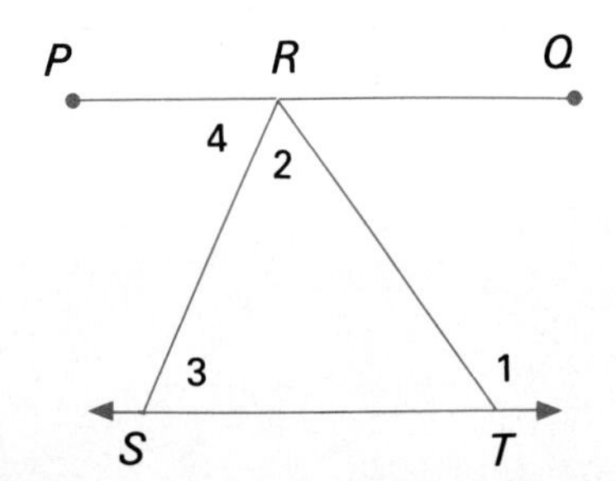

18. Given: $\overline{CE}$ bisects $\angle DCB$; m $\angle ECD$ = m $\angle B$
Prove: $\overline{CE} \parallel \overline{AB}$

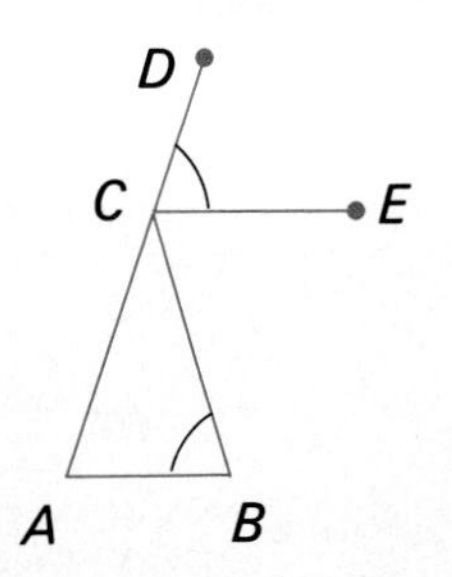

19. Given: $\overline{AD}$ bisects $\angle FAC$.
Find m $\angle BGD$.

20. Given: m $\angle GFB = 110$, $\overline{GF} \parallel \overline{AD}$, $\overline{FB} \parallel \overline{ED}$, $\angle 1 \cong \angle 2$
Find m $\angle 3$.

21. In triangle PQR, m $\angle P = \frac{4}{5}$ m $\angle Q$ and the measure of an exterior angle at R is 144. Find m $\angle Q$.
22. Find the measure of each angle of a triangle whose exterior angles, one at each vertex, have the measures $2x + 10$, $x + 20$, and $x + 30$.
23. In $\triangle ABC$, D is on $\overline{CB}$ such that m $\angle CAD$ = m $\angle CDA$. m $\angle CAB$ − m $\angle ABC$ = 30. Find m $\angle BAD$.
24. Each exterior angle of $\triangle ABC$ has the same measure. Find the measure of each exterior angle.

Mixed Review

True or false? If false, explain why. 1.7

1. A line has two endpoints *or* a plane is flat.
2. Vertical angles are congruent *and* parallel lines intersect.

Graph and name the resulting geometric figure. 1.8

3. $x \geq 5$ *or* $x \leq 7$
4. $x \geq 5$ *and* $x \leq 7$

Application: *Light Rays*

The figure at the right shows a ray of light, called the incident ray, and its reflection, called the reflection ray. In physics, the *Law of Reflection of Light* is stated as below.

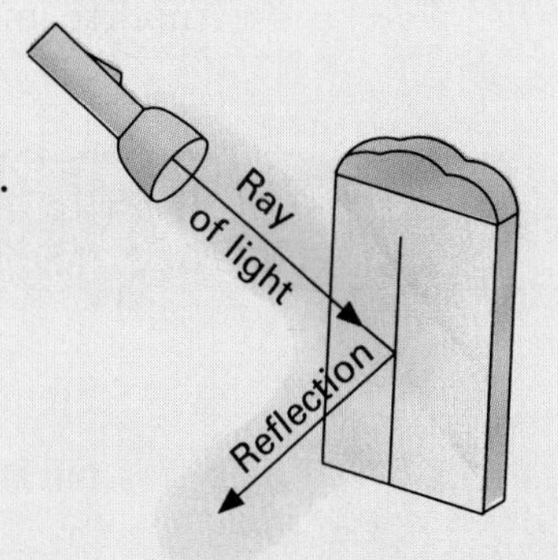

If an incident ray of light along $\overline{EC}$ is reflected from a mirror ($\overline{AB}$), the angle between the incident ray and the perpendicular or *normal* $\overline{CN}$ to the mirror is congruent to the angle between the reflected light ray along $\overline{CD}$ and the normal $\overline{CN}$.

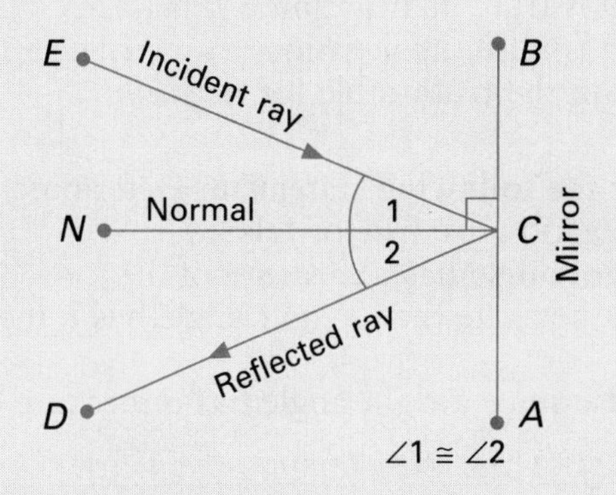

Exercise

$\overline{SR}$ and $\overline{VR}$ represent mirrors which are at right angles. An incident light ray along $\overline{PQ}$ is reflected from $\overline{VR}$ at Q, then from $\overline{SR}$ at T, and then emerges along the path $\overline{TU}$. Prove $\overline{UT} \parallel \overline{PQ}$.

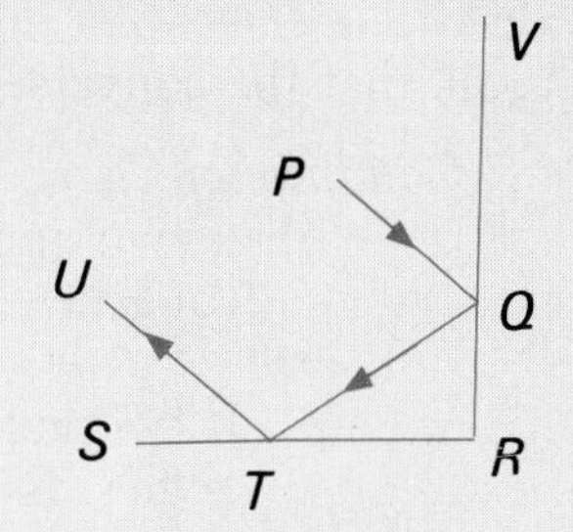

3.8 Negation, Contrapositive, Inverse, Biconditional

Objectives To write the negation, the contrapositive, and the inverse of a statement and to determine whether they are true

To write a biconditional and to determine whether it is true

The following logic statements have already been presented.

- conditional: $p \rightarrow q$
- converse of conditional $p \rightarrow q$: $q \rightarrow p$
- disjunction: $p \vee q$
- conjunction: $p \wedge q$

Another important statement in logic is the negation.

Definition If p is a statement, then the statement "not p" is called the **negation** of p. The negation of p is written as $\sim p$.

Statement	**Negation**
p: l is parallel to m	$\sim p$: l is not parallel to m; or it is not true that l is parallel to m.

When a statement p is true, its negation $\sim p$ is false.
When a statement p is false, its negation $\sim p$ is true.
This is summarized in the truth table for *negation*.

p	$\sim p$
T	F
F	T

EXAMPLE 1 Write a negation of the following statement. Determine whether the statement and its negation are true or false.
Statement (p): A right angle has a measure of 30.
Negation ($\sim$p): It is not true that a right angle has a measure of 30.

Solution The statement is false since a right angle has a measure of 90. The negation is therefore true.

Recall that the converse of a true conditional need not be true.

Conditional
($p \rightarrow q$): If two angles are right angles, then the two angles are congruent. (True)

Converse
($q \rightarrow p$): If two angles are congruent, then the two angles are right angles. (Not necessarily true)

Statements are said to be logically equivalent if they have the same truth values. A conditional and its converse are not logically equivalent, as shown in the truth table.

p	q	$p \to q$	$q \to p$
T	T	T	T
T	F	F	T
F	T	T	F
F	F	T	T

← The conditional ($p \to q$) and its converse ($q \to p$) have different truth values.

A conditional and its converse can be written as one statement using the expression "if and only if." This is illustrated below.

Conditional: If two lines are parallel, then alternate interior angles are congruent.

Converse: If alternate interior angles are congruent, then lines are parallel.

Single Statement: Two lines are parallel if and only if alternate interior angles are congruent; *or*,
Alternate interior angles are congruent if and only if lines are parallel.

Definition

When a conditional statement and its converse are combined by "if and only if," the resulting statement is called a **biconditional.**

The biconditional "p if and only if q" is also written $p \leftrightarrow q$.

The biconditional $p \leftrightarrow q$ is true only when the conditional $p \to q$ and its converse $q \to p$ are both true.

EXAMPLE 2 Write the converse of the given statement. Then write a biconditional. Determine whether the biconditional is true and give a reason.

Statement: If an angle has a measure of 90, then the sides of the angle are perpendicular.

Solution

Converse: If the sides of an angle are perpendicular, then the angle has a measure of 90. (True)

Biconditional: An angle has a measure of 90 if and only if the sides of the angle are perpendicular (*or*: The sides of an angle are perpendicular if and only if the angle has a measure of 90).

The biconditional is true because both the statement and its converse are true.

Another statement that can be formed from a conditional is the contrapositive.

Definition

The **contrapositive** of a conditional is the statement formed by interchanging the hypothesis and the conclusion and negating each of them. The contrapositive of $p \rightarrow q$ is $\sim q \rightarrow \sim p$.

The converse of a true conditional may not be true. However, the *contrapositive* of a true conditional is *always* true.

Statement: If two angles are vertical, then the angles have equal measure. (True)

Contrapositive: If two angles do *not* have equal measure, then they are *not* vertical angles.

A truth table can be used to show that whenever a conditional is true, so is its contrapositive. Likewise, whenever a conditional is false, so is its contrapositive. A conditional and its contrapositive are therefore *logically equivalent*. The significance of this relationship is that a conditional may be proved by proving its contrapositive instead.

Another statement formed from a conditional is the *inverse*. The inverse of a true conditional may or may not be true.

Definition

The **inverse** of a conditional statement is the statement formed by negating the hypothesis and the conclusion. The inverse of $p \rightarrow q$ is $\sim p \rightarrow \sim q$.

EXAMPLE 3 Write the inverse of the following true statements. Is the inverse true or false?

a. *Statement*: If two angles are vertical, then the angles have equal measures.

Solution *Inverse*: If two angles are *not* vertical, then the angles do *not* have equal measure.

The inverse is false. Two angles may be non-vertical and yet have equal measure.

b. *Statement*: If two angles are congruent, then they have equal measures.

Solution *Inverse*: If two angles are *not* congruent, then they do *not* have equal measures.

The inverse is true.

Classroom Exercises

For statements p and q and the conditional $p \rightarrow q$, match the term on the left with its symbolic representation on the right.

1. biconditional	**a.** $\sim p$
2. conjunction	**b.** $p \vee q$
3. converse	**c.** $\sim p \rightarrow \sim q$
4. disjunction	**d.** $p \leftrightarrow q$
5. inverse	**e.** $p \wedge q$
6. negation	**f.** $q \rightarrow p$

Written Exercises

Write the negation of the statement. Is the negation true or false?

1. Parallel lines intersect.
2. An obtuse angle has a measure less than 90.
3. The sides of a right angle are perpendicular.

Write the inverse and the contrapositive of the statement. Label each as true or false.

4. If the sum of the measures of two angles is 90, then the angles are complementary.
5. If two angles are adjacent and their outer rays form a straight angle, then the two angles are supplementary.
6. If an angle is a straight angle, then its measure is 90.
7. If an angle is a right angle, then its measure is not 20.

Write the converse of the statement. Then write a biconditional of the statement. Determine if the biconditional is true.

8. If M is the midpoint of $\overline{AB}$, then M divides $\overline{AB}$ into two congruent segments.
9. If corresponding angles are congruent, then lines are parallel.
10. If an angle is not acute, then the angle is obtuse.
11. Verify by truth table that $\sim(\sim p)$ is logically equivalent to p.
12. Verify by truth table that a conditional is logically equivalent to its contrapositive.

Mixed Review

Indicate whether the statement is always true, sometimes true, or never true.

1. Two coplanar segments that do not intersect are parallel. *3.1*
2. Two lines in a plane that do not intersect are parallel. *3.3*
3. A ray and a segment that are coplanar intersect if they are not parallel. *3.3*

3.9 Parallel Lines and Planes

Objectives To determine whether statements about parallel lines and planes are true

To prove statements about parallel lines and planes

In space, two lines can be parallel, intersecting, or skew. However, two distinct planes either intersect or are parallel.

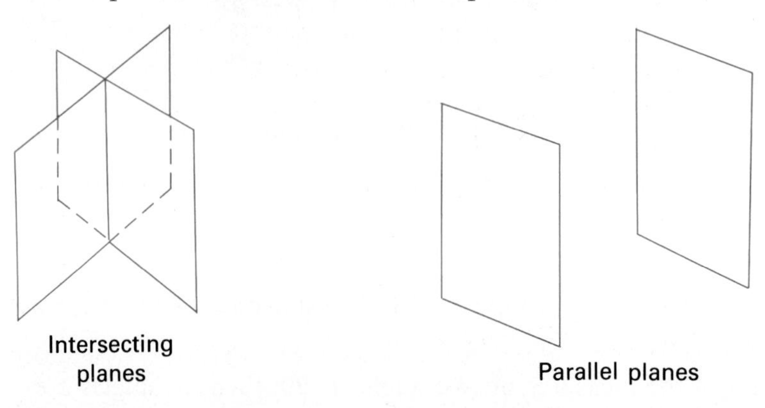

Definition **Parallel planes** are planes that do not intersect.

There are three possible relationships between a line and a plane.

(1) The line can be contained in the plane. The intersection of the line and the plane is the line itself.

(2) The line can intersect the plane in exactly one point.

(3) The line and plane can have no points in common. They do not intersect.

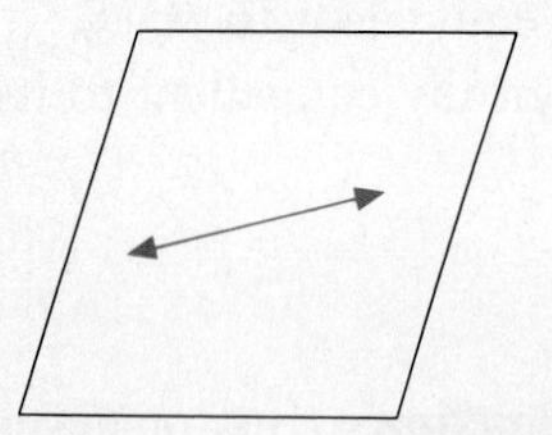

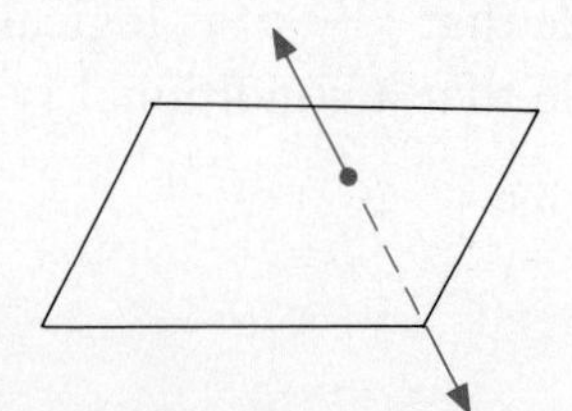

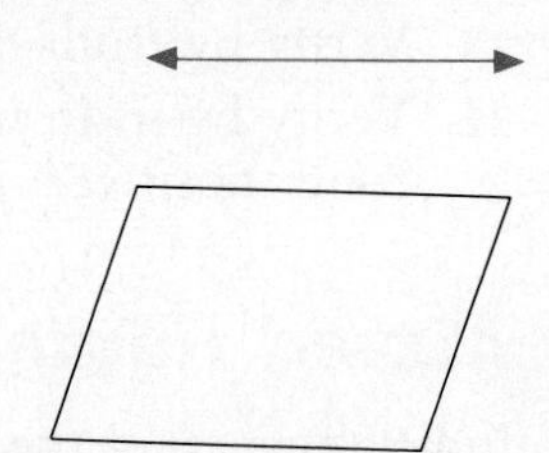

Definition A line and a plane are *parallel* if they do not intersect.

Theorem 3.11 If two parallel planes are intersected by a third plane, then the lines of intersection are parallel.

Given: Plane $\mathcal{P} \parallel$ plane $\mathcal{Q}$, $\mathcal{P}$ intersects $\mathcal{R}$ in $\overleftrightarrow{AB}$;
$\mathcal{Q}$ intersects $\mathcal{R}$ in $\overleftrightarrow{CD}$.
Prove: $\overleftrightarrow{AB} \parallel \overleftrightarrow{CD}$

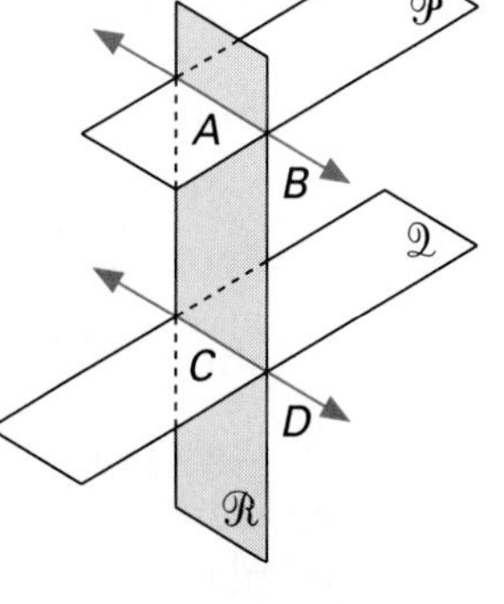

Indirect Proof
Given: $\mathcal{P} \parallel \mathcal{Q}$. Assume $\overleftrightarrow{AB} \nparallel \overleftrightarrow{CD}$. $\overleftrightarrow{AB}$ and $\overleftrightarrow{CD}$ are coplanar (both in plane $\mathcal{R}$), and must therefore intersect at some point. Call this point T. Because T is on $\overleftrightarrow{AB}$, the intersection of planes $\mathcal{P}$ and $\mathcal{R}$, it is in $\mathcal{P}$. Because T is on $\overleftrightarrow{CD}$, the intersection of planes $\mathcal{Q}$ and $\mathcal{R}$, it is also in $\mathcal{Q}$.

Therefore, planes $\mathcal{P}$ and $\mathcal{Q}$ have a point in common, T. Yet it is given that $\mathcal{P}$ and $\mathcal{Q}$ are *parallel.* This is a contradiction because parallel planes have no points of intersection. The assumption that $\overleftrightarrow{AB} \nparallel \overleftrightarrow{CD}$ must be false. Also, coplanar lines $\overleftrightarrow{AB}$ and $\overleftrightarrow{CD}$ cannot be skew.

Therefore, $\overleftrightarrow{AB} \parallel \overleftrightarrow{CD}$.

You have learned to prove a conditional by either a direct or indirect method of proof. Recall that a conditional is false only when the hypothesis is true and the conclusion is false. To prove a conditional false, you must find *one* example for which the conclusion is false when the hypothesis is true. Such an example is called a **counterexample.**

This counterexample can be a drawing which illustrates one case where the conditional is false.

EXAMPLE True or false? Give a reason.

Two lines parallel to the same plane are parallel.

Plan Try to find a counterexample that illustrates that when the *hypothesis*, "two lines are parallel to the same plane," is true, the *conclusion*, "the lines are parallel," is false.

Solution In the figure, $\overleftrightarrow{AC} \parallel$ plane $\mathcal{M}$ and $\overleftrightarrow{BD} \parallel$ plane $\mathcal{M}$. The *hypothesis* is true. But, $\overleftrightarrow{AC} \nparallel \overleftrightarrow{BD}$. The *conclusion* is false. Therefore, the statement, "two lines parallel to the same plane are parallel," is false. The drawing is a counterexample.

Classroom Exercises

1. What three possible relationships can exist between two distinct lines?
2. What two possible relationships can exist between two distinct planes?
3. Explain how to prove a conditional false using a counterexample.

Written Exercises

True or false? If false, draw a diagram to illustrate a counterexample.

1. Two lines parallel to the same line are parallel to each other.
2. Two planes parallel to the same plane are parallel to each other.
3. Two planes parallel to the same line are parallel to each other.
4. Two planes perpendicular to the same line are parallel.
5. If two lines are parallel, then every plane containing one of the lines is parallel to every plane containing the other line.
6. A line parallel to a plane is parallel to every line in that plane.
7. Given: planes $\mathcal{M}$ and $\mathcal{N}$ intersect in $\overleftrightarrow{AB}$; $\overleftrightarrow{CD} \parallel \mathcal{M}$, $\overleftrightarrow{CD} \parallel \mathcal{N}$, $\overleftrightarrow{CD}$ and $\overleftrightarrow{AB}$ are not skew.
   Prove: $\overleftrightarrow{CD} \parallel \overleftrightarrow{AB}$

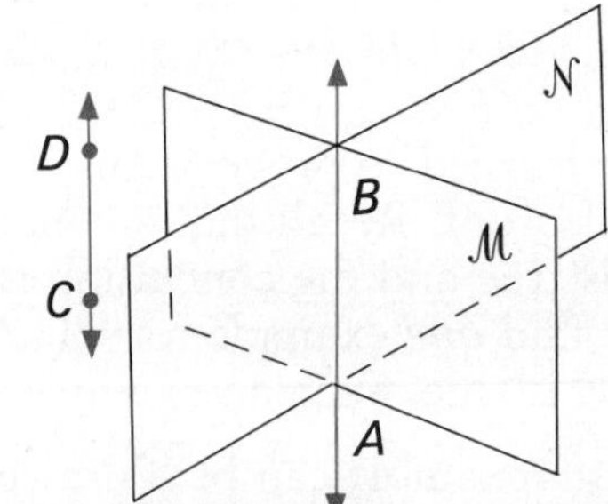

8. Prove: If a line intersects a plane in exactly one point, then it is not parallel to any line that is contained in the plane. (Use indirect proof.)
9. Assume the property that, through a point not on a plane, there is exactly one line and one plane parallel to the given plane. Prove: If a line intersects one of two parallel planes, then it also intersects the second plane.

Mixed Review

1. Given: m $\angle 6 = 75$, m $\angle 1 = 30$
   Find m $\angle 3$. *3.7*
2. Given: m $\angle 2 = 70$, m $\angle 1 = \frac{2}{3}$ m $\angle 6$
   Find m $\angle 1$ and m $\angle 6$. *3.6.*
3. Given: $\overleftrightarrow{AB} \parallel \overleftrightarrow{CD}$, m $\angle 4 = 3x + 20$, m $\angle 1 = x + 40$
   Find m $\angle 1$. *3.5.*

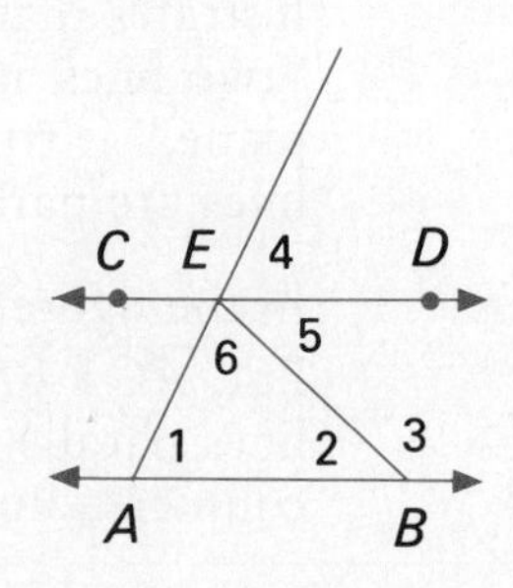

Problem Solving Strategies

Using an Alternate Approach

Many problems cannot be solved by a straightforward method. An alternate approach may have to be tried. For example, consider the drawing at the right. It is given that $p \parallel q$. You are asked to find m $\angle 1$ + m $\angle 2$.

A straightforward attempt might be trying to find the measure of both angles separately. This will not work since not enough information is given. This means that you have to look at the problem another way.

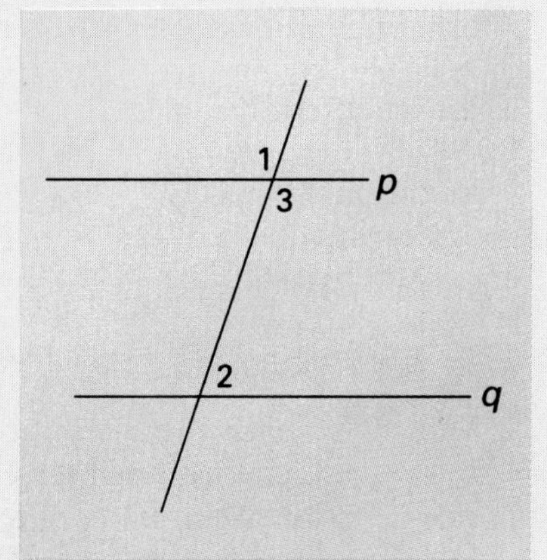

Example Given: $p \parallel q$
Find m $\angle 1$ + m $\angle 2$.

Solution m $\angle 1$ = m $\angle 3$ by the Vertical Angles Property.

Then m $\angle 2$ + m $\angle 3$ = 180 since these angles are on the same side of a transversal, interior angles of parallel lines are supplementary.
By substitution, m $\angle 1$ + m $\angle 2$ = 180.

Thus, we found m $\angle 1$ + m $\angle 2$ *without* knowing the measure of each angle separately.

1. In a later lesson you will prove the formula for the area of a triangle: $A = \frac{1}{2}h \cdot b$ where h is the length of the altitude (a perpendicular to a side of the triangle) and b is the length of the side to which the perpendicular is drawn, the **base.** In the figure at the right, if the area of triangle ABC is 30, then find the area of triangle BDC.

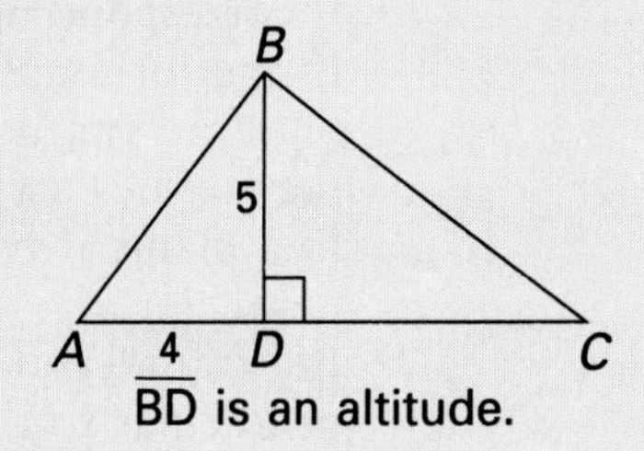

$\overline{BD}$ is an altitude.

2. If the triangle and rectangle at the right have equal areas, and if $\frac{pr}{2} = 60$, then xy = ?

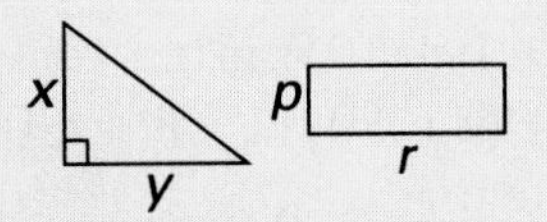

3. The formula for the volume of a cube is $V = s^3$, where s is the length of an edge. Without calculating any roots, find the length of the edge of a cube which has a volume = 238.328 cm^3 and the area of a face = 38.44 cm^2. (HINT: Write the formula in terms of s and s^2.)

Chapter 3 Review

Key Terms

alternate exterior angles (p. 98)
alternate interior angles (p. 97)
biconditional (p. 127)
contrapositive (p. 128)
converse (p. 109)
corresponding angles (p. 98)
exterior (p. 97)
exterior angles (p. 121)
indirect proof (p. 106)
inverse (p. 128)
negation (p. 126)
parallel lines (p. 93)
parallel planes (p. 130)
remote interior angles (p. 121)
skew lines (p. 93)
transversal (p. 97)
triangle (p. 115)

Key Ideas and Review Exercises

3.1 In space there are three possible relationships between two lines:
- Lines can be intersecting.
- Lines can be parallel.
- Lines can be skew, that is, noncoplanar.

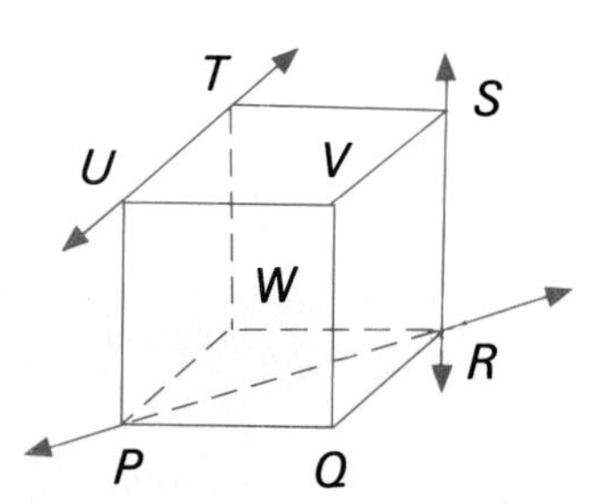

In the figure at the right:

1. Name a pair of lines that appear to be skew.
2. Name a segment that appears to be parallel to $\overleftrightarrow{UT}$.
3. Name a segment that appears to be skew to $\overleftrightarrow{PR}$.
4. What line can be drawn to be parallel to $\overleftrightarrow{PR}$?

3.2, 3.3, 3.5 Properties of parallel lines are summarized by the theorems below.
- If alternate interior angles are congruent, then lines are parallel.
- If corresponding angles are congruent, then lines are parallel.
- If interior angles on the same side of a transversal are supplementary, then lines are parallel.
- If two lines are parallel to the same line, then they are parallel.
- In a plane, if two lines are perpendicular to the same line, then they are parallel.
- If a transversal is perpendicular to one of two parallel lines, then it is perpendicular to the other.

Determine which lines are parallel. Give a reason.

5. $m\angle 1 = 105$, $m\angle 2 = 105$
6. $m\angle 3 = 80$, $m\angle 8 = 80$

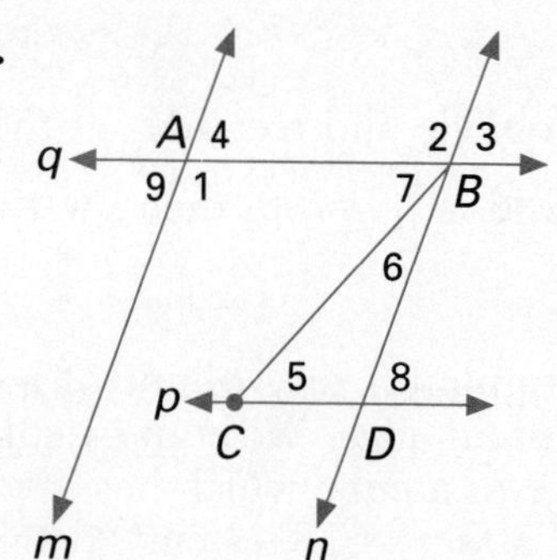

For Exercises 7–9, $m \parallel n$

7. Given: $m\angle 4 = 65$. Find $m\angle 2$. *3.5*
8. Given: $m\angle 9 = 70$. Find $m\angle 3$.
9. Given: $\angle 4 \cong \angle 8$
   Prove: $p \parallel q$

3.4 To write an indirect proof:

(1) Assume that the desired conclusion is false.
(2) Reason directly until a contradiction is reached.
(3) State that the assumption in the first step is false, so the original conclusion must be true.

10. Given: P is on $\overline{AB}$; P is not the midpoint of $\overline{AB}$.
Prove indirectly that $\overline{AP} \neq \overline{PB}$.

3.6, 3.7 To solve problems about angle measures of triangles, use the following properties of triangles:

- The sum of the measures of the angles of a triangle is 180.
- The measure of an exterior angle of a triangle is equal to the sum of the measures of its two remote interior angles.

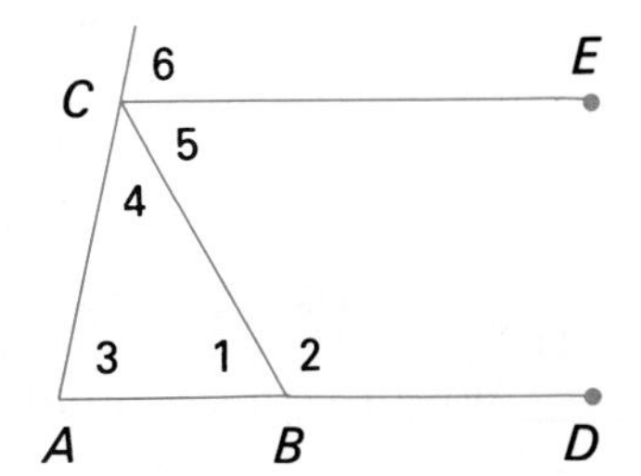

11. m $\angle 1 = 2x$, m $\angle 3 = 3x$, m $\angle 4 = 5x$. Find m $\angle 1$.

12. $\overline{AC} \perp \overline{AB}$, m $\angle 1 = \frac{4}{5}$ m $\angle 4$. Find m $\angle 1$ and m $\angle 4$.

13. m $\angle 3 = 4x + 20$, m $\angle 4 = x + 30$, m $\angle 2 = 2x + 110$. Find m $\angle 1$.

14. Given: $\overline{AD} \parallel \overline{CE}$
Prove: m $\angle 2$ = m $\angle 4$ + m $\angle 6$

3.8 To identify and determine whether certain statements are true:

- Negation of p: $\sim p$ (True when p is false)
- Converse of $p \rightarrow q$: $q \rightarrow p$ (Not necessarily true, even if $p \rightarrow q$ is true)
- Contrapositive of $p \rightarrow q$: $\sim q \rightarrow \sim p$ (True if $p \rightarrow q$ is true)
- Inverse of $p \rightarrow q$: $\sim p \rightarrow \sim q$ (Not necessarily true, even if $p \rightarrow q$ is true)
- Biconditional: $p \leftrightarrow q$ (True if $p \rightarrow q$ and $q \rightarrow p$ are both true)

Write the indicated expression. Determine if it is true. (Exercises 15–17)

15. Negation of "Parallel lines intersect."

16. Inverse of "If lines are parallel, then alternate interior angles are congruent."

17. Contrapositive of "If the sides of an angle are perpendicular, then the angle has a measure of 90."

18. Write the converse of "If lines are parallel, then corresponding angles are congruent." Then write a biconditional and determine if it is true.

3.9 **Determine whether the statements are true or false. Draw a diagram to support your answer. (Exercises 19–20)**

19. Two lines parallel to the same plane are parallel to each other.

20. A plane parallel to one of two skew lines contains the other.

Chapter 3 Test

True or false? If false, draw a figure to show why. (Exercises 1–5)

1. Lines which do not intersect are parallel.
2. Two planes parallel to the same line are parallel.
3. Skew lines do not intersect.
4. A line parallel to one of two skew lines is parallel to the other.
5. The measure of an exterior angle of a triangle equals the sum of the measures of its two remote interior angles.

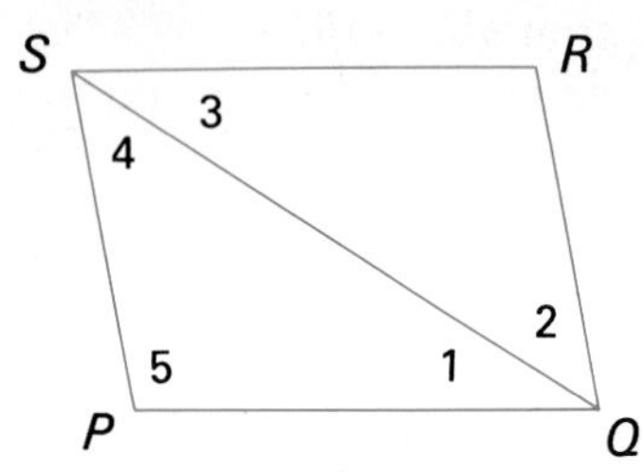

6. Given: $m\angle 4 = 40$, $m\angle 2 = 40$
   Which segments are parallel?
7. Given: $\overline{SP} \parallel \overline{QR}$
   Prove: $m\angle 2 + m\angle 3 = m\angle PSR$
*8. Given: $\overline{SP} \parallel \overline{RQ}$,
   $m\angle PSR = m\angle PQR$
   Prove: $\overline{PQ} \parallel \overline{RS}$
9. Given: $\angle 1 \cong \angle 4$, $m\angle 5 = 140$
   Find $m\angle 1$.

For Exercises 10–14, $m \parallel n$.

10. Given: $m\angle 8 = 100$. Find $m\angle 4$.
11. Given: $m\angle 1 = 70$. Find $m\angle 3$.
12. Given: $m\angle 2 = \frac{2}{7} m\angle 3$. Find $m\angle 1$.
13. Given: $m\angle 2$ is 60 less than twice $m\angle 3$. Find $m\angle 1$.
14. Given: $m\angle 10 = 3x + 30$, $m\angle 9 = x + 40$. Find $m\angle 9$.
15. Draw a line r and a point P not on r. Construct a line through P parallel to r.
16. Given: $m\angle A = 20$, $\angle 1 \cong \angle C$. Find $m\angle C$.
17. Given: $m\angle A = 2x + 10$, $m\angle 1 = 3x + 20$, $m\angle C = 15x - 50$. Find $m\angle C$.
18. Given: $m\angle A = 70$, $m\angle 2 = 120$. Find $m\angle C$.

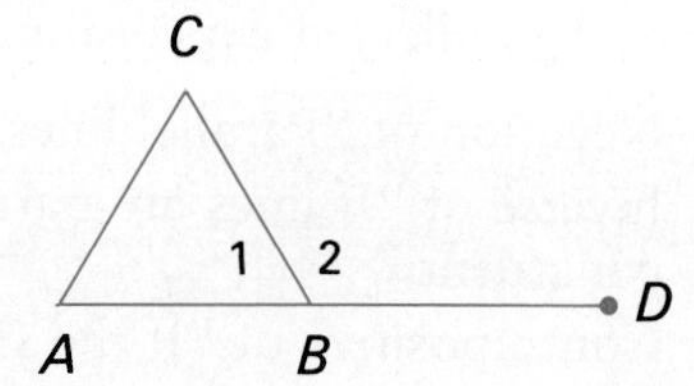

19. Write the negation of the statement "A right angle has a measure of 80." Is the negation true?

Use the following statement to answer Exercises 20–22.

"If two lines are parallel, then interior angles on the same side of a transversal are supplementary."

20. Write a biconditional for the statement and its converse. Is it true?
21. Write the inverse of the statement. Is it true?
22. Write the contrapositive of the statement. Is it true?

College Prep Test

In some cases, it may not be necessary to know the measures of two angles of a triangle to find the measure of the third. For example:

Given: m $\angle C = 50$, $\overline{AD}$ and $\overline{BD}$ are angle bisectors. Find x.

You don't have to find m $\angle 1$ and m $\angle 2$ separately to solve the problem. Rather, find their sum and subtract it from 180.

Since m $\angle C = 50$, m $\angle CAB$ + m $\angle CBA = 180 - 50 = 130$.

Then, from the definition of angle bisectors, m $\angle 1$ + m $\angle 2 = \frac{1}{2} \cdot 130 = 65$.

Now, $x = 180 - (\text{m } \angle 1 + \text{m } \angle 2) = 180 - 65 = 115$

1. Find the sum of the degree measures of the numbered angles.

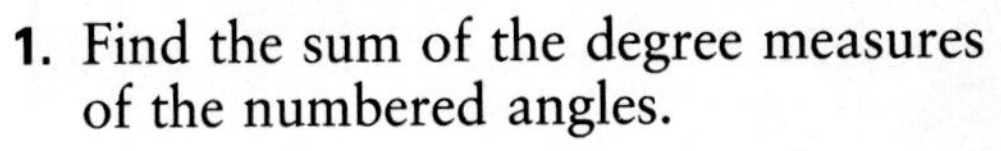

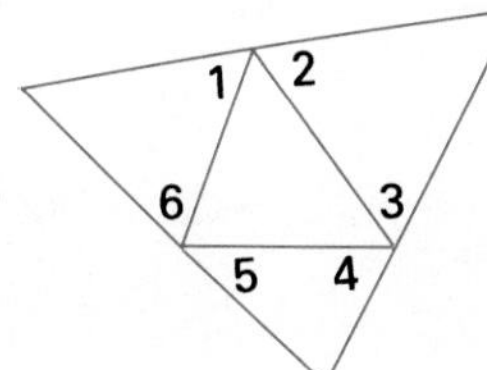

(A) 360 (B) 180 (C) 540 (D) 270 (E) It cannot be determined from the given information.

2. 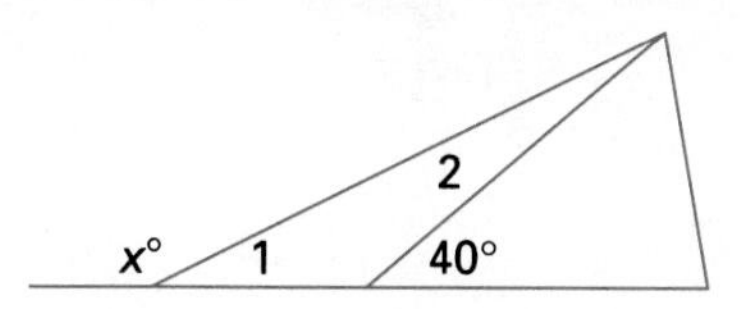

If $\angle 1 \cong \angle 2$ then x = ______.
(A) 60 (B) 160 (C) 40 (D) 140 (E) 120

3. 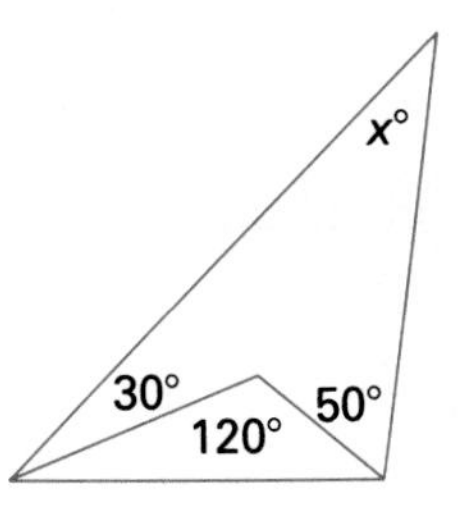

x = ______.
(A) 40 (B) 80 (C) 60 (D) 20 (E) 120

4. 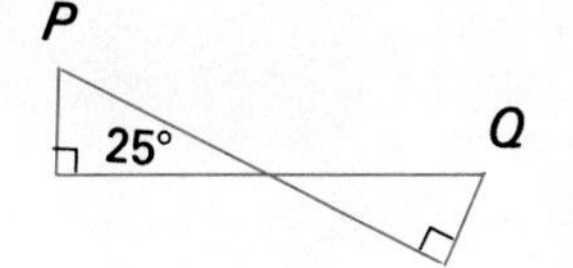

m $\angle P$ − m $\angle Q$ = ______.
(A) 115 (B) 65 (C) 40 (D) 0 (E) 50

5. 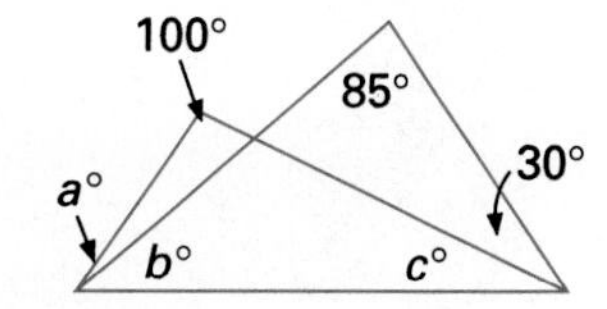

a = ______.
(A) 30 (B) 55 (C) 115 (D) 15 (E) 50

6. $\overline{PQ} \perp \overline{PR}$, m $\angle 1$ = m $\angle 2$

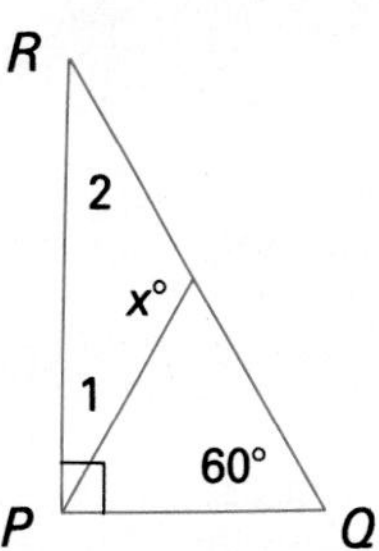

x = ______.
(A) 60 (B) 120 (C) 100 (D) 150 (E) None of these answers

4 CONGRUENT TRIANGLES

To recognize his achievements in mathematics and science, President George Washington appointed Benjamin Banneker a member of the commission which surveyed and laid out the streets for the District of Columbia. Banneker also wrote an Almanac, for which he did all the intricate mathematical calculations himself.

4.1 Triangle Classifications

Objectives To classify triangles according to measures of angles and lengths of sides

To find measures of angles and lengths of sides of a triangle

Triangles can be classified according to the measures of their angles. Recall the definition of acute and obtuse angles.

Acute angle:
$0 < m\ \angle A < 90$

Obtuse angle:
$90 < m\ \angle B < 180$

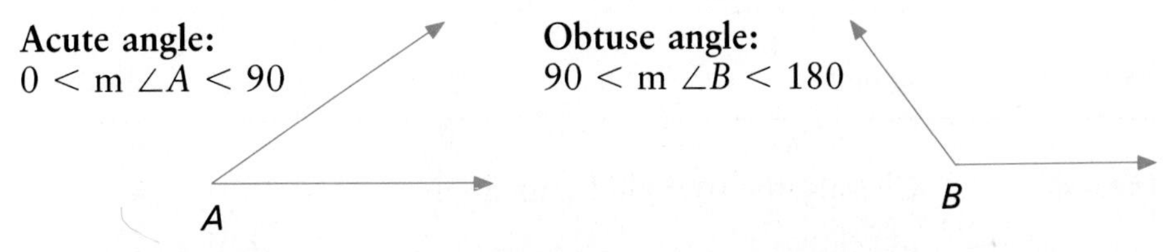

Definitions

An **acute triangle** is a triangle with three acute angles.
A **right triangle** is a triangle with one right angle.
An **obtuse triangle** is a triangle with one obtuse angle.
An **equiangular triangle** is a triangle with three congruent angles.

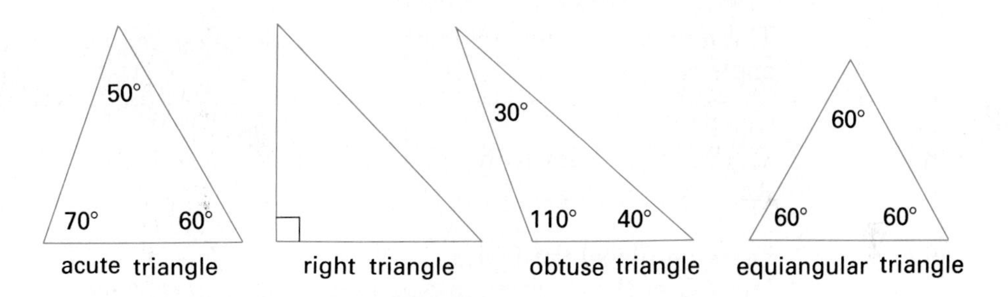

acute triangle　right triangle　obtuse triangle　equiangular triangle

Triangles can also be classified according to the lengths of their sides.

Definitions

An **equilateral triangle** is a triangle with three congruent sides.
An **isosceles triangle** is a triangle with at least two congruent sides.
A **scalene triangle** is a triangle with no congruent sides.

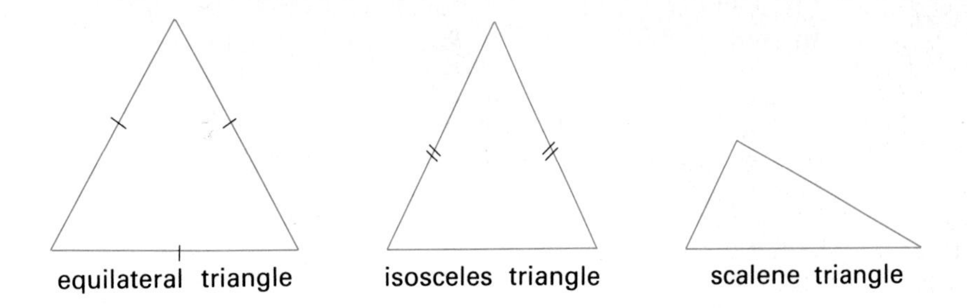

equilateral triangle　isosceles triangle　scalene triangle

EXAMPLE 1 Classify each triangle as indicated.

By sides:

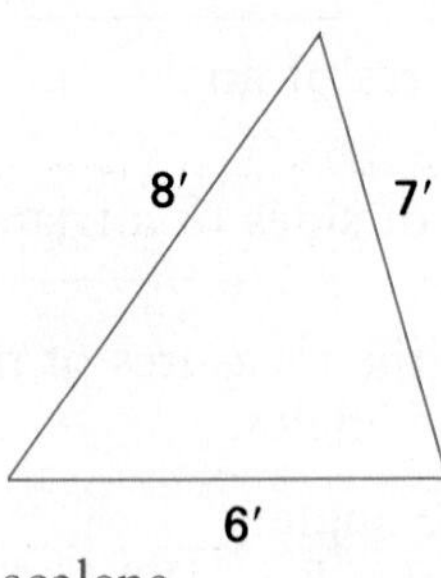

By angles:

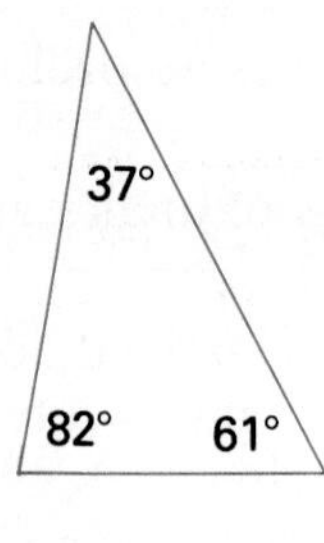

By sides and angles:

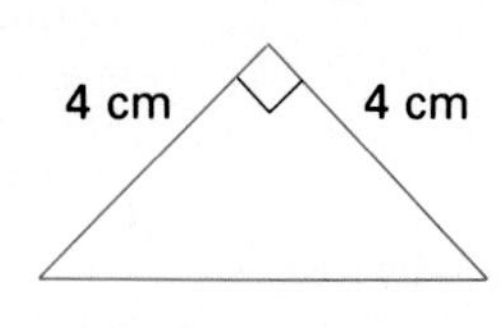

Solution scalene acute isosceles, right

EXAMPLE 2 Classify the triangle by its angles.

Plan Find the measure of the third angle.

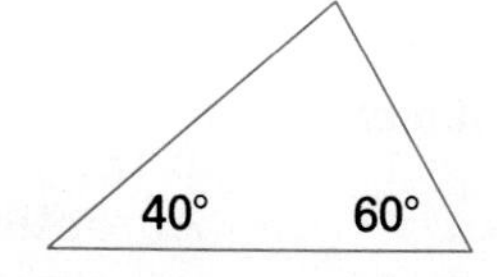

Solution Let x = measure of third angle.

$60 + 40 + x = 180$
$x = 80$

The measure of each angle of the triangle is less than 90. Thus, the triangle is acute.

EXAMPLE 3 $\triangle ABC$ is isosceles with $AC = BC$. $AC = 5x - 7$, $BC = 2x + 8$, and $AB = x + 2$. Find AB.

Plan Draw and label the figure.
Use $AC = BC$ to write an equation.

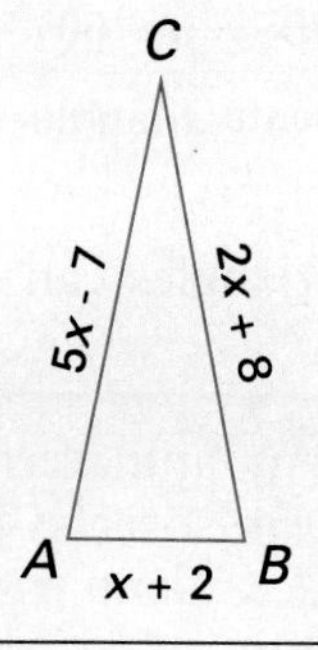

Solution

$5x - 7 = 2x + 8$
$3x - 7 = 8$
$3x = 15$
$x = 5$
$AB = x + 2 = 5 + 2 = 7$

In right $\triangle ABC$, $m\angle C = 90$

$$m\angle A + m\angle B + m\angle C = 180$$
$$m\angle A + m\angle B + 90 = 180$$
$$m\angle A + m\angle B = 90$$

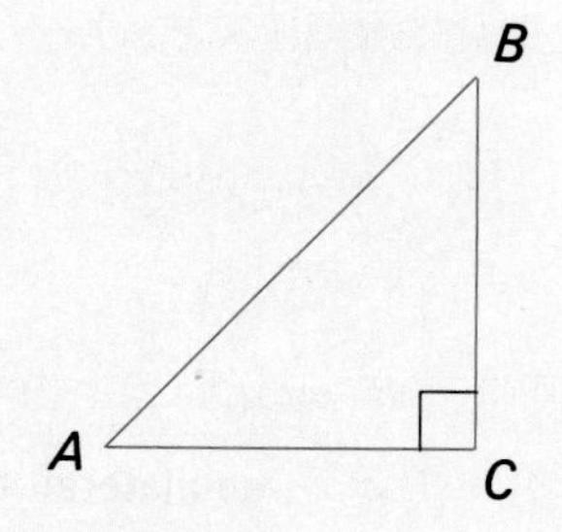

Therefore, $\angle A$ and $\angle B$ are complementary. Since the sum of the measures of $\angle A$ and $\angle B$ is 90, each angle is acute. This leads to Theorem 4.1.

Theorem 4.1 In a right triangle, the two angles other than the right angle are complementary and acute.

EXAMPLE 4 The measure of one acute angle of a right triangle is $\frac{2}{3}$ the measure of the other acute angle. Find the measure of the larger acute angle.

Plan Let x = larger angle
$\frac{2}{3}x$ = smaller angle

Solution

$$x + \frac{2}{3}x = 90 \quad \longleftarrow \text{Sum of measures of complementary } \angle\text{s is 90}$$
$$3x + 2x = 270$$
$$5x = 270$$
$$x = 54$$

The measure of the larger acute angle is 54.

Focus on Reading

Determine whether the statement is always true, sometimes true, or never true.

1. An equilateral triangle is isosceles.
2. An equiangular triangle is a right triangle.
3. A right triangle is a triangle with three acute angles.
4. An obtuse triangle is isosceles.
5. An obtuse triangle is a right triangle.

Classroom Exercises

Classify the triangle for the lengths of sides given.

1. 5 cm, 4 cm, 6 cm **2.** 4 in., 4 in., 4 in. **3.** 6 ft, 2 ft, 6 ft **4.** 7 in., 5 in., 6 in.

Classify the triangle for the angle measures given.

5. 90, 40, 50 **6.** 130, 20, 30 **7.** 60, 60, 60 **8.** 50, 50, 80

Written Exercises

Classify the triangle by its sides.

1. 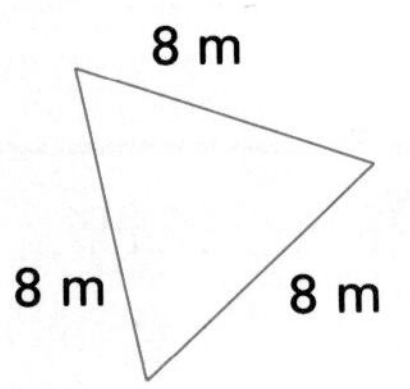

2. 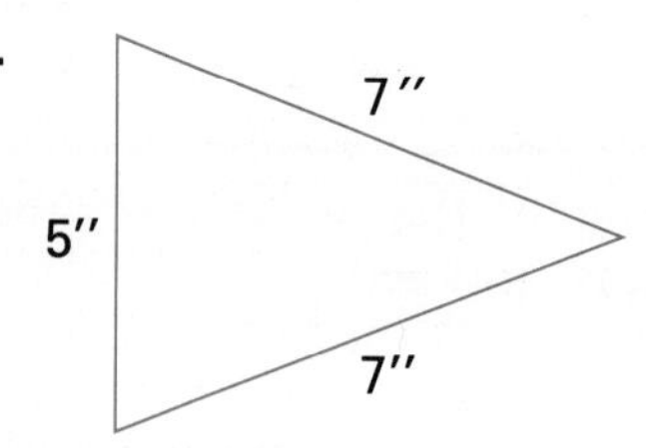

3. 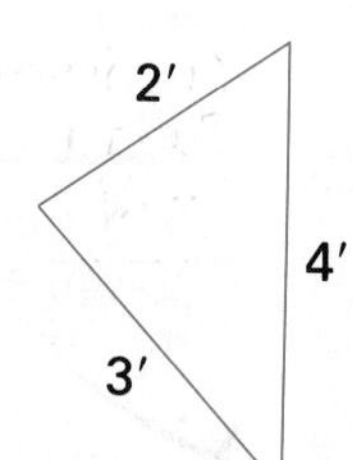

Classify the triangle by its angles.

4. The measures of two angles are 25 and 35.
5. The measures of two angles are 25 and 65.
6. The measures of two angles are 75 and 45.
7. $\triangle ABC$ is isosceles with $AC = BC$. $AC = 6x - 5$, $BC = 4x + 7$, and $AB = 5x - 2$. Find AB.
8. $\triangle ABC$ is equilateral with $BC = 3x + 7$ and $AC = 5x + 1$. Find AB.
9. The measure of an acute angle of a right triangle is $\frac{4}{5}$ the measure of the other acute angle. Find the measure of the smaller acute angle.
10. In a right triangle, the measure of one acute angle is 30 less than twice the measure of the other acute angle. Find the measure of each angle.
11. $\triangle PQR$ is isosceles with $PR = QR$. $PR = 10$, $QR = 3x - 11$, $PQ = x^2 - 25$. Find PQ.
12. The lengths of the sides of the triangle are given by $2n - 1$, $n + 7$, and $3n - 9$. Prove that if the triangle is isosceles, then the triangle is equilateral.
13. Prove Theorem 4.1.
14. Prove that the bisector of an exterior angle of an equiangular triangle is parallel to a side of a triangle.
15. Write an indirect proof that an obtuse triangle cannot be a right triangle.

Mixed Review

1. Given: The coordinates of the endpoints of a segment are -6 and 4. Find the coordinate of the midpoint. *1.3*
2. The measures of a pair of alternate interior angles of two parallel lines are $3x - 20$ and $x + 40$. Find the measure of each angle. *3.5*
3. If two adjacent angles are complementary, then their outer rays are ____. *1.6*

4.2 Congruence of Triangles

Objectives

To identify congruent triangles
To identify corresponding parts of congruent triangles
To determine missing measures in congruent triangles

Using the diagrams, trace Figure A and then place the tracing on top of Figure B. Since the two figures coincide, they are called congruent. Geometric figures are *congruent* if they have the same size and shape.

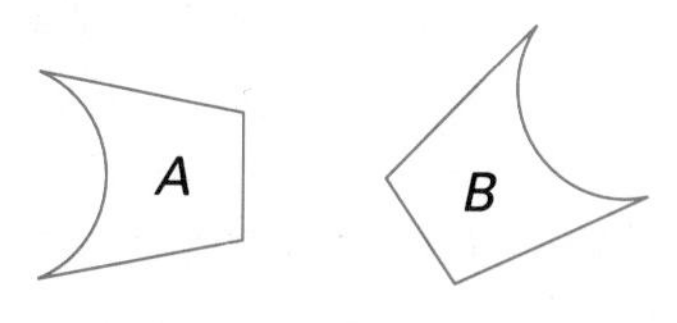

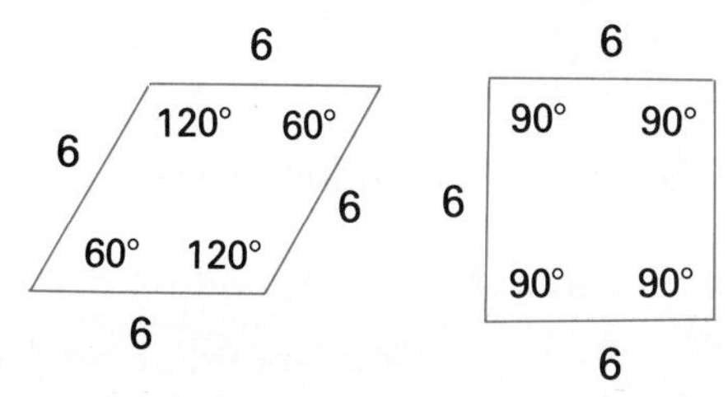

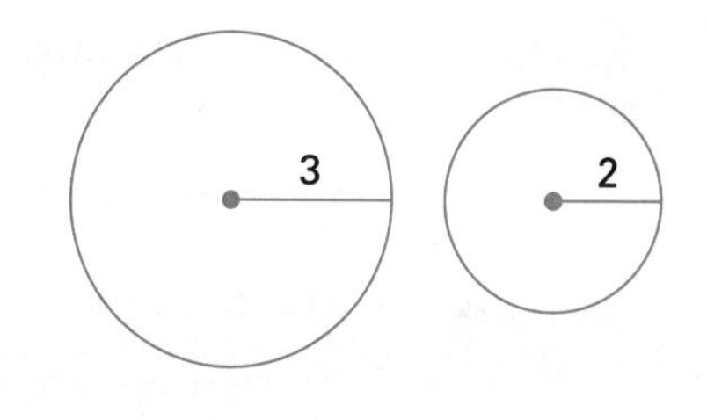

The lengths of the sides of the figures above are the same. But the figures differ in shape. The figures are *not* congruent.

The figures above have the same shape. But they differ in size. The figures are *not* congruent.

To determine whether two triangles are congruent, imagine overlaying one triangle on the other. The sides and angles of one triangle should be congruent to the corresponding sides and angles of the other triangle.

Definition

> Two triangles are **congruent** ($\cong$) if corresponding angles are congruent and corresponding sides are congruent.

The triangles at the right are congruent. The sets of arcs and tick marks indicate that the corresponding angles are congruent (equal in measure) and the corresponding sides are congruent (equal in length). To indicate that triangle ABC is congruent to triangle DEF, write $\triangle ABC \cong \triangle DEF$. The order of the letters always indicates the corresponding vertices.

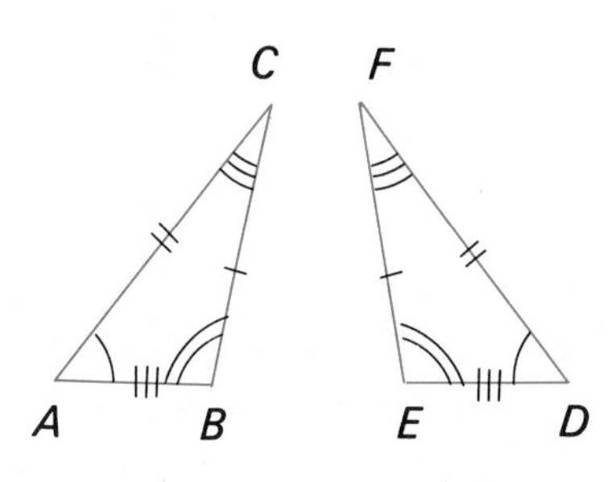

$\triangle ABC \cong \triangle DEF$

EXAMPLE 1 Using the markings shown, complete the congruence statement: $\triangle PQR \cong$ ______.

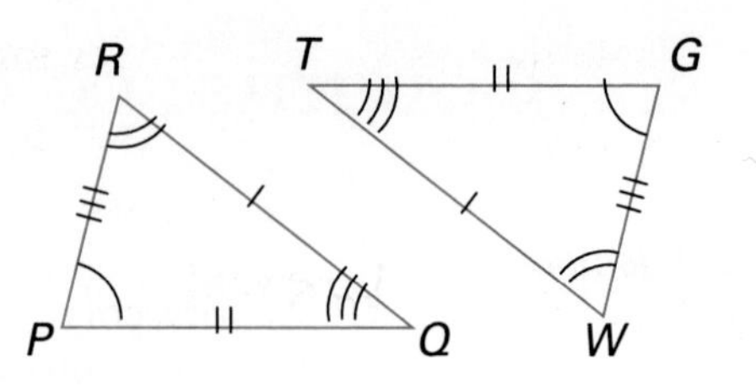

Solution From the markings, you know that

(1) $\angle P \cong \angle G$,
(2) $\angle Q \cong \angle T$, and
(3) $\angle R \cong \angle W$.

So, $\triangle PQR \cong \triangle GTW$.

EXAMPLE 2 Given: $\triangle PWG \cong \triangle SEM$

Identify the corresponding sides and the corresponding angles. Draw the two triangles and mark the corresponding parts.

Solution Use the order of letters to identify the corresponding parts.

Corresponding sides:
$\triangle PWG \cong \triangle SEM$
$\overline{PW} \cong \overline{SE}$
$\overline{WG} \cong \overline{EM}$
$\overline{PG} \cong \overline{SM}$

Corresponding angles:
$\triangle PWG \cong \triangle SEM$
$\angle P \cong \angle S$
$\angle W \cong \angle E$
$\angle G \cong \angle M$

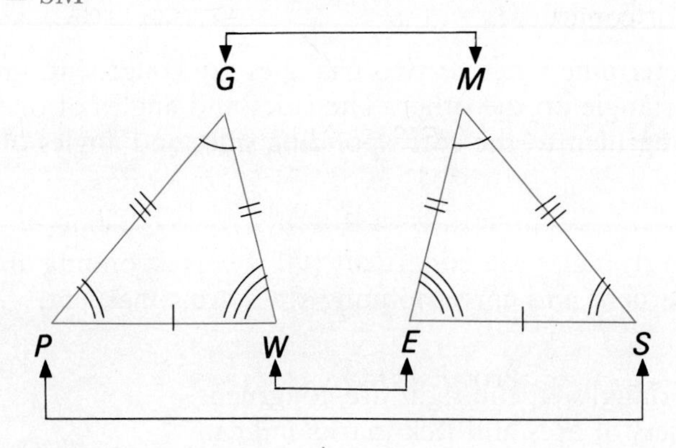

Classroom Exercises

Name the corresponding angles for the pair of congruent triangles.

1. $\triangle TYP \cong \triangle ERM$ **2.** $\triangle QWZ \cong \triangle UTA$ **3.** $\triangle PTC \cong \triangle DYH$

Name the corresponding sides for the pair of congruent triangles.

4. $\triangle GUM \cong \triangle PAT$ 5. $\triangle RUT \cong \triangle WDG$ 6. $\triangle SAD \cong \triangle LID$

Written Exercises

Use the markings on the figures to complete the congruence statement.

1.

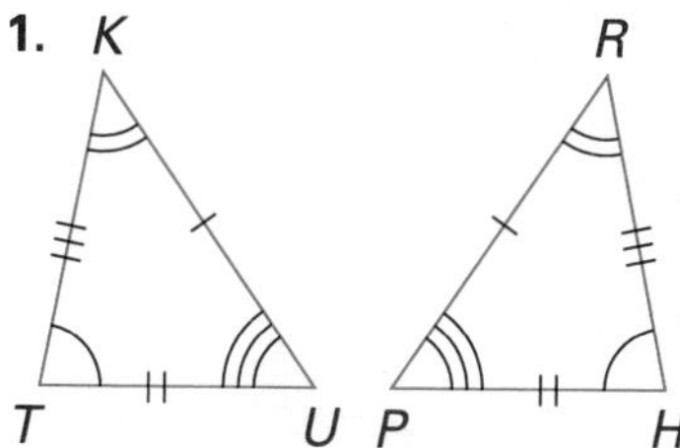

2.

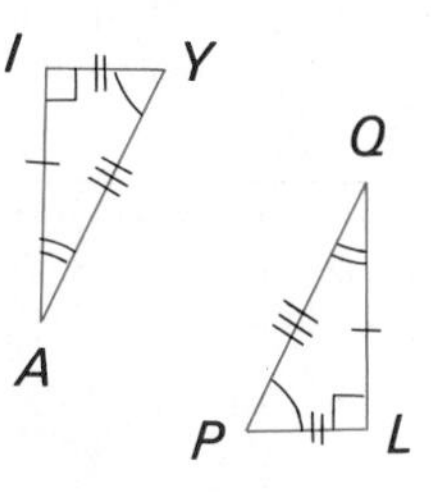

3.

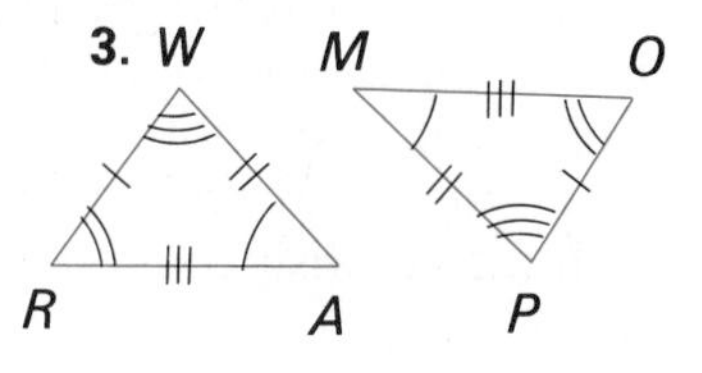

$\triangle TKU \cong$ ______ $\triangle LQP \cong$ ______ $\triangle RAW \cong$ ______

Complete the congruence statement for the given pair of congruent triangles.

4. $\triangle RTY \cong \triangle PUM$
   $\angle T \cong$ ______
5. $\triangle STY \cong \triangle APQ$
   $\angle Y \cong$ ______
6. $\triangle DFT \cong \triangle LKP$
   ______ $\cong \angle L$
7. $\triangle AFK \cong \triangle YWE$
   $\overline{AF} \cong$ ______
8. $\triangle KLB \cong \triangle PSM$
   $\overline{KB} \cong$ ______
9. $\triangle FAD \cong \triangle UKT$
   $\overline{AD} \cong$ ______

For Exercises 10–12, draw a diagram of the two congruent triangles. Mark the corresponding parts and identify the corresponding sides and the corresponding angles.

10. $\triangle TER \cong \triangle YUQ$ 11. $\triangle EJS \cong \triangle IPU$ 12. $\triangle DAS \cong \triangle HOP$
13. Given: $\triangle ABC \cong \triangle UKT$. $AB = 4$, $BC = 5$, $AC = 6$. Find UT.
14. Given: $\triangle GHT \cong \triangle MOW$. $m\angle G = 40$, $m\angle H = 70$. Find $m\angle M$ and $m\angle W$.
15. Given: $\triangle ABC \cong \triangle FGH$. $AB = 4x + 5$, $FG = 2x + 13$, $AC = 3x - 1$. Find FH.
16. Given: $\triangle PQR \cong \triangle RQP$. Prove: $\triangle PQR$ is isosceles.

Mixed Review

1. The measure of the exterior angle of a triangle is 100. The measure of one of its remote interior angles is 75. Find the measure of its other remote interior angle. *3.7*
2. Two angles are supplementary. The measure of one angle is twice the measure of the other. Find the measure of each angle. *1.6*
3. One of the acute angles of a right triangle measures 42. Find the measure of the other acute angle. *3.6*

Computer Investigation

Constructing Triangles

Use a computer software program that constructs triangles by definition of SSS, SAS, or ASA, and measures line segments and angles.

You have seen that two triangles are congruent if three pairs of corresponding angles are congruent and if three pairs of corresponding sides are congruent. The computer activities below will help you discover that congruence of certain combinations of three pairs of parts is enough to establish congruence of the triangles.

Activity 1

$\triangle ABC$ has side lengths 5, 6, and 7. If you construct a triangle with two side lengths 5 and 6 and measure of angle between the 5 and 6 sides equal to m $\angle A$, do you think the new triangle will be congruent to $\triangle ABC$?

C
6
7
A
5
B

1. Construct a triangle with sides of lengths 5, 6, and 7.
2. Find the measures of the angles of this triangle.
3. Construct a new triangle with two sides of lengths 5 and 6, and a measure of the angle between the 5 and 6 sides equal to the measure of the corresponding angle of the original triangle.
4. Find the measures of the remaining side and angles of this new triangle. Is this new triangle congruent to the original triangle?
5. Construct a new triangle with sides of lengths 4, 7, and 9.
6. Repeat Exercises 2–4 referring to this new triangle.

Summary
The results of Activity 1 suggest that congruence of triangles can be established by showing a certain combination of three pairs of corresponding parts congruent. Generalize from the above activity.

Activity 2

$\triangle ABC$ has side lengths 6, 7, and 8. If you construct a triangle with a side of length 8 and angles of measures equal to measures of the angles on both ends of the corresponding side, do you think the new triangle will be congruent to $\triangle ABC$?

7. Draw $\triangle ABC$. Then construct the triangle of measurements given above. Determine if the triangles are congruent.
8. Try to generalize the suggested pattern.

4.3 The SAS Postulate

Objectives

To supply missing reasons in proofs of triangle congruence
To prove triangles congruent using the SAS Postulate
To construct congruent triangles using the SAS correspondence

Two triangles are congruent if three pairs of corresponding angles are congruent and three pairs of corresponding sides are congruent. However, it is not necessary to show congruence of all six pairs of parts in order to guarantee congruence of two triangles.

One or two pairs of congruent parts will *not* guarantee congruence.

Two pairs of congruent sides:
The triangles are *not* congruent.

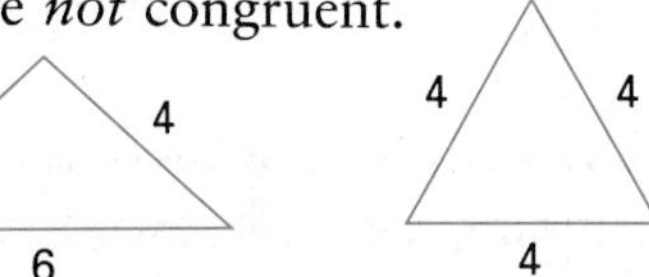

Congruence of triangles can be established by showing congruence of certain combinations of *three* pairs of parts. The combinations of pairs of parts that are required will be explained in this lesson and lessons that follow.

$\triangle DEF$ below was constructed using a compass and straightedge, so that $\overline{DE} \cong \overline{AB}$, $\angle E \cong \angle B$, and $\overline{EF} \cong \overline{BC}$.

Since both $\overline{AB}$ and $\overline{BC}$ are part of $\angle B$, $\angle B$ is the **included angle** for sides $\overline{AB}$ and $\overline{BC}$.

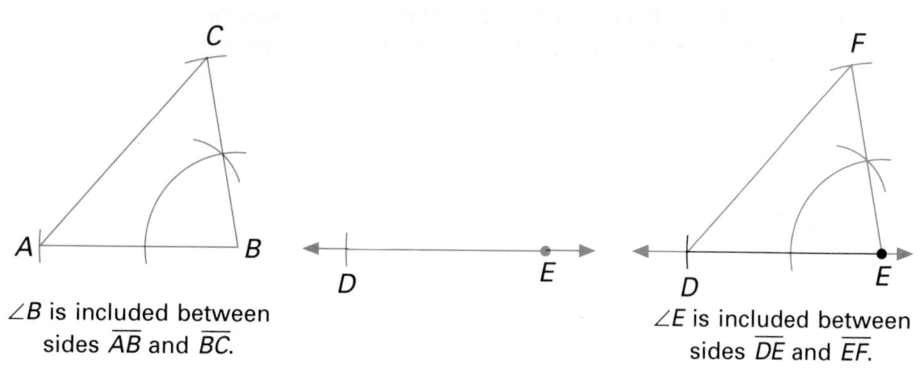

$\angle B$ is included between sides $\overline{AB}$ and $\overline{BC}$.

$\angle E$ is included between sides $\overline{DE}$ and $\overline{EF}$.

Now trace $\triangle DEF$ and fit it on top of $\triangle ABC$. You can see that they have the same size and shape. Therefore, $\triangle ABC \cong \triangle DEF$. This suggests the following postulate.

Postulate 15 **SAS Postulate for Congruence of Triangles:** If two sides and the included angle of one triangle are congruent to the corresponding two sides and included angle of a second triangle, then the triangles are congruent.

EXAMPLE 1 Which pairs of triangles are congruent by the SAS Postulate?

Solution

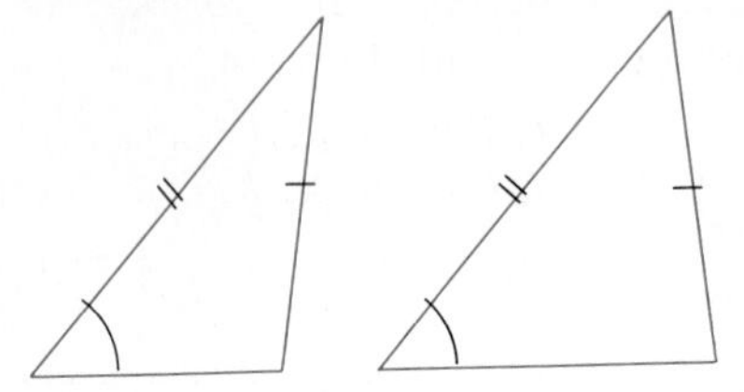

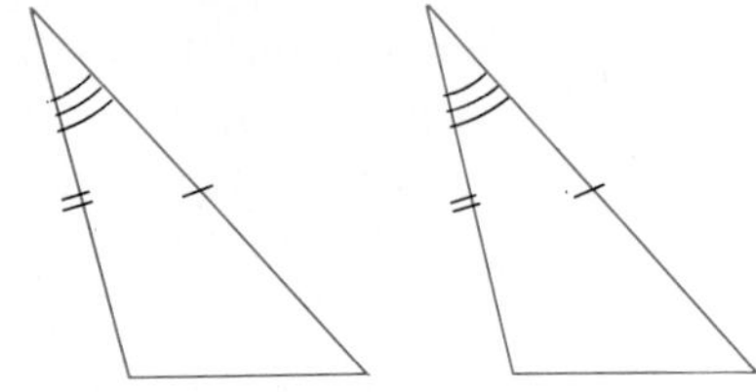

The angles marked congruent are not included between the sides marked congruent. The SAS Postulate does not apply. (In this case, the triangles are not congruent.)

The angles marked congruent are included between the sides marked congruent. The triangles are congruent.

EXAMPLE 2 To prove the triangles congruent by the SAS Postulate, determine which other pair of sides or angles needs to be congruent.

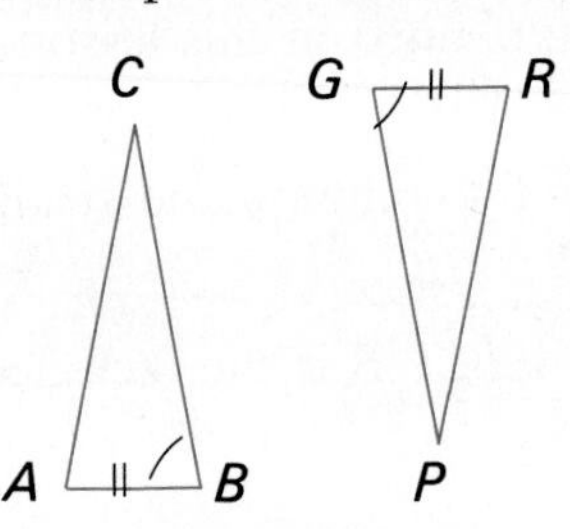

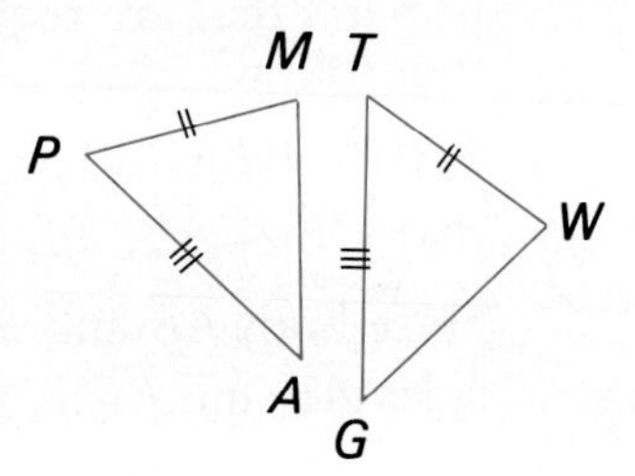

Solution Copy the figure and mark the congruent side and angle.

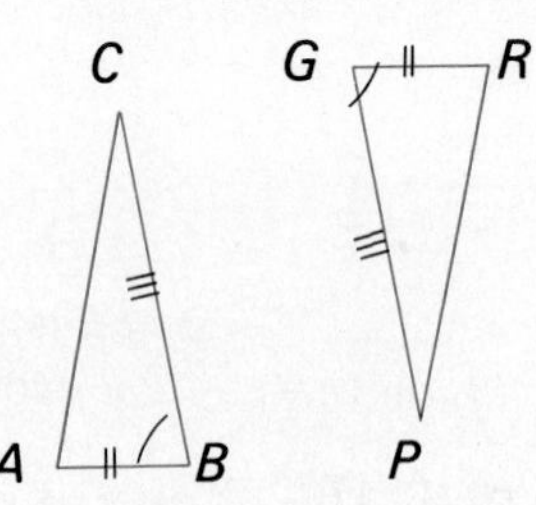

Needed: $\overline{BC} \cong \overline{GP}$

Copy the figure and mark the congruent sides.

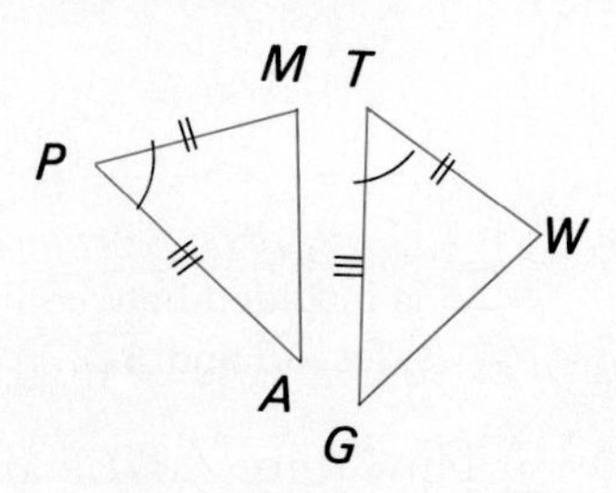

Needed: $\angle P \cong \angle T$

EXAMPLE 3 Given: $\overline{TS}$ and $\overline{PR}$ bisect each other.
Prove: $\triangle TPQ \cong \triangle SRQ$

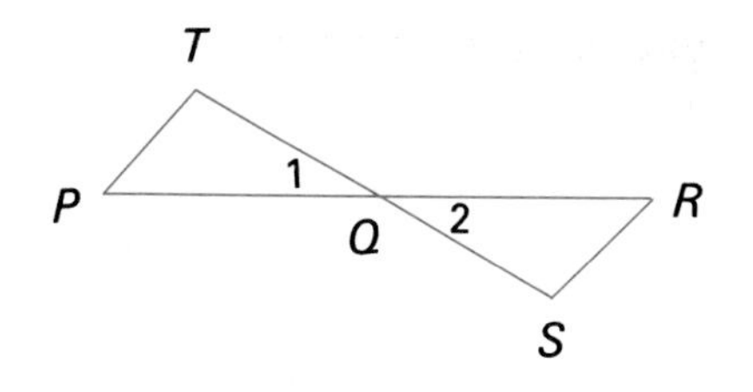

Plan Since $\overline{TS}$ and $\overline{PR}$ bisect each other, $\overline{PQ} \cong \overline{RQ}$ and $\overline{TQ} \cong \overline{SQ}$. $\angle 1$ and $\angle 2$ are vertical angles.

Proof

	Statement		Reason
1.	$\overline{TS}$ and $\overline{PR}$ bisect each other.		1. Given
2.	$\overline{PQ} \cong \overline{RQ}$	(S)	2. Def of bis
3.	$\overline{TQ} \cong \overline{SQ}$	(S)	3. Def of bis
4.	$\angle 1 \cong \angle 2$	(A)	4. Vert $\angle$s are $\cong$.
5.	$\therefore \triangle TPQ \cong \triangle SRQ$		5. SAS

EXAMPLE 4 Given: $\overline{AD}$ bisects $\angle A$; $\overline{AC} \cong \overline{AB}$.
Prove: $\triangle ACD \cong \triangle ABD$

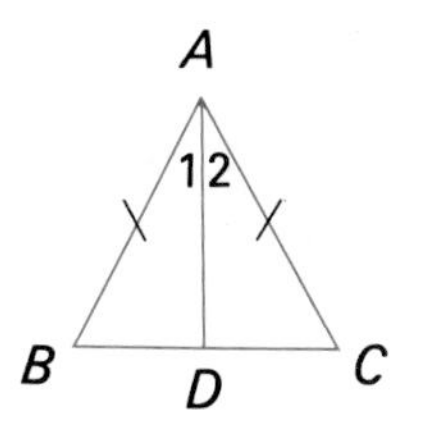

Plan Since $\overline{AD}$ bisects $\angle A$, $\angle 1 \cong \angle 2$. $\overline{AD}$ forms a side of both triangles.

Proof

	Statement		Reason
1.	$\overline{AD}$ bisects $\angle A$.		1. Given
2.	$\angle 1 \cong \angle 2$	(A)	2. Def of $\angle$ bis
3.	$\overline{AC} \cong \overline{AB}$	(S)	3. Given
4.	$\overline{AD} \cong \overline{AD}$	(S)	4. Reflex Prop
5.	$\therefore \triangle ACD \cong \triangle ABD$		5. SAS

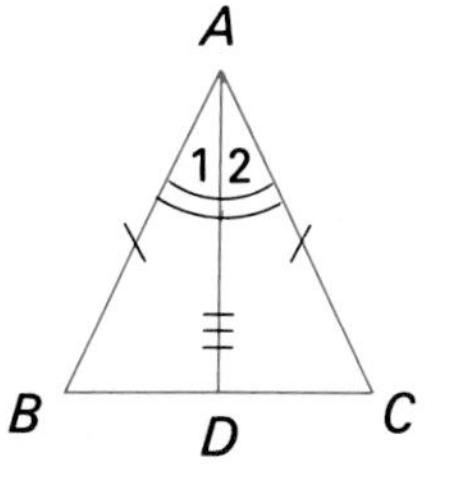

Notice that the Reflexive Property (Step 4) allows you to find a pair of congruent sides. This property is commonly used in geometric proofs.

Classroom Exercises

For each pair of sides of the triangle, identify the included angle.

1. $\overline{PW}$ and $\overline{WQ}$
2. $\overline{PQ}$ and $\overline{PW}$
3. $\overline{QP}$ and $\overline{WQ}$

4. $\overline{AK}$ and $\overline{AL}$
5. $\overline{KA}$ and $\overline{LK}$
6. $\overline{AL}$ and $\overline{LK}$

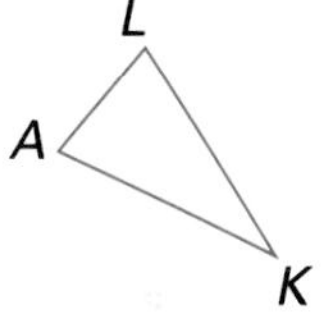

Use the markings to determine if there is enough information to show the two triangles are congruent. (Congruent or not enough information)

7.

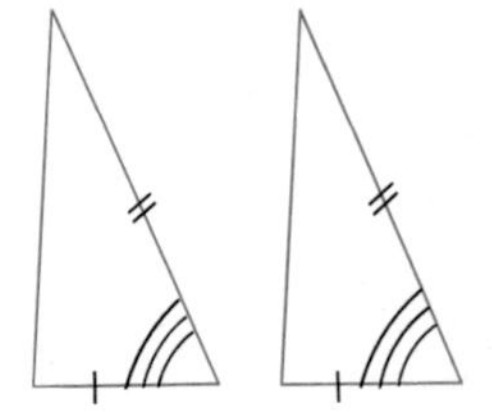

8.

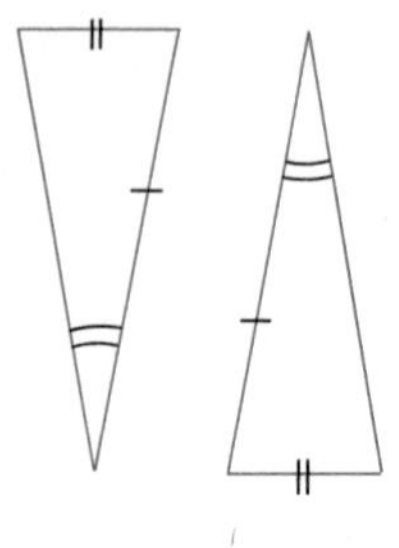

9.

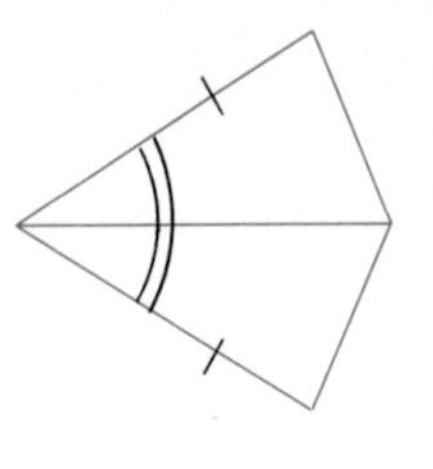

10.

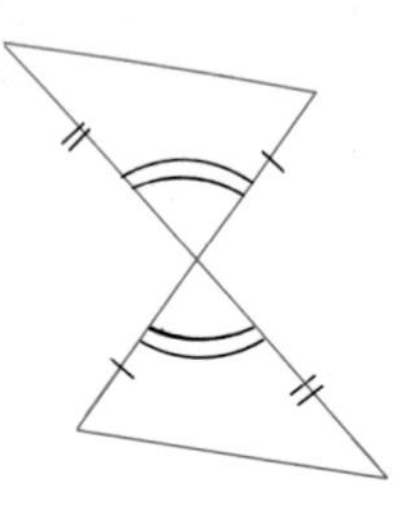

Written Exercises

To prove the two triangles congruent by the SAS Postulate, which additional pair of sides or angles must be congruent?

1.

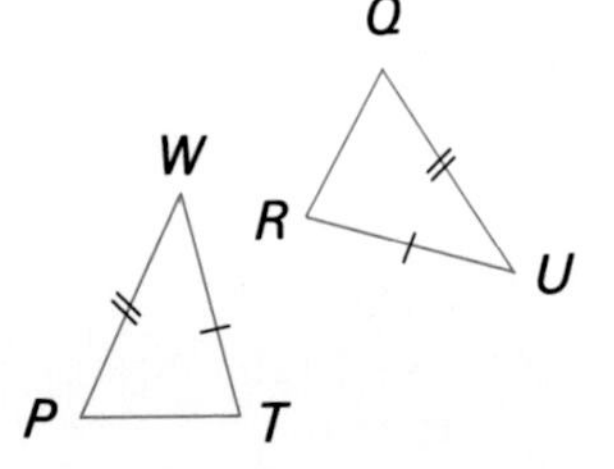

2.

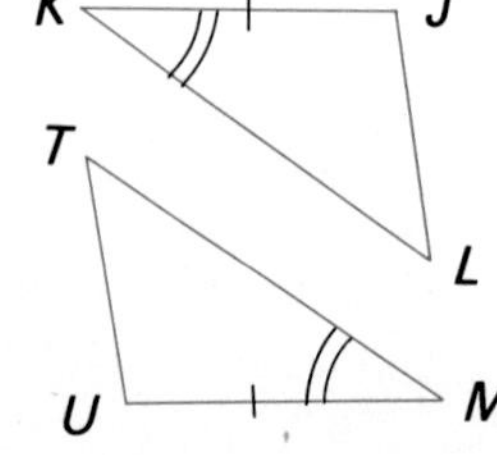

3.

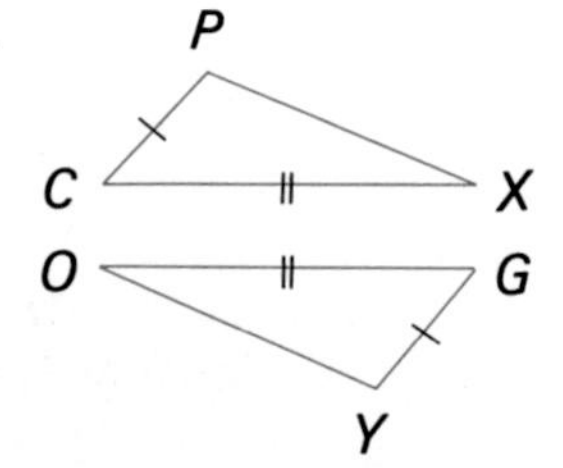

Supply the missing statements and reasons in the proof.

4. Given: B is the midpoint of $\overline{AC}$;
$\overline{AE} \perp \overline{AC}$, $\overline{CD} \perp \overline{CA}$, $\overline{EA} \cong \overline{DC}$.
Prove: $\triangle EAB \cong \triangle DCB$

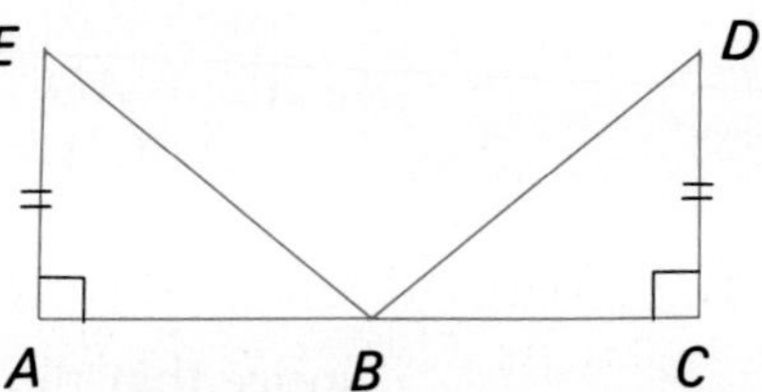

	Statement		Reason
	1. $\overline{AE} \perp \overline{AC}$, $\overline{CD} \perp \overline{CA}$		1. Given
	2. m $\angle A$ = 90, m $\angle C$ = 90		2. ________
	3. m $\angle A$ = m $\angle C$ ($\angle A \cong \angle C$)	(A)	3. ________
	4. B is the midpoint of $\overline{AC}$.		4. Given
	5. ____________	(S)	5. Def of midpt
	6. ____________	(S)	6. Given
	7. $\therefore \triangle EAB \cong \triangle DCB$		7. SAS

5. Given: $\overline{QP} \perp \overline{QR}$, $\overline{TS} \perp \overline{TU}$, $\overline{QR} \cong \overline{TU}$, $\overline{QP} \cong \overline{TS}$
Prove: $\triangle PQR \cong \triangle STU$

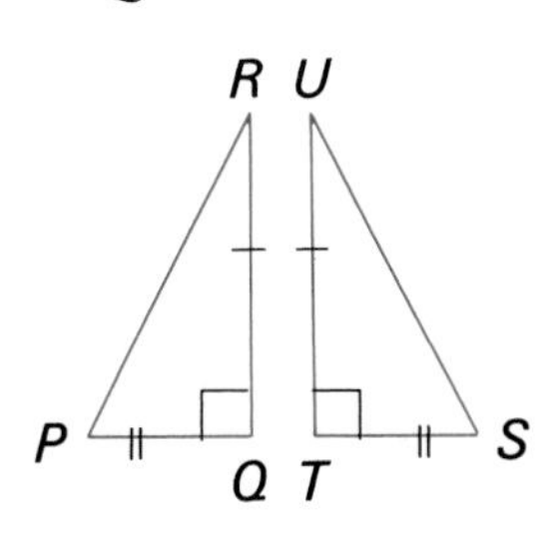

6. Given: $\overline{AD} \cong \overline{CD}$, $\angle 1 \cong \angle 2$
Prove: $\triangle ADB \cong \triangle CDB$

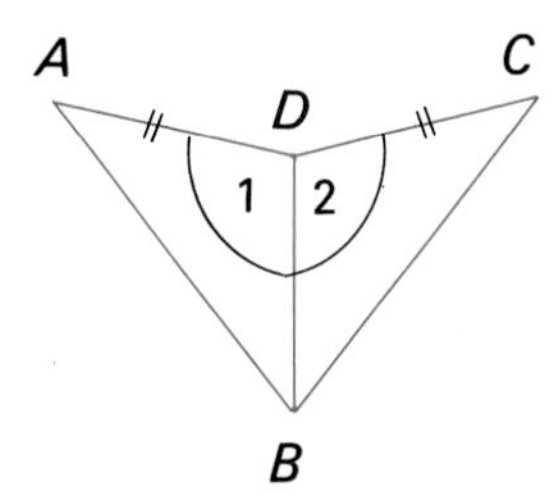

7. Given: $\angle 1 \cong \angle 2$, $\overline{EC} \cong \overline{ED}$, E is the midpoint of $\overline{AB}$.
Prove: $\triangle ACE \cong \triangle BDE$

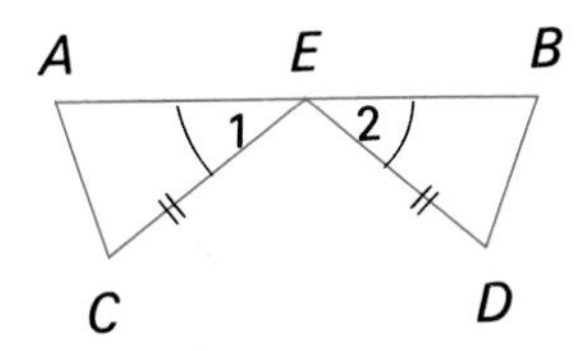

8. Given: $\overline{PO} \cong \overline{QO}$, $\overline{RO} \cong \overline{SO}$
Prove: $\triangle POR \cong \triangle QOS$

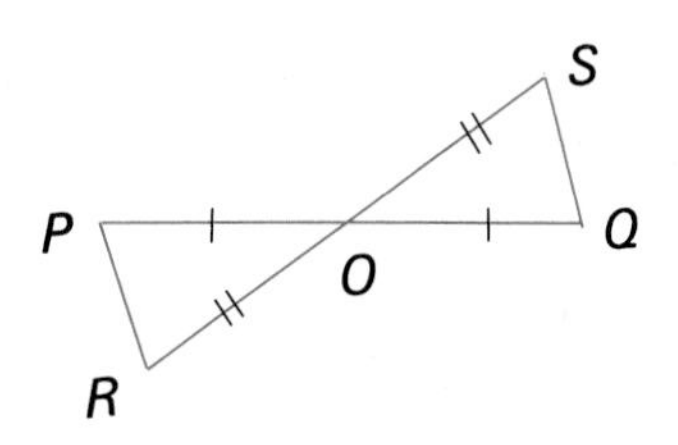

9. Given: $\overline{AB} \parallel \overline{FE}$, $\overline{AB} \cong \overline{EF}$, $\overline{AC} \cong \overline{ED}$
Prove: $\triangle ABC \cong \triangle EFD$

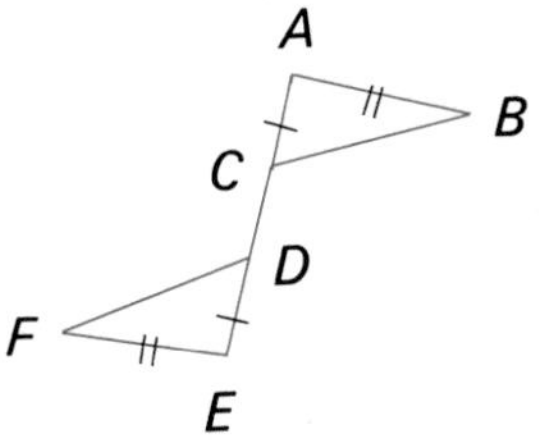

10. Given: $\overline{PQ} \parallel \overline{RS}$, $\overline{PQ} \cong \overline{RS}$
Prove: $\triangle PQS \cong \triangle RSQ$

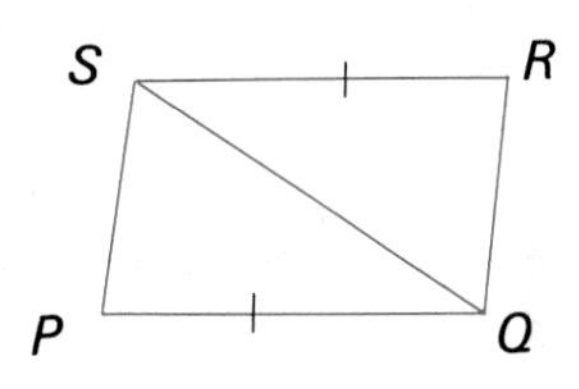

11. Given: $\overline{DB} \perp \overline{AC}$, $\overline{DB}$ bisects $\overline{AC}$.
Prove: $\triangle ABD \cong \triangle CBD$

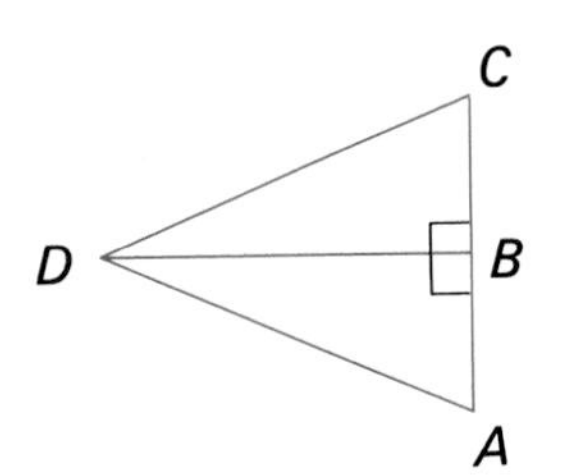

12. Given: $\overline{EA} \perp \overline{AB}$, $\overline{DC} \perp \overline{CB}$, $\overline{EA} \cong \overline{DC}$, B is the midpoint of $\overline{AC}$.
Prove: $\triangle EAB \cong \triangle DCB$

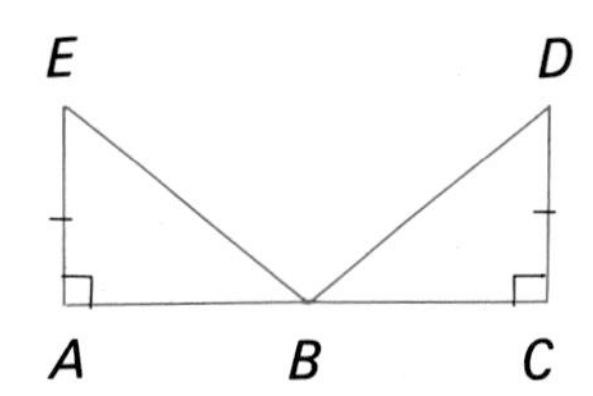

13. Given: $\triangle ABC$ with $\overline{BA} \cong \overline{CA}$, point D on $\overline{BC}$ such that $\overline{AD}$ bisects $\angle A$.
Prove: $\triangle BAD \cong \triangle CAD$

14. Draw an acute triangle. Construct a triangle congruent to it using SAS.

15. State the result of Example 4 as a theorem about isosceles triangles.
16. Given: $\overline{AD} \parallel \overline{BC}$, $\angle 1 \cong \angle 3$, $\overline{AC} \cong \overline{FB}$
Prove: $\triangle ABC \cong \triangle FCB$
17. Given: $\overline{CF} \parallel \overline{BE}$, $\angle 3 \cong \angle 4$, $\overline{BC} \cong \overline{BE}$
Prove: $\triangle CBF \cong \triangle EBF$

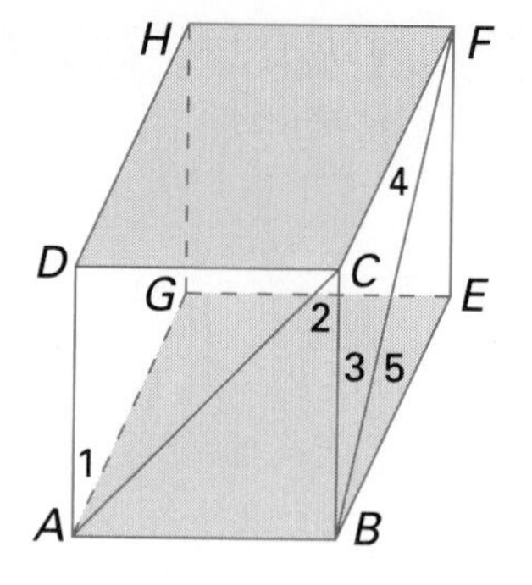

Mixed Review

In the figure at the right, $\overline{DE} \parallel \overline{AF}$.

1. Given: m $\angle 2 = 40$. Find m $\angle 4$. *3.5*
2. Given: m $\angle 2 = 65$. Find m $\angle DCB$. *3.5*
3. Given: m $\angle 1 = 60$, m $\angle 3$ is twice m $\angle 2$. Find m $\angle 2$. *3.6*
4. Given: m $\angle 6 = 120$, m $\angle 1 = 40$ Find m $\angle 3$. *3.7*

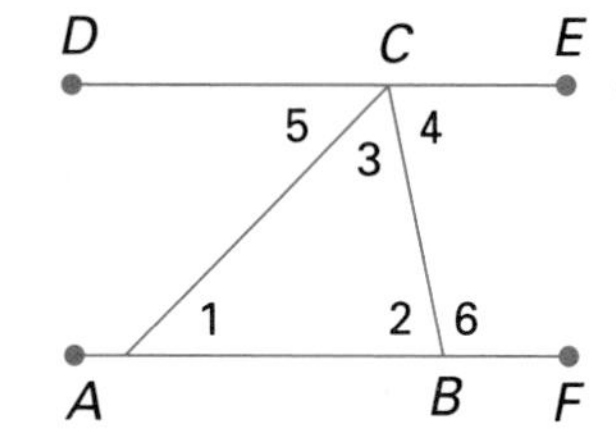

Application: *Angles in Sand Piles*

Materials such as sand, gravel, and coal are often stored in supply yards in loose piles. The angle made between the side of the pile and the (horizontal) ground is called the *angle of repose*. This angle varies depending on the characteristics of the material. For example, wet sand has a greater angle of repose than an equal volume of dry sand. The angle of repose can be used to estimate the relative sizes of piles, cross-sectional area, volume of a pile, ground area needed for storing given quantities, minimum amount of sand needed to bank a dam.

Material	Angle of repose (approximate)
Sand (dry)	30
Sand (wet)	38
Gravel	35
Salt	36
Coal	37
Cement	40
Crushed Stone	35

1. Which would form a higher pile—an equal volume of dry sand or wet sand?
2. If a pile of salt and a pile of coal had equal heights, which pile would have the greater volume?
3. Which would require a greater area for storage, a given volume of cement or of gravel?

4.4 SSS and ASA Congruence Proofs

Objectives

To supply missing reasons in proofs of triangle congruence
To construct congruent triangles using the SSS and ASA patterns
To prove triangles congruent using the SSS, ASA, and SAS postulates

Triangle congruence properties can be applied in various ways. Using triangles you can indirectly find the distance across a lake (see Exercise 15). Two new congruence postulates are developed in this lesson. $\triangle DEF$, below, was constructed so that $\overline{DE} \cong \overline{AB}$, $\overline{EF} \cong \overline{BC}$, and $\overline{FD} \cong \overline{CA}$.

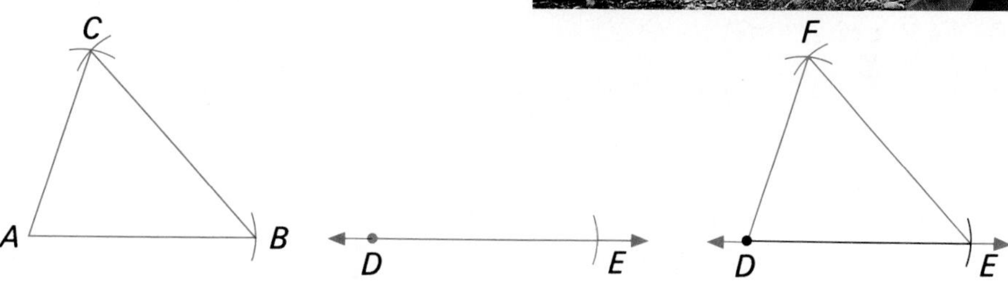

Trace $\triangle DEF$ and fit it on top of $\triangle ABC$. You can see that they have the same size and shape. Therefore, $\triangle ABC \cong \triangle DEF$. This suggests the following postulate.

Postulate 16

SSS Postulate for Congruence of Triangles: If the three sides of one triangle are congruent to the corresponding three sides of a second triangle, then the triangles are congruent.

A triangle can be constructed congruent to another triangle by an *angle-side-angle* method. Construct $\triangle DEF$ so that $\overline{DE} \cong \overline{AB}$, $\angle D \cong \angle A$, and $\angle E \cong \angle B$.

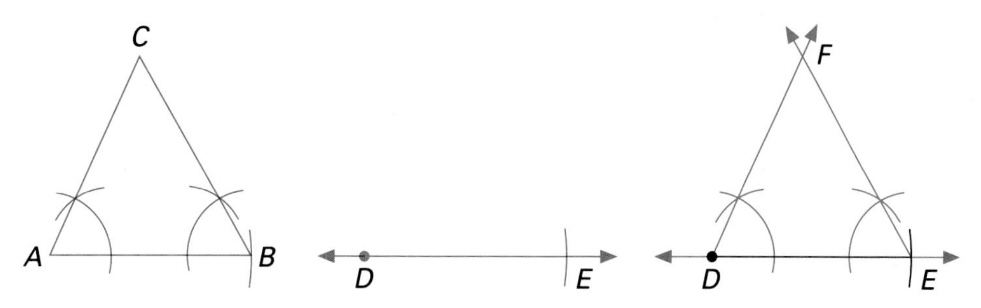

Postulate 17 **ASA Postulate for Congruence of Triangles:** If two angles and the included side of one triangle are congruent to the corresponding two angles and included side of a second triangle, then the triangles are congruent.

The *included side* mentioned in Postulate 17 is the side whose endpoints are the vertices of the two given angles. You now have three methods of establishing congruence of triangles: SSS, SAS, and ASA.

EXAMPLE 1 For each diagram, indicate whether the two triangles are congruent and, if so, write a statement of congruence. Justify your statement.

Solution

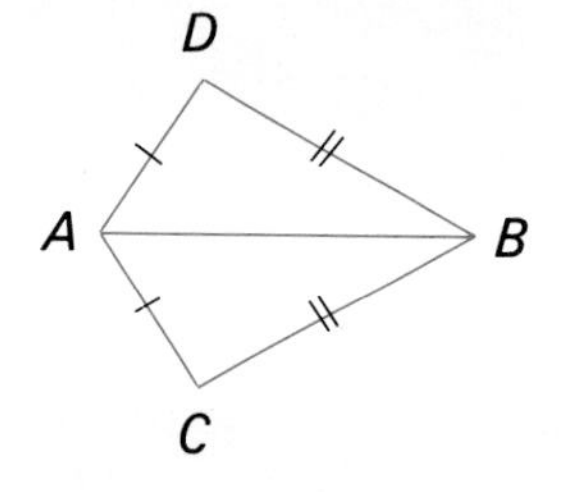

Yes. $\triangle ADB \cong \triangle ACB$
Reason: SSS

T S 2 O 1 P Q

Yes. $\triangle TOS \cong \triangle QOP$
Reason: SAS

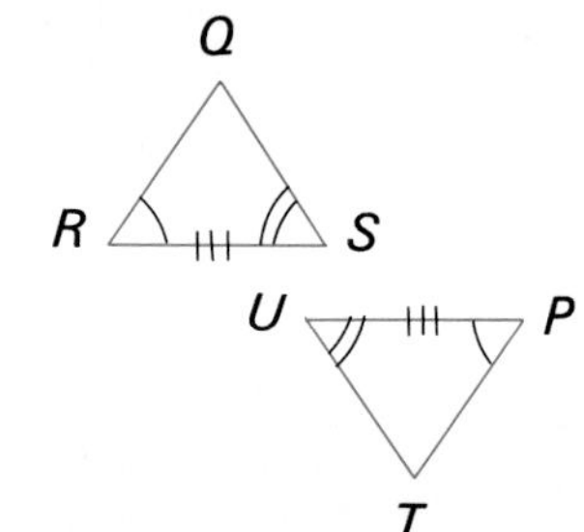

Yes. $\triangle QRS \cong \triangle TPU$
Reason: ASA

EXAMPLE 2 Given: $\overline{AB}$ bisects $\angle CAD$, $\angle 1 \cong \angle 2$.
Prove: $\triangle CAB \cong \triangle DAB$

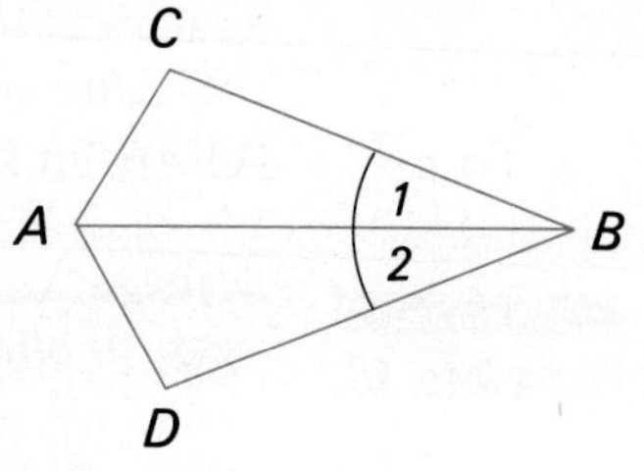

Proof

	Statement		Reason
1.	$\overline{AB}$ bisects $\angle CAD$.		1. Given
2.	$\angle CAB \cong \angle DAB$	(A)	2. Def of $\angle$ bis
3.	$\angle 1 \cong \angle 2$	(A)	3. Given
4.	$\overline{AB} \cong \overline{AB}$	(S)	4. Reflex Prop
5.	$\therefore \triangle CAB \cong \triangle DAB$		5. ASA

EXAMPLE 3 Given: $\overline{CD}$ bisects $\overline{AB}$, $\overline{AC} \cong \overline{BC}$.
Prove: $\triangle ADC \cong \triangle BDC$

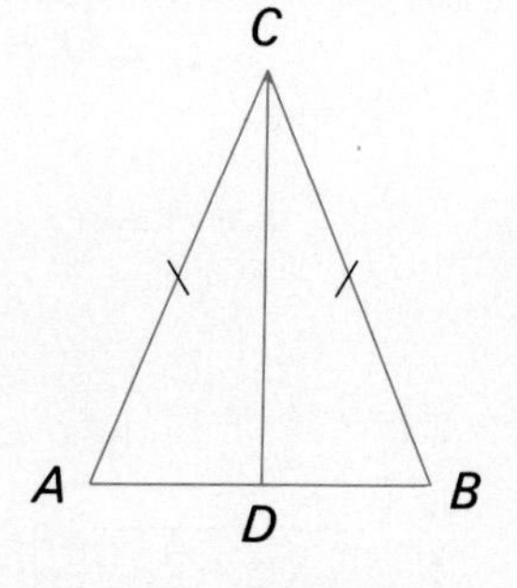

Proof

	Statement		Reason
1.	$\overline{CD}$ bisects $\overline{AB}$.		1. Given
2.	$\overline{AD} \cong \overline{BD}$	(S)	2. Def of bis
3.	$\overline{CD} \cong \overline{CD}$	(S)	3. Reflex Prop
4.	$\overline{AC} \cong \overline{BC}$	(S)	4. Given
5.	$\therefore \triangle ADC \cong \triangle BDC$		5. SSS

Classroom Exercises

For each pair of angles of the triangle, identify the included side.

1. $\angle P$ and $\angle Q$
2. $\angle P$ and $\angle R$
3. $\angle Q$ and $\angle R$

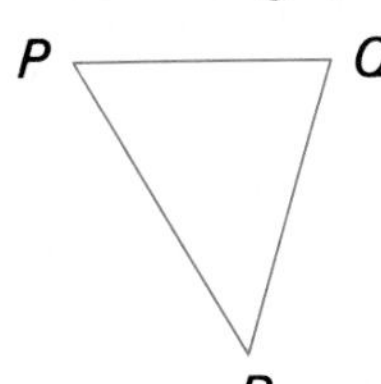

4. $\angle G$ and $\angle M$
5. $\angle G$ and $\angle X$
6. $\angle M$ and $\angle X$

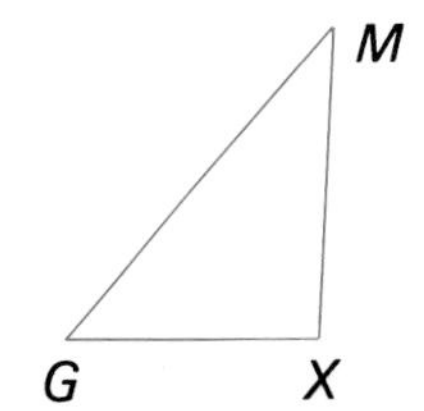

Indicate whether the pair of triangles is congruent, and, if so, state why.

7.

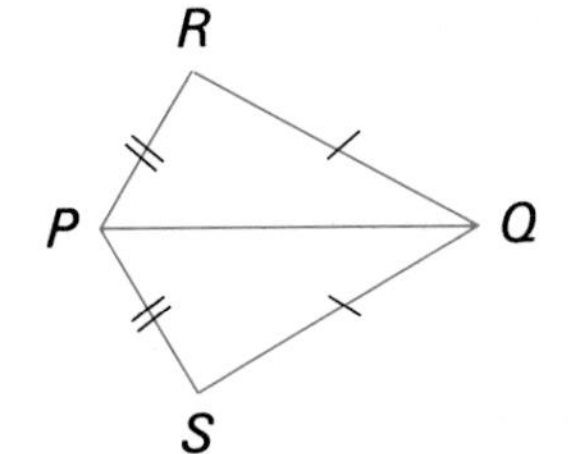

8.

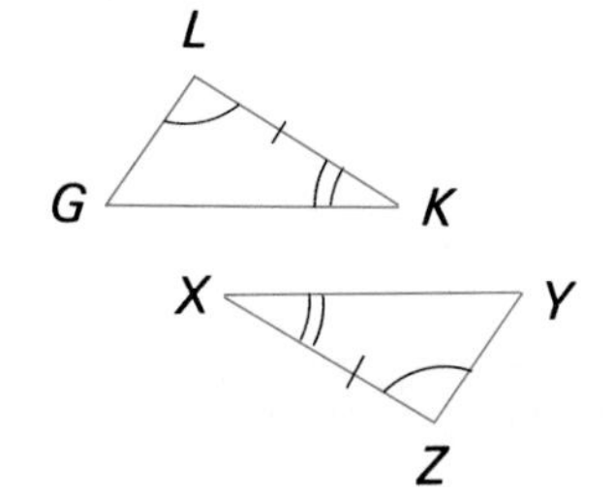

9.

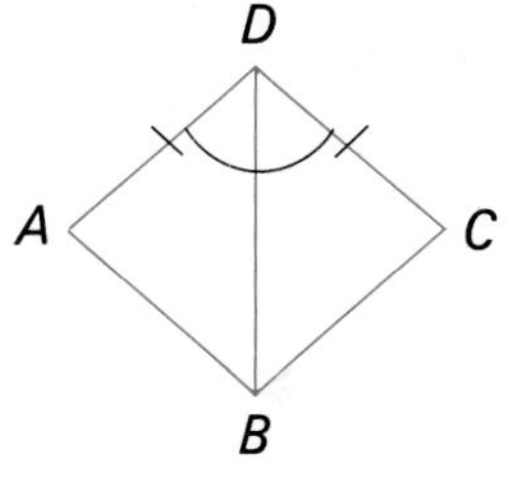

Written Exercises

Complete the statement.

1. To prove $\triangle TAH \cong \triangle YPM$ by ASA, show $\angle A \cong$ ______, $\angle H \cong$ ______, and ______ $\cong$ ______.
2. To prove $\triangle TAH \cong \triangle YPM$ by SAS, show $\overline{TH} \cong$ ______ $\overline{HA} \cong$ ______, and ______ $\cong$ ______.

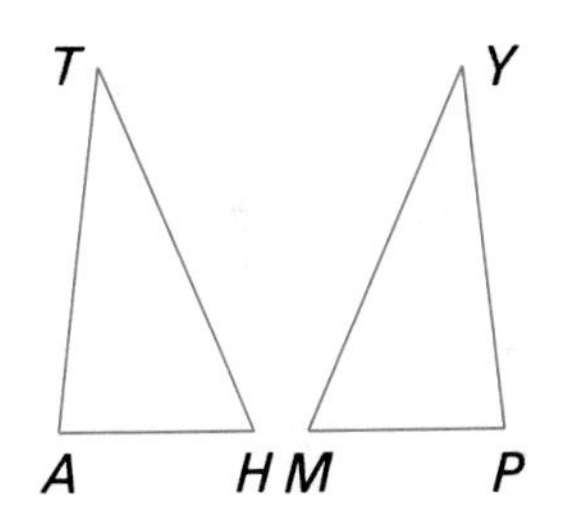

Supply the missing statements and reasons for the proof.

3. Given: $\overline{TS} \cong \overline{QR}$, $\overline{TU} \cong \overline{QP}$,
   $\overline{TU} \parallel \overline{QP}$
   Prove: $\triangle TSU \cong \triangle QRP$

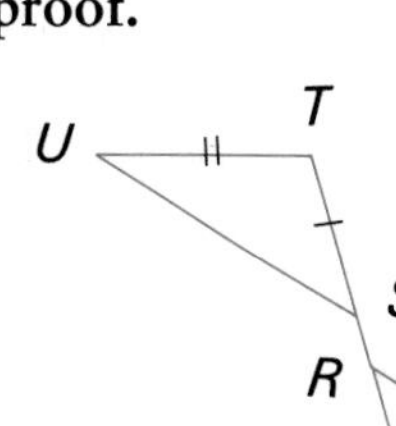
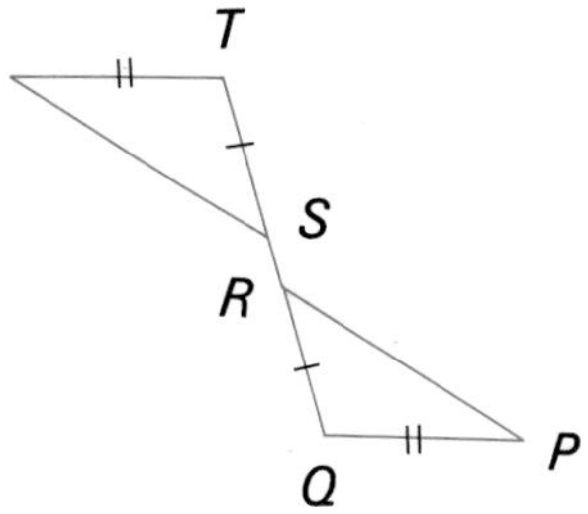

Statement	Reason
1. $\overline{TS} \cong \overline{QR}$	(S) 1. Given
2. ______	(S) 2. ______
3. $\overline{TU} \parallel \overline{QP}$	3. ______
4. ______	(A) 4. ______
5. $\therefore \triangle TSU \cong \triangle QRP$	5. ______

4. Given: $\overline{TR}$ and $\overline{MN}$ bisect each other.
Prove: $\triangle NTP \cong \triangle MRP$

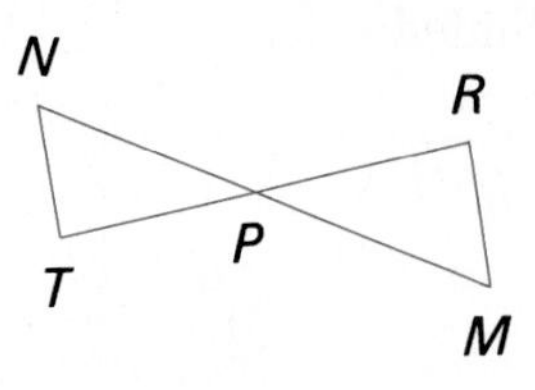

5. Given: $\overline{CD}$ bisects $\angle ACB$; $\angle 1 \cong \angle 2$.
Prove: $\triangle CDA \cong \triangle CDB$

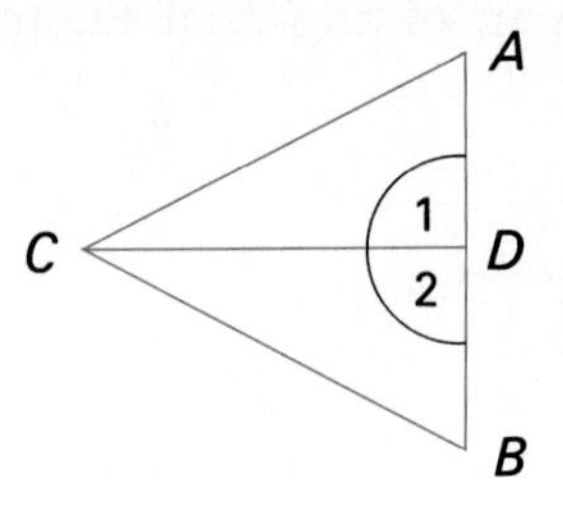

6. Given: $\overline{AB} \parallel \overline{CD}$, $\angle B \cong \angle D$,
$\overline{AB} \cong \overline{CD}$
Prove: $\triangle ABF \cong \triangle CDE$

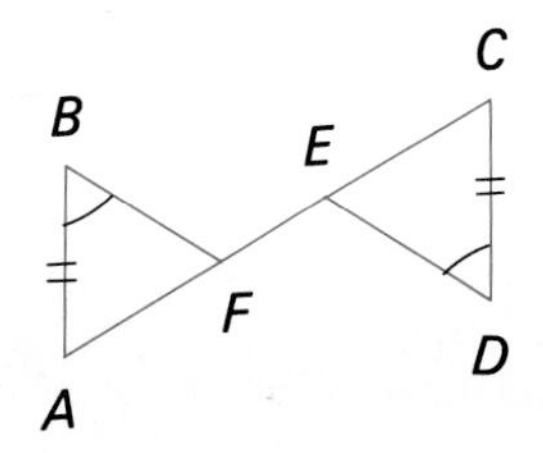

7. Given: $\overline{PG} \cong \overline{SG}$, $\overline{TP} \cong \overline{TS}$
Prove: $\triangle TPG \cong \triangle TSG$

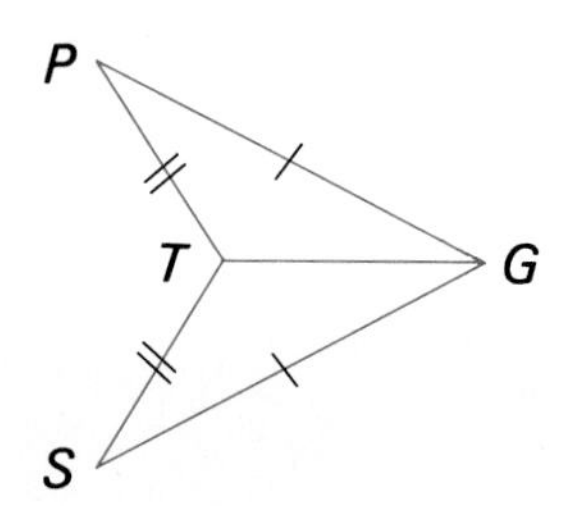

8. Draw an obtuse triangle. Construct a triangle congruent to it by SSS.

9. Draw a scalene triangle. Construct a triangle congruent to it by ASA.

10. Given: $\overline{OE} \perp \overline{MP}$, $\overline{OE}$ bisects $\angle MOP$.
Prove: $\triangle MOE \cong \triangle POE$

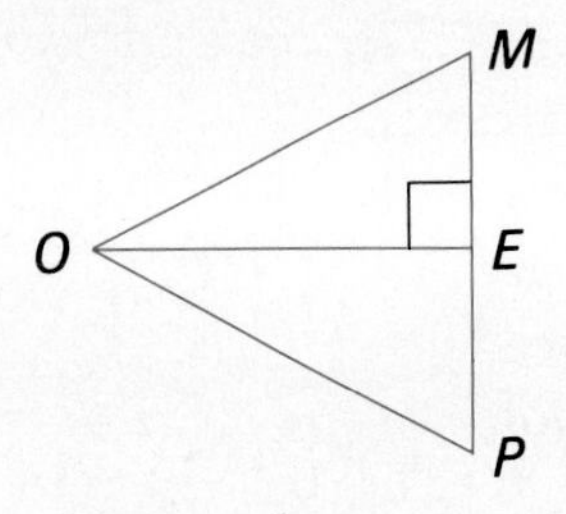

11. Given: $\overline{AD} \parallel \overline{BC}$, $\overline{DC} \parallel \overline{BA}$
Prove: $\triangle ADB \cong \triangle CBD$

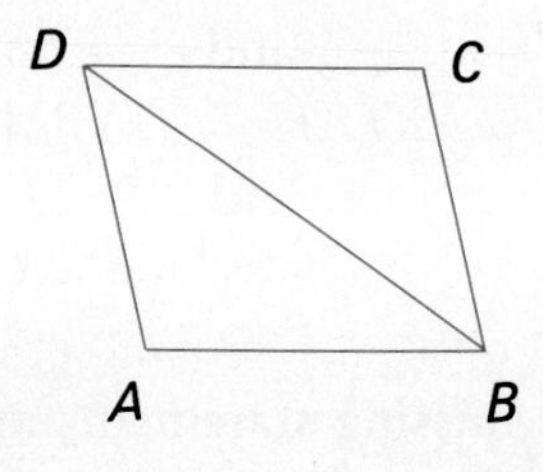

12. Given: Q is the midpoint of $\overline{PR}$; $\overline{TP} \perp \overline{PR}$, $\overline{SR} \perp \overline{RP}$,
$\angle TQP \cong \angle SQR$.
Prove: $\triangle TPQ \cong \triangle SRQ$

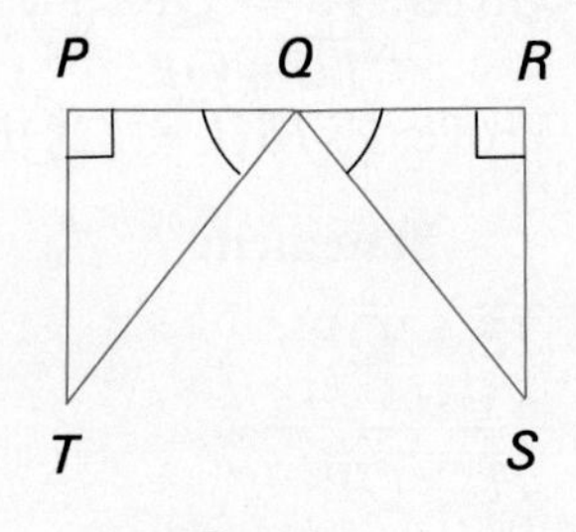

Ex. 12

13. It is given that $\overline{RS}$ bisects $\angle PRQ$ of $\triangle PRQ$. Also, $\overline{RS} \perp \overline{PQ}$ at point S. Prove that the two triangles formed are congruent.

14. A carpenter has to make a triangular brace for the gate at the right. He cuts two strips of wood of equal length, $\overline{PQ}$ and $\overline{RS}$. They are nailed to the gate so that $\overline{PQ} \parallel \overline{RS}$. Prove that the triangles formed are congruent. He checks his construction before driving the nails all the way in by measuring $\overline{PR}$ and $\overline{QS}$ to make sure they are equal. Why?

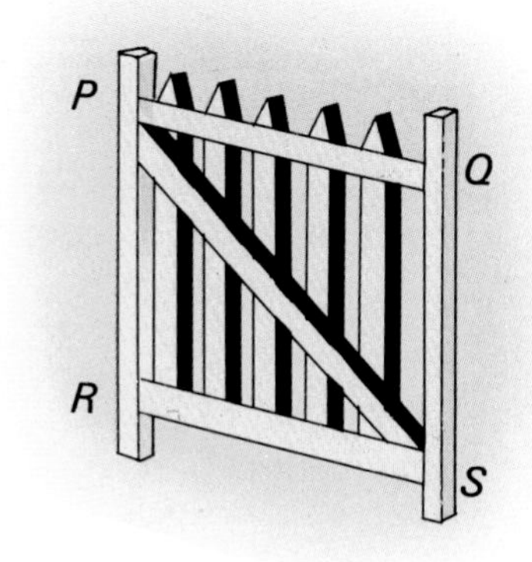

15. Jane wants to find the distance from her camp tent at G to a tree at M on the opposite side of a lake. She sets a stake at L so that $\overline{GM} \perp \overline{GL}$. She places a second stake at H in line with $\overline{GL}$, making $GL = HL$. She then stretches a tape from H, making $\overline{HN} \perp \overline{HL}$, and placing a stake at K, the intersection of $\overleftrightarrow{ML}$ with $\overleftrightarrow{HN}$. Prove $\triangle GLM \cong \triangle HLK$. Why will measuring $\overline{HK}$ now give the distance across the lake?

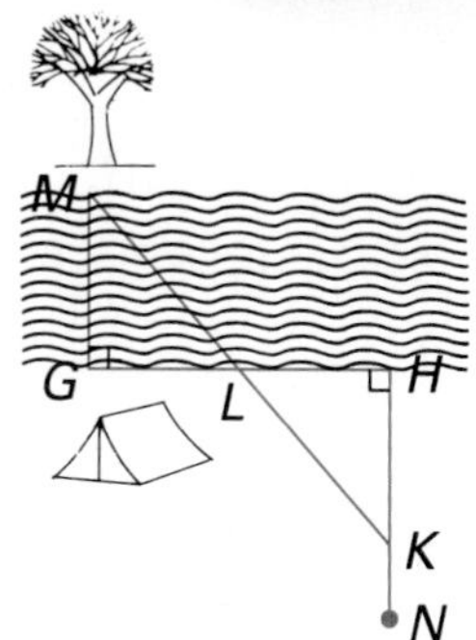

16. For what special case can you say that $\triangle ABC \cong \triangle BAC$?

Midchapter Review

Classify each triangle by its sides. *4.1*

1. 3 cm, 3cm, 3cm **2.** 5 in., 7 in., 6 in. **3.** 6 ft, 4 ft, 6 ft

Classify each triangle by its angles. *4.1*

4. The measures of two angles are 22 and 68.
5. The measures of two angles are 28 and 32.

Given: $\triangle ABC \cong \triangle XYZ$, $AC = 8$, $BC = 6$, m $\angle A = 37$, m $\angle B = 53$ *4.2*

6. Find the length of $\overline{YZ}$. **7.** Find the measure of $\angle Z$.

8. Given $\overline{AB}$ and $\overline{CD}$ bisect each other, complete the proof. *4.3*

Statement	Reason
1. $\overline{AB}$ and $\overline{CD}$ bisect each other.	1. Given
2. $AE = BE$, $CE = DE$	2. ______
3. m $\angle 1$ = m $\angle 2$	3. ______
4. ______	4. SAS

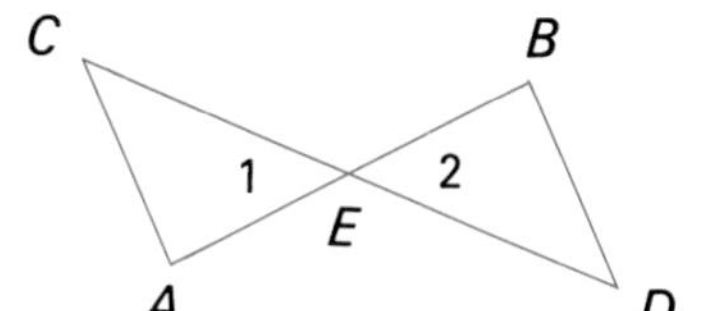

4.5 Congruence in Complex Figures

Objective To prove triangles are congruent when they are parts of complex figures

EXAMPLE Given: $\angle 1 \cong \angle 2$, $\overline{EA} \cong \overline{DC}$,
B is the midpoint of $\overline{AC}$.
Prove: $\triangle EAB \cong \triangle DCB$

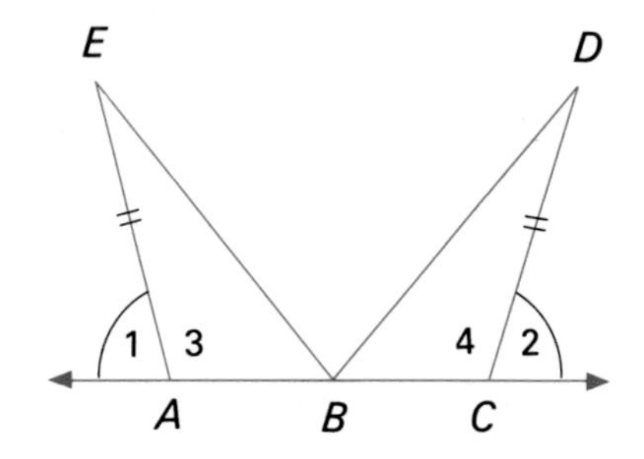

Proof

	Statement		Reason
1.	$\angle 1 \cong \angle 2$		1. Given
2.	$\angle 1$ and $\angle 3$ are supplementary. $\angle 2$ and $\angle 4$ are supplementary.		2. If the outer rays of two adj $\angle$s form a st $\angle$, then the $\angle$s are supp.
3.	$\angle 3 \cong \angle 4$	(A)	3. Supp of $\cong$ $\angle$s are $\cong$.
4.	B is the midpoint of $\overline{AC}$.		4. Given
5.	$\overline{AB} \cong \overline{CB}$	(S)	5. Def of midpt
6.	$\overline{EA} \cong \overline{DC}$	(S)	6. Given
7.	$\therefore \triangle EAB \cong \triangle DCB$		7. SAS

Classroom Exercises

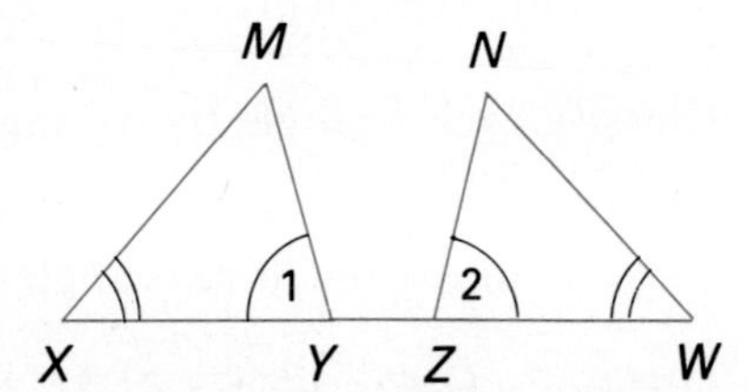

Supply the missing statements and reasons in the proof.

Given: $\overline{XZ} \cong \overline{WY}$, $\angle 1 \cong \angle 2$, $\angle X \cong \angle W$
Prove: $\triangle MXY \cong \triangle NWZ$

	Statement		Reason
1.	$\angle 1 \cong \angle 2$	(A)	1. __________
2.	$\angle X \cong \angle W$	(A)	2. __________
3.	__________		3. Given
4.	$XZ -$ ______ $= WY -$ ______		4. Subt Prop of Eq
5.	$XZ - YZ = XY$, $WY - YZ = ZW$		5. __________
6.	__________	(S)	6. Sub
7.	$\therefore \triangle MXY \cong \triangle NWZ$		7. ______

Written Exercises

1. Given: $\overline{AC} \cong \overline{CE}$, $\overline{AB} \cong \overline{DE}$, $\overline{BG} \cong \overline{DF}$, $\angle 1 \cong \angle 2$
Prove: $\triangle BGC \cong \triangle DFC$

2. Given: $\angle 3 \cong \angle 4$, $\angle 5 \cong \angle 6$, C is the midpoint of $\overline{BD}$.
Prove: $\triangle BGC \cong \triangle DFC$

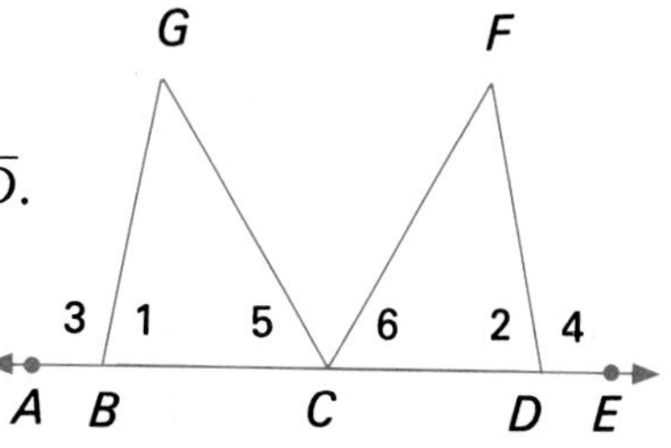

3. Given: $\overline{AB} \cong \overline{CD}$, $\angle 3 \cong \angle 2$, $\overline{EC} \cong \overline{FB}$
Prove: $\triangle AEC \cong \triangle DFB$

4. Given: $\angle 1 \cong \angle 4$, $\overline{EC} \cong \overline{FB}$, $\angle E \cong \angle F$
Prove: $\triangle AEC \cong \triangle DFB$

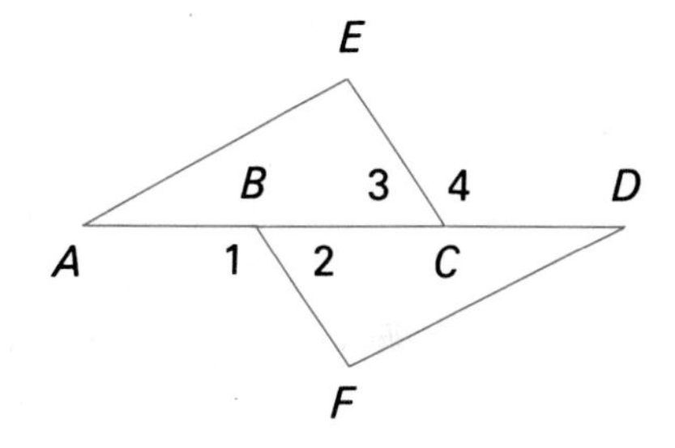

5. Given: $\overline{AG} \cong \overline{DG}$, $\overline{FG} \cong \overline{EG}$, $\overline{AB} \cong \overline{DC}$, $\angle A \cong \angle D$
Prove: $\triangle FAB \cong \triangle EDC$

6. Given: $\angle 1 \cong \angle 2$, $\overline{FB} \perp \overline{AD}$, $\overline{EC} \perp \overline{AD}$, $\overline{FB} \cong \overline{EC}$
Prove: $\triangle FAB \cong \triangle EDC$

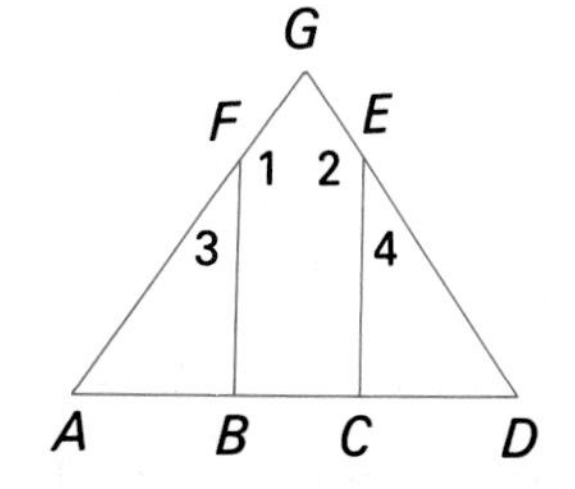

7. Given: $\overline{RP} \cong \overline{US}$, $\overline{RQ} \cong \overline{UT}$, $\overline{RP} \parallel \overline{US}$, $\overline{QW} \cong \overline{TV}$
Prove: $\triangle PQW \cong \triangle STV$

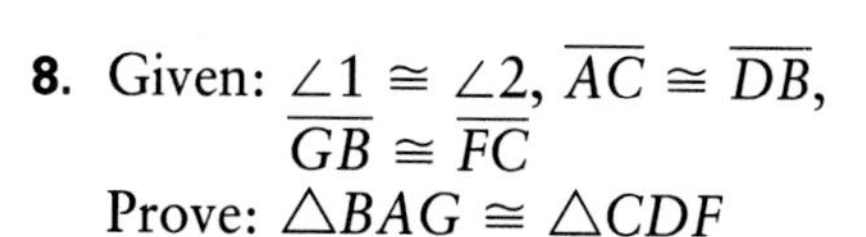
8. Given: $\angle 1 \cong \angle 2$, $\overline{AC} \cong \overline{DB}$, $\overline{GB} \cong \overline{FC}$
Prove: $\triangle BAG \cong \triangle CDF$

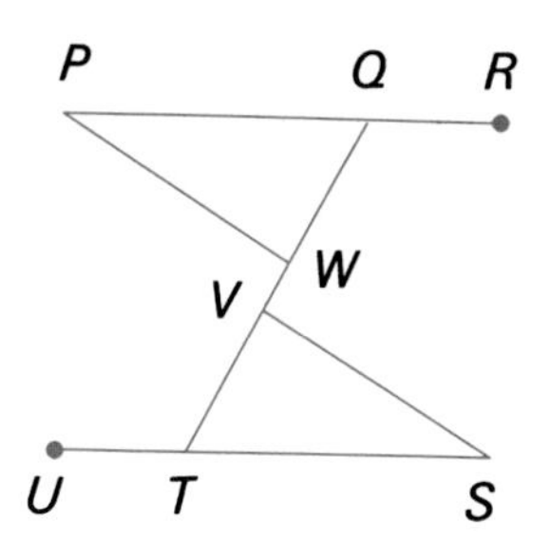

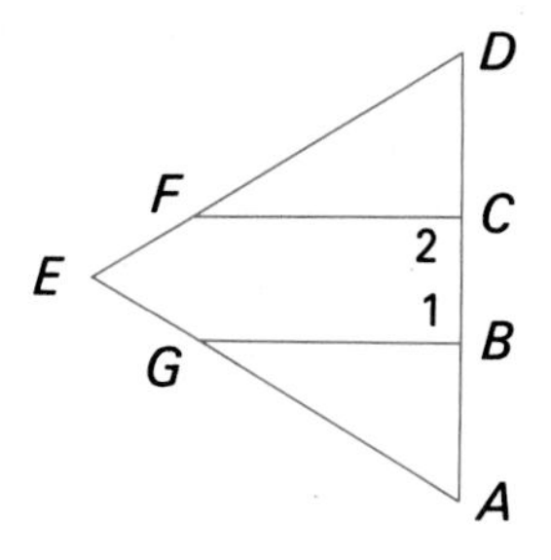

9. Given: $\overline{AD} \cong \overline{CB}$, $\overline{CE} \parallel \overline{BF}$, $\angle 1$ and $\angle 2$ are supplementary.
Prove: $\triangle CED \cong \triangle BFA$

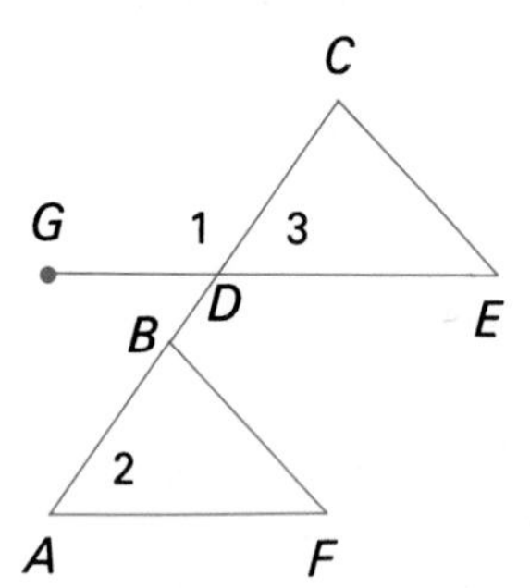

10. Given: $\overline{RV} \perp \overline{PT}$, $\angle 1 \cong \angle 2$, $\angle 3 \cong \angle 4$, R is the midpoint of $\overline{QS}$.
Prove: $\triangle QRW \cong \triangle SRU$

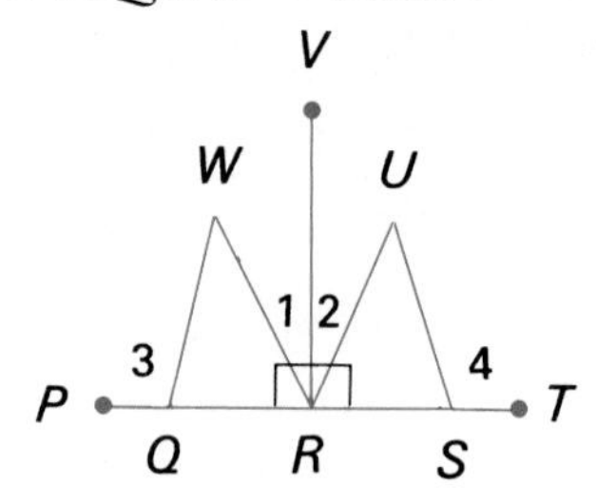

11. Given: $\overline{PQ} \cong \overline{RS}$, $\angle 1 \cong \angle 2$, $\angle 3 \cong \angle 4$, $\overline{WP} \perp \overline{PS}$, $\overline{SX} \perp \overline{PS}$
Prove: $\triangle PUR \cong \triangle STQ$

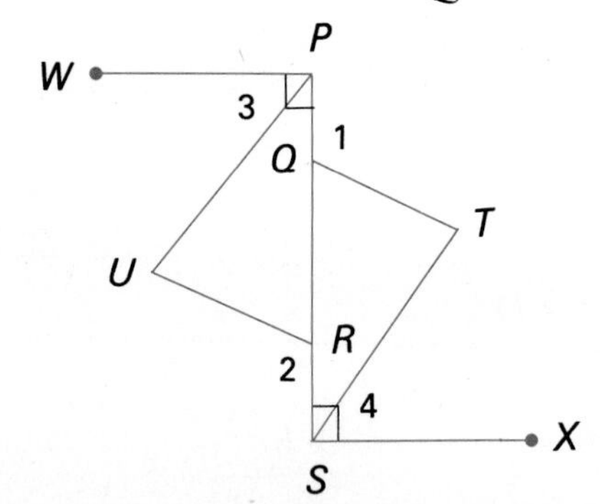

12. Given: $\overline{EB}$ bisects $\angle FBD$; $\angle 3 \cong \angle 4$, $\overline{FB} \cong \overline{DB}$, $\overline{EB} \perp \overline{AC}$.
Prove: $\triangle FAB \cong \triangle DCB$

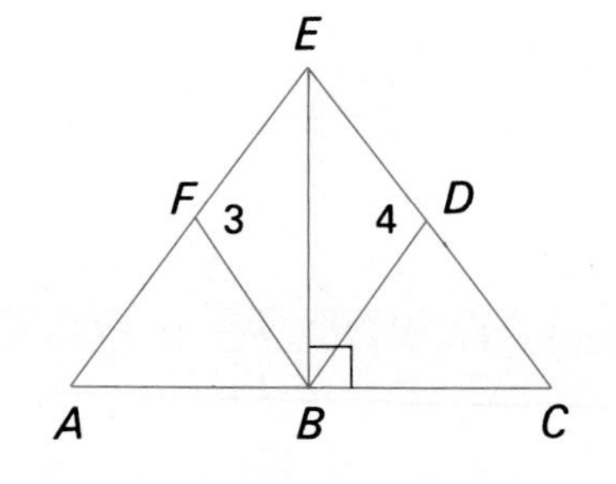

13. Given: $\overline{AB} \cong \overline{DC}$, $\overline{AE} \cong \overline{DF}$, and $\overline{CE} \cong \overline{BF}$
Prove: $\triangle ACE \cong \triangle DBF$

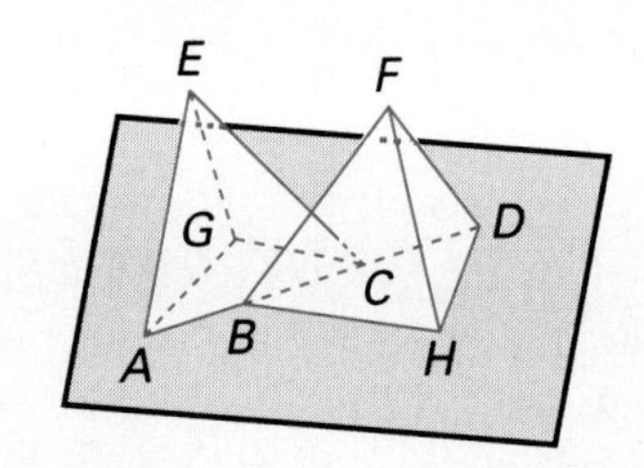

Mixed Review

1. Write the contrapositive of $p \rightarrow q$ *3.8*
2. The endpoints of $\overline{AB}$ have coordinates -4 and 6. Find the coordinate of M, the midpoint of $\overline{AB}$. *1.3*
3. Graph and name the resulting geometric figure: $x \geq 5$ *and* $x \leq 12$ *1.8*
4. On a number line, what is the distance between -2.5 and 1.6? *1.2*

4.6 The Third Angle and AAS Theorems

Objectives To apply the Third Angle Theorem
To apply the AAS Theorem
To recognize patterns for proving triangles congruent

Recall that the sum of the measures of the three angles of a triangle is 180. In the figure, the measures of two angles of one triangle are equal to the measures of two angles of the second triangle. Find the measures of the third angles. Repeat the process with another pair of triangles. You should notice a pattern.

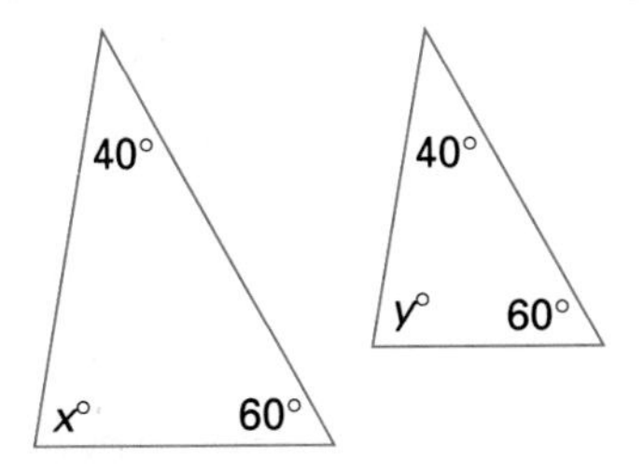

Theorem 4.2 **Third Angle Theorem:** If two angles of one triangle are congruent to two angles of a second triangle, then the third angles of the triangles are congruent.

Given: $\angle A \cong \angle D$, $\angle B \cong \angle E$
Prove: $\angle C \cong \angle F$

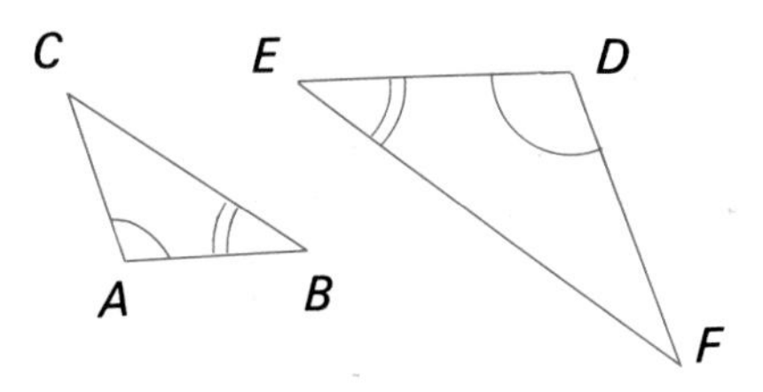

Proof

Statement	Reason
1. $m\angle A = m\angle D$, $m\angle B = m\angle E$	1. Given
2. $m\angle A + m\angle B + m\angle C = 180$	2. Sum of meas of $\angle$s of a $\triangle$ = 180
3. $m\angle D + m\angle E + m\angle F = 180$	3. Sum of meas of $\angle$s of a $\triangle$ = 180
4. $m\angle A + m\angle B + m\angle C = m\angle D + m\angle E + m\angle F$	4. Sub
5. $m\angle A + m\angle B + m\angle C = m\angle A + m\angle B + m\angle F$	5. Sub
6. $\therefore m\angle C = m\angle F$	6. Subt Prop of Eq
7. $\angle C \cong \angle F$	

EXAMPLE 1

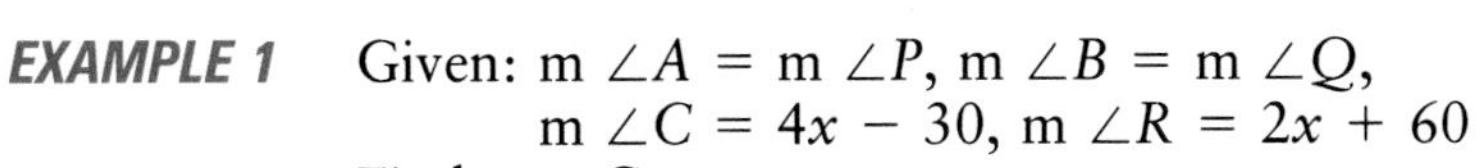
Given: $m\angle A = m\angle P$, $m\angle B = m\angle Q$,
$m\angle C = 4x - 30$, $m\angle R = 2x + 60$
Find $m\angle C$.

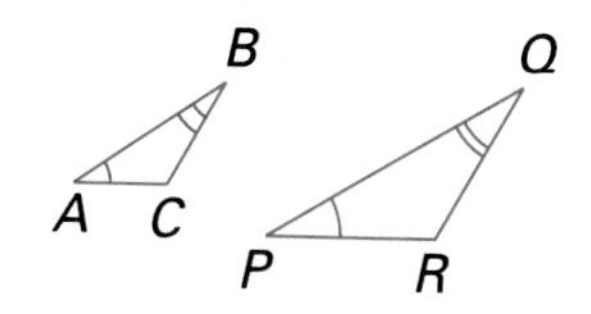

Solution

$m\angle C = m\angle R$ by the Third Angle Theorem

$$4x - 30 = 2x + 60$$
$$2x - 30 = 60$$
$$2x = 90$$
$$x = 45$$

$$m\angle C = 4x - 30$$
$$= 4 \cdot 45 - 30$$
$$= 150$$

So far, only three congruence patterns have been used to prove triangles congruent: SSS, SAS, and ASA. Other possibilities of side and angle combinations should be considered. Recall that a statement can be disproved by finding a counterexample.

Is AAA a congruence pattern? Consider the angles and sides in the figure. $\angle A \cong \angle X$, $\angle B \cong \angle Y$, and $\angle C \cong \angle Z$. Yet, corresponding sides are not congruent. The triangles have the same shape, but not the same size. Therefore, the AAA pattern *does not* guarantee triangle congruence.

Is SSA a congruence pattern? Once again, a counterexample can be used to show that this property is not always true. For the two triangles at the right, $\angle A \cong \angle D$, $\overline{AC} \cong \overline{DF}$, and $\overline{CB} \cong \overline{FE}$. Yet, the triangles have different shapes. Therefore, the triangles are not congruent. The SSA pattern *does not* guarantee congruence either.

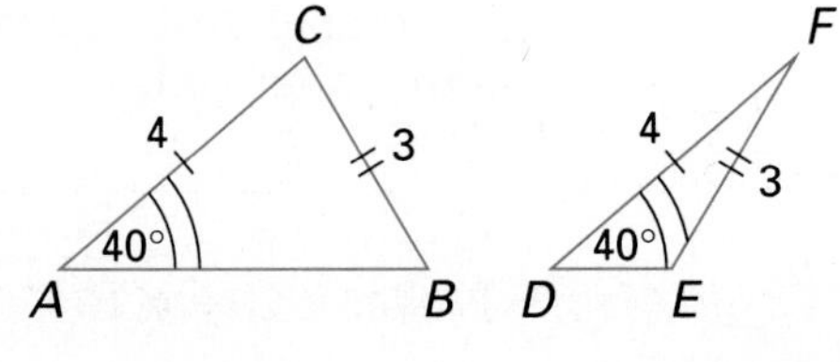

Is AAS a congruence pattern?

Since $\angle P \cong \angle A$ and $\angle Q \cong \angle B$, $\angle R \cong \angle C$ by the Third Angle Theorem. So the triangles *are* congruent, by ASA. Thus, the AAS pattern guarantees congruence, as stated in the next theorem.

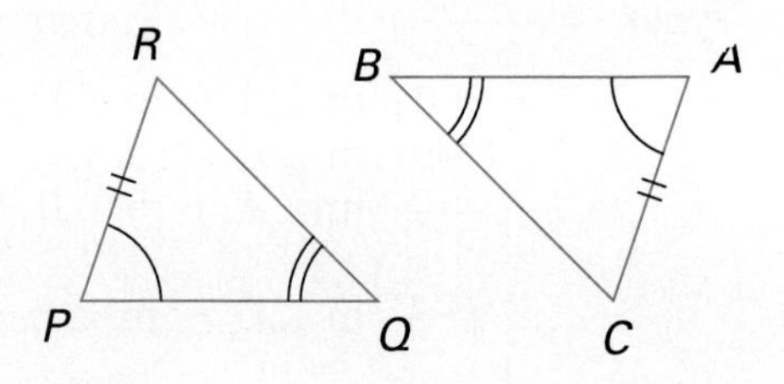

Theorem 4.3

AAS Theorem: If two angles and a non-included side of one triangle are congruent to the corresponding two angles and side of a second triangle, then the triangles are congruent.

You will be asked to prove this theorem in Exercise 17. Use the figure above for reference.

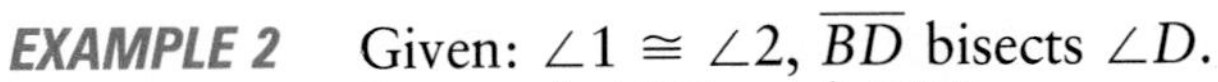

EXAMPLE 2 Given: $\angle 1 \cong \angle 2$, $\overline{BD}$ bisects $\angle D$.
Prove: $\triangle ABD \cong \triangle CBD$

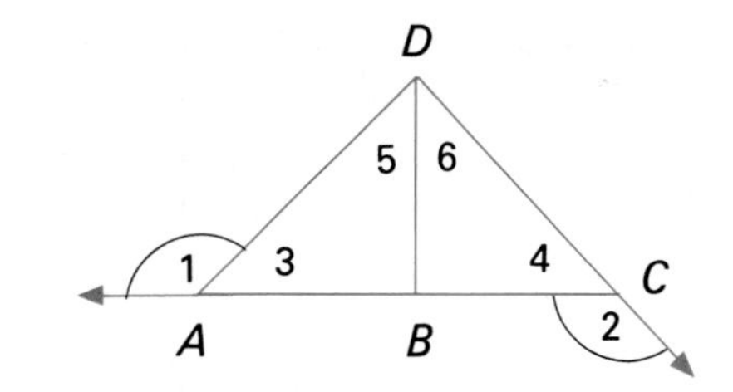

Proof

	Statement		Reason
	1. $\overline{BD}$ bisects $\angle D$.		1. Given
	2. $\angle 5 \cong \angle 6$	(A)	2. Def of $\angle$ bis
	3. $\overline{BD} \cong \overline{BD}$	(S)	3. Reflex Prop
	4. $\angle 1 \cong \angle 2$		4. Given
	5. $\angle 1$ and $\angle 3$ are supplementary; $\angle 2$ and $\angle 4$ are supplementary.		5. If outer rays of two adj $\angle$s form a st $\angle$, then the $\angle$s are supp.
	6. $\angle 3 \cong \angle 4$	(A)	6. Supp of $\cong$ $\angle$s are $\cong$.
	7. $\therefore \triangle ABD \cong \triangle CBD$		7. AAS

Summary

You *can* prove triangles are congruent by establishing these congruences.

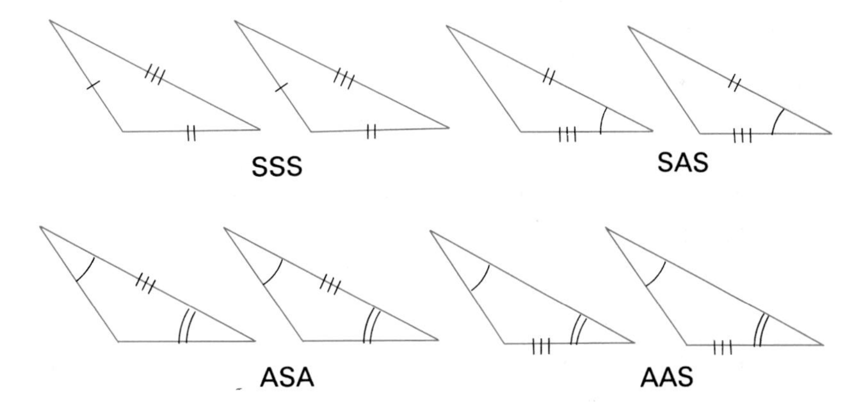

You *cannot* prove triangles congruent by establishing these congruences.

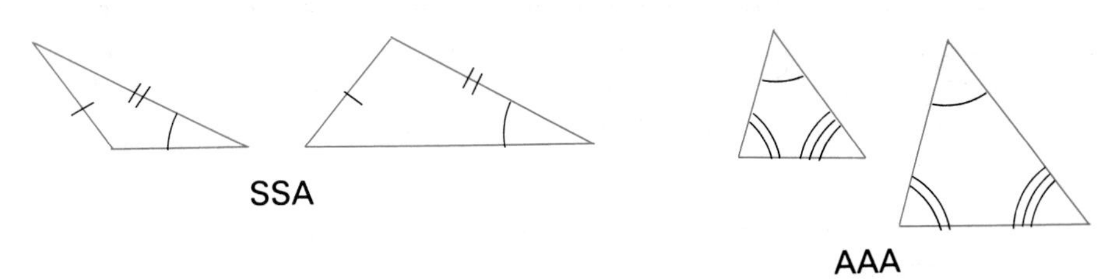

Focus on Reading

Indicate whether or not the set of given data guarantees congruence of two triangles.

1. congruence of three pairs of sides and three pairs of angles
2. congruence of three pairs of sides
3. congruence of three pairs of angles
4. congruence of two pairs of angles and any one pair of sides
5. congruence of two pairs of sides and any pair of non-included angles

Classroom Exercises

Based on the markings shown, determine whether the two triangles are congruent. State why or why not.

1. 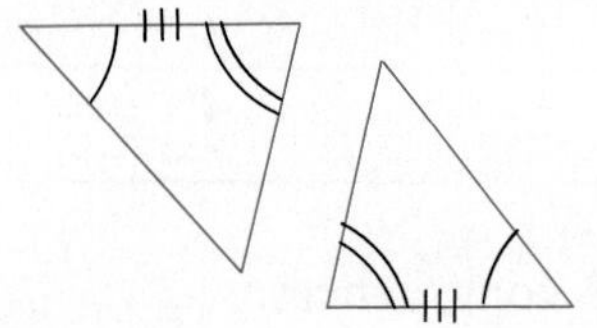

2. 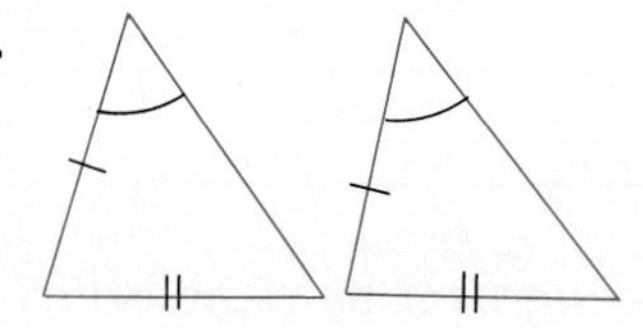

3. 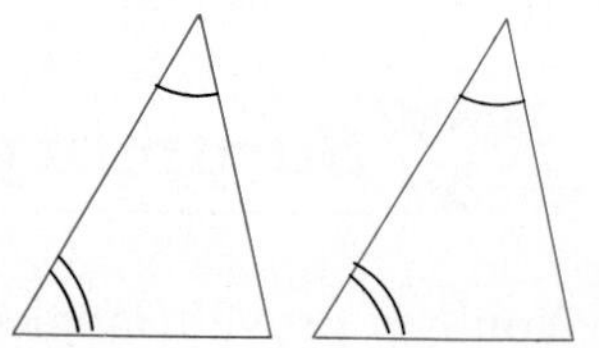

Written Exercises

Based on the given information, determine whether the triangles are congruent. State why or why not. (Exercises 1–4)

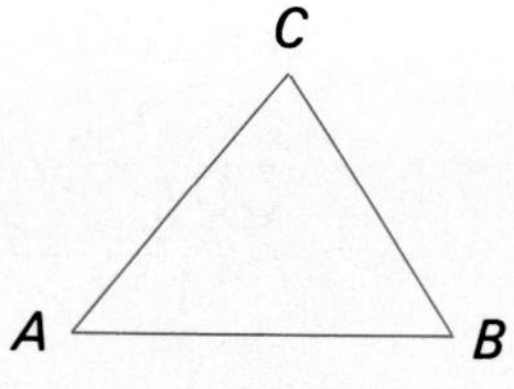

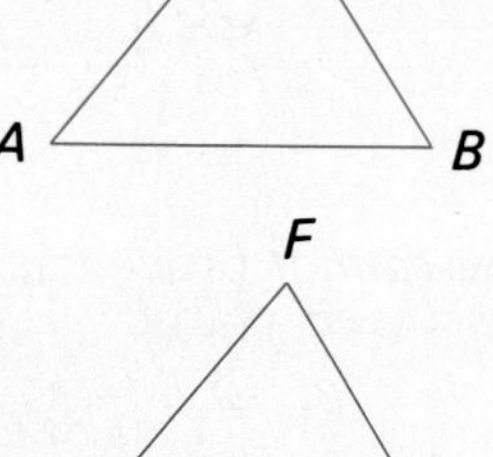

1. $\overline{AC} \cong \overline{DF}$, $\overline{BC} \cong \overline{EF}$, $\angle C \cong \angle F$
2. $\overline{FE} \cong \overline{CB}$, $\angle A \cong \angle D$, $\angle E \cong \angle B$
3. $\angle A \cong \angle D$, $\angle B \cong \angle E$, $\angle C \cong \angle F$
4. $\angle A \cong \angle D$, $\angle B \cong \angle E$, $\overline{AB} \cong \overline{DE}$

Find the indicated angle measure. (Exercises 5–7)

5. Given: $\angle A \cong \angle D$, $\angle B \cong \angle E$, $m\angle C = 4x - 20$, $m\angle F = 2x + 10$. Find $m\angle C$.
6. Given: $m\angle A = m\angle D$, $m\angle C = m\angle F$, $m\angle B = 4y + 10$, $m\angle E = y + 70$. Find $m\angle E$.
7. Given: $m\angle A = m\angle D$, $m\angle B = m\angle E$, $m\angle C = 2x - 30$, $m\angle F = x + 5$. Find $m\angle C$.

Complete the proof. (Exercises 8–9)

8. Given: $\overline{AD} \parallel \overline{CB}$, $\angle A \cong \angle C$
Prove: $\triangle ADB \cong \triangle CBD$

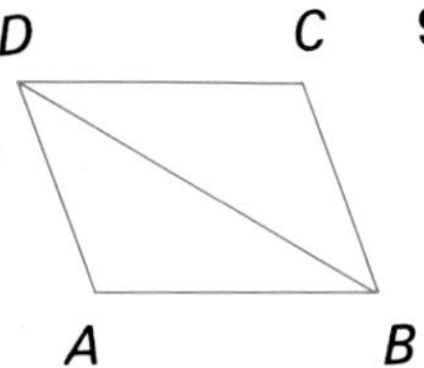

Statement	Reason
1. $\angle A \cong \angle C$	(A) 1. Given
2. $\overline{AD} \parallel \overline{CB}$	2. Given
3. ______ $\cong$ ______	(A) 3. Alt int $\angle$s of $\parallel$ lines are $\cong$.
4. $\overline{DB} \cong \overline{DB}$	(S) 4. __________
5. $\triangle ADB \cong \triangle CBD$	5. ______

9. Given: $\angle R \cong \angle S$, $\overline{PQ}$ bisects $\overline{RS}$; $\overline{PQ}$ bisects $\angle RPS$.
Prove: $\triangle PQR \cong \triangle PQS$

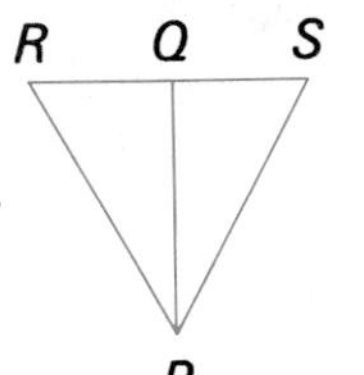

Statement	Reason
1. $\angle R \cong \angle S$	(A) 1. Given
2. $\overline{PQ}$ bisects $\overline{RS}$.	2. Given
3. $\overline{RQ} \cong \overline{SQ}$	(S) 3. __________
4. $\overline{PQ}$ bisects $\angle RPS$.	4. Given
5. $\angle RPQ \cong \angle SPQ$	(A) 5. __________
6. ______ $\cong$ ______	6. ______

10. Given: $\overline{AC}$ bisects $\overline{DE}$; $\angle A \cong \angle C$.
Prove: $\triangle ADB \cong \triangle CEB$

11. Given: $\overline{PQ} \perp \overline{RS}$, $\angle R \cong \angle S$
Prove: $\triangle PQR \cong \triangle PQS$

12. Given: $\overline{AB} \parallel \overline{DC}$, $\overline{AD} \parallel \overline{CB}$
Prove: $\triangle ADB \cong \triangle CBD$

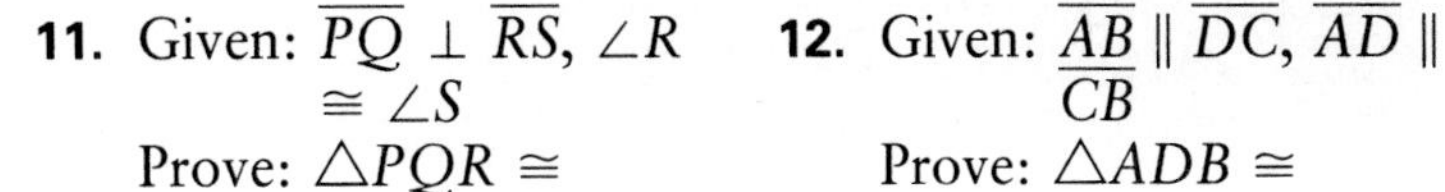

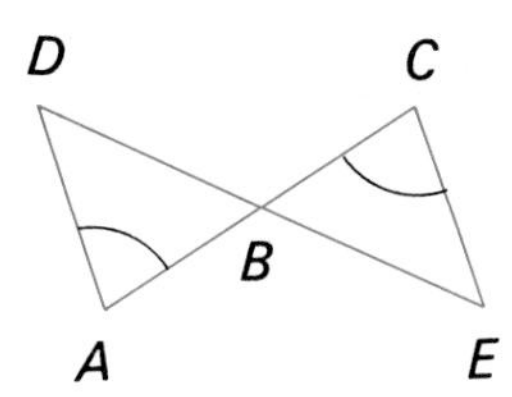

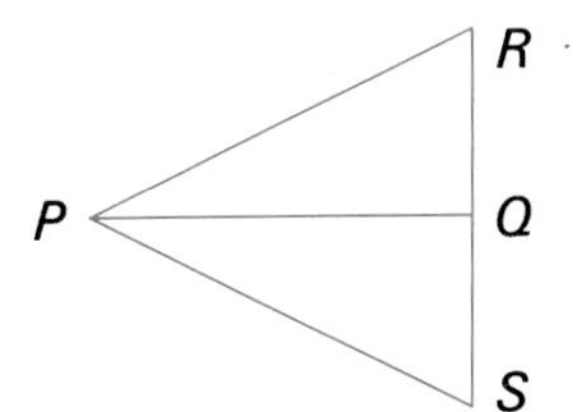

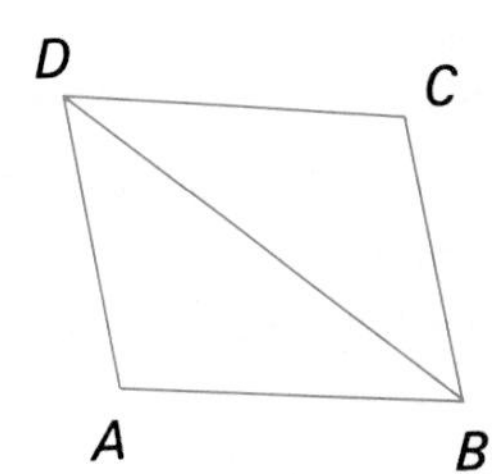

13. Given: $\angle P \cong \angle S$, $\overline{PU} \cong \overline{SR}$, $\overline{PR} \parallel \overline{US}$, $\overline{QU} \parallel \overline{TR}$
Prove: $\triangle PUQ \cong \triangle SRT$

14. Given: $\overline{PR} \cong \overline{US}$, $\overline{QR} \cong \overline{UT}$, $\angle 3 \cong \angle 4$, $\overline{RT} \parallel \overline{QU}$, $\angle P \cong \angle S$
Prove: $\triangle PUQ \cong \triangle SRT$

15. Given: $\overline{AB} \perp \overline{BE}$, $\overline{EF} \perp \overline{BE}$, $\angle 1 \cong \angle 2$, $\angle 3 \cong \angle 4$, $\overline{BD} \cong \overline{EC}$
Prove: $\triangle BCG \cong \triangle EDH$

16. Given: $\overline{CG} \parallel \overline{DH}$, $\overline{AB} \parallel \overline{FE}$, $\angle 1 \cong \angle 2$, $\overline{HE} \cong \overline{BG}$
Prove: $\triangle BCG \cong \triangle EDH$

17. Prove Theorem 4.3, the AAS Theorem.
18. Given: $\triangle ABC$ with points D and E on $\overline{AB}$ such that D is between A and E, $\overline{DC}$ and $\overline{EC}$ drawn such that $\angle CDE \cong \angle CED$, $\overline{AE} \cong \overline{BD}$, $\angle ACD \cong \angle BCE$.
Prove: $\triangle ADC \cong \triangle BEC$

Mixed Review

1. Draw an angle. Construct the bisector of this angle. *1.5*
2. The measures of two angles formed by the bisector of a given angle are $4x - 20$ and $x + 40$. Find the measures of the angles formed. *1.5*
3. The measure of an angle is three times the measure of its complement. Find the measures of the two angles. *1.6*
4. The measure of one acute angle of a right triangle is $\frac{2}{3}$ the measure of the other acute angle. Find the measure of each acute angle. *4.1*

Algebra Review

To solve a system of two equations with two variables by substitution:
(1) Solve for one variable in one equation.
(2) Substitute the expression for that variable in the second equation.
(3) Solve the resulting equation.
(4) Substitute the result from (3) in either one of the original equations and solve for the remaining variable.
(5) Check by substituting the values into both original equations.

Example: Solve and check. $a - 2b = 11$
$5a + 4b = 27$

(1) $a - 2b = 11$ — Solve for a.
$a = 2b + 11$
(2) $5a + 4b = 27$ — Substitute $(2b + 11)$ for a.
$5(2b + 11) + 4b = 27$
(3) $10b + 55 + 4b = 27$ — Solve for b.
$b = -2$
(4) $a - 2(-2) = 11$ — Substitute (-2) for b.
$a = 7$ — Solve for a.
(5) Check. $7 - (2)(-2) \stackrel{?}{=} 11$ — $5(7) + 4(-2) \stackrel{?}{=} 27$
$7 + 4 = 11$ ✓ — $35 - 8 = 27$ ✓

Solve and check.

1. $2a + b = 12$
$a + 2b = 9$
2. $-3p - q = 13$
$2p + 3q = -4$
3. $x - 2y = 8$
$2y = 3x - (-16)$

4.7 Applying Congruence: Corresponding Parts

Objectives

To prove congruence of sides and angles by first proving congruence of triangles

To prove that segments are parallel by first proving that triangles are congruent

Recall from the definition of triangle congruence that Corresponding Parts of Congruent Triangles are Congruent. This is abbreviated as CPCTC. If two triangles can be proved congruent, then any pair of corresponding sides or pair of corresponding angles are congruent.

EXAMPLE 1 Given: $\overline{EF}$ bisects $\angle BEC$; $\overline{BE} \cong \overline{CE}$.
Prove: $\angle 1 \cong \angle 2$

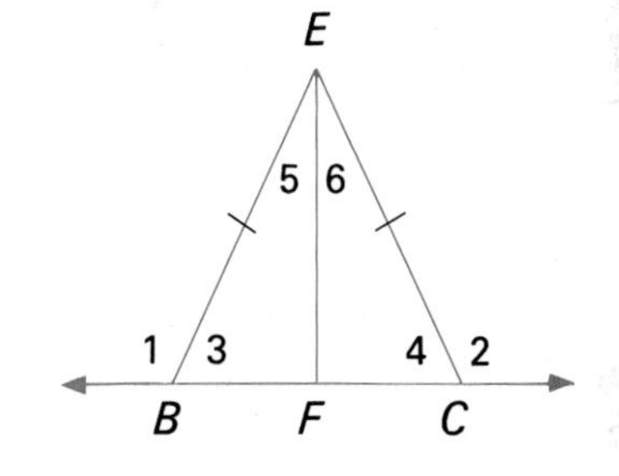

Plan $\angle 1$ and $\angle 2$ are supplements, respectively, of $\angle 3$ and $\angle 4$. Use SAS to prove the triangles congruent. Then, $\angle 3 \cong \angle 4$ by CPCTC. Finally, show that $\angle 1 \cong \angle 2$.

Proof

	Statement		Reason
1.	$\overline{BE} \cong \overline{CE}$	(S) 1.	Given
2.	$\overline{EF}$ bisects $\angle BEC$.	2.	Given
3.	$\angle 5 \cong \angle 6$	(A) 3.	Def of $\angle$ bis
4.	$\overline{EF} \cong \overline{EF}$	(S) 4.	Reflex Prop
5.	$\triangle BFE \cong \triangle CFE$	5.	SAS
6.	$\angle 3 \cong \angle 4$	6.	CPCTC
7.	$\angle 1$ and $\angle 3$ are supplementary; $\angle 2$ and $\angle 4$ are supplementary.	7.	If the outer rays of two adj $\angle$s form a st $\angle$, then the $\angle$s are supp.
8.	$\therefore \angle 1 \cong \angle 2$	8.	Supp of $\cong$ $\angle$s are $\cong$.

Lines can be proved parallel by showing that a pair of alternate interior angles is congruent. Sometimes this can be done by showing that the angles are corresponding parts of congruent triangles.

EXAMPLE 2 Given: $\overline{AD} \parallel \overline{CB}$, $\overline{AD} \cong \overline{CB}$
Prove: $\overline{CD} \parallel \overline{AB}$

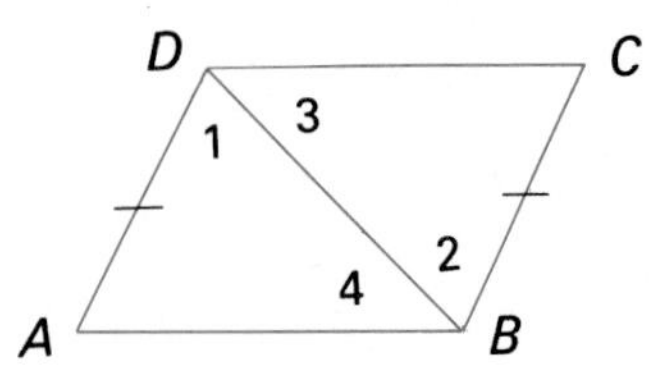

Proof

	Statement		Reason
	1. $\overline{AD} \cong \overline{CB}$	(S)	1. Given
	2. $\overline{AD} \parallel \overline{CB}$		2. Given
	3. $\angle 1 \cong \angle 2$	(A)	3. Alt int $\angle$s of $\parallel$ lines are $\cong$.
	4. $\overline{DB} \cong \overline{BD}$	(S)	4. Reflex Prop
	5. $\triangle DAB \cong \triangle BCD$		5. SAS
	6. $\angle 3 \cong \angle 4$		6. CPCTC
	7. $\therefore \overline{CD} \parallel \overline{AB}$		7. If alt int $\angle$s are $\cong$, then lines are $\parallel$.

Classroom Exercises

Provide a reason for each step in the proof below.
Given: $\overline{AB} \parallel \overline{CD}$, $\overline{AB} \cong \overline{CD}$
Prove: E is the midpoint of $\overline{AD}$.

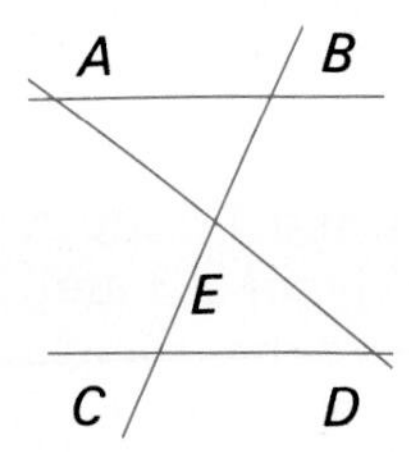

Statement		Reason
1. $\overline{AB} \parallel \overline{CD}$		**1.** ________
2. $\angle BAE \cong \angle CDE$	(A)	**2.** ________
3. $\overline{AB} \cong \overline{CD}$	(S)	**3.** ________
4. $\angle ABE \cong \angle DCE$	(A)	**4.** ________
5. $\triangle ABE \cong \triangle DCE$		**5.** ________
6. $\overline{AE} \cong \overline{DE}$		**6.** ________
7. E is the midpoint of $\overline{AD}$.		**7.** ________

Written Exercises

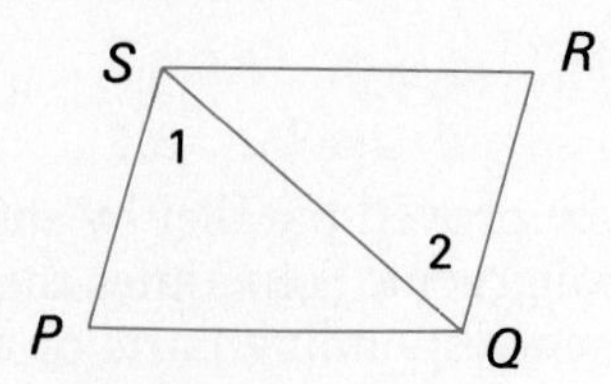

1. Given: $\overline{PS} \cong \overline{RQ}$, $\overline{SR} \cong \overline{QP}$
Prove: $\angle 1 \cong \angle 2$

2. Given: $\overline{PS} \parallel \overline{RQ}$, $\overline{PQ} \parallel \overline{RS}$
Prove: $\angle P \cong \angle R$

3. Given: $\overline{AC}$ and $\overline{DE}$ bisect each other.
Prove: $\overline{AD} \parallel \overline{CE}$

4. Given: $\overline{AD} \parallel \overline{CE}$, B is the midpoint of $\overline{AC}$.
Prove: B is the midpoint of $\overline{DE}$.

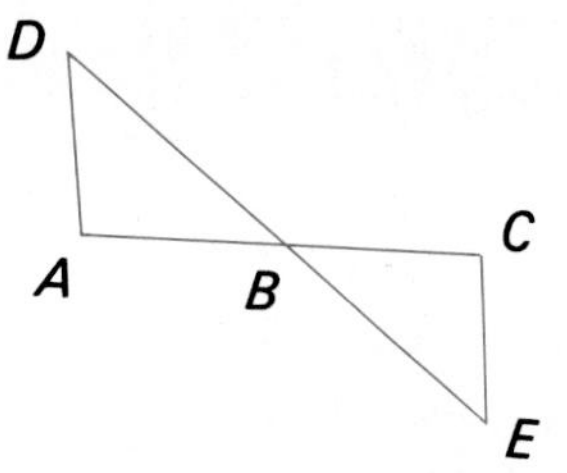

5. Given: $\overline{FS} \cong \overline{UQ}$,
$\overline{FP} \cong \overline{UR}$,
$\overline{FP} \parallel \overline{UR}$
Prove: $\overline{PQ} \parallel \overline{RS}$

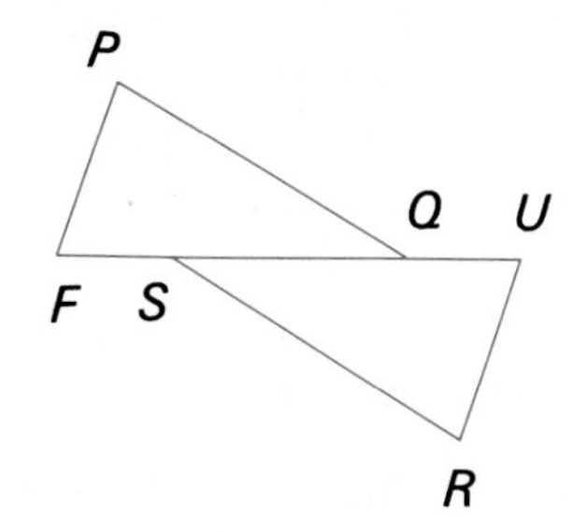

6. Given: $\overline{FP} \parallel \overline{UR}$,
$\overline{PQ} \parallel \overline{RS}$,
$\overline{PQ} \cong \overline{RS}$
Prove: $\overline{FQ} \cong \overline{US}$

7. Given: $\overline{TQ} \perp \overline{PR}$
$\overline{TQ} \perp \overline{US}$,
$\angle 1 \cong \angle 2$
Prove: $\overline{QU} \cong \overline{QS}$

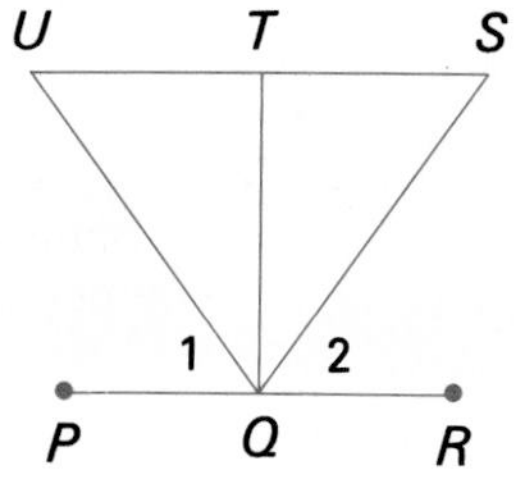

8. Given: $\overline{TQ}$ bisects $\angle UQS$;
$\overline{UQ} \cong \overline{SQ}$.
Prove: $\overline{TQ} \perp \overline{US}$

9. Given: $\overline{CB} \perp$ every line in plane $\mathcal{M}$ that passes through B,
$\angle 1 \cong \angle 2$, $\overline{DB} \cong \overline{IB}$
Prove: $\angle 5 \cong \angle 6$

10. Given: $\overline{BC} \perp$ every line in plane $\mathcal{N}$ that passes through C,
$\angle 3 \cong \angle 4$, $\overline{ED} \cong \overline{GI}$, $\overline{EC} \cong \overline{GC}$.
Prove: $\overline{BD} \cong \overline{BI}$

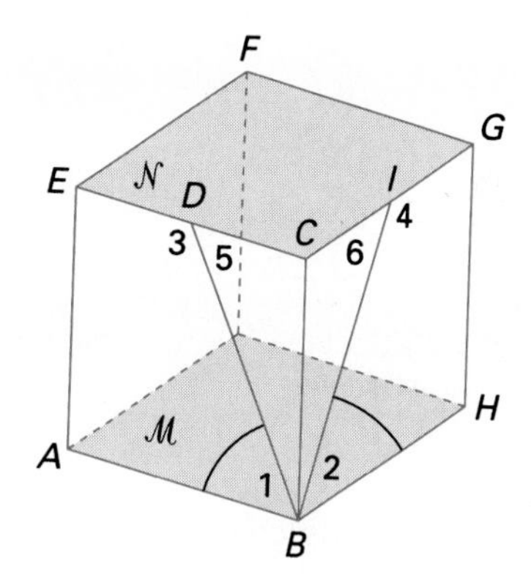

Mixed Review

In the figure, $\overline{QT} \parallel \overline{RS}$

1. Given: m $\angle 3 = 60$. Find m $\angle 5$. *3.5*

2. Given: m $\angle 2 = 70$. Find m $\angle QRS$. *3.5*

3. Given: m $\angle 3 = 75$, m $\angle 1 = 140$. Find m $\angle 4$. *3.6*

4. Given: m $\angle 2 = 40$, m $\angle 3$ is 4 times m $\angle 4$.
Find m $\angle 3$. *3.7*

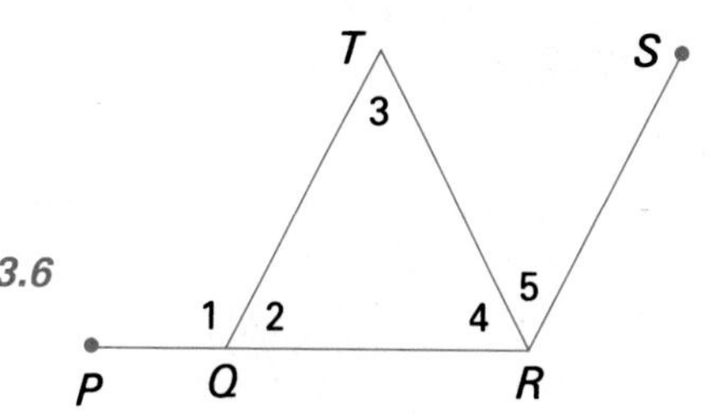

Chapter 4 Review

Key Terms

acute triangle (p. 139)
congruent triangles (p. 143)
corresponding parts of
congruent angles (p. 143)
equiangular triangle (p. 139)
equilateral triangle (p. 139)
included angle (p. 147)
included side (p. 154)
isosceles triangle (p. 139)
obtuse triangle (p. 139)
right triangle (p. 139)
scalene triangle (p. 139)

Key Ideas and Review

4.1 To classify triangles by angles, remember the following.

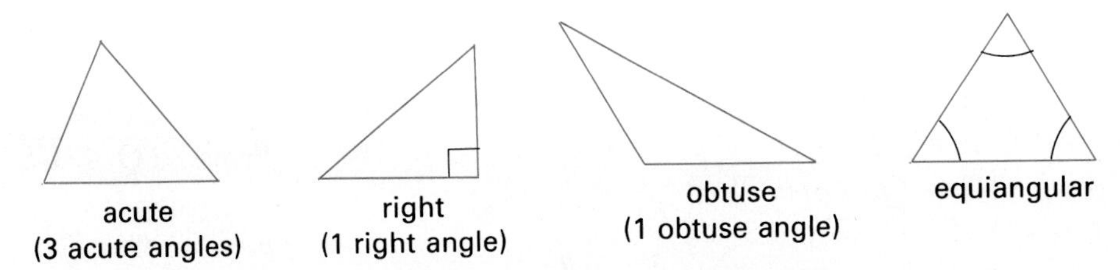

To classify triangles by sides, remember the following.

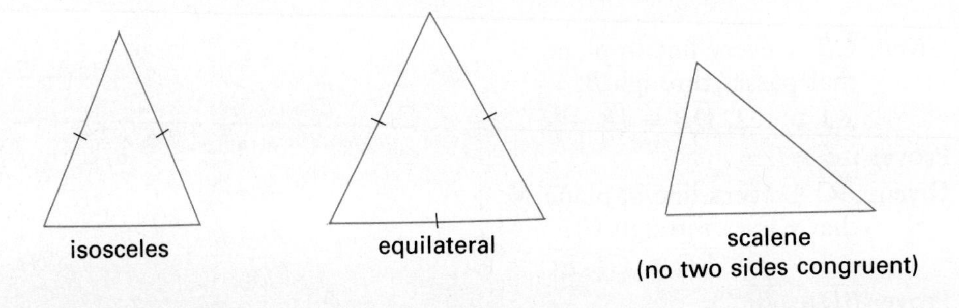

Classify each triangle by its sides.

1. 3 cm, 2 cm, 2.3 cm **2.** 3 ft, 3 ft, 3 ft **3.** 3 cm, 2 cm, 3 cm

Classify the triangle by its angles. (Exercises 4–5)

4. The measures of two angles are 35 and 55.
5. The measures of two angles are 70 and 50.
6. The measure of one acute angle of a right triangle is $\frac{3}{2}$ the measure of the other acute angle. Find the measure of each acute angle.
7. $\triangle ABC$ is isosceles with $AC = AB$. $AC = 4x - 2$, $AB = 2x + 6$, $BC = 3x - 1$. Find BC.

4.2, 4.6 To prove triangles congruent, use one of the following four congruence patterns.

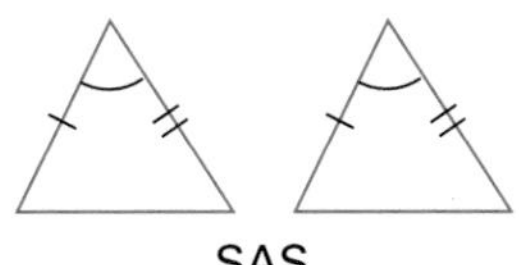

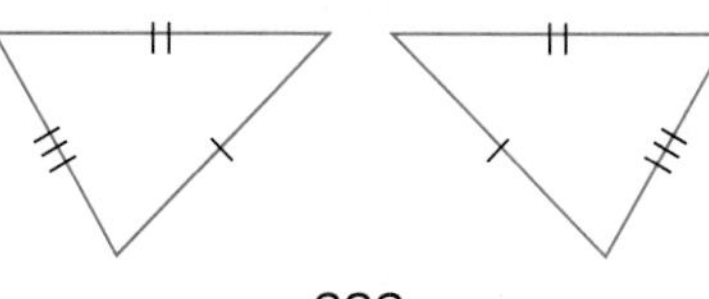

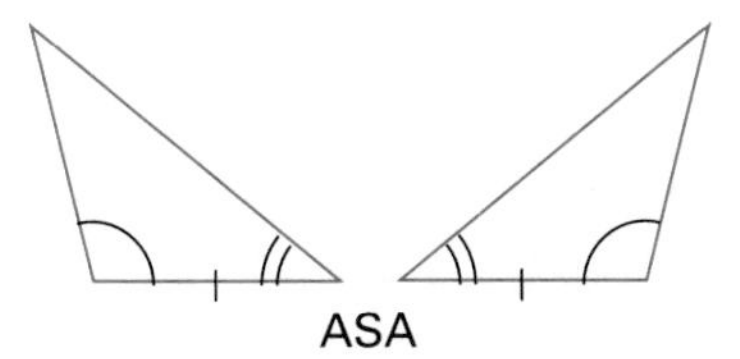

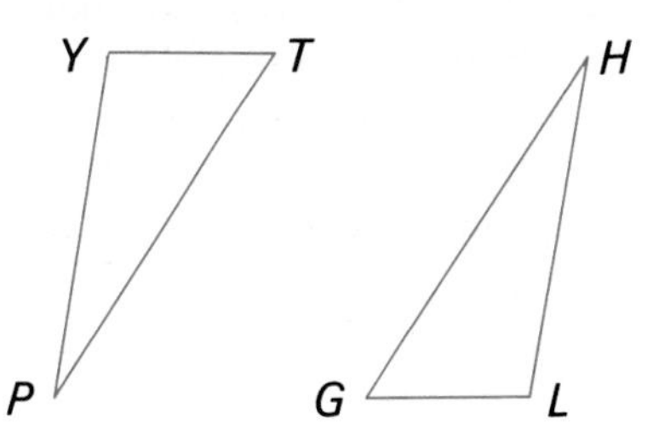

4.7 To prove corresponding parts of triangles congruent, first show that the triangles are congruent and then use CPCTC.

Based on the given information, determine whether the two triangles are congruent. State the reason. (Exercises 8–10)

8. $\angle Y \cong \angle L$, $\angle P \cong \angle H$, $\angle T \cong \angle G$

9. $\overline{YT} \cong \overline{LG}$, $\overline{PT} \cong \overline{HG}$, $\angle T \cong \angle G$

10. $\angle P \cong \angle H$, $\angle T \cong \angle G$, $\overline{PT} \cong \overline{HG}$

11. Given: $\angle Y \cong \angle L$, $\angle T \cong \angle G$, $m\angle P = 3x - 40$, $m\angle H = x + 20$
Find $m\angle P$.

12. Given: $\angle S \cong \angle R$, $\overline{PQ}$ bisects $\angle SQR$.
Prove: $\triangle SPQ \cong \triangle RPQ$

13. Given: $\overline{TU} \cong \overline{GY}$, $\overline{KY} \parallel \overline{HU}$, $\overline{KT} \perp \overline{TG}$, $\overline{HG} \perp \overline{TG}$
Prove: $\angle K \cong \angle H$

14. Given: $\overline{MQ} \parallel \overline{WL}$, $\overline{MQ} \cong \overline{WL}$
Prove: $\overline{ML} \parallel \overline{WQ}$

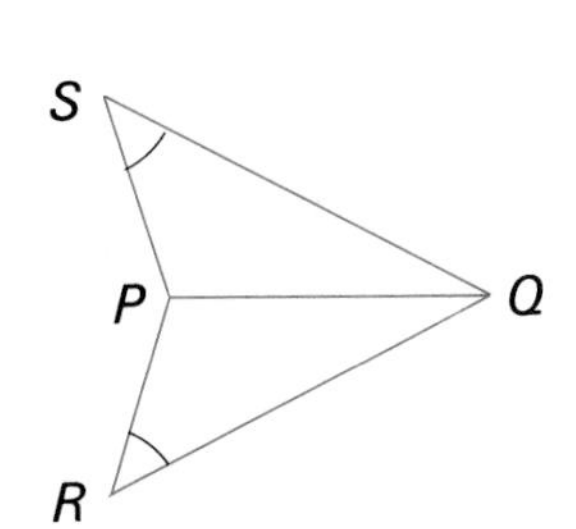

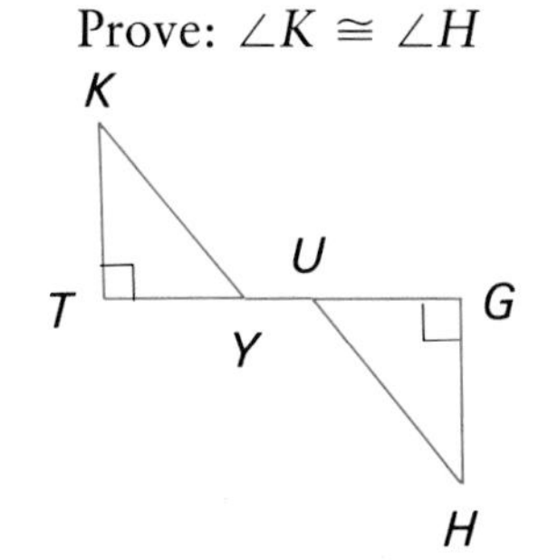

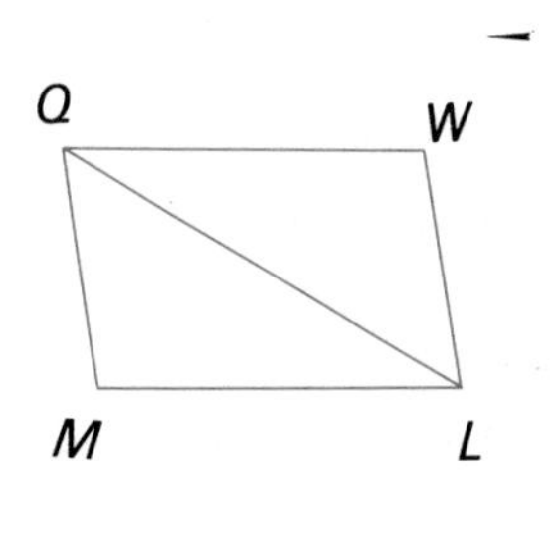

4.3, 4.4 To construct a triangle congruent to a given triangle, see pages 147 and 153.

15. Draw an acute triangle. Construct a triangle congruent to it, using the SSS pattern.

Chapter 4 Test

Classify the triangle by its sides and by its angles.

1. 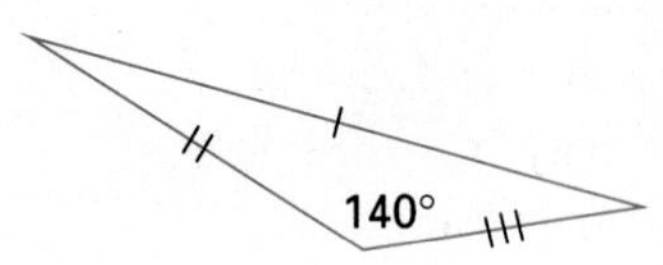

2. 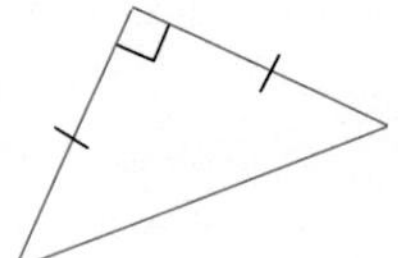

3. The measure of one acute angle of a right triangle is $\frac{3}{7}$ of the measure of the other acute angle. Find the measure of each acute angle.
4. Given: $\triangle ABC$ is isosceles with $AB = BC$, $AB = 12$, $BC = 3x + 6$, $AC = x^2 + 3x$. Find AC.
5. Given: $\angle A \cong \angle D$, $\angle B \cong \angle E$, $m\ \angle C = 4y + 20$, $m\ \angle F = y + 80$. Find $m\ \angle C$.
6. Draw an obtuse triangle. Construct a triangle congruent to this triangle, using an SAS pattern.

Based on the markings shown, determine whether the triangles are congruent. Provide a reason.

7. 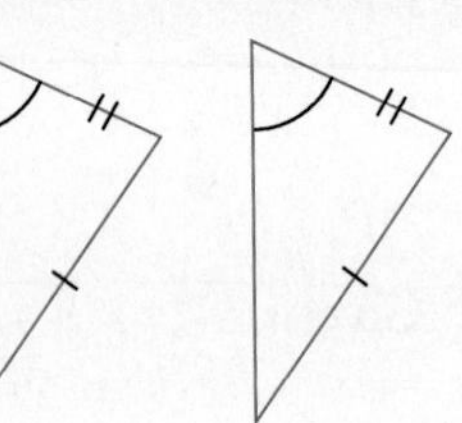

8. 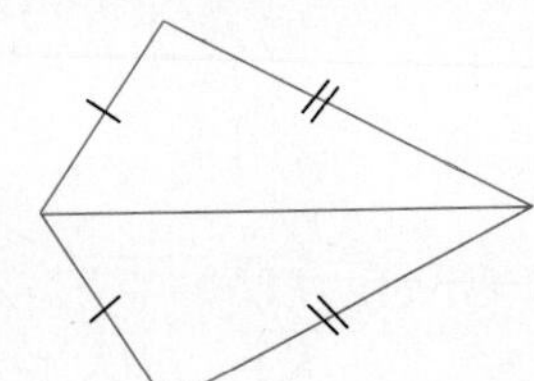

9. 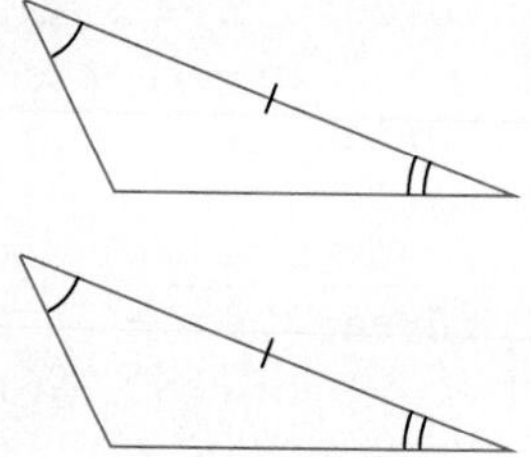

10. Given: $\overline{PQ}$ bisects $\angle MPT$; $\overline{PQ}$ bisects $\angle MQT$.
Prove: $\triangle PMQ \cong \triangle PTQ$

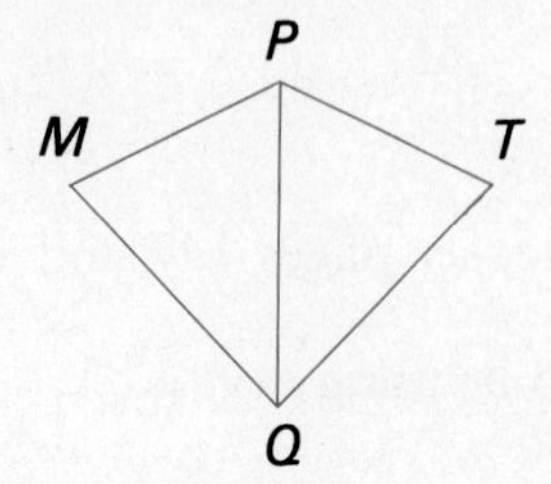

11. Given: $\overline{PT} \cong \overline{RT}$; $\overline{ST}$ bisects $\overline{PR}$.
Prove: $\overline{ST}$ bisects $\angle PTR$.

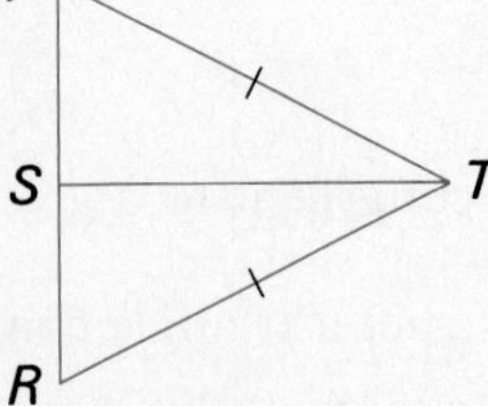

*12. Given: $\overline{TW} \parallel \overline{NM}$; $\angle 1 \cong \angle 2$; $\overline{WY} \cong \overline{MX}$
Prove: $\angle T \cong \angle N$

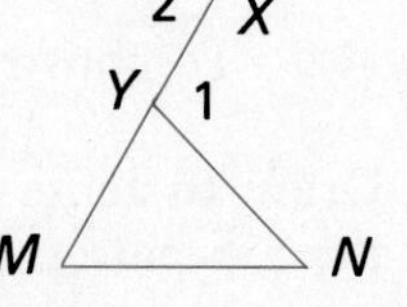

College Prep Test

In each item you are to compare a quantity in Column 1 with a quantity in Column 2. Write the letter of the correct answer from these choices:

A—The quantity in Column 1 is greater than the quantity in Column 2.
B—The quantity in Column 2 is greater than the quantity in Column 1.
C—The quantity in Column 1 is equal to the quantity in Column 2.
D—The relationship cannot be determined from the given information.

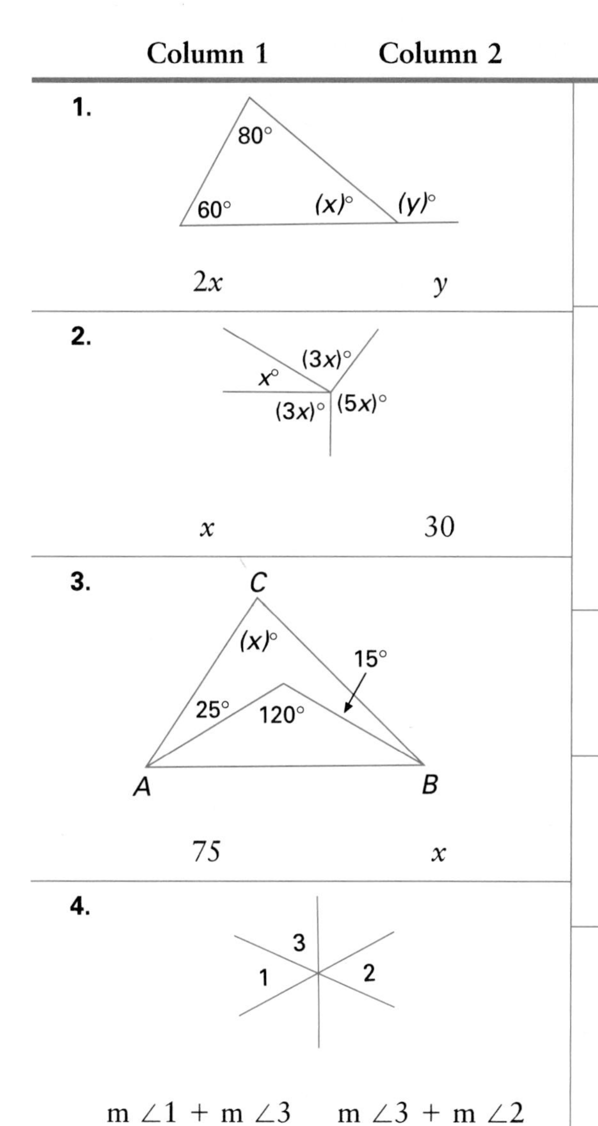

	Column 1	Column 2
1.	$2x$	y
2.	x	30
3.	75	x
4.	m ∠1 + m ∠3	m ∠3 + m ∠2

5.

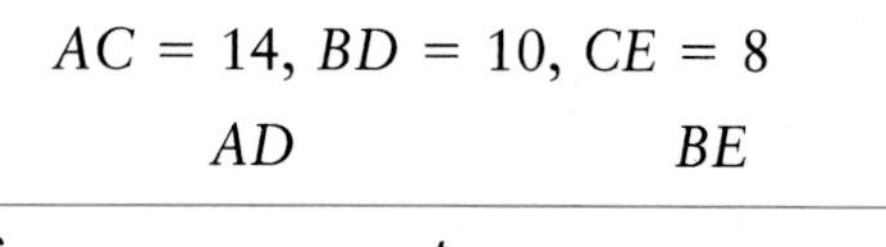

$AC = 14$, $BD = 10$, $CE = 8$

Column 1	Column 2
AD	BE

6.

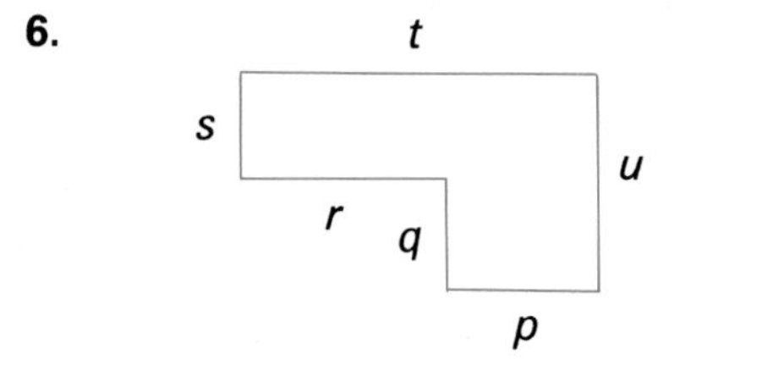

Column 1	Column 2
$p + q + r + s$	$t + u$

7. $k \neq 0$

Column 1	Column 2
$\frac{1}{k^2}$	$\left(\frac{1}{k}\right)^2$

8. x is a positive number.

Column 1	Column 2
x plus an increase of 30% of x	$0.3x$

9. On a certain day, 90% of the girls and 80% of the boys were present in geometry class.

Column 1	Column 2
Number of boys absent	Number of girls absent

Cumulative Review (Chapters 1–4)

Choose the best possible answer from A–E. (Exercises 1–4)

1. Given: Coordinates of the endpoints of a segment are -8 and 12. Find the coordinate of the midpoint. *1.3*
(A) 2 (B) 20 (C) 4
(D) -20 (E) None of these

2. If $\angle A$ is acute and m $\angle A = 2x$, then the measure of its complement is ______. *1.6*
(A) 180 (B) 90
(C) $90 - 2x$ (D) $180 - 2x$
(E) $2x - 90$

3. Given: lines m and n are coplanar. Which of the following relationships is not possible? *3.1*
(A) $m \parallel n$ (B) m and n intersect
(C) $m \perp n$ (D) m and n skew
(E) m and n form vertical angles.

4. The contrapositive of $p \rightarrow q$ is: *3.8*
(A) $q \rightarrow p$ (B) $\sim p \rightarrow q$
(C) $p \rightarrow \sim q$ (D) $\sim q \rightarrow \sim p$
(E) None of these

5. Graph and name the resulting geometric measure. *1.8*
$x \geq 6$ *and* $x \leq 10$

6. Draw a segment. Construct its midpoint. *1.3*

7. Given: $AB = CD$, *2.3*
$BP = DQ$
Prove: $AP = CQ$

8. Given: $AB = CD$
Prove: $CD + BP = AP$

9. Given: m $\angle 1 = 5x + 35$, *2.8*
m $\angle 3 = 70$
Find x.

10. Given: m $\angle 1 = 7x - 10$,
m $\angle 2 = 4x + 50$
Find m $\angle 2$.

11. Write the negation of the statement "Parallel lines are skew." Is the negation true? *3.8*

12. Write the converse of the statement "If two angles are vertical, then the angles are congruent." Is this converse true? *3.5*

Use the figure below for Exercises 13–15.

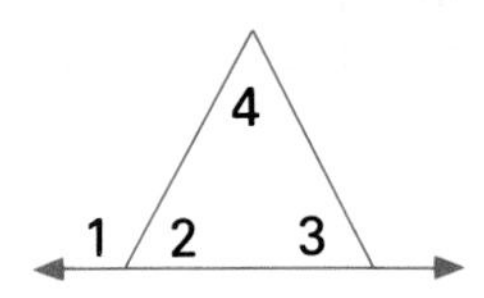

13. Given: m $\angle 1 = 140$, m $\angle 3 = 85$ *3.7*
Find m $\angle 4$.

14. Given: m $\angle 2 = 3x$, m $\angle 3 = 4x$, m $\angle 4 = 5x$ *3.6*
Find m $\angle 1$.

15. m $\angle 4 = 3x + 20$, m $\angle 3 = x + 10$, m $\angle 1 = 110$ *3.7*
Find m $\angle 3$.

For Exercises 16–18, use the diagram below and assume $\overleftrightarrow{CD} \parallel \overleftrightarrow{BA}$.

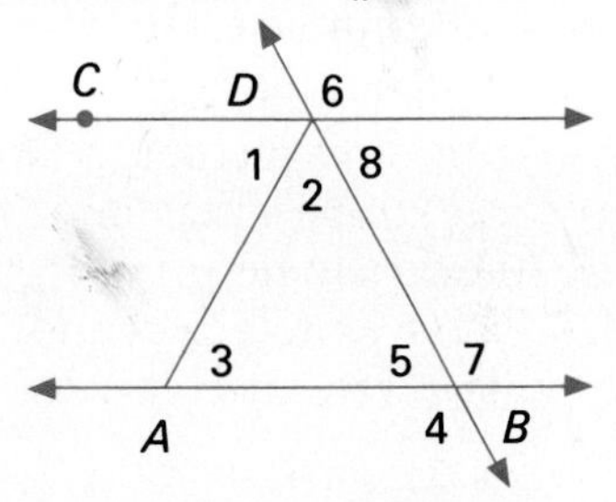

16. Given: m $\angle 4 = 100$. Find m $\angle 6$. *3.5*

17. Given: m $\angle 8 = \frac{4}{5}$ m $\angle 7$. Find m $\angle 8$. *3.5*

18. Given: $\overline{DA}$ bisects $\angle CDB$. *3.5*
Prove: $\angle 3 \cong \angle 2$

19. $AB = \frac{2}{3}BC$ and $AC = 25$. Find AB. *1.2*

20. Draw an angle. Construct the bisector of this angle. *1.5*

21. Write the conjunction and disjunction of the statements and determine if each is true or false. *1.7*
   a. Vertical angles are congruent.
   b. The sum of the measures of two supplementary angles is 90.

22. The measure of an angle is 3 times the measure of its supplement. Find the measure of each angle. *1.6*

23. How many different planes can contain one line? *3.1*

24. How many different planes can contain two intersecting lines?

State whether the statement is true or false. If false, draw a figure to show why. (Exercises 25–26)

25. Two planes parallel to the same line are parallel. *3.9*

26. Two planes intersected by a third plane are parallel.

27. Write the converse of "If two lines are parallel, then alternate interior angles are congruent." Then write a biconditional and determine if it is true. *3.8*

28. Given: $m \nparallel n$ *3.4*
   Prove: $\angle 1 \ncong \angle 2$ (Use an indirect proof.)

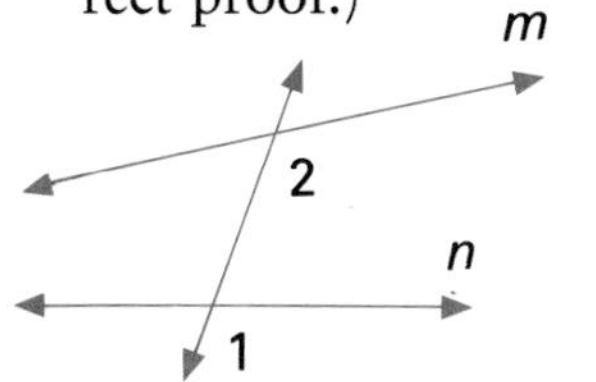

29. The measure of one acute angle of a right triangle is $\frac{1}{2}$ the measure of the other. Find the measure of each acute angle. *4.1*

Determine from the markings whether the two triangles are congruent. State why or why not. *4.3* *4.6*

30.

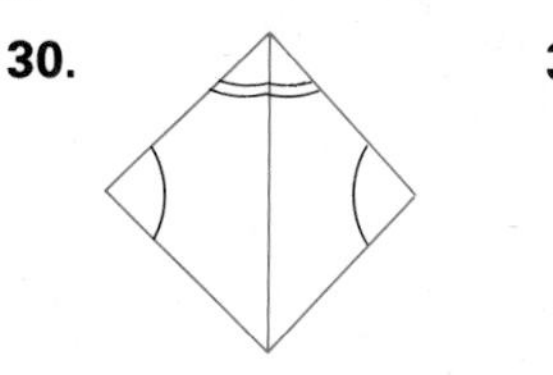

31.

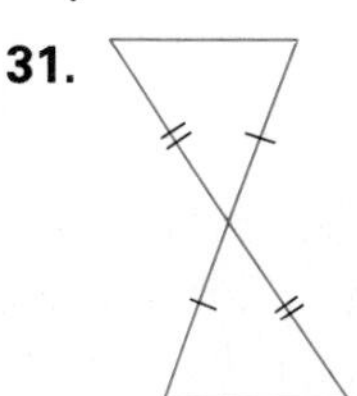

32. 33.

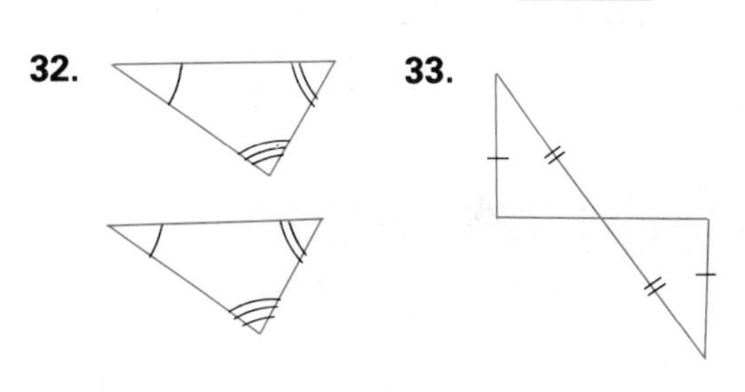

34. Given: $\overline{RM}$ and $\overline{LK}$ bisect each other. *4.7*
   Prove: $\overline{RL} \parallel \overline{MK}$

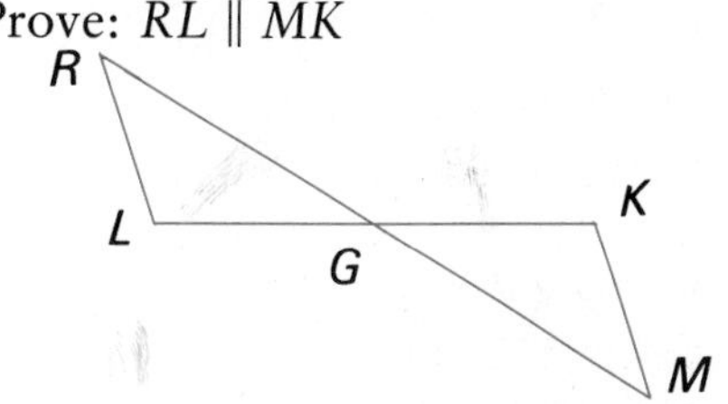

35. Given: $\angle 1 \cong \angle 2$, $\overline{ZT} \perp \overline{TY}$, $\overline{QY} \perp \overline{TY}$, $\overline{TR} \cong \overline{YR}$, $\overline{HR} \cong \overline{GR}$, W is the midpoint of $\overline{TY}$. *4.7*
   Prove: $\overline{WH} \cong \overline{WG}$

5 CONGRUENCE

Albert Einstein (1879–1955)—Although he failed algebra, Einstein developed the Theory of Relativity that forms the foundation for modern physics and space exploration. He played the violin to relax.

5.1 Isosceles Triangles

Objectives To prove and apply theorems about isosceles triangles
To solve problems involving isosceles triangles

Recall that an isosceles triangle has at least two congruent sides. Each of these sides is called a leg. The angle formed by the legs is called the **vertex angle.** The side opposite the vertex angle is called the **base.** The angles opposite the legs are called the **base angles.**

It can be proved that base angles are congruent. The drawing of an auxiliary segment within an isosceles triangle allows the proof to be easily completed.

Theorem 5.1 If two sides of a triangle are congruent, then the angles opposite these sides are congruent. (The base angles of an isosceles triangle are congruent.)

Given: $\triangle ABC$, $\overline{AC} \cong \overline{BC}$
Prove: $\angle A \cong \angle B$

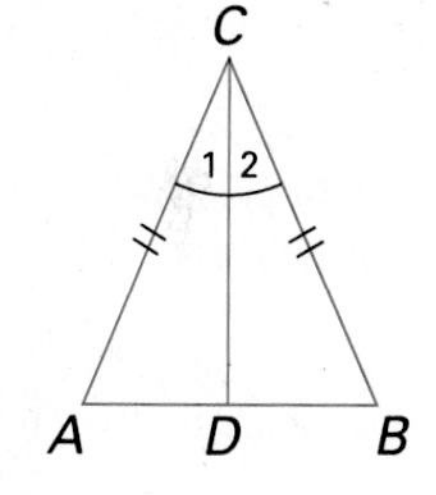

Plan Construct $\overline{CD}$, the angle bisector of $\angle ACB$. Then prove that $\triangle ADC \cong \triangle BDC$.

Proof

Statement		Reason
1. $\overline{AC} \cong \overline{BC}$	(S)	1. Given
2. Draw $\overline{CD}$, the bisector of $\angle ACB$, intersecting $\overline{AB}$ at D.		2. Every $\angle$, except a st $\angle$, has exactly one bis.
3. $\angle 1 \cong \angle 2$	(A)	3. Def of $\angle$ bis
4. $\overline{CD} \cong \overline{CD}$	(S)	4. Reflex Prop
5. $\triangle ADC \cong \triangle BDC$		5. SAS
6. $\therefore \angle A \cong \angle B$		6. CPCTC

Corollary If a triangle is equilateral, then it is also equiangular, and the measure of each angle is 60.

EXAMPLE 1 The measure of a base angle of an isosceles triangle is 10 less than twice the measure of the vertex angle. Find the measure of a base angle.

Plan Draw a diagram. Let x = measure of vertex angle. Then $2x - 10$ = measure of a base angle.

Solution

$$\begin{aligned} m\angle A + m\angle B + m\angle C &= 180 \\ x + (2x - 10) + (2x - 10) &= 180 \\ 5x - 20 &= 180 \\ 5x &= 200 \\ x &= 40 \\ 2x - 10 = (2 \cdot 40) - 10 &= 70 \end{aligned}$$

Thus, the measure of a base angle is 70.

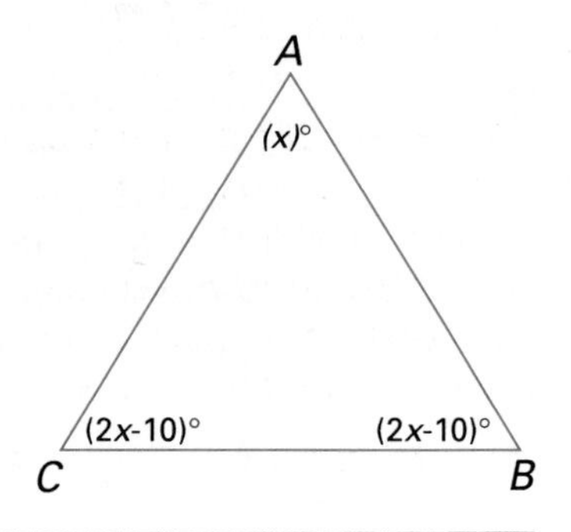

The converse of Theorem 5.1 is given below as Theorem 5.2.

Theorem 5.2 If two angles of a triangle are congruent, then the sides opposite these angles are congruent. (A triangle with two congruent sides is isosceles.)

The following corollary can be proved by using Theorem 5.2.

Corollary If a triangle is equiangular, then it is also equilateral.

The **perimeter** of a triangle is the sum of the lengths of its three sides. Example 2 applies the properties of isosceles triangles to a perimeter problem.

EXAMPLE 2 In $\triangle ABC$, $AC = BC$, $AB = 6$, and the perimeter is 20. Find AC.

Plan Draw $\triangle ABC$. Let x = length of $\overline{AC}$ and of $\overline{BC}$. Use the given perimeter to write an equation.

Solution

$$\begin{aligned} AC + BC + AB &= 20 \\ x + x + 6 &= 20 \\ 2x + 6 &= 20 \\ x &= 7 \end{aligned}$$

Thus, $AC = 7$.

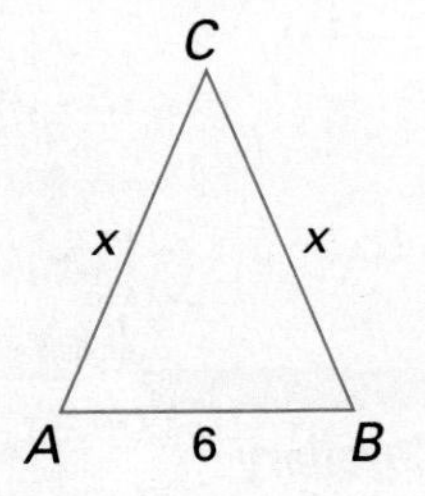

EXAMPLE 3 Given: $\angle 3 \cong \angle 4$, D is the midpoint of $\overline{EC}$; $\angle 1 \cong \angle 2$

Prove: $\overline{EA} \cong \overline{CB}$

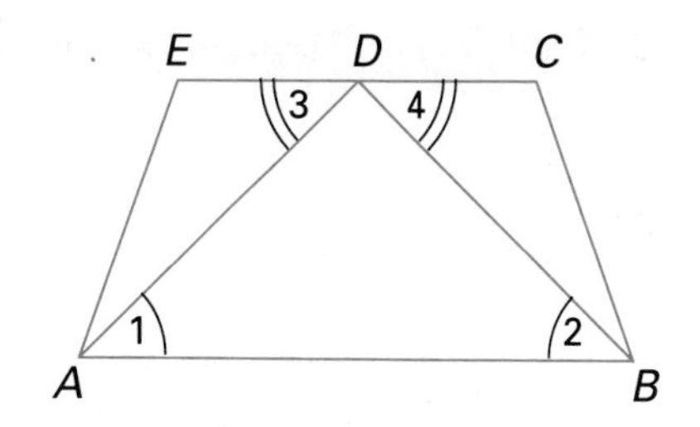

Plan Show $\overline{EA} \cong \overline{CB}$ by proving $\triangle EDA \cong \triangle CDB$. $\angle 1$ and $\angle 2$ are *not* angles of those triangles; however, since $\angle 1 \cong \angle 2$, $\overline{AD} \cong \overline{BD}$.

Proof

Statement		Reason
1. $\angle 3 \cong \angle 4$	(A)	1. Given
2. D is midpt of $\overline{EC}$.		2. Given
3. $\overline{ED} \cong \overline{CD}$	(S)	3. Def of midpt
4. $\angle 1 \cong \angle 2$		4. Given
5. $\overline{AD} \cong \overline{BD}$	(S)	5. Sides opp $\cong$ $\angle$s of a $\triangle$ are $\cong$.
6. $\triangle EDA \cong \triangle CDB$		6. SAS
7. $\therefore \overline{EA} \cong \overline{CB}$		7. CPCTC

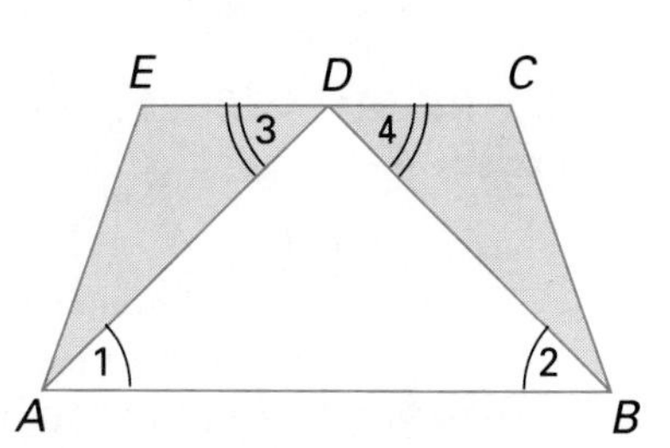

Classroom Exercises

1. Name the vertex angle, base, legs, and base angles of the isosceles triangle.

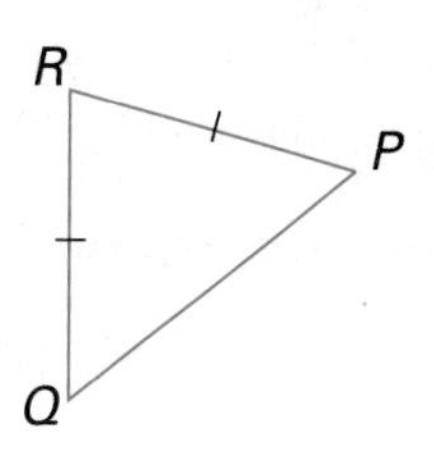

Complete.

2. If $\overline{FG} \cong \overline{CG}$, then _____ $\cong$ _____

3. If $\overline{EG} \cong \overline{ED}$, then _____ $\cong$ _____

4. If $\angle 6 \cong \angle 5$, then _____ $\cong$ _____

5. If $\angle 2 \cong \angle 7$, then _____ $\cong$ _____

6. If $\overline{EG} \cong \overline{DG}$, then _____ $\cong$ _____

7. If $\overline{AG} \cong \overline{BG}$, then _____ $\cong$ _____

8. If $\angle 1 \cong \angle 4$, then _____ $\cong$ _____

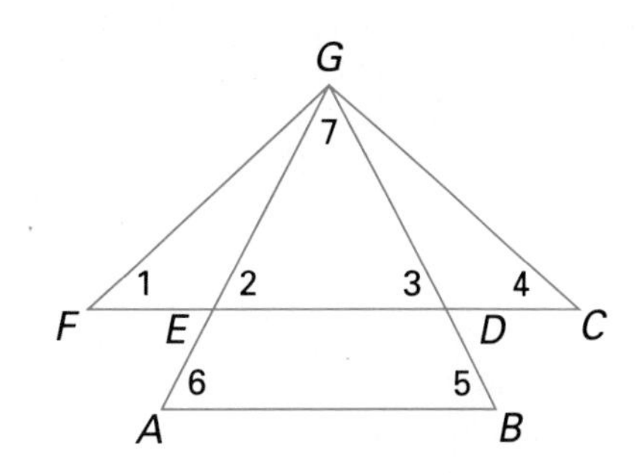

Written Exercises

Find the measure of each angle of the triangle.

1.

2.

3.

In $\triangle ABC$, $AC = BC$. Find the measure of each angle of the triangle. (Exercises 4–5)

4. $m\angle C = x$, $m\angle A = 2x - 20$

5. $m\angle C = 2x + 20$, $m\angle A = 6x + 10$

6. The measure of the vertex angle of an isosceles triangle is twice the measure of a base angle. Find the measure of a base angle.

7. The measure of the vertex angle of an isosceles triangle is 20 less than twice the measure of a base angle. Find the measure of the vertex angle.

8. Given: $\overline{BF} \cong \overline{CF}$, $m\angle 5 = 130$. Find $m\angle 4$.

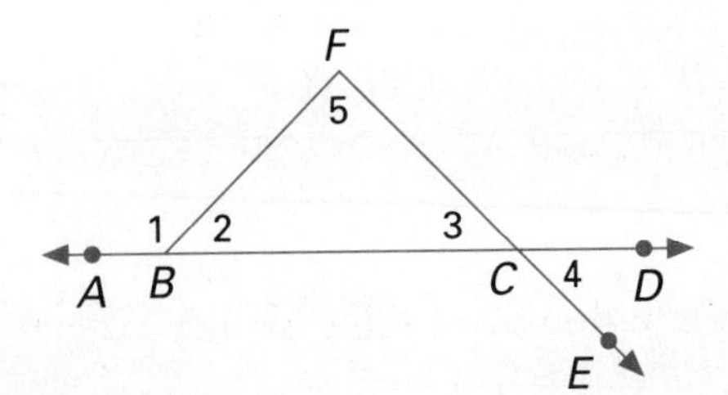

9. Given: $\overline{BF} \perp \overline{FC}$, $\overline{BF} \cong \overline{FC}$. Find $m\angle 2$.

10. Given: $\overline{BC} \cong \overline{FC}$, $m\angle 4 = 30$. Find $m\angle 1$.

11. Given: $\overline{FB} \cong \overline{FC}$, $FB = 10$, $BC = 15$. Find the perimeter of $\triangle FBC$.

12. Given: $\angle 2 \cong \angle 3$, $BF = 7\frac{1}{2}$, $BC = 11$. Find the perimeter of $\triangle FBC$.

13. Given: $\overline{BF} \cong \overline{CF}$, $BC = 20$, the perimeter of $\triangle BFC$ is 48. Find FC.

14. Given: $\overline{AB} \cong \overline{BC}$
Prove: $\angle 3 \cong \angle 2$

15. Given: $\angle 4 \cong \angle 2$
Prove: $\overline{AB} \cong \overline{AC}$

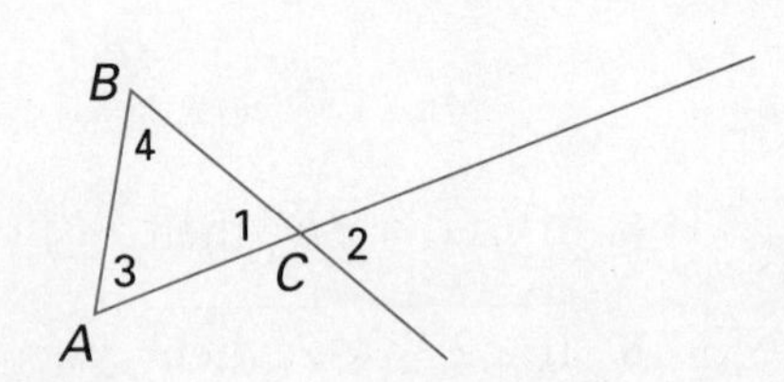

16. The perimeter of an equiangular triangle is 45 cm. Find the measure of each side.

17. $\triangle XYZ$ is equiangular and XY is 27 cm. Find the perimeter of $\triangle XYZ$.

18. Given: $\overline{BA} \parallel \overline{CD}$,
$\overline{CB} \cong \overline{DB}$
Prove: $\angle 1 \cong \angle 2$

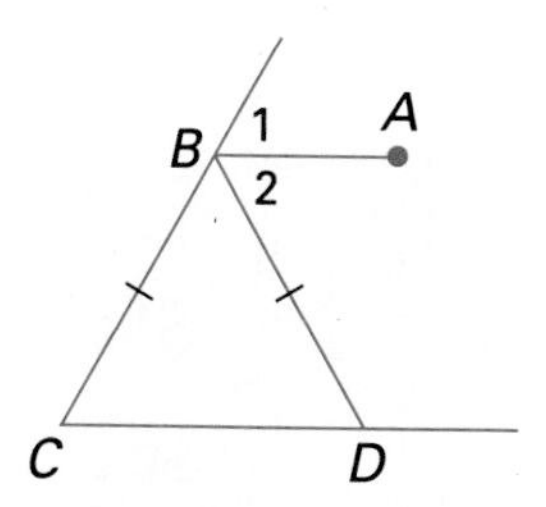

19. Given: $\overline{PQ} \cong \overline{PR}$,
$\overline{PS} \cong \overline{PT}$,
$\overline{QV} \cong \overline{RW}$
Prove: $\angle 1 \cong \angle 2$

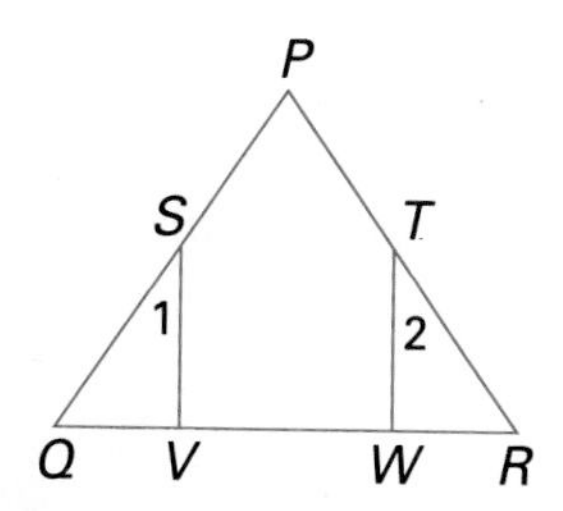

20. Given: $\overline{DA} \cong \overline{DB}$,
$\angle 1 \cong \angle 2$
Prove: $\overline{AC} \cong \overline{BC}$

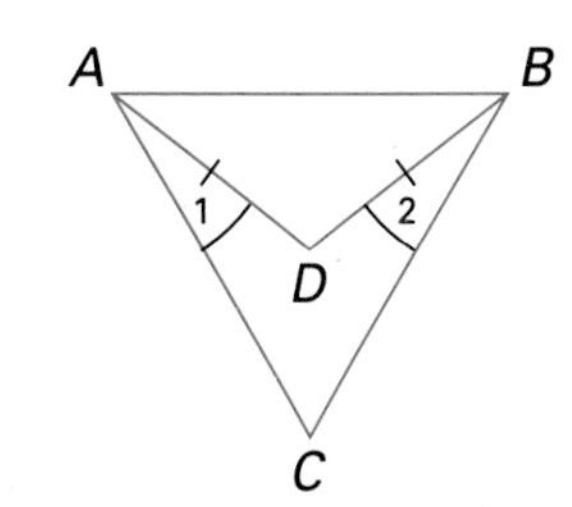

21. Prove the corollary to Theorem 5.1.

22. Prove Theorem 5.2. (HINT: Construct an auxiliary segment.)

23. $\triangle GHK$ is isosceles, with $\overline{GH}$ as the base. Point L is on $\overline{GH}$ such that $\overline{KL}$ bisects $\angle K$. m $\angle G = 40$. Find m $\angle GKL$ and m $\angle GLK$.

24. Prove the corollary to Theorem 5.2.

25. Prove: In an isosceles triangle, the bisector of an exterior angle of the vertex angle is parallel to the base.

26. Given: $\overline{BA}$ is perpendicular to every line in plane $\mathcal{M}$ passing through A; $\angle ACB \cong \angle ADB$
Prove: $\triangle BCD$ is isosceles.

27. Given: $\overline{AE} \perp \overline{CD}$, E is the midpoint of $\overline{CD}$; $\overline{BA}$ is perpendicular to every line in plane $\mathcal{M}$ passing through A.
Prove: $BC = BD$

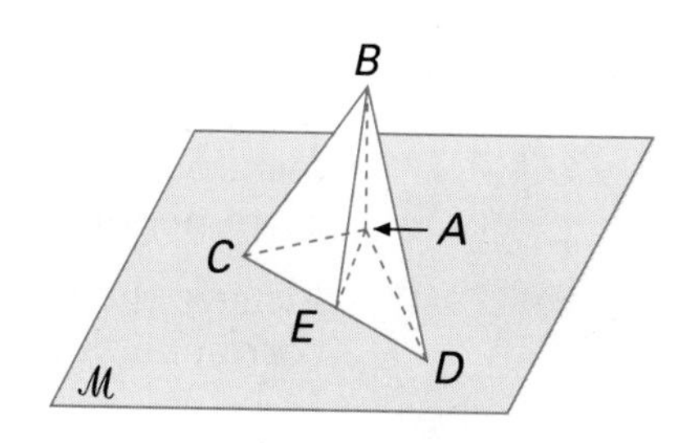

Mixed Review

1. The measure of an angle is twice the measure of its complement. Find the measure of the angle. *1.6*

2. $\angle A$ and $\angle B$ are vertical angles. m $\angle A = 4x - 40$, m $\angle B = 2x + 60$. Find m $\angle A$. *2.7*

3. The measure of an exterior angle of a triangle is 150. The measure of one of the two remote interior angles is $\frac{2}{3}$ the measure of the other. Find the measure of each remote interior angle. *3.7*

5.2 Introduction to Overlapping Triangles

Objective To identify and name overlapping triangles

When triangles overlap, redrawing them separately may help identify those that are required in a problem or proof.

EXAMPLE Identify the triangles that contain $\angle 1$ and $\angle 2$ as corresponding angles.

Identify the triangles that contain $\overline{QS}$ and $\overline{RT}$ as corresponding sides.

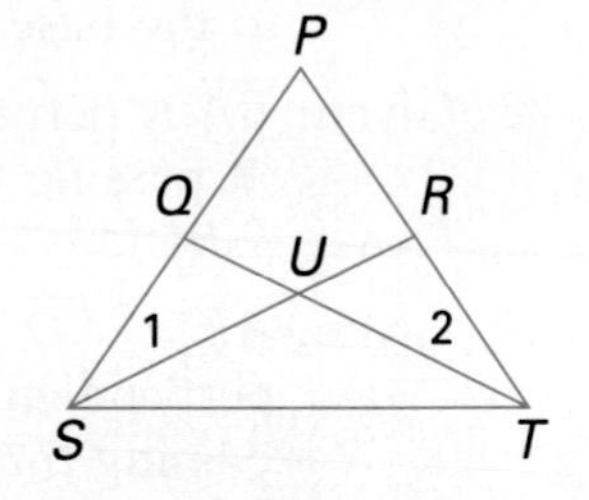

Solution Draw the three pairs of triangles separately.

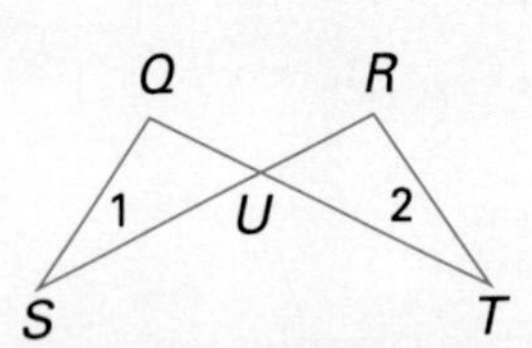

$\triangle QSU$ and $\triangle RTU$ contain $\overline{QS}$ and $\overline{RT}$, *and* $\angle 1$ and $\angle 2$.

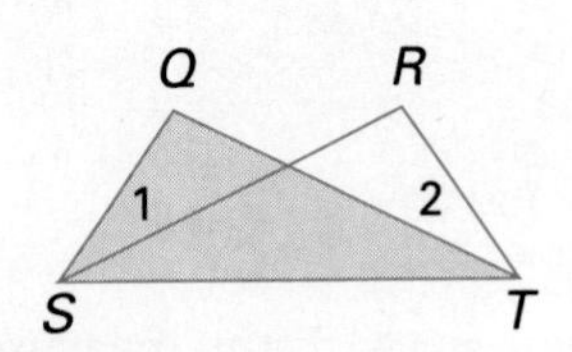

$\triangle QST$ and $\triangle RTS$ contain $\overline{QS}$ and $\overline{TR}$. ($\angle 1$ is *not* an angle of $\triangle QST$. $\angle 2$ is *not* an angle of $\triangle RTS$.)

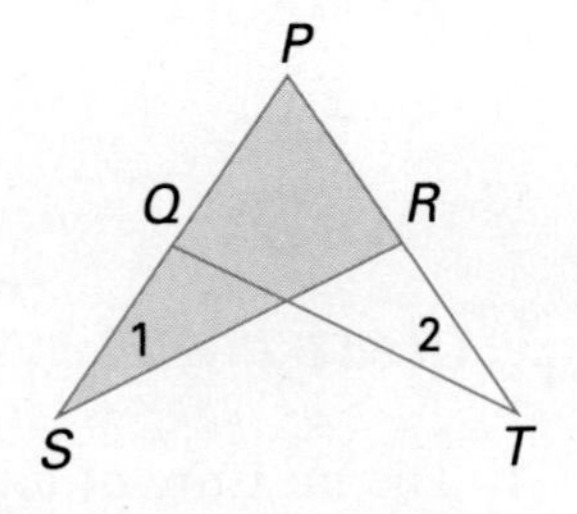

$\triangle PSR$ and $\triangle PTQ$ contain $\angle 1$ and $\angle 2$. ($\overline{QS}$ is *not* a side of $\triangle PSR$. $\overline{RT}$ is *not* a side of $\triangle PTQ$.)

Classroom Exercises

Identify a triangle in the figure that appears to be congruent to the given triangle.

1. $\triangle ATB$
2. $\triangle SAR$
3. $\triangle TAO$
4. $\triangle BAR$
5. $\triangle ORB$

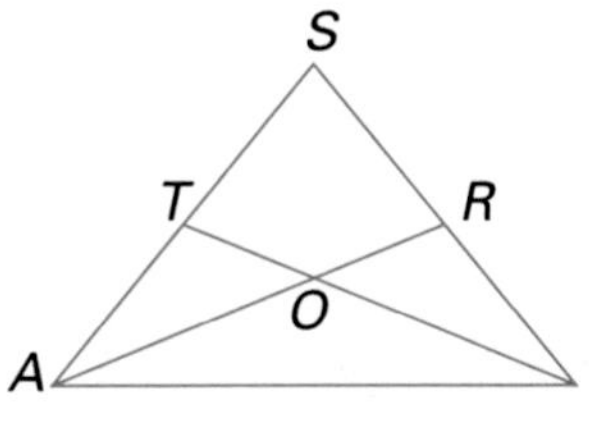

6. $\triangle YTW$
7. $\triangle YTM$
8. $\triangle MUY$
9. $\triangle WYU$
10. $\triangle MTY$

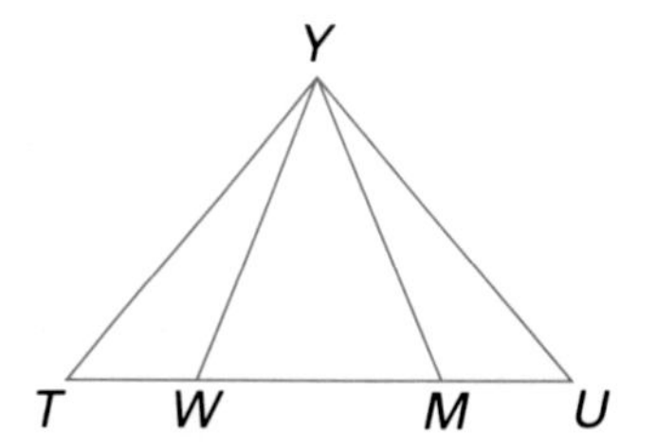

Written Exercises

Name each pair of triangles of which the given sides or angles are corresponding parts. Draw each pair of triangles separately.

1. $\angle D$ and $\angle C$
2. $\overline{AD}$ and $\overline{BC}$
3. $\overline{ED}$ and $\overline{EC}$
4. $\angle 1$ and $\angle 2$
5. $\angle 3$ and $\angle 4$

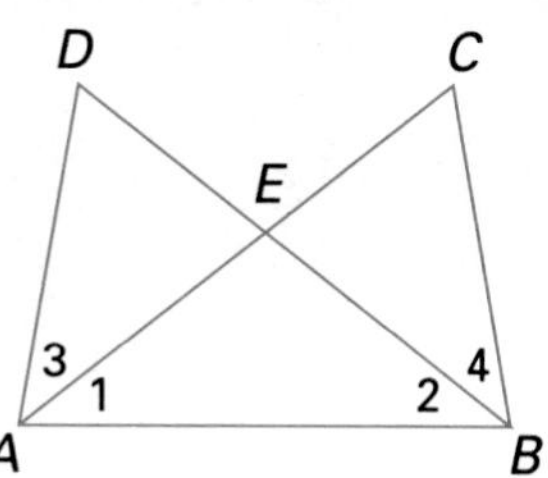

6. $\angle 1$ and $\angle 2$
7. $\angle 5$ and $\angle 6$
8. $\overline{PQ}$ and $\overline{RQ}$
9. $\angle 3$ and $\angle 4$
10. $\overline{UV}$ and $\overline{SV}$
11. $\overline{VP}$ and $\overline{VR}$
12. $\overline{PS}$ and $\overline{RU}$
13. $\angle UPR$ and $\angle SRP$

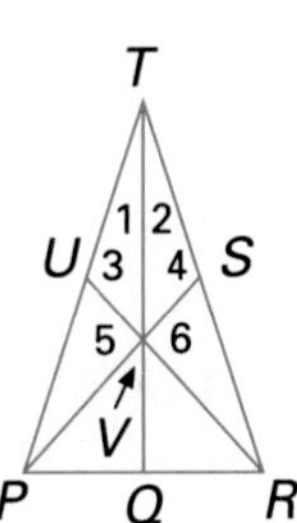

14. $\overline{IB}$ and $\overline{JA}$
15. $\overline{ID}$ and $\overline{JD}$
16. $\overline{EA}$ and $\overline{CB}$
17. $\overline{EB}$ and $\overline{CA}$
18. $\angle 1$ and $\angle 2$
19. $\overline{EH}$ and $\overline{CH}$
20. $\angle AEC$ and $\angle BCE$
21. $\angle DAJ$ and $\angle DBI$

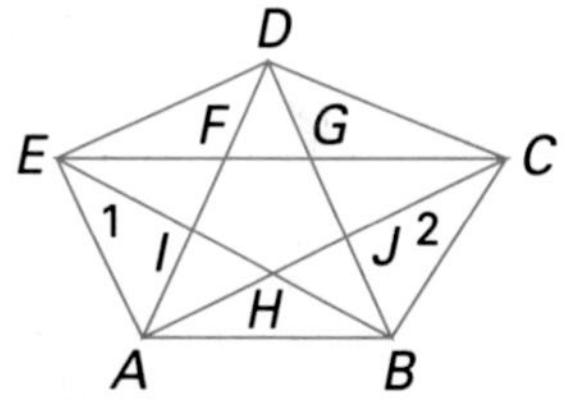

Mixed Review

1. Given: $\overline{AE} \cong \overline{AD}$, $\overline{AB}$ bisects $\angle EAD$.
   Prove: $\angle E \cong \angle D$ **4.3**
2. Given: $\angle EBC \cong \angle DBC$, $\angle 1 \cong \angle 2$
   Prove: $\overline{AE} \cong \overline{AD}$ **4.6**
3. Given: $\overline{AE} \parallel \overline{BD}$, $\overline{AE} \cong \overline{BD}$
   Prove: $\overline{AD} \cong \overline{BE}$ **4.3**

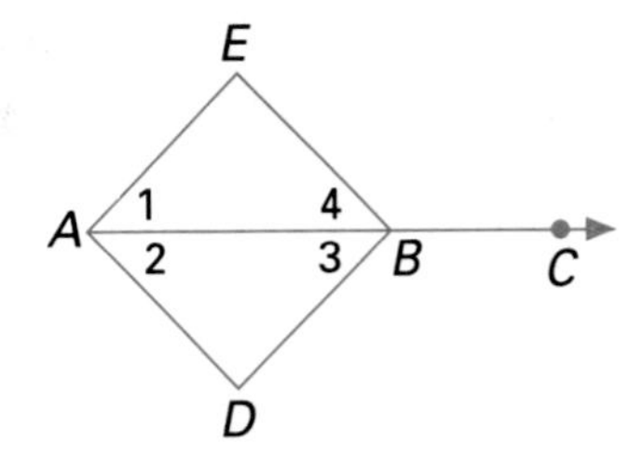

5.3 Congruence and Overlapping Triangles

Objective To write proofs involving congruent overlapping triangles

EXAMPLE 1 Given: $\overline{BD}$ bisects $\angle ADE$; $\angle 1 \cong \angle 2$
Prove: $\overline{AD} \cong \overline{ED}$

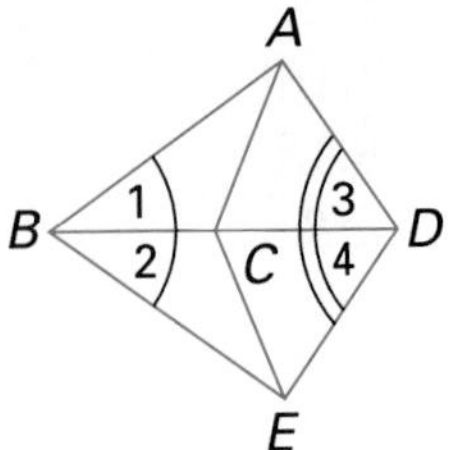

Plan Consider the three pairs of triangles separately. Mark congruent parts.

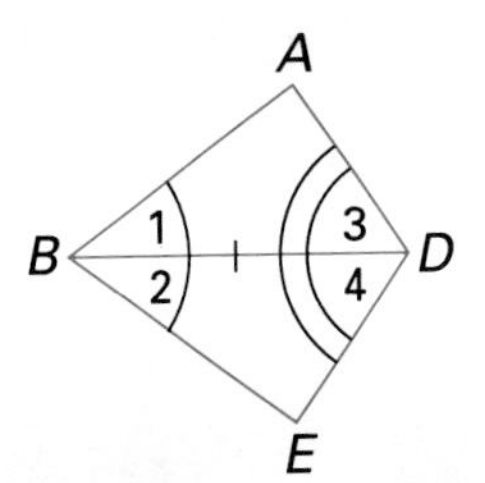

Three pairs of parts are marked. The triangles are congruent by ASA.

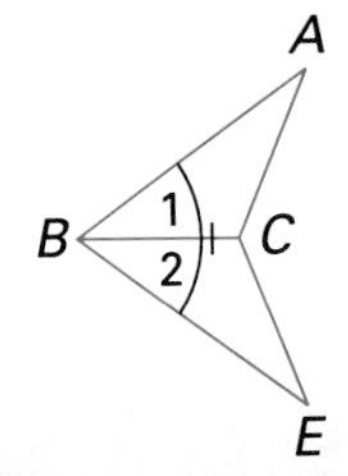

Only two pairs of parts are marked. This does not establish triangle congruence.

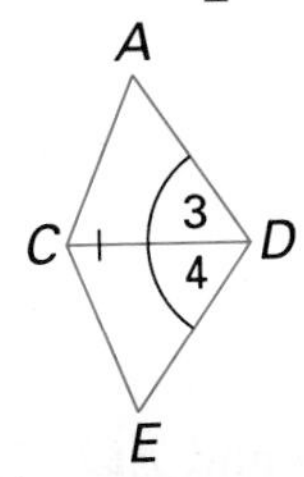

Only two pairs of parts are marked. This does not establish triangle congruence.

So, use $\triangle BAD \cong \triangle BED$ to prove $\overline{AD} \cong \overline{ED}$.

Proof

Statement		Reason
1. $\angle 1 \cong \angle 2$	**(A)**	1. Given
2. $\overline{BD}$ bisects $\angle ADE$.		2. Given
3. $\angle 3 \cong \angle 4$	**(A)**	3. Def of $\angle$ bis
4. $\overline{BD} \cong \overline{BD}$	**(S)**	4. Reflex Prop
5. $\triangle BAD \cong \triangle BED$		5. ASA
6. $\therefore \overline{AD} \cong \overline{ED}$		6. CPCTC

The Reflexive Property of Congruence can be used to state that angles common to two triangles are congruent. In the figure below, $\angle A$ is common to $\triangle ACE$ and $\triangle AFB$.

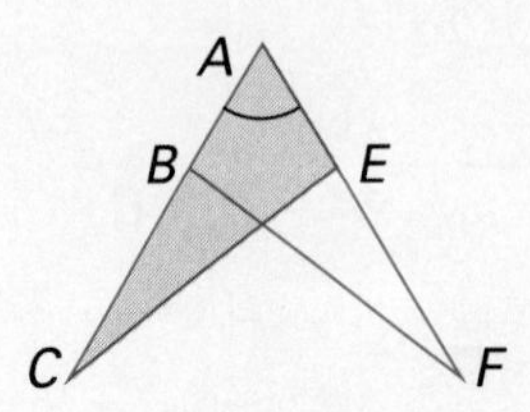

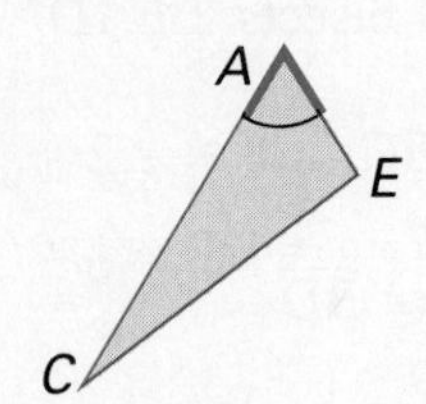

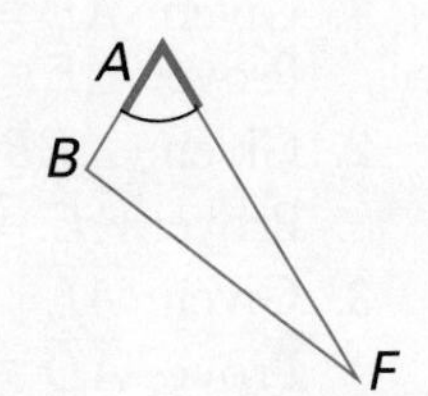

Sometimes it may be necessary to prove a pair of triangles congruent so that their corresponding parts can be used to prove a *second* pair of triangles congruent. These procedures are illustrated in the next example.

EXAMPLE 2 Given: $\overline{DB}$ bisects $\angle ADC$; $\angle 3 \cong \angle 4$
Prove: $\angle 5 \cong \angle 6$

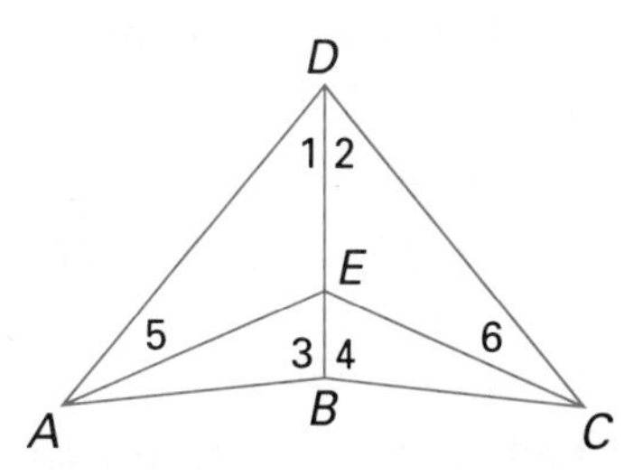

Plan Consider the pairs of triangles separately. Determine which pairs of triangles must be proved congruent to prove that $\angle 5 \cong \angle 6$.

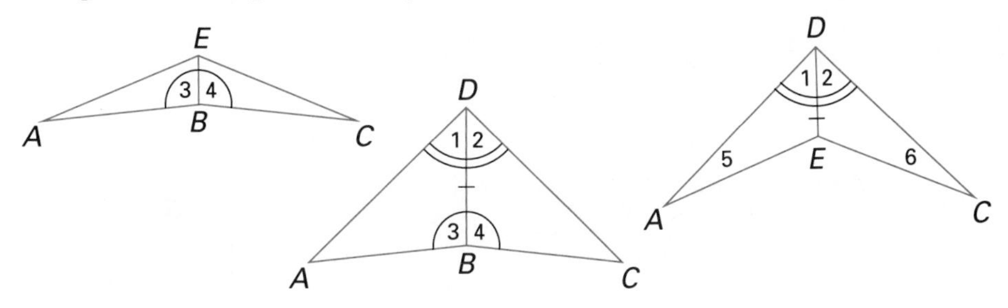

First prove $\triangle ADB \cong \triangle CDB$. Use this result to prove $\triangle ADE \cong \triangle CDE$. Then, it follows that $\angle 5 \cong \angle 6$ by CPCTC.

Proof

Statement		Reason
1. $\overline{DB}$ bisects $\angle ADC$.		1. Given
2. $\angle 1 \cong \angle 2$	(A)	2. Def of $\angle$ bis
3. $\angle 3 \cong \angle 4$	(A)	3. Given
4. $\overline{DB} \cong \overline{DB}$	(S)	4. Reflex Prop
5. $\triangle ADB \cong \triangle CDB$		5. ASA
6. $\overline{AD} \cong \overline{CD}$	(S)	6. CPCTC
7. $\overline{DE} \cong \overline{DE}$	(S)	7. Reflex Prop
8. $\angle 1 \cong \angle 2$	(A)	8. Given
9. $\triangle ADE \cong \triangle CDE$		9. SAS
10. $\angle 5 \cong \angle 6$		10. CPCTC

In more complex proofs such as the one above, it is sometimes helpful to draw a flow diagram before actually writing out the steps.

Flow Diagram

$\overline{DB}$ bisects $\angle ADC$. $\rightarrow$ $\angle 1 \cong \angle 2$
$\angle 3 \cong \angle 4 \rightarrow \triangle ADB \cong \triangle CDB \rightarrow \overline{AD} \cong \overline{CD}$
$\overline{DB} \cong \overline{DB}$
$\angle 1 \cong \angle 2 \rightarrow \triangle ADE \cong \triangle CDE \rightarrow \angle 5 \cong \angle 6$
$\overline{DE} \cong \overline{DE}$

Classroom Exercises

Name pairs of triangles that can be proved congruent from the given data and state the congruence pattern used as a reason.

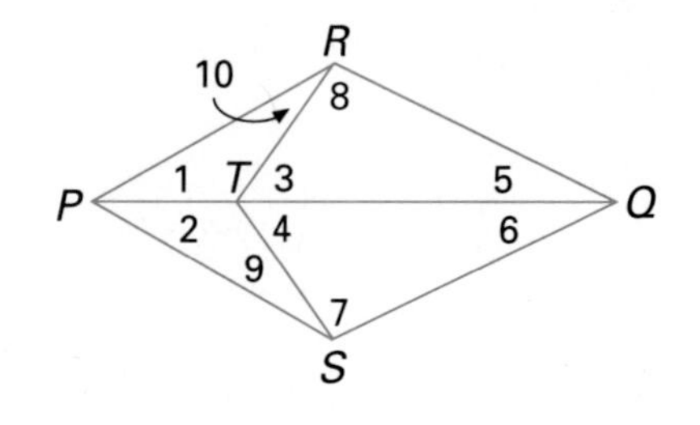

1. Given: $\overline{PR} \cong \overline{PS}$, $\angle 1 \cong \angle 2$
2. Given: $\angle 3 \cong \angle 4$, $\angle 5 \cong \angle 6$
3. Given: $\angle 7 \cong \angle 8$, $\angle 5 \cong \angle 6$
4. Given: $\overline{PR} \cong \overline{PS}$, $\overline{TR} \cong \overline{TS}$
5. Given: $\overline{TR} \cong \overline{TS}$, $\overline{QR} \cong \overline{QS}$
6. Given: $\angle 9 \cong \angle 10$, $\overline{PR} \cong \overline{PS}$, $\overline{TR} \cong \overline{TS}$
7. Given: $\overline{TR} \cong \overline{TS}$, $\angle PTR \cong \angle PTS$

Written Exercises

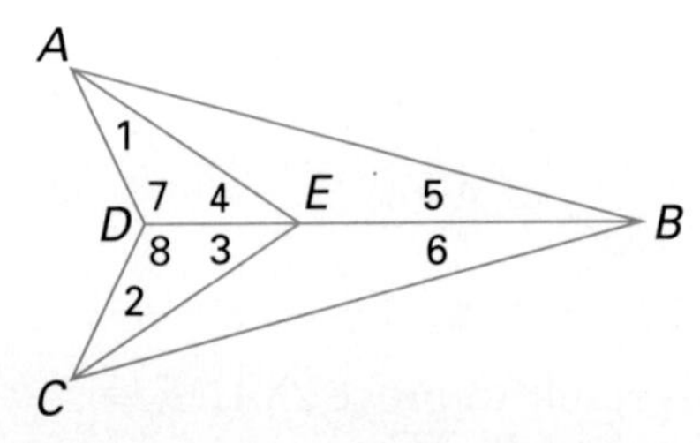

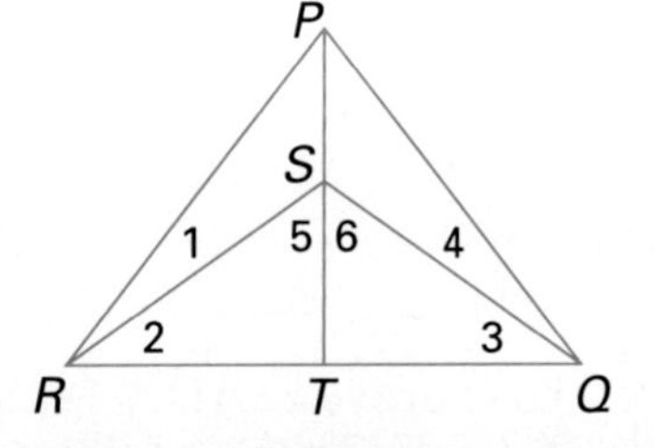

1. Given: $\angle 3 \cong \angle 4$, $\overline{AE} \cong \overline{CE}$
   Prove: $\overline{AD} \cong \overline{CD}$
2. Given: $\angle 7 \cong \angle 8$, $\overline{BD}$ bisects $\angle ABC$.
   Prove: $\angle DAB \cong \angle DCB$
3. Given: $\overline{PS}$ bisects $\angle RPQ$; $\overline{PR} \cong \overline{PQ}$.
   Prove: $\overline{RS} \cong \overline{QS}$
4. Given: $\overline{PT} \perp \overline{RQ}$, $\overline{PT}$ bisects $\overline{RQ}$.
   Prove: $\overline{PT}$ bisects $\angle RPQ$.

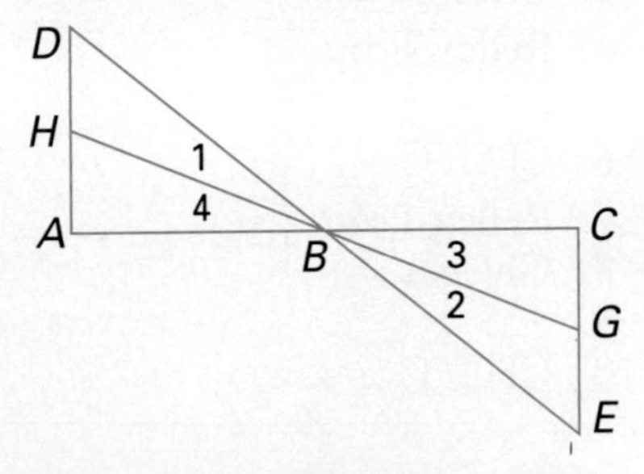

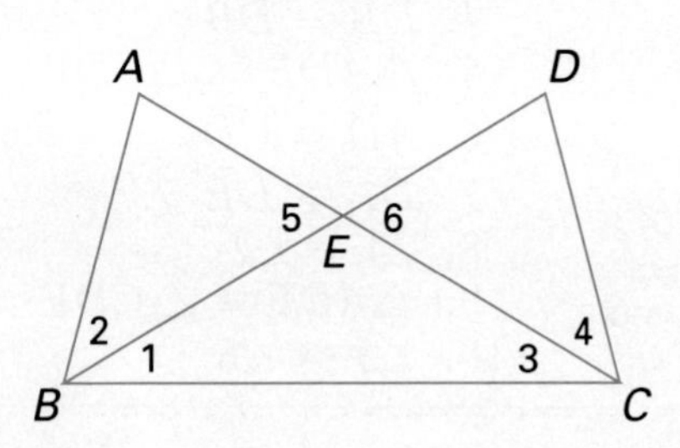

5. Given: $\overline{AD} \parallel \overline{CE}$, $\overline{HA} \cong \overline{GC}$
   Prove: $\overline{HB} \cong \overline{GB}$
6. Given: $\overline{HG}$ and $\overline{DE}$ bisect each other.
   Prove: $\overline{HD} \cong \overline{GE}$
7. Given: B is the midpoint of $\overline{AC}$;
   $\overline{AD} \perp \overline{AC}$, $\overline{EC} \perp \overline{AC}$
   Prove: B is the midpoint of $\overline{DE}$.
8. Given: $\angle 1 \cong \angle 3$, $\angle ABC \cong \angle DCB$
   Prove: $\overline{AB} \cong \overline{DC}$
9. Given: $\angle A \cong \angle D$, $\overline{AE} \cong \overline{DE}$
   Prove: $\overline{AB} \cong \overline{DC}$
10. Given: $\overline{AB} \cong \overline{DC}$, $\angle 1 \cong \angle 3$,
    $\angle 2 \cong \angle 4$
    Prove: $\overline{AC} \cong \overline{DB}$

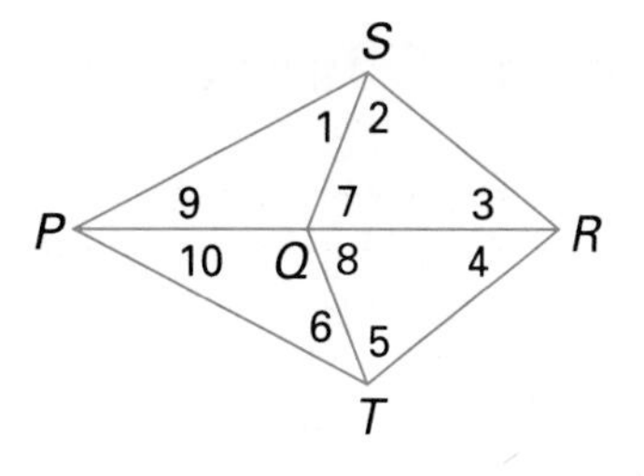

11. Given: $\overline{PR}$ bisects $\angle SPT$; $\angle 3 \cong \angle 4$
Prove: $\overline{QS} \cong \overline{QT}$

12. Given: $\overline{PR}$ bisects $\angle SRT$; $\overline{SR} \cong \overline{TR}$
Prove: $\angle 1 \cong \angle 6$

13. Given: $\overline{PS} \cong \overline{PT}$, $\overline{SR} \cong \overline{TR}$
Prove: $\angle 2 \cong \angle 5$

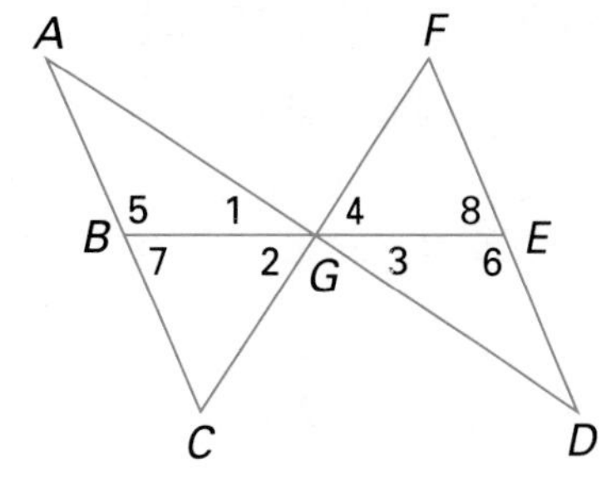

14. Given: $\overline{AD}$ and $\overline{CF}$ bisect each other.
Prove: $\angle 5 \cong \angle 6$

15. Given: $\overline{AD}$ bisects $\overline{BE}$; $\angle 5 \cong \angle 6$
Prove: $\overline{AC} \cong \overline{DF}$

16. Given: $\overline{AC} \parallel \overline{FD}$, $\overline{BC} \cong \overline{EF}$
Draw a flow diagram to prove $\overline{AC} \cong \overline{FD}$.

17. Write the proof for the flow diagram of Exercise 16.

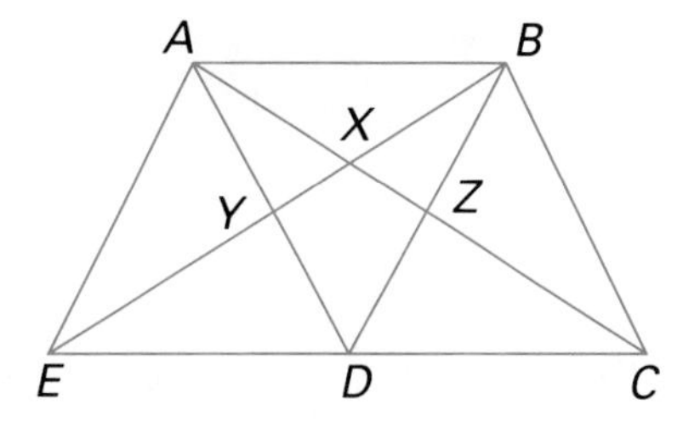

18. Given: $\triangle AED$ and $\triangle BCD$ are equilateral; D is the midpoint of $\overline{EC}$.
Prove: $XE = XC$

19. Given: $AX = BX$, $CX = EX$
Prove: $\overline{AB} \parallel \overline{EC}$

Mixed Review

Assume $\overline{PR} \parallel \overline{SU}$. 3.5, 3.7

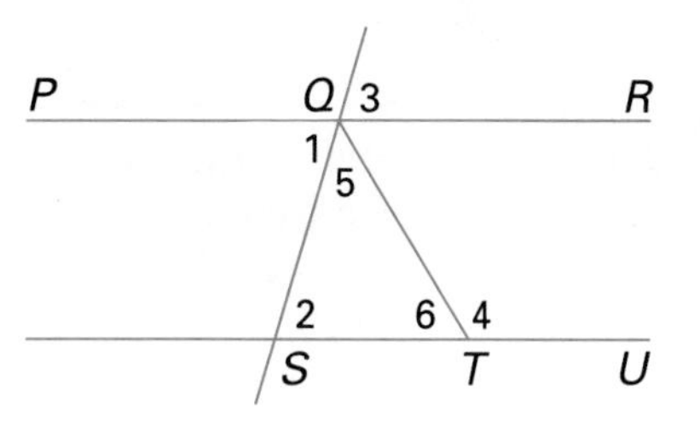

1. Given: m $\angle 2 = 40$
   Find m $\angle 3$.
2. Given: m $\angle 2 = 60$, m $\angle 5 = 50$
   Find m $\angle 4$.
3. Given: m $\angle 5 = 7x$, m $\angle 2 = 3x$, m $\angle 6 = 2x$
   Find m $\angle 2$.

Brainteaser

$\triangle ABC$ is equilateral. M, N, and P are the midpoints of the sides of $\triangle ABC$. X, Y, and Z are the midpoints of the sides of $\triangle MNP$. D, E, and F are the midpoints of the sides of $\triangle XYZ$. What can you conclude about $\triangle DEF$? Explain your reasoning.

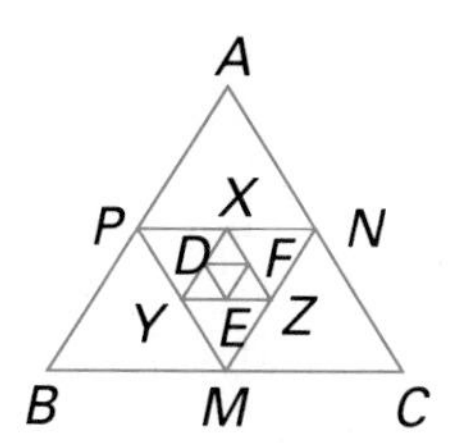

5.4 Congruence of Right Triangles

Objectives

To identify the parts of a right triangle
To write proofs involving congruence of right triangles

Definition

The sides forming the right angle of a right triangle are called **legs.** The side opposite the right angle is the **hypotenuse.**

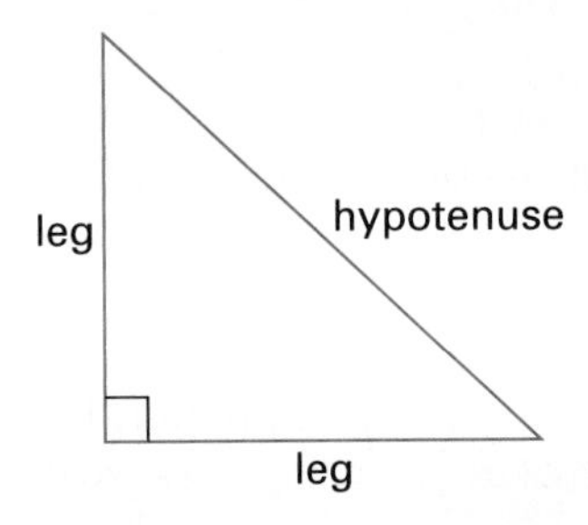

Right triangles can be proved congruent by any of the methods developed for other triangles.

Recall that the SAS congruency pattern requires that the corresponding angle be *included between* the sides. SSA does not, in general, guarantee the congruence of two triangles, as we have seen. However, the SSA pattern *can* be used for the special case of right triangles.

Theorem 5.3

Hypotenuse-Leg (HL) Theorem: Two right triangles are congruent if the hypotenuse and a leg of one are congruent, respectively, to the hypotenuse and corresponding leg of the other.

Given: Rt $\triangle$s ABC and DEF with rt $\angle$s B and E, $\overline{AC} \cong \overline{DF}$, $\overline{BC} \cong \overline{EF}$
Prove: $\triangle ABC \cong \triangle DEF$

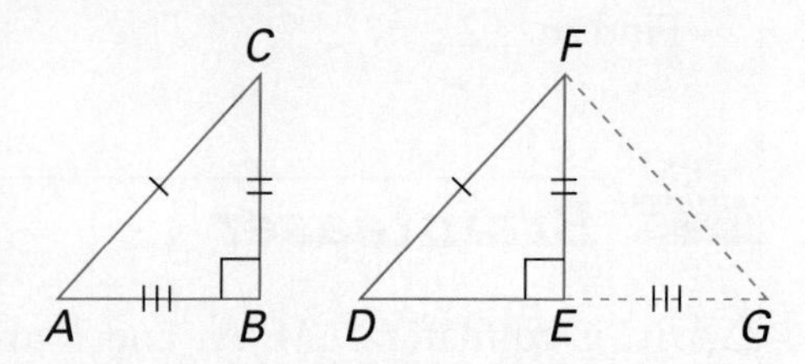

Plan

Extend $\overline{DE}$ to point G so that $\overline{AB} \cong \overline{EG}$. Prove $\triangle ABC \cong \triangle GEF$. Then by CPCTC, (1) $\angle G \cong \angle A$ and (2) $\overline{AC} \cong \overline{GF}$. Show that $\triangle DGF$ is isosceles. Then, $\angle D \cong \angle G$. From this and (1), $\angle D \cong \angle A$. Then, $\triangle ABC \cong \triangle DEF$ by SAS.

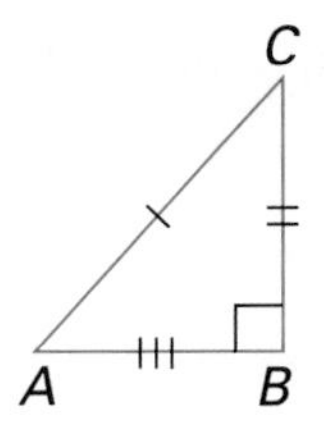

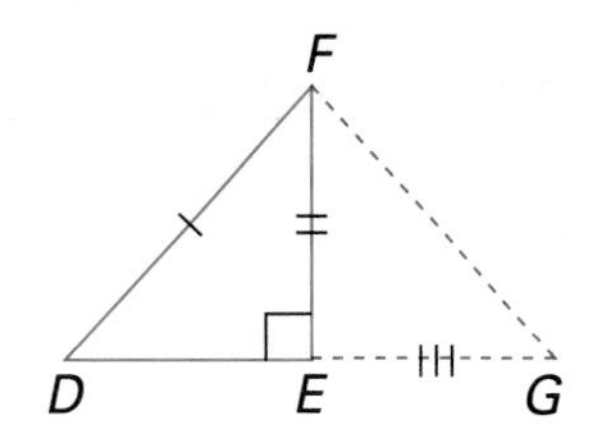

Proof	**Statement**	**Reason**
	1. Rt $\triangle$s ABC and DEF with rt $\angle$s B and E	1. Given
	2. Extend $\overline{DE}$ so that $\overline{EG} \cong \overline{BA}$.	(S) 2. Construction
	3. Draw $\overline{FG}$.	3. Any two points determine a line.
	4. $\overline{FE} \perp \overline{DG}$	4. Def of rt $\angle$
	5. $\angle FEG$ is a rt $\angle$.	5. Def of rt $\angle$
	6. $\angle ABC \cong \angle GEF$	(A) 6. All rt $\angle$s are $\cong$.
	7. $\overline{EF} \cong \overline{BC}$	(S) 7. Given
	8. $\triangle ABC \cong \triangle GEF$	8. SAS
	9. $\overline{AC} \cong \overline{GF}$	9. CPCTC
	10. $\overline{AC} \cong \overline{DF}$	(S) 10. Given
	11. $\overline{GF} \cong \overline{DF}$	11. Trans Prop Cong
	12. $\angle D \cong \angle G$	12. If 2 sides of a $\triangle$ are $\cong$, then $\angle$s opp $\cong$ sides are $\cong$.
	13. $\angle A \cong \angle G$	13. CPCTC
	14. $\angle A \cong \angle D$	(A) 14. Trans Prop Cong
	15. $\angle B \cong \angle DEF$	(A) 15. All rt $\angle$s are $\cong$.
	16. $\therefore \triangle ABC \cong \triangle DEF$	16. AAS

EXAMPLE 1 Given: $\overline{PS} \perp \overline{QS}$, $\overline{PR} \perp \overline{QR}$, $\overline{PS} \cong \overline{QR}$
Prove: $\overline{PR} \cong \overline{QS}$

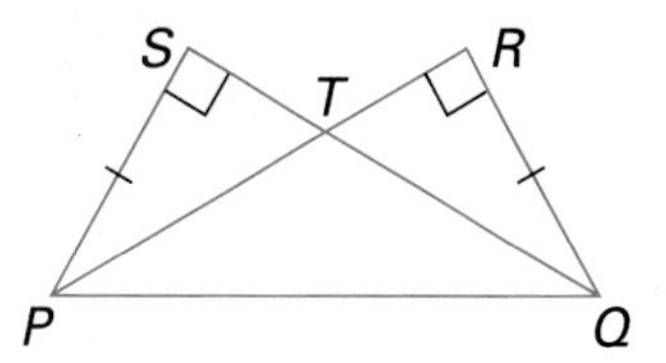

Proof	**Statement**	**Reason**
	1. $\overline{PS} \perp \overline{QS}$, $\overline{PR} \perp \overline{QR}$	1. Given
	2. $\angle S$ and $\angle R$ are rt $\angle$s.	2. $\perp$s form rt $\angle$s.
	3. $\triangle PSQ$ and $\triangle QRP$ are rt $\triangle$s.	3. Def of rt $\triangle$
	4. $\overline{PQ} \cong \overline{PQ}$	(H) 4. Reflex Prop
	5. $\overline{PS} \cong \overline{QR}$	(L) 5. Given
	6. $\triangle PSQ \cong \triangle QRP$	6. HL
	7. $\therefore \overline{PR} \cong \overline{QS}$	7. CPCTC

EXAMPLE 2 Given: $\overline{VP} \perp \overline{VQ}$, $\overline{TR} \perp \overline{TS}$, $\overline{PR} \cong \overline{SQ}$, $\overline{QV} \cong \overline{RT}$

Prove: $\angle P \cong \angle S$

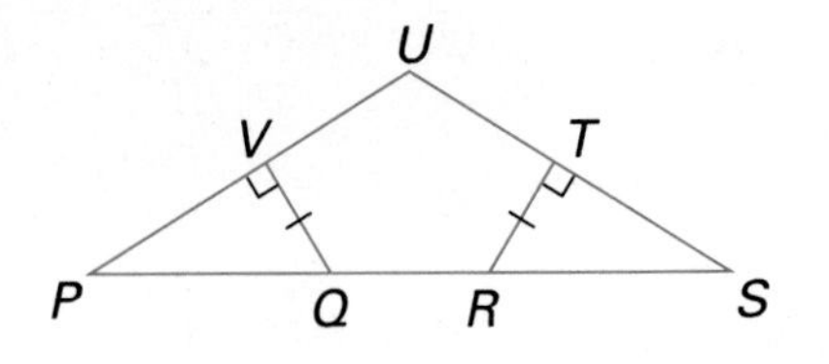

Plan Show $\overline{PQ} \cong \overline{SR}$. Then prove rt $\triangle PQV$ congruent to rt $\triangle SRT$.

Proof

	Statement		Reason
1.	$\overline{VP} \perp \overline{VQ}$, $\overline{TR} \perp \overline{TS}$		1. Given
2.	$\overline{QV} \cong \overline{RT}$	(L)	2. Given
3.	$\angle PVQ$ and $\angle STR$ are rt $\angle$s.		3. $\perp$s form rt $\angle$s.
4.	$\triangle PVQ$ and $\triangle STR$ are rt $\triangle$s.		4. Def of rt $\triangle$
5.	$PR = SQ$ ($\overline{PR} \cong \overline{SQ}$)		5. Given
6.	$PR - QR = SQ - QR$		6. Subt Prop of Eq
7.	$PR - QR = PQ$, $SQ - QR = SR$		7. Seg Add Post
8.	$PQ = SR$ ($\overline{PQ} \cong \overline{SR}$)	(H)	8. Sub
9.	$\triangle PVQ \cong \triangle STR$		9. HL for $\cong$ rt $\triangle$s
10.	$\therefore \angle P \cong \angle S$		10. CPCTC

Classroom Exercises

Based on the markings, state whether the pair of right triangles can be proved congruent. If congruent, state one of the triangle-congruence theorems as a reason.

1.

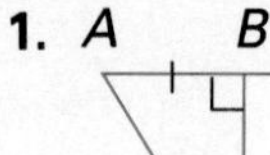

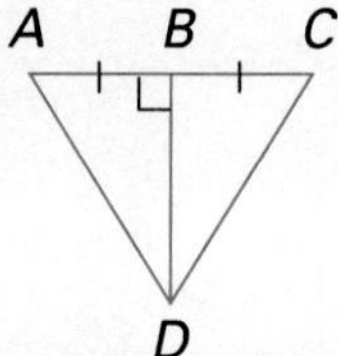

2.

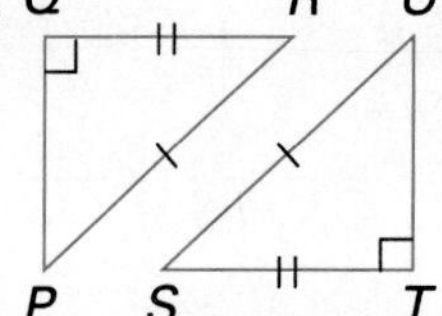

3.

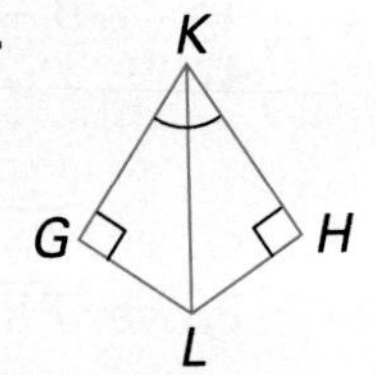

Written Exercises

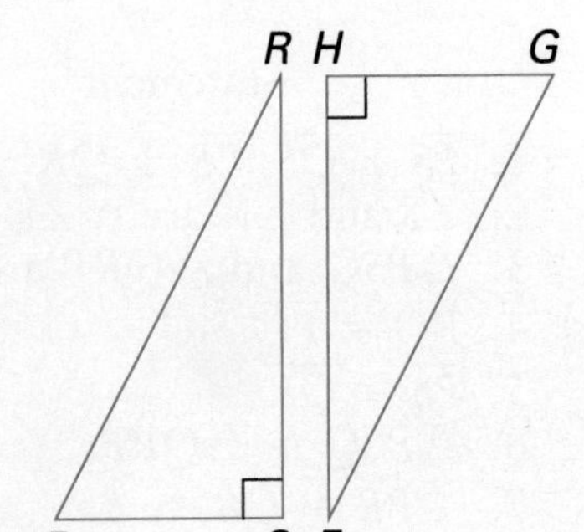

1. Given: $\overline{QR} \perp \overline{QP}$, $\overline{HF} \perp \overline{HG}$, $\overline{QR} \cong \overline{HF}$, $\overline{QP} \cong \overline{HG}$

Prove: $\triangle RQP \cong \triangle FHG$

2. Given: $\overline{QR} \perp \overline{QP}$, $\overline{HF} \perp \overline{HG}$, $\overline{RQ} \cong \overline{FH}$, $\overline{RP} \cong \overline{FG}$

Prove: $\angle P \cong \angle G$

3. Given: $\overline{QR} \perp \overline{QP}$, $\overline{HF} \perp \overline{HG}$, $\angle R \cong \angle F$, $\overline{PR} \cong \overline{GF}$

Prove: $\overline{RQ} \cong \overline{FH}$

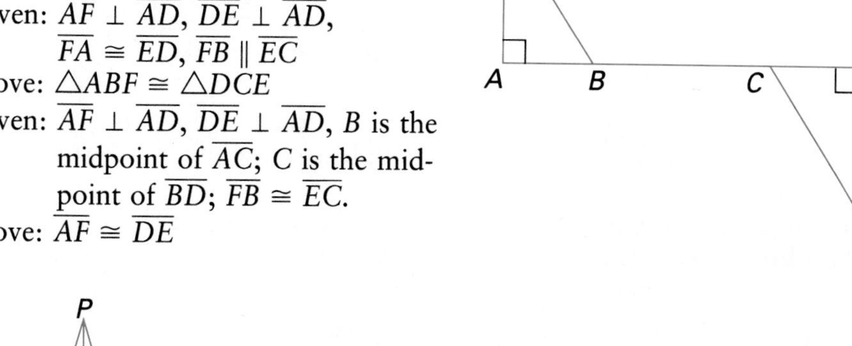

4. Given: $\overline{AF} \perp \overline{AD}$, $\overline{DE} \perp \overline{AD}$,
$\angle F \cong \angle E$, $\overline{FB} \cong \overline{EC}$
Prove: $\triangle ABF \cong \triangle DCE$

5. Given: $\overline{AF} \perp \overline{AD}$, $\overline{DE} \perp \overline{AD}$,
$\overline{FA} \cong \overline{ED}$, $\overline{FB} \parallel \overline{EC}$
Prove: $\triangle ABF \cong \triangle DCE$

6. Given: $\overline{AF} \perp \overline{AD}$, $\overline{DE} \perp \overline{AD}$, B is the midpoint of $\overline{AC}$; C is the midpoint of $\overline{BD}$; $\overline{FB} \cong \overline{EC}$.
Prove: $\overline{AF} \cong \overline{DE}$

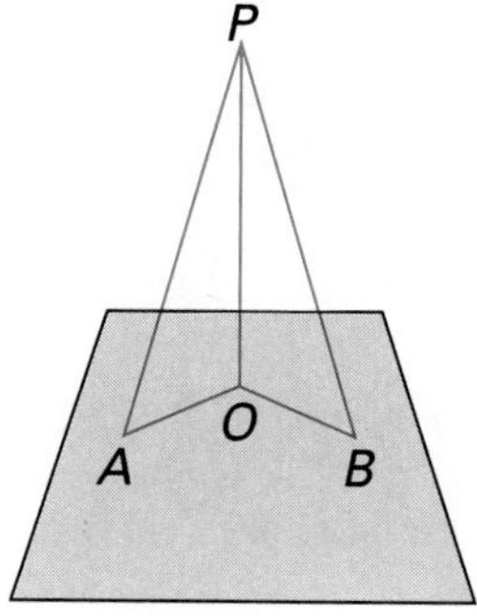

7. $\overline{OP}$ represents a TV antenna pole perpendicular to the plane of a roof (perpendicular to every line in the roof passing through O). If guy wires $\overline{AP}$ and $\overline{BP}$ are congruent, prove that A and B must be the same distance from O.

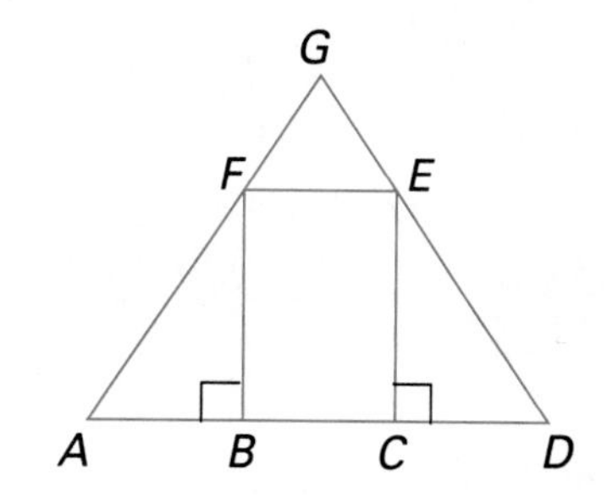

8. Rafters $\overline{AG}$ and $\overline{DG}$ are congruent. At points F and E, equidistant from G, braces $\overline{FB}$ and $\overline{EC}$ are constructed perpendicular to $\overline{AD}$. Prove that the braces must be congruent.

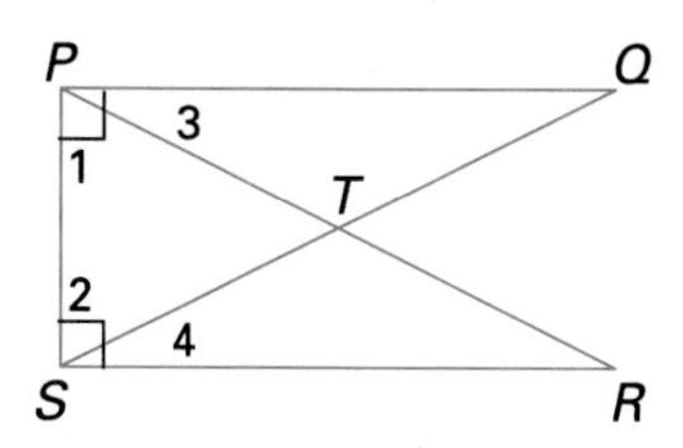

9. Given: $\overline{PQ} \perp \overline{PS}$, $\overline{SR} \perp \overline{SP}$, $\overline{PT} \cong \overline{ST}$
Prove: $\angle Q \cong \angle R$

10. Given: $\overline{PQ} \perp \overline{PS}$, $\overline{SR} \perp \overline{SP}$, $\overline{PR} \cong \overline{SQ}$
Prove: $\overline{PT} \cong \overline{ST}$

11. Given: $\overline{PQ} \perp \overline{PS}$, $\overline{SR} \perp \overline{SP}$, $\angle 3 \cong \angle 4$
Prove: $\angle Q \cong \angle R$

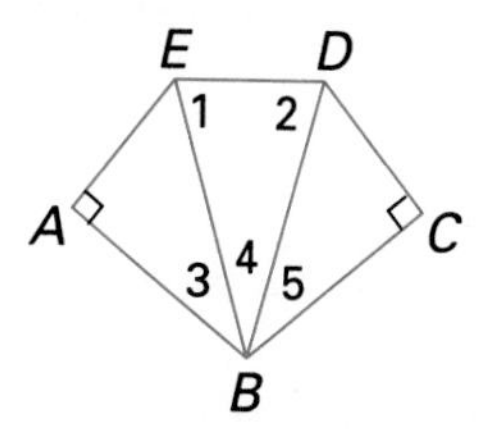

12. Given: $\overline{AB} \perp \overline{AE}$, $\overline{CB} \perp \overline{CD}$,
$\angle 1 \cong \angle 2$, $\overline{AB} \cong \overline{CB}$
Prove: $\overline{AE} \cong \overline{CD}$

13. Given: $\overline{AB} \perp \overline{AE}$, $\overline{CB} \perp \overline{CD}$,
$\angle ABD \cong \angle EBC$,
$\overline{AB} \cong \overline{CB}$
Prove: $\triangle EBD$ is isosceles.

The methods stated below are used to prove right triangles congruent. Prove each one. (Exercises 14–16)

14. Hypotenuse-Acute Angle Method (HA): Two right triangles are congruent if the hypotenuse and an acute angle of one are congruent to the hypotenuse and the corresponding acute angle of the other.

15. Leg-Leg Method (LL): Two right triangles are congruent if the two legs of one are congruent to the corresponding legs of the other.

16. Based on the pattern of Exercises 14–15, write the meaning of the LA method for proving right triangles congruent. Prove it.

17. Draw a flow diagram for the proof in Exercise 18.

18. Given: $\overline{SP} \cong \overline{SQ}$, $\overline{PR} \perp \overline{SQ}$, $\overline{QT} \perp \overline{SP}$
Prove: $\triangle UPQ$ is isosceles.

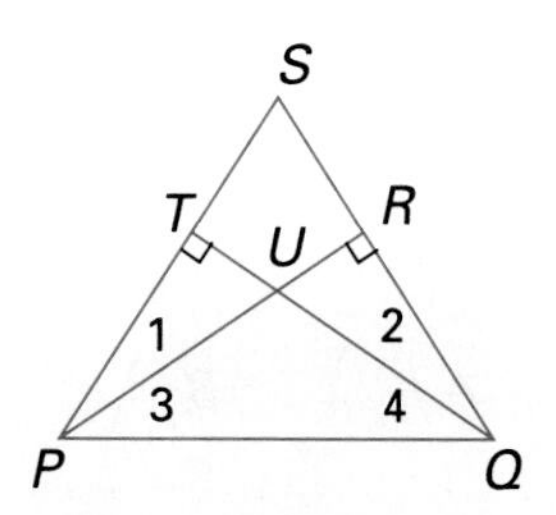

19. Draw a flow diagram for the proof in Exercise 20.

20. Given: $\angle 3 \cong \angle 4$, $\overline{PR} \perp \overline{SQ}$, $\overline{QT} \perp \overline{SP}$
Prove: $\triangle SPQ$ is isosceles.

21. Given: $\overline{PR} \perp \overline{SQ}$, $\overline{QT} \perp \overline{SP}$, $\overline{UT} \cong \overline{UR}$
Prove: $\overline{SR} \cong \overline{ST}$

Midchapter Review

1. Given: $\overline{PR} \cong \overline{QR}$, m $\angle R = 50$
Find m $\angle Q$. **5.1**

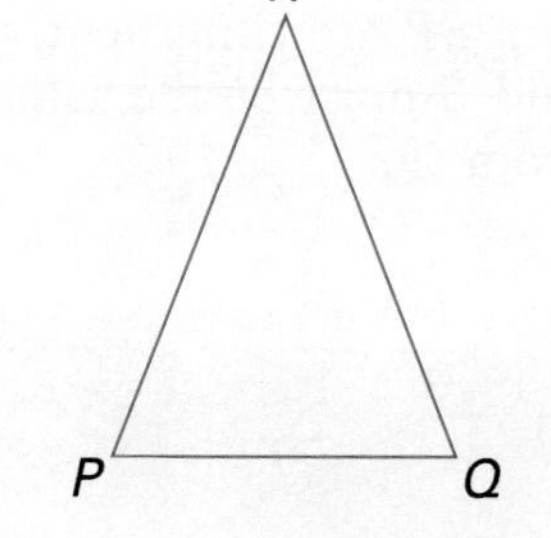

2. Given: $\angle P \cong \angle Q$, $PR = 4x - 2$, $QR = 2x + 8$, $PQ = 12$. Find the perimeter of $\triangle PQR$. **5.1**

3. Given: $\angle P \cong \angle Q$, m $\angle R$ is 3 times m $\angle P$.
Find m $\angle Q$. **5.1**

4. Name the triangles in the second figure on the right. **5.2**

5. Given: $\overline{QS} \perp \overline{PS}$, $\overline{QR} \perp \overline{PR}$, $\angle 3 \cong \angle 4$
Prove: $\triangle QSP \cong \triangle QRP$ **5.3, 5.4**

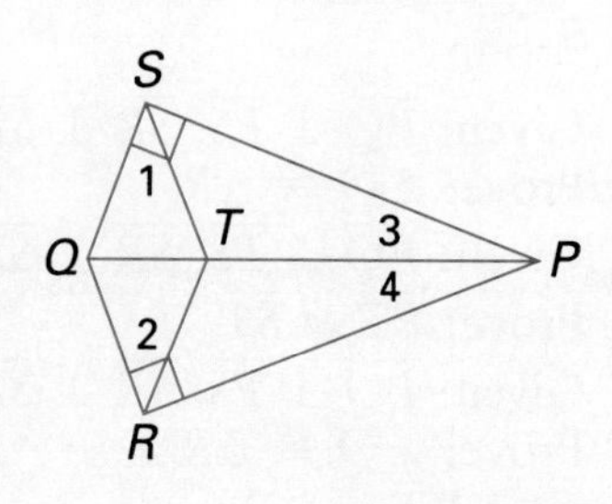

6. Given: $\overline{SP} \cong \overline{RP}$, m $\angle QSP = 90$,
m $\angle QRP = 90$
Prove: $\angle 3 \cong \angle 4$ **5.3, 5.4**

7. Given: $\overline{QP}$ bisects $\angle SQR$; $\angle 3 \cong \angle 4$,
$\overline{ST} \cong \overline{RT}$
Prove: $\angle 1 \cong \angle 2$ **5.2, 5.3**

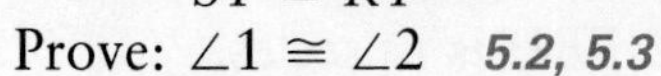

Computer Investigation

Altitudes and Medians of a Triangle

Use a computer to construct altitudes and medians of triangles by classification, and to measure the resulting figures.

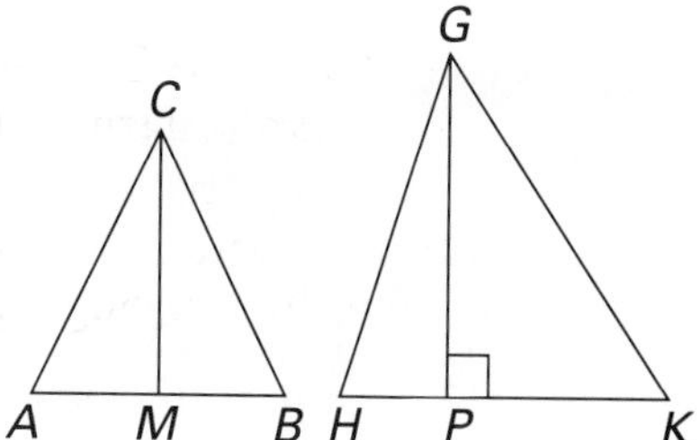

In $\triangle ABC$, $\overline{CM}$ is a segment from a vertex C to M, the midpoint of the opposite side $\overline{AB}$. $\overline{CM}$ is a **median** of $\triangle ABC$.

In $\triangle GHK$, $\overline{GP}$ is a segment from a vertex G, perpendicular to the opposite side $\overline{HK}$. $\overline{GP}$ is an **altitude** of $\triangle GHK$.

Activity 1

Draw an acute triangle.

1. Draw a median from each of the vertices.
2. Are the medians in the interior of the triangle? Do they intersect?
3. Are the medians congruent?
4. Determine if the median bisects the vertex angle from which it is drawn.

Answer questions 1–4 above for the following types of triangles which you are to draw using the computer.

5. obtuse
6. isosceles
7. equilateral

Activity 2

Draw an acute triangle.

8. Draw an altitude from each of the vertices.
9. Are the altitudes in the interior of the triangle? Do they intersect?
10. Is any altitude also a median?
11. Are the altitudes congruent?

Answer questions 8–11 above for the following types of triangles.

12. obtuse
13. isosceles
14. equilateral

Summary

Make some generalizations from the exercises above.

15. Are medians always in the interior of a triangle?
16. When do medians bisect the vertex angles from which they are drawn?
17. When will a median and an altitude be the same?

5.5 Altitudes and Medians of Triangles

Objectives To identify altitudes and medians of triangles
To prove and apply theorems about altitudes and medians of triangles

Definition

A **median** of a triangle is a segment whose endpoints are a vertex of the triangle and the midpoint of the opposite side.

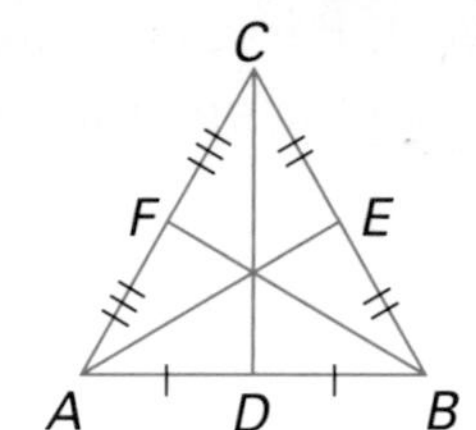

Every triangle has three medians, one from each vertex. As shown in the illustration, each median bisects the opposite side.

Notice that each median, except for its endpoints, is in the interior of the triangle. For convenience, this fact may be stated as "a median is in the interior of a triangle."

Definition

An **altitude** of a triangle is a segment from a vertex of the triangle perpendicular to the line containing the opposite side.

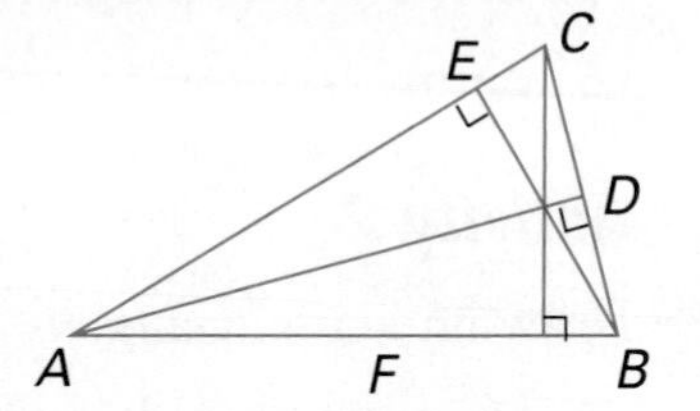

Every triangle has three altitudes, one from each vertex. In acute triangle ABC, as shown at the right, all three altitudes are in the interior of the triangle.

Unlike medians, altitudes are not necessarily in the interior of a triangle. The cases for right and obtuse triangles are seen below.

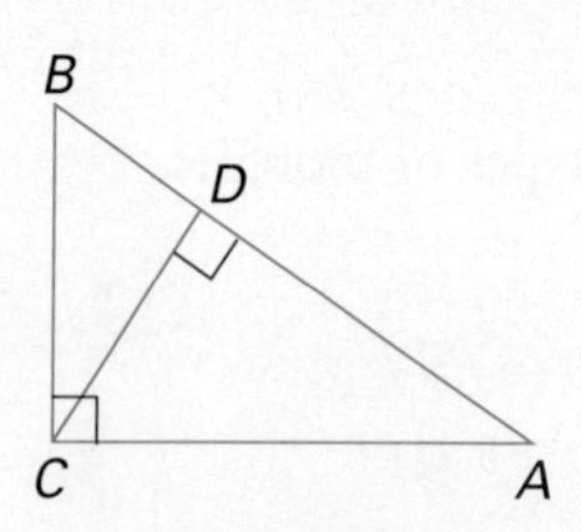

Right Triangle
Altitudes $\overline{BC}$ and $\overline{CA}$ are legs of the triangle.

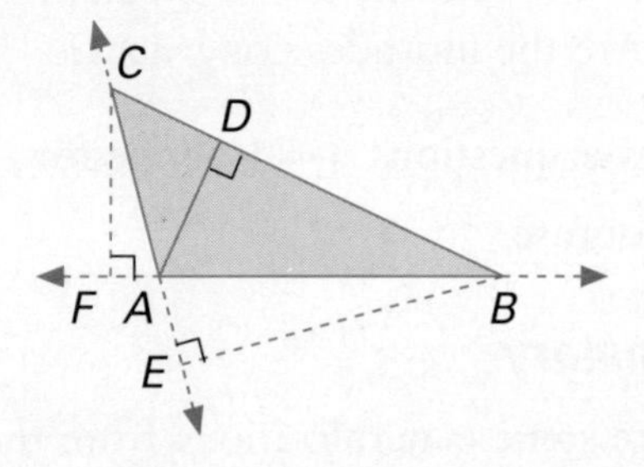

Obtuse Triangle
Altitudes $\overline{CE}$ and $\overline{BE}$ are in the exterior of the triangle.

EXAMPLE 1 Which segments are altitudes of $\triangle ABC$? Which are medians?

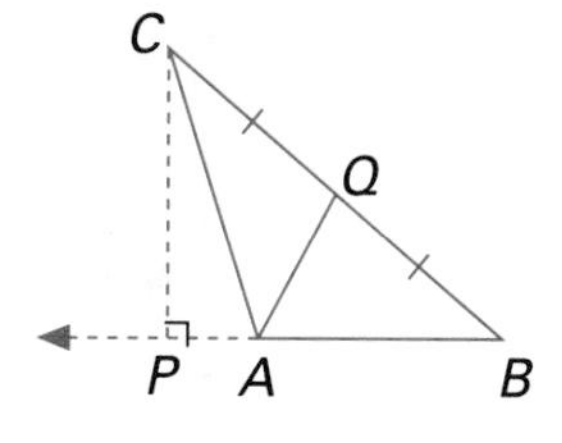

Solutions

$\overline{CP}$ is an altitude.
$\overline{AQ}$ is a median.

$\overline{BR}$ is a median.
$\overline{SB}$ is an altitude.

Sometimes an altitude can also be a median. Such is the case for the altitude from the vertex angle of an isosceles triangle. This is demonstrated by the next theorem. Notice that the proof is presented in *paragraph form.* In a paragraph proof, reasons that are expected to be clear to the reader may be omitted for the sake of brevity.

Theorem 5.4 The altitude from the vertex angle to the base of an isosceles triangle is a median (the altitude bisects the base).

Given: $\triangle ABC$ is isosceles, with $\overline{AC} \cong \overline{BC}$;
$\overline{CD}$ is an altitude.
Prove: $\overline{CD}$ is a median.

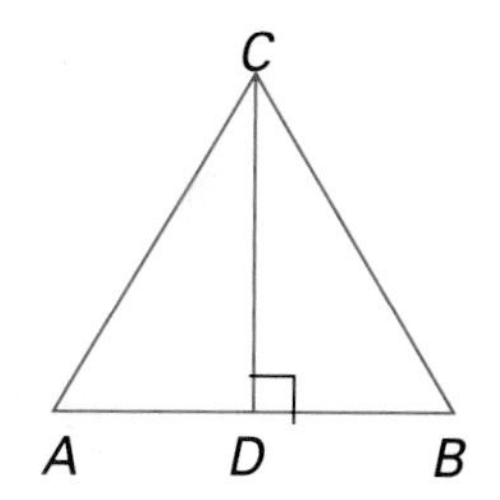

Proof (Paragraph Form) Since $\overline{CD}$ is an altitude, $\overline{CD} \perp \overline{AB}$ and two right triangles are formed. $\overline{AC} \cong \overline{BC}$ because $\triangle ABC$ is isosceles. Also, $\overline{CD} \cong \overline{CD}$ by the Reflexive Property. Therefore, $\triangle ADC \cong \triangle BDC$ by HL. Then $\overline{AD} \cong \overline{BD}$ by CPCTC. So, D is the midpoint of $\overline{AB}$, and altitude $\overline{CD}$ is also a median.

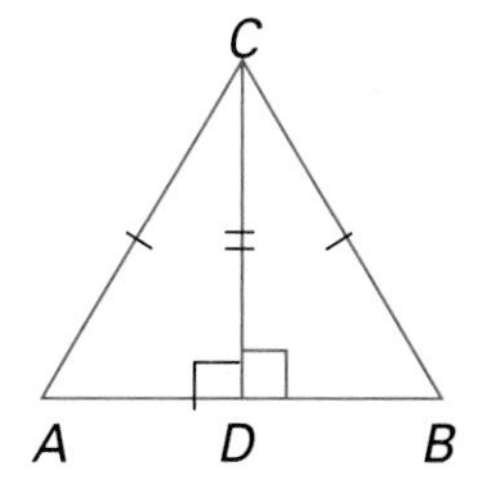

Medians that are drawn to corresponding sides of congruent triangles are called **corresponding medians.**

Theorem 5.5 Corresponding medians of congruent triangles are congruent.

Given: $\triangle ABC \cong \triangle EDF$, $\overline{CP}$ is a median of $\triangle ABC$; $\overline{FQ}$ is a median of $\triangle EDF$.
Prove: $\overline{CP} \cong \overline{FQ}$

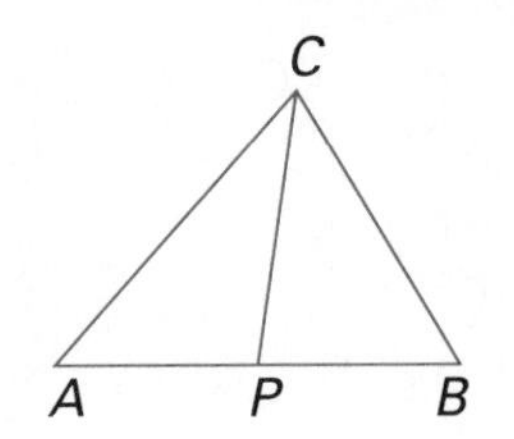

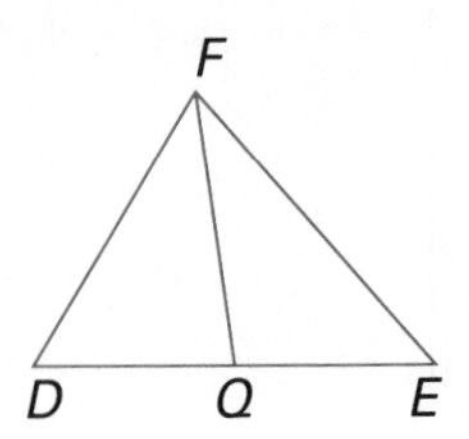

Proof

Statement	Reason
1. $\triangle ABC \cong \triangle EDF$, $\overline{CP}$ and $\overline{FQ}$ are medians.	1. Given
2. $AC = EF$ ($\overline{AC} \cong \overline{EF}$)	(S) 2. CPCTC
3. $\angle A \cong \angle E$	(A) 3. CPCTC
4. $AB = ED$ ($\overline{AB} \cong \overline{ED}$)	4. CPCTC
5. $AP = \frac{1}{2}AB$, $EQ = \frac{1}{2}ED$	5. Def of median
6. $\frac{1}{2}AB = \frac{1}{2}ED$	6. Mult Prop of Eq
7. $AP = EQ$	(S) 7. Sub
8. $\triangle APC \cong \triangle EQF$	8. SAS
9. $\therefore \overline{CP} \cong \overline{FQ}$	9. CPCTC

EXAMPLE 2 Given: $\triangle ABC \cong \triangle PQR$, $\overline{CM}$ and $\overline{RN}$ are corresponding medians; $CM = 9a - 2$, $RN = 4(a + 2)$
Find RN.

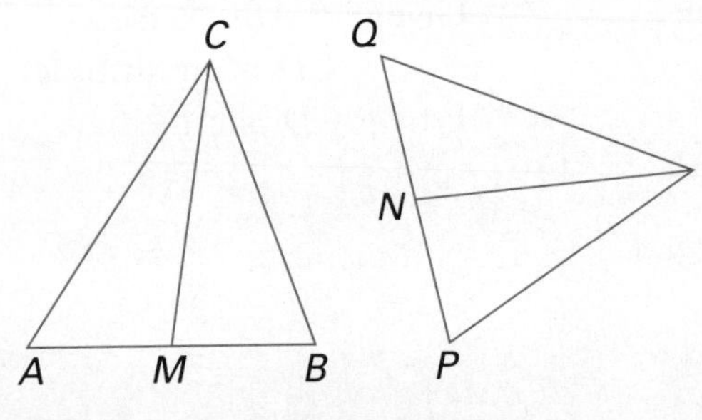

Solution $CM = RN$ by (Theorem 5.5)

$9a - 2 = 4(a + 2)$
$9a - 2 = 4a + 8$
$5a - 2 = 8$
$5a = 10$
$a = 2$

$RN = 4(a + 2) = 4(2 + 2) = 16$

Theorem 5.6 Corresponding altitudes of congruent triangles are congruent.

Classroom Exercises

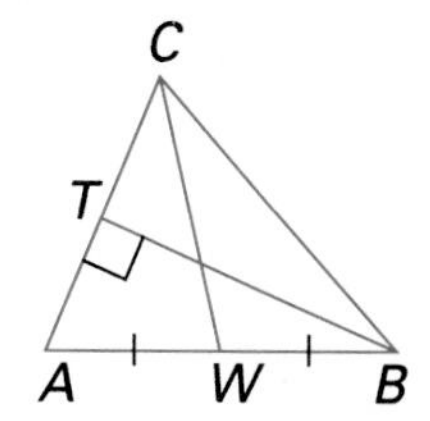

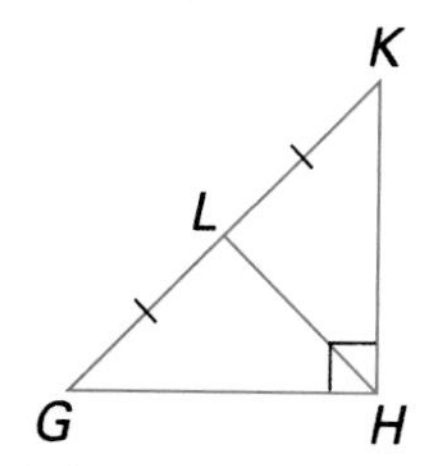

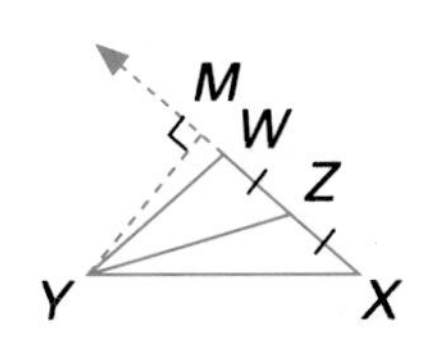

1. Name an altitude.
2. Name a median.
3. Name an altitude.
4. Name a median.
5. Name an altitude.
6. Name a median.
7. Is a median always in the interior of a triangle? Explain.
8. Is an altitude always in the interior of a triangle? Explain.

Written Exercises

Complete the statement by referring to the diagram at right.

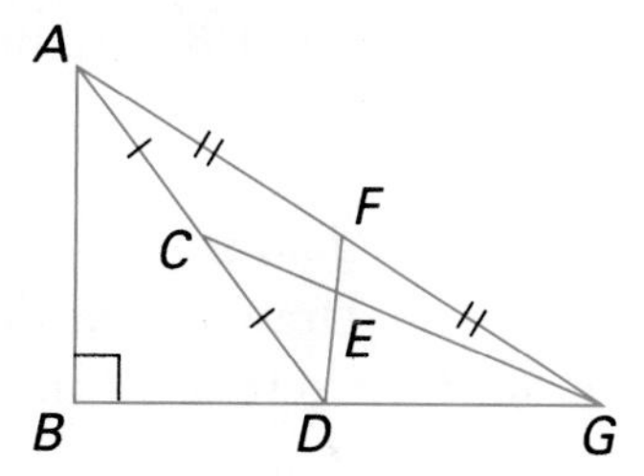

1. $\overline{DF}$ is a median of ______.
2. $\triangle ADG$ has altitude ______.
3. $\triangle ABG$ has altitudes ______ and ______.
4. $\triangle ABD$ has altitudes ______ and ______.
5. The median to $\overline{AD}$ in $\triangle ADG$ is ______

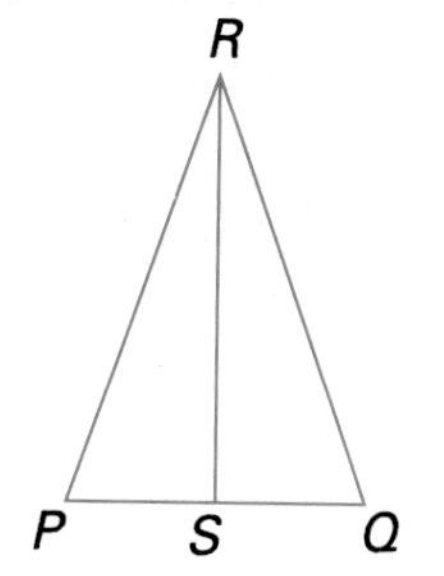

6. Given: $\overline{PR} \cong \overline{QR}$, $\overline{RS}$ is an altitude.
   Prove: $\triangle PSR \cong \triangle QSR$
7. Given: $\overline{RS}$ is a median; $\overline{PR} \cong \overline{QR}$.
   Prove: $\triangle PSR \cong \triangle QSR$
8. Given: $\overline{RS}$ is a median and an altitude.
   Prove: $\triangle PRQ$ is isosceles.

Find the indicated length. (Exercises 9–10)

9. Given: $\triangle TVS \cong \triangle XZW$, $\overline{SU}$ and $\overline{WY}$ are corresponding medians; $SU = 4x - 3$ and $WY = 2x + 7$. Find SU.

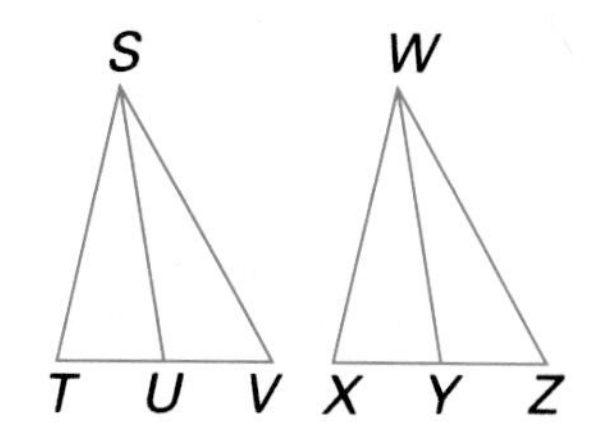

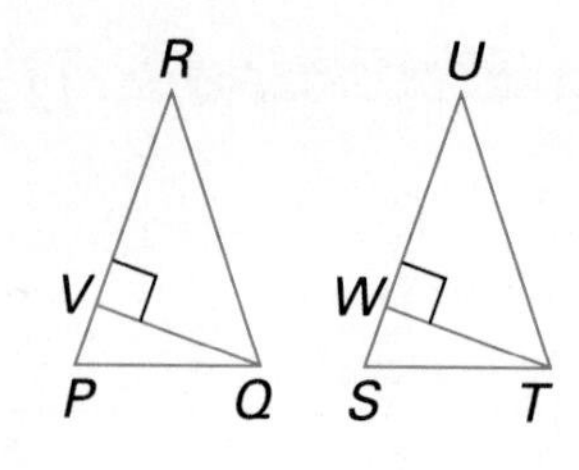

10. Given: $\triangle RPQ \cong \triangle UST$, $\overline{QV}$ and $\overline{TW}$ are corresponding altitudes; $QV = 2(t + 1)$ and $TW = 4(2t - 1)$. Find TW.

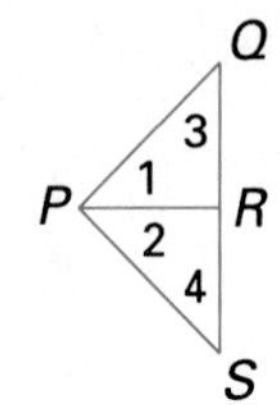

11. Given: $\overline{PR}$ is an altitude of $\triangle PQS$.
Prove: $m\angle 4 + m\angle 2 = m\angle 1 + m\angle 3$

12. Given: $\overline{PQ} \perp \overline{PS}$, $\angle 4 \cong \angle 1$
Prove: $\overline{PR}$ is an altitude of $\triangle PQS$.

13. Given: $\angle 3 \cong \angle 4$, $\overline{PR}$ is a median.
Prove: $\overline{PR}$ bisects $\angle QPS$.

14. Prove Theorem 5.6.

15. Given: $\triangle ACD$, $\overline{DB}$ bisects $\angle ADC$; $\overline{BE}$ and $\overline{BF}$ are altitudes of $\triangle$s ABD and CBD, respectively.
Prove: $\overline{BE} \cong \overline{BF}$

16. Given: $\overline{DB}$ is an altitude from the vertex angle of isosceles $\triangle ADC$; $\overline{BE}$ and $\overline{BF}$ are medians of $\triangle$s ABD and CBD, respectively.
Prove: $\overline{BE} \cong \overline{BF}$

17. Given: $\overline{DB}$ is a median from the vertex of isosceles $\triangle ADC$; $\angle 1 \cong \angle 2$. Draw a flow diagram showing that $\overline{BE} \cong \overline{BF}$.

18. Write a proof for Exercise 17 in two-column form.

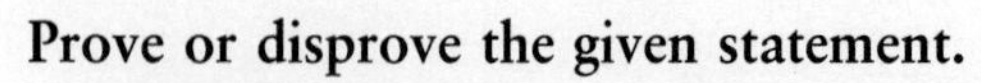

Prove or disprove the given statement.

19. If an altitude of a triangle bisects an angle of the triangle, then the triangle is isosceles.

20. The two medians drawn to the congruent sides of an isosceles triangle are congruent.

21. The angle bisector of a base angle of an isosceles triangle is also an altitude.

Mixed Review

1. Draw an angle. Construct the bisector of the angle. *1.5*

2. The measure of one of two complementary angles is $\frac{2}{3}$ the measure of the other. Find the measure of each angle. *1.6*

3. In $\triangle ABC$, $\overline{AC} \cong \overline{AB}$ and $m\angle A = 30$. Find $m\angle B$. *5.1*

5.6 Perpendicular Bisectors

Objectives

To prove and apply theorems about perpendicular bisectors
To construct perpendicular lines and segments

There are two segments and one line that are of special importance in the study of triangles. They are shown in the figures below.

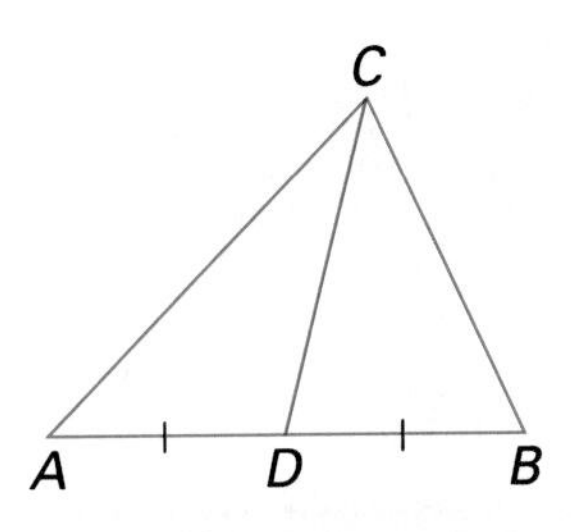

Recall that a *median* is a segment that has a vertex and the midpoint of the opposite side as endpoints.

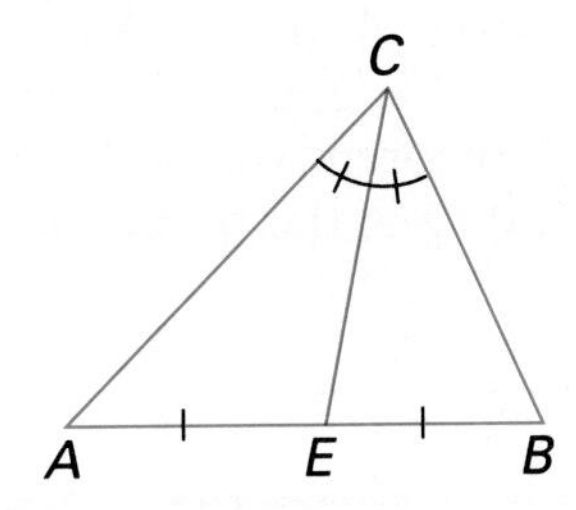

An *angle bisector* bisects an angle of the triangle.

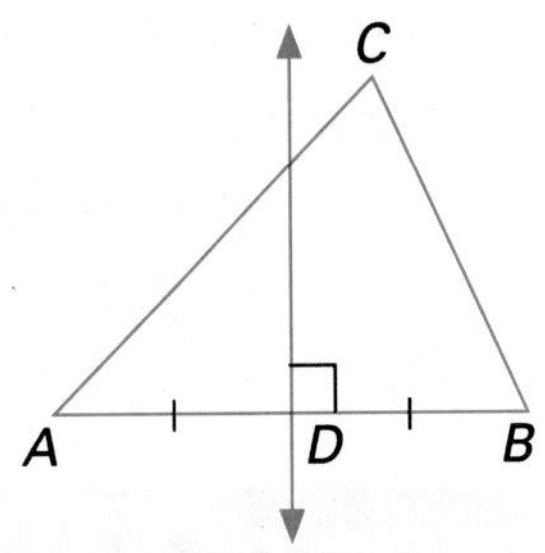

A *perpendicular bisector* of a side is perpendicular to that side at its midpoint.

The median, the angle bisector, and the perpendicular bisector are not necessarily contained in the same line. However, if a triangle is isosceles, then the bisector of the vertex angle is both a median and a perpendicular bisector of the base.

Theorem 5.7

The bisector of the vertex angle of an isosceles triangle is the perpendicular bisector of the base.

Given: $\overline{AC} \cong \overline{BC}$, $\overline{CD}$ bisects $\angle ACB$.
Prove: $\overline{CD}$ bisects $\overline{AB}$ and $\overline{CD} \perp \overline{AB}$.

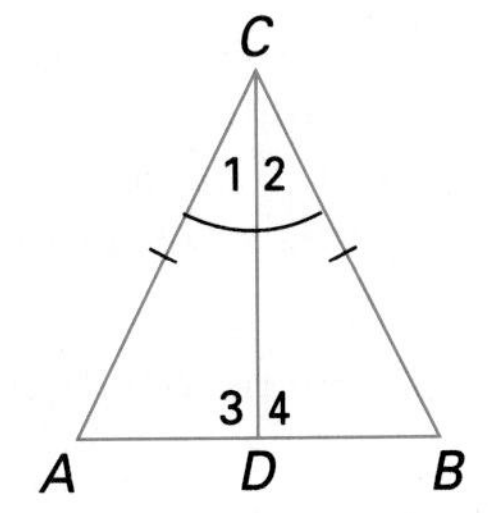

Plan

Prove $\triangle ADC \cong \triangle BDC$. Then $\overline{AD} \cong \overline{BD}$ and $\angle 3 \cong \angle 4$. Note then that m $\angle 3$ + m $\angle 4 = 180$. So, m $\angle 3$ = m $\angle 4 = 90$.

Corollary The bisector of the vertex angle of an isosceles triangle is also a median and an altitude of the triangle.

Construction Construct the perpendicular bisector of a segment.

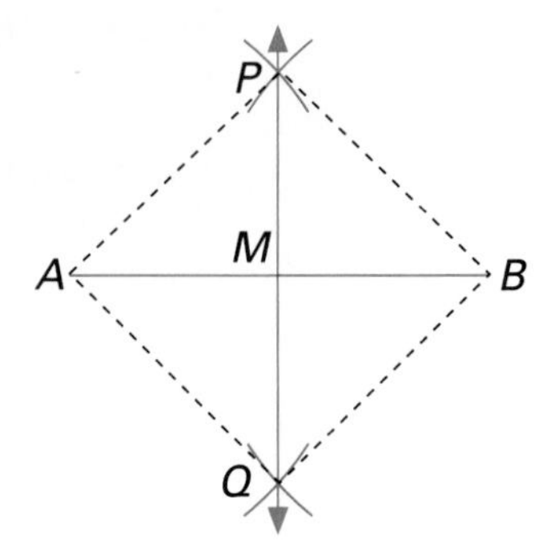

Follow the same steps as for constructing its midpoint (see Lesson 1.3). Notice in this construction that $PA = PB$—that is, P is equidistant from A and B. Also, $QA = QB$—that is, Q is equidistant from A and B. The next theorem states that the line containing P and Q is the perpendicular bisector of $\overline{AB}$.

Theorem 5.8 A line containing two points, each equidistant from the endpoints of a given segment, is the perpendicular bisector of the segment.

Given: $\overline{PA} \cong \overline{PB}$, $\overline{QA} \cong \overline{QB}$
Prove: $\overleftrightarrow{PQ}$ is the perpendicular bisector of $\overline{AB}$.

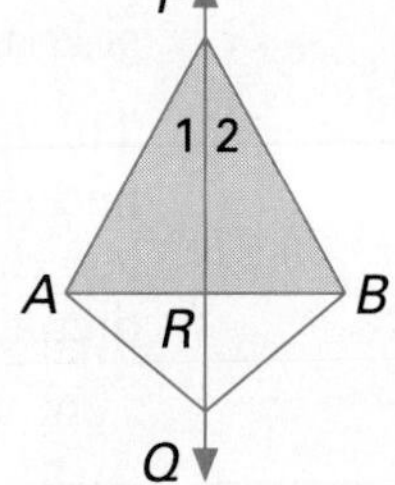

Plan Notice that $\triangle APB$ is isosceles. Prove $\triangle APQ \cong \triangle BPQ$ to get $\angle 1 \cong \angle 2$. Thus, $\overleftrightarrow{PR}$ is the bisector of the vertex angle of an isosceles triangle. Apply Theorem 5.7.

In the figure used for Theorem 5.8, points P and Q were placed on opposite sides of segment $\overline{AB}$. Points P and Q can also be placed on the same side of $\overline{AB}$ in the proof of this theorem.

The important idea in Theorem 5.8 is that points equidistant from the endpoints of the segment are on the perpendicular bisector of the segment. This idea is useful for constructing a perpendicular to a line through any point.

Construction A line perpendicular to a given line through a point not on the line.

Given: Line l and point P not on l
Construct: $\overleftrightarrow{PQ} \perp l$

•P

(See following page for construction.)

Draw an arc with center P, intersecting l at two points. Label the points A and B.

Using A and B as centers, swing equal arcs. Label the point of intersection Q.

Draw $\overleftrightarrow{PQ}$.

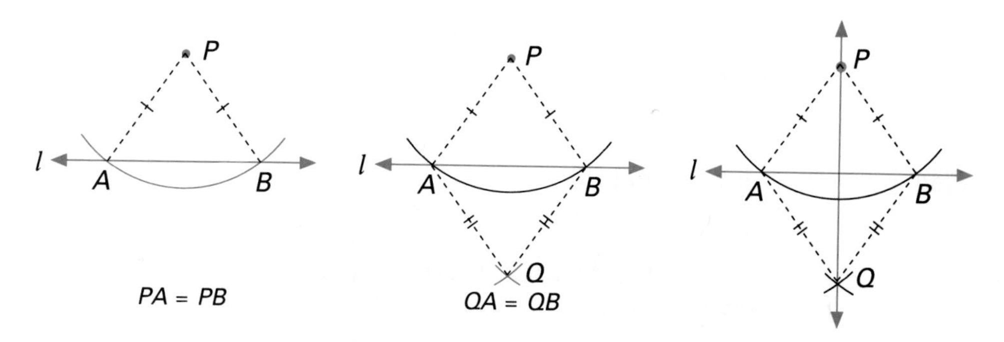

Points P and Q are equidistant from A and B. Result: $\overleftrightarrow{PQ} \perp l$ by Theorem 5.8

Construction A line perpendicular to a given line through a point on the line.

Given: Line l and point P on l

Construct: $\overleftrightarrow{QP} \perp l$

Draw an arc with center P intersecting l at two points. Label the points A and B.

Using A and B as centers, swing equal arcs. Label their intersection Q.

Draw $\overleftrightarrow{QP}$.

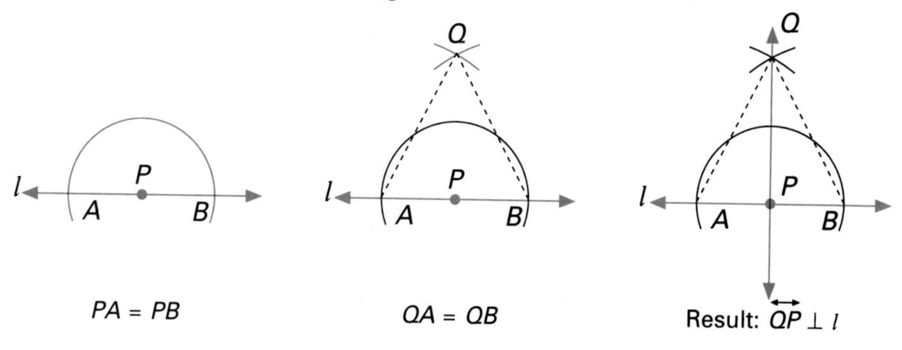

Result: $\overleftrightarrow{QP} \perp l$

You will be asked to prove the result of this construction in Exercise 13.

Theorem 5.9 Any point on the perpendicular bisector of a segment is equidistant from the endpoints of the segment.

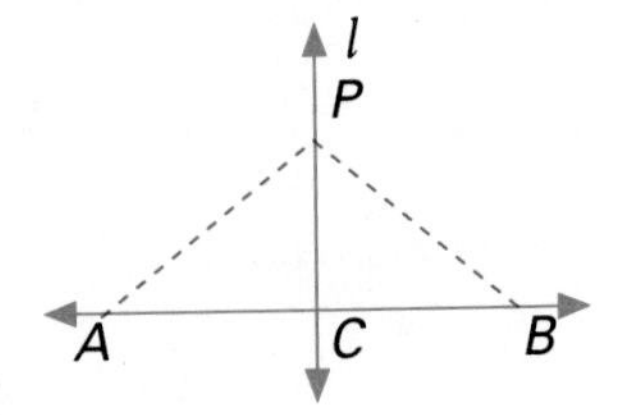

Given: l is the perpendicular bisector of $\overline{AB}$; P is any point on l.
Prove: P is equidistant from the endpoints of $\overline{AB}$.

Plan Draw $\overline{PA}$ and $\overline{PB}$, and prove $\triangle PAC \cong \triangle PBC$.

EXAMPLE 1 Given: $\overline{CD} \perp \overline{AB}$, $\overline{AD} \cong \overline{BD}$
Prove: $\angle A \cong \angle B$

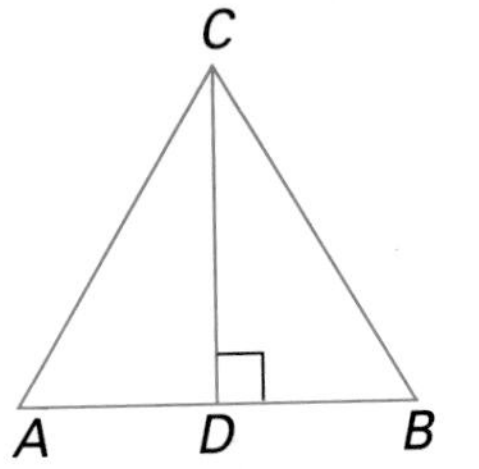

Plan $\overline{CD}$ is the $\perp$ bisector of $\overline{AB}$. Thus, C is equidistant from A and B, that is, $AC = BC$.

Proof

Statement	Reason
1. $\overline{CD} \perp \overline{AB}$, $\overline{AD} \cong \overline{BD}$	1. Given
2. D is the midpoint of $\overline{AB}$.	2. Def of midpt
3. $\overline{CD}$ is the $\perp$ bisector of $\overline{AB}$.	3. Def of $\perp$ bis
4. $AC = BC$ ($\overline{AC} \cong \overline{BC}$)	4. Any point on $\perp$ bis of a seg is equidistant from seg endpts.
5. $\therefore \angle A \cong \angle B$	5. $\angle$s opp $\cong$ sides of a $\triangle$ are $\cong$.

Focus on Reading

Use the figure and its markings to complete the sentence.

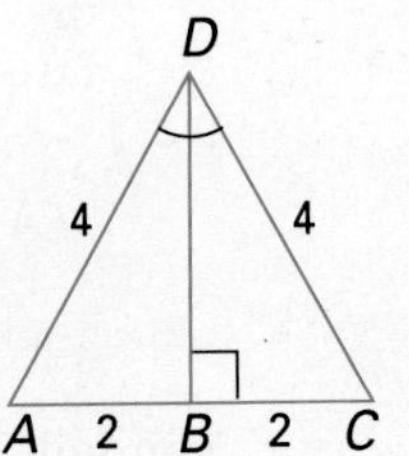

1. $\overline{DB}$ is the __________ of $\overline{AC}$.
2. Point D is __________ from the __________ of $\overline{AC}$.
3. $\angle ADC$ is the __________ of the isosceles triangle.
4. __________ is a median of $\triangle DAC$.
5. Any point on $\overline{DB}$ is equidistant from points __________ and __________.

Classroom Exercises

Based on the markings in the figure, state which segment is a perpendicular bisector of another segment. Justify your answer.

1.

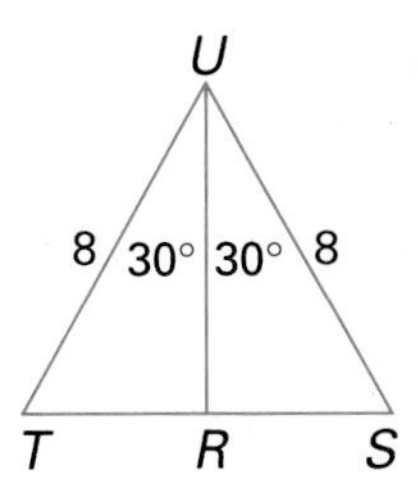

2.

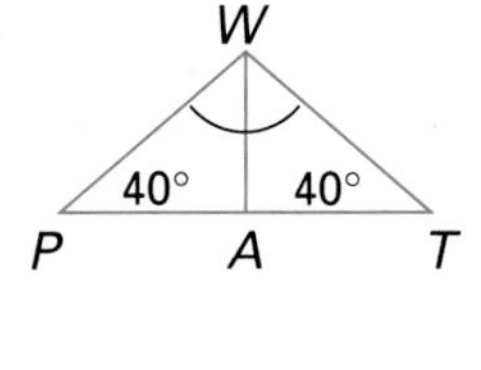

3.

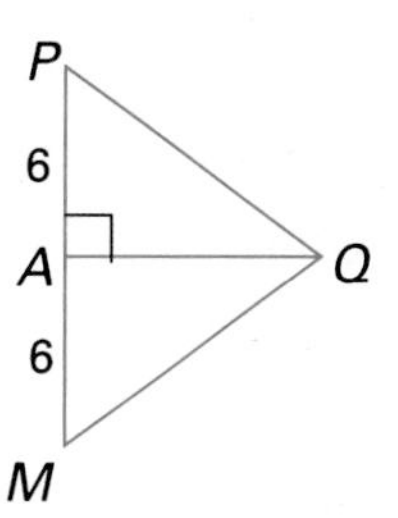

4.

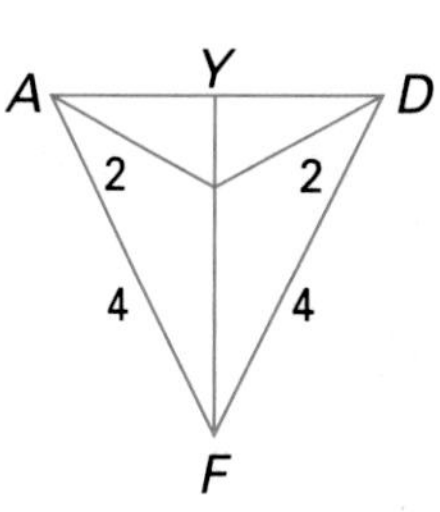

Written Exercises

1. Draw a line and a point not on the line. Construct the perpendicular to the line from the point.

2. Draw a line. Mark a point on the line. Construct the perpendicular to the line at the point.

Write the proof without showing triangles congruent. (Exercises 3–8)

3. Given: $\overline{TQ}$ bisects $\angle PTR$; $\overline{TP} \cong \overline{TR}$.
   Prove: $\overline{TQ}$ is $\perp$ bis of $\overline{PR}$.
4. Given: $\overline{TQ} \perp \overline{PR}$, $\overline{PQ} \cong \overline{RQ}$
   Prove: $\angle 5 \cong \angle 6$
5. Given: $\angle 3 \cong \angle 4$, $\overline{TQ}$ bisects $\angle PTR$.
   Prove: $\overline{PQ} \cong \overline{RQ}$

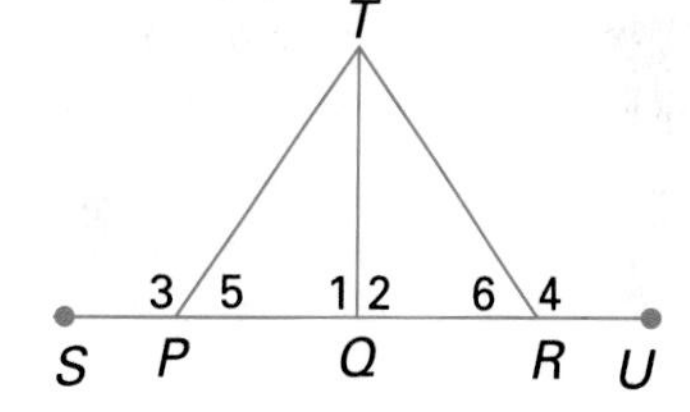

6. Given: $\angle 3 \cong \angle 4$, $\angle 1 \cong \angle 2$
   Prove: $\overline{CD}$ is $\perp$ bis of $\overline{AB}$.
7. Given: $\overline{CD}$ is $\perp$ bis of $\overline{AB}$.
   Prove: $\angle 1 \cong \angle 2$
8. Given: $\angle 3 \cong \angle 4$, $\angle CAD \cong \angle CBD$
   Prove: $\overline{CD}$ is $\perp$ bis of $\overline{AB}$.

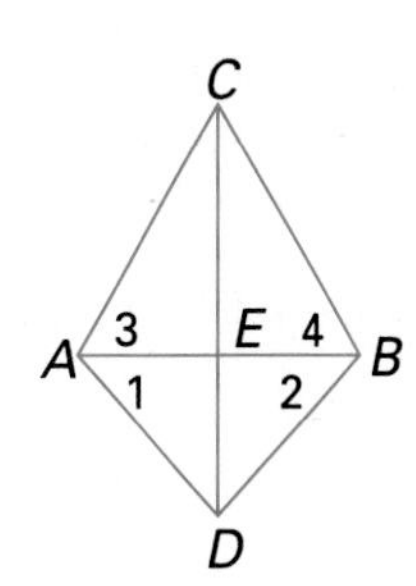

9. Prove that there is exactly one perpendicular to a line from a point not on that line. (HINT: Use an indirect proof.)
10. Prove that there is exactly one perpendicular to a line at a point on that line. (HINT: Use an indirect proof.)

Mixed Review

1. List the four congruence patterns for triangles. *4.3–4.6*
2. Draw an angle. Construct a second angle congruent to it. *1.4*

Find the complement and supplement of angle A, where possible. *1.6*

3. $m \angle A = 40$
4. $m \angle A = 140$
5. $m \angle A = x$
6. $m \angle A = 3x - 10$

Application: *Congruent Relationships*

The congruent relationships in an isosceles triangle and its altitude, a perpendicular bisector, are used often in construction. Following are different examples of how these properties can be applied.

1. Electric wires are to be strung between two poles across a ravine. The power company needs to know how much wire is needed, but it is impossible to measure directly across the ravine. How could this distance be found using the given diagram? Why does this work?

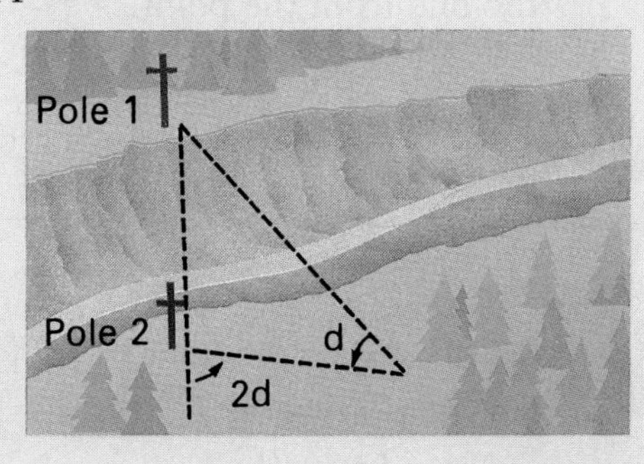

2. A homemade carpenter's level has an isosceles triangular frame, with a plumb bob hanging from the vertex. The plumb bob should always hang vertically. How can this be used to show whether or not a surface is horizontal? Why does this work?

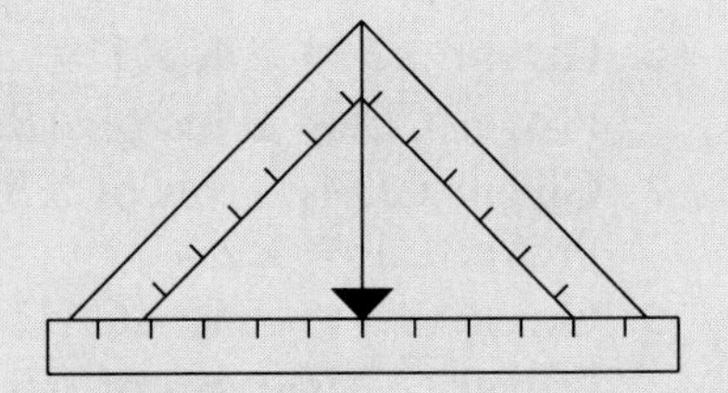

3. An artist is painting a large design in which an angle must be bisected. A carpenter's square is the only measuring device available. How can the square be used to bisect the angle? Why does this work?

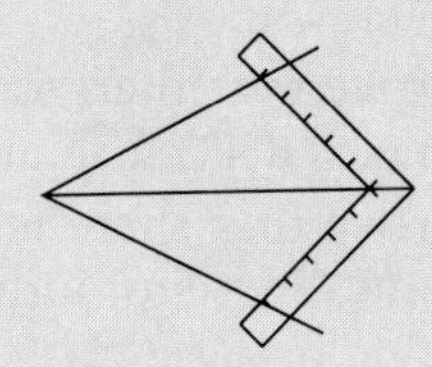

5.7 Inequalities in a Triangle

Objective To prove and apply inequality relationships in a triangle

Algebraic inequalities are frequently used in geometric proofs. The formal definitions of “$>$” and “$<$” are suggested by the following: $8 > 5$ since there exists a positive number, 3, such that $5 + 3 = 8$.

Definition

$a > b$ if there exists a positive number c such that $b + c = a$.
$a < b$ if there exists a positive number c such that $a + c = b$.

The definition of “$>$” can be used in proving the following theorem.

Theorem 5.10

Exterior Angle Inequality Theorem: The measure of an exterior angle of a triangle is greater than the measure of either of its remote interior angles.

Given: $\triangle ABC$ with exterior $\angle 1$
Prove: $m\angle 1 > m\angle 3$ and $m\angle 1 > m\angle 2$

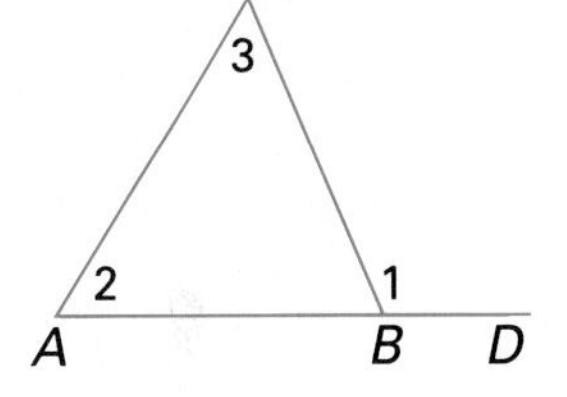

Proof

Statement	Reason
1. $\triangle ABC$ with exterior $\angle 1$	1. Given
2. $m\angle 1 = m\angle 2 + m\angle 3$	2. Ext $\angle$ Thm
3. $\therefore m\angle 1 > m\angle 3$, $m\angle 1 > m\angle 2$	3. Def of $>$

The following are properties of inequalities:

(1) If $a > b$ and $b > c$, then $a > c$ (Trans Prop of Ineq)
(2) If $a > b$ and $c = d$, then $a + c > b + d$ (Add Prop of Ineq)
(3) If $a > b$ and $c > d$, then $a + c > b + d$
(4) If $a > b$ and $b = c$, then $a > c$ (Sub Prop of Ineq)

Any triangle with sides of unequal length also has angles of unequal measure. The orders of inequality for the sides and the angles are related. In the figure, $AB > AC$.

It appears that $m\angle C > m\angle B$. The larger angle appears to be opposite the longer side. This is proved in the following theorem.

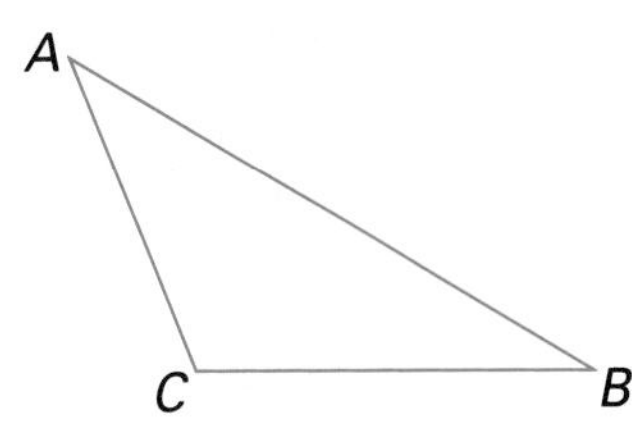

Theorem 5.11 If one side of a triangle is longer than another side, then the measure of the angle opposite the longer side is greater than the measure of the angle opposite the shorter side.

Given: $\triangle RPQ$ with $RQ > RP$
Prove: $m\angle RPQ > m\angle Q$

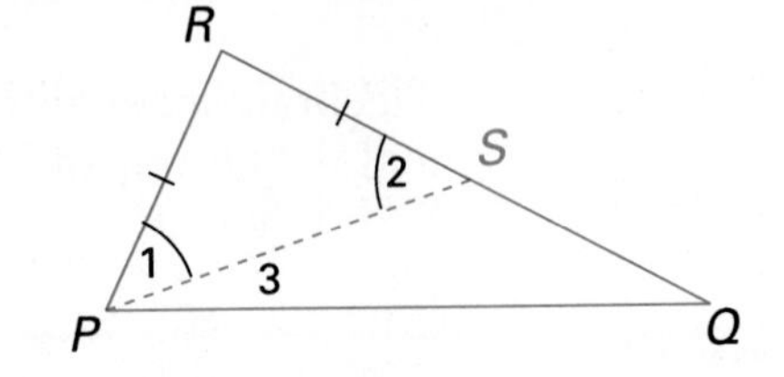

Plan Locate point S on $\overline{RQ}$ so that $\overline{RP} \cong \overline{RS}$. Then $\triangle RPS$ is isosceles.

Proof

Statement	Reason
1. $\triangle RPQ$ with $RQ > RP$	1. Given
2. On $\overline{RQ}$, choose S so that $RS = RP$.	2. Ruler Post
3. Draw $\overline{PS}$.	3. Two points determine a line.
4. $\triangle RPS$ is isosceles.	4. Def of isos $\triangle$
5. $\angle 1 \cong \angle 2$ ($m\angle 1 = m\angle 2$)	5. Base $\angle$s of isos $\triangle$ are $\cong$.
6. $m\angle RPQ = m\angle 1 + m\angle 3$	6. $\angle$ Add Post
7. $m\angle RPQ > m\angle 1$	7. Def of $>$
8. $m\angle RPQ > m\angle 2$	8. Sub Prop of Ineq
9. $m\angle 2 > m\angle Q$	9. Ext $\angle$ Ineq Thm
10. $\therefore m\angle RPQ > m\angle Q$	10. Trans Prop of Ineq

The converse of Theorem 5.11 can be proved by using the following property in an indirect proof.

Trichotomy Property
For any two real numbers a and b, only one of three possible relationships exists: $a < b$ or $a = b$ or $a > b$.

Theorem 5.12 If one angle of a triangle has a greater measure than a second angle, then the side opposite the greater angle is longer than the side opposite the smaller angle.

Given: $\triangle RPQ$ with $m\angle P > m\angle Q$
Prove: $RQ > RP$

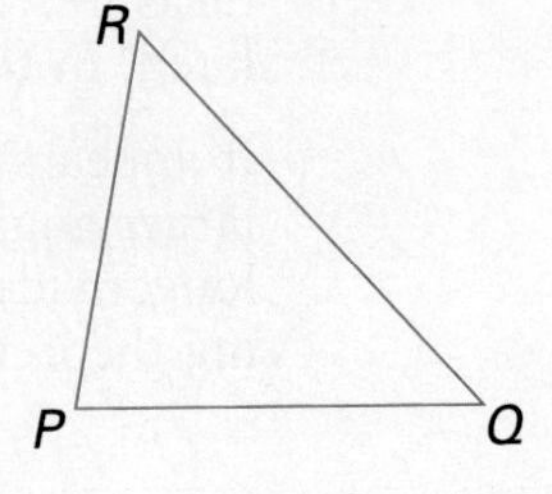

Plan According to the Trichotomy Property, one of three cases holds: $RQ < RP$, $RQ = RP$, or $RQ > RP$. Prove that each of the first two cases leads to a contradiction. Then the third case must hold.

Theorem 5.13 In a scalene triangle, the longest side is opposite the largest angle and the largest angle is opposite the longest side.

You will be asked to prove Theorem 5.13 in Exercise 9.

EXAMPLE 1 Name the longest side of the given triangle.

Solution

$$\begin{aligned} m\angle A + m\angle B + m\angle C &= 180 \\ (x - 30) + (2x + 10) + x &= 180 \\ 4x - 20 &= 180 \\ 4x &= 200 \\ x &= 50 \end{aligned}$$

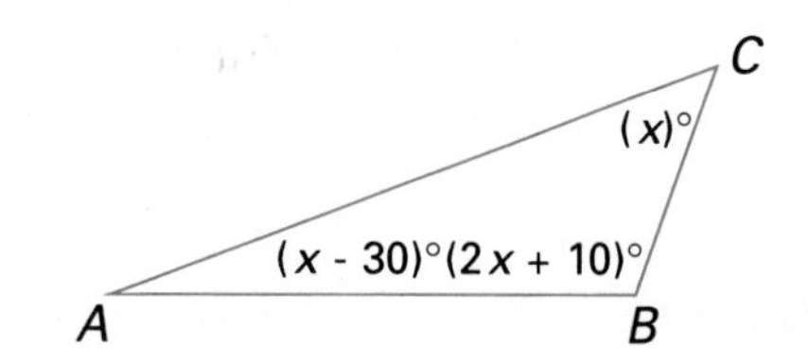

$m\angle A = x - 30 = 50 - 30 = 20$

$m\angle B = 2x + 10 = 2 \cdot 50 + 10 = 110$

$m\angle C = x = 50$

Thus, the longest side is $\overline{AC}$.

EXAMPLE 2 Name the longest segment in the figure.

Solution In $\triangle ABD$, $\overline{DB}$ is the longest side since it is opposite the largest angle. In $\triangle BCD$, $\overline{DC}$ is longer than any other side, including $\overline{DB}$. Therefore, $\overline{DC}$ is the longest segment in the figure.

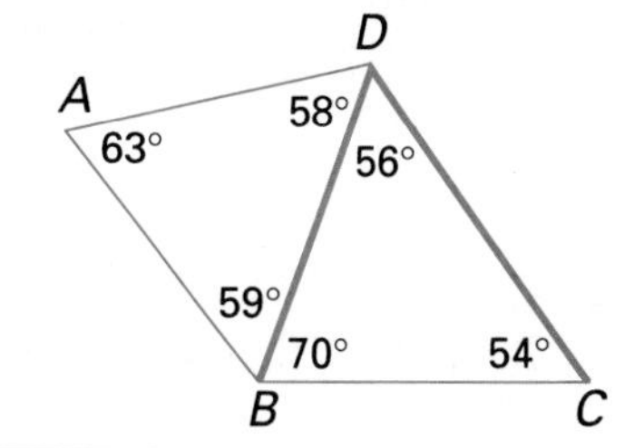

EXAMPLE 3 Given: $m\angle Q < m\angle P$, $RP > RS$
Prove: $m\angle Q < m\angle 1$

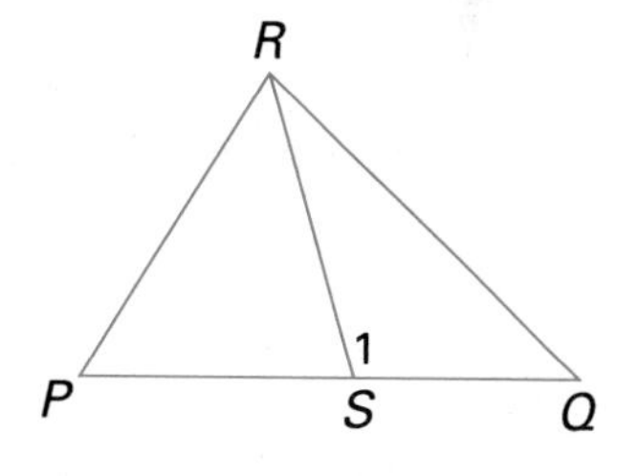

Proof

Statement	Reason
1. $m\angle Q < m\angle P$, or $m\angle P > m\angle Q$	1. Given
2. $RQ > RP$	2. The longer side is opp the larger $\angle$ in a $\triangle$.
3. $RP > RS$	3. Given
4. $RQ > RS$	4. Tran Prop of Ineq
5. $\therefore m\angle 1 > m\angle Q$, or $m\angle Q < m\angle 1$	5. The larger $\angle$ is opp the longer side in a $\triangle$.

Classroom Exercises

Name the angles in order from smallest to largest.

1. 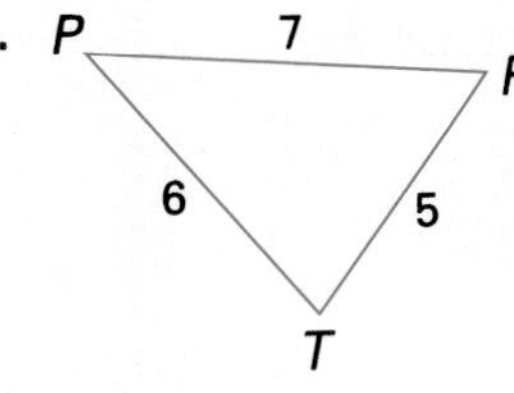

2. 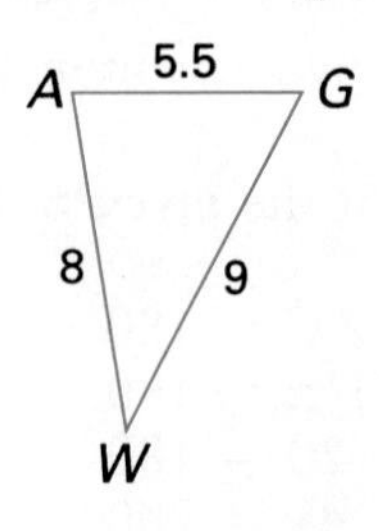

3. 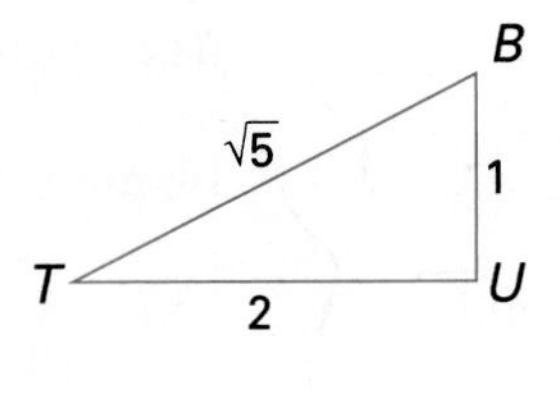

Written Exercises

In Exercises 1–3, determine the longest segment.

1. 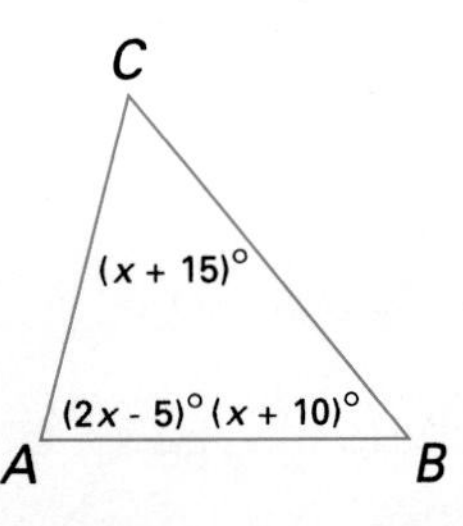

2. 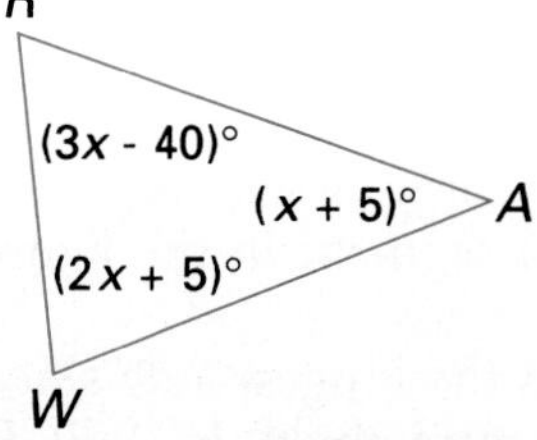

3. 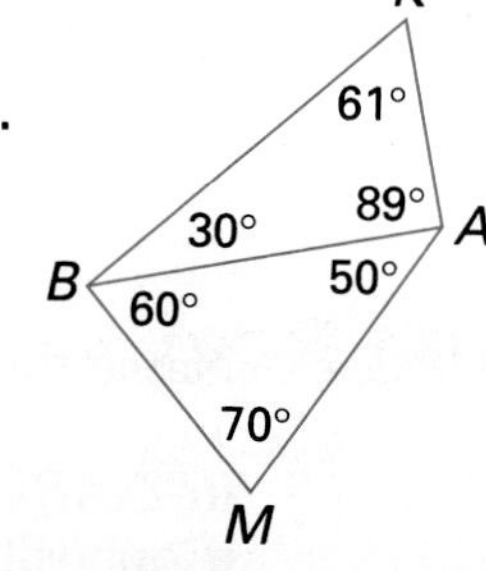

4. Given: $AD = BD$
   Prove: $DC > AD$

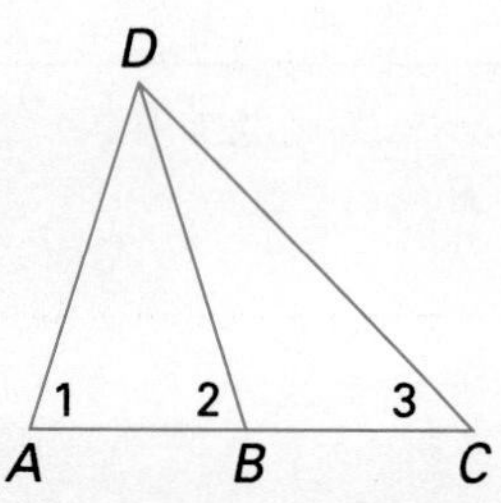

5. Given: $m\angle 2 > m\angle P$
   Prove: $PT > QT$

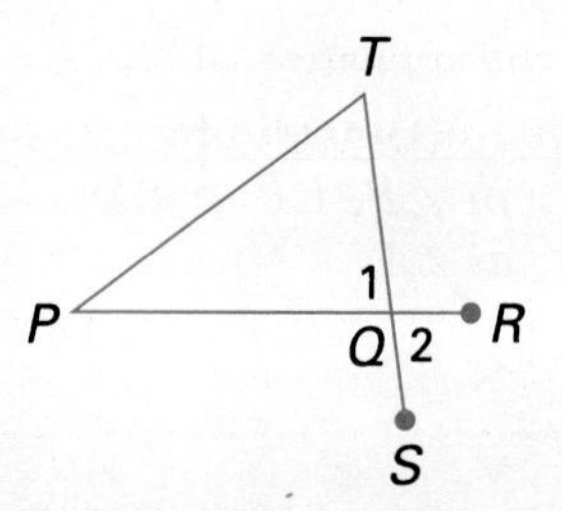

6. Given: $XZ > YZ$,
   $m\angle X > m\angle Z$
   Prove: $XZ > XY$

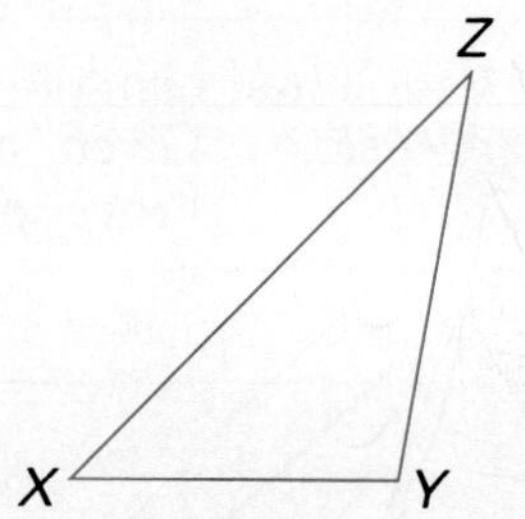

7. Given: $\overline{SQ}$ bisects $\angle PSR$.
   Prove: $RS > RQ$

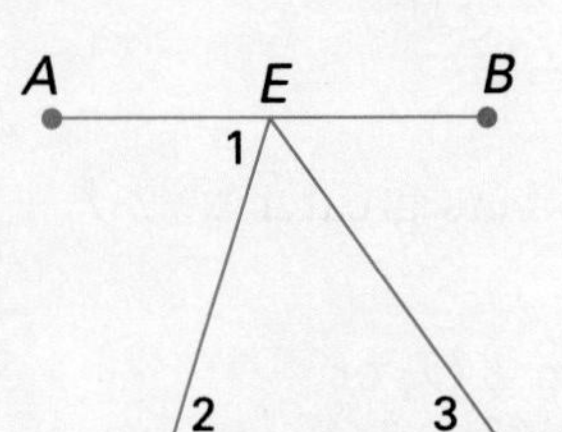

8. Given: $\overline{AB} \parallel \overline{CD}$,
   $m\angle 1 > m\angle 3$
   Prove: $ED > EC$

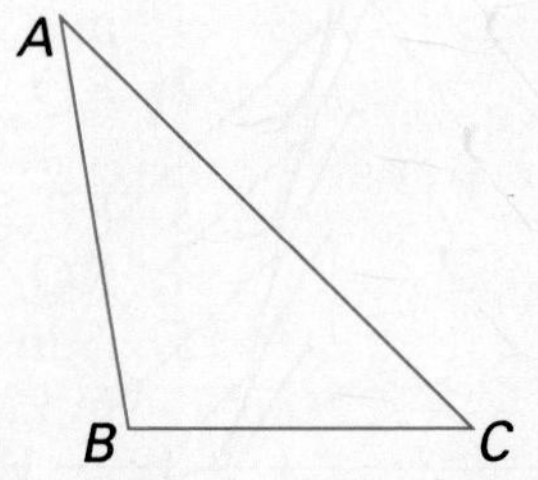

9. Prove Theorem 5.13.

Complete each statement for $\triangle ABC$.

10. If $AB > BC$, then ______ $>$ ______

11. If $AC < AB$, then ______ $<$ ______

12. If m $\angle A <$ m $\angle B$, then ______ $<$ ______

13. If m $\angle C >$ m $\angle B$, then ______ $<$ ______

14. Prove Theorem 5.12.

15. Given: $SQ > SP$, $SR = SQ$
Prove: m $\angle P$ + m $\angle R$ $>$ m $\angle PQR$

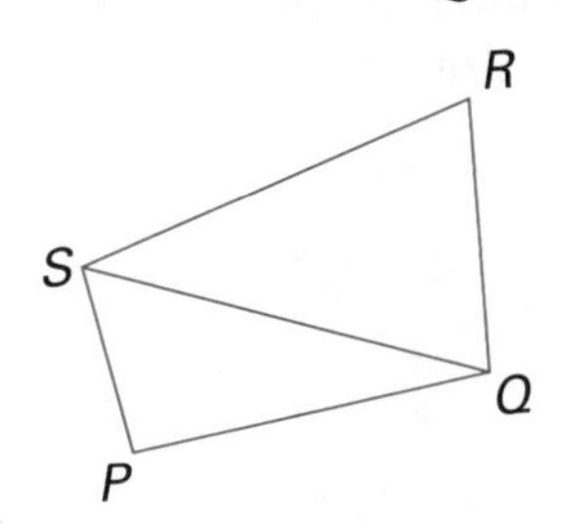

16. Given: $YW = YZ$, $YZ > YX$
Prove: m $\angle XWZ <$ m $\angle X$ + m $\angle Z$

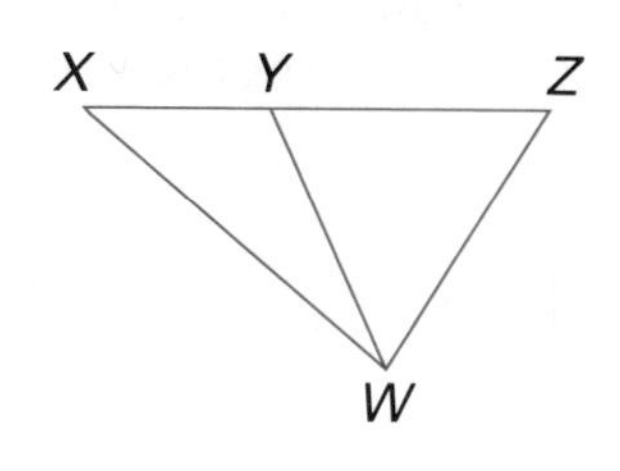

17. Given: $AB = BC$
Prove: $AB + BC > AC$

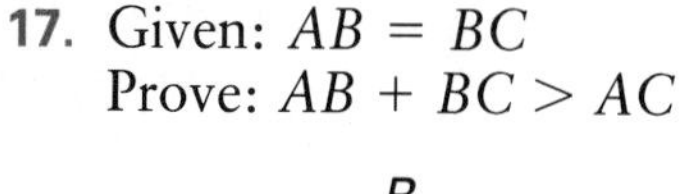

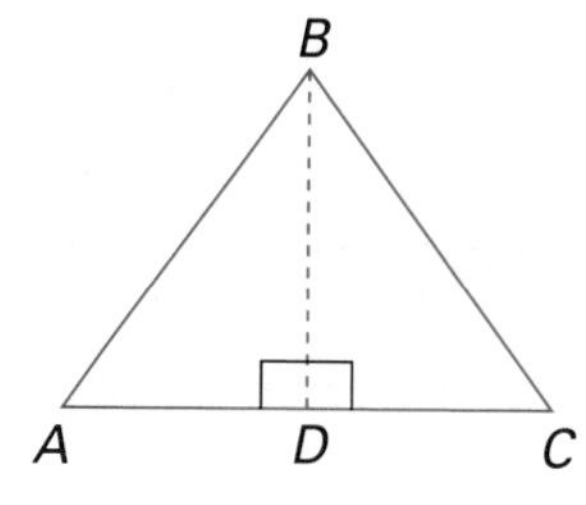

Mixed Review

1. The coordinates of the endpoints of a segment are -8 and 12. Find the coordinate of the midpoint of the segment. ***1.3***
2. The measures of two angles formed by an angle bisector of an angle are $3x - 20$ and $x + 40$. Find the measure of each angle. ***1.5***
3. The measure of one acute angle of a right triangle is 4 times the measure of the other. Find the measure of the larger acute angle. ***4.1***
4. Graph and name the resulting geometric figure: $x \geq 5$ *and* $x \geq 7$. ***1.8***

Algebra Review

Solve an inequality as you would solve an equation. However, when you multiply or divide either side of an inequality by a negative number, reverse the inequality symbol.

Example: Solve.
$$-3x + 14 < 8$$
$$-3x < -6$$
$$x > 2$$

The solution consists of all numbers greater than 2.

Solve.

1. $-15 < 9y - 6$

2. $3a + 8 < -19$

3. $-4x + 12 > -16$

4. $4x + 13 \leq 3x - 10$

5. $\frac{2a}{3} + 9 \leq 11$

6. $-4 + \frac{3m}{2} < m$

5.8 The Triangle Inequality Theorem

Objectives

To apply the Triangle Inequality Theorem

To determine whether a triangle can be constructed, given lengths for three sides

To write paragraph proofs

The first figure below shows a segment from point P perpendicular to line l. The second figure shows several segments drawn from P to l.

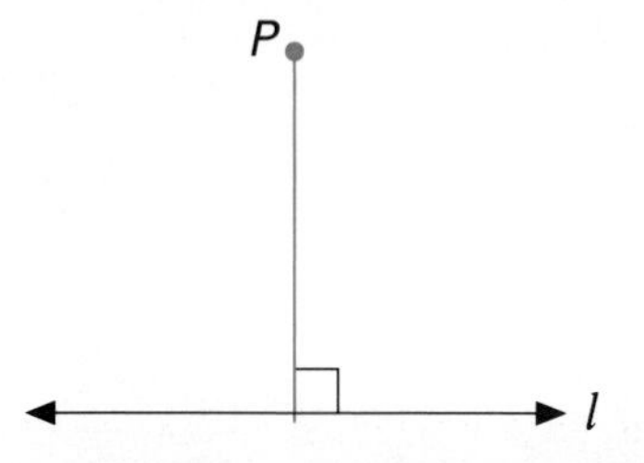

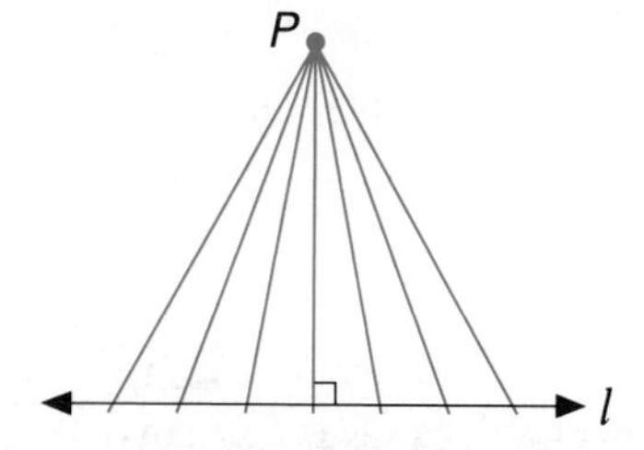

Notice that of all the segments drawn from point P to line l, the shortest one appears to be the perpendicular segment. This leads to the following theorem. The proof of the theorem is presented in paragraph form.

Theorem 5.14 The perpendicular segment from a point to a line is the shortest segment from the point to the line.

Given: $\overline{PA} \perp l$

Prove: $\overline{PA}$ is the shortest segment from P to l.

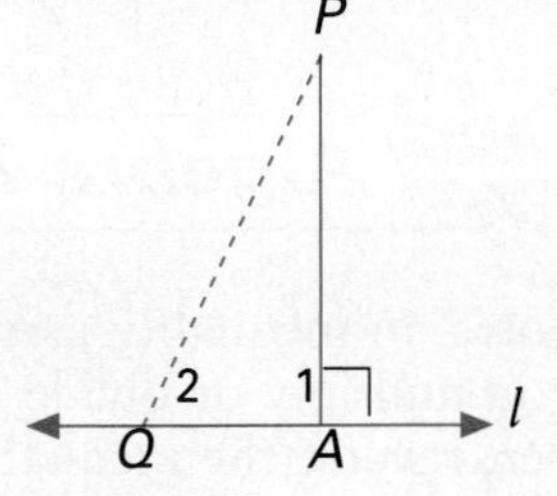

Proof (Paragraph Form)

Draw any segment, $\overline{PQ}$ from P to l, along with $\overline{PA} \perp l$. Then show that $PQ > PA$, as follows:

$\triangle PAQ$ is a right triangle. Thus, $\angle 1$ is a right angle. This makes $\angle 2$ an acute angle and m $\angle 1 >$ m $\angle 2$. Therefore, $PQ > PA$, since in a triangle the longer side is opposite the larger angle.

Corollary The longest side of a right triangle is the hypotenuse.

Given: Rt $\triangle ABC$ with hypotenuse $\overline{AB}$

Prove: $\overline{AB}$ is the longest side of the triangle.

EXAMPLE 1 If possible, draw a triangle with the given lengths for the three sides.

2 cm, 3 cm, 6 cm | 2 cm, 4 cm, 6 cm

Solution

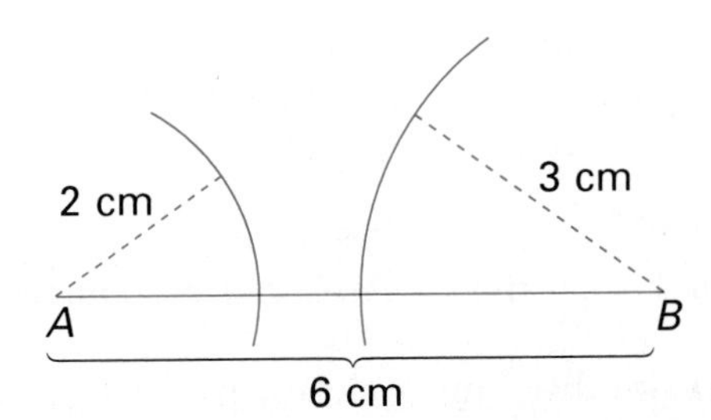

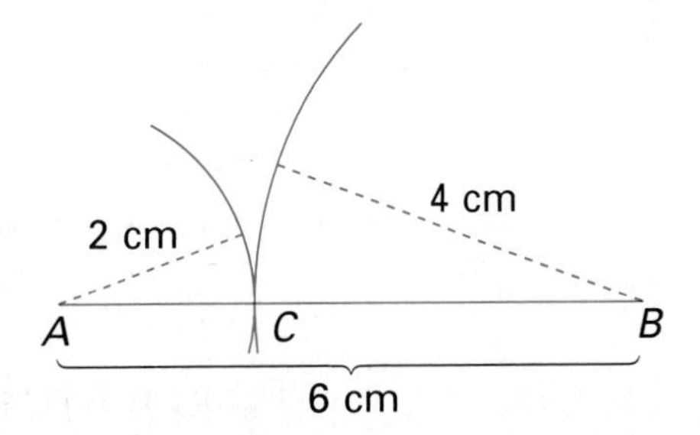

The 2-cm and 3-cm segments will not meet to form a third vertex, C. No triangle is formed.

$2 + 3 < 6$

The 2-cm and 4-cm segments meet on $\overline{AB}$. So, the third vertex lies on $\overline{AB}$. No triangle is formed.

$2 + 4 = 6$

Example 1 suggests that certain restrictions exist for the lengths of the sides of a triangle.

Theorem 5.15 **The Triangle Inequality Theorem:** The sum of the lengths of any two sides of a triangle is greater than the length of the third side.

Given: $\triangle ABC$
Prove: 1. $AB + BC > AC$
2. $AC + AB > BC$
3. $AC + BC > AB$

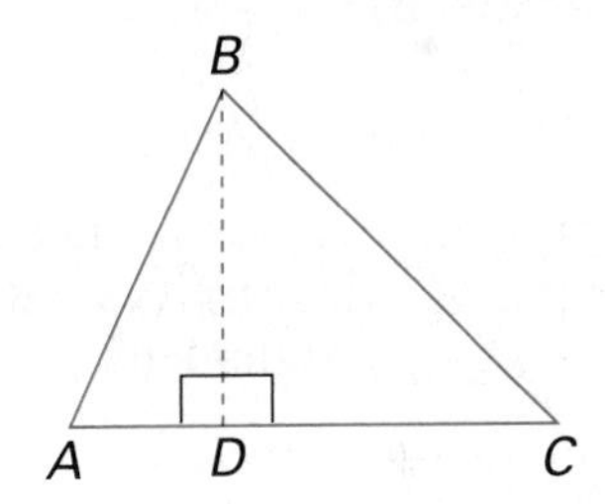

Proof

Statement	Reason
1. From B, construct $\overline{BD} \perp \overline{AC}$.	1. From a pt not on a line, exactly one ⊥ seg can be constructed to that line.
2. $AB > AD$; $BC > DC$	2. The longest side of a rt △ is the hyp.
3. $AB + BC > AD + DC$	3. Add Prop of Ineq
4. $AC = AD + DC$	4. Seg Add Post
5. $\therefore AB + BC > AC$	5. Sub

The proofs of parts 2 and 3 are done similarly.

EXAMPLE 2 Can a triangle be constructed with sides of the given lengths?

4 m, 8 m, 3 m   7 in, 2 in, 6 in

Solution Check the sum of the two smallest lengths.

$3 + 4 = 7$, and $7 < 8$. $2 + 6 = 8$, and $8 > 7$.

A triangle cannot be constructed.   A triangle can be constructed.

EXAMPLE 3 Find the restrictions on x such that $\triangle ABC$ can be constructed.

Solution Use Theorem 5.15 to write three inequalities.
Sum of two lengths > third length

(1) $4 + 6 > x$
(2) $4 + x > 6$
(3) $x + 6 > 4$

The third inequality is true in any case because $6 > 4$ and x is positive.

Solve the first two inequalities.

$4 + 6 > x$ and $4 + x > 6$
$10 > x$ and $x > 2$

So, $x > 2$ and $x < 10$, or $2 < x < 10$

Therefore, x can be any number between 2 and 10.

EXAMPLE 4 Given: $m\angle 1 = m\angle 2$
Prove: $BD + DC > AC$

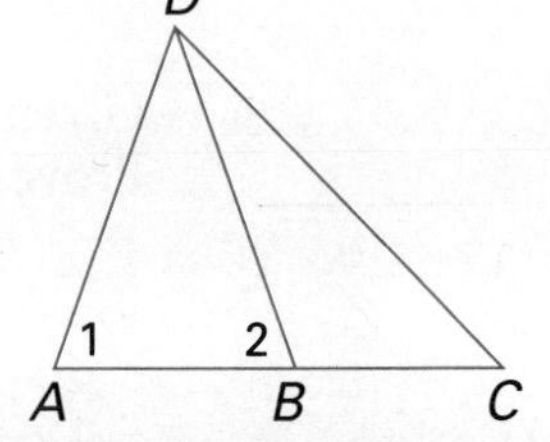

Plan $AD + DC > AC$.

Substitute BD for AD since $m\angle 1 = m\angle 2$.

Proof

Statement	Reason
1. $m\angle 1 = m\angle 2$ ($\angle 1 \cong \angle 2$)	1. Given
2. $AD = BD$ ($\overline{AD} \cong \overline{BD}$)	2. Sides opp $\cong$ $\angle$s of a $\triangle$ are $\cong$.
3. $AD + DC > AC$	3. $\triangle$ Ineq Thm
4. $\therefore BD + DC > AC$	4. Sub

Classroom Exercises

Tell whether or not it is possible to form a triangle with these sides.

1. 3, 5, 7   **2.** 2, 6, 3   **3.** 7, 5, 9   **4.** 4 m, 6 m, 1 m

5. 4.3, 6, 0.9   **6.** 3, 10, 13   **7.** $1\frac{1}{2}, 2\frac{1}{2}, 4\frac{1}{2}$ **8.** 5.1 ft, 7 ft, 2.3 ft

Written Exercises

Determine whether a triangle can be formed having the given lengths for sides. If not, indicate why.

1. 4 cm, 5 cm, 2 cm
2. 4 in, 7 in, 3 in
3. 9 cm, 3 cm, 7 cm

Complete each statement.

4. The longest side of a right triangle is the ________.
5. The shortest segment from a point to a line is the ________.
6. Determine the shortest segment from P to $\overline{RS}$. Explain why it is shortest.
7. From P, construct the shortest segment to l.

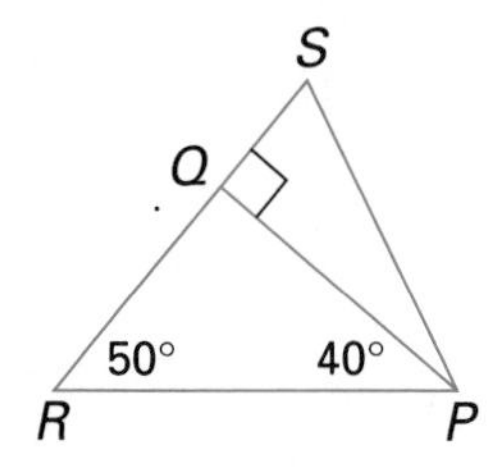

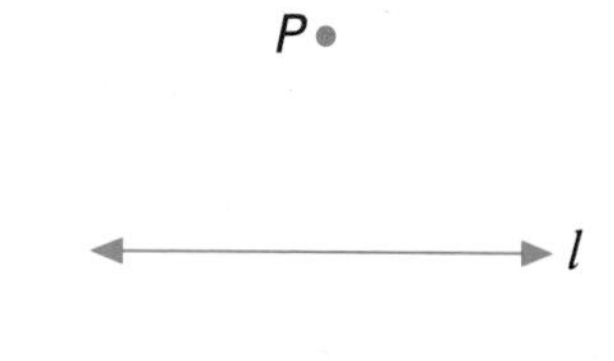

Find the restrictions on x for constructing the triangle.

8.

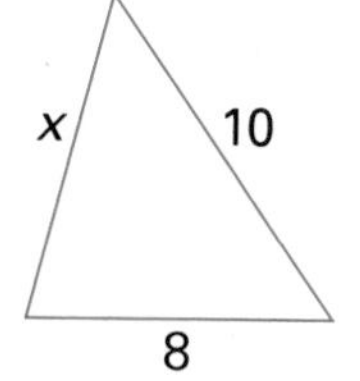

9.

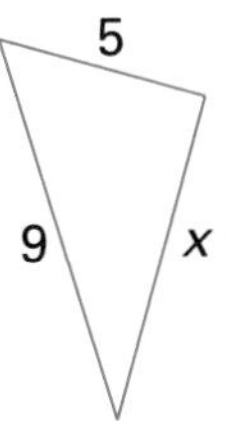

10.

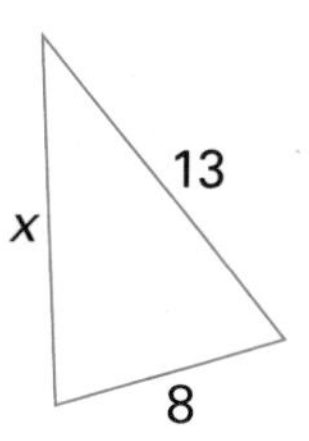

11.

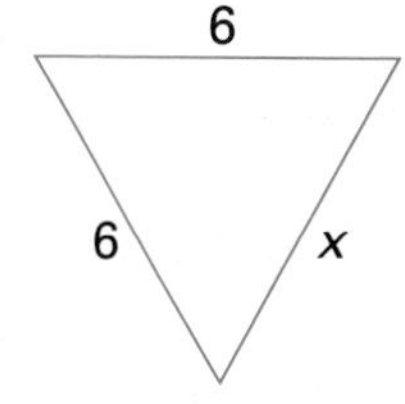

12. Two sides of a triangle have lengths 20 and 30. The length of the third side can be any number between ______ and ______.
13. The lengths of two sides of a triangle are 15 and 8. Use an inequality to express the range of the length of the third side.
14. The length of the base of an isosceles triangle is 12. What can be said about the lengths of the legs of the triangle?

The map at the right shows air distances from New York to Chicago, Dallas, and Miami. Use the Triangle Inequality Theorem to find the range of distances between:

15. Chicago and Dallas.
16. Dallas and Miami.
17. Chicago and Miami.

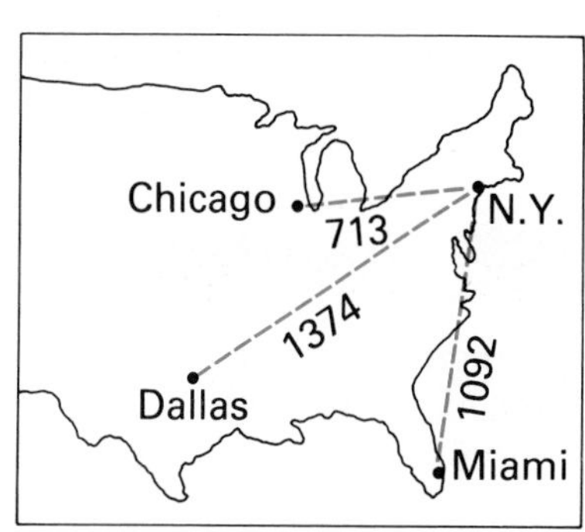

18. In the figure at the right, $\angle A \cong \angle C$.
Prove: $BD + DC > AB$

19. Write a paragraph proof of the Corollary to Theorem 5.14.

20. Prove Theorem 5.15 for a right triangle.

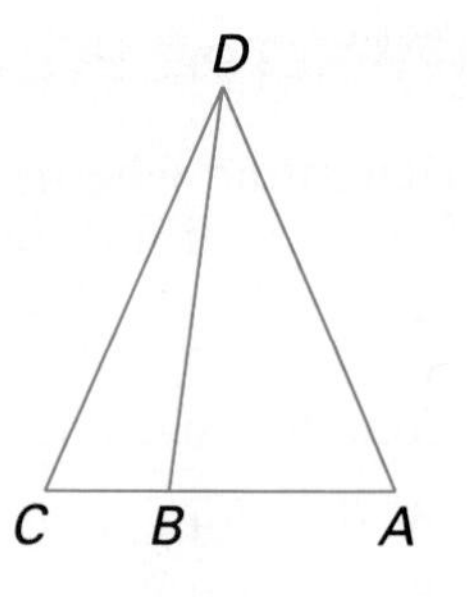

In $\triangle ABC$, AB and BC are given, find the range of values for AC.

21. $AB = 27.61$, $BC = 65.09$

22. $AB = 94.83$, $BC = 101.48$

23. Prove Theorem 5.15 for an obtuse triangle.

24. Prove that if Q is a point on $\overline{XY}$ of $\triangle XYZ$, then $2 \cdot ZQ < XZ + ZY + XY$.

25. Prove: $PR + RQ > PT + TQ$

26. Prove: $AC + BD > AD + BC$

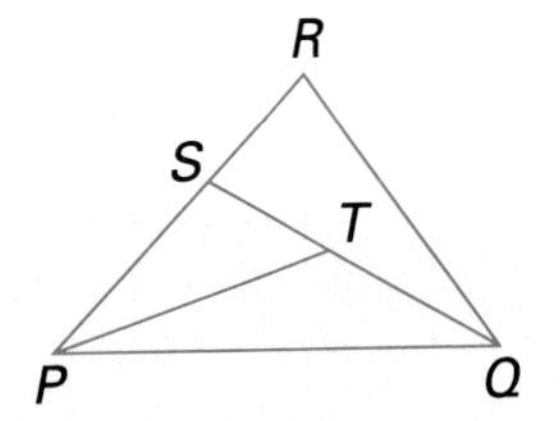

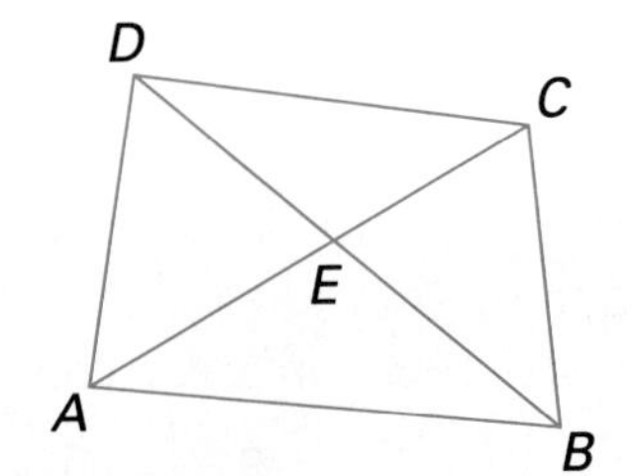

Mixed Review

1. The measure of the vertex angle of an isosceles triangle is 110. Find the measure of each base angle. 5.1

2. Given: $\overline{QS} \cong \overline{QW}$, $\overline{PS} \cong \overline{PW}$
Prove: $\angle 1 \cong \angle 2$ 5.3

3. Given: $\overline{AE} \perp \overline{AC}$, $\overline{CD} \perp \overline{AC}$, $\overline{AC}$ bisects $\overline{ED}$.
Prove: $\angle E \cong \angle D$ 4.6

4. Determine the longest segment. 5.7

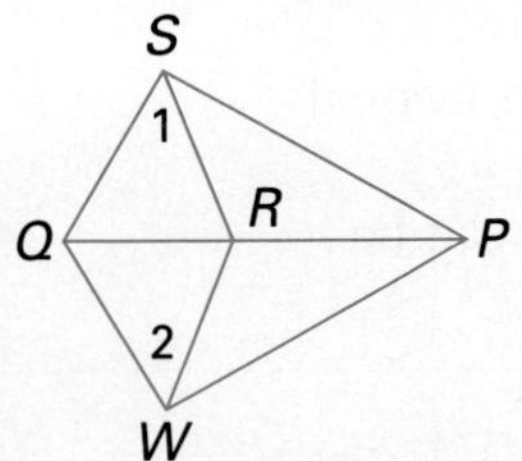

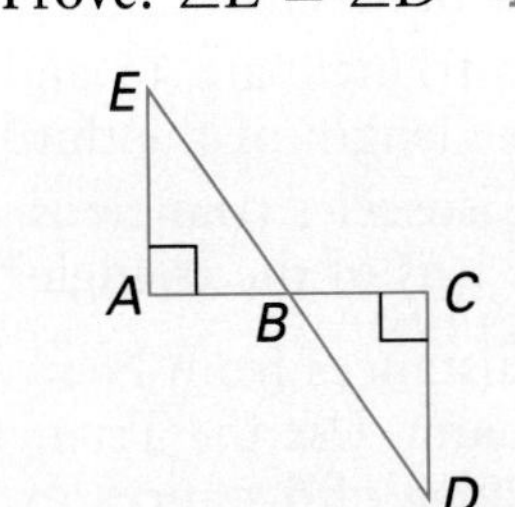

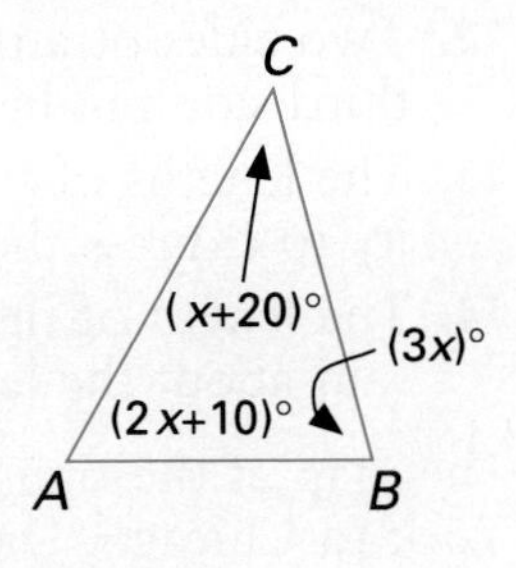

Brainteaser

The figure to the right shows a triangle with three medians. How many triangles in all are there in the figure?

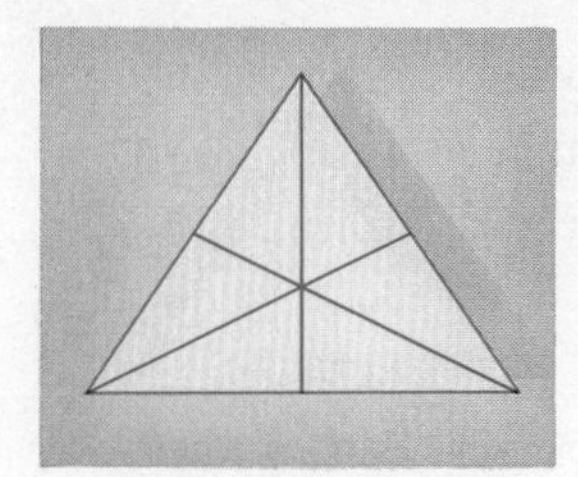

5.9 Inequalities for Two Triangles

Objective To apply the SAS and SSS Inequality Theorems

Think of two sticks hinged together and connected with an elastic band. As the angle between the sticks becomes greater, the length of the band increases. This suggests Theorem 5.16.

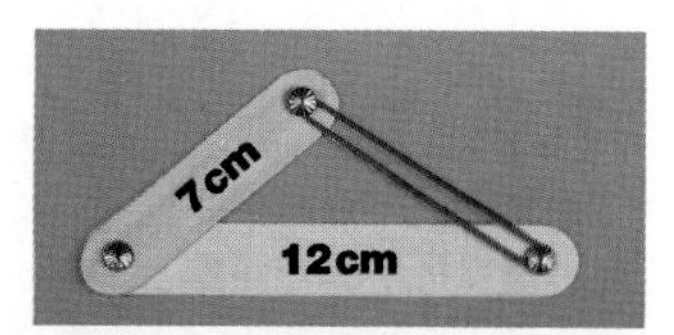

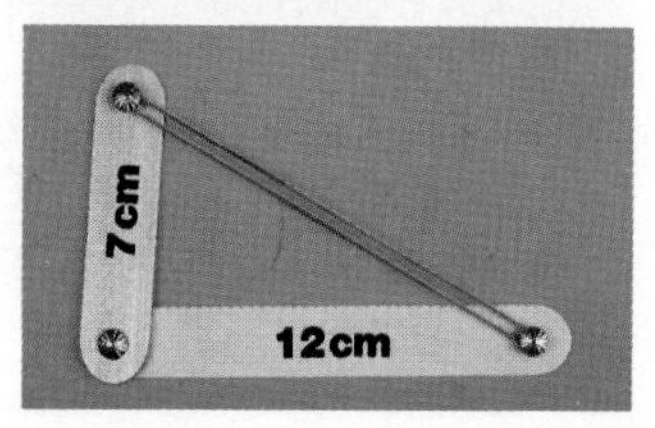

Theorem 5.16 **The SAS Inequality Theorem:** If two sides of one triangle are congruent, respectively, to two sides of a second triangle, and the included angle of the first triangle has a greater measure than the included angle of the second triangle, then the third side of the first triangle is longer than the third side of the second triangle.

Given: $\triangle ABC$ and $\triangle DEF$, with $\overline{CA} \cong \overline{FD}$, $\overline{CB} \cong \overline{FE}$, and $m\angle C > m\angle F$

Prove: $AB > DE$

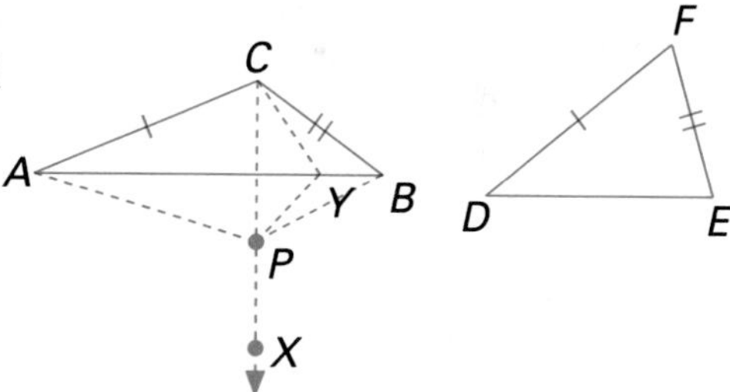

Proof

Statement	Reason
1. $\overline{CA} \cong \overline{FD}$	(S) 1. Given
2. Construct $\overrightarrow{CX}$ such that $m\angle ACX = m\angle F$.	(A) 2. Protractor Post
3. Locate point P on $\overrightarrow{CX}$ such that $CP = FE$ ($\overline{CP} \cong \overline{FE}$).	(S) 3. Ruler Post
4. $\triangle APC \cong \triangle DEF$	4. SAS
5. Construct bis of $\angle PCB$.	5. Protractor Post
6. Locate point Y, the intersection of $\overline{AB}$ and the bis of $\angle PCB$.	6. Ruler Post

7. $\overline{CY} \cong \overline{CY}$	(S) 7. Reflex Prop
8. $\angle PCY \cong \angle BCY$	(A) 8. Def of ∠ bis
9. $FE = CB$ ($\overline{FE} \cong \overline{CB}$)	9. Given
10. $CP = CB$ ($\overline{CP} \cong \overline{CB}$)	(S) 10. Trans Prop
11. $\triangle PCY \cong \triangle BCY$	11. SAS
12. $PY = BY$ ($\overline{PY} \cong \overline{BY}$)	12. CPCTC
13. $AY + PY > AP$	13. △ Ineq Thm
14. $AY + BY > AP$	14. Sub
15. $AY + BY = AB$	15. Seg Add Post
16. $AP = DE$	16. CPCTC
17. $\therefore AB > DE$	17. Sub Prop of Ineq

The SSS Inequality Theorem can be proved indirectly by using the Trichotomy Property.

Theorem 5.17 **The SSS Inequality Theorem:** If two sides of one triangle are congruent, respectively, to two sides of a second triangle, and the length of the third side of the first triangle is greater than the length of the third side of the second triangle, then the angle opposite the third side of the first triangle has a greater measure than the angle opposite the third side of the second triangle.

Given: $\triangle ABC$ and $\triangle DEF$ with $\overline{AC} \cong \overline{DF}$, $\overline{CB} \cong \overline{FE}$, and $AB > DE$

Prove: $m \angle C > m \angle F$

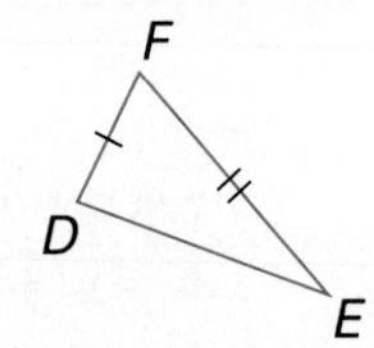

Plan Use an indirect proof. By the Trichotomy Property, $m \angle C < m \angle F$, or $m \angle C = m \angle F$, or $m \angle C > m \angle F$. Show that the first two assumptions lead to a contradiction.

EXAMPLE 1 Given: $\overline{AD} \cong \overline{DC}$, $AB > BC$

Prove: $m \angle 1 > m \angle 2$

Proof

Statement	Reason
1. $\overline{AD} \cong \overline{DC}$	1. Given
2. $\overline{DB} \cong \overline{DB}$	2. Reflex Prop
3. $AB > BC$	3. Given
4. $\therefore m \angle 1 > m \angle 2$	4. SSS Ineq Thm

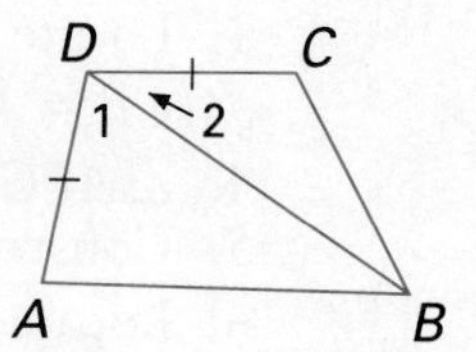

EXAMPLE 2 Given: $\overline{DB}$ bisects $\overline{AC}$; m $\angle 1 >$ m $\angle 2$
Prove: m $\angle A >$ m $\angle C$

Plan Use the SAS Inequality Theorem to prove $DC > DA$. Then, in $\triangle ACD$, m $\angle A >$ m $\angle C$.

Proof

Statement	Reason
1. $\overline{DB}$ bisects $\overline{AC}$.	1. Given
2. $\overline{AB} \cong \overline{BC}$	2. Def of seg bis
3. $\overline{DB} \cong \overline{DB}$	3. Reflex Prop
4. m $\angle 1 >$ m $\angle 2$	4. Given
5. $DC > DA$	5. SAS Ineq Thm
6. $\therefore$ m $\angle A >$ m $\angle C$	6. In a $\triangle$, larger $\angle$ is opp longer side.

Focus on Reading

Indicate whether the statement is sometimes true, always true, or never true.

1. The lengths of two sides of a triangle are equal, respectively, to the lengths of two sides of a second triangle, and the third side of the first triangle is longer than the third side of the second triangle.
2. The lengths of two sides of a triangle are equal, respectively, to the lengths of two sides of a second triangle, the included angle of the first triangle has a greater measure than the included angle of the second triangle, and the third sides are equal in length.
3. The legs of a right triangle are congruent, respectively, to the legs of a second right triangle, and the hypotenuse of the first right triangle is longer than the hypotenuse of the second right triangle.

Classroom Exercises

Write an inequality comparing the indicated side lengths or the indicated angle measures.

1. $\overline{AB}$ and $\overline{BC}$

2. $\angle A$ and $\angle B$

3. 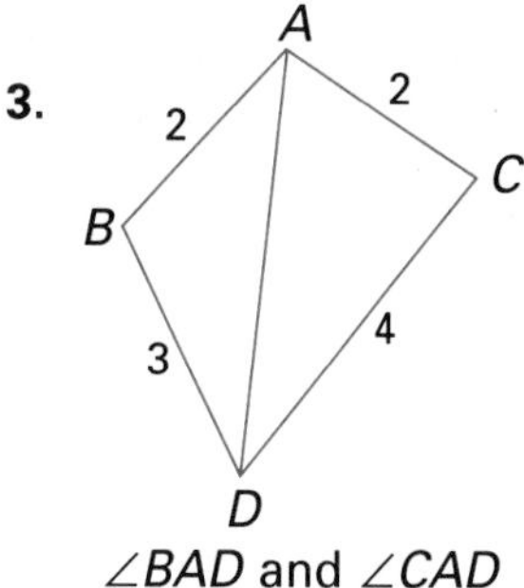

$\angle BAD$ and $\angle CAD$

Written Exercises

Insert the correct inequality symbol. (Exercises 1–3)

1. $AB = BC$, m $\angle 1 =$ 30, m $\angle 2 = 20$
AD ___ DC.

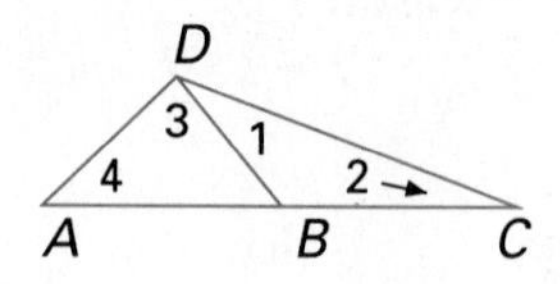

2. $DA = CB$, m $\angle DAB =$ 115, $\overline{DA} \parallel \overline{CB}$
DB ___ AC.

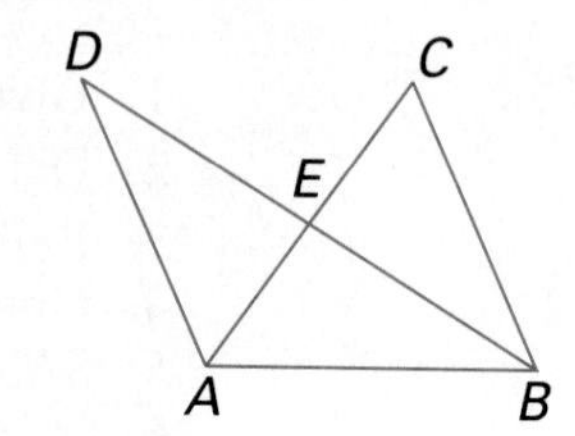

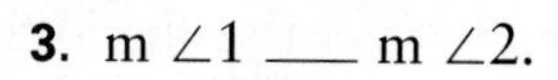

3. m $\angle 1$ ___ m $\angle 2$.

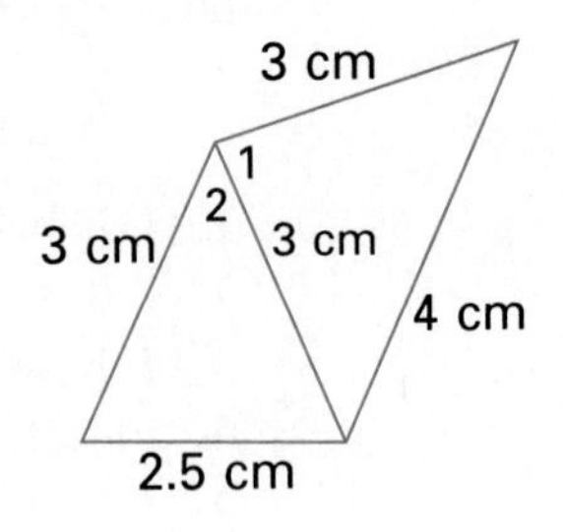

4. Given: $\overline{SQ}$ bisects $\overline{PR}$; $SR > SP$
Prove: m $\angle 2 >$ m $\angle 1$

5. Given: m $\angle P >$ m $\angle R$, $PQ = QR$
Prove: m $\angle 2 >$ m $\angle 1$

6. Given: m $\angle P =$ m $\angle R$, m $\angle 3 >$ m $\angle 4$
Prove: $PQ > QR$

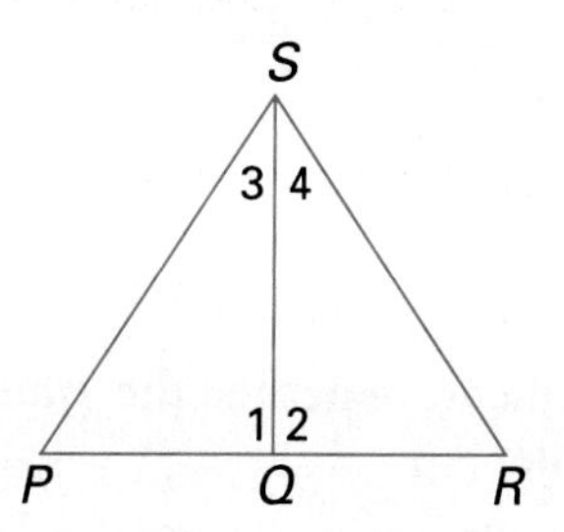

7. Given: $QR = PS$, $QS < PR$
Prove: m $\angle PQR >$ m $\angle SPQ$

8. Given: T is the midpoint of $\overline{QS}$; m $\angle PTQ >$ m $\angle PTS$
Prove: $PQ > PS$

9. Given: $TP > TQ$, $SQ = PR$
Prove: $PS > QR$

10. Given: m $\angle 4 >$ m $\angle 1$, m $\angle 3 >$ m $\angle 2$, $PS = QR$
Prove: $QS > PR$

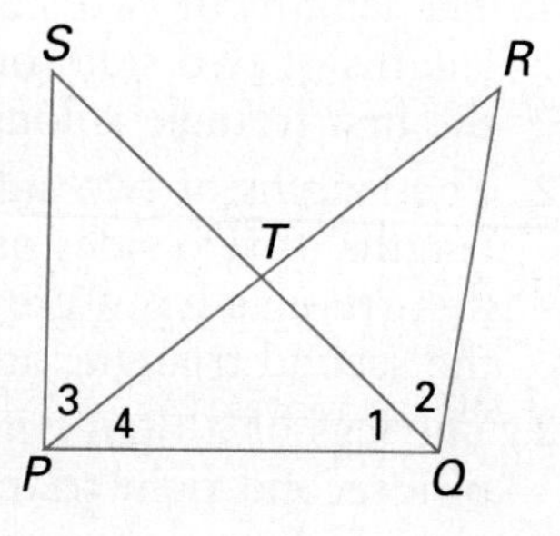

Essay

11. Write an explanation of why Theorem 5.16 may be referred to as the Hinge Theorem.

12. Given: $QS = TR$, $SR > TQ$
Prove: $PR > PQ$

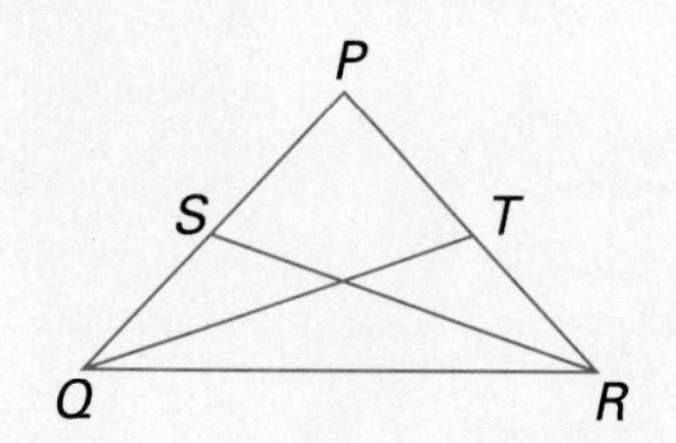

13. Given: $\overline{ED} \cong \overline{DF}$, m $\angle 1 >$ m $\angle 2$,
D is midpoint of $\overline{CB}$;
$\overline{AE} \cong \overline{AF}$
Prove: $AC > AB$

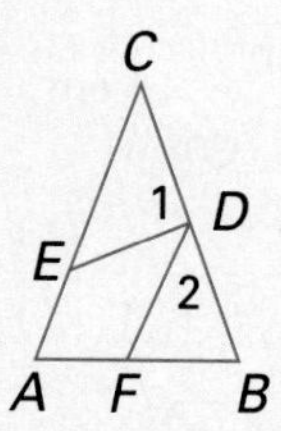

14. Given: m $\angle DBC$ = m $\angle DCB$,
m $\angle ADB$ < m $\angle ADC$
Prove: m $\angle ACB$ < m $\angle ABC$

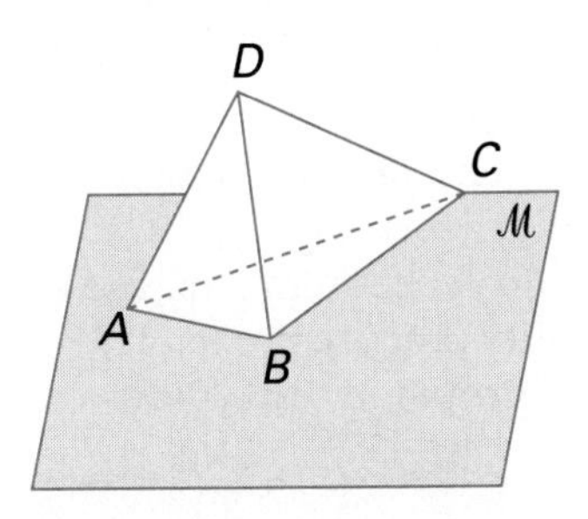

15. Given: $AE = AB = ED = DC$,
$EB < EC$
Prove: m $\angle AEB$ > m $\angle DCE$

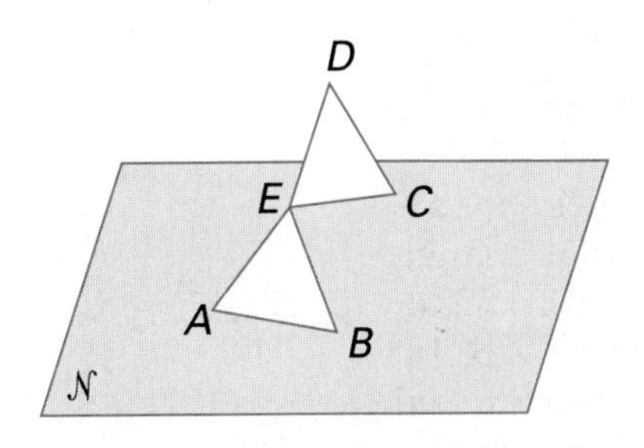

16. Prove the SSS Inequality Theorem.

Mixed Review

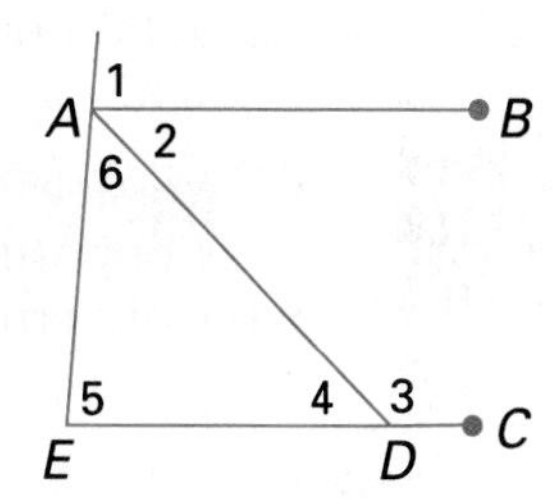

In the figure, $\overline{AB} \parallel \overline{EC}$ 3.5, 3.7

1. Find m $\angle 1$ if m $\angle 5 = 40$. 3.5
2. Find m $\angle 5$ if m $\angle EAB = 150$. 3.5
3. Given: m $\angle 5 = 70$, m $\angle 6 = 40$. Find m $\angle 3$. 3.7
4. Given: m $\angle 4 = 30$, m $\angle 6$ is twice m $\angle 5$. Find m $\angle 5$. 3.7

Algebra Review

To multiply or divide fractions:

Use $\frac{a}{b} \cdot \frac{c}{d} = \frac{a \cdot c}{b \cdot d}$ or $\frac{a}{b} \div \frac{c}{d} = \frac{a \cdot d}{b \cdot c}$.

Example: Divide $\frac{x^2 - 10x + 16}{x^2 - 49} \div \frac{3x - 24}{x - 7}$.

$$\frac{x^2 - 10x + 16}{x^2 - 49} \div \frac{3x - 24}{x - 7} = \frac{(x^2 - 10x + 16)(x - 7)}{(x^2 - 49)(3x - 24)}$$

$$= \frac{\overset{1}{\cancel{(x - 8)}}(x - 2)\overset{1}{\cancel{(x - 7)}}}{\underset{1}{\cancel{(x - 7)}}(x + 7)(3)\underset{1}{\cancel{(x - 8)}}} = \frac{x - 2}{3(x + 7)}$$

Multiply or divide.

1. $\frac{x^2 - 4}{4x + 12} \cdot \frac{2x + 6}{4x + 8}$

2. $\frac{x^2 - 2x - 8}{x^2 - 25} \div \frac{3x - 9}{x - 5}$

3. $\frac{m^2 - 4}{a + 1} \cdot \frac{7a + 7}{2m + 4}$

Chapter 5 Review

Key Terms

altitude (p. 194)
base (p. 177)
base angles (p. 177)
equiangular triangle (p. 177)
equilateral triangle (p. 177)
hypotenuse (p. 188)
leg of isosceles triangle (p. 177)
leg of right triangle (p. 188)
median (p. 194)
perimeter of equilateral triangle (p. 178)
perpendicular bisector (p. 199)
SAS Inequality Theorem (p. 215)
SSS Inequality Theorem (p. 216)
trichotomy property (p. 206)
vertex angle (p. 177)

Key Ideas and Review Exercises

5.1 To write proofs and solve problems about isosceles triangles, use these properties of an isosceles triangle:

- Two sides (legs) are $\cong$: $\overline{AB} \cong \overline{AC}$
- $\angle$s opposite $\cong$ legs are $\cong$: $\angle B \cong \angle C$
- Bis of vertex $\angle$ is $\perp$ bis of base.

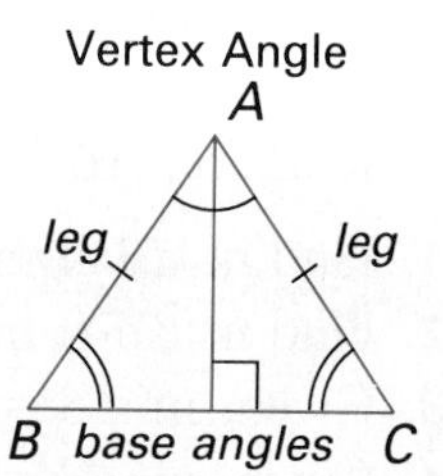

1. Given: $PS = RS$, m $\angle 3 = 80$. Find m $\angle PSR$.
2. Given: m $\angle PSR = 2x$, m $\angle P = 3x - 10$, $PS = RS$. Find m $\angle PSR$.
3. Given: $\angle 4 \cong \angle 1$
   Prove: $\overline{PS} \cong \overline{RS}$
4. Given: $\overline{PS} \cong \overline{RS}$, $\overline{QS}$ bisects $\angle PSR$.
   What conclusion about $\overline{SQ}$ and $\overline{PR}$ can be drawn? Why?

S, 5, 6, 4, 3, 2, 1, P, Q, R

5.3, 5.4 To prove right triangles congruent, use the Hypotenuse-Leg congruence pattern or any of the patterns established for congruence of triangles.

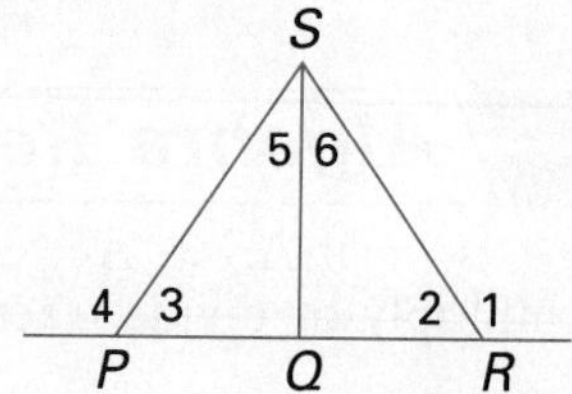

To write proofs involving overlapping triangles:

(1) Separate the pairs of triangles.
(2) Determine which pair of triangles you must prove congruent. If necessary, prove a pair of triangles congruent so that their corresponding sides or angles can be used to prove a second pair of triangles congruent.

5. Given: $\overline{BC} \cong \overline{BA}$, $\overline{CD} \cong \overline{AD}$
   Prove: $\angle 3 \cong \angle 4$
6. Given: $\overline{CA} \perp \overline{BD}$, $\angle 1 \cong \angle 2$
   Prove: $\angle 5 \cong \angle 6$

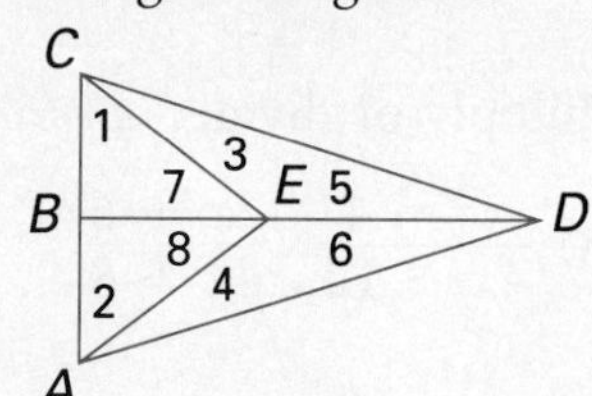

5.5 To work with altitudes and medians, use the following theorems.

- Corr alt of $\cong$ $\triangle$s are $\cong$.
- Corr medians of $\cong$ $\triangle$s are $\cong$.
- The alt to the base of an isos $\triangle$ is a median.

7. Prove that the altitude to the base of an isosceles triangle bisects the vertex angle.

5.6 To apply properties of perpendicular bisectors

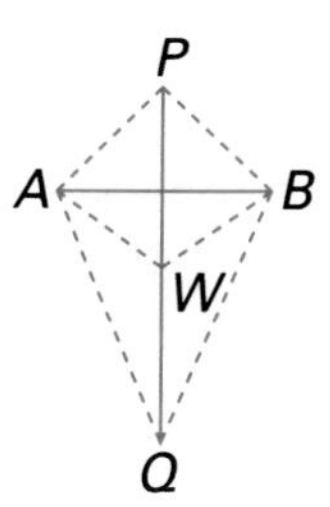

(1) If P and Q are equidistant ($PA = PB$, $QA = QB$) from the endpoints of a segment (A and B), then $\overline{PQ}$ is the perpendicular bisector of $\overline{AB}$.

(2) Any point W on the perpendicular bisector of a segment $\overline{AB}$ is equidistant ($WA = WB$) from the endpoints of the segment.

Based on the given information, what conclusions can be drawn?

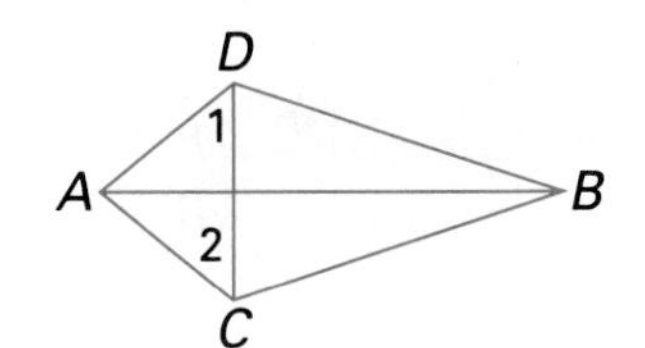

8. Given: $\overline{AD} \cong \overline{AC}$, $\overline{BD} \cong \overline{BC}$

9. Given: $\overline{AB}$ is the perpendicular bisector of $\overline{DC}$.

5.7–5.9 To work with inequalities in triangles, use

Triangle Inequality Theorem.

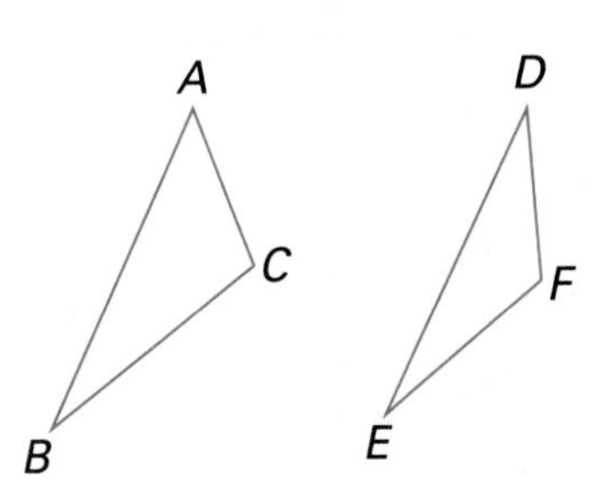

- $BA + AC > BC$
- If $AB > AC$, then m $\angle C >$ m $\angle B$
- If m $\angle A >$ m $\angle B$, then $CB > CA$

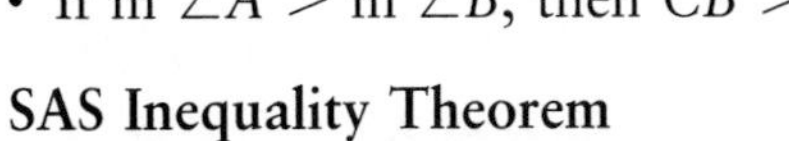

SAS Inequality Theorem

- If $AB = DE$, $AC = DF$, m $\angle A >$ m $\angle D$, then $BC > EF$

SSS Inequality Theorem

- If $AB = DE$, $AC = DF$, $BC > EF$, then m $\angle A >$ m $\angle D$

10. Name the longest side.

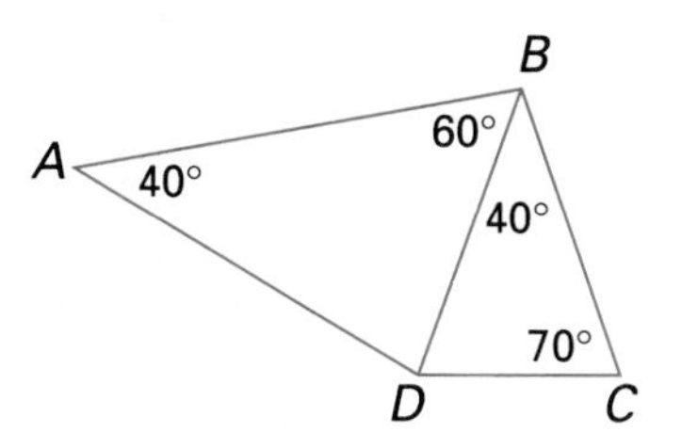

11. Write an inequality for the restrictions on x.

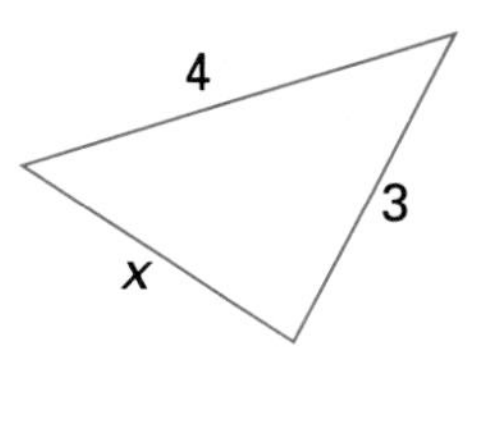

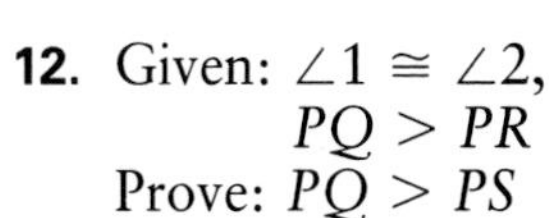

12. Given: $\angle 1 \cong \angle 2$, $PQ > PR$
Prove: $PQ > PS$

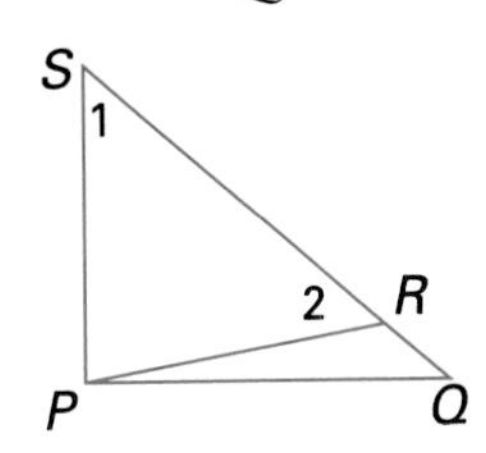

Chapter 5 Test

1. The measure of a base angle of an isosceles triangle is 80. Find the measure of the vertex angle.

Draw a conclusion from the diagram. (Exercises 2–3)
Name the longest side. (Exercise 4)

2. 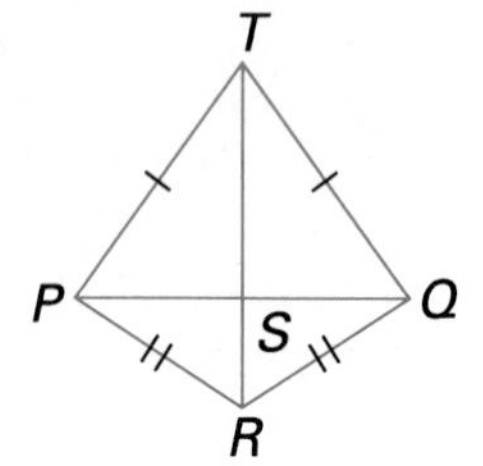

3. 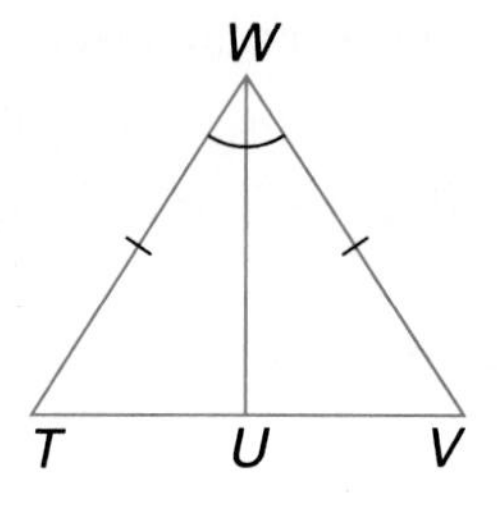

4. 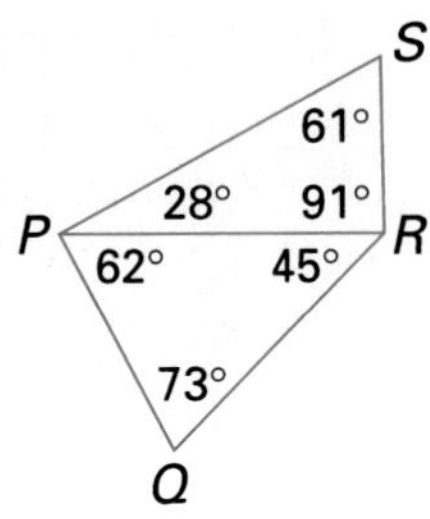

5. In $\triangle ABC$, m $\angle A = 2x - 10$, m $\angle B = x + 40$, m $\angle C = x + 30$ Name the longest side.
6. In $\triangle ABC$, $AB = 6$, $BC = 7$, $AC = 5$. Name the largest angle.
7. In $\triangle ABC$, $AB = 4$, $BC = 6$. What are the restrictions on the length of $\overline{AC}$?
8. Draw a line l with point Q on l. Construct a perpendicular to l at Q.

Complete the statement with the symbol =, >, or <.

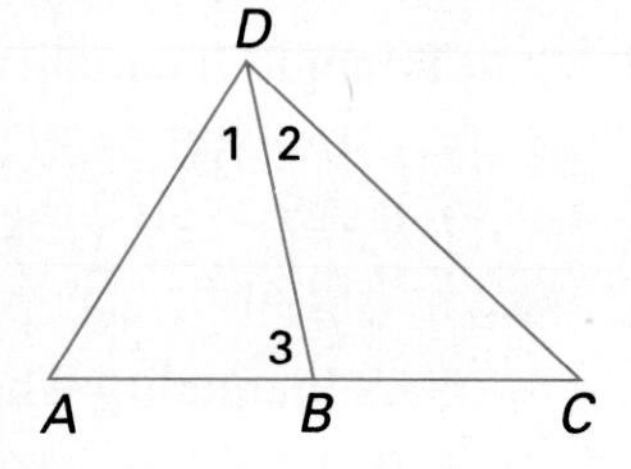

9. B is the midpoint of $\overline{AC}$. m $\angle 1 = 40$, m $\angle 2 = 70$.
   AB ___ BC
10. $AB = BC = 6$, m $\angle 1 = 50$, m $\angle A = 70$.
    AD ___ CD
11. $AD = BD$, m $\angle C <$ m $\angle 3$
    AD ___ DC

*12. Given: $\angle 1 \cong \angle 2$, $\overline{AB} \cong \overline{CD}$, $\angle 3 \cong \angle 4$
Prove: $\overline{FE} \cong \overline{GE}$

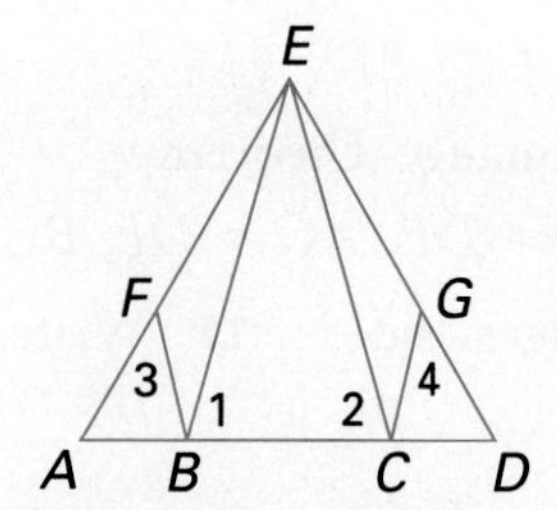

13. Identify the hypotenuse and legs of $\triangle RST$.
14. Identify an altitude and a median of $\triangle RST$.

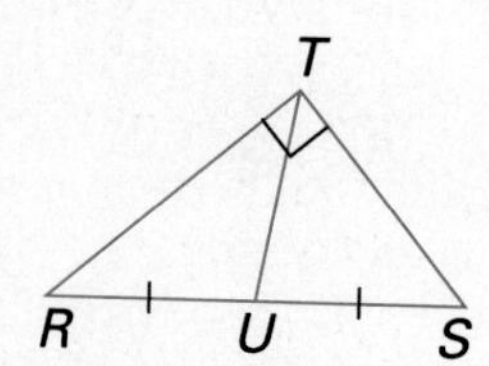

College Prep Test

Indicate the one correct answer for each question.

1.

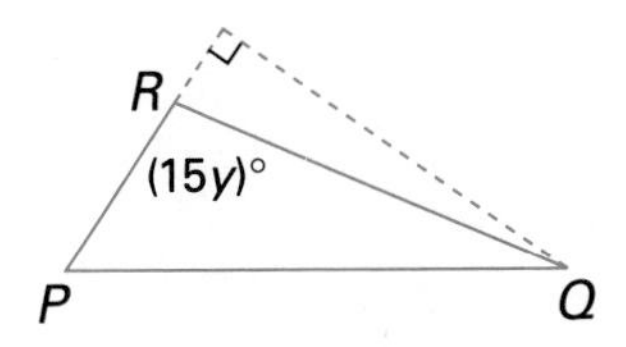

In $\triangle PQR$ above, which of the following could be a value of y?

(A) 4 (B) 12 (C) 2 (D) 6
(E) 8

2.

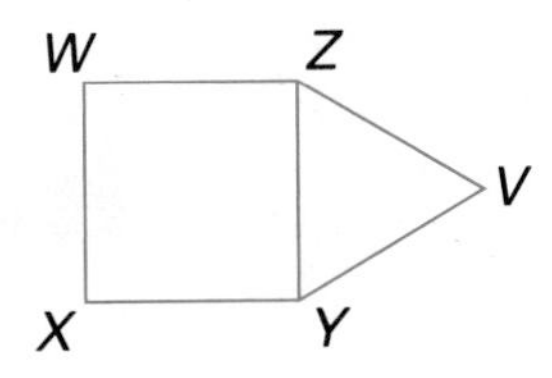

In the figure above, $XYZW$ is a square (all sides congruent and four right angles). Also, $\overline{VY} \cong \overline{VZ} \cong \overline{XY}$ Then, $m\angle V =$ ____.

(A) 15 (B) 45 (C) 60 (D) 90
(E) 30

3.

$$\begin{array}{r} 6\square 2 \\ \times\ 8 \\ \hline 5{,}1\triangle 6 \end{array}$$

In the correctly calculated product above, if □ and △ are replaced with different digits, then □ = ____.

(A) 6 (B) 2 (C) 5 (D) 4
(E) 7

4. In a senior class there are 400 boys and 500 girls. If 60% of the boys and 50% of the girls bought class rings, how many seniors did not buy class rings?

(A) 490 (B) 240 (C) 250
(D) 310 (E) 410

5.

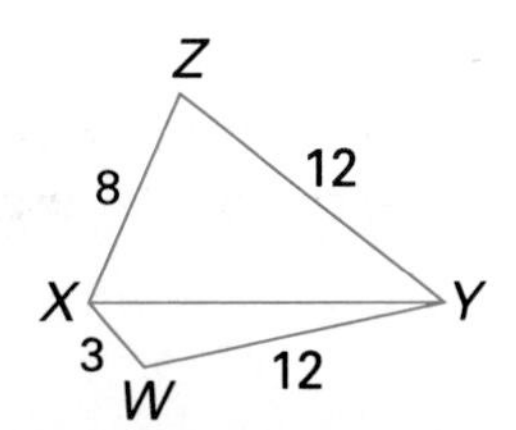

In the figure above, the perimeter of $\triangle XYZ$ is how much greater than the perimeter of $\triangle XYW$?

(A) 5 (B) 4 (C) 15 (D) 7
(E) 20

6.

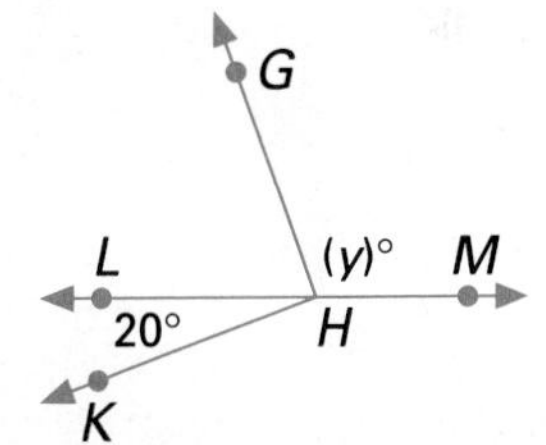

In the figure above, $\overline{GH} \perp \overline{KH}$ Find y.

(A) 70 (B) 110 (C) 200
(D) 160 (E) 100

7.

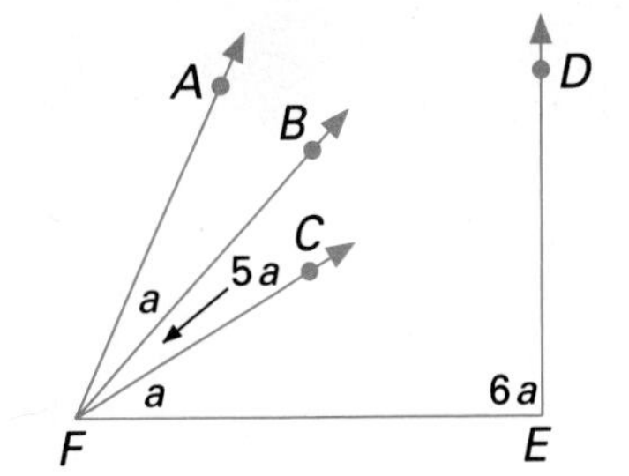

If the figure above were redrawn to scale so that $a = 15$, how many points of intersection would there be in addition to points E and F?

(A) 1 (B) 2 (C) 3 (D) 4
(E) 5

6 POLYGONS

Lewis Latimer (1848–1928)—By inventing a durable carbon filament for the electric light bulb, Latimer solved the problem of changing electric current into light. As an associate of Alexander Graham Bell, he also drew plans for the first telephone patent.

6.1 Introducing Polygons

Objectives To identify and name polygons and their parts
To identify and draw convex and concave polygons
To identify regular polygons

Some geometric figures have curved parts. Others, including polygons, consist entirely of segments.

Definition

A **polygon** is the union of three or more coplanar segments such that:
1. each endpoint is shared by exactly two segments;
2. segments intersect only at their endpoints; and
3. intersecting segments are noncollinear.

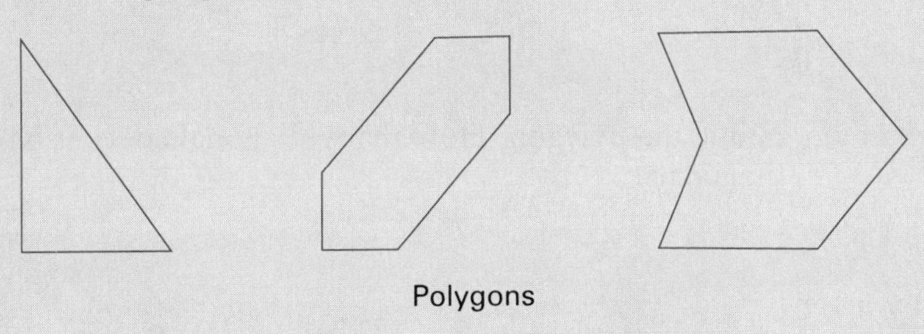

Polygons

Each segment of a polygon is called a **side.** Each endpoint of a side is called a **vertex** of the polygon. For each polygon, the number of sides is equal to the number of vertices. A polygon is usually named by listing its vertices in order.

Adjacent sides of a polygon intersect at a vertex. *Nonadjacent* sides do not intersect. The vertices of *consecutive angles* are endpoints of the same side of the polygon. If two angles are not consecutive, then they are called **nonconsecutive**.

EXAMPLE 1 Name two sides of polygon $ABCDE$ that are adjacent to $\overline{AB}$. Name two angles that are nonconsecutive with $\angle C$. Name two angles that are consecutive with $\angle C$.

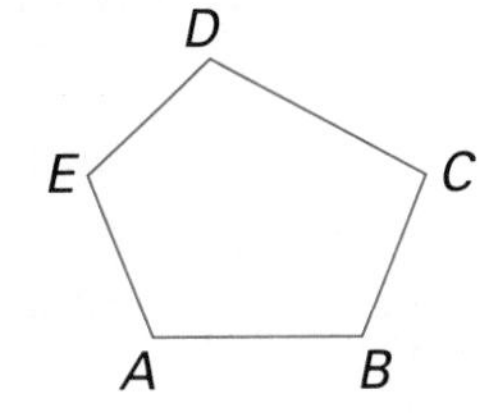

Solution $\overline{BC}$ and $\overline{EA}$ are adjacent to $\overline{AB}$.
$\angle E$ and $\angle A$ are nonconsecutive with $\angle C$.
$\angle B$ and $\angle D$ are consecutive with $\angle C$.

Polygons are usually classified according to the number of sides they have. A polygon with 13 sides is called a 13-gon; a polygon with 29 sides, a 29-gon; and a polygon with *n* sides, an *n*-gon. Some polygons, however, are given special names, as shown in the table.

Number of sides	Name	Number of sides	Name
3	Triangle	8	Octagon
4	Quadrilateral	9	Nonagon
5	Pentagon	10	Decagon
6	Hexagon	12	Dodecagon
7	Heptagon		

Definition

A **diagonal** of a polygon is a segment joining two nonconsecutive vertices.

EXAMPLE 2 Name the polygon. How many diagonals does it have? Name the diagonals.

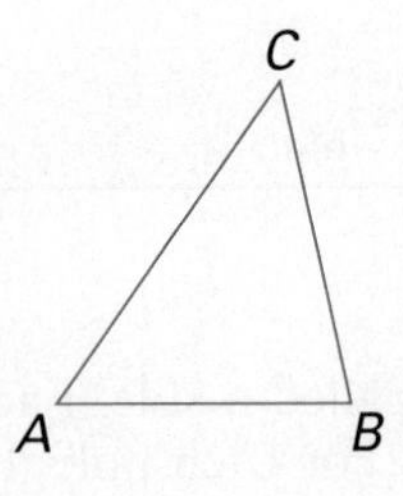

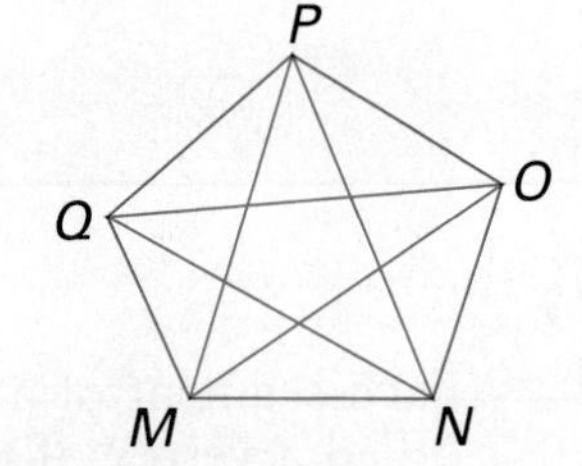

Solution

Triangle ABC
Number of Diagonals: None

Pentagon $MNOPQ$
Number of Diagonals: 5
($\overline{MO}$, $\overline{NP}$, $\overline{OQ}$, $\overline{PM}$, $\overline{QN}$)

A polygon separates the plane that contains it into three distinct sets of points, the *exterior*, the *polygon* itself, and the *interior*.

In the figure, point P is in the interior of pentagon $ABCDE$, and point Q is in the exterior.

Notice that the exterior contains lines, while the interior does not.

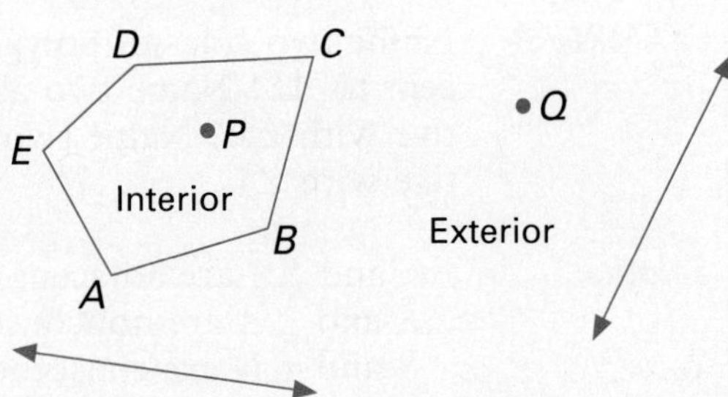

Definition

A polygon is **convex** if the segment $\overline{XY}$ joining any two interior points of the polygon is in the interior of the polygon. If a polygon is not convex, then it is **concave**.

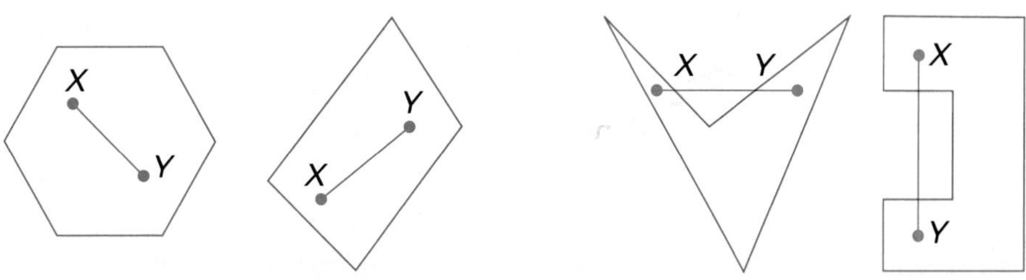

Convex Polygons

Concave Polygons

Some polygons are equilateral. All of the sides of an **equilateral** polygon are congruent. Some polygons are equiangular. All of the angles of an **equiangular** polygon are congruent.

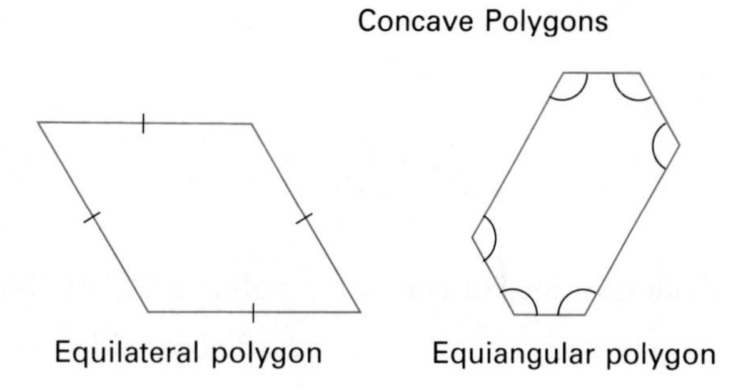

Equilateral polygon

Equiangular polygon

Definition

A **regular polygon** is a convex polygon that is both equilateral and equiangular.

Regular polygons

EXAMPLE 3 State whether the polygon is equilateral or equiangular. State whether it is regular or not regular.

Solution Equilateral; not regular — Equilateral and equiangular; regular

Classroom Exercises

State whether the figure is a polygon. If it is not, explain why.

1.

2.

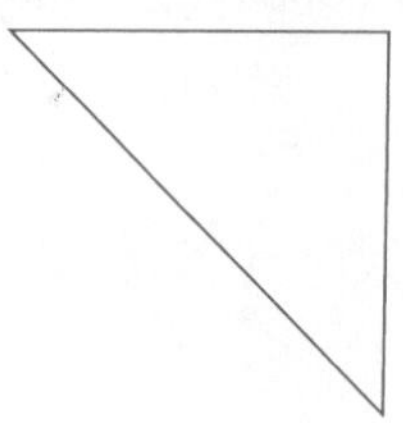

3.

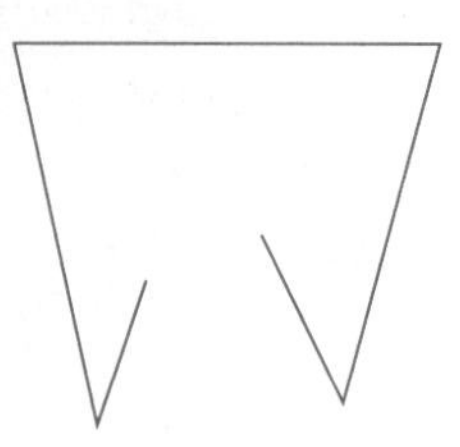

4.

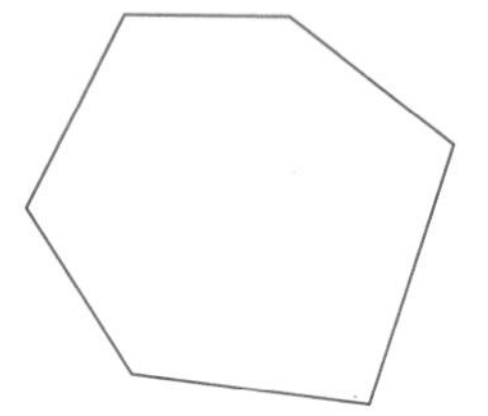

5.

6.

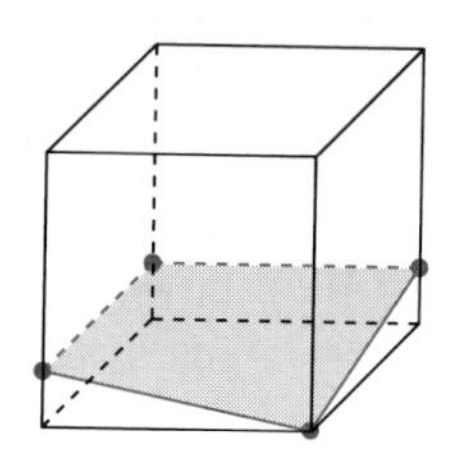

Complete the statement for polygon *PQRST*.

7. $\overline{RS}$ and __________ are adjacent sides.

8. $\overline{TP}$ and __________ are nonadjacent sides.

9. $\angle S$ and __________ are consecutive angles.

10. $\angle P$ and __________ are nonconsecutive angles.

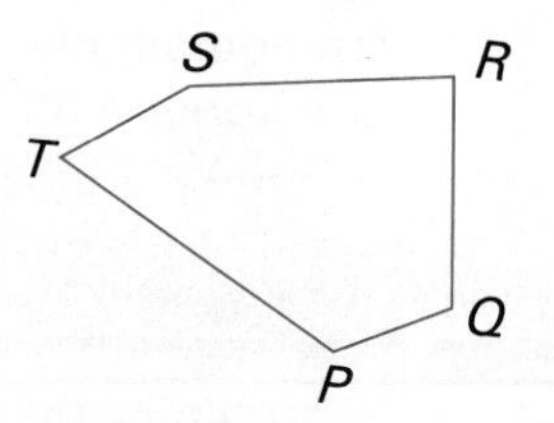

Written Exercises

Name the polygon. State whether it is convex or concave.

1.

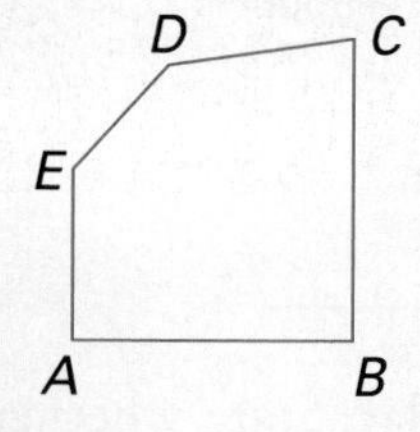

2.

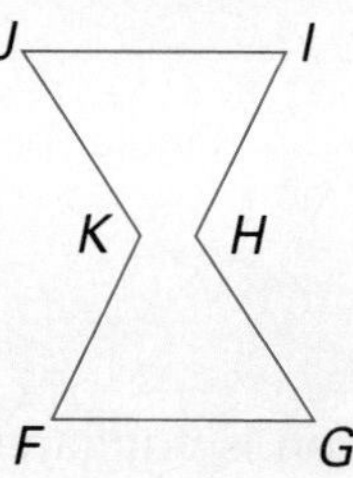

3.

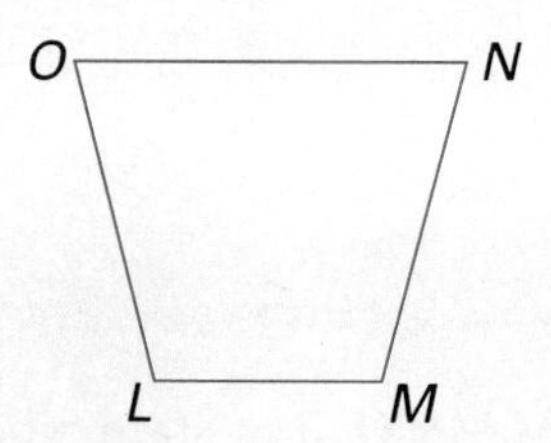

Determine whether the polygon is equilateral, equiangular, or regular.

4.

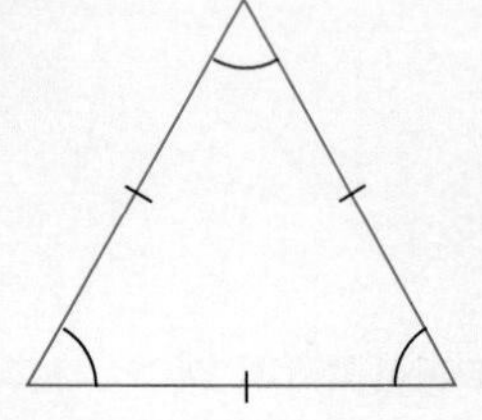

5.

6.

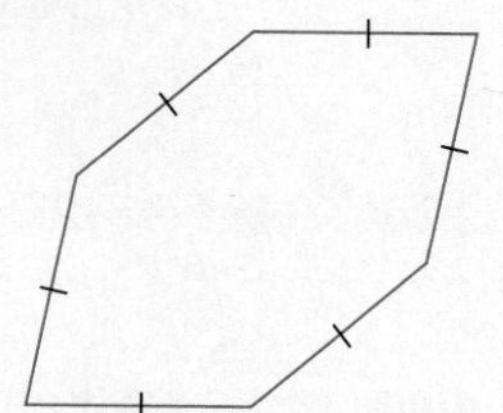

Draw the figure.

7. a convex pentagon
8. a concave hexagon
9. a concave pentagon
10. a convex octagon
11. an equiangular quadrilateral
12. an equilateral hexagon
13. a regular triangle
14. a regular quadrilateral
15. a regular pentagon
16. a regular octagon

Complete the table.

	Number of sides of polygon	Number of diagonals from one vertex	Number of diagonals
17.	3	______	______
18.	4	______	______
19.	5	______	______
20.	6	______	______
21.	7	______	______
22.	8	______	______
23.	50	______	______
24.	n	______	______

25. Draw an equilateral concave polygon.

26. The edges of the opening of a wrench are parallel. Standard nuts are regular polygons in shape. Will the wrench fit nuts with any given number of sides? Explain.

Mixed Review

1. Write the converse of "If a polygon is regular, then it is equiangular." Is the converse true or false? *3.5*

2. Write the inverse of "If a polygon is regular, then it is equilateral." Is the inverse true or false? *3.8*

3. Write the contrapositive of "If a polygon is regular, then it is equilateral." Is the contrapositive true or false? *3.8*

4. Write the inverse if "If a polygon is convex, then it is not concave." Is the inverse true or false? *3.8*

6.2 Interior Angles of Polygons

Objectives

To find the sum of the angle measures of a convex polygon
To find angle measures and the number of sides of polygons
To find angle measures and the number of sides of regular polygons

To store honey, bees build honeycombs with hundreds of *hexagonal* compartments. Why do you think a hexagon is better for this purpose than a circle, a square, or a triangle? (The computer storage shown in the photo is a modern industrial "honeycomb".)

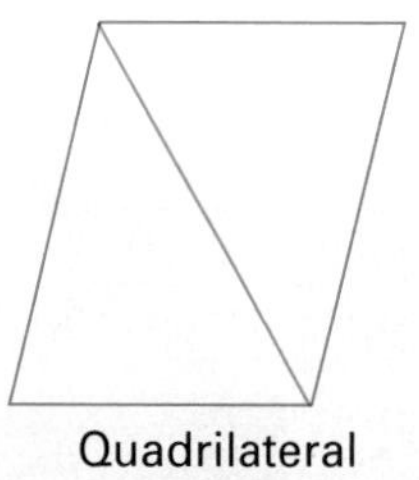

Quadrilateral

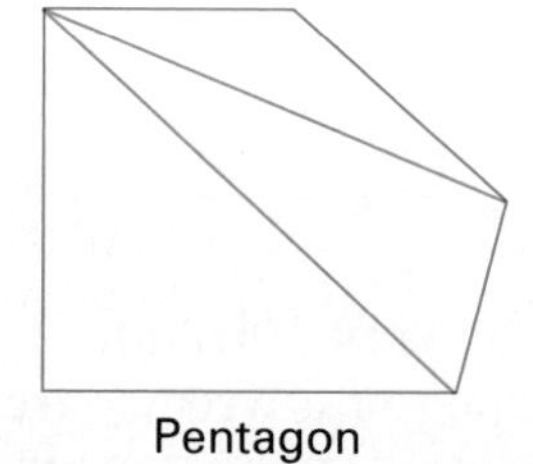

Pentagon

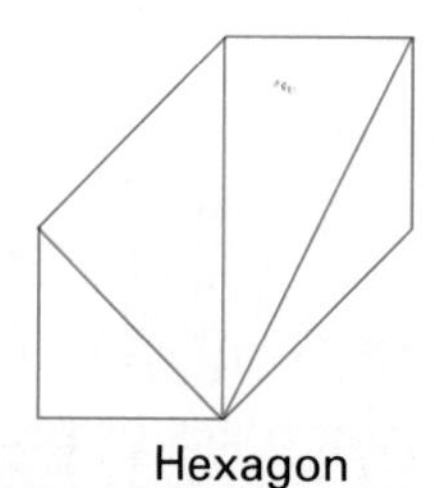

Hexagon

In each figure above the diagonals *from one vertex* to each of the other vertices form triangles. In each case, the total number of triangles formed is 2 less than the number of sides of the polygon. If a polygon has n sides, then $n - 2$ triangles are formed. The sum of the measures of the interior angles of the polygon is the sum of the measures of the angles of these triangles.

Polygon	Number of Sides	Number of Triangles	Sum of Angle Measures
quadrilateral	4	2	$2 \cdot 180 = 360$
pentagon	5	3	$3 \cdot 180 = 540$
hexagon	6	4	$4 \cdot 180 = 720$
heptagon	7	5	$5 \cdot 180 = 900$
octagon	8	6	$6 \cdot 180 = 1{,}080$

Theorem 6.1 The sum of the measures of the interior angles of a convex polygon with n sides is $(n - 2)180$.

Corollary 1 The sum of the measures of the interior angles of a convex quadrilateral is 360.

In this book, the *interior angles* of a polygon will simply be called the **angles of the polygon**.

EXAMPLE 1 Find the sum of the measures of the angles of a 22-gon.

Solution $n = 22$
Sum $= (22 - 2)180 = 20 \cdot 180 = 3{,}600$

EXAMPLE 2 In a hexagon, the measure of one angle is twice that of a second angle. The remaining angles are congruent, each with a measure three times that of the second angle. Find the measure of each angle.

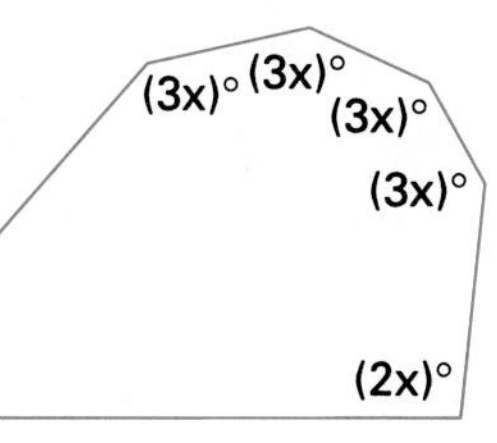

Plan By Theorem 6.1, the sum of the measures of the angles is $(6 - 2)180 = 720$. Let x = measure of the second angle.

Solution

$$x + 2x + 3x + 3x + 3x + 3x = 720$$
$$15x = 720$$
$$x = 48$$

The measure of the second angle is 48.

The measure of the original angle is $2 \cdot 48 = 96$.

The measure of each of the remaining angles is $3 \cdot 48 = 144$.

If you know the sum of the angle measures of a polygon, you can use the formula $S = (n - 2)180$ to find the number of sides of the polygon.

EXAMPLE 3 The sum of the measures of the angles of a convex polygon is 3,960. Find the number of sides of the polygon.

Solution

$$(n - 2)180 = 3{,}960$$
$$n - 2 = \frac{3{,}960}{180}$$ ← It is easier to divide by 180 than to expand the left-hand side first.
$$n - 2 = 22$$
$$n = 24$$

The polygon has 24 sides.

A regular polygon was defined to be an equilateral and equiangular convex polygon. Using the definition and Theorem 6.1, you can find the measure of each angle of a regular polygon if the number of sides is known.

EXAMPLE 4 Find the measure of each angle of a regular hexagon.

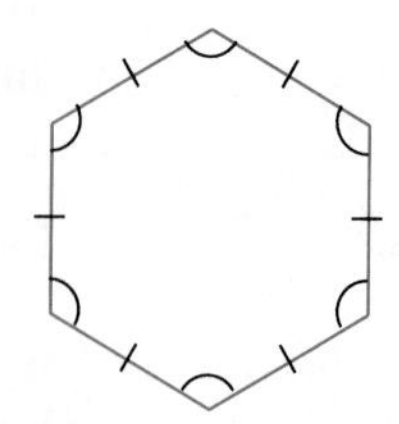

Solution The sum of the measures is $(6 - 2)180 = 720$. Each angle has the same measure.

Therefore, the measure of each angle is $\frac{720}{6} = 120$.

Example 4 can be generalized as a corollary to Theorem 6.1.

Corollary 2 The measure of an angle of a regular polygon with n sides is $\frac{(n-2)180}{n}$.

EXAMPLE 5 The measure of each angle of a regular polygon is 150. How many sides does the polygon have?

$$150 = \frac{(n-2)180}{n}$$
$$150n = (n-2)180$$
$$150n = 180n - 360$$
$$-30n = -360$$
$$n = 12$$

← Here, it is easier to expand first.

The polygon has 12 sides.

Classroom Exercises

How many sides does a polygon have if the measures of its angles have the given sum?

1. 180 **2.** $3 \cdot 180$ **3.** $6 \cdot 180$ **4.** 720
5. 360 **6.** 1,800 **7.** 18,000 **8.** 180,000

Written Exercises

Find the sum of the measures of the angles of the convex polygon.

1. a pentagon **2.** a hexagon **3.** a decagon
4. a 30-gon **5.** a 62-gon **6.** a 100-gon

Can the three angle measures given belong to a convex quadrilateral? If so, find the measure of the fourth angle.

7. 80, 50, 90
8. 100, 70, 120
9. 60, 60, 60
10. 95, 97, 83
11. 132, 112, 145
12. 38, 43, 57

The sum of the measures of the angles of a convex polygon is given. Find the number of sides of the polygon.

13. 900
14. 1,260
15. 1,980
16. 3,600
17. 4,500
18. 7,560

Find the measure of an angle of the regular polygon.

19. a square
20. a pentagon
21. a decagon
22. a 20-gon
23. a 30-gon
24. a 100-gon

Each angle of a regular polygon has the given measure. How many sides does the polygon have?

25. 60
26. 135
27. 108

Find the measure of each angle of quadrilateral *ABCD*.

28. $m\ \angle A = 10x$, $m\ \angle B = 6x + 10$, $m\ \angle C = 12x - 10$, $m\ \angle D = 8x$

29. $m\ \angle A = 8x + 5$, $m\ \angle B = 10x + 5$, $m\ \angle C = 10x - 8$, $m\ \angle D = 13x - 11$

30. The sum of the measures of four angles of a pentagon is 498. What is the measure of the unknown angle?

31. The sum of the measures of nine angles of a decagon is 1,320. What is the measure of the unknown angle?

32. Three angles of a hexagon are congruent. The other three angles are also congruent. Each of the first three angles has a measure twice that of one of the second three angles. What is the measure of each angle of the hexagon?

33. In a pentagon, the measure of one angle is twice that of a second angle. The remaining angles are congruent, each having a measure of three times that of the second angle. What is the measure of each angle of the pentagon?

Can the number be the sum of the angle measures of a polygon?

34. 10,180
35. 15,660
36. 18,180

Can the number be the measure of an angle of a regular polygon? If the measure is possible, how many sides does the polygon have?

37. 90
38. 100
39. 120
40. 140
41. 160
42. 175

Use the figures at the right for Exercises 43–44.

43. Prove Theorem 6.1 for convex quadrilaterals (Corollary 1).

44. Prove Theorem 6.1 for pentagons.

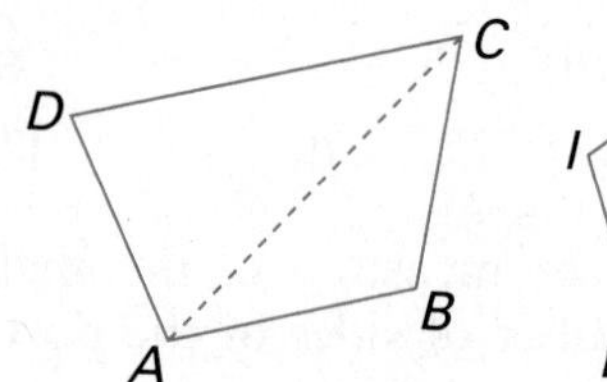

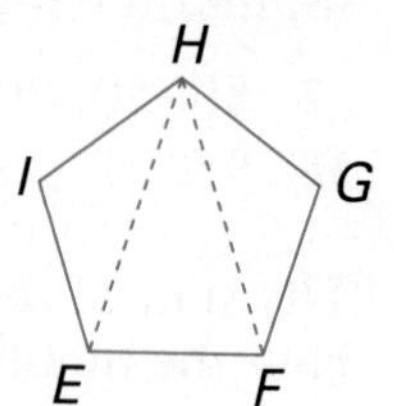

Another approach can be used to prove Theorem 6.1. Draw triangles with a common vertex in the interior of the polygon, as shown below. Any interior point may be chosen for the vertex. Then use the fact that the sum of the measures of the angles with the common vertex is 360.

45. Prove Theorem 6.1 for convex quadrilaterals.

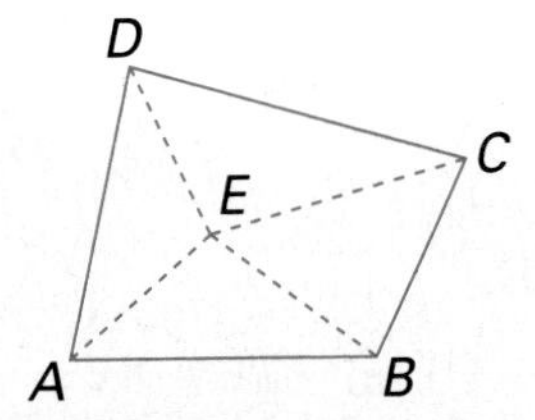

46. Prove Theorem 6.1 for convex hexagons.

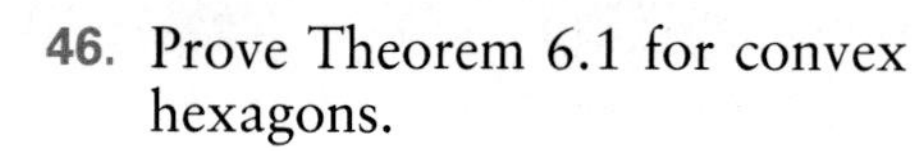

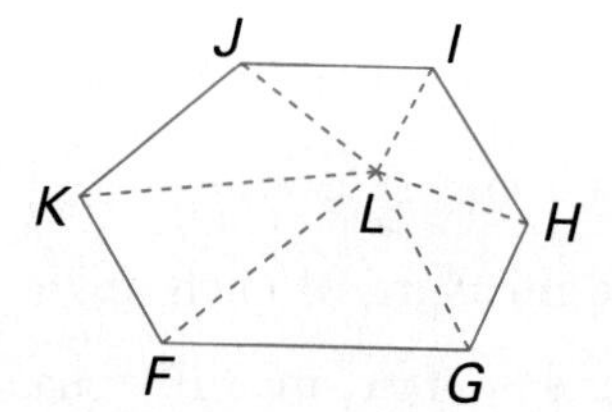

47. Prove Corollary 2.

48. Prove: Opposite sides of a regular hexagon are parallel. (HINT: Draw a transversal through vertices of the sides to be proven parallel such that it is $\perp$ to one side.)

49. Using a protractor, measure each angle of the given concave quadrilateral. Does the formula of Theorem 6.1 seem to apply?

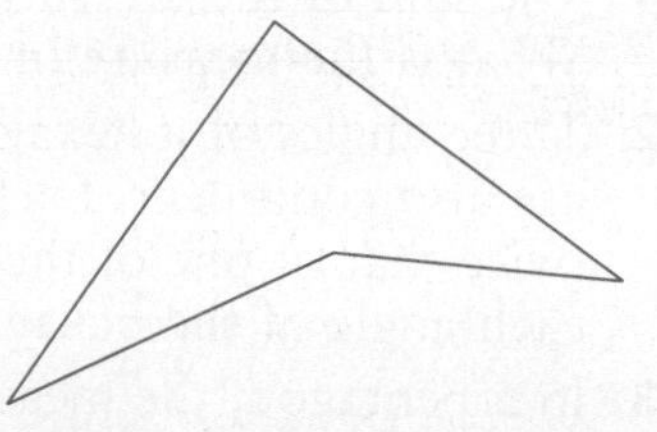

Mixed Review

Tell whether the statement is true or false.

1. Alternate interior angles of parallel lines are congruent. *3.5*
2. Exterior angles of parallel lines on the same side of a transversal are supplementary. *3.7*
3. If corresponding angles are supplementary, then lines are parallel. *3.3*
4. Any point on the perpendicular bisector of a segment is equidistant from the endpoints of the segment. *5.6*

Application: *Tiling and Soccer Ball*

Different tile patterns can be arranged by fitting together various types of regular polygons. For example, octagons and squares can be put together, as shown below. Can the following combinations of tiles be fitted together? If so, make paper cutouts or drawings to demonstrate.

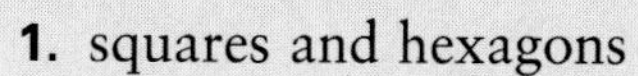

1. squares and hexagons
2. pentagons and triangles
3. pentagons and squares
4. hexagons and triangles
5. squares and triangles
6. hexagons and pentagons
7. octagons and triangles

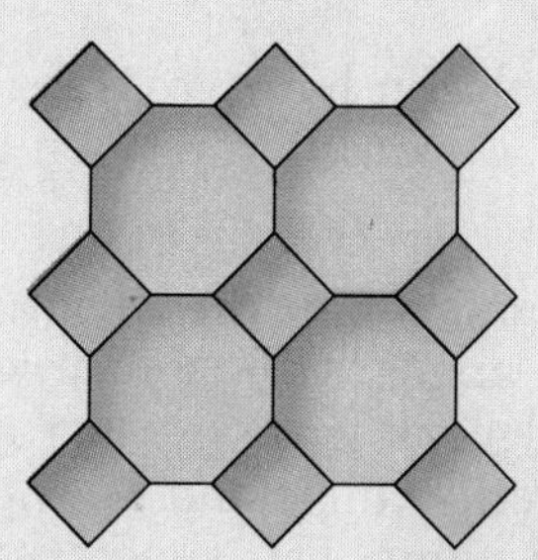

8. The cover of a soccer ball is made of two types of regular polygons sewn together. What are these polygons? Why was this combination not possible when tile patterns were considered in Exercises 1–7 above?

Algebra Review

To solve a fractional equation of the form $\frac{ax + b}{dx} = c$ for x, where $d \neq 0$, $x \neq 0$:

(1) Multiply both sides of the equation by dx.
(2) Solve for x.

Example: Solve $\frac{5x - 16}{3x} = 3$.

Multiply by $3x$.

$$\begin{aligned} 5x - 16 &= 9x \\ -4x &= 16 \\ x &= -4 \end{aligned}$$

Solve the equation.

1. $\frac{2x - 5}{3x} = 1$
2. $\frac{3x + 4}{2x} = 2$
3. $\frac{5x - 28}{3x} = 3$
4. $\frac{4x - 10}{2x} = 4$
5. $\frac{6x - 3}{5x} = 3$
6. $\frac{10x + 2}{2x} = 7$

6.3 Exterior Angles of Polygons

Objectives To identify the exterior angles of a polygon
To find the measures of exterior angles of polygons
To explain why the exterior angles of a regular polygon are congruent

A convex polygon has two exterior angles at each vertex. A pentagon has ten exterior angles.

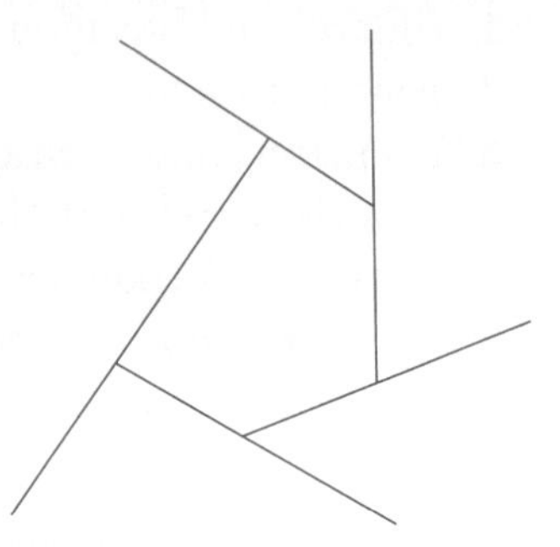

Trace the pentagon at the right as well as the exterior angles, one at each vertex. Cut out the wedges of tracing paper determined by each exterior angle and arrange them to fit around a point. This suggests the sum of these angle measures is 360.

Theorem 6.2 The sum of the measures of the exterior angles, one at each vertex, of any convex polygon is 360.

Plan Sketch a polygon from which to generalize. Use Theorem 6.1, which concerns the sum of the measures of *interior* angles.

Proof (Paragraph Form) At each vertex of a polygon, the sum of the measures of the interior angle and an exterior angle is 180. For example, in the figure at right, $m\angle 1 + m\angle 6 = 180$.

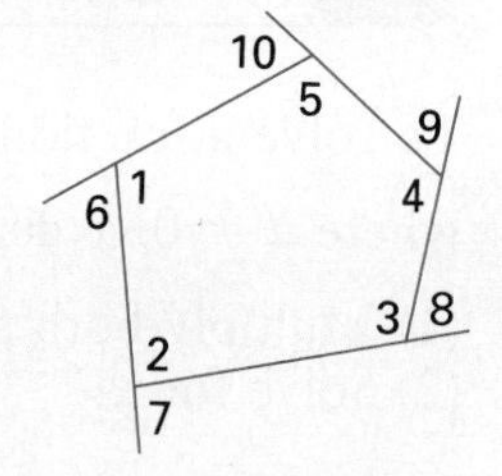

In a polygon of n sides, there are n vertices. Therefore, the sum of the measures of all pairs of interior angles and exterior angles is $180n$. It was stated in Theorem 6.1 that the sum of the measures of the interior angles of a polygon with n sides is $(n - 2)180$. So, the sum of the measures of the exterior angles of a polygon with n sides is the difference of the sum of interior angles from 180.

$$\begin{aligned} 180n - (n - 2)180 &= 180n - (180n - 360) \\ &= 180n - 180n + 360 \\ &= 360 \end{aligned}$$

EXAMPLE 1 What is the sum of the measures of the exterior angles, one at each vertex, of a convex dodecagon?

Solution Using Theorem 6.2, the sum is 360.

EXAMPLE 2 The measures of the exterior angles, one at each vertex, of a quadrilateral are m $\angle 1 = t + 4$, m $\angle 2 = t + 6$, m $\angle 3 = t + 8$, and m $\angle 4 = t + 10$. Find the measure of each exterior angle.

Solution

$$\begin{aligned} m\angle 1 + m\angle 2 + m\angle 3 + m\angle 4 &= 360 \\ (t + 4) + (t + 6) + (t + 8) + (t + 10) &= 360 \\ 4t + 28 &= 360 \\ 4t &= 332 \\ t &= 83 \end{aligned}$$

Therefore, m $\angle 1 = 83 + 4 = 87$, m $\angle 2 = 83 + 6 = 89$, m $\angle 3 = 83 + 8 = 91$, and m $\angle 4 = 83 + 10 = 93$.

All exterior angles of a regular polygon are congruent since they are supplements of congruent angles. The following corollary to Theorem 6.2 is based on this fact.

Corollary The measure of an exterior angle of a regular polygon with n sides is $\frac{360}{n}$.

EXAMPLE 3 The measure of an exterior angle of a regular polygon is $3x + 1$. The measure of a second exterior angle of the same polygon is $4x - 12$. Identify the polygon.

Solution Exterior angles of a regular polygon have the same measure.

$$\begin{aligned} 4x - 12 &= 3x + 1 \\ x &= 13 \\ 3x + 1 &= 3 \cdot 13 + 1 = 40 \end{aligned}$$

Each exterior angle has a measure of 40. By the corollary to Theorem 6.2,

$$\begin{aligned} 40 &= \frac{360}{n} \\ 40n &= 360 \\ n &= 9 \end{aligned}$$

Therefore, the polygon is 9-sided (a nonagon).

Classroom Exercises

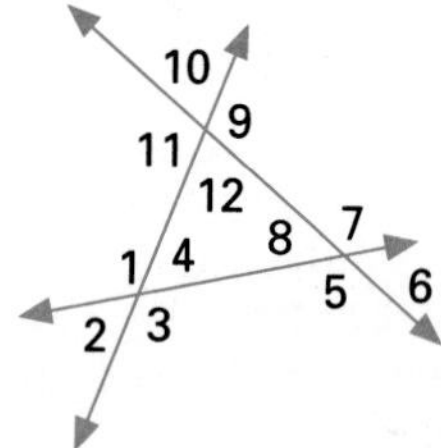

1. Name the interior angles.
2. Name the exterior angles.
3. Find m $\angle 1$ + m $\angle 9$ + m $\angle 5$.
4. Find m $\angle 4$ + m $\angle 12$ + m $\angle 8$.
5. Find m $\angle 2$ + m $\angle 6$ + m $\angle 10$.
6. Name the exterior angles.
7. Find m $\angle 6$ + m $\angle 12$ + m $\angle 16$ + m $\angle 4$.
8. Find m $\angle 15$ + m $\angle 11$ + m $\angle 7$ + m $\angle 3$.
9. Find m $\angle 13$ + m $\angle 9$ + m $\angle 5$ + m $\angle 1$.
10. Find m $\angle 14$ + m $\angle 16$ + m $\angle 10$ + m $\angle 12$ + m $\angle 6$ + m $\angle 8$ + m $\angle 2$ + m $\angle 4$.

Find the measure of each exterior angle of the given regular polygon.

11. a pentagon
12. an octagon
13. a decagon
14. a 15-gon

Written Exercises

Copy the polygon. Draw an exterior angle at each vertex. Use a protractor to measure each of these exterior angles. What is the sum of the measures? Does this sum agree with Theorem 6.2?

1.

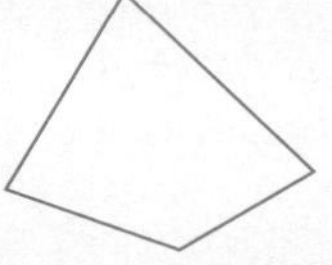

2.

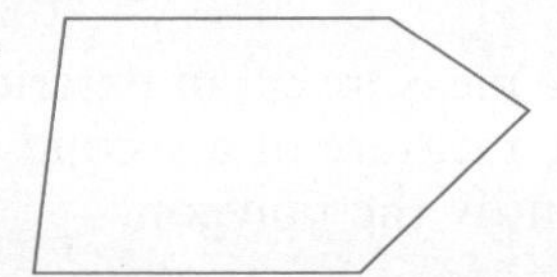

3. The measures of the exterior angles of a quadrilateral, one at each vertex, are x, $2x$, $3x$ and $4x$. Find the measure of each exterior angle.
4. The measures of the exterior angles of a hexagon, one at each vertex, are $5x + 8$, $3x + 13$, $5x - 3$, $6x + 10$, $4x - 5$, and $5x + 1$. Find x.

Find the measure of each exterior angle of the given regular polygon.

5. triangle
6. quadrilateral
7. hexagon
8. dodecagon
9. 20-gon
10. 100-gon

The measure of an exterior angle of a regular polygon is given. Find the number of sides of the polygon.

11. 60 **12.** 45 **13.** 30 **14.** 24

Essay.

15. The exterior angles of a regular polygon are congruent. Explain why.

16. Prove the corollary to Theorem 6.2.

17. The measure of an exterior angle of a regular polygon is $2x + 21$. The measure of a second exterior angle of the same polygon is $4x - 3$. Identify the polygon.

18. The measure of an exterior angle of a regular polygon is $12x - 6$. The measure of a second exterior angle of the same polygon is $8x + 6$. Identify the polygon.

19. A regular polygon has obtuse exterior angles. How many sides does the polygon have?

20. Of the regular polygons with obtuse interior angles, which has the largest exterior angle?

21. Is there a regular polygon with an interior angle congruent to an exterior angle? If so, draw a figure to illustrate.

Mixed Review

Refer to the figure to complete the statement.

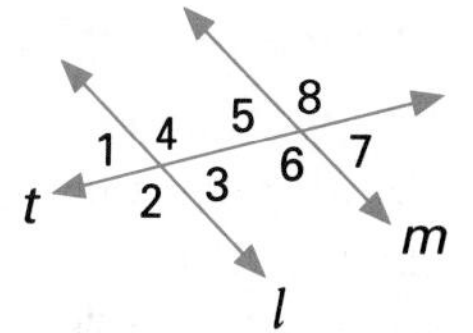

1. $\angle 3$ and $\angle$______ are alternate interior angles. *3.2*

2. $\angle 1$ and $\angle$______ are corresponding angles. *3.2*

3. $\angle 2$ and $\angle$______ are alternate exterior angles. *3.2*

4. $\angle 4$ and $\angle$______ are interior angles on the same side of the transversal. *3.2*

5. Given $\angle 8 \cong \angle 4$, prove $m \parallel l$. *3.3*

Application: *Mirrors*

Two mirrors can be arranged so that a regular polygon is shown. Draw a segment on a sheet of paper and arrange the two mirrors as shown. Move the mirrors until a regular hexagon is formed. Measure the angle formed by the mirrors. Move the mirrors to form other regular polygons. What is the relationship between the measure of the angle formed by the mirrors and the number of sides of the of the regular polygon?

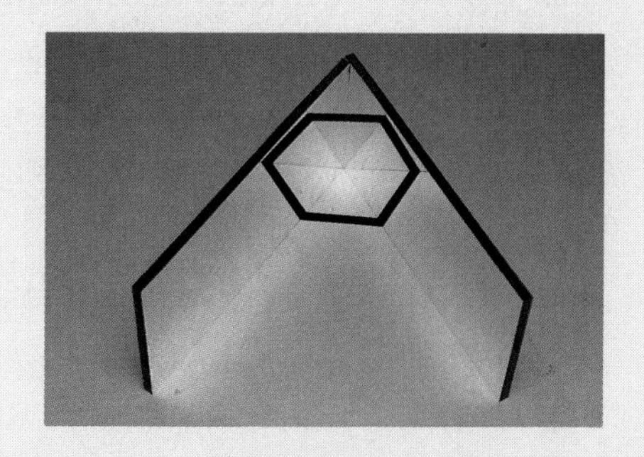

Computer Investigation

Properties of a Parallelogram

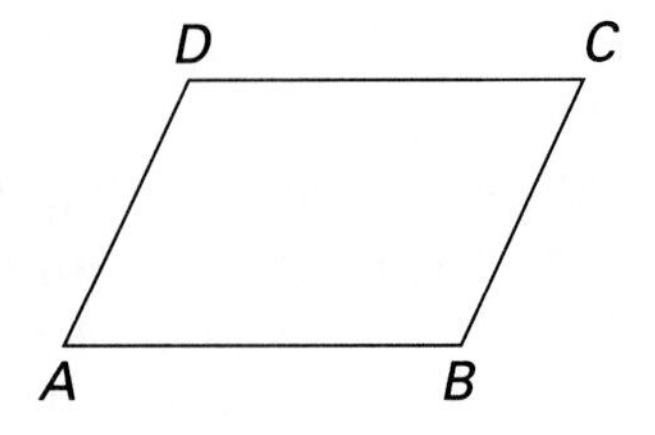

Use a computer software program that draws random parallelograms, labels points, and measures line segments or angles.

Quadrilateral *ABCD* has two pairs of opposite sides parallel: $\overline{CB} \parallel \overline{DA}$, $\overline{CD} \parallel \overline{BA}$. Such a figure is called a **parallelogram.** Opposite angles are angles *A* and *C* and angles *B* and *D*. The following activities are designed to help you discover some properties of parallelograms formally proved in the next lesson.

Activity 1

Draw a random parallelogram.

1. Find the lengths of all four sides.
2. Draw a new random parallelogram. Measure its sides.
3. Repeat the activity of Exercise 2 for a third random parallelogram.

Activity 2

What relationship seems to exist between opposite angles?

4. Use the measuring tool of the program to test your hypothesis for at least three different parallelograms.

Activity 3

What relationship seems to exist between consecutive angles?

5. Use the measuring tool of the program to test your hypothesis for at least three different parallelograms.

Activity 4

Draw a random parallelogram and label it *ABCD*. Draw the two diagonals $\overline{DB}$ and $\overline{AC}$. Label their point of intersection *E*.

6. Are the diagonals congruent? Measure to verify.
7. Do the diagonals bisect each other? Measure to verify.
8. Repeat the activities of Exercises 6 and 7 for two other random parallelograms.

Summary

What generalizations appear to be true of a parallelogram for: opposite sides? opposite angles? consecutive angles? diagonals?

6.4 Quadrilaterals and Parallelograms

Objectives

To identify parts and investigate properties of quadrilaterals
To prove and apply theorems about parallelograms

The figure to the right shows two patterns for building a structure such as a bridge, tower, or scaffold. Which do you think gives greater strength and why?

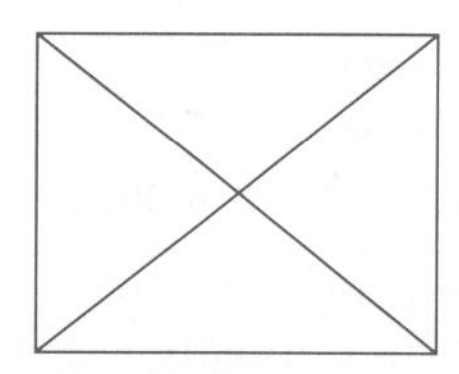

A **quadrilateral** is a polygon with four sides. *ABCD* is a quadrilateral.

$\overline{AB}$ and $\overline{BC}$ are adjacent sides.
$\overline{AB}$ and $\overline{CD}$ are opposite sides.
$\overline{AC}$ and $\overline{BD}$ are diagonals.
$\angle ABC$ and $\angle BAD$ are consecutive angles.
$\angle ABC$ and $\angle ADC$ are opposite angles.

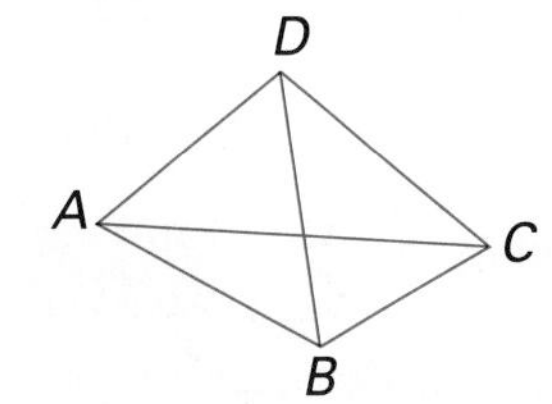

Definition

A **parallelogram** is a quadrilateral with both pairs of opposite sides parallel. ($\square EFGH$ means parallelogram *EFGH*.)

Theorem 6.3

A diagonal of a parallelogram forms two congruent triangles.

Given: $\square ABCD$ with diagonal $\overline{AC}$
Prove: $\triangle ABC \cong \triangle CDA$

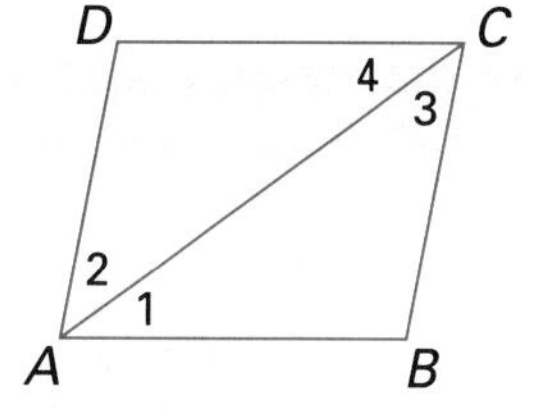

Plan

Use ASA. $\angle 1$ and $\angle 4$, and $\angle 2$ and $\angle 3$ are alternate interior angles.

Proof

	Statement		Reason
	1. $\square ABCD$ with diag $\overline{AC}$		1. Given
	2. $\overline{AB} \parallel \overline{DC}$, $\overline{AD} \parallel \overline{BC}$		2. Def of $\square$
	3. $\angle 1 \cong \angle 4$, $\angle 3 \cong \angle 2$	(A, A)	3. Alt int $\angle$s of $\parallel$ lines are $\cong$.
	4. $\overline{AC} \cong \overline{CA}$	(S)	4. Reflex Prop
	5. $\therefore \triangle ABC \cong \triangle CDA$		5. ASA

Note that in the proof of this theorem, it would be incorrect to state that $\triangle ABC \cong \triangle DCA$.

EXAMPLE 1 $WXYZ$ is a parallelogram. Which of the following pairs of triangles are congruent?

$\triangle XYZ$ and $\triangle YXW$ $\quad$ $\triangle XYZ$ and $\triangle ZWX$

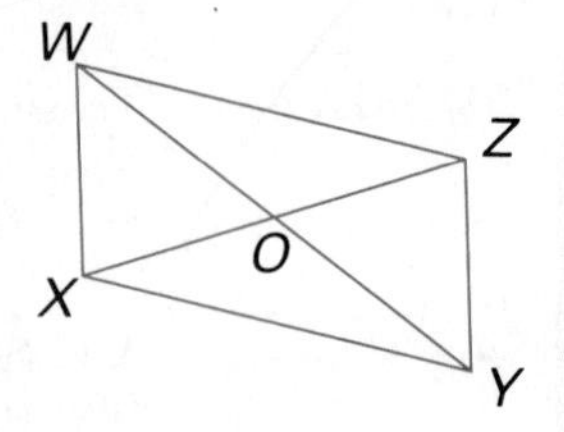

Solution $\triangle XYZ$ and $\triangle YXW$ are not necessarily congruent. $\quad$ $\triangle XYZ$ and $\triangle ZWX$ are congruent by Theorem 6.3.

In the proof of Theorem 6.3, you saw that $\triangle ABC \cong \triangle CDA$. Therefore, the opposite sides of $\square ABCD$ are congruent (CPCTC). The corresponding angles, B and D, are also congruent. By using diagonal $\overline{BD}$ you can show that $\angle DAB \cong \angle DCB$. Therefore, the opposite angles of $\square ABCD$ are congruent. This suggests the following corollaries.

Corollary 1 Opposite sides of a parallelogram are congruent.

Corollary 2 Opposite angles of a parallelogram are congruent.

EXAMPLE 2 In $\square ABCD$, $AB = 3x + 15$, $CD = 5x - 3$
Find AB.

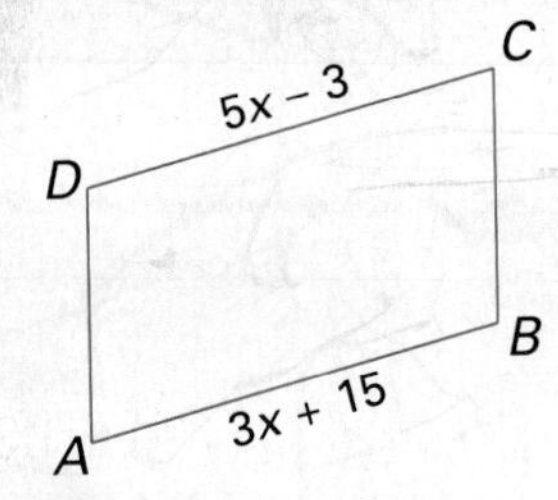

Plan Since opposite sides of a parallelogram are congruent, $AB = CD$.

Solution

$$AB = CD$$
$$3x + 15 = 5x - 3$$
$$x = 9$$
$$AB = 3(9) + 15 = 42$$

If two parallel lines are intersected by a transversal, then the interior angles on the same side of the transversal are supplementary. This can be used to prove Theorem 6.4.

Theorem 6.4 Consecutive angles of a parallelogram are supplementary.

EXAMPLE 3 In $\square PQRS$, $PQ = 23$ and $m\angle Q = 81$. Find SR and $m\angle R$.

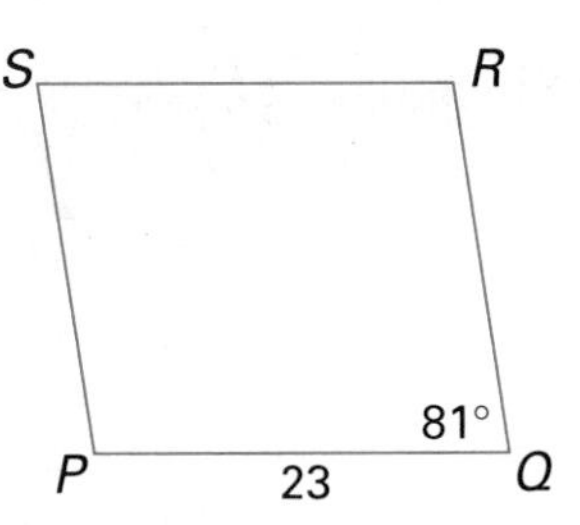

Solution Since $PQ = 23$, $SR = 23$

$$m\angle Q + m\angle R = 180$$
$$m\angle R = 180 - m\angle Q$$
$$m\angle R = 180 - 81 = 99$$

EXAMPLE 4 Given: $\square ABCD$ with $AB = AC$
Prove: $\angle ACB \cong \angle CDA$

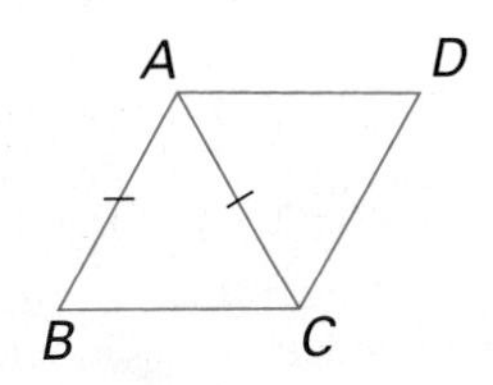

Proof

Statement	Reason
1. $\square ABCD$ with $AB = AC$	1. Given
2. $\angle ACB \cong \angle ABC$	2. In a $\triangle$, $\angle$s opp $\cong$ sides are $\cong$.
3. $\angle ABC \cong \angle CDA$	3. Opp $\angle$s of a $\square$ are $\cong$.
4. $\therefore \angle ACB \cong \angle CDA$	4. Trans Prop

You can use corresponding parts of congruent triangles to prove the following theorem about the diagonals of a parallelogram.

Theorem 6.5 The diagonals of a parallelogram bisect each other.

Given: $\square ABCD$ with diagonals $\overline{AC}$ and $\overline{BD}$ intersecting at O
Prove: $\overline{AC}$ and $\overline{BD}$ bisect each other.

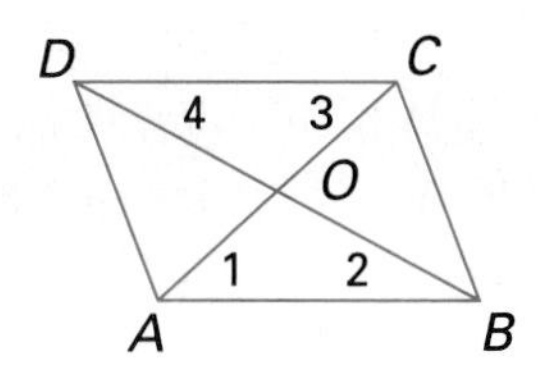

Proof

Statement	Reason
1. $\square ABCD$ with diags $\overline{AC}$ and $\overline{BD}$ intersecting at O	1. Given
2. $\overline{AB} \parallel \overline{CD}$	2. Opp sides of a $\square$ are $\parallel$.
3. $\angle 1 \cong \angle 3$, $\angle 2 \cong \angle 4$	(A, A) 3. Alt int $\angle$s of $\parallel$ lines are $\cong$.
4. $\overline{AB} \cong \overline{CD}$	(S) 4. Opp sides of a $\square$ are $\cong$.
5. $\triangle ABO \cong \triangle CDO$	5. ASA
6. $\overline{AO} \cong \overline{CO}$, $\overline{BO} \cong \overline{DO}$	6. CPCTC
7. $\therefore \overline{AC}$ and $\overline{BD}$ bisect each other.	7. Def of a bis

Summary

Properties of a Parallelogram

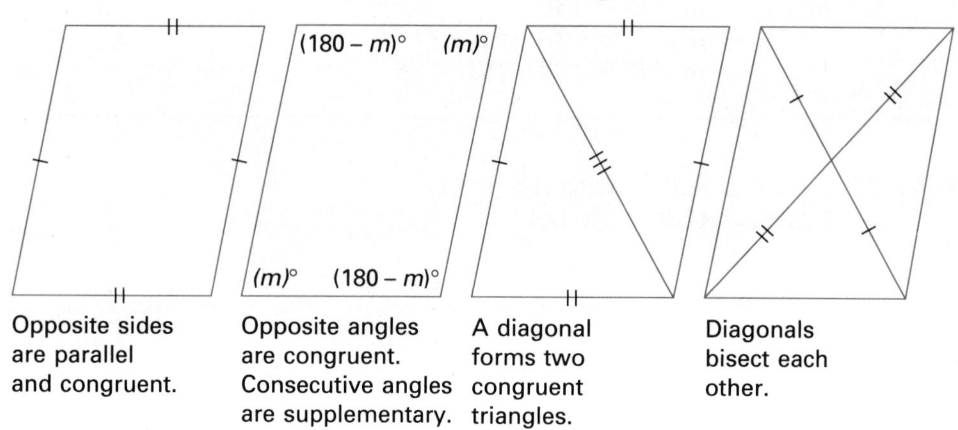

Opposite sides are parallel and congruent.

Opposite angles are congruent. Consecutive angles are supplementary.

A diagonal forms two congruent triangles.

Diagonals bisect each other.

Classroom Exercises

Complete the statement for parallelogram *TUVW*.

1. $\overline{WV}$ is opposite __________.
2. $\angle U$ and __________ are supplementary.
3. $\angle T \cong$ __________
4. $\overline{TU}$ and __________ are adjacent.
5. $\angle W$ is opposite __________.
6. $\angle T$ and __________ are consecutive angles.
7. Which sides of ▱ $TUVW$ are parallel?
8. Which sides of ▱ $TUVW$ are congruent?

Written Exercises

***ABCD* is a parallelogram. State whether the two triangles are congruent. If so, explain why.**

1. $\triangle ABC$ and $\triangle BCD$
2. $\triangle ABC$ and $\triangle CDA$
3. $\triangle CDB$ and $\triangle ABD$
4. $\triangle CDB$ and $\triangle BAD$

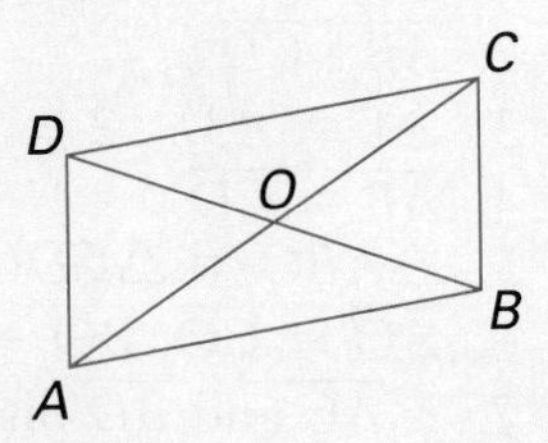

***EFGH* is a parallelogram. Find the unknown measure.**

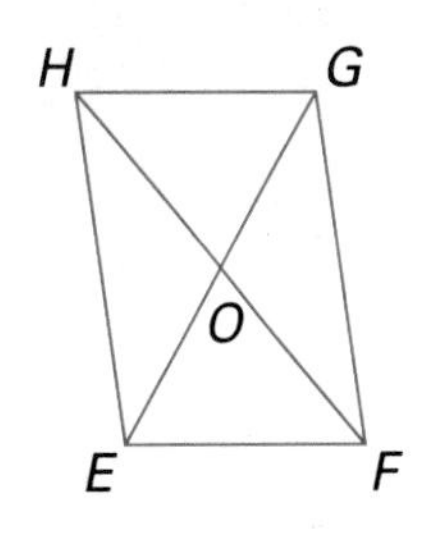

5. $EF = 17$, $GH =$ ______

6. m $\angle EFG = 67$, m $\angle GHE =$ ______

7. m $\angle HEF = 119$, m $\angle GHE =$ ______

8. m $\angle HGF = 125$, m $\angle EHG =$ ______

9. $EF = 12x$, $GH = 10x + 12$, $GH =$ ______

10. $EF = 5x - 7$, $GH = 3x + 1$, $EF =$ ______

11. $EH = 2x + 2$, $FG = 3x - 5$, $FG =$ ______

12. $EO = 3x + 2$, $GO = 5x - 8$, $EO =$ ______

13. $FO = 4x + 13$, $HO = 5x + 1$, $FH =$ ______

14. m $\angle EFG = 6x + 6$, m $\angle FGH = 3x + 3$, m $\angle FGH =$ ______

15. m $\angle EFG = 12x - 24$, m $\angle GHE = 9x + 12$, m $\angle EFG =$ ______

16. $HO = EO$, m $\angle HOE = 100$, m $\angle OHE =$ ______

17. Given: $\square ABCD$, $\angle A \cong \angle DBA$
Prove: $DB = BC$

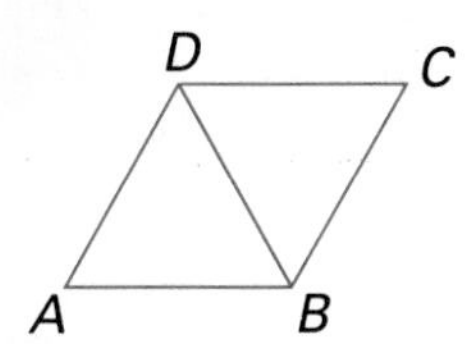

18. Given: $\square RSTU$, $\overline{RT}$ bisects $\angle URS$.
Prove: $\triangle RST$ is isosceles.

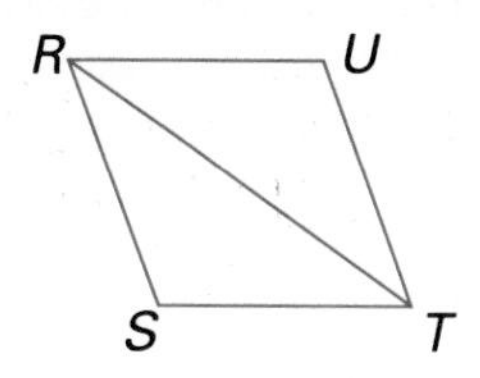

Explain why the figure cannot be a parallelogram.

19.

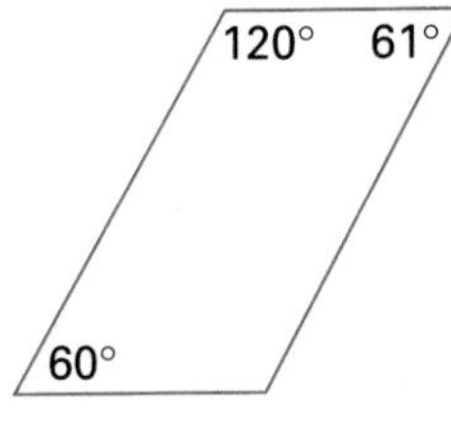

20.

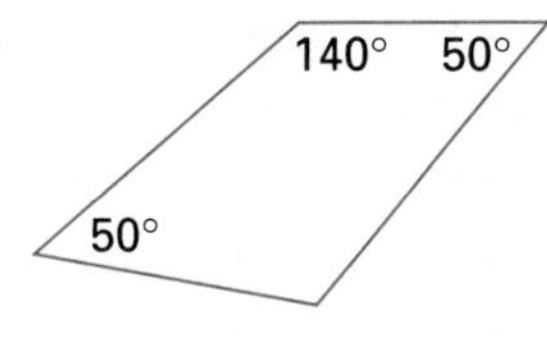

21.

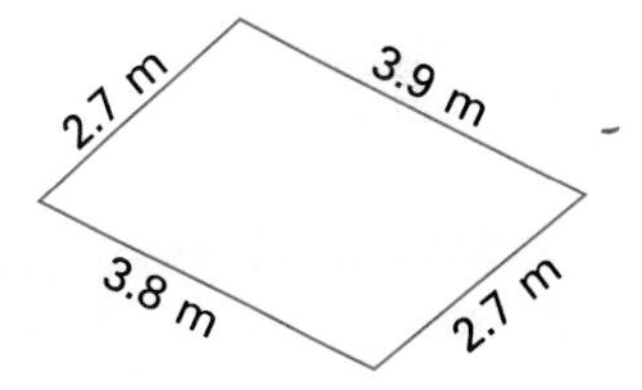

22. Prove Corollary 1.

23. Prove Corollary 2.

24. Prove Theorem 6.4.

25. Given: $\square ACDF$, $\square BCEF$
Prove: $\triangle ABF \cong \triangle DEC$

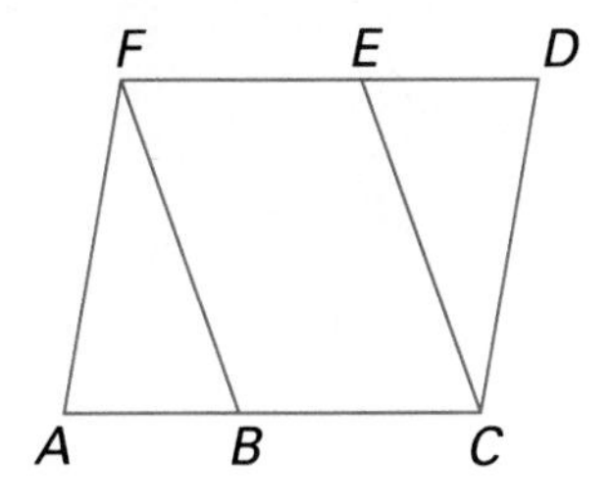

26. Given: $\square ABCD$, diagonals $\overline{AC}$ and $\overline{BD}$ intersect at O.
Prove: O is the midpoint of $\overline{EF}$.

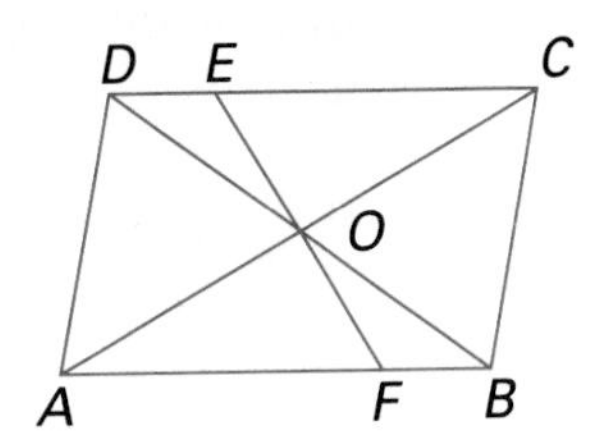

27. Given: $\square JLMP$, K is the midpoint of $\overline{JL}$; N is the midpoint of $\overline{PM}$.
Prove: O is the midpoint of JM.

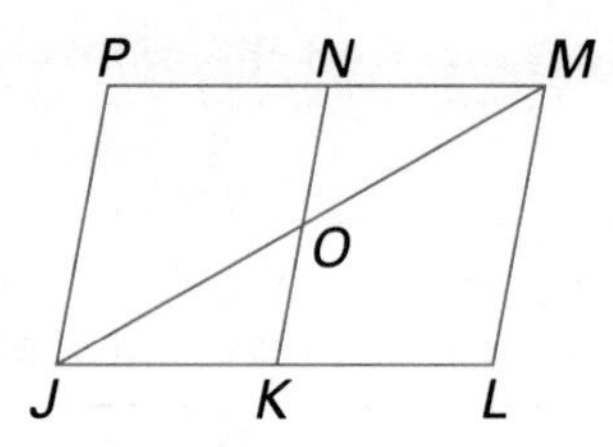

28. In $\square XYZW$, the sum of the measures of $\angle X$ and $\angle Z$ is 30 greater than the measure of $\angle Y$. Find the measure of $\angle X$.

29. Prove: The bisectors of consecutive angles of a parallelogram are perpendicular.

30. Prove: The bisectors of opposite angles of a parallelogram are parallel or concurrent.

Midchapter Review

Name each polygon; tell whether it is concave or convex; then tell whether it is equilateral, equiangular, or regular. *6.1*

1.

2.

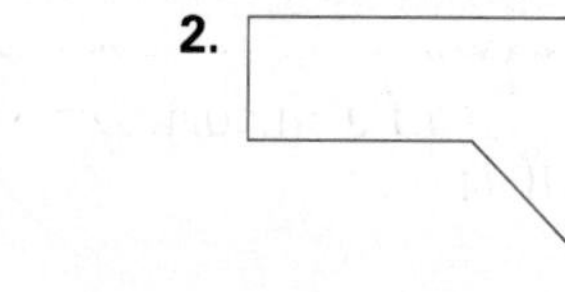

3.

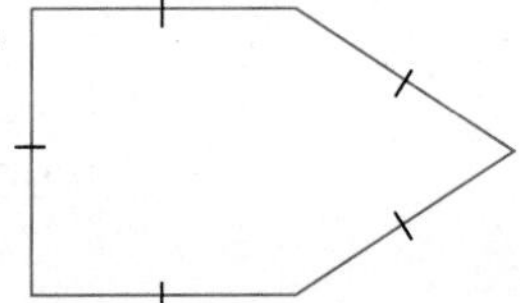

4. Find the sum of the measures of the angles of a pentagon. *6.2*

5. What is the measure of each angle of a convex regular octagon? *6.2*

6. What is the measure of each exterior angle of a regular pentagon? *6.3*

7. The measures of two exterior angles of a regular polygon are $2x + 10$ and $3x - 15$. Identify the polygon. *6.3*

Brainteaser

The design shown was made from a set of wooden tiles. Only three different kinds of tiles were used. Each tile is a parallelogram with all sides congruent. Without measuring, find the angle measures of the three different parallelograms.

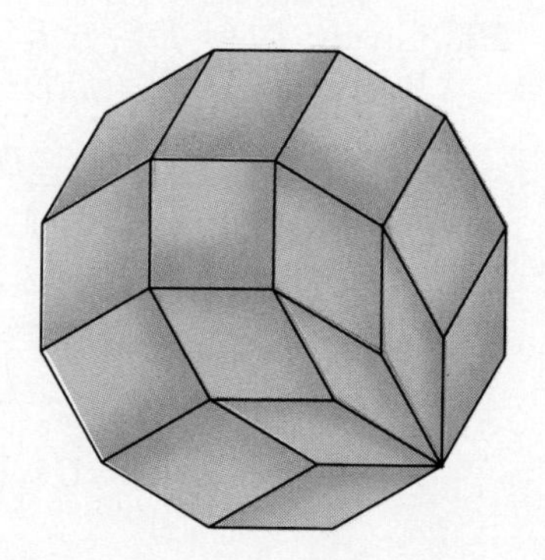

6.5 Quadrilaterals That Are Parallelograms

Objectives To determine if a given quadrilateral is a parallelogram
To prove that certain quadrilaterals are parallelograms

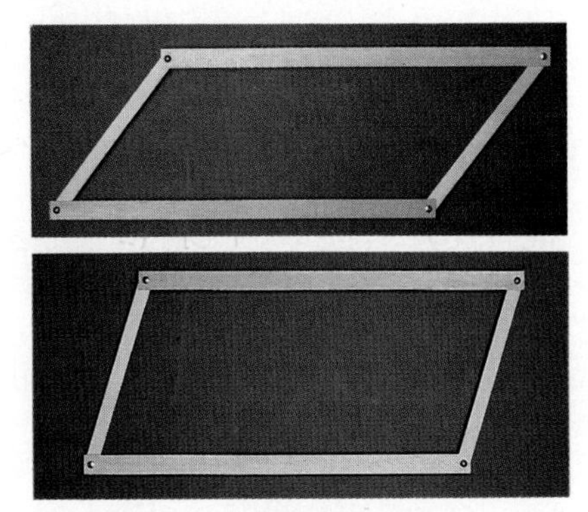

Experimenting with quadrilaterals suggests that there may be other ways to prove that a quadrilateral is a parallelogram. Fasten four sticks, two of one length and two of another length, at their ends to form a quadrilateral with opposite sides of equal length. The quadrilateral formed appears to be a parallelogram. If the sticks are moved so that the measures of the angles between them are changed, then the quadrilateral remains a parallelogram.

Theorem 6.6 If both pairs of opposite sides of a quadrilateral are congruent, then the quadrilateral is a parallelogram.

Given: Quad $ABCD$ with $\overline{AB} \cong \overline{CD}$, $\overline{CB} \cong \overline{AD}$
Prove: $ABCD$ is a parallelogram.

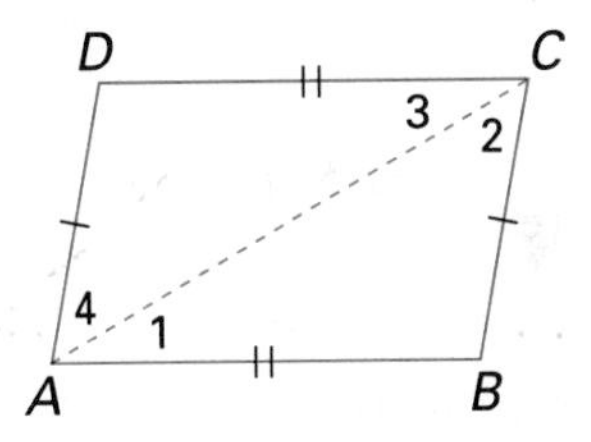

Plan Draw $\overline{AC}$ and prove $\triangle ABC \cong \triangle CDA$. Use congruent pairs of alternate interior angles to prove opposite sides parallel.

Proof

Statement		Reason
1. Quad $ABCD$ with $\overline{AB} \cong \overline{CD}$, $\overline{CB} \cong \overline{AD}$	(S, S)	1. Given
2. Draw $\overline{AC}$.		2. Two points determine a line.
3. $\overline{AC} \cong \overline{AC}$	(S)	3. Reflex Prop
4. $\triangle ABC \cong \triangle CDA$		4. SSS
5. $\angle 1 \cong \angle 3$, $\angle 2 \cong \angle 4$		5. CPCTC
6. $\overline{AB} \parallel \overline{CD}$, $\overline{CB} \parallel \overline{AD}$		6. If alt int $\angle$s are $\cong$, then lines are $\parallel$.
7. $\therefore ABCD$ is a $\square$.		7. Def of a parallelogram

Theorem 6.7 If the diagonals of a quadrilateral bisect each other, then the quadrilateral is a parallelogram.

Given: Diagonals $\overline{EG}$ and $\overline{FH}$ of quad $EFGH$ bisect each other.
Prove: $EFGH$ is a parallelogram.

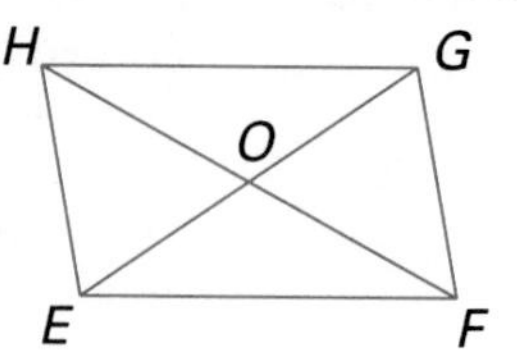

Plan Use SAS to prove $\triangle EOF \cong \triangle GOH$ and $\triangle GOF \cong \triangle EOH$. Then opposite sides of $EFGH$ are congruent.

EXAMPLE 1 Is the quadrilateral a parallelogram? Explain why.

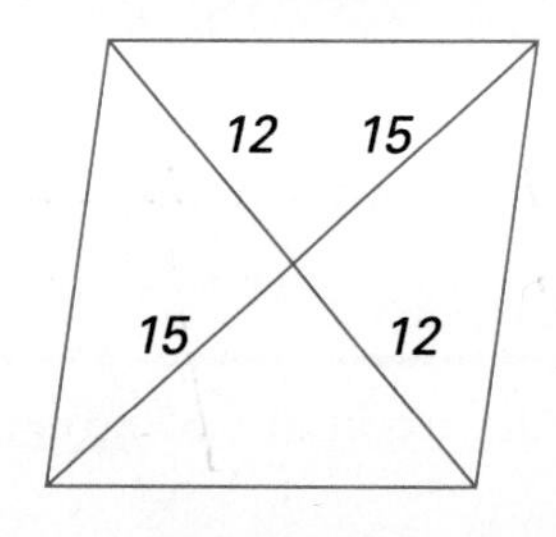

10
6
6
10

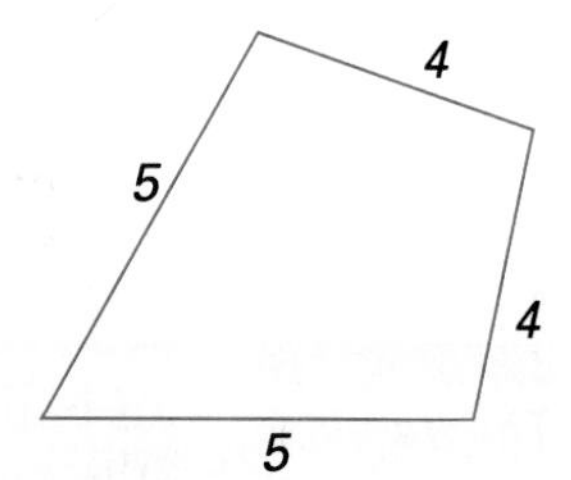

Solution It is a parallelogram, since diagonals bisect each other.

It is a parallelogram, since both pairs of opposite sides are congruent.

The opposite sides are not congruent. The figure is not a parallelogram.

The two following theorems are also useful when proving that a quadrilateral is a parallelogram.

Theorem 6.8 If two sides of a quadrilateral are parallel and congruent, then the quadrilateral is a parallelogram.

Given: Quad $ABCD$ with $\overline{AB} \parallel \overline{CD}$, $\overline{AB} \cong \overline{CD}$
Prove: $ABCD$ is a parallelogram.

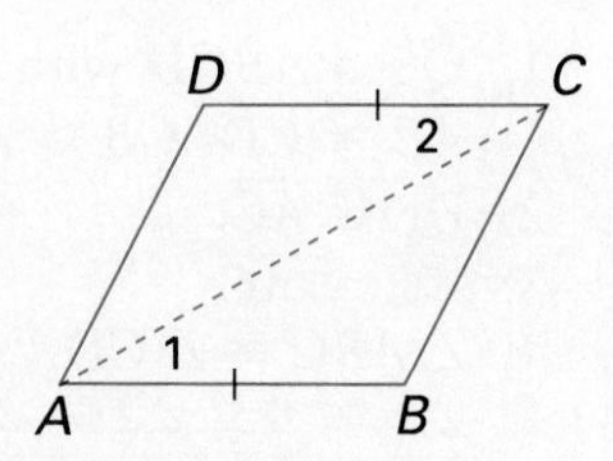

Plan Draw $\overline{AC}$. Use SAS to prove $\triangle ABC \cong \triangle CDA$.

Proof

	Statement		Reason
1.	Quad $ABCD$ with $\overline{AB} \cong \overline{CD}$	(S) 1.	Given
2.	Draw $\overline{AC}$.	2.	Two points determine a line.
3.	$\overline{AB} \parallel \overline{CD}$	3.	Given
4.	$\angle 1 \cong \angle 2$	(A) 4.	Alt int $\angle$s of $\parallel$ lines are $\cong$.
5.	$\overline{AC} \cong \overline{CA}$	(S) 5.	Reflex Prop
6.	$\triangle ABC \cong \triangle CDA$	6.	SAS
7.	$\overline{BC} \cong \overline{DA}$	7.	CPCTC
8.	$\therefore ABCD$ is a parallelogram.	8.	If both pairs of opp sides are $\cong$, then quad is a $\square$.

Theorem 6.9 If both pairs of opposite angles of a quadrilateral are congruent, then the quadrilateral is a parallelogram.

EXAMPLE 2 Given: Quad $ABCD$ with $\angle A \cong \angle C$, $\overline{AB} \parallel \overline{DC}$
Prove: $ABCD$ is a parallelogram.

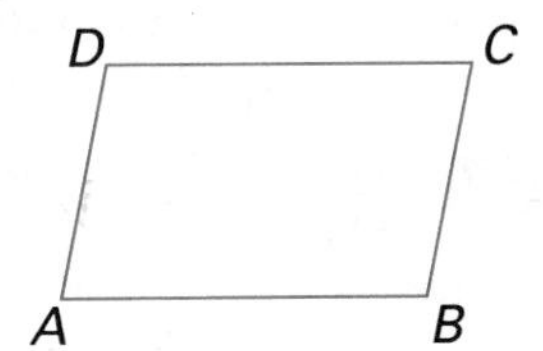

Proof

	Statement		Reason
1.	Quad $ABCD$ with $\overline{AB} \parallel \overline{DC}$	1.	Given
2.	$\angle D$ and $\angle A$ are supp; $\angle B$ and $\angle C$ are supp.	2.	If lines are $\parallel$, then int $\angle$s on same side of transv are supp.
3.	$\angle A \cong \angle C$	3.	Given
4.	$\angle D \cong \angle B$	4.	Supp of $\cong$ $\angle$s are $\cong$.
5.	$\therefore ABCD$ is a parallelogram.	5.	If a quad has both pairs of opp $\angle$s $\cong$, then it is a $\square$.

EXAMPLE 3 True or false? If a diagonal of a quadrilateral forms two congruent triangles, then the quadrilateral is a parallelogram.

Solution False. $EFGH$ is a parallelogram, but $ABCD$ is not.

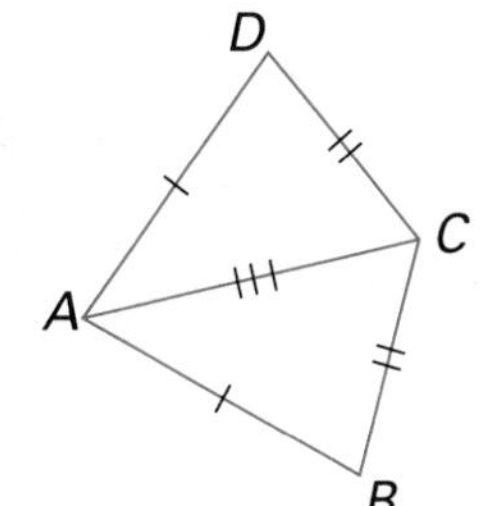

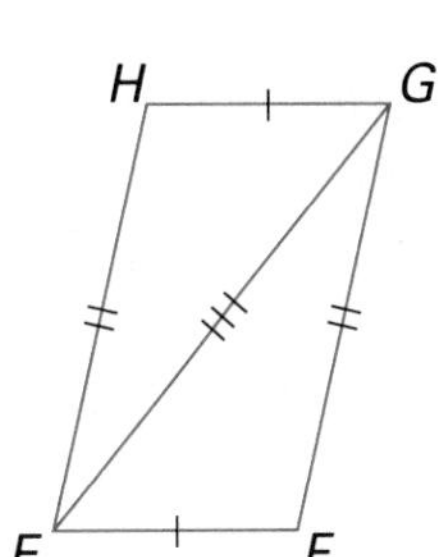

Summary

Sufficient Conditions for Proving a Quadrilateral is a Parallelogram

When proving that a quadrilateral is a parallelogram, it is enough to show that the quadrilateral satisfies any *one* of the five conditions below. Each is a *sufficient* condition.

1. Both pairs of opposite sides are parallel. (Definition)
2. Both pairs of opposite sides are congruent.
3. Diagonals bisect each other.
4. Two sides are parallel and congruent.
5. Both pairs of opposite angles are congruent.

Classroom Exercises

Form sufficient conditions for proving that *ABCD* is a parallelogram by completing the statement.

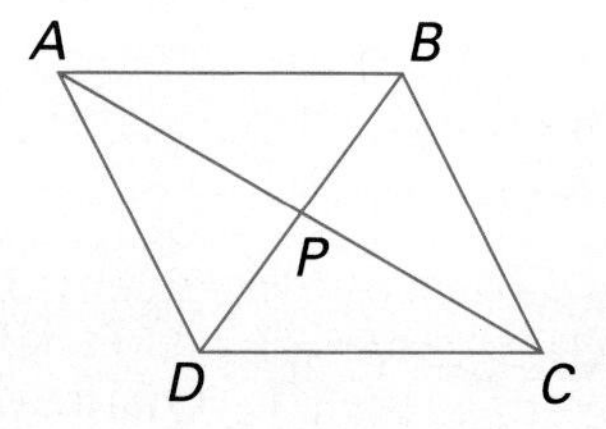

1. $\overline{AB} \cong$ ______ and $\overline{AD} \cong$ ______
2. $\angle BAD \cong$ ______ and $\angle ADC \cong$ ______
3. $\angle ABD \cong$ ______ and $\angle DAC \cong$ ______
4. $\overline{AP} \cong$ ______ and $\overline{DP} \cong$ ______
5. $\overline{AB} \parallel$ ______ and $\overline{BC} \parallel$ ______
6. $\overline{AB} \cong$ ______ and $\overline{AB} \parallel$ ______
7. $\overline{AD} \cong$ ______ and $\angle ADC$ is supplementary to ______.
8. $\triangle ABD \cong$ ______

Written Exercises

Using only the given information, determine whether *ABCD* is a parallelogram. Give a reason for your answer.

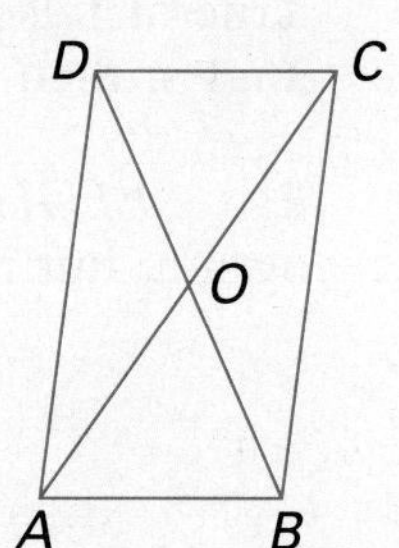

1. $\overline{AB} \parallel \overline{DC}$ and $\overline{BC} \parallel \overline{AD}$
2. $\overline{AB} \cong \overline{BC} \cong \overline{CD}$
3. $\overline{AB} \cong \overline{DC}$ and $\overline{BC} \cong \overline{AD}$
4. $\overline{AB} \cong \overline{DC}$ and $\overline{AB} \parallel \overline{DC}$
5. $\overline{AB} \cong \overline{DC}$ and $\overline{BC} \parallel \overline{AD}$
6. $\overline{AC} \cong \overline{DB}$
7. $\angle DAB \cong \angle BCD$
8. $\overline{AO} \cong \overline{CO}$ and $\overline{BO} \cong \overline{DO}$
9. $\triangle ABC \cong \triangle ADC$
10. $\overline{AO} \cong \overline{BO}$ and $\overline{CO} \cong \overline{DO}$

11. m $\angle DAB = 42$, m $\angle ABC = 138$, m $\angle BCD = 42$

12. m $\angle ABD = 37$, m $\angle CDB = 37$, $AB = CD = 12$

13. m $\angle ABD = 37$, m $\angle CDB = 37$, $AB = BC$

14. Prove Theorem 6.7.

15. Prove Theorem 6.9.

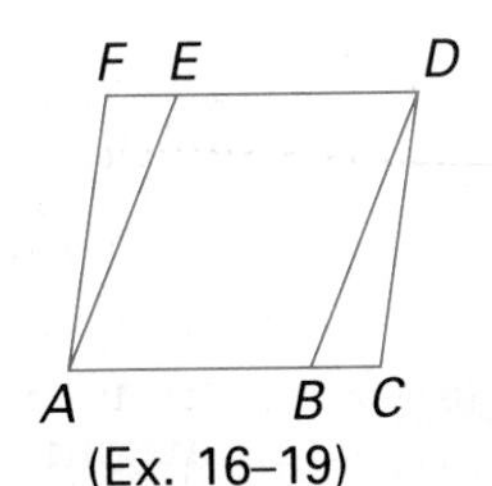

(Ex. 16–19)

16. Given: $\overline{ED} \cong \overline{BA}$, $\triangle AEF \cong \triangle DBC$
Prove: *ABDE* is a parallelogram.

17. Given: $\square ACDF$, $\overline{FE} \cong \overline{CB}$
Prove: *ABDE* is a parallelogram.

18. Given: $\square ABDE$, $\overline{FE} \cong \overline{CB}$
Prove: $\triangle AEF \cong \triangle DBC$

19. Given: $\overline{AF} \parallel \overline{CD}$, $\triangle AFE \cong \triangle DCB$
Prove: *ABDE* is a parallelogram.

20. Given: Equilateral $\triangle$s *GHJ* and *IHJ*
Prove: *GHIJ* is a parallelogram.

G
H J
I

True or false? Give a proof or a counterexample to defend your answer.

21. If consecutive angles of a quadrilateral are supplementary, then the quadrilateral is a parallelogram.
22. If the diagonals of a quadrilateral are perpendicular, then the quadrilateral is a parallelogram.
23. If one pair of opposite sides and one pair of opposite angles of a quadrilateral are congruent, then the quadrilateral is a parallelogram.
24. If the sum of the lengths of any two adjacent sides of a quadrilateral is constant, then the quadrilateral is a parallelogram.
25. If the sum of the measures of any two consecutive angles of a quadrilateral is constant, then the quadrilateral is a parallelogram.

Mixed Review

Find the measure of each angle of $\triangle ABC$.

1. m $\angle A = 36$, $\angle B$ is a right angle. *3.6*
2. $\angle A$ and $\angle C$ are congruent; $\angle B$ is a right angle. *3.6*
3. $\angle A$ and $\angle C$ are complementary; m $\angle A = 2$(m $\angle C$). *3.6*
4. $\overline{AB} \cong \overline{BC}$ and m $\angle A = 27$ *5.1*
5. m $\angle A = 90$, the exterior angle at *B* has measure 140. *3.7*
6. The exterior angles at *A* and *B* each have measure 120. *3.6*

6.6 The Midsegment Theorem

Objective To apply the Midsegment Theorem

A **triangle midsegment** joins the midpoints of two sides of a triangle. Draw several triangles, and construct midsegments. Measure the midsegment and the third side of each triangle.

Theorem 6.10 **Midsegment Theorem:** The segment joining the midpoints of two sides of a triangle is parallel to the third side, and its length is half the length of the third side.

Given: $\triangle ABC$, $\overline{CD} \cong \overline{AD}$, $\overline{CE} \cong \overline{BE}$
Prove: $\overline{DE} \parallel \overline{AB}$, $DE = \frac{1}{2}AB$

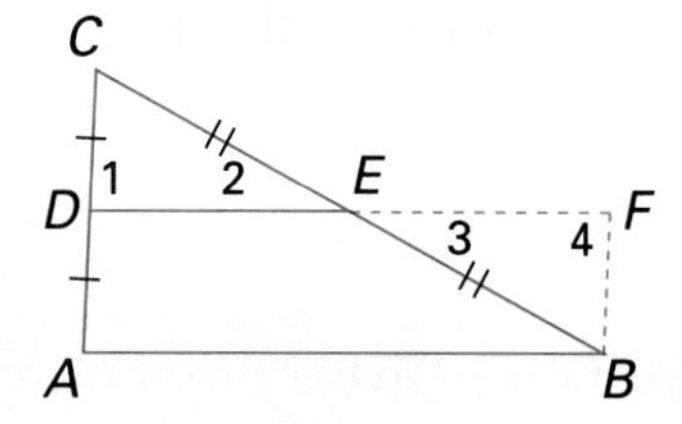

Proof

Statement	Reason
1. $\triangle ABC$, $\overline{CD} \cong \overline{AD}$, $\overline{CE} \cong \overline{BE}$	**(S)** 1. Given
2. Extend $\overline{DE}$ to point F such that $\overline{FE} \cong \overline{DE}$.	**(S)** 2. Ruler Post
3. E is the midpt of $\overline{DF}$.	3. Def of midpt
4. Draw $\overline{BF}$.	4. Two points determine a line.
5. $\angle 2 \cong \angle 3$	**(A)** 5. Vert $\angle$s are $\cong$.
6. $\triangle CED \cong \triangle BEF$	6. SAS
7. $\angle 1 \cong \angle 4$	7. CPCTC
8. $\overline{AD} \parallel \overline{BF}$	8. If alt int $\angle$s are $\cong$, then lines are $\parallel$.
9. $\overline{CD} \cong \overline{BF}$	9. CPCTC
10. $\overline{AD} \cong \overline{BF}$	10. Sub (Steps 1 and 9)
11. $ABFD$ is a $\square$.	11. If one pair of opp sides is $\parallel$ and $\cong$, then quad is a $\square$.
12. $\overline{DF} \cong \overline{AB}$ $(DF = AB)$	12. Opp sides of a $\square$ are $\cong$.
13. $DE = \frac{1}{2}DF$	13. Def of midpt
14. $\therefore DE = \frac{1}{2}AB$	14. Sub
15. $\therefore \overline{DE} \parallel \overline{AB}$	15. Opp sides of a $\square$ are $\parallel$.

EXAMPLE 1 Find RQ if S is the midpoint of $\overline{PR}$, T is the midpoint of $\overline{PQ}$, and $ST = 7$.

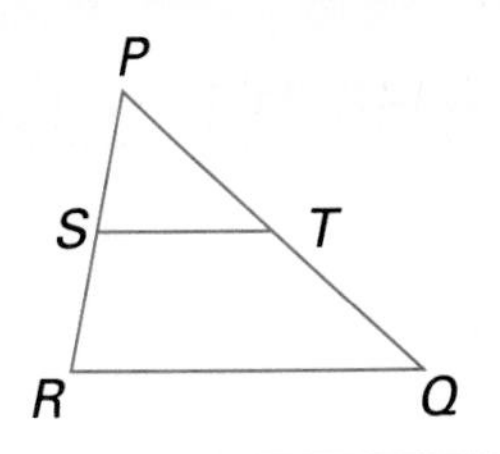

Solution By Theorem 6.10, $ST = \frac{1}{2}RQ$.

Therefore, $RQ = 14$.

EXAMPLE 2 Prove that the segments joining the midpoints of adjacent sides of a convex quadrilateral form a parallelogram.

Given: Quad $ABCD$ with E, F, G, and H the midpts of $\overline{AB}$, $\overline{BC}$, $\overline{CD}$, and $\overline{DA}$, respectively

Prove: $EFGH$ is a parallelogram.

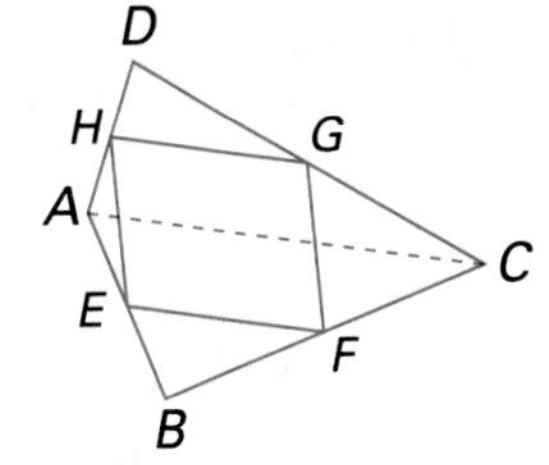

Plan Make a flow diagram.

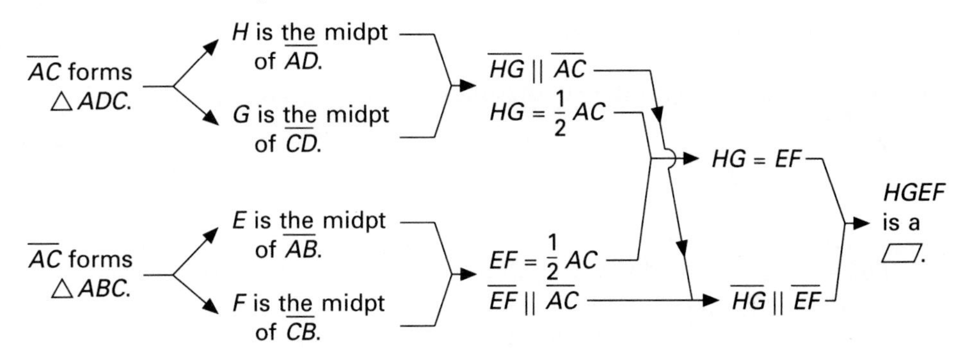

Proof

Statement	Reason
1. Draw $\overline{AC}$ to form $\triangle ABC$ and $\triangle ADC$.	1. Two pts determine a line; def of $\triangle$.
2. E is midpt of $\overline{AB}$; F is midpt of $\overline{BC}$.	2. Given
3. $EF = \frac{1}{2}AC$ and $\overline{EF} \parallel \overline{AC}$	3. Midseg Thm
4. G is midpt of $\overline{CD}$; H is midpt of $\overline{DA}$.	4. Given
5. $HG = \frac{1}{2}AC$ and $\overline{HG} \parallel \overline{AC}$	5. Midseg Thm
6. $EF = HG$ ($\overline{EF} \cong \overline{HG}$)	6. Sub
7. $\overline{EF} \parallel \overline{HG}$	7. Lines $\parallel$ to same lines are $\parallel$.
8. $\therefore$ $EFGH$ is a ▱.	8. If one pair of opp sides is $\parallel$ and $\cong$, then quad is a ▱.

EXAMPLE 3 Given: D and E are the midpoints of $\overline{AC}$ and $\overline{BC}$, respectively; $DE = x + 8$, $AB = 3x - 4$. Find DE and AB.

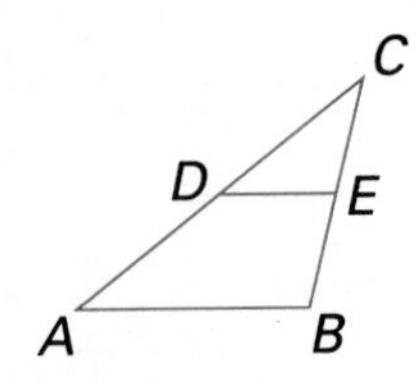

Solution

$$\begin{aligned} DE &= \tfrac{1}{2}AB \text{ by the Midsegment Theorem} \\ x + 8 &= \tfrac{1}{2}(3x - 4) \\ 2x + 16 &= 3x - 4 \\ -x &= -20 \\ x &= 20 \end{aligned}$$

$AB = 3x - 4 = 3 \cdot 20 - 4 = 56$

$DE = \frac{1}{2}AB = \frac{1}{2} \cdot 56 = 28$

Classroom Exercises

Find the indicated length if D is the midpoint of $\overline{AC}$ and E is the midpoint of $\overline{BC}$.

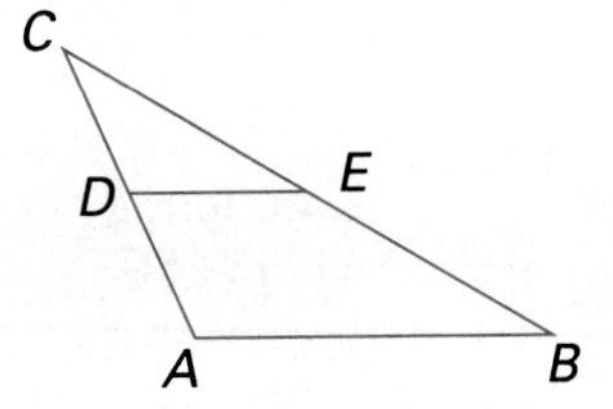

1. If $AB = 12$, then $DE =$ ______
2. If $AB = 15$, then $DE =$ ______
3. If $AB = 3.7$, then $DE =$ ______
4. If $DE = 12$, then $AB =$ ______
5. If $DE = 15$, then $AB =$ ______
6. If $DE = 3.7$, then $AB =$ ______

Written Exercises

In $\triangle FGH$, I, J, and K are the midpoints of $\overline{FG}$, $\overline{GH}$, and $\overline{HF}$, respectively. Using this information, complete the statements.

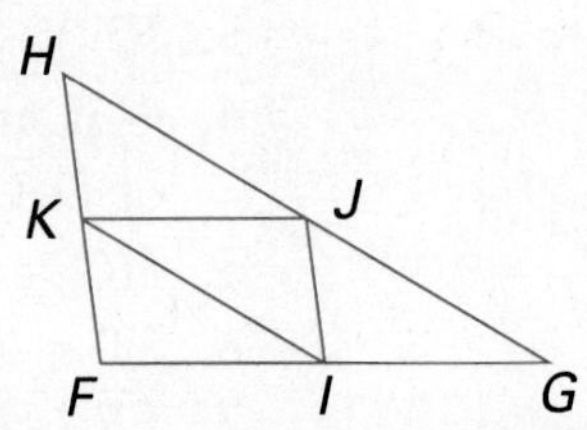

1. $\overline{KJ} \parallel$ ______
2. $\overline{FH} \parallel$ ______
3. If $HG = 12$, then $KI =$ ______
4. If $FH = 18$, then ______ $= 9$
5. If $KJ = 3$, then $FG =$ ______
6. If $IJ = 7$, then ______ $= 14$
7. If $FG = 6x - 4$ and $KJ = 2x + 1$, then $FG =$ ______
8. If $GH = 3x - 8$ and $KI = x + 4$, then $KI =$ ______

In the $\triangle$s shown, P, S, Q, T, G, and E are the midpoints of $\overline{CA}$, $\overline{AT}$, $\overline{CT}$, $\overline{AB}$, $\overline{AD}$, and $\overline{BD}$, respectively. Complete the statements.

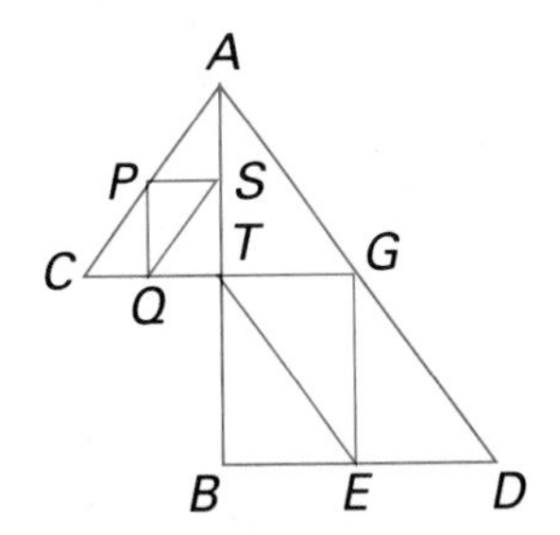

9. $\overline{TE} \parallel$ ______

10. $\overline{CA} \parallel$ ______

11. If $QS = 5$, then $CA =$ ______

12. If $AG = 7.5$, then $TE =$ ______

13. If $AB = 12$, then $AS =$ ______

14. If the perimeter of $\triangle ACT$ is 24, then the perimeter of $\triangle PSQ$ is ______.

O, P, and Q are the midpoints of $\overline{LM}$, $\overline{MN}$, and $\overline{NL}$, respectively.

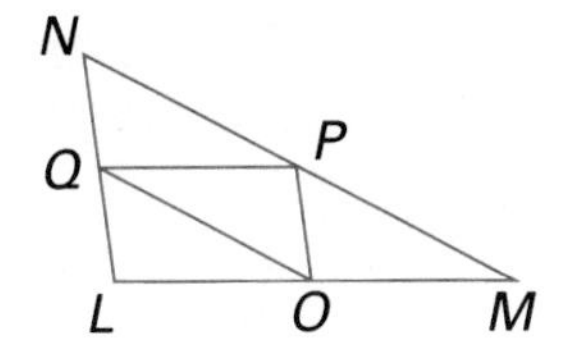

15. Prove: $OPNQ$ is a parallelogram.

16. Prove: $\triangle LOQ \cong \triangle OMP$

17. Prove: $\triangle LOQ \cong \triangle PQO$

18. Prove: $\triangle LOQ \cong \triangle QPN$

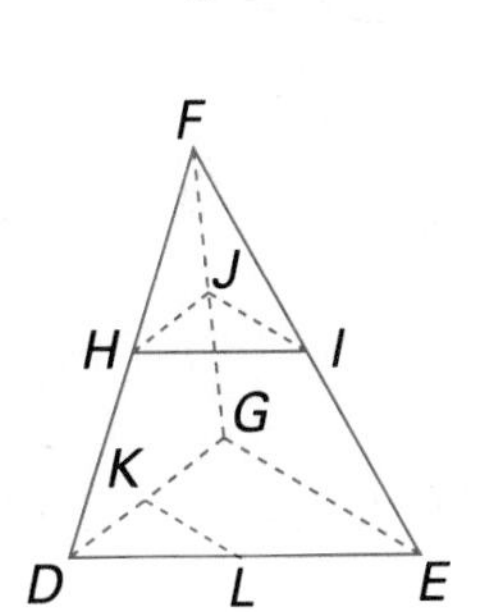

19. Given: U and V are the midpoints of $\overline{RS}$ and $\overline{ST}$, respectively. $\triangle RST$ is isosceles with $\overline{RT} \cong \overline{ST}$.
Prove: $\triangle USV$ is isosceles.

20. Given: U and V are the midpoints of $\overline{RS}$ and $\overline{ST}$, respectively. $\triangle USV$ is isosceles with $\overline{UV} \cong \overline{SV}$.
Prove: $\triangle RST$ is isosceles.

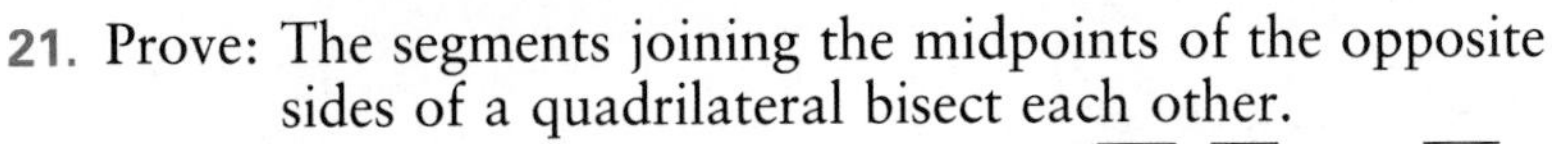

21. Prove: The segments joining the midpoints of the opposite sides of a quadrilateral bisect each other.

22. Given: H, I, and J are the midpoints of $\overline{DF}$, $\overline{EF}$, and $\overline{GF}$, respectively.
Prove: $\angle JHI \cong \angle KDL$

23. Given: H, I, J, K, and L are the midpoints of $\overline{DF}$, $\overline{EF}$, $\overline{GF}$, $\overline{DG}$, and $\overline{DE}$, respectively.
Prove: $\triangle JHI \cong \triangle KDL$

F
J
H
I
G
K
D
L
E

Mixed Review

Referring to the figure at the right, indicate whether the conditional statement is true or false.

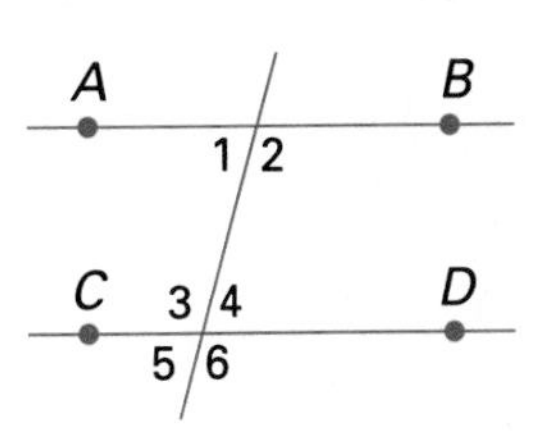

1. If $\overline{AB} \parallel \overline{CD}$, then $\angle 1 \cong \angle 4$. *3.5*

2. If $\angle 2 \cong \angle 4$, then $\overline{AB} \parallel \overline{CD}$.

3. If $\angle 2 \cong \angle 3$, then $\overline{AB} \parallel \overline{CD}$.

4. If $\overline{AB} \parallel \overline{CD}$, then $\angle 2 \cong \angle 6$. *3.5*

Problem Solving Strategies

Making a Table

A frequently used strategy for solving problems is to make a table of known data, and then look for a pattern. The technique is illustrated in the example below.

Example Five points are marked on a circle. All possible segments are drawn, joining each pair of points. How many segments are drawn?

Solution Draw three circles. Experiment with 2, 3, and 4 points. Look for a pattern. Make a table showing the number of segments drawn in each case.

Number of points	2	3	4	5
Number of segments	1	3	6	?

The pattern appears to be that each time a point is added, the number of *new* segments added is one less than the total number of points. If this pattern holds, then for 5 points there should be $6 + (5 - 1) = 10$ segments.

Exercises

1. Draw a circle with 6 points on it. Show that 15 segments are formed when all pairs of points are joined.

2. Extend the table above to predict the number of segments for 8 points.

3. Develop a formula for predicting the number of segments for any number of points on a circle for the example above. Then use the formula to predict the number of segments for 20 points.

6.7 Lines Parallel to Many Lines

Objectives To apply theorems about transversals to three or more parallel lines
To find the distance between two parallel lines

In Chapter 5, it was proven that the shortest segment from a point to a line is the segment perpendicular to the line. Therefore, **the distance from a point to a line** is defined as the length of the perpendicular segment from the point to the line.

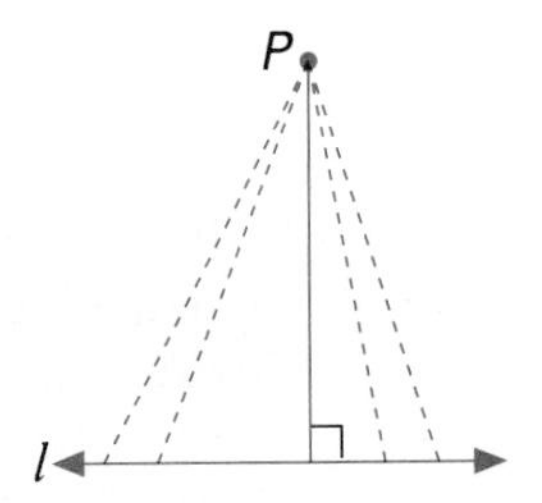

Recall also that the distance between two points is measured along a line. In Chapter 5, the triangle inequality allowed you to conclude that $AC < AB + BC$. In other words, the shortest path along line segments from point A to point C is the direct path $\overline{AC}$.

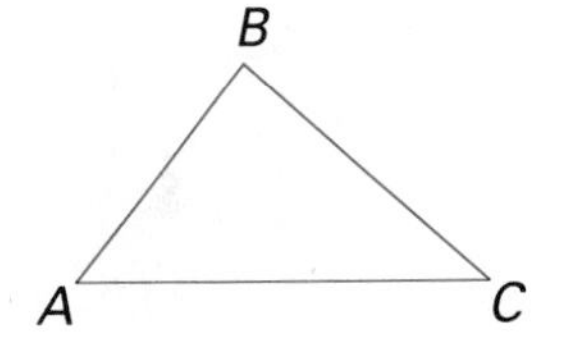

This notion of distance being measured along shortest paths can now be extended to a discussion of parallel lines. First, the following theorem assures that there is a unique distance between parallel lines.

Theorem 6.11 If two lines are parallel, then all points of each line are equidistant from the other line.

Given: $l \parallel m$, P and Q are on l; $\overline{PA} \perp m$, $\overline{QB} \perp m$.
Prove: $PA = QB$

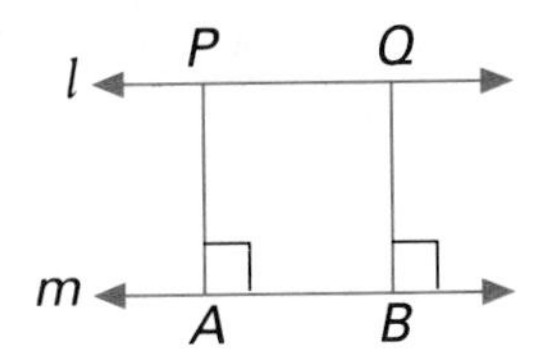

Plan Make a flow diagram.

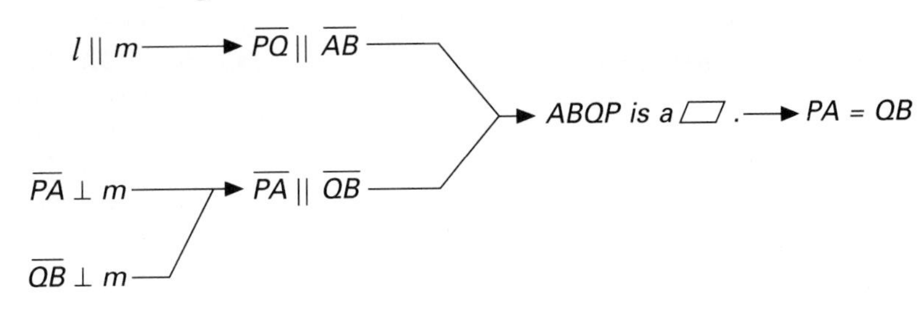

Definition

The **distance between two parallel lines** is the distance from a point on one line to the other line.

A typical ruler is not always helpful for dividing certain distances equally. For example, it is difficult to divide a two-inch segment into thirds with a standard ruler. As shown at right, a sheet of ruled notebook paper can be used to do this. The equally spaced lines on the paper help to locate equally spaced marks on the two-inch segment. This suggests the following theorem.

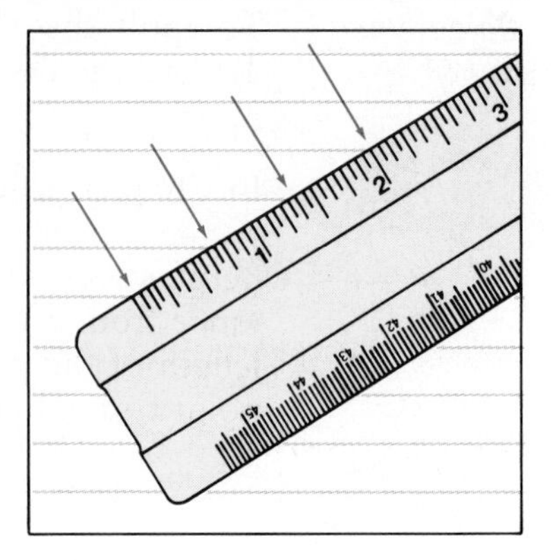

Theorem 6.12

If three parallel lines cut off congruent segments on one transversal, then they cut off congruent segments on every transversal.

Given: $l \parallel m \parallel n$ with transversals $\overleftrightarrow{AE}$ and $\overleftrightarrow{BF}$, $\overline{AC} \cong \overline{CE}$

Prove: $\overline{BD} \cong \overline{DF}$

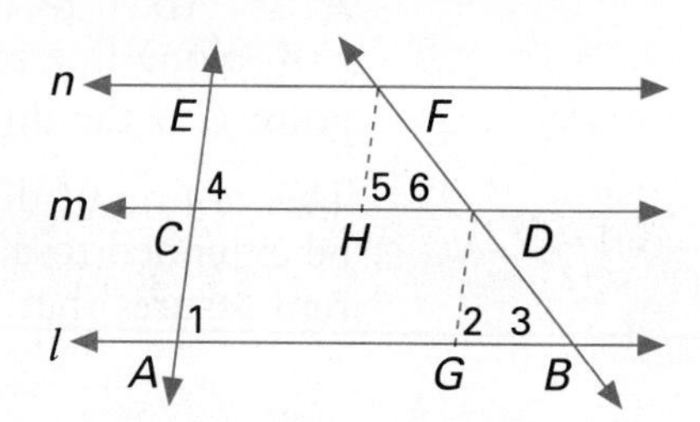

Plan

Draw $\overline{FH} \parallel \overline{EC}$ and $\overline{DG} \parallel \overline{AC}$. Prove that $AGDC$ and $CHFE$ are parallelograms and that $\triangle GBD \cong \triangle HDF$.

Proof

Statement		Reason
1. $l \parallel m \parallel n$ with transv $\overleftrightarrow{AE}$ and $\overleftrightarrow{BF}$		1. Given
2. Draw $\overline{DG} \parallel \overline{CA}$ and $\overline{FH} \parallel \overline{EC}$.		2. Through a given point, there is exactly one line $\parallel$ to a given line.
3. $AGDC$ and $CHFE$ are $\square$s.		3. If 2 pairs of opp sides of a quad are $\parallel$, then it is a $\square$.
4. $\overline{CE} \cong \overline{HF}$, $\overline{GD} \cong \overline{AC}$		4. Opp sides of $\square$ are $\cong$.
5. $\overline{AC} \cong \overline{CE}$		5. Given
6. $\overline{GD} \cong \overline{HF}$	(S)	6. Trans Prop Cong
7. $\angle 1 \cong \angle 2$, $\angle 4 \cong \angle 5$		7. Corr $\angle$s of $\parallel$ lines are $\cong$.
8. $\angle 1 \cong \angle 4$		8. Corr $\angle$s of $\parallel$ lines are $\cong$.
9. $\angle 2 \cong \angle 5$	(A)	9. Trans Prop Cong
10. $\angle 3 \cong \angle 6$	(A)	10. Corr $\angle$s of $\parallel$ lines are $\cong$.
11. $\triangle GBD \cong \triangle HDF$		11. AAS
12. $\therefore \overline{BD} \cong \overline{DF}$		12. CPCTC

Corollary If any number of parallel lines cut off congruent segments on one transversal, then they cut off congruent segments on every transversal.

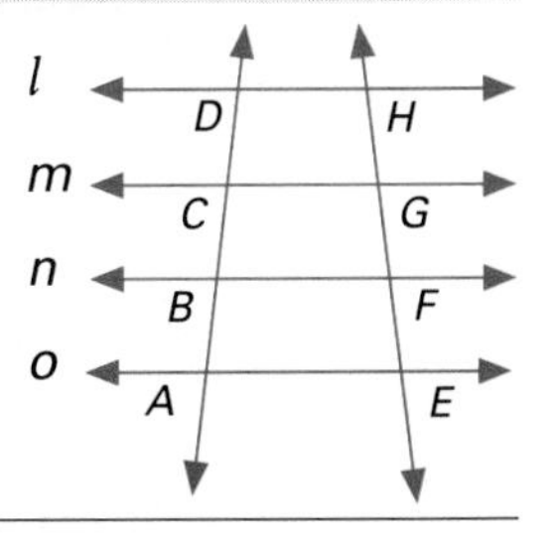

EXAMPLE 1 $l \parallel m \parallel n \parallel o$, $AB = BC = CD$, $EF = 5$. Find FH.

Solution $EF = FG = GH = 5$, $FG + GH = FH$
$FH = 5 + 5 = 10$

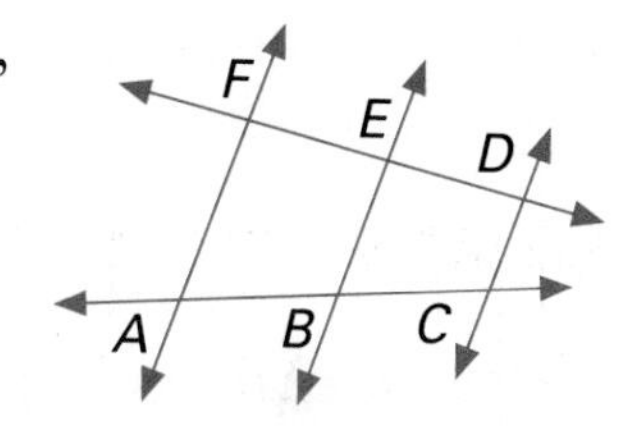

EXAMPLE 2 $\overleftrightarrow{AF} \parallel \overleftrightarrow{BE} \parallel \overleftrightarrow{CD}$, $FE = 7$, $ED = 7$, $AB = 7x - 5$, $BC = 3x + 3$. Find AB.

Solution
$AB = BC$ by Theorem 6.12
$7x - 5 = 3x + 3$
$4x = 8$
$x = 2$

Thus, $AB = 7x - 5 = 7 \cdot 2 - 5 = 9$.

EXAMPLE 3 Given: $\angle 1 \cong \angle 2$, $\angle 2 \cong \angle 3$, $\overline{BD} \cong \overline{DF}$
Prove: $\overline{AC} \cong \overline{CE}$

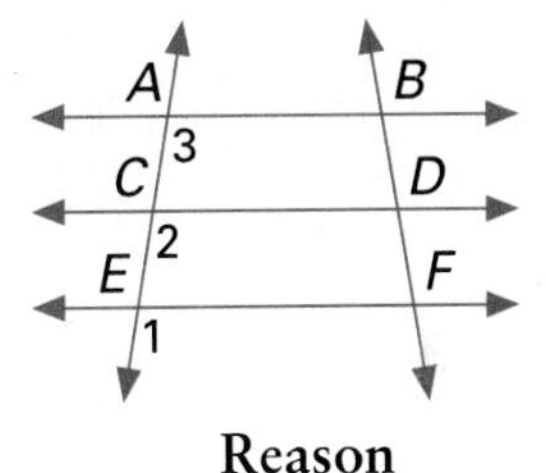

Proof

Statement	Reason
1. $\angle 1 \cong \angle 2$, $\angle 2 \cong \angle 3$	1. Given
2. $\overline{AB} \parallel \overline{CD}$, $\overline{CD} \parallel \overline{EF}$	2. If corr $\angle$s are $\cong$, lines are $\parallel$.
3. $\overline{AB} \parallel \overline{CD} \parallel \overline{EF}$	3. Lines $\parallel$ to the same line are $\parallel$.
4. $\overline{BD} \cong \overline{DF}$	4. Given
5. $\therefore \overline{AC} \cong \overline{CE}$	5. If $\parallel$ lines cut off $\cong$ segs on one transv, they cut off $\cong$ segs on every transv.

Theorem 6.13 If a segment is parallel to one side of a triangle and contains the midpoint of a second side, then this segment bisects the third side.

Using Theorem 6.12 and its corollary, you can divide a segment into any number of congruent segments.

Construction Divide a segment into three congruent segments.

Given: $\overline{AB}$
Choose any point not on $\overline{AB}$. Label it C. Draw $\overrightarrow{AC}$.

Use a compass with any convenient opening to construct $\overline{AX} \cong \overline{XY} \cong \overline{YZ}$.

Draw $\overline{BZ}$. Construct $\overline{XP} \parallel \overline{BZ}$ and $\overline{YQ} \parallel \overline{BZ}$.

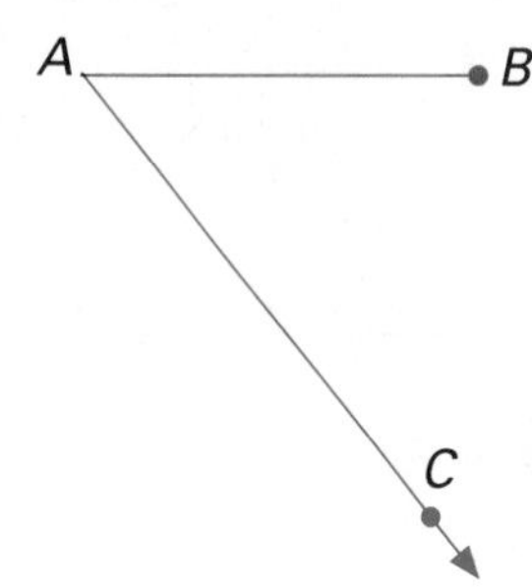

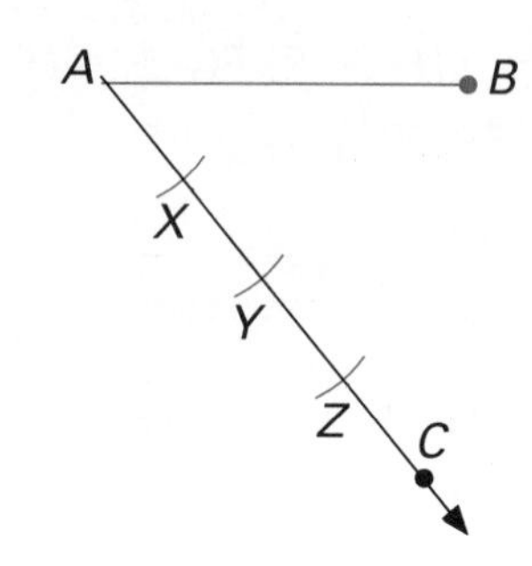

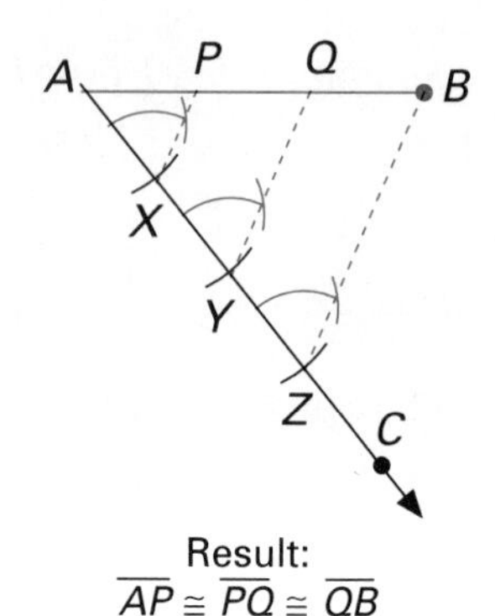

Result:
$\overline{AP} \cong \overline{PQ} \cong \overline{QB}$

Classroom Exercises

Given: $l \parallel m \parallel n \parallel o$; $\overline{AD} \cong \overline{DG} \cong \overline{GJ}$. Using this information, complete the statement.

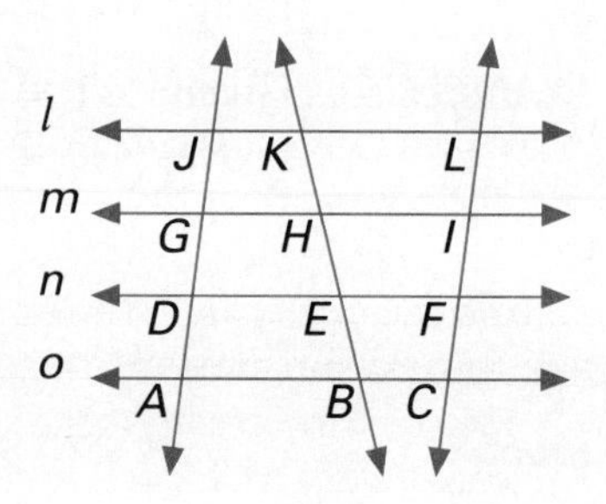

1. If $KH = 7$, then $HE =$ ______
2. If $LI = 9$, then $FC =$ ______
3. If $KE = 14$, then $KH =$ ______
4. If $LC = 12$, then $IF =$ ______
5. If $KE = 15$, then $HB =$ ______
6. If $IC = 20$, then $LF =$ ______
7. If $EB = 1$, then $KB =$ ______
8. If $LF = 16$, then $LC =$ ______
9. If $HE = 4$, then $KE =$ ______
10. If $IF = 9$, then $LC =$ ______

Written Exercises

Given: $\overleftrightarrow{AB} \parallel \overleftrightarrow{CD} \parallel \overleftrightarrow{EF}$, $\overline{AC} \cong \overline{CE}$. Using this information, complete the statement.

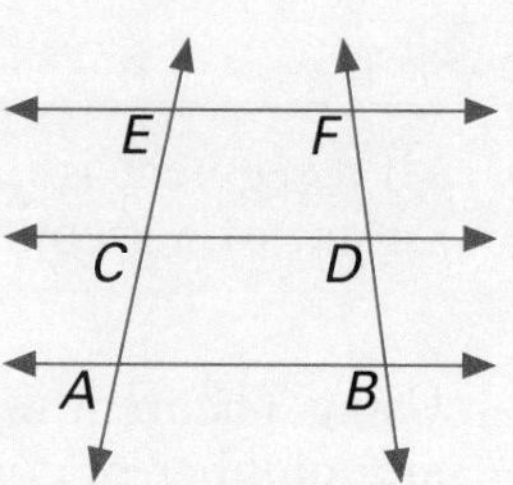

1. $BD = 5x + 1$, $DF = 2x + 13$, $BD =$ ______
2. $BD = 2x + 1$, $DF = 4x - 7$, $DF =$ ______
3. $BD = 3x + 4$, $DF = 4x - 5$, $DF =$ ______
4. $BD = 3x + 4$, $BF = 7x - 2$, $DF =$ ______
5. $DF = 2x + 5$, $BF = 5x - 2$, $BD =$ ______
6. $DF = 3x + 2$, $BF = 12x + 1$, $BD =$ ______

7. Draw a segment 15 cm long. Divide it into three congruent segments.
8. Draw a segment 17 cm long. Divide it into five congruent segments.
9. Given: $\angle 1 \cong \angle 5$, $\angle 5 \cong \angle 9$, $\overline{GH} \cong \overline{HI}$
Prove: $\overline{LK} \cong \overline{KJ}$
10. Given: $\angle 3 \cong \angle 6$, $\angle 8 \cong \angle 9$, $\overline{LK} \cong \overline{KJ}$
Prove: $\overline{GH} \cong \overline{HI}$
11. Given: $\angle 3 \cong \angle 10$, $\overline{KH} \parallel \overline{JI}$, $\overline{GH} \cong \overline{HI}$
Prove: $\overline{LK} \cong \overline{KJ}$
12. Given: $\overline{LG} \parallel \overline{KH}$, $\angle 5 \cong \angle 9$, $\overline{LK} \cong \overline{KJ}$
Prove: $\overline{GH} \cong \overline{HI}$

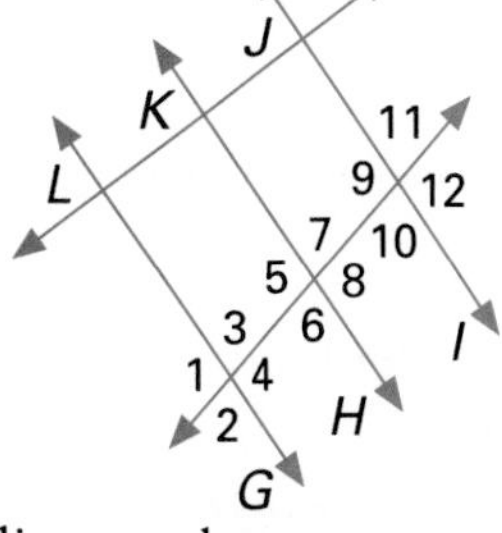

13. Construct a pair of parallel lines. Find the distance between the lines.
14. Prove Theorem 6.11.
15. Prepare a flow diagram for the proof of Theorem 6.12.
16. Prove the corollary to Theorem 6.12 for four parallel lines.
17. Prove Theorem 6.13.

Provide a counterexample to show that the conditional is false.

18. If three lines cut off congruent segments on each of two transversals, then the lines are parallel.
19. If a segment joining a point on each of two sides of a triangle is half as long as the third side, then it is parallel to the third side.
20. The yard lines of a football field must be marked off for the first game of the season. A permanent marker is at the end of each goal line. There are no other markers on the field. How can the maintenance crew lay off the yard lines so that they are parallel to the goal line and parallel to each other?

Mixed Review

Complete the statement, using $\square MNOP$. 6.5

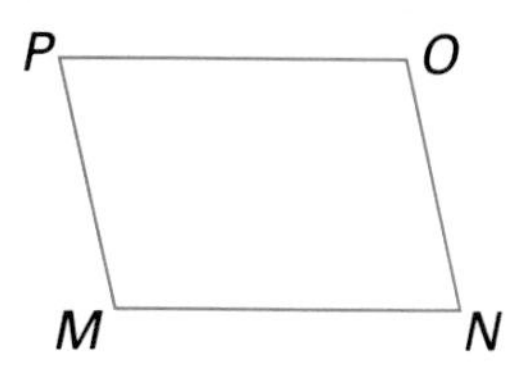

1. If $MN = 6$, then $PO =$ ______
2. If $PM = 8.7$, then $ON =$ ______
3. If $MN + NO = 37$, then $OP + PM =$ ______
4. If $PM + MN = 4.9$, then $MN + NO =$ ______
5. If $m\angle M = 78$, then $m\angle O =$ ______
6. If $m\angle N = 83$, then $m\angle O =$ ______

Chapter 6 Review

Key Terms

adjacent sides (p. 225)
concave (p. 227)
consecutive angles (p. 225)
convex (p. 227)
diagonal (p. 226)
distance between parallel lines (p. 258)
equiangular polygon (p. 227)
equilateral polygon (p. 227)
opposite angles (p. 241)
opposite sides (p. 241)
parallelogram (p. 241)
polygon (p. 225)
quadrilateral (p. 241)
regular polygon (p. 227)
triangle midsegment (p. 252)

Key Ideas and Review Exercises

6.1 To identify convex and concave polygons, determine if a segment connecting any two points is in the interior of the polygon. If so, the polygon is convex. If not, it is concave.

To identify a regular polygon, determine if all sides and all angles of the polygon are congruent. If angles are congruent, it is equiangular. If sides are congruent, it is equilateral.

1. Name the polygon. Is it convex or concave?

2. Is the given polygon equilateral, equiangular, or regular?

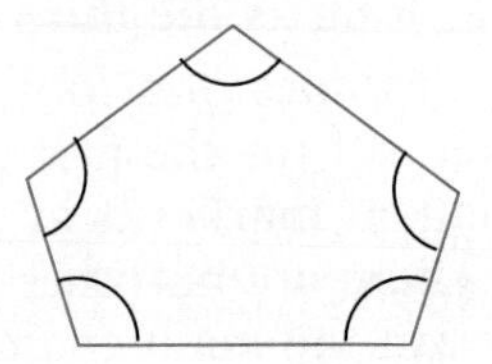

6.2 To find the sum of the angle measures of a polygon with n sides, use: Sum $= (n - 2)180$.

To find the measure of each angle of a regular polygon with n sides, use: Measure $= \dfrac{(n - 2)180}{n}$.

3. Find the sum of the measures of the angles of a dodecagon.

4. Find the measure of each angle of a regular hexagon.

6.3 The sum of the measures of the exterior angles, one at each vertex, of a polygon is 360.

5. The measures of three exterior angles of a quadrilateral are 37, 58, and 92. Find the measure of the exterior angle at the fourth vertex.

6. Find the measure of an exterior angle of a regular octagon.

6.4 To prove theorems or to determine properties of parallelograms, use one or more of the following:

(1) Opposite sides of a parallelogram are parallel.
(2) Opposite sides of a parallelogram are congruent.
(3) Opposite angles of a parallelogram are congruent.

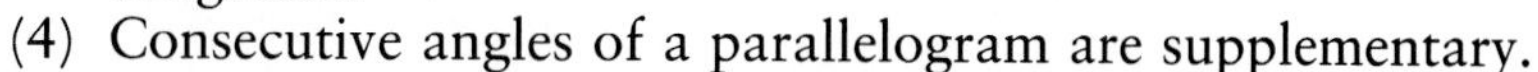

(4) Consecutive angles of a parallelogram are supplementary.
(5) The diagonals of a parallelogram bisect each other.

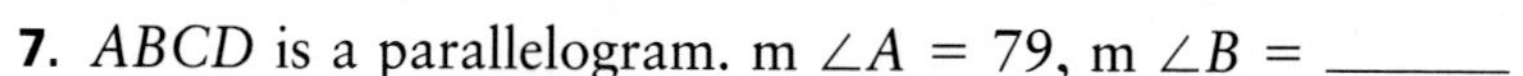

7. $ABCD$ is a parallelogram. m $\angle A = 79$, m $\angle B =$ ______

8. $ABCD$ is a parallelogram. $AO = 2x + 1$, $AC = 5x - 5$, $AO =$ ______

9. Given: $\square ABCD$
Prove: $\triangle AOD \cong \triangle COB$

6.5 To prove that a quadrilateral is a parallelogram, use one of the following:

(1) Both pairs of opposite sides are parallel.
(2) Both pairs of opposite sides are congruent.
(3) The diagonals bisect each other.
(4) One pair of opposite sides is parallel and congruent.
(5) Both pairs of opposite angles are congruent.

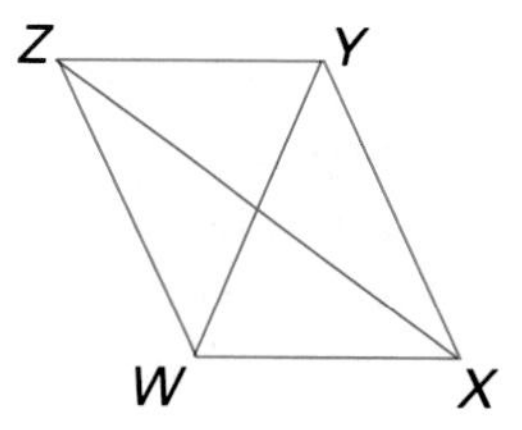

10. $\overline{XY} \parallel \overline{WZ}$, $\overline{XY} \cong \overline{WZ}$. Is $WXYZ$ a parallelogram? Why?

11. Given: $\triangle ZYX \cong \triangle XWZ$
Prove: $XWZY$ is a parallelogram.

6.6 Midsegment Theorem: The segment joining the midpoints of two sides of a triangle is parallel to the third side, and its length is half the length of the third side.

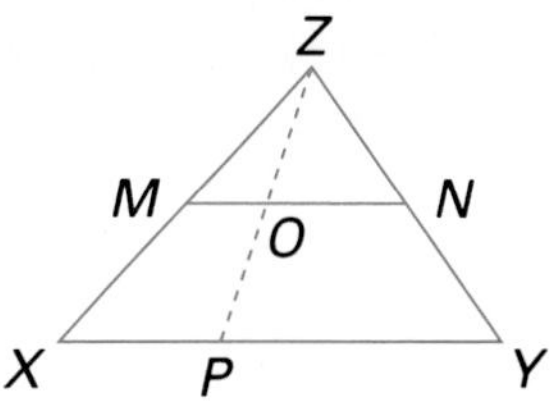

12. M and N are the midpoints of $\overline{XZ}$ and $\overline{YZ}$, respectively; $MN = 12$. Find XY.

13. Given: M and N are the midpoints of $\overline{XZ}$ and $\overline{YZ}$, respectively.
Prove: m $\angle OPY = 180 -$ m $\angle MOZ$

6.7 Three or more parallel lines that cut off congruent segments on one transversal cut off congruent segments on every transversal they intersect.

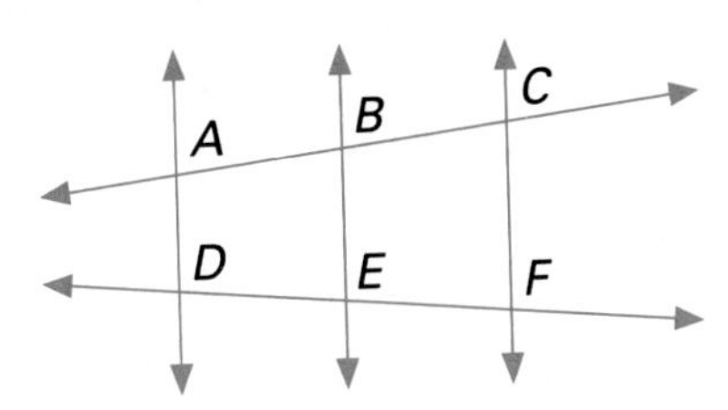

14. $\overleftrightarrow{AD} \parallel \overleftrightarrow{BE} \parallel \overleftrightarrow{CF}$, $AB = 14$, $BC = 14$, $DE = 3x + 2$, $DF = 7x - 1$. Find EF.

15. Draw a segment 12 cm long. Using a straightedge and compass, divide it into five congruent segments.

Chapter 6 Test

Name the polygon. Is the polygon equilateral, equiangular, or regular?

1.

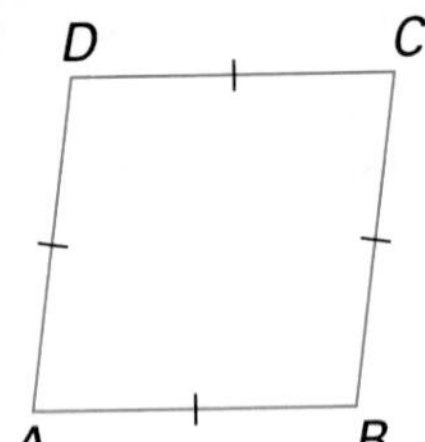

2.

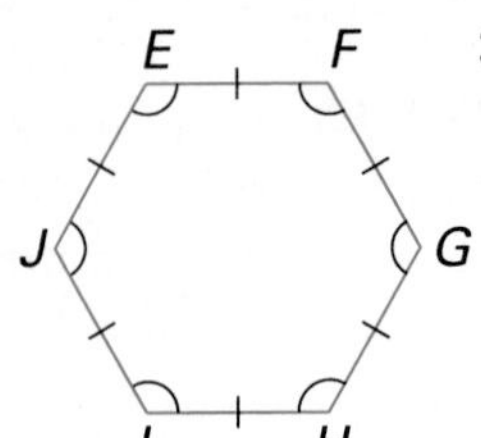

3.

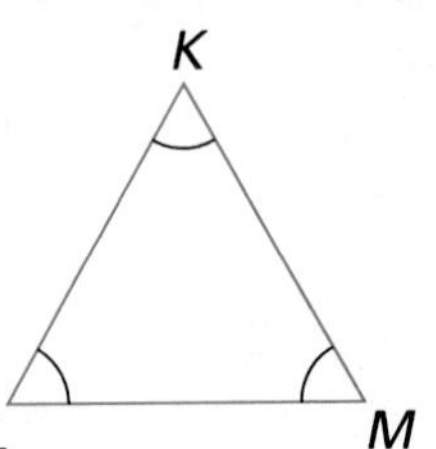

4.

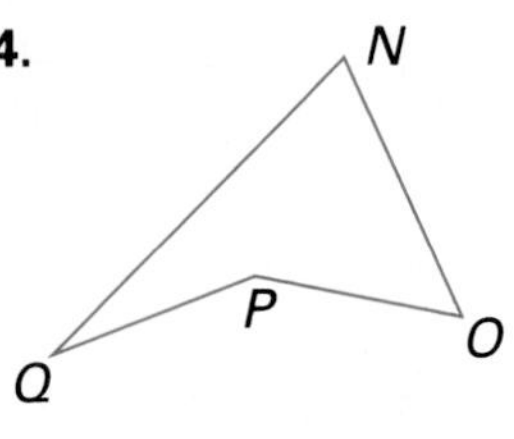

5. Which of the polygons in Items 1—4 are concave?

6. Find the sum of the measures of the angles of a 100-gon.

7. Find the measure of each angle of a regular octagon.

8. Find the measure of each exterior angle of a regular pentagon.

9. In quad *ABCD*, m $\angle A = x - 10$, m $\angle B = x$, m $\angle C = x + 10$ and m $\angle D = x + 20$. Find the measure of each angle.

10. The measures of the exterior angles of a pentagon, one at each vertex, are 45, 86, 135, and 54. Find the measure of the fifth exterior angle.

11. Find the number of sides in a regular polygon if the measure of each exterior angle is 12.

12. Name a pair of consecutive angles of quad *RSTU*.

True or false?

13. Adjacent sides of a parallelogram are congruent.

14. Opposite angles of a parallelogram are supplementary.

15. If each pair of opposite sides is congruent in a quadrilateral, then it is a parallelogram.

16. Diagonals of a parallelogram are perpendicular.

Given: *WXYZ* is a parallelogram.

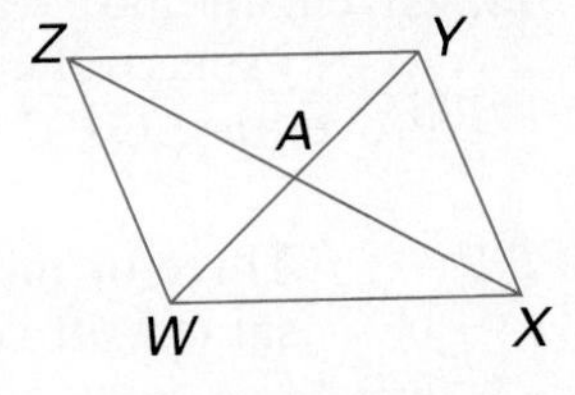

17. m $\angle XWZ$ + m $\angle YXW$ = ______

18. $WA = x + 5$, $YA = 2x - 7$. Find WA.

19. Draw a segment four inches long. Using a straightedge and a compass, divide it into five congruent segments.

***20.** Given: $\square ABCD$, $BM = CN$
Prove: *AMND* is a parallelogram.

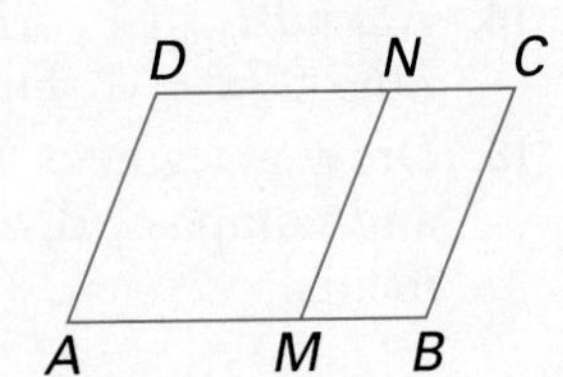

College Prep Test

Strategy for Achievement in Testing

Questions on standardized achievement tests often involve geometric figures. If a diagram is not given, you may want to draw one. Try to make a reasonably accurate diagram, but it is important to avoid special cases. For example, if a question refers to a quadrilateral, do not draw a parallelogram. A good diagram uses all of the given information, and only that information.

Choose the best answer to each question or problem.

1. Which of the following is true of a line, but not true of a segment?
   (A) has exactly one endpoint
   (B) contains an infinite number of points
   (C) is named by two points
   (D) has no endpoints
   (E) has a midpoint

2. Which number is closest to 3?
   (A) $\frac{10}{3}$ (B) 2.9 (C) 11
   (D) $\sqrt{8}$ (E) $\frac{11}{4}$

3. An acute angle can have a measure of:
   I. 89.999 II. 0.0001 III. 90.0001
   (A) I only (B) II only (C) III only
   (D) I and II only (E) I and III only

4. If $l \parallel m$, then
   I. $\angle 3$ and $\angle 5$ are supplementary.
   II. $m\angle 7 = m\angle 8$
   III. $m\angle 3 = m\angle 7 + m\angle 6$
   (A) I only (B) II only (C) III only
   (D) I and II only (E) I, II, and III

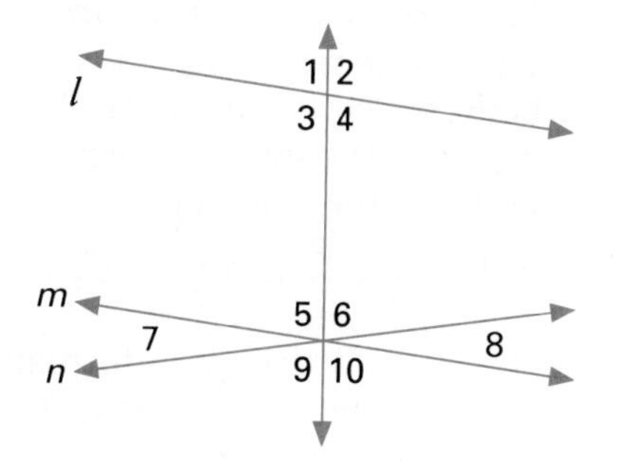

5. $m\angle A = 68$ and $m\angle C = 73$

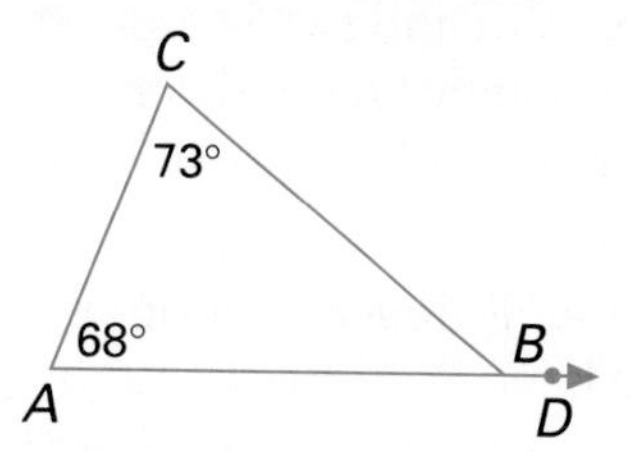

   Which of these is not possible?
   (A) $m\angle CBD$ is acute.
   (B) $m\angle ABC$ is acute.
   (C) $m\angle CBD = 141$
   (D) $m\angle ABC < 141$
   (E) None of these

6. If $\overline{AB} \cong \overline{TP}$, $\overline{BC} \cong \overline{PS}$, and $\overline{CA} \cong \overline{ST}$, then:
   (A) $\triangle ABC \cong \triangle TSP$
   (B) $\triangle ABC \cong \triangle PTS$
   (C) $\triangle ABC \cong \triangle SPT$
   (D) $\triangle ABC \cong \triangle STP$
   (E) None of these

7. Each angle of a regular polygon with r sides has a measure of ______.
   (A) $\frac{360}{r}$ (B) $(r - 2)180$
   (C) 360 (D) $\frac{(r - 2)180}{r}$ (E) 60

8. To construct an angle of 150, you should first construct a(n) ______.
   (A) angle of 45 (B) angle bisector
   (C) isosceles triangle
   (D) pair of parallel lines
   (E) equilateral triangle

Cumulative Review (Chapters 1–6)

1. If C is the midpoint of AB, find the coordinates of A and B for $A:(p-6)$, $B:(4p+2)$, and $C:(8)$. *1.3*

2. The measure of the supplement of an angle is $2\frac{1}{2}$ times the measure of its complement. Find the angle. *1.6*

3. Write the conjunction and the disjunction of the two statements: Triangles have three sides. Triangles have three angles. *1.7*

4. Graph $x \leq 8$ *and* $x \geq 3$ on a number line. *1.8*

5. Are $\angle AOB$ and $\angle CBO$ adjacent? If not, why not? *1.5*

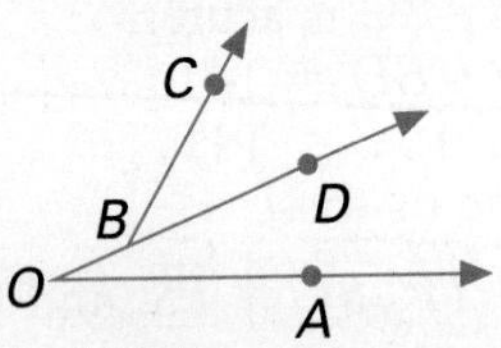

6. Draw an obtuse angle and label it $\angle WXY$. Construct $\angle ABC$ congruent to $\angle WXY$. *1.4*

7. Construct the bisector of $\angle ABC$ in Exercise 5. *1.5*

8. Solve for x below and find all the angles. *2.7*

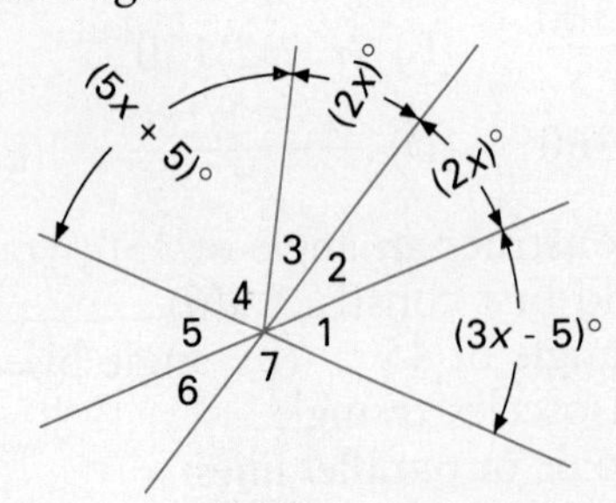

9. Given: Adjacent angles ABP and PBC which are $\ncong$. To write an indirect proof that $\overrightarrow{BP}$ is not the bisector of $\angle ABC$, what assumption must be made? *3.4*

10. Two parallel lines are intersected by a transversal. Which nonadjacent pairs of angles are supplementary? *3.5*

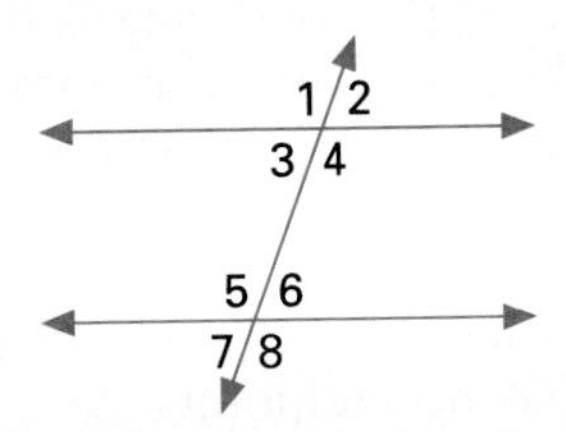

11. Given: Isosceles triangle ABC with $\overline{AC} \cong \overline{BC}$, and exterior $\angle DCB$. If m $\angle A = 72$, what is m $\angle DCB$? *3.6*

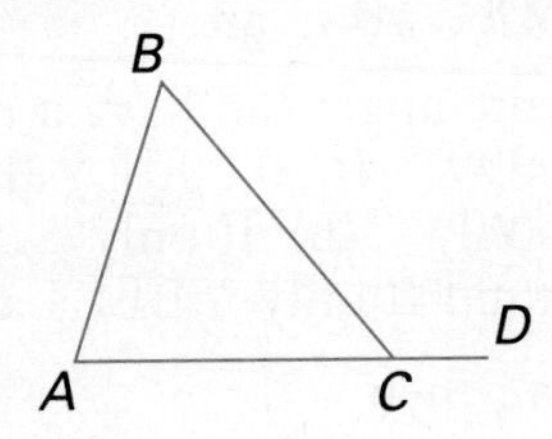

12. In $\triangle ABC$, m $\angle A = 2x$, m $\angle B = 3x - 60$, and m $\angle C = x$. Find m $\angle B$. *3.6*

13. Write the converse, inverse, and contrapositive of the following statement and tell whether each is true or false: If points are coplanar, then they lie in the same plane. *3.5, 3.8*

14. Given: $\triangle ABC$ and $\triangle XYZ$, $\overline{AC} \cong \overline{XZ}$, $\overline{CB} \cong \overline{ZY}$, $\angle B \cong \angle Y$. Is enough information given to prove $\triangle ABC \cong \triangle XYZ$? Why or why not? *4.6*

15. The measure of one acute angle of a right triangle is $\frac{2}{3}$ the measure of the other. Find the measures of the two acute angles. *4.1*

16. Given: Isosceles triangle ABC with $AC = BC$. If $AC = x + 8$, $BC = 3x - 2$, and $AB = 2x + 1$, find AB. *5.1*

17. Given: Isosceles triangle ABC with $AC = BC$, and $\overline{AD}$ bisecting $\angle CAB$, and $\overline{BD}$ bisecting $\angle CBA$. If m $\angle C = 40$, find m $\angle ADB$. *5.2*

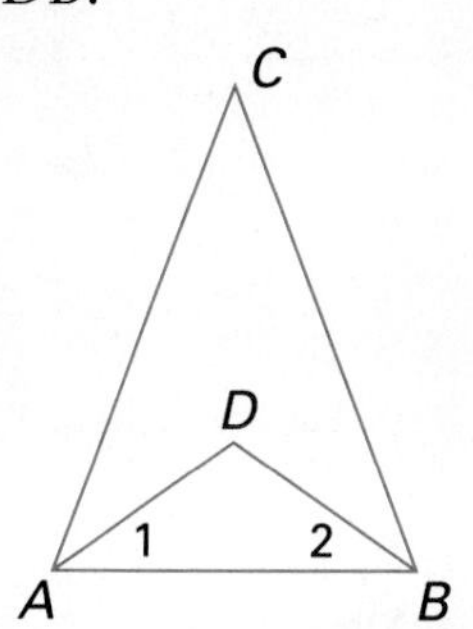

18. Given: P and Q are equidistant from X and Y. What conclusion can be drawn? *5.6*

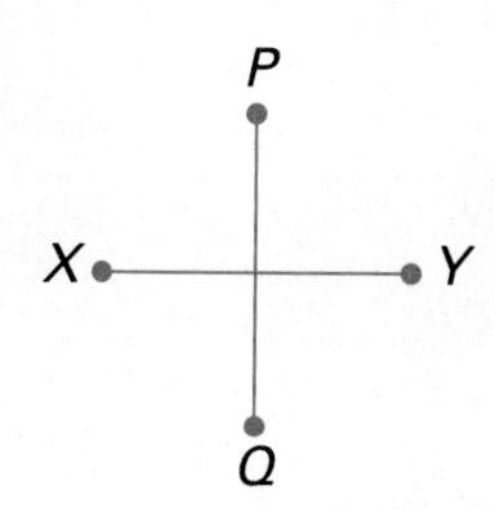

19. In $\triangle DEF$, $\overline{DE} \perp \overline{EF}$. Which side of the triangle is the hypotenuse? *4.1*

20. In $\triangle ABC$, if $AB = 5$, and $BC = 10$, what restrictions are there on the length of $\overline{AC}$? *5.7*

21. If the measure of each interior angle of a polygon is 135, how many sides does the polygon have? *6.2*

22. Is Quad $ABCD$ a $\square$ if *6.5*
(a) $\overline{AB} \parallel \overline{CD}$ and $\overline{AD} = \overline{BC}$?
(b) $\overline{AB} \parallel \overline{CD}$ and $\overline{AD} \parallel \overline{BC}$?
(c) $\overline{AM} \cong \overline{MC}$ and $\overline{DM} \cong \overline{BM}$?
(d) m $\angle DAB$ = m $\angle BCD$ and $\angle DAB$ is supplement of $\angle ABC$?

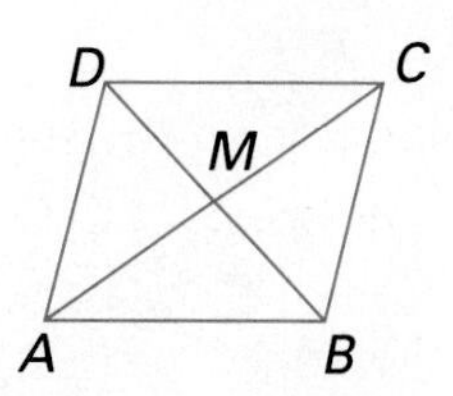

23. Given: $\overline{AB}$ is the perpendicular bisector of $\overline{CD}$; $\overline{AC} \cong \overline{BD}$. *5.6*
Prove: $\overline{AM} \cong \overline{BM}$

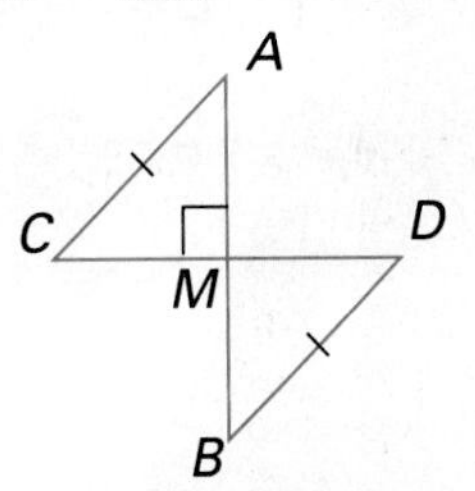

24. What is the sum of the measures of the angles of a decagon? *6.2*

25. Given: $m \parallel n$ and angle measures as indicated. Find x. *2.7*

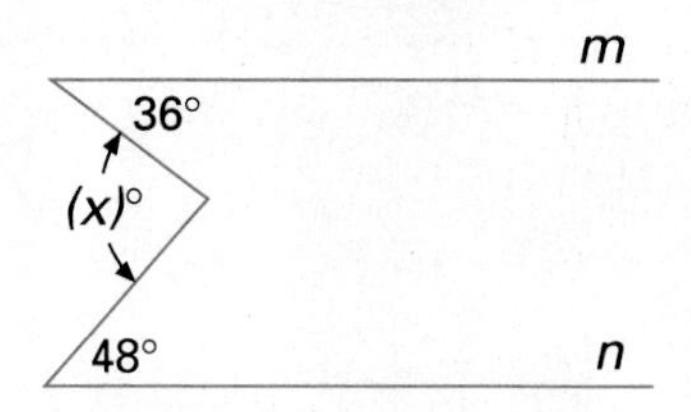

26. Given: the solid figure as shown *3.1*
(a) Name two skew segments.
(b) Name two coplanar segments.

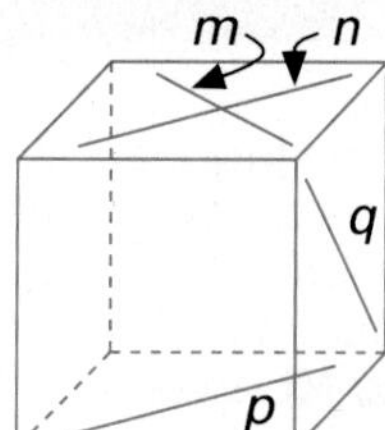

7 SPECIAL QUADRILATERALS

Stephen Hawking (Jan. 8, 1942)—Hawking developed a new model for black holes (stars that have collapsed). By means of mathematical calculations, he demonstrated that there can be "mini" black holes too small to see that weigh as much as Mt. Everest.

7.1 Proving Quadrilaterals Congruent

Objectives To identify and name congruent quadrilaterals and their parts
To prove quadrilaterals congruent

Congruent quadrilaterals have the same size and shape. In this lesson, postulates and theorems for congruent triangles will be used to prove theorems concerning congruence of convex quadrilaterals.

Definition **Congruent quadrilaterals** are quadrilaterals whose corresponding sides are congruent and whose corresponding angles are congruent.

Quad $ABCD \cong$ quad $EFGH$

$\angle A \cong \angle E$	$\overline{AB} \cong \overline{EF}$
$\angle B \cong \angle F$	$\overline{BC} \cong \overline{FG}$
$\angle C \cong \angle G$	$\overline{CD} \cong \overline{GH}$
$\angle D \cong \angle H$	$\overline{DA} \cong \overline{HE}$

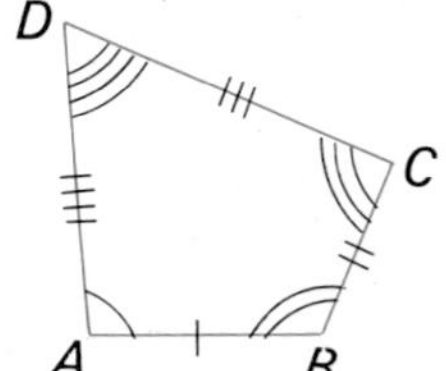

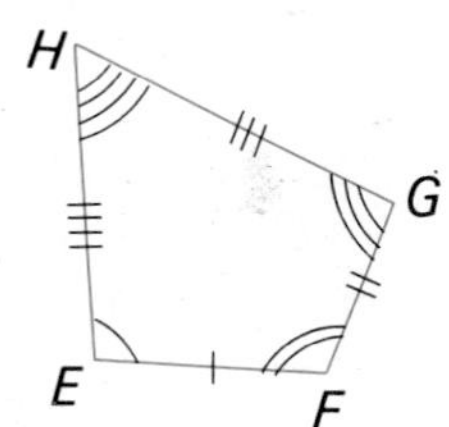

EXAMPLE 1 Using the markings below, write a congruence statement.

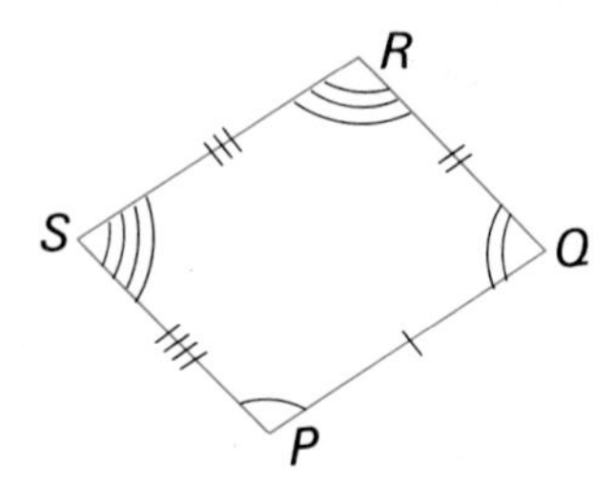

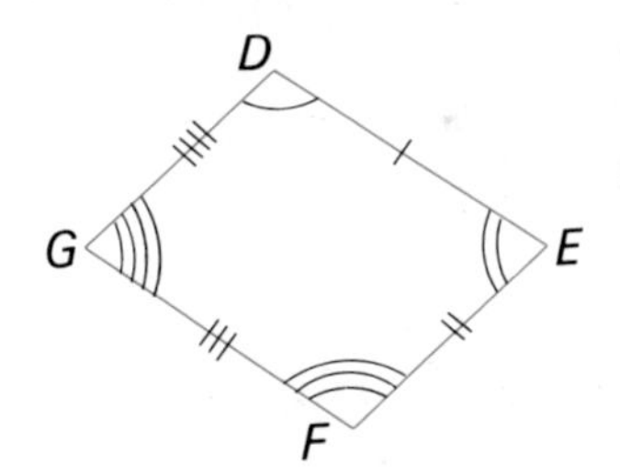

Solution Quad $PQRS \cong$ quad $DEFG$

In the case of triangles, three pairs of congruent parts were needed to establish congruent triangles. It does *not* follow, however, that four pairs of congruent parts can establish congruent quadrilaterals.

$\overline{AB} \cong \overline{EF}$
$\overline{BC} \cong \overline{FG}$
$\overline{CD} \cong \overline{GH}$
$\overline{DA} \cong \overline{HE}$

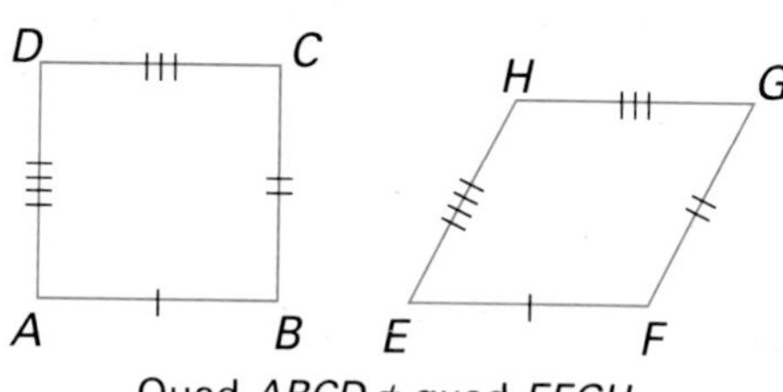

Quad $ABCD \not\cong$ quad $EFGH$

Four pairs of congruent sides do not establish congruence of quadrilaterals because a quadrilateral, unlike a triangle, is not rigid. In the previous figure, the sides of quadrilateral $ABCD$ collapse inward at angles B and D to form quadrilateral $EFGH$. In general, it will be necessary to establish *five* pairs of congruences to prove quadrilaterals congruent.

Theorem 7.1 Two quadrilaterals are congruent if any three sides and the included angles of one are congruent, respectively, to three sides and the included angles of the other. (SASAS for congruent quadrilaterals)

Given: $\overline{AB} \cong \overline{EF}$, $\overline{BC} \cong \overline{FG}$, $\overline{CD} \cong \overline{GH}$, $\angle B \cong \angle F$, $\angle C \cong \angle G$

Prove: Quad $ABCD \cong$ quad $EFGH$

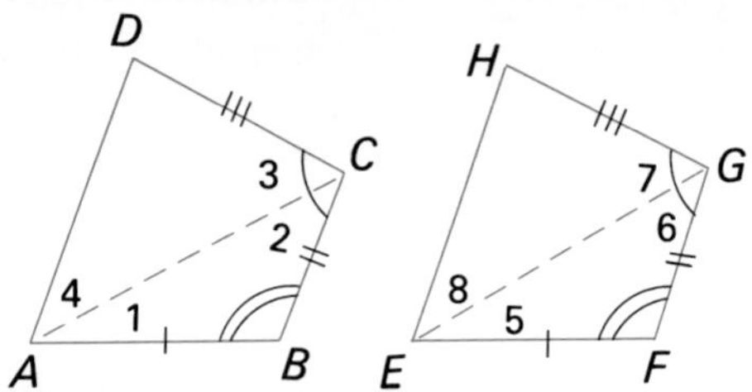

Plan Draw $\overline{AC}$ and $\overline{EG}$. $\triangle ABC \cong \triangle EFG$ by SAS. Using corresponding parts of these triangles, prove that $\triangle ACD \cong \triangle EGH$. Use CPCTC to complete the proof.

EXAMPLE 2 Given: $\square ABCD$ and $\square EFGH$, $\overline{AB} \cong \overline{EF}$, $\overline{AD} \cong \overline{EH}$, $\angle A \cong \angle E$

Prove: $\square ABCD \cong \square EFGH$

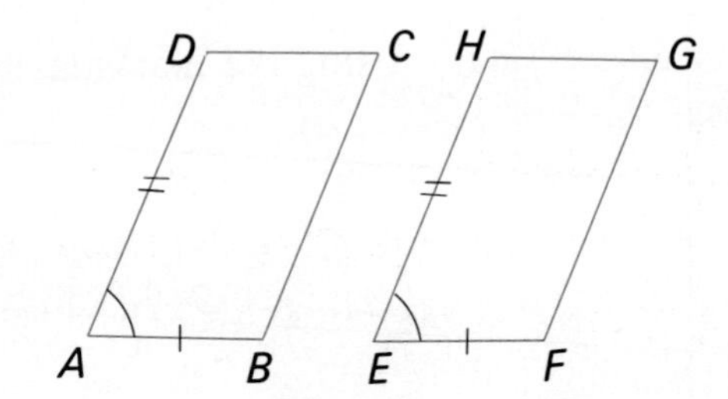

Proof

Statement	Reason
1. $\overline{AB} \cong \overline{EF}$	(S) 1. Given
2. $\angle A \cong \angle E$	(A) 2. Given
3. $\overline{AD} \cong \overline{EH}$	(S) 3. Given
4. $\square ABCD$, $\square EFGH$	4. Given
5. $\angle B$ and $\angle A$ are supp; $\angle F$ and $\angle E$ are supp.	5. Adj $\angle$s of a $\square$ are supp.
6. $\angle B \cong \angle F$	(A) 6. Supp of $\cong$ $\angle$s are $\cong$.
7. $\overline{BC} \cong \overline{AD}$, $\overline{FG} \cong \overline{EH}$	7. Opp sides of a $\square$ are $\cong$.
8. $\overline{BC} \cong \overline{FG}$	(S) 8. Sub (Steps 3 and 7)
9. $\therefore \square ABCD \cong \square EFGH$	9. SASAS

An angle-side-angle-side-angle congruence pattern can also be proved.

Theorem 7.2 Two quadrilaterals are congruent if any three angles and the included sides of one are congruent, respectively, to three angles and the included sides of the other. (ASASA for congruent quadrilaterals)

Given: $\angle F \cong \angle J$, $\angle G \cong \angle K$,
$\angle H \cong \angle L$, $\overline{FG} \cong \overline{JK}$,
$\overline{GH} \cong \overline{KL}$

Prove: Quad $EFGH \cong$ quad $IJKL$

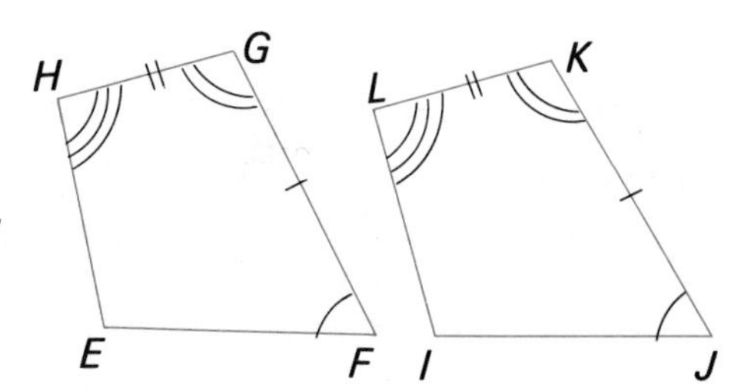

Classroom Exercises

Based on the markings shown, write a correct congruence statement.

1.

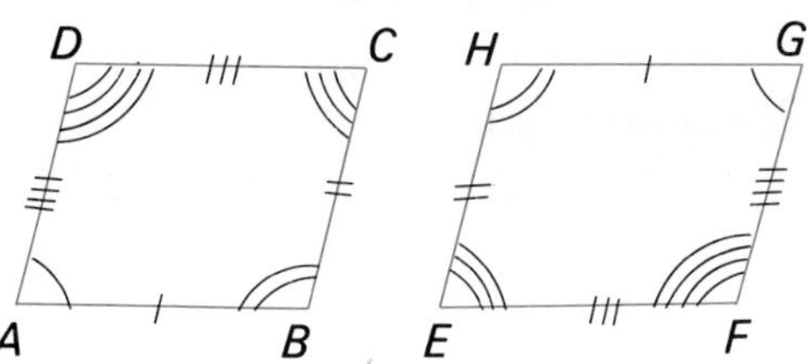

2.

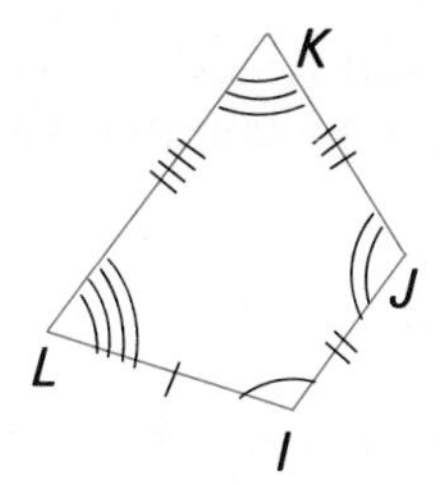

Complete the congruence statement.

Given: Quad $QRST \cong$ quad $XWVU$

3. $\angle R \cong$ ______
4. $\overline{TS} \cong$ ______
5. $\angle W \cong$ ______

Given: Quad $YZAB \cong$ quad $CDEF$

6. $\overline{FE} \cong$ ______
7. $\angle E \cong$ ______
8. $\overline{DE} \cong$ ______

Written Exercises

Name the pairs of congruent sides and the pairs of congruent angles for the congruence statement.

1. Quad $ABCD \cong$ quad $WXYZ$
2. Quad $LMVQ \cong$ quad $XCRN$
3. Given: $\overline{EB}$ is the $\perp$ bisector of $\overline{AC}$; $\overline{AF} \cong \overline{CD}$, $\angle A \cong \angle C$.
   Prove: Quad $ABEF \cong$ quad $CBED$
4. Given: $\overline{EB}$ is the $\perp$ bisector of $\overline{AC}$; $\overline{BE}$ bisects $\angle FED$;
   $\overline{FE} \cong \overline{DE}$.
   Prove: Quad $ABEF \cong$ quad $CBED$

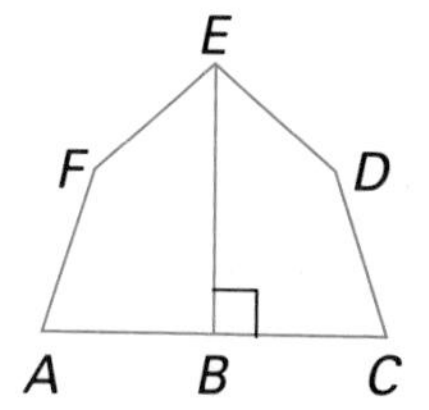

5. Given: $\overline{AF} \cong \overline{BD}$, $\overline{AG} \cong \overline{BC}$, $\angle A \cong \angle B$
   Prove: Quad $ABDG \cong$ quad $BAFC$
6. Given: $\triangle GEF \cong \triangle CED$, $\overline{AF} \cong \overline{BD}$
   Prove: Quad $ABDG \cong$ quad $BAFC$

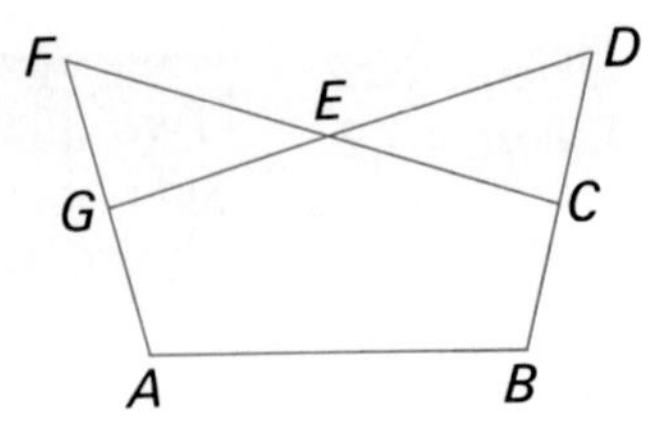

7. Given: $\overline{TX} \cong \overline{WU}$, $\triangle TZY \cong \triangle WZV$
   Prove: Quad $TUVZ \cong$ quad $WXYZ$
8. Given: $\triangle TUW \cong \triangle WXT$, $\overline{TY} \cong \overline{XY}$, $\overline{UV} \cong \overline{WV}$, $\overline{TZ} \cong \overline{WZ}$
   Prove: Quad $TUVZ \cong$ quad $WXYZ$
9. Given: $\square TUWX$, $\overline{YV}$ and $\overline{TW}$ bisect each other at Z.
   Prove: Quad $TUVZ \cong$ quad $WXYZ$

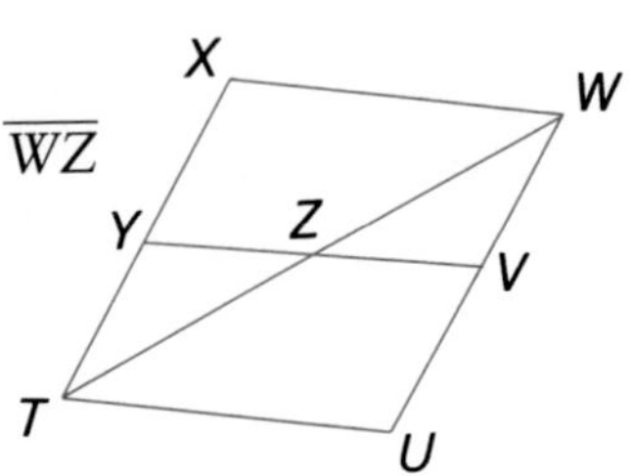

10. Prove Theorem 7.1.
11. Prove Theorem 7.2.
12. There are several patterns to be considered for proving convex quadrilaterals congruent. Consider these patterns: SSAAS, SSASS, AASAA, AASSA, and ASSAS. Choose one. Prove that it establishes congruence or show by counterexample that it does not.

Mixed Review

Determine whether the statement is true or false.

1. Opposite sides of a parallelogram are congruent. **6.4**
2. If both pairs of opposite sides of a quadrilateral are congruent, then the quadrilateral is a parallelogram. **6.5**
3. Diagonals of a parallelogram are congruent. **6.4**
4. If the diagonals of a quadrilateral are congruent, then the quadrilateral is a parallelogram. **6.5**

Algebra Review

To divide a polynomial by a monomial, divide each term of the polynomial by the monomial.

Example Divide $18p^5 + 9p^3 - 12p^2 \div 3p^2$.

$$\frac{18p^5 + 9p^3 - 12p^2}{3p^2} = \frac{18p^5}{3p^2} + \frac{9p^3}{3p^2} - \frac{12p^2}{3p^2} = 6p^3 + 3p - 4$$

Divide.

1. $(t^9 + t^7 - t^5) \div t^2$
2. $(m^6 - m^4 + m^2) \div m^2$
3. $(25x^6 - 15x^4 + 10x^2) \div 5x^2$
4. $(14y^7 + 28y^6 - 35y^5) \div 7y^3$

7.2 Necessary Conditions: Rectangles, Rhombuses, and Squares

Objectives

To identify rectangles, rhombuses, and squares
To prove and apply properties of rectangles, rhombuses, and squares
To identify necessary conditions for special quadrilaterals

A parallelogram is a quadrilateral with both pairs of opposite sides parallel. The pairs of opposite sides and opposite angles of a parallelogram are congruent. Some special parallelograms and their properties will be considered in this lesson. (The overlapping of classes of parallelograms is shown by the diagram on the right.)

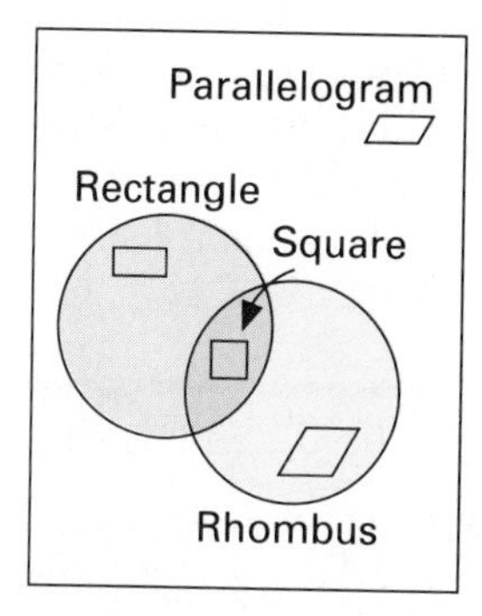

Definitions

A **rectangle** is a parallelogram with four right angles.
A **rhombus** is a parallelogram with four congruent sides.
A **square** is a rectangle with four congruent sides.

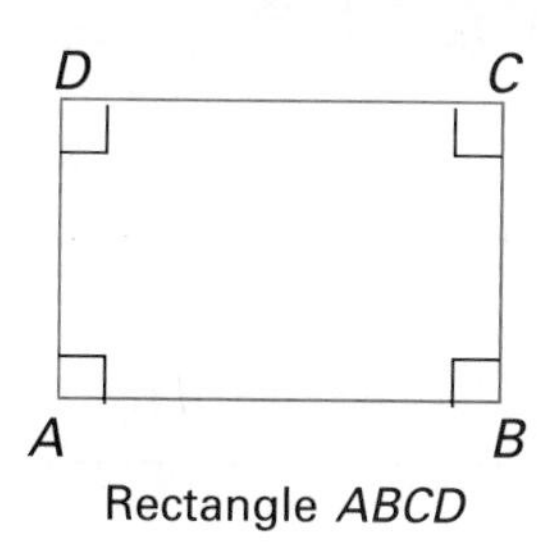

Rectangle *ABCD*

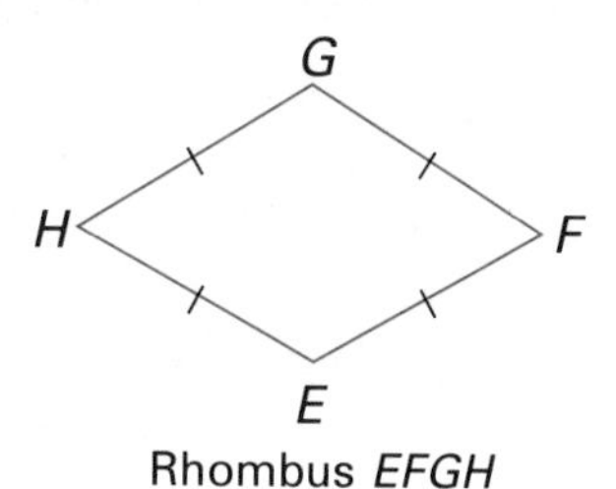

Rhombus *EFGH*

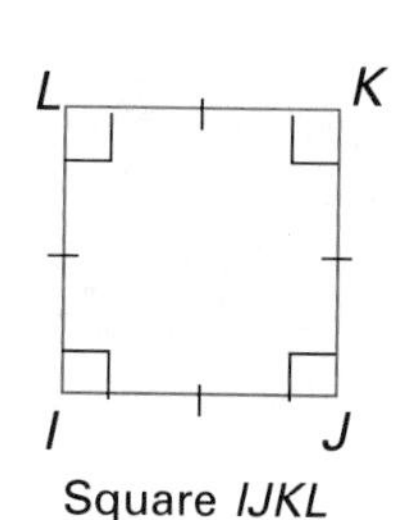

Square *IJKL*

From these definitions, you can see that figures may be classified in more than one way:

All rectangles, rhombuses, and squares are parallelograms.
All squares are rectangles.
All squares are rhombuses.

A **necessary condition** is a property or characteristic that must be satisfied in order to achieve a desired result. For example, for a given quadrilateral to be a parallelogram, it must satisfy the necessary condition that the opposite sides be parallel. If the opposite sides are not parallel, then the quadrilateral is not a parallelogram.

EXAMPLE 1 Which of the following are necessary conditions for a parallelogram to be a square? Why?

a. The parallelogram has four congruent sides.
b. The parallelogram has at least two right angles.

Solution Both conditions are necessary.
a. Since a square is a rhombus, it must have four congruent sides.
b. Since a square is also a rectangle, it must have four right angles.

From Lesson 6.4, you know certain properties of a parallelogram. These properties are necessary conditions for a quadrilateral to be a parallelogram. Necessary conditions for rhombuses and rectangles will now be established.

Theorem 7.3 The diagonals of a rhombus are perpendicular.

Given: Rhom $ABCD$ with diagonals $\overline{AC}$ and $\overline{BD}$
Prove: $\overline{AC} \perp \overline{BD}$

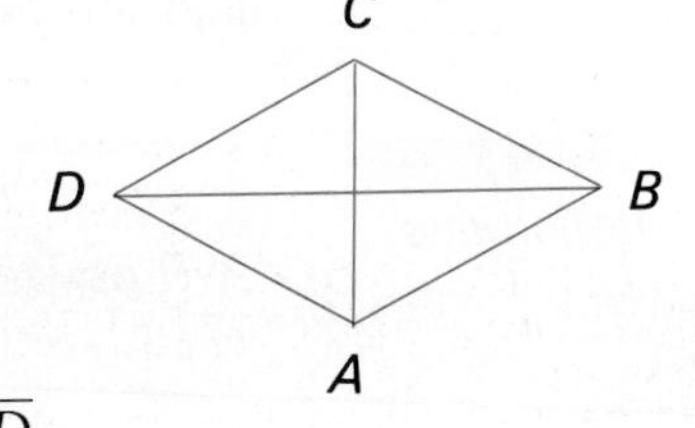

Plan All sides of a rhombus are congruent, so $\overline{AD} \cong \overline{AB}$ and $\overline{CD} \cong \overline{CB}$. Points A and C are each equidistant from the endpoints of $\overline{BD}$.

Proof

Statement	Reason
1. Rhomb $ABCD$ with diags $\overline{AC}$ and $\overline{BD}$	1. Given
2. $\overline{AD} \cong \overline{AB}$, $\overline{CD} \cong \overline{CB}$	2. Def of rhom
3. $\therefore \overline{AC} \perp \overline{BD}$	3. A line containing 2 pts, each equidist from endpts of a seg, is $\perp$ bis of the seg (Thm 5.8).

Theorem 7.4 The diagonals of a rectangle are congruent.

Given: Rect $EFGH$ with diagonals $\overline{HF}$ and $\overline{GE}$
Prove: $\overline{HF} \cong \overline{GE}$

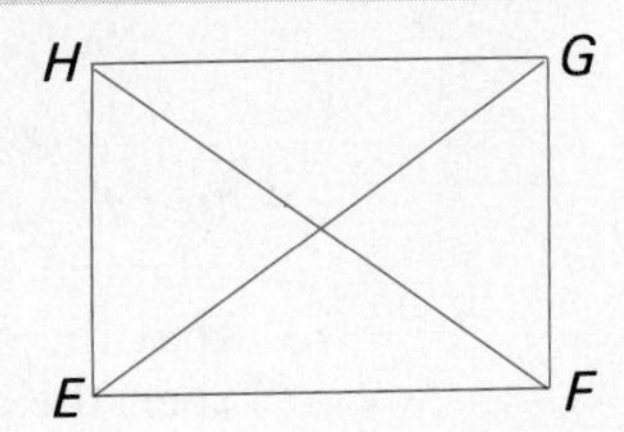

Plan Use the definition of a rectangle and properties of a parallelogram to prove $\triangle HEF \cong \triangle GFE$.

Theorem 7.5 Each diagonal of a rhombus bisects two angles of the rhombus.

Given: Rhom $IJKL$, with diagonals $\overline{IK}$ and $\overline{JL}$
Prove: $\overline{IK}$ bisects $\angle JIL$ and $\angle JKL$; $\overline{JL}$ bisects $\angle ILK$ and $\angle IJK$.

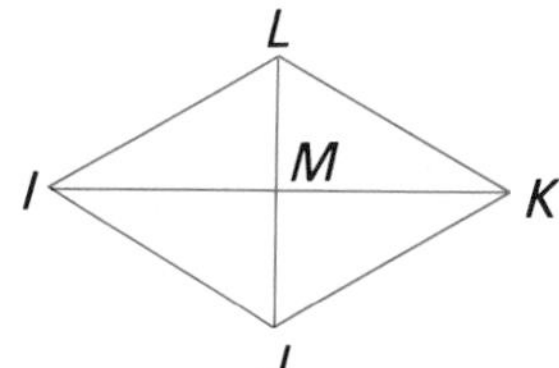
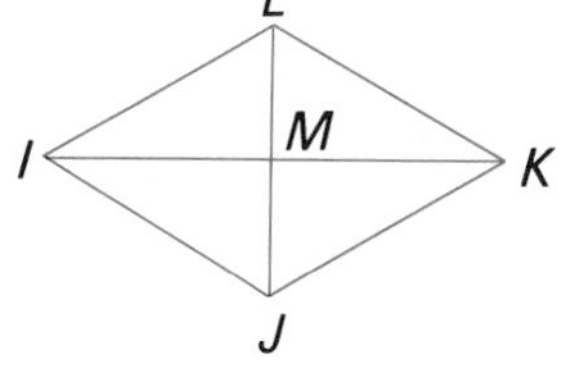

Plan Use the fact that diagonals of a rhombus are perpendicular and that diagonals of a parallelogram bisect each other.

EXAMPLE 2 Given: Rect $IJKL$, $JL = 5x - 1$, $IK = 3x + 5$
Find JL.

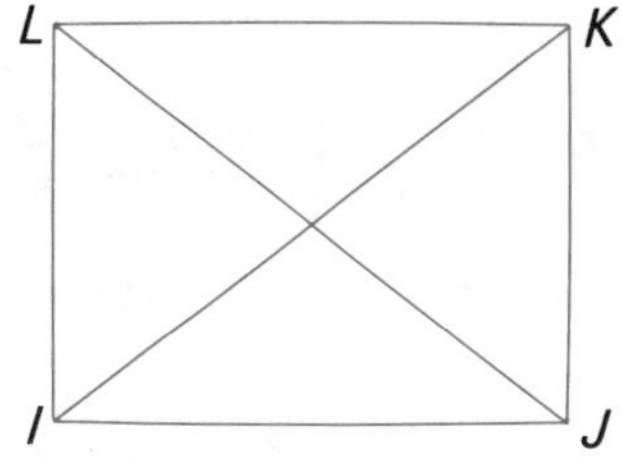

Solution

$$JL = IK$$
$$5x - 1 = 3x + 5$$
$$2x = 6$$
$$x = 3$$
$$JL = 5 \cdot 3 - 1 = 14$$

Summary

The table below shows necessary conditions for the given figures.

Condition	Rectangle	Rhombus	Square
All sides are ≅.		X	X
4 right ∠s	X		X
≅ diagonals	X		X
⊥ diagonals		X	X
Diagonals bisect ∠s.		X	X

Focus on Reading

Suppose you were proving each statement. State the hypothesis and the conclusion.

1. If a parallelogram has four congruent sides, then it is a rhombus.
2. If a parallelogram is a rectangle, then the diagonals are congruent.
3. The angles of a rectangle are right angles.
4. The adjacent sides of a rhombus are congruent.

Classroom Exercises

Based only on the markings shown, identify the parallelogram as a rectangle, a rhombus, a square, or none of these. A figure may be correctly identified in more than one way.

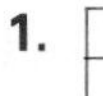

1. 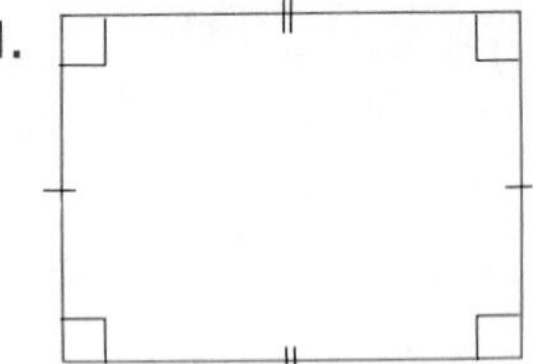

2. 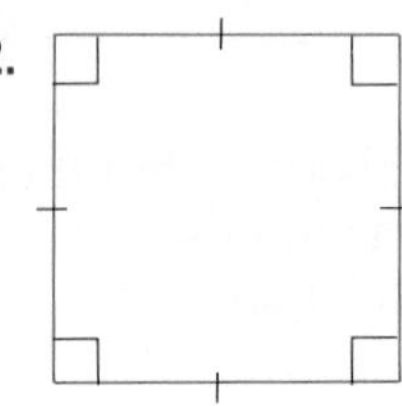

3. 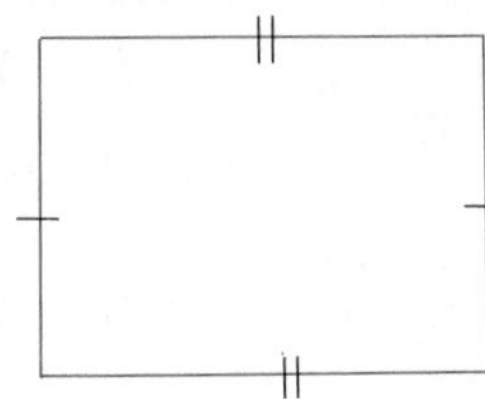

Written Exercises

True or false? If false, draw a picture to defend your answer.

1. All squares are rhombuses.
2. All rhombuses are rectangles.
3. Some rhombuses are rectangles.
4. Some parallelograms are squares.
5. Some parallelograms are rectangles.
6. No rectangles are squares.
7. All squares are rectangles.
8. All rhombuses are parallelograms.
9. All rectangles are squares.
10. Some rhombuses are squares.

Is the statement a necessary condition for a quadrilateral to be a rectangle?

11. Opposite angles are congruent.
12. Adjacent angles are congruent.
13. Diagonals are congruent.
14. Diagonals are perpendicular.

Find the indicated measure.

15. Given: Square $ABCD$,
    $AB = 23$
    Find BC.

16. Given: Rect $ABCD$,
    $BD = 13$
    Find AC.

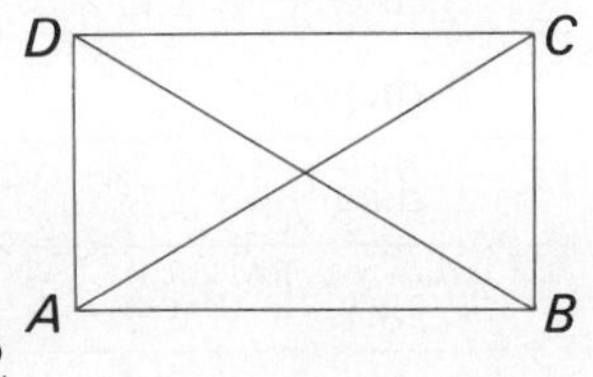

17. Given: Rhom $ABCD$
    Find m $\angle AEB$.

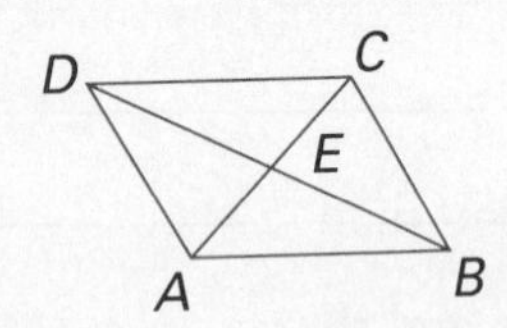

18. Given: Rect $ABCD$,
    $AC = 7x + 5$,
    $BD = 14x - 2$
    Find AC.

If the statement is true, explain why. If false, draw a counterexample.

19. The diagonals of a rhombus are congruent.
20. The diagonals of a square are congruent.
21. The diagonals of a square are perpendicular.
22. A diagonal of a rectangle bisects opposite angles.

Find the indicated measure.

23. Given: $\square ABCD$, $AC = 9x + 1$, $AE = 5x - 1$
Find CE.

24. Given: $\square ABCD$, $BD = 4x + 4$, $ED = 3x - 2$
Find BE.

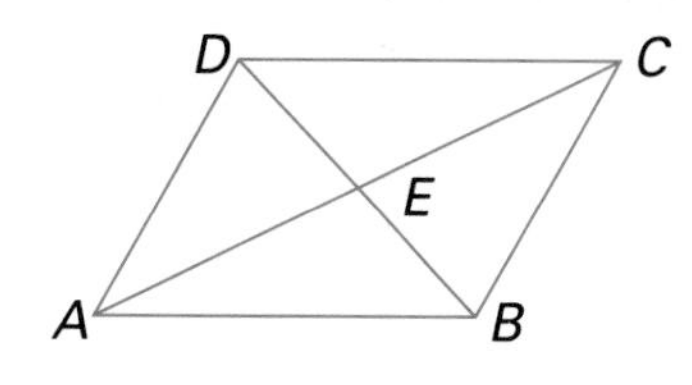

25. Given: Square $ABCD$
Find m $\angle EAB$.

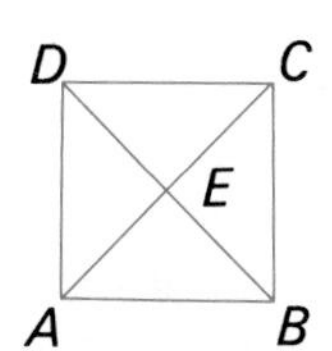

26. Given: Rhom $XYZW$, m $\angle XYW = 3x + 10$, m $\angle ZYW = 4x - 5$
Find m $\angle XYW$.

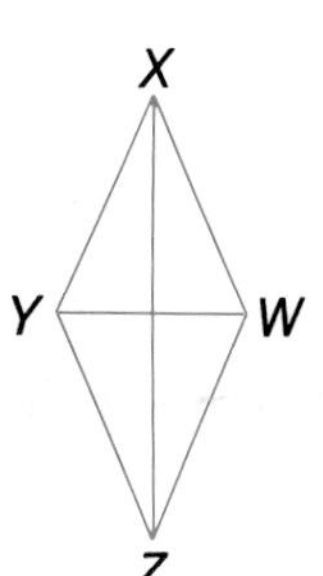

27. Prove Theorem 7.4.

28. Prove Theorem 7.5.

29. Prove: The midpoint of the hypotenuse of a right triangle is equidistant from the three vertices of the triangle. (HINT: Construct a rectangle.)

30. Given: Regular pentagon $EFGHI$, diagonals $\overline{FH}$ and $\overline{GI}$ intersect at J.
Prove: $EFJI$ is a parallelogram.

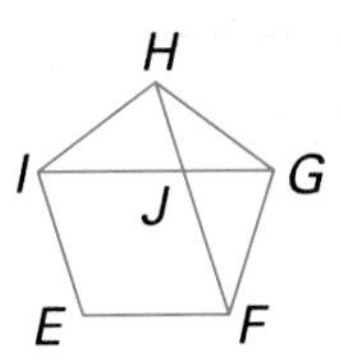

31. Given: Square $KLMN$, $\overline{NO} \cong \overline{KP} \cong \overline{LQ} \cong \overline{MR}$
Prove: $OPQR$ is a square.

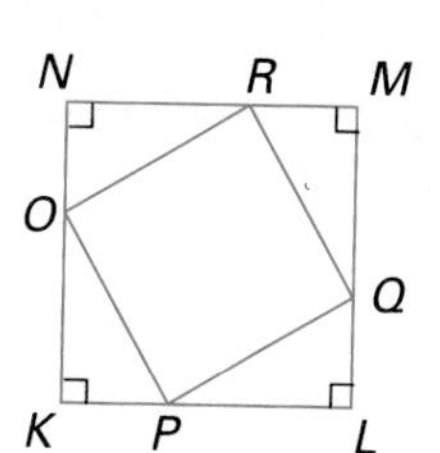

Mixed Review

Is the statement true or false?

1. Adjacent sides of a parallelogram are congruent. *6.4*
2. If exactly two adjacent sides of a quadrilateral are congruent, then the quadrilateral is a parallelogram. *6.5*
3. If all consecutive angles of a quadrilateral are congruent, then the quadrilateral is a parallelogram. *6.5*
4. Consecutive angles of a parallelogram are congruent. *6.4*
5. Diagonals of a parallelogram bisect each other. *6.4*
6. If the diagonals of a quadrilateral bisect each other, then the quadrilateral is a parallelogram. *6.5*

7.3 Sufficient Conditions: Rectangles, Rhombuses, and Squares

Objectives

To identify sufficient conditions for special quadrilaterals
To prove that a figure is a rectangle, rhombus, or square

Think about the following statements.

(1) If a quadrilateral has right angles, then it is a rectangle.
(2) If a quadrilateral has four right angles, then it is a rectangle.
(3) If a quadrilateral is a rhombus, then it is a rectangle.

Statement 1 is false. Having a right angle is a *necessary* condition for a rectangle, but it does not assure that a quadrilateral is a rectangle. Statement 2 is true. Every quadrilateral with four right angles is a rectangle. This is a **sufficient** condition for a rectangle because it guarantees that the figure is a rectangle. Statement 3 is false. Being a rhombus is *neither* necessary nor sufficient to be a rectangle.

Theorem 7.6 A parallelogram with one right angle is a rectangle.

Theorem 7.7 A parallelogram with two adjacent congruent sides is a rhombus.

Theorem 7.8 A parallelogram with perpendicular diagonals is a rhombus.

Given: $\square ABCD$, $\overline{AC} \perp \overline{BD}$
Prove: $ABCD$ is a rhombus.

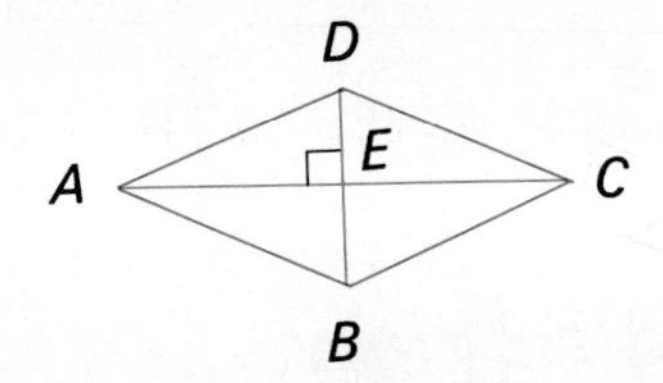

Plan Show that $\overline{AD}$ and $\overline{AB}$ are $\cong$ corresponding sides of $\cong$ $\triangle$s AED and AEB.

Proof

Statement	Reason
1. $ABCD$ is a parallelogram.	1. Given
2. $\overline{DE} \cong \overline{BE}$	(S) 2. Diags of a $\square$ bis each other.
3. $\overline{AC} \perp \overline{BD}$	3. Given
4. $\angle DEA \cong \angle BEA$	(A) 4. $\perp$ lines form $\cong$ rt $\angle$s.
5. $\overline{AE} \cong \overline{AE}$	(S) 5. Reflex Prop
6. $\triangle AED \cong \triangle AEB$	6. SAS
7. $\overline{AD} \cong \overline{AB}$	7. CPCTC
8. $\therefore ABCD$ is a rhombus.	8. A $\square$ with two adj $\cong$ sides is a rhom.

EXAMPLE 1 Consider a quadrilateral with perpendicular diagonals. Is this property a sufficient condition for the quadrilateral to be a rhombus? If it is not sufficient, provide a counterexample.

Solution The quadrilateral at the right has perpendicular diagonals, but it does not have four congruent sides. Having perpendicular diagonals is *not* a sufficient condition for a quadrilateral to be a rhombus.

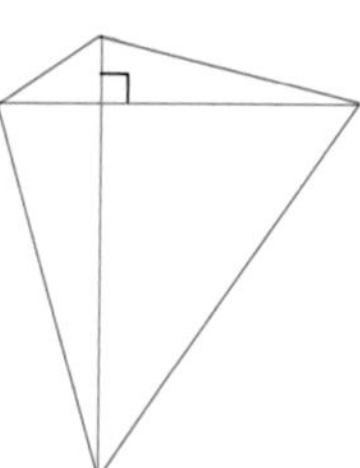

Two other sufficient conditions for a parallelogram to be a rectangle or a rhombus are given as theorems below.

Theorem 7.9 A parallelogram with congruent diagonals is a rectangle.

Theorem 7.10 A parallelogram with a diagonal that bisects opposite angles is a rhombus.

There are sufficient conditions for proving that a quadrilateral not specifically identified as a parallelogram is a rectangle, a rhombus, or a square.

Theorem 7.11 A quadrilateral with four congruent sides is a rhombus.

EXAMPLE 2 True or false? A quadrilateral with congruent diagonals is a rectangle. Draw a figure to support your answer.

Solution False. It is possible for a quadrilateral to have congruent diagonals and not be a rectangle. Congruent diagonals are necessary, but *not* sufficient for proving that a quadrilateral is a rectangle.

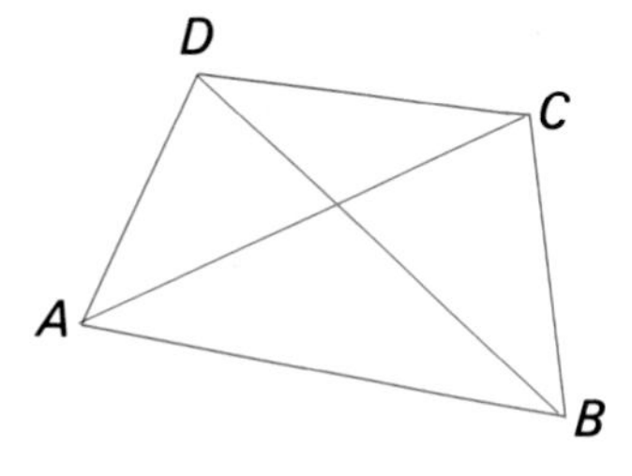

EXAMPLE 3 Is the following a sufficient condition for a quadrilateral to be a square?

One pair of opposite angles are right angles.

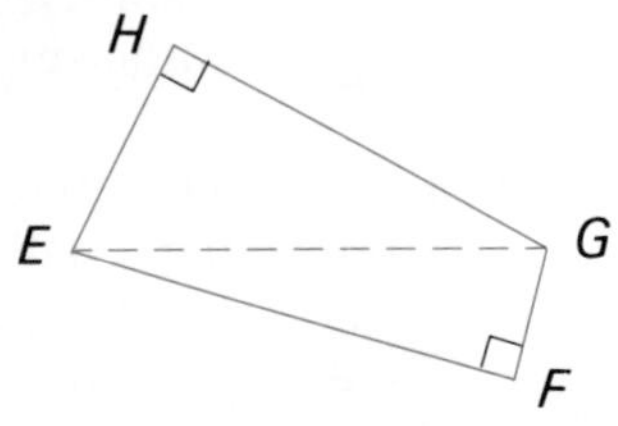

Solution This is *not* a sufficient condition for a quadrilateral to be a square. A counterexample can be devised by drawing two noncongruent right triangles with a common hypotenuse.

Summary

The following chart shows sufficient conditions for a *parallelogram* to be a rectangle, rhombus, or square.

Condition	Rectangle	Rhombus	Square
2 adj ≅ sides		X	
1 right ∠	X		
≅ diagonals	X		
⊥ diagonals		X	
1 right ∠ and 2 adj ≅ sides			X
diags bis opp ∠s.		X	

Classroom Exercises

Based only on the markings, identify the parallelogram as a rectangle, a rhombus, a square, or none of these. A figure may be correctly identified in more than one way.

1. 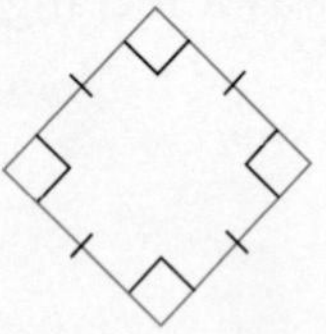

2. 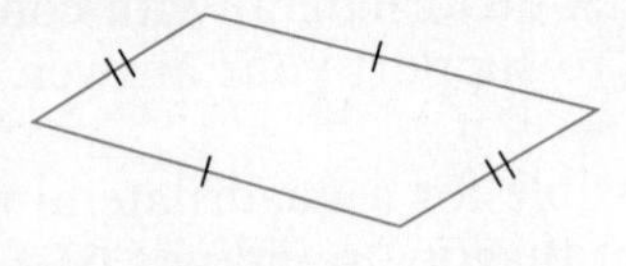

3. 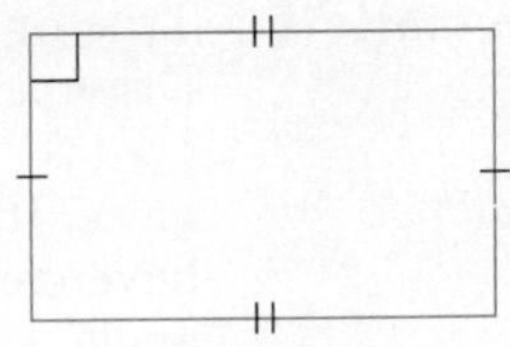

4. 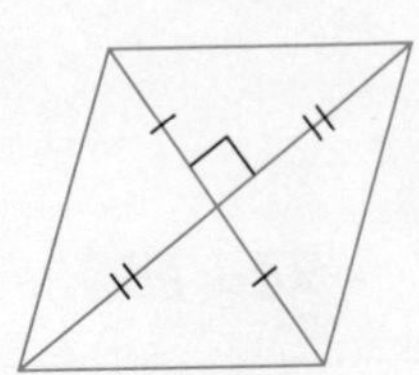

5. 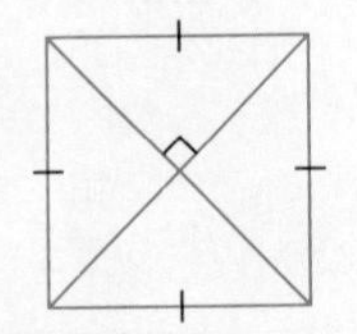

6. 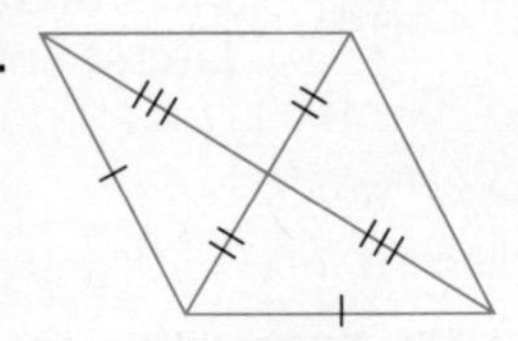

Written Exercises

True or false? Draw a picture to defend your answer.

1. A quadrilateral with two consecutive right angles is a rectangle.
2. A parallelogram with two consecutive right angles is a rectangle.
3. A rhombus with two consecutive right angles is a square.
4. A quadrilateral with perpendicular diagonals is a rhombus.
5. A parallelogram with perpendicular diagonals is a rhombus.
6. A quadrilateral with four congruent sides is a rhombus.

Determine whether the stated condition is sufficient for the figure to be a square. If not sufficient, draw a diagram of a counterexample.

7. a quadrilateral with at least one right angle
8. a quadrilateral with at least two right angles
9. a quadrilateral with at least three right angles
10. a quadrilateral with four right angles
11. a parallelogram with at least one right angle
12. a rhombus with at least one right angle
13. Given: $\overline{AB} \cong \overline{DC}$, $\overline{AB} \parallel \overline{DC}$, $\overline{AC} \cong \overline{BD}$
    Prove: $ABCD$ is a rectangle.
14. Given: $\overline{AE} \cong \overline{BE} \cong \overline{CE} \cong \overline{DE}$
    Prove: $ABCD$ is a rectangle.

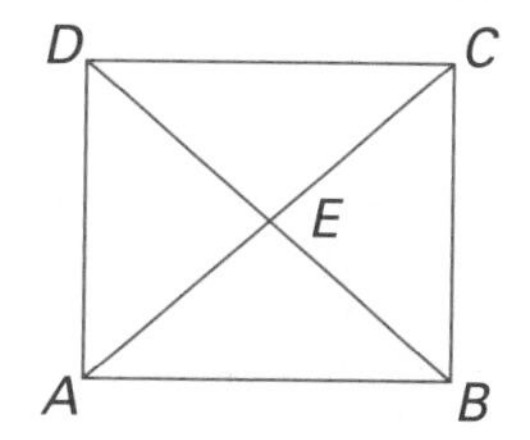

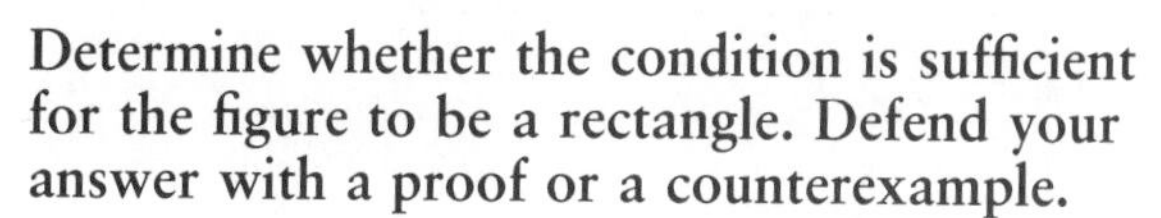

Determine whether the condition is sufficient for the figure to be a rectangle. Defend your answer with a proof or a counterexample.

15. a quadrilateral with congruent diagonals
16. a quadrilateral with at least one right angle
17. a quadrilateral with congruent diagonals that bisect each other
18. a parallelogram with opposite angles supplementary

Determine whether the condition is sufficient for the figure to be a rhombus. Provide a proof or a counterexample.

19. a quadrilateral with diagonals that are perpendicular bisectors of each other
20. a quadrilateral with two pairs of adjacent congruent sides

Determine whether the condition is sufficient for the figure to be a square. Provide a proof or a counterexample.

21. a parallelogram with one right angle
22. a rectangle with two consecutive congruent angles
23. a rectangle with perpendicular diagonals
24. a rhombus with congruent diagonals

Essay.

25. Explain in your own words the difference between a necessary condition for a quadrilateral to be a rectangle and a sufficient condition for a quadrilateral to be a rectangle.

26. Prove Theorem 7.6.

27. Prove Theorem 7.7.

28. Prove Theorem 7.9.

29. Prove Theorem 7.10.

30. Prove Theorem 7.11.

Is the condition sufficient for the two given figures to be congruent? Give a proof or a counterexample.

31. two squares with a diagonal of the first congruent to a diagonal of the second

32. two rhombuses with a diagonal of the first congruent to a diagonal of the second

33. two rectangles with two adjacent sides of the first congruent, respectively, to two adjacent sides of the second

34. two rectangles with a side and a diagonal of the first congruent to a side and a diagonal of the second

Midchapter Review

1. Given: Rectangles $ABCD$ and $WXYZ$, $AB = WX$
   Prove: Quad $ABCD \cong$ quad $WXYZ$

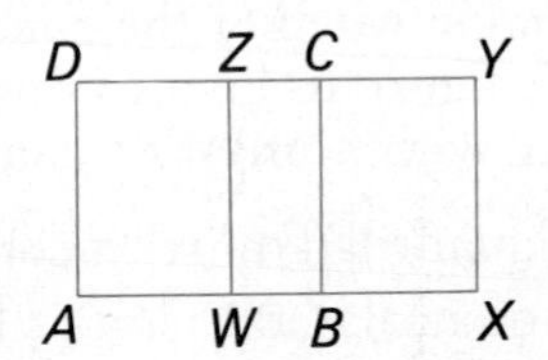

Given: Rhom $ABCD$, $DB = 10$, m $\angle ABC = 120$

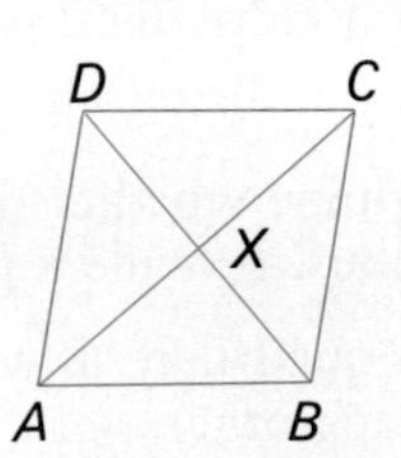

2. Find XB.

3. Find m $\angle AXB$.

4. Find m $\angle ABX$.

True or false? If false, give a counterexample.

5. A rhombus with one right angle is a square.
6. A parallelogram with two congruent sides is a rhombus.
7. Every square is a rhombus.

Determine whether the given condition is only necessary or sufficient for the figure to be a rectangle.

8. a quadrilateral with one right angle
9. a parallelogram with one right angle

Application: *Conditions for Quadrilaterals*

We use necessary and sufficient conditions in everyday life to minimize our efforts. Determine the conditions for each of the following situations.

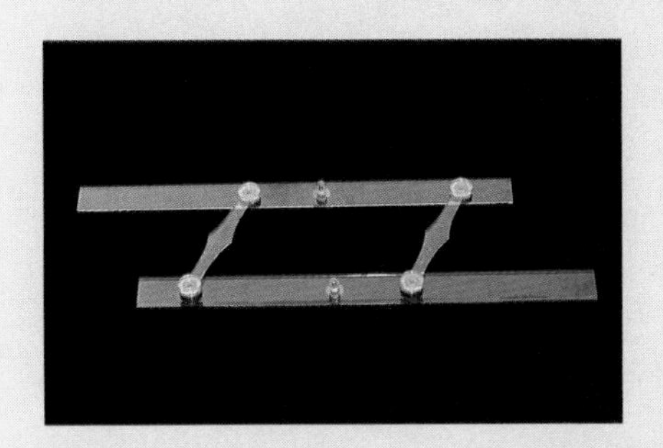

1. Parallel rulers are used to draw parallel lines. Consider the photo at the left. How are the parallel rulers made? What conditions will insure that the rulers are parallel in all positions?

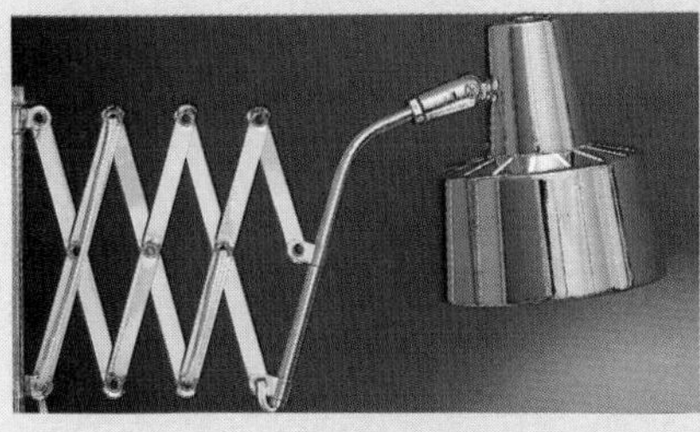

2. What type of quadrilateral is used in making the lamp shown above? What insures that the lamp moves to and from the wall in a straight line?

3. The lamp shown above can be pivoted into many different positions. The light itself can be raised or lowered while pointing in the same direction. Why is this so?

4. A shortwave radio antenna is stretched between a pole on one side of a house and a tree on the other. Its length cannot be measured directly. What kind of geometric figure could be used on the ground in front of the house to determine the distance? What would be necessary and sufficient conditions for that figure to achieve the desired result?

Computer Investigation

Necessary Conditions

Use a computer software program that draws random trapezoids, labels points, constructs segments between two given points, and measures line segments.

The quadrilateral *ABCD* at the right has only one pair of parallel sides: $\overline{AB} \parallel \overline{CD}$. Such a quadrilateral is defined formally in the next lesson as a **trapezoid.** The computer activity below will help you discover some of the properties of trapezoids.

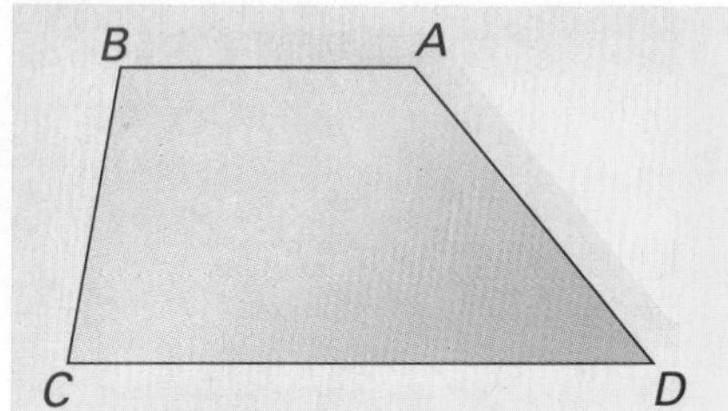

Activity 1

Draw a random trapezoid. The nonparallel sides $\overline{BC}$ and $\overline{AD}$ do not have to be congruent.

1. Find *CD* and *BA*.
2. Do you think that the parallel sides of a trapezoid could ever be congruent? Draw three more random trapezoids, measuring the lengths of the parallel sides of each. Are the parallel sides of a trapezoid ever congruent?
3. Recall that if one pair of sides of a quadrilateral are both *parallel* and *congruent*, the quadrilateral is a **parallelogram** and both pairs of sides are parallel. How does this explain the conclusion of Exercise 2?

Activity 2

Draw a random trapezoid *ABCD*. Draw the diagonals $\overline{AC}$ and $\overline{BD}$.

4. Are the diagonals congruent?
5. Do the diagonals bisect each other?
6. Repeat Exercises 4–5 for two more random trapezoids.
7. Can the diagonals of a trapezoid ever bisect each other? HINT: Recall that if the diagonals of a quadrilateral bisect each other, then the quadrilateral must be a ______.
8. Can you draw a trapezoid with congruent diagonals?

Summary

Can a trapezoid have two pairs of congruent sides?
Can the diagonals of a trapezoid bisect each other?
Can the diagonals of a trapezoid ever be congruent? When?

7.4 Trapezoids

Objectives To identify bases, legs, medians, and altitudes of trapezoids
To apply theorems about medians and altitudes of trapezoids

A parallelogram has both pairs of opposite sides parallel. Another special quadrilateral, the *trapezoid,* has only one set of parallel sides.

The photo shows a table in the shape of a trapezoid. What are some advantages of this shape? How might a number of such tables be arranged for a conference?

Definition

A **trapezoid** is a quadrilateral with exactly one pair of parallel sides.

The sides of a trapezoid are given specific names. The *bases* are the two parallel sides. The *legs* are the two nonparallel sides.

EXAMPLE 1 Which sides of each trapezoid are the bases? Which sides are the legs?

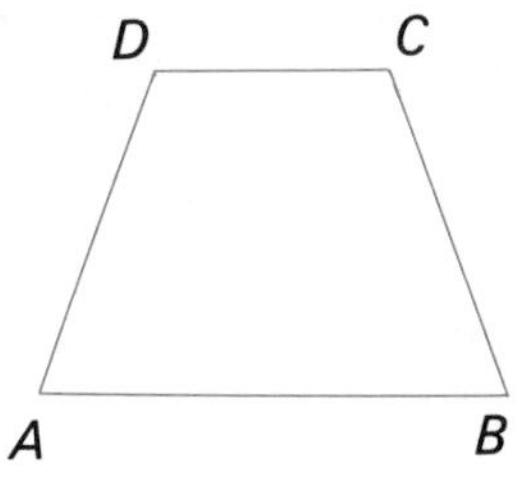

$\overline{AB} \parallel \overline{DC}$

H G E F

$\overline{FG} \parallel \overline{EH}$

Solution $\overline{AB}$ and $\overline{DC}$ are the bases. $\overline{AD}$ and $\overline{BC}$ are the legs.

$\overline{FG}$ and $\overline{EH}$ are the bases. $\overline{EF}$ and $\overline{HG}$ are the legs.

EXAMPLE 2 Determine whether the diagonals of a trapezoid bisect each other. Use a drawing to illustrate your answer.

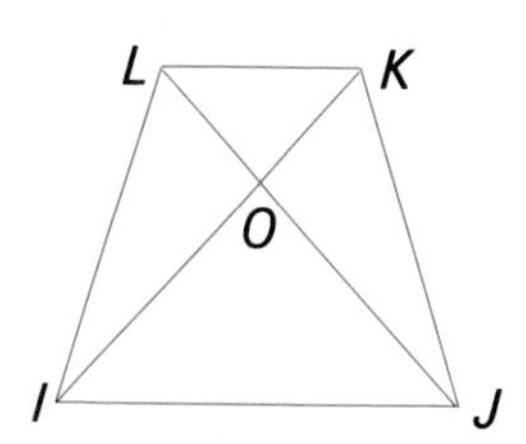

Solution Since $IO \neq KO$ and $JO \neq LO$, the diagonals $\overline{IK}$ and $\overline{JL}$ do not bisect each other.

Definition

An **altitude** of a trapezoid is a perpendicular segment from any point on one base to the line containing the other base.

A triangle has exactly three altitudes. A trapezoid, however, has infinitely many altitudes.

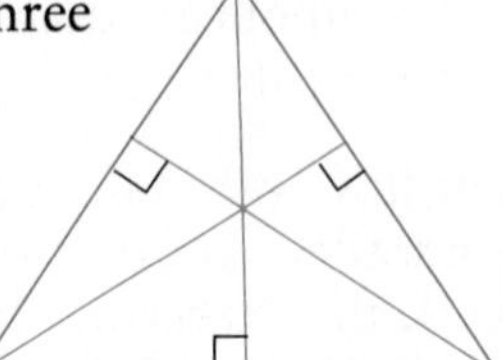

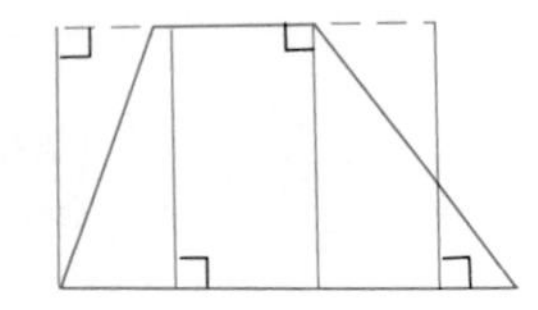

Theorem 7.12

All altitudes of a trapezoid are congruent.

Given: Trap $ABCD$ with $\overline{DC} \parallel \overline{AB}$ and altitudes $\overline{EH}$ and $\overline{FG}$

Prove: $\overline{EH} \cong \overline{FG}$

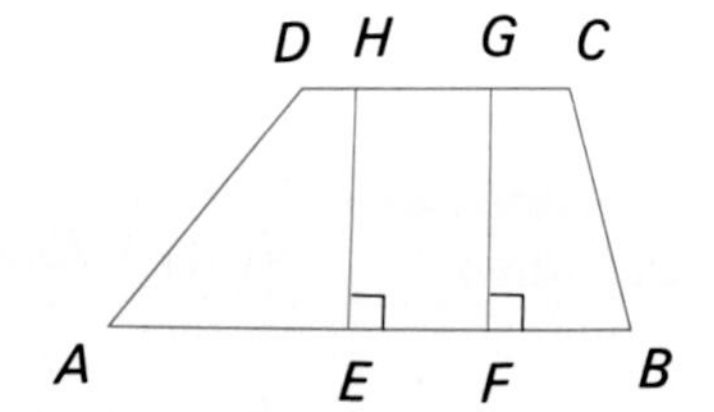

Proof

Statement	Reason
1. Trap $ABCD$, alt $\overline{EH}$ and $\overline{FG}$	1. Given
2. $\overline{AB} \parallel \overline{DC}$	2. Given
3. $\overline{EH} \perp \overline{AB}$, $\overline{FG} \perp \overline{AB}$	3. Def of alt of trap
4. $\therefore \overline{EH} \cong \overline{FG}$ $(EH = FG)$	4. If 2 lines are $\parallel$, all pts of each are equidist from other line.

Another special segment in a trapezoid is the median.

Definition

The **median** of a trapezoid is the segment that joins the midpoints of the legs.

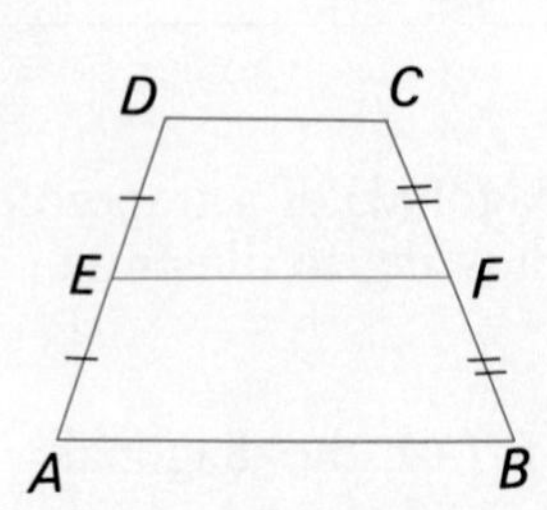

Median $\overline{EF}$ of trapezoid $ABCD$

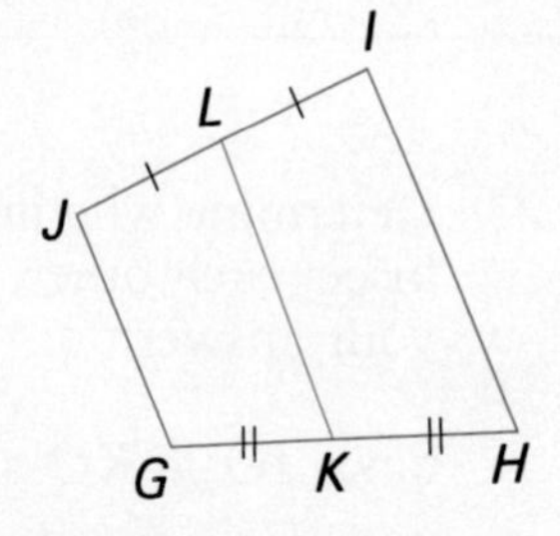

Median $\overline{LK}$ of trapezpoid $GHIJ$

Theorem 7.13 The median of a trapezoid is parallel to its bases. Its length is half the sum of the lengths of the two bases.

Given: Trap $ABCD$ with median $\overline{XY}$,
$\overline{AB} \parallel \overline{DC}$

Prove: $\overline{DC} \parallel \overline{XY} \parallel \overline{AB}$,
$XY = \frac{1}{2}(DC + AB)$

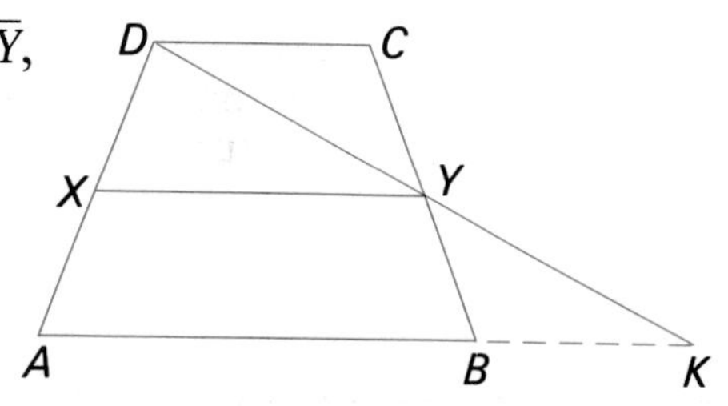

Proof

Statement	Reason
1. Trap $ABCD$ with median $\overline{XY}$, $\overline{AB} \parallel \overline{DC}$	1. Given
2. Draw $\overleftrightarrow{DY}$ intersecting $\overleftrightarrow{AB}$ at K.	2. Two pts determine a line.
3. $\angle DYC \cong \angle KYB$	(A) 3. Vert $\angle$s are $\cong$.
4. $CY = BY$	(S) 4. Def of median
5. $\angle DCB \cong \angle KBC$	(A) 5. Alt int $\angle$s of $\parallel$ lines are $\cong$.
6. $\triangle DCY \cong \triangle KBY$	6. ASA
7. $DC = KB$	7. CPCTC
8. $AB + KB = AK$	8. Seg Add Post
9. $AB + DC = AK$	9. Sub
10. $DY = KY$	10. CPCTC
11. $DX = AX$	11. Def of median
12. $\overline{XY} \parallel \overline{AB}$, $XY = \frac{1}{2}AK$	12. $\triangle$ Midseg Thm
13. $\therefore \overline{XY} \parallel \overline{DC} \parallel \overline{AB}$	13. 2 lines $\parallel$ to same line are $\parallel$.
14. $\therefore XY = \frac{1}{2}(AB + DC)$	14. Sub

EXAMPLE 3 Given: Trapezoid $ABCD$, median $\overline{EF}$, $AB = 20$, $EF = 15$
Find DC.

Solution

$$15 = \tfrac{1}{2}(20 + DC)$$
$$30 = 20 + DC$$
$$DC = 10$$

Both sides of the equation were multiplied by 2 to simplify computations.

EXAMPLE 4 Given: Trap $GHIJ$, median $\overline{KL}$,
$JI = 4x - 4$, $KL = 3x + 6$,
$GH = 5x - 5$
Find KL.

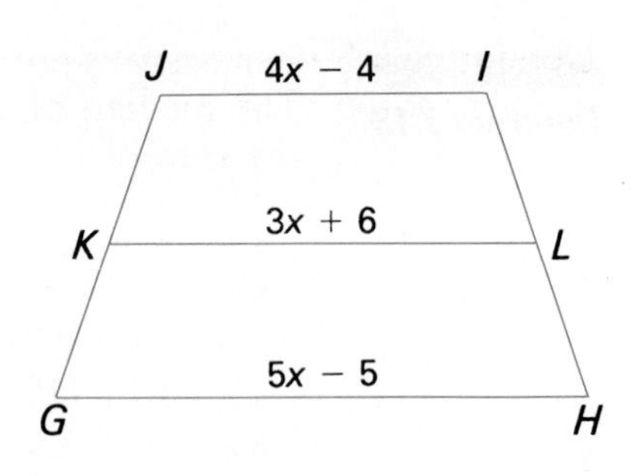

Solution

$$3x + 6 = \frac{1}{2}[(4x - 4) + (5x - 5)]$$
$$6x + 12 = (4x - 4) + (5x - 5)$$
$$21 = 3x$$
$$7 = x$$
$$KL = 3 \cdot 7 + 6 = 27$$

Classroom Exercises

Use the diagram to name the parts of the figure.

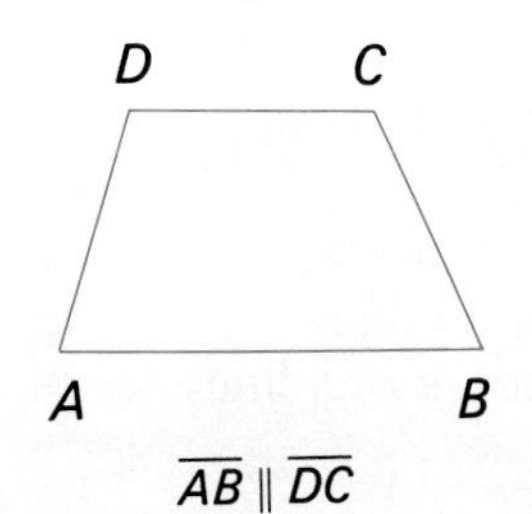

$\overline{AB} \parallel \overline{DC}$

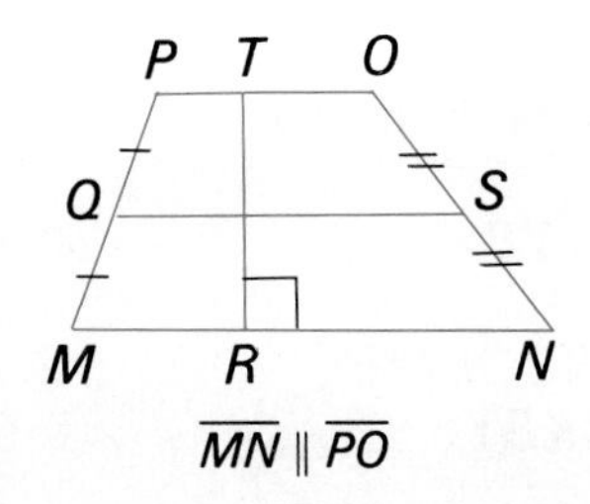

$\overline{MN} \parallel \overline{PO}$

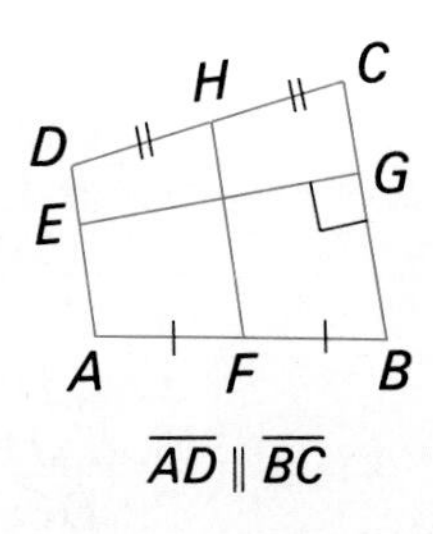

$\overline{AD} \parallel \overline{BC}$

1. The bases are ______.
2. The legs are ______.
3. ______ is an altitude.
4. ______ is a median.
5. ______ is an altitude.
6. ______ is a median.

Written Exercises

$WXYZ$ is a trapezoid with $\overline{WX} \parallel \overline{ZY}$. Find the length of the median.

1.

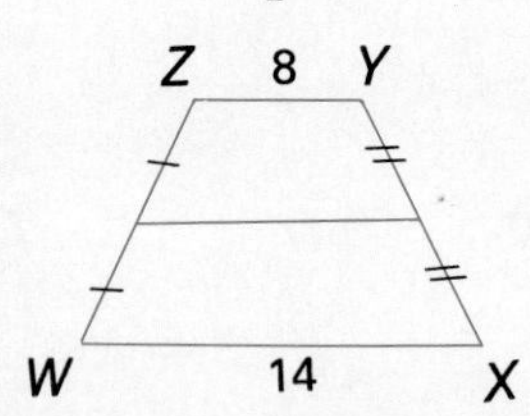

2.

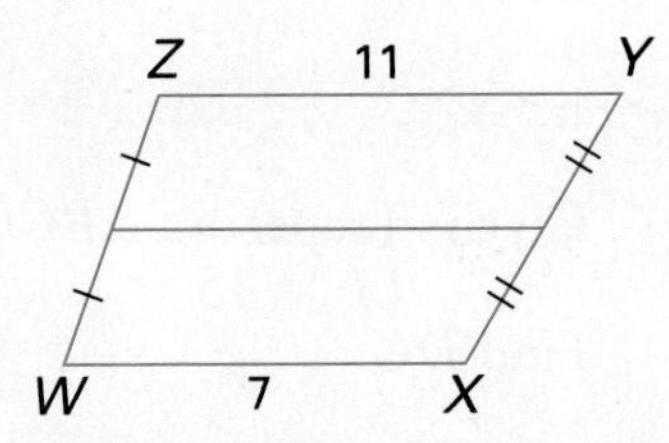

$ABCD$ is a trapezoid, with median $\overline{EF}$. Find the indicated length.

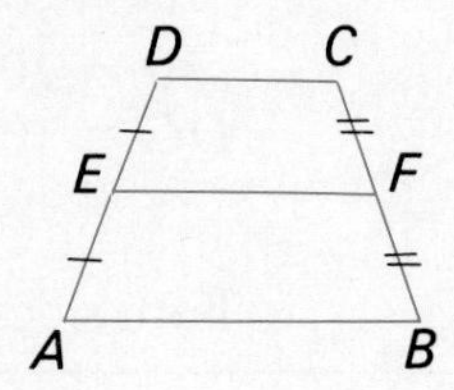

3. $AB = 15$, $DC = 5$, $EF =$ ______
4. $AB = 28$, $DC = 22$, $EF =$ ______
5. $AB = 24$, $DC =$ ______, $EF = 20$
6. $AB = 31$, $DC =$ ______, $EF = 26$
7. $AB =$ ______, $DC = 3$, $EF = 11$

Determine whether the trapezoid described is possible. If it is possible, use a drawing to illustrate your answer.

8. a trapezoid with two congruent sides
9. a trapezoid with three congruent sides
10. a trapezoid with four congruent sides
11. a trapezoid with two congruent angles
12. a trapezoid with three congruent angles
13. a trapezoid with four congruent angles
14. a trapezoid with congruent diagonals
15. a trapezoid with diagonals that are perpendicular

***GHIJ* is a trapezoid, with median $\overline{KL}$. Find the indicated length.**

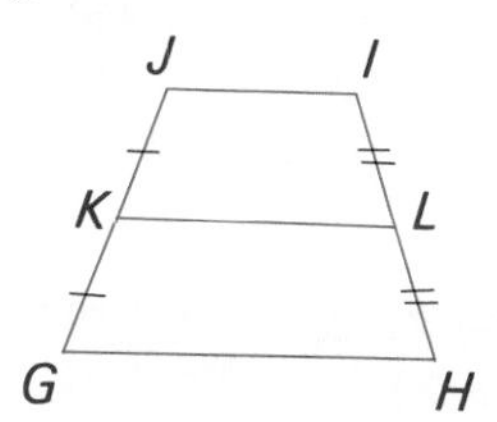

16. $GH = 3x - 5$, $JI = x + 3$, $KL = x + 7$
    Find KL.
17. $GH = 4x + 3$, $JI = 2x - 1$, $KL = 3x + 1$
    Find GH.
18. $GH = 4x - 3$, $JI = x + 5$, $KL = 3x$
    Find JI.

Construct a counterexample for each statement.

19. Consecutive angles of a trapezoid are congruent.
20. Consecutive angles of a trapezoid are supplementary.
21. The legs of a trapezoid are congruent.

Prove the conclusion, or disprove it by giving a counterexample.

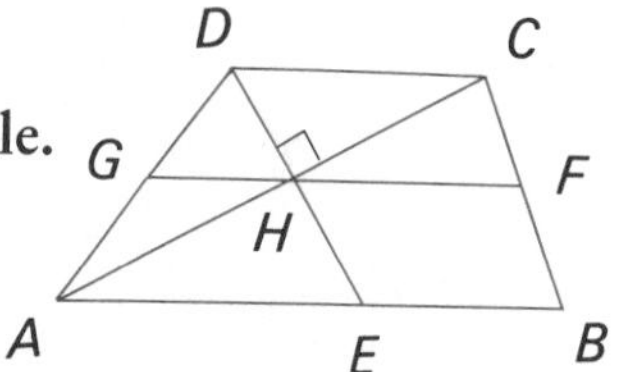

Given: Trap $ABCD$, with median $\overline{GF}$; $\overline{AC}$ intersects $\overline{GF}$ at H; $\overline{DE} \perp \overline{AC}$ at H.

22. Conclusion: $EBCD$ is a trapezoid.
23. Conclusion: $\triangle AED$ is isosceles.
24. Conclusion: $\triangle AHG$ is isosceles.
25. Conclusion: $\triangle HFC$ is isosceles.

Mixed Review

1. Prove: The base angles of an isosceles triangle are congruent. *5.1*
2. Prove: Opposite angles of a parallelogram are congruent. *6.4*

Determine whether the stated condition is necessary or not, and whether it is sufficient or not for a quadrilateral to be a parallelogram. *7.2, 7.3*

3. Two opposite angles are congruent.
4. Four angles are right angles.
5. Both pairs of opposite sides are congruent.

7.5. Isosceles Trapezoids

Objectives

To identify isosceles trapezoids
To prove and apply theorems about isosceles trapezoids
To identify necessary and sufficient conditions for isosceles trapezoids

Definition

An **isosceles trapezoid** is a trapezoid with congruent legs.

It was proved earlier that the base angles of an isosceles triangle are congruent. A similar theorem for isosceles trapezoids can now be proved.

Theorem 7.14

The base angles of an isosceles trapezoid are congruent.

Given: Trap $ABCD$, $\overline{AB} \parallel \overline{DC}$, $\overline{AD} \cong \overline{BC}$
Prove: $\angle A \cong \angle B$ and $\angle ADC \cong \angle BCD$

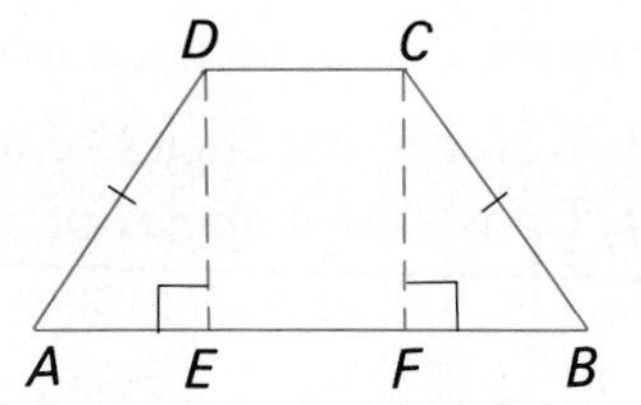

Plan

Draw perpendiculars $\overline{DE}$ and $\overline{CF}$ to form congruent right triangles AED and BFC. $\angle A$ and $\angle B$ are corresponding parts of these triangles.

Proof

Statement		Reason
1. Trap $ABCD$, $\overline{AB} \parallel \overline{DC}$		1. Given
2. Draw $\overline{DE} \perp \overline{AB}$, $\overline{CF} \perp \overline{AB}$.		2. There is exactly one $\perp$ from a pt to a line.
3. $\angle DEA$ and $\angle CFB$ are rt $\angle$s.		3. Def of $\perp$
4. $\triangle AED$ and $\triangle BFC$ are rt $\triangle$s.		4. Def of rt $\triangle$
5. $\overline{AD} \cong \overline{BC}$	(H)	5. Given
6. $\overline{DE} \cong \overline{CF}$ ($DE = CF$)	(L)	6. $\parallel$ lines are equidistant.
7. $\triangle AED \cong \triangle BFC$		7. HL
8. $\angle A \cong \angle B$		8. CPCTC
9. $\angle A$ and $\angle ADC$ are supp; $\angle B$ and $\angle BCD$ are supp.		9. $\parallel$ lines form supp int $\angle$s on same side of transv.
10. $\therefore \angle ADC \cong \angle BCD$		10. Supp of $\cong$ $\angle$s are $\cong$.

The next theorem is the converse of Theorem 7.14.

Theorem 7.15 If the base angles of a trapezoid are congruent, then the trapezoid is isosceles.

EXAMPLE 1 Find m $\angle W$ and m $\angle Y$ in isosceles trapezoid $WXYZ$.

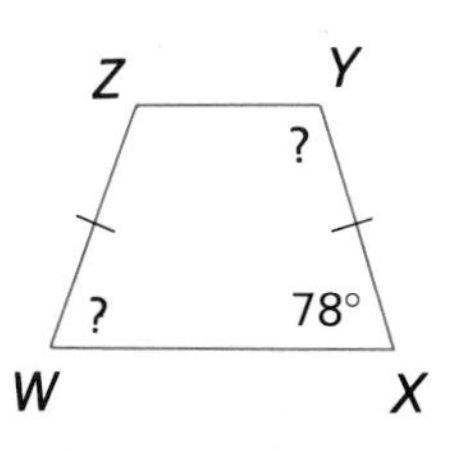

Plan m $\angle W$ = m $\angle X$ by Theorem 7.14. $\angle X$ and $\angle Y$ are supplementary since they are interior angles on the same side of transversal $\overline{XY}$, and $\overline{WX} \parallel \overline{ZY}$.

Solution

m $\angle W$ = m $\angle X$
m $\angle W$ = 78

m $\angle X$ + m $\angle Y$ = 180
m $\angle Y$ = 180 − m $\angle X$
m $\angle Y$ = 180 − 78 = 102

EXAMPLE 2 Given: Trap $MNOP$, $\overline{MN} \parallel \overline{PO}$, $\angle M \cong \angle N$, $MP = 5x - 12$, $NO = x + 8$
Find NO.

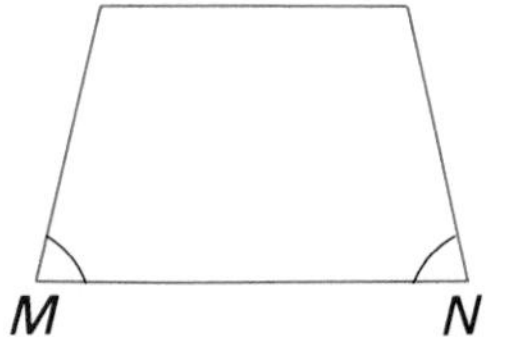

Plan $MNOP$ is an isosceles trapezoid by Theorem 7.15. Therefore, $MP = NO$.

Solution

$5x - 12 = x + 8$
$4x = 20$
$x = 5$
$NO = 5 + 8$, or 13

Theorem 7.16 The diagonals of an isosceles trapezoid are congruent.

Given: Isosceles trap $ABCD$, $\overline{AD} \cong \overline{BC}$, $\overline{AB} \parallel \overline{DC}$
Prove: $\overline{AC} \cong \overline{BD}$

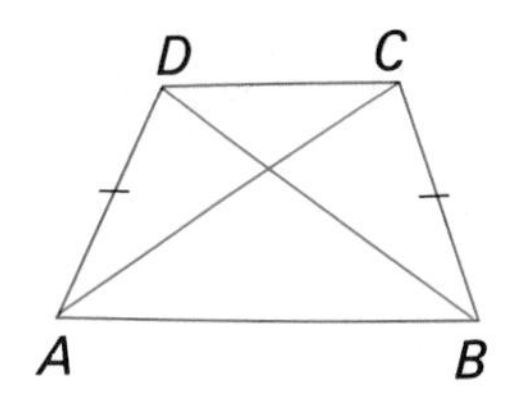

Some quadrilaterals with congruent diagonals are isosceles trapezoids, while others are not. However, every trapezoid with congruent diagonals is isosceles, as stated in Theorem 7.17.

Theorem 7.17 If the diagonals of a trapezoid are congruent, then the trapezoid is isosceles.

Given: Trap $ABCD$, $\overline{DC} \parallel \overline{AB}$, $\overline{AC} \cong \overline{BD}$
Prove: $\overline{BC} \cong \overline{AD}$

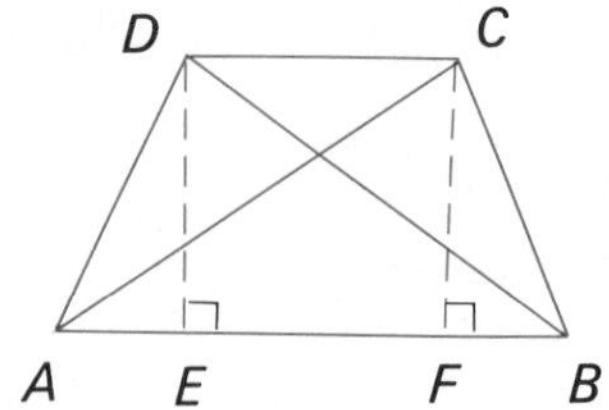

Plan Draw $\overline{DE} \perp \overline{AB}$ and $\overline{CF} \perp \overline{AB}$.
Prove $\triangle AFC \cong \triangle BED$ and then prove $\triangle BAC \cong \triangle ABD$.

Theorems 7.16 and 7.17 allow the conclusion that the property of congruent diagonals is *both* a necessary and a sufficient condition for a trapezoid to be isosceles. Every isosceles trapezoid has congruent diagonals, and every trapezoid with congruent diagonals is isosceles.

EXAMPLE 3 Is the property of one pair of congruent opposite sides a necessary condition for a quadrilateral to be an isosceles trapezoid? Is it a sufficient condition? Why?

Solution It is a necessary condition. By definition, an isosceles trapezoid *must* have congruent legs.

It is *not* a sufficient condition. A quadrilateral can have a pair of congruent opposite sides and not be an isosceles trapezoid.

Classroom Exercises

Refer to the trapezoids for Exercises 1–9.

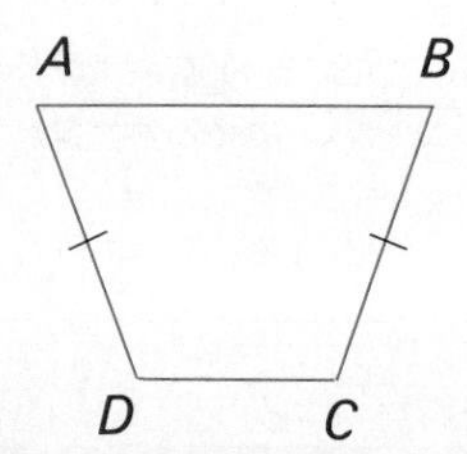

1. Name two congruent angles.
2. Name two pairs of supplementary angles.
3. Name two parallel sides.
4. Name two nonparallel sides.

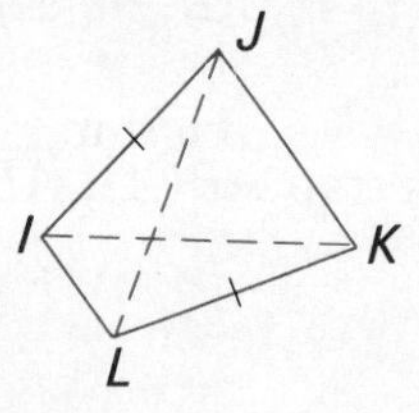

5. If trapezoid $IJKL$ is isosceles, then $\angle IJK \cong$ __________.
6. If trapezoid $IJKL$ is isosceles, then $\overline{JL} \cong$ __________.

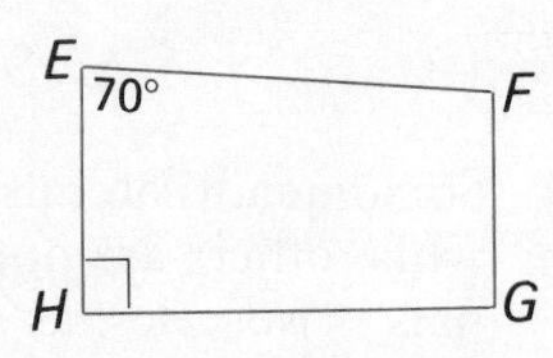

7. Name two supplementary angles.
8. Name two nonsupplementary angles.
9. Why is trapezoid $EFGH$ not isosceles?

Written Exercises

Find the indicated angle measure in the isosceles trapezoid.

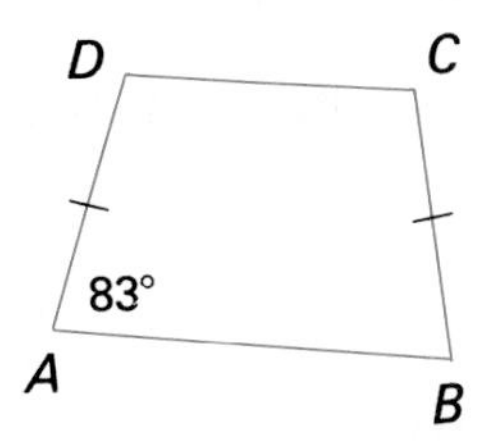

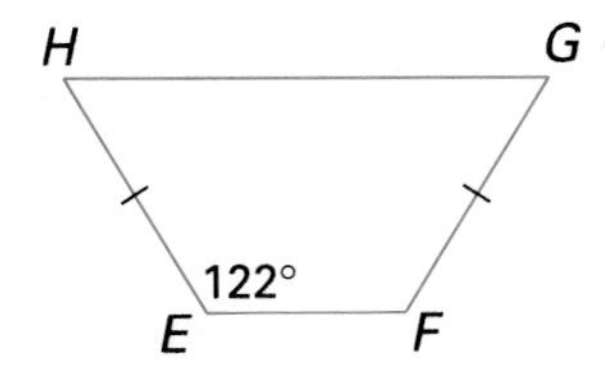

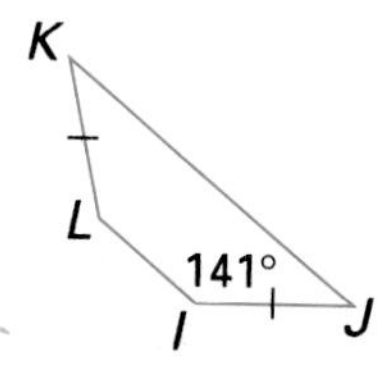

1. m $\angle B$ = ______
2. m $\angle C$ = ______
3. m $\angle H$ = ______
4. m $\angle G$ = ______
5. m $\angle K$ = ______
6. m $\angle J$ = ______

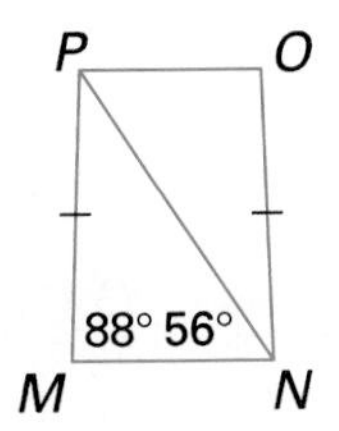

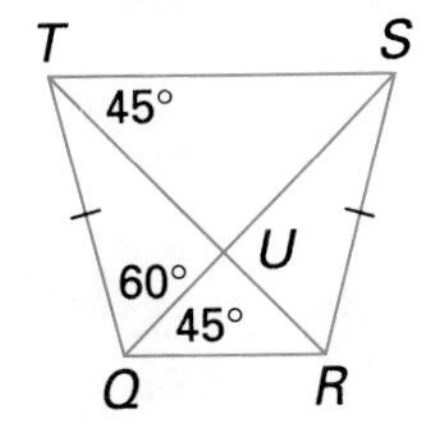

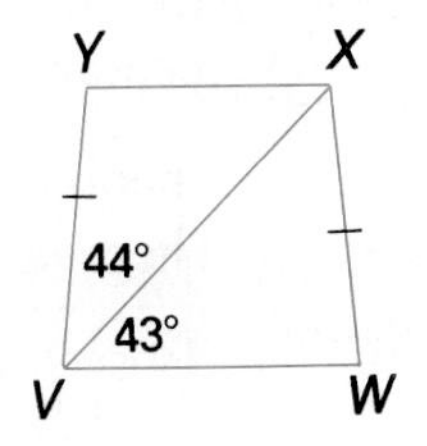

7. m $\angle MPN$ = ______
8. m $\angle PNO$ = ______
9. m $\angle QTU$ = ______
10. m $\angle TSU$ = ______
11. m $\angle W$ = ______
12. m $\angle VXW$ = ______

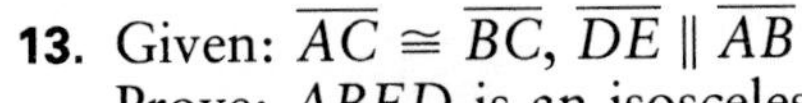

13. Given: $\overline{AC} \cong \overline{BC}$, $\overline{DE} \parallel \overline{AB}$
    Prove: $ABED$ is an isosceles trapezoid.
14. Given: Isosceles trap $ABED$, $\overline{DE} \parallel \overline{AB}$
    Prove: $\triangle ABC$ is isosceles.

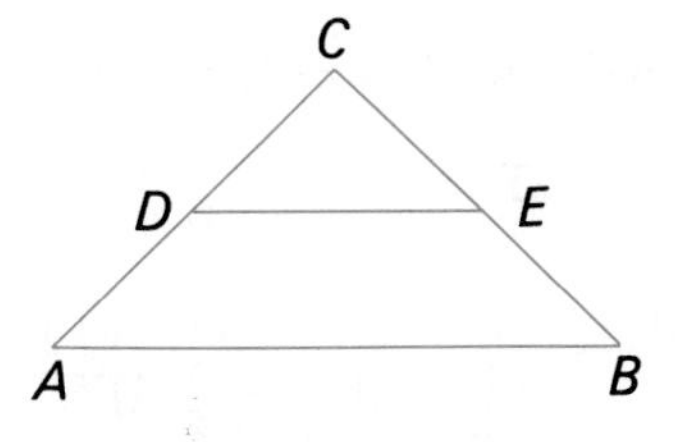

Use isosceles trapezoid $XYZW$ with $\overline{WZ} \parallel \overline{XY}$ for Exercises 15–18.

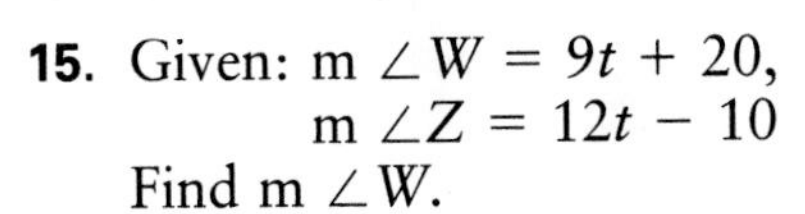

15. Given: m $\angle W = 9t + 20$, m $\angle Z = 12t - 10$
    Find m $\angle W$.
16. Given: m $\angle X = 12t$, m $\angle Y = 11t + 4$
    Find m $\angle X$.
17. Given: m $\angle W = 5m + 6$, m $\angle X = 2m + 6$
    Find m $\angle X$.
18. Given: m $\angle Z = 7n - 6$, m $\angle Y = 4n - 12$
    Find m $\angle Z$.

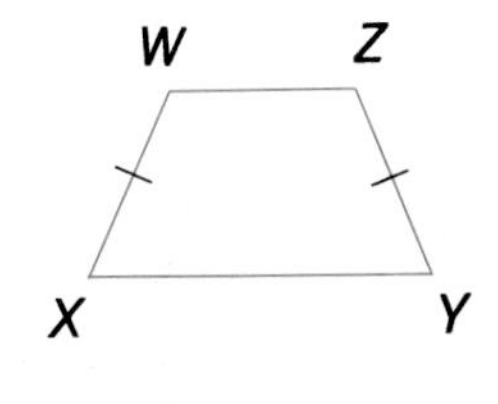

19. Given: $\overline{RS} \parallel \overline{UT}$, $\angle R \cong \angle S$, $RU = 4x - 6$, $ST = x + 15$
    Find ST.

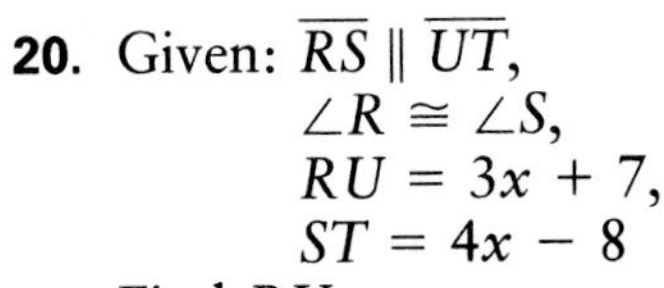

20. Given: $\overline{RS} \parallel \overline{UT}$, $\angle R \cong \angle S$, $RU = 3x + 7$, $ST = 4x - 8$
    Find RU.

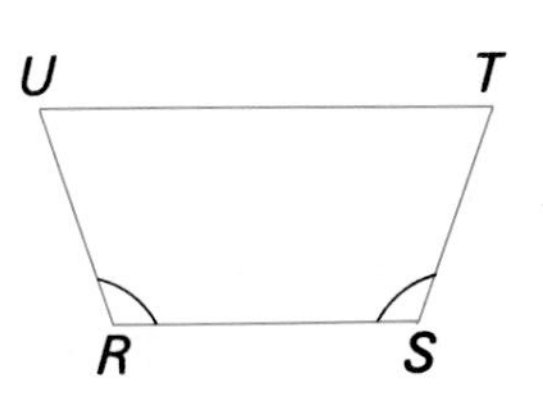

Is the property a necessary condition for a trapezoid to be an isosceles trapezoid? Is it a sufficient condition?

21. Diagonals are perpendicular to each other.

22. One pair of opposite sides is congruent.

23. Opposite angles are supplementary.

24. Two pairs of consecutive angles are supplementary.

Is the property a necessary condition for a quadrilateral to be an isosceles trapezoid? Is it a sufficient condition? If not sufficient, give a counterexample.

25. Opposite angles are supplementary.

26. Two pairs of consecutive angles are supplementary.

27. Two pairs of consecutive angles are congruent.

28. One pair of opposite sides is congruent.

29. Prove Theorem 7.15.

30. Prove Theorem 7.16.

31. Prove Theorem 7.17.

32. Given: Isosceles trap $ABCD$, $\overline{DC} \parallel \overline{AB}$
Prove: $\triangle ABD \cong \triangle BAC$

33. Given: Isosceles trap $ABCD$, $\overline{DC} \parallel \overline{AB}$
Prove: $\triangle AED \cong \triangle BEC$

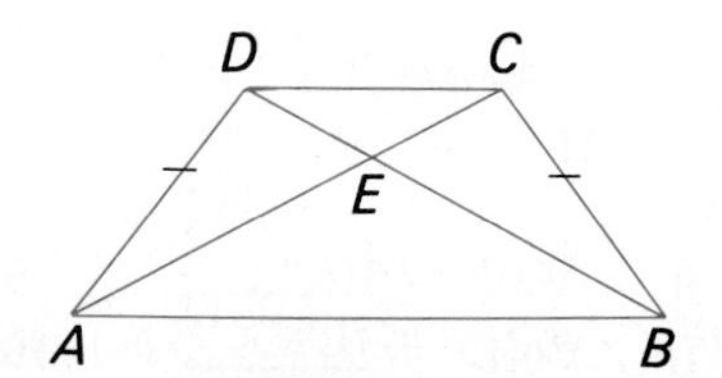

34. Given: $\overline{FB} \perp \overline{AB}$, $\overline{FB} \perp \overline{BC}$, $\overline{AB} \cong \overline{BC}$, M is the midpoint of $\overline{AF}$; N is the midpoint of $\overline{FC}$.
Prove: $ACNM$ is an isosceles trapezoid.

35. Given: M is the midpoint of $\overline{AF}$; N is the midpoint of $\overline{FC}$; $\triangle AFE \cong \triangle CFG$.
Prove: $ACNM$ is an isosceles trapezoid.

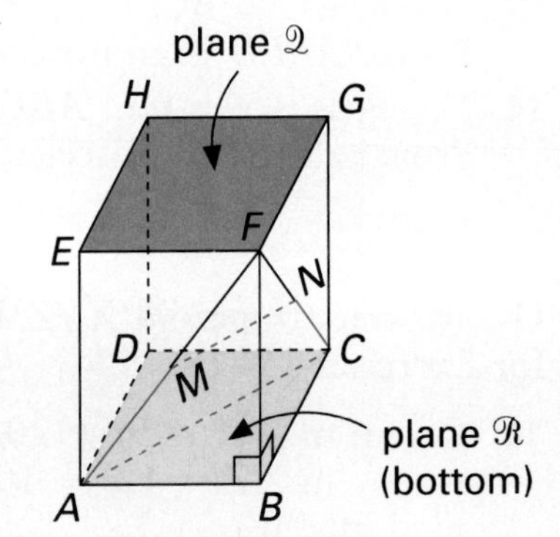

Mixed Review

Is the statement true or false?

1. Two quadrilaterals can be proved congruent by a side-side-side-side pattern. *7.1*
2. A rhombus has congruent diagonals. *7.2*
3. Draw an acute angle. Construct an angle congruent to it. *1.4*
4. Draw an obtuse triangle. Construct a triangle congruent to it. *4.3*

7.6 Constructing Quadrilaterals

Objective To construct quadrilaterals with given properties

Special quadrilaterals can be constructed by using their unique properties. To do so efficiently, make use of the *sufficient conditions* developed in the preceding lessons.

EXAMPLE 1 Construct a rectangle with adjacent sides congruent to two given segments.

a b

Plan A parallelogram with one right angle is a rectangle. Construct one right angle. Then construct a quadrilateral with opposite sides congruent.

Solution Construct a perpendicular at E. Construct $\overline{EF}$ and $\overline{EH}$ so that $EF = a$ and $EH = b$.

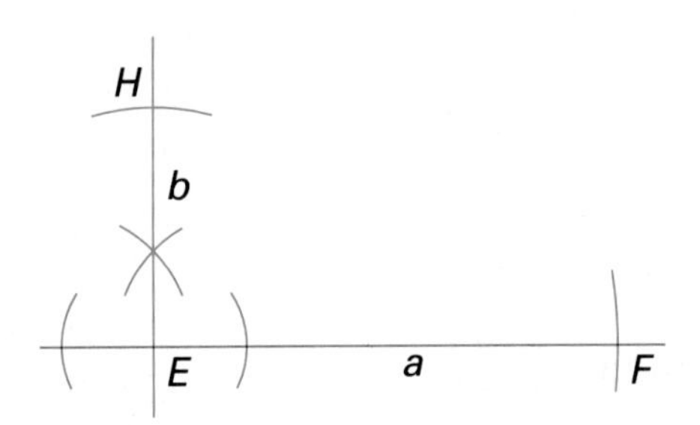

Swing arcs of length a from H and length b from F. Label the point of intersection G and draw $\overline{HG}$ and $\overline{FG}$.

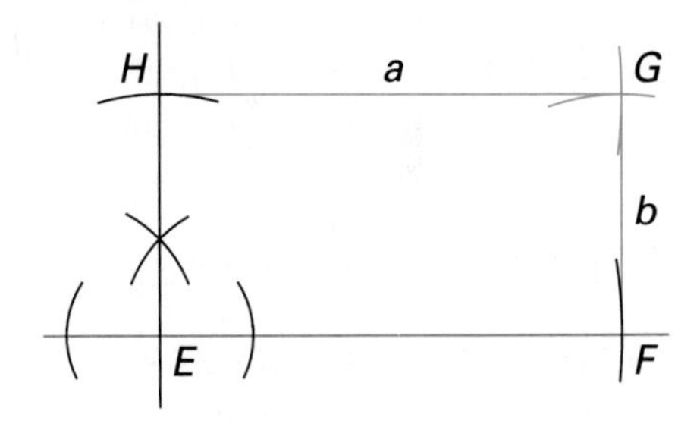

Because the opposite sides in the construction are congruent, the result is a parallelogram. Then, since $\angle E$ is a right angle, $EFGH$ is the required rectangle.

EXAMPLE 2 Construct a rhombus with diagonals congruent to two given segments.

a b

Plan A parallelogram with diagonals that are perpendicular is a rhombus. Construct diagonals that are perpendicular bisectors of each other.

Solution Construct $\overline{AC}$ with $AC = a$. Construct the perpendicular bisector of $\overline{AC}$.

Construct a segment of length b. Bisect the segment to find the length of $\frac{1}{2}b$.

On the perpendicular bisector of $\overline{AC}$, mark off segments of length $\frac{1}{2}b$ to determine B and D. Connect points A, B, C, and D.

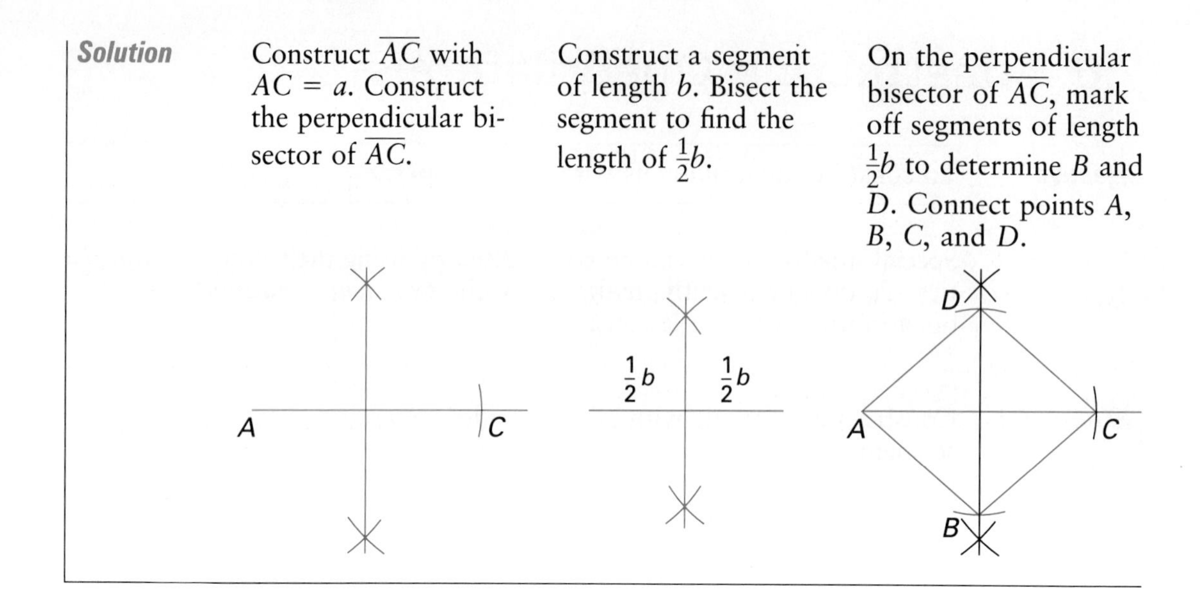

Classroom Exercises

Is the construction sufficient or not sufficient to result in a rhombus?

1. Construct a quadrilateral with two adjacent congruent sides.
2. Construct a quadrilateral with four congruent sides.
3. Construct a quadrilateral with perpendicular diagonals.
4. Construct a parallelogram with perpendicular diagonals.
5. Construct a parallelogram with two adjacent congruent sides.

Written Exercises

1. Draw two segments of two different lengths. Construct a rectangle with sides congruent to these segments.
2. Draw a segment. Construct a square with sides congruent to this segment.
3. Construct a parallelogram with adjacent sides of lengths a and b and an angle congruent to the given angle.
4. Construct a rhombus with sides of length c and an angle congruent to the given angle.

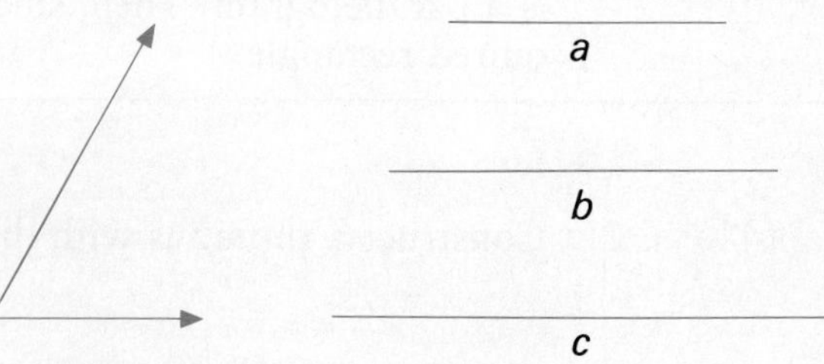

5. Draw two segments. Construct a rhombus with diagonals congruent to these two segments.

6. Draw a segment. Construct a parallelogram with one side congruent to this segment, another side half as long, and an angle of measure 60.
7. Construct a quadrilateral congruent to quad $ABCD$ using an SASAS pattern.
8. Construct a quadrilateral congruent to quad $ABCD$ using an ASASA pattern.
9. Is it possible to construct a quadrilateral congruent to quad $ABCD$ using an SSSS pattern?

D C A B

10. Construct a parallelogram with diagonals of lengths a and b and one side of length c.

a

b

c

11. Construct an isosceles trapezoid with three sides of length s and a diagonal of length d.

s d

12. Construct a trapezoid with bases of lengths a and b, a leg of length c, and a diagonal of length d.

a c

b d

Mixed Review

Is the statement true or false? ***5.6, 7.4***

1. All altitudes of a triangle are congruent.
2. All altitudes of a trapezoid are congruent.

State whether the congruence pattern is sufficient to establish congruence of two triangles. ***4.3, 4.4, 4.6***

3. SAS
4. SSA
5. AAA
6. SSS
7. AAS

Brainteaser

The figure to the right is a square with an equilateral triangle on top of it. Can you find the value of x?

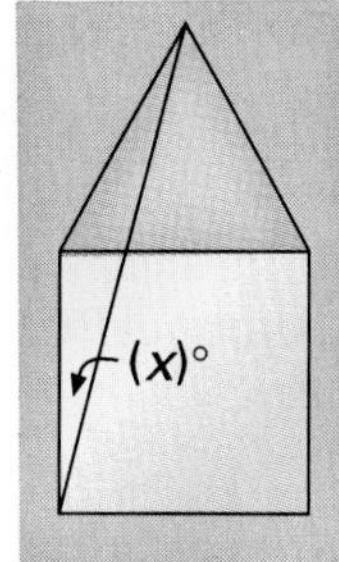

Chapter 7 Review

Key Terms

altitude of a trapezoid (p. 286)
congruent quadrilaterals (p. 269)
isosceles trapezoid (p. 290)
median of a trapezoid (p. 286)
necessary condition (p. 273)
rectangle (p. 273)
rhombus (p. 273)
square (p. 273)
sufficient condition (p. 278)
trapezoid (p. 285)

Key Ideas and Review Exercises

7.1 To identify and name congruent quadrilaterals, determine corresponding sides and angles.

1. If quad $LMNO \cong$ quad $ZYXW$, name the pairs of congruent sides and the pairs of congruent angles.

2. Based on the marks shown in the figure, complete the congruence statement: Quad $JKLM \cong$ ______.

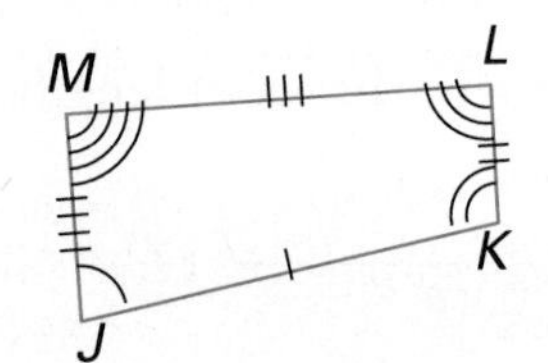

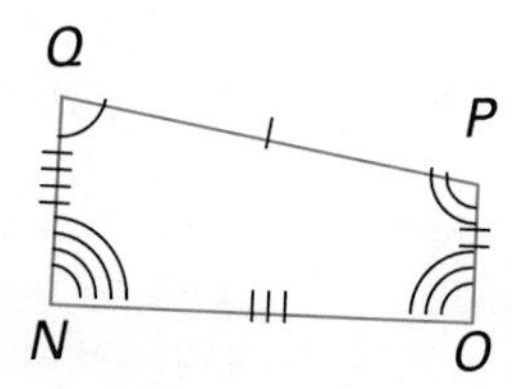

3. Identify two congruence patterns that can be used to prove that two given quadrilaterals are congruent.

7.2 To prove and apply properties of rectangles, rhombuses, and squares, and to identify necessary conditions for them, use one or more of the following:
- A rectangle is a parallelogram with four right angles.
- A rhombus is a parallelogram with four congruent sides.
- A square is a rectangle with four congruent sides.
- The diagonals of a rhombus are perpendicular.
- The diagonals of a rectangle are congruent.
- Each diagonal of a rhombus bisects two opposite angles of the rhombus.

4. Is the property that diagonals are congruent a necessary condition for a parallelogram to be a square?
5. In rhombus $ABCD$, $AB = 5x + 3$, $BC = 9x - 5$ Find AB.

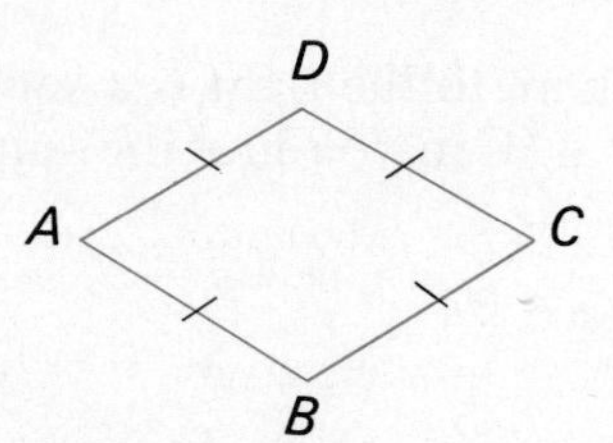

6. Given: Rect $XYZW$, M is the midpoint of $\overline{XY}$.
Prove: $WM = ZM$

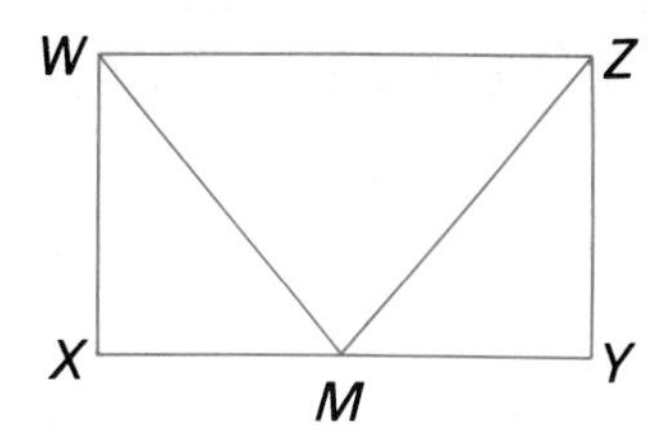

7.3 To prove that a figure is a rectangle, a rhombus, or a square, and to identify sufficient conditions for them, use one or more of the following:

- A parallelogram with perpendicular diagonals is a rhombus.
- A parallelogram with at least one right angle is a rectangle.
- A parallelogram with congruent diagonals is a rectangle.
- A parallelogram with a diagonal that bisects its opposite angles is a rhombus.
- A quadrilateral with four congruent sides is a rhombus.
- A parallelogram with two adjacent congruent sides is a rhombus.
- A parallelogram with two adjacent congruent sides and one right angle is a square.

7. Prove: A parallelogram with two consecutive congruent angles is a rectangle.

8. Prove: A rectangle with two adjacent congruent sides is a square.

7.4 A trapezoid is a quadrilateral with exactly one pair of parallel sides. The median of a trapezoid has length one-half the sum of the lengths of its bases. The median is parallel to the bases.

9. Can a trapezoid have exactly one right angle? Provide a drawing to support your answer.

10. The lengths of the two bases of a trapezoid are 9 and 17. What is the length of the median?

7.5 The base angles of an isosceles trapezoid are congruent.
If the base angles of a trapezoid are congruent, then the trapezoid is isosceles.
The diagonals of an isosceles trapezoid are congruent.
If the diagonals of a trapezoid are congruent, then the trapezoid is isosceles.

11. In quadrilateral $PQRS$, if $\overline{PQ} \parallel \overline{RS}$, and $\angle P \cong \angle Q$, is $PQRS$ an isosceles trapezoid? Why or why not?

12. In isosceles trapezoid $LMNO$, $\overline{LO} \cong \overline{MN}$, $m\angle L = 5x + 10$, and $m\angle M = 6x$. Find $m\angle L$.

7.6 To construct a special quadrilateral (rhombus, rectangle, square, or trapezoid), use one or more of the unique properties of that quadrilateral.

13. Draw two segments. Construct a rectangle with sides congruent to the segments.

14. Draw two segments. Construct a rhombus with diagonals congruent to the segments.

Chapter 7 Test

If quad $ABCD \cong$ quad $PQRS$, which angle or side is congruent to the given angle or side?

1. $\overline{BC} \cong$ ______
2. $\overline{SP} \cong$ ______
3. $\angle B \cong$ ______
4. $\angle S \cong$ ______
5. Based on the markings shown, write a correct congruence statement. Which congruence pattern can be used to prove the two quadrilaterals congruent?

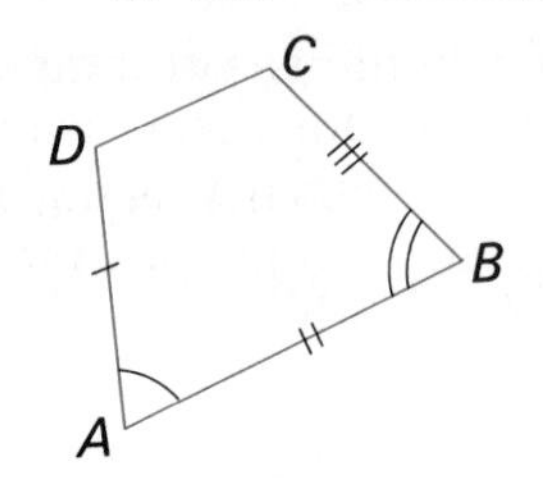

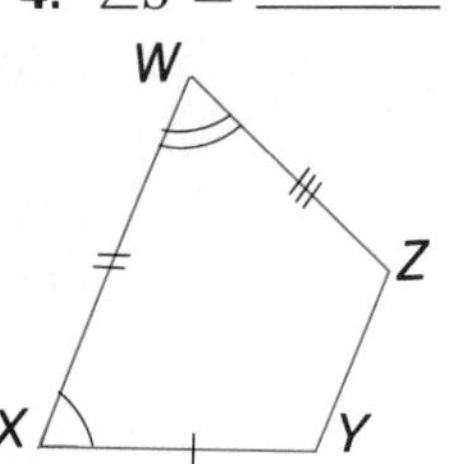

Based only on the markings shown, identify the parallelogram as a rectangle, rhombus, square, or none of these.

6. 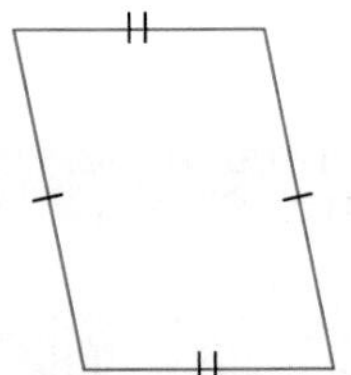

7. 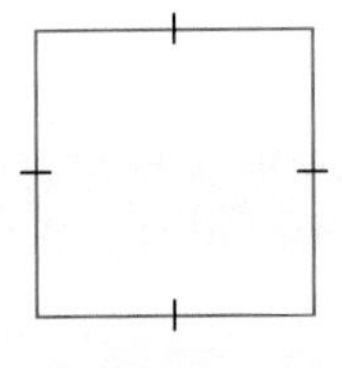

8. 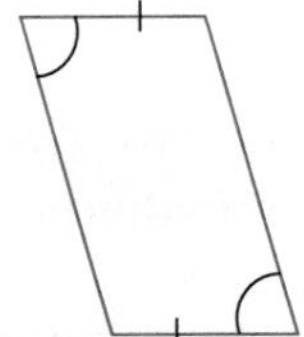

9. 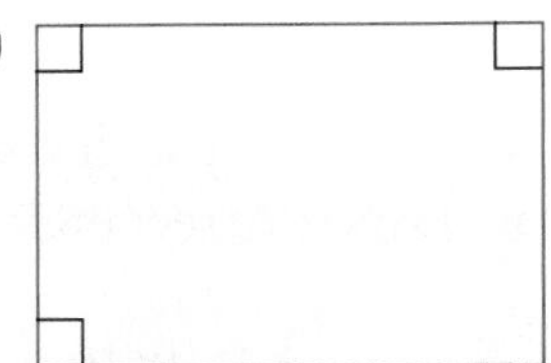

Is the stated condition *necessary* for a quadrilateral to be a rhombus?

10. a quadrilateral with perpendicular diagonals
11. a quadrilateral with four congruent sides

Is the stated condition *sufficient* for a quadrilateral to be a rectangle?

12. a quadrilateral with three right angles
13. a quadrilateral with congruent diagonals

True or false? Draw a picture to defend your answer.

14. A parallelogram with diagonals that bisect each other is a rhombus.
15. A quadrilateral with congruent diagonals is an isosceles trapezoid.
16. Given: Trap $ABCD$, median $\overline{EF}$, $AB = x + 2$, $DC = 2x + 1$, $EF = 2x - 3$. Find EF.

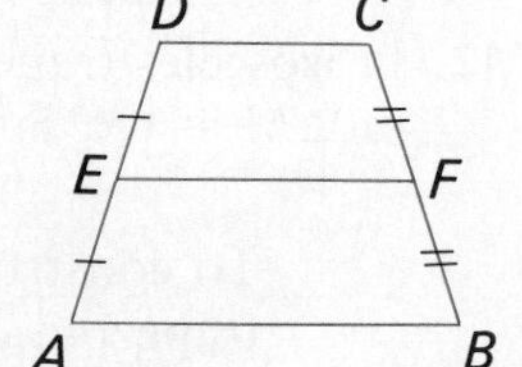

17. Construct a rectangle with sides of lengths d and s.

______ d ______ s

* 18. Prove: The diagonals of a rhombus bisect the opposite angles.

* 19. Prove: The midpoint of the hypotenuse of a right triangle is equidistant from the three vertices of the triangle.

College Prep Test

In each item you are to compare a quantity in Column 1 with a quantity in Column 2. Write the letter of the correct answer from these choices.

A—The quantity in Column 1 is greater than the quantity in Column 2.
B—The quantity in Column 2 is greater than the quantity in Column 1.
C—The quantity in Column 1 is equal to the quantity in Column 2.
D—The relationship cannot be determined from the given information.

	Column 1	Column 2
1. A prime factor of 77 is 11.	11	the other prime factor
2. $x = 5$	$5x + 5$	$7x - 5$
3. $\angle A$ and $\angle B$ are complementary. $\angle A$ is acute.	m $\angle A$	m $\angle B$
4. $\angle C$ and $\angle D$ are vertical angles. $\angle C$ is obtuse.	m $\angle C$	m $\angle D$
5. m $\angle G <$ m $\angle F <$ m $\angle E$	m $\angle E$	m $\angle G$
6. $\angle A$ and $\angle B$ are the base angles of isosceles trapezoid $ABCD$; m $\angle A = 75$.	m $\angle B$	m $\angle C$
7. $\angle C$ is an obtuse angle in $\triangle ABC$.	AB	BC
8. In $\triangle XYZ$, $XY = 9$, $YZ = 10$, and $ZX = 11$	m $\angle X$	m $\angle Z$
9. $ABCDEF$ is a regular hexagon. $\angle 1$ is an angle of $ABCDEF$. $\angle 2$ is an exterior angle of $ABCDEF$.	m $\angle 1$	m $\angle 2$
10. $\angle A$ is the vertex angle of isosceles $\triangle ABC$. m $\angle A = 74$	m $\angle A$	m $\angle B$
11. $\angle A$ is a base angle of isosceles $\triangle ABC$. m $\angle A = 74$.	AB	BC
12. $ABCDE$ is an equilateral pentagon.	m $\angle C$	m $\angle E$
13. $x = 0$	$3x$	$5x$
14. Quad $ABCD \cong$ quad $EFGH$. $AB = 12$ and $FG = 5$.	BC	EF

8 SIMILARITY

Pythagoras (about 569 B.C.)—Pythagoras founded a secret society of mathematicians. One secret the members were sworn not to reveal was the newly-discovered fact that the square root of two cannot be expressed as a fraction.

8.1 Ratio and Proportion

Objectives

To solve proportions
To find the geometric mean of two numbers

When a slide is projected, the screen image has the same shape as the image on the slide, but its size differs. An important question then arises: How do measures of the slide and of its image compare?

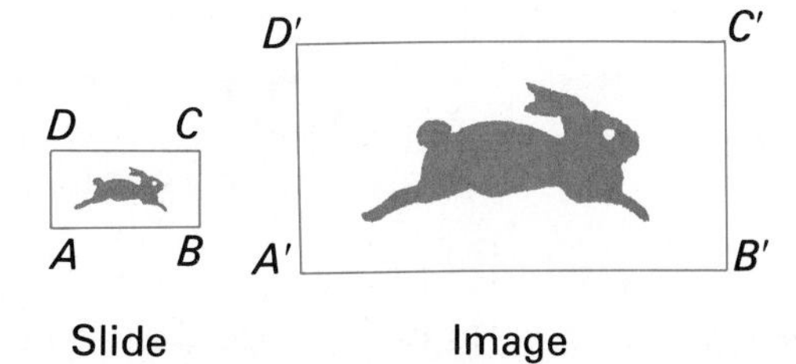

Slide   Image

Definition

A **ratio** is a comparison of two numbers by division. The ratio of a to b $(b \neq 0)$ may be written: a to b, $\frac{a}{b}$, or $a:b$.

As with fractions, it often is best to write a ratio in simplest form. For example, $\frac{14}{56}$ in simpler form is $\frac{1}{4}$.

Look at the illustration. Measure the sides. Note that $\frac{AB}{BC} = \frac{A'B'}{B'C'}$.

Definition

A **proportion** is an equation of the form $\frac{a}{b} = \frac{c}{d}$ $(b \neq 0, d \neq 0)$. a and d are the **extremes** of the proportion; b and c are the **means.**

Theorem 8.1

In a proportion the product of the extremes equals the product of the means.

Plan

Prove that $\frac{a}{b} = \frac{c}{d}$ $(b \neq 0, d \neq 0)$ implies $ad = bc$. This is called the **cross product.**

Proof

Statement	Reason
1. $\frac{a}{b} = \frac{c}{d}$ $(b \neq 0, d \neq 0)$	1. Given
2. $\frac{a}{b}(bd) = \frac{c}{d}(bd)$	2. Mult Prop of Eq
3. $a \cdot \frac{1}{b}(bd) = c \cdot \frac{1}{d}(bd)$	3. Def of division
4. $a \cdot (bd)\frac{1}{b} = c \cdot (bd)\frac{1}{d}$	4. Comm Prop of Mult

5. $a \cdot (db) \frac{1}{b} = (cb)\, d \cdot \frac{1}{d}$	5. Comm and Assoc Prop of Mult
6. $(ad)\, b \cdot \frac{1}{b} = (bc)\, d \cdot \frac{1}{d}$	6. Assoc and Comm Prop of Mult
7. $b \cdot \frac{1}{b} = 1,\ d \cdot \frac{1}{d} = 1$	7. Prop of Mult Inverses
8. $ad \cdot 1 = bc \cdot 1$	8. Sub
9. $\therefore ad = bc$	9. Identity Prop of Mult

Corollary If the product of the extremes equals the product of the means, then a proportion exists.

In other words, $ad = bc$ implies $\frac{a}{b} = \frac{c}{d}$ $(b \neq 0, d \neq 0)$.

EXAMPLE 1 Solve the proportion $\frac{6}{x} = \frac{9}{12}$.

Solution

$$\frac{6}{x} = \frac{9}{12}$$
$$6 \cdot 12 = 9 \cdot x$$
$$72 = 9x$$
$$8 = x$$

Theorem 8.2 If $\frac{a}{b} = \frac{c}{d}$ (assume no denominator equals zero), then:

(1) $\frac{b}{a} = \frac{d}{c}$

(2) $\frac{a}{c} = \frac{b}{d}$

(3) $\frac{a+b}{b} = \frac{c+d}{d}$

(4) $\frac{a-b}{b} = \frac{c-d}{d}$

(5) $\frac{a}{b} = \frac{a+c}{b+d}$

Plan To prove Part 3, show that $(a + b)d = b(c + d)$. If the product of the extremes equals the product of the means, then a proportion exists.

EXAMPLE 2 State three proportions that follow when Theorem 8.2 is applied to $\frac{2}{3} = \frac{6}{9}$. Check by using Theorem 8.1.

Solution

(1) $\frac{3}{2} = \frac{9}{6}$ True, since $3 \cdot 6 = 9 \cdot 2 = 18$

(2) $\frac{2}{6} = \frac{3}{9}$ True, since $2 \cdot 9 = 3 \cdot 6 = 18$

(3) $\frac{5}{3} = \frac{15}{9}$ $\frac{2+3}{3} = \frac{6+9}{9}$, true; since $5 \cdot 9 = 3 \cdot 15 = 45$

One special proportion has the form $\frac{a}{m} = \frac{m}{b}$. For example, if $\frac{3}{m} = \frac{m}{27}$, then $81 = m^2$ and $m = 9$ *or* $m = -9$.

The number 9 is called the **geometric mean** of 3 and 27.

Definition

The **geometric mean** of two positive numbers, a and b, is the positive number m, such that $\frac{a}{m} = \frac{m}{b}$.

EXAMPLE 3

a. Find the geometric mean m of 6 and 15.

b. Find the geometric mean m of a and $9a$ $(a > 0)$.

Solutions

a.

$$\frac{6}{m} = \frac{m}{15}$$
$$m^2 = 90$$
$$m = \sqrt{90}$$
$$= \sqrt{9} \cdot \sqrt{10}$$
$$= \pm 3\sqrt{10}$$

$3\sqrt{10}$ is the geometric mean.

b.

$$\frac{a}{m} = \frac{m}{9a}$$
$$m^2 = 9a^2$$
$$m = \pm 3a$$

$3a$ is the geometric mean.

EXAMPLE 4 The measures of three angles of a triangle are in the ratio 2:3:5. Find the measure of the largest angle.

Solution Draw a diagram. Let the angle measures be $2x$, $3x$, and $5x$ (ratio 2:3:5). Write an equation using the fact that the sum of the measures of the angles of a triangle is 180.

$$2x + 3x + 5x = 180$$
$$10x = 180$$
$$x = 18$$

$$5x = 5 \cdot 18 = 90$$

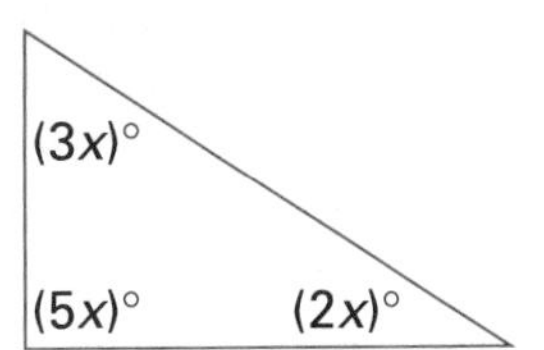

Therefore, 90 is the measure of the largest angle.

Classroom Exercises

Write the ratio in simplest form.

1. $\frac{16}{64}$ **2.** 18 to 54 **3.** 28:52 **4.** 20 to 85 **5.** 1.5:2

Written Exercises

Is the proportion true or false?

1. $\frac{3}{2} = \frac{18}{12}$ **2.** $\frac{5}{6} = \frac{6}{7}$ **3.** $\frac{5}{7} = \frac{25}{49}$ **4.** $\frac{7}{11} = \frac{77}{121}$

5. $\frac{2}{3} = \frac{2 + 2}{3 + 3}$ **6.** $\frac{19}{23} = \frac{19^2}{23^2}$ **7.** $\frac{6}{11} = \frac{6 \cdot 13}{11 \cdot 13}$ **8.** $\frac{12}{18} = \frac{12 - 9}{18 - 9}$

Solve the proportion for x.

9. $\frac{3}{4} = \frac{x}{24}$ **10.** $\frac{x}{7} = \frac{35}{49}$ **11.** $\frac{5}{9} = \frac{x}{108}$ **12.** $\frac{6}{15} = \frac{3}{x}$

13. $\frac{1}{x} = \frac{x}{4}$ **14.** $\frac{2}{x} = \frac{x}{8}$ **15.** $\frac{x}{4} = \frac{64}{x}$ **16.** $\frac{x}{1} = \frac{25}{x}$

Use Theorem 8.2 to state five proportions that follow from the given proportion. Use Theorem 8.1 to check.

17. $\frac{1}{4} = \frac{2}{8}$ **18.** $\frac{6}{12} = \frac{3}{6}$

Find the value of x in the proportion.

19. If $\frac{a}{b} = \frac{3}{4}$, then $\frac{a + b}{b} = \frac{3 + x}{4}$ **20.** If $\frac{p}{q} = \frac{9}{13}$, then $\frac{p - q}{q} = \frac{9 - x}{13}$

Find the geometric mean m of the pair of positive numbers.

21. 1 and 9 **22.** 5 and 20 **23.** 12 and 3 **24.** 6 and 7

25. 9 and 5 **26.** p and q **27.** p^2 and q^2 **28.** c and $9c$

Is the statement true for all values of the variables? (Assume each denominator $\neq$ 0.) (Exercises 29–32)

29. $\frac{cx}{cy} = \frac{dx}{dy}$ **30.** $\frac{c + x}{c + y} = \frac{d + x}{d + y}$ **31.** $\frac{ab}{bc} = \frac{ad}{dc}$ **32.** $\frac{p - q}{r - q} = \frac{p + q}{r + q}$

33. The ratio of the measures of two complementary angles is 4 to 5. What are the measures of the angles?

34. The ratio of the measures of two supplementary angles is 3 to 6. What are the measures of the angles?

35. The measures of three angles of a triangle are in the ratio 5:6:9. What are the measures of the angles?

36. The measures of three angles of a triangle are in the ratio 2:4:9. What are the measures of the angles?

37. The perimeter of a rectangle is 60 cm. The ratio of the length of a base to the length of a corresponding altitude of the rectangle is 5 to 7. Find the lengths of the base and altitude.

38. The perimeter of an isosceles triangle is 99 cm. The ratio of the length of the base to the length of a leg is 3:4. What are the lengths of the base and leg of the triangle?

39. If two boxes of cereal cost $1.38, what will seven boxes cost?

40. A baseball player has a batting average of .320. If he has been at bat 225 times, how many hits does he have?

41. Pat answered correctly 80% of the questions on a test. If there were 15 questions, how many were answered correctly?

Tell whether the proportion is true for all values of the variables. If it is true, give a proof. If false, give a numerical counterexample. (Assume each denominator is not equal to zero.)

42. If $\frac{x}{y} = \frac{r}{s}$, then $\frac{x + c}{y} = \frac{r + c}{s}$

43. If $\frac{x}{y} = \frac{r}{s}$, then $\frac{x + a}{y + a} = \frac{r + a}{s + a}$

44. If $\frac{x}{y} = \frac{r}{s}$, then $\frac{x}{x + y} = \frac{r}{r + s}$

45. If $\frac{x}{y} = \frac{r}{s} = \frac{m}{n}$, then $\frac{x}{y} = \frac{x + r + m}{y + s + n}$

46. Prove the corollary to Theorem 8.1.

47. Prove Property 1 of Theorem 8.2.

48. Prove Property 2 of Theorem 8.2.

49. Prove Property 3 of Theorem 8.2.

50. Prove Property 4 of Theorem 8.2.

51. Prove Property 5 of Theorem 8.2.

Mixed Review

Assume $l \parallel m$. Determine whether the two angles are congruent. *3.5*
(Exercises 1–4)

1. $\angle 3$ and $\angle 5$
2. $\angle 1$ and $\angle 8$
3. $\angle 2$ and $\angle 7$
4. $\angle 1$ and $\angle 5$

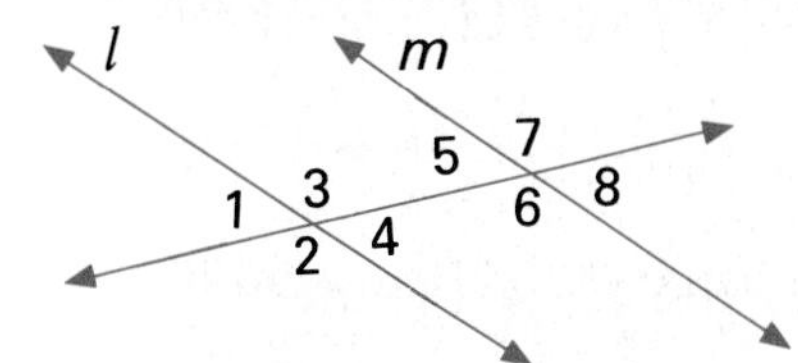

5. If the ratio of the measures of $\angle 4$ and $\angle 6$ is 2 to 3, what are the measures of the angles? *8.1*

Application: *Pitch of a Roof*

The pitch of a roof is the ratio of the rise of the roof to the run of the roof. For example a 4-in-12 roof rises 4 inches for every 12 inches of run.

If the building is 24 ft wide, what is the total rise for a 4-in-12 roof? For a 6-in-12 roof?

8.2 Similar Polygons

Objectives To identify similar polygons
To find the measures of parts of similar polygons
To explain the difference between similar and congruent polygons

The figures below are not the same size, but they have the same shape. Such figures are called similar polygons.

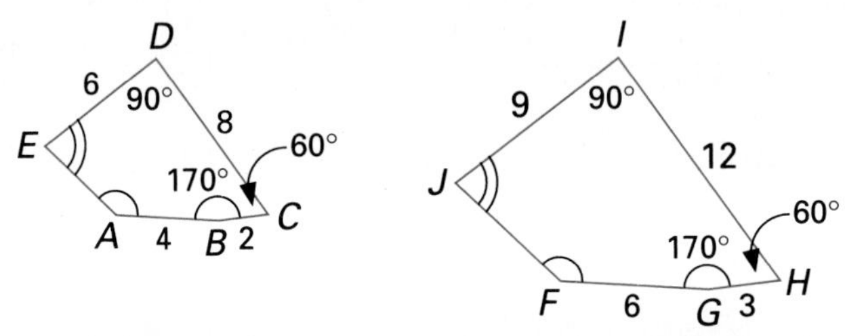

Polygon *ABCDE* is similar to polygon *FGHIJ*.

Definition Two **polygons are similar** (~) if corresponding angles are congruent and the ratios of the lengths of corresponding sides are equal.

Polygon *ABCDE* ~ polygon *FGHIJ* means that "polygon *ABCDE is similar to* polygon *FGHIJ*." Note that the corresponding vertices must be named in the same order. The symbol $\nsim$ means *is not similar to.*

EXAMPLE 1 Is square *KLMN* similar to rectangle *OPQR*? Why or why not?

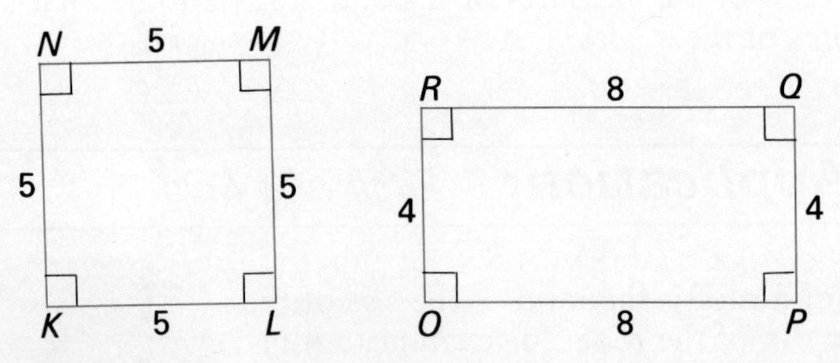

Solution Corresponding angles are congruent, since all the angles are right.

Corresponding sides are not proportional: $\frac{KL}{OP} = \frac{5}{8}$ and $\frac{LM}{PQ} = \frac{5}{4}$.

Therefore, square *KLMN* $\nsim$ rectangle *OPQR*.

Having congruent corresponding angles is not a sufficient condition to prove the similarity of two polygons. Nor is having corresponding proportional sides sufficient. Both congruent corresponding angles *and* corresponding proportional sides must be true. In the two quadrilaterals,

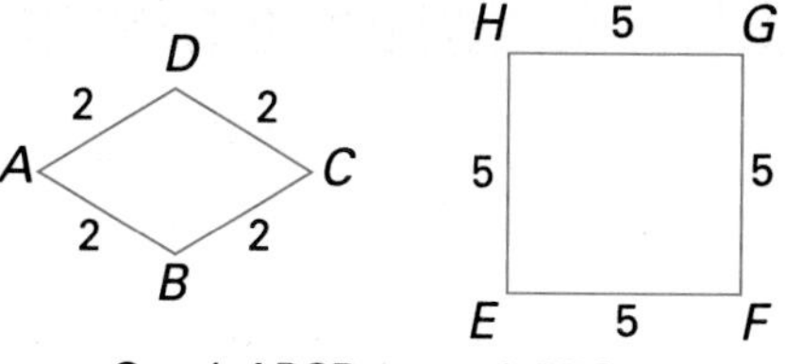

Quad $ABCD \not\sim$ quad $EFGH$.

$$\frac{AB}{EF} = \frac{BC}{FG} = \frac{CD}{GH} = \frac{DA}{HE} = \frac{2}{5}.$$

Given that the lengths of corresponding sides of similar polygons are proportional, use a proportion to find the unknown lengths of sides.

EXAMPLE 2 Polygon $ABCDE \sim$ polygon $FGHIJ$. $AB = 16$, $BC = 20$, $CD = 14$, $DE = 15$, $EA = 25$, and $FG = 24$. Find GH.

Solution

$$\frac{AB}{FG} = \frac{BC}{GH}$$

$$\frac{16}{24} = \frac{20}{GH}$$

$$16(GH) = 20 \cdot 24 = 480$$

$$GH = 30$$

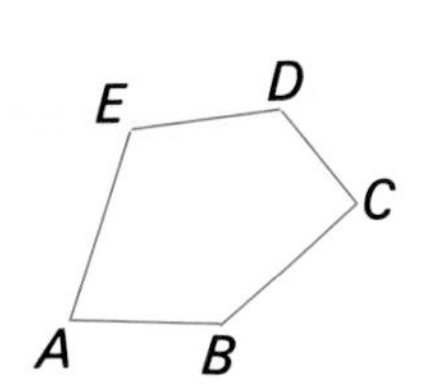

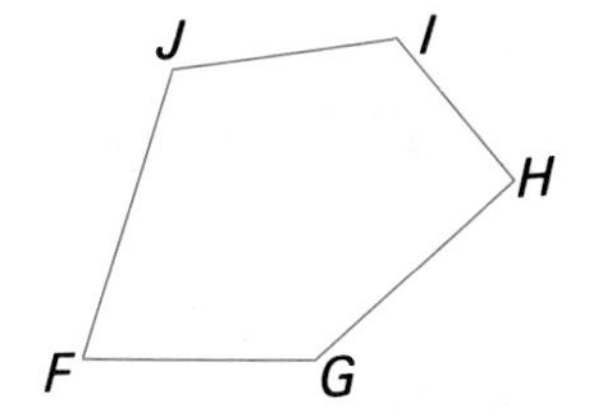

Similarity provides a basis for important applications of geometry.

EXAMPLE 3 A blueprint for a house was drawn on a scale in which $\frac{1}{8}$ in. on the plan represents 1 ft of actual length. Part of the blueprint is shown. What is the actual length of the living room?

2 in.

Plan Write a proportion using the ratio $\frac{\text{scale length}}{\text{actual length}}$.

Solution The length of the room on the blueprint is 2 in.

$$\frac{\frac{1}{8}\text{ in.}}{1\text{ ft}} = \frac{2\text{ in.}}{x\text{ ft}}$$

$$\frac{1}{8}x = 2, \text{ so } x = 16$$

Thus, the living room is 16 ft long.

Classroom Exercises

Determine whether the polygons are similar on the basis of the marking alone. Explain your answer.

1. 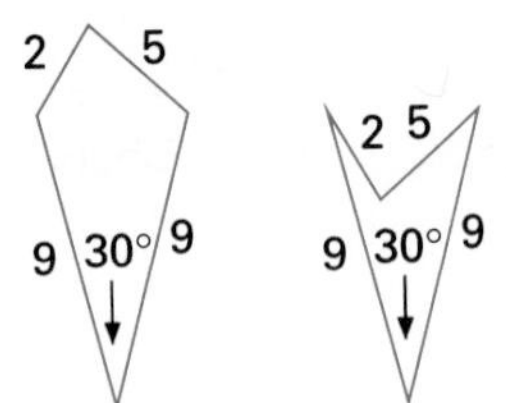

2. 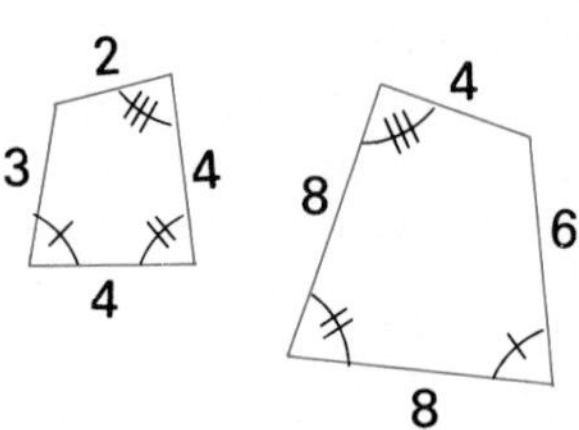

3. 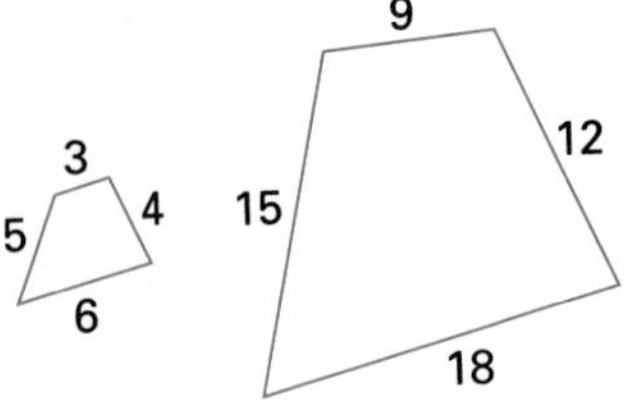

Use the properties of proportions of Theorem 8.2 to complete the following statements. Quad $RSTU \sim$ quad $JKLM$.

4. $\frac{RS}{ST} = \frac{?}{KL}$

5. $LM \cdot RS = TU \cdot$ ______

6. $\frac{RS + JK}{JK} = \frac{TU + ?}{LM}$

Written Exercises

Find the unknown lengths for the pair of similar polygons.

1. $ABCD \sim EFGH$

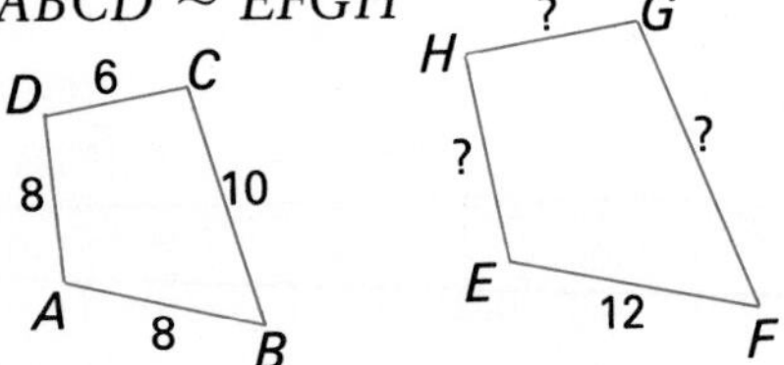

2. Square $IJKL \sim$ square $MNOP$

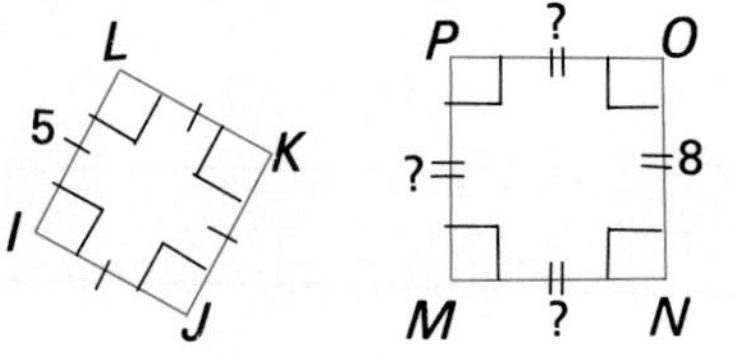

3. $\square QRST \sim \square UVWX$

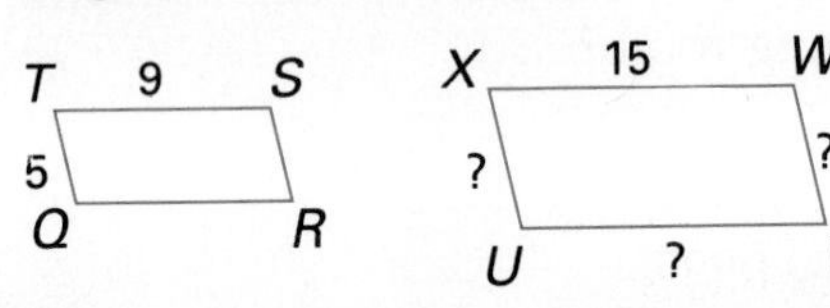

4. $ABCDE \sim FGHIJ$

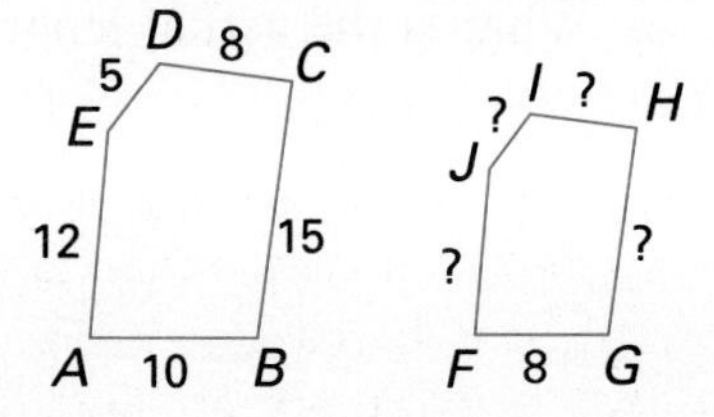

For Exercises 5–8, find the missing measures.

5. Given: Rhomb $ABCD \sim$ rhomb $EFGH$, $AB = 7$, $EF = 12$, $m\angle A = 55$
Find FG and $m\angle F$.

6. Given: Pentagon $IJKLM \sim$ pentagon $NOPQR$, $IJ = 24$, $JK = 20$, $KL = 28$, $NO = 32$
Find OP and PQ.

7. Given: Polygon $STUVW \sim$ polygon $XYZAB$, $XY = 32$, $YZ = 36$, $ST = 4c + 2$, $TU = 5c + 2$
Find ST and TU.

8. Given: Polygon $CDEFG \sim$ polygon $HIJKL$, $DE = 13$, $EF = 15$, $IJ = 5x + 2$, $JK = 7x - 2$
Find IJ and JK.

9. Given: Square $ABCD$, square $XYZW$
Prove: $ABCD \sim XYZW$

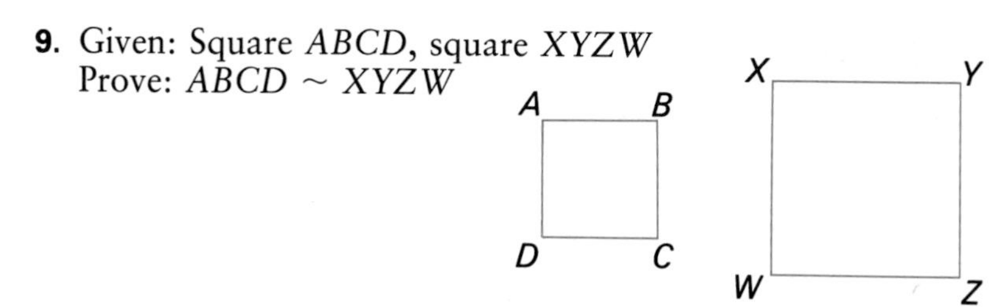

10. If two polygons are similar, are they necessarily congruent? Explain.
11. If two polygons are congruent, are they necessarily similar? Explain.

A portion of a blueprint for a house is shown. The scale used is $\frac{1}{8}$ of an inch to the foot. In Exercises 12–15, find the indicated measure.

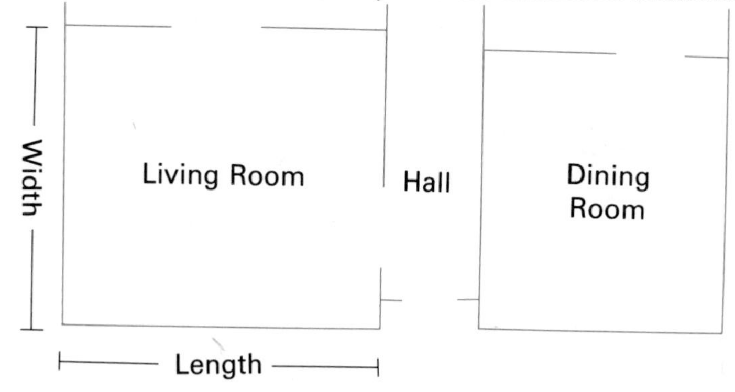

12. How wide is the living room?
13. How long is the living room?
14. How wide is the dining room?
15. How wide is the hall?
16. Draw two equilateral pentagons that are not similar.
17. Draw two equiangular pentagons that are not similar.
18. Refer to an encyclopedia or book of art history to find uses of the **golden ratio** in painting, architecture, or photography. Look also under Golden Section, Divine Proportion, and Golden Rectangle.

Mixed Review

1. List four necessary conditions for a quadrilateral to be a parallelogram. *6.6*
2. List four sufficient conditions for a quadrilateral to be a parallelogram. *6.6*
3. If a parallelogram has congruent diagonals, is the parallelogram necessarily a square? If not, show a counterexample. *7.3*

Problem Solving Strategies

Using a Formula to Find the Golden Ratio

Look at the three rectangles. Which do you think is most pleasing to the eye? Most people choose the one in the middle. Why? We don't really know, but this shape has been used throughout history to create works of art and beautiful buildings such as the Parthenon.

In the golden rectangle the long side is the geometric mean between the width and the long side plus the width. To solve this proportion you must use the quadratic formula.

$$\frac{w}{l} = \frac{l}{w + l}$$

Let $l = 1$ and cross multiply.

$$w(w + 1) = 1$$
$$w^2 + w - 1 = 0$$
$$w = \frac{-1 \pm \sqrt{5}}{2} \approx 0.618$$

The **golden ratio** in mathematical literature is the reciprocal $\frac{1}{w}$. See Exercise 3.

A golden rectangle can be constructed as follows: Draw square $ABCD$; find the midpoint of $\overline{AB}$ and draw segment $\overline{TC}$ as shown. Extend the segment $\overline{AB}$ to R by using the length of the segment $\overline{TC}$ from T. Complete the rectangle by drawing $\overline{RS} \perp$ to $\overline{AR}$.

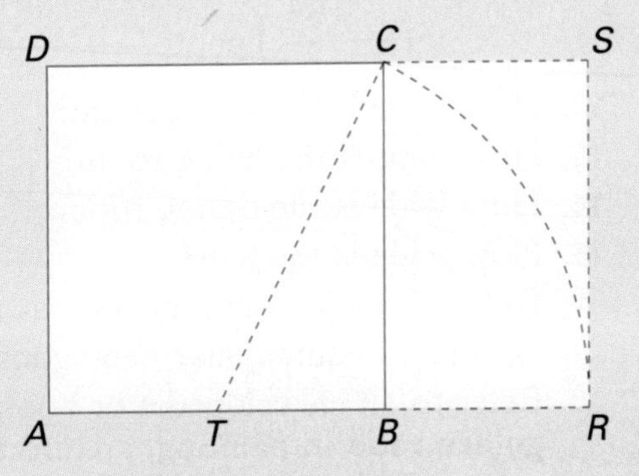

Exercises

1. Since all golden rectangles are similar, we can choose any convenient lengths. Let a side of the square used for the construction above have length 2. Use the Pythagorean Theorem to find the length of $\overline{TC}$. Now set up the proportion used to define the golden ratio and prove that it is a true proportion.
2. Using a calculator, find the ratio $\frac{AB}{AR}$ to the nearest thousandth.
3. Using a calculator, find the ratio $\frac{AR}{AB}$ to the nearest thousandth.

8.3 Similar Triangles

Objectives To write and use proportions relating to similar triangles
To apply the AA Similarity Postulate

A printer can quickly tell if two rectangular designs are similar by placing one on top of the other so that two corners coincide. If the diagonal of the larger rectangle contains the diagonal of the smaller, forming similar triangles, then the two rectangles are similar. Many other practical applications of geometry depend upon properties of similar triangles.

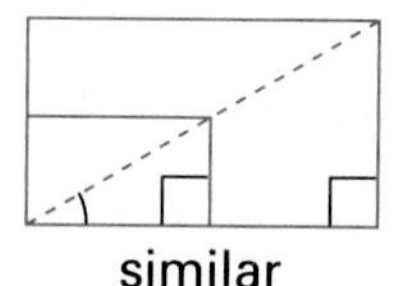

similar

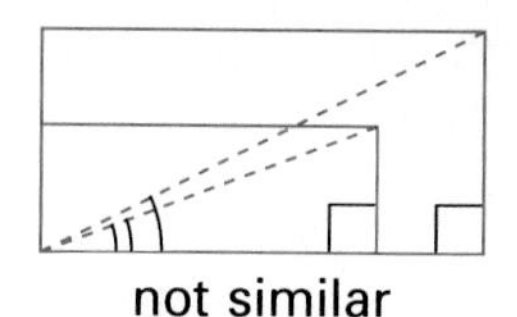

not similar

Definition Two **triangles are similar** if corresponding angles are congruent and corresponding sides are proportional.

EXAMPLE 1 Given: $\triangle XYZ \sim \triangle TUV$, $XY = 12$, $YZ = 15$, $ZX = 18$, $TU = 16$
Find UV and VT.

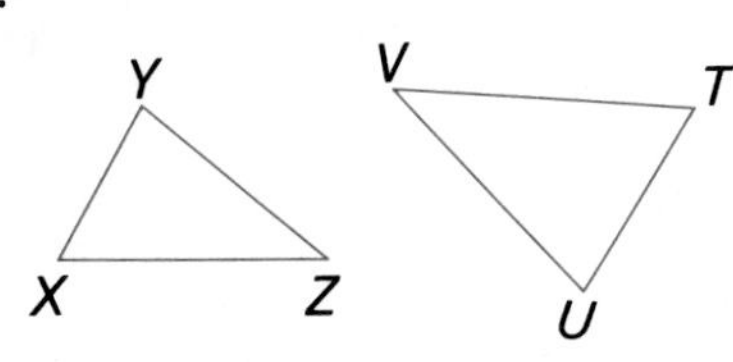

Solution

$$\frac{XY}{TU} = \frac{YZ}{UV} = \frac{ZX}{VT}$$

$$\frac{XY}{TU} = \frac{12}{16} = \frac{3}{4}$$

So, $\frac{15}{UV} = \frac{3}{4}$ and $\frac{18}{VT} = \frac{3}{4}$

$3 \cdot UV = 60$ $\quad$ $3 \cdot VT = 72$

$UV = 20$ $\quad$ $VT = 24$

Theorem 8.3 Congruent triangles are similar.

Theorem 8.4 **Transitive Property of Triangle Similarity:** If $\triangle ABC \sim \triangle DEF$ and $\triangle DEF \sim \triangle GHI$, then $\triangle ABC \sim \triangle GHI$.

Postulate 18 **AA Similarity Postulate:** If two angles of a triangle are congruent to two angles of another triangle, then the two triangles are similar.

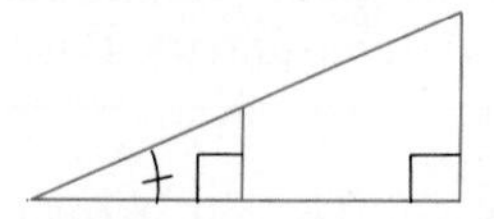

EXAMPLE 2 Write a correct similarity statement for the triangles if m $\angle X = 32$, m $\angle Z = 87$, m $\angle R = 87$, and m $\angle Q = 32$.

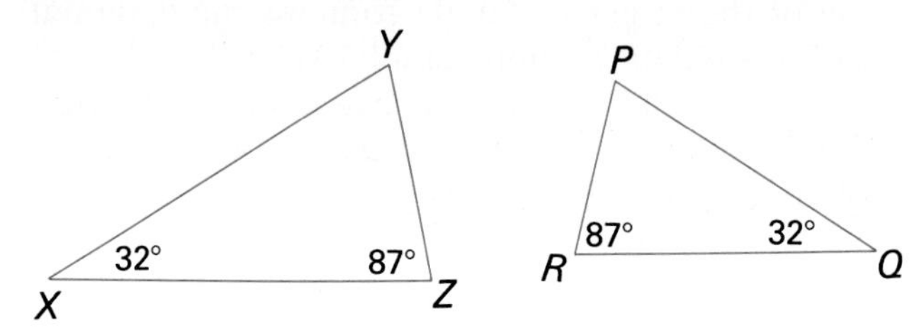

Solution By the AA Similarity Postulate, $\triangle XZY \sim \triangle QRP$.

EXAMPLE 3 Given: $\overleftrightarrow{RS} \parallel \overline{XY}$
Prove: $\triangle ZRS \sim \triangle ZXY$

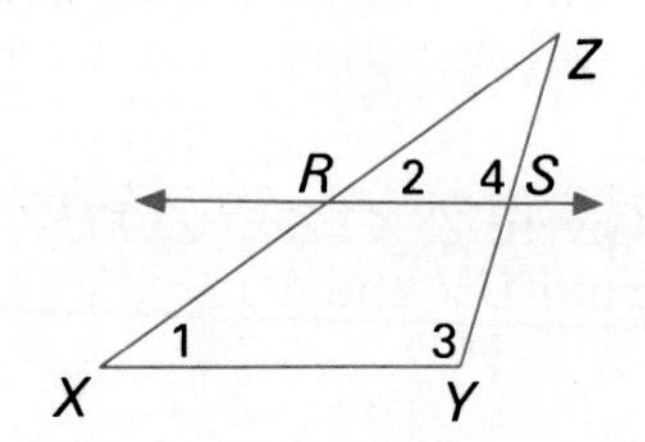

Proof

Statement	Reason
1. $\overleftrightarrow{RS} \parallel \overline{XY}$	1. Given
2. $\angle 1 \cong \angle 2$, $\angle 3 \cong \angle 4$	2. Corr $\angle$s of $\parallel$ lines are $\cong$.
3. $\therefore \triangle ZRS \sim \triangle ZXY$	3. AA $\sim$

Every proportion has a related pair of **cross products**: the product of the extremes and the product of the means. For example, in $\frac{AB}{CD} = \frac{EF}{GH}$, $AB \cdot GH$ and $CD \cdot EF$ are cross products. Given the cross products, you can write the corresponding proportion(s).

EXAMPLE 4 Write two proportions from $ZY \cdot AB = CD \cdot XW$.

Plan Choose the factors of one of the cross products to be the means of the proportion; the factors of the other cross product are the extremes.

Solution $\frac{ZY}{CD} = \frac{XW}{AB}$, or $\frac{XW}{ZY} = \frac{AB}{CD}$ (Other solutions are possible.)

EXAMPLE 5 Given: $\square XYZW$

Prove: $XQ \cdot PQ = ZQ \cdot RQ$

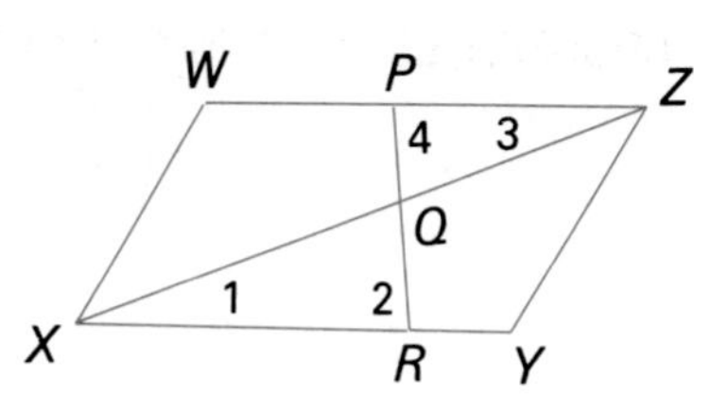

Plan Draw a flow diagram.

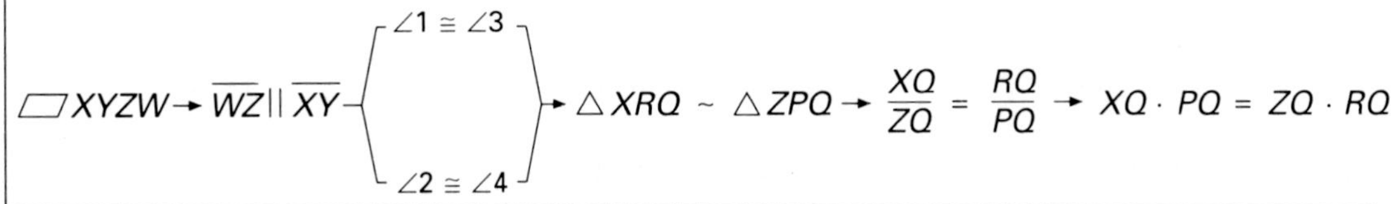

Classroom Exercises

State whether the two triangles are similar.

1.

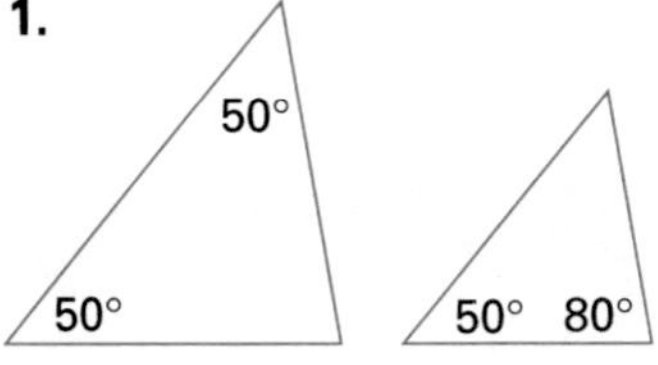

2.

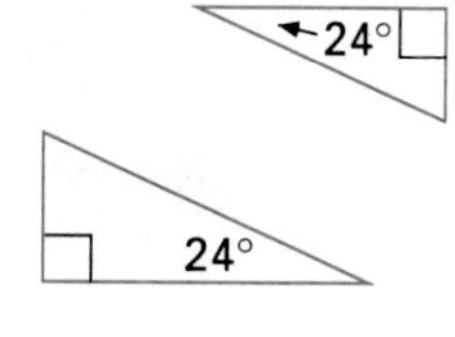

3.

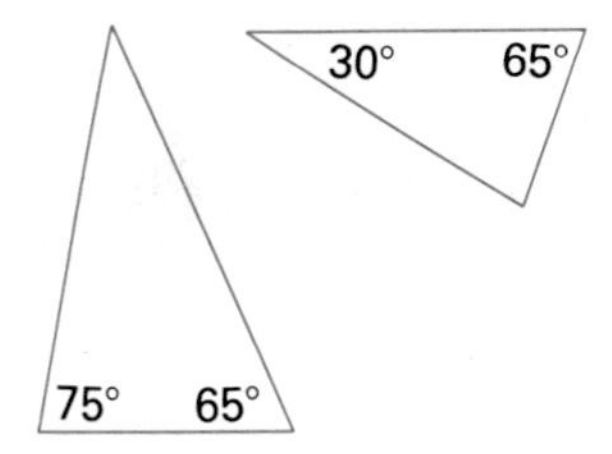

Complete the statement. $\triangle ABC \sim \triangle DEF$

4. $\angle A = \angle$ ____, $\angle B = \angle$ ____, $\angle C = \angle$ ____.

5. $\dfrac{AB}{?} = \dfrac{BC}{?} = \dfrac{CA}{?}$

Written Exercises

For Exercises 1–6, assume $\triangle ABC \sim \triangle DEF$. Find the missing measure.

1. $AB = 8$, $BC = 10$, $DE = 12$, $EF =$ ______

2. $AB = 12$, $CA = 16$, $DE = 9$, $FD =$ ______

3. $m\angle A = 80$, $m\angle B = 40$, $m\angle D =$ ______, $m\angle E =$ ______

4. $m\angle B = 65$, $m\angle C = 35$, $m\angle E =$ ______, $m\angle F =$ ______

5. $m\angle A = 40$, $m\angle B = 60$, $m\angle D =$ ______, $m\angle F =$ ______

6. $m\angle B = 35$, $m\angle C = 85$, $m\angle D =$ ______, $m\angle E =$ ______

For Exercises 7–10, use the diagram at right.

7. Given: $\overline{CD} \parallel \overline{AB}$
Prove: $\triangle ABE \sim \triangle DCE$

8. Given: $\angle A \cong \angle D$
Prove: $\triangle ABE \sim \triangle DCE$

9. Given: $\overline{CD} \parallel \overline{AB}$
Prove: $\dfrac{AB}{DC} = \dfrac{AE}{DE}$

10. Given: $\overline{CD} \parallel \overline{AB}$
Prove: $AE \cdot CE = BE \cdot DE$

Based on the given information, is $\triangle ABC \sim \triangle XYZ$?

11. $m\angle A = 60$, $m\angle B = 50$, $m\angle X = 60$, $m\angle Z = 70$

12. $m\angle A = 55$, $m\angle C = 45$, $m\angle Y = 80$, $m\angle Z = 45$

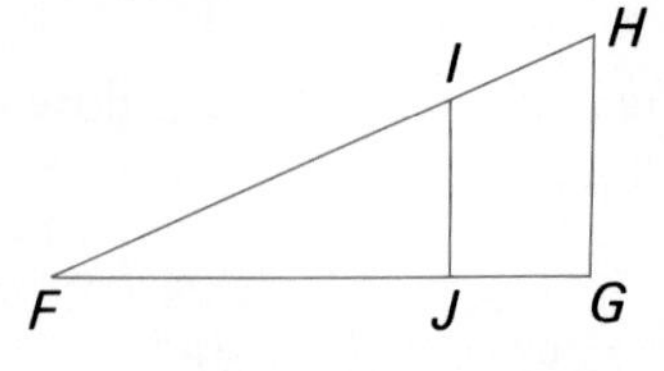

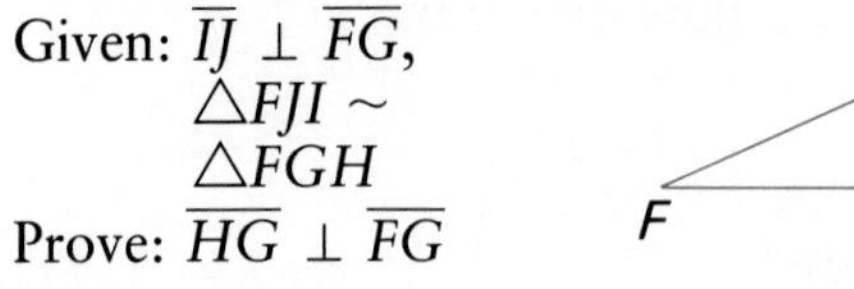

13. Given: $\overline{IJ} \perp \overline{FG}$, $\overline{HG} \perp \overline{FG}$
Prove: $\triangle FJI \sim \triangle FGH$

14. Given: $\overline{IJ} \perp \overline{FG}$, $\triangle FJI \sim \triangle FGH$
Prove: $\overline{HG} \perp \overline{FG}$

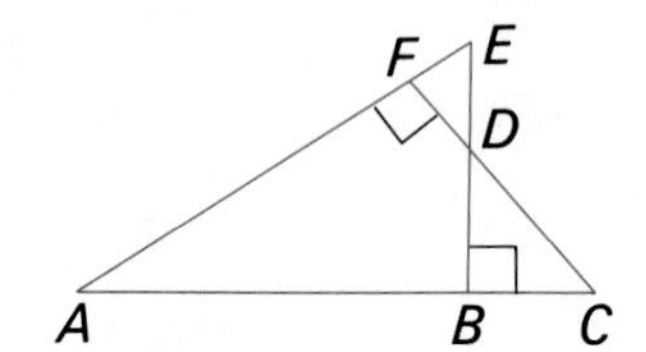

15. Given: $\overline{EB} \perp \overline{AC}$, $\overline{CF} \perp \overline{AE}$
Prove: $\triangle ABE \sim \triangle AFC$

16. Given: $\overline{EB} \perp \overline{AC}$, $\overline{CF} \perp \overline{AE}$
Prove: $\triangle BCD \sim \triangle FED$

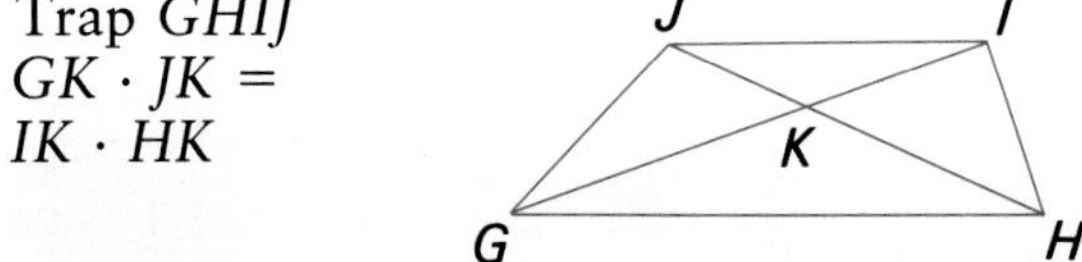

17. Given: Trap $GHIJ$
Prove: $GH \cdot JK = IJ \cdot HK$

18. Given: Trap $GHIJ$
Prove: $GK \cdot JK = IK \cdot HK$

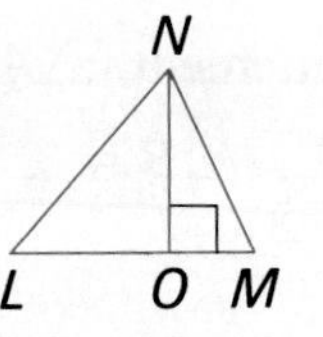

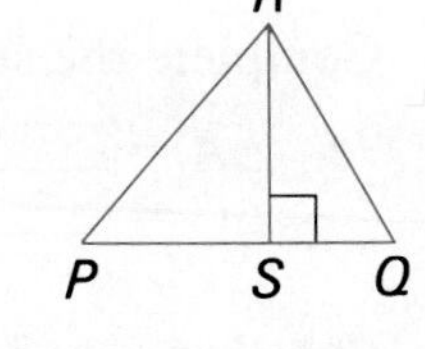

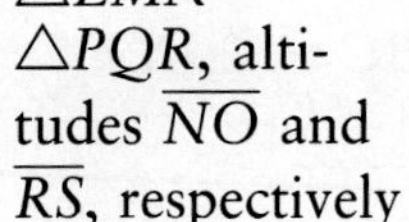

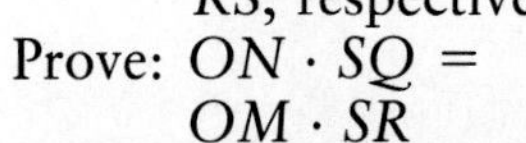

19. Given: $\triangle LMN \sim \triangle PQR$, altitudes $\overline{NO}$ and $\overline{RS}$, respectively
Prove: $\dfrac{NL}{RP} = \dfrac{NO}{RS}$

20. Given: $\triangle LMN \sim \triangle PQR$, altitudes $\overline{NO}$ and $\overline{RS}$, respectively
Prove: $ON \cdot SQ = OM \cdot SR$

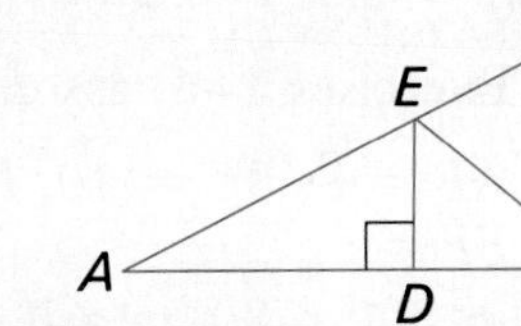

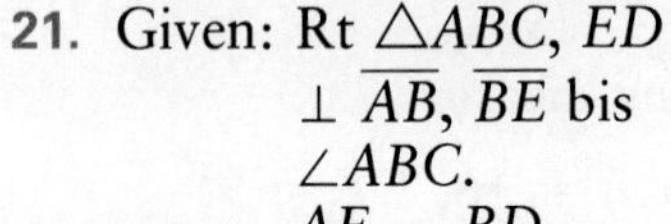

21. Given: Rt $\triangle ABC$, $\overline{ED} \perp \overline{AB}$, $\overline{BE}$ bis $\angle ABC$.
Prove: $\dfrac{AE}{AC} = \dfrac{BD}{BC}$

22. Given: Rt $\triangle ABC$, $\overline{ED} \perp \overline{AB}$, $\angle DEB \cong \angle EBA$
Prove: $\dfrac{1}{BC} + \dfrac{1}{AB} = \dfrac{1}{ED}$

23. Prove Theorem 8.3.

24. Prove Theorem 8.4.

25. Prove Example 5 using a two-column proof.

Midchapter Review

Is the proportion true or false? *8.1*

1. $\dfrac{2}{6} = \dfrac{9}{27}$

2. $\dfrac{3}{4} = \dfrac{3+4}{4+3}$

3. $\dfrac{5}{9} = \dfrac{2 \cdot 5}{2 \cdot 9}$

Solve the proportion for x. *8.1*

4. $\frac{3}{x} = \frac{6}{12}$

5. $\frac{1}{3} = \frac{x}{21}$

6. $\frac{2}{x} = \frac{3}{20}$

Find the geometric mean of the pair of positive numbers. *8.1*

7. 1 and 16

8. 5 and 6

9. p and q

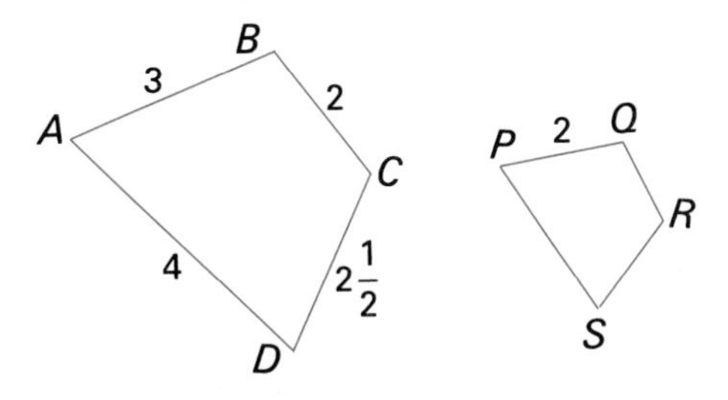

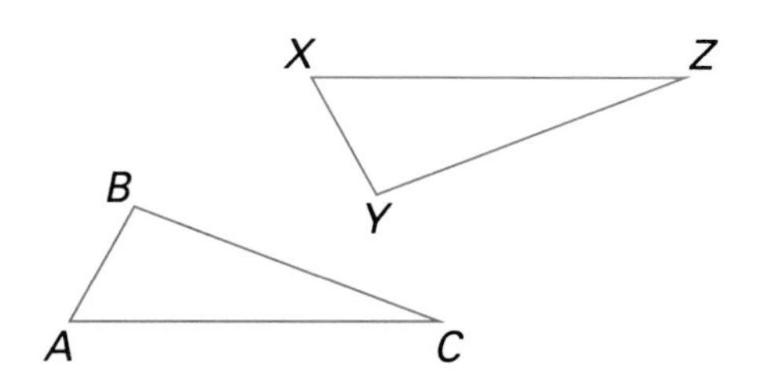

10. Find the missing lengths for the pair of similar quadrilaterals above.

11. Write a similarity statement for the triangles above. *8.3*
$m\angle B = 100$, $m\angle Y = 100$,
$m\angle A = 60$, $m\angle X = 60$

12. Given: $\overline{AB} \parallel \overline{RL}$. Write a similarity statement. Prove triangle similarity. *8.3*

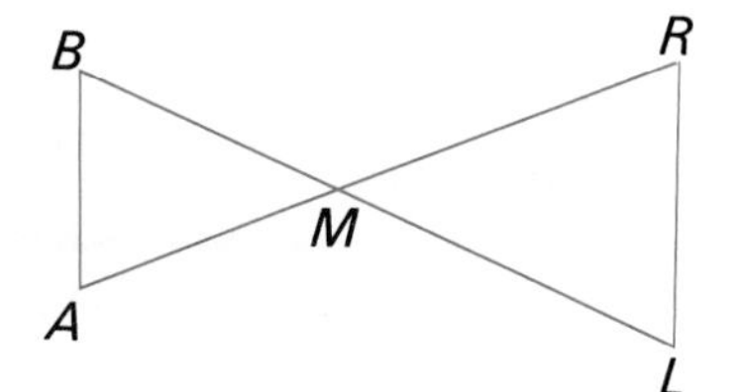

Application: *Finding the Height of a Tree*

One method of estimating the height of a tall tree involves properties of similar triangles. Place a mirror on level ground, as shown in the diagram, and stand where you can see the top of the tree in the mirror. Since the light bounces off a mirror at an angle equal to that at which it strikes the mirror, $\angle HRF \cong \angle TRB$. Assuming that the tree and you are both perpendicular to the ground, $\angle TBR \cong \angle HFR$; therefore, $\triangle TBR \sim \triangle HFR$ by AA Similarity. Measurements can then be made and the proportion solved for the height of the tree.

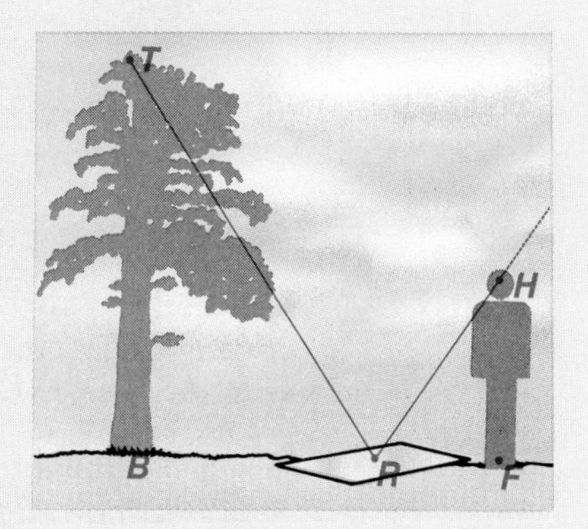

1. Suppose a person 6 ft tall places the mirror so that the top of the tree is visible. The mirror sits 20 ft from the tree and 1 ft from the person. How tall is the tree?

2. Suppose the person repeats the procedure with another tree and finds $BR = 20$, $RF = 2$. How tall is this tree?

3. Using a person's height is actually a simplification. What precisely should $\overline{HF}$ represent in the diagram?

8.4 The Triangle Proportionality Theorem

Objectives To prove theorems involving triangle proportionality
To apply the Triangle Proportionality Theorem and its converse

Two segments are divided proportionally when the lengths of the parts of one segment form a proportion with the lengths of the parts of the other segment.

Points P and Q divide the two segments below proportionally, since $\frac{2}{3} = \frac{4}{6}$. $\overline{AB}$ and $\overline{CD}$ are divided proportionally if $\frac{AP}{PB} = \frac{CQ}{QD}$.

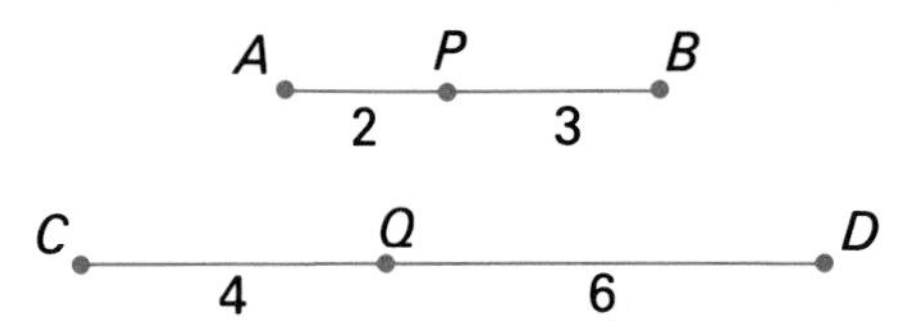

Theorem 8.5 **Triangle Proportionality Theorem:** If a line is parallel to one side of a triangle and intersects the other two sides, then it divides the two sides proportionally.

Given: $\triangle ABC$, with $\overleftrightarrow{DE} \parallel \overline{AB}$

Prove: $\frac{AD}{DC} = \frac{BE}{EC}$

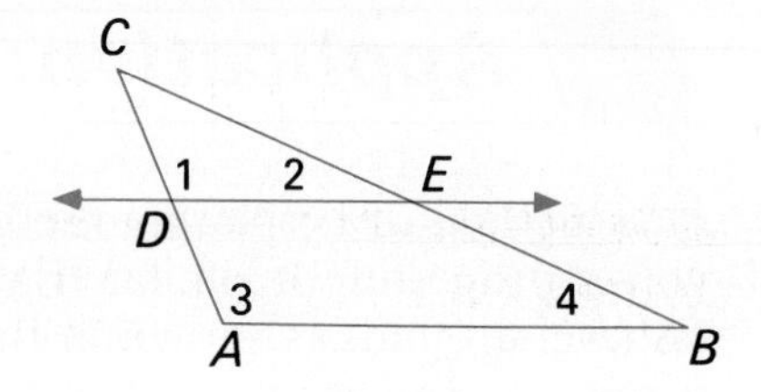

Proof

Statement	Reason
1. $\triangle ABC$ with $\overleftrightarrow{DE} \parallel \overline{AB}$	1. Given
2. $\angle 1 \cong \angle 3$, and $\angle 2 \cong \angle 4$	2. Corr $\angle$s of $\parallel$ lines are $\cong$.
3. $\triangle ABC \sim \triangle DEC$	3. AA $\sim$
4. $\frac{AC}{DC} = \frac{BC}{EC}$	4. Def of $\sim$ $\triangle$s
5. $\frac{AC - DC}{DC} = \frac{BC - EC}{EC}$	5. If $\frac{a}{b} = \frac{c}{d}$, then $\frac{a - b}{b} = \frac{c - d}{d}$
6. $AC - DC = AD$, $BC - EC = BE$	6. Seg Add Post
7. $\therefore \frac{AD}{DC} = \frac{BE}{EC}$	7. Sub

Theorem 8.2 leads to the following additional properties for $\triangle ABC$.

$$\frac{DC}{AC} = \frac{EC}{BC} \qquad \frac{DC}{AD} = \frac{EC}{BE} \qquad \frac{DC}{EC} = \frac{AD}{BE}$$

EXAMPLE 1 If $\overline{ST} \parallel \overline{RQ}$, which of the following are true?

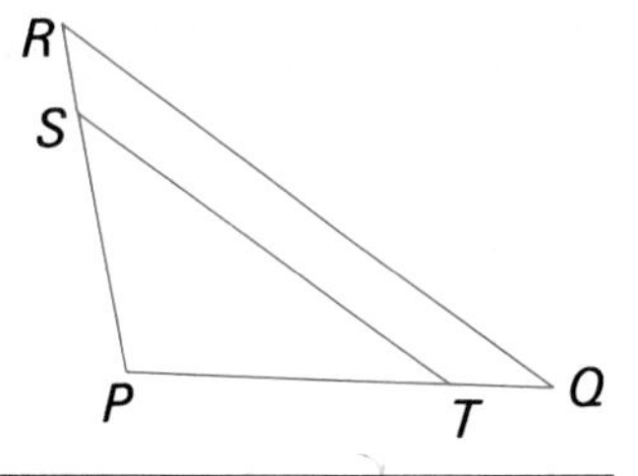

a. $\frac{PS}{SR} = \frac{PT}{TQ}$ b. $\frac{PS}{PR} = \frac{TQ}{PQ}$

Solution a. True, by Triangle Proportionality Theorem b. False (unless $\overline{ST}$ is midsegment)

EXAMPLE 2 $\overline{ED} \parallel \overline{AB}$, $AE = 16$, $EC = 12$

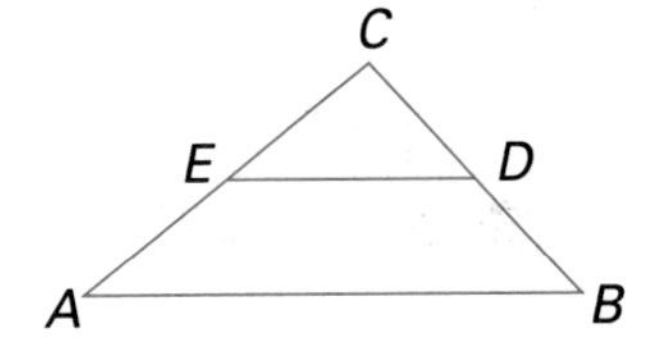

a. If $DC = 15$, find BD. b. If $BD = 9$, find BC.

Solutions

a.
$$\frac{AE}{EC} = \frac{BD}{DC}$$
$$\frac{16}{12} = \frac{BD}{15}$$
$$12 \cdot BD = 16 \cdot 15$$
$$BD = 20$$

b.
$$\frac{AE}{AC} = \frac{BD}{BC}$$
$$\frac{16}{28} = \frac{9}{BC}$$
$$16 \cdot BC = 28 \cdot 9$$
$$BC = 15\tfrac{3}{4}$$

The converse of the Triangle Proportionality Theorem can be proved by using an auxiliary line.

Theorem 8.6 If a line divides two sides of a triangle proportionally, then the line is parallel to the third side of the triangle.

Given: $\triangle ABC$ with D between A and C, and E between B and C, $\frac{AD}{DC} = \frac{BE}{EC}$

Prove: $\overleftrightarrow{DE} \parallel \overline{AB}$

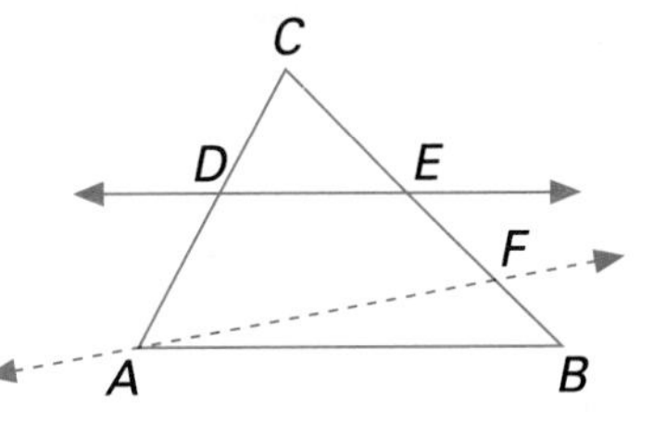

Plan Through point A, draw a line $\overleftrightarrow{AF}$ assumed parallel to $\overleftrightarrow{DE}$. Show that $BE = FE$.

EXAMPLE 3 Based on the information given, is $\overline{JK} \parallel \overline{HI}$?

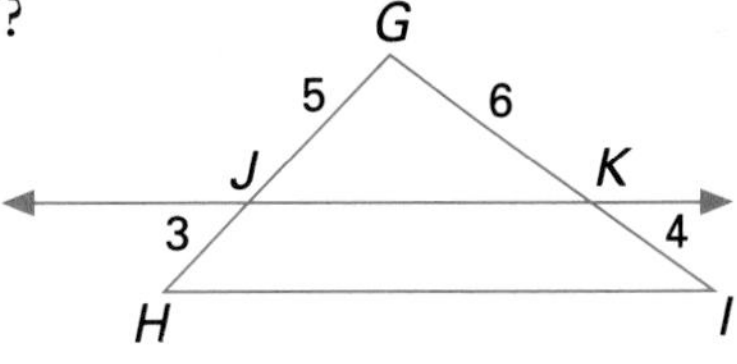

Solution If $\overline{JK} \parallel \overline{HI}$, then $\frac{JH}{JG} = \frac{KI}{KG}$

But this is not the case: $\frac{3}{5} \neq \frac{4}{6}$.

Therefore, $\overline{JK} \nparallel \overline{HI}$.

Classroom Exercises

For Exercises 1–4, $\overline{DE} \parallel \overline{AB}$. Complete the proportion.

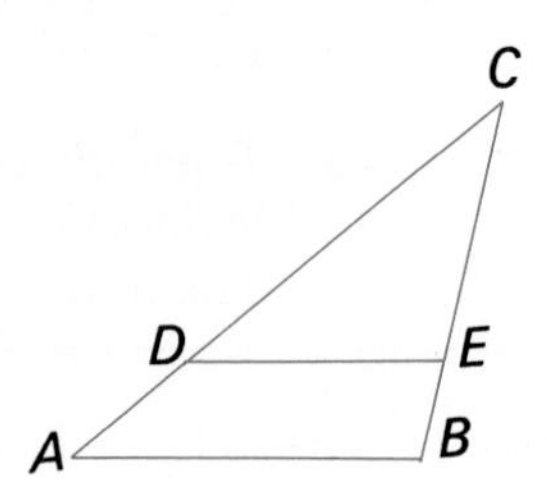

1. $\frac{CD}{CA} = \frac{CE}{?}$

2. $\frac{CE}{CB} = \frac{?}{CA}$

3. $\frac{?}{DC} = \frac{BC}{EC}$

4. $\frac{CD}{?} = \frac{CE}{EB}$

Written Exercises

Assume that the line intersecting the two sides of the triangle is parallel to the third side. Find the missing measure.

1.

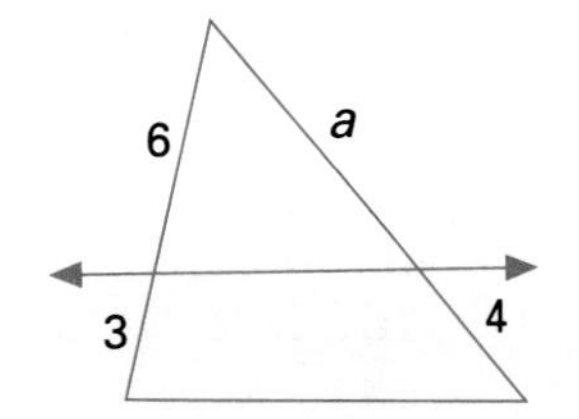

2.

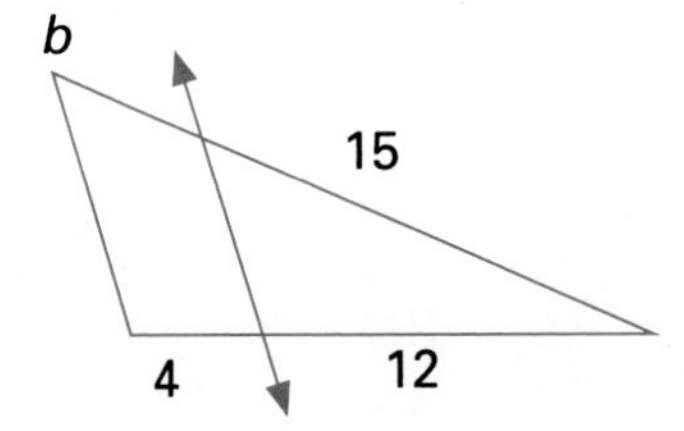

3.

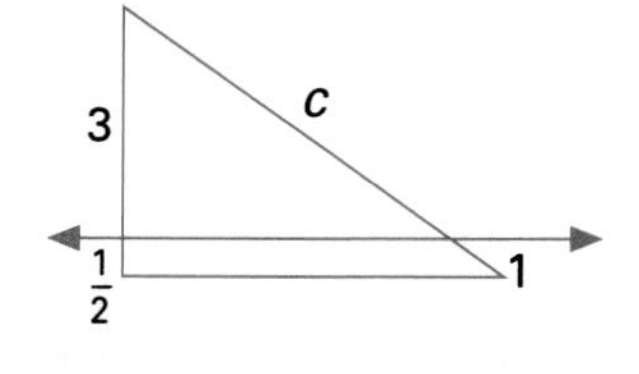

4.

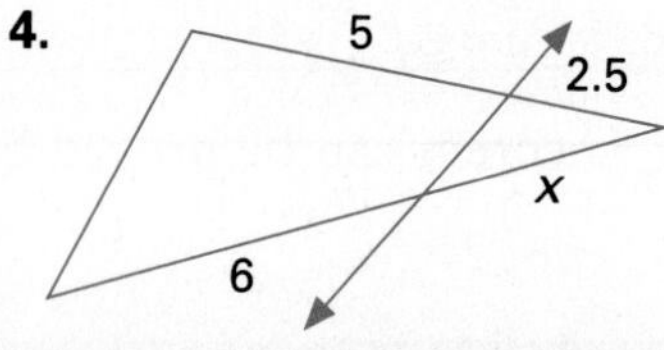

5.

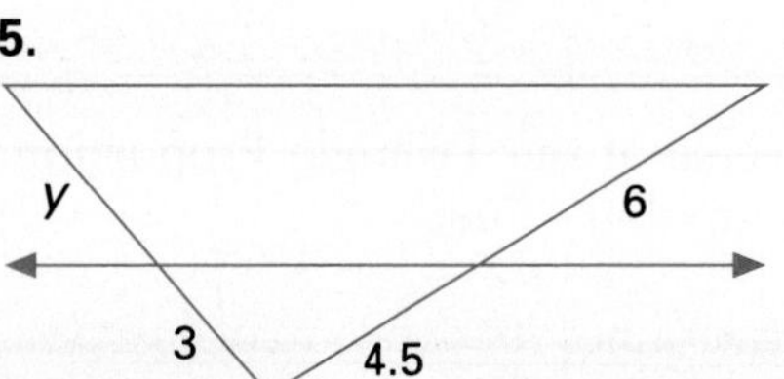

6.

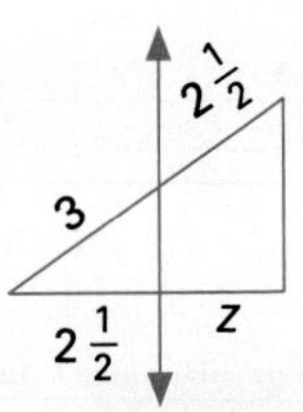

Based on the information given, is $\overline{IJ} \parallel \overline{FG}$?

7. $FI = 12, IH = 8, GJ = 15, JH = 10$
8. $FI = 15, IH = 20, GJ = 18, JH = 25$
9. $FH = 20, FI = 12, GH = 30, GJ = 18$
10. $FH = 18, FI = 10, GH = 27, JH = 15$

Determine whether the line(s) intersecting two of the sides of the triangle is(are) parallel to the third side.

11.

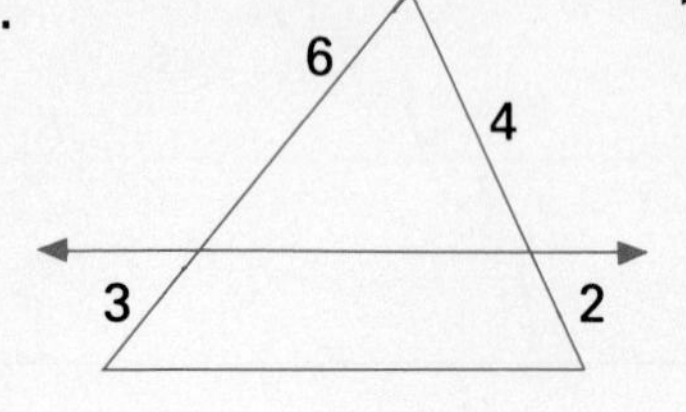

12.

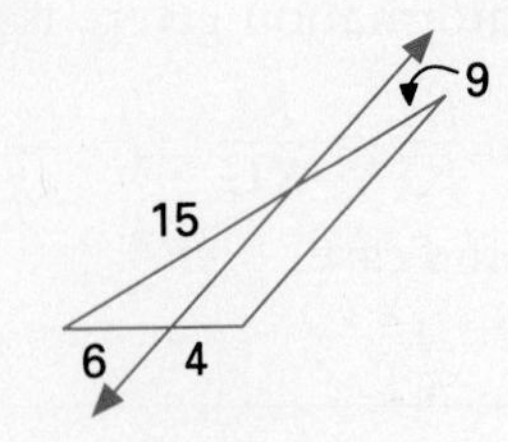

13.

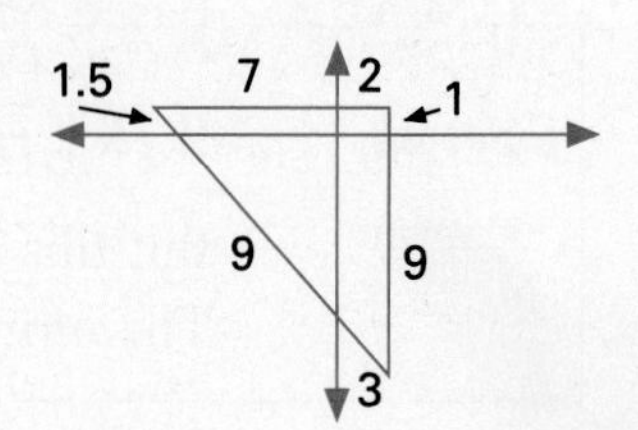

$\overline{IJ} \parallel \overline{FG}$. Find the indicated length.

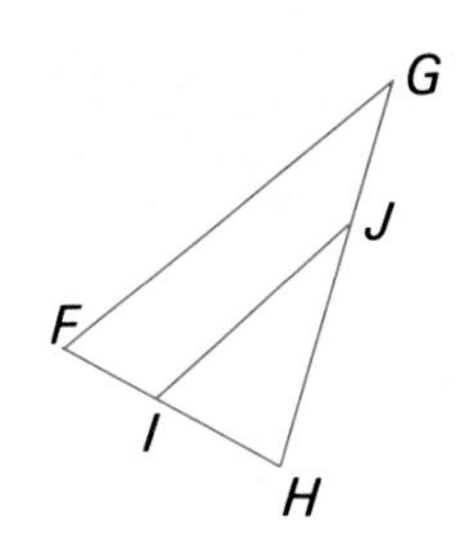

14. $HJ = 4, HG = 6, HI = 6, HF =$ ______

15. $HJ = 12, HG = 15, HF = 20, HI =$ ______

16. $HJ = 12, JG = 24, HI = 8, IF =$ ______

17. $HI = 15, IF = 10, JG = 12, HJ =$ ______

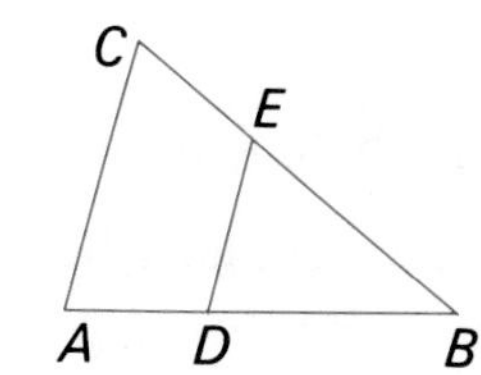

18. Given: $\frac{AD}{DB} = \frac{CE}{EB}$, $\angle BDE \cong \angle BED$

Prove: $\angle A \cong \angle C$

19. Given: $\angle A \cong \angle C$, $\angle BDE \cong \angle BED$

Prove: $\frac{AD}{DB} = \frac{CE}{EB}$

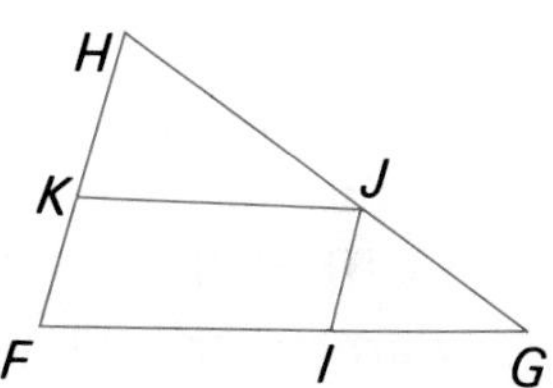

20. Given: $\overline{KJ} \parallel \overline{FG}$, $\overline{IJ} \parallel \overline{FH}$

Prove: $\frac{FK}{FH} = \frac{IG}{FG}$

21. Complete the proof of Theorem 8.6 in a two-column form.

Solve the following proportions, correct to the nearest tenth, using a calculator.

22. $\frac{25}{x} = \frac{49}{127}$ **23.** $\frac{17}{55} = \frac{x}{9}$ **24.** $\frac{101}{36} = \frac{74}{x}$

25. Prove: If three parallel lines intersect two transversals, then they divide the transversals proportionally.

Prove or disprove with a counterexample.

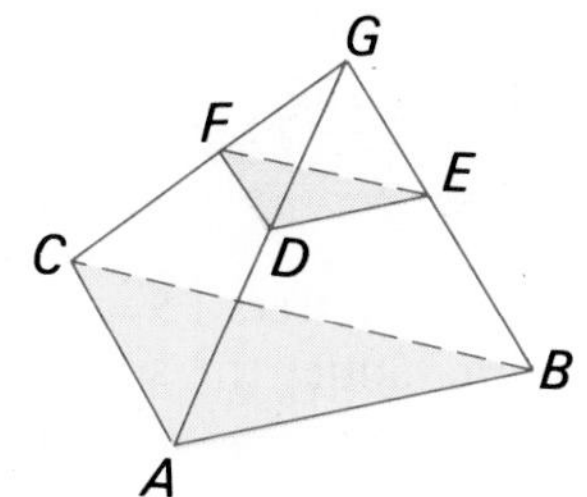

26. Given: $\overline{DE} \parallel \overline{AB}$, $\overline{FE} \parallel \overline{CB}$

Prove or disprove: $\overline{FD} \parallel \overline{CA}$

27. Given: $\overline{FD} \parallel \overline{CA}$, $\overline{DE} \parallel \overline{AB}$

Prove or disprove: $\frac{GF}{FC} = \frac{GE}{EB}$

28. Give a proof of the Midsegment Theorem based on Theorem 8.6, the converse of the Triangle Proportionality Theorem.

Mixed Review

1. State the Symmetric Property of Equality. *2.3*
2. State the Transitive Property of Congruence. *2.3*
3. Draw a concave pentagon. *6.1*
4. Draw a convex polygon. *6.1*

Computer Investigation

Similar Triangles

Use a computer software program that constructs triangles by definition of SAS and SSS, labels points, and measures line segments and angles.

Activity 1

Draw the following triangles.

1. Construct and label $\triangle ABC$ with $AC = 2$, AB = 1, and m $\angle A = 40$.
2. Use the measuring tool of the software program to find CB, m $\angle B$, and m $\angle C$.
3. Construct a new triangle and label it $A'B'C'$ with $A'C' = 4$, $A'B' = 2$, and m $\angle A = 40$.
4. Find $C'B'$, m $\angle B'$, and m $\angle C'$.
5. Find $C'B'$: CB.
6. Why are the triangles now similar?

Activity 2

7. Construct $\triangle ABC$ with $AC = 3$, $AB = 2$, and m $\angle A = 30$.
8. Find the measure of each of the remaining parts.
9. Construct a new triangle and label it $\triangle A'B'C'$ with $A'B' = 6$, $A'C' = 9$, and m $\angle A = 30$.
10. Find the measures of each of the remaining parts.
11. Why are the two triangles now similar?

Activity 3

Lower case letters refer to the sides of the triangle opposite the angle of the same name; the side can also be named by the adjacent vertices. (Side a is opposite $\angle A$, also referred to as $\overline{BC}$.)

12. Construct and label $\triangle ABC$ with sides: $a = 2$, $b = 4$, and $c = 3$.
13. Find the measure of each angle.
14. Construct and label a new $\triangle A'B'C'$ with each side measuring twice the corresponding side lengths of $\triangle ABC$.
15. Find the measure of each angle of $\triangle A'B'C'$.
16. Why are the triangles now similar?
17. Repeat the steps of Exercises 13–17 for two new triangles with sides: $a = 3$, $b = 4$, $c = 5$; and $a' = 6$, $b' = 8$, $c' = 10$.
18. Try to generalize a new way to show triangles similar.

8.5 SAS and SSS Similarity Theorems

Objective To apply the SAS and SSS Similarity Theorems

Given $\triangle ABC$, you can construct $\triangle XYZ$, so that $\angle X \cong \angle A$ and the sides $\overline{XY}$ and $\overline{XZ}$ are twice as long as sides $\overline{AB}$ and $\overline{AC}$, respectively. The two triangles will appear to be similar. This suggests the following theorem.

Theorem 8.7

SAS Similarity Theorem: If an angle of one triangle is congruent to an angle of another triangle and the corresponding sides that include these angles are proportional, then the triangles are similar.

Given: $\angle C \cong \angle F$, $\frac{AC}{DF} = \frac{BC}{EF}$

Prove: $\triangle ABC \sim \triangle DEF$

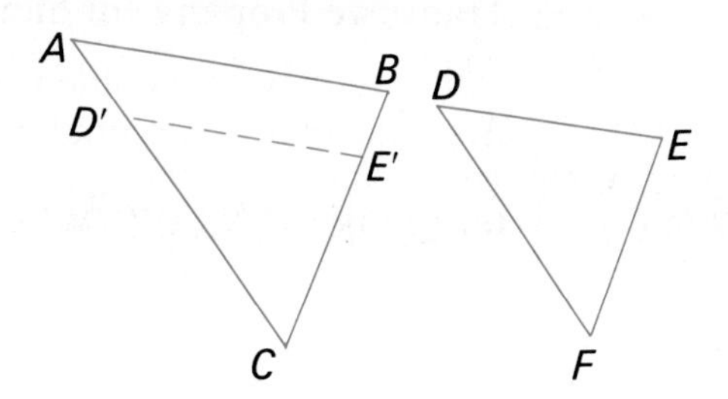

Proof

Statement	Reason
1. $\angle C \cong \angle F$	(A) 1. Given
2. Choose D' on $\overline{AC}$ so that $\overline{D'C} \cong \overline{DF}$ $(D'C = DF)$	(S) 2. Ruler Post
3. Choose E' on $\overline{BC}$ so that $\overline{E'C} \cong \overline{EF}$ $(E'C = EF)$	(S) 3. Ruler Post
4. $\triangle D'E'C \cong \triangle DEF$	4. SAS
5. $\triangle D'E'C \sim \triangle DEF$	5. $\cong$ $\triangle$s are similar.
6. $\frac{AC}{DF} = \frac{BC}{EF}$	6. Given
7. $\frac{AC}{D'C} = \frac{BC}{E'C}$	7. Sub
8. $\overline{D'E'} \parallel \overline{AB}$	8. If line divides two sides of $\triangle$ proport, it is $\parallel$ to third side.
9. $\angle CD'E' \cong \angle A$ and $\angle CE'D' \cong \angle B$	9. Corr $\angle$s of $\parallel$ lines are $\cong$.
10. $\triangle ABC \sim \triangle D'E'C$	10. AA $\sim$
11. $\therefore \triangle ABC \sim \triangle DEF$	11. Trans Prop of $\triangle$ $\sim$

This SAS similarity pattern is suggested by the SAS congruence theorem.

Another similarity pattern suggested by congruence is an SSS pattern.

Theorem 8.8 **SSS Similarity Theorem:** If all three pairs of corresponding sides of two triangles are proportional, then the two triangles are similar.

Given: $\frac{CA}{FD} = \frac{AB}{DE} = \frac{CB}{FE}$
Prove: $\triangle ABC \sim \triangle DEF$

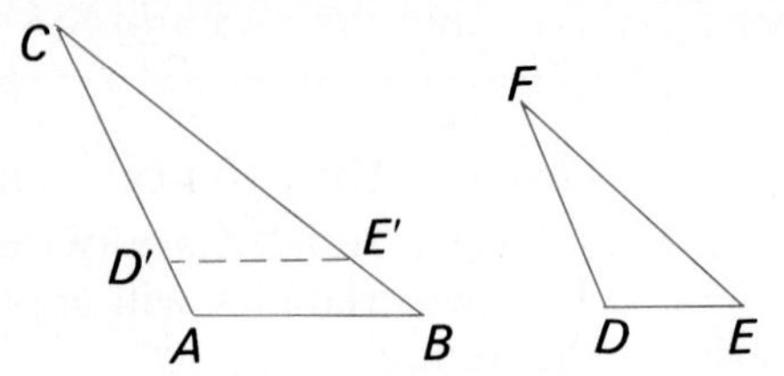

Plan Assume $CA > FD$. Locate D' on $\overline{CA}$ and E' on $\overline{CB}$ so that $\overline{CD'} \cong \overline{FD}$ and $\overline{CE'} \cong \overline{FE}$. Use SAS Similarity to prove $\triangle ABC \sim \triangle D'E'C$. Prove $D'E' \cong DE$ by using these similar triangles and the given. Then $\triangle D'E'C \cong \triangle DEF$ by SSS. Since congruent triangles are similar, the Transitive Property for Similar Triangles completes the proof.

You now have available three methods for proving triangles similar: AA Similarity, SAS Similarity, SSS Similarity.

EXAMPLE 1 Is $\triangle PQR \sim \triangle XYZ$? Why or why not?

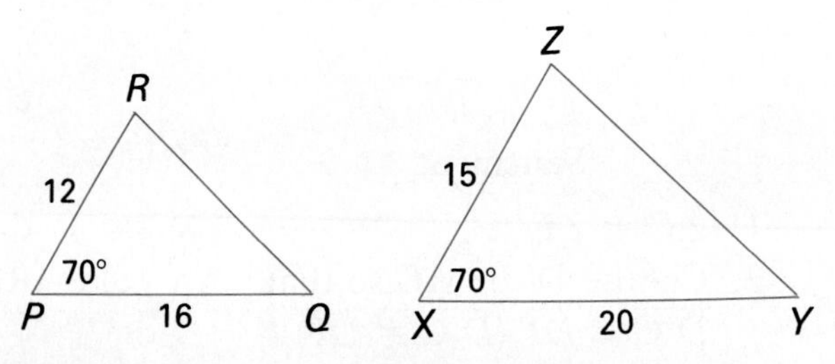

Solution m $\angle P = 70$ and m $\angle X = 70$, so $\angle P \cong \angle X$

$\frac{12}{15} = \frac{4}{5}$ and $\frac{16}{20} = \frac{4}{5}$, so $\frac{PR}{XZ} = \frac{PQ}{XY}$

Thus, $\triangle PQR \sim \triangle XYZ$ by SAS Similarity.

Some special triangles are easily shown to be similar. All angles of equiangular triangles have measure of 60. Therefore, all equiangular triangles are similar by the AA Similarity Postulate. Isosceles right triangles can be proved similar by the SAS Similarity Theorem.

EXAMPLE 2 Prove all equilateral triangles are similar.
Given: Equilateral $\triangle$s ABC and DEF
Prove: $\triangle ABC \sim \triangle DEF$

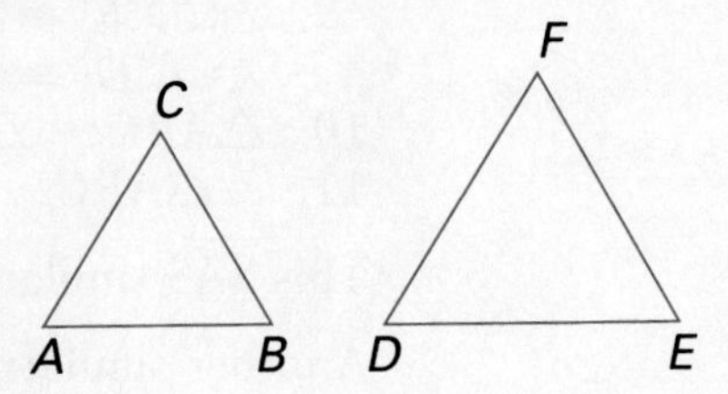

Proof

Statement	Reason
1. $\triangle ABC$ and $\triangle DEF$ are equilateral.	1. Given
2. $AB = BC = CA$, $DE = EF = FD$	2. Def of equil $\triangle$
3. $\frac{AB}{DE} = \frac{BC}{DE} = \frac{CA}{DE}$	3. Div Prop of Eq
4. $\frac{AB}{DE} = \frac{BC}{EF} = \frac{CA}{FD}$	4. Sub
5. $\therefore \triangle ABC \sim \triangle DEF$	5. SSS $\sim$

EXAMPLE 3 Given: $\frac{KI}{HI} = \frac{JI}{GI}$

Prove: $KJ \cdot HI = HG \cdot KI$

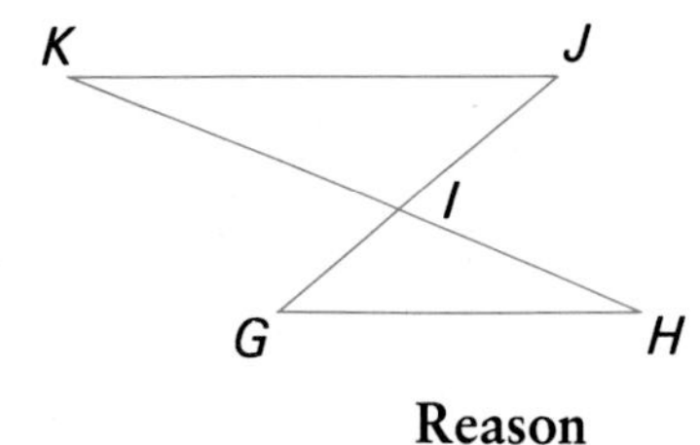

Proof

Statement	Reason
1. $\frac{KI}{HI} = \frac{JI}{GI}$	1. Given
2. $\angle KIJ \cong \angle HIG$	2. Vert $\angle$s are $\cong$.
3. $\triangle KIJ \sim \triangle HIG$	3. SAS $\sim$
4. $\frac{KJ}{HG} = \frac{KI}{HI}$	4. Def of similar $\triangle$s
5. $\therefore KJ \cdot HI = HG \cdot KI$	5. In a proportion, the prod of extremes = the prod of means.

Focus On Reading

Select the correct spelling of each word.

1.	allitude	altatude	altitude
2.	congruant	congruent	congerant
3.	coresponding	coressponding	corresponding
4.	extremes	extreems	extreemes
5.	similiar	simular	similar

Classroom Exercises

Determine whether the two triangles are similar. Explain why or why not.

1.

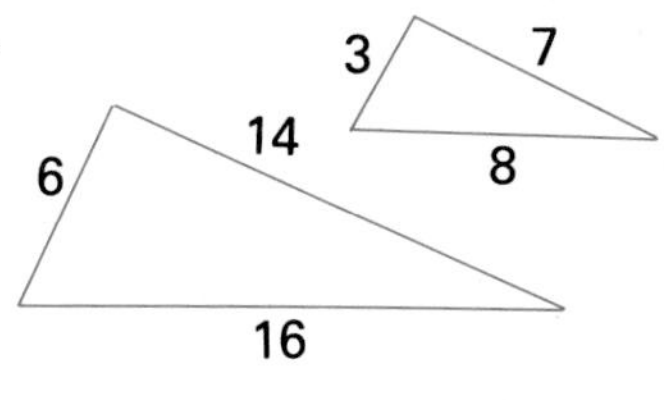

2.

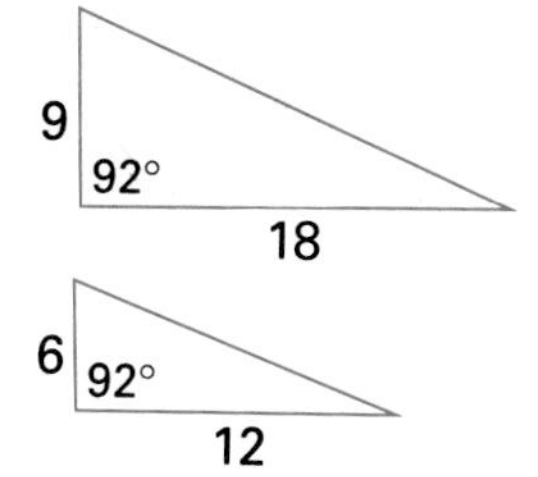

3.

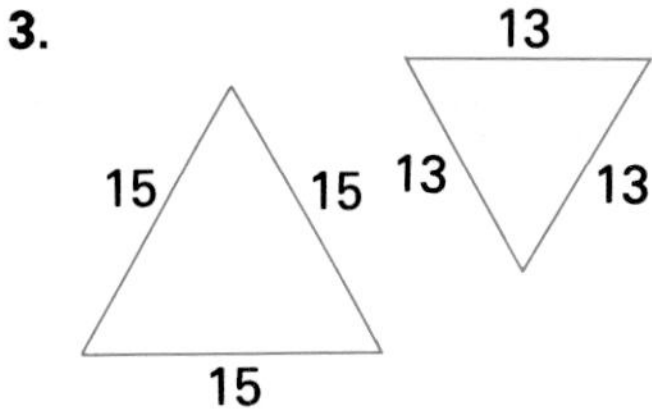

4.

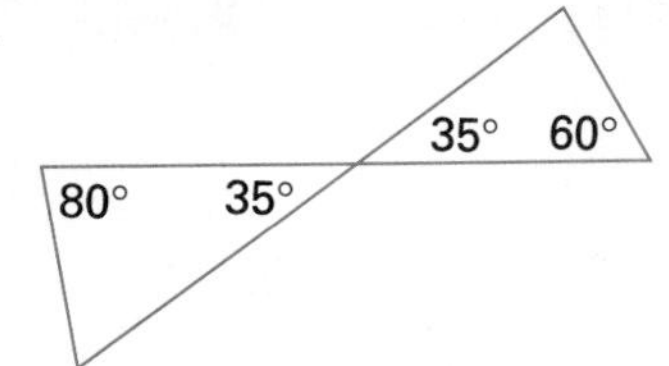

5.

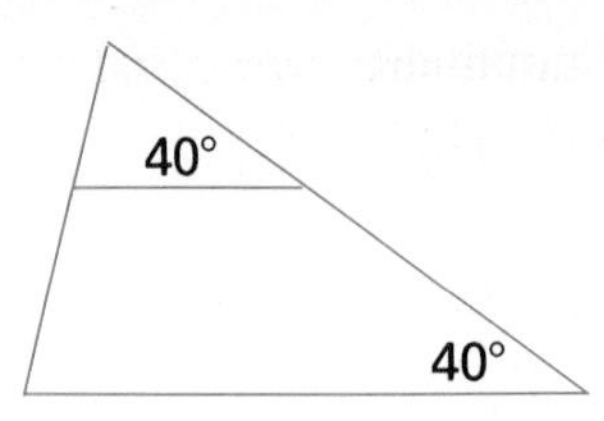

6.

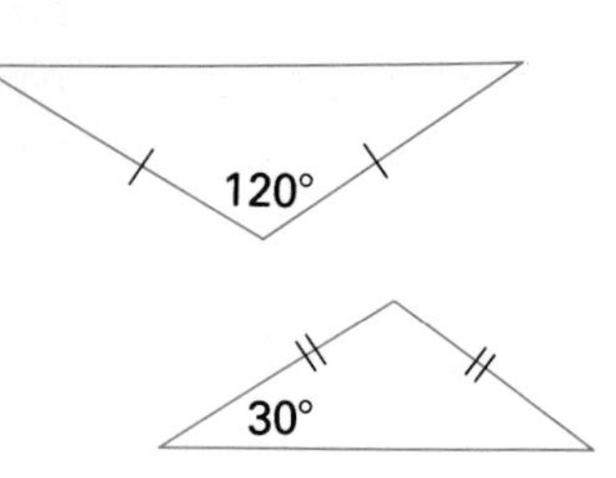

Written Exercises

On the basis of the given information, is $\triangle ABC \sim \triangle DEF$? Explain. (Exercises 1–8)

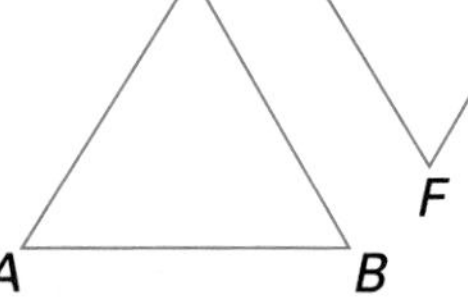

1. $\angle A \cong \angle D, \angle B \cong \angle E$

2. $\frac{AB}{DE} = \frac{BC}{EF} = \frac{CA}{FD}$

3. $\angle A \cong \angle D \ \frac{AB}{DE} = \frac{BC}{EF}$

4. $\frac{BC}{EF} = \frac{CA}{FD}$

5. m $\angle A = 40$, m $\angle B = 40$, m $\angle D = 40$, m $\angle F = 100$

6. $\frac{AC}{DF} = \frac{AB}{DE}$, $\angle A \cong \angle D$

7. $\frac{AC}{DE} = \frac{AB}{DF} = \frac{AB}{EF}$

8. $\triangle ABC$ is equilateral, $\angle D \cong \angle F$

9. Prove that all equiangular triangles are similar.

10. Prove that all isosceles right triangles are similar.

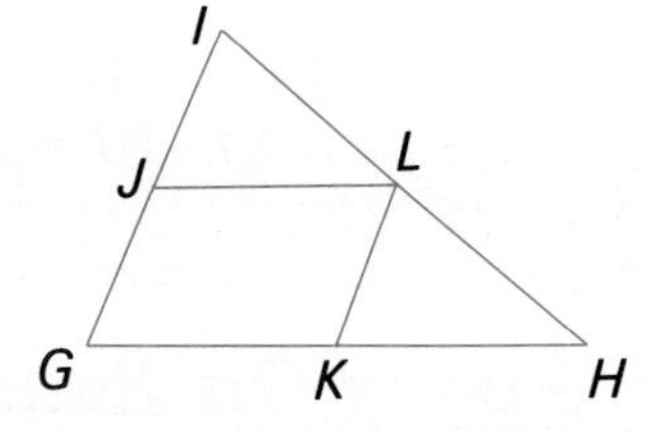

11. Given: $\square GKLJ$
Prove: $\triangle JLI \sim \triangle KHL$

12. Given: $\triangle KHL \sim \triangle JLI$
Prove: $\triangle KHL \sim \triangle GHI$

13. Given: $\frac{NQ}{NO} = \frac{NP}{NM}$
Prove: $\frac{NQ}{NO} = \frac{QP}{OM}$

14. Given: $\frac{NP}{NM} = \frac{QN}{ON}$
Prove: $QP \cdot NM = OM \cdot NP$

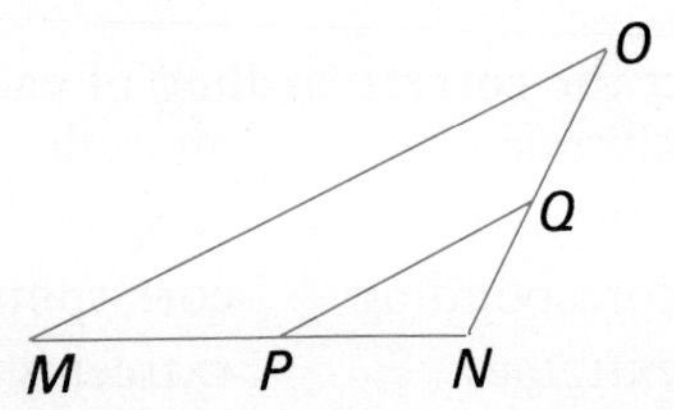

15. Prove that two isosceles triangles are similar if the vertex angle of one is congruent to the vertex angle of the other.

16. Prove that two isosceles triangles are similar if the base and a leg of one are proportional to the base and a leg of the other.

17. Prove Theorem 8.8 for $CA > FD$.

18. Prove or disprove: The triangle formed by connecting the midpoints of the sides of a triangle is similar to the given triangle.

19. Prove or disprove: The triangle formed by perpendiculars to the sides of a triangle is similar to the given triangle $\triangle ABC \sim \triangle EDF$.

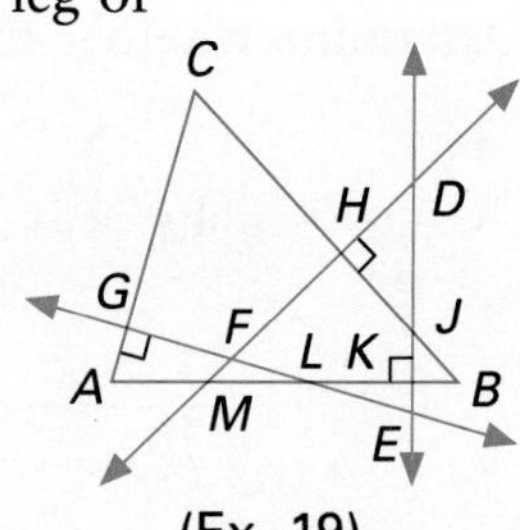

(Ex. 19)

For which of the following patterns can two quadrilaterals be proved similar? Explain your reasoning step-by-step.

20. SASAS Similarity

21. ASASA Similarity

22. The steepness of a road is measured by its *grade*. A three-percent grade means that the road rises 3 ft for every 100 ft of horizontal length. If you drive up a three-percent grade for 2 mi, how many ft higher are you than when you started (1 mi = 5,280 ft)?

Mixed Review

True or false?

1. A median of a triangle, except for its endpoints, is always in the interior of a triangle. 5.5
2. An altitude of a triangle, except for its endpoints, is always in the interior of a triangle. 5.5
3. An angle bisector of a triangle cannot be an altitude of the triangle. 5.5
4. A median of a triangle cannot be an altitude of the triangle. 5.5
5. A triangle cannot be constructed with sides 2, 3, and 5. 5.5

Algebra Review

To graph the solution set of an inequality, locate the points on the number line that correspond to the solutions.

Example: Graph the solution set of $x < 2$ *and* $x \geq -1$.

Graph each part separately, then determine the intersection.

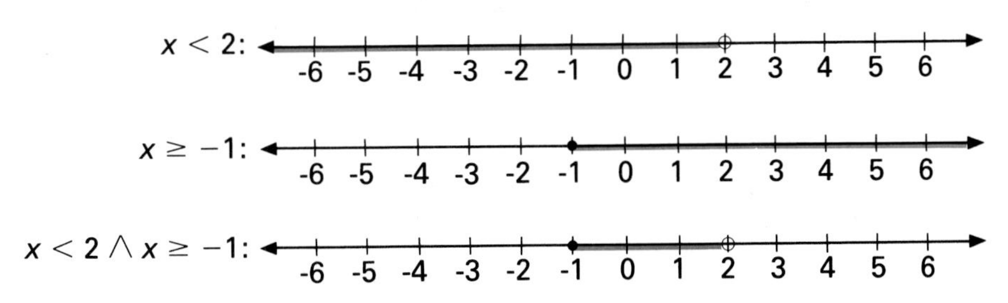

Graph the solution set of the inequality.

1. $x > 3$
2. $x < 5$
3. $x > -5$
4. $x < -3$
5. $a < 6$
6. $t > 0$
7. $s > -1$
8. $m < -3$
9. $x > -2$ *and* $x < 1$
10. $x < 0$ *and* $x > -2$
11. $x < 6$ *and* $x > 1$
12. $x > 0$ *and* $x < 4$
13. $x < -2$ *and* $x < 3$
14. $x > -2$ *and* $x > 0$

8.6 Segments in Similar Triangles

Objectives To apply theorems concerning proportional parts in triangles
To construct segments with lengths in a given ratio

Theorem 8.9 Corresponding altitudes of similar triangles are proportional to corresponding sides.

Given: $\triangle ABC \sim \triangle DEF$, altitudes $\overline{CG}$ and $\overline{FH}$

Prove: $\frac{CG}{FH} = \frac{AC}{DF}$

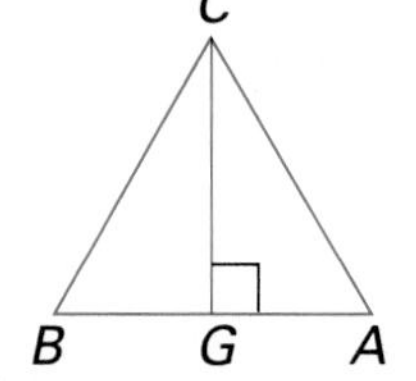

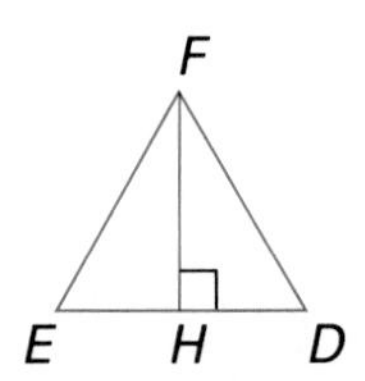

Proof

Statement	Reason
1. $\triangle ABC \sim \triangle DEF$, altitudes $\overline{CG}$ and $\overline{FH}$	1. Given
2. $\angle CGA$ and $\angle FHD$ are rt $\angle$s.	2. Def of alt
3. $\angle CGA \cong \angle FHD$	(A) 3. All rt $\angle$s are $\cong$.
4. $\angle A \cong \angle D$	(A) 4. Def of similar $\triangle$s
5. $\triangle CGA \sim \triangle FHD$	5. AA $\sim$
6. $\therefore \frac{CG}{FH} = \frac{AC}{DF}$	6. Def of similar $\triangle$s

Theorem 8.10 Corresponding medians of similar triangles are proportional to corresponding sides.

EXAMPLE 1 $\triangle LMN \sim \triangle PQR$, $\overline{NW}$ and $\overline{RO}$ are medians. Find NW and RO.

Solution

$$\frac{9}{12} = \frac{x + 4}{2x + 2}$$
$$9(2x + 2) = 12(x + 4)$$
$$18x + 18 = 12x + 48$$
$$6x = 30$$
$$x = 5 \text{ (So, } x + 4 = 9 \text{ and } 2x + 2 = 12)$$

Thus, $NW = 9$ and $RO = 12$.

Theorem 8.11 The bisector of an angle of a triangle divides the opposite side of the triangle into segments proportional to the other two sides.

Given: $\triangle ABC$, $\overline{AD}$ bisects $\angle CAB$.

Prove: $\dfrac{DC}{DB} = \dfrac{AC}{AB}$

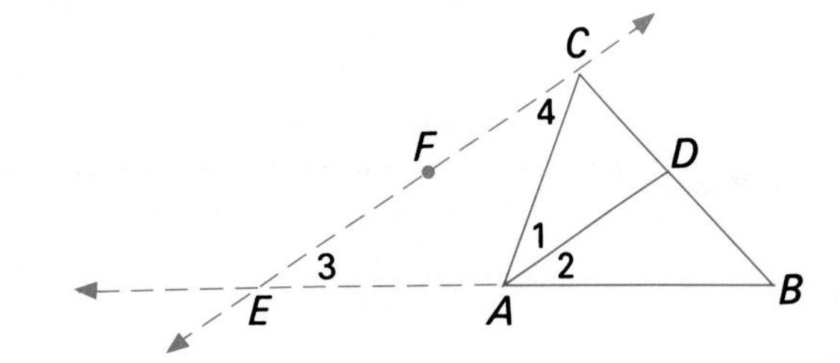

Plan

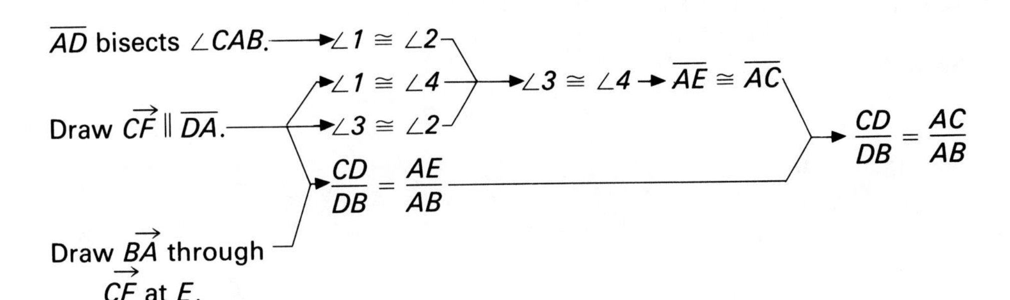

EXAMPLE 2 Given: $\overline{CD}$ bisects $\angle ACB$.
$AC = 12$, $BC = 28$, $AB = 20$
Find AD and DB.

Plan Let $x = AD$. Then $20 - x = DB$

Solution

$$\frac{AD}{DB} = \frac{12}{28}$$
$$\frac{x}{20 - x} = \frac{12}{28}$$
$$28x = 12(20 - x)$$
$$28x = 240 - 12x$$
$$40x = 240$$
$$x = 6$$

Therefore, $AD = 6$ and $DB = 14$.

Construction To divide a segment into two segments whose lengths have a ratio $\frac{2}{3}$.

Given: $\overline{AB}$

Construct P on $\overline{AB}$ so that $\dfrac{AP}{PB} = \dfrac{2}{3}$.

Draw any segment $\overline{XY}$.
Construct segments of lengths $2(XY)$ and $3(XY)$.

Construct $\triangle ABC$ so that $AC = 2(XY)$ and $BC = 3(XY)$. Bisect $\angle ACB$. Draw the angle bisector intersecting $\overline{AB}$ at P.

Result: $\frac{AP}{PB} = \frac{2}{3}$

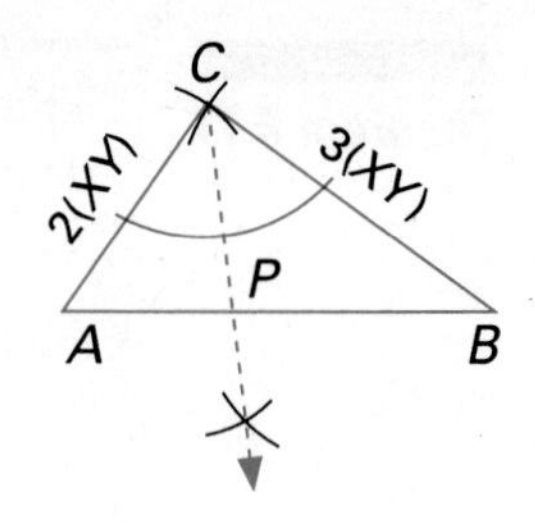

Classroom Exercises

Solve for x.

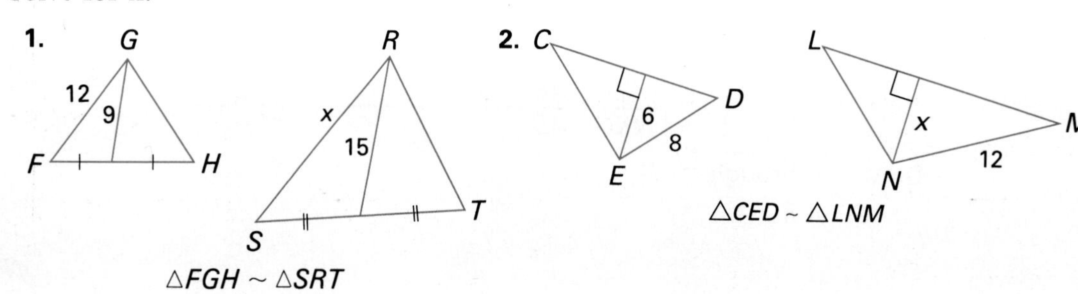

Written Exercises

$\triangle ABC \sim \triangle EFG$, $\overline{CD}$ and $\overline{GH}$ are altitudes. Find the indicated measure.

1. $AC = 16$, $EG = 12$, $CD = 8$, $GH =$ ______
2. $AC = 18$, $EG = 21$, $CD =$ ______, $GH = 14$
3. $AC = 14$, $EG = 21$, $CD = 3x + 2$, $GH = 7x - 2$, $CD =$ ______
4. $AC = 10$, $EG = 15$, $CD = x + 3$, $GH = 2x + 2$, $GH =$ ______

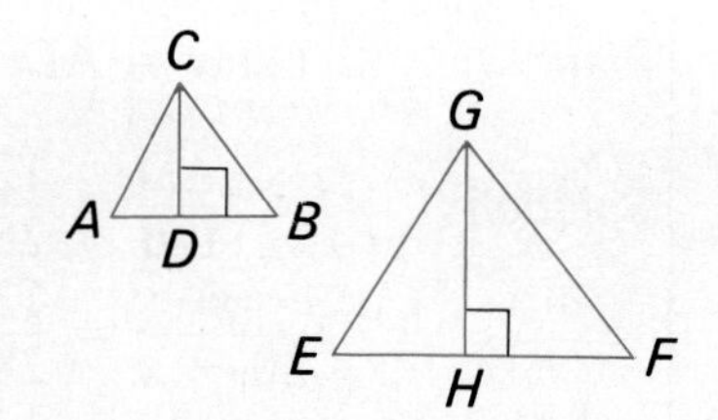

$\triangle IJK \sim \triangle MNO$, $\overline{KL}$ and $\overline{OP}$ are medians. Find the indicated measure.

5. $KI = 25$, $OM = 30$, $KL = 20$, $OP =$ ______
6. $KI = 25$, $OM = 35$, $OP = 14$, $KL =$ ______
7. $KI = 24$, $OM = 30$, $KL = 3x + 4$, $OP = 6x - 4$, $KL =$ ______
8. $KI = 6$, $OM = 9$, $KL = x + 3$, $OP = 2x + 1$, $OP =$ ______

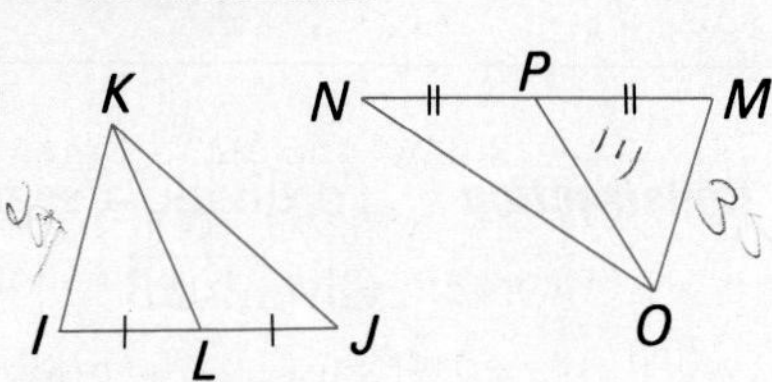

Given: $\overline{ST}$ bisects $\angle RSQ$ of $\triangle QRS$. Find the indicated measure.

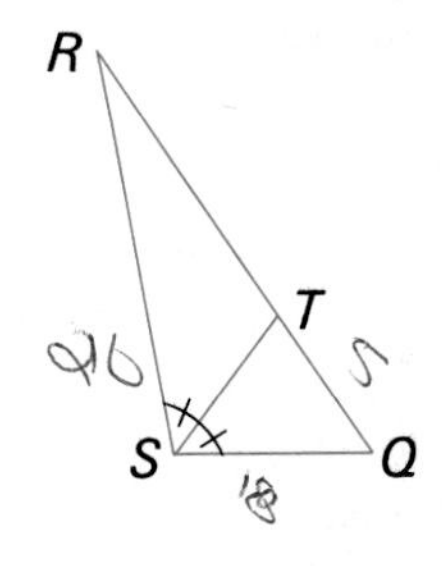

9. $QS = 6$, $RS = 9$, $QT = 2$, $TR =$ ______

10. $QS = 8$, $RS = 10$, $TR = 5$, $QT =$ ______

11. $QS = 8$, $RS = 12$, $QR = 10$, $TR =$ ______

12. $QS = 12$, $RS = 16$, $QR = 14$, $QT =$ ______

13. $QS = 3$, $RS = 5$, $QT = 2x + 1$, $TR = 4x - 5$, $QT =$ ______

14. $QS = 16$, $RS = 20$, $QT = x + 3$, $TR = 2x - 3$, $TR =$ ______

15. Perimeter of $\triangle QRS = 27$, $QT = 4$, $TR = 5$, $QS =$ ______

16. Perimeter of $\triangle QRS = 80$, $QT = 14$, $TR = 18$, $SR =$ ______

Given: $\triangle EFG \sim \triangle KLM$, $\overline{GH}$ and $\overline{MN}$ are altitudes; $\overline{EI}$ and $\overline{KO}$ are medians; and $\overline{FJ}$ and $\overline{LP}$ are angle bisectors. Find the indicated measure.

17. $KL = 3.6$, $EF = 2.4$, $KO = 2.1$, $EI =$ ______

18. $KM = 4$, $GE = 3.2$, $MN =$ ______, $GH = 2.4$

19. $MN = 2$, $GH =$ ______, $KO = 2.8$, $EI = 2.1$

20. $PL = 4.9$, $JF = 5.6$, $KO =$ ______, $EI = 2.4$

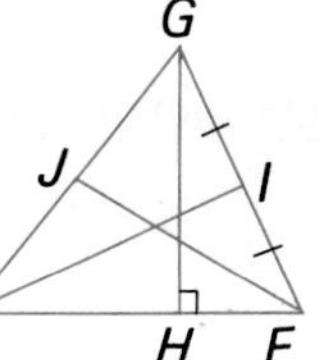

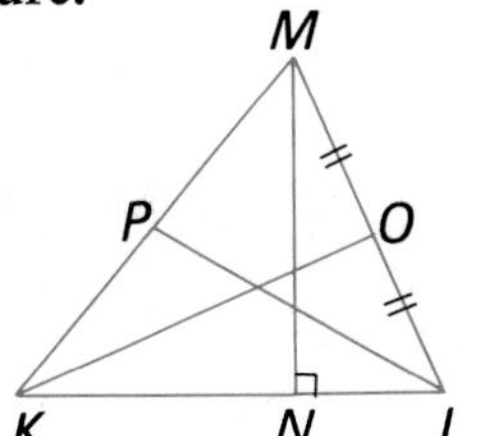

21. Given: $\overline{AE}$ bisects $\angle DAB$;
$\overline{CE}$ bisects $\angle DCB$.
Prove: $\dfrac{AD}{AB} = \dfrac{DC}{CB}$

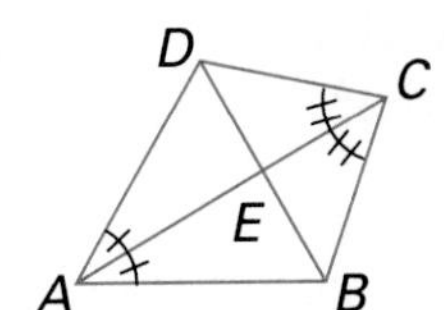

22. Draw a segment $\overline{XY}$. Construct P so that $\dfrac{XP}{PY} = \dfrac{3}{4}$.

23. Prove Theorem 8.10.

24. Prove Theorem 8.11.

25. Prove: If two triangles are similar, then corresponding altitudes are proportional to corresponding medians.

26. Prove: Diagonals of a trapezoid divide each other into proportional segments.

27. Prove or disprove the converse of Theorem 8.9.

Mixed Review

1. What is the geometric mean of 8 and 50? *8.1*

2. Give an example to show that the AAA pattern is insufficient for congruence of triangles. *4.6*

3. State the Triangle Inequality Theorem. *5.7*

4. How many diagonals does a convex hexagon have? *6.1*

Chapter 8 Review

Key Terms

AA Similarity p. 314
cross products p. 314
extremes p. 303
geometric mean p. 305
means p. 303
proportion p. 303
ratio p. 303
SAS Similarity p. 323
SSS Similarity p. 324
similar polygons p. 308
similar triangles p. 313
Transitive Property of Triangle Similarity p. 313
Triangle Proportionality Theorem p. 318

Key Ideas and Review Exercises

8.1 To solve a proportion, equate the product of the extremes and the product of the means.

To find the geometric mean of two positive numbers a and b, solve the proportion $\frac{a}{m} = \frac{m}{b}$ for the positive number m.

1. Solve the proportion: $\frac{8}{15} = \frac{x}{20}$.
2. Find the geometric mean of 12 and 75.
3. The ratio of the measures of two complementary angles is 2 to 7. What are the measures of the angles?

8.2 To show that two polygons are similar, show that corresponding angles are congruent and corresponding sides are proportional.

4. Based on the information given, is quad $ABCD \sim$ quad $EFGH$?

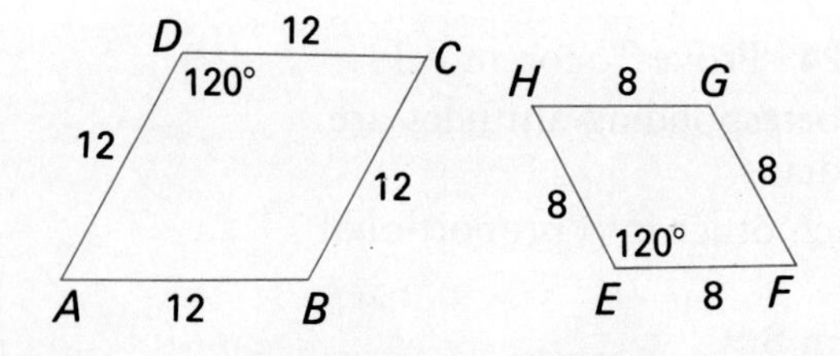

5. Quad $IJKL \sim$ quad $MNOP$. Find the missing lengths.

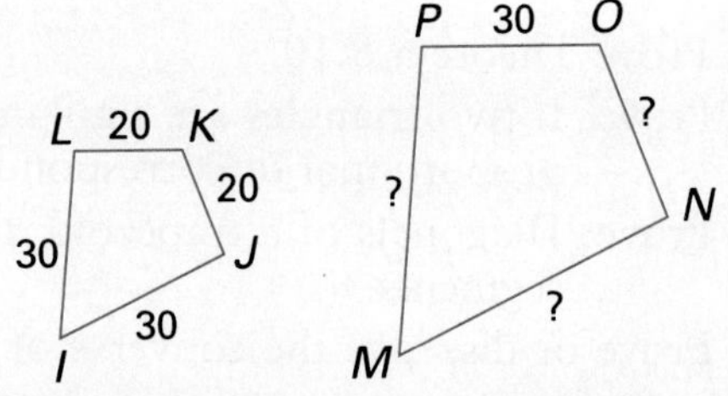

8.3 Two triangles are similar if two angles of one triangle are congruent to two angles of the other.

6. If m $\angle X = 65$, m $\angle Y = 55$, m $\angle L = 65$, and m $\angle N = 60$, is $\triangle XYZ \sim \triangle LMN$?
7. For similar triangles XYZ and LMN, write two proportions that give $YZ \cdot LN = MN \cdot XZ$.

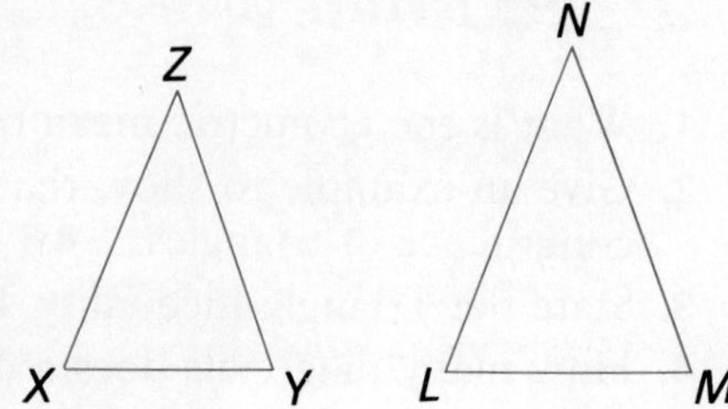

8.4 The Triangle Proportionality Theorem states that if a line is parallel to one side of a triangle and intersects the other two sides, then it divides the two sides proportionally.

The *converse* states that if a line divides two sides of a triangle proportionally, then the line is parallel to the third side of the triangle.

8. If $\overline{TV} \parallel \overline{XY}$, is $\frac{XT}{TZ} = \frac{ZV}{VY}$?

9. $\overline{TV} \parallel \overline{XY}$, $XT = 9$, $TZ = 12$, $ZV = 15$. Find VY.

10. If $XT = 54$, $TZ = 42$, $YV = 63$, and $VZ = 48$, is $\overline{TV} \parallel \overline{XY}$?

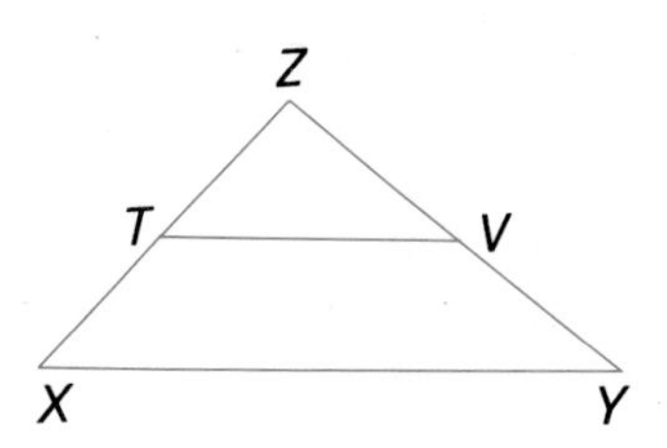

8.5 To prove two triangles similar, the SAS Similarity and the SSS Similarity theorems can be used.

11. If $\frac{GH}{IH} = \frac{JK}{LK}$, is $\triangle GHI \sim \triangle JKL$? Why?

12. Given: $\overline{GI} \cong \overline{IH}$, $\overline{JL} \cong \overline{KL}$, $\angle I \cong \angle L$
Prove: $\triangle GHI \sim \triangle JKL$

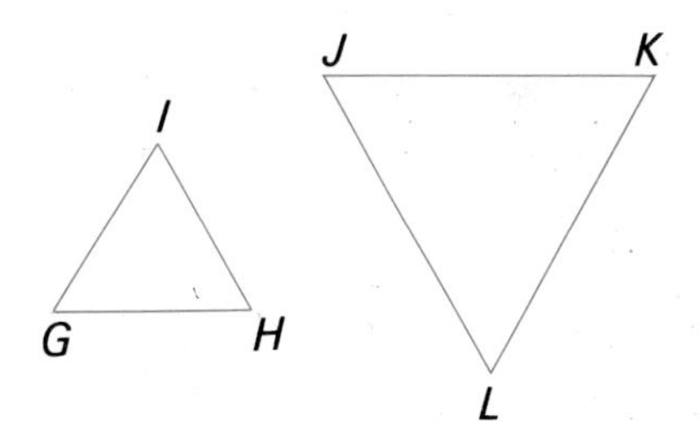

8.6 Corresponding altitudes (and medians) of similar triangles are proportional to corresponding sides.

The bisector of an angle divides the opposite side of a triangle into segments proportional to the other two sides.

Proportional segments can be constructed using an angle bisector.

13. Explain how to determine whether the medians of similar triangles are corresponding medians.

14. $\triangle XYZ \sim \triangle PQR$, $\overline{ZA}$ and $\overline{RB}$ are altitudes; $XY = 40$, $PQ = 60$, $ZA = 18$, $RB =$ ______.

15. $\overline{AD}$ bisects $\angle CAB$; $AC = 24$, $AB = 32$, $CD = 18$, $BD =$ ______.

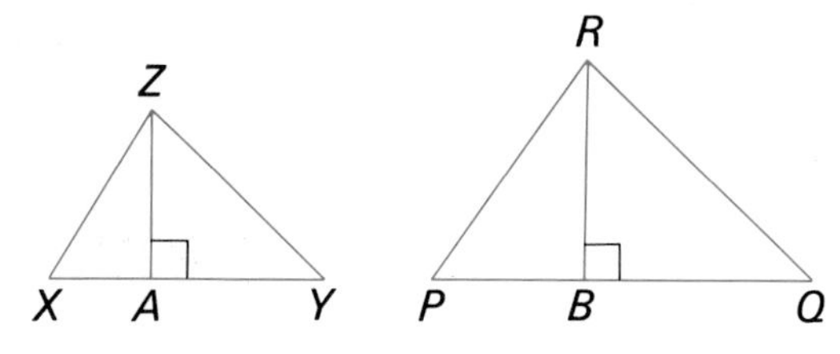

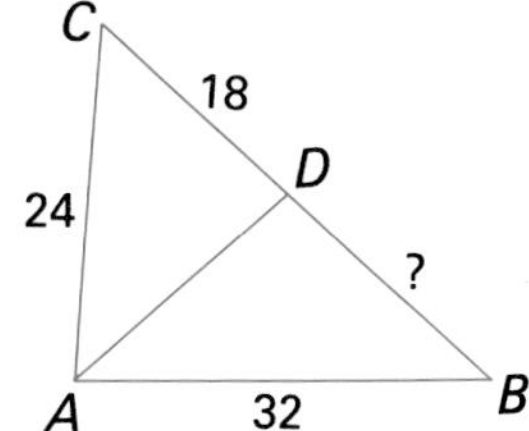

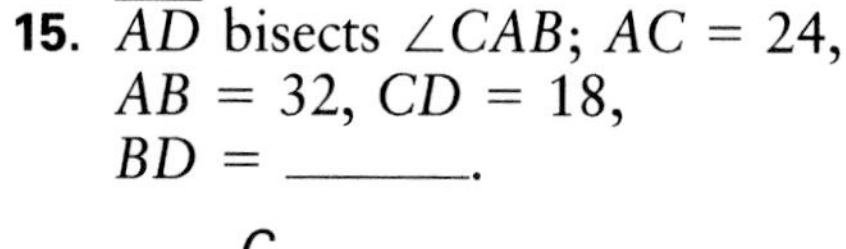

16. Draw a segment $\overline{AB}$. Construct C so that $\frac{AC}{CB} = \frac{3}{5}$.

Chapter 8 Test

Solve the proportion for x. (Exercises 1–2)

1. $\frac{x}{15} = \frac{32}{40}$

2. $\frac{21}{27} = \frac{x}{45}$

3. Find the geometric mean of 8 and 2.

$\overline{DE} \parallel \overline{AB}$. Find the indicated length.

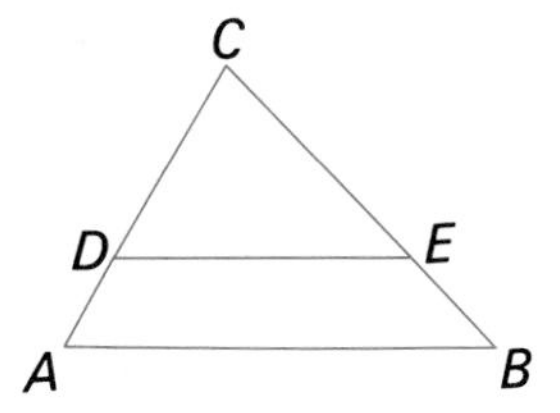

4. $DA = 12$, $CE = 12$, $EB = 15$, $CD =$ ______

5. $CA = 35$, $CE = 25$, $EB = 10$, $AD =$ ______

Based on the information given, is $\triangle ABC \sim \triangle DEF$? If so, why?

6. $\angle A \cong \angle D$, $\angle C \cong \angle F$

7. $\angle B \cong \angle E$, $\frac{AC}{DF} = \frac{AB}{DE}$

8. $\frac{AB}{DE} = \frac{BC}{EF} = \frac{CA}{FD}$

9. $\triangle GHI \sim \triangle PQR$. Find GF.

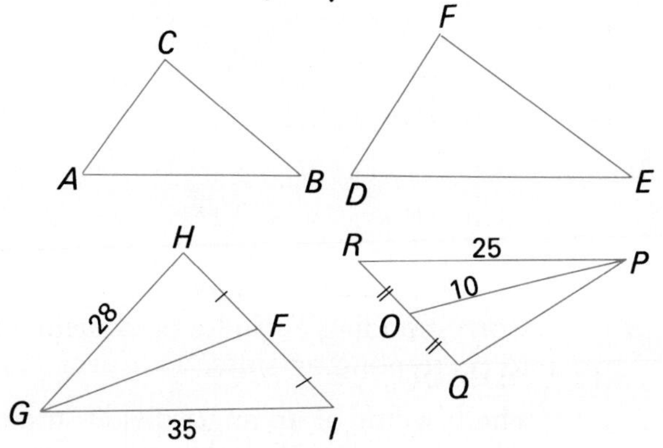

True or false? (Exercises 10–12)

10. If two triangles are similar, then their corresponding altitudes are congruent.

11. Congruent triangles are similar.

12. Two quadrilaterals are similar if their corresponding angles are congruent.

13. Given: $\overline{IJ}$ bisects $\angle GIH$; $IG = 45$, $IH = 25$, $GH = 42$

Find GJ.

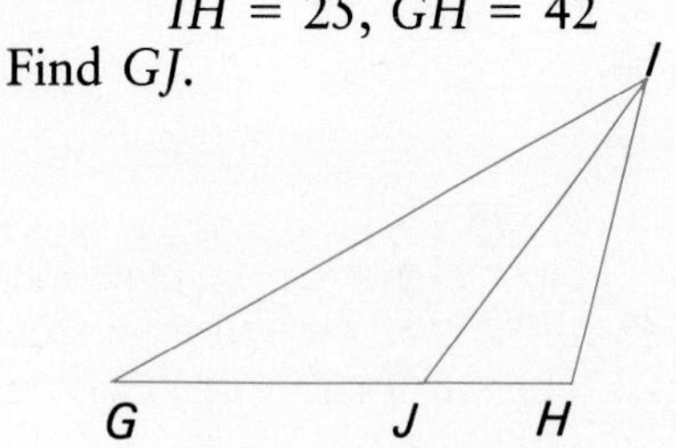

14. Given: $\overline{MN}$ and $\overline{KO}$ are altitudes of $\triangle KLM$.

Prove: $\triangle MOP \sim \triangle MNL$

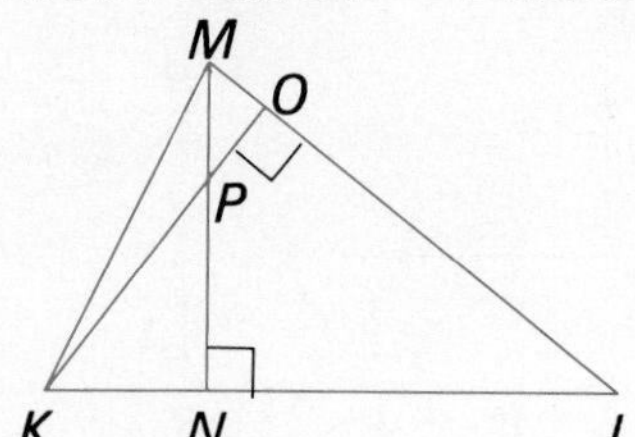

***15.** Prove: if $\frac{m}{n} = \frac{p}{q}$, then $\frac{m - n}{n} = \frac{p - q}{q}$

College Prep Test

Choose the one best answer to each question or problem.

1. In $\triangle ABC$, $\overline{DE} \parallel \overline{BC}$, $AE = 5$, $EB = 3$, and $AD = 4$. Find DC.

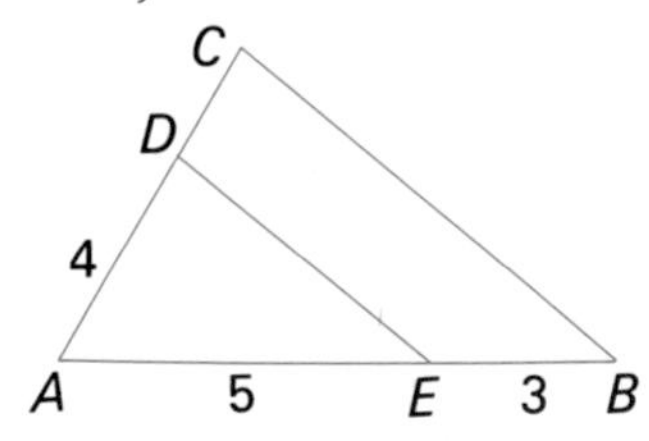

(A) 2.4 (B) 3.75 (C) 4
(D) 6.4 (E) Cannot be determined

2. In $\triangle FGH$, $FG = 6$, $GH = 7$, and $HF = 8$. Find IG.

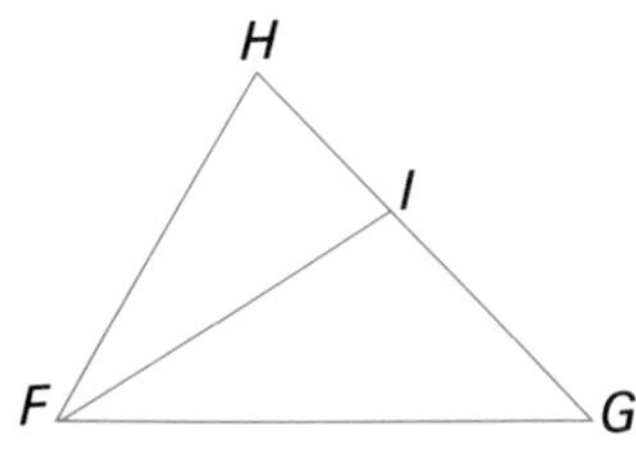

(A) 3 (B) 3.5 (C) 4 (D) 6
(E) Cannot be determined

3. If one dozen pencils cost one dollar, how many dollars will n pencils cost?
(A) $\frac{12n}{1.00}$ (B) $\frac{1.00n}{12}$ (C) $\frac{12}{1.00n}$
(D) $\frac{1.00}{12n}$ (E) Cannot be determined

4. Given: Trap $JKLM$, median $\overline{NO}$, $JK = 6x + 1$, $NO = 7x - 3$, $ML = 4x + 3$. Find NO.

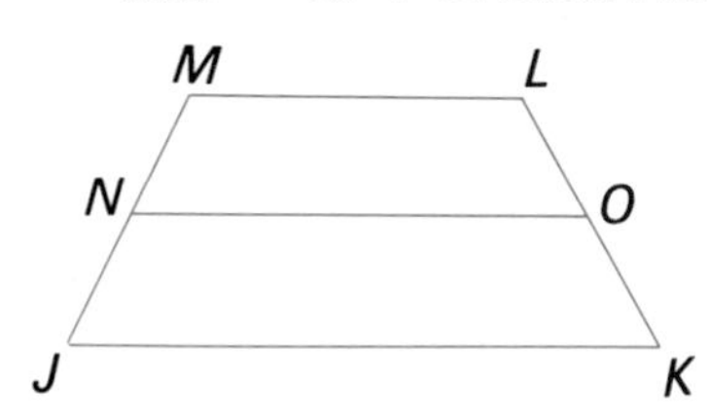

(A) 6.5 (B) 13 (C) 14.5
(D) 16 (E) Cannot be determined

5. If it takes Sam 15 min to type two pages, how many hours will it take him to type 50 pages?
(A) 6.25 (B) 12.5
(C) 15 (D) 25
(E) Cannot be determined

6. On a map drawn to scale 0.125 in. = 10 mi, what is the actual distance in miles between two cities which are 2.5 in. apart on the map?
(A) 12.5 (B) 25
(C) 200 (D) 1,250
(E) Cannot be determined

7. In the figure below, rect $PQRS$ is composed of five congruent rectangles; $SP = 75$. Find TU.

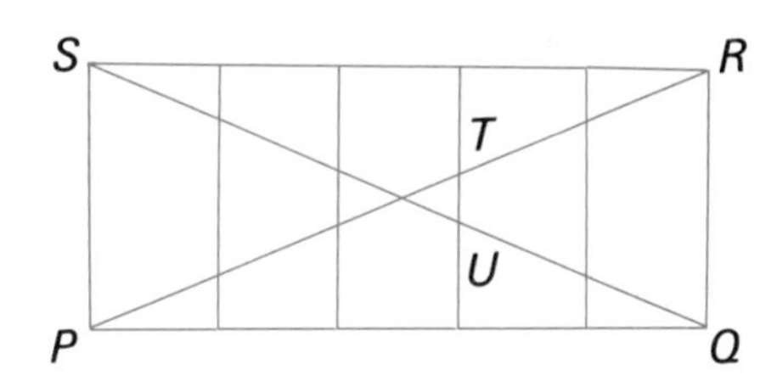

(A) 7.5 (B) 15
(C) 20 (D) 30
(E) Cannot be determined

8. A five foot-tall person casts a shadow 8 ft long. At the same time, a tree casts a shadow 64 ft long. How many feet tall is the tree?
(A) 40 (B) 51.2
(C) 80 (D) 102.4
(E) Cannot be determined

Cumulative Review (Chapters 1–8)

1. Identify the following points as collinear, coplanar but noncollinear, or noncoplanar. *1.1*
   (a) E, B (b) A, B, C
   (c) A, F, D (d) A, B, C, D
   (e) A, D, C, E (f) A, B, C, F

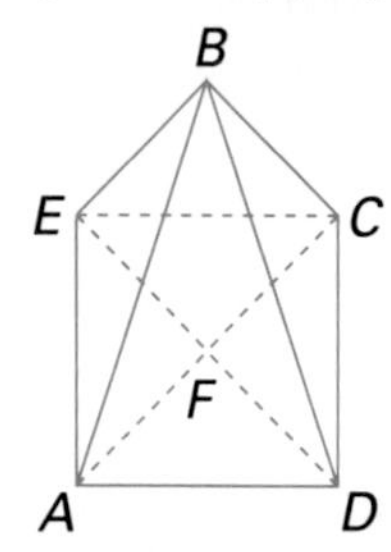

2. Draw an acute angle. Construct the bisector of the angle. *1.5*

3. State a conclusion from this statement: $\overrightarrow{OP}$ bisects $\angle AOB$. *2.1*

4. Write the following statement in "If . . ., then . . ." form: A right angle has perpendicular sides.

5. True or false? Skew lines are noncoplanar lines. *3.1*

6. Give a proof of the theorem: The sum of the measures of the angles of a triangle is 180. *3.6*

7. Given: Parallel lines l and m with transversal t. $\overline{BC}$ bisects $\angle ABD$ and $\overline{DC}$ bisects $\angle EDB$. What is the measure of $\angle DCB$? *3.5*

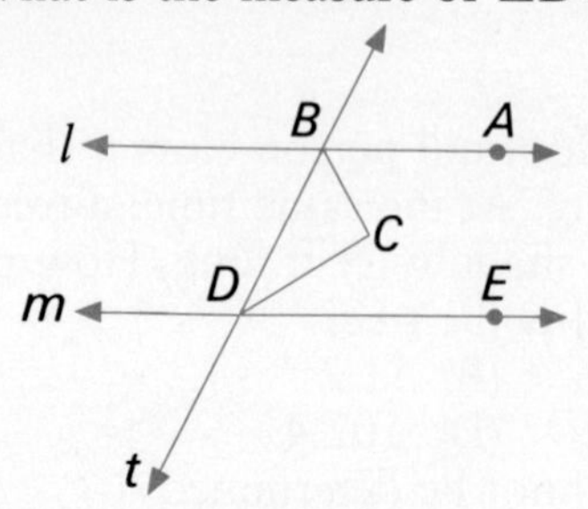

8. True or false? If two planes intersect, then they intersect in exactly one line. *3.1*

9. What is the measure of the angle formed by the bisectors of two adjacent complementary angles? *1.6*

10. Find the measure of all the angles of $\triangle ABC$ if the m $\angle CBD$ = 150, and m $\angle A$ is one-half the m $\angle C$. *3.6*

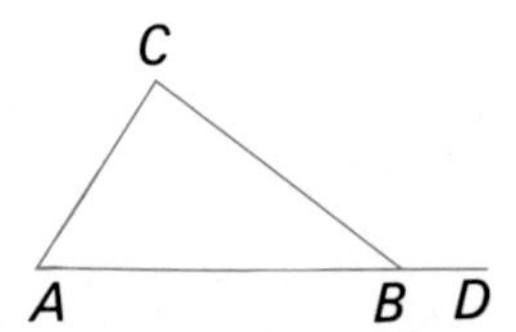

11. $\triangle ABC$ is isosceles with $\angle A \cong \angle B$; $AC = x^2 - 10$, $AB = x + 2$, $CB = 3x$. Find AB. *5.1*

12. True or false? If the measures of two angles of a triangle are 30 and 40, then the triangle is obtuse. *4.1*

13. Given: $\angle A \cong \angle E$, $\overline{AC} \cong \overline{EF}$, $\overline{CB} \cong \overline{FD}$. Is enough information given to prove $\triangle ABC \cong \triangle EDF$? Why or why not? *4.4*

14. The measure of one of the base angles of an isosceles triangle is 20 less than the measure of the vertex angle. Find the measure of the vertex angle. *5.1*

15. Draw a segment approximately 5 cm long. Construct the perpendicular bisector of the segment. *5.6*

16. Is there a regular polygon in which an exterior angle has a measure of 20? *6.3*

17. Is the property, a quadrilateral has congruent opposite angles, a necessary condition for a quadrilateral to be a parallelogram? *6.4, 7.2*

18. Is the property, consecutive angles are supplementary, a sufficient condition for a quadrilateral to be a parallelogram? *6.5, 7.3*

19. Given: D and E are the midpoints of $\overline{AC}$ and $\overline{BC}$, respectively. $DE = 3x - 2$, $AB = 2x + 10$. Find AB. *6.6*

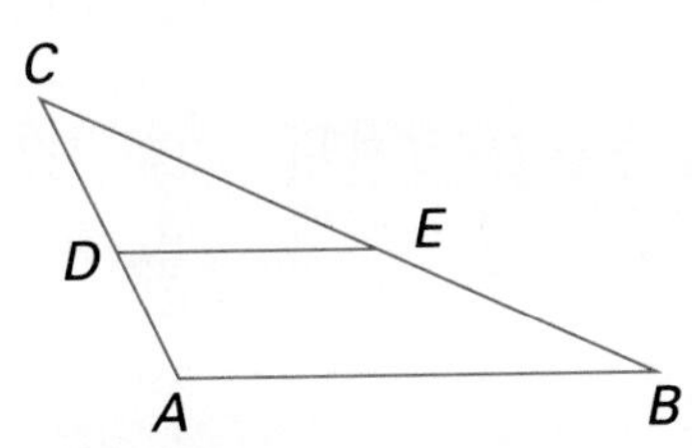

20. Is the property, a parallelogram has congruent diagonals, a necessary condition for a parallelogram to be a rhombus? *7.2*

21. Is it possible to prove $\triangle ABC \cong \triangle DEF$ if $\angle A \cong \angle D$, $\overline{AB} \cong \overline{DE}$, and $\overline{BC} \cong \overline{EF}$? *4.5*

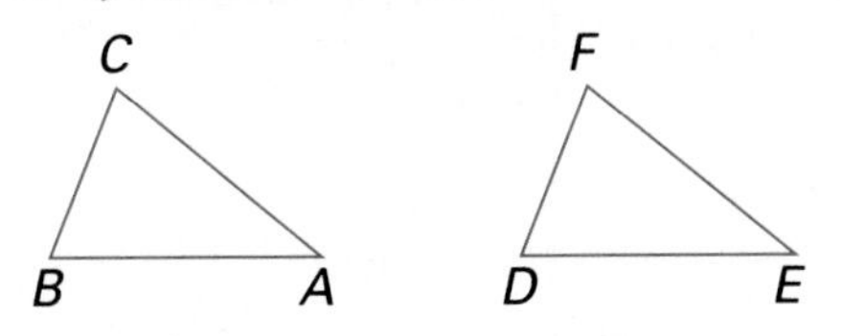

22. Construct an angle of 30° using straight edge and compass. *1.5, 4.1*

23. Point D is the midpoint of $\overline{AB}$ and point P is located so that $PA = PB$. What is m $\angle PDA$? *5.4*

24. Find the values of x and y in $\square ABCD$ if $AB = 3x + 2$, $BC = y + 12$, $CD = 5x - 8$, and $DA = 4y + 3$. *6.4*

25. Given: $\triangle ABC$ with M the midpoint of $\overline{AC}$, and N the midpoint of $\overline{BC}$. If $MN = 2x + 5$, and $AB = 2x + 15$. Find MN. *6.6*

26. A regular polygon has each interior angle half as large as each exterior angle. How many sides does the polygon have? *6.3*

27. If $4x = 2y$, what is the ratio of x to y? *8.1*

28. Is the property, a parallelogram has at least one right angle, a sufficient condition for a parallelogram to be a rectangle? *7.3*

29. Is the property, a trapezoid has congruent diagonals, a necessary condition for a trapezoid to be isosceles? *7.5*

30. Find the geometric mean of 2 and 18. *8.1*

31. Are any two regular pentagons similar? If so, prove it. If not, provide a counterexample. *8.2*

32. True or false? If $\triangle LMN \sim \triangle TUV$, and $\triangle TUV \sim \triangle XYZ$, then $\frac{MN}{YZ} = \frac{NL}{ZX}$. *8.3*

33. In $\triangle ABC$, $AB = 3$, $BC = 6$, and m $\angle B = 60$. In $\triangle EFG$, $EF = 8$, $FG = 16$, and m $\angle F = 60$. Are $\triangle ABC$ and $\triangle EFG$ similar? Why or why not? *8.5*

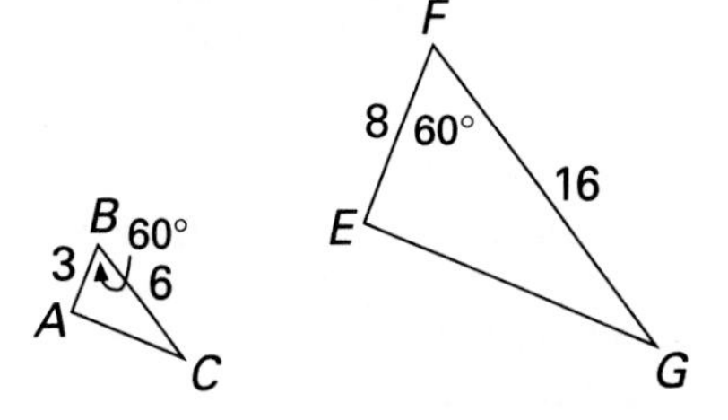

9 RIGHT TRIANGLES

Maria Agnesi (1718–1799)—Maria Agnesi, the eldest of 21 children, could speak five languages before age 15. She is said to have written complete solutions to puzzling math problems while walking in her sleep.

9.1 Right Triangle Similarity Properties

Objective To find the lengths of sides and altitudes of right triangles

In right triangle PRQ, the altitude $\overline{RS}$ from R to the hypotenuse forms two right triangles, $\triangle PSR$ and $\triangle QSR$. Suppose that m $\angle 1 = 50$. Then the measure of each of the other numbered angles can be found by using Theorem 4.1, which states that the acute angles of a right triangle are complementary.

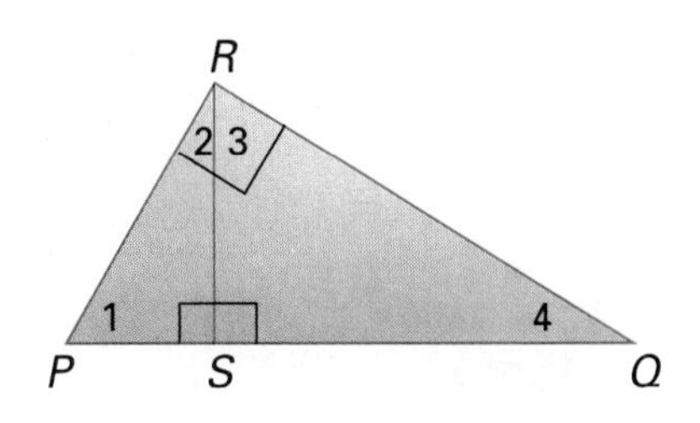

Note that in each of the three triangles above ($\triangle PSR$, $\triangle QSR$, $\triangle RPQ$), the measures of the angles are the same: 40, 50, and 90. Thus, by AA Similarity, the triangles are similar. The general case for this result is stated in the following theorem.

Theorem 9.1 In a right triangle, the altitude to the hypotenuse forms two similar right triangles, each of which is also similar to the original triangle.

Given: Right $\triangle PRQ$, with right $\angle PRQ$ and altitude $\overline{RS}$

Prove: $\triangle PRQ \sim \triangle PSR$, $\triangle PRQ \sim \triangle RSQ$, $\triangle PRS \sim \triangle RSQ$

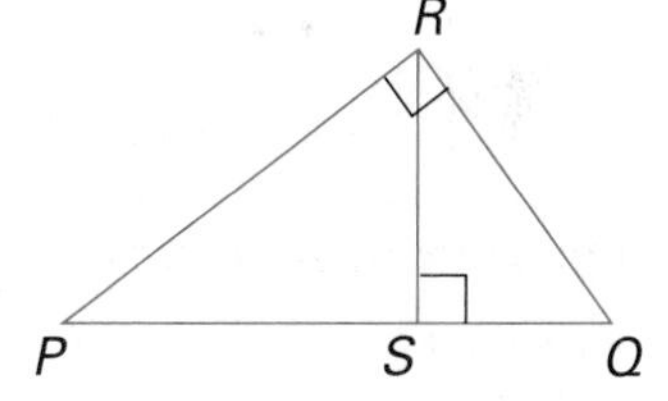

Proof

	Statement		Reason
	1. Rt $\triangle PRQ$ with rt $\angle PRQ$, altitude $\overline{RS}$		1. Given
	2. $\angle PSR$ and $\angle QSR$ are rt $\angle$s.		2. Def of alt
	3. $\angle PRQ \cong \angle PSR$	(A)	3. All rt $\angle$s are $\cong$.
	4. $\angle P \cong \angle P$	(A)	4. Reflex Prop
	5. $\therefore \triangle PSR \sim \triangle PRQ$		5. AA $\sim$
	6. $\angle PRQ \cong \angle RSQ$	(A)	6. All rt $\angle$s are $\cong$.
	7. $\angle Q \cong \angle Q$	(A)	7. Reflex Prop
	8. $\therefore \triangle PRQ \sim \triangle RSQ$		8. AA $\sim$
	9. $\therefore \triangle PSR \sim \triangle RSQ$		9. Trans Prop of Similarity

In $\triangle PRQ$ at the right, h, m, and n represent the lengths of sides of the triangles formed by the altitude to the hypotenuse. Notice that since $\triangle PRS \sim \triangle RQS$ by Theorem 9.1, their corresponding sides are proportional.

$$\frac{m}{h} = \frac{h}{n}, \text{ or } h^2 = mn$$

Corollary 1

In a right triangle, the square of the length of the altitude to the hypotenuse equals the product of the lengths of the segments formed on the hypotenuse.

Corollary 1 is often stated as follows:

In a right triangle, the length of the altitude to the hypotenuse is the geometric mean of the lengths of the segments formed on the hypotenuse.

EXAMPLE 1 Given: Right $\triangle ABC$ with altitude $\overline{BD}$ to hypotenuse $\overline{AC}$, $BD = 12$, $AD = 8$

Find CD.

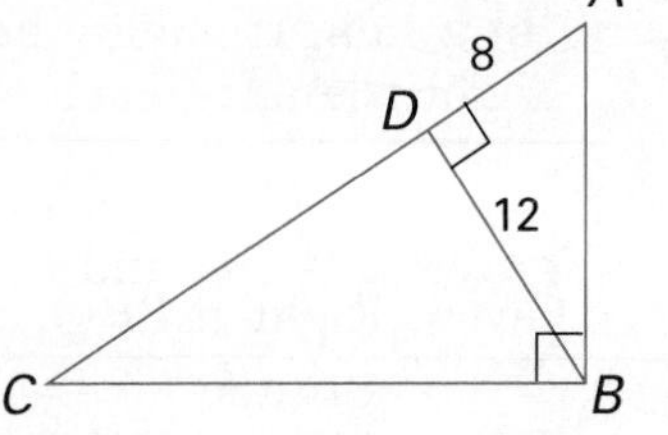

Solution

$(BD)^2 = AD \cdot CD$ (Corollary 1)

$12^2 = 8 \cdot CD$

$\frac{144}{8} = CD$

$CD = 18$

EXAMPLE 2 Given: Right $\triangle KHG$ with altitude $\overline{HT}$ to hypotenuse $\overline{KG}$, $KG = 14$, $KT = 6$

Find HT.

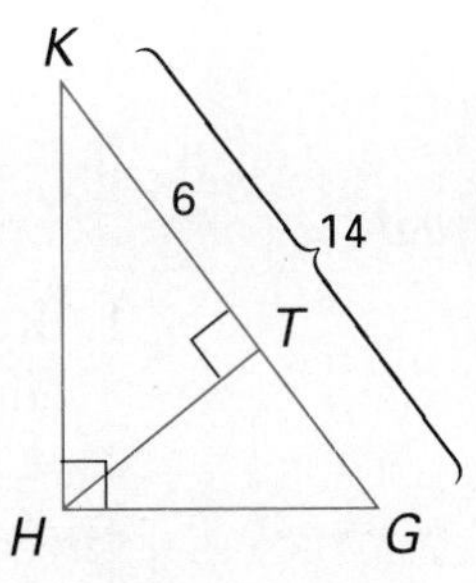

Plan Find the length of the other segment of the hypotenuse, TG.

$$TG = KG - KT$$
$$= 14 - 6$$
$$= 8$$

Then apply Corollary 1.

Solution (Length of altitude)^2 = product of lengths of segments of hypotenuse

$$(HT)^2 = 8 \cdot 6 = 48$$
$$HT = \sqrt{48} = \sqrt{16} \cdot \sqrt{3} = 4\sqrt{3}$$

EXAMPLE 3 Right $\triangle ABC$ has its right $\angle$ at C. The altitude from C to the hypotenuse meets the hypotenuse at point D. $CD = 4$ and $AD = 2$. Find DB.

Plan First draw a right triangle. Label the vertex of the right angle C. Draw the altitude from that vertex to $\overline{AB}$. The segments of the hypotenuse are then $\overline{AD}$ and $\overline{DB}$. $AD = 2$ and $CD = 4$. Let $DB = x$.

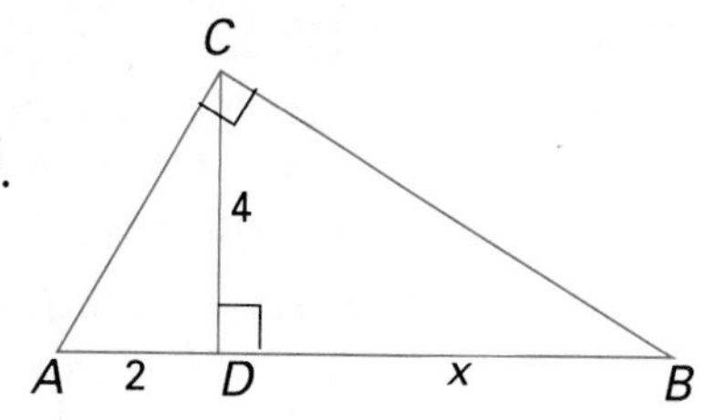

Solution (Length of altitude)^2 = product of lengths of segments of hypotenuse

$$4^2 = 2 \cdot x$$
$$16 = 2x$$
$$8 = x$$

So, $DB = 8$

Corollary 2 If the altitude is drawn to the hypotenuse of a right triangle, then the square of the length of either leg equals the product of the lengths of the hypotenuse and the segment of the hypotenuse adjacent to that leg.

Given: Right $\triangle RTS$ with right $\angle RTS$ and altitude $\overline{TU}$

Prove: $s^2 = mt$ and $r^2 = nt$

Plan $\triangle RTS \sim \triangle RUT$, and $\triangle RTS \sim \triangle TUS$. Use corresponding sides of similar triangles to write proportions.

T
s
r
R m U n S
t

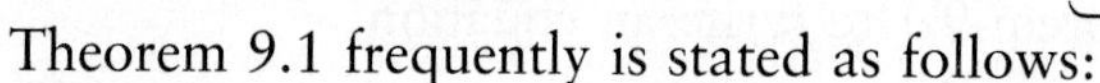

Theorem 9.1 frequently is stated as follows:

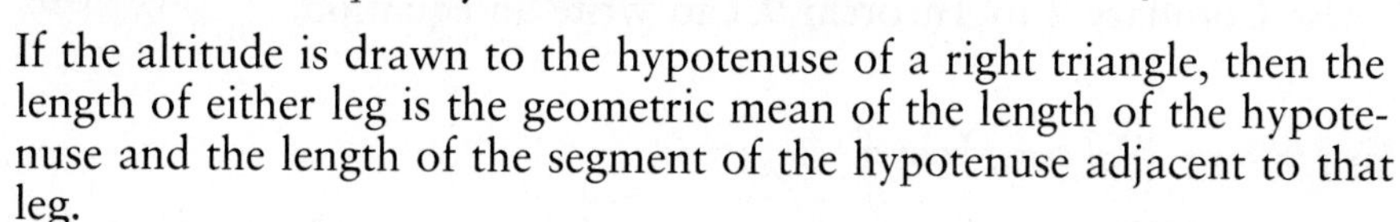

If the altitude is drawn to the hypotenuse of a right triangle, then the length of either leg is the geometric mean of the length of the hypotenuse and the length of the segment of the hypotenuse adjacent to that leg.

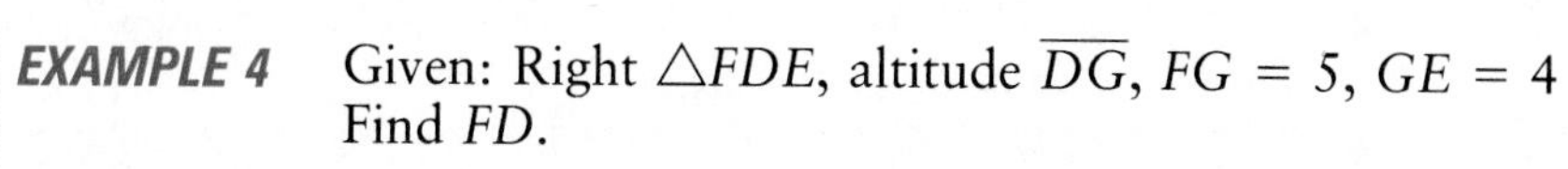

EXAMPLE 4 Given: Right $\triangle FDE$, altitude $\overline{DG}$, $FG = 5$, $GE = 4$ Find FD.

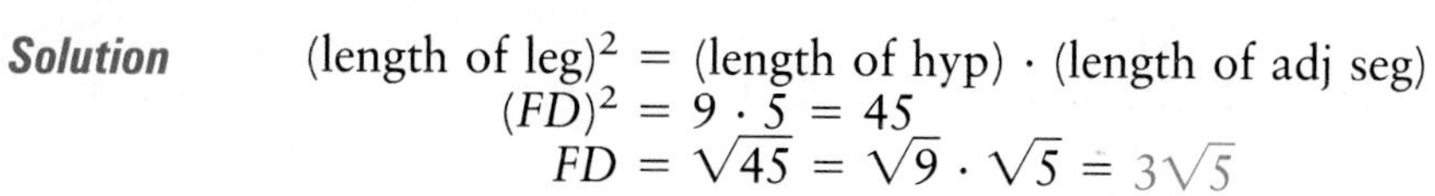

Solution (length of leg)^2 = (length of hyp) $\cdot$ (length of adj seg)

$$(FD)^2 = 9 \cdot 5 = 45$$
$$FD = \sqrt{45} = \sqrt{9} \cdot \sqrt{5} = 3\sqrt{5}$$

F
5
G
4
D E

EXAMPLE 5 Given: Right $\triangle DEF$, $\overline{FG} \perp \overline{DE}$, $DF = 8$, $EG = 12$. Find DG.

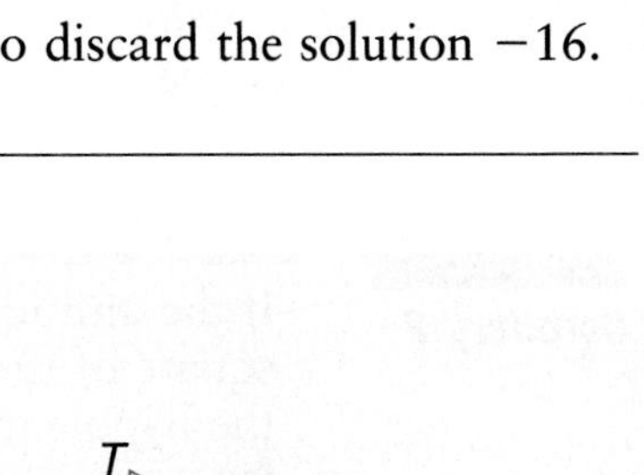

Plan Let $DG = x$. [Length of segment adjacent to leg $\overline{DF}$]
Then $DE = x + 12$. [length of hypotenuse]

Solution

$$(DF)^2 = DE \cdot DG$$
$$8^2 = (x + 12)x$$
$$64 = x^2 + 12x$$
$$0 = x^2 + 12x - 64$$
$$0 = (x + 16)(x - 4)$$
$$x + 16 = 0 \quad or \quad x - 4 = 0$$
$$x = -16 \quad or \quad x = 4$$

A segment cannot have a negative length, so discard the solution -16. Thus, $DG = 4$.

Classroom Exercises

Identify the following using the given figure.

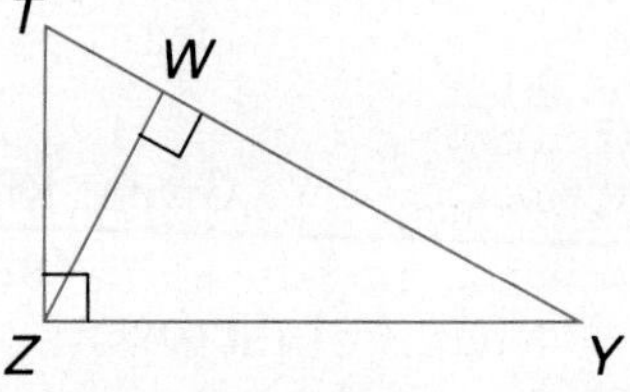

1. hypotenuse of $\triangle TZY$
2. hypotenuse of $\triangle WZY$
3. legs of $\triangle TYZ$
4. legs of $\triangle TZW$
5. legs of $\triangle WZY$
6. altitude to hypotenuse of $\triangle TYZ$
7. segment of hypotenuse of $\triangle TYZ$ adjacent to $\overline{ZY}$
8. segment of hypotenuse of $\triangle TYZ$ adjacent to $\overline{ZT}$
9. Use Corollary 1 to Theorem 9.1 to write an equation.
10. Use Corollary 2 to Theorem 9.1 to write an equation.

Written Exercises

In right $\triangle SVU$, $\angle SVU$ is a right angle and $\overline{VT}$ is an altitude. Find the length.

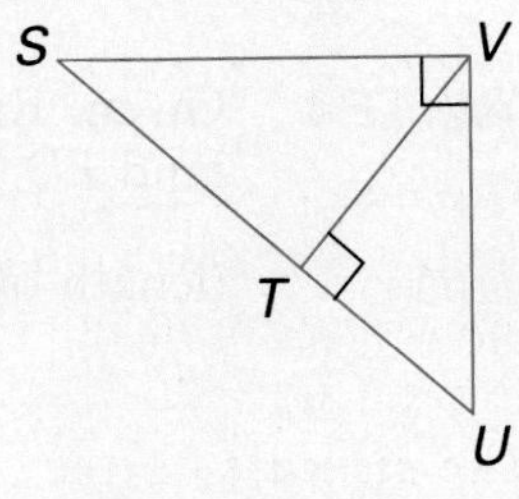

1. $TV = 4$, $ST = 2$, $TU =$ ______
2. $ST = 12$, $TU = 3$, $TV =$ ______
3. $ST = 6$, $TU = 4$, $VT =$ ______
4. $TV = 4$, $TU = 6$, $ST =$ ______
5. $ST = 8$, $SU = 10$, $TV =$ ______
6. $SU = 12$, $TU = 8$, $TV =$ ______
7. $ST = 2$, $SU = 8$, $SV =$ ______
8. $TU = 3$, $SU = 8$, $UV =$ ______

In right $\triangle QMP$, $\angle QMP$ is a right angle and $\overline{MN}$ is an altitude. Find the length.

9. $QM = 6$, $QN = 2$, $QP =$ ______
10. $QN = 8$, $NP = 4$, $MP =$ ______
11. $MP = 8$, $QP = 16$, $NP =$ ______
12. $QN = 4$, $NP = 8$, $MP =$ ______
13. $QN = 12$, $NP = 2$, $MP =$ ______
14. $QN = 3$, $QP = 12$, $MN =$ ______
15. $QM = 6$, $PN = 5$, $NQ =$ ______
16. $QN = 6$, $PM = 4$, $QP =$ ______
17. $QP = 10$, $MN = 4$, $QN =$ ______
18. $PQ = 13$, $MN = 6$, $NP =$ ______
19. $QN = NP$, $QM = 4$, $QN =$ ______
20. $QP = 20$, $MN = 8$, $NP =$ ______
21. Right triangle RST has its right angle at S. The altitude from S meets the hypotenuse at U. $RU = 4$ and $UT = 6$. Find RS, SU, and ST.
22. The right angle of right triangle KGA is at A. The altitude from A meets the hypotenuse at P. $PK = 32$ and $GA = 12$. Find GP and PA.
23. If the altitude to the hypotenuse of a right triangle bisects the hypotenuse and the length of the hypotenuse is 6, find the length of each leg.
24. Prove Corollary 1 to Theorem 9.1.
25. Prove Corollary 2 to Theorem 9.1.
26. Prove: In a right triangle, the product of the lengths of the hypotenuse and the altitude to the hypotenuse equals the product of the lengths of the legs.
27. The lengths of the two legs and hypotenuse of a right triangle are a, b, and c respectively. Find the length of the altitude to the hypotenuse in terms of a, b, and c.

In the figure, m $\angle EAC = 90$, $\overline{DB} \perp \overline{AC}$, $\overline{AD} \perp \overline{EC}$

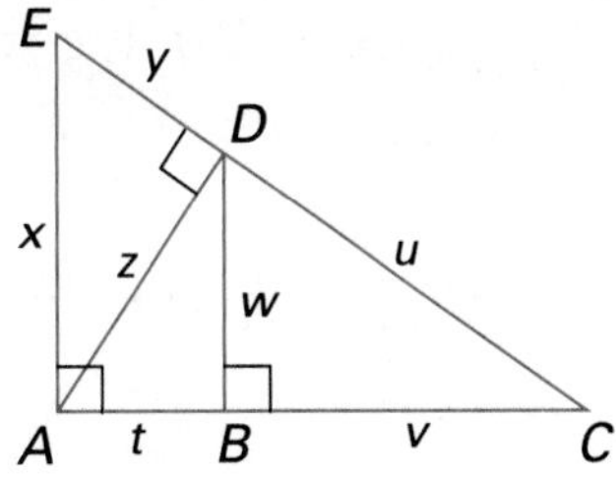

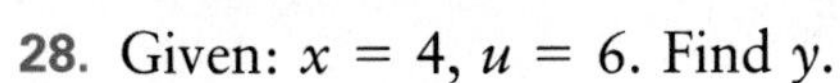

28. Given: $x = 4$, $u = 6$. Find y.
29. Prove: $w(u + y) = xu$
30. Given: $t = 9$, $v = 4$. Find x.
31. Given: $u = 6$, $t = 5$. Find x.

Mixed Review

1. The measures of the acute angles of a right triangle are in the ratio 3:7. Find the measure of each acute angle. *8.1*
2. Find the sum of the measures of the interior angles of a pentagon. *6.2*
3. The measure of one of the angles of a parallelogram is 140. Find the measure of each of the other three angles. *6.4*

9.2 The Pythagorean Theorem

Objective To apply the Pythagorean Theorem

In this lesson, you will explore a method to find the length of the hypotenuse of a right triangle, given the lengths of the two legs.

Theorem 9.2

Pythagorean Theorem: In any right triangle, the sum of the squares of the lengths of the legs is equal to the square of the length of the hypotenuse.

Given: Right $\triangle ACB$ with right $\angle ACB$
Prove: $a^2 + b^2 = c^2$

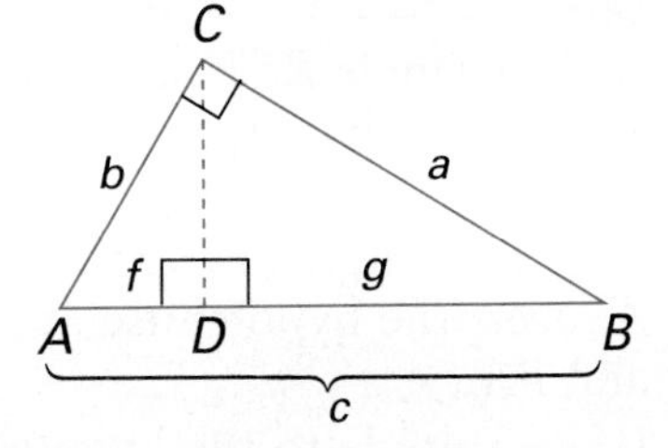

Plan Draw $\overline{CD} \perp \overline{AB}$. $\overline{CD}$ is the altitude to hypotenuse $\overline{AB}$. Write two equations using corollary 2 to Theorem 9.1.

Proof

Statement	Reason
1. Rt $\triangle ACB$ with rt $\angle ACB$	1. Given
2. Draw $\overline{CD} \perp$ to $\overline{AB}$.	2. From a pt not on a line, there is exactly one line $\perp$ to the line.
3. $\overline{CD}$ is an altitude.	3. Def of alt
4. $b^2 = fc$, $a^2 = gc$	4. Square of length of leg = prod of lengths of hyp and adj seg.
5. $a^2 + b^2 = fc + gc$	5. Add Prop of Eq
6. $a^2 + b^2 = (f + g)c$	6. Distr Prop
7. $f + g = c$	7. Seg Add Post
8. $\therefore a^2 + b^2 = c \cdot c$ or $a^2 + b^2 = c^2$	8. Sub

EXAMPLE 1

Find c.

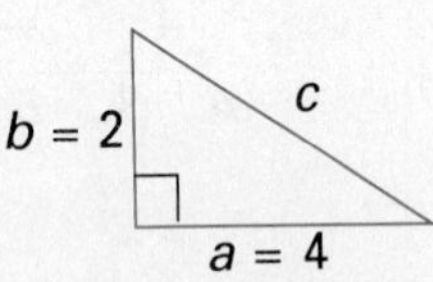

Find b.

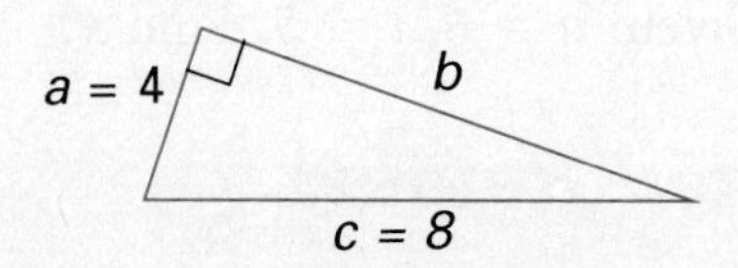

Solutions

$$a^2 + b^2 = c^2$$
$$4^2 + 2^2 = c^2$$
$$16 + 4 = c^2$$
$$20 = c^2$$
$$c = \sqrt{20}, \text{ or } 2\sqrt{5}$$

$$a^2 + b^2 = c^2$$
$$4^2 + b^2 = 8^2$$
$$16 + b^2 = 64$$
$$b^2 = 48$$
$$b = \sqrt{48}, \text{ or } 4\sqrt{3}$$

EXAMPLE 2 The length of a diagonal of a square is 6. Find the length of a side.

Plan Draw a square with a diagonal. The diagonal is the hypotenuse of each right triangle formed.

Let s = length of a side.

Solution

$$\begin{aligned} s^2 + s^2 &= 6^2 \\ 2s^2 &= 36 \\ s^2 &= 18 \\ s &= \sqrt{18} = \sqrt{9} \cdot \sqrt{2}, \text{ or } 3\sqrt{2} \end{aligned}$$

Thus, the length of a side of the square is $3\sqrt{2}$.

EXAMPLE 3 The length of a side of a rhombus is 13 cm, and the length of one diagonal is 10 cm. Find the length of the other diagonal.

Plan Draw a rhombus with its diagonals. Each side is a hypotenuse of a right triangle. The diagonals bisect each other.

Let $a = \frac{1}{2}$(length of the other diagonal).

Solution

$$\begin{aligned} 13^2 &= a^2 + 5^2 \\ 169 &= a^2 + 25 \\ 144 &= a^2 \\ 12 &= a \end{aligned}$$

Thus, the length of the other diagonal is $2 \cdot 12$, or 24 cm.

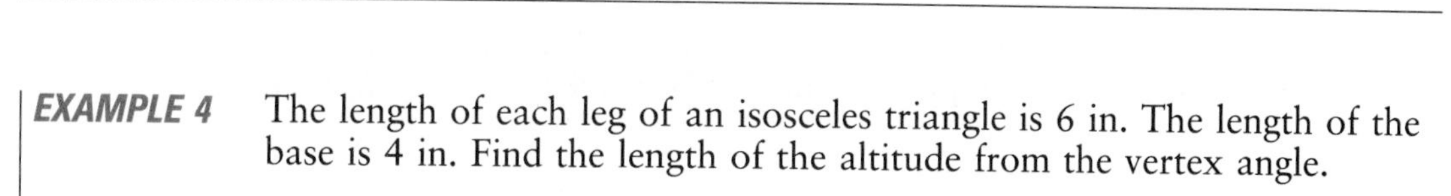

EXAMPLE 4 The length of each leg of an isosceles triangle is 6 in. The length of the base is 4 in. Find the length of the altitude from the vertex angle.

Plan Draw an isosceles triangle with an altitude from the vertex angle. Since by Theorem 5.7 such an altitude bisects the base, $CD = \frac{1}{2} \cdot 4$, or 2. Let h equal the length of the altitude. By the Pythagorean Theorem:

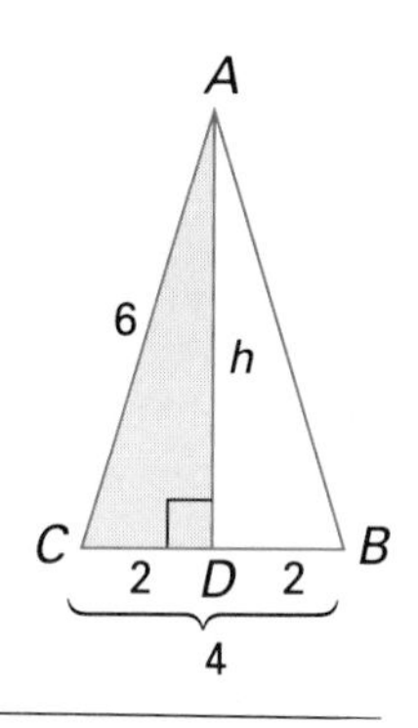

Solution

$$\begin{aligned} 2^2 + h^2 &= 6^2 \\ 4 + h^2 &= 36 \\ h &= \sqrt{32}, = \sqrt{16} \cdot \sqrt{2}, \text{ or } 4\sqrt{2} \end{aligned}$$

The length of the altitude is $4\sqrt{2}$ in.

EXAMPLE 5 Find the length of the altitude to the hypotenuse of a right triangle with legs of lengths 6 and 8.

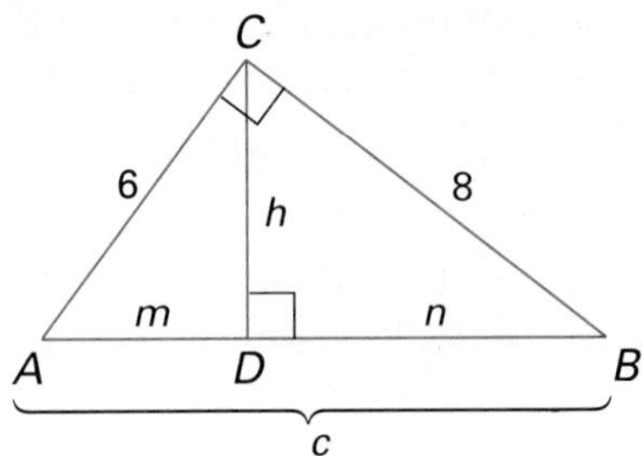

Solution Use the Pythagorean Theorem to find c.

$$c^2 = 6^2 + 8^2$$
$$c^2 = 36 + 64$$
$$c^2 = 100$$
$$c = 10$$

According to Theorem 9.1, $\triangle CBD \sim \triangle ABC$

$$\frac{CD}{AC} = \frac{CB}{AB}$$
$$\frac{h}{6} = \frac{8}{10}$$
$$h = \frac{48}{10}, \text{ or } 4.8$$

Thus, the length of the altitude is 4.8.

Classroom Exercises

For Exercises 1–8, write an equation that can be solved to find a, b, or c.

1.

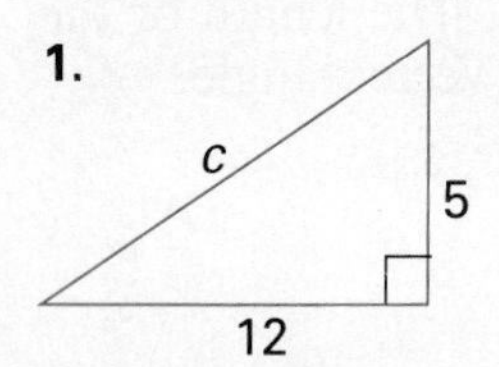

2.

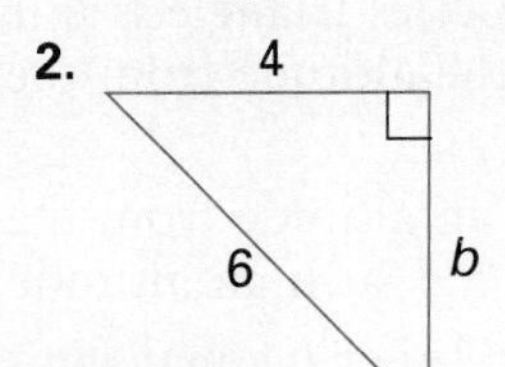

3.

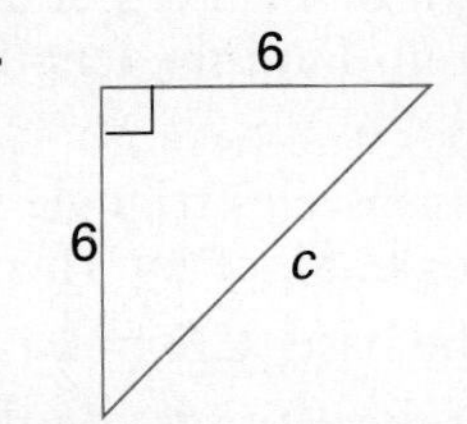

4.

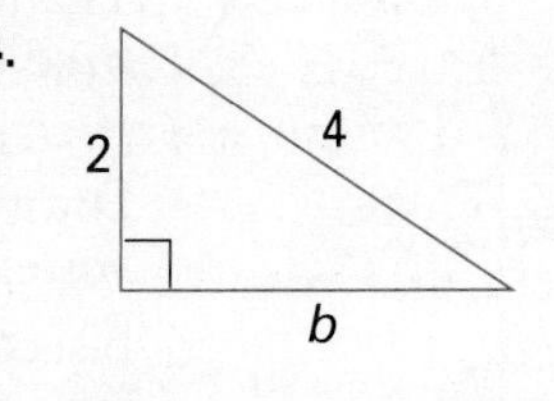

5.

6.

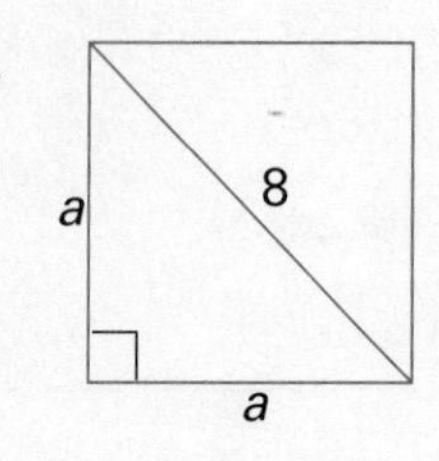

7.

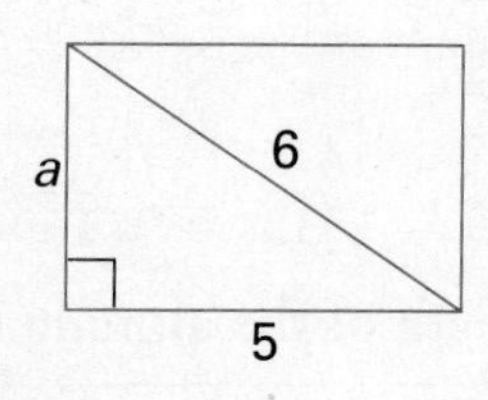

8.

Written Exercises

For Exercises 1–8, find *a*, *b*, or *c* for Classroom Exercises 1–8. For Exercises 9–14, *p* and *q* are lengths of legs, and *r* is the length of the hypotenuse of a right triangle.

9. If $p = 4$, $q = 2$, find r.

10. If $r = 8$, $p = 2$, find q.

11. If $r = 8$, $p = 6$, find q.

12. If $q = 15$, $r = 17$, find p.

13. If $p = 10$, $q = 10$, find r.

14. If $r = 10$, $p = 5$, find q.

Find the indicated length to the nearest tenth.

15. $a = 12$, $b = 14$, $c =$ ______

16. $a = 18$, $b = 23$, $c =$ ______

17. $a = 18$, $c = 32$, $b =$ ______

18. $c = 37$, $b = 14$, $a =$ ______

19. $a = 13.6$, $b = 11.8$, $c =$ ______

20. $c = 23.5$, $b = 11.3$, $a =$ ______

21. The figure at the right shows a shelf 12″ wide supported by 20″ braces like $\overline{AB}$. Find the distance $\overline{AC}$ from shelf to the brace.

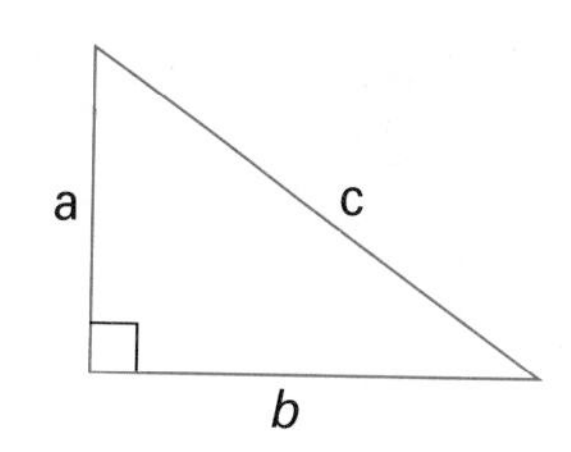

22. The lengths of two adjacent sides of a rectangle are 8 and 6. Find the length of a diagonal of the rectangle.

23. The length of each side of a rhombus is 8 ft. The length of one of its diagonals is 4 ft. Find the length of the other diagonal.

24. The lengths of the diagonals of a rhombus are 16 km and 8 km. Find the length of a side of the rhombus.

25. The lengths of the legs of an isosceles triangle are each 6 in. The length of the base is 8 in. Find the length of an altitude from the vertex angle to the base.

26. Find the length of the altitude to the hypotenuse of a right triangle if the lengths of the legs are 5 ft and 12 ft.

27. Find the length of the altitude to the hypotenuse of a right triangle if the lengths of the legs are 15 cm and 8 cm.

Find the value of *x* in each figure.

28.

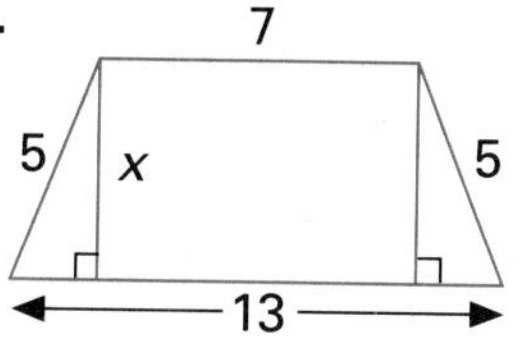

29.

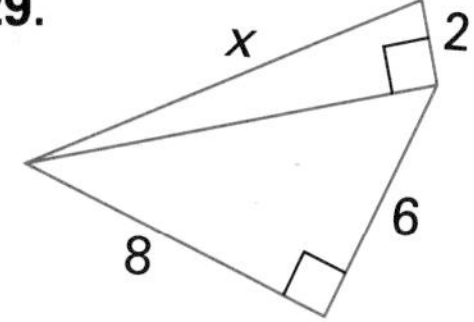

30.

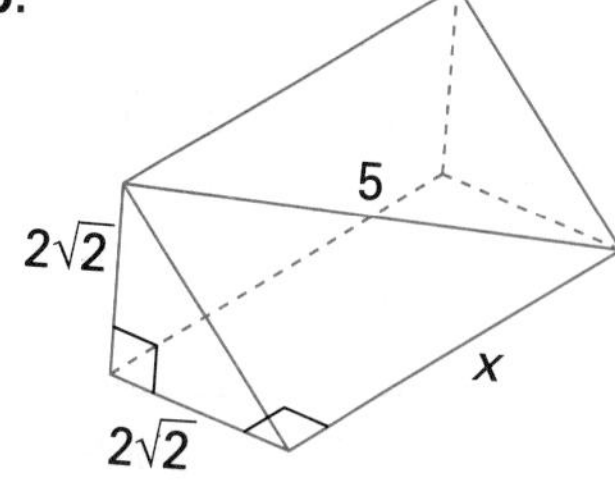

31. The base of an aerial fire truck ladder is 5 ft above the ground and 25 ft from the bottom of the wall of a burning building. How long must a ladder be to reach the roof, which is known to be 65 ft above the ground? (Fire companies find this out in advance for tall buildings.)

32. Find d, the length of the diagonal of a rectangular solid with dimensions 2, 4, and 6. (HINT: Use separate triangles, as shown.)

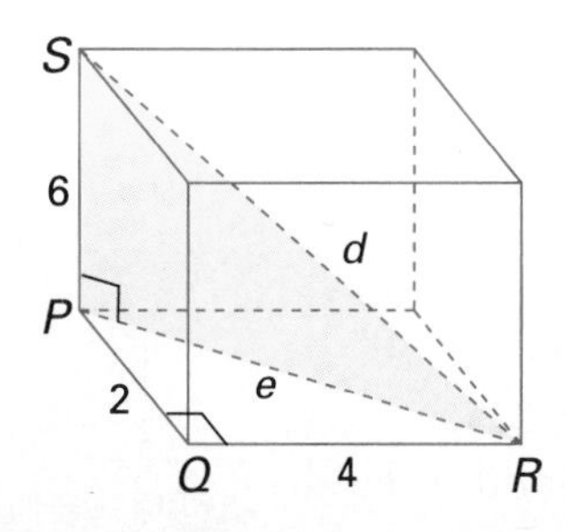

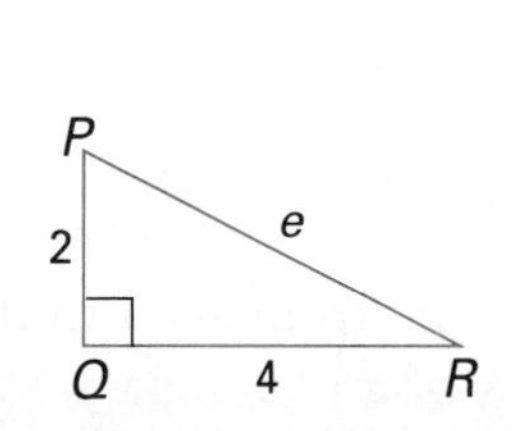

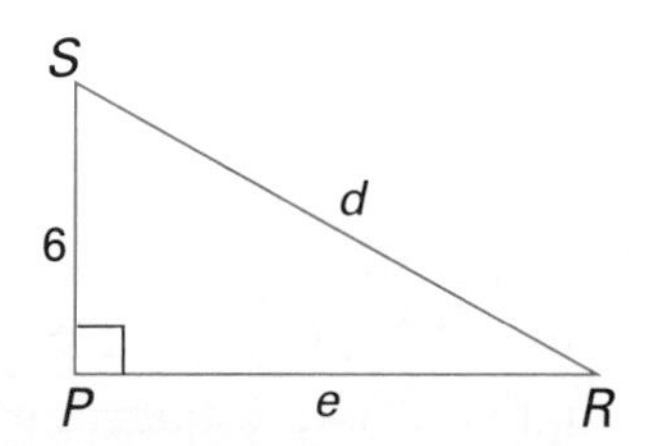

Find a formula for the length of a diagonal of a rectangular solid with sides of the given dimensions. (Exercises 33–35)

33. ℓ, w, and h

34. e, e, e (a cube)

35. x, $2\sqrt{3x}$, 6

36. In $\triangle ABC$, $AB = 9$, $BC = 11$, $AC = 10$. Find the length of the altitude $\overline{BN}$. $\triangle ABC$ is *not* a right triangle. (HINT: Apply the Pythagorean Theorem twice, using the diagram at the right.)

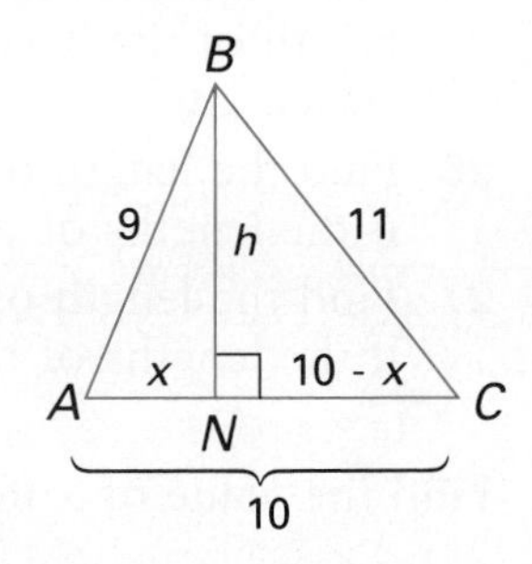

Mixed Review

1. Find the coordinate of the midpoint of a segment if the coordinates of its endpoints are -4 and 10. *1.3*
2. The measure of the vertex angle of an isosceles triangle is 40. Find the measure of each base angle. *4.1*
3. Given $\triangle ABC$ with m $\angle A = 40$, m $\angle B = 60$. Name the longest side of the triangle. *5.8*

Application: *Baseball "Diamond"*

The so-called baseball "diamond" is actually a square with sides of length 90 ft. The pitcher's mound is located near the center of the square, with the pitching rubber 60 ft 6 in. from home plate. (In the example and problems that follow, assume that the pitching rubber lies on the plane of the playing field. In fact, the rubber is slightly elevated by the mound.)

Example Show that the pitching rubber is situated closer to home plate than to second base.

Solution According to the Pythagorean Theorem,

$(HS)^2 = (HF)^2 + (FS)^2$
$(HS)^2 = 90^2 + 90^2$
$(HS)^2 = 2 \cdot 90^2$
$HS = 90\sqrt{2} \approx 127.3$ ft

S
T
F
H

Since the rubber is 60.5 ft from home plate, it must be approximately $127.3 - 60.5$, or 66.8 ft from second base.

Exercises

1. A third baseman fields a ground ball, steps on third to force a runner, and throws to first to get the batter. How far is his throw to first base?

It sometimes surprises baseball fans when a fast runner is unable to beat a slow pitcher to first base on a ball hit to the first baseman. In the next four Exercises compute the distances related to this problem.

2. Compute the distance from the pitching rubber to the center of the diamond.
3. Compute the distance from the center of the "diamond" to first base.
4. Using the results of Exercises 2 and 3, compute the distance from the pitching rubber to first base.
5. How much further is it from home plate to first base than from the pitching rubber to first base?

9.3 Converse of the Pythagorean Theorem

Objective To determine whether a triangle is right, acute, or obtuse

A person bought a triangular piece of riverfront property measuring 150′ by 200′ by 250′ and wondered if it formed a right triangle. This lesson will show you how to answer this question.

Theorem 9.3

Converse of the Pythagorean Theorem: If the sum of the squares of the lengths of two sides of a triangle is equal to the square of the length of the third side, then the triangle is a right triangle.

Given: $\triangle PQR$ with $p^2 + q^2 = r^2$
Prove: $\triangle PQR$ is a right triangle.

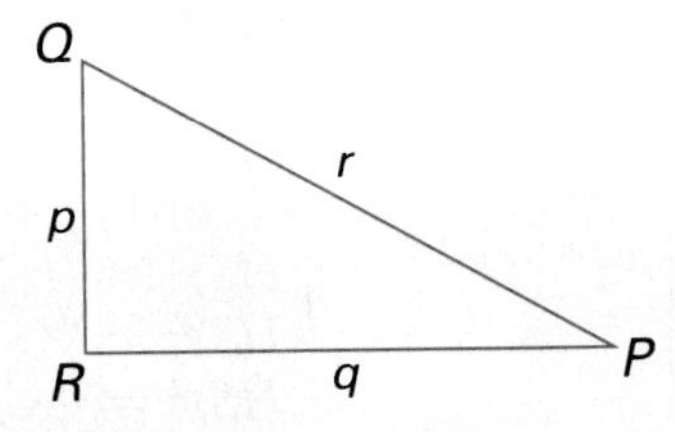

Proof (Paragraph form)

Construct $\triangle GHK$, with rt $\angle K$, $HK = p$, $GK = q$. By the Pythagorean Theorem,

$$(GH)^2 = p^2 + q^2$$

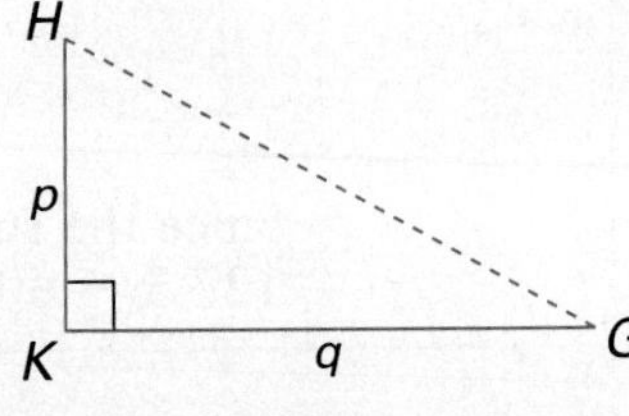

It is given that $r^2 = p^2 + q^2$.
Then $(GH)^2 = r^2$ and $GH = r$.
Because of the way $\triangle GHK$ was constructed and the fact that $GH = r$, $\triangle PQR \cong \triangle GHK$ by SSS. Then, by CPCTC, $\angle R$ is a right angle. Therefore, $\triangle PQR$ is a right triangle.

EXAMPLE 1 Given: $AB = 18$, $BC = 82$, $AC = 80$
Determine whether the triangle is a right triangle. If so, name the right angle.

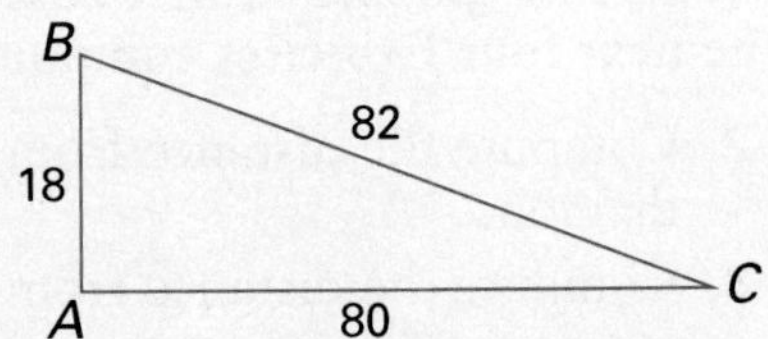

Solution

$$(AB)^2 + (AC)^2 \stackrel{?}{=} (BC)^2$$
$$18^2 + 80^2 \stackrel{?}{=} 82^2$$
$$324 + 6{,}400 \stackrel{?}{=} 6{,}724$$
$$6{,}724 = 6{,}724 \checkmark$$

The triangle is a right triangle since $(AB)^2 + (AC)^2 = (BC)^2$.
The right angle is $\angle A$, the angle opposite the longest side $\overline{BC}$.

Theorem 9.4 If the square of the longest side of a triangle is greater (less) than the sum of the squares of the lengths of the other two sides, then the triangle is obtuse (acute).

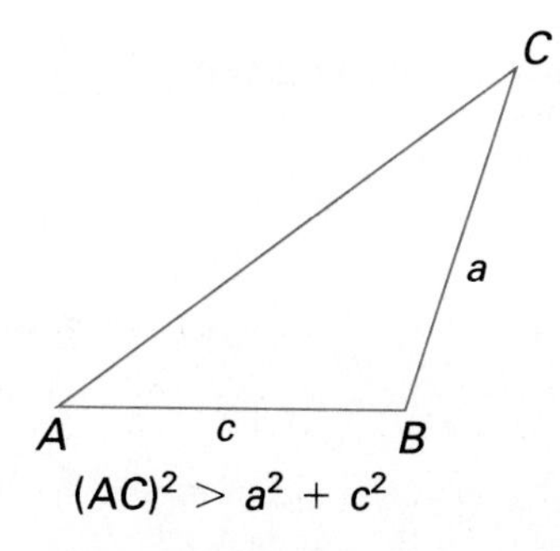

$(AC)^2 > a^2 + c^2$

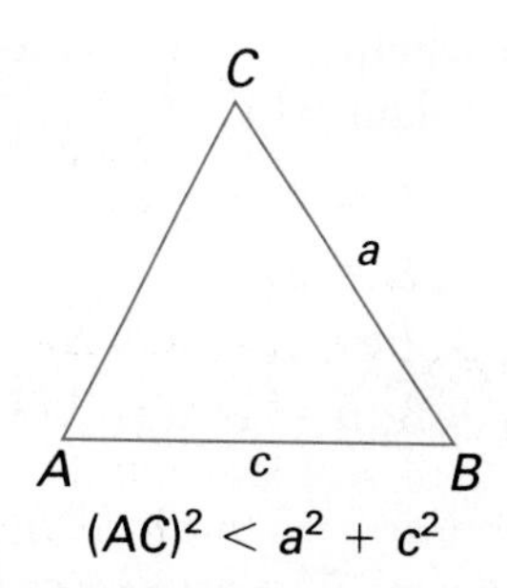

$(AC)^2 < a^2 + c^2$

Theorems 9.3 and 9.4 can be used to determine whether a triangle is acute, obtuse, or right. It may be necessary to use the algebraic property that $(\sqrt{a})^2 = \sqrt{a} \cdot \sqrt{a} = a$, for all $a \geq 0$.

EXAMPLE 2 Determine whether the triangle is acute, obtuse, or right for three sides of the given lengths.

a. 10, 2, 9 **b.** 15, 36, 39 **c.** $\sqrt{6}, \sqrt{3}, \sqrt{5}$

Solutions Compare the square of the greatest length with the sum of the squares of the other two lengths.

a.
$10^2 \overset{?}{=} 2^2 + 9^2$
$100 \overset{?}{=} 4 + 81$
$100 > 85$

The triangle is an obtuse triangle.

b.
$39^2 \overset{?}{=} 15^2 + 36^2$
$1{,}521 \overset{?}{=} 225 + 1{,}296$
$1{,}521 = 1{,}521$

The triangle is a right triangle.

c.
$(\sqrt{6})^2 \overset{?}{=} (\sqrt{3})^2 + (\sqrt{5})^2$
$6 \overset{?}{=} 3 + 5$
$6 < 8$

The triangle is an acute triangle.

EXAMPLE 3 Sketch parallelogram $ABCD$ with diagonals intersecting at O, $DC = 10$, $AC = 16$, and $BD = 12$. Determine whether $ABCD$ is a rhombus.

Solution $OD = 6$, $OC = 8$ since diagonals of a ▱ bisect each other.

$(OC)^2 + (OD)^2 \overset{?}{=} (DC)^2$
$8^2 + 6^2 \overset{?}{=} 10^2$
$64 + 36 \overset{?}{=} 100$
$100 = 100$ ✓

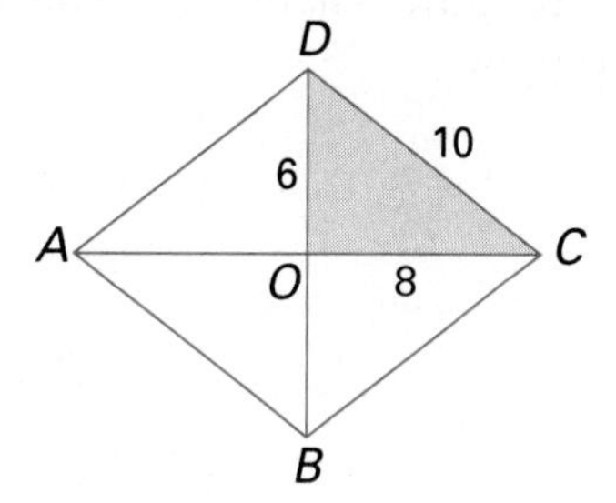

Therefore, $\triangle DOC$ is a right triangle, and the diagonals of ▱$ABCD$ are perpendicular to each other.

Thus, ▱$ABCD$ is a rhombus.

Classroom Exercises

Simplify.

1. $(\sqrt{7})^2$ 2. $(\sqrt{3})^2$ 3. $(\sqrt{y})^2$ 4. $(\sqrt{2y})^2$ 5. $(3\sqrt{2})^2$

Indicate whether $\triangle ABC$ is a right triangle for the given lengths of a, b, and c. Explain why.

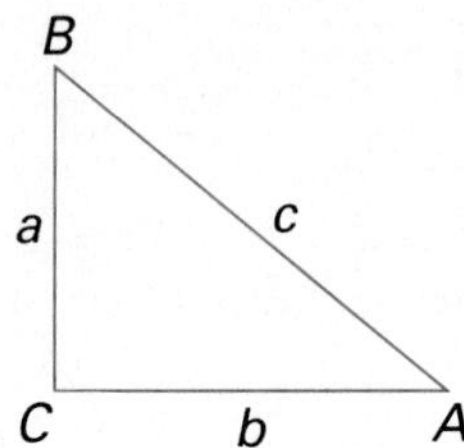

6. $a = 3, b = 4, c = 5$
7. $a = 5, b = 2, c = 6$
8. $a = 2, b = 1, c = \sqrt{5}$
9. $a = \sqrt{3}, b = 2, c = \sqrt{7}$

Written Exercises

The lengths of three sides of $\triangle ABC$ are given. Determine whether $\triangle ABC$ is a right triangle. If so, name the right angle.

1. $AC = 18, BC = 24, AB = 30$
2. $BC = 30, AB = 72, AC = 78$
3. $AC = 17, AB = 16, BC = 14$
4. $AC = 1.5, AB = 2, BC = 2.5$

Determine whether the triangle formed with given lengths of the three sides is acute, obtuse, or right.

5. 24, 32, 50
6. 9, 16, 19
7. 18, 24, 28
8. 0.6, 0.6, 1
9. $\frac{5}{13}, 1, \frac{12}{13}$
10. 8, 5, 5
11. 7, 6, 9
12. 7, 3, 5
13. $8x, 17x, 15x$
14. a, a, a
15. $2\sqrt{3}, \sqrt{30}, 3\sqrt{2}$
16. $a, a, a\sqrt{2}$

Determine whether the parallelogram $ABCD$ is a rhombus, using the given data.

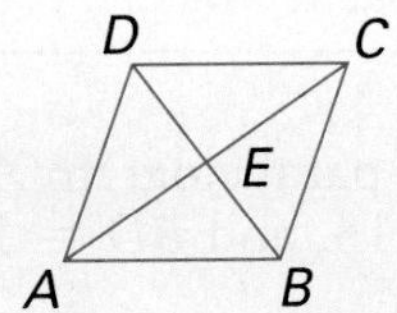

17. $AD = 10, AC = 16, DB = 12$
18. $DC = 4, DB = 4\sqrt{3}, AE = 2$
19. $DB = 2\sqrt{3}, AC = 2\sqrt{2}, AB = \sqrt{5}$
20. $AC = 2\sqrt{3}, DB = 2, AD = 2$

21. Determine whether $\triangle BCD$ is a right triangle.

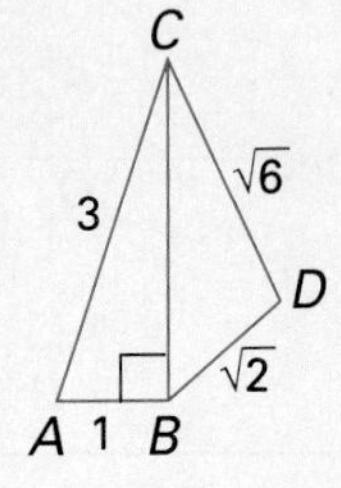

22. Given: $\overline{PQ} \perp$ plane $\mathcal{M}$. Determine whether $\triangle RQS$ is acute, obtuse, or right.

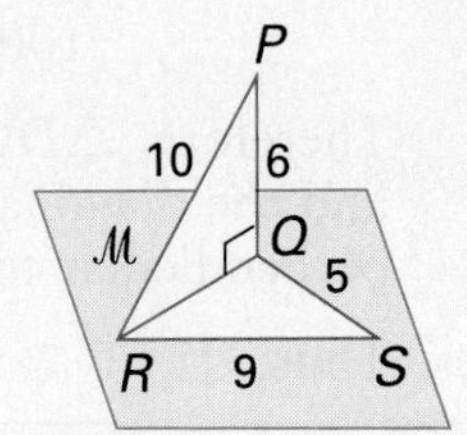

Essay.

23. Describe in a short paragraph how to determine whether a triangle is right, acute, or obtuse, given the lengths of the three sides.

24. Given: $\overline{AD}$ is the altitude to $\overline{BC}$; $(AD)^2 = (DC) \cdot (BD)$
Prove: $\angle BAC$ is a right angle.

25. Given: $\overline{AD}$ is the altitude to $\overline{BC}$; $(AC)^2 = (BC) \cdot (DC)$
Prove: $\angle BAC$ is a right angle.

26. Prove Theorem 9.4 for an obtuse triangle.

27. The medians to the legs of a right triangle have lengths $2\sqrt{13}$ and $\sqrt{73}$. Find the lengths of the three sides of the triangle.

28. Given: x and y are the lengths of the legs of a right triangle with hypotenuse of length z. Show that the triangle with sides of lengths $x + 2$, $y + 2$, and $z + 2$ is an acute triangle.

29. In $\triangle ABC$, $\overline{AB}$ is the longest side. AC is 2 more than twice CB, and AC is 1 less than AB. Find the length of CB to guarantee that $\triangle ABC$ will be obtuse.

Mixed Review

In the figure at the right, $m \parallel n$.

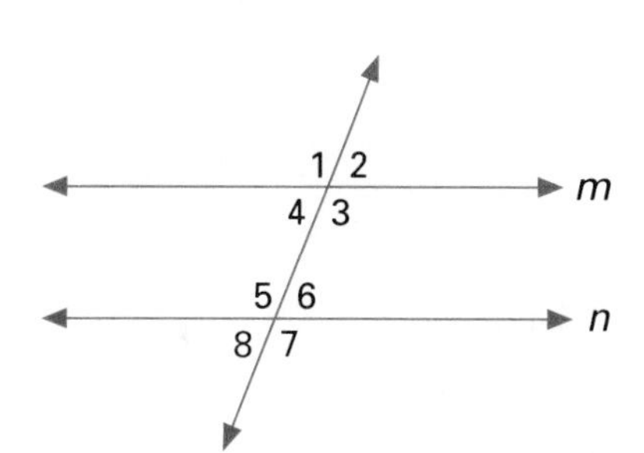

1. Find m $\angle 7$ if m $\angle 1 = 105$. *3.5*

2. m $\angle 4 = \frac{2}{3}$m $\angle 5$. Find m $\angle 5$. *3.5*

3. m $\angle 4 = 3x + 20$, m $\angle 6 = x + 60$. Find m $\angle 4$. *3.5*

4. What is the corresponding angle to $\angle 2$? *3.1*

Algebra Review

Example: To simplify an expression with a radical in the denominator: Rationalize the denominator.

$$\frac{4}{\sqrt{2}} = \frac{4 \cdot \sqrt{2}}{\sqrt{2} \cdot \sqrt{2}} = \frac{4\sqrt{2}}{2} = 2\sqrt{2}$$

Simplify the expression by rationalizing the denominator.

1. $\frac{1}{\sqrt{2}}$ **2.** $\frac{6}{\sqrt{3}}$ **3.** $\frac{4}{\sqrt{2}}$ **4.** $\frac{28}{\sqrt{7}}$ **5.** $\frac{10}{\sqrt{5}}$ **6.** $\frac{4}{\sqrt{6}}$

9.4 Two Special Types of Right Triangles

Objectives To apply the properties of the 45-45-90 triangle
To apply the properties of the 30-60-90 triangle

The isosceles right triangle, shown at the right, occurs frequently in the study of mathematics. Notice that the measure of each base angle is 45. This special triangle is often called a *45-45-90 triangle*.

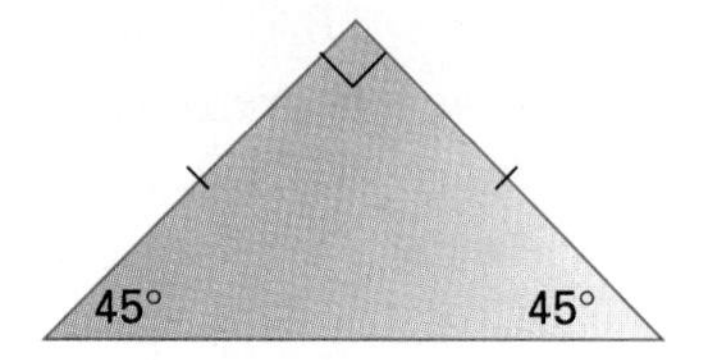

EXAMPLE 1 Find the length of the hypotenuse of the right triangle.

Solution

$h^2 = 4^2 + 4^2$
$h^2 = 16 + 16$
$h^2 = 32$
$h = \sqrt{32}$, or $\sqrt{16} \cdot \sqrt{2}$
$h = 4\sqrt{2}$

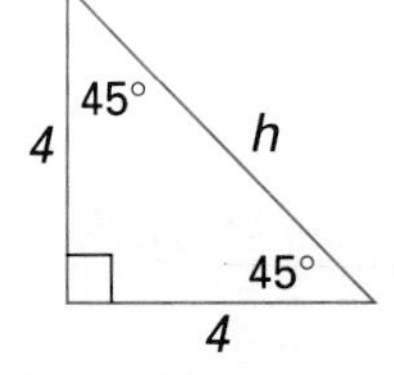

Notice a pattern for the 45-45-90 triangle.

Theorem 9.5 In a 45-45-90 triangle, the hypotenuse is $\sqrt{2}$ times as long as a leg.

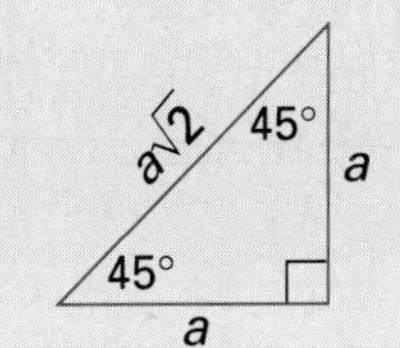

EXAMPLE 2 $\triangle XYZ$ is a 45-45-90 triangle. Find the length of a leg if the length of the hypotenuse is 12.

Solution

$12 = a\sqrt{2}$

$a = \frac{12}{\sqrt{2}}$

$a = \frac{12 \cdot \sqrt{2}}{\sqrt{2} \cdot \sqrt{2}}$

$= \frac{12\sqrt{2}}{2}$, or $6\sqrt{2}$

X
a
12
Y
a
Z

Thus, the length of a leg of $\triangle XYZ$ is $6\sqrt{2}$.

Another special type of right triangle is the 30-60-90 triangle.

Theorem 9.6 In a 30-60-90 triangle:
(1) the hypotenuse is twice as long as the leg opposite the angle of measure 30; (2) the leg opposite the angle of measure 60 is $\sqrt{3}$ times as long as the leg opposite the angle of measure 30.

Given: Right $\triangle PQR$ with right $\angle PRQ$, m $\angle Q = 60$,
m $\angle P = 30$
Prove: $QP = 2 \cdot QR$, $PR = \sqrt{3} \cdot QR$

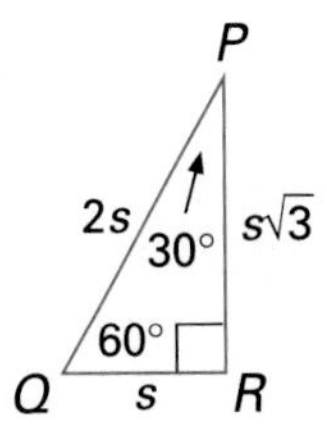

Plan Extend $\overline{QR}$ to T so that $RT = RQ$. Draw $\overline{PT}$. Prove that $\triangle PRQ \cong \triangle PRT$. Show that m $\angle Q$ = m $\angle T$ = m $\angle QPT = 60$. Let $QR = s$. Then QP must equal $2s$. Use the Pythagorean Theorem to find PR in terms of s.

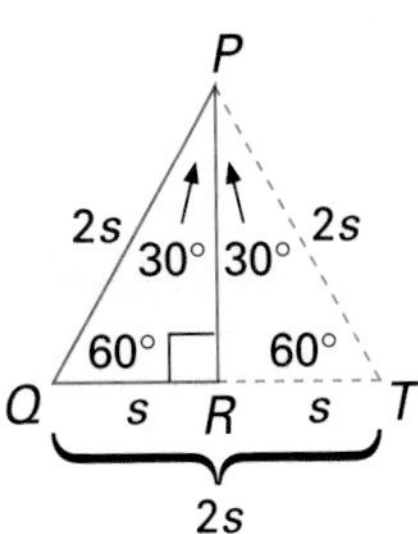

EXAMPLE 3 Find the lengths of the legs of a 30-60-90 triangle if the length of the hypotenuse is 8.

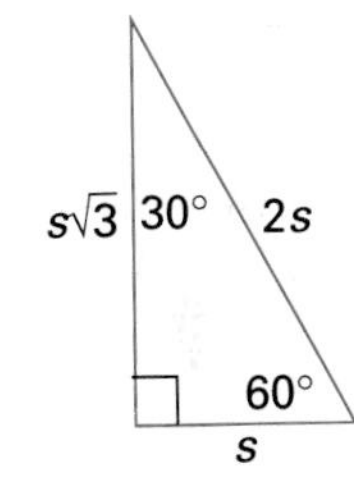

Solution

$$2s = 8$$
$$s = 4$$

Then, $s\sqrt{3} = 4\sqrt{3}$

Thus, the lengths of the legs are 4 and $4\sqrt{3}$.

EXAMPLE 4 $\triangle ABC$ is a 30-60-90 triangle. $AC = 12$. Find AB and BC.

Solution $AC = s\sqrt{3} = 12$

$s = \dfrac{12}{\sqrt{3}}$

$s = \dfrac{12 \cdot \sqrt{3}}{\sqrt{3} \cdot \sqrt{3}}$

$s = \dfrac{12\sqrt{3}}{3}$

$AB = s = 4\sqrt{3}$
$BC = 2s = 8\sqrt{3}$

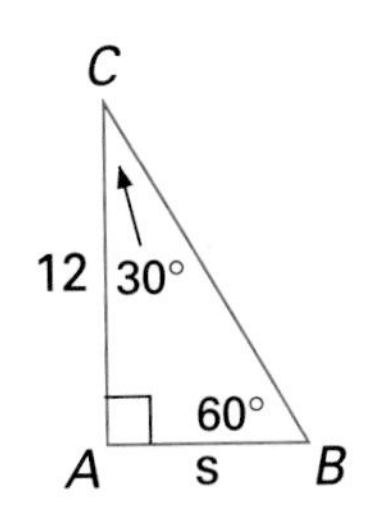

Thus, $AB = 4\sqrt{3}$ and $BC = 8\sqrt{3}$.

EXAMPLE 5 The length of an altitude of an equilateral triangle is $4\sqrt{3}$. Find the length of a side of the triangle.

Plan Draw an equilateral triangle with an altitude. Two 30-60-90 triangles are formed. Label the sides of one of the triangles as s, $2s$, and $s\sqrt{3}$.

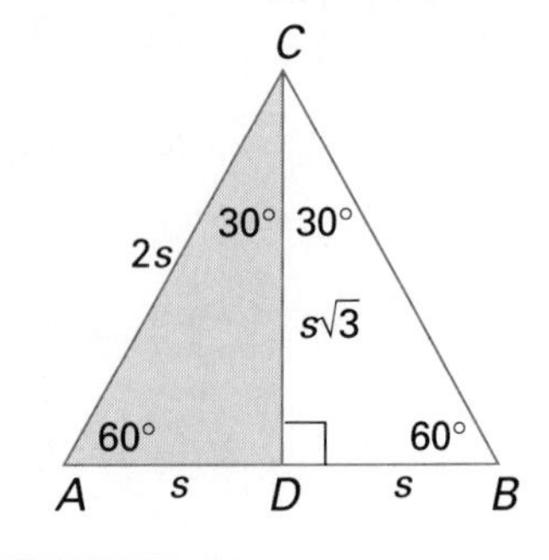

Solution

$$CD = s\sqrt{3} = 4\sqrt{3}$$
$$s = 4$$
$$AB = 2s, \text{ or } 8$$

Thus, the length of a side is 8.

EXAMPLE 6 Given: $\square ABCD$, altitude $\overline{DE}$, $AD = 6$, $m\angle B = 120$

Find the length of the altitude, $\overline{DE}$.

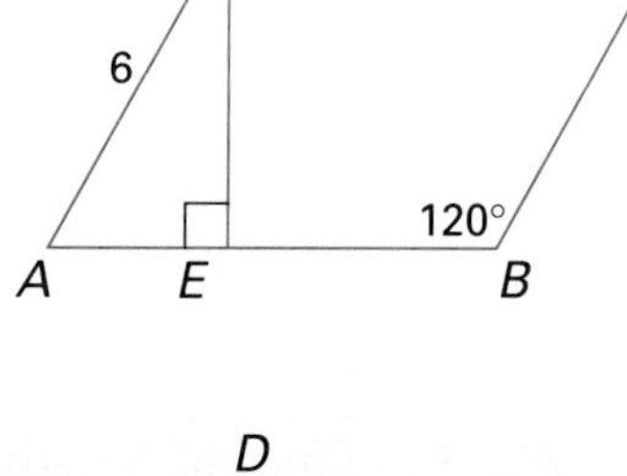

Plan Since $ABCD$ is a parallelogram, the consecutive angles A and B are supplementary. $m\angle A = 180 - 120 = 60$. Use the 30-60-90 triangle.

Solution Use the 30-60-90 triangle at the right.

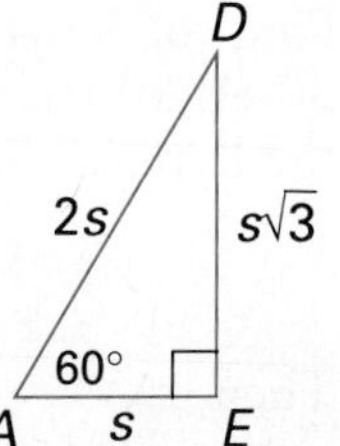

$$AD = 2s = 6$$
$$s = 3$$

Thus, $DE = s\sqrt{3} = 3\sqrt{3}$.

Summary

Special Right Triangle Properties

45-45-90 Triangle

30-60-90 Triangle

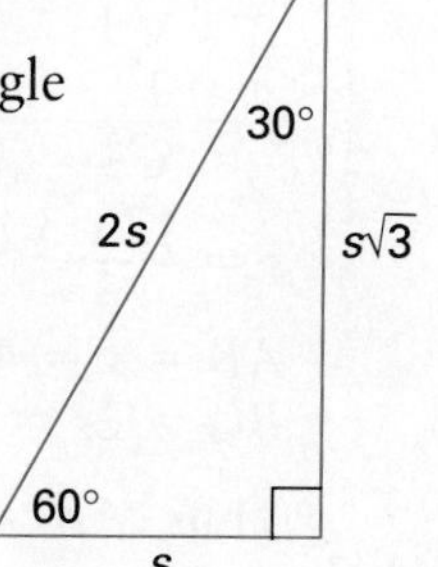

Focus on Reading

Copy and complete the sentence to make a true statement.

1. In a 30-60-90 triangle, the length of the leg opposite the angle measuring ______ is $\frac{1}{2}$ the length of the hypotenuse.
2. An altitude of an equilateral triangle divides the triangle into two ______-______-______ triangles.
3. In a 45-45-90 triangle, the length of the hypotenuse is ______.
4. For a 30-60-90 triangle, the lengths of the three sides can be represented as s, $2s$ and ______.

Classroom Exercises

Find the indicated lengths.

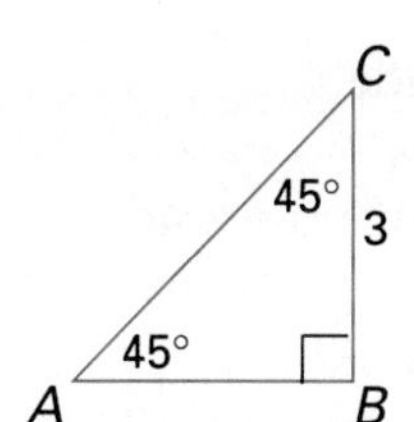

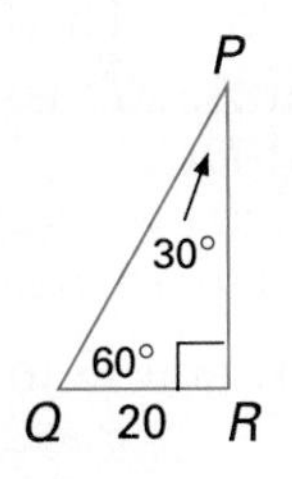

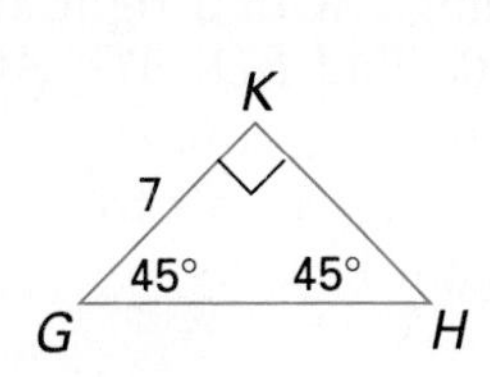

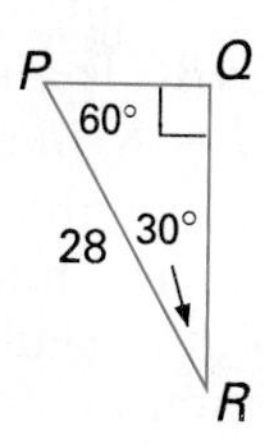

1. $AB =$ ______
2. $AC =$ ______
3. $QP =$ ______
4. $PR =$ ______
5. $KH =$ ______
6. $GH =$ ______
7. $RQ =$ ______
8. $PQ =$ ______

Written Exercises

For the given length, find each of the remaining two lengths.

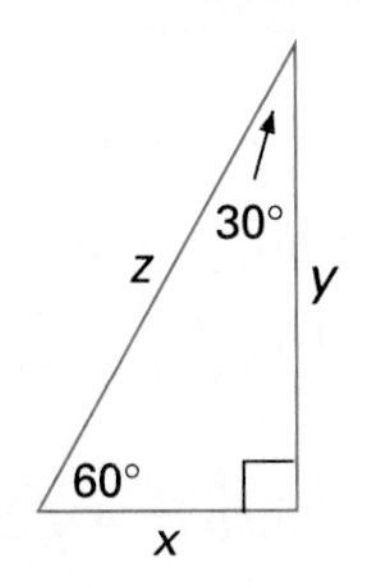

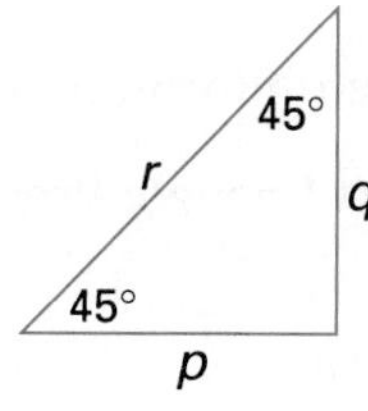

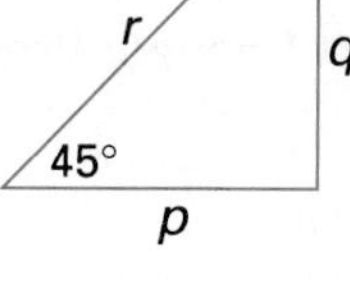

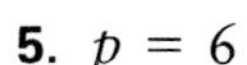

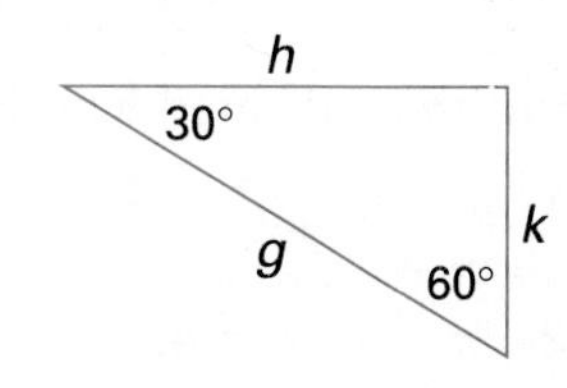

1. $x = 6$
2. $y = 6$
3. $z = 14$
4. $y = 4\sqrt{3}$
5. $p = 6$
6. $r = 6$
7. $q = 4\sqrt{2}$
8. $r = 10$
9. $k = 3.4$
10. $k = 6\sqrt{3}$
11. $g = 17$
12. $h = 2\sqrt{3}$

Find the indicated lengths using the given information.

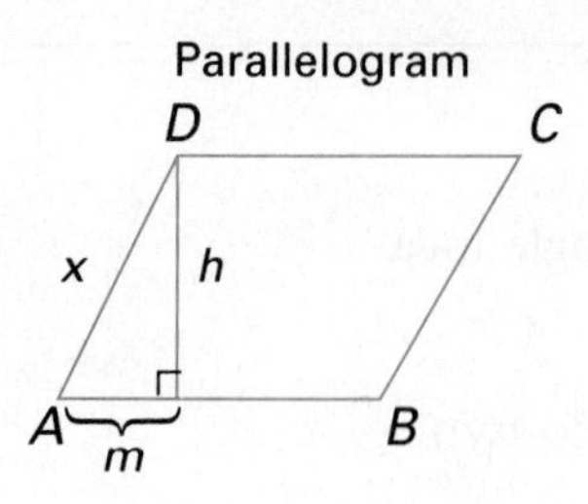

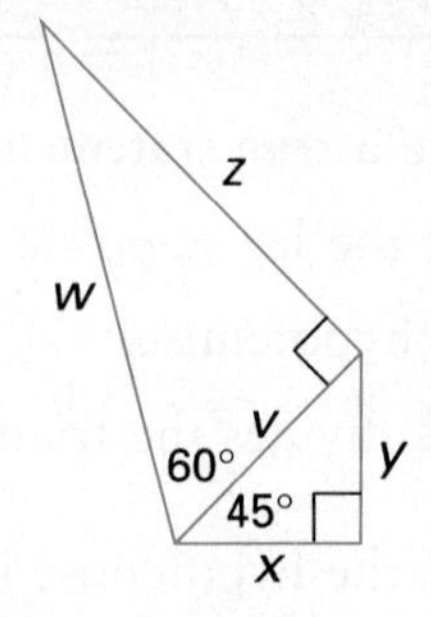

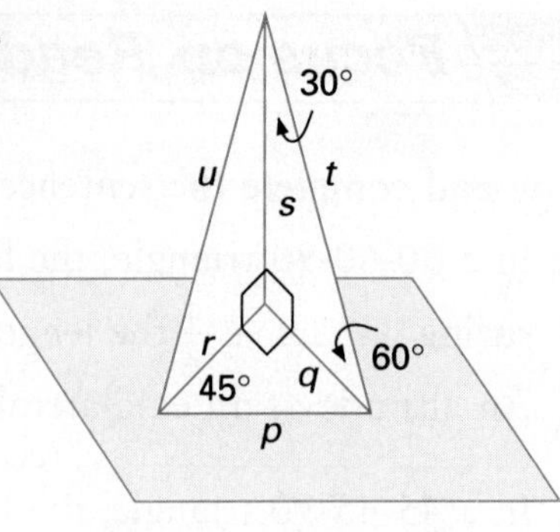

13. m $\angle B = 120$, $x = 18$, $h =$ ______

14. m $\angle D = 150$, $x = 4$, $h =$ ______

15. m $\angle B = 135$, $x = 6$, $m =$ ______

16. $x = 4$, $w =$ ______

17. $y = \sqrt{3}$, $z =$ ______

18. $w = 6$, $x =$ ______

19. $z = 4\sqrt{3}$, $y =$ ______

20. $t = 10$, $q =$ ______

21. $s = 6\sqrt{3}$, $q =$ ______

22. $p = 10$, $q =$ ______

23. $t = 8$, $p =$ ______

24. Prove Theorem 9.5

25. Prove Theorem 9.6

26. $\triangle ABC$ is a right triangle with a right angle at A. $\overline{AD}$ is an altitude $AB = 8$, m $\angle B = 30$. Find BD, AD, AC, and DC.

27. The length of each side of a square is 8. The four corners are cut off to form a regular octagon. Find its perimeter.

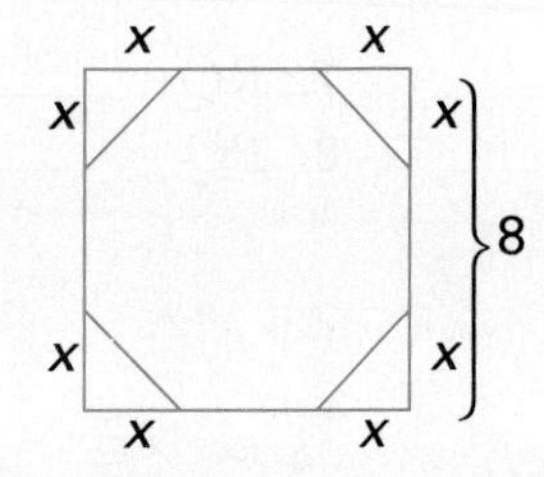

28. Find the perimeter of trapezoid $ABCD$. (HINT: Draw altitudes from C and D to the base $\overline{AB}$.)

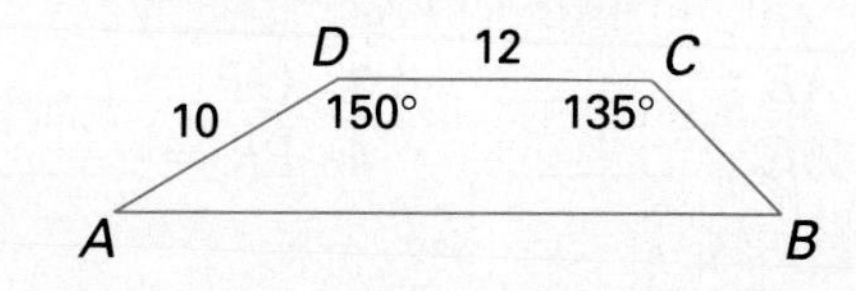

Midchapter Review

Find the indicated lengths using the given information. *9.1, 9.2, 9.4*

1. $p = 4$, $QP = 8$, $m =$ ______

2. $m = 6$, $r = 4$, $h =$ ______

3. $m = 4$, $r = 5$, $p =$ ______

4. m $\angle Q = 60$, $p = 6$, $q =$ ______

5. $p = 4$, $q = 4\sqrt{2}$, $QP =$ ______

6. m $\angle P = 45$, $p = 4\sqrt{2}$, $m + r =$ ______

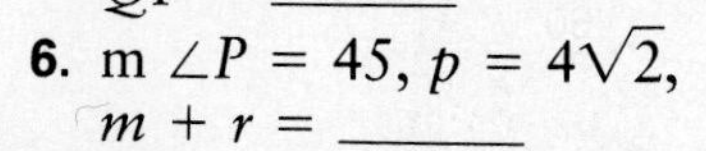

Determine whether the triangle with given side lengths is acute, obtuse, or right. If the triangle is right, indicate which side is the hypotenuse. *9.3*

7. 4, 6, 7

8. 3, 7, 9

9. $\sqrt{6}, \sqrt{2}, 2\sqrt{2}$

Computer Investigation

A Special Ratio in Right Triangles

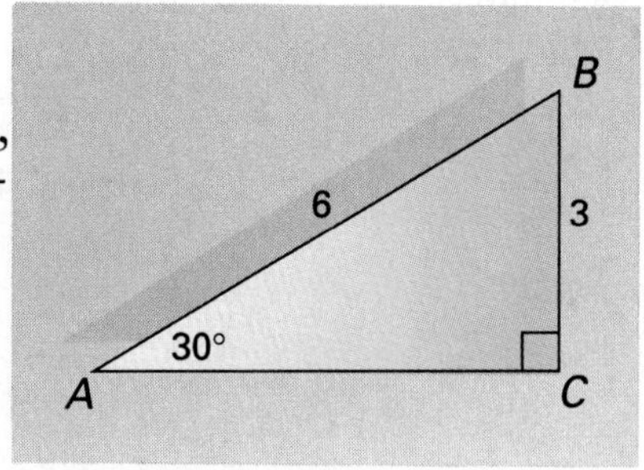

Use a computer software program that draws angles of a given measure, constructs perpendicular segments, extends segments, labels points, and measures line segments and angles. In right $\triangle ACB$ $\overline{AB}$ is the hypotenuse. $\overline{BC}$ is the leg of the right $\triangle$, *opposite* the angle with measure 30. Notice the ratio

$$\frac{\text{length of leg opposite } \angle \text{ of } 30}{\text{length of hyp}} = \frac{3}{6}, \text{ or } \frac{1}{2}.$$

This ratio for a right triangle is very important in mathematics and will be formally defined in the next lesson.

Activity 1

Draw a right triangle as directed below.

1. Draw $\angle CBA$ with measure 30.
2. From C draw a segment $\overline{CD}$ perpendicular to $\overline{BA}$ intersecting $\overline{BA}$ at E; $\triangle BEC$ is a right triangle.
3. Measure to find CB (the length of the hypotenuse) and CE (the length of the leg opposite $\angle B$).
4. Find the ratio $\frac{CE}{CB}$ to the nearest hundredth of a unit.

Activity 2

Repeat the steps above for a new angle of measure 30. Draw the angle with its sides extended so that the new right triangle will not be necessarily congruent to the first triangle of Activity 1.

5. How does the ratio $\frac{CE}{CB}$ compare with the ratio of Activity 1?

Activity 3

Repeat the steps of Activities 1 and 2 above for right triangles each containing an acute angle with the given measure and the given name.

6. m $\angle CBA = 25$
7. m $\angle CBA = 45$

Summary

Generalize the results of Activities 1–3 above.

In a right triangle with an acute angle of a given measure, what ratio is always the same, regardless of the lengths of the sides?

9.5 The Sine Ratio

Objective To find side lengths and angle measures in right triangles using the sine ratio

Notice a pattern that applies to the three triangles below.

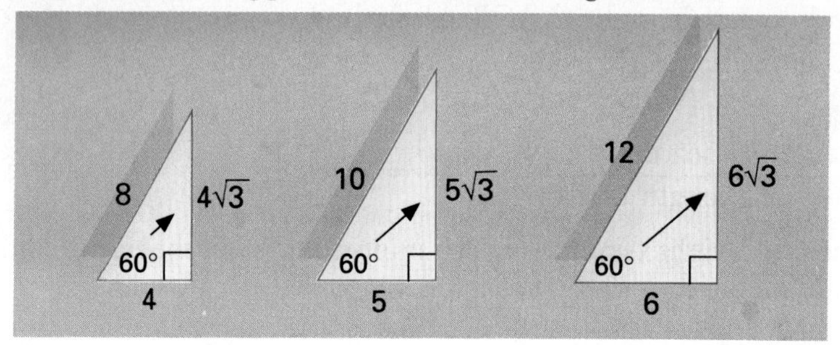

The ratio of the length of the side opposite the angle measuring 60 to the length of the hypotenuse is the same for each triangle.

$$\frac{\text{length of side opposite angle of 60}}{\text{length of hypotenuse}} = \frac{4\sqrt{3}}{8} = \frac{5\sqrt{3}}{10} = \frac{6\sqrt{3}}{12} = \frac{\sqrt{3}}{2}$$

Using a calculator or square root table, the ratio $\frac{\sqrt{3}}{2} \approx \frac{1.732}{2}$, or 0.8660. By the properties of similar triangles, it can be shown that this ratio is the same for any angle measuring 60, regardless of the lengths of the sides. This ratio is called *sine* of the angle measuring 60. So, the sine of 60 $\approx$ 0.8660. This is written and abbreviated as sin 60 $\approx$ 0.8660. For convenience, we agree to use = rather than the approximation symbol, $\approx$.

The sine of an angle is an example of a trigonometric ratio. The word *trigonometry* is derived from two Greek words meaning *triangle measurement.*

Definition For right triangle ABC with angle A as shown,

sine of angle $A = \dfrac{\text{length of opposite leg}}{\text{length of hypotenuse}}$;

$\sin A = \dfrac{a}{c}$.

hypotenuse → c; a ← leg opposite $\angle A$; A, B, C

You can use the trigonometric table on page 643 to find decimal approximations for the sines of angles from 0 to 90.

EXAMPLE 1 Use the table of ratios to find sin 16 and sin 60.

Solution Locate 16 and 60 in the angle measure column. Then read across to the sine column.

Angle Measure	Sin	Cos	Tan	Angle Measure	Sin	Cos	Tan
12°	.2079	.9781	.2126	58°	.8480	.5299	1.600
13°	.2250	.9744	.2309	59°	.8572	.5150	1.664
14°	.2419	.9703	.2493	→60° →	.8660	.5000	1.732
15°	.2588	.9659	.2679	61°	.8746	.4848	1.804
→16° →	.2756	.9613	.2867	62°	.8829	.4695	1.881
17°	.2924	.9563	.3057	63°	.8910	.4540	1.963

Thus, sin 16 = 0.2756 and sin 60 = 0.8660.

EXAMPLE 2 Find m $\angle A$ to the nearest degree if sin A = 0.9394.

Solution Read down the sine column. The value closest to 0.9394 is 0.9397. From this number read across to the angle column.

Angle Measure	Sin	Cos	Tan	Angle Measure	Sin	Cos	Tan
22°	.3746	.9272	.4040	68°	.9272	.3746	2.475
23°	.3907	.9205	.4245	69°	.9336	.3584	2.605
24°	.4067	.9135	.4452	70° ←	.9397 ←	.3420	2.747
25°	.4226	.9063	.4663	71°	.9455	.3256	2.904
26°	.4384	.8988	.4877	72°	.9511	.3090	3.077

So, m $\angle A$ = 70 to the nearest degree.

Equations can now be formed with the sine ratio in order to calculate side lengths or angle measures of given right triangles.

EXAMPLE 3 If m $\angle A$ = 25 and c = 7, find a to the nearest tenth.

A
25°
b
c = 7
B
a
C

Solution Write an equation using the sine ratio.

$$\sin A = \frac{\text{length of leg opposite } \angle A}{\text{length of hypotenuse}} = \frac{a}{c}$$

$$\sin 25 = \frac{a}{7}$$ [Use the table in Example 2 to find sin 25.]

$$0.4226 = \frac{a}{7}$$

$$0.4226(7) = \left(\frac{a}{7}\right)7$$

$$2.9582 = a$$ Thus, rounded to the nearest tenth, a is 3.0.

EXAMPLE 4 The length of the hypotenuse of a right triangle is 9. The length of a leg is 6. Find the measure of each acute angle to the nearest degree.

Solution First sketch a right triangle. Write an equation using the sine ratio.

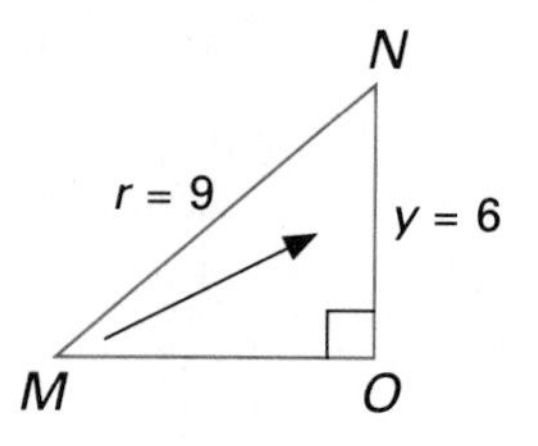

$$\sin M = \frac{\text{length of leg opposite angle}}{\text{length of hypotenuse}}$$

$$\sin M = \frac{y}{r} = \frac{6}{9}, \text{ or } \frac{2}{3}$$

Divide to 5 decimal places: $\sin M = 0.66666$
Round to 4 decimal places: $\sin M = 0.6667$
Find closest value in sine column: $m\angle M = 42$ to the nearest degree.
$m\angle N = 90 - 42$, or 48.

Thus, the measures of the acute angles are approximately 42 and 48.

EXAMPLE 5 The measure of the vertex angle of an isosceles triangle is 40. The length of the altitude to the base is 12. Find the length of a leg to the nearest tenth.

Plan Draw the triangle. The altitude bisects the vertex angle. Find $m\angle B$.

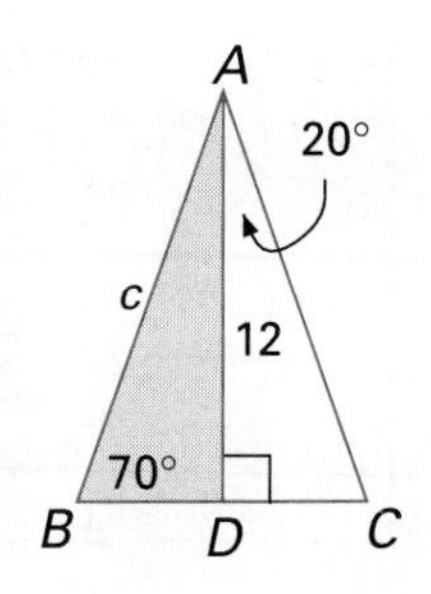

$$m\angle B = 90 - \frac{1}{2} \cdot 40 = 90 - 20 = 70$$

Write an equation using the sine ratio for right $\triangle BDA$.

Solution

$$\sin 70 = \frac{12}{c}$$

$$c(\sin 70) = 12$$

$$c = \frac{12}{\sin 70} = \frac{12}{0.9397} = 12.77$$

Thus, the length of a leg is 12.8, rounded to the nearest tenth.

Classroom Exercises

For the right triangle, give sin *A* and sin *B* as a ratio of lengths of sides.

1. 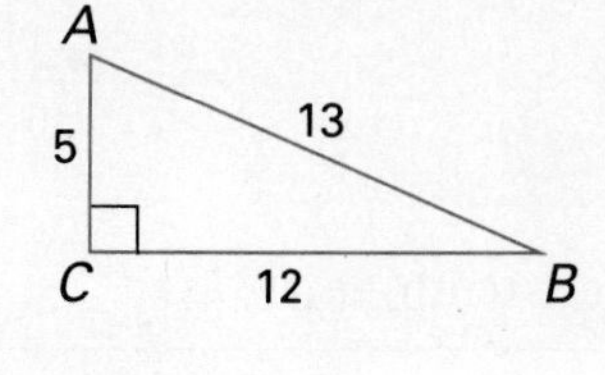

2. 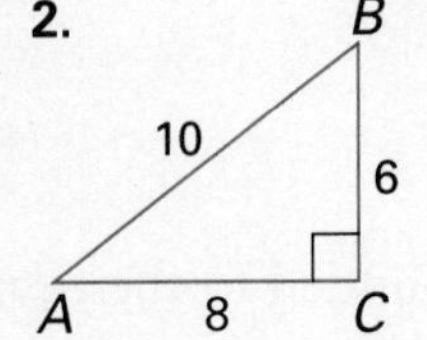

3. 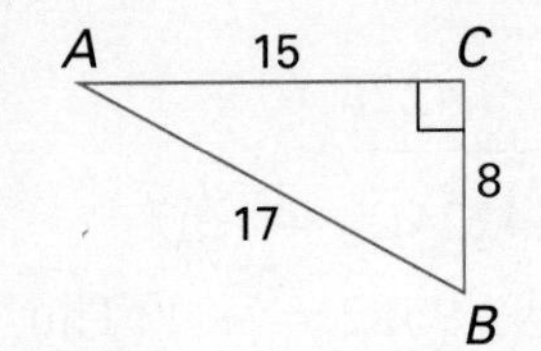

4. 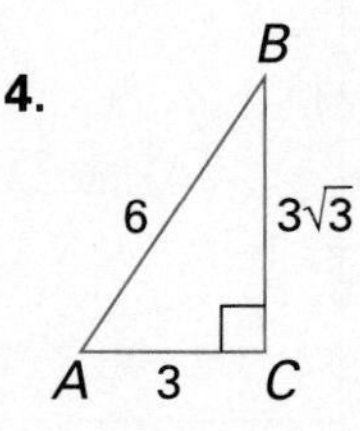

Write an equation that can be used to find the indicated side length or angle measure.

5.

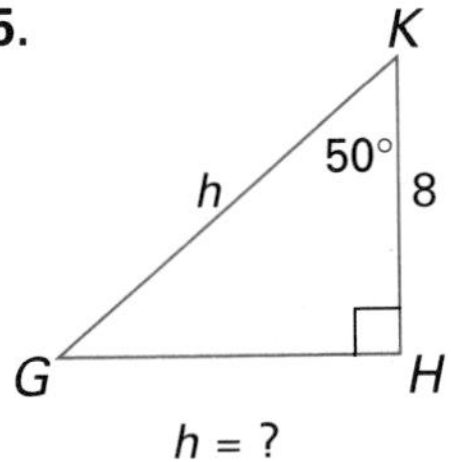

6.

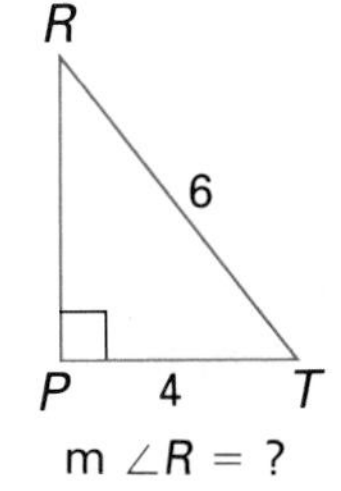

7.

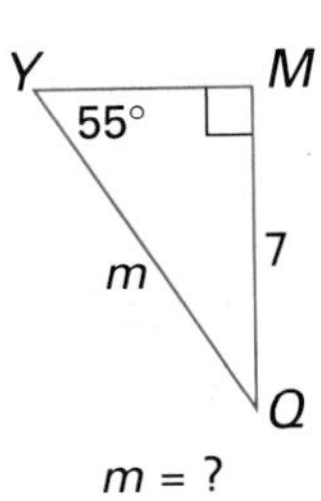

8.

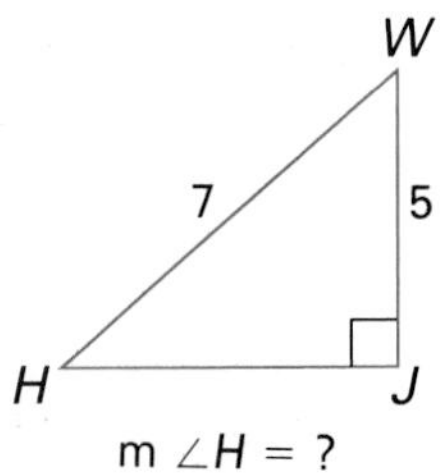

Written Exercises

Use the table on page 643 to find the indicated value.

1. sin 49 **2.** sin 79 **3.** sin 19 **4.** sin 84 **5.** sin 62 **6.** sin 88

Use the table on page 643 to find m $\angle A$ to the nearest degree.

7. $\sin A = 0.6293$ **8.** $\sin A = 0.9976$ **9.** $\sin A = 0.5446$
10. $\sin A = 0.6020$ **11.** $\sin A = 0.9562$ **12.** $\sin A = 0.3018$

Find the indicated measure to the nearest tenth (sides) or to the nearest degree (angles).

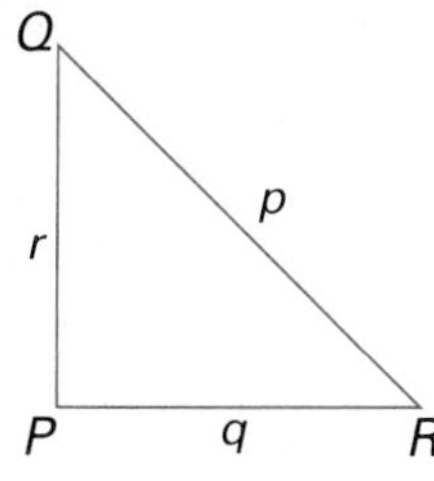

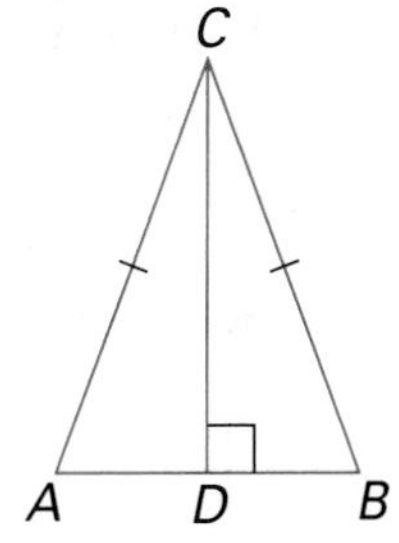

13. m $\angle R = 42$, $p = 18$, $r =$ ______
14. m $\angle Q = 65$, $p = 19$, $q =$ ______
15. $q = 10$, $p = 12$, m $\angle Q =$ ______
16. $q = 15$, $p = 20$, m $\angle Q =$ ______

17. m $\angle ACB = 50$, $AB = 20$, $AC =$ ______
18. m $\angle ACB = 40$, $AC = 16$, $AD =$ ______
19. $AC = 30$, $AB = 20$, m $\angle A =$ ______

20. A jet plane is 10 km above the ground in the figure below. Find the distance d from the runway.

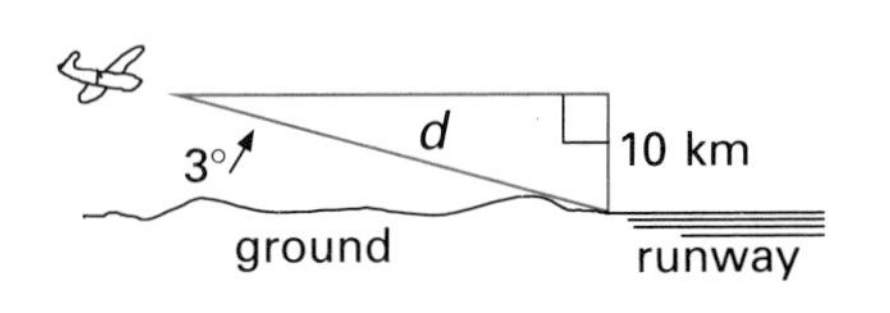

21. Find the property boundary from A to B.

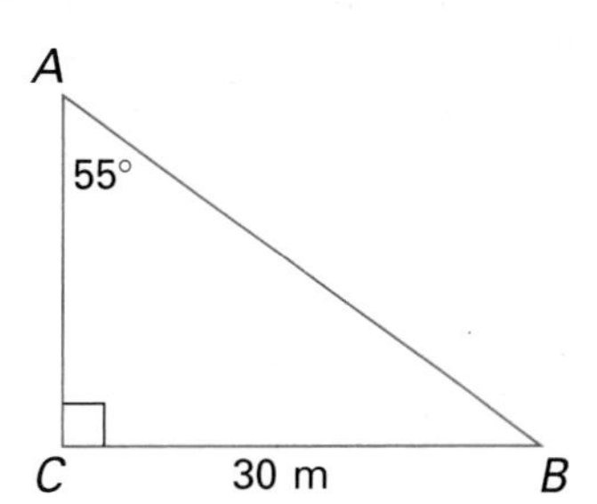

In Exercises 22–26, round answers to the nearest tenth (sides) or nearest degree (angles).

22. The length of the hypotenuse of a right triangle is 8. The length of a leg is 6. Find the measures of each acute angle.

23. Triangle ABC has a right angle at A. $AB = 5$, m $\angle C = 40$. Find BC.

24. The measure of an acute angle of a right triangle is 20. The length of the leg opposite this angle is 10. Find the length of the hypotenuse.

25. The measure of the vertex angle of an isosceles triangle is 32. The length of each leg is 16. Find the length of the altitude drawn to the base.

26. The length of a leg of an isosceles triangle is 14, and the length of the base is 12. Find the measure of the vertex angle.

27. An airplane takes off with an airspeed of 290 ft/s and climbs at an angle that measures 9 with the horizontal. How long will it take the plane to reach a height of 6,000 ft?

28. A student solved the following problem. Triangle ABC has a right angle at C. $AB = 49.4$ and m $\angle A = 72$. Find CB. The student's answer was 64.3. Was this a reasonable answer? Why?

In Exercises 29–31, find y to the nearest tenth or degree.

29.

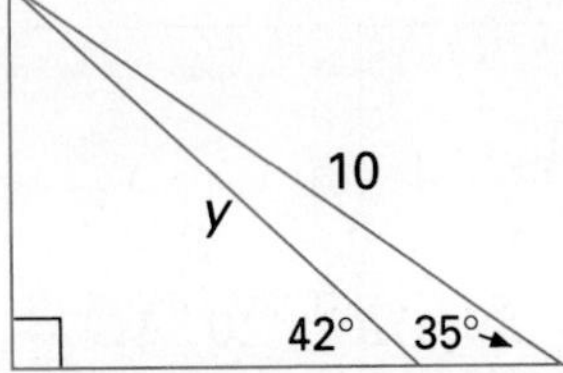

30.

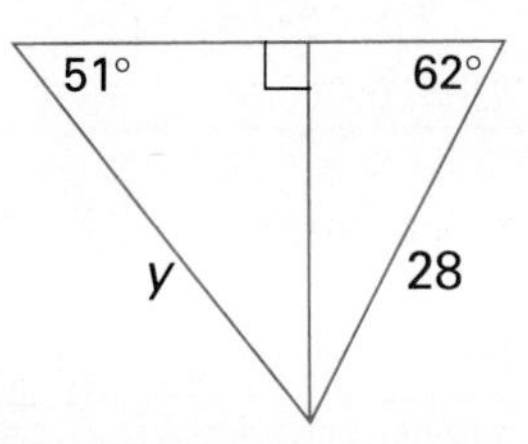

31.

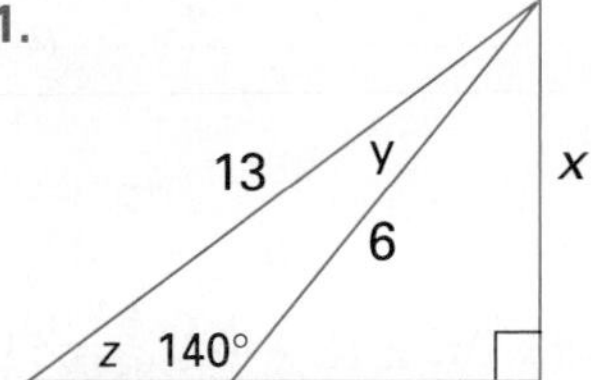

32. Explain why for any acute angle A, $0 < \sin A < 1$.

Mixed Review

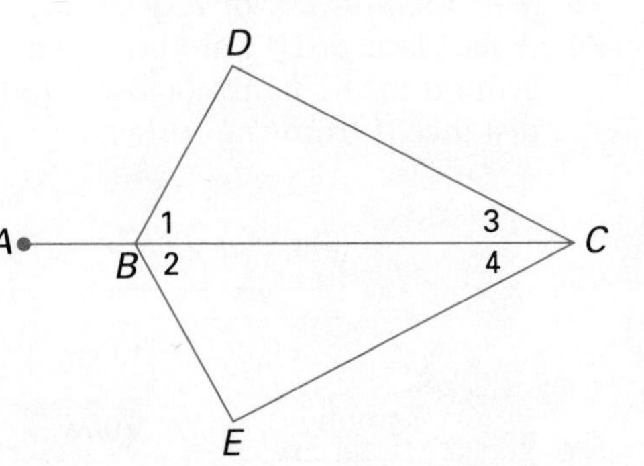

1. Given: $\angle 3 \cong \angle 4$, $\overline{DC} \cong \overline{EC}$
Prove: $\angle 1 \cong \angle 2$ *4.7*

2. Given: $\angle ABD \cong \angle ABE$, $\overline{BC}$ bisects $\angle DCE$
Prove: $\angle D \cong \angle E$ *4.7*

3. Given: m $\angle D = 60$, m $\angle 3 = 20$
Find m $\angle ABD$. *3.7*

4. Given: m $\angle 1$: m $\angle D$: m $\angle 3 = 3:4:5$.
Find m $\angle ABD$. *8.1*

Using the Calculator

A scientific calculator with keys for computing trigonometric ratios can be used to find the sine of a given angle or the acute angle with a given sine ratio.

To find sin 65 to four decimal places:

(1) Make sure the calculator is in the degree mode. DEG must appear in the display window.
(2) Enter the angle measure 65.
(3) Press the [sin] key.
(4) The calculator will display 0.906307787.
(5) Round to four decimal places, 0.9063.

The calculator steps for finding sin 65 are as follows.

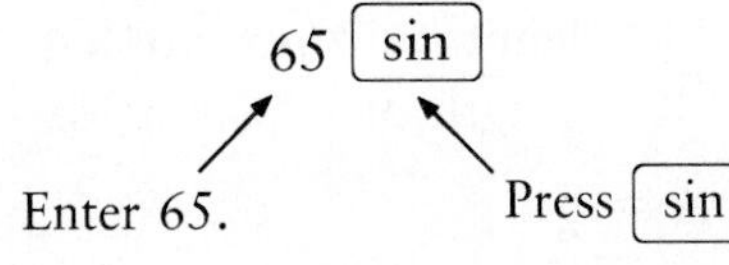

Example 1 Using the figure at the right, find a to the nearest tenth.

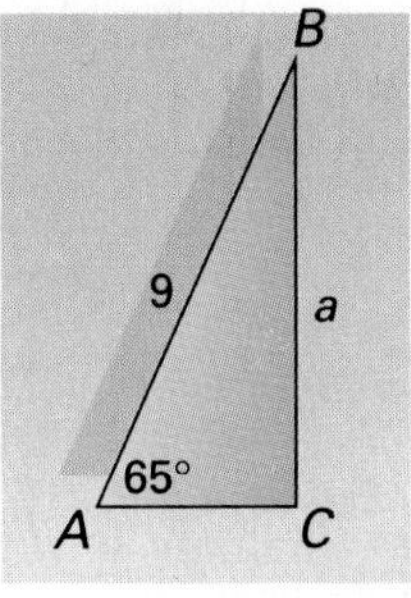

Solution

$\sin 65 = \frac{a}{9}$

$9 \sin 65 = a$

Calculator Steps:
9 × 65 sin = 8.156770083

Thus, a = 8.2 to the nearest tenth.

Example 2 Suppose you are given $\sin A = 0.8774$. Find m $\angle A$.

Solution Finding m $\angle A$ is the *inverse* of finding the sine $\angle A$. To do this, the [INV] (inverse) key or the [2nd f] (second-function) key is used. For example, if $\sin A = 0.8774$, you can find m $\angle A$ to the nearest degree by using the calculator as follows.

Calculator Steps:

0.8774 [INV] [sin] or 0.8774 [2nd f] [sin]

Display: 61.33030001 Display: 61.33030001

Thus, m $\angle A$ = 61, to the nearest degree.

Exercises

Do Written Exercises 1–12 of lesson 9.5 using a calculator.

9.6 Other Trigonometric Ratios

Objective To find side lengths and angle measures using sine, cosine, and tangent ratios

In this lesson, you will study two more ratios that will help you find the lengths of sides and the measures of angles.

Definition

In right triangle ABC with acute angle A:

$$\text{cosine of angle } A = \frac{\text{length of adjacent leg}}{\text{length of hypotenuse}}$$

$$\cos A = \frac{b}{c}$$

$$\text{tangent of angle } A = \frac{\text{length of opposite leg}}{\text{length of adjacent leg}}$$

$$\tan A = \frac{a}{b}$$

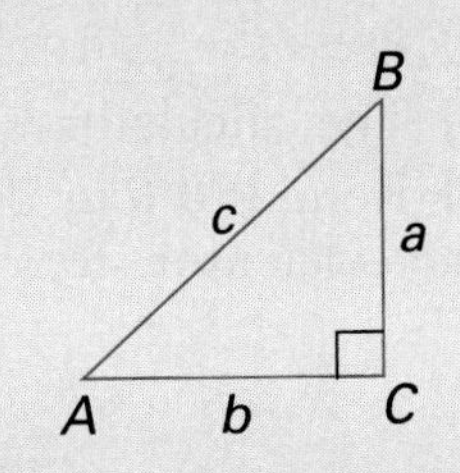

When solving problems involving right triangles, you must choose which trigonometric ratio to use.

EXAMPLE 1 For each triangle, find x to the nearest tenth.

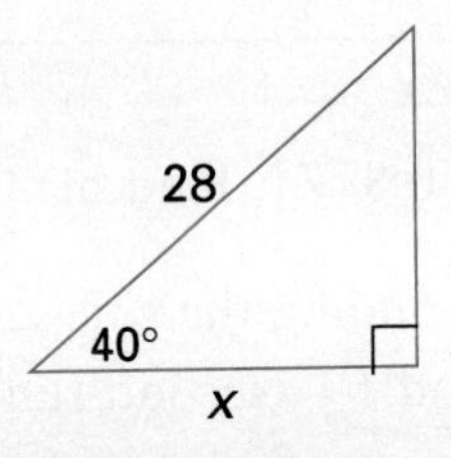

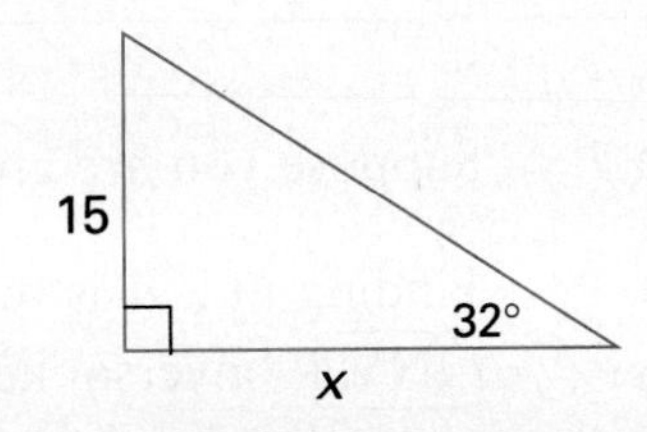

Plan

28 = length of hypotenuse
x = length of adjacent leg
Use the cosine ratio.

15 = length of opposite leg
x = length of adjacent leg
Use the tangent ratio.

Solution

$$\cos 40 = \frac{x}{28} = \frac{\text{adj}}{\text{hyp}}$$

$$0.7660 = \frac{x}{28}$$

$$0.7660(28) = x$$

$$21.4480 = x$$

Thus, $x = 21.4$.

$$\tan 32 = \frac{15}{x} = \frac{\text{opp}}{\text{adj}}$$

$$0.6249 = \frac{15}{x}$$

$$0.6249x = 15$$

$$x = \frac{15}{0.6249} = 24.00384$$

Thus, $x = 24.0$.

EXAMPLE 2 Find x to the nearest degree.

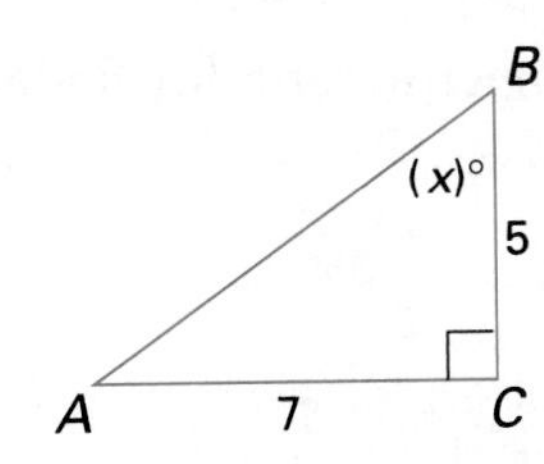

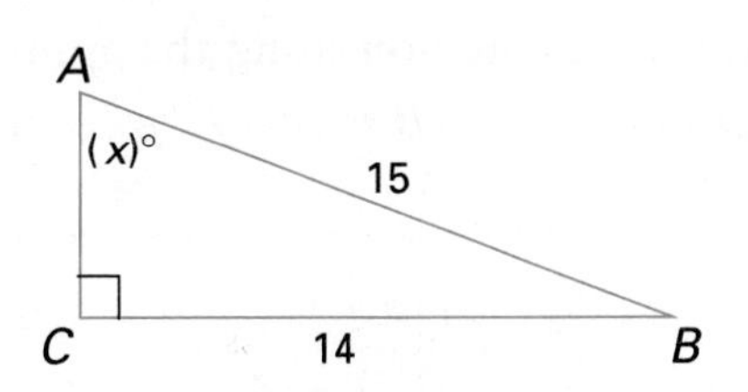

Plan

Given: lengths of both legs
Use the tangent ratio.

Solutions

$\tan B = \frac{7}{5} = \frac{\text{opp}}{\text{adj}}$

$\tan B = 1.4000$

m $\angle B = 54$ [1.4000 is closer to 1.3764 than 1.4281.]

Given: length of opp leg and hypotenuse
Use the sine ratio.

$\sin A = \frac{14}{15} = \frac{\text{opp}}{\text{hyp}}$

$\sin A = 0.93333$ ← Divide to 5 decimal places.

$\sin A = 0.9333$ ← Round to 4 decimal places.

m $\angle A = 69$

EXAMPLE 3 The length of the altitude to the base of an isosceles triangle is 16. The measure of a base angle is 55. Find the length of the base to the nearest tenth.

Solution

$\tan 55 = \frac{16}{x}$

$1.428 = \frac{16}{x}$

$1.428x = 16$

$x = 11.204$

$AC = 2x = 2(11.204) = 22.408$

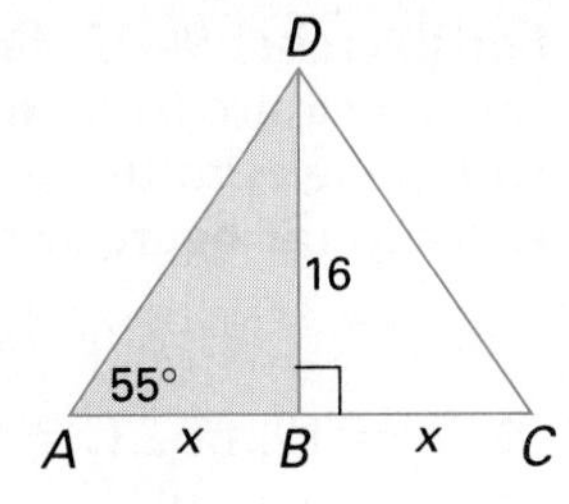

Thus, the length of the base is 22.4 to the nearest tenth.

Focus On Reading

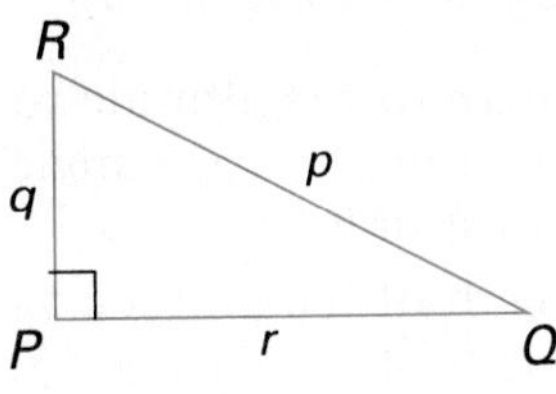

1. Name the hypotenuse.
2. Name the leg adjacent to $\angle R$.
3. Name the leg opposite $\angle Q$.
4. Given p and q, which ratio enables you to find the measure of $\angle R$?
5. Given the measure of $\angle Q$ and r, which ratio enables you to find p?

Classroom Exercises

Write an equation using the appropriate trigonometric ratio for finding *a*.

1.

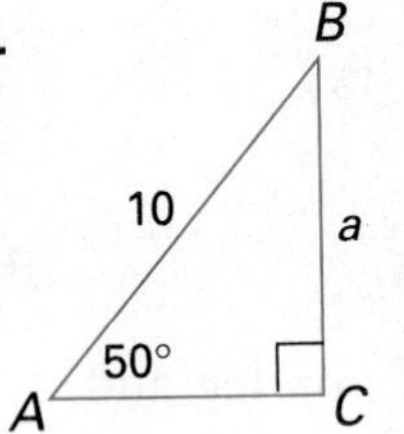

2.

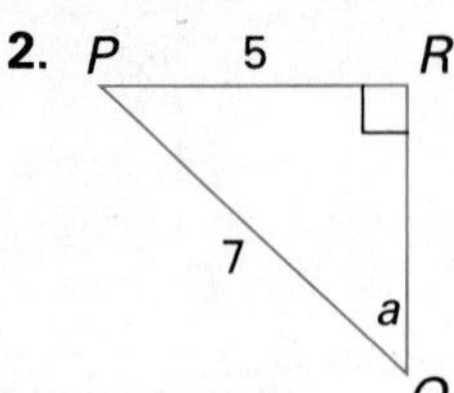

3.

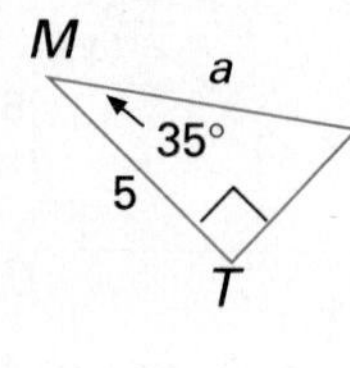

4.

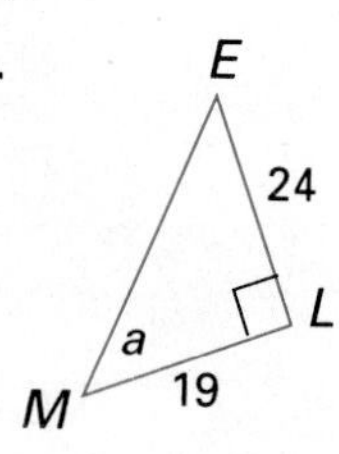

5.

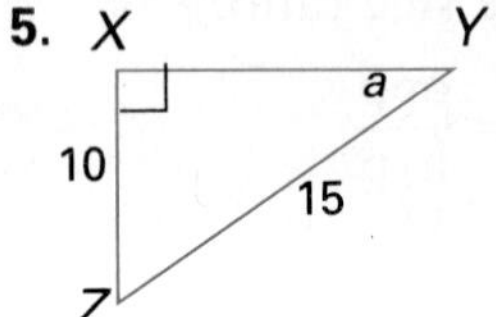

6.

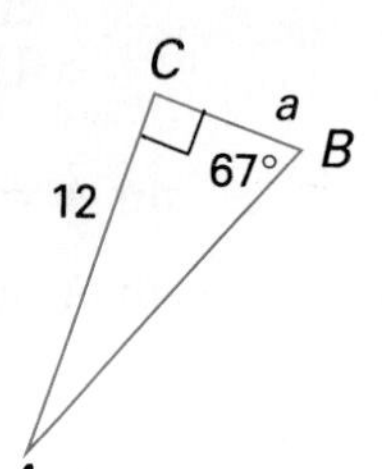

7.

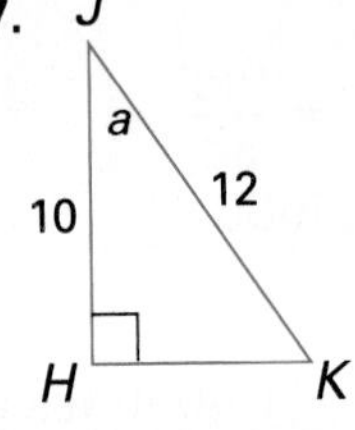

8.

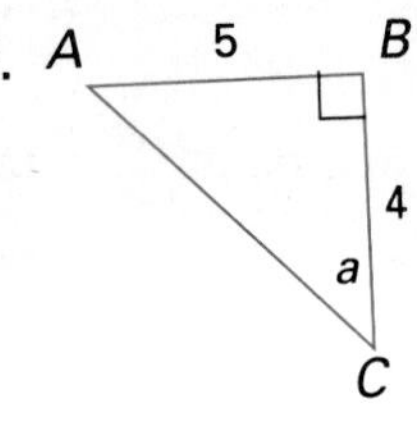

Written Exercises

For Exercises 1–8, find the indicated measures for Classroom Exercises 1–8 above. (Give lengths to the nearest tenth of a unit and angle measures to the nearest degree.)

For Exercises 9–22, find the indicated measure (sides to the nearest tenth, angles to the nearest degree). Refer to the figure at the right.

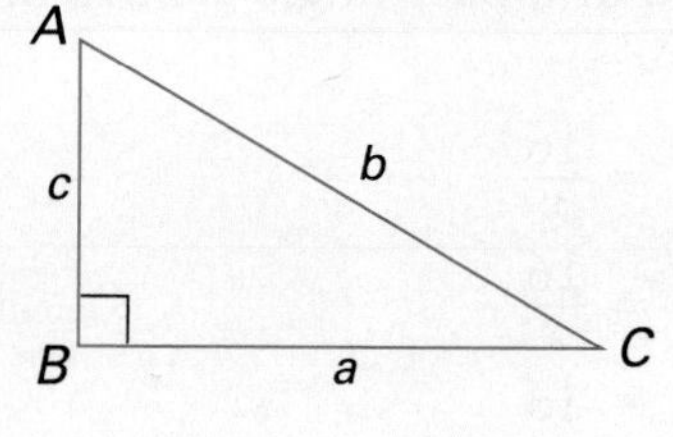

9. $b = 19$, $m\angle C = 48$, $a =$ ______

10. $m\angle A = 29$, $b = 14$, $c =$ ______

11. $m\angle A = 49$, $a = 19$, $c =$ ______

12. $c = 13$, $m\angle C = 26$, $b =$ ______

13. $c = 32$, $m\angle A = 54$, $b =$ ______

14. $b = 35$, $m\angle C = 75$, $a =$ ______

15. $c = 12$, $b = 16$, $m\angle C =$ ______

16. $b = 18$, $a = 8$, $m\angle C =$ ______

17. $a = 14$, $b = 16$, $m\angle A =$ ______

18. $c = 20$, $a = 18$, $m\angle A =$ ______

19. $c = 40$, $m\angle A = 70$, $b =$ ______

20. $b = 30$, $m\angle C = 85$, $c =$ ______

21. $c = 124$, $a = 275$, $m\angle A =$ ______

22. $m\angle A = 65$, $a = 2{,}345$, $c =$ ______

23. The length of the altitude to the base of an isosceles triangle is 14. The measure of a base angle is 62. Find the length of the base to the nearest unit.

24. The length of the altitude to the base of an isosceles triangle is 17. The length of the base is 18. Find the measure of a base angle to the nearest degree.

25. Right triangle ABC has right angle at C, m $\angle A = 42$, and $AC = 24$. Find the length of the hypotenuse to the nearest unit.

26. Find, to the nearest unit, the length of the longer leg of a right triangle if the length of the hypotenuse is 8, and the measure of an acute angle is 22.

27. The lengths of the sides of a rectangle are 12 and 22. Find, to the nearest degree, the measure of the angle formed by a diagonal and the longer side.

28. The measure of the vertex angle of an isosceles triangle is 110. The length of a leg is 18. Find the length, to the nearest unit, of the altitude to the base.

29. The length of each side of a rhombus is 12. The measure of an angle between two adjacent sides is 40. Find, to the nearest unit, the length of each diagonal.

30. Right triangle ABC has right angle at B. $\overline{BD}$ is an altitude to the hypotenuse. $DC = 5$ and $AD = 16$. Find m $\angle ABD$ to the nearest degree.

31. Given: $\overline{EB} \perp \overline{AD}$, $\overline{AC} \perp \overline{CD}$, $AB = BD$, m $\angle ABE = 62$, $CD = 8$. Find EB to the nearest unit.

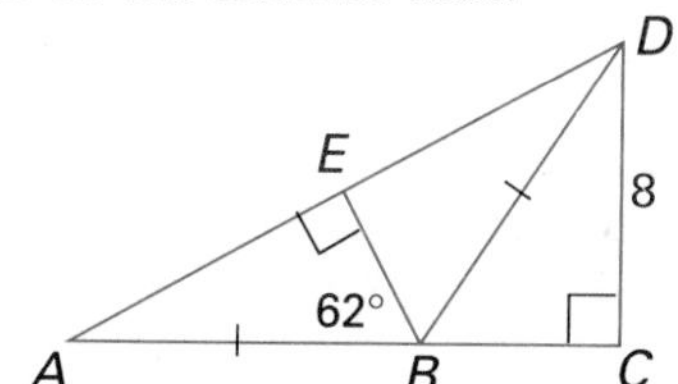

32. Given: length of each side of the cube is 10. Find m $\angle ACB$ to the nearest degree.

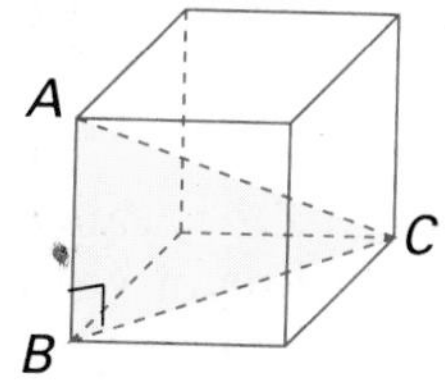

Mixed Review

In the figure at the right, $\overline{AC} \perp \overline{AB}$ and $\overline{AD} \perp \overline{CB}$. Find the indicated measure.

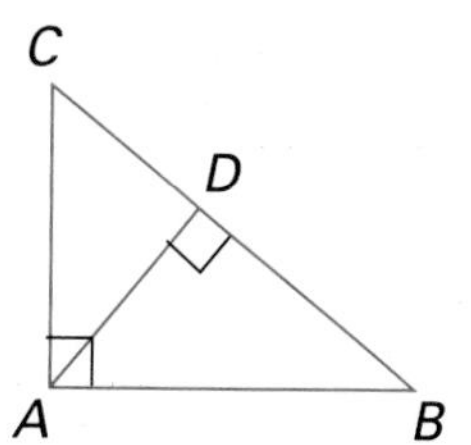

1. m $\angle B = 60$, $BC = 8$, $AC =$ ______ *9.4*
2. m $\angle C = 45$, $BC = 6$, $AC =$ ______ *9.4*
3. m $\angle B = 30$, $AB = 4$, $BC =$ ______ *9.4*
4. $CD = 4$, $CB = 9$, $AC =$ ______ *9.1*
5. $CD = 6$, $BD = 8$, $AD =$ ______ *9.1*

Brainteaser

The practice field at a high school is a rectangle whose length is twice as long as its width. The shorter ends have semicircles on them. The distance around the inside of the track that encloses the field is 100 m. Find the area of the practice field to the nearest meter.

9.7 Applying Trigonometric Ratios

Objective To solve word problems using trigonometric ratios

Surveyors, pilots, and navigators frequently apply trigonometric ratios to problems involving *angles of elevation* or *depression.*

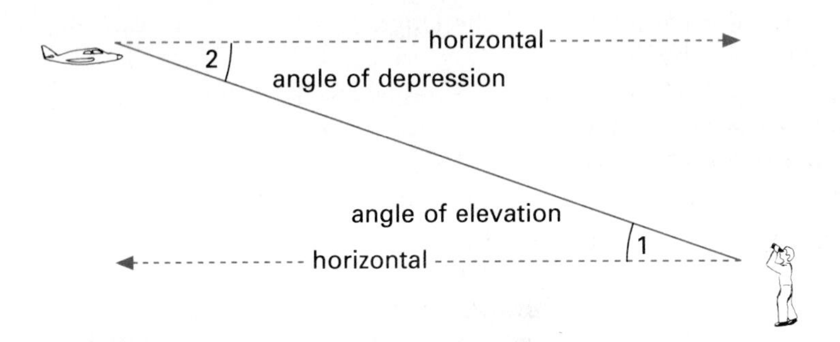

In the diagram above, $\angle 1$ indicates the *angle of elevation.* It is the angle by which the ground observer's line of vision must be raised, or elevated, with respect to the horizontal, to sight the plane.

$\angle 2$ is called the *angle of depression.* It is the angle by which the pilot's line of vision must be lowered, or depressed, with respect to the horizontal, to sight the ground observer.

EXAMPLE 1 A plane is flying over level ground at an altitude of 900 m. When the pilot sights a landing field, the measure of the angle of depression is 27. Find the distance, to the nearest meter, from the point on the ground directly under the pilot to the landing field.

This is known as the *ground distance.*

Plan Draw a diagram with the plane at P flying to the field at F. Then use the tangent ratio to find GF.

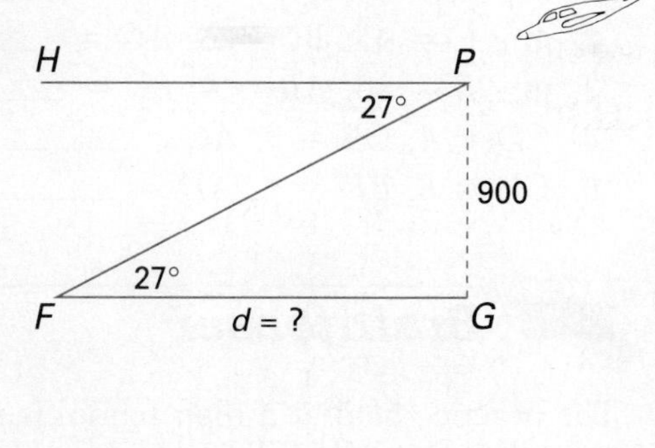

Solution

$$\tan 27 = \frac{PG}{GF}$$

$$0.5095 = \frac{900}{d}$$

$$0.5095d = 900$$

$$d = 1{,}766.4$$

So the ground distance, to the nearest meter, is 1,766 m.

EXAMPLE 2 A tree 50 ft high casts a 35 ft shadow. Find, to the nearest degree, the measure of the angle of elevation of the sun.

Solution Draw a sketch.

$$\tan A = \frac{50}{35}$$
$$\tan A = 1.4285$$
$$m\angle A \approx 55$$

Thus, the measure of the angle of elevation is 55 to the nearest degree.

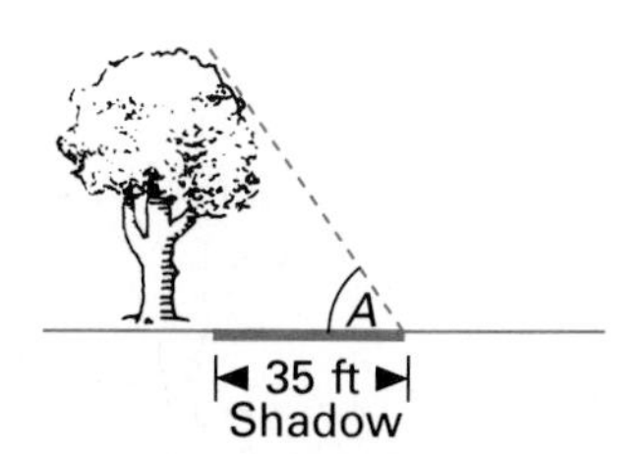

EXAMPLE 3 A 20-ft ladder is leaning against a wall. The foot of the ladder forms an angle of measure 65 with the ground. How far, to the nearest foot, is the top of the ladder from the ground?

Solution Draw a sketch.

$$\sin 65 = \frac{h}{20}$$
$$0.9063 = \frac{h}{20}$$
$$h = 20(0.9063) = 18.126$$

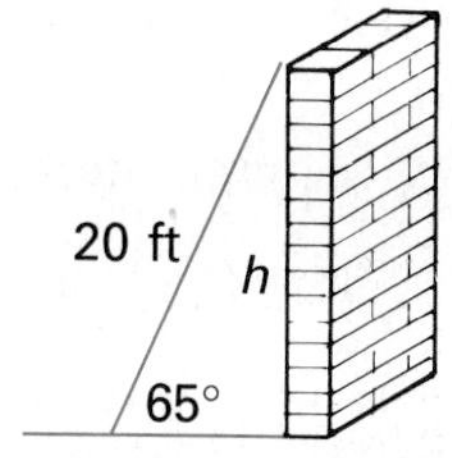

Thus, the top of the ladder is about 18 ft above the ground.

Classroom Exercises

Give the number of the angle and tell whether it is an angle of elevation or depression if a person at:

1. R sights the point T
2. P sights the point R
3. T sights the point R
4. R sights the point P

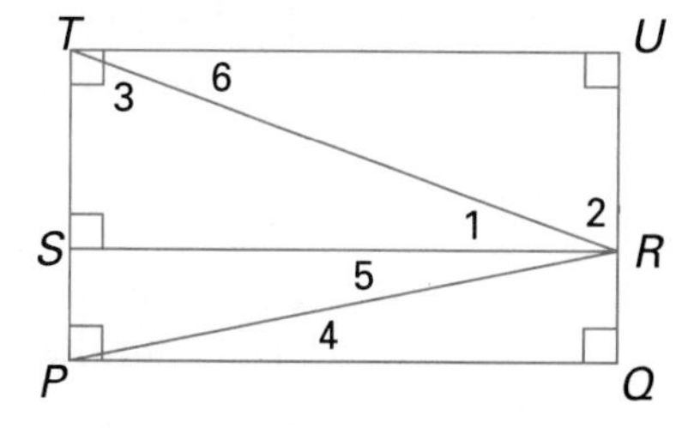

Written Exercises

Find the indicated measure to the nearest unit or the nearest degree.

1. The measure of the angle of depression is 36. The altitude of the plane is 9,000 ft. Find the ground distance d from the plane to the runway.
2. Find the flight distance s from the plane to the runway.

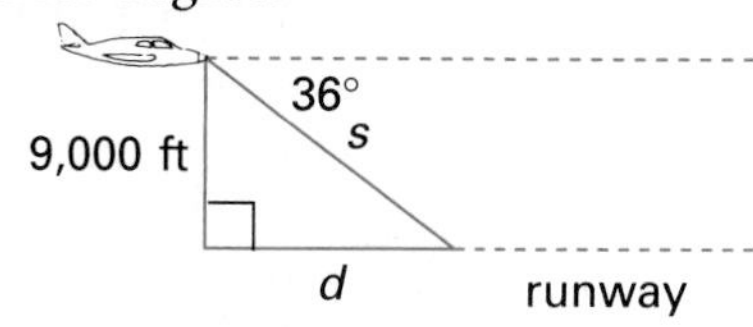

3. A 45-ft ladder makes an angle of measure 55 with the ground. How high up the wall does the ladder reach?

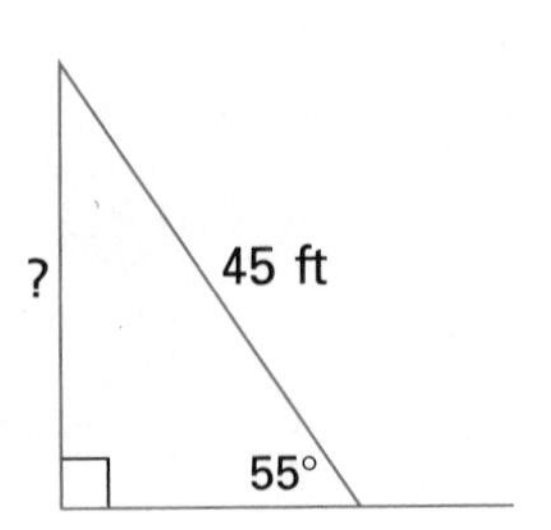

4. A building 200 ft tall casts a 155-ft shadow. Find the angle of elevation of the sun.

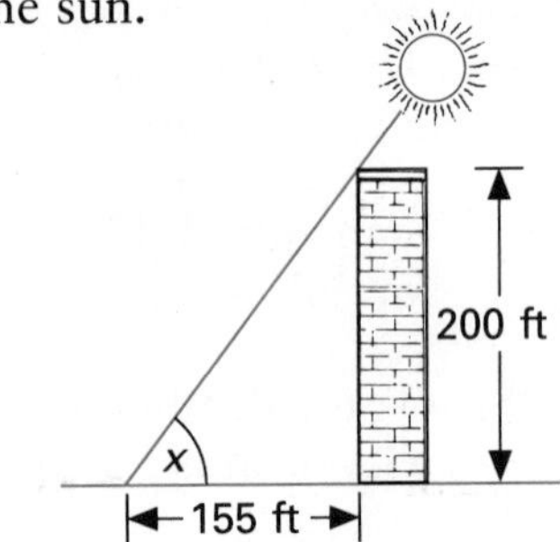

5. At a point 125 ft from the base of a tower, the angle of elevation of the top of the tower has a measure of 38. How high is the tower?

38°
125 ft

6. A 15-ft ladder leans against a wall. The foot of the ladder is 5 ft from the bottom of the wall. At what angle is the ladder to the ground?

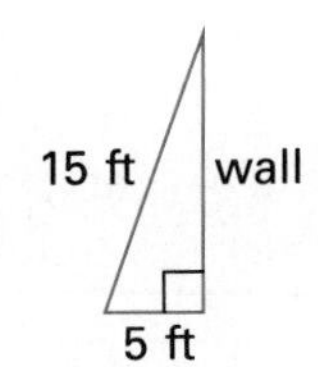

7. To estimate the width of a lake, Bob stood directly opposite a large tree. He then walked 140 ft along the bank and approximated the measure of the angle between his line of sight to the tree and the lake's edge as 32. Find the approximate width of the lake. 87 ft

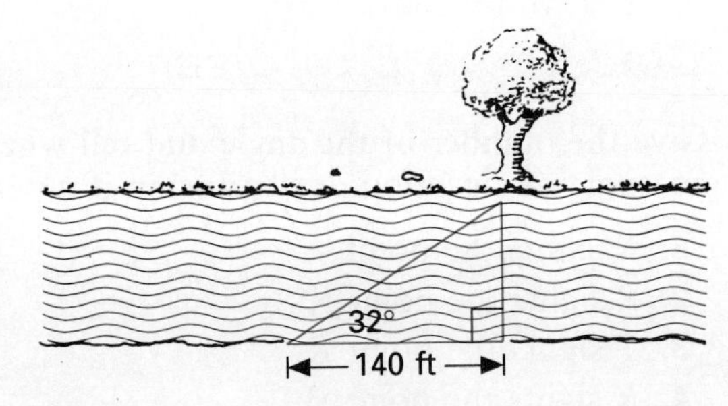

8. A tree casts a 60-ft shadow. Find the height of the tree if the angle of elevation of the sun has a measure of 51.

9. The top of a lighthouse is 250 ft above sea level. From the top, the measure of the angle of depression to a boat is 48. How far is the boat from the bottom of the lighthouse?

10. The height of a cloud cover (called the "ceiling") at an airport was found the following way. A searchlight was pointed straight up. From a point on the ground 1,500 ft from the searchlight, the measure of the angle of elevation to the illumination of the clouds was 72. Find the height of the cloud cover.

11. A pilot flying over level ground at an altitude of 2,400 ft sights a building. The angle of depression measures 6. Find the ground distance between the building and the point directly below the pilot.
12. The surface of a ramp is 475 m long. It rises a vertical distance of 28 m. Find the measure of the angle of elevation.
13. A man in a sailboat is directly opposite the base of a 750-m high cliff. The angle of elevation measures 31. How far from land is the sailor?

Use the diagram to answer Exercises 14–16.

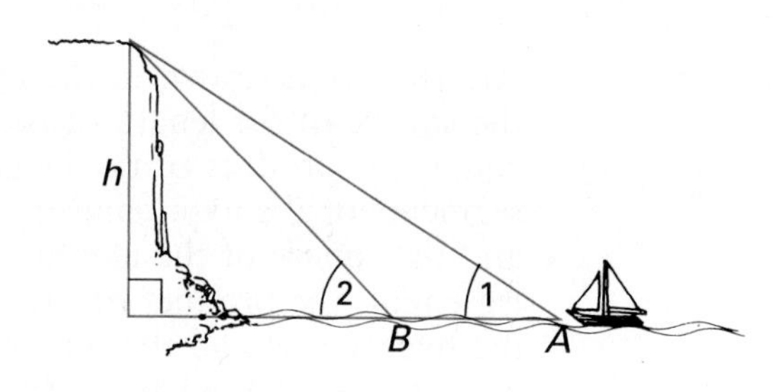

14. A navigator at A observes the measure of the angle of elevation to the top of the cliff to be 15. This measure changes to 26 at B. If the height of the cliff is known to be 750 m, how far has the boat moved to get from point A to point B?

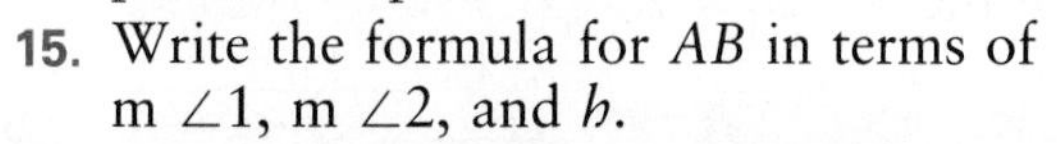

15. Write the formula for AB in terms of m $\angle 1$, m $\angle 2$, and h.
16. If the distance $AB = 1{,}500$ m, m $\angle 1 = 19$, and m $\angle 2 = 28$, how high is the cliff?

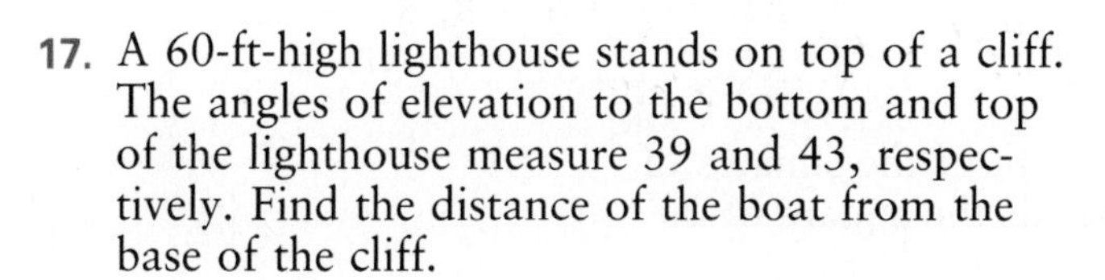

17. A 60-ft-high lighthouse stands on top of a cliff. The angles of elevation to the bottom and top of the lighthouse measure 39 and 43, respectively. Find the distance of the boat from the base of the cliff.

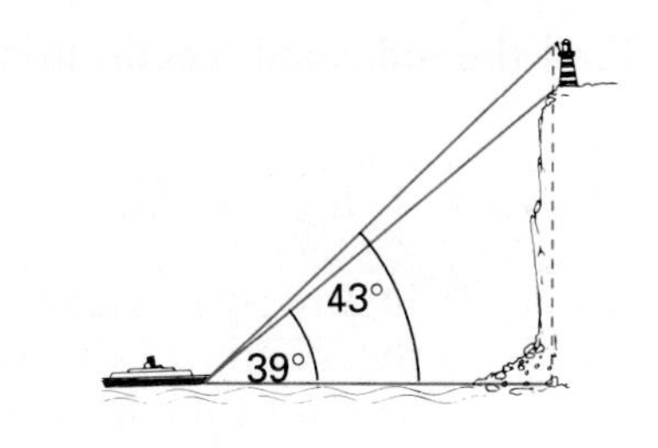

18. A surveyor wants to measure the distances from points A and B to point C on the opposite side of a stream. Point C can be sighted from both A and B. She measures $\overline{AB}$ and angles A and B with the results indicated. Find AC and BC to nearest meter.

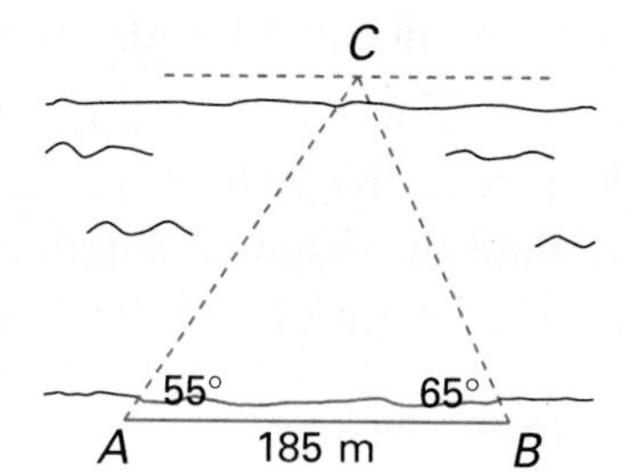

Mixed Review

1. Find the sum of the measures of the interior angles of a regular decagon. *6.2*
2. Points A and B on a number line have coordinates -8 and 6. Find the coordinate of M, the midpoint of $\overline{AB}$. *1.3*
3. Graph the conjunction $x \geq 3$ and $x \leq 8$ and describe the geometric figure. *1.8*
4. The ratio of the measures of the two acute angles of a right triangle is 2:7. Find the measure of each angle. *8.1*

Chapter 9 Review

Key Terms

angle of depression (p. 370)
angle of elevation (p. 370)
cosine (p. 366)
Pythagorean Theorem (p. 344)
sine (p. 360)
tangent (p. 366)
trigonometric ratio (p. 360)

Key Ideas and Review Exercises

9.1 If the altitude is drawn to the hypotenuse of a right triangle, then

- the square of the length of the altitude equals the product of the lengths of the segments of the hypotenuse: $h^2 = mn$,
- and the square of the length of either leg equals the product of the lengths of the hypotenuse and the segment of the hypotenuse adjacent to that leg: $r^2 = nt$ and $s^2 = mt$.

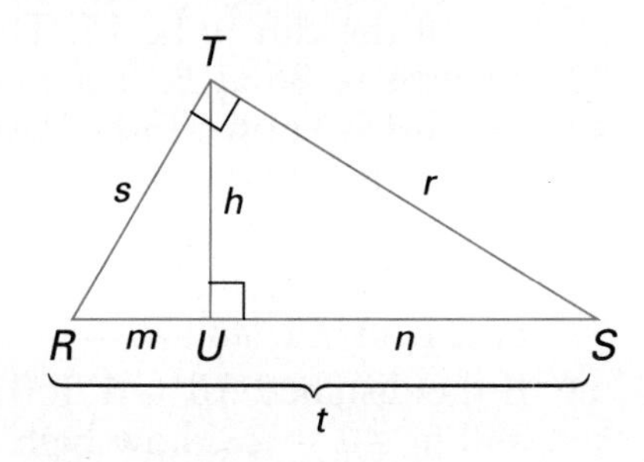

Find the indicated length, using the figure above.

1. $TU = 4$, $RU = 8$, $US =$ ______
2. $RU = 2$, $US = 8$, $RT =$ ______
3. $TS = 4$, $RS = 8$, $US =$ ______
4. $RU = 12$, $ST = 8$, $SU =$ ______

9.2 In any right triangle, the sum of the squares of the lengths of the legs equals the square of the length of the hypotenuse: $a^2 + b^2 = c^2$.

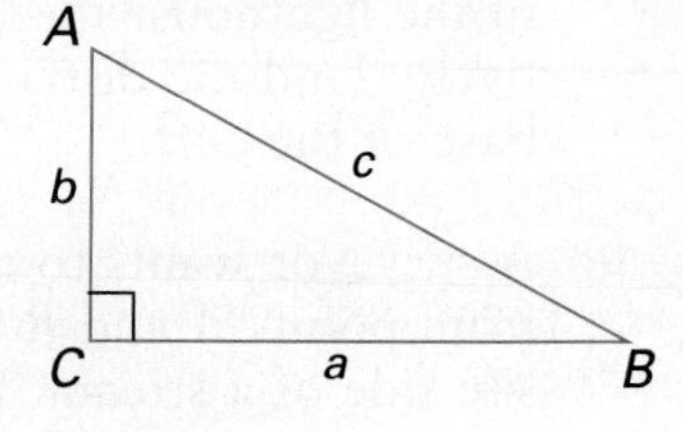

Find the indicated length, using the figure at the right.

5. $a = 4$, $b = 6$, $c =$ ______
6. $a = 2$, $c = 7$, $b =$ ______
7. Find the diagonal length of a square if the length of a side is 6 m.
8. The length of each leg of an isosceles triangle is 4 ft. The length of the base is 6 ft. Find the length of the altitude from the vertex angle to the base.

9.3 $\triangle XYZ$, with $\overline{XY}$ the longest side, is:
a *right* triangle if $z^2 = x^2 + y^2$,
an *obtuse* triangle if $z^2 > x^2 + y^2$, or
an *acute* triangle if $z^2 < x^2 + y^2$.

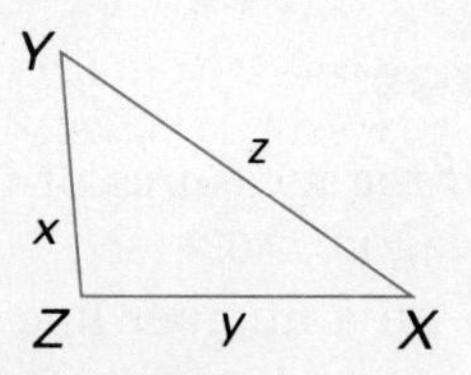

Determine whether the triangle formed with the given lengths of the three sides is acute, obtuse, or right. If the triangle is right, indicate which side is the hypotenuse.

9. 4, 3, 6
10. 5, 6, 7
11. $\sqrt{2}$, 1, $\sqrt{3}$
12. a, a, $a\sqrt{3}$

9.4 Properties of Special Right Triangles

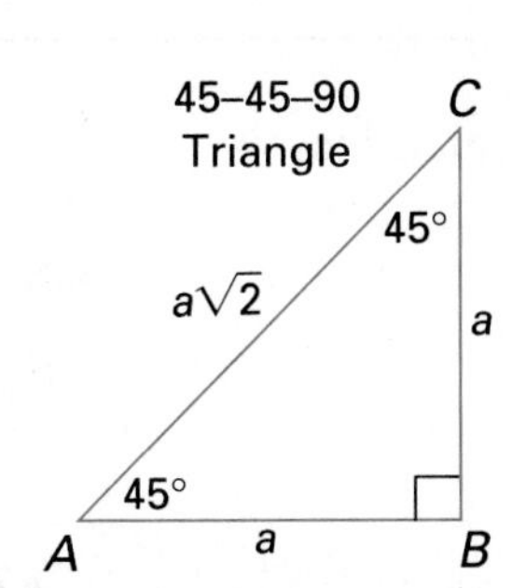

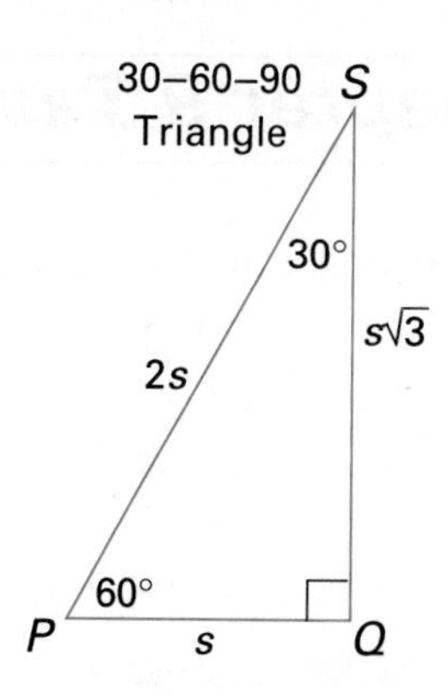

Use the figures above to find the indicated lengths.

13. $AC = 8$, $AB =$ ____, $BC =$ ____

14. $BC = 3\sqrt{2}$, $AB =$ ____, $AC =$ ____

15. $PS = 12$, $PQ =$ ____, $SQ =$ ____

16. $SQ = 6$, $PQ =$ ____, $PS =$ ____

Find the indicated length for an equilateral triangle with given lengths.

17. side: 8; altitude: ____

18. altitude: 4, side: ____

9.5, 9.6 Trigonometric Ratios for a Right Triangle

$\sin A = \frac{a}{c}$ $\quad \sin B = \frac{b}{c} \leftarrow \frac{\text{opp}}{\text{hyp}}$

$\cos A = \frac{b}{c}$ $\quad \cos B = \frac{a}{c} \leftarrow \frac{\text{adj}}{\text{hyp}}$

$\tan A = \frac{a}{b}$ $\quad \tan B = \frac{b}{a} \leftarrow \frac{\text{opp}}{\text{adj}}$

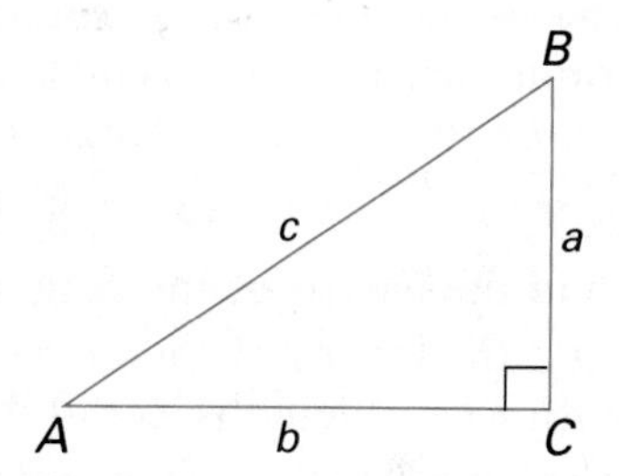

Find the indicated measure (sides to the nearest unit, angles to the nearest degree). Use the figure above.

19. $b = 8$, $a = 4$, $m\angle A =$ ____

20. $m\angle A = 40$, $a = 5$, $c =$ ____

21. $a = 20$, $m\angle B = 25$, $c =$ ____

22. $b = 5$, $c = 15$, $m\angle A =$ ____

9.7

Trigonometric ratios can be applied to word problems involving angles of elevation and angles of depression.

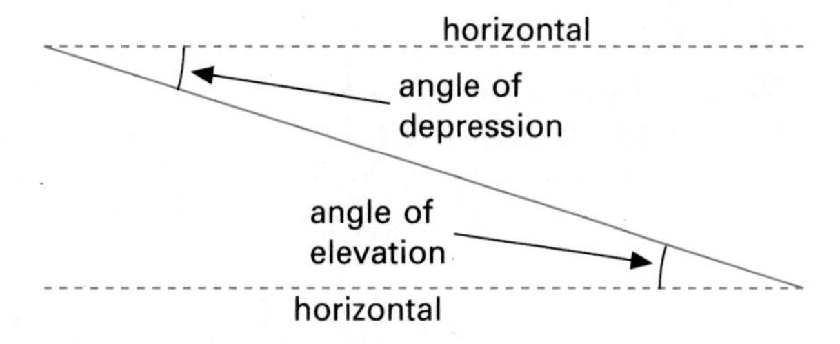

Find the required measure to the nearest unit.

23. A tree casts a 150-ft shadow. Find the height of the tree if the angle of elevation of the sun has a measure of 49.

Chapter 9 Test

Use the right triangle at the right to find the indicated length.

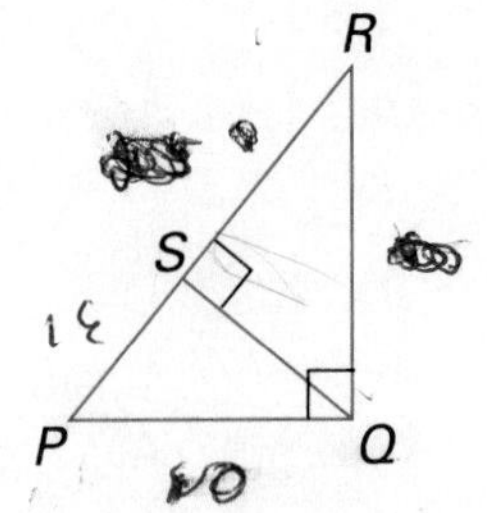

1. $PR = 25$, $SP = 16$, $PQ =$ ______
2. $SQ = 4$, $SP = 2$, $SR =$ ______
3. $SR = 5$, $PQ = 6$, $SP =$ ______
4. $SR = 6$, $SP = 2$, $SQ =$ ______

Use the right triangle at the right to find the indicated length(s).

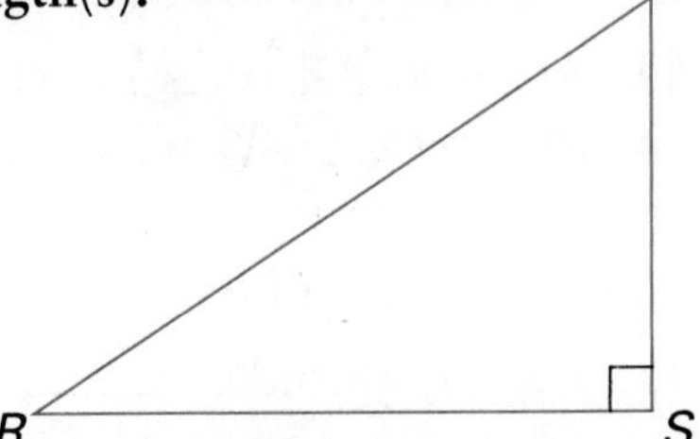

5. $RS = 6$, $ST = 4$, $RT =$ ______
6. $RT = 17$, $RS = 15$, $ST =$ ______
7. m $\angle R = 45$, $RT = 6$, $ST =$ ______
8. m $\angle R = 60$, $RS = 6$, $RT =$ ___, $ST =$ ___
9. m $\angle T = 60$, $TS = 4$, $RT =$ ___, $RS =$ ___
10. m $\angle T = 45$, $RS = 8$, $ST =$ ___, $RT =$ ___

Determine whether the triangle with the given lengths of sides is acute, obtuse, or right. If the triangle is right, tell which side is the hypotenuse.

11. 5, 7, 9
12. 7, 8, 5
13. $4\sqrt{2}$, 2, 6
14. a, $a\sqrt{5}$, $a\sqrt{7}$
15. Find the length of the diagonal of a square with a side of length 4.
16. Find the length of the altitude to the hypotenuse of a right triangle with legs of lengths 2 and 4.
17. The lengths of the diagonals of a rhombus are 6 and 12. Find the length of a side of the rhombus.
18. Find the length of a side of an equilateral triangle with altitude of length 6.

Use the right triangle at the right to find the indicated measure (sides to the nearest unit, angles to the nearest degree).

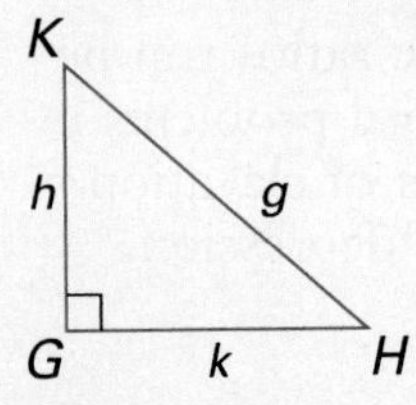

19. $h = 6$, $g = 8$, m $\angle H =$ ______
20. $h = 12$, m $\angle K = 62$, $k =$ ______
21. $g = 15$, m $\angle H = 40$, $h =$ ______

22. A tree casts a shadow of 140 ft. Find the height of the tree to the nearest foot if the angle of elevation of the sun measures 6.

* 23. An angle of a rhombus measures 118. The length of the diagonal opposite that angle is 16 ft. Find the length of the other diagonal to the nearest foot.

College Prep Test

Indicate the one correct answer for each question.

1. In the figure below, $x + y =$ ______
(A) 130 (B) 50 (C) 80
(D) 0 (E) 40

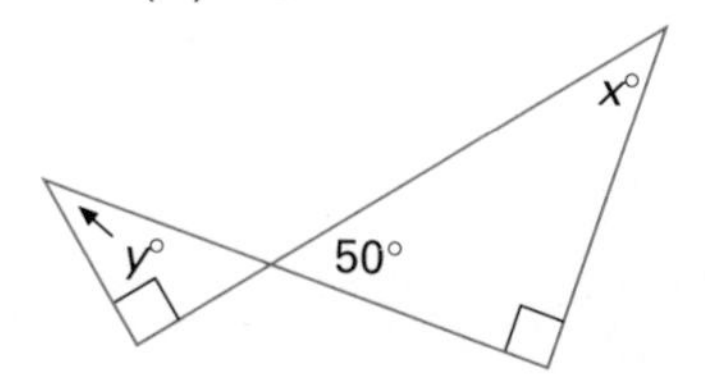

2. If the measure of one acute angle of a right triangle is 25, find the measure of the supplement of the other.
(A) 65 (B) 155 (C) 115
(D) 35 (E) None of these

3. In the diagram below, $\angle RSQ$, $\angle QTS$, and $\angle QTP$ are right angles. Find QR.

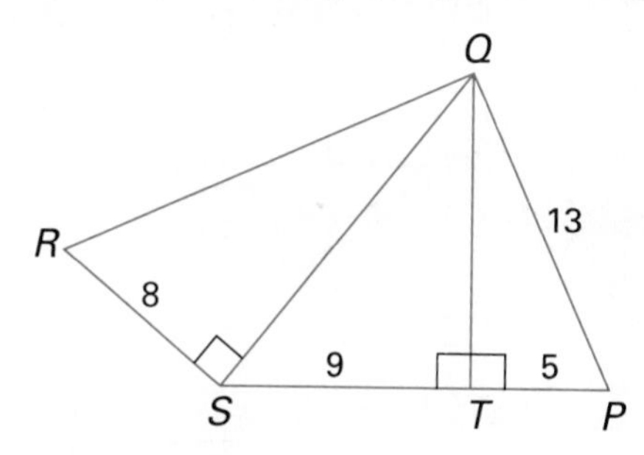

(A) 12 (B) 13 (C) 15
(D) 25 (E) 17

4. In the figure of the square and rectangle below, the value of x is.
(A) 120 (B) 100 (C) 150
(D) 135 (E) Not enough information is given to answer the question.

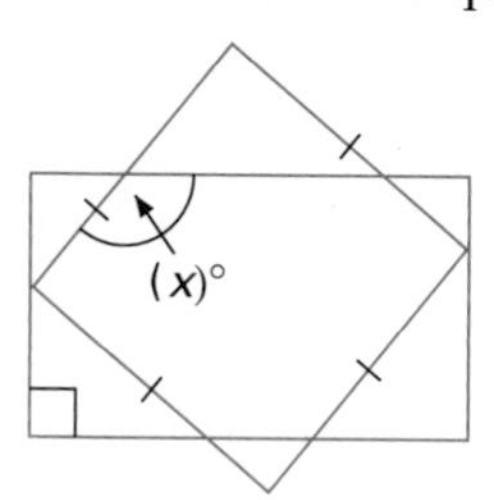

5.

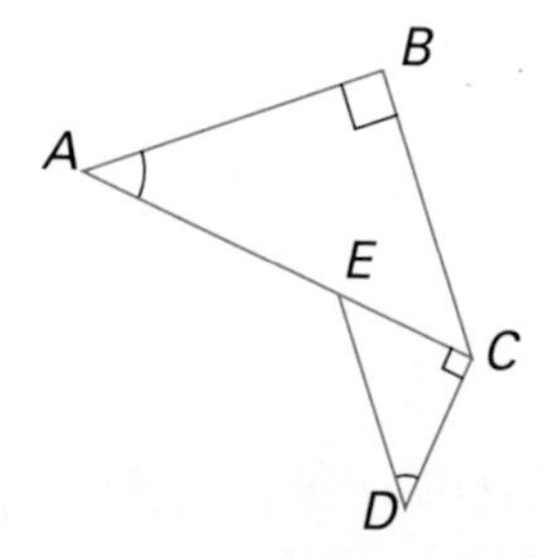

In the figure above, $DE = 5$, $DC = 4$, $AE = 7$. $\angle A \cong \angle D$. Find AB.
(A) 8 (B) 6 (C) 7 (D) 5
(E) 10

6. For the three squares below, find the ratio $\frac{PQ}{RS}$.

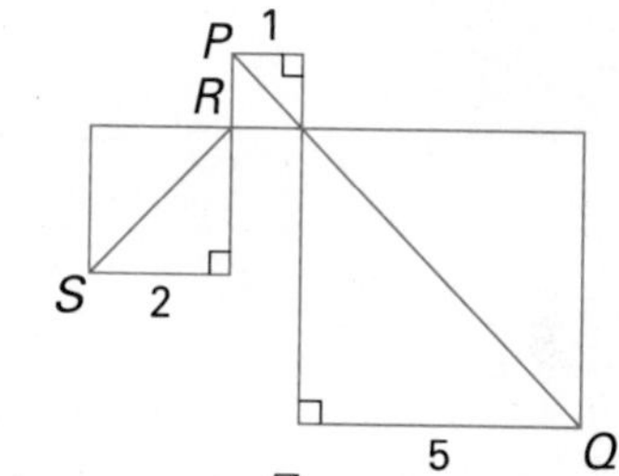

(A) $\frac{1}{2}$ (B) $\frac{3\sqrt{2}}{2}$ (C) $\frac{3}{1}$ (D) $\frac{3}{2}$
(E) $\frac{18}{\sqrt{2}}$

7. In the figure below, $AC = 4\sqrt{2}$, $AD = 2\sqrt{2}$, m $\angle DAB = 105$. Find BA.

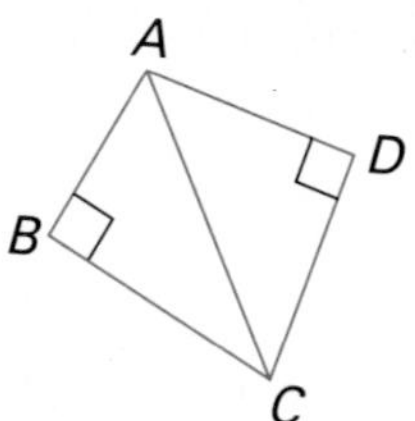

(A) 8 (B) 4 (C) $2\sqrt{2}$
(D) $\sqrt{2}$ (E) None of these

10 COORDINATE GEOMETRY

René Descartes (1596–1650)—Descartes developed the idea that any point on a plane can be named by a pair of numbers. Today this idea is used in street maps, lines of longitude and latitude on a globe, and graph paper.

10.1 Coordinate Systems and Distance

Objectives To graph ordered pairs of numbers
To find the distance between two points in a coordinate plane

Every point on a number line can be named by exactly one real number.

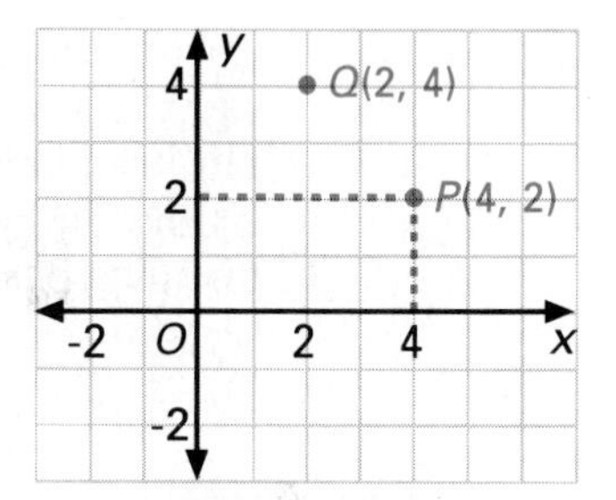

To name points in a plane, a system based on two perpendicular number lines is often used.

The horizontal line is called the **x-axis**; the vertical line, the **y-axis**. The **origin** is the point of intersection of the two axes.

For every point in a plane, there is a corresponding ordered pair of numbers (x, y) called the **coordinates** of the point. For every ordered pair of numbers, there is a corresponding point, called the **graph** of the ordered pair. The graphs of points $P(4, 2)$ and $Q(2, 4)$ are shown above. Observe that the order of the coordinates is important. The x-coordinate, or **abscissa**, is the horizontal distance from the point to the y-axis. The y-coordinate, or **ordinate**, is the vertical distance from the point to the x-axis. The abscissa of P is 4 and the ordinate of P is 2.

A plane where points are associated with ordered pairs in this way is called a **coordinate plane**.

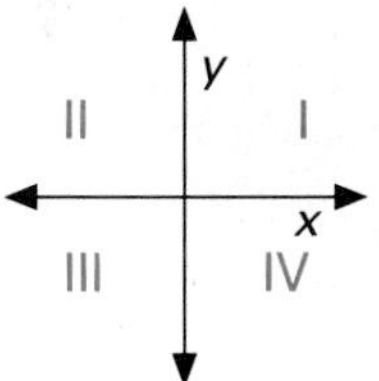

The axes separate the plane into four quadrants, numbered I, II, III, and IV in a counterclockwise manner. Both coordinates of a point in Quadrant I are positive. Abscissas are negative in Quadrants II and III, while ordinates are negative in Quadrants III and IV. Points on the axes are not contained in any quadrant.

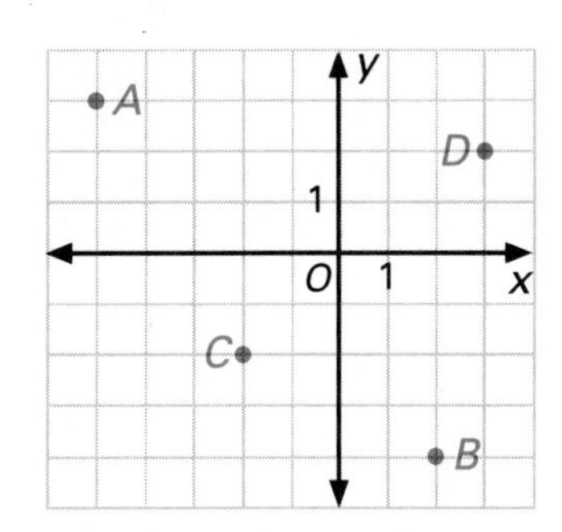

EXAMPLE 1 Write the coordinates of points A, B, C, and D. Tell which quadrant contains each point.

Solution
$A(-5, 3)$ Quadrant II
$B(2, -4)$ Quadrant IV
$C(-2, -2)$ Quadrant III
$D(3, 2)$ Quadrant I

The distance AB between the two points A and B on a number line is $|m - n|$ or $|n - m|$ where m and n are the coordinates of A and B, respectively. For example, AB below is $|4 - (-2)| = 6$.

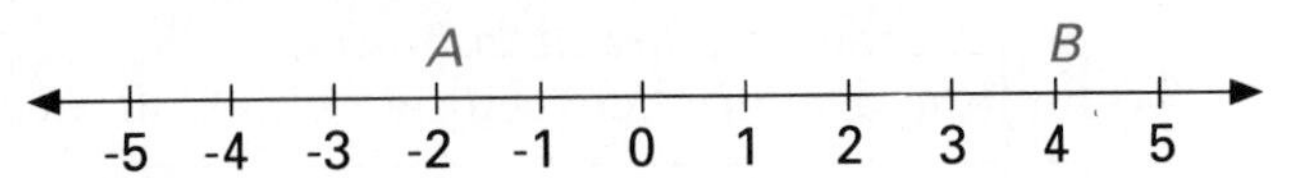

To find the distance between two points in a coordinate plane, the Pythagorean Theorem can be used.

EXAMPLE 2 Locate the points $A(-2,-1)$ and $B(2, 2)$ in a coordinate plane. Draw a right triangle which has the two points as vertices of the acute angles. Find the lengths of the legs of this triangle. Using the Pythagorean Theorem, find the distance between the two points.

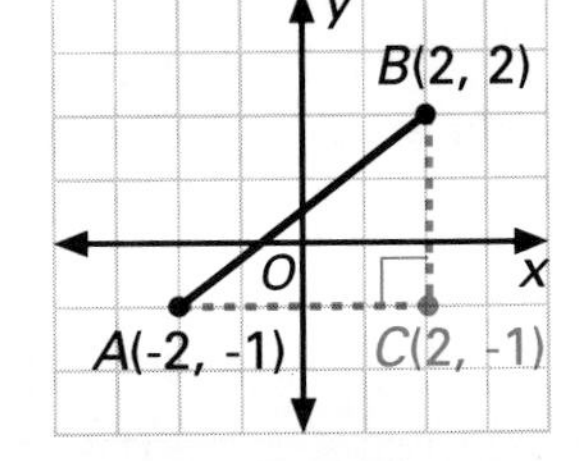

$AC = |2 - (-2)| = 4$ ← difference of abscissas

$BC = |2 - (-1)| = 3$ ← difference of ordinates

$AB = \sqrt{(AC)^2 + (BC)^2}$

$AB = \sqrt{4^2 + 3^2} = 5$

Example 2 suggests a general formula for finding the distance between any two points on a plane.

Theorem 10.1 **The Distance Formula:** The distance d between $P_1(x_1,y_1)$ and $P_2(x_2,y_2)$ is given by the formula $d = \sqrt{(x_2 - x_1)^2 + (y_2 - y_1)^2}$.

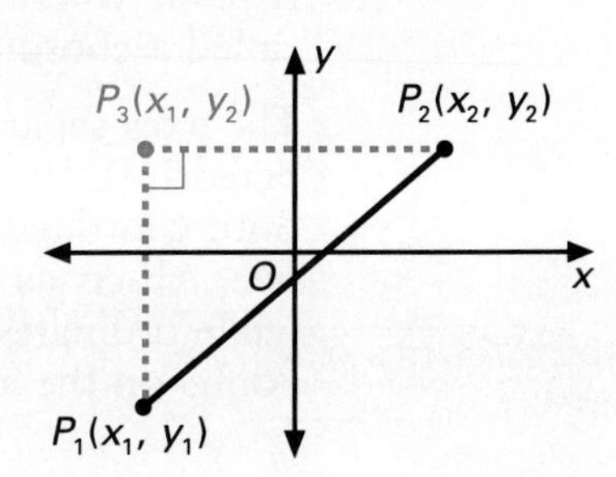

Given: $P_1(x_1, y_1)$, $P_2(x_2, y_2)$

Prove: $d = P_1P_2$

$= \sqrt{(x_2 - x_1)^2 + (y_2 - y_1)^2}$

Outline of Proof

(1) Locate $P_3(x_1, y_2)$ and draw $\triangle P_1P_2P_3$.

(2) $\overline{P_1P_3}$ is vertical. (Why?) Therefore, $P_1P_3 = |y_2 - y_1|$

(3) $\overline{P_2P_3}$ is horizontal. (Why?) Therefore, $P_2P_3 = |x_2 - x_1|$

(4) $\triangle P_1P_2P_3$ is a right triangle. (Why?) So, by the Pythagorean Theorem,

$$d^2 = (P_1P_3)^2 + (P_2P_3)^2$$
$$d^2 = |y_2 - y_1|^2 + |x_2 - x_1|^2$$
$$d^2 = (y_2 - y_1)^2 + (x_2 - x_1)^2$$
$$d = \sqrt{(x_2 - x_1)^2 + (y_2 - y_1)^2}$$

EXAMPLE 3 Use the Distance Formula to find the distance between $A(-1,-2)$ and $B(3, 3)$.

Solution If $A = P_1$, and $B = P_2$, then $x_1 = -1$, $y_1 = -2$, $x_2 = 3$, and $y_2 = 3$

$$\begin{aligned} d &= \sqrt{(x_2 - x_1)^2 + (y_2 - y_1)^2} \\ &= \sqrt{[3 - (-1)]^2 + [3 - (-2)]^2} \\ &= \sqrt{4^2 + 5^2} \\ &= \sqrt{16 + 25}, \text{ or } \sqrt{41} \end{aligned}$$

The Distance Formula can sometimes be used to prove properties of geometric figures.

EXAMPLE 4 $\triangle ABC$ has vertices $A(-2, 6)$, $B(6, 4)$, and $C(0,-2)$. Is $\triangle ABC$ scalene, isosceles, or equilateral?

Plan Find the lengths of the three sides of $\triangle ABC$.

Solution

$$\begin{aligned} AB &= \sqrt{[6 - (-2)]^2 + (4 - 6)^2} \\ &= \sqrt{8^2 + (-2)^2} = \sqrt{64 + 4} \\ &= \sqrt{68} = 2\sqrt{17} \end{aligned}$$

$$\begin{aligned} AC &= \sqrt{[0 - (-2)]^2 + (-2 - 6)^2} \\ &= \sqrt{2^2 + (-8)^2} \\ &= \sqrt{4 + 64} = \sqrt{68} = 2\sqrt{17} \end{aligned}$$

$$\begin{aligned} BC &= \sqrt{(0 - 6)^2 + (-2 - 4)^2} \\ &= \sqrt{(-6)^2 + (-6)^2} \\ &= \sqrt{36 + 36} = \sqrt{72} = 6\sqrt{2} \end{aligned}$$

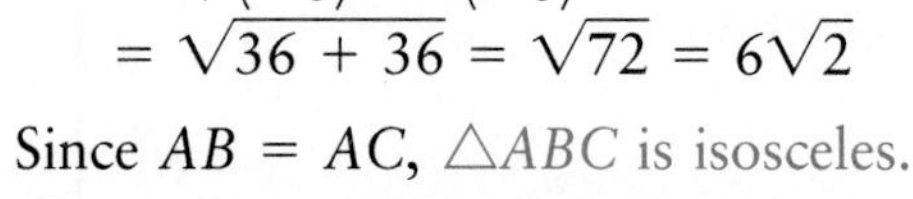

Since $AB = AC$, $\triangle ABC$ is isosceles.

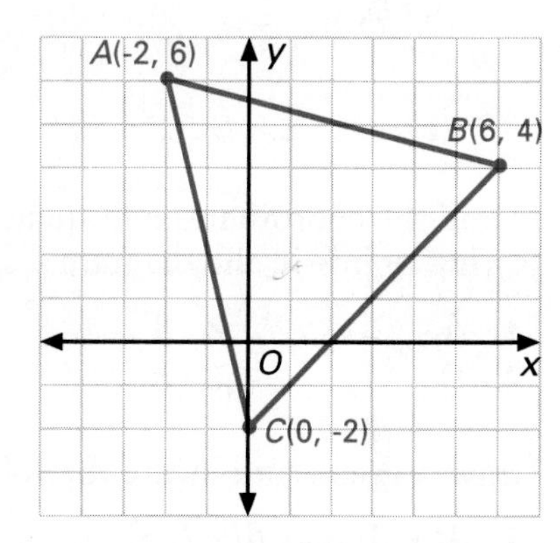

Focus on Reading

Translate each sentence into symbols.

1. The length of segment $\overline{PQ}$ is 7.
2. The distance between two points P and Q on a number line is the absolute value of the difference of the coordinates of the two points.
3. The distance between the two points A and B with coordinates (x, y) and (u, v) is 5.
4. The line containing the points D and E is perpendicular to the line containing points F and G.

Classroom Exercises

The figure at the right shows a street map. In some ways it is like the coordinate axes we have been studying. North is to the top and east is to the right. (Exercises 1–5)

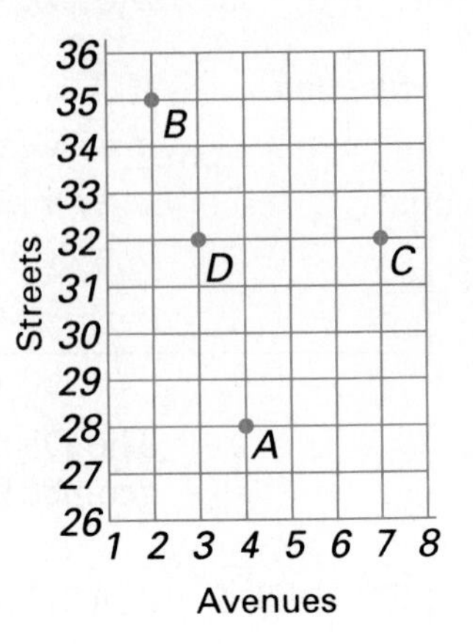

1. What would be your direction and distance in walking from A to B?
2. How would you walk from B to C?
3. How would you walk from D to A?
4. What is the address of point C?
5. What is the address of point D?

6. In what quadrants is the abscissa positive?
7. In what quadrants is the ordinate negative?
8. In what quadrant is the point $P(3,-2)$?

Written Exercises

Name the coordinates and quadrant of the following points in the diagram.

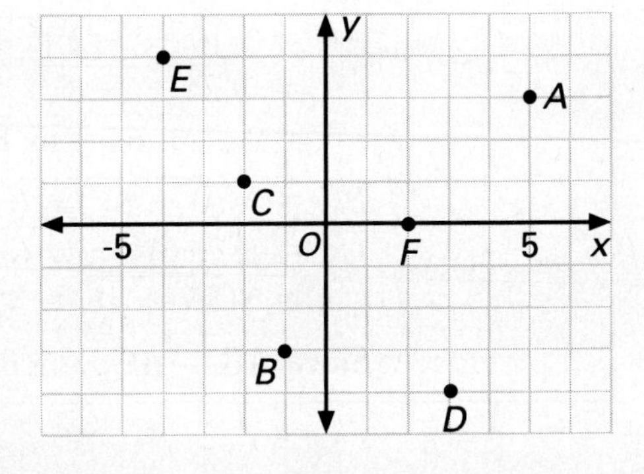

1. A
2. B
3. C
4. D
5. E
6. F

Draw a graph and locate the following points.

7. $A(4, 3)$ and $B(-2, 3)$
8. $P(4, 0)$ and $Q(3, -2)$
9. $S(-1, -1)$ and $T(0, 0)$
10. $G(-3, 4)$ and $H(-3, -3)$

Locate the pair of given points in a coordinate plane. Draw a right triangle that has the two points as vertices of the acute angles. Find the lengths of the legs and the length of the hypotenuse of the triangle.

11. $A(2, 1)$, $B(4, 5)$
12. $C(3, -2)$, $D(0, 4)$
13. $E(-1, -3)$, $F(4, 2)$
14. $G(0, -4)$, $H(-3, 2)$

Use the Distance Formula to find the distance between the given points.

15. $P_1(0, 0)$, $P_2(4, 3)$
16. $P_1(1, 1)$, $P_2(4, 5)$
17. $P_1(-1, 2)$, $P_2(11, 7)$
18. $P_1(2, -3)$, $P_2(7, 9)$
19. $P_1(1, 2)$, $P_2(5, 6)$
20. $P_1(6, 2)$, $P_2(1, 7)$

Find the length of $\overline{AB}$.

21. $A(0, 9), B(-5, -3)$

22. $A(-1, 2), B(2, -2)$

23. $A(2, -3), B(-3, 1)$

24. $A(-2, -3), B(3, 1)$

***ABC* is a triangle. Use the lengths *AB*, *BC*, and *AC* to tell whether or not $\triangle ABC$ is isosceles.**

25. $A(-2, 0), B(2, 0), C(0, 5)$

26. $A(0, 3), B(5, 1), C(0, -1)$

27. $A(0, 5), B(5, 0), C(7, 7)$

28. $A(1, 5), B(-4, 3), C(-1, 0)$

29. $A(4, 4), B(-6, 2), C(5, -1)$

30. $A(1, -6), B(-3, -3), C(-2, -2)$

***ABC* is a triangle. Use the lengths *AB*, *BC*, and *AC* to tell whether or not $\triangle ABC$ is a right triangle.**

31. $A(0, 0), B(2, 4), C(4, -2)$

32. $A(1, -2), B(-5, 4), C(5, 2)$

Using the Distance Formula to find the lengths of the sides and tell whether or not *ABCD* is a parallelogram.

33. $A(0, -2), B(5, -3), C(7, 1), D(2, 2)$

34. $A(-4, -2), B(2, 0), C(3, 3), D(-2, 1)$

35. $A(-2, -1), B(3, -2), C(4, 1), D(-1, 2)$

***ABCD* is a parallelogram. Using Theorem 7.4, tell whether or not it is a rectangle. (Exercises 36–37)**

36. $A(1, -2), B(7, 0), C(5, 3), D(-1, 1)$

37. $A(-4, -2), B(-2, -6), C(8, -1), D(6, 3)$

38. $A(3, y)$ is a distance of 10 units from $B(-3, -1)$. Find all possible values of y.

39. $C(-1, 2)$ is a distance of $\sqrt{17}$ units from $D(x, -2)$. Find all possible values of x.

40. $(-2, -2)$ and $(2, -1)$ are the coordinates of opposite vertices of a square. What are the coordinates of the other vertices?

41. $(-1, -3)$, $(2, -2)$, and $(1, 1)$ are the coordinates of three vertices of a square. What are the coordinates of the fourth vertex?

42. $(-4, -1)$, $(-3, 1)$, and $(1, -1)$ are the coordinates of three vertices of a rectangle. What are the coordinates of the fourth vertex?

43. Find the coordinates of all points with integer coordinates that are a distance of 5 units from the origin.

Mixed Review

1. Draw a segment 5 cm long. Construct the perpendicular bisector of the segment. *5.6*

Find the coordinate of the midpoint of each segment. *1.3*

2. $\overline{AG}$ **3.** $\overline{EK}$ **4.** $\overline{DJ}$

5. $\overline{LF}$ **6.** $\overline{KG}$ **7.** $\overline{JB}$

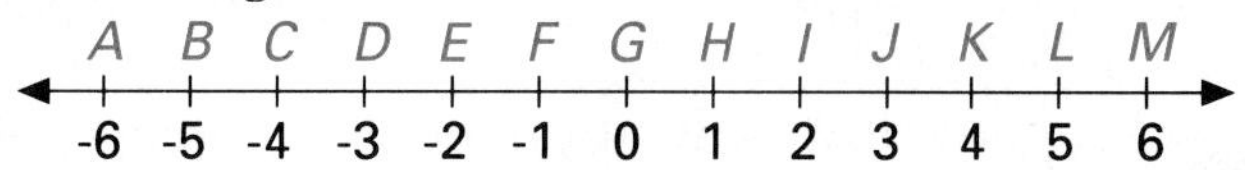

10.2 The Midpoint Formula

Objective To find the coordinates of the midpoint of a segment

The coordinate x_m of the midpoint of a segment on a number line is the average of the coordinates of the endpoints. On the number line below, the midpoint of $\overline{CM}$ is 2, since $\frac{-3 + 7}{2} = 2$.

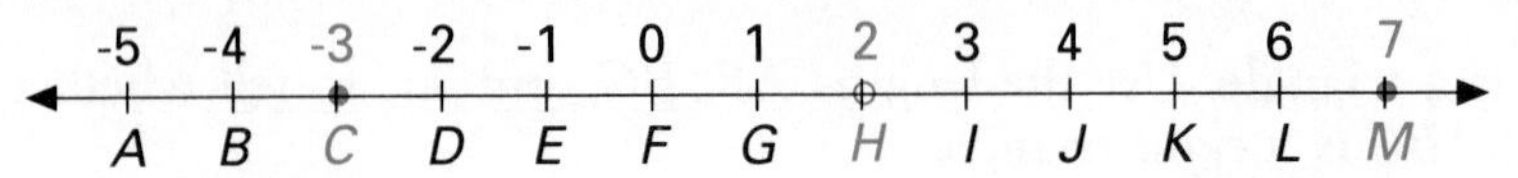

The coordinates of the midpoint of a segment in a coordinate plane can be found in a similar way. $\overline{HJ}$, with endpoints $H(1, 5)$ and $J(3, 7)$, has the midpoint $M(2, 6)$. The abscissa 2 is the average of the abscissas 1 and 3. The ordinate 6 is the average of the ordinates 5 and 7.

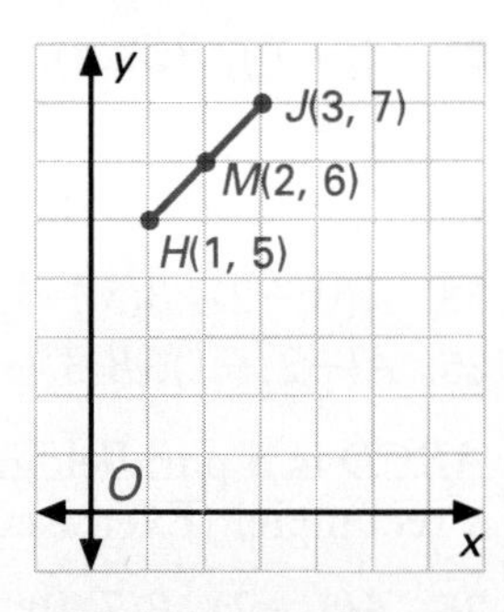

This result is stated in the following theorem.

Theorem 10.2 **The Midpoint Formula:** Given $P(x_1, y_1)$ and $Q(x_2, y_2)$, the coordinates (x_m, y_m) of M, the midpoint of $\overline{PQ}$, are $\left(\frac{x_1 + x_2}{2}, \frac{y_1 + y_2}{2}\right)$.

Given: $P(x_1, y_1)$, $Q(x_2, y_2)$

Prove: $x_m = \frac{x_1 + x_2}{2}$, $y_m = \frac{y_1 + y_2}{2}$

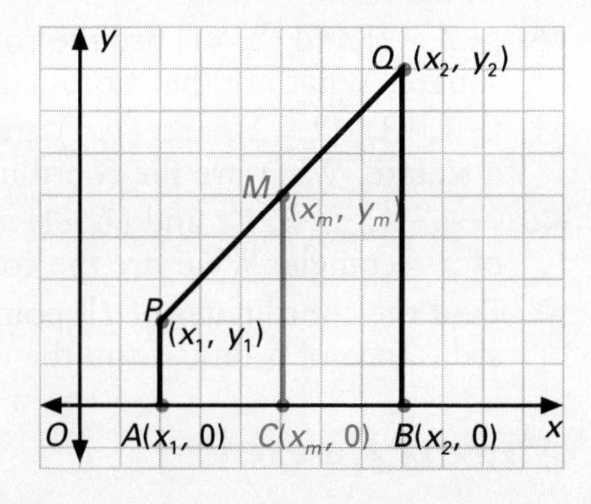

Proof Draw $\overline{PA}$, $\overline{MC}$, and $\overline{QB}$ all ⊥ to the x-axis. These are all ∥ to the y-axis and therefore vertical. A, B, and C have the same x-coordinates as P, Q, and M, respectively. C is the midpoint of $\overline{AB}$. By Theorem 6.12, if three parallel lines cut congruent segments on one transversal, they cut congruent segments on all transversals. But the coordinates of the midpoint of two points on a number line is the average.

Therefore, $x_m = \frac{x_1 + x_2}{2}$ for all points on $\overline{MC}$.

Similarly, it can be shown that $y_m = \frac{y_1 + y_2}{2}$.

EXAMPLE 1 Given $P(3, 2)$ and $Q(-5,-8)$, find the midpoint of $\overline{PQ}$.

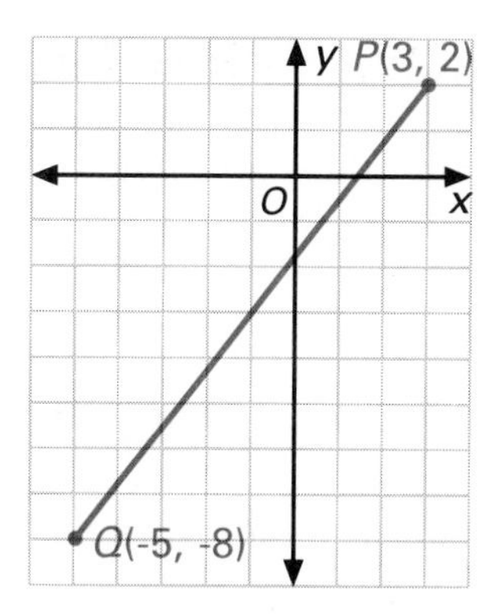

Solution

$$x_m = \frac{x_1 + x_2}{2} \qquad y_m = \frac{y_1 + y_2}{2}$$

$$= \frac{3 + (-5)}{2} = -1 \qquad = \frac{2 + (-8)}{2} = -3$$

Therefore, the coordinates of the midpoint are $(-1,-3)$.

EXAMPLE 2 Given: $M(1, 2)$ is the midpoint of $\overline{AB}$. The coordinates of A are $(-3, 8)$. Find the coordinates of B.

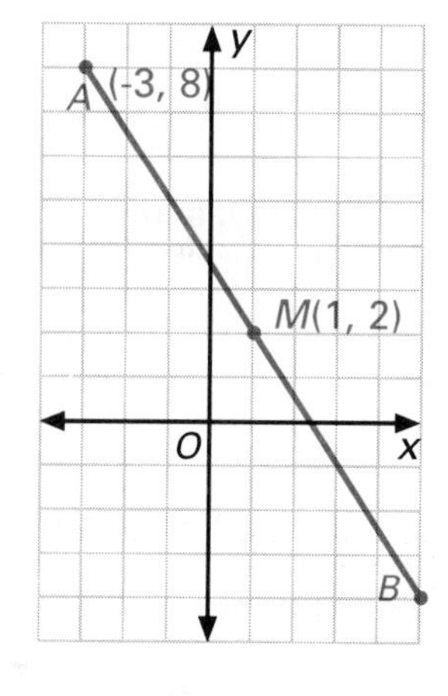

Solution Let A have the coordinates (x_1,y_1) and B have the coordinates (x_2,y_2).

$$x_m = \frac{x_1 + x_2}{2} \qquad y_m = \frac{y_1 + y_2}{2}$$

$$1 = \frac{-3 + x_2}{2} \qquad 2 = \frac{8 + y_2}{2}$$

$$2 = -3 + x_2 \qquad 4 = 8 + y_2$$

$$x_2 = 5 \qquad y_2 = -4$$

Therefore, the coordinates of B are $(5,-4)$.

EXAMPLE 3 $ABCD$ is a quadrilateral with vertices $A(-1,-3)$, $B(9,-3)$, $C(11, 5)$, and $D(1, 5)$. Show that $ABCD$ is a parallelogram using the midpoint formula.

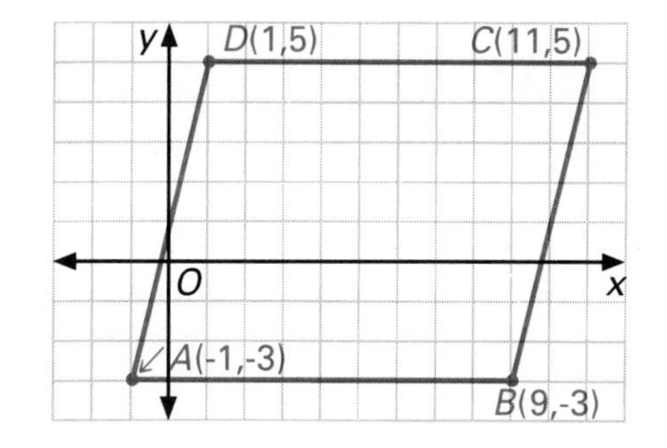

Plan Show that diagonals $\overline{AC}$ and $\overline{BD}$ bisect each other by showing that they have the same midpoint.

Solution

For diagonal $\overline{AC}$,

$$x_m = \frac{-1 + 11}{2},\ y_m = \frac{-3 + 5}{2}$$

$$x_m = 5,\ y_m = 1$$

For diagonal $\overline{BD}$,

$$x_m = \frac{9 + 1}{2},\ y_m = \frac{-3 + 5}{2}$$

$$x_m = 5,\ y_m = 1$$

Since the diagonals have the same midpoint, they bisect each other, and the quadrilateral is a parallelogram.

Classroom Exercises

Find the coordinate of the midpoint of each given segment.

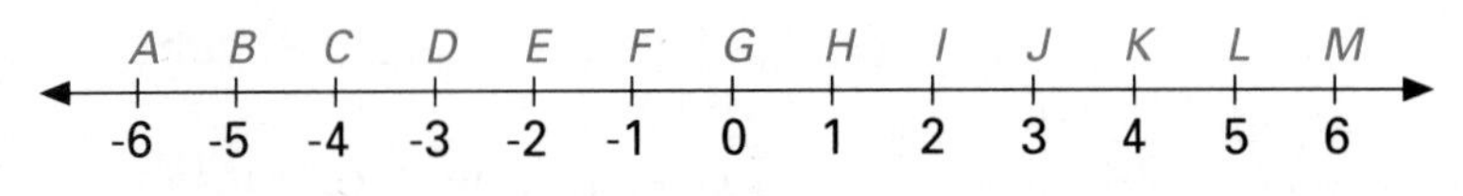

1. $\overline{GK}$
2. $\overline{AG}$
3. $\overline{DJ}$
4. $\overline{AI}$
5. $\overline{DL}$
6. $\overline{FL}$

For each pair of points A and B, find the coordinates of the midpoint of $\overline{AB}$.

7. A(0, 0), $B(8, 8)$
8. $A(-6, 6)$, $B(0, 0)$
9. $A(0, 0)$, $B(-4, 10)$

Written Exercises

For each pair of points A and B, find the coordinates of the midpoint of $\overline{AB}$.

1. $A(0, 0)$, $B(4, 4)$
2. $A(3, 5)$, $B(1, 7)$
3. $A(-4, 3)$, $B(6, -5)$
4. $A(7, -6)$, $B(-1, 2)$
5. $A(2, 9)$, $B(-8, -1)$
6. $A(-1, 0)$, $B(-9, -4)$
7. $A(7, 5)$, $B(0, 0)$
8. $A(2, 5)$, $B(1, 2)$
9. $A(-1, 3)$, $B(6, -2)$
10. $A(3, -4)$, $B(-8, 1)$
11. $A(2, -7)$, $B(-9, 1)$
12. $A(-5, -4)$, $B(-1, -9)$

M is the midpoint of $\overline{AB}$. Find the coordinates of B.

13. $A(0, 0)$, $M(3, 3)$
14. $A(1, 2)$, $M(3, 5)$
15. $A(4, 2)$, $M(1, -2)$
16. $A(1, 5)$, $M(2, -2)$
17. $A(-3, 3)$, $M(2, -4)$
18. $A(-3, -1)$, $M(-1, -4)$

ABCD is a quadrilateral with the given vertices. Using the fact that a quadrilateral is a parallelogram if its diagonals bisect each other, show that *ABCD* is a parallelogram.

19. $A(0, 1)$, $B(6, 1)$, $C(6, 5)$, $D(0, 5)$
20. $A(-3, 0)$, $B(-1, -3)$, $C(5, 1)$, $D(3, 4)$
21. $A(-2, -1)$, $B(1, -1)$, $C(4, 4)$, $D(1, 4)$
22. $A(-3, 1)$, $B(6, -1)$, $C(7, 1)$, $D(-2, 3)$
23. $\triangle XYZ$ has vertices $X(-3, -5)$, $Y(1, 7)$, and $Z(0, 4)$. Find the length of median $\overline{ZM}$.
24. $\triangle RST$ has vertices $R(-5, 0)$, $S(5, 0)$, and $T(0, 6)$. Find the length of median $\overline{TM}$.

ABC is a triangle with given vertices. Find the coordinates of M, the midpoint of $\overline{AB}$, and N, the midpoint of $\overline{AC}$. How does MN relate to BC?

25. $A(-3, 6)$, $B(-3, 0)$, $C(1, 4)$
26. $A(4, -3)$, $B(2, 1)$, $C(-4, -5)$

Find the coordinates of the midpoint M of $\overline{CD}$, correct to the nearest hundredth.

C **27.** $C(7.3, 4.2)$, $D(5.8, -1.9)$

C **28.** $C(-1.1, 5.6)$, $D(-3.8, -8.7)$

29. Find the coordinates of two points that trisect the segment with endpoints $A(-3, 8)$ and $B(9, -1)$.

30. Find the coordinates of the point on the segment with endpoints $A(0, 0)$ and $B(3, 6)$ that is twice as far from A as from B.

31. Find the coordinates of the point on the segment with endpoints $P_1(x_1, y_1)$ and $P_2(x_2, y_2)$ that is twice as far from P_1 as from P_2.

32. Develop a formula for finding the coordinates of a point on the segment with endpoints $P_1(x_1, y_1)$ and $P_2(x_2, y_2)$ that is n times as far from P_1 as from P_2.

33. Use the formula from Exercise 32 to find the coordinates of a point on the segment with endpoints $A(-6, -3)$ and $B(9, 7)$ that is four times as far from A as from B.

Mixed Review

1. Draw a line. Through a given point not on the line, construct a line parallel to the given line. *3.3*
2. Draw a line. Through a given point not on the line, construct a line perpendicular to the given line. *5.6*
3. Draw a line. Through a given point on the line, construct a line perpendicular to the given line. *5.6*

Algebra Review

To solve a fractional equation of the form $\frac{ax + b}{c} = \frac{dx + e}{f}$, where $c \neq 0$, $f \neq 0$:
(1) Multiply each side of the equation by cf.
(2) Solve the resulting equation.

Example Solve. $\frac{4x + 2}{3} = \frac{3x - 5}{2}$

$$2(4x + 2) = 3(3x - 5) \quad \leftarrow \text{result of multiplying both sides by } 3 \cdot 2$$
$$8x + 4 = 9x - 15$$
$$x = 19$$

Solve each equation.

1. $\frac{x - 2}{2} = \frac{x + 1}{3}$
2. $\frac{x + 4}{3} = \frac{x - 3}{4}$
3. $\frac{3x + 1}{4} = \frac{2x - 2}{3}$
4. $\frac{5x + 4}{4} = \frac{4x - 2}{3}$

10.3 Slope of a Line

Objectives To find the slope of a segment or a line
To draw a line given the slope and a point

In the photograph at the right, the roads are steeper at certain places than at others. The steepest parts of a road have the greatest *slope*. In mathematics, slope is the measure of the steepness of a segment or a line.

A segment determines a special right triangle having a horizontal and a vertical leg. The length of the vertical leg, sometimes called the *rise*, is $|y_2 - y_1|$. The length of the horizontal leg, sometimes called the *run*, is $|x_2 - x_1|$. The slope is the ratio of the lengths of these two legs, or the ratio of rise to run. It is necessary, however, to distinguish between a segment which rises from left to right, from one which descends from left to right. If a segment rises from left to right, then it will be assigned a positive slope. If it descends, then it will be assigned a negative slope. The formula in the following definition does this.

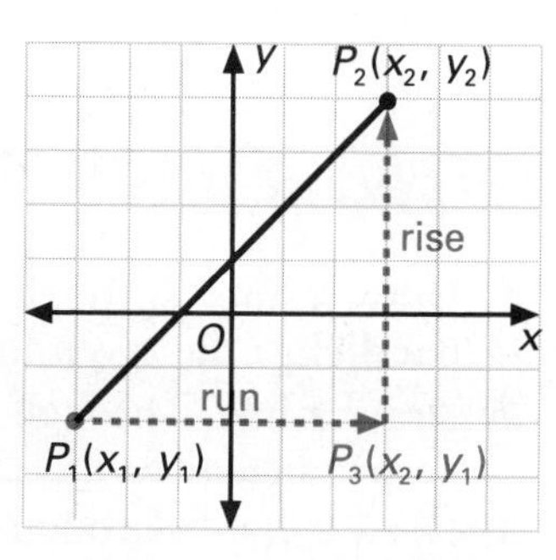

Definition The **slope** of a segment $\overline{P_1P_2}$, with endpoints $P_1(x_1, y_1)$ and $P_2(x_2, y_2)$, $x_2 \neq x_1$, is the ratio $\frac{y_2 - y_1}{x_2 - x_1}$.

EXAMPLE 1 Find the slope m of $\overline{AB}$ for $A(-1,-2)$ and $B(3, 5)$. Draw the graph of $\overline{AB}$ to check.

Solution

$$m = \frac{y_2 - y_1}{x_2 - x_1}$$

$$= \frac{5 - (-2)}{3 - (-1)} = \frac{7}{4}$$

Therefore, the slope of $\overline{AB}$ is $\frac{7}{4}$.

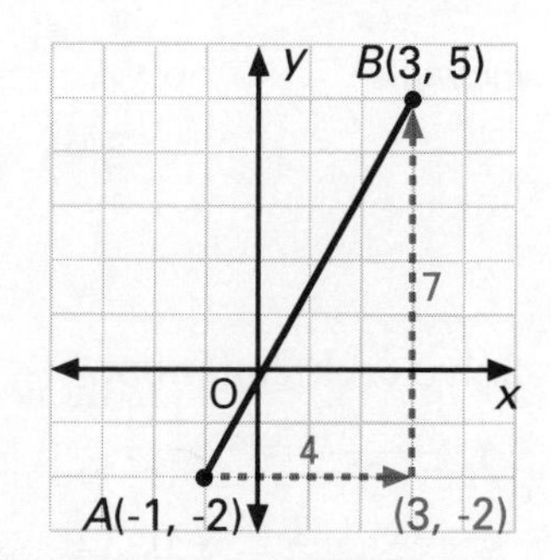

A horizontal segment neither rises nor descends. Since the y-coordinates of all points on a horizontal segment are equal, the slope of a horizontal segment is 0. For example, the slope of $\overleftrightarrow{PQ}$ is $\frac{4-4}{4-1} = 0$. The x-coordinates of a vertical segment are all equal. Therefore, the slope of a vertical segment is undefined, because division by 0 is undefined. The slope of $\overleftrightarrow{MN}$, for example, is $\frac{-1-(-6)}{-2-(-2)} = \frac{5}{0}$, which is undefined.

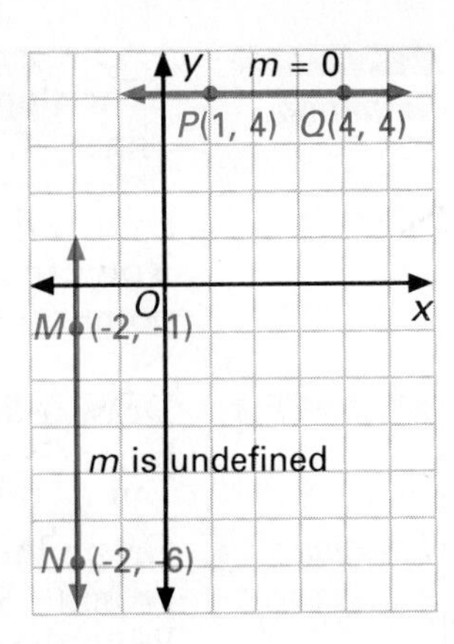

EXAMPLE 2 Is the segment with the given endpoints horizontal, vertical, or neither?

$P_1 = (4,-7)$, $P_2 = (4, 6)$

Solution The slope is $\frac{-7-6}{4-4}$, or $\frac{-13}{0}$, which is undefined.
Therefore, the segment is vertical.

Theorem 10.3 All segments of a non-vertical line have equal slopes.

Given: $\overline{AB}$ and $\overline{DE}$ on line l
Prove: Slope of $\overline{AB}$ = slope of $\overline{DE}$

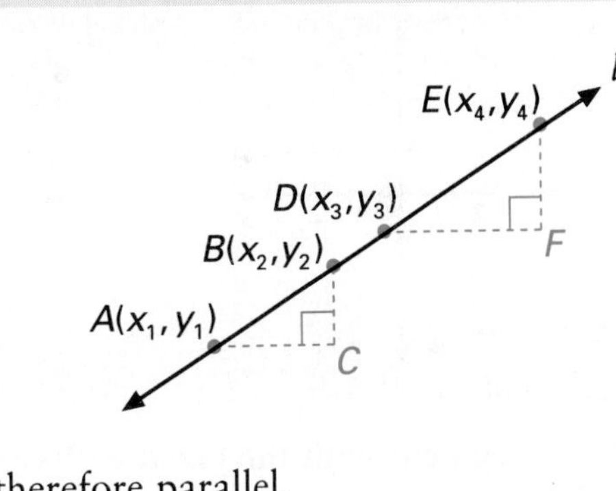

Outline of Proof:

1. Assign coordinates as follows: $A(x_1, y_1)$, $B(x_2, y_2)$, $D(x_3, y_3)$, $E(x_4, y_4)$.
2. Locate $C(x_2, y_1)$ and $F(x_4, y_3)$, and draw $\triangle ABC$ and $\triangle DEF$.
3. $\overline{AC}$ and $\overline{DF}$ are horizontal and therefore parallel.
4. $\angle BAC \cong \angle EDF$ (Corr $\angle$s of $\parallel$ lines are $\cong$.)
5. $\angle C \cong \angle F$, since each is a right angle formed by the horizontal and vertical segments.
6. $\triangle ABC \sim \triangle DEF$ (AA$\sim$)
7. $\frac{BC}{AC} = \frac{EF}{DF}$ ($\sim\triangle$s and properties of proportion)
8. If the line rises to the right, slope of $\overline{AB} = \frac{BC}{AC}$, slope of $\overline{DE} = \frac{EF}{DF}$.

   If the line descends to the right, slope of $\overline{AB} = -\frac{BC}{AC}$, slope of $\overline{DE} = -\frac{EF}{DF}$.
9. Slope of $\overline{AB}$ = slope of $\overline{DE}$ (Substitution)

Definition The **slope** m of a non-vertical line is the slope of any segment of the line.

You can draw a line if you are given its slope and the coordinates of a point on the line.

EXAMPLE 3 Draw a line with slope $= \frac{1}{2}$, containing the point $(-3,-4)$.

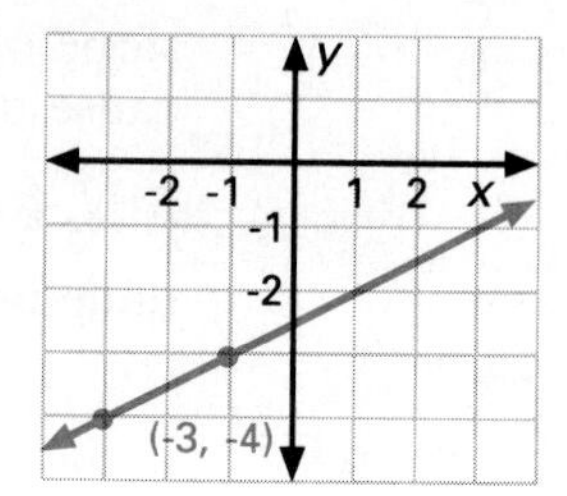

Solution Graph the point $(-3,-4)$.
From $(-3,-4)$, move up 1 unit.
From that point, move to the right 2 units.
Mark that point.
Draw the line through the two points.

Classroom Exercises

Tell whether the slope of the line is positive, negative, zero, or undefined.

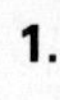

1.

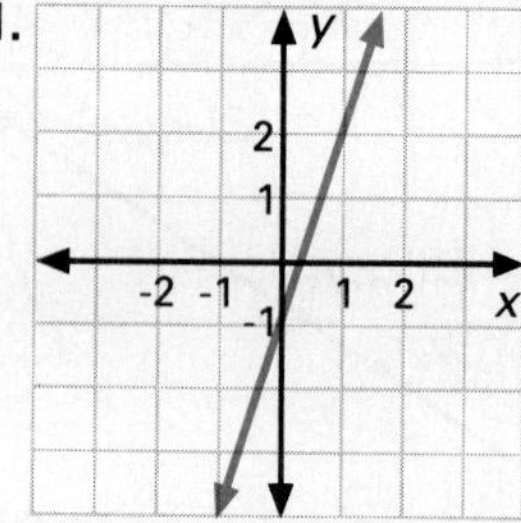

2.

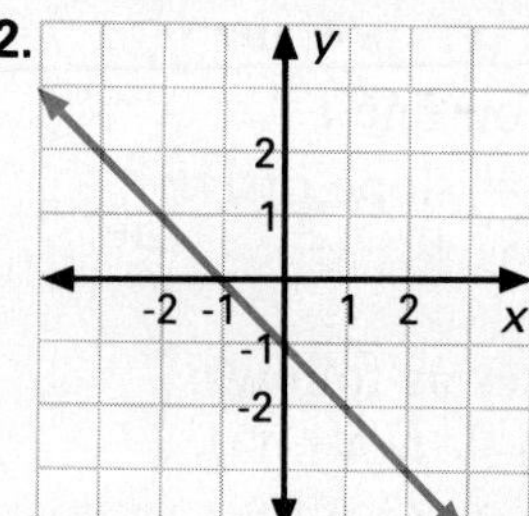

3.

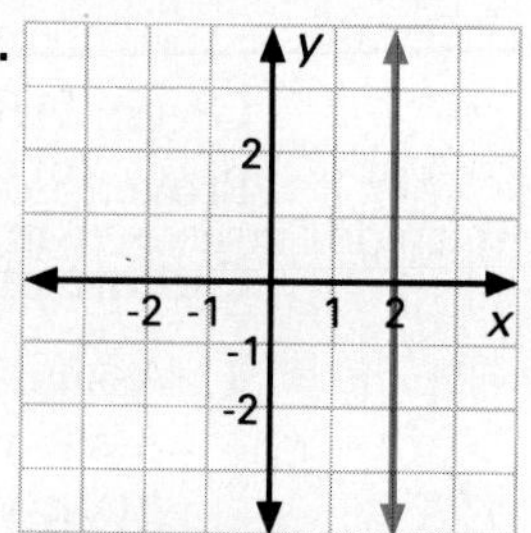

Is the segment with the given endpoints horizontal, vertical, or neither?

4. $A(0, 1)$, $B(0,-3)$ **5.** $C(-3, 5)$, $D(2,-4)$ **6.** $E(5,-3)$, $F(-2,-3)$
7. $G(5,-1)$, $H(4, 1)$ **8.** $I(-2, a)$, $J(5, a)$ **9.** $K(b, 4)$, $L(b,-1)$

Written Exercises

Find the slope of the segment with the given endpoints.

1. $A(3, 5)$, $B(6, 9)$ **2.** $C(-3, 5)$, $D(6, 5)$ **3.** $E(0, 0)$, $F(-3,-5)$
4. $G(-1, 4)$, $H(-1,-4)$ **5.** $I(3,-3)$, $J(2,-1)$ **6.** $K(4,-4)$, $L(3,-5)$
7. $M(5,-2)$, $N(-3, 4)$ **8.** $O(-6, 1)$, $P(-7, 5)$ **9.** $Q(-2,-1)$, $R(3,-4)$

Find the slope of the line containing the given points.

10. $P_1(4, 2), P_2(6, 4)$
11. $P_1(9, 2), P_2(3, 0)$
12. $P_1(-2, 4), P_2(-8, 2)$
13. $P_1(-4, 5), P_2(-2, 5)$
14. $P_1(3, 6), P_2(-3,-3)$
15. $P_1(0, 2), P_2(-2, 1)$

Draw a line with the given slope containing the given point.

16. $m = \frac{2}{3}, (1,-3)$
17. $m = \frac{1}{2}, (-2, 4)$
18. $m = \frac{-3}{2}, (-3, 1)$
19. $m = \frac{-2}{5}, (1, 4)$

Use slopes to show whether the three given points are in the same line.

20. $A(0, 0), B(2, 4), C(6, 12)$
21. $D(-3,-2), E(0, 0), F(6, 4)$
22. $G(2,-5), H(-5,-5), I(0,-5)$
23. $J(0, 7), K(4, 1), L(6, 5)$

Find the unknown coordinate so that $\overline{PQ}$ will have the given slope.

24. $P(3, 5), Q(x, 1), m = 4$
25. $P(4, y), Q(9, 2)\ m = -5$
26. $P(8, 8), Q(3, y), m = 0$
27. $P(6, 9), Q(x,-2)$, m is undefined.
28. $P(4, 7), Q(-2, y), m = \frac{2}{3}$
29. $P(x,-1), Q(-1, 3), m = \frac{-4}{5}$

30. A carpenter must know the "pitch" or "slope" of a roof when building it. This is the number of inches of vertical rise in 12 in. of horizontal run. A slope of 3-in-12 means a vertical rise of 3 in. in a horizontal run of 12 in. How many inches would the roof rise in a run of 15 ft if it had a slope of 3-in-12?

31. If a roof rises 3 ft in a run of 15 ft, what is the slope of the roof?

32. $\triangle ABC$ has vertices $A(1,-2)$, $B(4,-5)$, and $C(2, 6)$. What is the slope of the median to side $\overline{BC}$?

33. The coordinates of the midpoints of the sides of a triangle are $(1, 3)$, $(4,-2)$, and $(5, 1)$. Find the coordinates of the vertices.

34. Discuss the relationship between the slope of a line and the trigonometric ratios of the angle that the line makes with the x-axis.

Midchapter Review

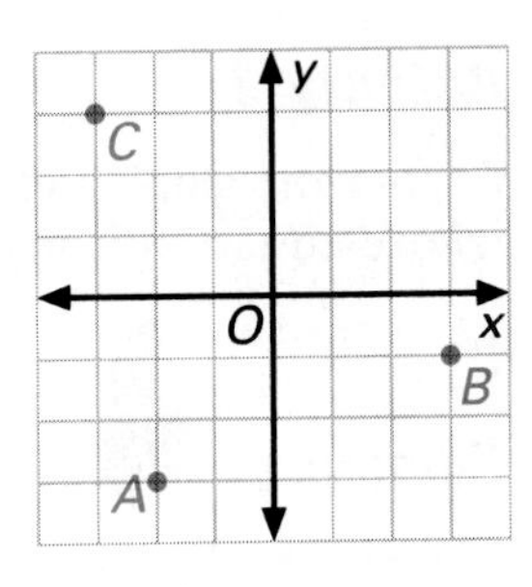

Name the coordinates and the quadrant for each of the points.

1. A
2. B
3. C

4. Is the triangle formed by points A, B, and C isosceles? *10.1*

5. Find the coordinate of the midpoint of $\overline{AB}$.

6. Find the slope of $\overleftrightarrow{AB}$.

Computer Investigation

Equation of a Line

Use a computer software program that graphs equations.

Activity 1

Draw the graphs below. If necessary, adjust the boundaries of the grid to accommodate the given lines.

1. Draw the graph of $y = \frac{2}{3}x$. NOTE: Be sure to graph $y = \frac{2}{3} \cdot x$, not $y = \frac{2}{3x}$.
2. Use the graph to find the slope of this line. Start wherever the line crosses the y-axis: (0, 0). Count 3 blocks to the right along the x-axis. Then, count up the number of blocks until you are on the line: 2. The slope of the line is $\frac{\text{rise}}{\text{run}} = \frac{2}{3}$.
3. Draw the graph of $y = 3x + 1$. Use the graph to find the slope of the line. Start wherever the line crosses the y-axis: (0,1). Move 1 block to the right along the x-axis. Then, count up the number of blocks until you are on the line: 3. The slope of this line is $\frac{\text{rise}}{\text{run}} = \frac{3}{1}$, or 3.
4. Predict the slope of the line $y = \frac{4}{5}x - 2$ before graphing it. Then verify your conclusion by graphing the line.

Activity 2

Draw the graph of each equation below on the same set of axes.

5. $y = \frac{3}{4}x$
6. $y = \frac{3}{4}x + 2$
7. $y = \frac{3}{4}x + 6$
8. $y = \frac{3}{4}x - 4$

Use the graphs of Exercises 5–8 above to answer Exercises 9–11 below.

9. What is the slope of each line above?
10. How are all the lines related to each other?
11. For each line of Exercises 5–8, where does the line cross the y-axis?

Activity 3

Predict the slope and the point of crossing the y-axis for each line whose equation is given below. Verify by graphing.

12. $y = 4x -$ [illegible]
13. $y = \frac{3}{5}x - 8$
14. $y = -\frac{1}{5}x + 3$

Summary

The graph of an equation of the form $y = mx + b$ has a slope of ______ and crosses the y-axis at ______.

10.4 Equation of a Line

Objectives To draw the graph of an equation of a line
To write an equation of a line in three forms

A **graph of an equation** is the set of points whose coordinates satisfy the given equation. An equation whose graph is a line is called a **linear equation.**

All points of a *vertical* line have the same x-coordinate or abscissa. All points with this x-coordinate lie on the given vertical line. The equation $x = a$, where a is any real number, is an equation of the vertical line which intersects the x-axis at $(a, 0)$. Similar reasoning leads to the conclusion that an equation of a *horizontal* line which intersects the y-axis at $(0, b)$ is $y = b$.

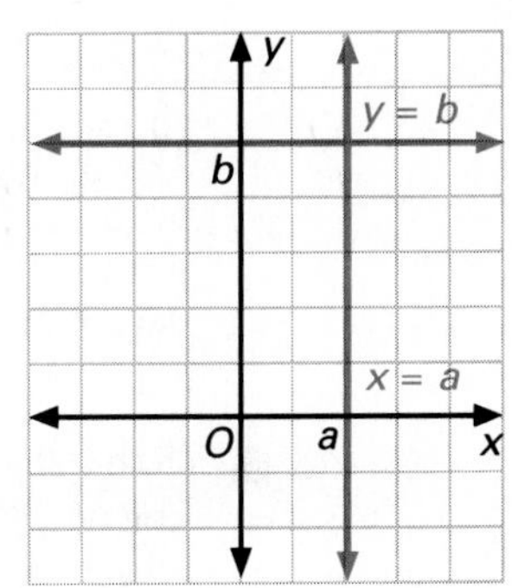

The formula for the slope of a line can be used to obtain an equation of a line through a given point with a given slope. If $P_1(x_1, y_1)$ is a specific point on a line and $P(x, y)$ is any other point on the line, then $m = \frac{y - y_1}{x - x_1}$ is an equation of the given line with slope m and containing $P_1(x_1, y_1)$. Using the Multiplication Property of Equality with this equation essentially proves the following theorem.

Theorem 10.4 An equation of a line with slope m containing the point $P_1(x_1, y_1)$ is $y - y_1 = m(x - x_1)$.

This is the **point-slope** form of the equation of a line.

EXAMPLE 1 Write an equation in point-slope form of the line with slope 2 containing the point $(3, -4)$.

Solution Let $x_1 = 3$, $y_1 = -4$, and $m = 2$.
$y - (-4) = 2(x - 3)$

The **standard form** of an equation of a line is an equation of the form $ax + by = c$, where a and b are not both zero, and where $a \geq 0$. (In some texts $ax + by + c = 0$ is considered to be the standard form.)

EXAMPLE 2 Write an equation in standard form of the line containing points (2, 1) and (4, −4).

Plan Find the slope. Use the point-slope form of the equation of a line.

Solution

$$m = \frac{-4 - 1}{4 - 2} = \frac{-5}{2}$$

$$\begin{aligned} (y - 1) &= \tfrac{-5}{2}(x - 2) \\ 2(y - 1) &= -5(x - 2) \\ 2y - 2 &= -5x + 10 \\ 5x + 2y &= 12 \end{aligned} \qquad \text{or} \qquad \begin{aligned} [y - (-4)] &= \tfrac{-5}{2}(x - 4) \\ 2(y + 4) &= -5(x - 4) \\ 2y + 8 &= -5x + 20 \\ 5x + 2y &= 12 \end{aligned}$$

The **y-intercept** of a line is the y-coordinate of the point at which the line intersects the y-axis. If the y-intercept is 6, for example, then it intersects the y-axis at the point (0, 6). In general, if the y-intercept is b, then the line crosses the y-axis at $(0, b)$.

If the slope m and the y-intercept b of a line are known, then the equation is $(y - b) = m(x - 0)$, which simplifies to $y = mx + b$. This proves the following theorem.

Theorem 10.5 If a line has slope m and y-intercept b, then an equation of the line is $y = mx + b$.

$y = mx + b$ is the **slope-intercept** form of the equation of a line.

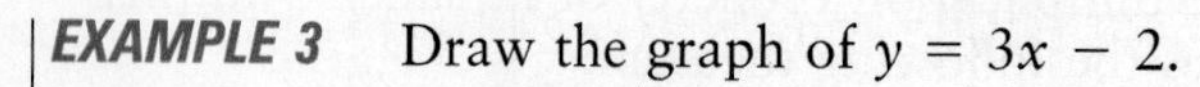

EXAMPLE 3 Draw the graph of $y = 3x - 2$.

Plan The slope m is 3. The y-intercept b is −2.

Solution Plot the point (0, −2). From that point, go up three units and to the right one unit. Draw a point at (1, 1). Draw a line through the two points.

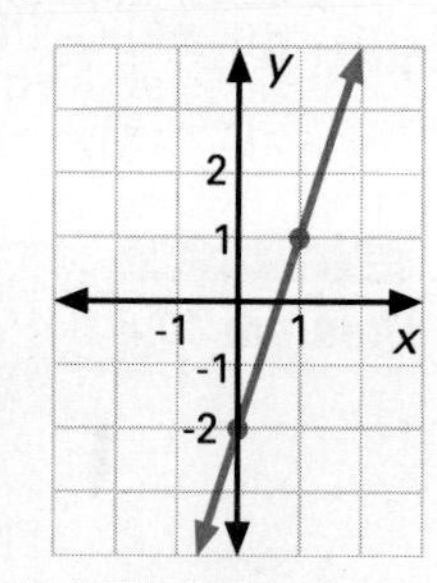

EXAMPLE 4 Write an equation in slope-intercept form of the line pictured below.

Solution The y-intercept b is −3. Since the line passes through (0, −3) and (4, 0), the slope m is $\frac{0 - (-3)}{4 - 0} = \frac{3}{4}$.

Therefore, $y = \frac{3}{4}x - 3$ is the equation of the line.

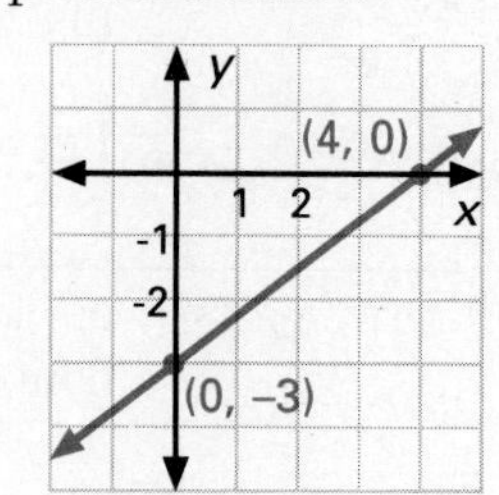

Classroom Exercises

Find the equation of the line with the given slope and y-intercept.

1. $m = 5, b = 2$ **2.** $m = -1, b = 4$ **3.** $m = \frac{4}{3}, b = -2$

Find the slope and y-intercept of the line. Then draw the graph.

4. $y = 4x + 3$ **5.** $y = \frac{-2}{3}x - 2$ **6.** $y = -x + 1$

Written Exercises

What is an equation of a horizontal line that contains the given point?

1. $(3, 4)$ **2.** $(-4, 3)$ **3.** $(5, -2)$

What is an equation of a vertical line that contains the given point?

4. $(3, 4)$ **5.** $(-4, 3)$ **6.** $(5, -2)$

Write an equation in slope-intercept form of the line with the given slope and the given y-intercept.

7. $m = -3, b = 2$ **8.** $m = \frac{5}{4}, b = -6$ **9.** $m = \frac{1}{3}, b = -\frac{4}{3}$

10. $m = 0, b = 5$ **11.** $m = -2, b = 0$ **12.** $m = 1, b = 1$

13. Write an equation in point-slope form of the line having slope -3 and containing $(1, 2)$.

14. Write an equation in standard form of the line in Exercise 13.

15. Find the y-intercept of the line with equation $2x - y = 3$.

Write an equation in point-slope form of the line with the given slope containing the given point.

16. $m = 4, (2, 3)$ **17.** $m = 3, (-1, 4)$ **18.** $m = -2, (4, 1)$

19. $m = -1, (-3, -1)$ **20.** $m = \frac{-2}{5}, (-2, -1)$ **21.** $m = \frac{-4}{3}, (4, 0)$

Write an equation in standard form of the line containing the given points.

22. $(2, 0), (4, 4)$ **23.** $(1, 3), (5, 2)$ **24.** $(-1, 3), (5, -2)$

25. $(-2, 0), (0, -5)$ **26.** $(-2, -7), (3, -1)$ **27.** $(4, -1), (-6, -2)$

Draw a graph of each equation.

28. $y = 2x + 1$ **29.** $y = 3x - 2$ **30.** $y = -4x - 4$

31. $y = -x - 2$ **32.** $y = \frac{-3}{2}x + 2$ **33.** $y = \frac{-3}{4}x - 3$

Write an equation for each line.

34.
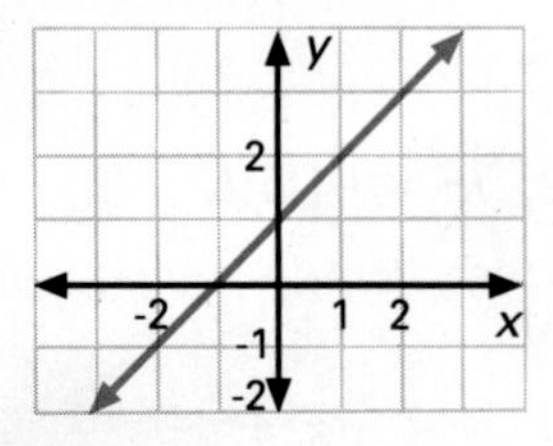

35.
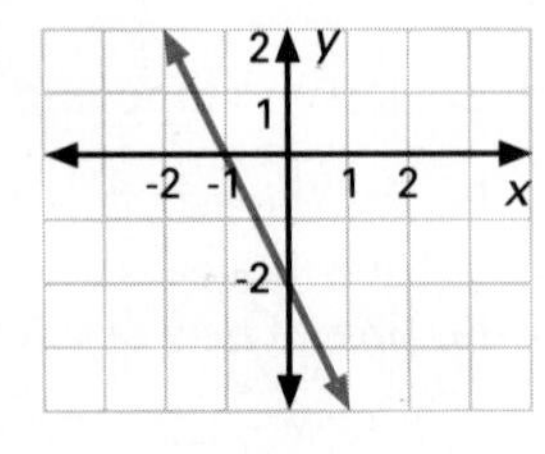

36.
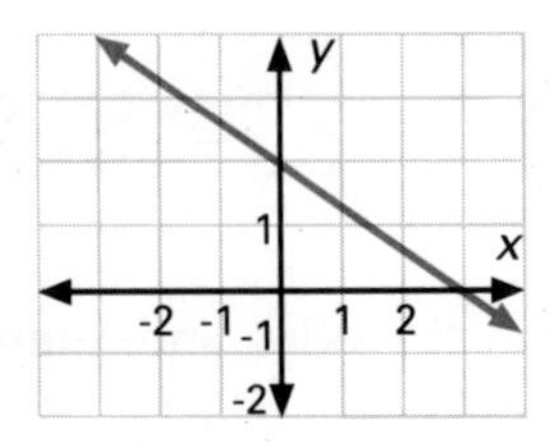

37.
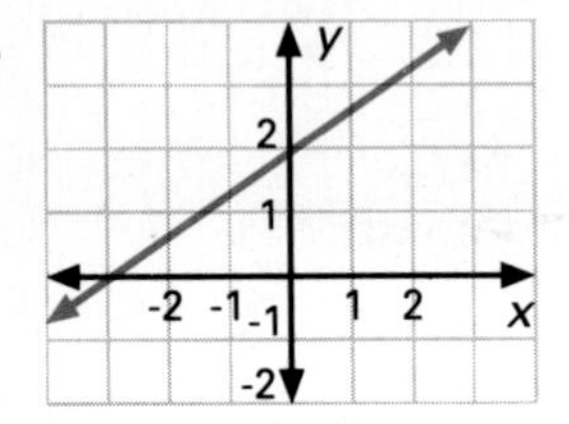

38.
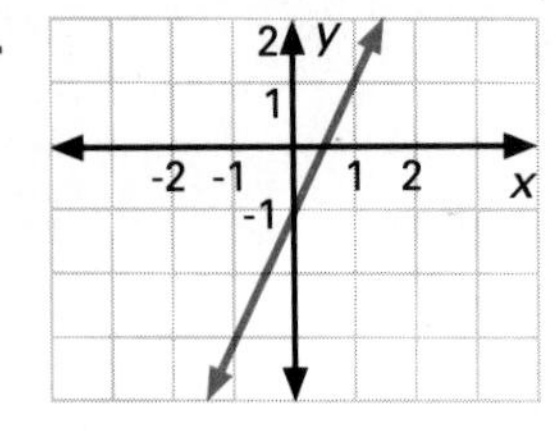

39.
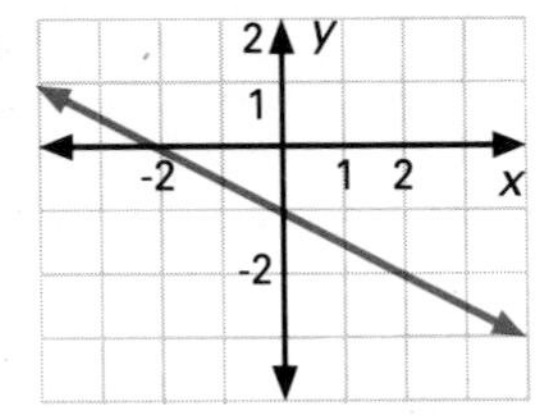

40. What are the possible slopes of lines that make an angle of measure 30 with the x-axis?

41. What are the possible slopes of lines that make an angle of measure 45 with the x-axis?

42. What are the possible slopes of lines that make an angle of measure 60 with the x-axis?

43. Write an equation of a line that makes an angle of measure 30 with the x-axis and which contains the point (2, 1).

44. Write an equation of a line that makes an angle of measure 45 with the x-axis and which contains the point $(-3, 4)$.

Mixed Review

1. Write the converse of "If two angles are congruent, then they are equal in measure." *3.5*
2. Write the converse of "All right angles are congruent." *3.5*
3. Write the inverse of "If two lines are parallel, then they are coplanar." *3.8*
4. Write the inverse of "Two congruent triangles have equal areas." *3.8*
5. Write the contrapositive of "If a triangle is scalene, then its sides are of different lengths." *3.8*
6. Write the contrapositive of "An obtuse triangle has an obtuse angle." *3.8*

10.5 Parallel or Perpendicular Lines

Objectives

To determine if lines are parallel or perpendicular
To find the slope of a line parallel or perpendicular to a given line
To write an equation of a line parallel or perpendicular to a given line

The roofs of the buildings on the left have the same pitch or slope. It appears that the corresponding lines on the roofs are parallel.

In the figure, perpendicular segments have been constructed from points on lines l_1 and l_2 to the x-axis. If $\triangle ABP \sim \triangle CDQ$, then $\frac{PB}{AB} = \frac{QD}{CD}$. These ratios are the slopes of the lines l_1 and l_2. Since $\triangle ABP \sim \triangle CDQ$, $\angle PAB \cong \angle QCD$, and $\overline{AP} \parallel \overline{CQ}$. This suggests the following two theorems, which are given without proof.

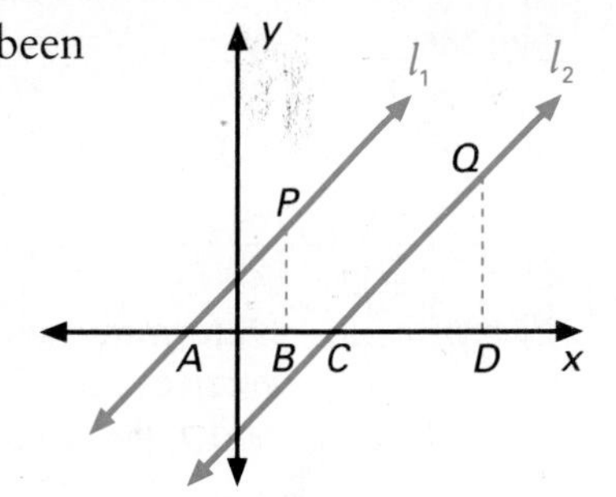

Theorem 10.6 If two non-vertical lines have equal slopes, then they are parallel.

Theorem 10.7 If two non-vertical lines are parallel, then they have equal slopes.

Theorems 10.6 and 10.7 are converses of each other. Remember that the proof of a theorem does not guarantee that its converse is true. The combination of Theorems 10.6 and 10.7 allows us to state that parallelism and equal slopes are necessary and sufficient conditions for each other.

Notice in Example 1 that it is not necessary to graph the lines before determining that they are parallel.

EXAMPLE 1 Do the given equations describe parallel lines?

$$2x - 3y = -12 \qquad 9y = 6x - 9$$

Plan Rewrite the equations in slope-intercept form. Then compare the slopes.

Solution

$$\begin{aligned} 2x - 3y &= -12 \\ -3y &= -2x - 12 \\ y &= \tfrac{2}{3}x + 4 \longrightarrow m = \tfrac{2}{3} \end{aligned} \qquad \begin{aligned} 9y &= 6x - 9 \\ y &= \tfrac{2}{3}x - 1 \longrightarrow m = \tfrac{2}{3} \end{aligned}$$

Since the slopes are equal, the lines are parallel.

EXAMPLE 2 Write an equation in slope-intercept form of a line containing $P(2, 1)$ parallel to the line whose equation is $y = 3x - 1$.

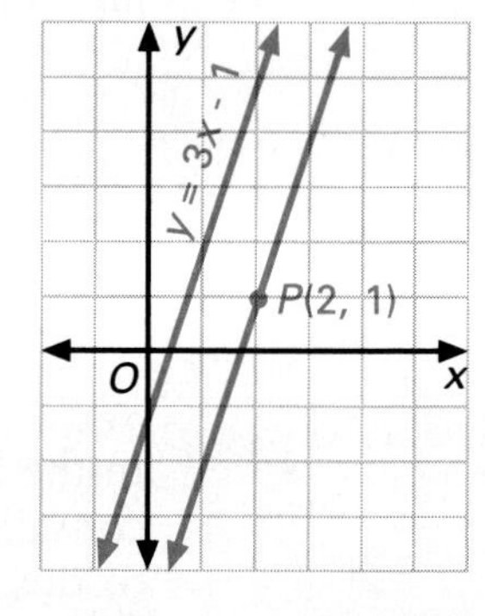

Plan Since the parallel lines have the same slope, use the slope of 3 and the point (2, 1) in the point-slope form.

Solution

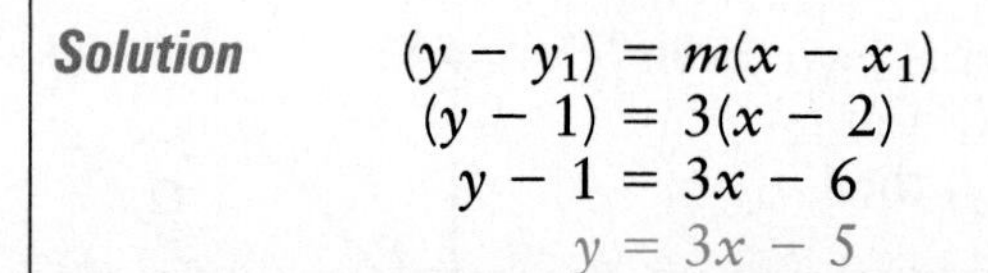

$$\begin{aligned} (y - y_1) &= m(x - x_1) \\ (y - 1) &= 3(x - 2) \\ y - 1 &= 3x - 6 \\ y &= 3x - 5 \end{aligned}$$

EXAMPLE 3 Determine whether $ABCD$ with vertices $A(-2, 2)$, $B(3, -1)$, $C(0, -6)$, and $D(-5, -3)$ is a parallelogram.

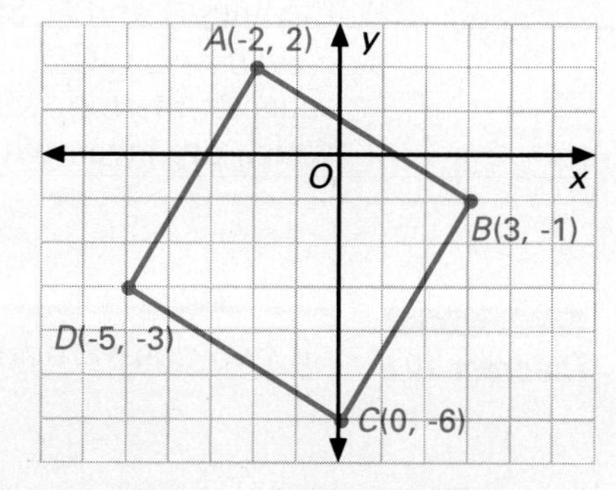

Plan Show that opposite sides are parallel.

Solution

$$\text{Slope of } \overline{AB} = \frac{2 - (-1)}{-2 - 3} = \frac{3}{-5} = -\frac{3}{5} \qquad \text{Slope of } \overline{CB} = \frac{-6 - (-1)}{0 - 3} = \frac{-5}{-3} = \frac{5}{3}$$

$$\text{Slope of } \overline{CD} = \frac{-3 - (-6)}{-5 - 0} = \frac{3}{-5} = -\frac{3}{5} \qquad \text{Slope of } \overline{DA} = \frac{-3 - 2}{-5 - (-2)} = \frac{-5}{-3} = \frac{5}{3}$$

Since their slopes are equal, $\overline{AB} \parallel \overline{CD}$ and $\overline{CB} \parallel \overline{DA}$. Therefore, $ABCD$ is a parallelogram.

The relationship between perpendicularity and slope is not as obvious as that between parallelism and slope. Two perpendicular lines are shown in the diagram. The equation of one of the lines is $y = \frac{3}{4}x - 2$. The equation of the other is $y = -\frac{4}{3}x + 3$. Notice that the slopes of the lines are negative reciprocals of each other. In other words, the product $(\frac{3}{4})(-\frac{4}{3})$ of the two slopes is -1. This property is true for all non-vertical perpendicular lines.

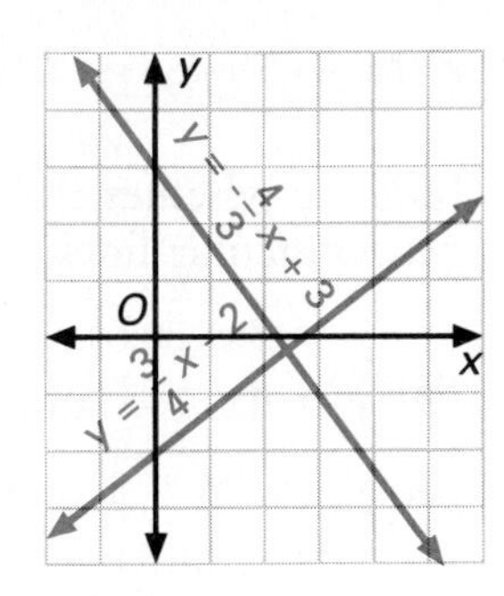

Theorem 10.8 If the product of the slopes of two non-vertical lines is -1, then the lines are perpendicular.

EXAMPLE 4 For the points $A(0, 0)$, $B(5, 5)$, $C(-1, 1)$, and $D(4, 6)$ determine whether $\overleftrightarrow{AB}$ and $\overleftrightarrow{CD}$ are parallel, perpendicular, or neither of these.

Solution Slope of $\overleftrightarrow{AB} = \frac{5-0}{5-0} = \frac{5}{5} = 1$ Slope of $\overleftrightarrow{CD} = \frac{6-1}{4-(-1)} = \frac{5}{5} = 1$

Since the slopes are equal, $\overleftrightarrow{AB} \parallel \overleftrightarrow{CD}$.

EXAMPLE 5 Write an equation in slope-intercept form of a line containing $(-2, 0)$, perpendicular to the line whose equation is $y = -\frac{2}{3}x + 3$.

Solution

$mn = -1$ ← Let m and n be the slopes.

$-\frac{2}{3}n = -1$ ← m = slope of given line

$n = \frac{3}{2}$

$(y - 0) = \frac{3}{2}[x - (-2)]$ ← line containing $(-2, 0)$

$y = \frac{3}{2}x + 2$

Theorem 10.9 The product of the slopes of two non-vertical perpendicular lines is -1.

EXAMPLE 6 Determine whether $ABCD$ from Example 3 is a rectangle.

Solution $ABCD$ is a parallelogram as shown in the solution of Example 3.

Since slope of $\overline{AB}$ · slope of $\overline{BC} = -\frac{3}{5} \cdot \frac{5}{3} = -1$, $\overline{AB} \perp \overline{BC}$.

Therefore, $ABCD$ is a rectangle.

Classroom Exercises

Indicate whether the pair of equations describes parallel lines, perpendicular lines, or neither.

1. $y = 3x + 2$ and $y = -3x - 2$

2. $y = 2x + 1$ and $y = 2x - 1$

3. $y = 4x + 5$ and $y = 4x - 3$

4. $y = 2x + 1$ and $y = -\frac{1}{2}x - 1$

5. $y = -x + 1$ and $y = x - 1$

6. $y = \frac{2}{3}x - 5$ and $y = \frac{2}{3}x - 1$

7. $y = 6$ and $x = 4$

8. $y = 7$ and $y = -7$

Give the slope of a line parallel to the line with the given equation. Then give the slope of a line perpendicular to the line with the given equation.

9. $y = 2x + 2$

10. $y = -3x - 2$

11. $y = \frac{1}{2}x - 3$

12. $y = \frac{-2}{3}x + 5$

Written Exercises

Indicate whether the pair of equations describes parallel lines, perpendicular lines, or neither.

1. $2x + 3y = 7,\ 3x + 2y = 9$

2. $3x - 4y = -2,\ 3x + 4y = 6$

3. $-4x + 3y = 6,\ -12x + 9y = 9$

4. $2x + 5y = -4,\ -5x + 2y = 3$

Write an equation in slope-intercept form of the line containing the given point and parallel to the line whose equation is given.

5. $y = 2x + 1,\ (0, 0)$

6. $y = -3x - 5,\ (1, 1)$

7. $y = 5x - 2,\ (3, -4)$

8. $y = -4x + 6,\ (-2, 1)$

9. $y = \frac{3}{4}x,\ (4, -2)$

10. $y = \frac{3}{2}x + 1,\ (-2, -5)$

Write an equation in slope-intercept form of the line containing the given point and perpendicular to the line whose equation is given.

11. $y = 2x,\ (0, 0)$

12. $y = -3x + 1,\ (1, 1)$

13. $y = x - 3,\ (-3, 2)$

14. $y = -x + 7,\ (-1, -1)$

15. $y = \frac{3}{4}x,\ (-1, 3)$

16. $y = -\frac{5}{3}x - 2,\ (2, -2)$

For the given points A, B, C, and D, determine whether $\overleftrightarrow{AB}$ and $\overleftrightarrow{CD}$ are parallel, perpendicular, or neither of these.

17. $A(0, 0)$, $B(3, 3)$, $C(-1, 1)$, $D(2, -2)$

18. $A(-3, 2)$, $B(-2, 5)$, $C(-4, -1)$, $D(2, 1)$

19. $A(3, 9)$, $B(-2, 4)$, $C(-1, 0)$, $D(4, 5)$

20. $A(6, -1)$, $B(-2, -2)$, $C(-6, 3)$, $D(0, 4)$

Determine whether or not $ABCD$ is a parallelogram.

21. $A(-3, 2)$, $B(-1, 4)$, $C(0, -3)$, $D(-2, -5)$

22. $A(2, -7)$, $B(7, -2)$, $C(6, 1)$, $D(1, -3)$

Determine whether or not $ABCD$ is a rectangle.

23. $A(-6,-3)$, $B(-1, 2)$, $C(1, 0)$, $D(-4,-5)$

24. $A(-1,-4)$, $B(-7, 0)$, $C(-9,-3)$, $D(-3,-7)$

Determine whether or not the diagonals of $ABCD$ are perpendicular.

25. $A(0, 5)$, $B(4, 2)$, $C(0,-2)$, $D(-4,-2)$

26. $A(5, 4)$, $B(-1, 3)$, $C(0,-3)$, $D(6,-2)$

27. The slope of a line is 3. A line parallel to this given line contains the point $(-2, 5)$. The abscissa of a second point on this line is 0. Find the ordinate of this point.

28. The slope of a line is $-\frac{3}{4}$. A line perpendicular to this line contains the point $(2,-7)$. The ordinate of a second point on this line is 1. Find the abscissa of this point.

29. The slope of a line is -2. A line perpendicular to this line contains the point $(-1,-3)$. The abscissa of a second point on this line is 7. Find the ordinate of this point.

30. The slope of a line is $\frac{4}{5}$. A line perpendicular to this line contains the point $(5, 2)$. The ordinate of a second point on this line is -3. Find the abscissa of this point.

Classify the quadrilateral in all appropriate categories: rhombus, trapezoid, parallelogram, rectangle, square, and "kite". A kite has exactly one of its diagonals as a perpendicular bisector of the other.

31. $A(-2, 1)$, $B(-1, 4)$, $C(2, 3)$, $D(3,-4)$

32. $E(-1, 0)$, $F(2, 3)$, $G(6, 1)$, $H(3,-2)$

33. $M(-4, 0)$, $N(-1, 2)$, $O(1,-1)$, $P(-2,-3)$

34. $R(-1, 1)$, $S(2, 2)$, $T(6, 1)$, $V(-3, 2)$

Triangle TRI has vertices $T(3, 2)$, $R(-3,-1)$, and $I(1,-4)$.

35. Find an equation of the line containing the median from T to $\overline{RI}$.

36. Find an equation of the line containing the altitude from T to $\overline{RI}$.

37. Find an equation of the line containing the midsegment parallel to $\overline{RI}$.

Mixed Review

Which property is a property of a trapezoid, an isosceles trapezoid, a parallelogram, a rectangle, a rhombus, or a square? *6.5, 6.6*

1. Diagonals are congruent.

2. Diagonals are perpendicular.

3. Opposite angles are congruent.

4. Opposite sides are congruent.

5. At least one pair of consecutive angles is congruent.

6. Diagonals bisect each other.

Application: *Vector Methods*

Anything that has both *magnitude* and *direction* can be represented by a vector. Vectors are drawn as arrows, because arrows have both magnitude (length) and direction.

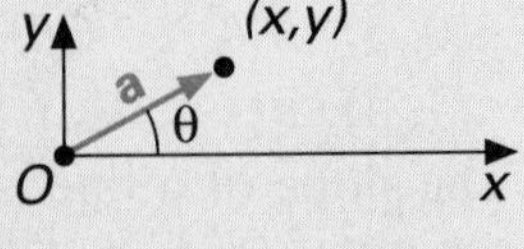

Vector **a** is in **standard position**, since it begins at the origin. Only one distinct vector can be drawn from the origin to (x, y), so we may refer to **a** simply as (x, y).

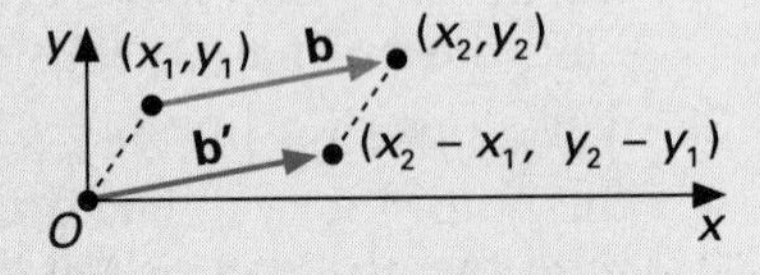

Any **free vector** such as **b**, with its **tail** at (x_1, y_1) and its **head** at (x_2, y_2), can be standardized: Subtract $(x_2, y_2) - (x_1, y_1) = (x_2 - x_1, y_2 - y_1)$ and relocate as shown. Vector **b′** has the same length and direction as **b**. (Why?)

We can multiply a vector in standard position by a real number c: $c(x, y) = (cx, cy)$. When c does not equal 0 or 1, the result is a new vector with a different length, but the same *or opposite* direction. Vectors that have the same or opposite direction, even collinear vectors, are called **parallel**.

We can use vector methods to determine if line segments are perpendicular or parallel. In the following examples we will think of line segments as vectors. (It won't matter which endpoints we call heads or tails.)

EXAMPLE 1 Given segment $\overline{AC}$ with endpoints $(-5, 1)$ and $(-1, 3)$ and segment $\overline{BD}$ with endpoints $(2, 1)$ and $(8, 4)$, determine whether the two are parallel.

Plan Regard $\overline{AC}$ and $\overline{BD}$ as free vectors. Standardize them as vectors **a** and **b**. If **a** $\parallel$ **b**, then $\overline{AC} \parallel \overline{BD}$. Vectors **a** and **b** are parallel if **a** times some real number c equals **b**: $c\mathbf{a} = \mathbf{b}$.

Solution $\mathbf{a} = (-5, 1) - (-1, 3) = (-4, -2)$ $\quad\quad$ $\mathbf{b} = (8, 4) - (2, 1) = (6, 3)$
$-1.5(-4, -2) = (6, 3)$, so $\mathbf{a} \parallel \mathbf{b}$. We conclude that $\overline{AC} \parallel \overline{BD}$.

EXAMPLE 2 Given segment $\overline{AC}$ with endpoints $(3, 7)$ and $(4, 9)$ and segment $\overline{BD}$ with endpoints $(1, 3)$ and $(-1, 4)$, determine whether the two are perpendicular.

Plan Regard $\overline{AC}$ and $\overline{BD}$ as free factors. Standardize them as vectors **a** and **b**. If $\mathbf{a} \perp \mathbf{b}$, then $\overline{AC} \perp \overline{BD}$. Vectors **a** and **b** are perpendicular if their **dot product** $\mathbf{a} \cdot \mathbf{b} = 0$. For vectors $\mathbf{a} = (x_1, y_1)$ and $\mathbf{b} = (x_2, y_2)$, $\mathbf{a} \cdot \mathbf{b}$ is defined as $x_1x_2 + y_1y_2$. (Note: this is a real number, not a new vector.)

Solution $\mathbf{a} = (4, 9) - (3, 7) = (1, 2)$ $\quad\quad$ $\mathbf{b} = (-1, 4) - (1, 3) = (-2, 1)$
$\mathbf{a} \cdot \mathbf{b} = (1 \cdot -2 + 2 \cdot 1) = 0$, so $\mathbf{a} \perp \mathbf{b}$. We conclude that $\overline{AC} \perp \overline{BD}$.

Try using a vector approach to solve problems 17–20, page 400.

10.6 Proofs with Coordinates

Objectives To assign coordinates to the vertices of a polygon in a plane
To use coordinate geometry to prove statements

Coordinate geometry can be used to prove many theorems. In all such cases the coordinate proof is not the only proof, but it is often the simpler proof. The following theorem was proved in this book as Theorem 6.10. Compare this coordinate proof to the earlier proof.

Theorem 10.10 **Midsegment Theorem:** The segment joining the midpoints of two sides of a triangle is parallel to the third side, and its length is half the length of the third side.

Given: $\triangle RST$, with $R(0, 0)$, $S(2a, 0)$, and $T(2b, 2c)$, M is the midpoint of $\overline{RT}$; N is the midpoint of $\overline{ST}$.
Prove: $\overline{MN} \parallel \overline{RS}$, $MN = \frac{1}{2}(RS)$

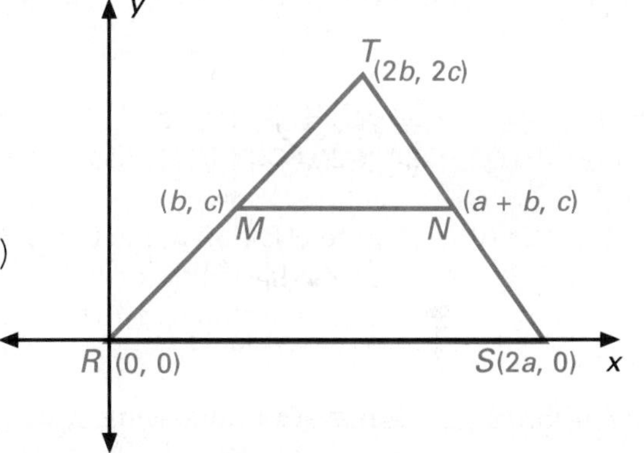

Outline of Proof:

1. By the Midpoint Formula, the coordinates of M are (b, c) and of N are $(a + b, c)$.
2. The slope of $\overline{MN}$ = the slope of $\overline{RS}$, which is zero ($\overline{MN} \parallel \overline{RS}$).
3. $MN = |a + b - b| = a$
4. $RS = 2a$
5. $MN = \frac{1}{2}(RS)$

A first step in setting up a coordinate proof involving a polygon is to place the figure in a coordinate plane.

(1) If possible, place the polygon so that a side is on one of the axes with a vertex at the origin.
(2) If a polygon contains a right angle, place the polygon so that two sides lie on the x- and y-axes.

(3) If the proof involves the Midpoint Formula, use coordinates such as $(2a, 2b)$ to avoid fractions in the proof.

(4) If the proof involves an isosceles triangle or trapezoid, it is sometimes convenient to place the figure with its axis of symmetry on the y-axis.

(5) If the proof involves perpendicular diagonals of a square or rhombus, it is sometimes convenient to place the figure so that the diagonals coincide with the x-axis and y-axis.

As long as you do not assume more than is given in the hypothesis of a theorem, the placement of the figure does not affect the validity of the proof. It may, however, simplify the computation necessary to complete the proof.

EXAMPLE 1 Which placement of the figure is preferable to prove that the median to the base of an isosceles right triangle is perpendicular to the base? Why?

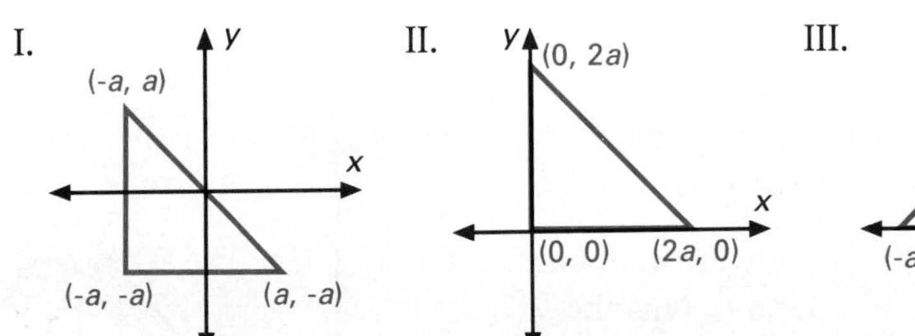

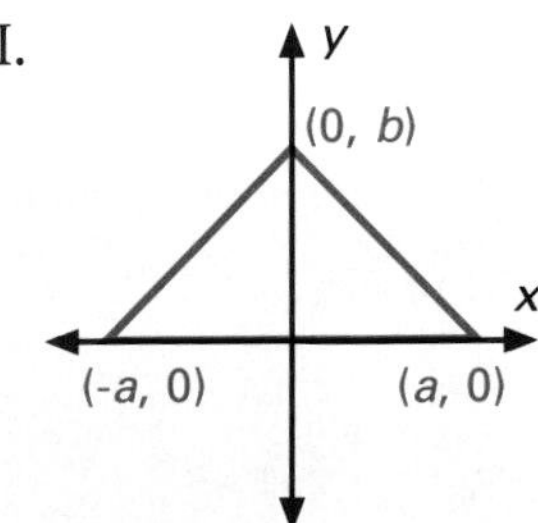

Solution Placement II. The use of zero coordinates simplifies computation. The use of a coordinate such as $2a$ avoids fractions when midpoints must be found. Placement III is isosceles but not necessarily right.

EXAMPLE 2 What are the missing coordinates in parallelogram $ABCD$ if $AB = 2a$?

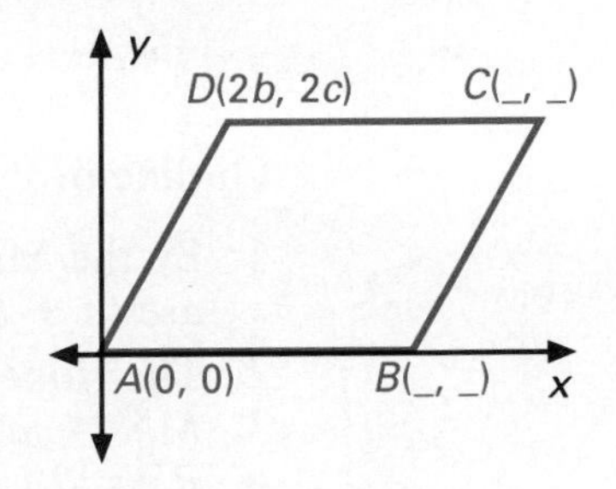

Solution Side $\overline{AB}$ of the parallelogram with length $2a$ was placed along the x-axis with one vertex at $(0, 0)$. If $AB = 2a$, then B must have the coordinates $(2a, 0)$.

Since opposite sides of a parallelogram are congruent, $DC = 2a$. To show this, if the abscissa of D is $2b$, then the abscissa of C must be $2b + 2a$.

Since opposites sides of a parallelogram are parallel, the side opposite $\overline{AB}$ is horizontal. If the ordinate of D is $2c$, then the ordinate of C is $2c$.

The coordinates of C are $(2a + 2b, 2c)$.

EXAMPLE 3 Using a coordinate proof, prove that the diagonals of a parallelogram bisect each other.

Given: $\square ABCD$, with diagonals $\overline{AC}$ and $\overline{BD}$

Prove: $\overline{AC}$ and $\overline{BD}$ bisect each other.

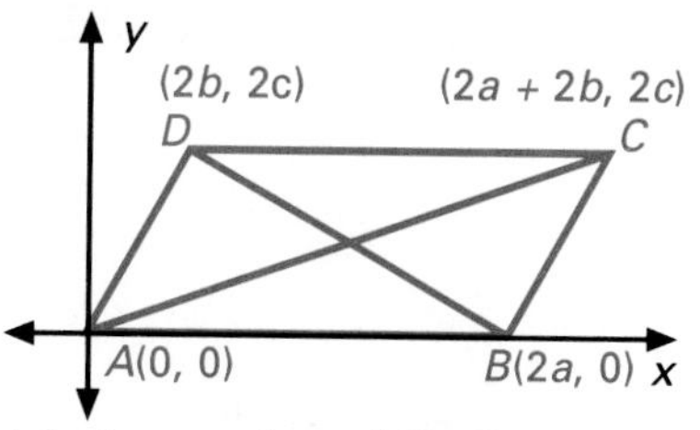

Proof

1. Place $\square ABCD$ in a coordinate plane with the vertices $A(0, 0)$, $B(2a, 0)$, $C(2a + 2b, 2c)$, and $D(2b, 2c)$.
2. $M_{\overline{AC}}$, the midpoint of $\overline{AC}$, has coordinates $(\frac{0 + 2a + 2b}{2}, \frac{0 + 2c}{2})$, or $(a + b, c)$.
3. $M_{\overline{BD}}$, the midpoint of $\overline{BD}$, has the coordinates $(\frac{2a + 2b}{2}, \frac{0 + 2c}{2})$, or $(a + b, c)$.
4. Since $M_{\overline{AC}} = M_{\overline{BD}}$, $\overline{AC}$ and $\overline{BD}$ have the same midpoint, $(a + b, c)$, and therefore bisect each other.

Classroom Exercises

What are the missing coordinates for the given figures?

1.

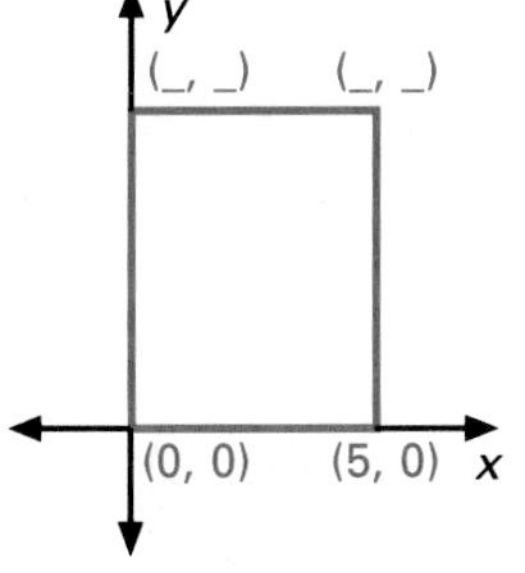

Square

2.

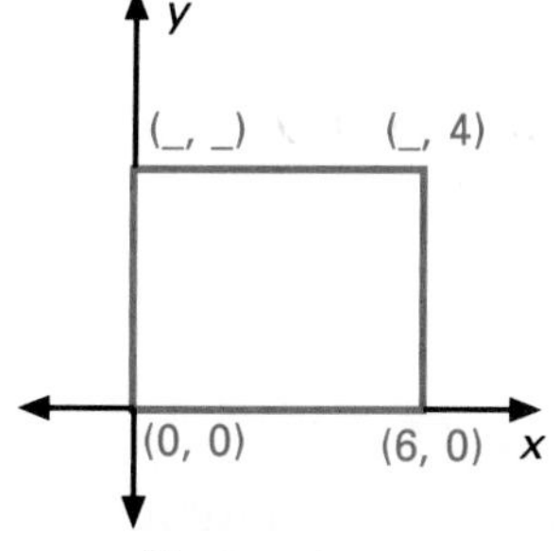

Rectangle

3.

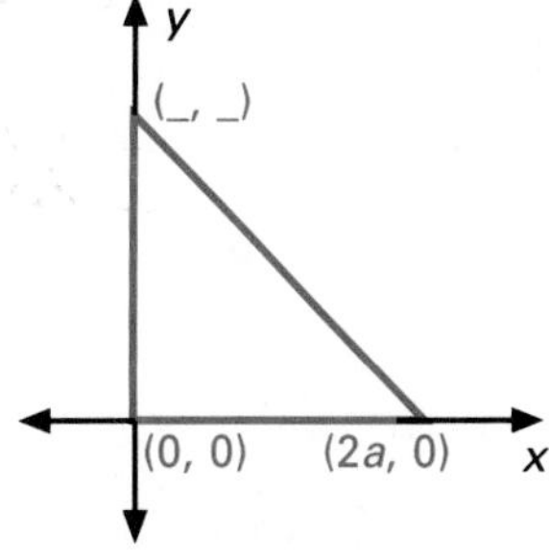

Isosceles right triangle

4.

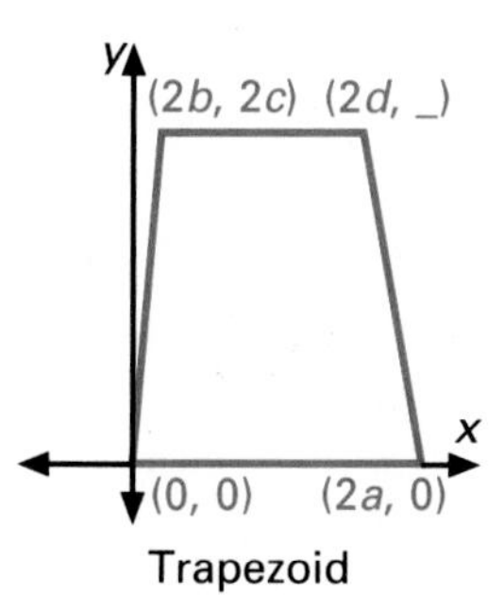

Trapezoid

5.

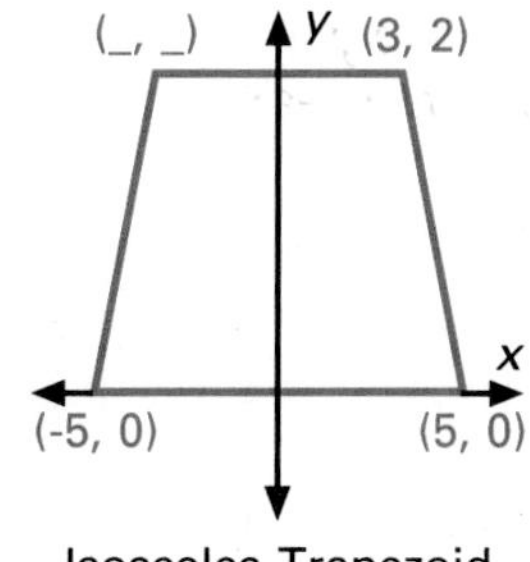

Isosceles Trapezoid

6.

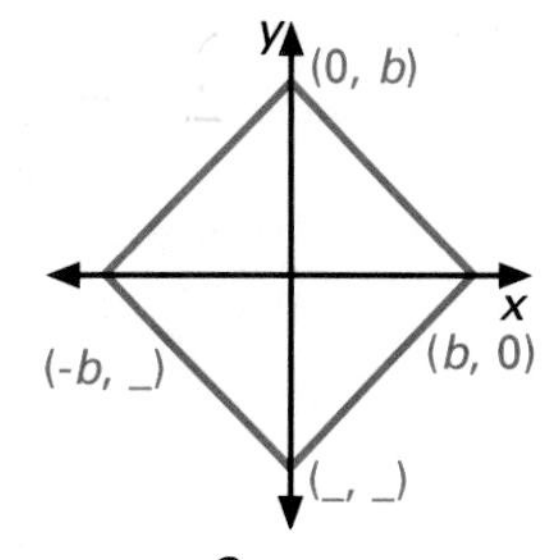

Square

7. Discuss a plan for a proof that the median of a trapezoid is parallel to the base.

Written Exercises

What are the missing coordinates for the given figures?

1.

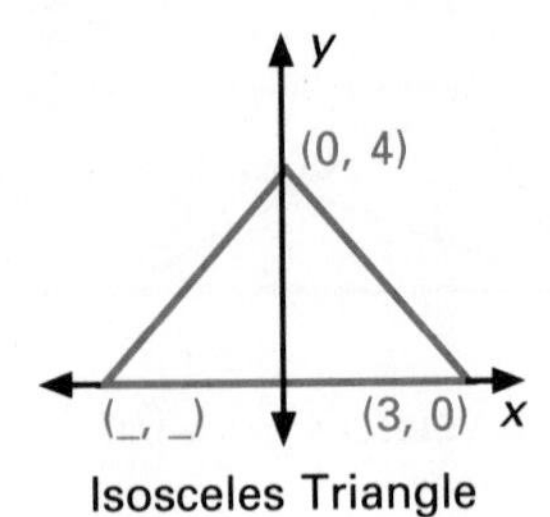

Isosceles Triangle

2.

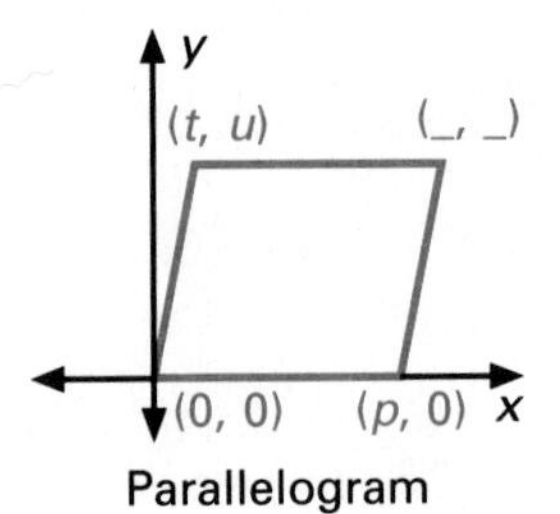

Parallelogram

3.

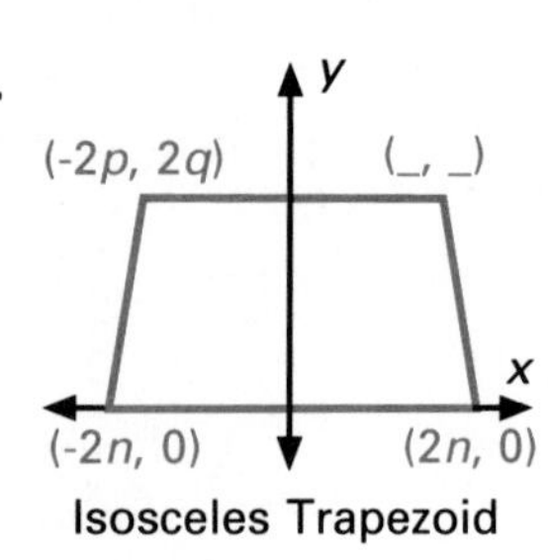

Isosceles Trapezoid

Place each given figure in a coordinate plane. What are the coordinates of each vertex? (Exercises 4–7)

4. a right triangle

5. an isosceles triangle

6. a scalene triangle

7. a quadrilateral

8. Explain why the given coordinates are those of a rectangle.

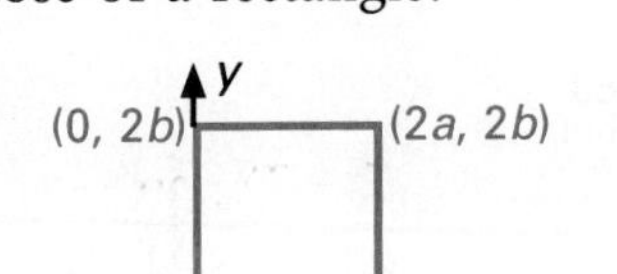

9. Explain why the given coordinates are those of a trapezoid.

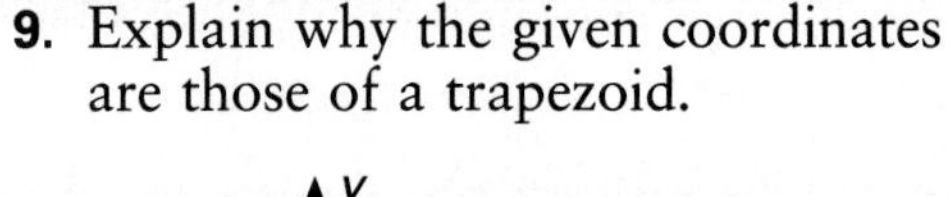

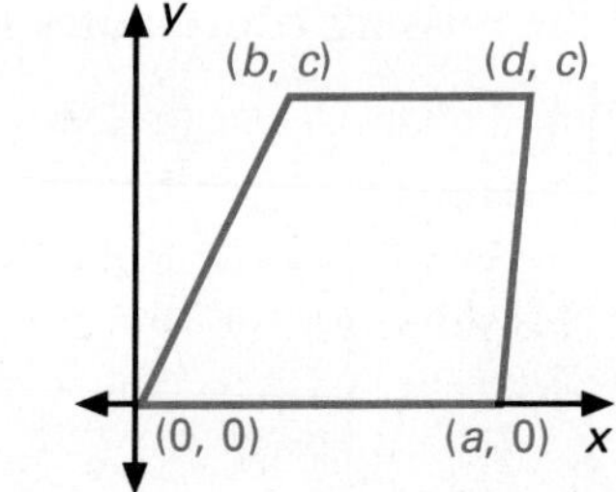

Prove the following statements using a coordinate proof.

10. The diagonals of a rectangle are congruent.

11. The diagonals of a square have the same midpoint.

12. The diagonals of a rectangle bisect each other.

Draw a set of coordinate axes. Place the given figure in the coordinate plane so that the coordinates of the vertices are as simple as possible.

13. a rhombus

14. a kite

Prove the following statements using a coordinate proof.

15. The diagonals of an isosceles trapezoid are congruent.

16. The midpoint of the hypotenuse of a right triangle is equidistant from the vertices.

17. The median of a trapezoid is parallel to the bases.
18. The length of the median of a trapezoid is one-half the sum of the lengths of the two bases.
19. The segments joining the midpoints of the opposite sides of a quadrilateral bisect each other.
20. The segments determined by successive midpoints of any quadrilateral form a parallelogram.

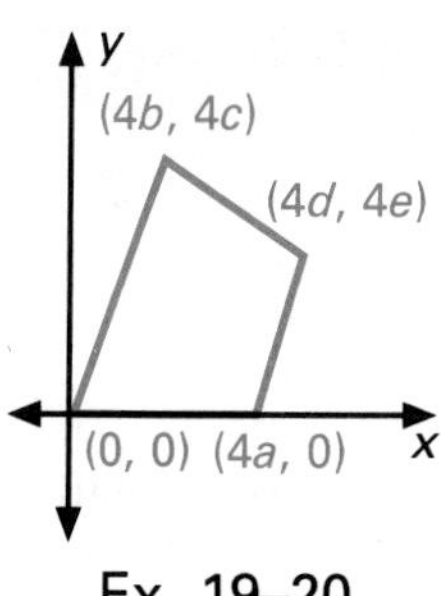

Ex. 19–20

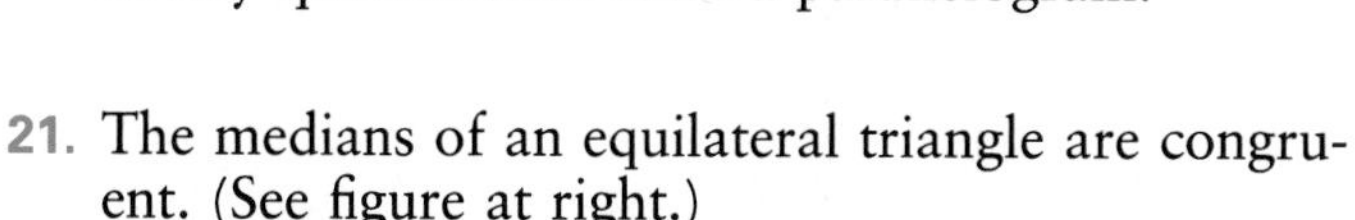

21. The medians of an equilateral triangle are congruent. (See figure at right.)

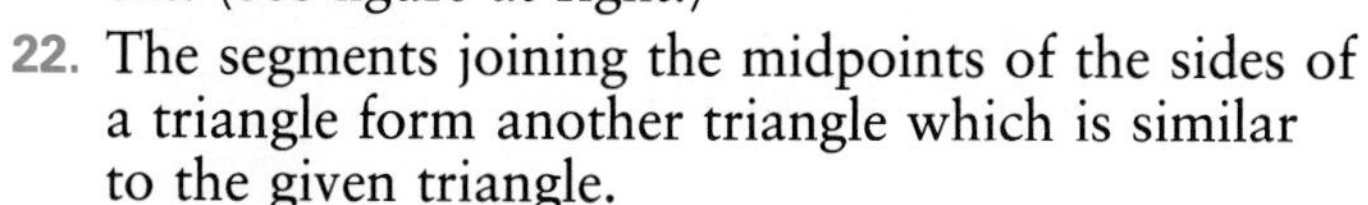

22. The segments joining the midpoints of the sides of a triangle form another triangle which is similar to the given triangle.
23. The segments joining the midpoints of consecutive sides of a rectangle form a rhombus.
24. If the diagonals of a parallelogram are congruent, then the parallelogram is a rectangle.
25. If the diagonals of a quadrilateral bisect each other, then the quadrilateral is a parallelogram.

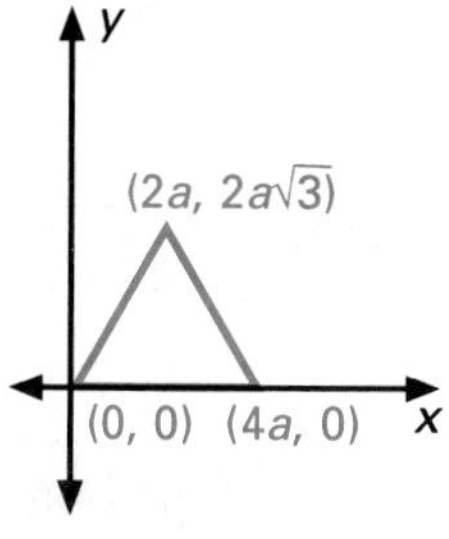

Ex. 21

Mixed Review

1. Find the midpoint of the segment joining the points $A(-1, -3)$ and $B(3, 5)$. 10.2
2. Find the length of the segment given in Exercise 1. 10.1
3. Find the slope and y-intercept of the line whose equation is $2y = 4x - 6$. 10.4
4. What is the slope of a line perpendicular to the line whose equation is given in Exercise 3? 10.5

Algebra Review

To square a binomial, use $(a + b)^2 = a^2 + 2ab + b^2$.

Example: Simplify $(2x + 3)^2$.

$$(a + b)^2 = a^2 + 2ab + b^2$$
$$(2x + 3)^2 = (2x)^2 + 2(2x)(3) + 3^2$$
$$= 4x^2 + 12x + 9$$

Simplify.

1. $(x + 2)^2$
2. $(x - 3)^2$
3. $(y + 5)^2$
4. $(a - 8)^2$
5. $(2x + 1)^2$
6. $(3y - 2)^2$
7. $(2c - 5)^2$
8. $(4x + 1)^2$
9. $(3a + b)^2$
10. $(x + 3y)^2$
11. $(2c - d)^2$
12. $(5x - 4y)^2$

Chapter 10 Review

Key Terms

abscissa (p. 379)
coordinates (p. 379)
coordinate plane (p. 379)
Distance Formula (p. 380)
linear equation (p. 393)
Midpoint Formula (p. 384)
ordinate (p. 379)
origin (p. 379)
point-slope (p. 393)
quadrant (p. 379)
slope (p. 388)
slope-intercept (p. 394)
standard form (p. 393)
x-axis (p. 379)
y-axis (p. 379)
y-intercept (p. 394)

Key Ideas and Review Exercises

10.1 To graph ordered pairs and find the distance between two points
The distance between two points $P_1(x_1, y_1)$ and $P_2(x_2, y_2)$ is given by the formula $d = \sqrt{(x_2 - x_1)^2 + (y_2 - y_1)^2}$.

1. On the coordinate system, graph the points $A(0, 5)$, $B(-3,-2)$, $C(4,-1)$, $D(3, 0)$, and $E(-2, 4)$.
2. Find the distance between the points $P(6, 1)$ and $Q(-1,-5)$.
3. $\triangle ABC$ has coordinates $A(4, 1)$, $B(1,-6)$, $C(-2, 1)$. Is the triangle scalene, isosceles, or equilateral?

10.2 To find the coordinates of the midpoint of a segment
The Midpoint Formula: Given $P(x_1, y_1)$ and $Q(x_2, y_2)$, the coordinates (x_m, y_m) of M, the midpoint, are $\left(\frac{x_1 + x_2}{2}, \frac{y_1 + y_2}{2}\right)$.

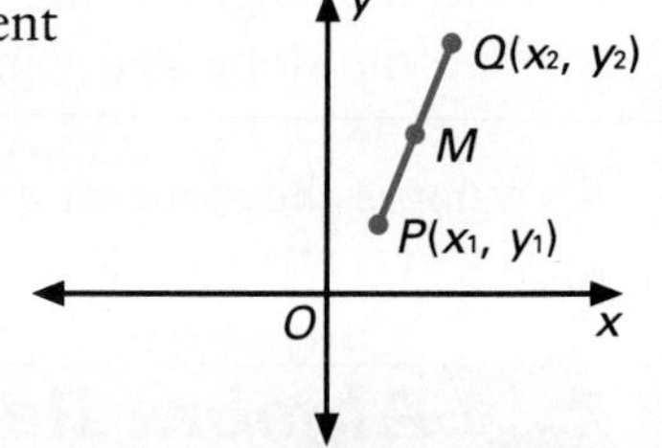

4. Given $P(-1, 2)$ and $Q(3,-5)$ find the midpoint of $\overline{PQ}$.
5. Given $M(2, 3)$ is the midpoint of $\overline{AB}$. The coordinates of A are $(-1, 0)$. Find the coordinates of B.
6. $ABCD$ is a quadrilateral with vertices $A(0, 0)$, $B(5, 1)$, $C(6, 6)$, $D(1, 5)$. Show that $ABCD$ is a rhombus.

10.3 To find the slope of a line and to draw a line with a given slope
The slope of a line with points $P_1(x_1, y_1)$ and $P_2(x_2, y_2)$ is the ratio $\frac{y_2 - y_1}{x_2 - x_1}$, $x_1 \neq x_2$.

7. Find the slope of a line with points $(4,-1)$ and $(-3, 6)$.

8. Tell whether the slope of each of the following is positive, negative, zero, or undefined.

a. **b.** **c.** **d.**

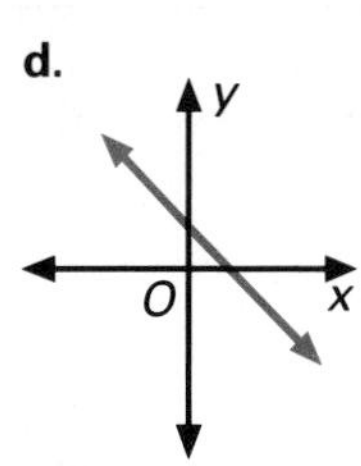

9. Draw a line with slope $-\frac{1}{2}$ containing the point $(1,-1)$.

10.4 To draw the graph of an equation of a line and to write the equation of a line in three forms

A graph of an equation is the set of points whose coordinates satisfy the given equation.

10. Write an equation in standard form of the line containing the points $(-1, 1)$ and $(3, 5)$.

11. Find the slope and the y-intercept of the line whose equation is $y = -2x + 3$. Then draw the graph.

12. Write an equation in slope-intercept form of the line pictured at the right.

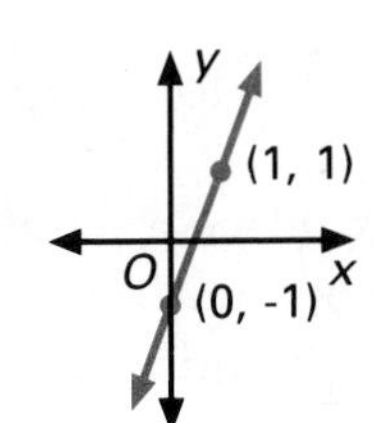

10.5 If two non-vertical lines have the same slope, then they are parallel. The product of the slope of two non-vertical perpendicular lines is -1.

13. Write an equation in slope-intercept form of a line containing $(2,-1)$ and perpendicular to the line whose equation is $y = \frac{3}{4}x + 1$.

14. Determine whether the diagonals are perpendicular for the quadrilateral with vertices $A(3, 0)$, $B(8, 4)$, $C(4, 9)$, and $D(-1, 5)$.

10.6 To assign coordinates to the vertices of a polygon in a plane

To use coordinate geometry to prove statements

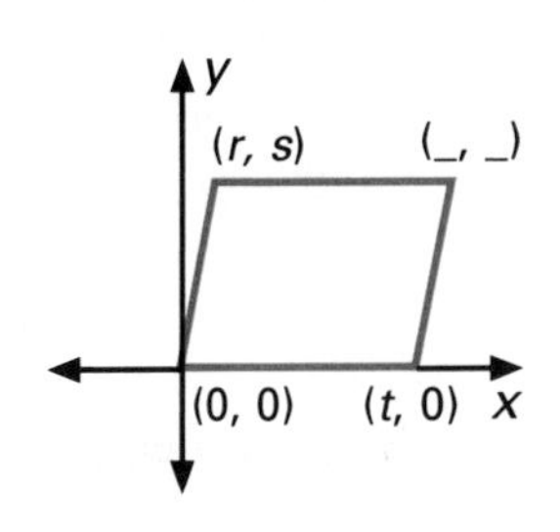

15. Find the missing coordinate for the parallelogram.

16. Use a coordinate proof to prove that the diagonals of a square are perpendicular.

Chapter 10 Test

Write the coordinates of each point.

1. A
2. B
3. C
4. D

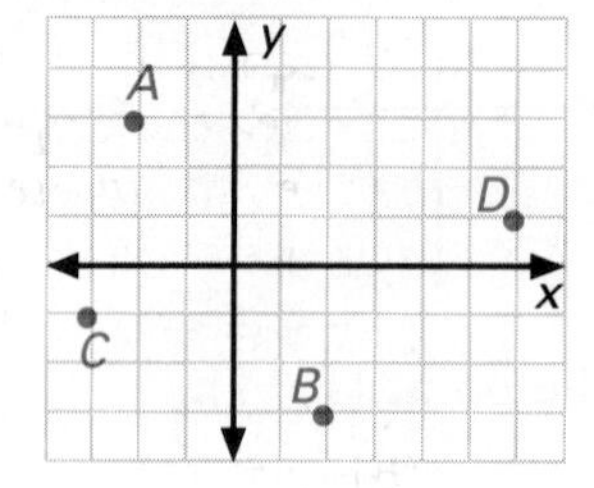

Find the length of $\overline{AB}$.

5. $A(-1, 0), B(2, -4)$

6. $A(3, -2), B(7, 6)$

ABC is a triangle. Use the lengths AB, BC, and AC to tell whether or not $\triangle ABC$ is isosceles. (Exercises 7 and 8)

7. $A(1, -3)$ $B(-3, 1)$ $C(1, 1)$

8. $A(2, 4)$ $B(-1, 3)$ $C(1, 0)$

9. Find the coordinates of the midpoint of $\overline{AB}$ for $A(7, -6)$ and $B(-1, 2)$
10. If $M(-3, 4)$ is the midpoint of $\overline{XY}$ for $X(5, -2)$, find the coordinates of Y.
11. Draw a set of coordinate axes. Place a parallelogram on these axes so that the simplest coordinates are assigned to the vertices.
12. Find the slope of a segment with endpoints $(-2, 1)$ and $(3, 5)$.
13. Draw a set of coordinate axes. Draw a line through $(0, 3)$ with a slope of $-\frac{2}{3}$.
14. Write an equation in slope-intercept form of a line with a slope of 2 containing the point $(0, -3)$.
15. Write an equation in standard form of a line containing the points $(-1, 3)$ and $(3, 1)$.
16. Draw a set of coordinate axes. Draw a graph of $y = 2x - 1$.
17. Write an equation in slope-intercept form of a line parallel to $y = -5x + 3$ through the point $(1, 0)$.
18. Write an equation in slope-intercept form of a line perpendicular to $y = 2x + 1$ through the point $(-1, 3)$.
19. Give a coordinate proof that the diagonals of a rectangle are congruent.
20. Give a coordinate proof that the slope of the midsegment of a triangle is equal to the slope of the base.

*21. Give a coordinate proof that the medians of an equilateral triangle are congruent.

College Prep Test

Choose the one best answer to each question or problem.

1. A line segment has one endpoint at (3,−2) and its midpoint at (2,−5). What are the coordinates of the other endpoint?
(A) (1,−8) (B) (5,−7)
(C) (1,−3) (D) (1, 8)
(E) (−1, 8)

2. What is the length of a line segment joining the points whose coordinates are (−2,−7) and (6, 8)?
(A) 4 (B) 5 (C) $7\frac{1}{2}$
(D) $8\frac{1}{2}$ (E) 17

3. Find the slope of the line whose equation is $3x + 2y = 6$.
(A) $\frac{2}{3}$ (B) $-\frac{2}{3}$ (C) $\frac{3}{2}$
(D) $-\frac{3}{2}$ (E) 6

4. Which point lies at the greatest distance from the origin?
(A) (0,−9) (B) (−2, 9)
(C) (−7,−6) (D) (8, 5) (E) (0, 0)

5. $\triangle AOB$ and $\triangle PCB$ are isosceles right triangles with equal areas. What are the coordinates of point P?

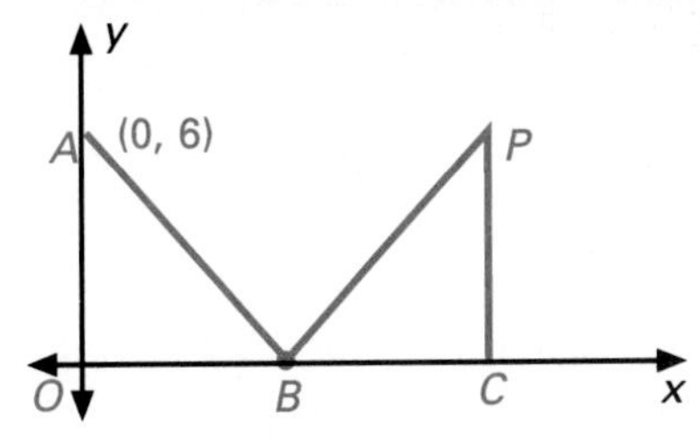

(A) (6, 0) (B) (6, 12) (C) (12, 0)
(D) (0, 12) (E) (12, 6)

6. An equation of the line parallel to the x-axis and passing through point (4, 2) is ______.
(A) $x = -4$ (B) $y = 2$
(C) $x = 2$ (D) $x = 4$
(E) $y = 0$

7. An equation of the line passing through the origin and perpendicular to the line whose equation is $x - y = 2$ is ______.
(A) $y = -x$ (B) $y = \frac{1}{2}x + 2$
(C) $y = -2x$ (D) $y = x$
(E) $y = -x + 1$

8. The coordinates of A are (4, 0) and of C are (15, 0). Find the area of $\triangle ABC$ if the equation of $\overline{AB}$ is $y = 2x - 8$.

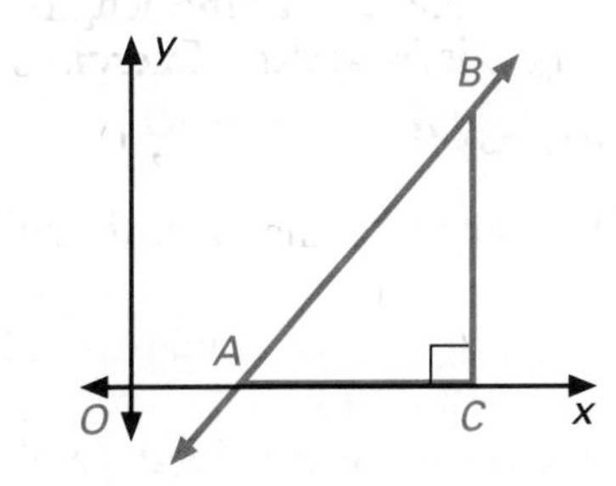

(A) 100 (B) 121 (C) 144
(D) 169 (E) 132

9. What is an equation of the line $\overleftrightarrow{AB}$?

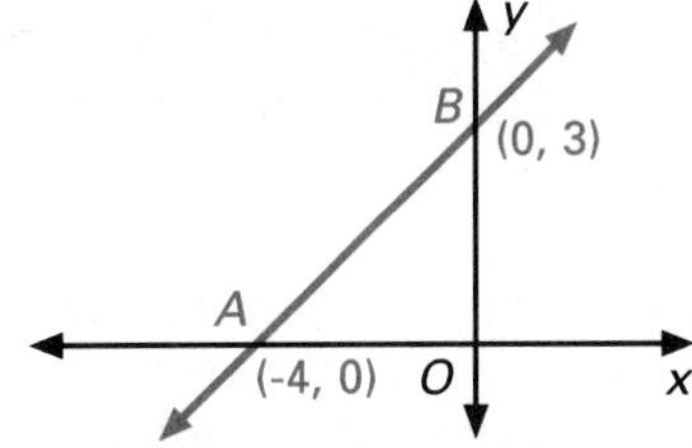

(A) $y = \frac{3}{4}x + 3$ (B) $y = \frac{3}{4}x - 4$
(C) $y = \frac{4}{3}x + 3$ (D) $y = -\frac{4}{3}x + 3$
(E) $y = -\frac{3}{4}x + 3$

10. The vertices of rectangle $ABCD$ are the points $A(0, 0)$, $B(8, 0)$, $C(8, k)$, $D(0, 5)$. The value of k is ______.
(A) 4 (B) 5 (C) 6 (D) 3 (E) 2

Cumulative Review (Chapters 1–10)

1. Identify each of the following as line segment, ray, line, or the measure of a segment. *1.1*
(A) $\overrightarrow{AB}$ (B) $\overline{AB}$ (C) $\overleftrightarrow{AB}$
(D) AB

2. Given: $\overline{RP}$ bisects $\angle SPQ$; $m\angle 1 = 4x - 5$, $m\angle 2 = 2x + 6$. Find $m\angle SPQ$. *1.5*

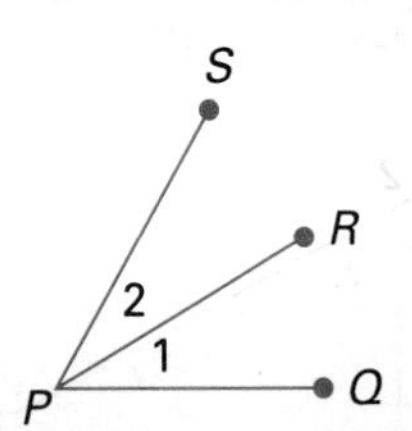

3. Graph the conjunction and the disjunction. Identify the resulting figure. *1.7*
(A) $x \geq 3$ *and* $x \leq 7$
(B) $x \geq 4$ *or* $x \leq 3$

4. Find $m\angle 1$. *3.6*

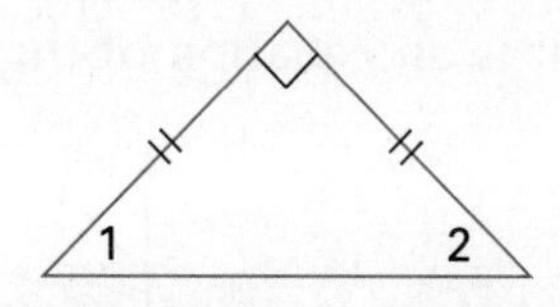

Use the following statement to answer Exercises 6–8: "If two lines are parallel, then corresponding angles are congruent." *3.8*

5. Write a biconditional for the statement. Is it true?

6. Write the inverse of the statement. Is it true?

7. Write the contrapositive of the statement. Is it true?

8. Based on the markings, determine whether you can prove the triangles are congruent. *4.6*

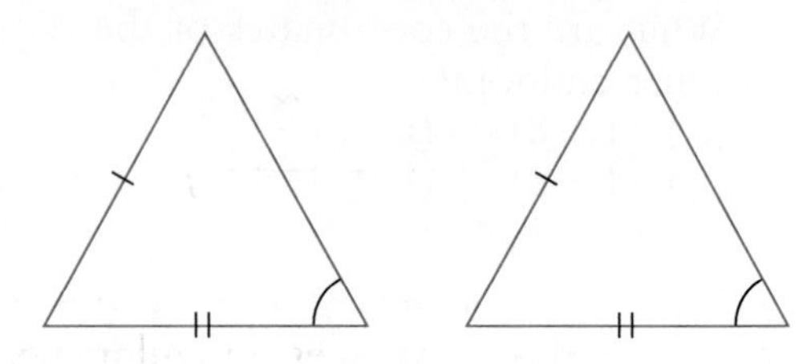

9. Given: $\overline{DC}$ bisects $\angle ACB$; $AC = CB$. *4.7*
Prove: $\overline{AD} \cong \overline{DB}$

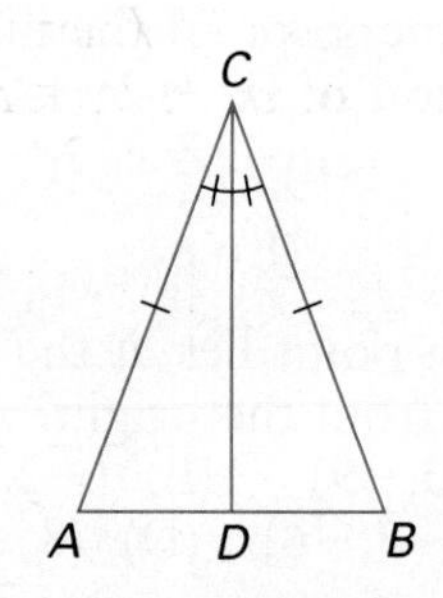

10. In $\triangle ABC$, $AB = 5$, $BC = 8$, $AC = 7$ *5.7*
Name the largest angle.

Complete the statement with =, <, or >. *5.7*

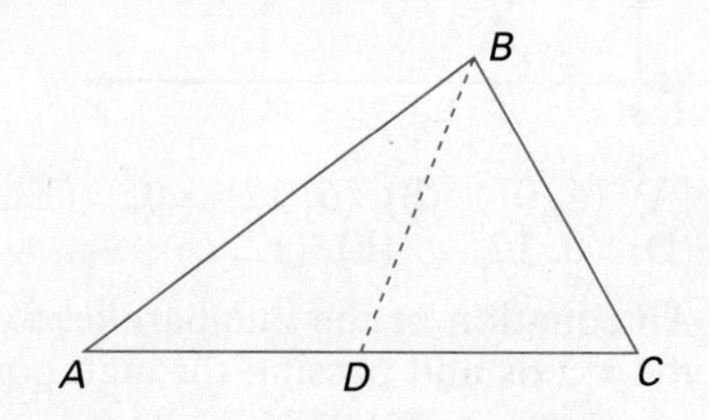

11. $\angle A < \angle C$, BC ______ AB

12. $\overline{BD}$ is a median; AD ______ DC.

13. Identify the hypotenuse, legs, median, and altitude(s) of right triangle TRS. *5.6*

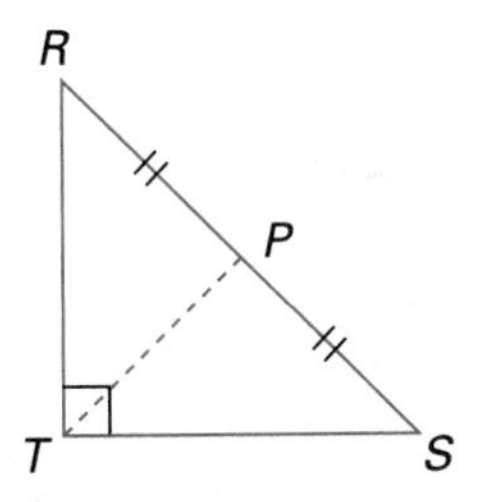

14. How many sides does a regular polygon have if each exterior angle has a measure of 72? *6.2*

15. The measure of three exterior angles of a quadrilateral are 75, 85, and 95. What is the measure of the exterior angle at the fourth vertex? *6.3*

16. Given: $\square ABCD$ *6.4*
Prove: $\triangle ABC \cong \triangle CDA$

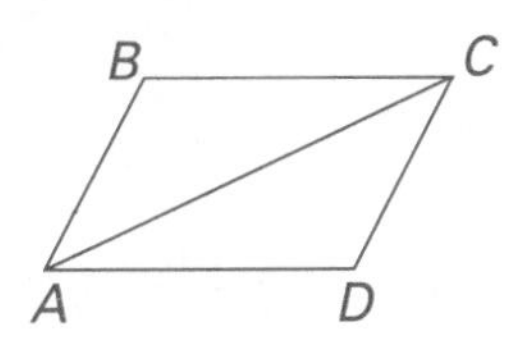

True or False? *7.2, 7.3*

17. The diagonals of a parallelogram bisect each other.

18. The diagonals of a rectangle are perpendicular.

19. The diagonals of a square are congruent.

20. The diagonals of a rhombus are congruent.

21. The diagonals of a square are perpendicular.

22. The diagonals of a rhombus bisect the vertex angles.

23. The diagonals of a parallelogram are congruent.

24. Given: Isos trap $ABED$, C is the midpoint of $\overline{DE}$. *7.5*
Prove: $\triangle ABC$ is isosceles.

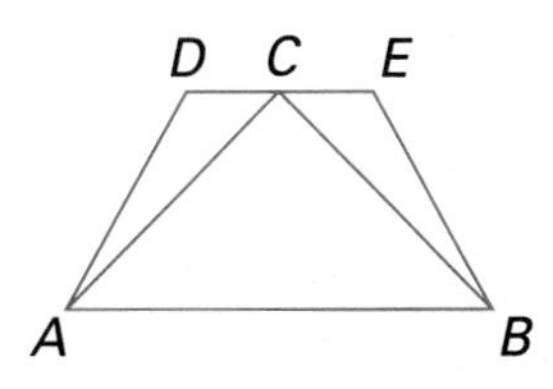

25. Find the geometric mean of 10 and 40. *8.1*

26. Given: $\triangle ARB \sim \triangle DRC$ *8.3*
Find RD.

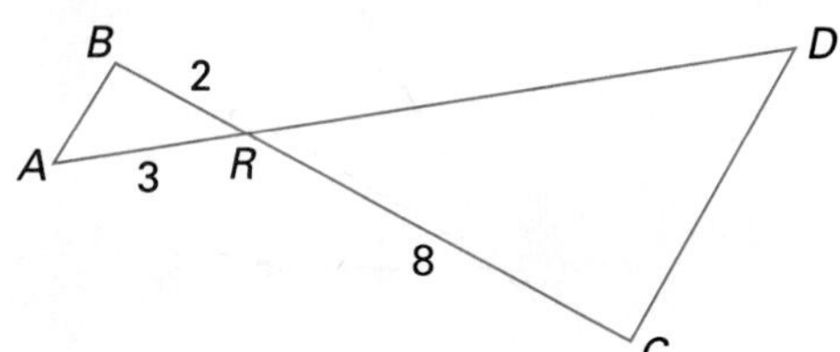

27. Prove: *8.1*
If $\frac{a}{b} = \frac{c}{d}$, then $\frac{a+b}{b} = \frac{c+d}{d}$

28. Find the length of the diagonal of a square with sides measuring 8 in. *9.2*

29. A tree casts a 30-ft shadow. Find the height of the tree if the angle of elevation to the top of the tree has a degree measure of 51. *9.7*

30. Find the distance AB for points $A(-1,-2)$, and $B(5,-3)$. *10.1*

31. Find the slope and y-intercept of the line whose equation is $3y - 2x = 9$. *10.3, 10.4*

32. Write an equation, in standard form, of a line perpendicular to the line with equation $y = \frac{1}{2}x - 1$ and containing the point (0, 0). *10.5*

33. Use a coordinate proof to prove that the diagonals of a square bisect each other. *10.6*

11 CIRCLES

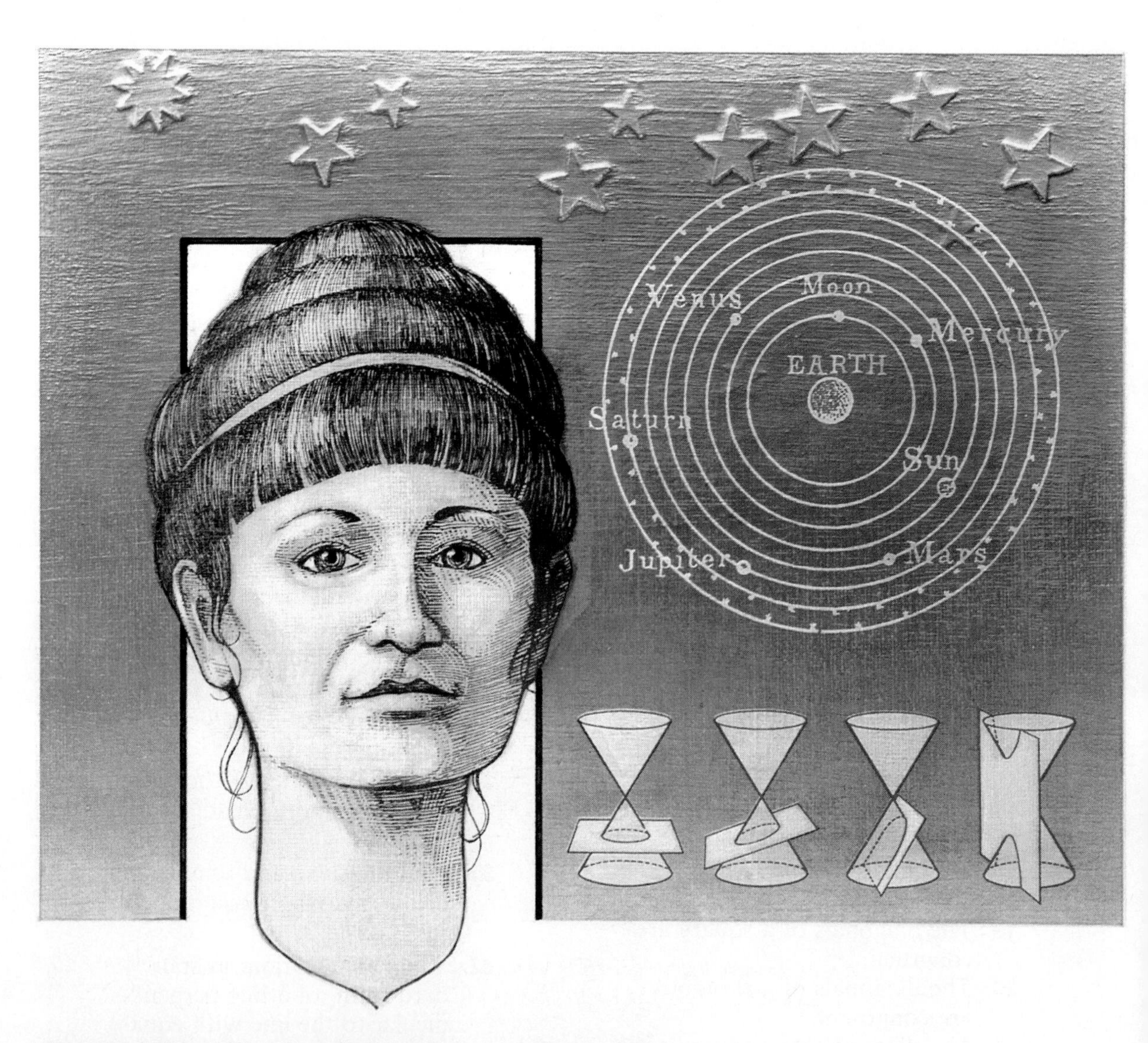

Hypatia (370?–415)—Hypatia lectured on mathematics, philosophy, and astronomy, as well as on simple mechanics. She designed scientific instruments such as the astrolabe, which measures the positions of planets and stars.

11.1 Circles and Spheres: Basic Definitions

Objectives

To identify special segments, lines, and other sets of points related to circles and spheres

To find lengths of radii and diameters

The photograph at the right shows a bicycle wheel, which is a physical model of a *circle* in geometry. Every point of the rim of this wheel is the same distance, 13 in., from the hub or center of the wheel. A, B, C, and D are four such points.

Definitions

A **circle** is the set of all points in a plane that are a given distance from a fixed point in that plane. The fixed point is the **center** of the circle. A segment from the center to any point on the circle is a **radius** (plural: **radii**).

A circle is named by its center. The circle to the right is named Circle O or $\odot O$, where the symbol for circle is $\odot$.

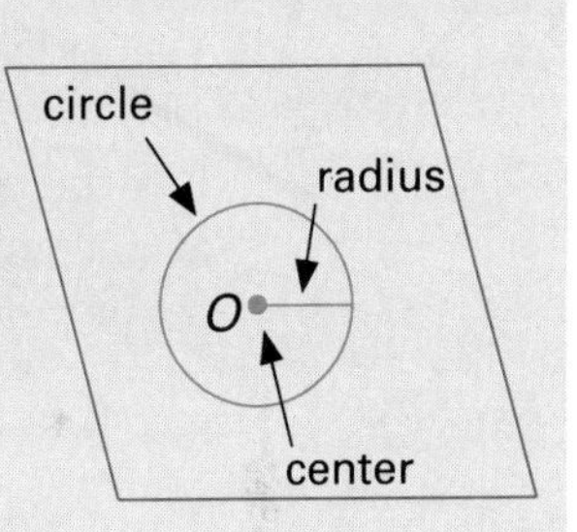

It follows from the definition of a circle that all radii of the same circle are congruent. It is now possible to define *congruent circles* in terms of their radii.

Definition

Congruent circles are circles whose radii are congruent.

A circle in a plane separates the plane into three sets of points: the circle itself, the interior of the circle, and the exterior of the circle.

Definitions

The **interior** of a circle is the set of all points in the plane of the circle whose distance from the center is *less* than the length of the radius. The **exterior** of a circle is the set of all points in the plane whose distance from the center is *greater* than the length of the radius.

EXAMPLE 1 Given: $\odot O$, $OR = 2x - 10$, $OP = 8$
Find all values of x for which point R is in the exterior of the circle.

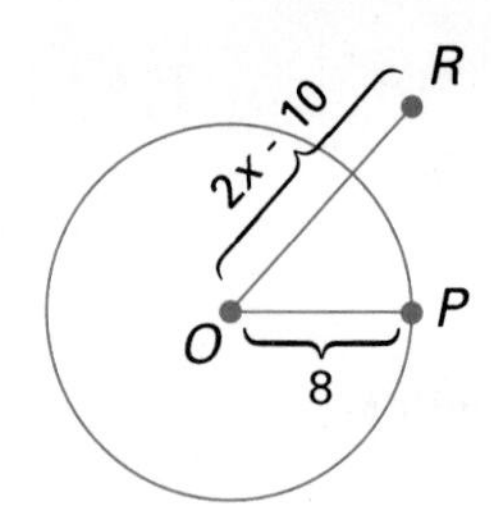

Plan To meet the conditions, set $OR > OP$.

Solution

$$OR > 8$$
$$2x - 10 > 8$$
$$2x > 18, \text{ or } x > 9$$

Definitions

A **chord** of a circle is a segment whose endpoints are on the circle. A **diameter** is a *chord* that contains the center of the circle.

If the length of the radius is r and the length of the diameter is d, then $d = 2r$ and $r = \frac{d}{2}$.

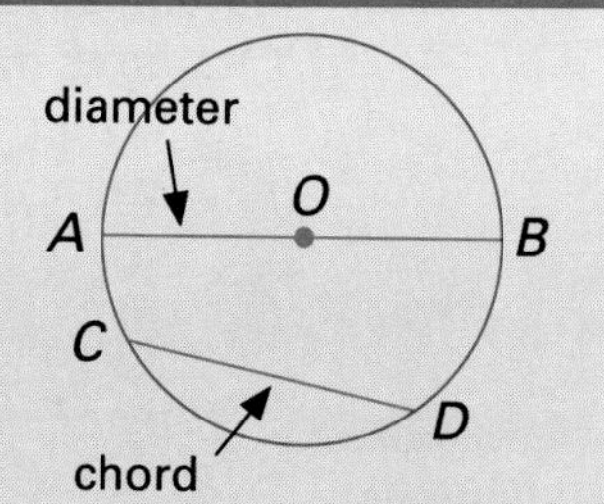

EXAMPLE 2 For the given radius or diameter, find the indicated length.

a. Given: $d = 18$
Find r.

b. Given: $r = 2a + 4$
Find d.

c. Given: $d = 4x - 12$
Find r.

Solutions

a. $r = \frac{d}{2} = \frac{18}{2}$
Thus, $r = 9$.

b. $d = 2r = 2(2a + 4)$
Thus, $d = 4a + 8$.

c. $r = \frac{d}{2} = \frac{4x - 12}{2}$
Thus, $r = 2x - 6$.

Given a line and a circle that are coplanar, the possible number of points in which the line can intersect the circle are indicated by the following diagrams.

0 points 1 point 2 points

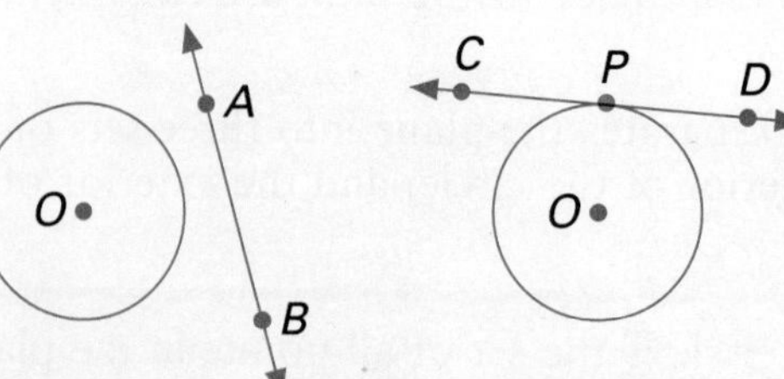

E P O Q F

$\overleftrightarrow{AB}$ does not intersect the circle.

$\overleftrightarrow{CD}$ is a *tangent* to the circle; P is the point of *tangency*.

$\overleftrightarrow{EF}$ is a *secant* of the circle.

Definitions

A **tangent** to a circle is a line that is coplanar with the circle and intersects the circle in exactly one point, called the **point of tangency.** A **secant** of a circle is a line that intersects the circle in two points.

EXAMPLE 3 Identify all radii, diameters, chords, secants, and tangents of $\odot O$ at the right.

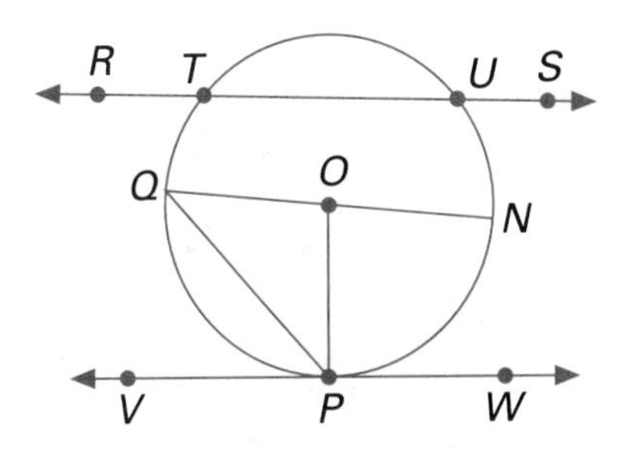

Solution

Radii: $\overline{OQ}, \overline{ON}, \overline{OP}$
Diameter: $\overline{QN}$
Secant: $\overleftrightarrow{TU}$ (or $\overleftrightarrow{RS}$)
Chords: $\overline{TU}, \overline{QN}$ (also diameter), $\overline{QP}$
Tangent: $\overleftrightarrow{VW}$

The definition of circle can be extended to describe the three-dimensional counterpart of the circle, the *sphere.*

Definitions

A **sphere** is the set of all points in space that are a given distance from a fixed point. The fixed point is the **center** of the sphere. A segment from the center of the sphere to any point on the sphere is a **radius** of the sphere.

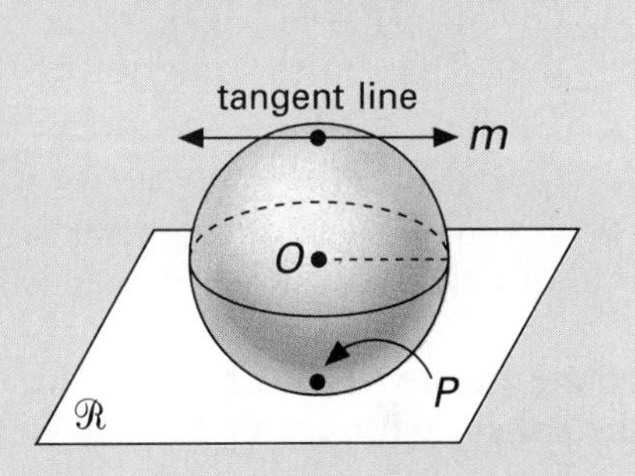

A **chord of a sphere** has endpoints on the sphere. A **secant of a sphere** is a line that intersects the sphere in two points.

A line or plane that intersects a sphere in exactly one point is **tangent** to that sphere.

Focus on Reading

1. Rewrite the definition of a circle using the word *equidistant.*

Complete to make a true statement.

2. A tangent intersects a circle in ____________.
3. A secant intersects a circle in ____________.
4. The length of the radius of a circle is ____________ the length of the diameter.
5. Given: $\overline{OP}$ is the radius of a circle with center O; $OP = 5$; Q is such that $OQ = 9$. Point Q lies in the ____________ of the circle.

Classroom Exercises

Indicate whether P is in the interior, the exterior, or on the circle.

1. P is 5 in. from the center of a circle with radius of length 4 in.
2. P is 7 in. from the center of a circle with diameter of length 10 in.
3. P is 4 ft from the center of a circle with radius of length 6 ft.
4. P is 7 in. from the center of a circle with diameter of length 14 in.

For the given length of the radius or diameter, find the indicated length.

5. $r = 6$, $d =$ ______
6. $d = 14$, $r =$ ______
7. $r = 2.5$, $d =$ ______
8. $d = 6.2$, $r =$ ______

Written Exercises

Use the figure at the right for Exercises 1–3.

1. Given: $OP = 12$, $OR = 3x - 6$. Find all values of x such that R lies in the exterior of the circle.
2. Given: $OP = 13$, $OS = 2x - 7$. Find all values of x such that S lies in the interior of the circle.
3. Given: Length of diameter is 30; $OR = 5x - 10$. Find all values of x such that R is in the exterior of the circle.

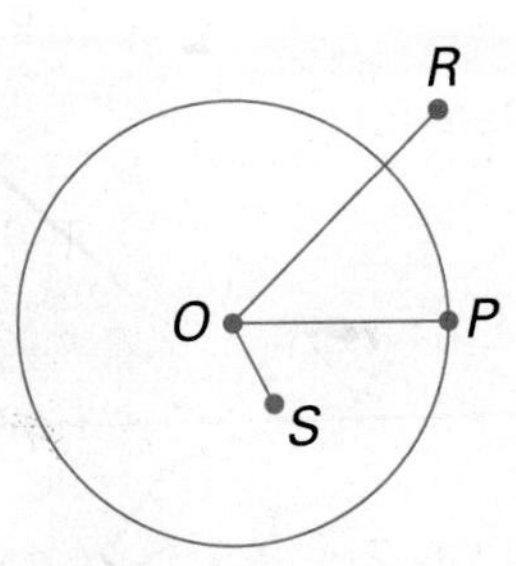

Name the given set of points, first for the circle with center O, then for the sphere with center O.

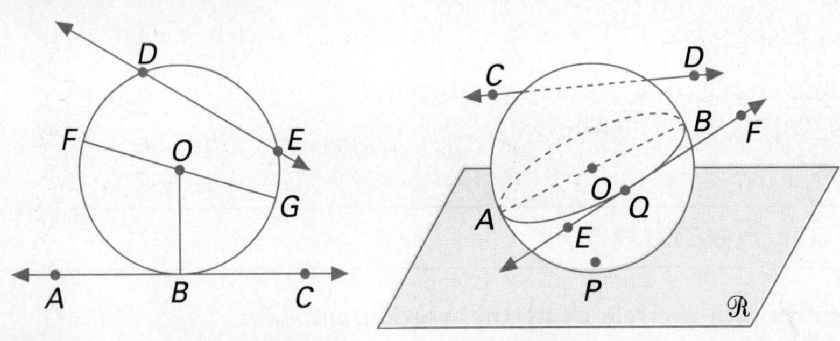

4. radius
5. diameter
6. chord
7. secant
8. tangent

Find the length of a diameter for the indicated length of a radius.

9. 3.7
10. $5\frac{1}{2}$
11. $3x$
12. $2a + 3$
13. $6x - 4$

Find the length of a radius for the indicated length of a diameter.

14. 17.8
15. $10\sqrt{3}$
16. $6x$
17. $4a - 12$
18. $8x + 10$

R and S are centers of the two intersecting circles. Identify the given set of points as a radius, diameter, chord, secant, or tangent, either for $\odot R$ or $\odot S$ or both.

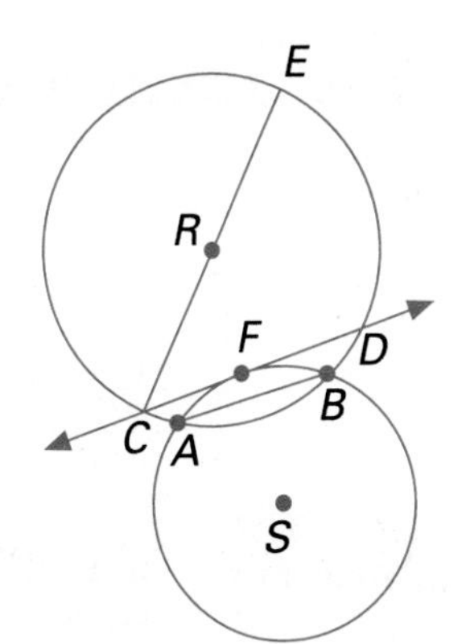

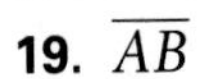

19. $\overline{AB}$

20. $\overline{CE}$

21. $\overleftrightarrow{CD}$

22. $\overline{RE}$

23. $\overline{CD}$

Use Circle O at the right, for Exercises 24–27.

24. $QR = 4x + 6$, $OQ = x + 8$. Find QR.

25. $OP = 6x - 4$, $OR = -x + 10$. Find OQ.

26. $PQ = 10$. Would 9 be a reasonable length of the diameter? Why or why not?

27. Given: m $\angle POQ = 60$
Prove: $\triangle POQ$ is equilateral.

28. Write a definition of the interior of a sphere.

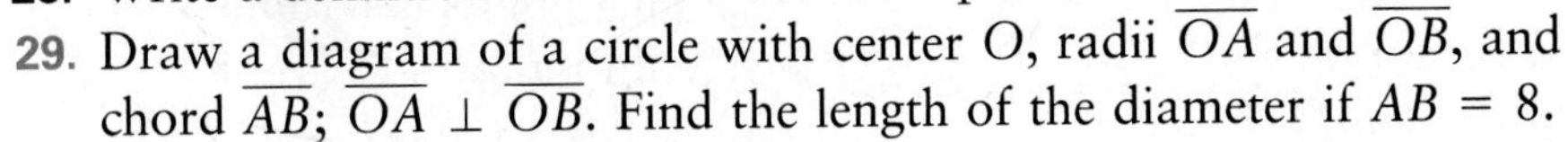

29. Draw a diagram of a circle with center O, radii $\overline{OA}$ and $\overline{OB}$, and chord $\overline{AB}$; $\overline{OA} \perp \overline{OB}$. Find the length of the diameter if $AB = 8$.

30. The length of the diameter of a circle is 28. The length of the radius is the square of some number, increased by 5 times that number. Find the number.

31. The distance from a point P to a plane is 4. A sphere with center P intersects the plane in a circle with diameter of length 6. Find the length of a radius of the sphere.

32. Consider again the definitions of tangent and secant given in this lesson. Why was it not necessary to state that a secant is coplanar with the circle?

Mixed Review

1. Find the length of an altitude of an equilateral triangle with a side of length 4. *9.4*

2. The length of the hypotenuse of a 45-45-90 triangle is 20. Find the length of each leg of the right triangle. *9.4*

3. The measure of the vertex angle of an isosceles triangle is 70. Find the measure of a base angle. *5.1*

11.2 Properties of Chords

Objective To prove and apply theorems about chords

Recall that the distance from a point to a line is the length of a perpendicular segment from the point to the line.

Theorem 11.1 If a line or segment contains the center of a circle and is perpendicular to a chord, then it *bisects* the chord.

Given: $\odot O$ with $\overline{OD} \perp$ chord $\overline{AB}$
Prove: $\overline{OD}$ bisects $\overline{AB}$.

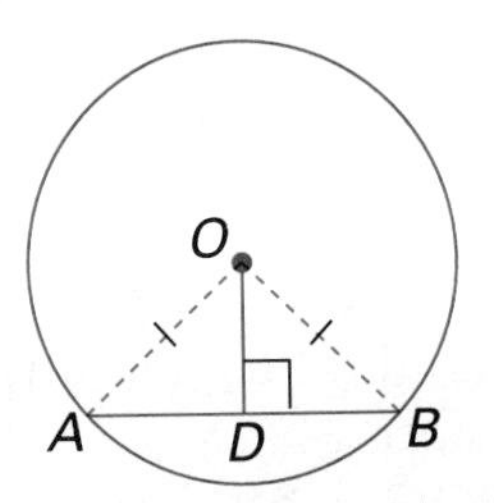

Proof

Statement	Reason
1. $\odot O$ with $\overline{OD} \perp$ chord $\overline{AB}$	1. Given
2. Draw radii $\overline{OA}$ and $\overline{OB}$.	2. Two points determine a line.
3. $\overline{OA} \cong \overline{OB}$	(H) 3. All radii of a $\odot$ are $\cong$.
4. $\overline{OD} \cong \overline{OD}$	(L) 4. Reflex Prop
5. $\angle ODA$ and $\angle ODB$ are rt $\angle$s.	5. Def of $\perp$ lines
6. $\triangle ODA$ and $\triangle ODB$ are rt $\triangle$s.	6. Def of rt $\triangle$s
7. $\triangle ODA \cong \triangle ODB$	7. HL
8. $\overline{AD} \cong \overline{BD}$	8. CPCTC
9. $\therefore \overline{OD}$ bisects $\overline{AB}$.	9. Def of bis

EXAMPLE 1 A chord of length 8 is 2 units from the center of a circle. Find the length of a diameter.

Plan Draw a diagram. $\overline{OD} \perp \overline{AB}$. Therefore, $\overline{OD}$ bisects $\overline{AB}$. $BD = \frac{1}{2} \cdot 8 = 4$; $OD = 2$. Let $OB = r$ and use the Pythagorean Theorem.

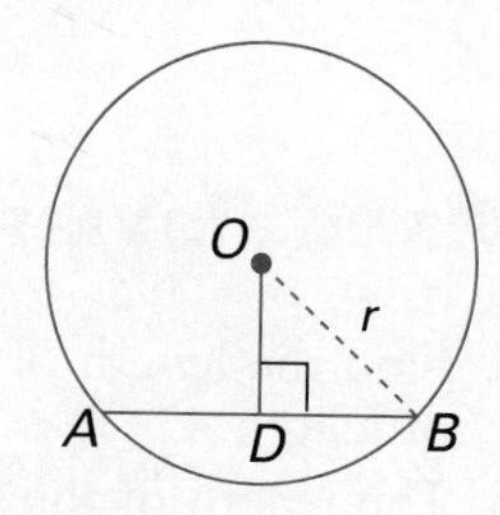

Solution

$$r^2 = 2^2 + 4^2$$
$$r^2 = 20$$
$$r = \sqrt{20}$$
$$= \sqrt{4} \cdot \sqrt{5} = 2\sqrt{5}$$

Thus, since $d = 2r$, the length of the diameter is $4\sqrt{5}$.

Applications of Theorem 11.1 may involve properties of special right triangles such as the 30-60-90 or 45-45-90 triangles.

EXAMPLE 2 Given: $\overline{OA}$ and $\overline{OB}$ are radii of circle O; $\overline{AB}$ is a chord; m $\angle AOB = 120$; $AB = 12$. Find the distance from O to $\overline{AB}$.

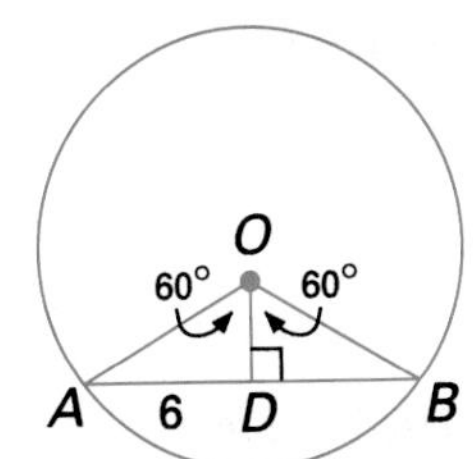
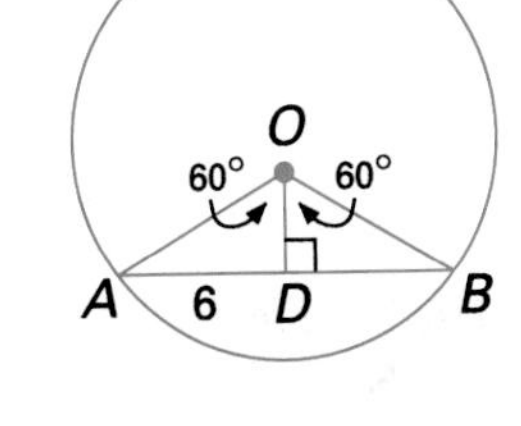

Plan Draw a diagram. Since m $\angle AOB = 120$, m $\angle A$ + m $\angle B = 60$. Then, since $\triangle AOB$ is isosceles, m $\angle A$ = m $\angle B = 30$. With $\overline{OD} \perp \overline{AB}$, there are two 30-60-90 triangles.

Solution Let $OD = s$.

Then $s\sqrt{3} = AD$

$s\sqrt{3} = 6$ [$\overline{OD}$ bisects $\overline{AB}$, $AD = 6$.]

$$s = \frac{6}{\sqrt{3}} = \frac{6 \cdot \sqrt{3}}{\sqrt{3} \cdot \sqrt{3}} = \frac{6\sqrt{3}}{3} = 2\sqrt{3}$$

Thus, the chord is $2\sqrt{3}$ units from the center O.

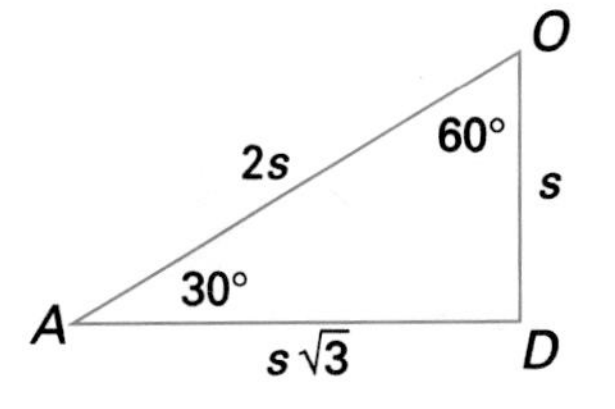

Theorem 11.2 In the same circle or in congruent circles, congruent chords are equidistant from the center(s).

Given: $\odot D \cong \odot S$,
$\overline{AC} \cong \overline{PR}$
Prove: $\overline{BD} \cong \overline{QS}$

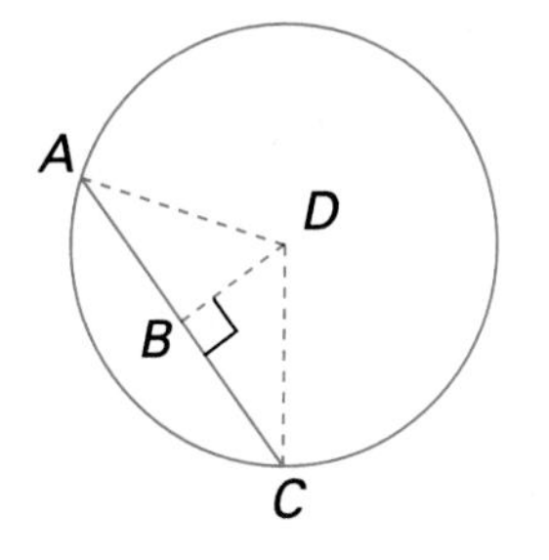

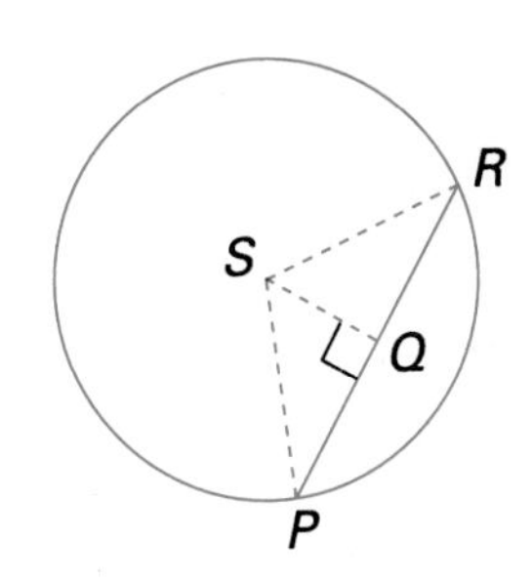

Plan Prove $\triangle ADC \cong \triangle PSR$. Then the corresponding altitudes $\overline{BD}$ and $\overline{QS}$ are congruent.

Proof

Statement		Reason
1. $\odot D \cong \odot S$, $\overline{AC} \cong \overline{PR}$	(S)	1. Given
2. $\overline{AD} \cong \overline{PS}$, $\overline{CD} \cong \overline{RS}$	(S, S)	2. Def of $\cong$ $\odot$s
3. $\triangle ADC \cong \triangle PSR$		3. SSS
4. $\therefore \overline{BD} \cong \overline{QS}$ ($BD = QS$)		4. Corr alt of $\cong$ $\triangle$s are $\cong$.

The converse of Theorem 11.2 also is true.

Theorem 11.3 In the same circle or in congruent circles, chords that are equidistant from the center(s) are congruent.

EXAMPLE 3 Given: $\odot O$, $\overline{OP} \cong \overline{OQ}$, $\overline{OP} \perp \overline{RT}$, $\overline{OQ} \perp \overline{ST}$, $m\angle T = 40$

Find $m\angle R$.

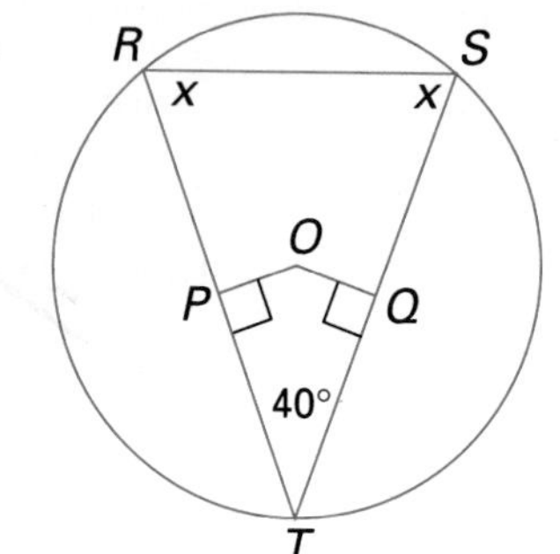

Plan Chords $\overline{TR}$ and $\overline{TS}$ are equidistant from the center O. So by Theorem 11.3, the chords are congruent. Then $\triangle RTS$ is isosceles and $m\angle R = m\angle S$.

Solution Let $m\angle R = m\angle S = x$.

Then,
$$x + x + 40 = 180$$
$$2x + 40 = 180$$
$$2x = 140$$
$$x = 70$$
Thus, $m\angle R = 70$

EXAMPLE 4 Given: $\odot O$ has radius of length 13; chords $\overline{AB}$ and $\overline{DF}$ are 5 and 12 units from O, respectively.

Show that $AB > DF$.

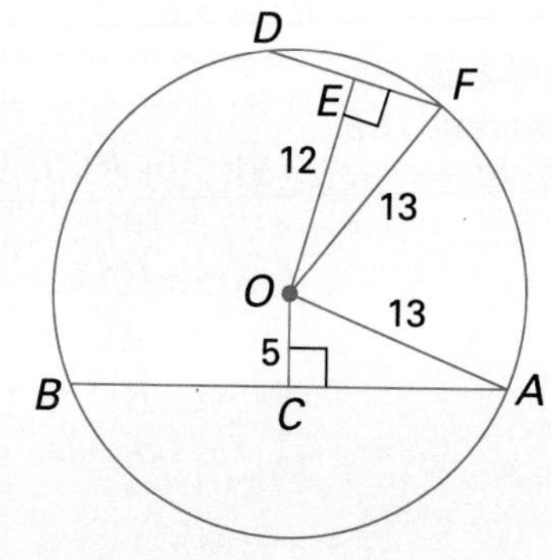

Solution Use the Pythagorean Theorem to compute the lengths.

$$(AC)^2 + 5^2 = 13^2$$
$$(AC)^2 + 25 = 169$$
$$(AC)^2 = 144$$
$$AC = \sqrt{144} = 12$$
$$AB = 2 \cdot AC = 2 \cdot 12 = 24$$

$$(EF)^2 + 12^2 = 13^2$$
$$(EF)^2 + 144 = 169$$
$$(EF)^2 = 25$$
$$EF = \sqrt{25} = 5$$
$$DF = 2 \cdot EF = 2 \cdot 5 = 10$$

Thus, $AB > DF$.

It is seen in Example 4 that the chord nearer the center of the circle is the longer chord. This suggests the following theorem and its converse.

Theorem 11.4 In the same circle or congruent circles, if two chords are unequally distant from the center(s), then the chord nearer its corresponding center is the longer chord.

Theorem 11.5 In the same circle or congruent circles, if two chords are unequal in length, then the longer chord is nearer the center of its circle.

EXAMPLE 5 Determine which side of the triangle is farthest from the center.

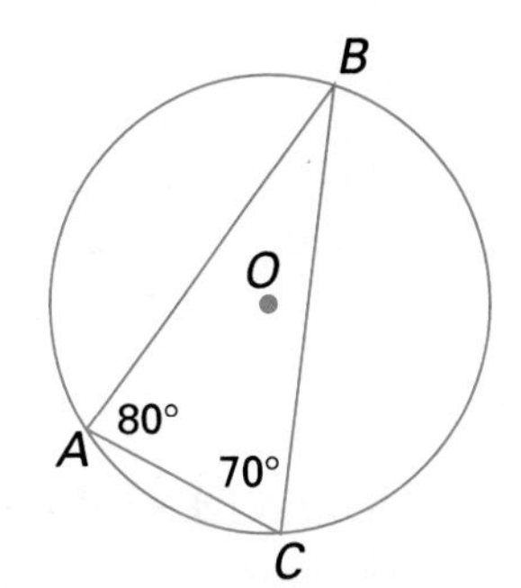

Solution

$$70 + 80 + m\,\angle B = 180$$
$$150 + m\,\angle B = 180$$
$$m\,\angle B = 30$$

$\overline{AC}$ is the shortest side of $\triangle ABC$, or the shortest chord of $\odot O$. It follows then from Theorem 11.5, that $\overline{AC}$ is the chord farthest from the center.

In the figure above, $\triangle ABC$ is said to be *inscribed* in the circle, because its sides are chords of the circle.

Definitions A polygon whose sides are chords of a circle is an **inscribed polygon.** The polygon is **inscribed** in the circle; the circle is **circumscribed** about the polygon.

Classroom Exercises

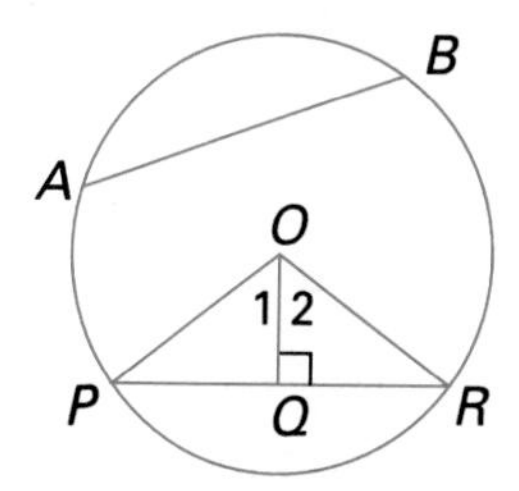

1. What length represents the distance from chord $\overline{PR}$ to O?
2. Find m $\angle 1$ if m $\angle POR = 100$.
3. Given: $PQ = 2a$; $PR =$ ___
4. Given: $PR = 16$; $PQ =$ ___

Written Exercises

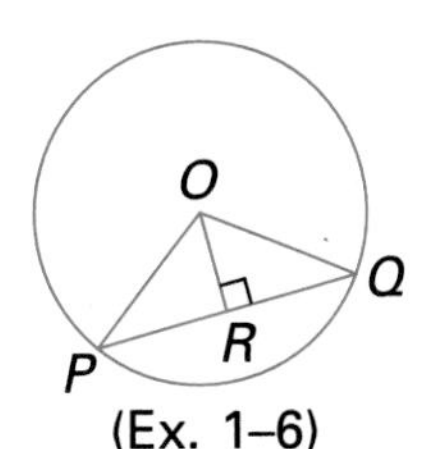

(Ex. 1–6)

1. Given: $PQ = 24$, $OR = 5$
   Find PO.
2. Given: $PQ = 8$, $OR = 4$
   Find PO.

3. Given: $OR = 4$, $PO = 8$
Find PQ.

4. Given: $PO = 6$, $OR = 4$
Find PQ.

5. m $\angle POQ = 120$, $PQ = 10$
Find OP.

6. Given: m $\angle POQ = 60$,
$OR = 6$
Find OQ.

In circle O, $\overline{OU} \perp \overline{TS}$ and $\overline{OV} \perp \overline{RS}$.

7. Given: $OU = 4$, $OV = 4$,
$RV = 8$
Find TS.

8. Given: $TS = 12$, $RV = 6$,
and $OV = 4$
Find OU.

9. Given: $OU = OV$,
m $\angle S = 70$
Find m $\angle T$.

10. Given: $OU = OV$,
m $\angle T = 20$
Find m $\angle S$.

11. Given: m $\angle R = 25$,
m $\angle T = 55$
Name the shortest chord.

12. Given: m $\angle T = 60$,
m $\angle S = 64$
Which chord is closest to the center?

13. A chord of length 12 is 4 units from the center of a circle. Find the length of a diameter to the nearest tenth.

14. A chord of a circle is 12 units from the center. Draw a picture and estimate the length of the chord if the length of a radius is 14. Then find the length to the nearest hundredth.

15. A diameter of a circle is of length 16. The length of a chord is 12. Find the distance of the chord from the center of the circle.

16. In circle O, $\overline{AB}$ is a chord. $\overline{OA}$ and $\overline{OB}$ are radii. m $\angle AOB = 120$, $OA = 4$. Find the distance of the chord from the center of the circle.

17. Write a biconditional for Theorem 11.2 and its converse.

18. Prove Theorem 11.3.

19. Prove Theorem 11.4.

20. Prove Theorem 11.5.

21. An archaeologist finds a piece of what appears to be a circular pipe. Holding the ends of a meter stick against the inside of the pipe, he finds that the midpoint of the 100-cm stick is 8 cm from the pipe wall. Find the diameter of the pipe.

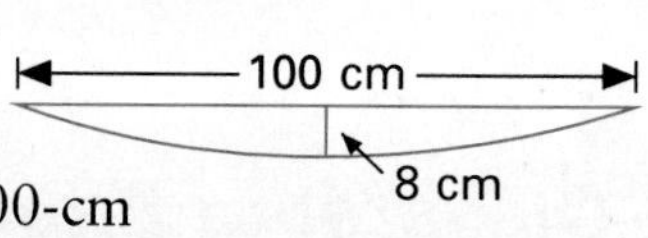

Mixed Review

1. Given: $\overline{DE} \parallel \overline{AB}$, m $\angle 2 = 30$, m $\angle 5 = 20$
Find m $\angle 4$. 3.2

2. Given: m $\angle 1 = 40$, m $\angle 2 = 45$
Find m $\angle 3$. 3.6

3. Given: m $\angle 1 = 60$, m $\angle 4 = 80$
Find m $\angle 2$. 3.7

4. Given: $\overline{AD} \perp \overline{AB}$, m $\angle 5 = \frac{2}{3}$m $\angle 6$
Find m $\angle 6$. 5.4

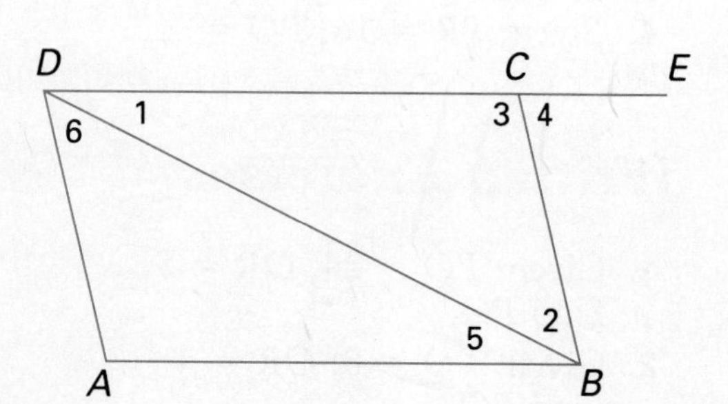

11.3 Special Properties of Tangents to Circles

Objective To prove and apply theorems about tangents to circles

This lesson develops basic relationships of tangents and circles.

Theorem 11.6 If a line is perpendicular to a radius at its endpoint on the circle, then the line is tangent to the circle.

Given: $\overline{OP} \perp \overleftrightarrow{AB}$
Prove: $\overleftrightarrow{AB}$ is tangent to $\odot O$ at P.

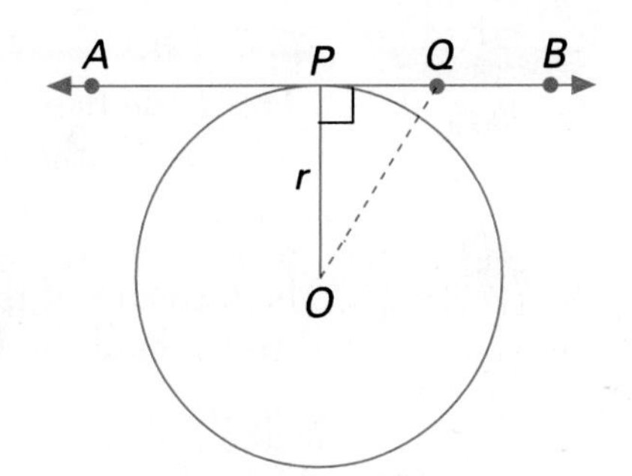

Proof Choose any point Q on $\overleftrightarrow{AB}$ other than P, and draw right $\triangle OPQ$. The hypotenuse of a right triangle is its longest side, so $OQ > r$. Therefore, Q is the exterior of the circle. This makes P the *only* point on $\overleftrightarrow{AB}$ lying on $\odot O$; thus, by definition $\overleftrightarrow{AB}$ is tangent to $\odot O$ at P.

Definition A **tangent segment** is a segment that contains a point of tangency and another point of a tangent line to a circle.

Theorem 11.7 If a line is tangent to a circle, then the line is perpendicular to the radius drawn to the point of tangency.

EXAMPLE 1 Given: $\overline{AB}$ tangent to $\odot O$, $\overline{OA}$ is a radius, $OA = 6$, $OB = 8$.

Find the length of segment $\overline{AB}$.

Plan Angle OAB is a right angle since the radius is perpendicular to the tangent at A.

Solution Let $AB = x$.

$$x^2 + 6^2 = 8^2$$
$$x^2 + 36 = 64$$
$$x^2 = 28$$
$$x = \sqrt{28} = \sqrt{4} \cdot \sqrt{7} = 2\sqrt{7}$$

Thus, the length of segment $\overline{AB}$ is $2\sqrt{7}$.

The figure at the right shows two segments tangent to a circle from an exterior point T. To prove that these segments are congruent segments, draw auxiliary lines $\overline{OA}$, $\overline{OB}$, and $\overline{OT}$. The right triangles formed can be proved congruent.

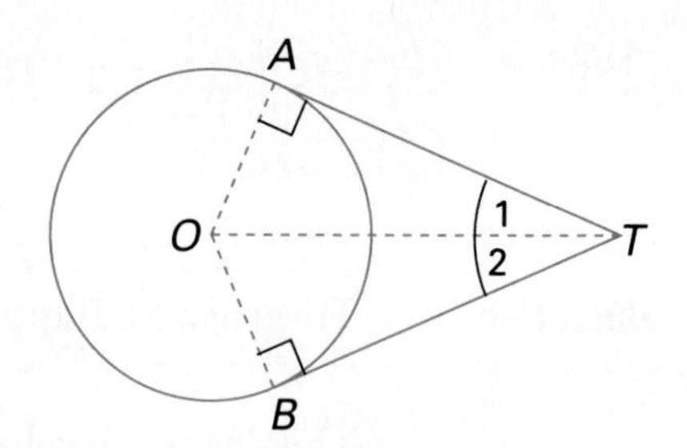

Theorem 11.8 Two segments drawn tangent to a circle from an exterior point are congruent.

Corollary The angle between two tangents to a circle from an exterior point is bisected by the segment joining its vertex to the center of the circle.

EXAMPLE 2 The measure of the angle between two tangents to a circle from a point P is 60. Find the length of a radius if the length of each tangent is 8.

Plan Make a sketch. By the corollary to Theorem 11.8, $\overline{OP}$ bisects $\angle APB$. So, m $\angle APO = 30$. $\triangle OAP$ is a 30-60-90 triangle.

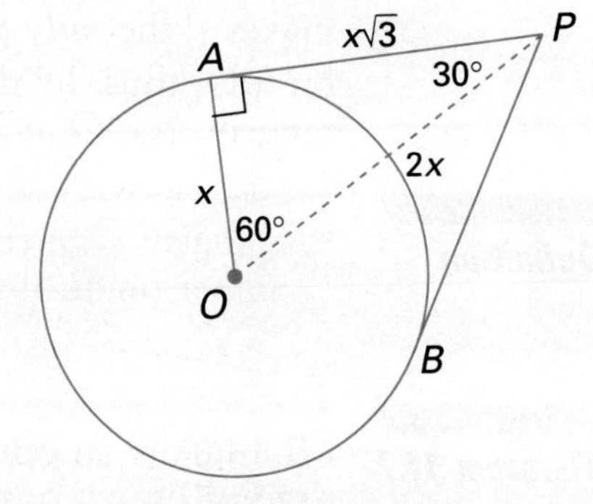

Solution $AP = x\sqrt{3} = 8$

$$x = \frac{8}{\sqrt{3}} = \frac{8 \cdot \sqrt{3}}{\sqrt{3} \cdot \sqrt{3}}$$

$$x = \frac{8\sqrt{3}}{3}$$

Thus, the length of a radius is $\frac{8\sqrt{3}}{3}$.

EXAMPLE 3 Find x in the figure at the right if each side of $\triangle ABC$ is tangent to $\odot O$.

Plan Use the property that tangent segments from an exterior point are congruent. Label each segment in terms of x.

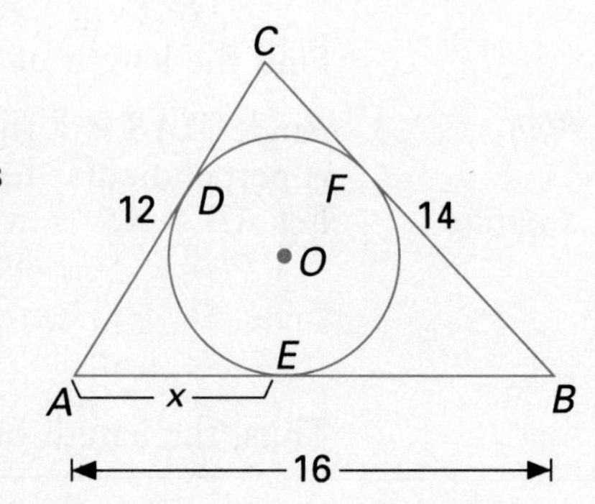

Solution Let $AD = AE = x$.
Then, $EB = 16 - x$ and $DC = 12 - x$
Also, $FB = 16 - x$ and $CF = 12 - x$

By the Segment Addition Postulate,

$$CB = CF + FB$$
$$14 = (12 - x) + (16 - x)$$
$$14 = 28 - 2x$$
$$-14 = -2x$$

Thus, $x = 7$.

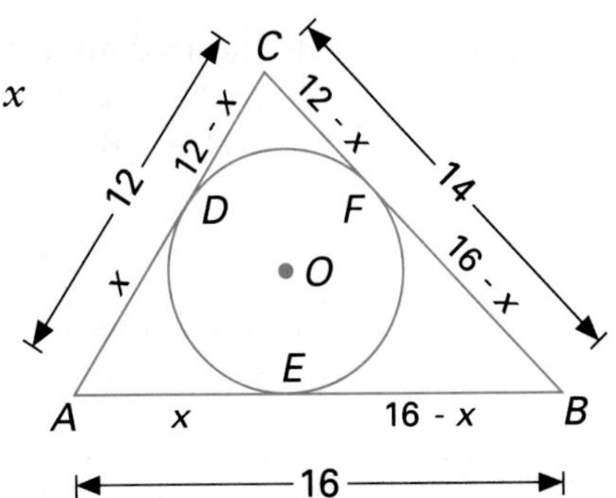

The figure for Example 3 above shows a circle inscribed in a polygon.

Definition

A polygon whose sides are tangent to a circle is a **circumscribed polygon**. The polygon is circumscribed about the circle, and the circle is *inscribed* in the polygon.

Definition

A line tangent to each of two coplanar circles is a **common tangent**.

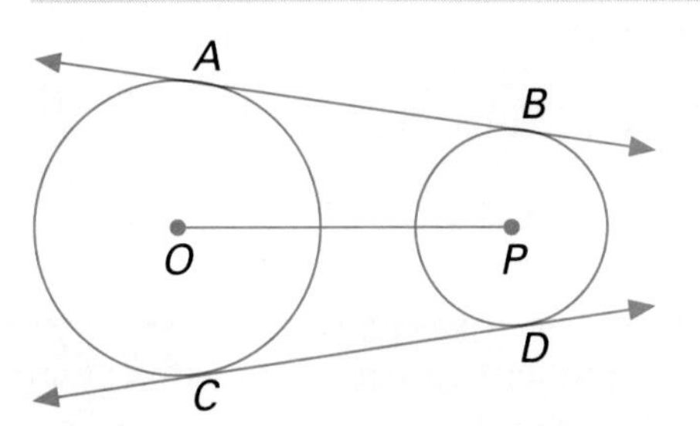

Common *external* tangents $\overleftrightarrow{AB}$ and $\overleftrightarrow{CD}$ do not intersect the segment $\overline{OP}$ joining the centers.

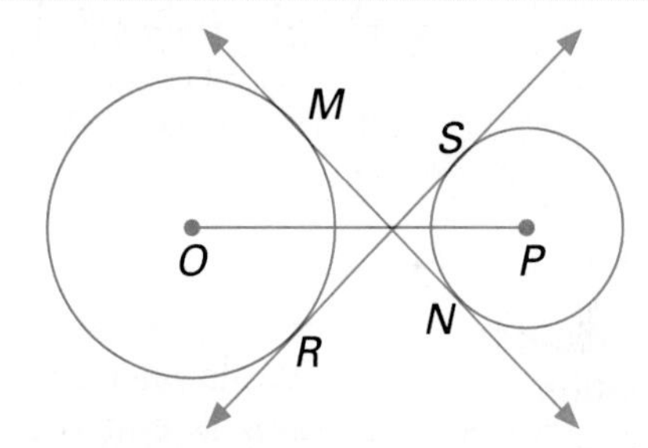

Common *internal* tangents $\overleftrightarrow{MN}$ and $\overleftrightarrow{RS}$ intersect the segment $\overline{OP}$ joining the centers.

EXAMPLE 4 The centers of two circles of radii of lengths 3 and 8 are 13 units apart. Find the length of a common external tangent segment.

Plan Draw two circles with centers O and P. Draw $\overline{OP}$, common external tangent $\overline{AB}$, and radii $\overline{OA}$ and $\overline{PB}$. Draw $\overline{PQ} \perp \overline{OA}$ so that $ABPQ$ forms a rectangle. Since the opposite legs of a rectangle are congruent, $AQ = BP = 3$ and $AB = QP$. Note that $\overline{QP}$ is a leg of $\triangle OQP$. $OQ = 8 - 3 = 5$.

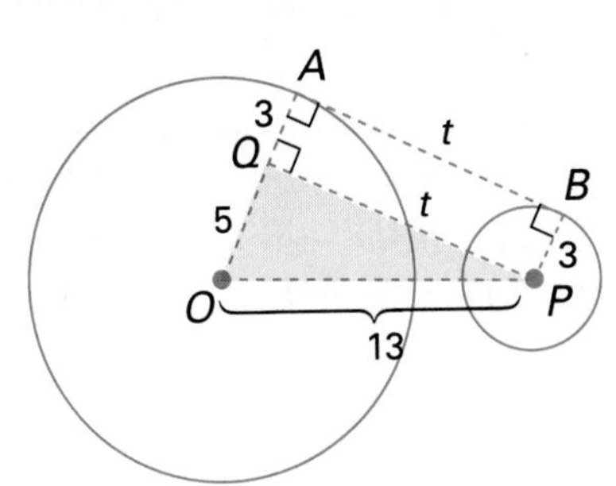

Solution Use the Pythagorean Theorem.

$$5^2 + t^2 = 13^2$$
$$25 + t^2 = 169$$
$$t^2 = 144$$
$$t = \sqrt{144} = 12$$

Thus, the length of tangent segment $\overline{AB}$ is 12.

Two circles can be tangent to the same line at the same point. At the right, circle O is in the interior of circle P, except for the point of tangency. The two circles are said to be *internally tangent* at point T.

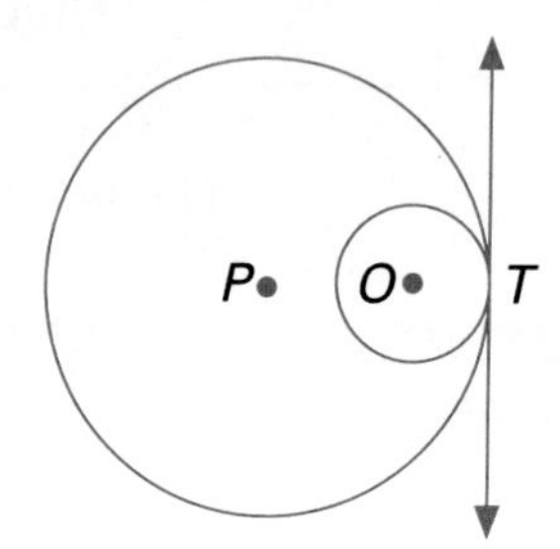

In the next diagram, circle S is in the exterior of circle R, except for the point of tangency V. The circles are said to be *externally tangent* at point V.

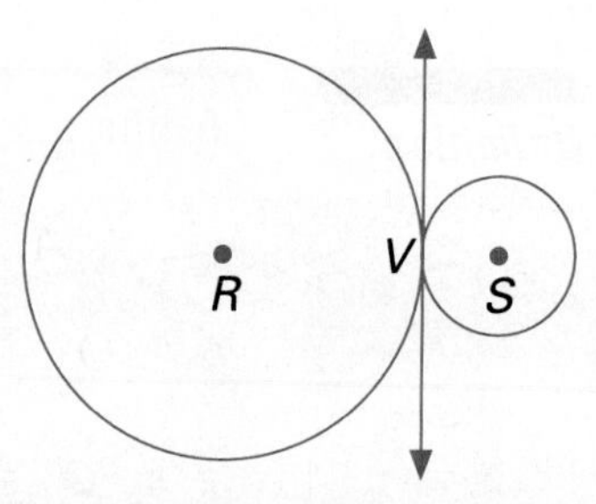

Definition Two coplanar circles are tangent to each other if they are tangent to the same line at the same point.

Classroom Exercises

Given: $\odot O$ is inscribed in $\triangle BFD$.

1. Which segments are perpendicular? Why?
2. $\overline{CD} \cong$ __________. Why?
3. $\overline{FE} \cong$ __________. Why?
4. $AB = 4$, $OC = 4$. Tell how to find OB.
5. m $\angle FBD = 80$; m $\angle 1 =$ __________. Why?
6. $CD = 5$, $FD = 12$, $FE =$ __________

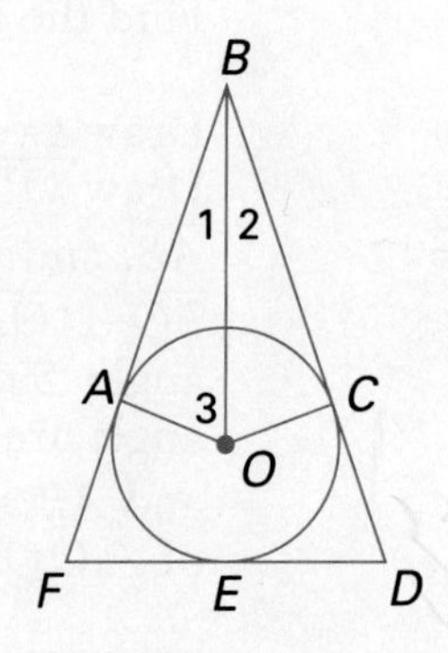

Written Exercises

Use the figures below for Exercises 1–6.

$\overline{PQ}$ is a tangent segment.

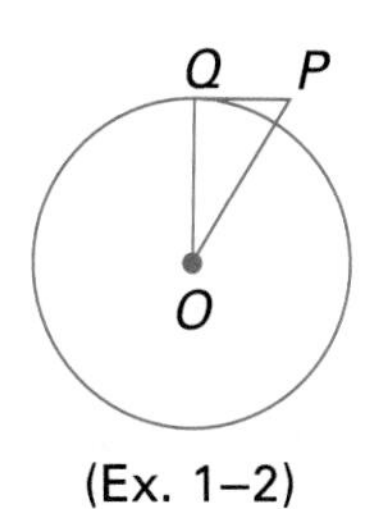

(Ex. 1–2)

$\overline{AB}$ and $\overline{AC}$ are tangent segments.

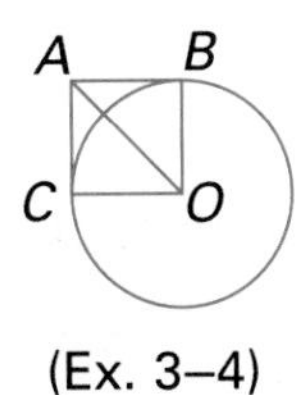

(Ex. 3–4)

Circle O is inscribed in $\triangle ABC$.

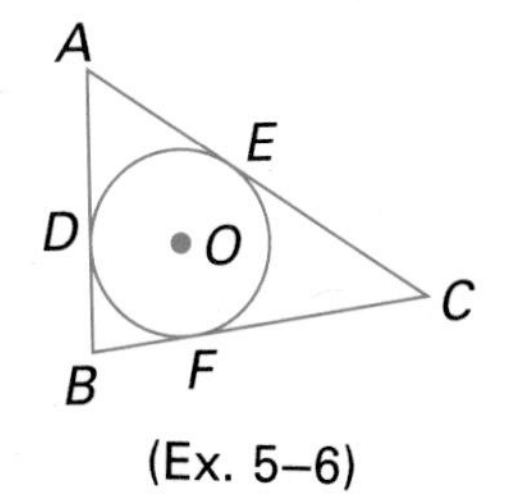

(Ex. 5–6)

1. $QP = 2$, $OP = 6$
Find OQ.

2. $m\angle P = 60$,
$OQ = 2$
Find PQ.

3. $m\angle CAB = 120$,
$OB = 6$
Find AB and AC.

4. $m\angle CAB = 90$,
$OA = 10$
Find OB and OC.

5. $AC = 12$, $CB = 10$,
$AB = 8$
Find BF.

6. $AB = 6$, $AC = 8$,
$BC = 9$
Find EC.

In the figure at the right, $\overline{RQ}$ is an external tangent segment to circles O and P; $\overline{SP} \perp \overline{OR}$.

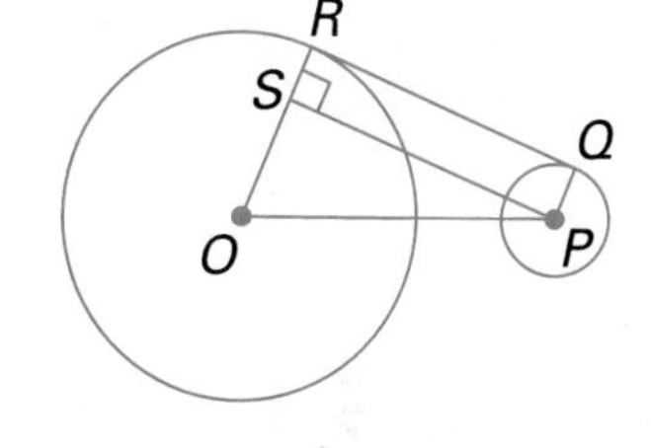

7. Given: $PQ = 6$, $OR = 10$, $OP = 20$
Find RQ.

8. Given: $RQ = 8$, $OP = 10$, $QP = 1$
Find OS and OR.

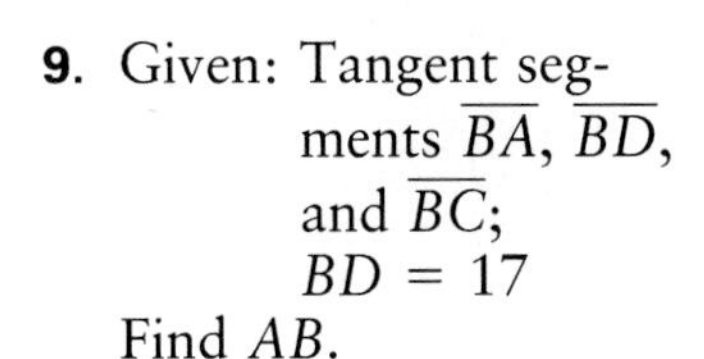

9. Given: Tangent segments $\overline{BA}$, $\overline{BD}$, and $\overline{BC}$;
$BD = 17$
Find AB.

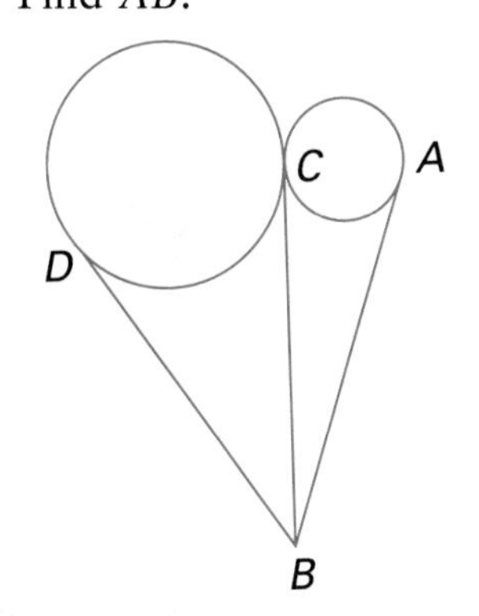

10. Given: Circumscribed $\triangle PQR$, $QU = 2$, $UR = 5$, and $PS = 4$
Find $QR + RP + QP$.

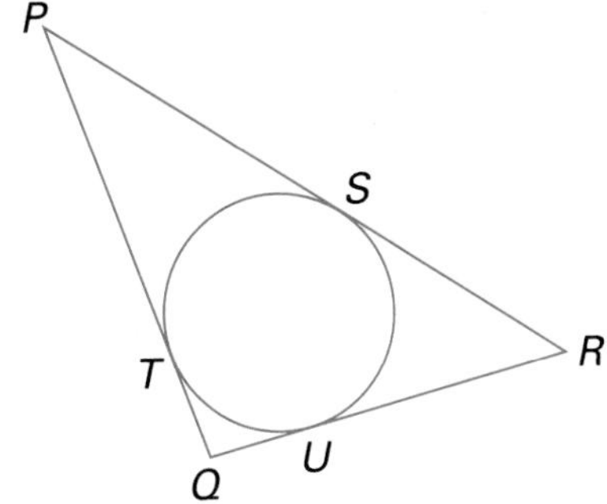

11. Given: Circumscribed polygon $ABCD$, $BE = 5$, $GD = 3$, $AH = 4$, and $CG = 6$
Find $AB + BC + CD + DA$.

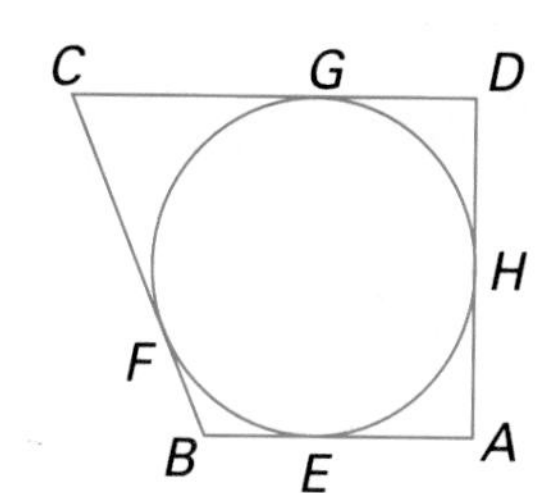

12. Prove Theorem 11.8.

13. Prove the corollary to Theorem 11.8.

14. Given: $\odot O \cong \odot Q$, $\odot O$ and $\odot Q$ are tangent to $\overline{SM}$ at M.
Prove: $\overline{OS} \cong \overline{QS}$

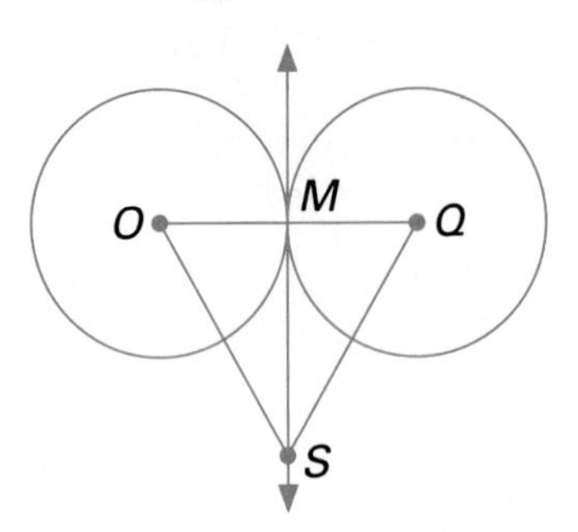

15. Given: Circles O and P are tangent to $\overline{AC}$ and $\overline{AD}$.
Prove: $CB = DE$

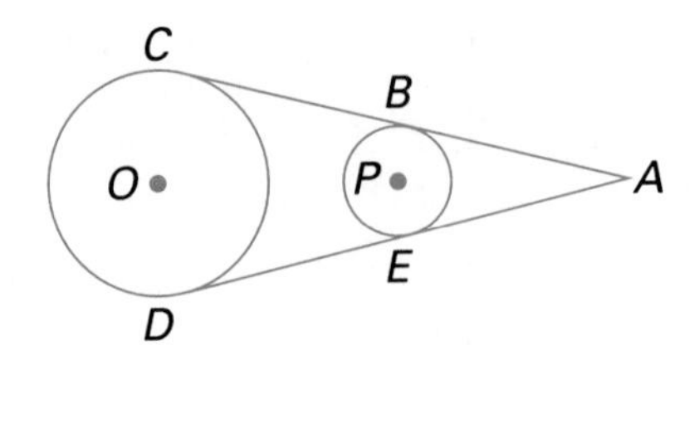

16. The measure of the angle between two tangent segments to a circle from a point R is 120. Find the length of a diameter if the length of each tangent segment is 10.
17. The centers of two circles of radii lengths 10 and 2 are 17 units apart. Find the length of a common external tangent segment.
18. The lengths of the radii of two circles are 4 and 2. The centers are 10 units apart. Find the length of the common internal tangent segment.
19. Prove Theorem 11.7. (HINT: Use an indirect proof.)

The figure shows a point C at a height h above the earth. d is the distance from C to the visual horizon. r is the radius of the earth. $\overline{AC}$ is a tangent segment.

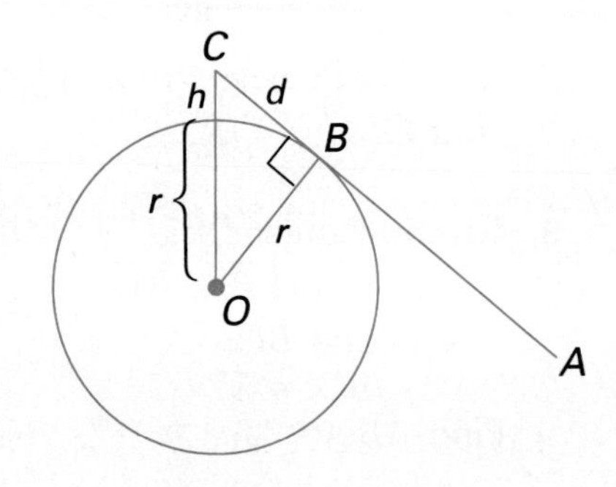

20. Write a formula for d in terms of h and r.

21. The formula of Exercise 20 can be used to calculate the range of VHF (very high frequency) ship-to-shore communications. A ship has a VHF transceiver on top of a mast 36 m tall. Find d, the maximum communication range of the ship to the nearest kilometer. (Use $r \approx 6{,}380$ km.)

Mixed Review

1. Find the number of sides of a regular polygon if the measure of each interior angle is 120. *6.2*
2. Find the measure of each of two complementary angles if their measures are in the ratio 2 to 3. *8.1*
3. Find the length of an altitude of equilateral triangle ABC if $AB = 8$. *9.4*

Computer Investigation

Arcs and Related Angles

Use a computer software program that draws circles of a given radius length, places random points on or in the exterior of the circle, labels points, constructs line segments between two given points, draws tangent segments to the circle at a given point, and measures angles.

The measure of a central angle equals the degree measure of its arc. Thus, in the figure, $\angle BAC$ is a central angle and m $\angle BAC$ = m$\overset{\frown}{BC}$. $\angle BDC$ is *not* a central angle. It is an angle formed by two chords, not radii. The computer activities below will help you discover a pattern for finding the measure of an angle such as $\angle BDC$, which intercepts $\overset{\frown}{BC}$, the arc of central angle BAC. This property will be formally stated and proved in the next lesson.

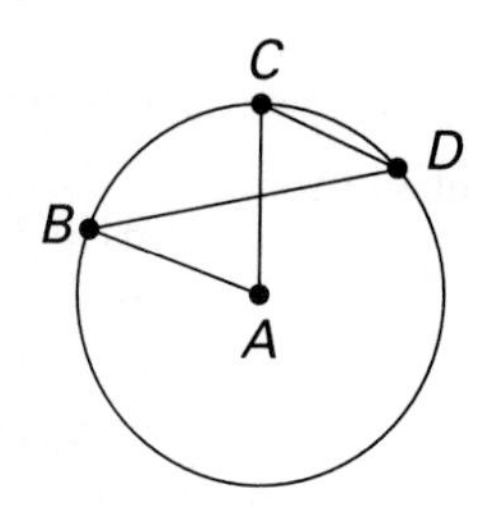

Activity 1

Draw a circle of radius 5, or some length that will fit on the screen. Label the center A. Draw three random points, B, C, and D on the circle as illustrated in the diagram above. Draw $\overline{AB}$, $\overline{AC}$, $\overline{DB}$, and $\overline{DC}$.

Find the following measures.

1. m $\angle BAC$ (measure of a central angle)
2. m $\angle BDC$

Repeat the steps above for the following three circles.

3. radius = 5, but with three new points B, C, and D on the circle
4. radius = 6
5. radius = 4
6. Draw $\odot A$. Place B and C randomly on the circle; draw $\overline{AB}$, $\overline{AC}$, and $\overline{BC}$, and a tangent segment $\overline{BD}$.
7. Find m $\angle BAC$ and m $\angle CBD$.
8. Repeat Exercises 6 and 7 for circles of radii 4 and 5.

Summary

What generalizations appear to hold true for the measures of:

1. an angle with its vertex on the circle and having its sides as chords of the circle?
2. an angle with its vertex on the circle and having its sides as a chord and tangent of the circle?

11.4 Arcs and Central Angles

Objective To find degree measures of arcs

You know that a segment may be considered part of a line. In the same way, an *arc* is an unbroken part of a circle. An arc is measured by referring to an angle whose vertex is the center of the circle.

Definitions

A **central angle** of a circle is an angle with measure less than 180 whose vertex is the center of the circle. The **minor arc** $\overset{\frown}{AB}$ of central angle AOB consists of points A, B, and all points on the circle that lie in the interior of the central angle. The **degree measure of a minor arc** is equal to the measure of its central angle: $m\overset{\frown}{AB} = m\angle AOB$.

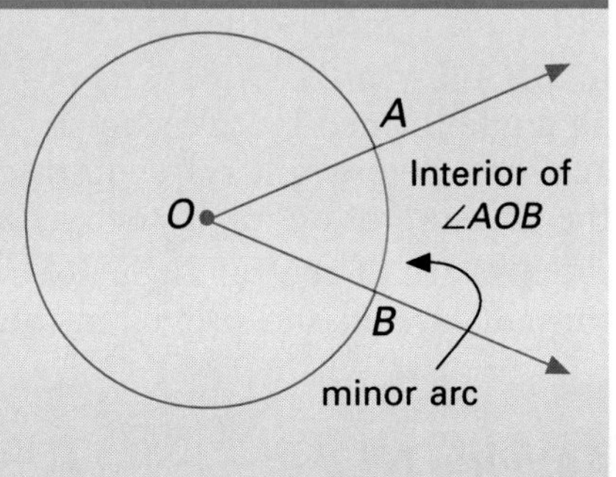

Since Chapter 1, you have assumed that measures of angles are given in degrees. Similarly, you may now assume that measures of arcs are given in degrees.

EXAMPLE 1 Given: $m\angle AOB = 20$, $m\angle BOC = 40$
Find $m\overset{\frown}{AC}$.

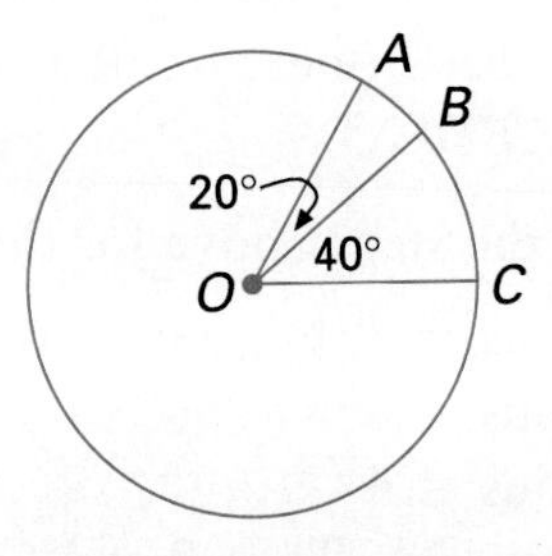

Solution $m\angle AOC = m\angle AOB + m\angle BOC$
by the Angle Addition Postulate.

$m\angle AOC = 20 + 40 = 60$

Thus, by definition of the measure of a minor arc, $m\overset{\frown}{AC} = 60$.

At the right, two adjacent central angles together form a straight angle. A straight angle separates the points of the circle into two special arcs, $\overset{\frown}{RTS}$ and $\overset{\frown}{RQS}$, called *semicircles*. The degree measure of a semicircle is 180.

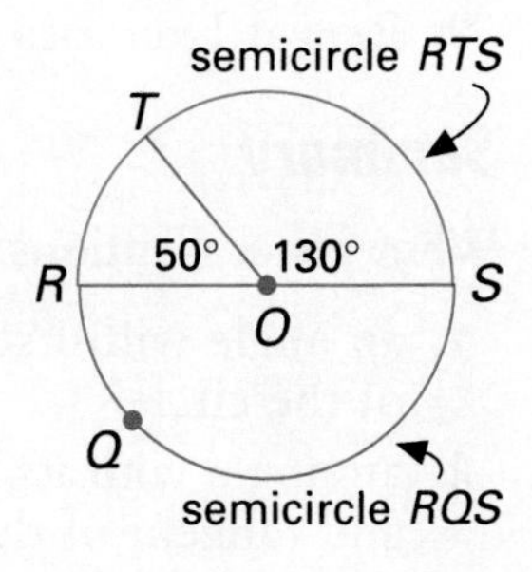

Definition

Semicircle $\overparen{RTS}$ consists of the endpoints R and S of diameter $\overline{RS}$ and all points of $\odot O$ that lie on the same side of $\overline{RS}$ as T.

Definition

The **major arc** $\overparen{ACB}$ consists of points A and B and all points of the circle in the *exterior* of central angle AOB. The degree measure of $\overparen{ACB}$ equals 360 minus the measure of central angle AOB.

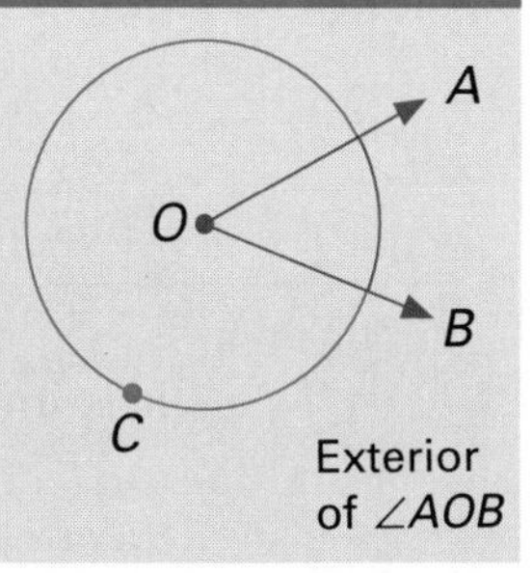

Degree measures of arcs can be added in a manner very similar to that in which degree measures of angles are added. The following postulate permits this. It applies to minor arcs, major arcs, and semicircles.

Postulate 19

Arc Addition Postulate
If P is a point on $\overparen{APB}$, then
$m\overparen{AP} + m\overparen{PB} = m\overparen{APB}$.

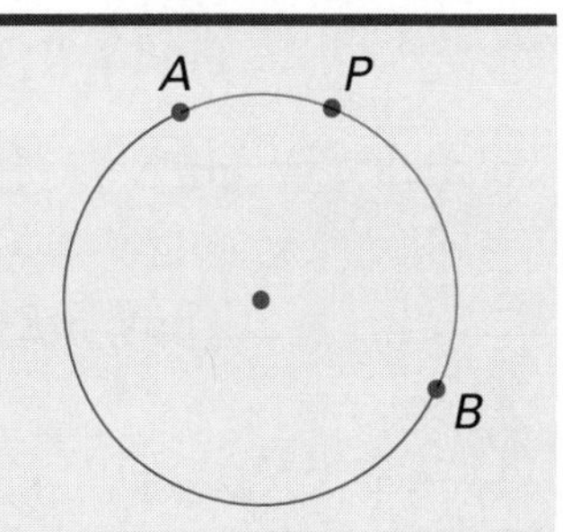

The sum of the degree measures of the two semicircles is $m\overparen{RTS} + m\overparen{RQS} = 180 + 180 = 360$. Thus, the degree measure of a circle is 360.

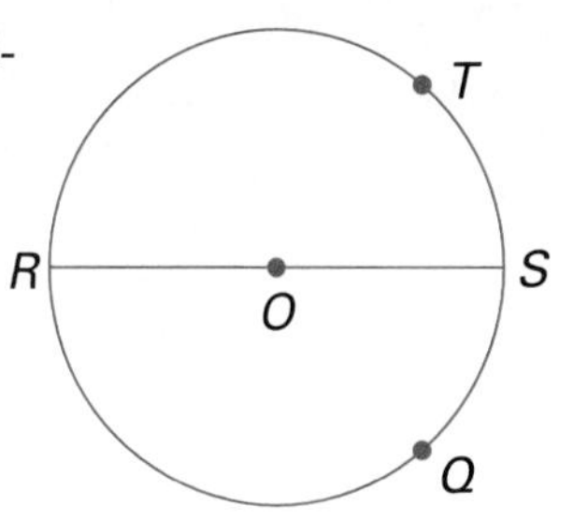

The following equations show addition of degree measures of arcs for the figure at the right.

$$m\overparen{APB} = m\overparen{AP} + m\overparen{PB}$$
$$= 25 + 80 = 105$$
$$m\overparen{CAP} = m\overparen{CA} + m\overparen{AP}$$
$$= 100 + 25 = 125$$

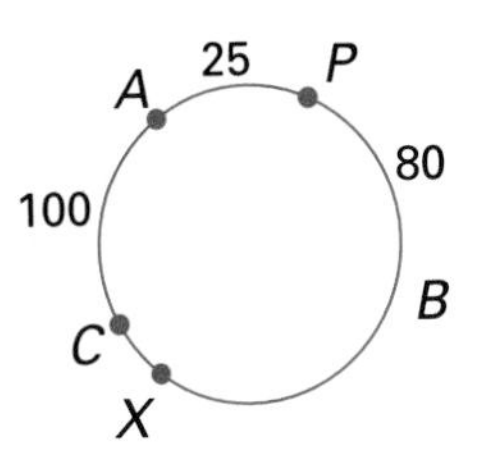

EXAMPLE 2 Given that $m\angle B = 20$, find $m\overset{\frown}{ACB}$.

Plan Since radii are congruent, $\triangle AOB$ is isosceles and $m\angle A = m\angle B$. Use the Triangle Sum Theorem.

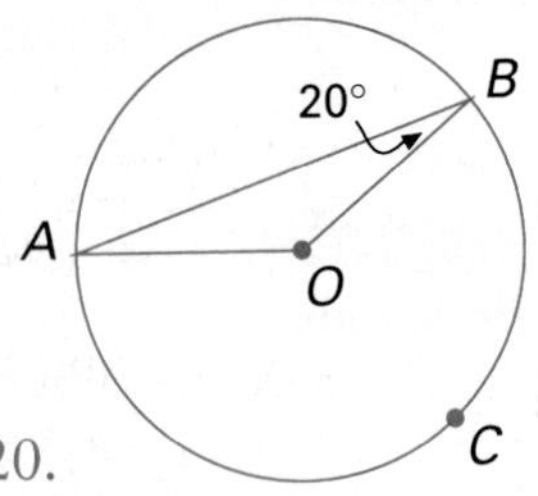

Solution

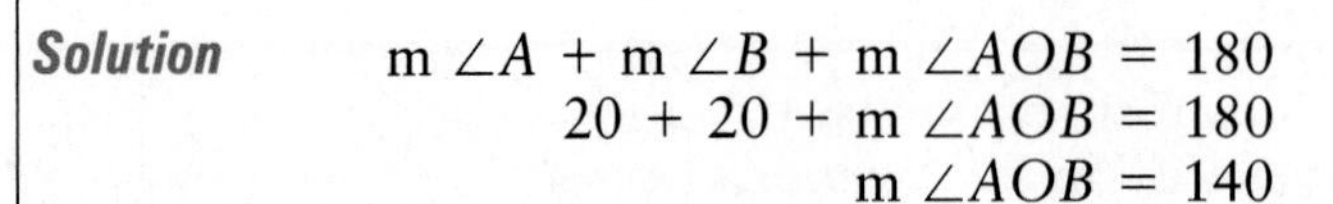

$$\begin{aligned} m\angle A + m\angle B + m\angle AOB &= 180 \\ 20 + 20 + m\angle AOB &= 180 \\ m\angle AOB &= 140 \end{aligned}$$

Thus, $m\overset{\frown}{AB} = 140$, and $m\overset{\frown}{ACB} = 360 - 140 = 220$.

It is important to distinguish between two such symbols as $\overset{\frown}{AB}$ and $m\overset{\frown}{AB}$. $\overset{\frown}{AB}$ is *a set of points* that make up an arc; $m\overset{\frown}{AB}$ is the degree *measure* of the arc.

In this text, two letters under an arc sign, such as $\overset{\frown}{AB}$, refer *only* to minor arcs. Three letters, such as $\overset{\frown}{ACB}$, may refer either to a minor arc (for clarity), a major arc, or a semicircle.

EXAMPLE 3 Given: $\odot O$ with diameter $\overline{PQ}$, $m\angle POS = 120$, $m\angle QOR = 50$
Find $m\overset{\frown}{RQS}$.

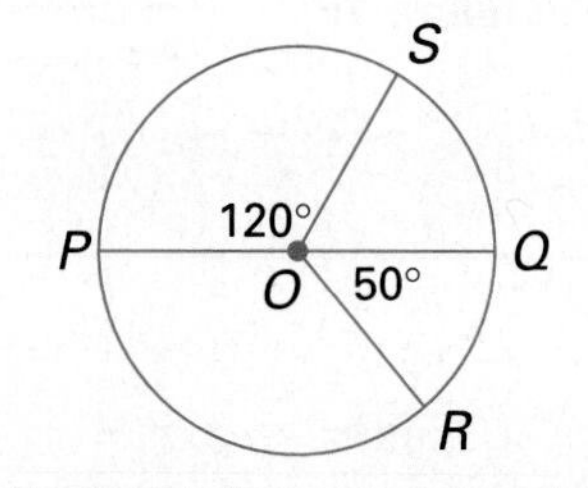

Solution $m\overset{\frown}{PS} = 120$ and $m\overset{\frown}{RQ} = 50$ (by definition)
$m\overset{\frown}{QS} = 180 - 120 = 60$ ($\overline{PQ}$ is a diameter.)

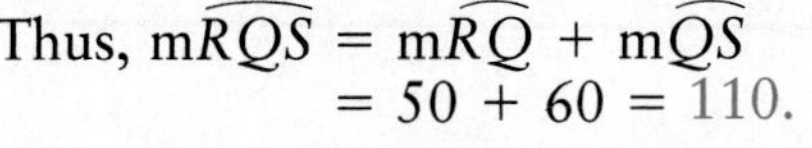

Thus, $m\overset{\frown}{RQS} = m\overset{\frown}{RQ} + m\overset{\frown}{QS}$
$= 50 + 60 = 110$.

EXAMPLE 4 Given: $\overline{AD}$ is a diameter; $m\angle BOC = 40$, $m\overset{\frown}{AB}:m\overset{\frown}{CD} = 5:2$.
Find $m\overset{\frown}{AC}$.

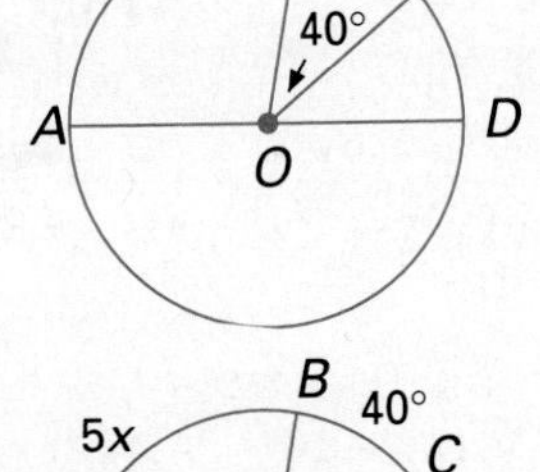

Plan $\overset{\frown}{ABD}$ is a semicircle since $\overline{AD}$ is a diameter.
Write an equation relating the three minor arcs.
Let $m\overset{\frown}{AB} = 5x$, $m\overset{\frown}{CD} = 2x$ and $m\overset{\frown}{BC} = 40$, since $m\angle BOC = 40$.

Solution

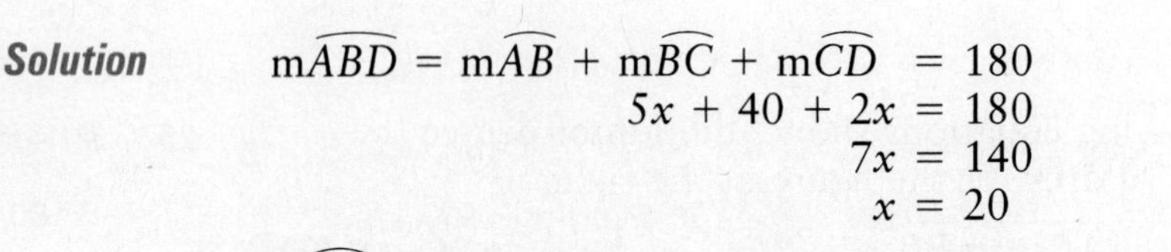

$$\begin{aligned} m\overset{\frown}{ABD} = m\overset{\frown}{AB} + m\overset{\frown}{BC} + m\overset{\frown}{CD} &= 180 \\ 5x + 40 + 2x &= 180 \\ 7x &= 140 \\ x &= 20 \end{aligned}$$

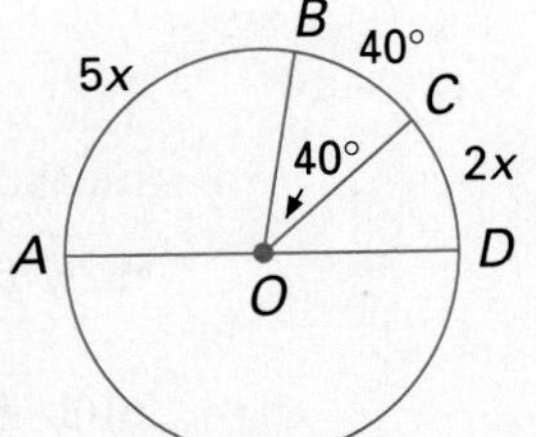

$m\overset{\frown}{AB} = 5x = 5 \cdot 20$, or 100
Thus, $m\overset{\frown}{AC} = m\overset{\frown}{AB} + m\overset{\frown}{BC} = 100 + 40$, or 140.

EXAMPLE 5 Find $m\overset{\frown}{ADC}$ for $\odot O$ at right.

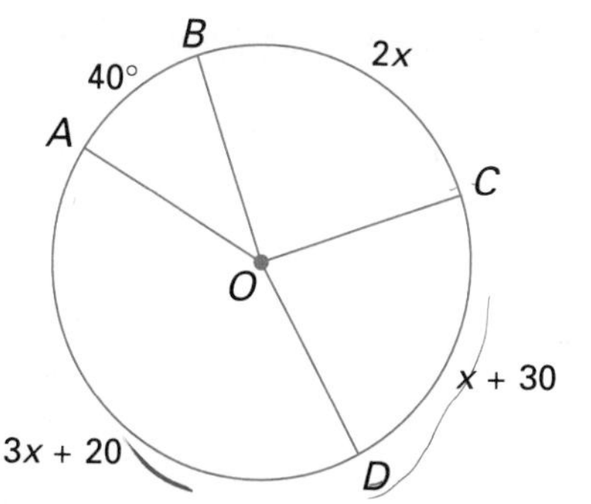

Solution The degree measure of a circle is 360.

$$\begin{aligned} m\overset{\frown}{AB} + m\overset{\frown}{BC} + m\overset{\frown}{CD} + m\overset{\frown}{DA} &= 360 \\ 40 + 2x + (x + 30) + (3x + 20) &= 360 \\ 6x + 90 &= 360 \\ 6x &= 270 \\ x &= 45 \end{aligned}$$

Find $m\overset{\frown}{AD}$ and $m\overset{\frown}{DC}$.

$m\overset{\frown}{AD} = 3x + 20$
$= 3 \cdot 45 + 20$, or $= 155$

$m\overset{\frown}{DC} = x + 30$
$= 45 + 30 = 75$

Thus, $m\overset{\frown}{ADC} = m\overset{\frown}{AD} + m\overset{\frown}{DC} = 155 + 75$, or 230.

Focus on Reading

Indicate whether the statement is sometimes true, always true, or never true.

1. The degree measure of a minor arc equals the degree measure of its central angle.
2. The degree measure of a semicircle is half the degree measure of the circle.
3. If the degree measures of two minor arcs are added, the result is the degree measure of a major arc.
4. Given an arc of a circle that is not a minor arc, the arc is a major arc.

Classroom Exercises

The measures of some central angles of a circle are given in the figure at the right. Find the measure of the indicated arc.

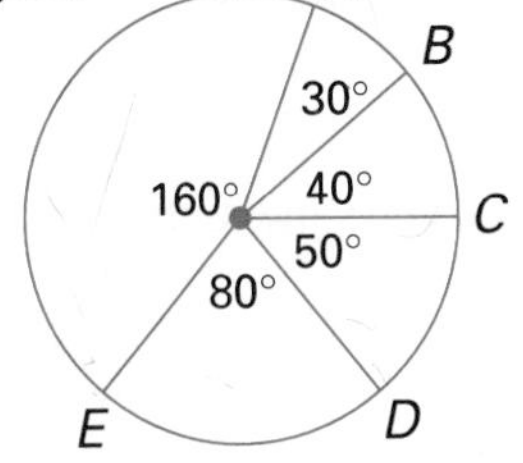

1. $m\overset{\frown}{AC}$ 2. $m\overset{\frown}{BD}$ 3. $m\overset{\frown}{EC}$ 4. $m\overset{\frown}{AD}$ 5. $m\overset{\frown}{BE}$
6. $m\overset{\frown}{EAB}$ 7. $m\overset{\frown}{ECA}$ 8. $m\overset{\frown}{DAE}$ 9. $m\overset{\frown}{EDA}$ 10. $m\overset{\frown}{CBD}$

Written Exercises

Given: $\odot O$ with diameter $\overline{AB}$. Find the measure of the indicated arc.

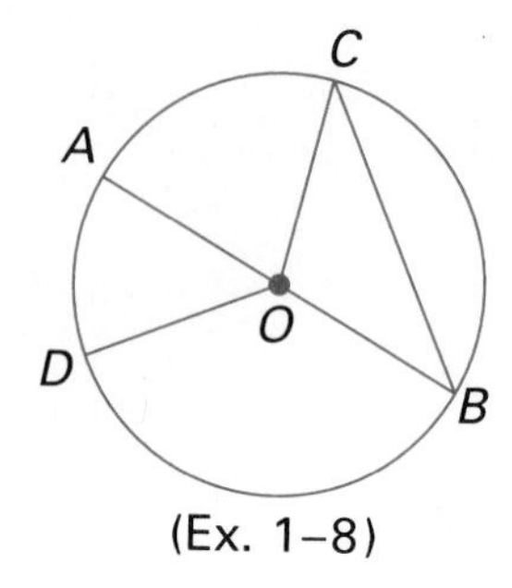

(Ex. 1–8)

1. $m\overset{\frown}{AC} = 60$, $m\overset{\frown}{CB} =$ ______

2. $m\overset{\frown}{AD} = 50$, $m\overset{\frown}{DB} =$ ______

3. $m\overset{\frown}{CB} = 100$, $m\overset{\frown}{AD} = 40$, $m\overset{\frown}{ABC} =$ ______

4. $m\overset{\frown}{DB} = 160$, $m\overset{\frown}{AC} = 30$, $m\overset{\frown}{BAC} =$ ______

5. $m\overset{\frown}{AD} = 50$, $m\overset{\frown}{CB} = 110$, $m\overset{\frown}{DCB} =$ ______

6. $m\ \angle B = 30$
$m\overset{\frown}{BAC} =$ ______

7. $m\overset{\frown}{BC} = 100$
$m\ \angle B =$ ______

8. $m\overset{\frown}{BAC} = 200$
$m\ \angle C =$ ______

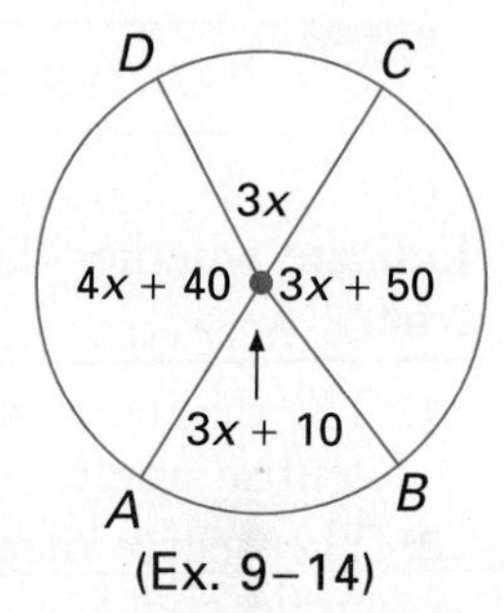

(Ex. 9–14)

9. $m\overset{\frown}{AD} =$ ______

10. $m\overset{\frown}{DC} =$ ______

11. $m\overset{\frown}{ADC} =$ ______

12. $m\overset{\frown}{CB} =$ ______

13. $m\overset{\frown}{ABC} =$ ______

14. $m\overset{\frown}{ADB} =$ ______

For Exercises 15–23, $\odot O$ has diameter $\overline{AB}$.

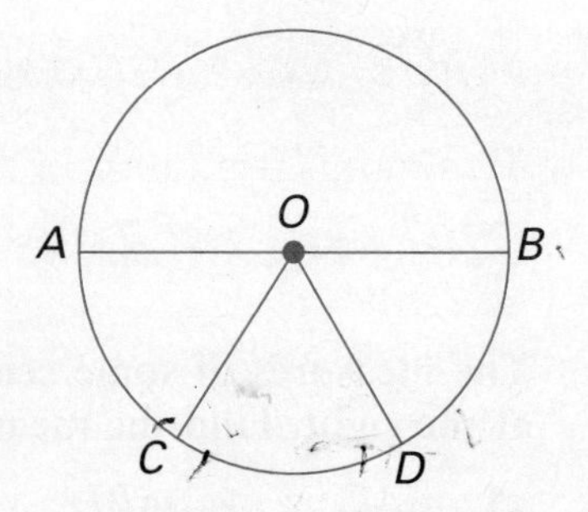

Given: $m\overset{\frown}{AC}:m\overset{\frown}{CD} = 3:2$, $m\overset{\frown}{DB} = 50$

15. $m\overset{\frown}{CD} =$ ______

16. $m\overset{\frown}{CB} =$ ______

17. $m\overset{\frown}{CBA} =$ ______

Given: $m\overset{\frown}{CD} = 90$, $m\overset{\frown}{AC}:m\overset{\frown}{DB} = 5:4$

18. $m\overset{\frown}{AC} =$ ______ **19.** $m\overset{\frown}{DB} =$ ______ **20.** $m\overset{\frown}{DAB} =$ ______

Given: $m\overset{\frown}{AC}:m\overset{\frown}{CD}:m\overset{\frown}{DB} = 3:2:4$

21. $m\overset{\frown}{AC} =$ ______

22. $m\overset{\frown}{DB} =$ ______

23. $m\overset{\frown}{DBC} =$ ______

24. Given: $\odot O$ with tangent $\overline{RS}$
Prove: $m\overarc{ST} = 90 - m\angle R$

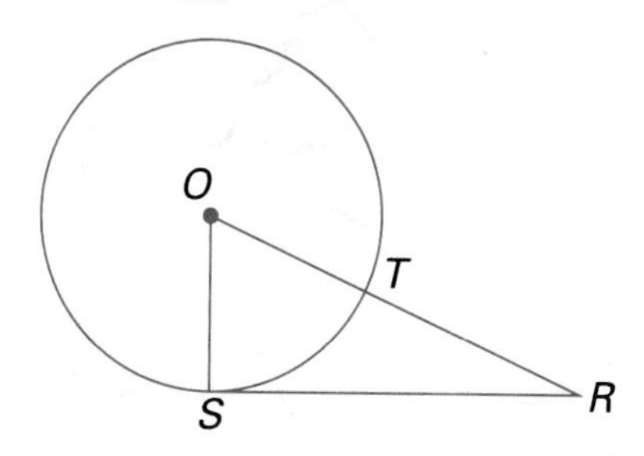

25. Given: $\overline{AD}$ and $\overline{CD}$ are tangent to $\odot O$; $m\overarc{ABC} = 270$.
Prove: $AOCD$ is a square.

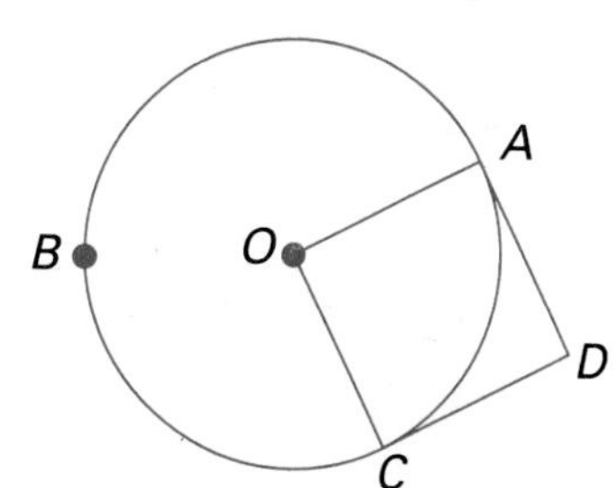

In $\odot O$ at the right, B and D are points of tangency, and $\overline{ED}$ is a diameter.

26. $m\overarc{EB}:m\overarc{BD} = 4:5$, $ED = 10$
Find AB to the nearest tenth. (HINT: Use a trig ratio.)

27. $AB = 4\sqrt{3}$, $OB = 4$
Find $m\overarc{BDE}$.

28. Prove: $m\angle C = m\overarc{EB}$

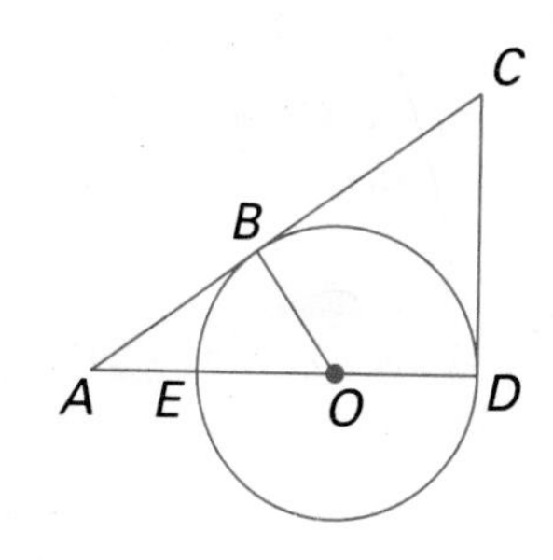

Mixed Review

1. The measure of a base angle of an isosceles triangle is 15 greater than the measure of the vertex angle. Find the measure of the vertex angle. **5.1**
2. In quadrilateral $ABCD$, $\overline{AB} \cong \overline{DC}$ and $\overline{AD} \parallel \overline{BC}$. Is $ABCD$ a parallelogram? Justify your answer. **6.4**
3. True or false? Skew lines are coplanar lines. **3.1**
4. Draw an acute angle. Construct the bisector of the angle. **1.5**

Algebra Review

To simplify real numbers and expressions involving square roots, the following properties are needed:

- $\sqrt{a^2} = |a|$, for all real numbers a
- $\sqrt{a} \cdot \sqrt{a} = a$, for all positive real numbers a
- $\sqrt{ab} = \sqrt{a} \cdot \sqrt{b}$, for all positive real numbers a and b

Simplify the expression.

1. $(3\sqrt{5})^2$	**2.** $\sqrt{27}$	**3.** $\sqrt{x^2y^2}$	**4.** $3\sqrt{32}$
5. $(\sqrt{3})^2 + (\sqrt{6})^2$	**6.** $\sqrt{9 + 16}$	**7.** $\sqrt{4x^2}$	**8.** $\sqrt{24a^2}$
9. $a^2\sqrt{a^2}$	**10.** $(2\sqrt{2})^2$	**11.** $\sqrt{5^2 + 12^2}$	**12.** $\sqrt{k^4}$

11.5 Arcs, Chords, and Central Angles

Objectives To identify congruent arcs and congruent chords
To prove relationships among arcs, chords, and central angles

Definition **Concentric circles** are coplanar circles with a common center.

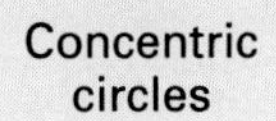

Definition In the same circle or congruent circles, **congruent arcs** are arcs that have the same degree measure.

In the diagram of concentric circles at the right, some arcs with the same degree measure are congruent and some are not.

$m\,\angle AOB = m\,\angle COD = m\,\angle EOF = 40$,

$\overset{\frown}{EF} \cong \overset{\frown}{AB}$ (same circle)

$\overset{\frown}{CD} \not\cong \overset{\frown}{AB}$ (circles are not congruent)

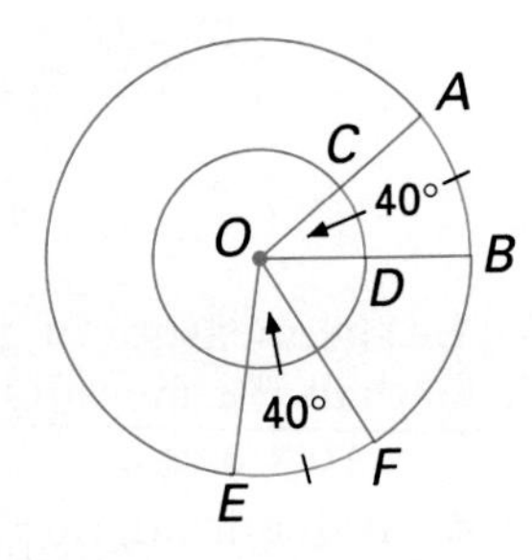

EXAMPLE 1 Given: $m\overset{\frown}{AE}:m\overset{\frown}{ED} = 4:5$, $m\overset{\frown}{AB} = 40$, $m\overset{\frown}{BC} = 80$, $m\overset{\frown}{CD} = 150$.
Determine which arcs are congruent.

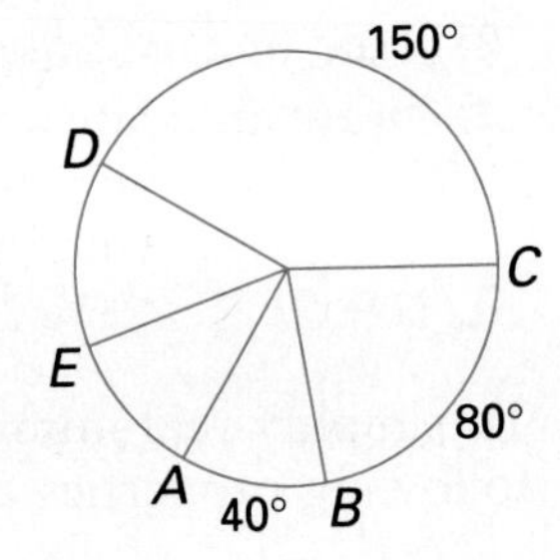

Plan Find $m\overset{\frown}{AE}$ and $m\overset{\frown}{ED}$, then compare all the arcs.
Let $m\overset{\frown}{AE} = 4x$ and $m\overset{\frown}{ED} = 5x$.

Solution

$$\begin{aligned} 4x + 5x + 150 + 80 + 40 &= 360 \\ 9x + 270 &= 360 \\ 9x &= 90 \\ x &= 10 \end{aligned}$$

Therefore, $m\overset{\frown}{AE} = 4x = 40$ and $m\overset{\frown}{ED} = 5x = 50$.

Since $m\overset{\frown}{AB} = m\overset{\frown}{AE} = 40$ in the same circle, $\overset{\frown}{AB} \cong \overset{\frown}{AE}$.

To avoid confusion, whenever reference is made to the *arc of a chord*, it will be assumed to be the *minor* arc.

Theorem 11.9 In the same circle or in congruent circles:

1. if chords are congruent, then their corresponding arcs and central angles are congruent;
2. if arcs are congruent, then their corresponding chords and central angles are congruent;
3. if central angles are congruent, then their corresponding arcs and chords are congruent.

The proof of part 1 is given below.

Given: $\odot O \cong \odot M$, $\overline{PR} \cong \overline{QS}$

Prove: $\angle O \cong \angle M$ and $\widehat{PR} \cong \widehat{QS}$

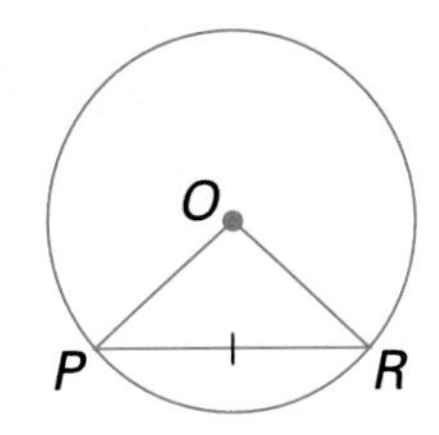

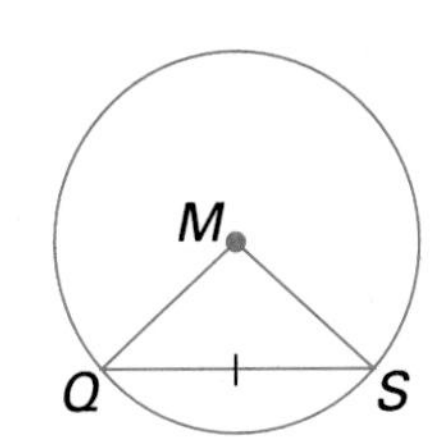

Proof

Statement	Reason
1. $\odot O \cong \odot M$, $\overline{PR} \cong \overline{QS}$	(S) 1. Given
2. $\overline{OP} \cong \overline{MQ}$, $\overline{OR} \cong \overline{MS}$	(S, S) 2. Radii of $\cong$ $\odot$s are $\cong$.
3. $\triangle POR \cong \triangle QMS$	3. SSS
4. $\angle POR \cong \angle QMS$ ($m\angle POR = m\angle QMS$)	4. CPCTC
5. $m\widehat{PR} = m\angle POR$, $m\widehat{QS} = m\angle QMS$	5. Def of meas of minor arc
6. $m\widehat{PR} = m\widehat{QS}$	6. Sub, Trans
7. $\widehat{PR} \cong \widehat{QS}$	7. Def of $\cong$ arcs

EXAMPLE 2 Which chords are congruent?

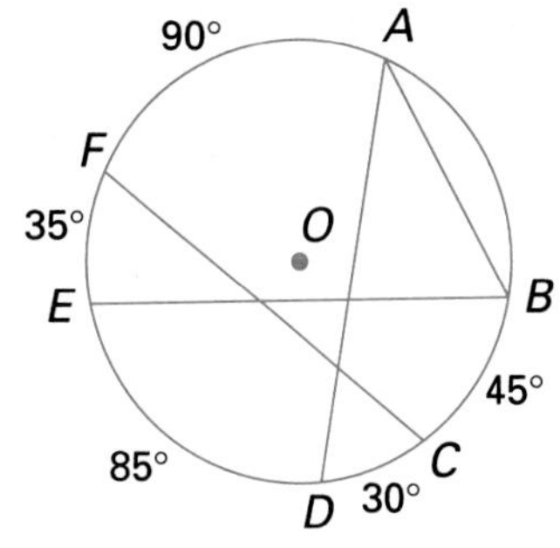

Plan Find the measure of the arc of each chord. First, find $m\widehat{AB}$. (Use *minor* arcs, *not* major arcs.)

Solution

$m\widehat{AB} = 360 - (45 + 30 + 85 + 35 + 90)$
$= 75$

$m\widehat{FC} = 35 + 85 + 30 = 150$

$m\widehat{EB} = 85 + 30 + 45 = 160$

$m\widehat{AD} = m\widehat{AB} + 45 + 30 = 75 + 45 + 30 = 150$

Thus, $\overline{FC} \cong \overline{AD}$, since $m\widehat{FC} = m\widehat{AD} = 150$.

Definition A point M is the **midpoint** of $\widehat{AMB}$ if $\widehat{AM} \cong \widehat{MB}$. A line, ray, or segment passing through point M *bisects* the arc.

EXAMPLE 3 Prove that in a circle, a radius perpendicular to a chord bisects the minor arc corresponding to the chord.

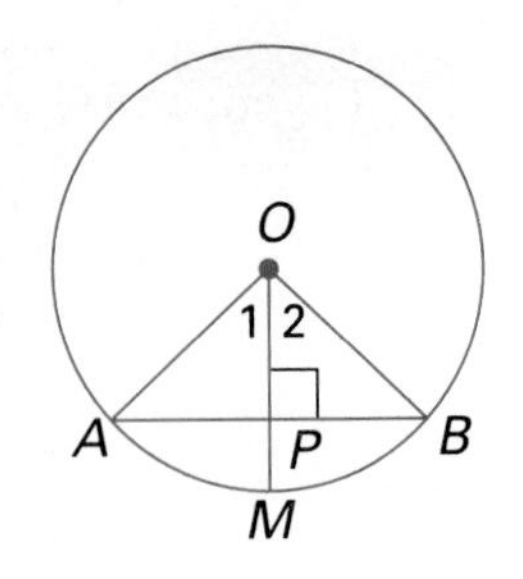

Given: $\odot O$ with radius $\overline{OM} \perp$ chord $\overline{AB}$ at P
Prove: $\overline{OM}$ bisects $\overset{\frown}{AB}$, the arc of chord $\overline{AB}$.

Proof

Statement	Reason
1. $\overline{OM} \perp \overline{AB}$ at P	1. Given
2. $\overline{OA} \cong \overline{OB}$	2. Radii of the same $\odot$ are $\cong$.
3. $\triangle AOB$ is isos.	3. Def of isos $\triangle$
4. $\overline{OP}$ is an altitude.	4. Def of alt
5. $\overline{OP}$ bis $\angle AOB$.	5. Alt to base of isos $\triangle$ bis vertex $\angle$.
6. $\angle 1 \cong \angle 2$	6. Def of $\angle$ bis
7. $\overset{\frown}{AM} \cong \overset{\frown}{MB}$	7. If central $\angle$s are $\cong$, then corr arcs are $\cong$.
8. $\therefore \overline{OM}$ bis $\overset{\frown}{AB}$.	8. Def of arc bis

Classroom Exercises

For Exercises 1–4, complete the proof of part 2 of Theorem 11.9.

Given: $\odot O \cong \odot M$, $\overset{\frown}{PR} \cong \overset{\frown}{QS}$
Prove: $\angle POR \cong \angle QMS$ and $\overline{PR} \cong \overline{QS}$

Statement		Reason
1. $\odot O \cong \odot M$, $\overset{\frown}{PR} \cong \overset{\frown}{QS}$ ($m\overset{\frown}{PR} = m\overset{\frown}{QS}$)		1. ________
2. $m\angle POR = m\overset{\frown}{PR}$, $m\angle QMS = m\overset{\frown}{QS}$		2. ________
3. $m\angle POR = m\angle QMS$ ($\angle POR \cong \angle QMS$)	**(A)**	3. ________
4. $\overline{OP} \cong \overline{MQ}$, $\overline{OR} \cong \overline{MS}$	**(S, S)**	4. ________
5. $\triangle POR \cong \triangle QMS$		5. ________
6. $\therefore \overline{PR} \cong \overline{QS}$		6. ________

7. Write the proof of part 3 of Theorem 11.9.

Written Exercises

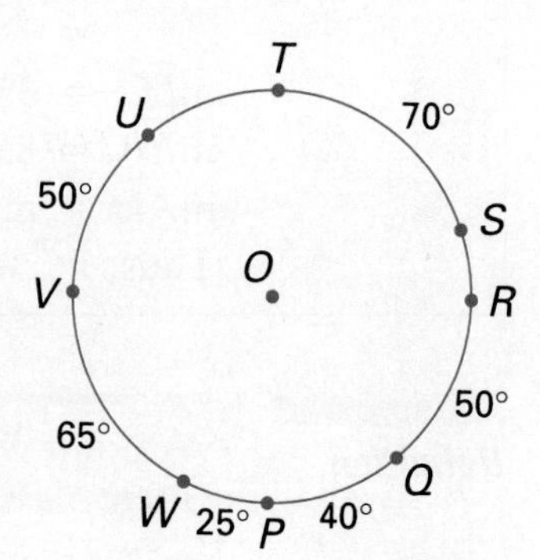

Given: $m\overset{\frown}{UT}:m\overset{\frown}{SR} = 2:1$, $m\overset{\frown}{TS} = 70$, $m\overset{\frown}{RQ} = 50$, $m\overset{\frown}{QP} = 40$, $m\overset{\frown}{PW} = 25$, $m\overset{\frown}{WV} = 65$, $m\overset{\frown}{VU} = 50$

1. Which arc is congruent to $\overset{\frown}{QP}$?

2. Which arc is congruent to $\overset{\frown}{QW}$?

3. Which arc is congruent to $\overset{\frown}{WR}$?

4. Which arc is congruent to $\overset{\frown}{PQS}$?

5. Which arcs are congruent to $\overset{\frown}{TR}$?

6. Given: $m\widehat{AC} = m\widehat{BA}$
Prove: $\triangle ABC$ is isosceles.

7. Given: $\angle B \cong \angle C$
Prove: $\widehat{AC} \cong \widehat{AB}$

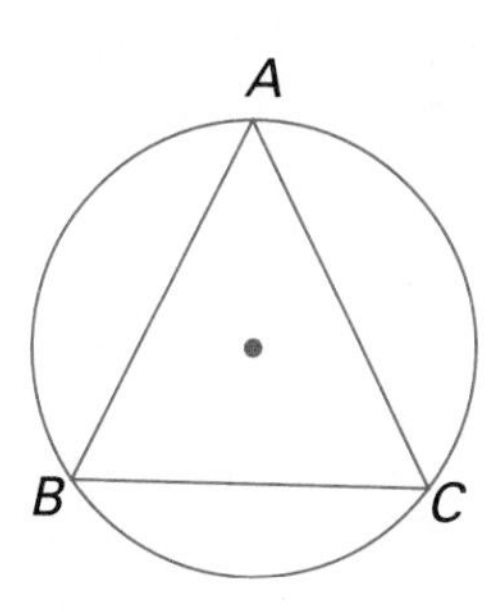

8. Given: $\odot O$ with $\widehat{AB} \cong \widehat{AD}$
Prove: $\angle 1 \cong \angle 2$

9. Given: $\odot O$, diameter $\overline{AC}$, $\overline{DC} \cong \overline{BC}$
Prove: $\widehat{AD} \cong \widehat{AB}$

10. Given: $\overline{AC}$ is a diameter of $\odot O$;
$\overline{AC}$ bisects $\angle DCB$.
Prove: $\widehat{AD} \cong \widehat{AB}$

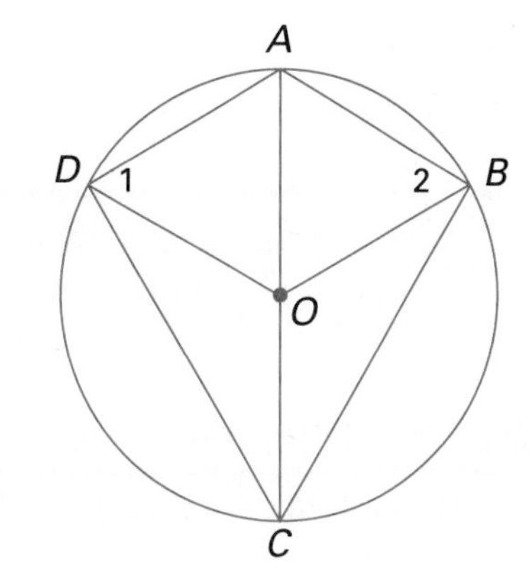

11. Given: $\overline{PQ}$ a diameter, trapezoid $PQRS$
Prove: The diagonals of the trapezoid are $\cong$.

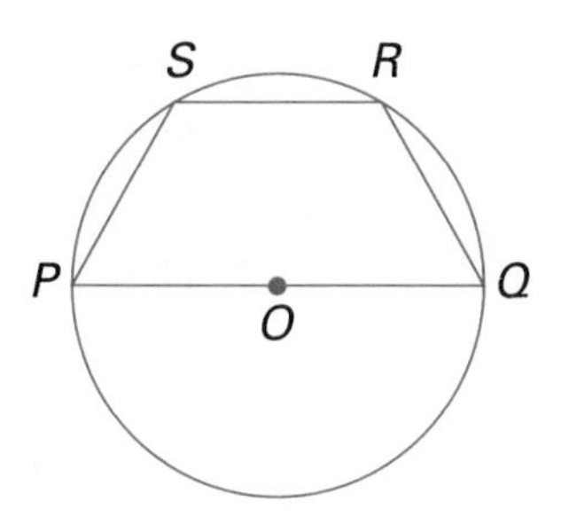

12. Given: $\odot O$, diameter $\overline{AB}$, $\overline{OD} \parallel \overline{BC}$
Prove: $\overline{OD}$ bisects $\widehat{AC}$.

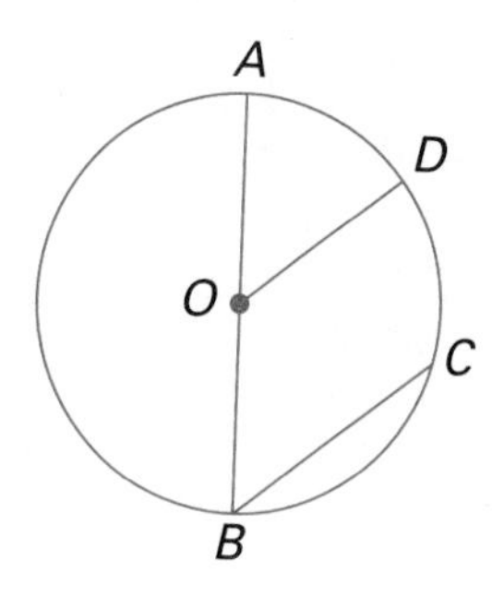

Mixed Review

Use the right triangle at the right to find each indicated length.

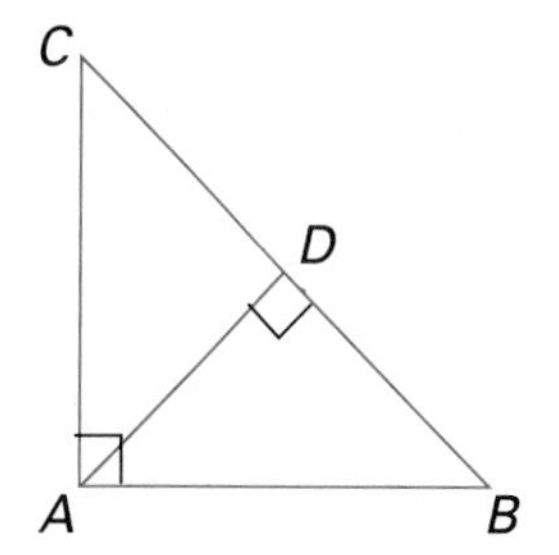

1. $AC = 6$, $AB = 4$, $BC =$ ______ *9.2*
2. $AC = 6$, $BC = 9$, $CD =$ ______ *9.1*
3. $CD = 4$, $BD = 9$, $AD =$ ______
4. $m\angle C = 30$, $AC = 4$, $BC =$ ______ *9.4*
5. $m\angle B = 45$, $BC = 8$, $AB =$ ______

11.6 Inscribed Angles

Objective To prove and apply theorems about inscribed angles

The angle illustrated at the right is not central. The sides of the angle contain chords of the circle, and the vertex is on the circle.

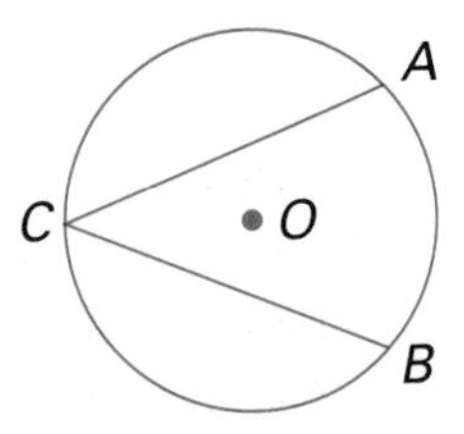

Definitions

An **inscribed angle** ACB is an angle whose vertex lies on a circle and whose sides contain chords of the circle.
The arc $\overparen{AB}$ is called the **intercepted arc** of inscribed angle ACB.

Theorem 11.10

The Inscribed Angle Theorem: The measure of an inscribed angle is one-half of the degree measure of its intercepted arc.

Given: $\odot O$ with inscribed $\angle ACB$

Prove: $m\angle ACB = \frac{1}{2} m\overparen{AB}$

Case 1
Point O lies *on* one of the sides of $\angle C$.

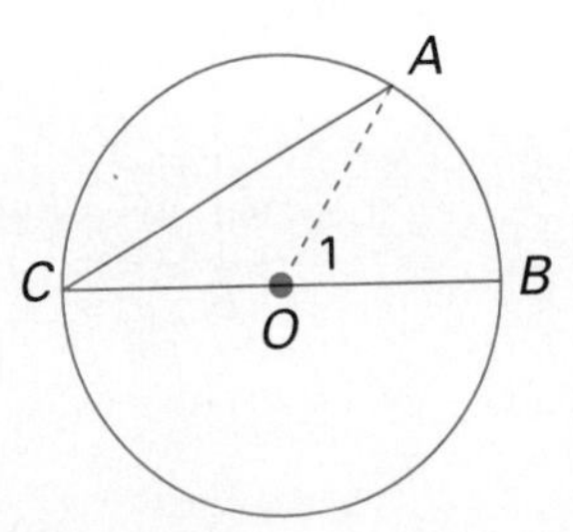

Case 2
Point O lies in the *interior* of $\angle C$.

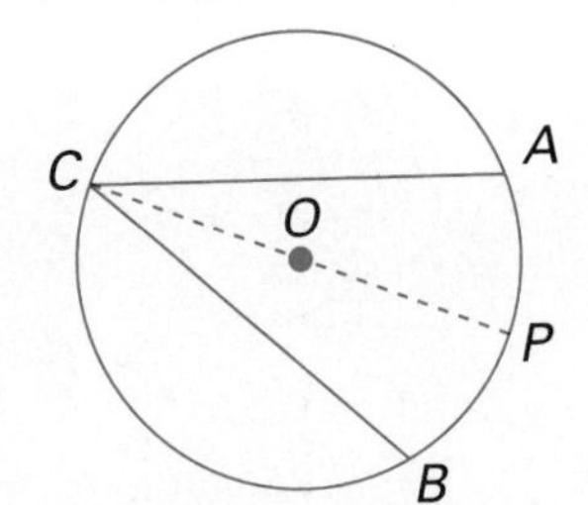

Case 3
Point O lies in the *exterior* of $\angle C$.

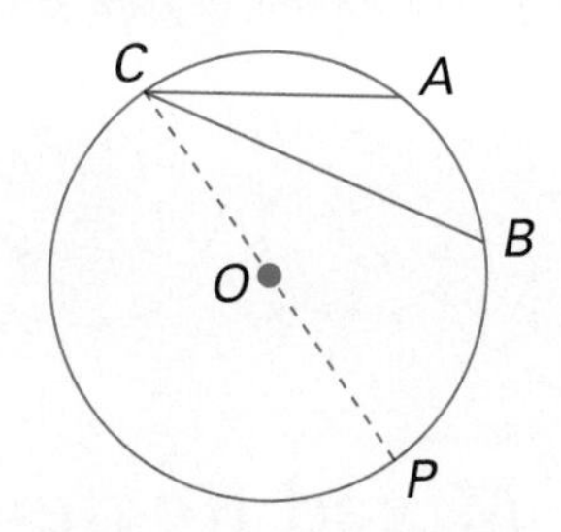

Proof Case 1

Statement	Reason
1. Draw radius $\overline{OA}$.	1. Two pts determine a line.
2. $\overline{OA} \cong \overline{OC}$	2. Radii of the same $\odot$ are $\cong$.
3. $\angle C \cong \angle A$ ($m\angle C = m\angle A$)	3. $\angle$s opp $\cong$ sides of a $\triangle$ are $\cong$.
4. $m\angle C + m\angle A = m\angle 1$	4. Ext Angle Thm

5. $m\angle C + m\angle C = 2(m\angle C) = m\angle 1$	5. Sub
6. $m\angle C = \frac{1}{2}m\angle 1$	6. Div Prop of Equality
7. $m\angle 1 = m\overset{\frown}{AB}$	7. Def meas of minor arc
8. $\therefore m\angle C = \frac{1}{2}m\overset{\frown}{AB}$	8. Sub

Corollary 1

If two inscribed angles intercept the same arc or congruent arcs, then the angles are congruent.

Given: $\odot O$ with inscribed angles A and E, $\overset{\frown}{BC} \cong \overset{\frown}{DF}$

Prove: $\angle A \cong \angle E$

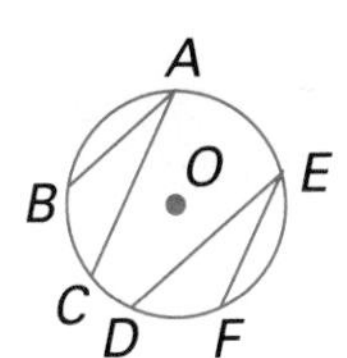

EXAMPLE 1 Given: $m\overset{\frown}{CA} = 50$, $m\overset{\frown}{CB} = 130$

Find $m\angle C$.

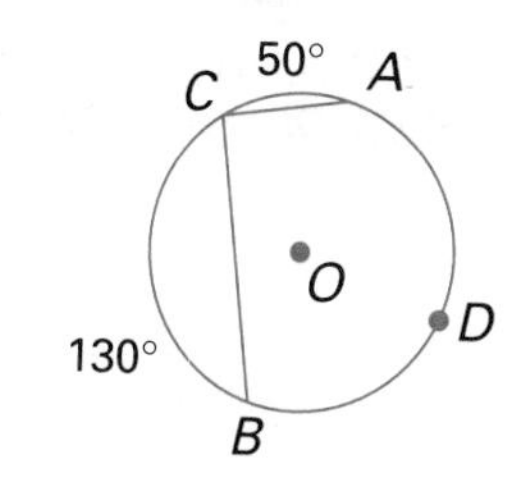

Solution

$$m\overset{\frown}{BC} + m\overset{\frown}{CA} + m\overset{\frown}{ADB} = 360$$
$$130 + 50 + m\overset{\frown}{ADB} = 360$$
$$180 + m\overset{\frown}{ADB} = 360$$
$$m\overset{\frown}{ADB} = 180$$

Therefore, $m\angle C = \frac{1}{2} \cdot 180$, or 90.

Corollary 2

An angle inscribed in a *semicircle* is a *right* angle.

EXAMPLE 2 Given: $\triangle ABC$ inscribed in $\odot O$,

$m\overset{\frown}{AB}:m\overset{\frown}{BC}:m\overset{\frown}{AC} = 3:4:2$

Find $m\angle B$.

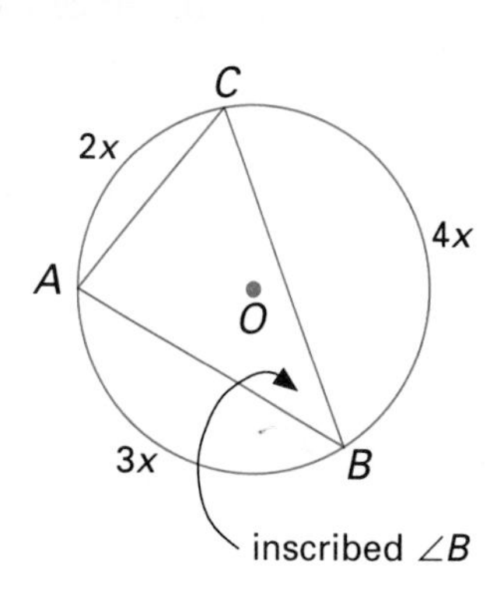

Plan Sketch $\triangle ABC$ inscribed in $\odot O$. Use the ratio of arc measures to find $m\overset{\frown}{AC}$.
Apply the Inscribed Angle Theorem.

Solution Let $m\overset{\frown}{AB} = 3x$, $m\overset{\frown}{BC} = 4x$, and $m\overset{\frown}{AC} = 2x$.

Then, $3x + 4x + 2x = 360$
$$9x = 360$$
$$x = 40$$

Then, $m\overset{\frown}{AC} = 2x = 80$. Thus, $m\angle B = \frac{1}{2} \cdot 80$, or 40.

Corollary 3 If two arcs of a circle are included between parallel chords or secants, then the arcs are congruent.

Proof

Statement	Reason
1. $\overline{SR} \parallel \overline{PQ}$	1. Given
2. Draw $\overline{PR}$.	2. Two pts determine a line.
3. $\angle 1 \cong \angle 2$, $(m\ \angle 1 = m\ \angle 2)$	3. If lines are $\parallel$, alt int $\angle$s are $\cong$.
4. $m\ \angle 1 = \frac{1}{2} m\widehat{SP}$, $m\ \angle 2 = \frac{1}{2} m\widehat{RQ}$	4. Inscr Angle Thm
5. $\frac{1}{2} m\widehat{SP} = \frac{1}{2} m\widehat{RQ}$	5. Sub, Trans
6. $\therefore m\widehat{SP} = m\widehat{RQ}$	6. Mult Prop of Eq

Corollary 4 The opposite angles of an inscribed quadrilateral are supplementary.

EXAMPLE 3 Given: Quadrilateral $ABCD$ inscribed in $\odot O$.
$\overline{AB} \parallel \overline{CD}$, $m\ \angle A = 100$, $m\widehat{DC} = 150$

Find (a) $m\ \angle C$, (b) $m\widehat{BC}$, and (c) $m\widehat{DA}$.

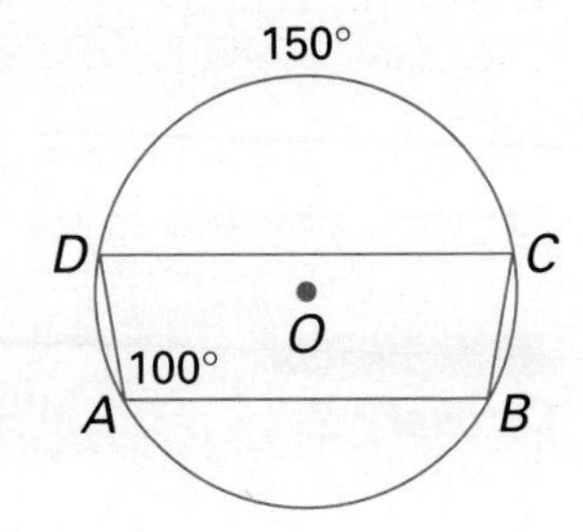

Solutions

a. $m\ \angle C = 180 - 100 = 80$ by Corollary 4.

b. Since $m\ \angle A = 100$, $m\widehat{BCD} = 200$ by the Inscr Angle Thm. Then $m\widehat{BC} = m\widehat{BCD} - m\widehat{DC} = 200 - 150 = 50$.

c. Finally, $m\widehat{DA} = 50$ by Corollary 3.

Thus, $m\ \angle C = 80$, $m\widehat{BC} = 50$ and $m\widehat{DA} = 50$.

Classroom Exercises

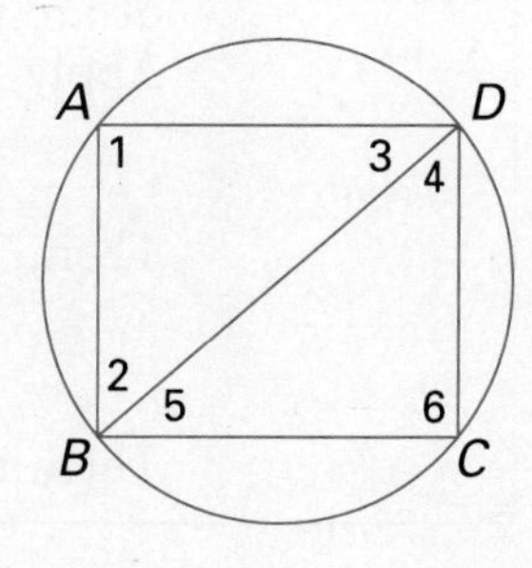

Name the arc intercepted by the indicated inscribed angle.

1. $\angle 1$ **2.** $\angle 2$

3. $\angle 3$ **4.** $\angle 4$

5. $\angle 5$ **6.** $\angle 6$

7. $\angle ABC$ **8.** $\angle ADC$

Written Exercises

Use the figure at the right to find the indicated measure for Exercises 1–9.

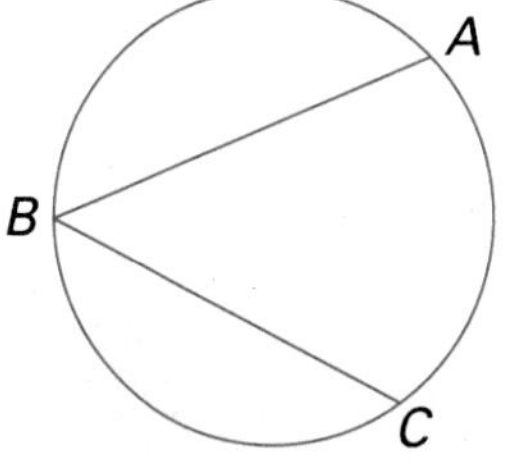

1. $m\widehat{AC} = 36$, $m\angle B =$ ____ 2. $m\widehat{AC} = 200$, $m\angle B =$ ____
3. $m\angle B = 65$, $m\widehat{AC} =$ ____ 4. $m\angle B = 40$, $m\widehat{AC} =$ ____
5. $m\widehat{AC} = x$, $m\angle B =$ ____ 6. $m\angle B = y$, $m\widehat{AC} =$ ____
7. $m\widehat{BA} = 100$, $m\widehat{BC} = 150$, $m\angle B =$ ______
8. $m\widehat{BA} = 110$, $m\widehat{BC} = 200$, $m\angle B =$ ______
9. $AB = CB$, $m\widehat{AB} = 150$, $m\angle B =$ ______

Use the figure at the right to find the indicated measure for Exercises 10–16.

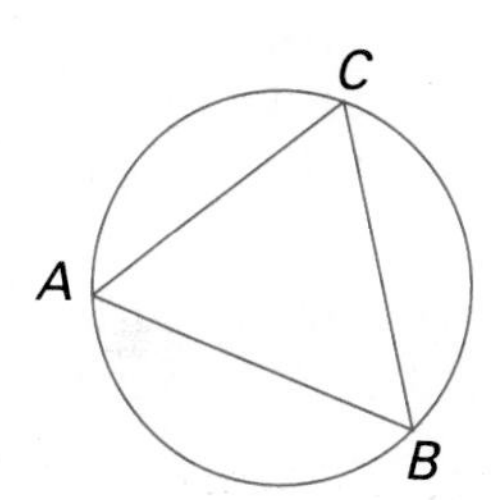

10. $m\widehat{AB}:m\widehat{BC}:m\widehat{AC} = 2:5:3$, $m\angle B =$ ______
11. $m\widehat{AB}:m\widehat{AC}:m\widehat{BC} = 3:2:1$, $m\angle C =$ ______
12. $\triangle ABC$ is equilateral; $m\widehat{AC} =$ ______.
13. $\overline{AB}$ is a diameter; $m\angle C =$ ______.
14. $\overline{AB}$ is a diameter; $m\widehat{AC} = m\widehat{BC}$, $m\angle A =$ ______.
15. $AC = BC$, $m\angle C = 70$, $m\widehat{AC} =$ ______
16. $m\widehat{AC} = 200$, $m\angle C = 30$, $m\widehat{CB} =$ ______

In the figure at the right, $\overline{PQ} \parallel \overline{RS}$. Find the indicated measure for Exercises 17–21.

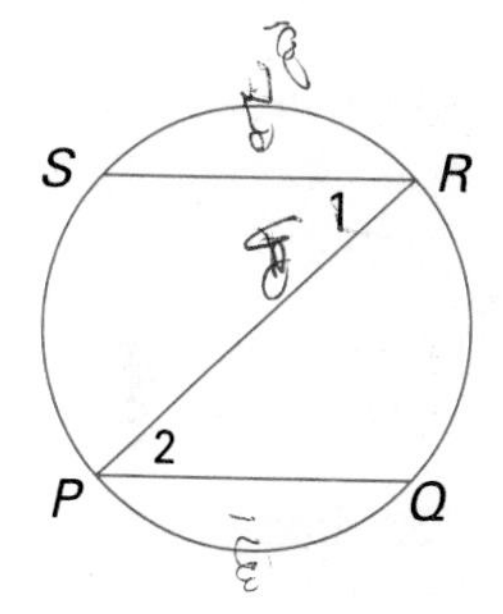

17. $m\widehat{SP} = 90$, $m\angle 2 =$ ______
18. $m\angle 1 = 60$, $m\widehat{RS} = 70$, $m\widehat{PQ} =$ ______
19. $m\widehat{SR} = 100$, $m\widehat{PQ} = 140$, $m\angle 1 =$ ______
20. $m\widehat{PS} = 140$, $m\widehat{SR}:m\widehat{PQ} = 3:2$, $m\widehat{PQ} =$ ______
21. $\overline{PR}$ is a diameter; $m\widehat{PS}:m\widehat{SR} = 2:1$, $m\angle 2 =$ ______.

In the figure at the right, *ABCD* is an inscribed quadrilateral. Find the indicated measure for Exercises 22–26.

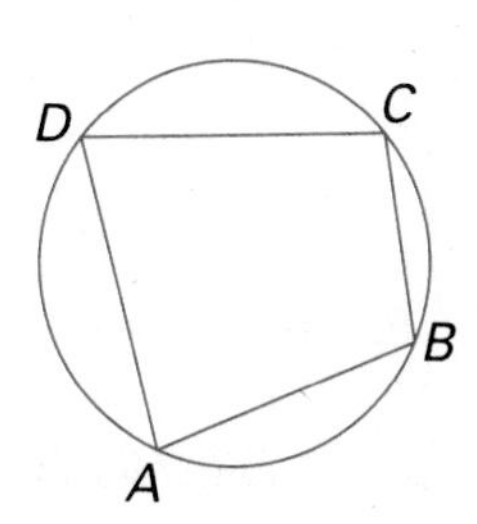

22. $m\angle D = 100$, $m\angle B =$ ______
23. $m\angle C = 80$, $m\angle A =$ ______
24. $m\widehat{DCB} = 200$, $m\angle C =$ ______
25. $m\angle B:m\angle D = 4:5$, $m\angle D =$ ______
26. $m\angle A$ is twice $m\angle C$; $m\angle C =$ ______.

27. Prove Case 2 of Theorem 11.10. (HINT: Apply Case 1 to $\angle ACP$ and $\angle BCP$. Then add the measures.)

28. Prove Case 3 of Theorem 11.10. (HINT: Apply Case 1 to $\angle ACP$ and $\angle BCP$. Then subtract the measures.)

29. Prove Corollary 1. 30. Prove Corollary 2. 31. Prove Corollary 4.

32. Given: $ABCD$ is a trapezoid with $\overline{DC}$ and $\overline{AB}$ as bases.
Prove: Diagonals are congruent.

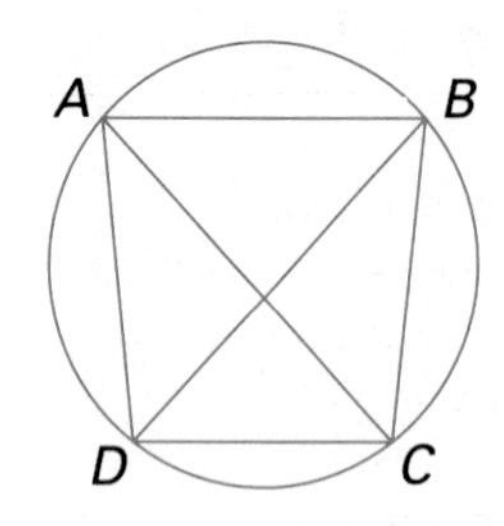

33. Given: m $\angle 1 = \frac{1}{2}$m$\widehat{CD}$, $\overline{BD}$ is a diameter.
Prove: m$\widehat{BA}$ = m$\widehat{BC}$

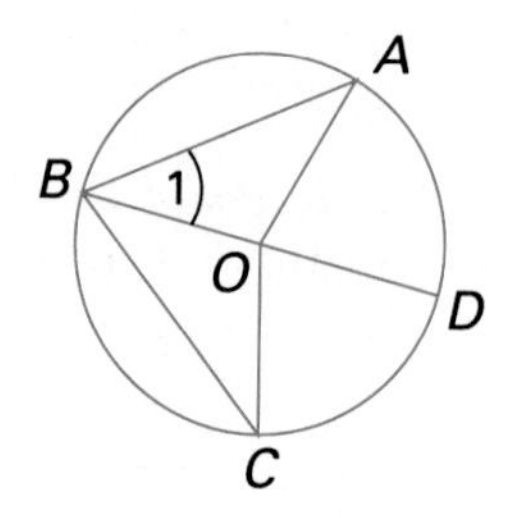

34. Given: $ABCD$ is a parallelogram inscribed in a circle.
Prove that the parallelogram is a rectangle.

35. Prove that a rhombus inscribed in a circle is a square.

36. Prove that if a pair of opposite sides of an inscribed quadrilateral are congruent, then the other pair of sides are parallel.

37. Given: $\triangle ABC$ is an equilateral triangle inscribed in a circle. R and S are midpoints of the arcs $\widehat{AB}$ and $\widehat{BC}$, respectively.
Prove: $ARSC$ is a rectangle.

38. Prove that an equilateral hexagon inscribed in a circle is regular. Does this result extend to all equilateral n-gons (polygons of n sides)?

Midchapter Review

Identify the following for $\odot O$. *11.1*

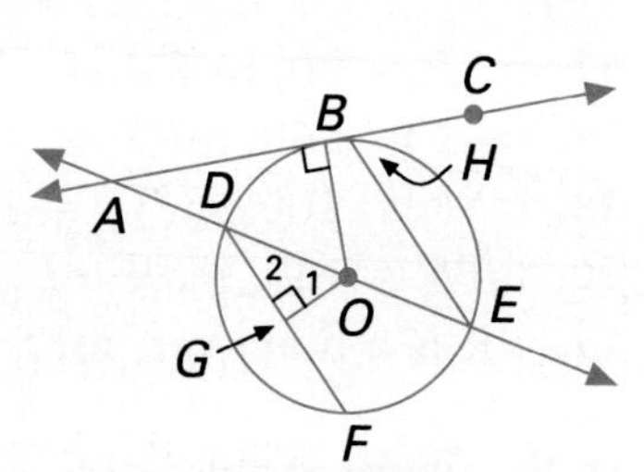

(Ex. 1–12)

1. radius
2. chord
3. diameter
4. secant
5. tangent

6. Given: $OB = 6x - 2$, $OD = 4x + 8$. Find OE. *11.1*

7. Given: $ED = 14$, $OA = 5x + 4$. Find all values of x for which A is in the exterior of the circle. *11.1*

8. Given: $\overline{HE} \parallel \overline{DF}$, m$\widehat{HE} = 70$, m$\widehat{DF} = 100$. Find m$\widehat{FE}$. *11.6*

9. Given: m $\angle 1 = 30$, $OG = 4$. Find the length of the radius. *11.2*

10. Given: m $\angle 2 = 25$. Find m$\widehat{FE}$. *11.6*

11. Given: m$\widehat{DF}$:m$\widehat{FE}$ = 2:3. Find m $\angle 1$. *11.6*

12. Given: $DF = 24$, $OD = 13$. Find OG. *11.2*

13. $\triangle ABC$ is inscribed in a circle. m $\angle A = 50$, m $\angle C = 70$. Name the longest chord of the circle. *11.5*

14. The measure of the angle between two tangents to a circle from P is 60. Find the length of each radius if the length of a tangent segment is 8. *11.3*

11.7 Angles Formed by Secants and Chords

Objective To prove and apply theorems about secants and chords

In this lesson the Inscribed Angle Theorem is used to explore various types of angles related to circles. In the figure at the right, notice that the vertical angles 1 and 2 are neither central nor inscribed. Yet their measures can be found by working from the given measures of their intercepted arcs. In the next figure, an auxiliary segment $\overline{BC}$ has been drawn.

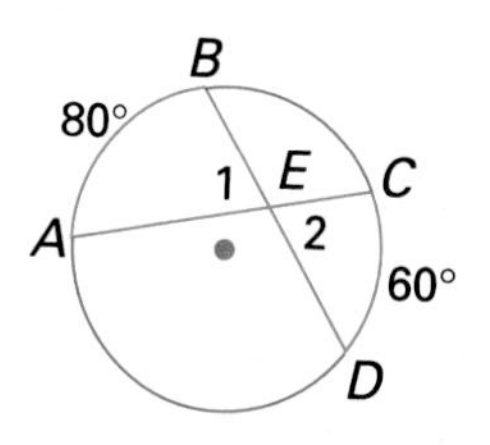

$$\begin{aligned} m\angle 1 &= m\angle B + m\angle C \\ &= \tfrac{1}{2}m\widehat{CD} + \tfrac{1}{2}m\widehat{AB} \\ &= \tfrac{1}{2}(m\widehat{CD} + m\widehat{AB}) \end{aligned}$$

Thus, $m\angle 1 = \frac{1}{2}(60 + 80) = 70$.

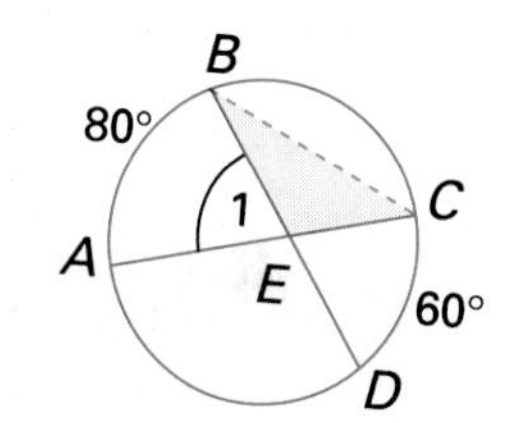

Theorem 11.11 The measure of an angle formed by two secants or chords intersecting in the interior of a circle is one-half the sum of the measures of the arcs intercepted by the angle and its vertical angle.

EXAMPLE 1 Given: $m\angle PTQ = 120$, $m\widehat{PQ} = 100$
Find $m\widehat{RS}$.

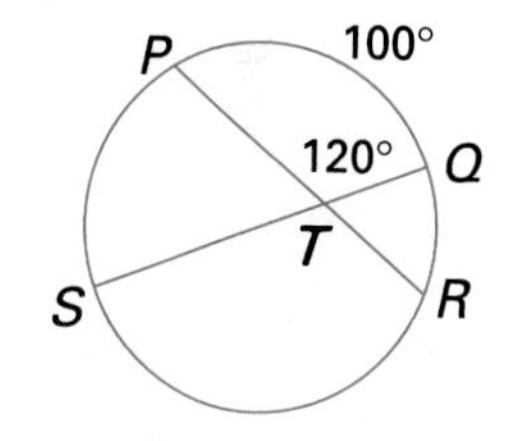

Solution Write an equation based on Theorem 11.11.

$$\begin{aligned} m\angle PTQ &= \tfrac{1}{2}(m\widehat{PQ} + m\widehat{RS}) \\ 120 &= \tfrac{1}{2}(100 + m\widehat{RS}) \\ 240 &= 100 + m\widehat{RS} \\ m\widehat{RS} &= 140 \end{aligned}$$

EXAMPLE 2 Given: $m\widehat{AB} = 3x + 40$, $m\widehat{BC} = 2x + 10$,
$m\widehat{CD} = 2x + 30$, $m\widehat{DA} = 3x - 20$
Find $m\angle DEC$.

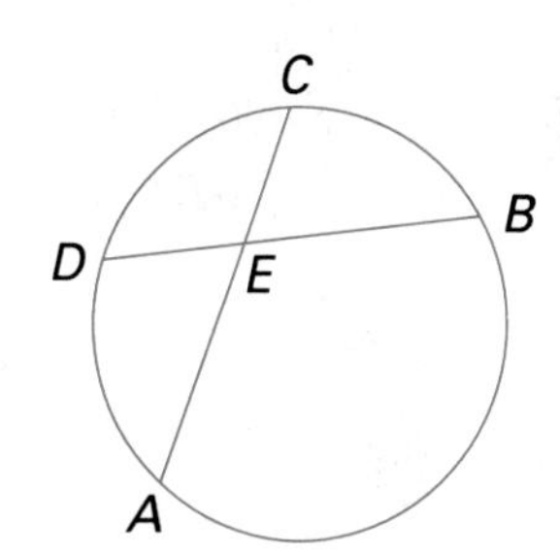

Plan Use the fact that a circle has measure of 360 to write and solve an equation for x. Find $m\widehat{CD}$ and $m\widehat{AB}$, then apply Theorem 11.11.

Solution $m\widehat{AB} + m\widehat{BC} + m\widehat{CD} + m\widehat{DA} = 360$

$$\begin{aligned}(3x + 40) + (2x + 10) + (2x + 30) + (3x - 20) &= 360\\ 10x + 60 &= 360\\ 10x &= 300\\ x &= 30\end{aligned}$$

$$\begin{aligned}m\widehat{CD} &= 2x + 30\\ &= (2 \cdot 30) + 30 = 90\end{aligned} \qquad \begin{aligned}m\widehat{AB} &= 3x + 40\\ &= (3 \cdot 30) + 40 = 130\end{aligned}$$

Thus, $m\angle DEC = \frac{1}{2}(m\widehat{CD} + m\widehat{AB})$
$= \frac{1}{2}(90 + 130) = 110$

In the figure at the right, two secant segments are drawn to $\odot O$ from a point A in the exterior of the circle. Angle A intercepts two arcs, $\widehat{EB}$ and $\widehat{CD}$. The relationship between $\angle A$ and these arcs can be seen by drawing chord $\overline{DB}$. Again, the property of exterior angles of a triangle is applicable.

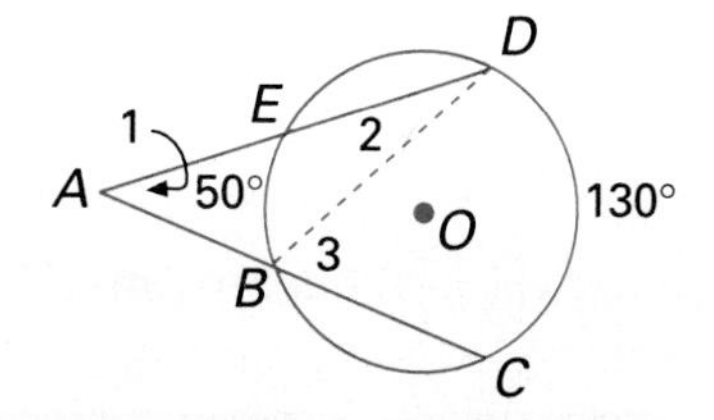

$$\begin{aligned}m\angle 1 + m\angle 2 &= m\angle 3\\ m\angle 1 &= m\angle 3 - m\angle 2\\ m\angle 1 &= \frac{1}{2}\,m\widehat{DC} - \frac{1}{2}\,m\widehat{EB}\\ &= \frac{1}{2}(m\widehat{DC} - m\widehat{EB})\end{aligned}$$

Thus, $m\angle 1 = \frac{1}{2}(130 - 50) = 40$,

or $m\angle 1 = \frac{1}{2}$(*difference* of the measures of the intercepted arcs).

The following theorem is suggested.

Theorem 11.12 The measure of an angle formed by two secants intersecting in the exterior of the circle is one-half the difference of the measures of the intercepted arcs.

EXAMPLE 3 Given: $m\widehat{BC} = 85$, $m\widehat{ED} = 95$,
$m\widehat{BE}:m\widehat{CD} = 4:5$
Find $m\angle A$.

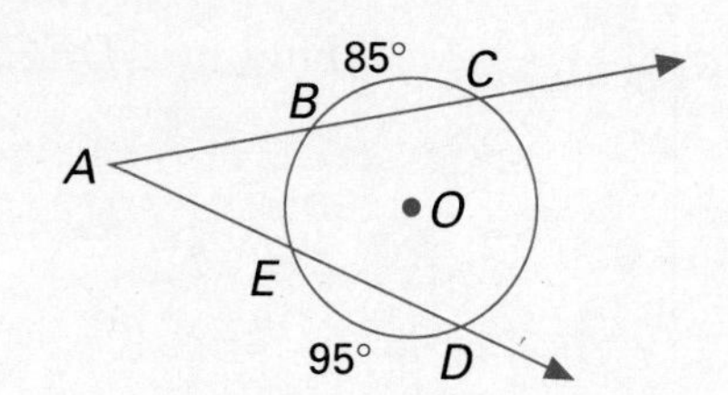

Solution Let $m\widehat{BE} = 4x$, $m\widehat{CD} = 5x$.

Then, $m\widehat{BC} + m\widehat{CD} + m\widehat{ED} + m\widehat{BE} = 360$

$$85 + 5x + 95 + 4x = 360$$
$$9x + 180 = 360$$
$$9x = 180$$
$$x = 20$$

$m\widehat{BE} = 4x = 4 \cdot 20 = 80$ and

$m\widehat{CD} = 5x = 5 \cdot 20 = 100$

$$m\angle A = \frac{1}{2}(m\widehat{CD} - m\widehat{BE})$$
$$= \frac{1}{2}(100 - 80) = 10$$

Classroom Exercises

Find each indicated measure.

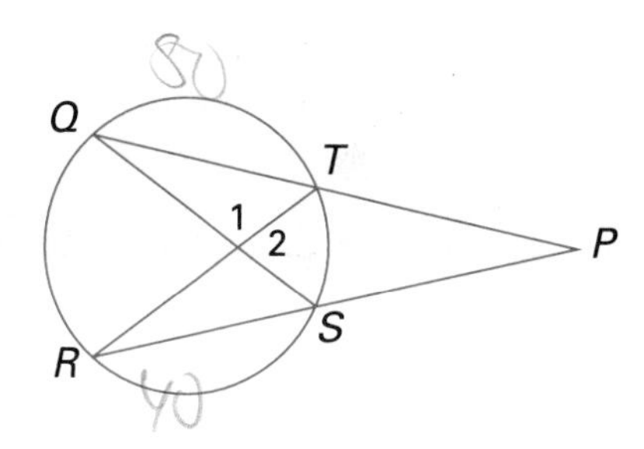

1. $m\widehat{QR} = 100$, $m\widehat{TS} = 60$, $m\angle 2 =$ ______
2. $m\widehat{QT} = 50$, $m\widehat{RS} = 40$, $m\angle 1 =$ ______
3. $m\widehat{QR} = 60$, $m\widehat{TS} = 40$, $m\angle P =$ ______
4. $m\widehat{QR} = 90$, $m\widehat{TS} = 40$, $m\angle P =$ ______
5. $m\widehat{RQ} = m\widehat{QT} = m\widehat{RS} = 100$, $m\angle P =$ ______

Written Exercises

Find each indicated measure.

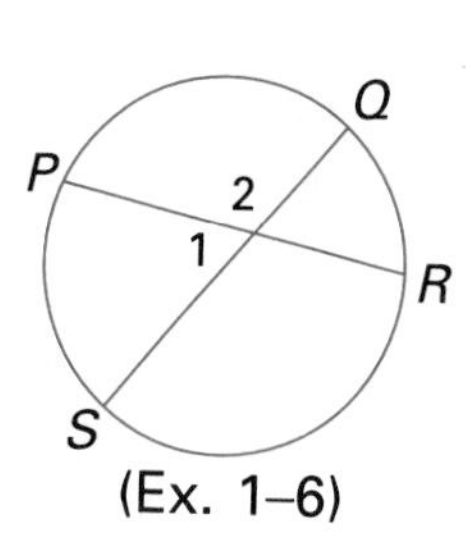

(Ex. 1–6)

1. $m\widehat{PS} = 138$, $m\widehat{QR} = 76$, $m\angle 1 =$ ______
2. $m\widehat{PQ} = 119$, $m\widehat{SR} = 65$, $m\angle 2 =$ ______
3. $m\widehat{PS} = 100$, $m\angle 1 = 80$, $m\widehat{QR} =$ ______
4. $m\widehat{SR} = 240$, $m\angle 2 = 175$, $m\widehat{PQ} =$ ______
5. $m\widehat{SP} = x + 30$, $m\widehat{PQ} = x + 50$, $m\widehat{QR} = 2x + 60$, $m\widehat{RS} = 5x - 50$, $m\angle 1 =$ ______
6. $m\widehat{SP}:m\widehat{PQ}:m\widehat{QR}:m\widehat{RS} = 5:4:3:6$, $m\angle 1 =$ ______

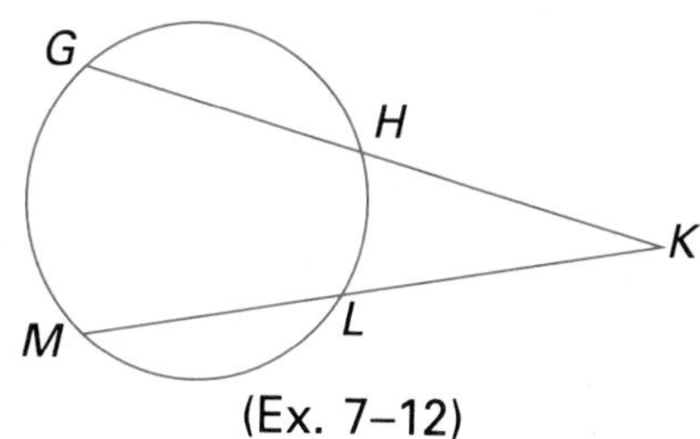

(Ex. 7–12)

7. $m\widehat{GM} = 163$, $m\widehat{HL} = 39$, $m\angle K =$ ______
8. $m\widehat{GM} = 103$, $m\widehat{HL} = 47$, $m\angle K =$ ______
9. $m\widehat{GH} = 90$, $m\widehat{HL} = 80$, $m\widehat{ML} = 100$, $m\angle K =$ ______
10. $m\widehat{GM} = 100$, $m\angle K = 30$, $m\widehat{HL} =$ ______
11. $m\widehat{GM}:m\widehat{HL} = 5:3$, $m\widehat{GH} = 150$, $m\widehat{ML} = 50$, $m\angle K =$ ______
12. $m\widehat{GM} = 3x + 40$, $m\widehat{GH} = 2x - 20$, $m\widehat{HL} = x - 30$, $m\widehat{LM} = 2x + 10$, $m\angle K =$ ______

13. Given: $\overline{AB} \parallel \overline{CD}$, $m\widehat{AD} = 60$
Find m $\angle 1$.

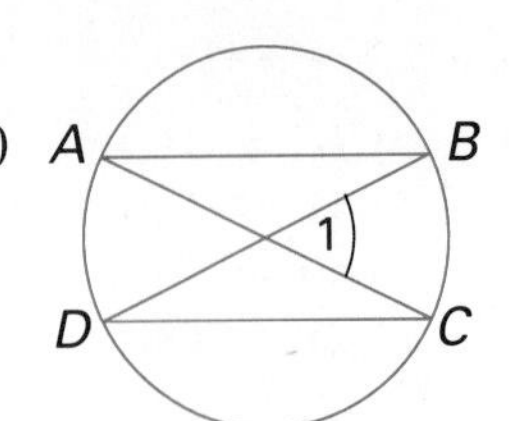

14. Given: diameter $\overline{SQ}$, $m\widehat{SP} = 140$, $m\widehat{RS} = 108$
Find m $\angle 1$.

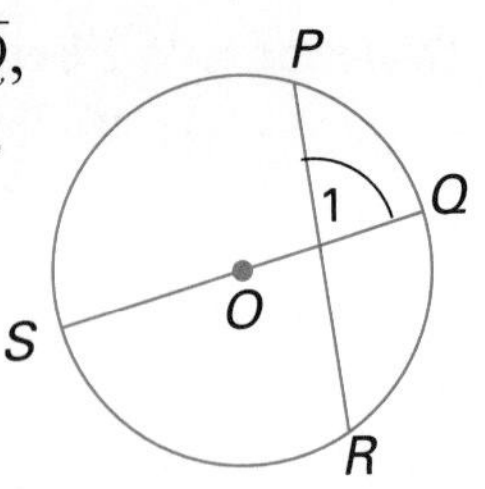

15. Prove Theorem 11.11.

16. Prove Theorem 11.12.

17. Given: $m\widehat{AE} = 140$, m $\angle C = 30$
Find m $\angle 1$.

18. Given: $m\widehat{AE} = 100$, m $\angle 1 = 70$
Find m $\angle C$.

19. Given: m $\angle BAD = 30$, m $\angle ABE = 40$
Find m $\angle C$.

20. Given: $m\widehat{AE}$ is 3 times $m\widehat{BD}$; $m\widehat{BA} = 130$, $m\widehat{ED} = 130$. Find m $\angle C$.

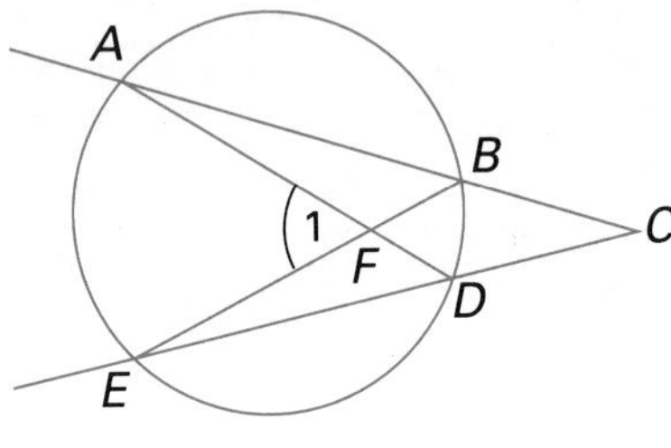

(Ex. 17–20)

Given: $\overline{AB}$ is a diameter; $m\widehat{AF}$:$m\widehat{FE}$:$m\widehat{EB}$ = 3:4:2, $m\widehat{BC} = 140$. Find the following.

21. m $\angle 6$
22. m $\angle 2$
23. m $\angle 5$
24. m $\angle 3$

Given: $\overline{AB}$ is a diameter; $m\widehat{BC} = \frac{2}{3}m\widehat{CA}$, $m\widehat{AF} = 60$, $m\widehat{EB} = 30$. Find the following.

25. m $\angle 7$
26. m $\angle 3$
27. m $\angle 1$
28. m $\angle 4$
29. Given: $m\widehat{AF} = 60$. Would it be reasonable to expect $\angle 4$ to have a measure of 30? Why?

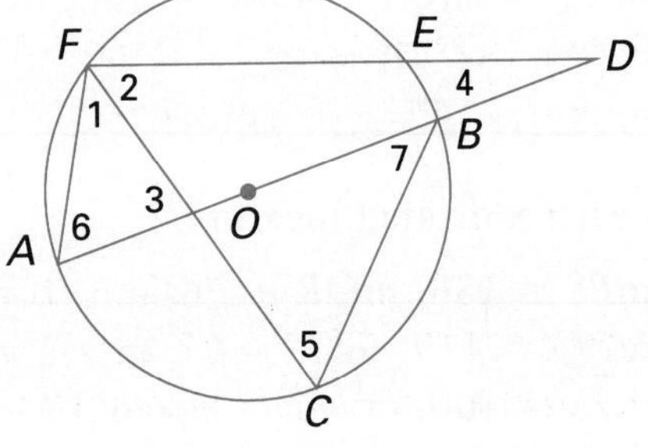

(Ex. 21–29)

30. Two chords $\overline{PQ}$ and $\overline{RS}$ meet at a point Y in the interior of a circle. m $\angle PYR = 54$ and $m\widehat{PR} = 58$. Find $m\widehat{SQ}$.

31. Given: chords $\overline{AB}$ and $\overline{CD}$ of $\odot O$ are perpendicular at point E.
Prove: $m\widehat{BD} + m\widehat{AC} = 180$

32. The photo at the right shows a carpenter's square and a metal disk. How can the carpenter's square be used to find the center of the circular disk? Explain why the method works.

33. Prove: If P is any point in the exterior of $\odot O$, then $m\angle P < m\angle C$.

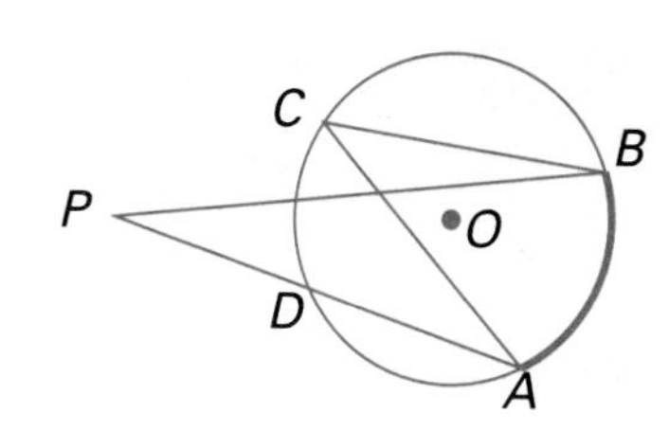

34. The figure at the right shows a ship at point F. The "horizontal danger angle" is used by navigators to chart a course that avoids rocks and shoals close to the shore. Points A and B represent two lighthouses. It is known that within the circle passing through A, B, and C, there are dangerous rocks. If no rocks are located outside the circle, why is the ship safe when $m\angle F < m\angle C$? ($\angle C$ is the "Horizontal danger angle.")

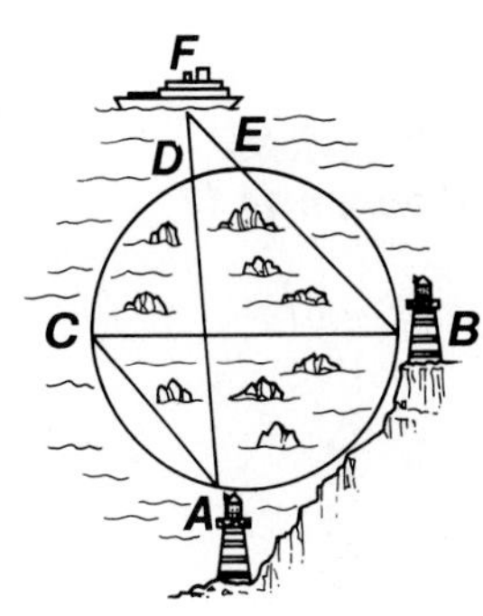

35. Given: $m\angle 1 = 20$, $m\angle 2 = 70$
Find $m\overset{\frown}{DE}$ and $m\overset{\frown}{CB}$. (HINT: Use simultaneous equations.)

36. Given: $\overline{EC}$ is the perpendicular bisector of chord $\overline{DB}$; $m\angle 1 = 40$.
Find $m\overset{\frown}{DE}$ and $m\overset{\frown}{CB}$.

37. Given: $\overline{CD}$ and $\overline{CE}$ are congruent chords in a circle. W is a point on $\overset{\frown}{CE}$; $\overleftrightarrow{CW}$ and $\overleftrightarrow{DE}$ meet at some point Y.
Prove: $\angle CYD \cong \angle CDW$

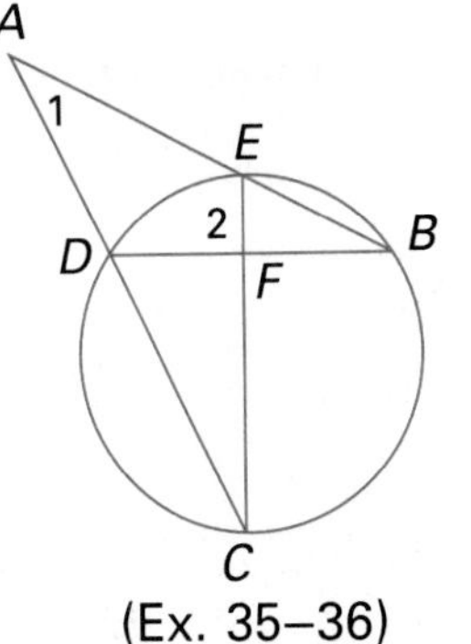

(Ex. 35–36)

Mixed Review

In $\odot O$ at the right, $\overline{AB}$ is a diameter.

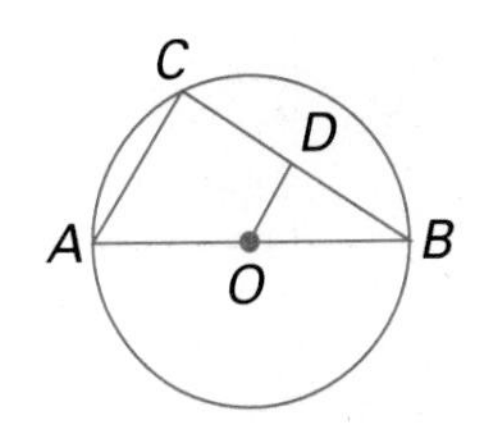

1. What kind of arc is $\overset{\frown}{ACB}$? ***11.4***
2. Find $m\angle C$. ***11.6***
3. $m\overset{\frown}{CB} = 70$, $m\angle A =$ ______ ***11.6***
4. Given: $\overline{OD} \perp \overline{BC}$, $AB = 10$, $BC = 8$. Find OD. ***11.2***

Brainteaser

In the figure at the right, $PQRST$ is an inscribed pentagon and $m\overset{\frown}{RQ} = 80$.
Find $m\angle S + m\angle P$.

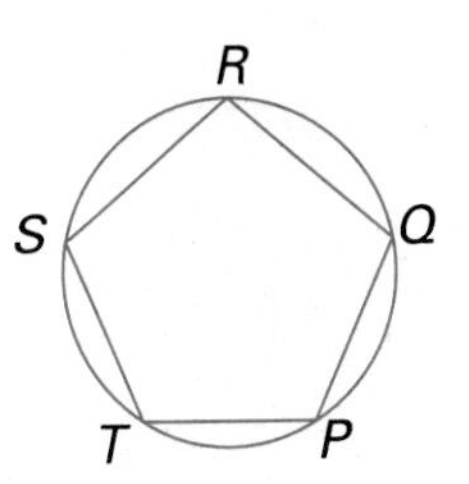

11.8 Angles Formed by Tangents and Secants

Objective To prove and apply theorems about angles formed by tangents and secants or by two tangents

Theorem 11.13 If a tangent and a secant (or a chord) intersect at the point of tangency on a circle, then the measure of the angle formed is one-half the measure of its intercepted arc.

Given: Secant $\overleftrightarrow{AC}$, tangent $\overleftrightarrow{AB}$ intersecting at point of tangency A on $\odot O$.
Prove: $m \angle 1 = \frac{1}{2} m\overset{\frown}{AC}$

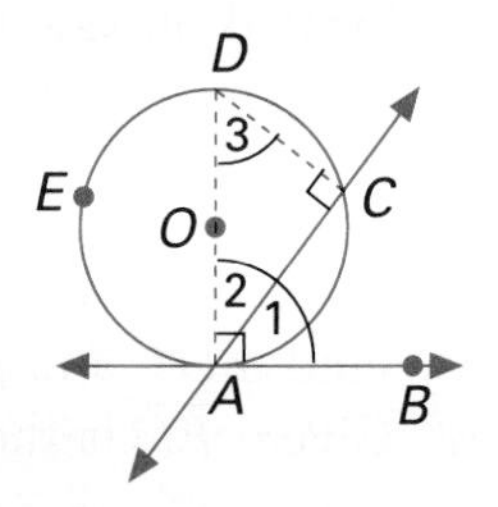

Proof

Statement	Reason
1. Sec $\overleftrightarrow{AC}$ and tan $\overleftrightarrow{AB}$ intersecting at A	1. Given
2. Draw diam $\overline{AD}$. Draw $\overline{DC}$.	2. Two pts determine a line.
3. $\overline{DA} \perp \overleftrightarrow{AB}$	3. Radius is $\perp$ to tan at pt of tangency.
4. $\overset{\frown}{DCA}$ is a semicircle.	4. Def of semicircle
5. $\angle DCA$ is a rt $\angle$.	5. $\angle$ inscr in semicircle is rt.
6. $\triangle DCA$ is a rt $\triangle$.	6. Def of rt $\triangle$
7. $\angle 3$ and $\angle 2$ are comp.	7. Acute $\angle$s of rt $\triangle$s are comp.
8. $\angle 1$ and $\angle 2$ are comp.	8. Acute adj $\angle$s are comp if outer rays are $\perp$.
9. $\angle 3 \cong \angle 1$ ($m \angle 3 = m \angle 1$)	9. Comps of the same $\angle$ are $\cong$.
10. $m \angle 3 = \frac{1}{2} m\overset{\frown}{AC}$	10. Inscr Angle Theorem
11. $\therefore m \angle 1 = \frac{1}{2} m\overset{\frown}{AC}$	11. Sub

EXAMPLE 1 Given: Tangent $\overleftrightarrow{QR}$ and chord $\overline{PQ}$, $m\overset{\frown}{PSQ} = 300$
Find $m \angle PQR$.

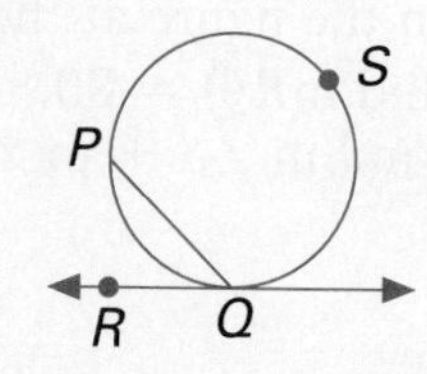

Solution $m\overset{\frown}{PQ} = 360 - 300 = 60$

Then, $m \angle PQR = \frac{1}{2} m\overset{\frown}{PQ} = \frac{1}{2} \cdot 60 = 30$.

The theorem for finding the measure of an angle formed by two secants intersecting at an exterior point to a circle can be extended to two similar cases.

Theorem 11.14 The measure of an angle formed either by

1. a tangent and a secant intersecting at a point exterior to a circle, or
2. two tangents intersecting at a point exterior to a circle equals one-half the *difference* of the measures of the intercepted arcs.

Proof

Case 1

Given: Tangent $\overleftrightarrow{QR}$ and secant $\overleftrightarrow{QS}$ intersecting at point Q

Prove: $m\angle 1 = \frac{1}{2}(m\widehat{SUR} - m\widehat{TR})$

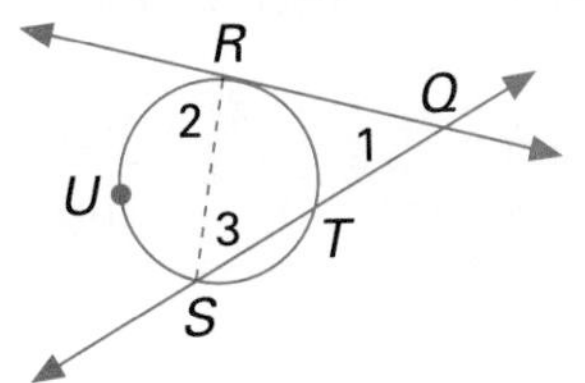

Case 2

Given: Tangents $\overleftrightarrow{QR}$ and $\overleftrightarrow{QS}$ intersecting at point Q

Prove: $m\angle 1 = \frac{1}{2}(m\widehat{RTS} - m\widehat{RS})$

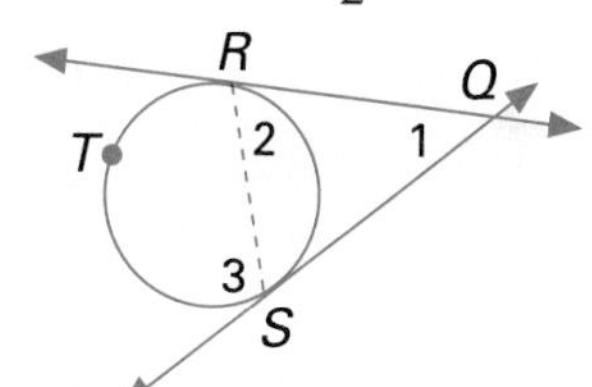

EXAMPLE 2 Given: Tangent $\overleftrightarrow{EA}$, secant $\overleftrightarrow{ED}$, $m\widehat{AB} = 60$, $m\widehat{BCD} = 200$ Find $m\angle E$.

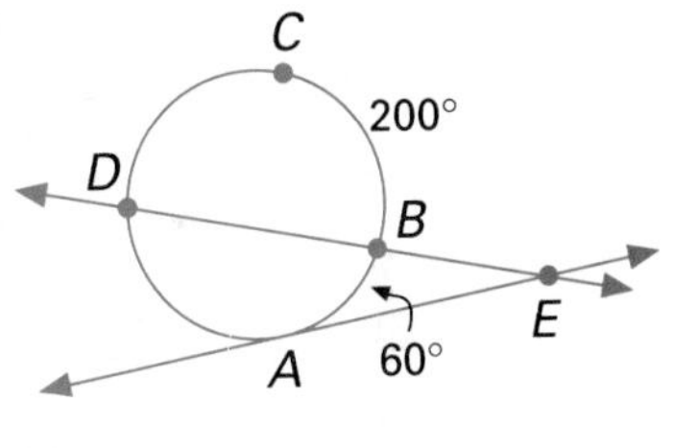

Solution

$m\angle E = \frac{1}{2}(m\widehat{AD} - m\widehat{AB}) = \frac{1}{2}(m\widehat{AD} - 60)$

$m\widehat{AD} = 360 - (200 + 60) = 100$

Therefore, $m\angle E = \frac{1}{2}(100 - 60) = 20$.

EXAMPLE 3 Two tangents intersect at point Q so that $m\angle Q = 40$. Find the measures of the intercepted arcs of $\angle Q$.

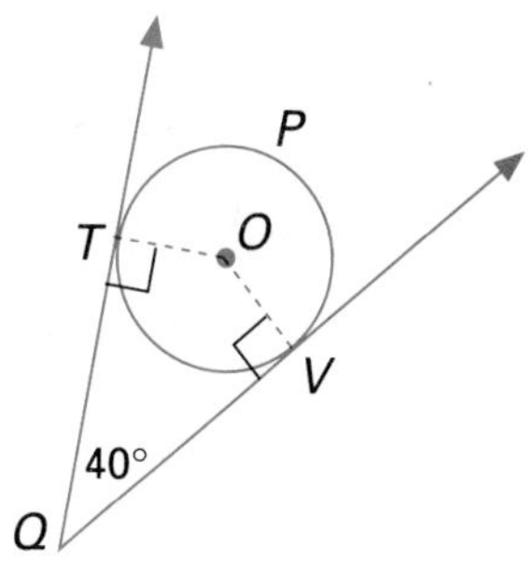

Solution

Let $m\widehat{TV} = x$. Then $m\widehat{TPV} = 360 - x$, since the measure of a circle is 360.

$$m\angle Q = \frac{1}{2}[(360 - x) - x]$$

$$40 = \frac{1}{2}(360 - 2x)$$

$$40 = 180 - x$$

$$-140 = -x, \text{ or } x = 140$$

Thus, $m\widehat{TV} = 140$, and $m\widehat{TPV} = 360 - 140 = 220$.

EXAMPLE 4 Given: Tangent $\overleftrightarrow{AB}$, secant $\overleftrightarrow{AC}$, diameter $\overline{DC}$, $m\overset{\frown}{CF} = 65$, $m\overset{\frown}{GF} = 40$, $m\overset{\frown}{BC} = 135$

Find $m\angle 1$, $m\angle 2$, and $m\angle 3$.

Solution First find the measures of $\overset{\frown}{DG}$ and $\overset{\frown}{BD}$.

$\overline{CD}$ is a diameter. So, $m\overset{\frown}{CGD} = m\overset{\frown}{CBD} = 180$.

$m\overset{\frown}{DG} = 180 - (m\overset{\frown}{CF} + m\overset{\frown}{GF})$

$m\overset{\frown}{DG} = 180 - (65 + 40)$

$m\overset{\frown}{DG} = 75$

$m\overset{\frown}{BD} = 180 - m\overset{\frown}{BC}$

$m\overset{\frown}{BD} = 180 - 135$

$m\overset{\frown}{BD} = 45$

$\angle 1$ is formed by a tangent and a secant with vertex in the exterior.

$m\angle 1 = \frac{1}{2}(m\overset{\frown}{BC} - m\overset{\frown}{BD})$

$m\angle 1 = \frac{1}{2}(135 - 45)$

$m\angle 1 = 45$

$\angle 2$ is an inscribed angle.

$m\angle 2 = \frac{1}{2}m\overset{\frown}{DG}$

$m\angle 2 = \frac{1}{2} \cdot 75$

$m\angle 2 = 37\frac{1}{2}$

$\angle 3$ is formed by two intersecting chords.

$m\angle 3 = \frac{1}{2}(m\overset{\frown}{DG} + m\overset{\frown}{CF})$

$m\angle 3 = \frac{1}{2}(75 + 65)$

$m\angle 3 = 70$

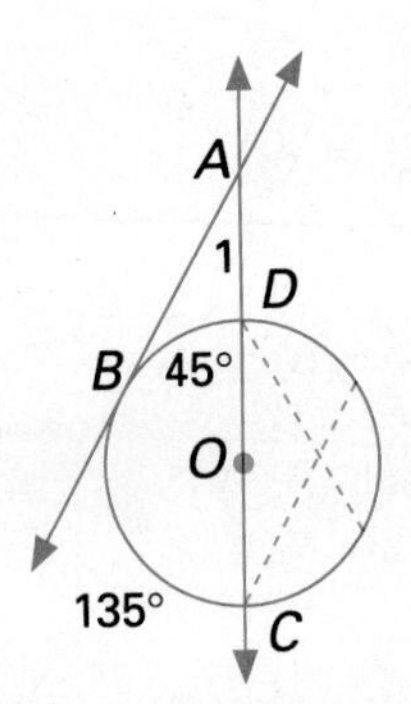

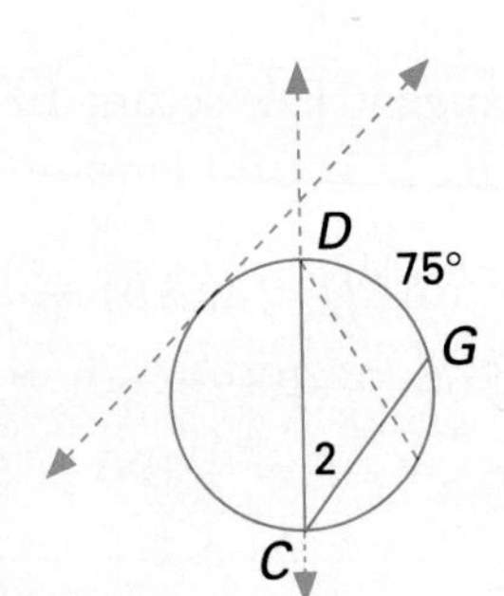

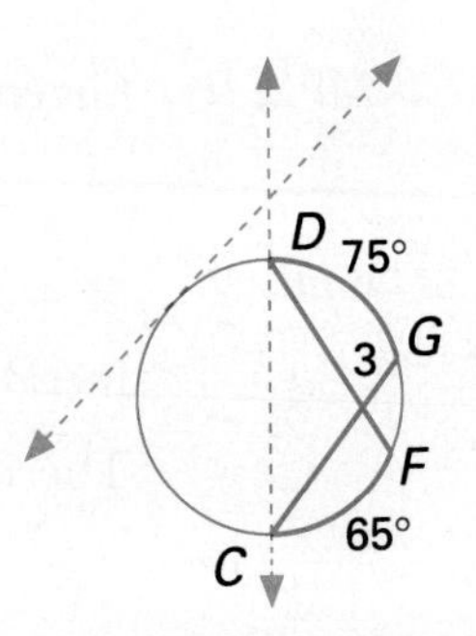

Classroom Exercises

Find the indicated measure.

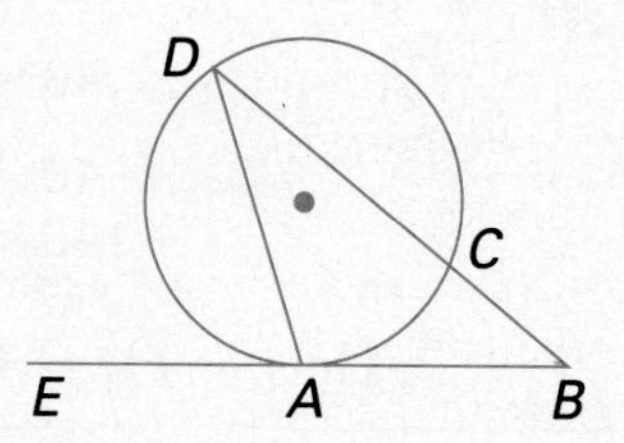

1. $m\overset{\frown}{AC} = 150$, $m\angle D =$ ______
2. $m\overset{\frown}{AD} = 200$, $m\overset{\frown}{AC} = 100$, $m\angle B =$ ______
3. $m\overset{\frown}{AD} = 40$, $m\angle EAD =$ ______
4. $m\overset{\frown}{AD} = x$, $m\overset{\frown}{AC} = y$, $m\angle B =$ ______

5. $m\overset{\frown}{PT} = 80$, $m\overset{\frown}{SQ} = 60$, $m\angle R =$ ______
6. $m\overset{\frown}{SQ} = 90$, $m\angle 1 =$ ______
7. $m\overset{\frown}{PT} = 70$, $m\overset{\frown}{SQ} = 30$, $m\angle 3 =$ ______

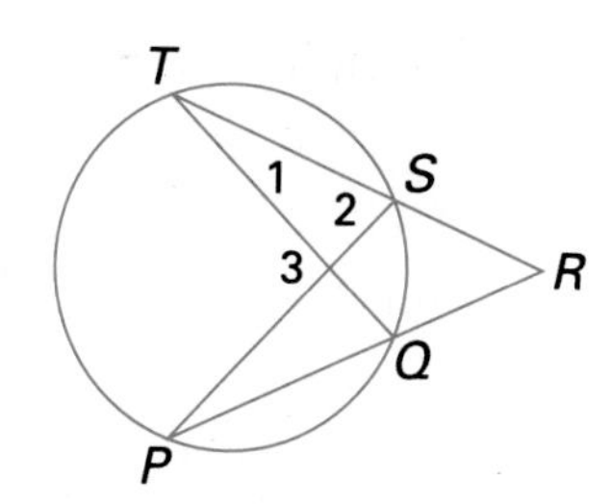

Written Exercises

1. Given: Tangent $\overleftrightarrow{AB}$, $m\overset{\frown}{BDC} = 240$
   Find $m\angle ABC$.

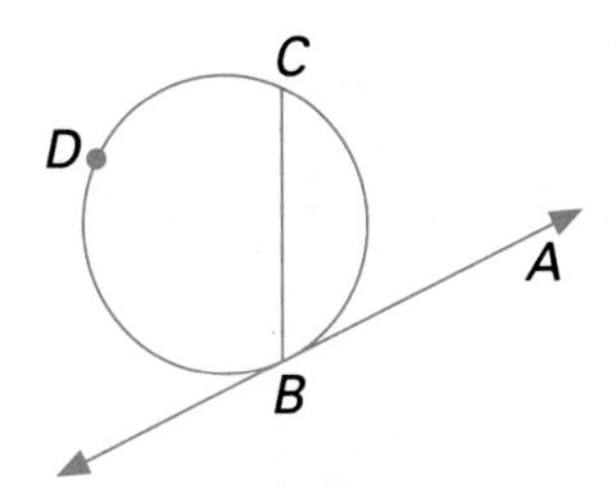

2. Given: Tangent $\overleftrightarrow{PS}$, $m\overset{\frown}{RQ} = 160$,
   $m\overset{\frown}{RS} = 60$
   Find $m\angle P$.

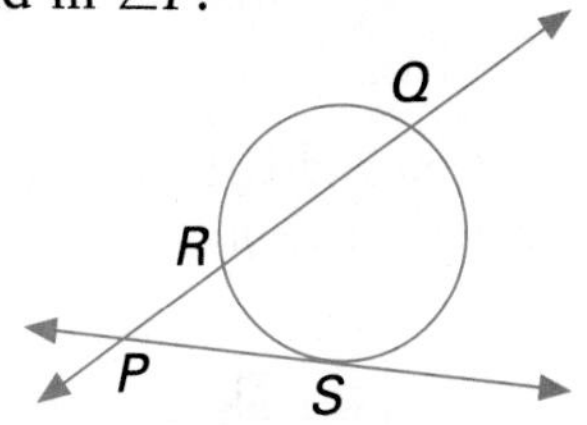

3. Given: Tangents $\overleftrightarrow{AB}$ and $\overleftrightarrow{AC}$,
   $m\overset{\frown}{BC}:m\overset{\frown}{BDC} = 1:2$
   Find $m\angle 1$.

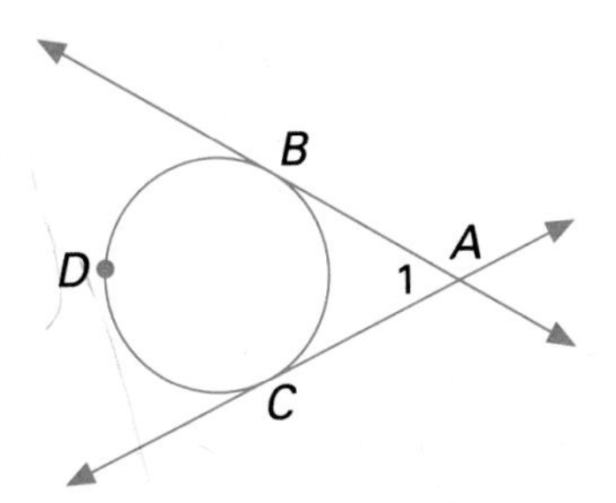

4. Given: Tangent $\overleftrightarrow{RS}$ and $\overleftrightarrow{RT}$,
   $m\overset{\frown}{SUT} = 235$
   Find $m\angle SRT$.

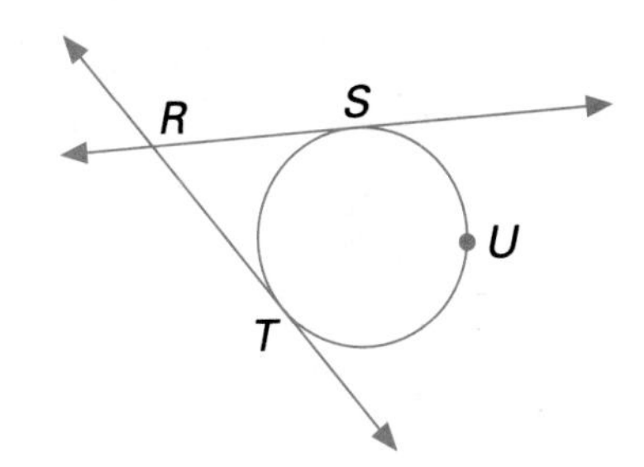

In the figure at the right, $\overleftrightarrow{AB}$ and $\overleftrightarrow{AC}$ are tangents.

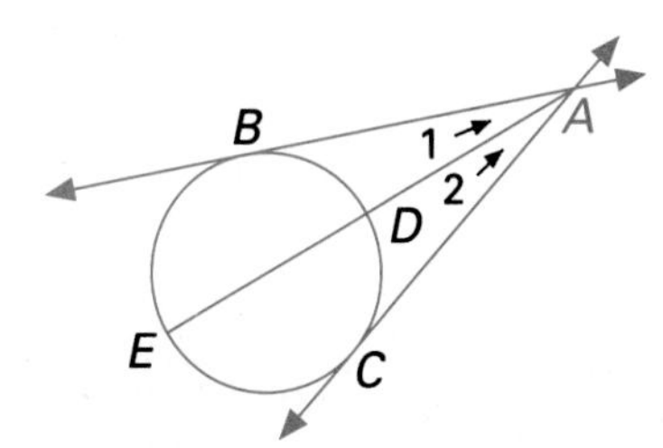

5. $m\overset{\frown}{EB} = 60$, $m\overset{\frown}{BD} = 40$, $m\angle 1 =$ ______
6. $m\overset{\frown}{EC} = 100$, $m\angle 2 = 35$, $m\overset{\frown}{DC} =$ ______
7. $m\overset{\frown}{EC} = 80$, $m\angle 2 = 5$, $m\overset{\frown}{DC} =$ ______
8. $m\overset{\frown}{BEC} = 250$, $m\angle BAC =$ ______
9. $m\overset{\frown}{EB}:m\overset{\frown}{BD}:m\overset{\frown}{DC}:m\overset{\frown}{CE} = 5:4:4:5$, $m\angle 1 =$ ______
10. The measure of an angle formed by two tangents to a circle is 70. Find the measure of the central angle formed by radii to the points of tangency.
11. A central angle measures 120. Find the measure of the angle formed by two tangents to the circle at the intersections of the central angle and the circle.

In the figure at the right, $\overleftrightarrow{AF}$ is a tangent.

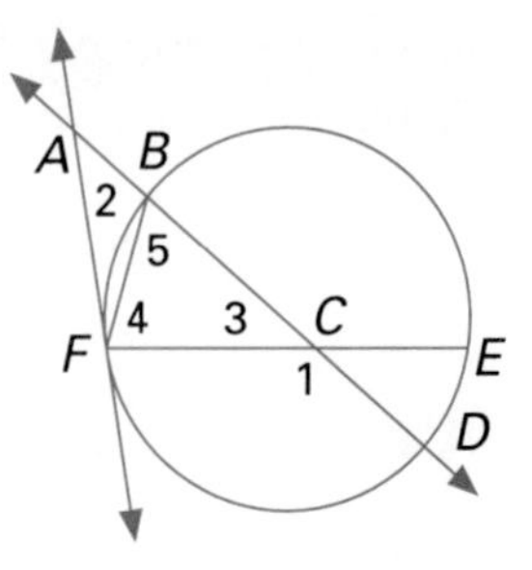

12. m $\angle 2 = 35$, m$\widehat{FB} = 45$, m$\widehat{FD}$ = ______

13. m$\widehat{FD} = 140$, m $\angle 1 = 125$, m$\widehat{BE}$ = ______

14. m$\widehat{FB} = 115$, m $\angle 3 = 95$, m$\widehat{DE}$ = ______

15. $\overline{BD}$ is a diameter; $\dfrac{m\widehat{FB}}{m\widehat{FD}} = \dfrac{4}{5}$, m $\angle 2$ = ______.

16. $\overline{FE}$ is a diameter; m$\widehat{BE} = 120$, m$\widehat{FD} = 80$, m $\angle 2$ = ______.

17. m $\angle 2 = 30$, m$\widehat{FD} = 140$, m$\widehat{DE} = 50$, m $\angle 3$ = ______

18. m $\angle 5 = 65$, m$\widehat{FB} = 60$, m $\angle 2$ = ______

19. $\overline{FE}$ is a diameter; m$\widehat{BE} = 160$, m $\angle 1 = 155$, m $\angle 2$ = ______.

20. Prove Case 1 of Theorem 11.14.

21. Prove Case 2 of Theorem 11.14.

In the figure at the right $\overleftrightarrow{AC}$ is a tangent, $\overline{BD} \parallel \overline{FG}$, $\overline{BD}$ is a diameter. Find the indicated measure.

22. m$\widehat{DG} = 50$, m $\angle 2$ = ______

23. m $\angle 5 = 70$, m $\angle 1$ = ______

24. m $\angle 1 = 30$, m $\angle 3 = 105$, m$\widehat{BF}$ = ______

25. m $\angle 1 = 50$, m $\angle 6 = 20$, m $\angle 7$ = ______

26. m $\angle 5 = 85$, m $\angle 1$ = ______

27. m $\angle 4 = 30$, m $\angle 1 = 60$, m$\widehat{FDE}$ = ______

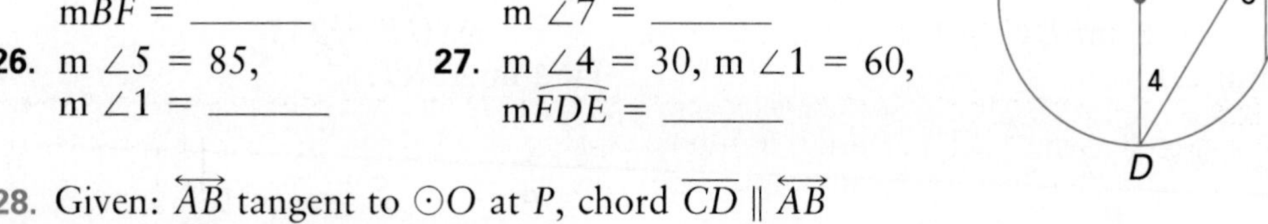

28. Given: $\overleftrightarrow{AB}$ tangent to $\odot O$ at P, chord $\overline{CD} \parallel \overleftrightarrow{AB}$
Prove: $\widehat{CP} \cong \widehat{DP}$

29. Prove: If two tangents to a circle are parallel, then the points of tangency divide the circle into two congruent arcs.

30. Triangle PQR is circumscribed about a circle. m $\angle P$:m $\angle Q$:m $\angle R = 3:2:1$. The three points of tangency are joined to form an inscribed triangle. Find the measure of each angle of the inscribed triangle.

31. From a point P exterior to $\odot O$, two tangents are drawn intersecting the circle at points A and B. Prove that m $\angle APB$ + m$\widehat{AB} = 180$.

Mixed Review

1. Find the sum of the measures of the interior angles of a decagon. ***6.2***
2. Given: Parallelogram $ABCD$, m $\angle A = 120$. Find m $\angle B$. ***6.4***
3. Find the length of a leg of an isosceles right triangle if the length of the hypotenuse is 6. ***9.2***
4. Find the measure of each base angle of an isosceles triangle if the measure of the vertex angle is 50. ***5.1***

11.9 Lengths of Segments Formed by Secants, Chords, and Tangents

Objective To find lengths of segments related to chords, secants, and tangents

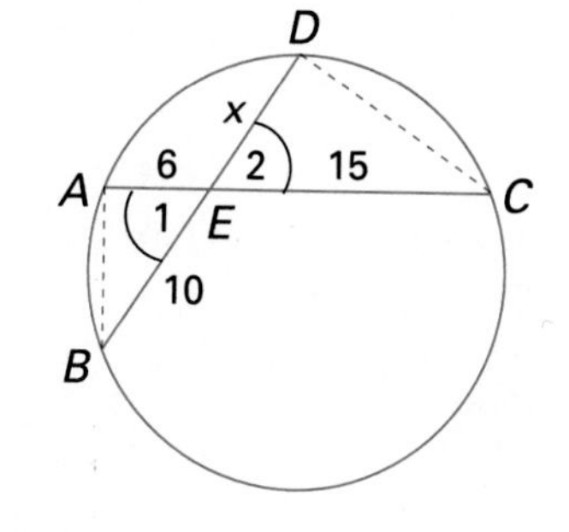

Besides the measures of angles and arcs, the lengths of various segments relating to circles can also be determined. In the figure at the right, it can be shown by AA Similarity that $\triangle ABE \sim \triangle DCE$. First, $\angle 1 \cong \angle 2$ by the vertical angles property; then, $\angle A \cong \angle D$ (or $\angle B \cong \angle C$), because inscribed angles intercepting the same arc are congruent. As a result:

$$\frac{BE}{EC} = \frac{AE}{ED} \longrightarrow \frac{10}{15} = \frac{6}{x}$$

$$BE \cdot ED = EC \cdot AE \longrightarrow 10x = 90$$

$$x = 9$$

Thus, $ED = 9$. It is also seen that the *product* of the lengths of the segments of one chord equals the *product* of the lengths of the segments of the other chord. This suggests the following theorem.

Theorem 11.15 If two chords of a circle intersect, then the product of the lengths of the segments of one chord equals the product of the lengths of the segments of the other chord.

Given: Chords $\overline{AC}$ and $\overline{BD}$ intersect at E.
Prove: $BE \cdot ED = AE \cdot EC$

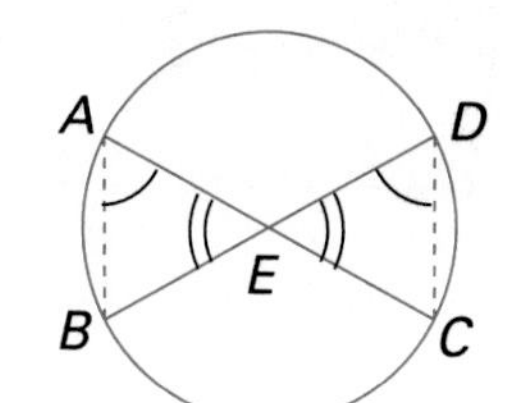

EXAMPLE 1 Given: $BE = 3(EA)$, $DC = 7$, $EC = 3$
Find EA.

Solution Let $EA = x$; then $BE = 3x$.

$EA \cdot BE = ED \cdot EC$ ⟵ (Theorem 11.15)

$x \cdot 3x = (7 - 3) \cdot 3$

$3x^2 = 12$

$x^2 = 4$

$x = \sqrt{4} = 2$

(Use the principal square root.)

Thus, $EA = 2$.

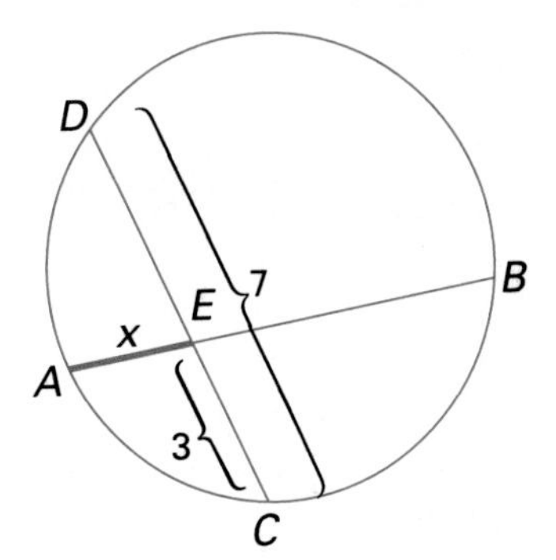

In the figure below, $\overline{PQ}$ is a *tangent* segment, $\overline{PS}$ is a *secant* segment, and $\overline{PR}$ is an *external secant* segment.

Theorem 11.16 If a tangent and a secant intersect in the exterior of a circle, then the square of the length of the tangent segment equals the product of the lengths of the secant segment and the external secant segment.

Given: $\odot O$, tangent $\overleftrightarrow{PQ}$ and secant $\overleftrightarrow{PS}$ intersecting at point P
Prove: $(PQ)^2 = PS \cdot PR$

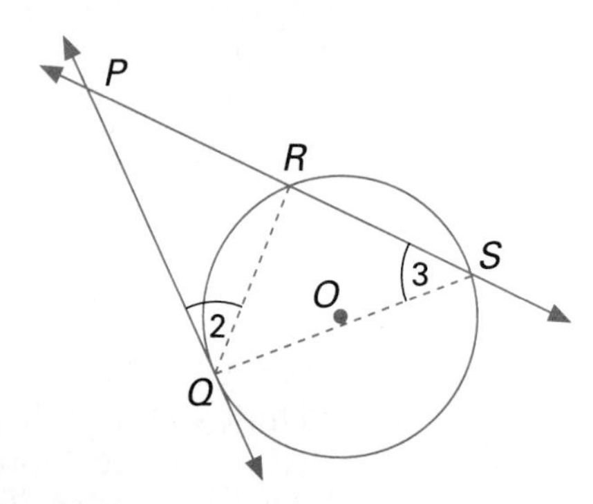

Plan m $\angle 2$ and m $\angle 3$ each equal $\frac{1}{2}\text{m}\overparen{QR}$, so m $\angle 2$ = m $\angle 3$. Also, $\angle P \cong \angle P$. Therefore $\triangle PQR \sim \triangle PSQ$ by AA Similarity. Then, by similar triangles, $\frac{PQ}{PS} = \frac{PR}{PQ}$, and by a proportion property $PQ \cdot PQ = PS \cdot PR$, or $(PQ)^2 = PS \cdot PR$.

In the figure at the right, $\overleftrightarrow{AC}$ and $\overleftrightarrow{AE}$ are *secants* and $\overleftrightarrow{AF}$ is a *tangent*. By Theorem 11.16:

$$(AF)^2 = AC \cdot AB,$$
$$\text{and } (AF)^2 = AE \cdot AD.$$

So, by substitution, $AC \cdot AB = AE \cdot AD$. This result is stated as a corollary to Theorem 11.16.

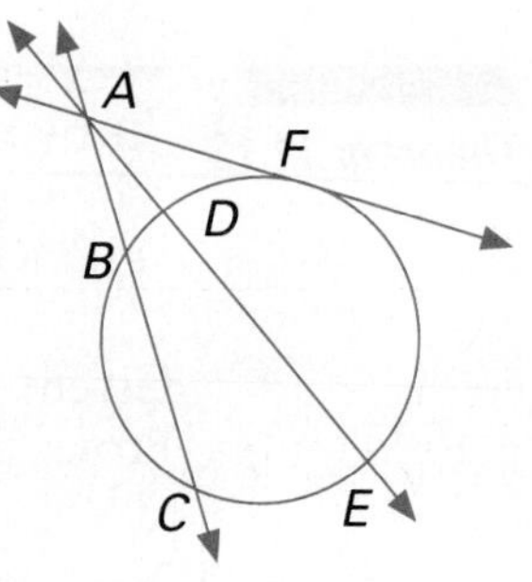

Corollary If two secants intersect in the exterior of a circle, then the product of the lengths of one secant segment and its external segment equals the product of the lengths of the other secant segment and its external segment.

Given: $\odot O$ with secants $\overleftrightarrow{BC}$ and $\overleftrightarrow{DE}$ intersecting at point A
Prove: $AC \cdot AB = AE \cdot AD$

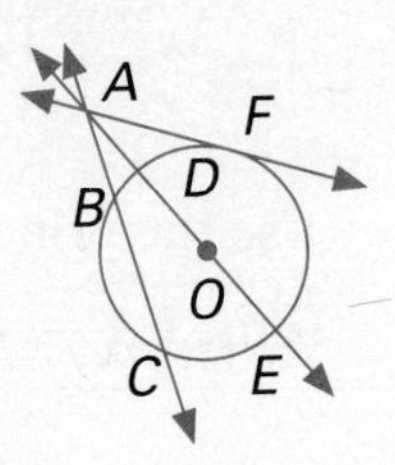

EXAMPLE 2 Given: Tangent $\overleftrightarrow{AB}$, secant $\overleftrightarrow{AD}$, $AC = 4$, $CD = 8$
Find AB.

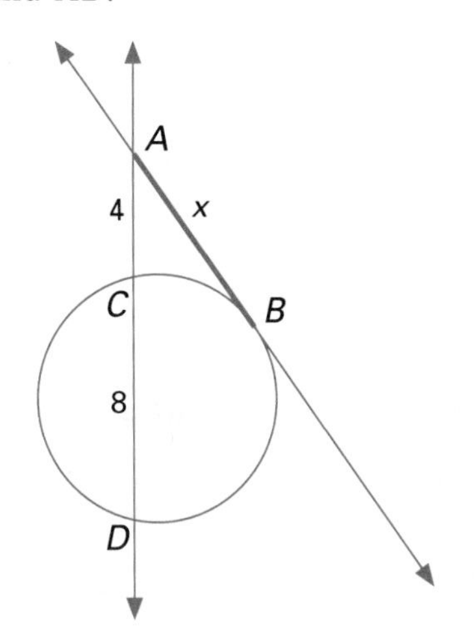

Given: Secants $\overleftrightarrow{PT}$ and $\overleftrightarrow{PR}$, $PQ = 6$, $PS = 4$, $TS = 10$
Find QR.

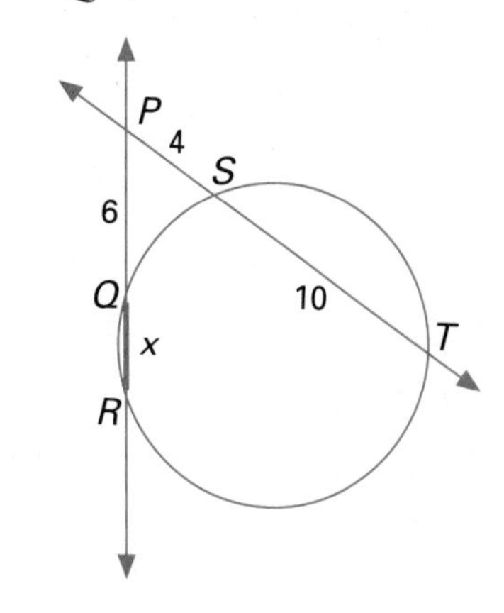

Solution Let $x = AB$.

Then by Theorem 11.16:

$$(AB)^2 = AD \cdot AC$$
$$x^2 = (4 + 8)4$$
$$x^2 = 48$$
$$x = \sqrt{48} = \sqrt{16} \cdot \sqrt{3} = 4\sqrt{3}$$

Thus, $AB = 4\sqrt{3}$.

Let $x = QR$.

Then by Corollary 1:

$$PR \cdot PQ = PT \cdot PS$$
$$(x + 6)6 = (4 + 10)4$$
$$6x + 36 = 56$$
$$6x = 20$$
$$x = 3\tfrac{1}{3}$$

Thus, $QR = 3\frac{1}{3}$.

EXAMPLE 3 From exterior point A, a tangent segment $\overline{AB}$ is drawn to $\odot O$ meeting the circle at B. A secant $\overline{AC}$ is drawn, meeting the circle at D and C. If $AB = 6$ and $DC = 5$, find AD.

Plan First draw the figure.
Let $AD = x$.

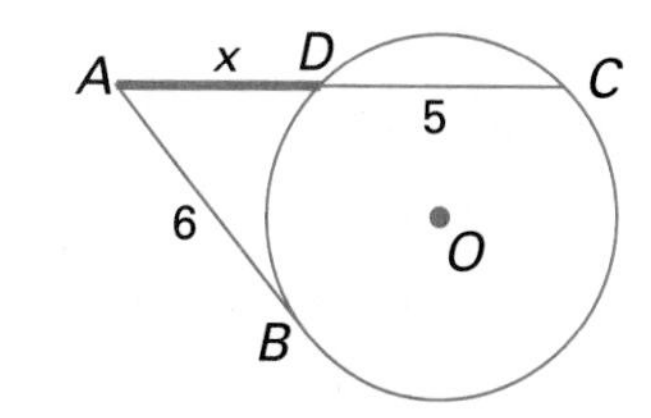

Solution

$$(AB)^2 = AD \cdot AC \text{ by Theorem 11.16}$$
$$6^2 = x(x + 5)$$
$$36 = x^2 + 5x$$

$$x^2 + 5x - 36 = 0$$
$$(x + 9)(x - 4) = 0$$
$$x + 9 = 0 \quad or \quad x - 4 = 0$$
$$x = -9 \quad or \quad x = 4$$

Since $x = AD$, a length, $x = 4$ is the only possible solution.
Therefore, $AD = 4$.

Classroom Exercises

Write an equation you could use to find the value of x.

1.

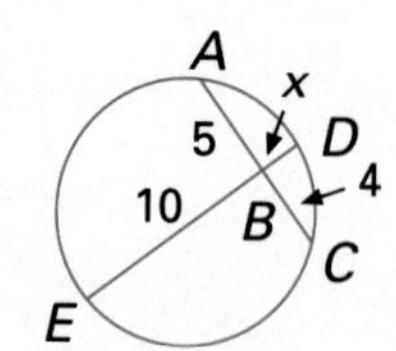

2.

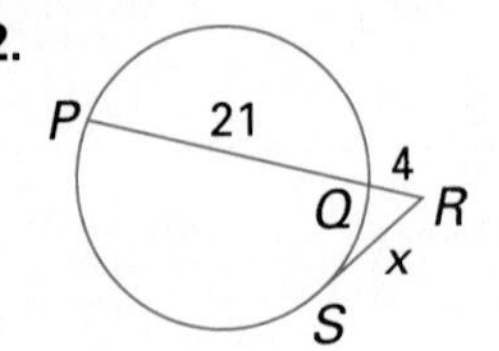

3.

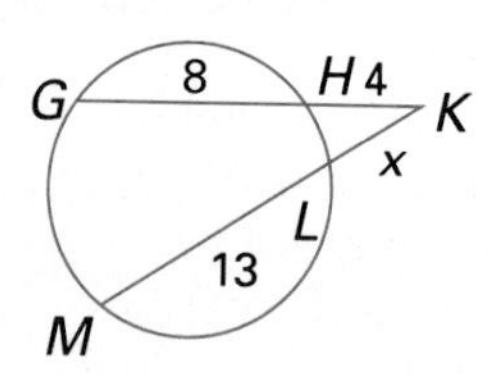

Written Exercises

For Exercises 1–10, assume that lines that appear to be tangent are tangent.

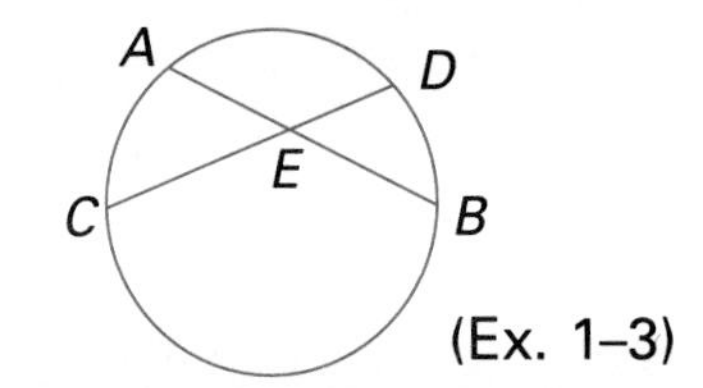

(Ex. 1–3)

1. Given: $AE = 8$, $EB = 10$, $DE = 5$
   Find EC.
2. Given: $DE = 4(EC)$, $AE = 4$, $EB = 25$
   Find DE.
3. Given: $DE = 2(EC)$, $AE = 2$, $BE = 10$
   Find DE.

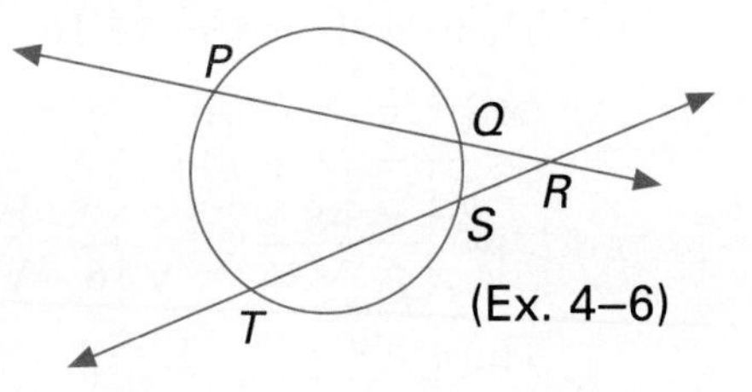

(Ex. 4–6)

4. Given: $PQ = 7$, $QR = 8$, $RS = 2$
   Find TS.
5. Given: $PR = 9$, $PQ = 1$, $TR = 24$
   Find TS.
6. Given: $PQ = 3$, $QR = 5$, $SR = 4$
   Find TS.

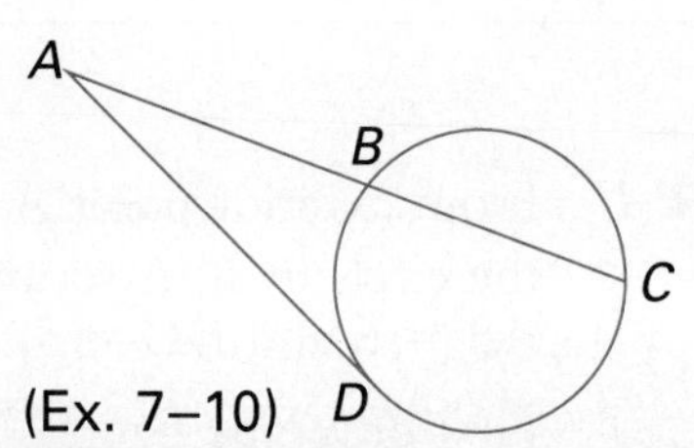

(Ex. 7–10)

7. Given: $AB = 4$, $AC = 10$
   Find AD.
8. Given: $AB = 4$, $BC = 16$
   Find AD.
9. Given: $AD = 4$, $AC = 2(AB)$
   Find AB.
10. $AD = 8$, $BC = 12$. Find AB.
11. Two chords $\overline{AB}$ and $\overline{CD}$ intersect at a point E in the interior of a circle. $BE = 15.4$, $DE = 8.5$, $EC = 3.1$. Find AE to the nearest tenth.
12. From exterior point P, tangent $\overline{PT}$ is drawn to $\odot O$ meeting the circle at T. A secant $\overline{PB}$ is drawn meeting the circle at A and B. $PT = 6$ and $AB = 9$. Find PB.
13. Two chords $\overline{PO}$ and $\overline{RS}$ of a circle meet, when extended through O and S, at point T. $PO = 8$, $RS = 22$, $ST = 2$. Find OT.
14. Two chords $\overline{AB}$ and $\overline{CD}$ intersect at a point E in the interior of a circle. $AB = 16$, $DE = 12$, $EC = 4$. Find AE.
15. Prove Theorem 11.15.
16. Prove Theorem 11.16.
17. Prove the corollary to Theorem 11.16.

18. From exterior point P, tangent $\overline{PA}$ is drawn to $\odot O$ meeting the circle at A. From P, secant $\overleftrightarrow{PR}$ is drawn meeting the circle at M and R and passing through O. $\odot O$ has a radius of length 4. $PA = 3$. Find PR.

19. A piece of a broken gear is brought to a machinist. In order to build a new gear, he must first compute the diameter of the original gear. He finds the distance AB to be 8 in. He then measures ED, which is the perpendicular distance from E (the midpoint of $\overline{AB}$) to $\widehat{AB}$. $DE = 2$ in. Find the diameter.

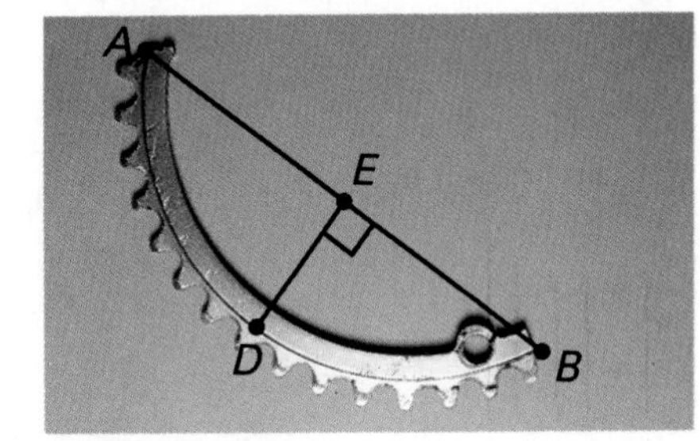

20. From exterior point T, tangent $\overline{TQ}$ and secant $\overline{TS}$ are drawn to a circle meeting the circle at Q, R, and S. The two segments are at right angles to each other. $TQ = 4$ and $TR = 2$. Find the length of a radius of the circle.

21. Given: $\overleftrightarrow{PR}$ is tangent to $\odot O$ at Q; $\overline{VU} \parallel \overline{PR}$.
Prove: $VQ \cdot QS = QU \cdot TQ$

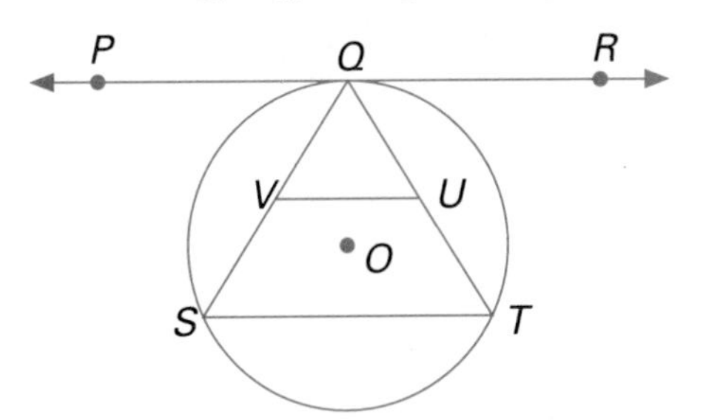

22. Given: $AB = 16$, $BC = 6$, $CD = 8$
Find BE.

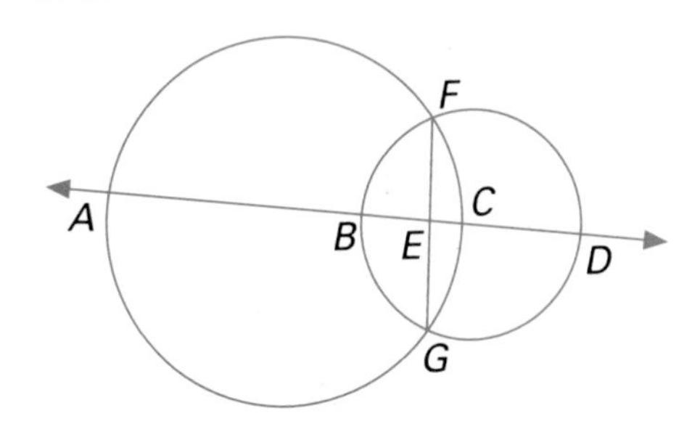

23. Given: Two noncongruent circles externally tangent at A with points C on the larger circle, D on the smaller circle such that $\overleftrightarrow{CD}$ passes through A, E on the larger circle, and F on the smaller circle such that $\overleftrightarrow{EF}$ passes through A.
Prove: $CA \cdot AF = EA \cdot AD$

Mixed Review

1. Given: A segment $\overline{AB}$. Construct the midpoint. *1.3*
2. A 14-ft ladder leans against a wall. The foot of the ladder is 5 ft from the bottom of the wall. At what angle (measured to the nearest degree) is the ladder to the ground? *9.7*
3. How many diagonals does a convex pentagon have? *6.1*
4. Given: $\triangle ABC$ with D on $\overline{AC}$ and E on $\overline{CB}$ such that $\overline{DE} \parallel \overline{AB}$, $DC = 4$, $AD = 5$, $CE = 12$. Find EB. *8.2*

11.10 Circles and Constructions

Objective To perform and explain constructions involving circles

Two properties of circles form the basis for constructing two tangents to a circle from an exterior point:

- The sides of an angle inscribed in a semicircle are perpendicular.
- A line perpendicular to a radius at its endpoint on the circle is tangent to the circle.

Construction Construct two tangents to a circle from a point in the exterior.

Procedure

Draw $\overline{OT}$. Construct M, the midpoint of $\overline{OT}$.

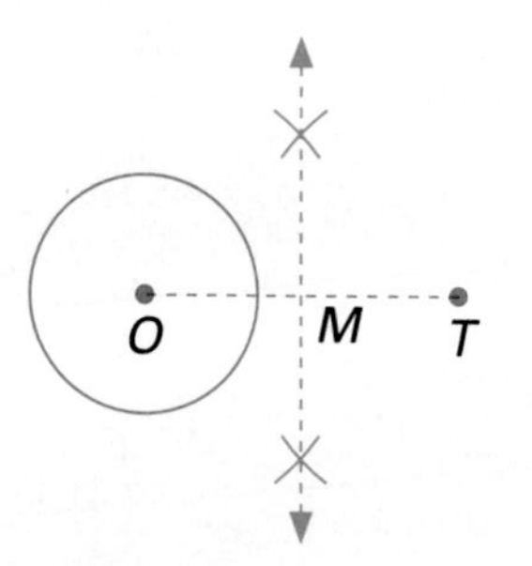

Use MT as the length of a radius to draw a circle intersecting $\odot O$ at points A and B.

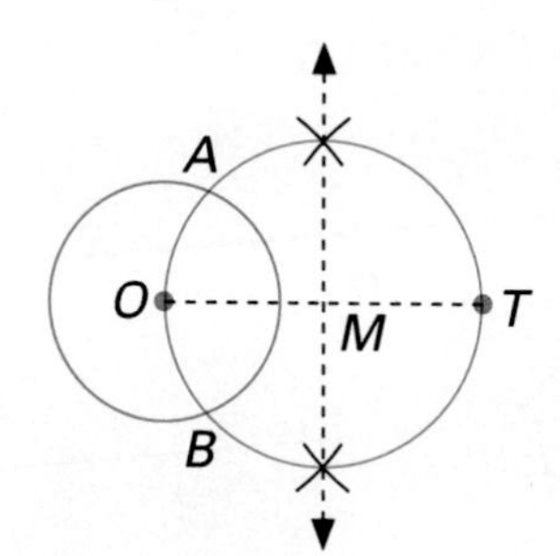

Draw $\overline{TA}$ and $\overline{TB}$. Result: $\overleftrightarrow{TA}$ and $\overleftrightarrow{TB}$ are tangents.

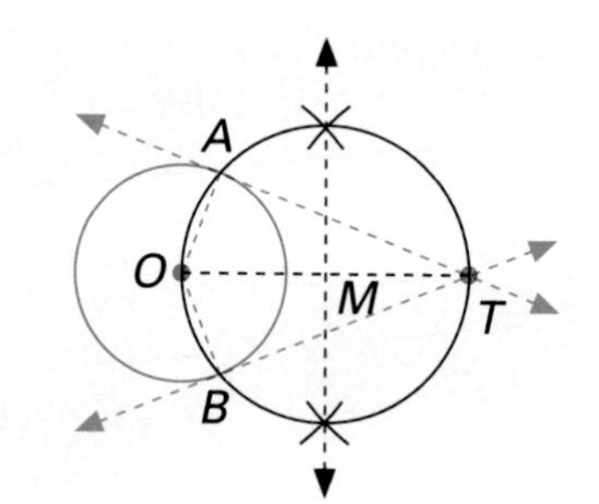

In the figure at the right, $\overline{AB}$ is constructed congruent to radius $\overline{OA}$. A radius $\overline{OB}$ is drawn; $OA = AB = OB$. Therefore, $\triangle OAB$ is equilateral. m $\angle O = 60$, and it follows that $m\widehat{AB} = 60$. Since the measure of a circle is 360, *six* of such arcs can be marked off around the circle. This suggests the following construction.

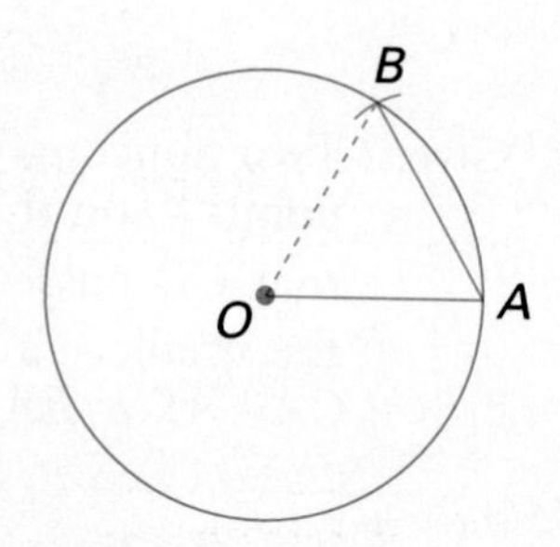

Construction Inscribe a regular hexagon in a circle.

Procedure Open compass to the length of a radius. Beginning at any point A on the circle, mark successive arcs on the circle. Draw chords to form a regular hexagon. Thus, $ABCDEF$ is a regular hexagon inscribed in $\odot O$.

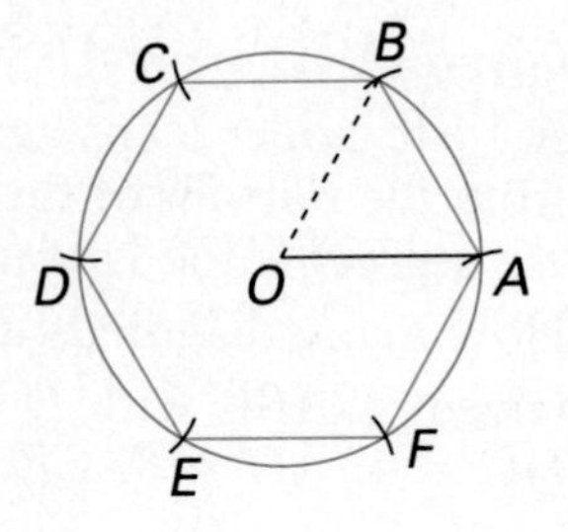

Once it is determined how to construct a central angle and arc of given measure, other arcs of equal measure may be marked off around the circle. Chords between them can then be drawn. After marking off the six points, a chord can be drawn between every other point: $\overline{AC}$, $\overline{CE}$, and $\overline{EA}$. This results in an equilateral triangle.

EXAMPLE Construct: A regular inscribed octagon.

Plan An octagon has eight sides, so the measure of each central angle of the inscribed regular octagon is $\frac{360}{8}$, or 45.

Solution

Draw radius $\overline{OA}$. At O, construct $\overline{OP} \perp \overline{OA}$.

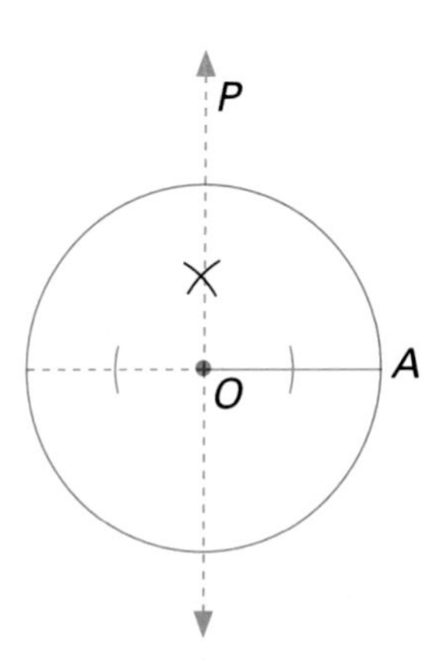

Construct $\overline{OB}$, the bisector of right $\angle POA$.

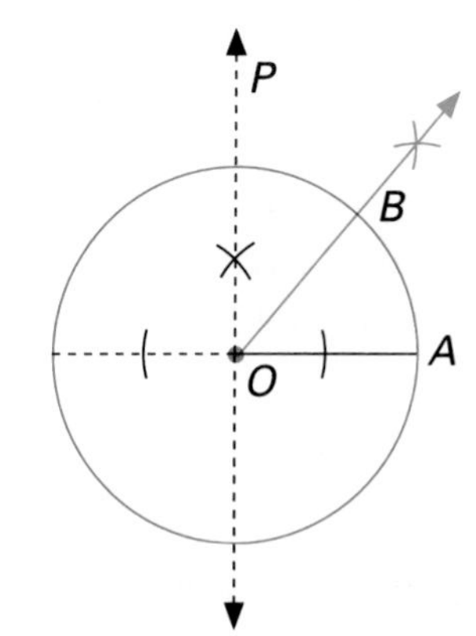

Starting at B, mark off successive arcs congruent to $\overset{\frown}{AB}$. Draw chords.

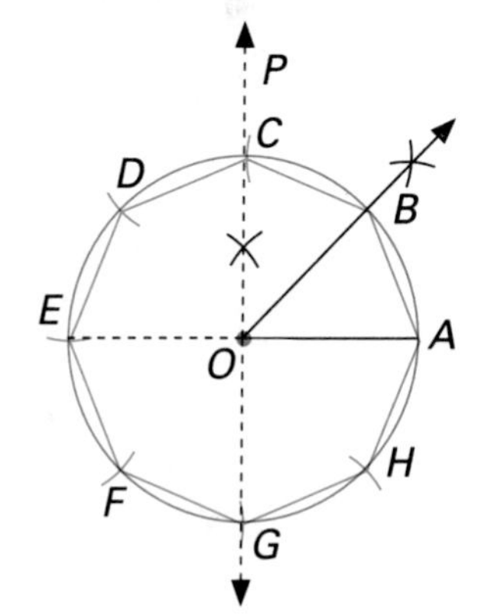

Result: $ABCDEFGH$ is a regular inscribed octagon.

Focus on Reading

List the following four steps in proper sequence for constructing two tangents to a circle O from an exterior point T.

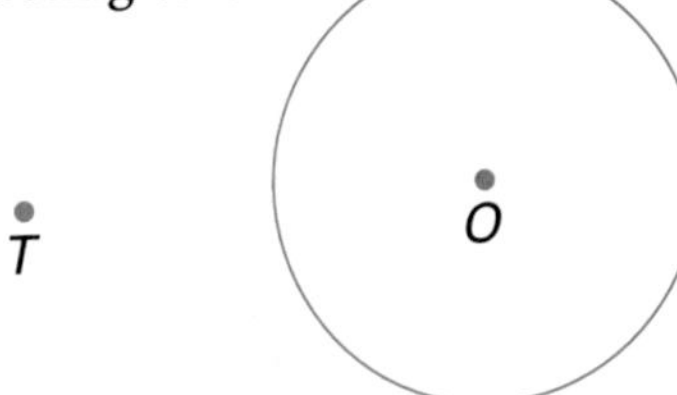

a. Using $\overline{TQ}$ as a radius, draw a circle intersecting $\odot O$ at points A and B.

b. Construct Q, the midpoint of $\overline{OT}$.

c. Draw $\overline{TA}$ and $\overline{TB}$.

d. Draw $\overline{OT}$.

Classroom Exercises

1. How is a tangent related to the radius of a circle drawn to the point of tangency?
2. Tell how to construct a tangent to a circle at a point on the circle.
3. Tell how to construct a tangent to a circle from an exterior point.
4. Tell how to inscribe a regular hexagon in a circle.
5. Tell how to inscribe an equilateral triangle in a circle.
6. Tell how to inscribe a regular octagon in a circle.

Written Exercises

1. Draw a segment $\overline{AB}$. Construct a circle with $\overline{AB}$ as diameter.
2. Draw a circle and a point P in the exterior. From P, construct two tangents to the circle.
3. Draw a segment $\overline{PQ}$ of length 4 units. Construct a circle with $\overline{PQ}$ as diameter. Construct a tangent at Q.
4. Draw a circle with a 6-cm diameter. From a point in the exterior, construct 2 tangents to the circle.
5. Inscribe a regular hexagon in a circle.
6. Inscribe an equilateral triangle in a circle. Justify the construction.
7. Inscribe a regular octagon in a circle.
8. Inscribe a regular 12-sided polygon in a circle.
9. Construct a circle with radius of 3 in. From a point 5 in. from the center of the circle, construct two tangents to the circle. Predict the length of each tangent segment using the properties of tangents and the Pythagorean Theorem. Then verify the length of each tangent with a ruler.
10. Draw a circle and a point A in the exterior of the circle. From A, construct two tangents to the circle. Construct the bisector of $\angle A$. Does this line pass through the center of the circle? Explain.
11. Draw a segment $\overline{PQ}$. Construct a circle with $\overline{PQ}$ as diameter. Construct a diameter $\overline{ST} \perp \overline{PQ}$. Draw the four chords $\overline{SP}$, $\overline{PT}$, $\overline{TQ}$, and $\overline{QS}$. Prove that the quadrilateral $PTQS$ is a square.
12. Use the results of Exercise 11 to construct a square inscribed in a circle.
13. At any point T on a circle, construct a tangent to the circle. At this point T, construct another circle externally tangent to the first circle. From any point on the common tangent to the two circles, construct two other tangents to the circles.
14. Construct a right triangle. Construct a circle that circumscribes it. Justify why the hypotenuse will be the diameter of the circle.

15. Consider the construction of tangents from an external point, as given in the lesson (see p. 462).
    a. Explain why $\overline{OT}$ is a diameter of $\odot M$.
    b. Explain why $\angle OAT$ is a right angle.
    c. Explain why $\overline{TA}$ is perpendicular to $\odot O$ at point A.
16. Consider the construction of a regular hexagon, as given in the lesson. Explain why the hexagon is
    a. equilateral; and b. equiangular.
17. Construct two circles with radii of lengths 3 units and 8 units, and centers 13 units apart. Then construct two external tangents to the circles. Predict the lengths of these two tangent segments. After performing the construction, confirm by measurement.
18. Construct $\overline{AB}$ with length of 4 units and point C between A and B such that $AC = 1$ and $CB = 3$. Construct a circle with $\overline{AB}$ as diameter. At C, construct a perpendicular to the diameter meeting the circle at points D and E. Prove that $CD = \sqrt{3}$. This provides a procedure for constructing the square root of a number, for example, $\sqrt{3}$.

Mixed Review

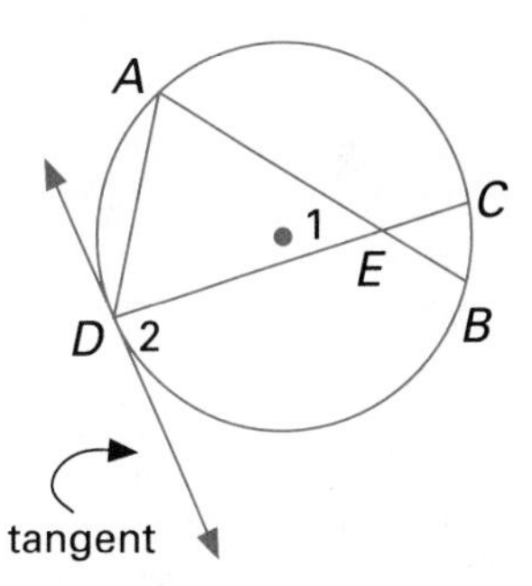

1. $m\widehat{AD} = 80$, $m\widehat{BC} = 40$, $m\angle 1 =$ ______ *11.7*
2. $\overline{DC}$ is a diameter; $m\widehat{CB} = 30$, $m\angle A =$ ______. *11.6*
3. $AE = 20$, $EB = 4$, $DE = 16$, $EC =$ ______ *11.9*
4. $m\widehat{DBC} = 150$, $m\angle 2 =$ ______ *11.8*

Application: *Geometric Designs*

Using the method of inscribing a regular hexagon in a circle, construct the following mathematical designs.

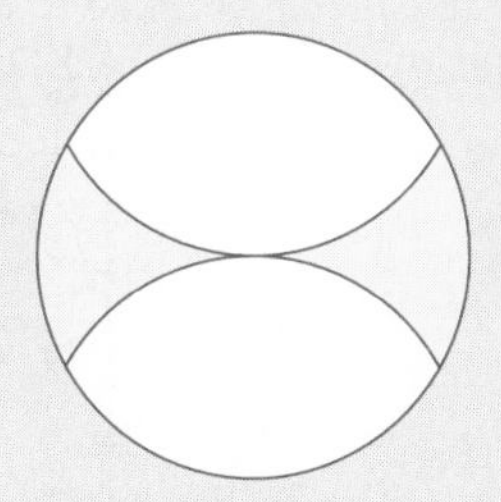

1. Explain your step-by-step procedure for constructing these designs.
2. Design another pattern using circles, central angles, and arcs. Write out a procedure for constructing it.

11.11 Equations of Circles

Objectives To write equations of circles
To graph equations of circles

A circle is the set of all points in a plane a given distance from a fixed point. Using this definition, an equation of a circle can be derived.

Suppose that a circle has its center at some point $P(h,k)$ and has a radius of length r. Let $P(x,y)$ be any point on the circle. The Distance Formula leads to this equation.

$$\sqrt{(x-h)^2 + (y-k)^2} = r, \text{ or}$$
$$(x-h)^2 + (y-k)^2 = r^2$$

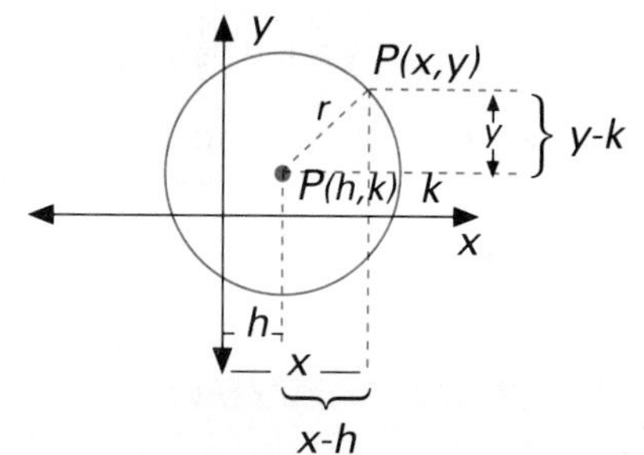

This proves the following theorem.

Theorem 11.17 The equation of a circle with the coordinates of the center (h,k) and a radius of length r is $(x-h)^2 + (y-k)^2 = r^2$.

When the equation of a circle is written in the form $(x-h)^2 + (y-k)^2 = r^2$, it is easy to identify the coordinates of the center and the length of the radius. The graph of the circle can then be drawn.

EXAMPLE 1 Draw the graph of $(x-1)^2 + (y+3)^2 = 16$.

Plan Rewrite the equation to show the center and radius.

Solution $(x-1)^2 + [y-(-3)]^2 = 4^2$
$(h,k) = (1,-3)$ $r = 4$

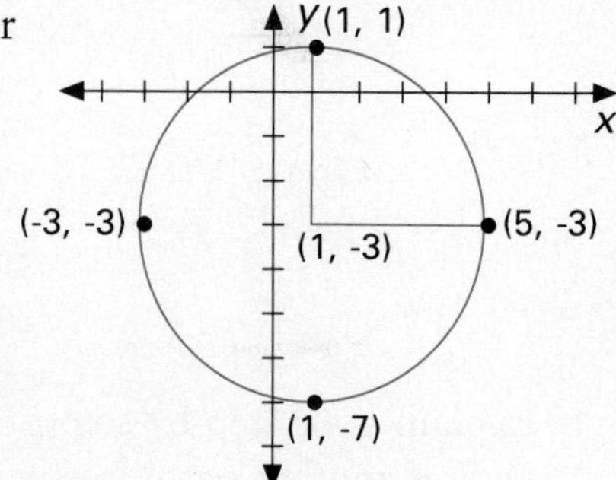

EXAMPLE 2 Find an equation of a circle with a radius of length 3 and center $(-2,3)$.

Solution

$$(x - h)^2 + (y - k)^2 = r^2$$
$$[x - (-2)]^2 + (y - 3)^2 = 3^2$$
$$(x + 2)^2 + (y - 3)^2 = 9$$

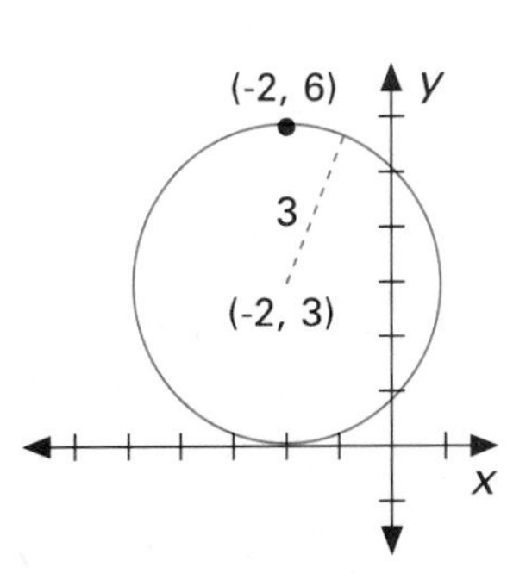

If the center of a circle is the origin, the equation $(x - 0)^2 + (y - 0)^2 = r^2$ simplifies to $x^2 + y^2 = r^2$. For example, the graph of $x^2 + y^2 = 16$ is the circle with the origin as its center and with a radius of length 4.

To simplify an equation of a circle means to write it in the form $x^2 + y^2 + dx + ey + f = 0$. Such an equation is a *quadratic equation.*

EXAMPLE 3 Simplify the equation in Example 2.

Solution

$$(x + 2)^2 + (y - 3)^2 = 9$$
$$x^2 + 4x + 4 + y^2 - 6y + 9 = 9$$
$$x^2 + y^2 + 4x - 6y + 4 = 0$$

EXAMPLE 4 Find the coordinates of the center and the length of a radius of a circle whose equation is $x^2 + y^2 + 2x - 4y + 1 = 0$.

Plan Change the given equation to form $(x - h)^2 + (y - k)^2 = r^2$.

Solution

$$x^2 + y^2 + 2x - 4y + 1 = 0$$
$$(x^2 + 2x + \underline{\qquad}) + (y^2 - 4y + \underline{\qquad}) = -1$$
$$(x^2 + 2x + 1) + (y^2 - 4y + 4) = -1 + 1 + 4$$
$$(x + 1)^2 + (y - 2)^2 = 4$$
$$[x - (-1)]^2 + (y - 2)^2 = 2^2$$

Therefore, the coordinates of the center are $(-1, 2)$, and the length of a radius is 2.

EXAMPLE 5 Write an equation of the form $x^2 + y^2 + dx + ey = 0$ for a circle with center $A(4,-3)$ if the circle passes through the origin.

Plan Find the length of a radius by finding the distance from the center to the origin.

Solution

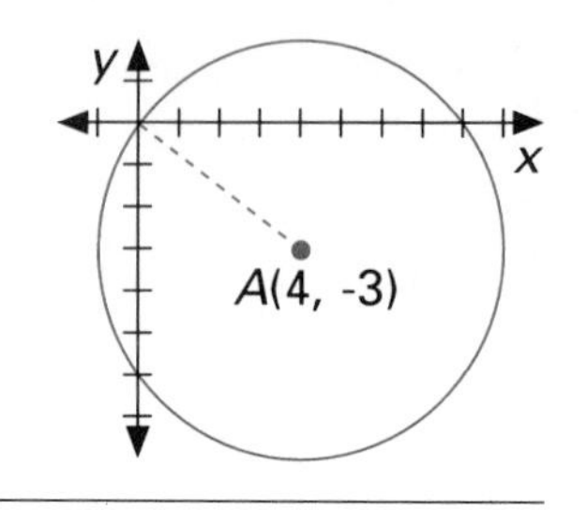

$$r = \sqrt{(4-0)^2 + (-3-0)^2}$$
$$= \sqrt{16 + 9}$$
$$= \sqrt{25}, \text{ or } 5$$

$$(x-4)^2 + [y-(-3)]^2 = 5^2$$
$$x^2 - 8x + 16 + y^2 + 6y + 9 = 25$$
$$x^2 + y^2 - 8x + 6y = 0$$

Classroom Exercises

Give the coordinates of the center and the length of a radius.

1. $(x-2)^2 + (y-5)^2 = 9$
2. $(x-4)^2 + (y-1)^2 = 4$
3. $(x+3)^2 + (y-9)^2 = 1$
4. $(x-7)^2 + (y+1)^2 = 16$
5. $(x+7)^2 + (y+2)^2 = 4$
6. $(x+1)^2 + (y+8)^2 = 25$

Written Exercises

Write an equation of the form $(x-h)^2 + (y-k)^2 = r^2$ for the circle with center O and radius of length r.

1. $O(0,0)$, $r = 5$
2. $O(0,0)$, $r = 2$
3. $O(1,1)$, $r = 2$
4. $O(-2,-2)$, $r = 1$
5. $O(1,-3)$, $r = 4$
6. $O(-5,3)$, $r = 9$
7. $O(-2,-4)$, $r = 1$
8. $O(-4,-2)$, $r = 4$

Write an equation of the form $x^2 + y^2 + dx + ey + f = 0$ for the circle with center Q and radius of length r.

9. $Q(0,0)$, $r = 2$
10. $Q(1,3)$, $r = 4$
11. $Q(-3,4)$, $r = 5$
12. $Q(4,-2)$, $r = 4$

Draw the graph of each circle.

13. $(x+2)^2 + (y-1)^2 = 4$
14. $(x-4)^2 + (y+2)^2 = 16$
15. $(x-2)^2 + y^2 = 1$
16. $x^2 + (y-2)^2 = 9$
17. $x^2 + y^2 = 4$
18. $x^2 + y^2 = 16$

Find the coordinates of the center and the length of a radius of each circle.

19. $x^2 + y^2 - 6x + 5 = 0$

20. $x^2 + y^2 - 4y - 4 = 0$

21. $x^2 + y^2 - 2x - 8y + 13 = 0$

22. $x^2 + y^2 - 16x - 6y + 72 = 0$

23. $x^2 + y^2 + 18x - 20y + 177 = 0$

24. $x^2 + y^2 + 4x + 2y - 145 = 0$

Write an equation of the circle with center A if the circle passes through the origin.

25. $A(3,4)$

26. $A(5,12)$

27. $A(-4,-3)$

28. $A(-12,-5)$

Tell whether point A is on the circle with the given equation.

29. $A(0,4)$, $x^2 + y^2 = 16$

30. $A(-1,-1)$, $(x - 2)^2 + (y + 1)^2 = 9$

31. $A(-2,3)$, $x^2 + y^2 + 4x = 0$

32. $A(2,5)$, $x^2 + y^2 + 2x - 8y + 13 = 0$

Write an equation of the circle containing the two given points as endpoints of a diameter.

33. $(1,2)$ and $(-3,-4)$

34. $(4,5)$ and $(-4,-1)$

Tell whether the two circles are congruent.

35. $x^2 + y^2 - 4x + 2y - 4 = 0$ and $x^2 + y^2 + 2x - 6y - 6 = 0$

36. $x^2 + y^2 + 10x + 8 = 0$ and $x^2 + y^2 - 8y - 1 = 0$

Write an equation of the line that is tangent to the given circle at the given point P.

37. $x^2 + y^2 + 4x - 6y - 12 = 0$, $P(-6,6)$

38. $x^2 + y^2 + 10x + 8y + 21 = 0$, $P(-7,-8)$

Which of the following sets of points is a circle?

39. the set of all points in a plane equidistant from two points

40. the set of all points in a plane twice as far from one point as from another

Mixed Review

1. Write an equation of a line containing the points $P(0,1)$ and $Q(-3,-5)$. ***10.4***
2. Find the distance between the points $R(2,-5)$ and $S(-3,0)$. ***10.1***
3. Find the midpoint of $\overline{MN}$ for $M(-1,-3)$ and $N(5,9)$. ***10.2***
4. Find the slope of the line $2x + y = 5$. ***10.4***

Chapter 11 Review

Key Terms

arc (p. 432)
central angle (p. 432)
chord (p. 416)
circle (p. 415)
circumscribed polygon (p. 427)
common tangent (p. 427)
concentric circles (p. 438)
diameter (p. 416)
external secant segment (p. 458)
inscribed angle (p. 442)
inscribed polygon (p. 423)
intercepted arc (p. 442)
major arc (p. 433)
minor arc (p. 432)
radius (p. 415)
secant (p. 417)
semicircle (p. 432)
tangent (p. 417)
tangent circles (p. 428)
tangent segment (p. 425)

Key Ideas and Review Exercises

11.1 To identify lines and segments related to circles

Identify the following for the circle.

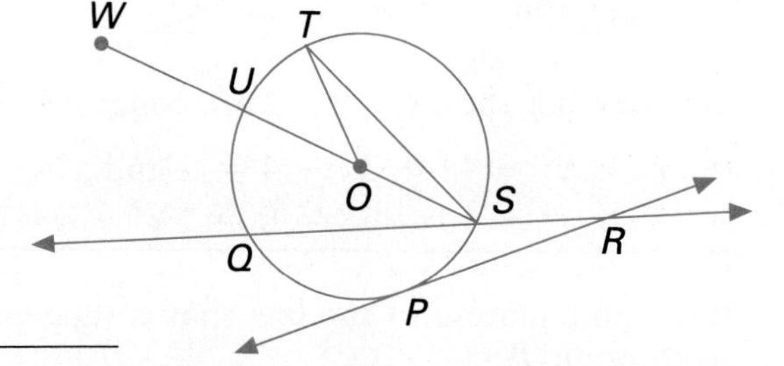

1. radius **2.** diameter **3.** chord
4. secant **5.** tangent
6. Given: $OU = 7$, $OW = 3x - 5$. Find all the values of x such that W lies in the exterior of the circle.
7. $OT = 2x - 8$, $OU = 4x - 20$, $OS =$ ______

11.2 About chords and their relation to the center of a circle:
- A perpendicular to a chord from the center of a circle bisects the chord.
- Congruent chords are equidistant from the center, and conversely.
- Chords of greater length are closer to the center of a circle.

8. $AB = 12$, $AO = 10$, $OF =$ ______
9. $m\angle AOB = 120$, $AB = 8$, $OA =$ ______
10. $OD = 5$, $OF = 4$. Name the longer of the two chords $\overline{AB}$ and $\overline{EC}$.

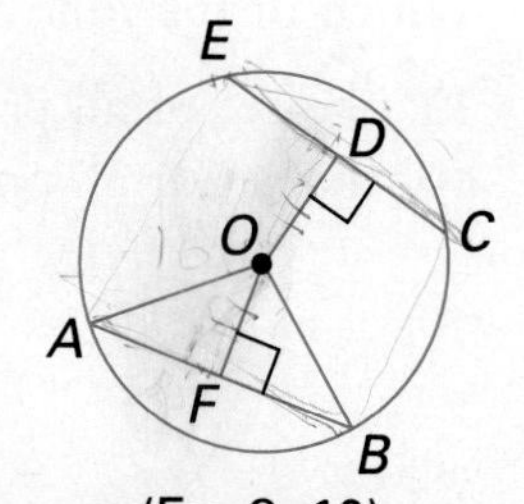

(Ex. 8–10)

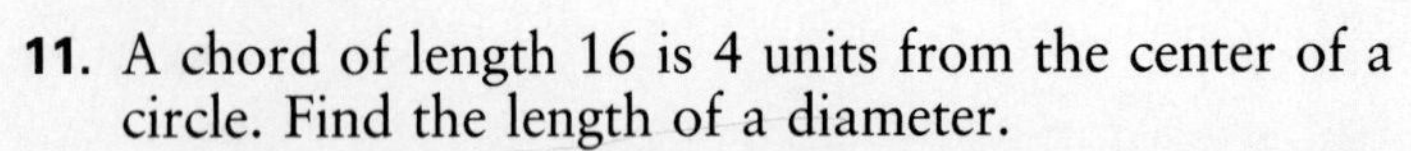

11. A chord of length 16 is 4 units from the center of a circle. Find the length of a diameter.
12. $\triangle ABC$ is inscribed in a circle. $m\angle A = 30$, $m\angle B = 70$. Name the longest chord.

11.3 Properties of tangents and radii: If $\overline{TA}$ and $\overline{TB}$ are tangent segments to $\odot O$ with radius $\overline{OA}$, then $\overline{OA} \perp \overline{AT}$, $AT = BT$, and $\overline{OT}$ bisects $\angle T$ (see p. 426).

13. $TB = 6$, $OT = 8$, $OA =$ ______
14. $m\angle ATB = 120$, $AT = 10$, $OA =$ ______

Arcs and Angles of Circles

In the diagrams below, $\overline{RV} \parallel \overline{PQ}$, $\overline{PQ}$ is a diameter, $\overline{UT}$ is tangent to $\odot O$, and $\overline{FE}$ is tangent to $\odot P$.

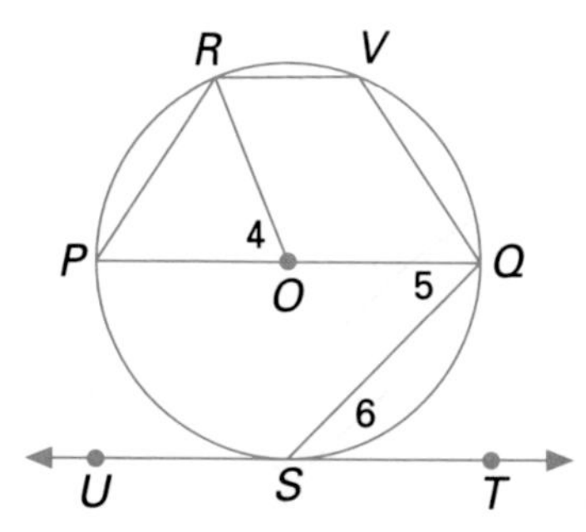

$m\angle 4 = m\widehat{PR}$ **11.4**

$m\angle 5 = \frac{1}{2}m\widehat{PS}$ **11.6**

$m\angle 6 = \frac{1}{2}m\widehat{SQ}$ **11.8**

$m\widehat{PR} = m\widehat{VQ}$ (since $\overline{RV} \parallel \overline{PQ}$) **11.6**

15. $m\widehat{RQ} = 160$, $m\angle 4 =$ ______

16. $m\widehat{PR}:m\widehat{RQ} = 5:4$, $m\angle 4 =$ ______

17. $m\widehat{PS} = 100$, $m\angle 6 =$ ______

18. $m\widehat{RV} = 60$, $m\angle 4 =$ ______

19. $m\angle V = 150$, $m\angle P =$ ______

$m\angle 1 = \frac{1}{2}(m\widehat{AB} + m\widehat{CD})$ **11.7**

$m\angle 2 = \frac{1}{2}(m\widehat{AB} - m\widehat{CD})$

$m\angle 3 = \frac{1}{2}(m\widehat{BF} - m\widehat{FC})$ **11.8**

20. $m\widehat{AB} = 100$, $m\angle 1 = 75$, $m\widehat{CD} =$ ______

21. $m\widehat{AB}:m\widehat{CD} = 4:1$, $m\widehat{AD} = 130 = m\widehat{BC}$, $m\angle 2 =$ ______

22. $m\widehat{AB} = 100$, $m\angle 1 = 60$, $m\angle 2 =$ ______

11.9 *Products of Segments*

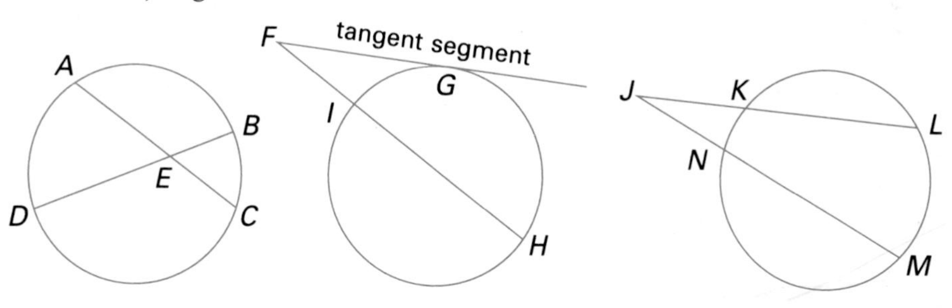

$AE \cdot EC = DE \cdot EB$ $\quad (FG)^2 = FH \cdot FI$ $\quad JM \cdot JN = JL \cdot JK$

23. $AE = 8$, $EC = 10$, $ED = 16$, $EB =$ ____

24. $FG = 4$, $IH = 6$, $FH =$ ____

25. $JK = 4$, $JL = 10$, $MN = 3$, $JN =$ ____

26. $FI = 4$, $IH = 6$, $FG =$ ____

27. Construct two tangents to a circle from a point exterior to the circle. **11.10**

11.11 The equation of a circle with center at point $P(h,k)$ and radius r is $(x - h)^2 + (y - k)^2 = r^2$.

28. Graph the equation $(x - 3)^2 + (y + 2)^2 = 9$.

Chapter 11 Test

Identify the following for $\odot O$ at the right.

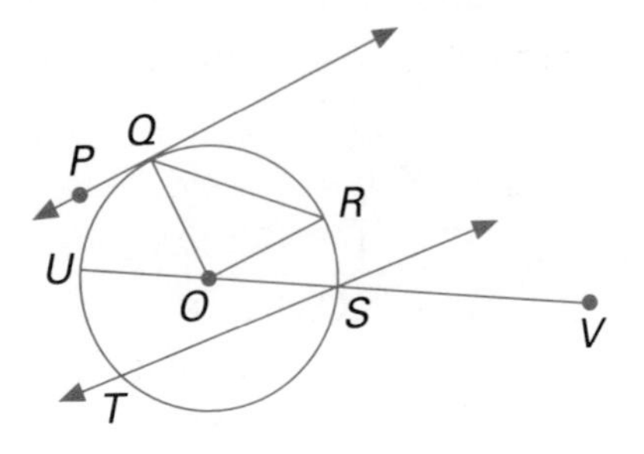

1. radius
2. chord
3. diameter
4. tangent
5. secant
6. Given: $OQ = 4x - 2$, $OS = 2x + 8$. Find OU.
7. Given: $SU = 12$, $OV = 3x - 9$. Find all values of x for which V is in the exterior of the circle.

Use $\odot O$ with radius $\overline{OD}$ for Exercises 8–10.

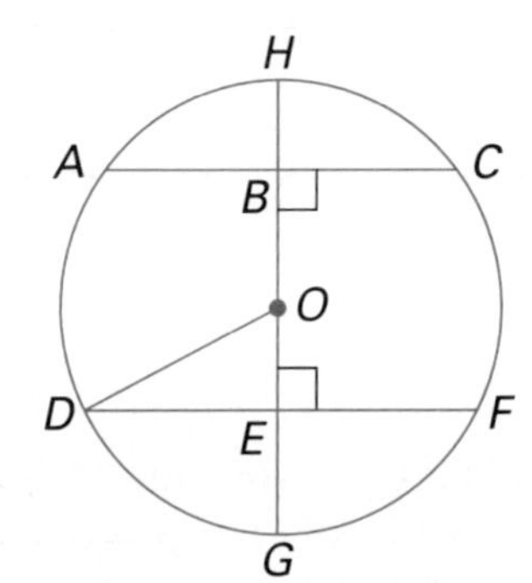

8. Given: Diameter $\overline{GH} \perp$ chords $\overline{AC}$ and $\overline{DF}$
   Prove: $m\widehat{AD} = m\widehat{CF}$
9. Given: $m\widehat{AHC} = 115$, $m\widehat{DGF} = 185$
   Find $m\widehat{AD}$.
10. Given: $m\angle DOE = 30$, $OE = 4$
    Find the length of the radius.

Use $\odot O$ with tangent $\overleftrightarrow{EF}$ for Exercises 11–14.

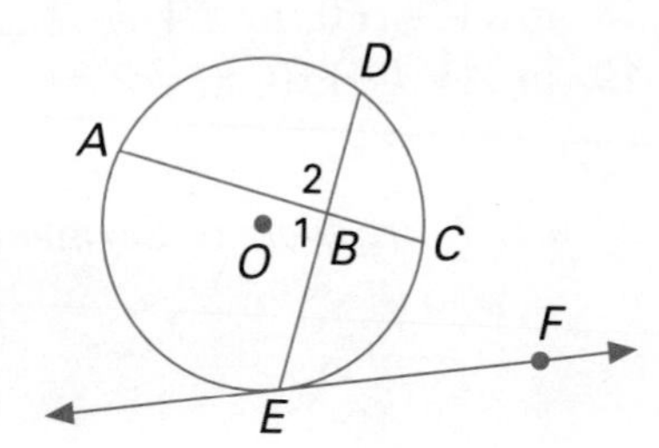

11. $m\widehat{AE} = 100$, $m\widehat{DC} = 70$, $m\angle 1 =$ ______
12. $AB = 24$, $BC = 4$, $EB = 6$, $BD =$ ______
13. $m\widehat{AE}:m\widehat{AD}:m\widehat{DC}:m\widehat{EC} = 4:3:2:1$,
    $m\angle 2 =$ ______
14. $m\widehat{AE} = 100$, $m\angle 1 = 70$, $m\widehat{DC} =$ ______

Use $\odot O$ with tangent segment $\overline{AB}$ for Exercises 15–18.

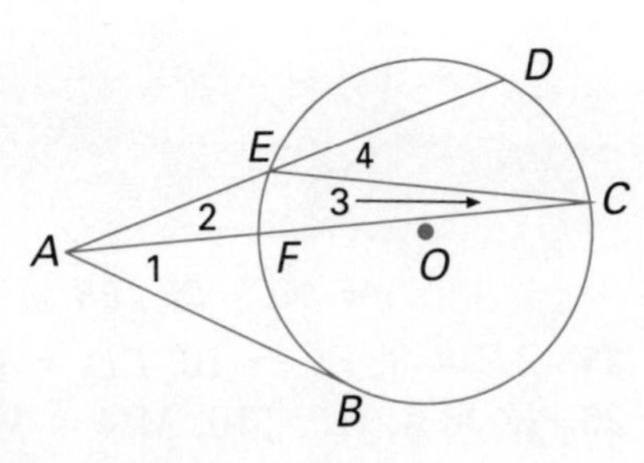

15. $m\widehat{DC} = 50$, $m\angle 2 = 5$, $m\widehat{FE} =$ ______
16. $m\widehat{FBC} = 160$, $m\widehat{BC}:m\widehat{FB} = 3:2$, $m\angle 1 =$ ______
17. $AF = 2$, $FC = 4$, $AB =$ ______
18. $ED = 1$, $AF = 2$, $FC = 8$, $AE =$ ______
19. Find the length of a chord of a circle if the chord is 6 units from the center and the length of the radius is 10 units.
20. The length of the common external tangent to two circles is 12. Find the distance between the centers if the lengths of the radii are 7 and 2.

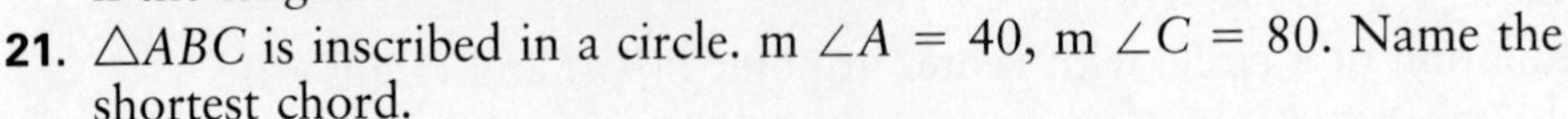

21. $\triangle ABC$ is inscribed in a circle. $m\angle A = 40$, $m\angle C = 80$. Name the shortest chord.
22. Find the coordinates of the center and the length of the radius of the circle. Draw the circle $x^2 + y^2 - 4x - 5 = 0$.

College Prep Test

Strategy for Achievement in Testing

Each item consists of two quantities, one in Column A and one in Column B. You are to compare the two quantities and choose an answer from the following list.

A—The quantity in Column A is greater.
B—The quantity in Column B is greater.
C—The two quantities are equal.
D—The relationship cannot be determined from the information given.

	Column 1	Column 2
1.	the length of the hypotenuse of a right triangle with legs of lengths 16 and 12	the length of the hypotenuse of a right triangle with legs of lengths 5 and 12

2. Lines a and b are parallel. Lines m and n are parallel.

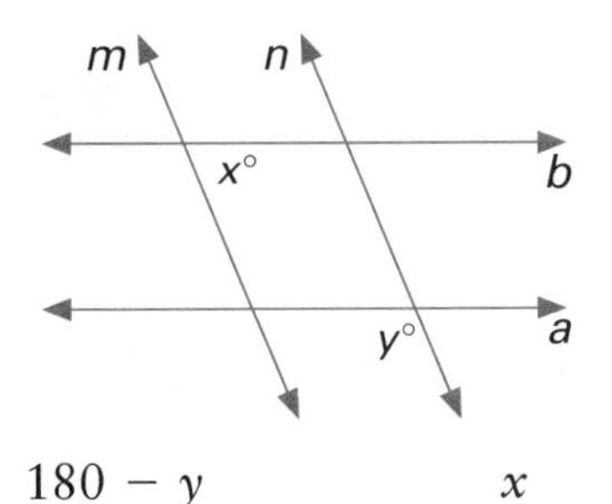

Column 1	Column 2
$180 - y$	x

3.

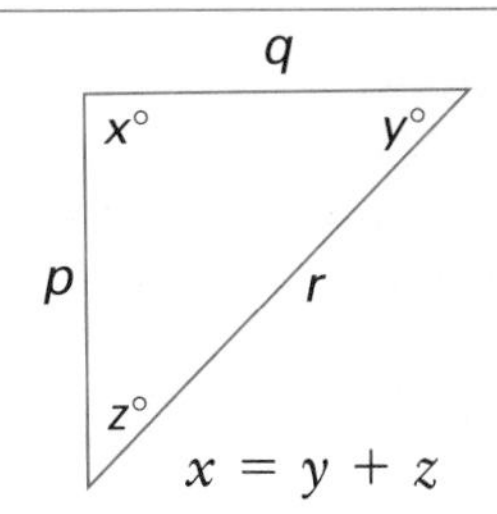

$x = y + z$

NOTE: The figure is not drawn to scale.

Column 1	Column 2
$p^2 + q^2$	r^2

4.

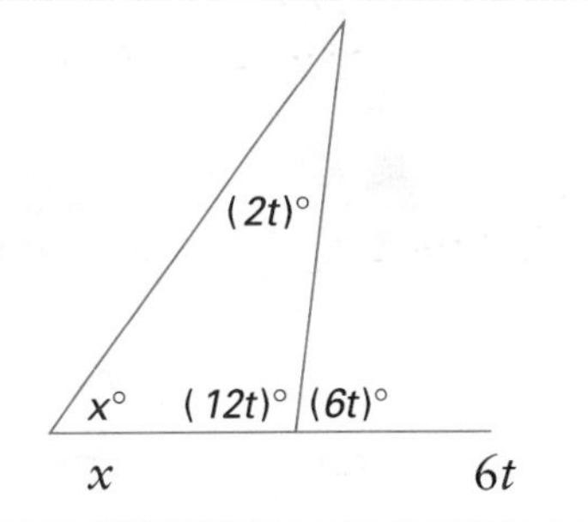

Column 1	Column 2
x	$6t$

5. Circle with center O.

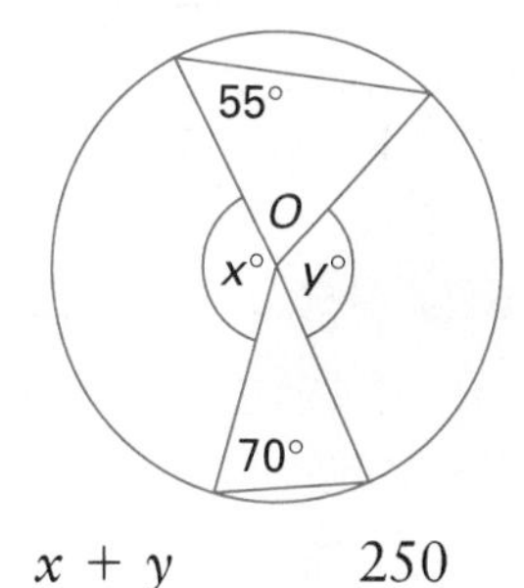

Column 1	Column 2
$x + y$	250

6. $a \parallel b \parallel c$

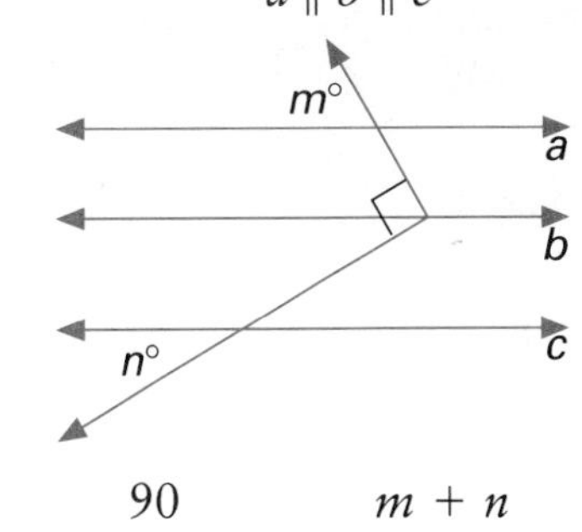

Column 1	Column 2
90	$m + n$

12 AREA

Kurt Gödel (1906–1978)—Gödel, at age 25, showed that some mathematical statements cannot be proved either true or false. This shook the foundations of mathematics and logic which had existed for thousands of years.

12.1 Standard Units

Objective To determine appropriate standard units of area measurement

All measurement systems use standard units. To measure distance or length, a unit such as the *centimeter* is used. If a segment is 5 cm long, the segment can be divided into five adjacent, congruent, nonoverlapping segments, each of which is one centimeter long.

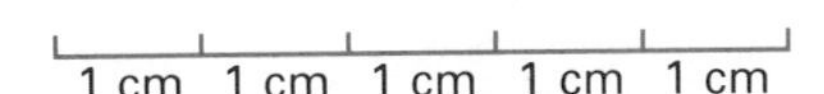

To measure area, a standard unit is also used. People have used many different units over the years. The square foot, the acre, and the hectare are three such units. Each has its own history. For example, an acre was originally the amount of land that could be plowed in a single day by a man and a team of oxen. Since different men and different oxen moved at different speeds, this was not a very accurate unit of measure.

Standard units of area today are based on congruent figures that will cover a surface without overlapping. The square is usually used. However, many other figures have these properties.

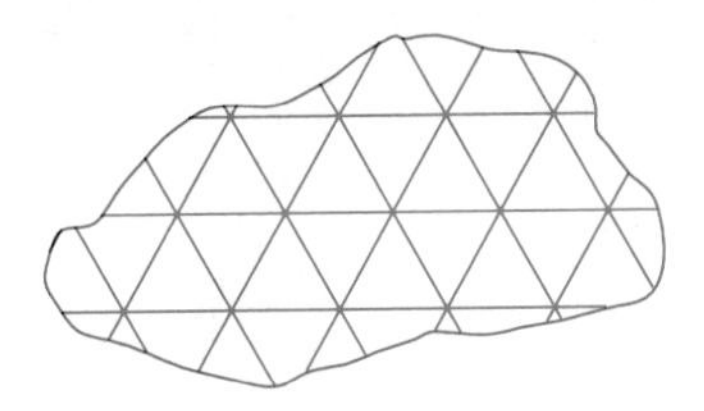

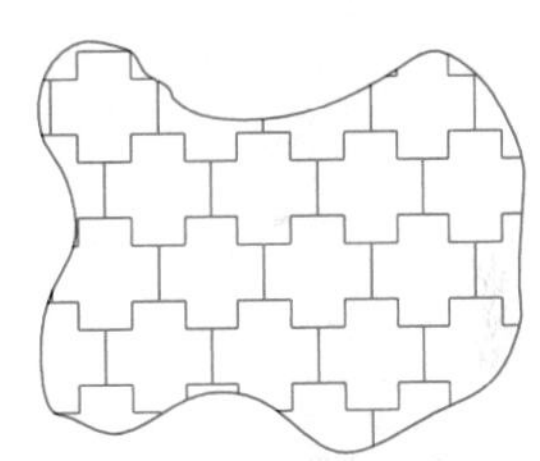

EXAMPLE Copy the figure. Show how it can be used as a standard unit of area that covers the surface with no overlapping.

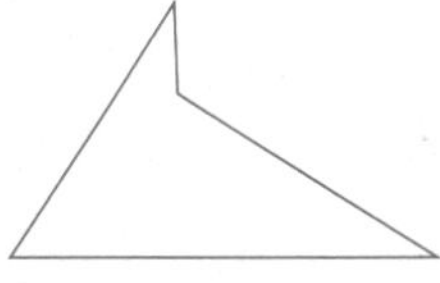

Solution

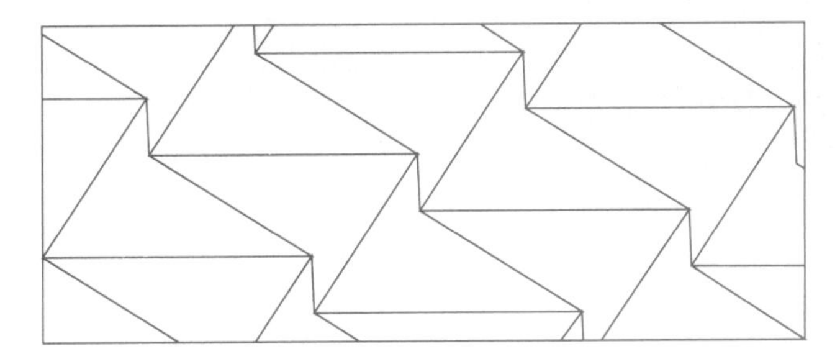

Not all shapes have these properties. For example, a circle cannot be used as a standard unit for area. The rule for covering a piece of surface with a given shape is that the shapes cannot overlap and yet must cover all of the surface. This is not possible with circles.

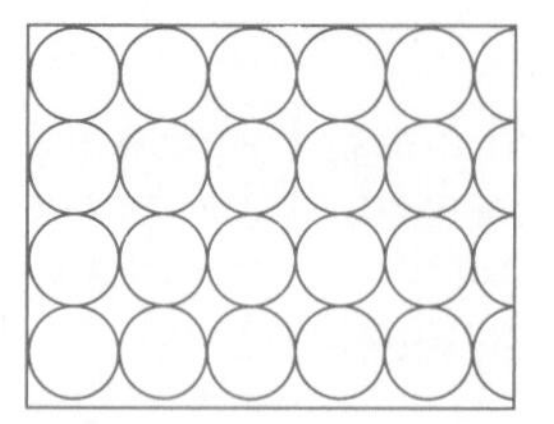
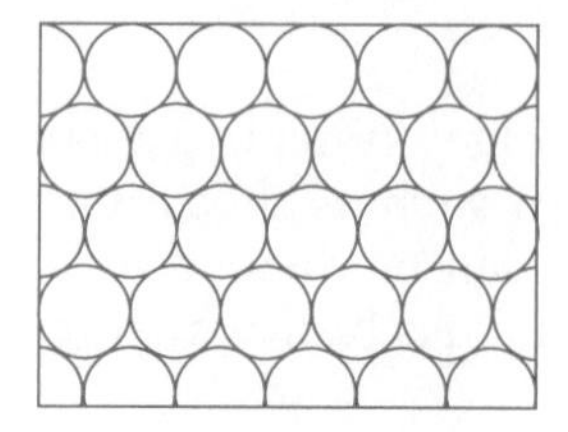
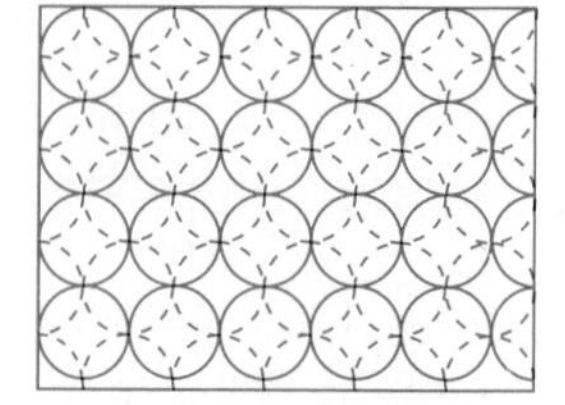

For a regular polygon to serve as a standard unit of area, the measure of each of its angles must be a factor of 360. Each angle of the square below measures 90. Since $4 \cdot 90 = 360$, four such angles fit around a common vertex without overlapping. Each angle of the regular pentagon measures 108. Since 108 is not a factor of 360, three pentagons leave space uncovered and four pentagons overlap.

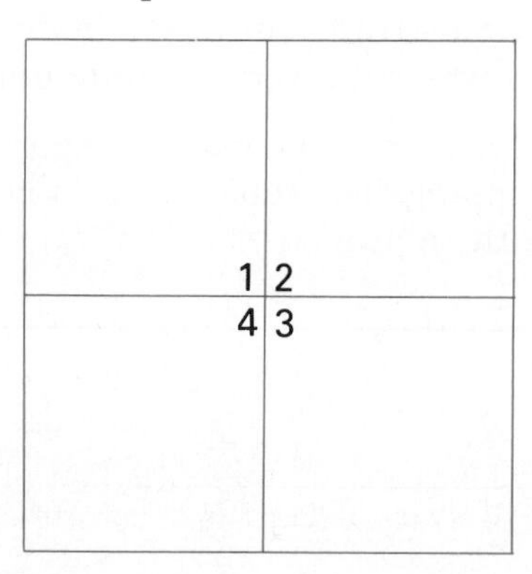

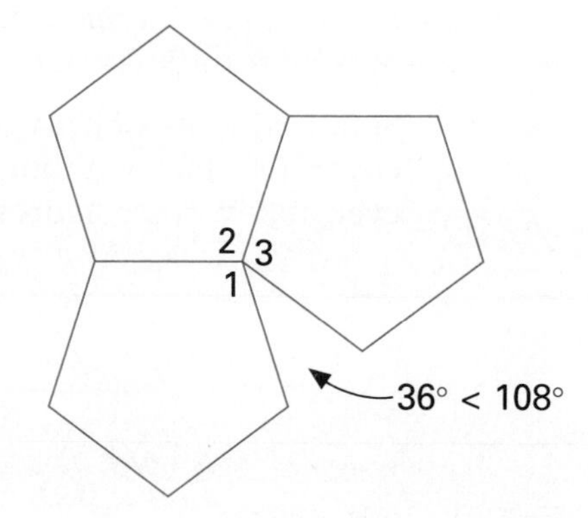

The concept of a standard unit is based on the assumption that standard units have constant areas.

Postulate 20 Congruent polygons have equal areas.

Focus on Reading

Choose the best answer to complete each statement.

1. The square mile is a unit of measure that might be used to indicate
   a. the distance from the earth to the moon.
   b. the area of Indiana.
   c. the area of your classroom.

2. The square centimeter might be used to give
   a. the area of this page.
   b. the area of a ball field.
   c. the area of the United States.

3. The area of this page is approximately
   a. 4 in^2.
   b. 4 ft^2.
   c. 63 in^2.

Classroom Exercises

Each angle of the given regular polygon has the given measure. Can the polygon be used as a standard unit of area?

1. triangle, 60
2. quadrilateral, 90
3. decagon, 144
4. pentagon, 108
5. hexagon, 120
6. octagon, 135

Written Exercises

Copy the figure. Show how it can be used as a standard unit that covers a surface with no overlapping.

1. 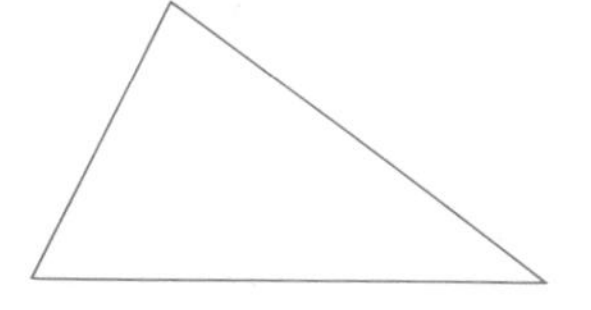

2. 

3. 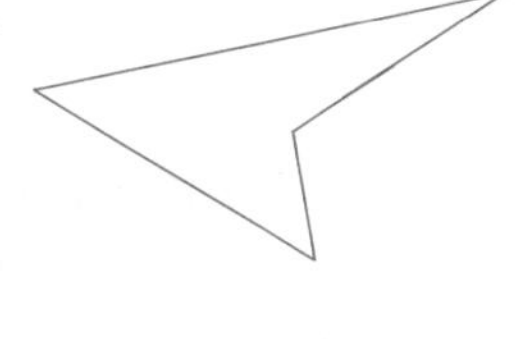

Can the shape described be used as a standard unit of area? If so, draw a picture to show how.

4. a right triangle
5. an acute triangle
6. an obtuse triangle
7. any rectangle
8. any parallelogram
9. a trapezoid
10. an isosceles trapezoid
11. any convex quadrilateral
12. a regular hexagon
13. any convex hexagon
14. any convex pentagon
15. a regular octagon
16. any concave quadrilateral
17. any equilateral polygon
18. any equiangular polygon
19. any regular polygon

Mixed Review

What is the measure of each interior angle and each exterior angle of the polygon? *6.2, 6.3*

1. a regular pentagon
2. a regular octagon
3. a regular 100-gon

12.2 Areas of Rectangles and Squares

Objective To apply the formulas for the areas of a rectangle and a square

A **polygonal region** is the union of a polygon and its interior. For example, the *triangular region* below is the union of a triangle and its interior.

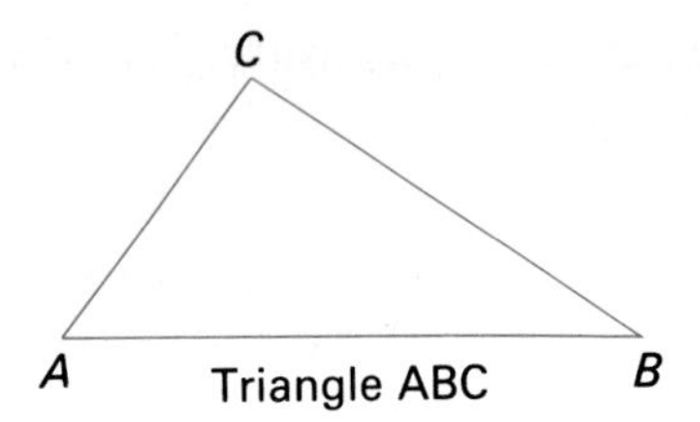

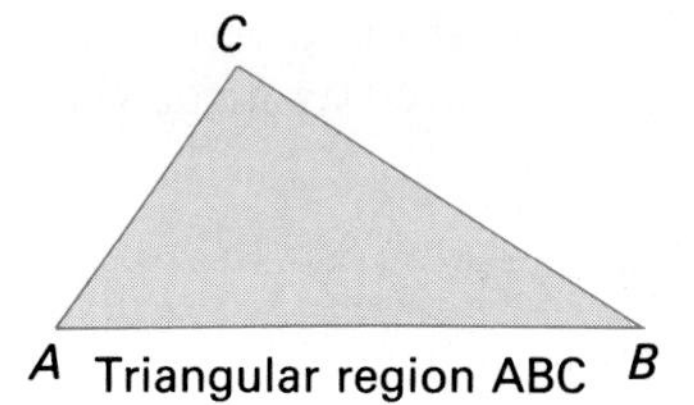

The relative sizes of some regions can be compared directly. If one region can fit in the interior of a second region, then the second is larger than the first.

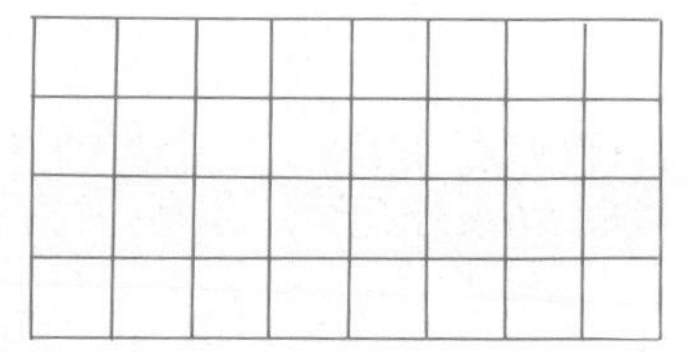

It is not practical to compare most regions in this manner. However, their areas can be used to compare their sizes. Using the small square as a standard unit, the area of the 4-by-8 rectangle at right is 32 square units. When the unit of length used is the inch, the unit of area is the square inch (in^2). When the unit of length is the centimeter, the unit of area is the square centimeter (cm^2).

Definition The **area of a polygon** is the number of square units in the region bounded by the polygon.

EXAMPLE 1 Using the small square region as the unit of area, find the area of the given rectangle.

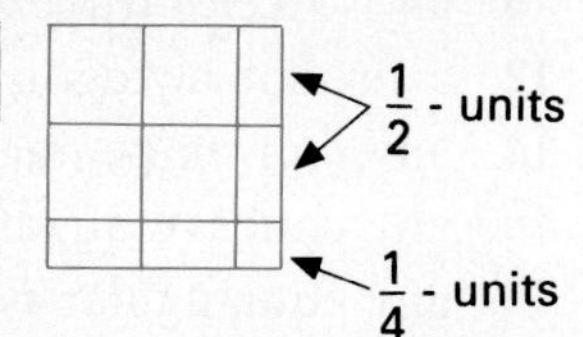

Solution The rectangle is $2\frac{1}{2}$ units wide and $2\frac{1}{2}$ units long.

$2\frac{1}{2} \cdot 2\frac{1}{2} = \frac{5}{2} \cdot \frac{5}{2} = \frac{25}{4} = 6\frac{1}{4}$

It contains $6\frac{1}{4}$ square units.

Postulate 21 The area of a rectangle is the product of the lengths of a base and a corresponding altitude (Area of rectangle $= bh$).

EXAMPLE 2 If the area of a rectangle is 42 cm^2, and the length of an altitude is 6 cm, what is the length of the base?

$A = 42$ $h = 6$

$b = ?$

Solution

$A = bh$

$42 = b \cdot 6$, or $6b$

$b = 7$ cm

Since a square is a rectangle, you can use the formula $A = bh$ to find its area. However, a special formula is often used.

Theorem 12.1 The area of a square is the square of the length s of a side ($A = s^2$).

EXAMPLE 3 What is the area of a square with sides 34.5 cm long?

Solution

$A = s^2$

$= (34.5)^2 = 1{,}190.25$ cm^2

Postulate 22 **Area Addition Postulate:** If a region is the union of two or more nonoverlapping regions, then its area is the sum of the areas of these nonoverlapping regions.

EXAMPLE 4 Find the area of the polygon. Each angle is a right angle.

Plan Divide the region into three rectangular regions. Apply Postulate 22.

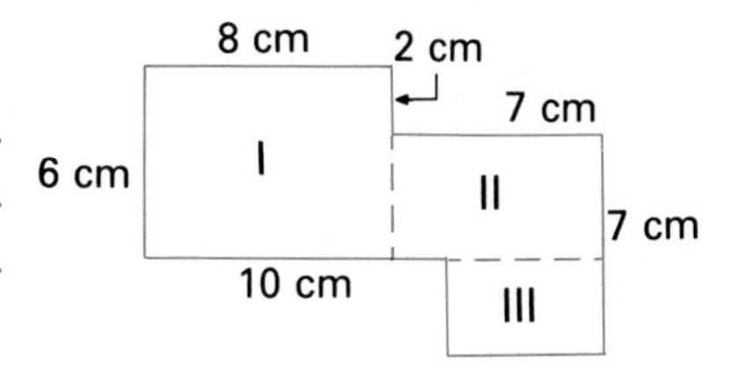

Solution

Area of rectangle I $= 8 \cdot 6$, or 48 cm^2

Area of rectangle II $= 7 \cdot 4$, or 28 cm^2

Area of rectangle III $= 5 \cdot 3$, or 15 cm^2

The total area is 91 cm^2.

Classroom Exercises

What is the area of the rectangle or square?

1. $b = 5$ cm, $h = 2$ cm
2. $b = 3$ in., $h = 7$ in.
3. $b = 9$ yd, $h = 7$ yd
4. $b = 8$ m, $h = 6$ m
5. $s = 8$ cm
6. $s = 6$ ft

Written Exercises

Copy the rectangle. Using the small square region as the unit of area, find the area.

1.

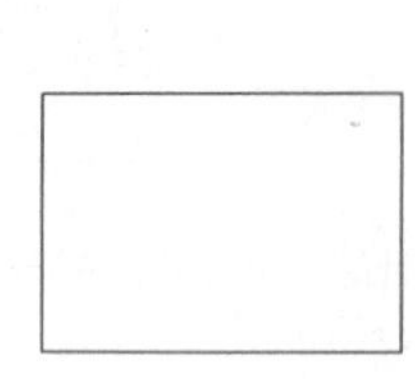

2.

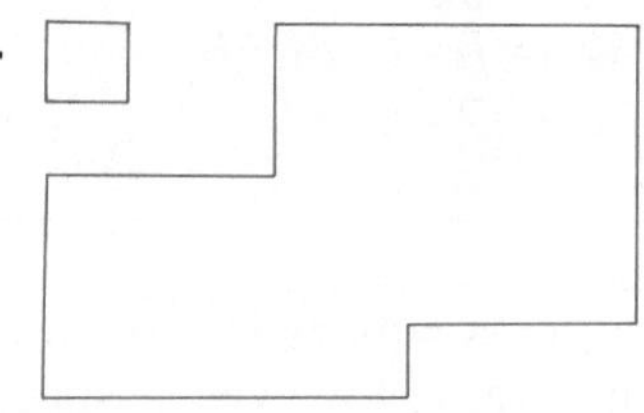

What is the area of the rectangle or square?

3. $b = 4.5$ m, $h = 3.2$ m

4. $b = 2.6$ cm, $h = 6.5$ cm

5. $b = 2.7$ in., $h = 4.6$ in.

6. $b = 8.3$ ft, $h = 1.9$ ft

7. $s = 2.5$ mm

8. $s = 4.5$ cm

9. $s = 8.7$ yd

10. $s = 9.4$ ft

Find the unknown length of a side of the rectangle.

11. $b = 9$ in., $h =$ ______, $A = 72$ in^2

12. $b =$ ______, $h = 13$ cm, $A = 65$ cm^2

13. $b = 1.5$ yd, $h =$ ______, $A = 6$ yd^2

14. Prove Theorem 12.1.

What is the area of the rectangle?

15. $b = x + 3, h = x - 3$

16. $b = y + 4, h = y + 4$

17. $b = y + z, h = y - z$

18. $b = 2x + y, h = 3x - 2y$

Find the unknown length of a side of the rectangle.

19. $h = 4.5$ cm, $A = 13.5$ cm^2

20. $b = 3.6$ m, $A = 18$ m^2

21. $b = \sqrt{6}$ m, $A = 6$ m^2

22. $h = \sqrt{3}$ cm, $A = \sqrt{18}$ cm^2

23. $b = 2x + y, A = 4x^2 - y^2$

24. $h = x - 2y, A = x^2 - 4yx + 4y^2$

Find the area of the polygon. Each angle is a right angle.

25.

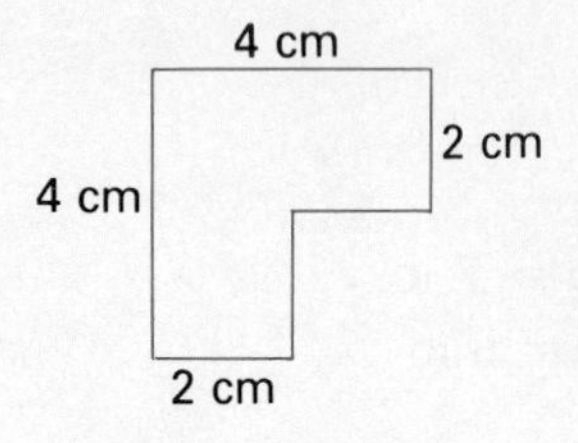

26.

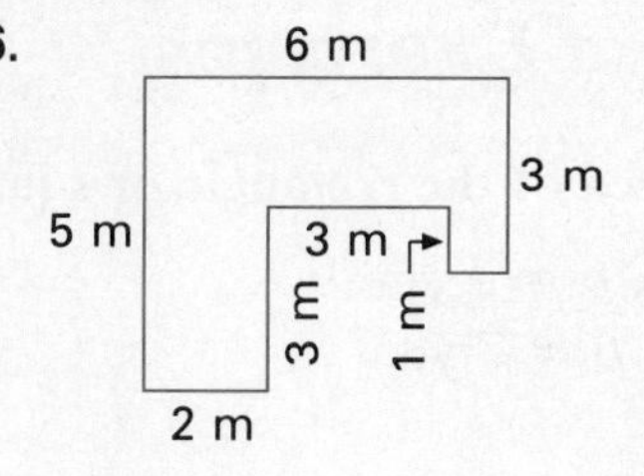

27.

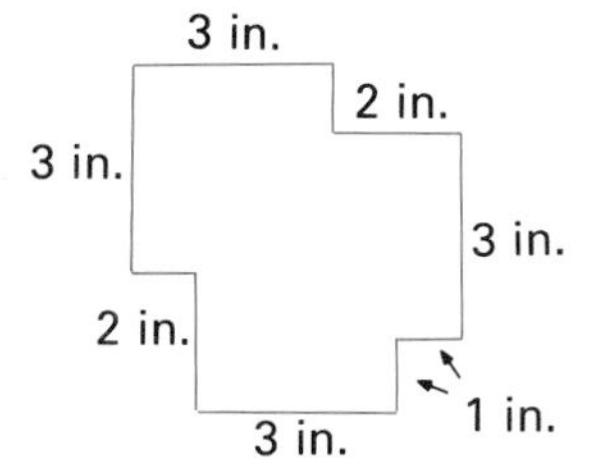

28.

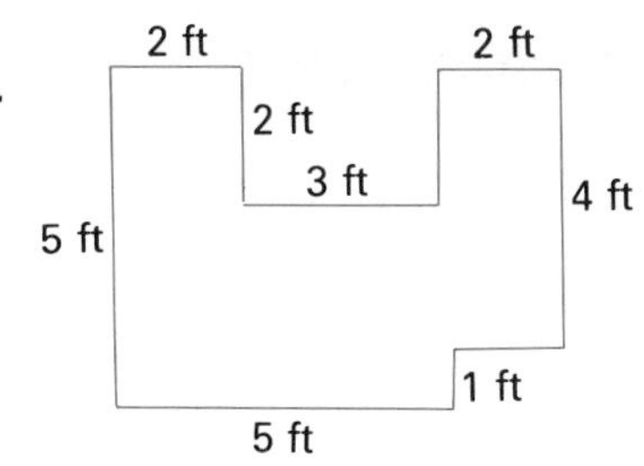

For Exercises 29–31, find the area of the rectangle. (HINT: **Use the Pythagorean Theorem.**)

29.

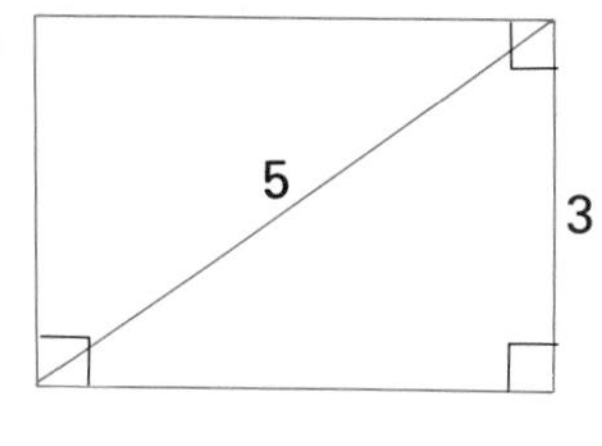

30.

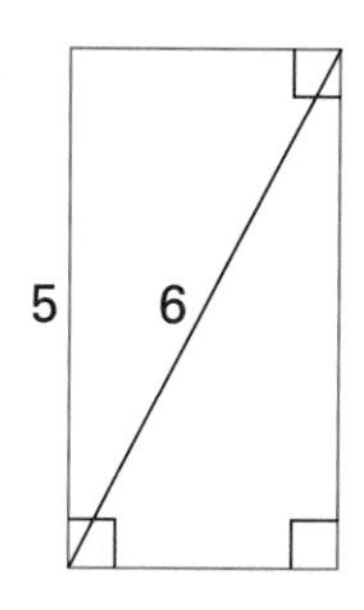

31.

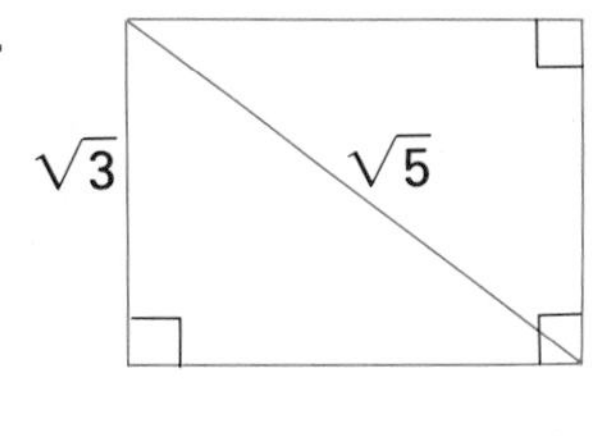

32. The length of a diagonal of a square is 20 in. What is the area of the square?

33. The length of a diagonal of a square is 16 ft. What is the area of the square?

34. Two sides of a rectangle have lengths of 5 cm and 6 cm. What is the length of a side of a square with the same area?

35. Two sides of a rectangle have lengths of 3.4 m and 2.6 m. What is the length, to the nearest hundredth of a meter, of a side of a square with the same area?

36. The lengths of two consecutive sides of a rectangle have the ratio of 3 to 4. The area is 108 cm^2. What are the lengths?

37. The lengths of two consecutive sides of a rectangle have the ratio of 3 to 2. The area is 216 mm^2. What are the lengths?

Prove or disprove.

38. The ratio of the areas of two rectangles with congruent bases is equal to the ratio of the lengths of their altitudes.

39. If the length of a base of a rectangle is doubled and the length of a corresponding altitude remains constant, then the area is doubled.

40. If the lengths of the sides of one square are twice the lengths of the sides of a second square, then the area of the first square is twice that of the second.

41. If the lengths of the sides of a rectangle are multiplied by any constant, then the area of the original rectangle is multiplied by the same constant.

42. A painter is buying paint for the walls and ceiling of a room of a house. The room is 14 ft wide by 16 ft long, with walls 8 ft tall. There are two doors in the room, each $2\frac{1}{2}$ ft wide and 6 ft tall. There are two windows, each 4 ft wide and 3 ft tall. If one gallon of paint covers approximately 500 ft^2 of space, how many gallons must be bought? How much paint will be left over?

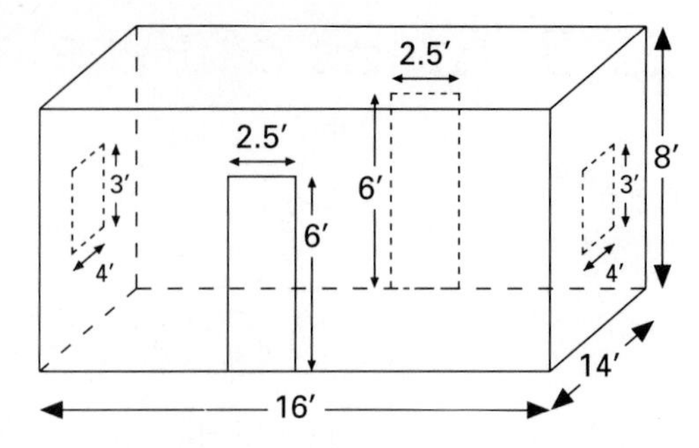

Mixed Review

1. Which of the following patterns can be used to prove that two quadrilaterals are congruent: SSSS, AAAA, SASAS, ASASA, AASAA? *7.1*

True or False?

2. A necessary condition for a quadrilateral to be a parallelogram is that opposite angles are congruent. *6.4*
3. A sufficient condition for a parallelogram to be a rectangle is for diagonals to be congruent. *7.3*

Brainteaser

The rectangular region shown has been divided into nine square regions, all different in size. The area of Square III is 64 cm^2. The area of Square IV is 81 cm^2. What is the area of the rectangle?

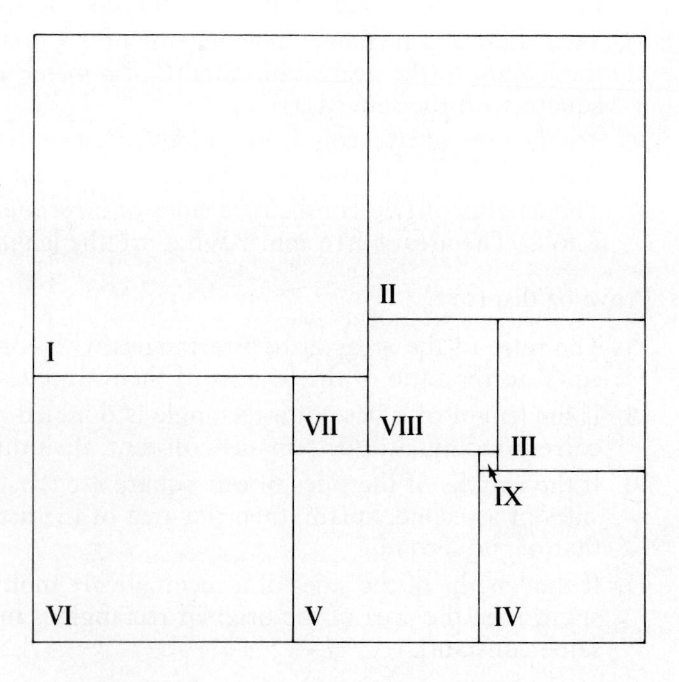

12.3 Areas of Parallelograms

Objective To apply the formula for the area of a parallelogram

If you formed a parallelogram out of paper, you could cut it and rearrange the pieces to form a rectangle. The base and the altitude of this rectangle are congruent to those of the original parallelogram. This fact can be used to prove the following theorem.

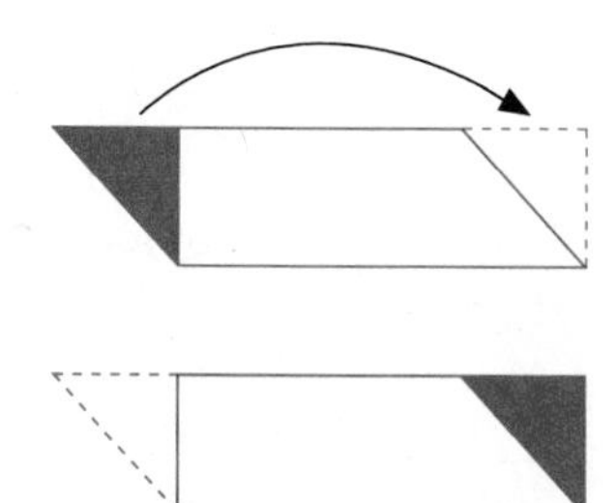

Theorem 12.2 The area of a parallelogram is the product of the length of a *base* and the length of a corresponding *altitude* (Area of parallelogram = bh).

Given: $\square ABCD$ with base $\overline{AB}$ and altitude $\overline{CE}$, $AB = b$, $CE = h$

Prove: Area($\square ABCD$) = bh

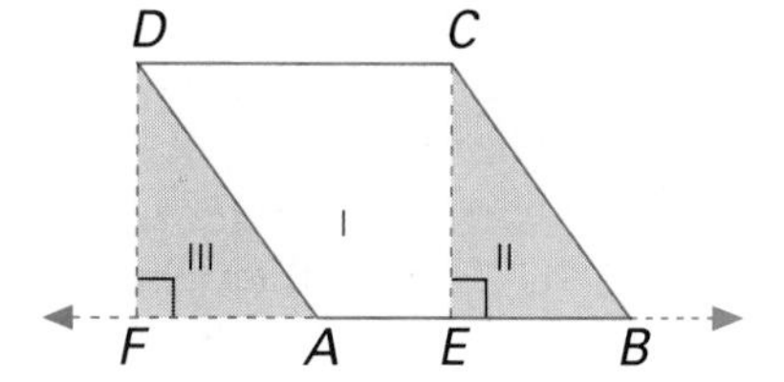

Plan for proof:

The region $ABCD$ can be separated into two parts, which can be rearranged to form a rectangular region $FECD$. By constructing $\overline{DF} \perp \overleftrightarrow{AB}$ and $\overline{CE} \perp \overleftrightarrow{AB}$, right triangles DFA and CEB are formed. $\triangle DFA \cong \triangle CEB$ by HL, and Area II = Area III. Therefore, Area I + Area II = Area I + Area III, or Area ($\square ABCD$) = Area(rect $FECD$) = bh.

EXAMPLE 1 Find the area of $\square ABCD$.

Solution $b = 3.4$ cm, $h = 2.5$ cm
Area($\square ABCD$) = $bh = 3.4 \cdot 2.5 = 8.5$ cm^2

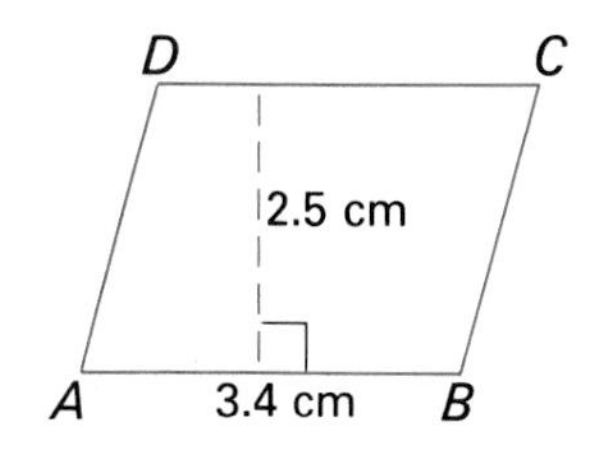

If the area of a parallelogram and the length of either a base or an altitude are known, then the other length can be found.

EXAMPLE 2 Given: Area($\square PQRS$) = 36 cm², length of altitude $\overline{HS}$ = 4 cm
Find the length of base $\overline{PQ}$.

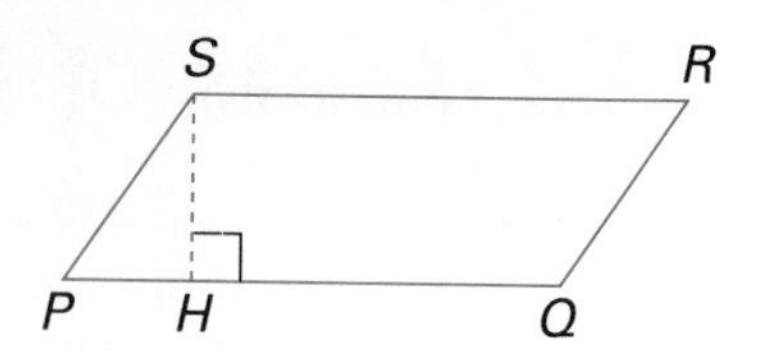

Solution
$A = bh$
$36 = 4 \cdot PQ$
$PQ = 9$ cm

EXAMPLE 3 Given: $\square ABCD$, $\overline{DE} \perp \overline{AB}$, $\overline{BF} \perp \overline{AD}$, $AB = 15$, $DE = 10$, $AD = 12$
Find BF.

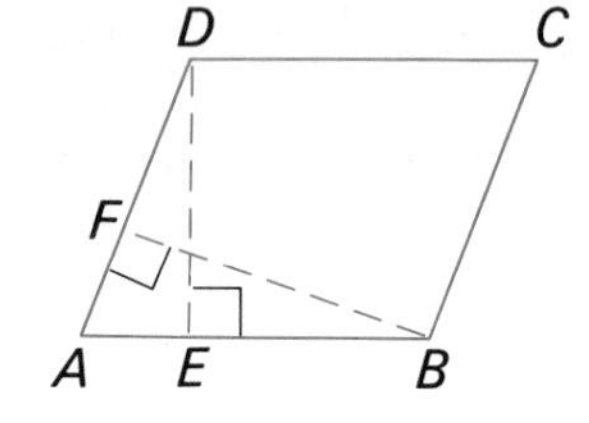

Plan Use both $\overline{AB}$ and $\overline{DA}$ as bases, with corresponding altitudes $\overline{DE}$ and $\overline{BF}$.

Solution Area($\square ABCD$) $= AB \cdot DE$
$= 15 \cdot 10 = 150$

Area($\square ABCD$) $= AD \cdot BF$
$= 12 \cdot BF$

Therefore, $150 = 12 \cdot BF$ and $BF = 12\frac{1}{2}$.

EXAMPLE 4 The lengths of two sides of a parallelogram are 12 cm and 10 cm. The measure of the angle between the two sides is 30. Find the area of the parallelogram.

Plan Draw altitude $\overline{JK}$ to make a 30-60-90 triangle. The length of the side opposite the angle of measure 30 is half that of the hypotenuse.

Solution $JK = \frac{1}{2} \cdot GJ = 5$
Therefore, Area($\square GHIJ$) $= 5 \cdot 12$, or 60 cm².

Classroom Exercises

Tell whether enough information is given to find the area.

1.
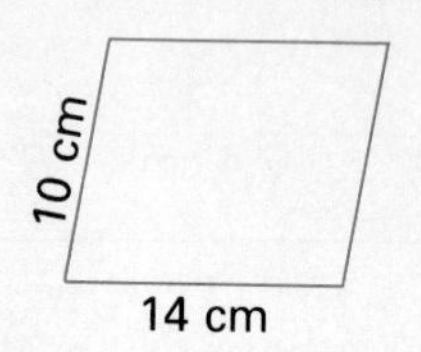

2.
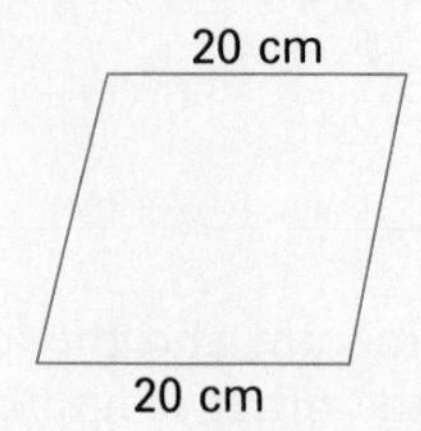

3.
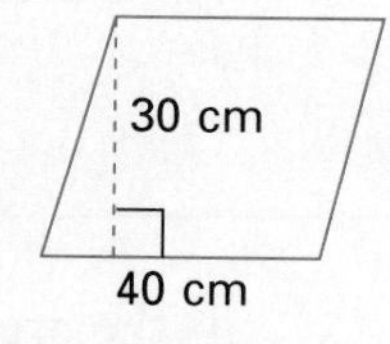

Find the area of the parallelogram.

4.

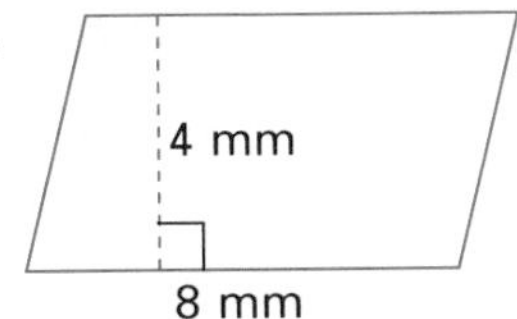

5.

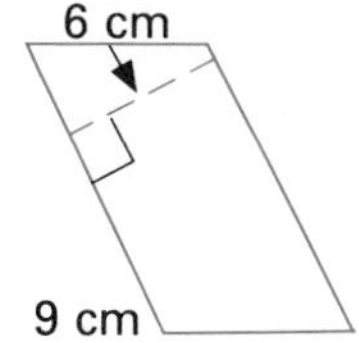

6.

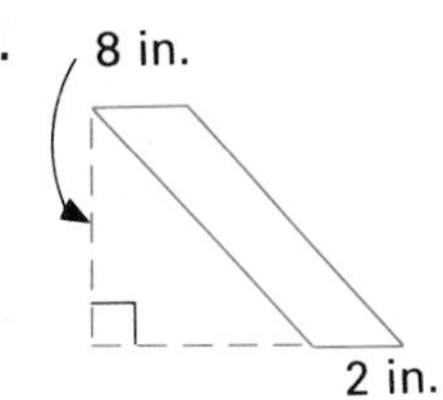

Written Exercises

Find the missing measurement for the indicated parallelogram.

	Length of base b	Length of altitude h	Area
1.	8 cm	7 cm	______
2.	9 m	3 m	______
3.	5.2 in.	3.4 in.	______
4.	2.7 ft	1.2 ft	______
5.	______	5 mm	30 mm^2
6.	8 cm	______	72 cm^2
7.	2.7 ft	3.4 ft	______
8.	6.9 in.	7.8 in.	______
9.	$\sqrt{5}$ m	______	5 m^2
10.	______	$\sqrt{7}$ mm	$\sqrt{14}$ mm^2
C **11.**	2.6 cm	______	8.84 cm^2
C **12.**	______	9.2 in.	76.36 in^2

In $\square ABCD$, $\overline{FD} \perp \overline{AB}$ and $\overline{EG} \perp \overline{DA}$. Find the missing measurement.

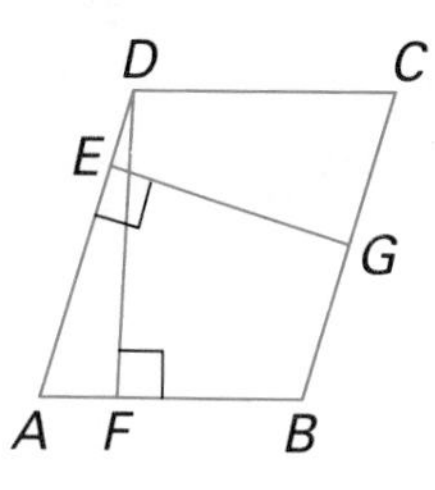

13. $AB = 12$, $FD = 6$, $DA = 8$, $EG =$ ______

14. $AB = 30$, $FD =$ ______, $DA = 12$, $EG = 5$

15. $AB =$ ______, $FD = 2.5$, $DA = 10$, $EG = 5$

16. $AB = 24$, $FD = 1.2$, $DA =$ ______, $EG = 3.6$

17. $AB = \sqrt{6}$, $FD = 2$, $DA =$ ______, $EG = \sqrt{3}$

18. $AB =$ ______, $FD = AB$, $DA = 4$, $EG = 9$

Find the area of $\square PQRS$.

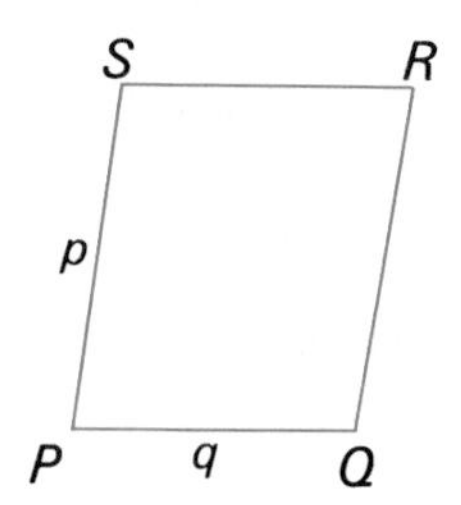

19. $p = 6$, $q = 12$, $m \angle P = 30$

20. $p = 10$, $q = 8$, $m \angle P = 30$

21. $p = 10$, $q = 8$, $m \angle P = 45$

22. $p = 10$, $q = 8$, $m \angle P = 60$

23. Solve Example 3 of this lesson using similar triangles instead of area formulas.

Find the area of the parallelogram.

24.

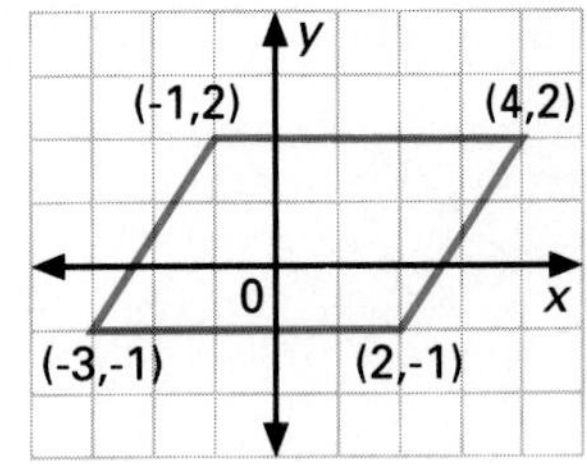

25.

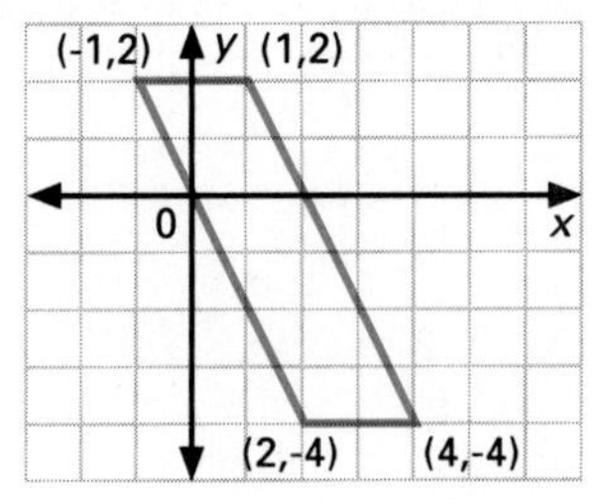

26.

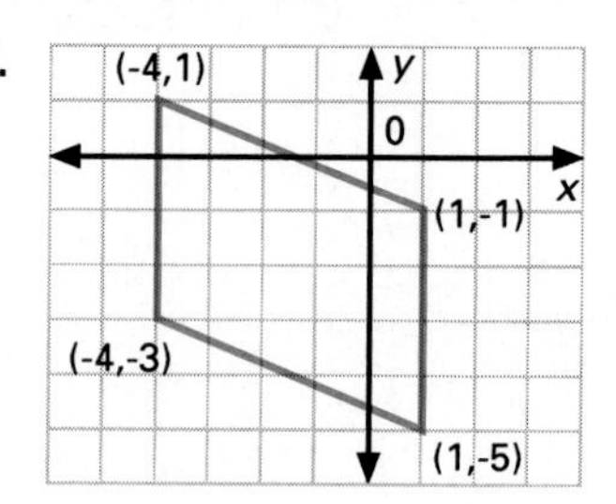

Use the appropriate trigonometric ratio to find the area, to the nearest tenth, of a parallelogram with the indicated measures.

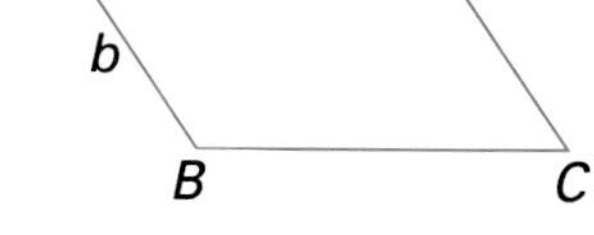

27. $a = 3.8$, $b = 2.7$, m $\angle A = 49$

28. $a = 5.9$, $b = 4.2$, m $\angle A = 67$

29. $a = 3.6$, $b = 4.7$, m $\angle A = 68$

30. $a = 5.8$, $b = 3.1$, m $\angle A = 31$

31. Prove Theorem 12.2.

32. Prove: Area($\square ABCD$) $= ab \cdot \sin A$, where a and b are the lengths of the sides that include $\angle A$, and A is an acute angle.

33. Several parallelograms are shown on the dot grid below. The horizontal or vertical distance between adjacent dots is 1 cm. Find the area of each parallelogram. Determine if there is any relationship between the area, the number of dots on the sides of the parallelogram, and the number of dots in the interior of the parallelogram.

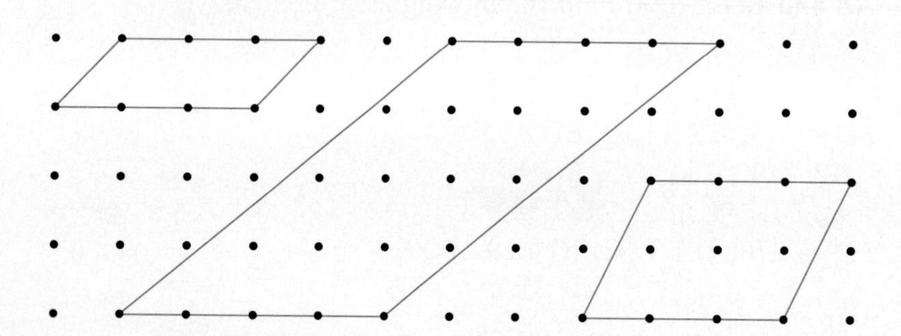

Mixed Review

Do the three numbers form a Pythagorean triple? *9.2*

1. 4, 5, 6
2. 8, 9, 10
3. 6, 8, 10
4. If the length of each leg of an isosceles right triangle is 5 in., find the measure of the hypotenuse. *9.4*

Problem Solving Strategies

More than One Way

The Pythagorean Theorem can be proved in a number of ways. Three different ways are suggested here. Each one requires that you compare different areas and use algebra. You will need to use the formulas for the areas of a square, a triangle, and a trapezoid. You will have to add or subtract areas in order to find areas in different ways and then compare them.

1. Four identical right triangles are cut out and arranged as shown. Prove the Pythagorean Theorem by finding the area of the large square in two ways—side squared and sum of five pieces. Set the two areas equal to each other.

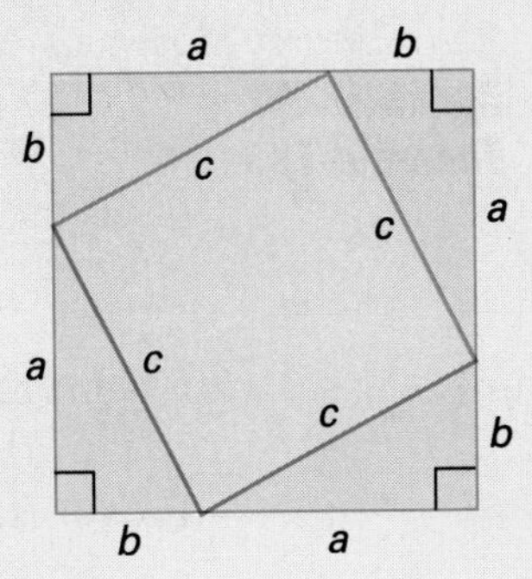

2. The figure to the right shows four right triangles arranged differently. This time there is a small square in the center. Again, use the area of the large square in two ways to prove the Pythagorean Theorem.

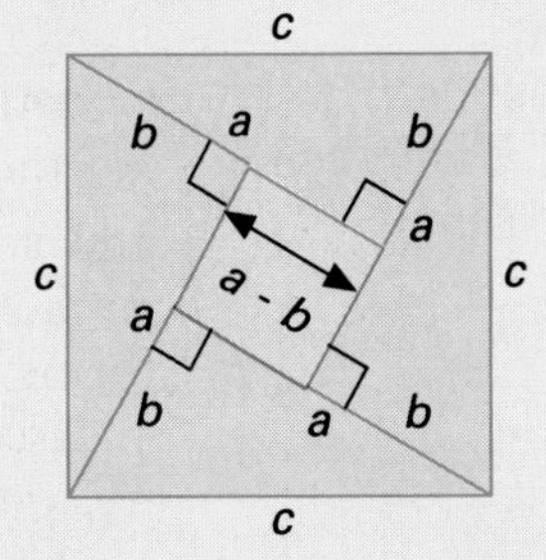

3. This method was discovered by President Garfield. Two identical right triangles are arranged as shown. Follow the steps below to prove the Pythagorean Theorem.
   a. Verify that the figure in the center is a right triangle.
   b. Verify that the large figure is a trapezoid with a and b the lengths of the bases and $(a + b)$ the length of an altitude.
   c. Find the area of the trapezoid in two different ways.

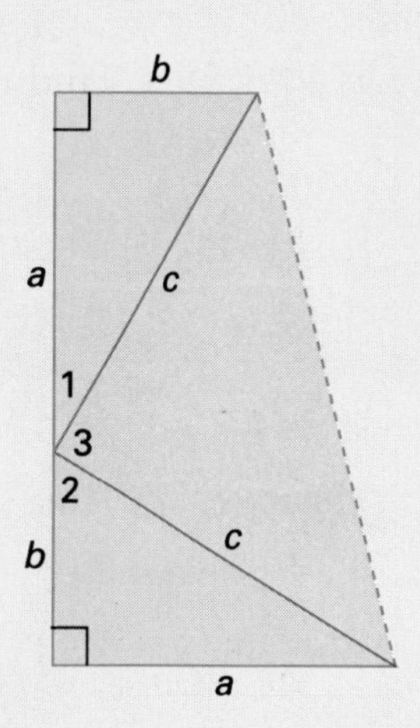

12.4 Areas of Triangles

Objectives To apply the formula for the area of a triangle
To apply the formulas for the areas of a rhombus and a kite

If a parallelogram region is divided by a diagonal, two congruent triangles are formed. This fact suggests a way of finding the formula for the area of any triangle.

Theorem 12.3 The area of a triangle is one-half the product of the length of a base and the length of a corresponding altitude (Area of triangle $= \frac{1}{2}bh$).

Given: $\triangle ABC$, altitude $\overline{CE} \perp \overleftrightarrow{AB}$,
$AB = b$, $CE = h$
Prove: Area($\triangle ABC$) $= \frac{1}{2}bh$

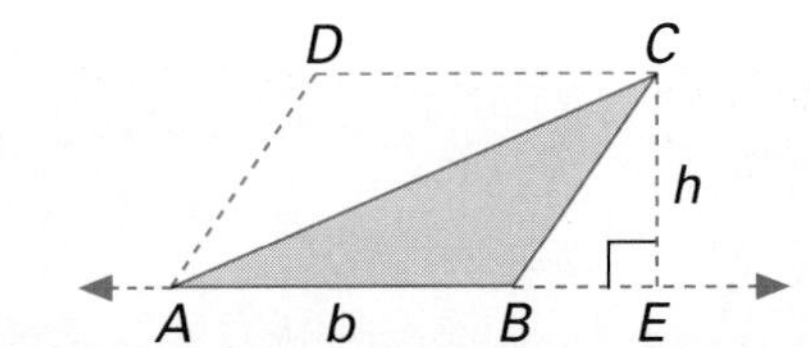

Proof: (Paragraph Form) Construct a parallelogram, the area of which is bh. This is done by using the Parallel Postulate: Through points A and C construct $\overline{AD} \parallel \overline{BC}$ and $\overline{CD} \parallel \overline{AB}$. By the Area Addition Postulate, Area($\square ABCD$) = Area($\triangle ABC$) + Area($\triangle CDA$). Since the diagonal of a parallelogram forms two congruent triangles, $\triangle ABC \cong \triangle CDA$ so that Area($\triangle ABC$) = Area($\triangle CDA$). Then Area($\square ABCD$) $= 2 \cdot$ Area($\triangle ABC$) $= bh$, or Area($\triangle ABC$) $= \frac{1}{2}bh$.

EXAMPLE 1 Find the area of the given triangles.

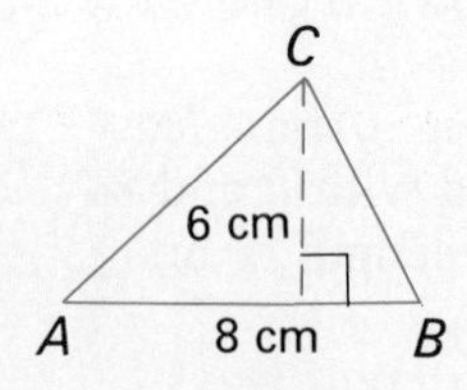

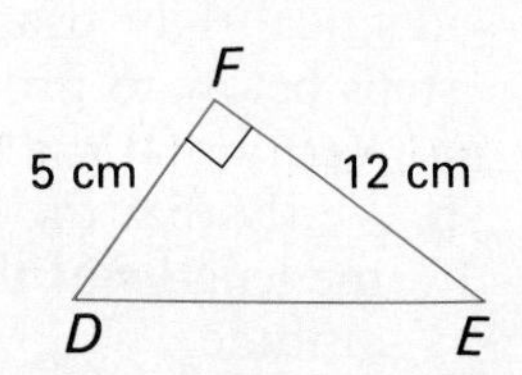

Solutions

Area $= \frac{1}{2}bh$

$= \frac{1}{2} \cdot 8 \cdot 6 = 24 \text{ cm}^2$

Area $= \frac{1}{2}bh$

$= \frac{1}{2} \cdot 5 \cdot 12 = 30 \text{ cm}^2$

EXAMPLE 2 Given: Area($\triangle GHI$) = 36 mm², IM = 6 mm
Find GH.

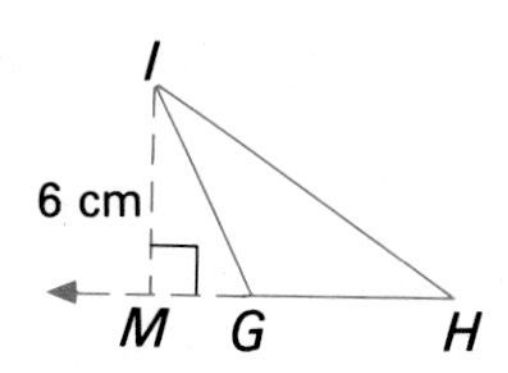

Solution

$$\text{Area}(\triangle GHI) = \tfrac{1}{2}bh$$
$$36 = \tfrac{1}{2} \cdot GH \cdot 6$$
$$36 = 3(GH)$$
$$GH = 12 \text{ mm}$$

A **kite** is a quadrilateral in which one diagonal is the perpendicular bisector of the other. The region bounded by a kite can be thought of as two triangular regions. This leads to a formula for the area of a kite.

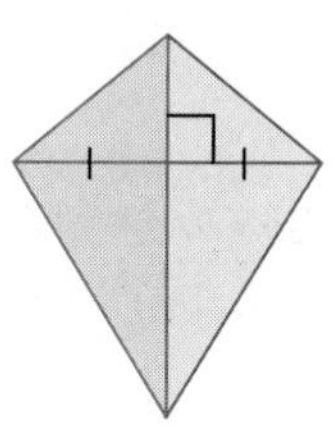

Theorem 12.4 The area of a kite is one-half the product of the lengths of the diagonals. (Area of Kite = $\frac{1}{2}d_1d_2$)

Given: Kite $ABCD$; diagonal $\overline{AC}$ is the perpendicular bisector of diagonal $\overline{BD}$, $AC = d_1$, $BD = d_2$
Prove: Area(Kite $ABCD$) = $\frac{1}{2}d_1d_2$

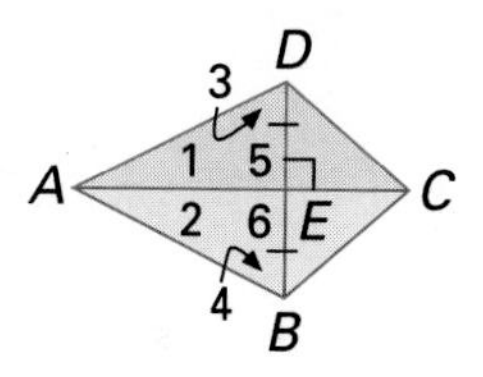

Plan

Flow Diagram

Kite $ABCD$ ⟶ $\overline{AC} \perp \overline{DB}$ ⟶ $\begin{cases} \text{Area}(\triangle ADC) = \frac{1}{2}\, d_1 \cdot DE \\ \text{Area}(\triangle ABC) = \frac{1}{2}\, d_1 \cdot EB \end{cases}$

$$\text{Area(Kite } ABCD) = \text{Area}(\triangle ADC) + \text{Area}(\triangle ABC)$$
$$= \tfrac{1}{2}\, d_1(DE + EB) = \tfrac{1}{2}\, d_1d_2$$

Corollary The area of a rhombus is one-half the product of the lengths of the two diagonals.

EXAMPLE 3 Find the area of a rhombus with diagonals 16 cm and 12 cm long.

Solution

$$\text{Area} = \tfrac{1}{2}d_1d_2$$
$$= \tfrac{1}{2} \cdot 16 \cdot 12 = 96 \text{ cm}^2$$

There is a special formula for the area of an equilateral triangle.

Theorem 12.5 If s is the length of a side of an equilateral triangle, then the area is $\frac{s^2}{4}\sqrt{3}$.

Given: Equilateral $\triangle MNO$, $MN = NO = OM = s$, altitude $\overline{OP}$ of length h

Prove: Area($\triangle MNO$) $= \frac{s^2}{4}\sqrt{3}$

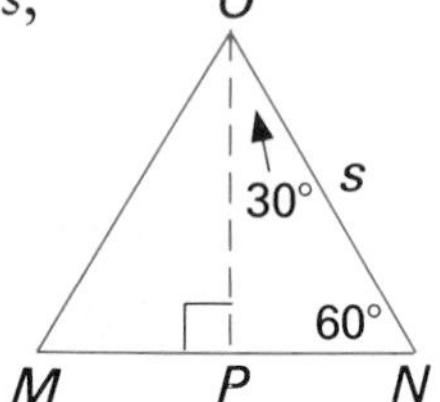

Plan The altitude of an equilateral triangle forms two congruent triangles. Use the properties of 30-60-90 triangles to find OP in terms of s.

Another famous formula is ascribed to the ancient Greek mathematician, Heron. *Heron's Formula* uses the lengths of the sides of a triangle to find the area. The formula is stated here without proof. Note that the formula uses half the perimeter, called the *semiperimeter*.

Theorem 12.6 **Heron's Formula:** If a, b, and c are the lengths of the sides of a triangle and s is the semiperimeter, such that $s = \frac{1}{2}(a + b + c)$, then Area(triangle) $= \sqrt{s(s - a)(s - b)(s - c)}$.

EXAMPLE 4 Find the area of a triangle with the sides 5 cm, 12 cm, and 13 cm long.

Solution Let $a = 5$, $b = 12$, $c = 13$.

$s = \frac{1}{2}(5 + 12 + 13) = 15$

$A = \sqrt{15(15 - 5)(15 - 12)(15 - 13)}$

$A = \sqrt{15 \cdot 10 \cdot 3 \cdot 2} = \sqrt{900} = 30$

Therefore, the area is 30 cm^2.

C
5 cm
12 cm
A 13 cm B

Classroom Exercises

Find the area of the triangle, rhombus, or kite.

1.

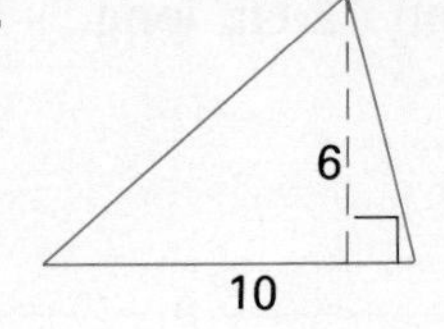

2.

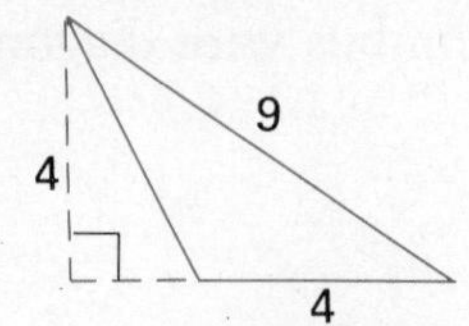

3.

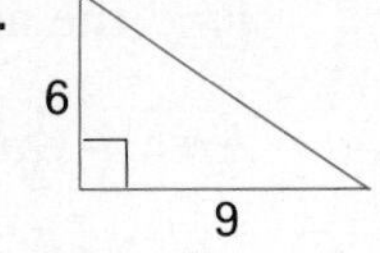

4.

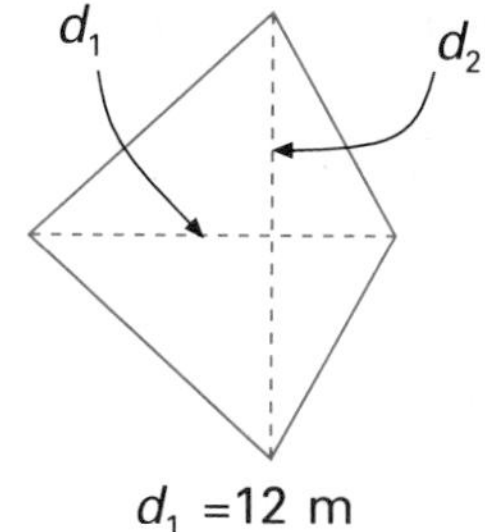

5.

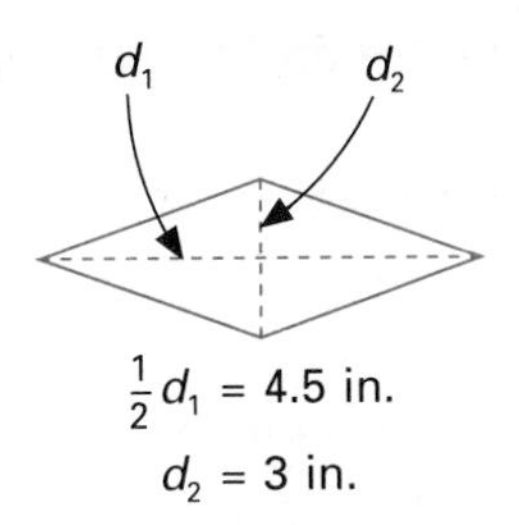

Written Exercises

Find the area of the triangle.

1.

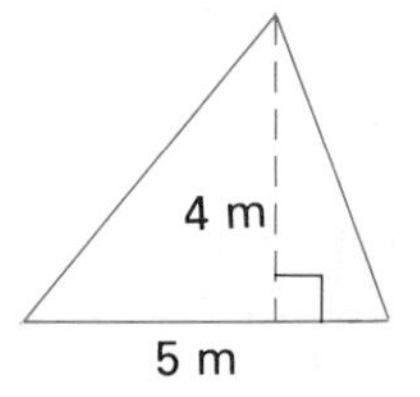

2.

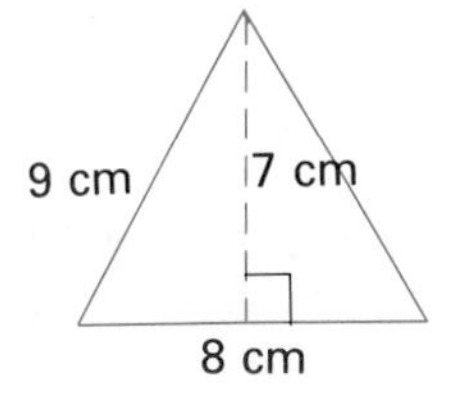

3.

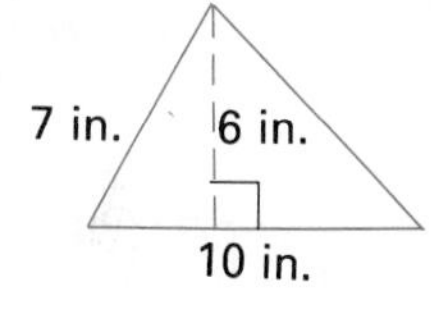

4.

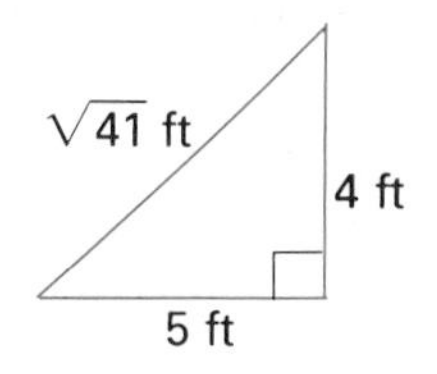

5.

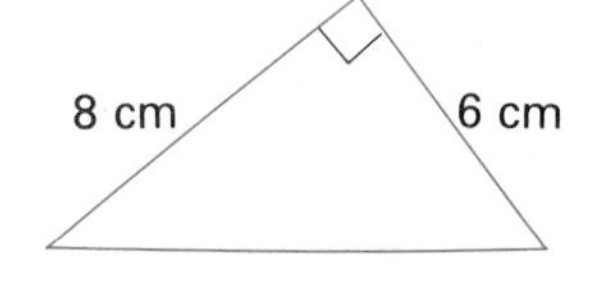

6.

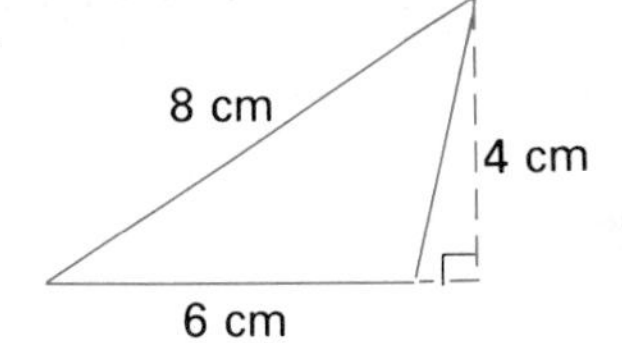

Find the missing measurement for the indicated triangle.

	b	*h*	Area
7.	8 cm	______	80 cm^2
8.	______	12 m	60 m^2
9.	3.4 mm	______	17 mm^2
10.	______	1.6 in.	4 in^2

11. Find the area of a rhombus with diagonals 9 mm and 14 mm long.

Find the area of each of the following polygons.

12. Find the area of a kite with diagonals 12 in. and 23 in. long.

13. Find the area of an equilateral triangle with sides 10 in. long.

In $\triangle ABC$, $\overline{CF} \perp \overline{AB}$, $\overline{AD} \perp \overline{BC}$, and $\overline{BE} \perp \overline{CA}$. Find the missing length.

14. $AB = 15$, $CF = 8$, $BC = 12$, $AD =$ ______

15. $AB = 24$, $CF = 18$, $CA =$ ______, $BE = 16$

16. $CA = 15$, $BE = 7$, $BC = 10$, $AD =$ ______

17. $CA = 6$, $BE = 9$, $BC =$ ______, $AD = 8$

Use Heron's Formula to find the area of a triangle with sides of the given lengths.

18. $a = 6, b = 8, c = 10$

19. $a = 5, b = 12, c = 13$

20. $a = 4, b = 5, c = 6$

21. $a = 6, b = 9, c = 12$

Find the length of an altitude and the area of an equilateral triangle with sides of each length. (Exercises 22–24)

22. $s = 10$

23. $s = 8$

24. $s = t$

25. The area of a rhombus is 25 cm^2. The length of one diagonal is 8 cm. What is the length of the second diagonal?

26. The area of a kite is 32 ft^2. The length of one diagonal is 12 ft. What is the length of the second diagonal?

Find the area of the triangle or rhombus.

27.

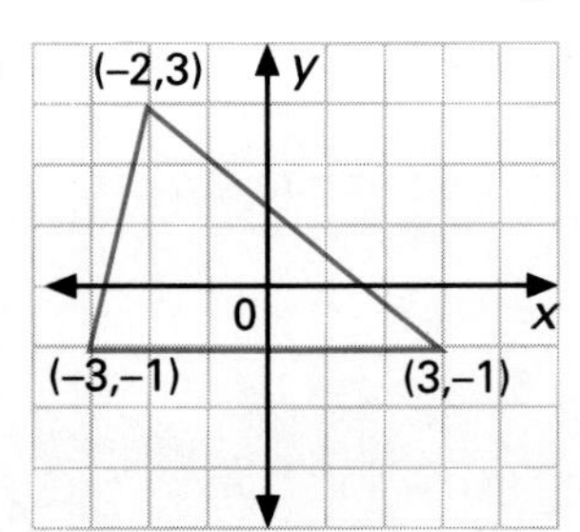

28.

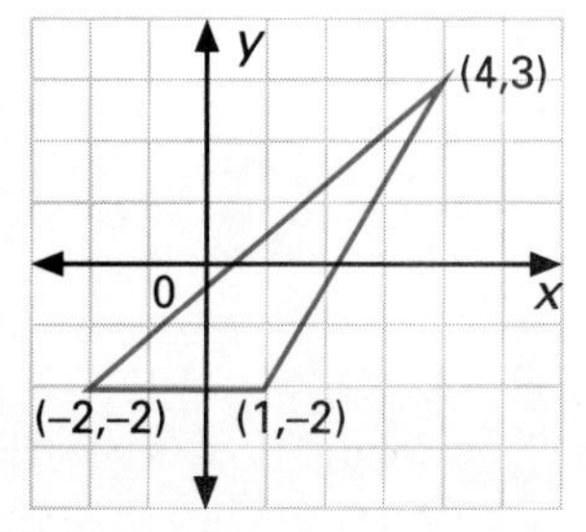

29.

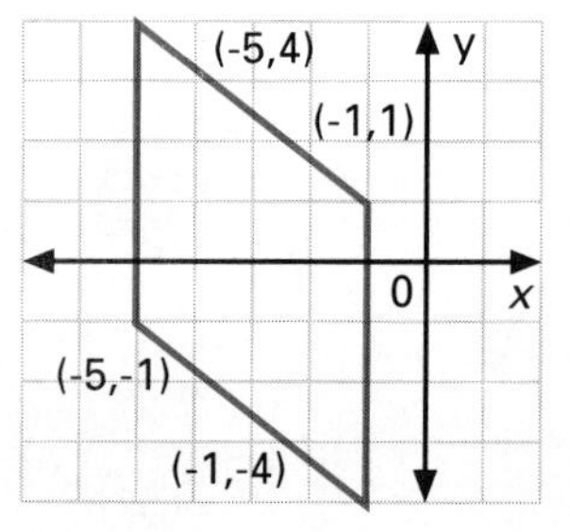

30. Write the proof of Theorem 12.4.

31. Prove: The area of a square is one-half the square of the length of the diagonal.

32. Write a plan for the proof of Theorem 12.5.

33. Prove Theorem 12.5.

Prove or disprove the statement.

34. If two triangles have equal areas, then the ratio of the lengths of their bases is equal to the ratio of the lengths of their altitudes.

Mixed Review

Which of the properties listed is a sufficient reason to prove that a quadrilateral is a parallelogram? To prove that it is a trapezoid? To prove that it is an isosceles trapezoid? *6.5, 7.4, 7.5*

1. Diagonals bisect each other.

2. A pair of opposite sides are congruent.

3. A pair of opposite sides are parallel.

4. Diagonals are congruent.

5. Opposite angles are congruent.

6. Opposite angles are supplementary.

12.5 Areas of Trapezoids

Objective To apply the formula for the area of a trapezoid

The roof of the New York Life Insurance Building is composed of trapezoidal sections. To find how much roofing material is needed for the building, the area of each section must be found.

Theorem 12.7 The area of a trapezoid is one-half the product of the length of an altitude and the sum of the lengths of the upper and lower bases [Area $= \frac{1}{2}h(b_1 + b_2)$].

Given: Trapezoid $ABCD$, $b_2 = DC$, $b_1 = AB$, length of altitude is h.
Prove: Area(trap $ABCD$) $= \frac{1}{2}h(b_1 + b_2)$

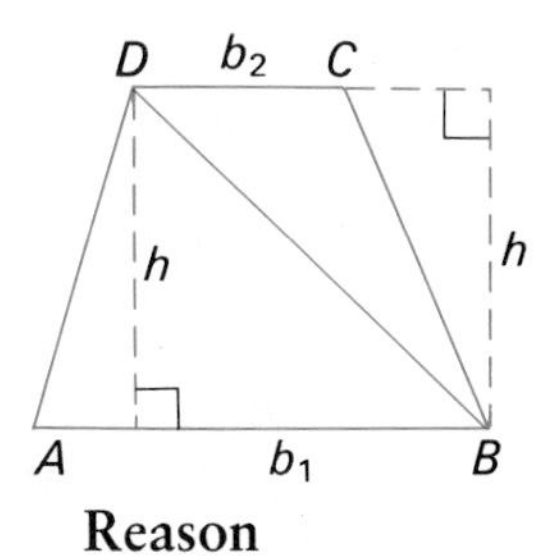

Statement	Reason
1. Trap $ABCD$, $AB = b_1$, $DC = b_2$, length of alt is h.	1. Given
2. Draw $\overline{DB}$.	2. Two points determine a line.
3. Area($\triangle ADB$) $= \frac{1}{2}b_1h$, Area($\triangle CBD$) $= \frac{1}{2}b_2h$	3. Area($\triangle$) $= \frac{1}{2}bh$
4. Area(trap $ABCD$) $=$ Area($\triangle ADB$) $+$ Area($\triangle CBD$)	4. Area Add Post
5. Area($\triangle ADB$) $+$ Area($\triangle CBD$) $= \frac{1}{2}b_1h + \frac{1}{2}b_2h$	5. Add Prop of Eq
6. Area($\triangle ADB$) $+$ Area($\triangle CBD$) $= \frac{1}{2}h(b_1 + b_2)$	6. Distr Prop
7. $\therefore$ Area(trap $ABCD$) $= \frac{1}{2}h(b_1 + b_2)$	7. Trans Prop of Eq

EXAMPLE 1 Find the area of trapezoid $ABCD$.

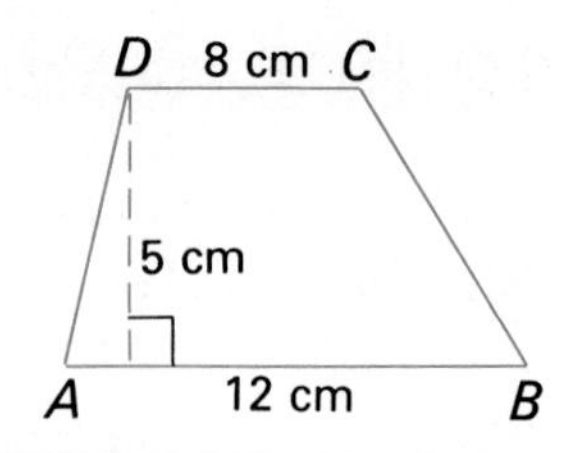

Solution $b_1 = 12$ cm, $b_2 = 8$ cm, $h = 5$ cm

Area $= \frac{1}{2} \cdot 5(12 + 8) = 50 \text{ cm}^2$

EXAMPLE 2 The area of a trapezoid is 80 mm^2. The length of one base is 6 mm. The length of an altitude is 8 mm. What is the length of the other base?

Solution Area $= 80 \text{ mm}^2$, $b_2 = 6$ mm, $h = 8$ mm

$$80 = \frac{1}{2} \cdot 8(b_1 + 6)$$
$$80 = 4b_1 + 24$$
$$56 = 4b_1$$
$$b_1 = 14$$

The length of the other base is 14 mm.

The lengths of the two nonparallel sides, or legs, are not used in the formula for the area of a trapezoid. However, for certain angle measures, the area can be found if one of the legs and the measure of an angle are known.

EXAMPLE 3 Find the area of trapezoid $IJKL$.

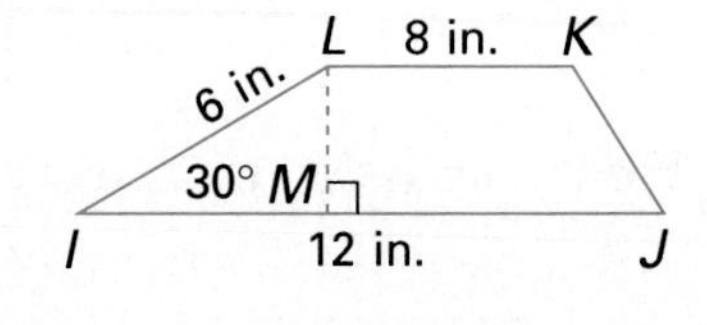

Solution $LM = 3$ in., since $\triangle ILM$ is 30-60-90.

Then, Area(trap $IJKL$) $= \frac{1}{2} \cdot 3(8 + 12)$
$= 30 \text{ in}^2$

The legs of an isosceles trapezoid are congruent. If the lengths of the bases are known, then the area of the isosceles trapezoid can be found.

EXAMPLE 4 Find the area of isosceles trapezoid $ABCD$.

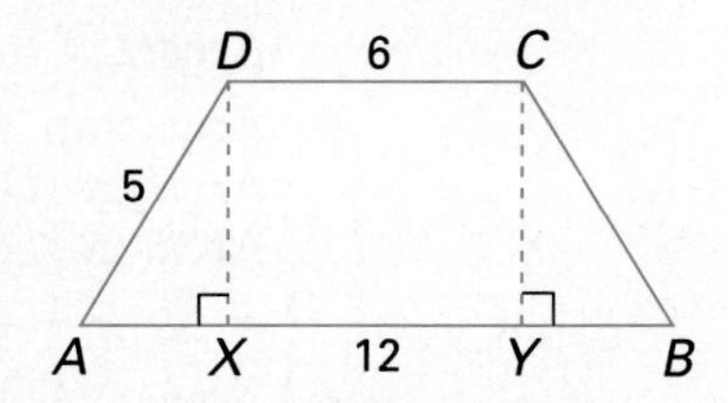

Solution When $\overline{DX}$ and $\overline{CY}$ are drawn perpendicular to $\overline{AB}$, rectangle $CDXY$ is formed. Therefore, $XY = DC = 6$. $\triangle AXD \cong \triangle BYC$ by HL, so that $AX = BY = 3$.

Then $h = 4$ by the Pythagorean Theorem.

Area($ABCD$) $= \frac{1}{2}h(b_1 + b_2)$
$= \frac{1}{2}(4)(18) = 36$

Classroom Exercises

State whether enough information is given to find the area.

1.
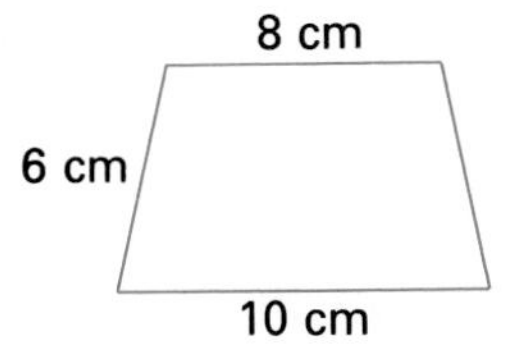

2.
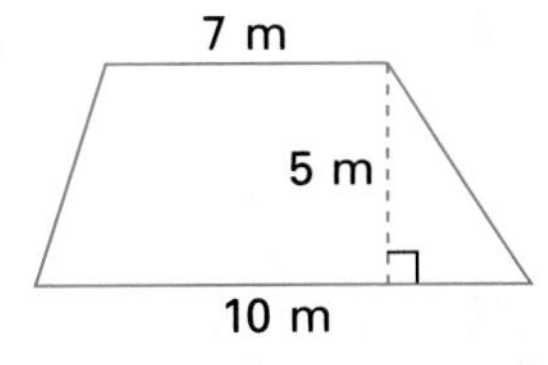

3.
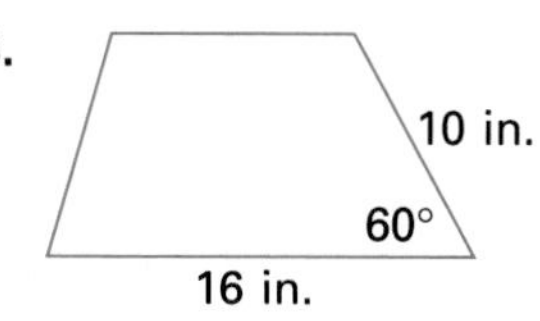

4.
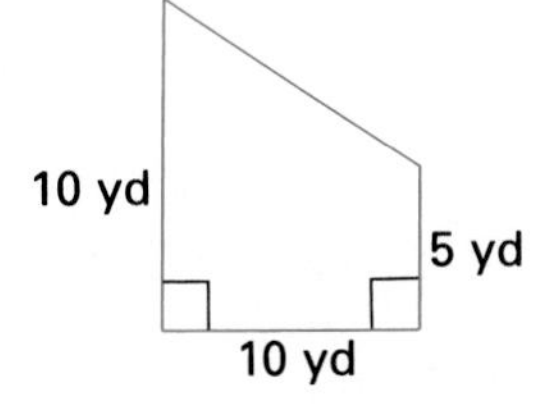

5.
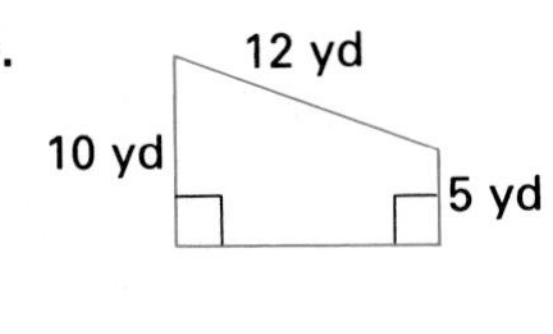

6.
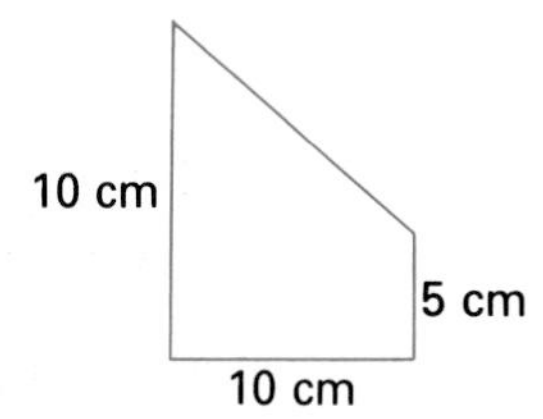

Written Exercises

Find the area of the trapezoid.

1.
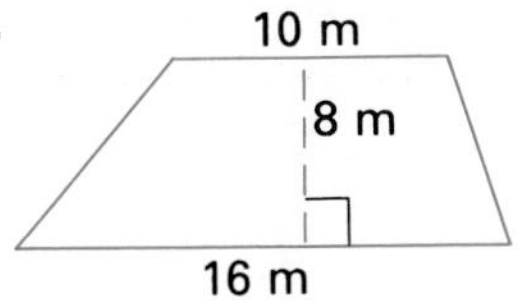

2.
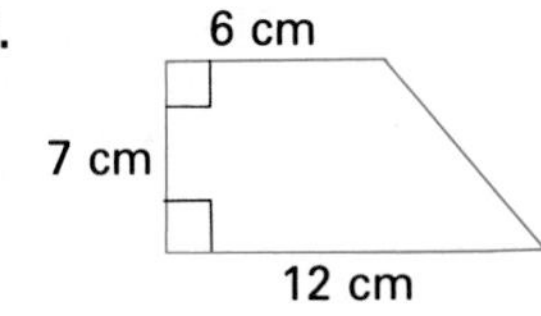

3.
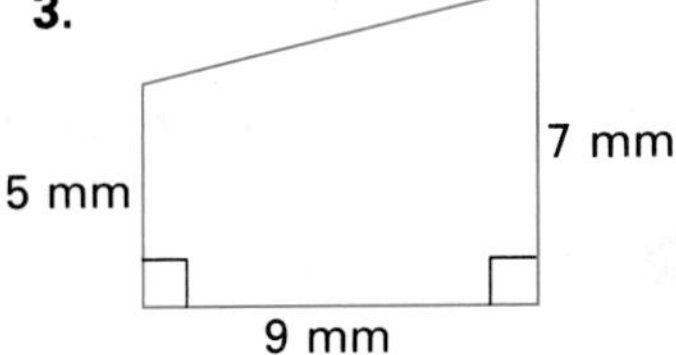

Find the missing measurement for the indicated trapezoid.

	b_1	b_2	h	Area
4.	8 cm	12 cm	5 cm	______
5.	10 mm	15 mm	9 mm	______
6.	8 in.	______	7 in.	70 in^2
7.	______	12 ft	5 ft	75 ft^2
8.	7 m	10 m	______	25.5 m^2
9.	16 cm	9 cm	______	87.5 cm^2

$\overline{AB}$ and $\overline{CD}$ are the bases of trapezoid $ABCD$. Find the area.

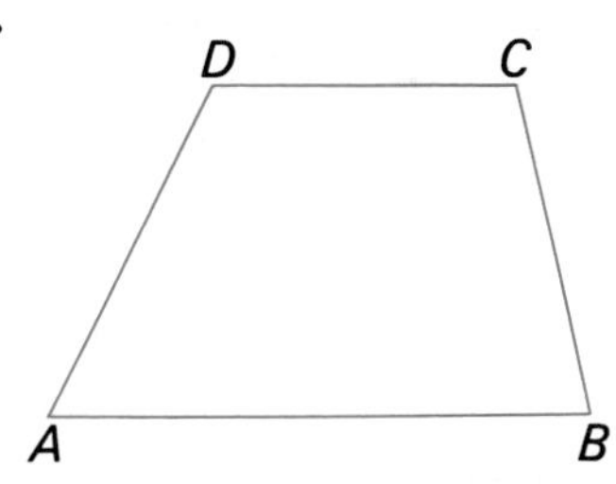

10. $AB = 16$, $CD = 12$, $AD = 8$, $m\angle A = 30$

11. $AB = 19$, $CD = 12$, $AD = 7$, $m\angle A = 30$

12. $AB = 12$, $CD = 8$, $AD = 5$, $m\angle A = 60$

13. $AB = 17$, $CD = 8$, $AD = 7$, $m\angle A = 60$

14. $AB = 20$, $CD = 14$, $AD = 10$, $m\angle A = 45$

15. $AB = 13$, $CD = 6$, $AD = 3$, $m\angle A = 45$

Find the area of the indicated isosceles trapezoid $ABCD$.

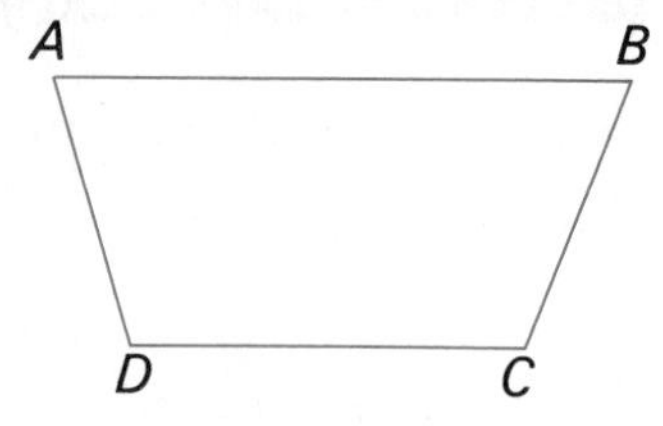

16. $AB = 12$, $CD = 6$, $AD = 5$
17. $AB = 40$, $CD = 30$, $AD = 13$
18. $AB = 16$, $CD = 8$, $m\angle A = 30$
19. $AB = 12$, $CD = 6$, $m\angle A = 60$
20. $AB = 20$, $CD = 12$, $m\angle A = 45$

Find the dimensions of a figure with the same altitude and the same area as trapezoid $PQRS$. Then draw the figure. (Exercises 21–23)

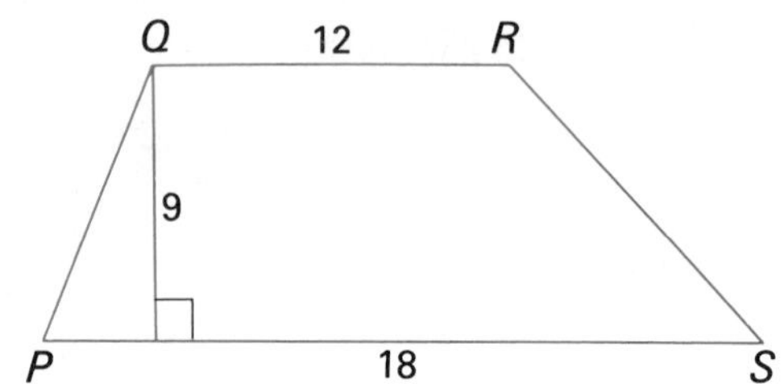

21. a triangle
22. a rectangle
23. a parallelogram

24. It has been suggested that the trapezoid formula is the only formula needed for finding the areas of all of the following: rectangle, square, parallelogram, triangle, and trapezoid. Give an argument for or against this.

Find the area of the trapezoid.

25.

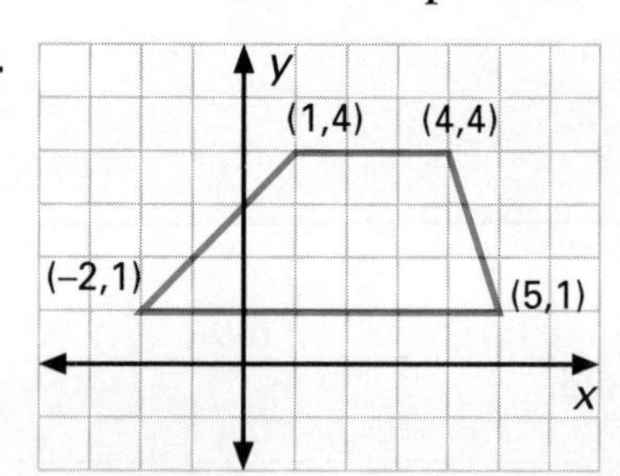

26.

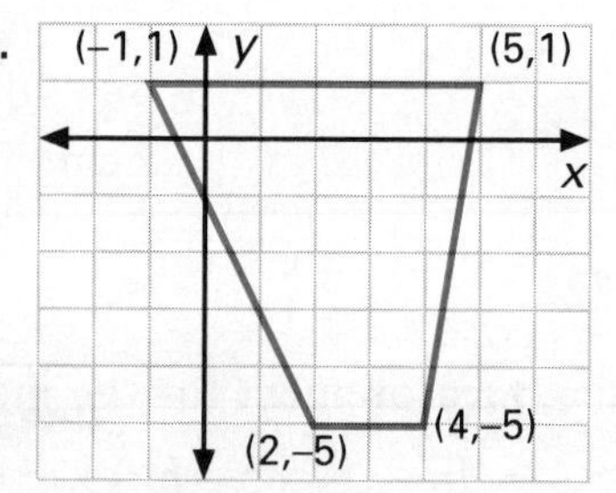

Find the area, to the nearest hundredth, of the indicated trapezoid using the appropriate trigonometric ratio.

	b_1	b_2	s	$m\angle A$
27.	10 cm	8 cm	5 cm	56
28.	16 mm	12 mm	6 mm	87
29.	17.6 m	12.3 m	11.7 m	77
30.	16.2 cm	13.7 cm	9.6 cm	58

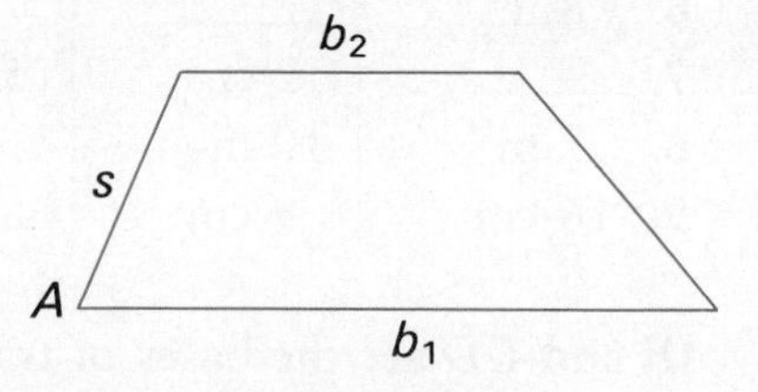

31. An altitude of a trapezoid is 8 cm long. The area is 168 cm^2. If one base is 12 cm longer than twice the other base, what is the length of each base?

32. An altitude of a trapezoid is 11 m long. The area is 143 m^2. If the lower base is 3 m shorter than three times the upper base, what is the length of each base?

Use the trapezoid from Exercises 27–30 to prove the following.

33. If b_1 and b_2 are the lengths of the bases of a trapezoid, and s is the length of a side adjacent to $\angle A$, then the area of the trapezoid is $\frac{1}{2} \cdot s(\sin A)(b_1 + b_2)$.

Midchapter Review

Find the area of each figure.

1.

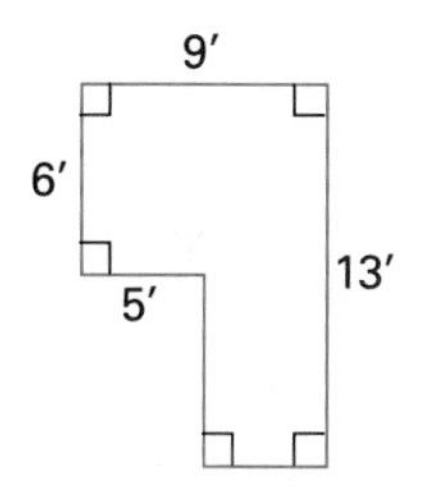

2. parallelogram

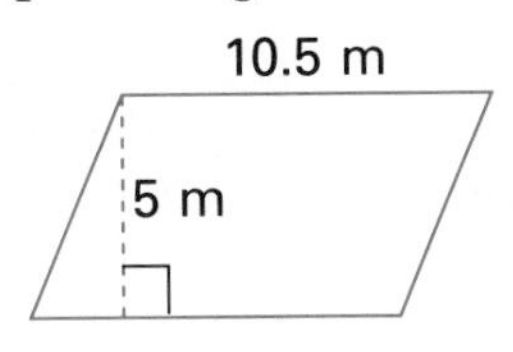

3.

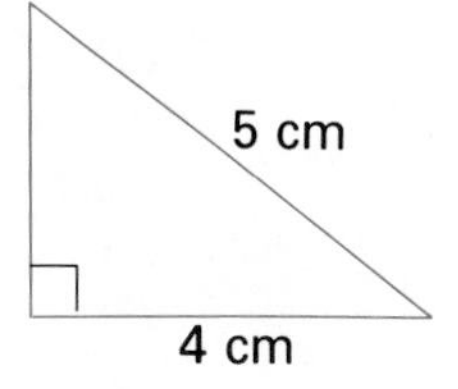

4. equilateral triangle

5.

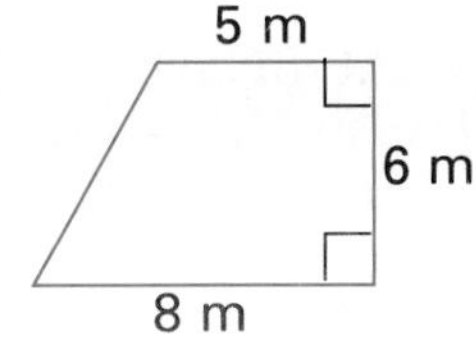

6.

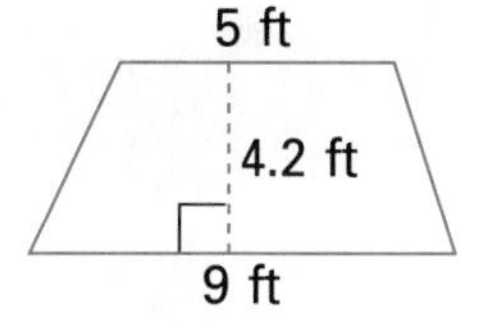

Application: *Areas of Irregularly-Shaped Regions*

The area of an irregularly-shaped region can be estimated by dividing the region into narrow strips, all the same width. Each strip is approximately trapezoidal in shape. A surveyor's map shows the region below. Each strip is 10 m wide. What is the approximate area of the region in square meters? If a hectare is 10,000 m^2, what is the approximate area in hectares?

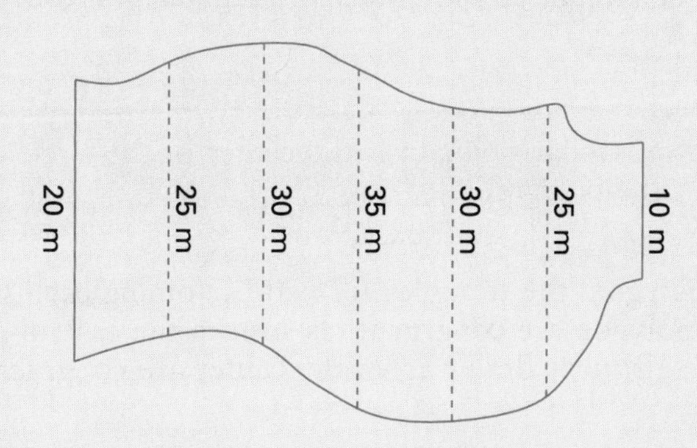

12.6 Measuring the Regular Polygons

Objectives To find the perimeter and area of a regular polygon
To find the apothem of a regular polygon

A regular polygon is both equilateral and equiangular. These properties allow you to find the circle that circumscribes the regular polygon.

Suppose $ABCDEF$ is a regular hexagon with $\overline{AO}$ and $\overline{BO}$ the angle bisectors of $\angle FAB$ and $\angle ABC$, respectively. Since the hexagon is equiangular, $\angle 2 \cong \angle 3$. Therefore, $OA = OB$.

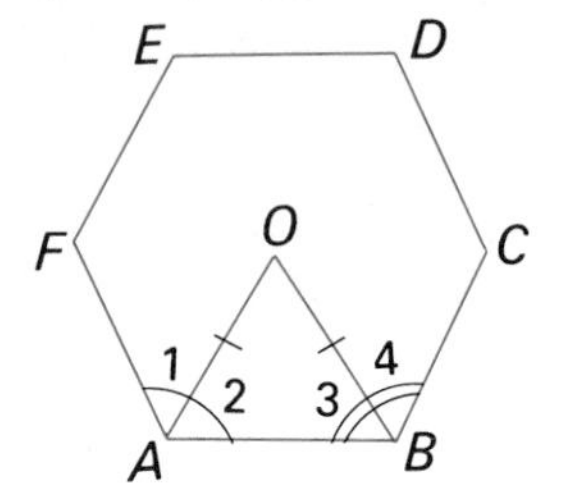

Draw $\overline{OC}$. $\triangle OBA \cong \triangle OBC$ by *SAS*. Therefore, $OA = OC$ by *CPCTC*. So $OA = OB = OC$. This process can be continued for each vertex of the hexagon so that $OA = OB = OC = OD = OE = OF$. As a result, O is the center of the circle containing points A, B, C, D, E, and F.

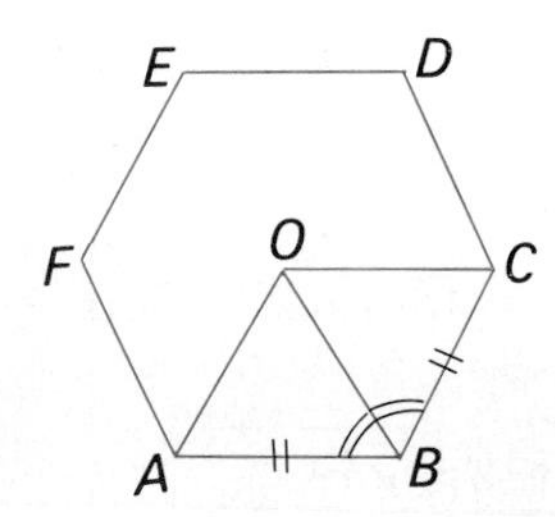

The above method of constructing the center of the circumscribed circle can be extended to any regular polygon. This essentially proves the following theorem.

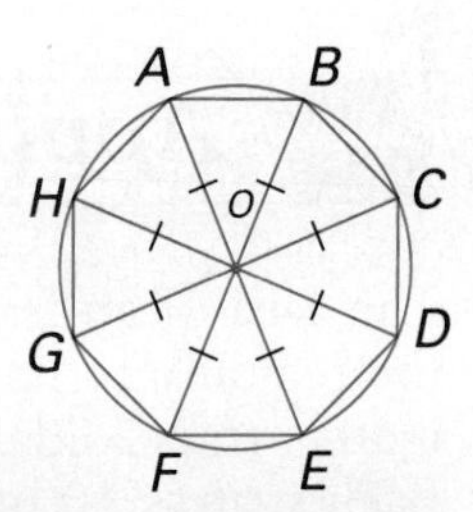

Theorem 12.8 A circle can be circumscribed about any regular polygon.

Definition The **center of a regular polygon** is the center of the circle circumscribed about the polygon.

In a regular polygon, the triangles formed by joining each vertex to the center of the polygon are congruent, as seen in the case of the hexagon. Therefore, the altitudes drawn from the center are congruent for each triangle.

Definitions

The **apothem of a regular polygon** is the length of a perpendicular segment from the center of the polygon to a side. A **radius of the regular polygon** is a segment from the center to a vertex of the polygon.

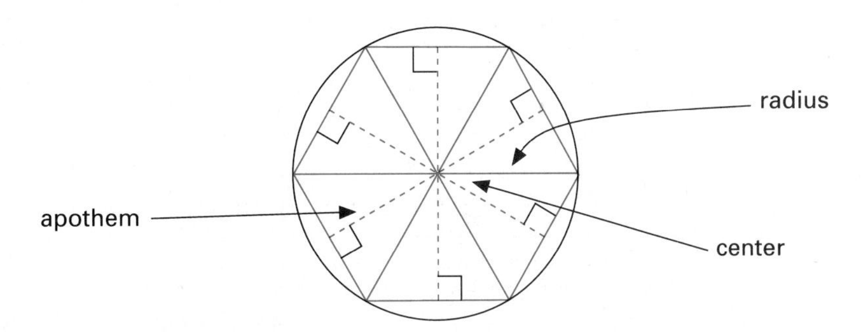

EXAMPLE 1 Find the apothem of a regular hexagon with a side 10 cm long.

Plan In the regular hexagon, $\triangle ABO$ is equiangular (equilateral) since $m \angle OAB = m \angle OBA = \frac{1}{2} \cdot 120 = 60$.
Find the length of the altitude of $\triangle ABO$.

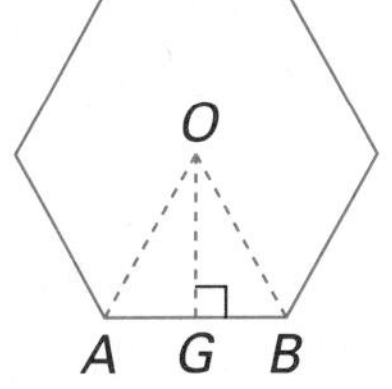

Solution The length h of an altitude of an equilateral triangle is $\frac{s\sqrt{3}}{2}$. In $\triangle ABO$, $s = AB = 10$.
Therefore, the apothem is $\frac{10\sqrt{3}}{2} = 5\sqrt{3}$ cm.

Definition

The **perimeter** of a polygon is the sum of the lengths of its sides.

The radii of a regular n-gon, or polygon with n sides, form n congruent triangles, where a is the apothem and s is the length of a side of the polygon.

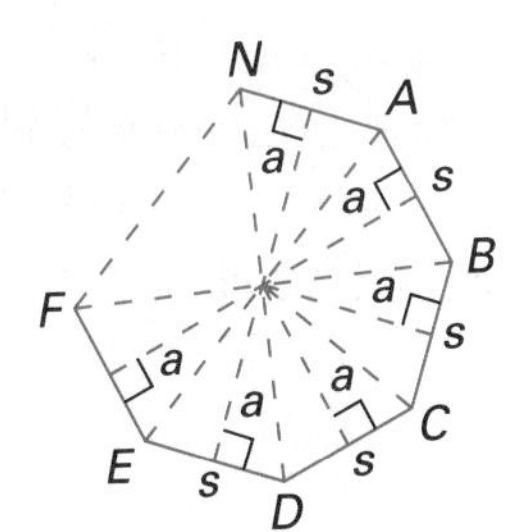

The area of the polygon is equal to the sum of the areas of all the triangles. Thus, Area(n-gon) $= n \cdot \frac{1}{2}a \cdot s$, where n is the number of sides. Since ns is equal to the perimeter p, Area(n-gon) $= \frac{1}{2}ap$.

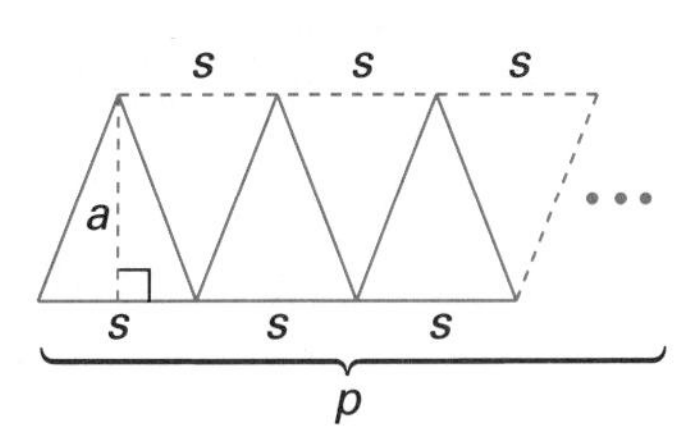

Theorem 12.9 The area of a regular polygon is one-half the product of the apothem and the perimeter [Area (n-gon) $= \frac{1}{2}ap$].

EXAMPLE 2 The perimeter of a regular pentagon is 21.8 mm. Its apothem is 3 mm. Find its area.

Solution

$$\text{Area} = \frac{1}{2}ap$$

$$\text{Area} = \frac{1}{2}(3 \cdot 21.8) = 32.7 \text{ mm}^2$$

EXAMPLE 3 The area of a regular octagon is 482.84 cm^2. The length of a side is 10 cm. Find the apothem.

Solution

$$\text{Area} = \frac{1}{2}ap = \frac{1}{2}a \cdot ns$$

$$482.84 = \frac{1}{2}a(8 \cdot 10)$$

$$482.84 = 40a$$

$$a = \frac{482.84}{40} \approx 12.071 \text{ cm}$$

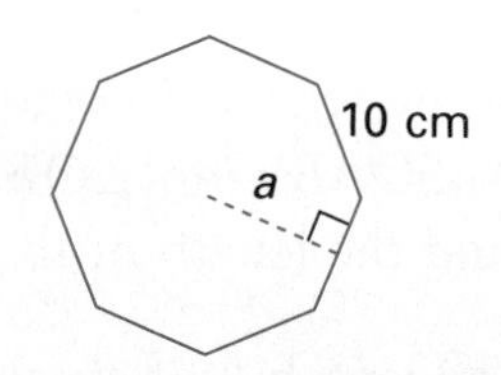

Other formulas relate the area of a regular polygon directly to the length of a side or to the length of a radius.

Theorem 12.10 The area of a regular polygon is $n\left[\sin\left(\frac{180}{n}\right)\right]\left[\cos\left(\frac{180}{n}\right)\right]r^2$, or $\dfrac{ns^2}{4\tan\left(\frac{180}{n}\right)}$, where n is the number of sides, s is the length of a side, and r is the length of a radius.

EXAMPLE 4 Find the area, to the nearest hundredth, of a regular hexagon with sides 5 cm long.

Solution

$$n = 6, s = 5$$

$$A = \frac{6 \cdot 5^2}{4\tan\left(\frac{180}{6}\right)}$$

$$A = \frac{150}{4 \cdot 0.5774} \approx 64.95 \text{ cm}^2$$

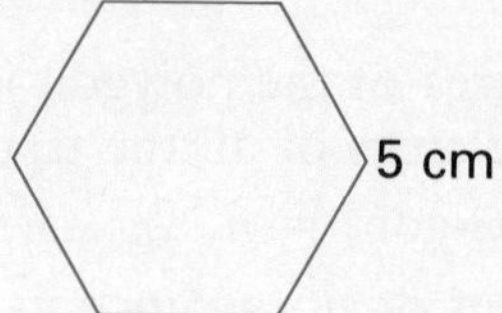

Classroom Exercises

Find the area of the regular polygon.

1. perimeter = 40 in., apothem = 5 in.
2. perimeter = 8 yd, apothem = 4 yd

Find the area of the regular polygon if the length of a side is 10 cm.

3. a square, apothem = 5 cm
4. a hexagon, apothem = 8.7 cm

Written Exercises

Find the apothem of the regular polygon.

1. a hexagon with a side 2 cm long
2. a hexagon with a side 10 cm long
3. a square with a side 6 mm long
4. a square with a side 10 mm long

Find, to the nearest tenth, the perimeter of the regular polygon.

5. a pentagon with a side 13 cm long
6. a dodecagon with a side 17 mm long
7. a 130-gon with a side 14 in. long
8. a 150-gon with a side 29 ft long

Find the area of the regular polygon.

	Polygon	Perimeter	Apothem	Area
9.	square	16 cm	2 cm	______
10.	pentagon	10 m	1.38 m	______
11.	pentagon	7.28 in.	1 in.	______
12.	hexagon	60 mm	8.66 mm	______

13. The area of a regular hexagon is 64.95 cm^2. The perimeter is 30 cm. Find the apothem.
14. The area of a regular octagon is 43.46 in^2. The apothem is 3.62 in. Find the perimeter.

Find the missing information for the regular polygon.

	Polygon	Length of side	Perimeter	Apothem	Area
15.	square	8 m	______	______	______
16.	hexagon	10 cm	______	______	______
17.	hexagon	______	______	$\sqrt{10}$ ft	______
18.	octagon	1 in.	______	______	______

Find, to the nearest tenth, the area of the regular polygon.

19. a pentagon with sides 10 cm long
20. a hexagon with sides 8 m long
21. an octagon with sides 12 yd long
22. a dodecagon with sides 20 ft long
23. a 30-gon with sides 10 mm long
24. a 100-gon with sides 50 cm long

Prove the following.

25. An apothem of a regular polygon bisects a side of the polygon.
26. A radius of a regular polygon bisects an angle of the polygon.
27. All triangles formed by the sides of a regular polygon and the radii of the polygon are congruent.

28. Prove that the area of a regular polygon is $n\left[\sin\left(\frac{180}{n}\right)\right]\left[\cos\left(\frac{180}{n}\right)\right]r^2$.

29. Prove that the area of a regular polygon is $\dfrac{ns^2}{4\tan\left(\frac{180}{n}\right)}$.

    $\left(\text{HINT: } \tan A = \dfrac{\sin A}{\cos A}.\right)$

30. Find the ratio of the area of an equilateral triangle inscribed in a circle to an equilateral triangle circumscribed about the same circle.
31. Find the ratio of the area of a square inscribed in a circle to a second square circumscribed about the same circle.
32. A surveyor was instructed to start at a given point, walk 10 ft due north, turn two degrees to the left, walk 10 ft, turn two more degrees to the left, and repeat the process until he was back at the starting point. How many acres of land did he walk around? (1 acre = 43,560 ft^2)
33. A 20-ft by 30-ft courtyard is to be surfaced with hexagonal concrete paving tiles. If each edge of each tile is 3 in. long, how many tiles will it take to cover the floor of the courtyard?

Mixed Review

Find the area. *12.2–12.5*

1.
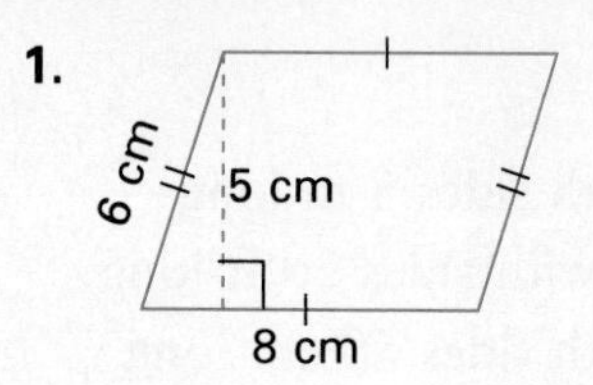

2.
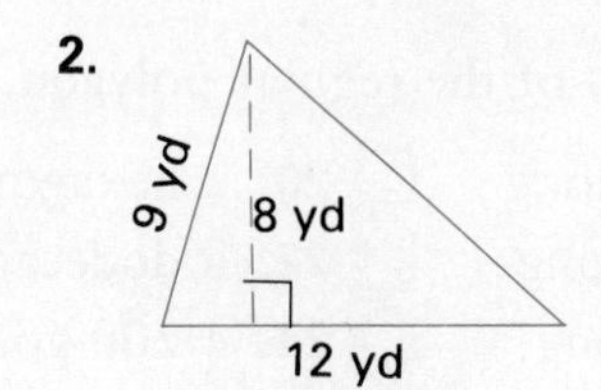

3.
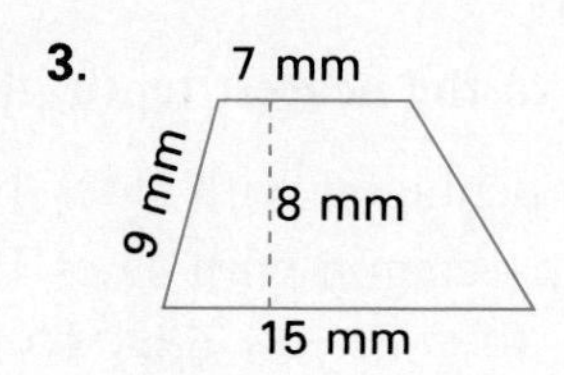

Algebra Review

To simplify radical expressions:
(1) Rationalize the denominator.
(2) Write in simplest form.

Example: Simplify $\frac{3}{2 + \sqrt{3}}$.

$$\frac{3}{2 + \sqrt{3}} = \frac{3}{2 + \sqrt{3}} \cdot \frac{2 - \sqrt{3}}{2 - \sqrt{3}} = \frac{3(2 - \sqrt{3})}{4 - 3} = 6 - 3\sqrt{3}$$

Simplify.

1. $\frac{4}{2 - \sqrt{5}}$

2. $\frac{3}{3 + \sqrt{2}}$

3. $\frac{5}{\sqrt{7} - 2}$

4. $\frac{-2}{\sqrt{5} - 4}$

5. $\frac{3\sqrt{2} - 1}{3\sqrt{2} + 1}$

6. $\frac{2\sqrt{8} - 3}{3\sqrt{12} + 1}$

Application: *Polygons in Structures*

A certain zoo constructed buildings and outdoor areas in polygonal shapes.

1. The area in the giraffe cage is made up of parts of hexagons. Each side is 8 ft long. How much fencing was needed for the enclosure? What is the area of the enclosure?

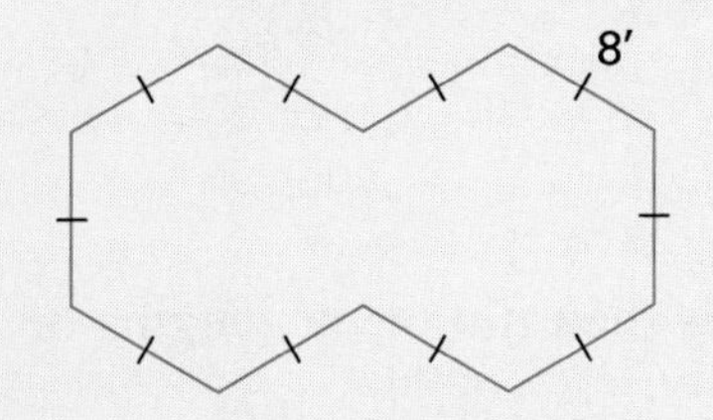

2. The gorilla cage is a partial regular octagon. Find the total area of the floor space if each of the shorter sides has a length of 10 ft.

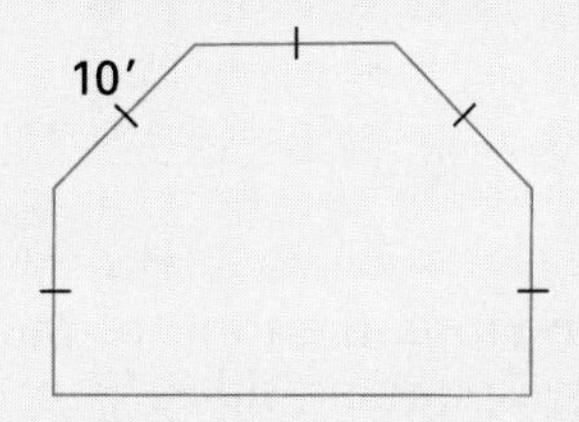

12.7 Areas and Perimeters of Similar Polygons

Objectives To find the ratio of the perimeters of similar polygons
To find the ratio of the areas of similar polygons

A window in an old house was originally a square with sides 1 yd long. The owner nailed boards covering half the window. The resulting window was still a square, 1 yd tall and 1 yd wide. The new window and the original window were similar shapes, although turned in different directions. What is the ratio of their perimeters?

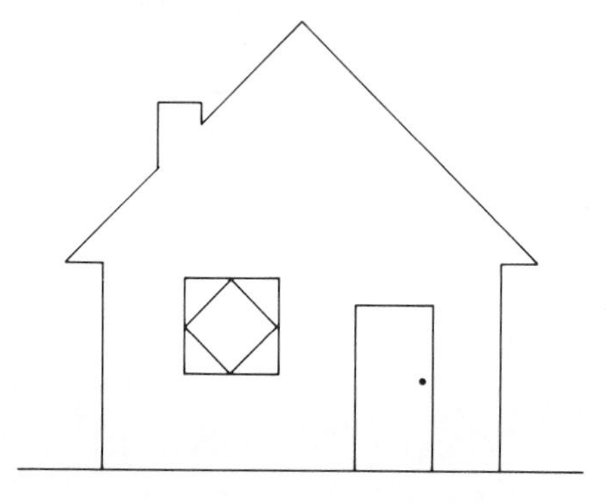

If two triangles are similar, then the ratios of the lengths of corresponding sides are equal. A triangle with sides of lengths 4 cm, 5 cm, and 6 cm is similar to a triangle with sides of lengths 8 cm, 10 cm, and 12 cm. The perimeters are 15 cm and 30 cm. Notice that the ratio of the perimeters of the triangles, 15:30, equals the ratio of the lengths of corresponding sides, 1:2.

Theorem 12.11 The ratio of the perimeters of two similar polygons is the same as the ratio of the lengths of any two corresponding sides.

Proof Suppose two similar polygons have sides of length $a_1, a_2, \cdots, a_n$ and $b_1, b_2, \cdots, b_n$ respectively. By the properties of similar triangles, $\frac{a_1}{b_1} = \frac{a_2}{b_2} = \cdots = \frac{a_n}{b_n} = s$, where s is the ratio of corresponding sides.

By multiplication, $a_1 = sb_1, a_2 = sb_2, \cdots, an = sb_n$.

The perimeter $a_1 + a_2 + \cdots + a_n = sb_1 + sb_2 + \cdots + sb_n$
$= s(b_1 + b_2 + \cdots + b_n)$

and $\frac{a_1 + a_2 + \cdots + a_n}{b_1 + b_2 + \cdots + b_n} = s$

Therefore, the ratio of the perimeters equals the ratio of the lengths of corresponding sides.

EXAMPLE 1 Given: $\triangle DEF \sim \triangle ABC$, $AB = 7$, $BC = 9$, $CA = 11$, $DE = 21$
Find the perimeters of $\triangle ABC$ and $\triangle DEF$.

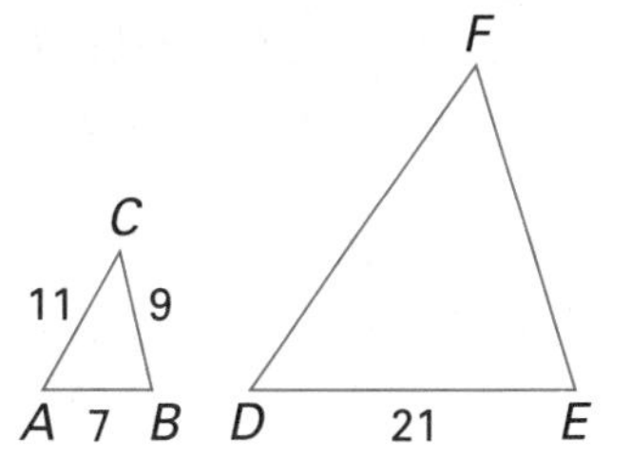

Plan $\frac{DE}{AB} = \frac{21}{7} = 3$. The ratio of the perimeters is 3:1.

Solution Perimeter of $\triangle ABC = 7 + 9 + 11 = 27$
Perimeter of $\triangle DEF = 3(\text{perimeter of } \triangle ABC)$
$= 3 \cdot 27 = 81$

EXAMPLE 2 The perimeter of hexagon $ABCDEF$ is 64 m. The length of a side of similar hexagon $GHIJKL$ is $\frac{3}{4}$ that of the corresponding side of hexagon $ABCDEF$. Find the perimeter of hexagon $GHIJKL$.

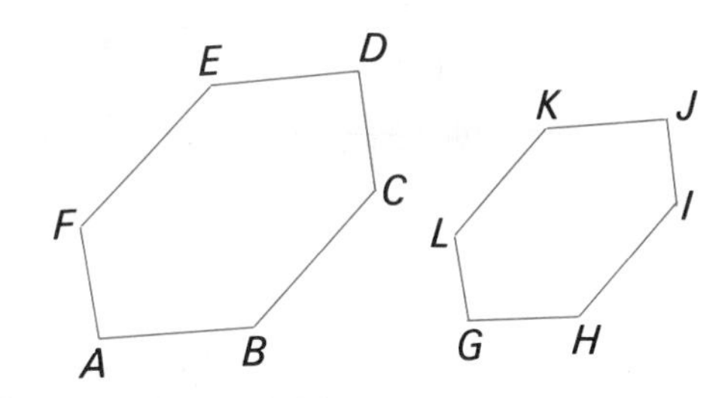

Solution Perimeter of hex $GHIJKL = \frac{3}{4}(\text{perimeter of hex } ABCDEF)$
$= \frac{3}{4} \cdot 64 = 48$ m

Theorem 12.12 The **ratio** of the areas of two similar triangles is the square of the ratio of the lengths of any two corresponding sides.

EXAMPLE 3 Find the ratios of the perimeters and of the areas of the similar triangles.

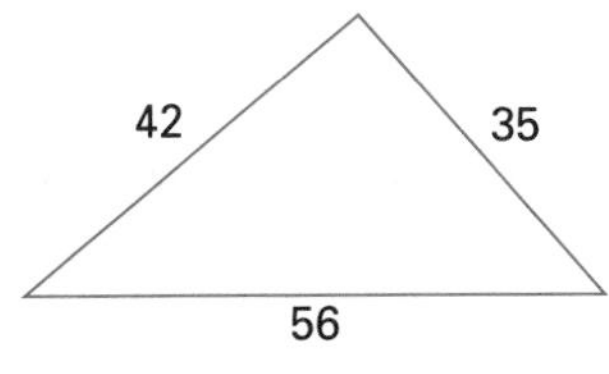

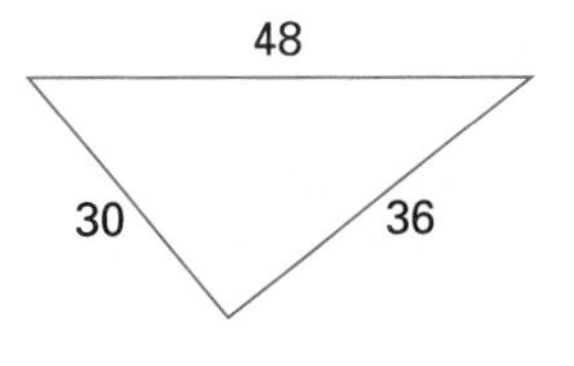

Solution The ratio of the lengths of two corresponding sides is $\frac{56}{48}$, or $\frac{7}{6}$.
Therefore, the ratio of the perimeters is $\frac{7}{6}$,
and the ratio of the areas is $\frac{7^2}{6^2} = \frac{49}{36}$.

Any polygonal region can be broken up into triangular regions. This fact suggests a generalization of Theorem 12.12.

Theorem 12.13 The ratio of the areas of two similar polygons is the square of the ratio of the lengths of any two corresponding sides.

EXAMPLE 4 The length of a side of square *ABCD* is 5mm. The length of a side of square *EFGH* is 8 mm. What is the ratio of the areas of the two squares?

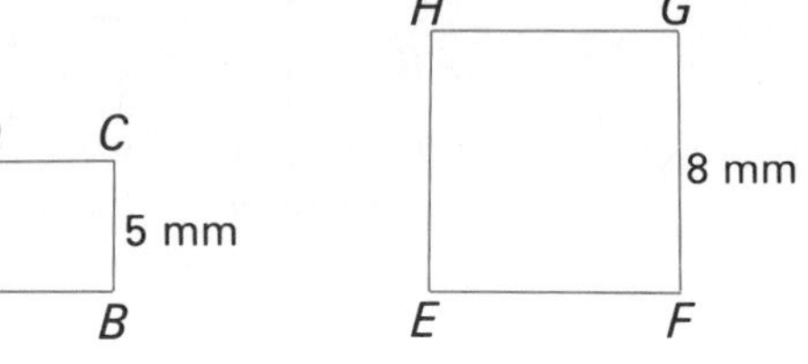

Solution $\frac{AB}{EF} = \frac{5}{8}$

Therefore, $\frac{\text{Area}(ABCD)}{\text{Area}(EFGH)} = \left(\frac{5}{8}\right)^2 = \frac{25}{64}$.

Classroom Exercises

Find the ratio of the perimeters of two similar polygons, given the ratio of the lengths of two corresponding sides.

1. $\frac{2}{3}$ **2.** $\frac{5}{9}$ **3.** $\frac{7}{4}$ **4.** $\frac{3}{11}$

Find the ratio of the areas of two similar polygons, given the ratio of the lengths of two corresponding sides.

5. $\frac{4}{9}$ **6.** $\frac{2}{9}$ **7.** $\frac{8}{5}$ **8.** $\frac{9}{11}$

Find the ratio of the lengths of corresponding sides of the two similar polygons, given the ratio of areas.

9. $\frac{16}{9}$ **10.** 4:4 **11.** $\frac{4}{25}$ **12.** 9:1 **13.** 49 to 100

Written Exercises

Find the ratio of perimeters and the ratio of areas for pairs of similar triangles, given the measures of their corresponding sides.

1.

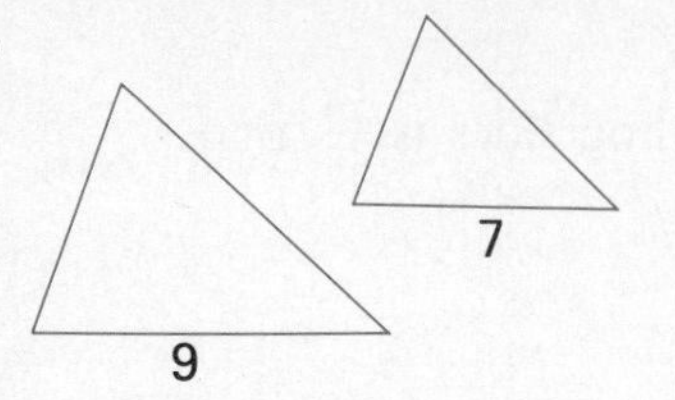

2.

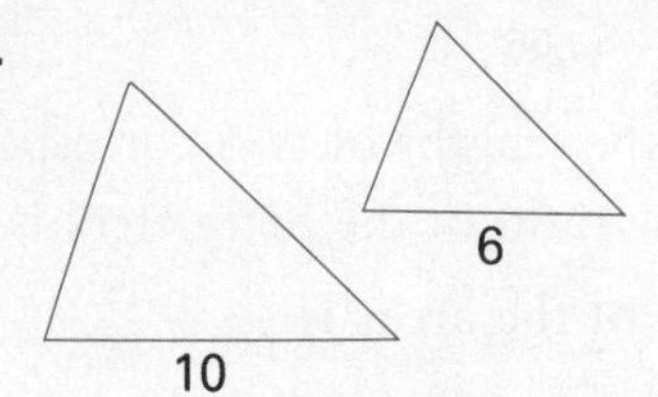

3.

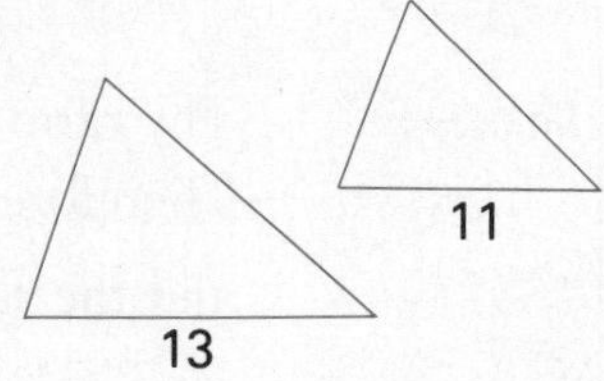

4.

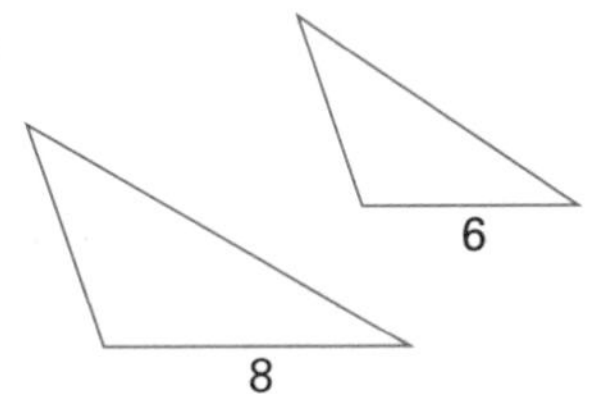

5.

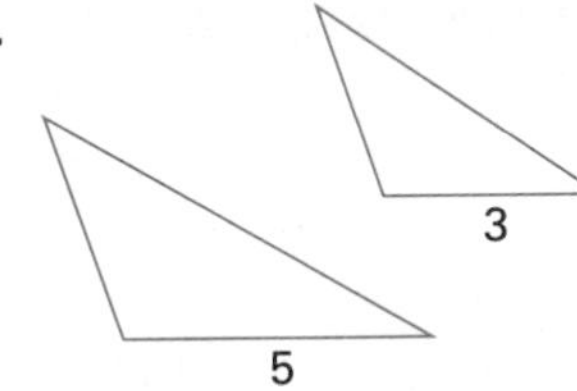

6.

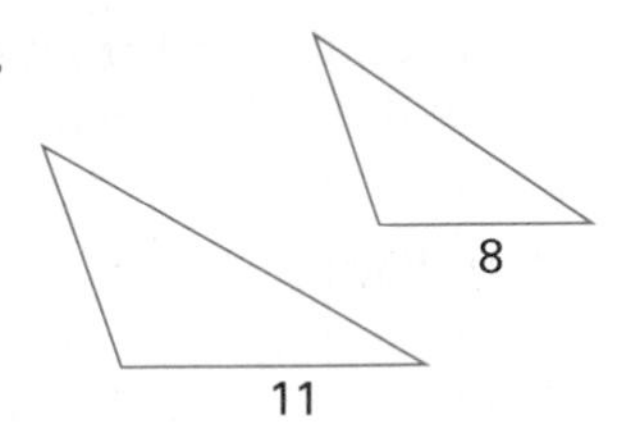

7. The perimeter of pentagon $ABCDE$ is 35 cm. The length of a side of similar pentagon $FGHIJ$ is $\frac{2}{5}$ that of the corresponding side of $ABCDE$. Find the perimeter of $FGHIJ$.

8. The perimeter of heptagon $KLMNOPQ$ is 33 in. The length of a side of similar heptagon $RSTUVWX$ is $\frac{9}{11}$ that of the corresponding side of $KLMNOPQ$. Find the perimeter of $RSTUVWX$.

9. The perimeter of octagon $ABCDEFGH$ is 54 ft. The length of a side of similar octagon $IJKLMNOP$ is $\frac{13}{9}$ that of the corresponding side of $ABCDEFGH$. Find the perimeter of $IJKLMNOP$.

For the given ratio of areas for two similar polygons, find the ratio of lengths of corresponding sides.

10. $\frac{25}{81}$ **11.** $\frac{49}{64}$ **12.** $\frac{121}{225}$ **13.** $\frac{289}{361}$

Find the ratios of the perimeters and the ratios of the areas for the similar triangles.

14.

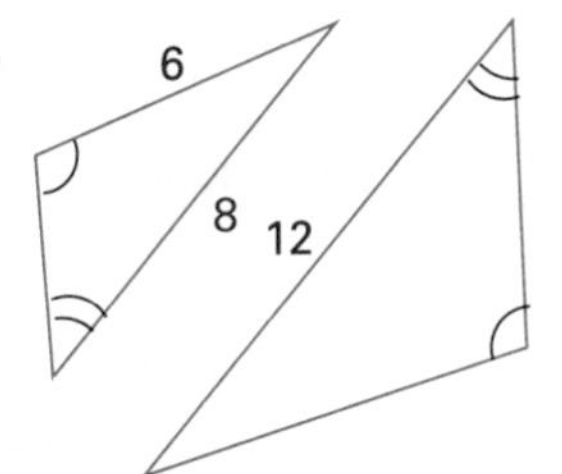

15.

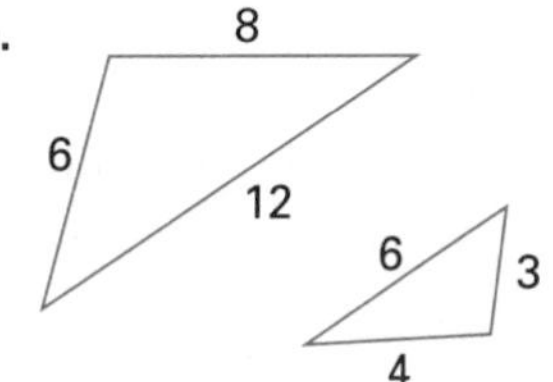

16.

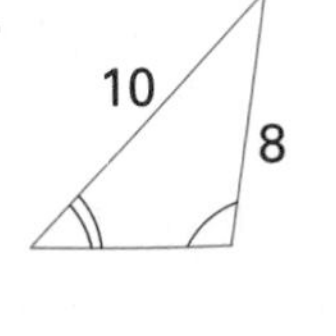

$\triangle ABC \sim \triangle DEF$. Find the missing length or ratio.

	$\frac{\text{Area}(\triangle ABC)}{\text{Area}(\triangle DEF)}$	AB	DE	Perimeter of $\triangle ABC$	Perimeter of $\triangle DEF$
17.	$\frac{16}{9}$	12 cm	______	36 cm	______
18.	$\frac{49}{36}$	______	18 mm	______	48 mm
19.	$\frac{2}{3}$	5 in.	______	12 in.	______
20.	$\frac{7}{3}$	______	7 ft	______	17 ft

21. An equilateral triangle has an area of 1 cm^2. What is the area of an equilateral triangle with sides twice as long?
22. Prove Theorem 12.12 using the standard formula for the area of a triangle.
23. Prove Theorem 12.13 for rectangles.
24. Prove or disprove: The ratio of the areas of two similar regular polygons is equal to the ratio of their apothems.
25. Area($\triangle ABC$) = 16 cm^2; Area($\triangle DEF$) = 25 cm^2; the perimeter of $\triangle ABC$ is $2x + 8$; the perimeter of $\triangle DEF$ is $9x - 3$. Find each perimeter.
26. Area($\square GHIJ$) = 18 mm^2; Area($\square KLMN$) = 50 mm^2; the perimeter of $\square GHIJ$ is $5x - 3$; the perimeter of $\square KLMN$ is $7x + 11$. Find each perimeter.
27. A side of an equilateral triangle is 1 cm long. What is the length of the side of an equilateral triangle with twice the area?
28. A home had a window in the shape of a regular hexagon 1 yd tall. The owner nailed boards over it in such a way that the new window was still a regular hexagon 1 yd tall. What is the ratio of the area of this new window to the original window?
29. Prove Theorem 12.12 using Heron's formula.

Mixed Review

Define the following.

1. coplanar lines *1.1*
2. parallel lines *3.1*
3. supplementary angles *1.6*
4. Transitive Property of Congruence *2.3*
5. alternate interior angles *3.2*

Application: *Enlarging Drawings*

A pantograph is a device used for enlarging (or reducing) drawings. The entire device is free to move, except for a fixed pivot point at P. The adjustment of the pivots determines the ratio of the lengths of corresponding parts in the two drawings. Explain how to set the pivots so that the area of the larger figure is twice that of the smaller.

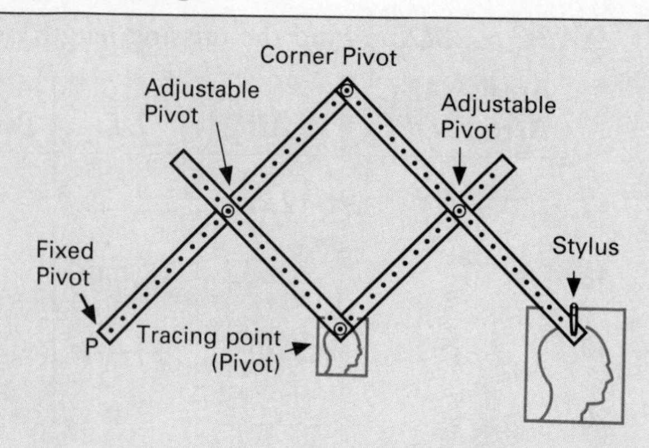

Computer Investigation

Estimating π

Use a computer software program that draws regular polygons of a given size, defined by the distance from the center to a vertex, and that measures its area.

Notice in the figure at the right, the area of the circle appears to be a little larger than 3 times the area of a square with side lengths equal to the length of the radius of the circle. Greek mathematicians used this intuitive idea to guess that the area of a circle, of radius length R, is R^2 times a number a little larger than 3. They decided to call this number pi, or π. The computer activity below illustrates a method for approximating the value of π.

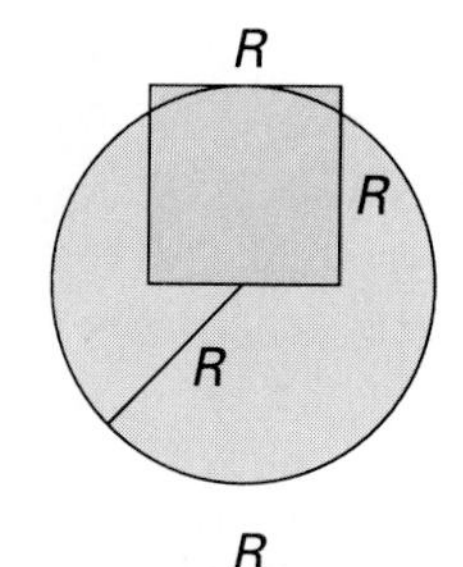

Activity 1

Draw several regular polygons, each with a distance of length 3 from the center to a vertex with the following number of sides.

1. 3 sides
2. 4 sides
3. 5 sides
4. 6 sides
5. 7 sides
6. 9 sides
7. 11 sides
8. 13 sides
9. 15 sides
10. 16 sides
11. 17 sides
12. 18 sides

Activity 2

The figure at the right shows an 8-sided regular polygon inscribed in a circle. The length of a radius is 3.

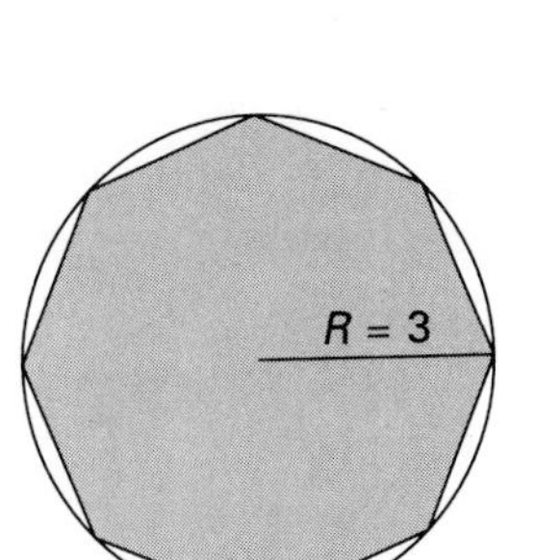

13. Construct the polygon shown at the right.
14. Find the area of this polygon.
15. Divide the area by R^2, in this case 3^2, or 9.
16. Repeat the steps of Exercises 13-15 for each of the polygons whose number of sides is given in Exercises 1-12. Keep a record of the results.

Summary

As the number of sides of a regular polygon of a given radius increases, the shapes of successive polygons approach a ______. The corresponding quotients, $\frac{\text{Area of polygon}}{R^2}$, approach the number ______.

12.8 Circumferences and Areas of Circles

Objectives To find the circumference of a circle
To find the area of a circle

Consider a circle with a radius 1 unit long. Regular polygons can be inscribed in it. As the number of sides of the polygon increases, the perimeter also increases, but not indefinitely.

Number of sides	Approximate perimeter
4	5.6569
8	6.1229
16	6.2429
32	6.2731
64	6.2807
128	6.2826
256	6.2830

As the number of sides of the inscribed polygons increases, the polygons more closely approximate a circle. The perimeters approach a value of approximately 6.2832. A good measure of the distance around the circle seems to be the perimeter of an inscribed regular polygon with a very large number of sides.

These perimeters are approaching a *limit.* A sequence of numbers approaches a limit if there is a number b to which the numbers of the sequence get closer and closer. The number b is called the *limit of the sequence.*

Definition The **circumference of a circle** is the limit of the perimeters of inscribed regular polygons of the circle as the number of sides of the polygon increases indefinitely.

If you wrapped a piece of string around a can, you would find that the string is slightly more than three times the length of a diameter of the can. This would be true for any can that is a circular cylinder, regardless of its size. It appears that the ratio of the distance around a circle to the length of the diameter is a constant.

Theorem 12.14 The ratio of the circumference to the length of a diameter is the same for all circles.

This constant ratio of the circumference to the length of a diameter is called **pi**, denoted by the symbol π. π is an irrational number that can be approximated to any degree of accuracy. To fifteen decimal places $\pi = 3.141592653589793$. Commonly, 3.14, 3.1416, or $\frac{22}{7}$ is used as an approximation. Unless otherwise specified, you may leave answers in terms of π rather than approximating.

Let C be the circumference of a circle, and d the length of its diameter. Since $\frac{C}{d} = \pi$, then $C = \pi d$. Since a diameter has twice the length of a radius, $C = 2\pi r$.

Corollary The circumference of a circle with radius of length r is $2\pi r$.

EXAMPLE 1 The diameter of a circle is 5 cm long. What is the circumference?

Solution

$C = \pi d$
$C = \pi \cdot 5$, or 5π

EXAMPLE 2 If the circumference of a circle is 30π, what is the length of a radius and the length of a diameter?

Solution

$C = 2\pi r$
$30\pi = 2\pi r$

Therefore, $r = \frac{30\pi}{2\pi} = 15$, and $d = 2r = 30$.

The area of a circle can be found through a process similar to that used for the circumference. As the number of sides of a regular polygon inscribed in a circle increases, the area also increases, but not indefinitely.

Sides (n)	Approximate area
4	2
8	2.8284
16	3.0615
32	3.1214
64	3.1365
128	3.1403
256	3.1413

1

radius = 1

These areas, like the perimeters earlier, approach a limit.

Definition

The **area of a circle** is the limit approached by the areas of inscribed regular polygons as the number of sides of the polygons increases indefinitely. The length of the apothem of a polygon with many sides approaches the length of a radius of the circumscribing circle.

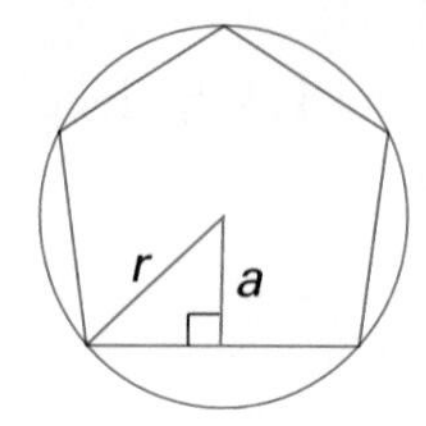

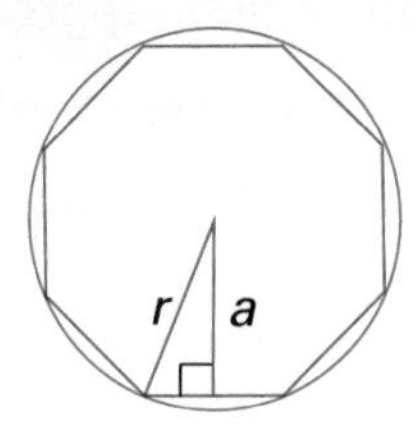

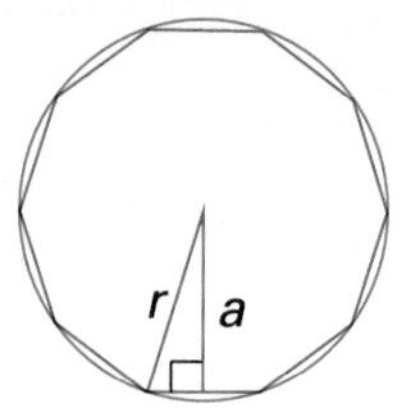

Area(regular polygon) $= \frac{1}{2}ap$. This suggests that as a approaches r,
Area($\odot$) $= \frac{1}{2}rC = \frac{1}{2}r(2\pi r)$, or πr^2.

Theorem 12.15 The area of a circle with radius of length r is πr^2.

EXAMPLE 3 Find the area of a circle with a radius of 6 mm. Use 3.14 for π.

Solution Area $= \pi r^2 = 3.14 \cdot 6^2 = 113.04$ mm^2

EXAMPLE 4 Find the area of a circle with a diameter 20 cm long.

Solution $A(\odot) = \pi r^2$
$A(\odot) = \pi(\frac{1}{2} \cdot 20)^2 = 100\pi$ cm^2

EXAMPLE 5 Find the area of the shaded region between $\odot O$ and the inscribed square $ABCD$. The radius of the circle is 2 cm long.

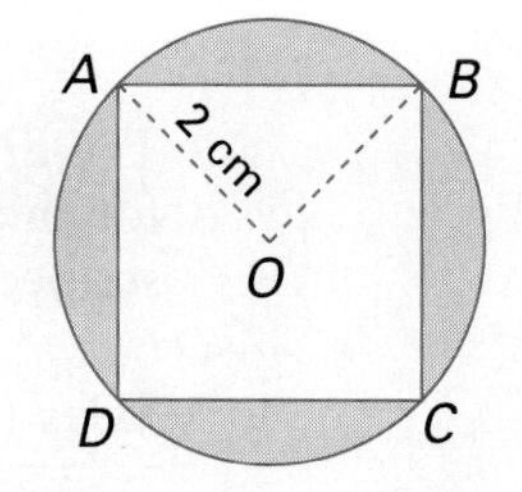

Solution Area($\odot O$) $= \pi \cdot 2^2 = 4\pi$ cm^2
$\triangle AOB$ is a 45-45-90 triangle,
so $AB = \sqrt{2} \cdot AO = 2\sqrt{2}$ cm.

The area of the square is $(2\sqrt{2})^2 = 8$ cm^2.
Therefore, the area of the shaded region is $(4\pi - 8)$ cm^2.

Classroom Exercises

Find the circumference of the circle in terms of π.

1. $d = 10$ m **2.** $d = 2$ cm **3.** $d = \frac{1}{4}$ ft **4.** $d = 2$ mi **5.** $d = 4$ yd

Find the area of the circle in terms of π.

6. $r = 10.1$ mm **7.** $r = 3$ cm **8.** $r = 12$ in. **9.** $r = 1\frac{1}{2}$ in. **10.** $r = 7$ ft

Written Exercises

Find, to the nearest tenth, the circumference of the circle.

1. $d = 20$ cm **2.** $d = 6.7$ m **3.** $d = 100$ yd **4.** $d = 30$ ft

Find, to the nearest tenth, the area of the circle.

5. $r = 2$ mm **6.** $r = 5$ cm **7.** $r = 10$ in. **8.** $r = 20\frac{1}{2}$ ft

Find the missing measurements.

	r	d	C	Area of circle
9.	5 cm	10 cm	10π cm	
10.	10 m			100π m^2
11.	4 in.			
12.	12 ft			
13.		10 m		
14.		16 cm		
15.			10π in.	
16.			16π yd	
17.				100π cm^2
18.				64π m^2
19.	2.5 m			
20.		6.8 cm		
21.			100 in.	
22.				100 in^2

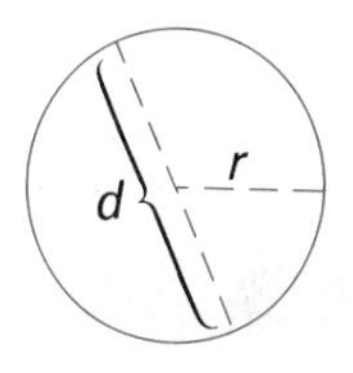

Find, to the nearest tenth, the area of the shaded region. The circle has a radius 10 cm long, and the polygon is a regular polygon.

23.

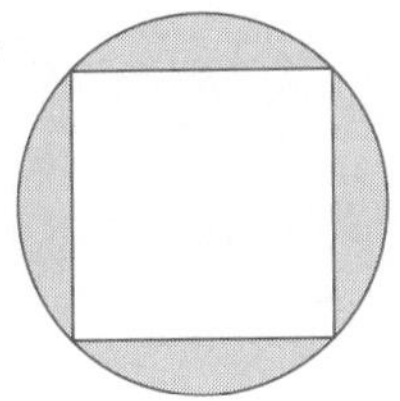

24.

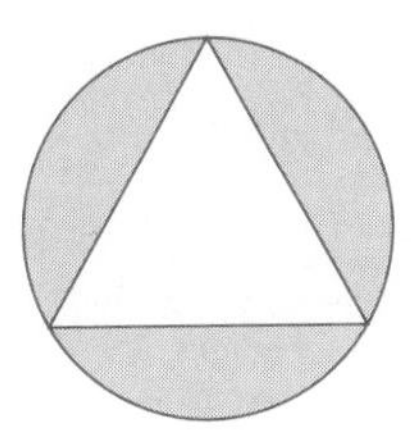

25.

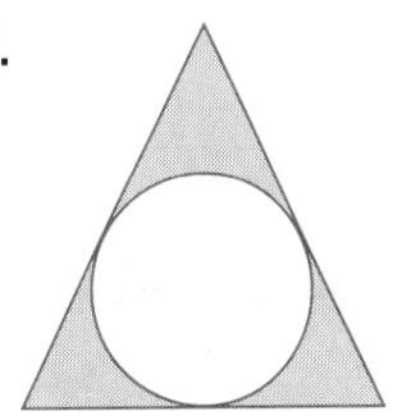

26. Two approximations, 3.14 and $\frac{22}{7}$, are often used for π. Which is closer to π?

27. π is known to be between $3\frac{1}{7}$ and $3\frac{10}{71}$. Which is closer to π?

For Exercises 28–30, *ABCD* is a square with a side of length 10. All circles in the square are congruent and tangent to adjacent circles and/or the square. Find the area of the shaded region in terms of π.

28.

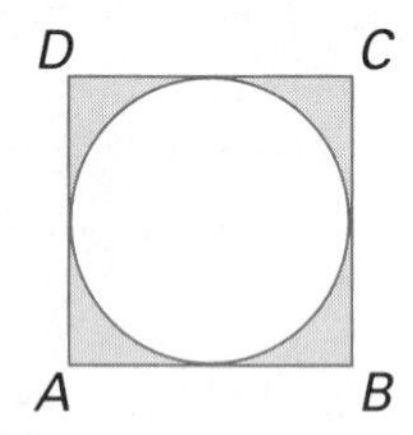

29.

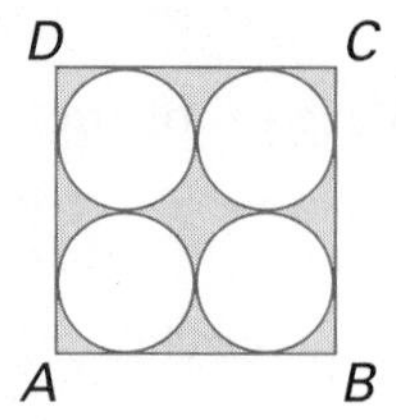

30.

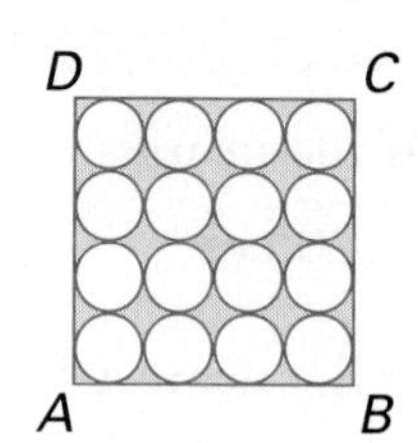

31. Find the ratio of the area of a circle inscribed in an equilateral triangle to the area of the circle circumscribed about the same triangle.
32. Find the ratio of the area of a circle inscribed in a square to the area of the circle circumscribed about the same square.
33. The radius of the earth is approximately 6,400 km. If a rope were stretched around the earth at the equator, how long would the rope be? Suppose the rope were lifted 3 km off the surface of the earth all the way around. Approximately how much longer would the rope have to be? The radius of the moon is approximately 1,700 km. If the same experiment were repeated on the moon, how much longer would the rope be?

Prove or disprove the statement.

34. The ratio of the areas of two circles equals the square of the ratio of the lengths of the radii.
35. The ratio of the areas of two circles equals the ratio of their circumferences.
36. The ratio of the areas of two circles equals the ratio of the areas of their inscribed squares.
37. The ratio of the areas of two circles equals the ratio of the areas of their circumscribed squares.

Mixed Review

Find the degree measure of the arc. ***11.4–11.7***

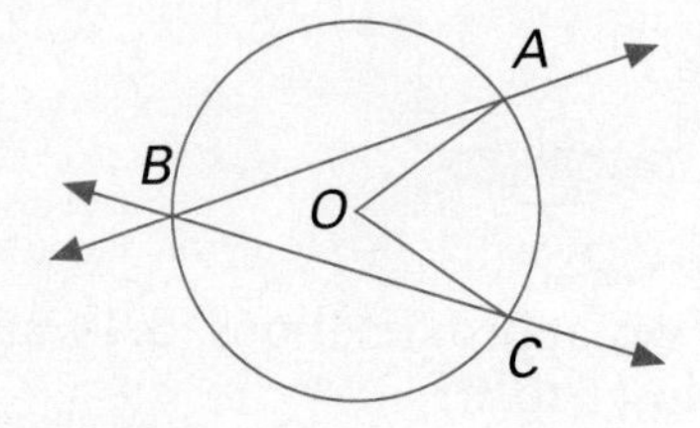

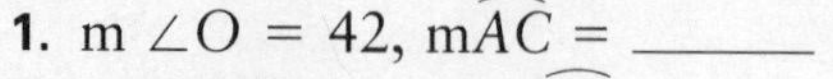

1. $m\angle O = 42$, $m\overset{\frown}{AC} =$ ______
2. $m\angle ABC = 31$, $m\overset{\frown}{AC} =$ ______
3. $m\overset{\frown}{AC} = 64$, $m\overset{\frown}{ABC} =$ ______
4. $m\overset{\frown}{AB} = 108$, $m\overset{\frown}{BC} = 113$, $m\overset{\frown}{AC} =$ ______
5. Find the area of an equilateral triangle with a side 4 cm long. ***12.4***

12.9 Measuring Arcs and Sectors of Circles

Objectives

To find the lengths of arcs
To find the areas of sectors and segments
To find the areas of annuli

The measure of an arc was defined as the measure of its corresponding central angle. The concept of arc measure may also serve to develop the meaning of the *length* of an arc of a circle. The diagram at right shows twelve arcs, each of which, like $\widehat{AB}$, has a degree measure of 30. One may think of $\widehat{AB}$ as having a length equal to $\frac{1}{12}$ the circumference of the circle.

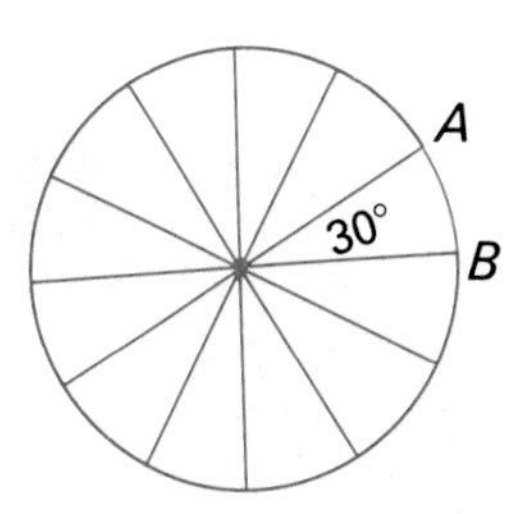

Definition

In a circle of radius r, an arc of degree measure m has **arc length** equal to $\frac{m}{360} \cdot 2\pi r$.

In other words, the length of the arc is that fraction $(\frac{m}{360})$ of the circumference $(2\pi r)$ that corresponds to the arc.

EXAMPLE 1 A circle has a circumference of 24 cm. The degree measure of an arc of the circle is 36. What is the length of the arc?

Solution Let l = length of the arc.

Then $l = \frac{m}{360} \cdot C$

$= \frac{36}{360} \cdot 24 = \frac{1}{10} \cdot 24 = 2.4$ cm

EXAMPLE 2 Given: Diameter $\overline{AB}$, $AB = 12$ cm, $m\widehat{BC} = 60$
Find the length of $\widehat{BC}$.

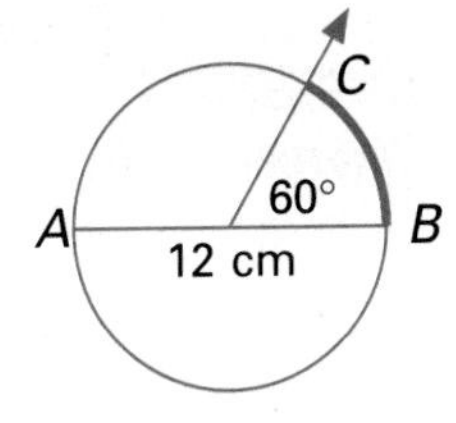

Solution Let l = length of $\widehat{BC}$.
If $AB = 12$, then $r = 6$

$l = \frac{60}{360} \cdot 2\pi \cdot 6 = 2\pi$ cm

Two arcs may have the same lengths, but not the same degree measure. For example, in a circle with diameter 8, an arc of degree measure 90 has a length of 2π. In a circle with diameter 16, an arc of degree measure 45 has the same length, 2π.

Definition

A **sector of a circle** is the region bounded by an arc of the circle and two radii whose endpoints are endpoints of that arc.

When you find the area of a sector, you actually are finding part of the area of the circle. For example, if two radii form an angle of 60, the area of the corresponding sector is $\frac{60}{360} = \frac{1}{6}$ of the area of the circle.

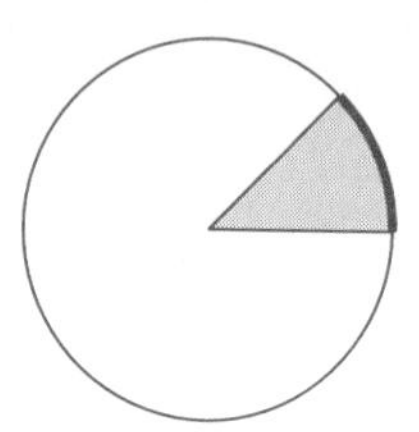

Definition

In a circle of radius r, where a sector has an arc of degree measure m, the **area of the sector** is $\frac{m}{360} \cdot \pi r^2$.

In other words, the area of a sector is that fraction $(\frac{m}{360})$ of the total area (πr^2) that corresponds to the sector.

EXAMPLE 3 A circle has circumference of 24π. Find the area of a sector with arc of degree measure 50.

Plan Use the circumference to compute the radius. Then use the formula for the area of a sector.

Solution

$$2\pi r = C$$
$$2\pi r = 24\pi$$
$$r = \frac{24\pi}{2\pi}$$
$$= 12$$

$$\text{area of sector} = \frac{m}{360} \cdot \pi r^2$$
$$= \frac{50}{360} \cdot \pi(12)^2$$
$$= \frac{7{,}200}{360}\pi = 20\pi$$

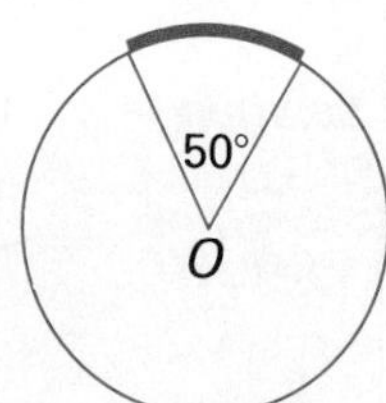

Theorem 12.16 The area of a sector of a circle is one-half the product of the length s of the arc and the length r of its radius (Area $= \frac{1}{2}rs$).

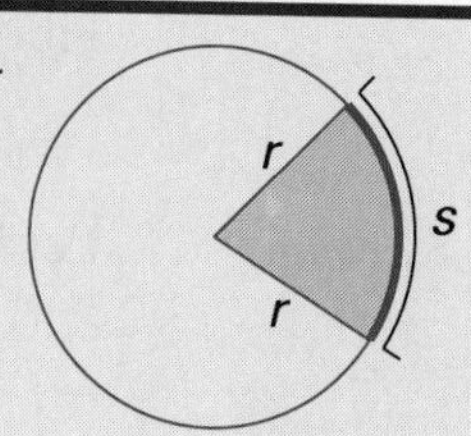

EXAMPLE 4 The length of a radius of a circle is 12 in. If the area of a sector of this circle is 36 in^2, what is the length of the arc of the sector?

Solution

$$\text{Area} = \frac{1}{2}rs$$
$$36 = \frac{1}{2} \cdot 12s$$
$$s = \frac{36}{6} = 6 \text{ in.}$$

Areas of several other regions can be found as parts of circles.

Definition

A **segment of a circle** is a region bounded by a minor arc of the circle and the chord determined by the arc.

EXAMPLE 5 The length of a radius of a circle is 6 cm. Find the area of a segment bounded by an arc of degree measure 60 and its corresponding chord.

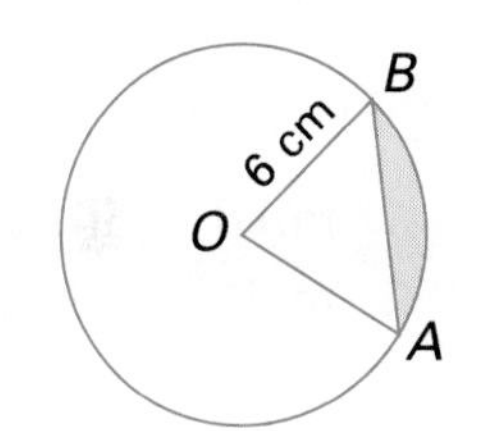

Plan From the area of the sector subtract the area of equilateral $\triangle AOB$. The sides of the triangle are the same length as the radius.

Solution

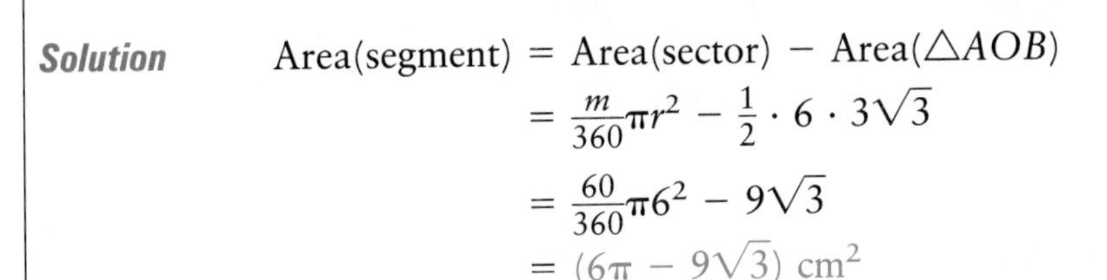

$$\text{Area(segment)} = \text{Area(sector)} - \text{Area}(\triangle AOB)$$
$$= \frac{m}{360}\pi r^2 - \frac{1}{2} \cdot 6 \cdot 3\sqrt{3}$$
$$= \frac{60}{360}\pi 6^2 - 9\sqrt{3}$$
$$= (6\pi - 9\sqrt{3}) \text{ cm}^2$$

Definition

An **annulus** is a region bounded by two concentric circles.

EXAMPLE 6 Find the area of the annulus bounded by concentric circles with radii 12 cm and 8 cm long.

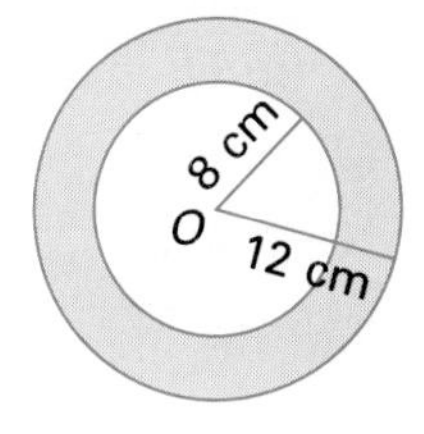

Plan Subtract the area of the smaller circle from the area of the larger circle.

Solution

$$\text{Area(annulus)} = \text{Area(outer circle)} - \text{Area(inner circle)}$$
$$\text{Area(annulus)} = \pi \cdot 12^2 - \pi \cdot 8^2$$
$$\text{Area(annulus)} = 144\pi - 64\pi$$
$$= 80\pi \text{ cm}^2$$

Classroom Exercises

What part of the area of a circle is a sector with the given degree measure of the arc?

1. 60 **2.** 12 **3.** 45 **4.** 120

What fractional part of the circumference is an arc with the given degree measure?

5. 30 **6.** 90 **7.** 120 **8.** 180

Written Exercises

$\odot O \cong \odot P$; $\odot O \ncong \odot Q$. Do the arcs have equal length? Why or why not?

1. $m\widehat{AB} = 50$, $m\widehat{BC} = 50$
2. $m\widehat{DE} = 58$, $m\widehat{EF} = 59$
3. $m\widehat{AC} = 112$, $m\widehat{DF} = 112$
4. $m\widehat{AB} = 50$, $m\widehat{IH} = 50$
5. $m\widehat{GI} = 115$, $m\widehat{DF} = 115$
6. $m\widehat{BC} = 53$, $m\widehat{ED} = 54$

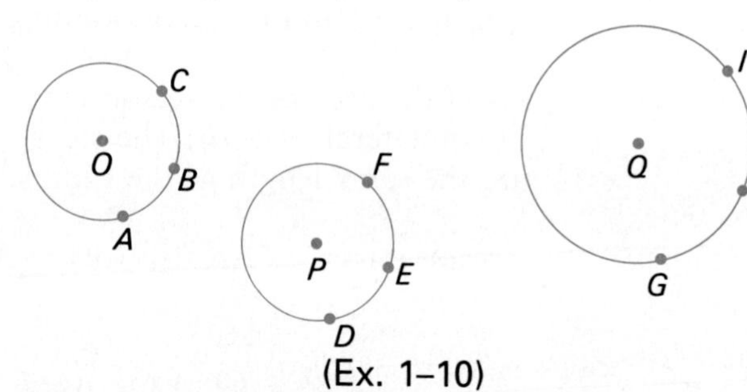

(Ex. 1–10)

Find the arc length.

7. Circumference = 24, $m\widehat{AB} = 60$
8. Circumference = 36, $m\widehat{CD} = 45$
9. Circumference = 18π, $m\widehat{EF} = 120$
10. Circumference = 25π, $m\widehat{GH} = 36$

Find the arc length and the area of the indicated sector.

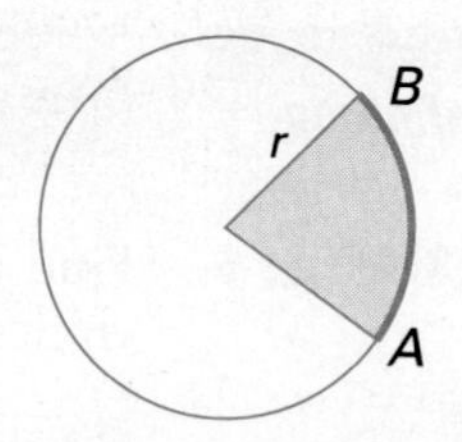

11. $m\widehat{AB} = 60$, $r = 10$ m
12. $m\widehat{AB} = 90$, $r = 4$ cm
13. $m\widehat{AB} = 120$, $r = 8$ ft
14. $m\widehat{AB} = 180$, $r = 9$ in.
15. $m\widehat{AB} = 210$, $r = 3$ mm
16. $m\widehat{AB} = 270$, $r = 12$ cm

Find the area of the annulus bounded by concentric circles with the given radii.

17. $r_1 = 10$ cm, $r_2 = 5$ cm
18. $r_1 = 12$ cm, $r_2 = 6$ cm
19. $r_1 = 9$ mm, $r_2 = 3$ mm
20. $r_1 = 5$ mm, $r_2 = 2$ mm

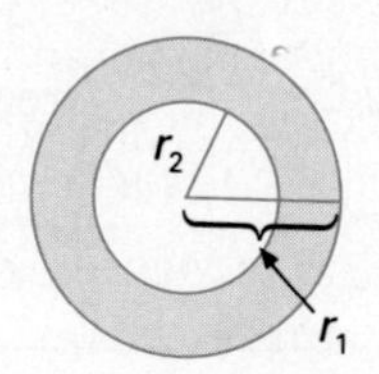

Find the missing length of arc s, length of radius r, or area of sector A for the indicated circle.

21. $s = 4$ ft, $r = 3$ ft, $A =$ ______

22. $s = 8$ m, $A = 48$ m^2, $r =$ ______

23. $A = 1$ in^2, $r = 8$ in., $s =$ ______

24. $A = 2$ ft^2, $r = 24$ in., $s =$ ______

$\overline{OA}$ and $\overline{OB}$ are radii of $\odot O$. Find the missing measures.

	m $\angle O$	r	m$\widehat{AB}$	Area (sector AB)	Area (segment AB)
25.	______	6	60	______	______
26.	120	12	______	______	______
27.	______	10	______	25π sq units	______
28.	______	______	120	24π sq units	______

29. Prove Theorem 12.16.

Find the area of the shaded region if the length of the side of the square is 10. (Curved lines are arcs of circles.)

30.

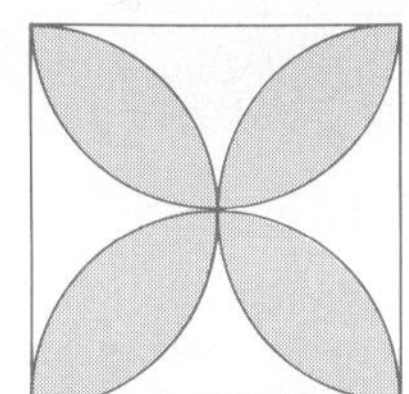

31.

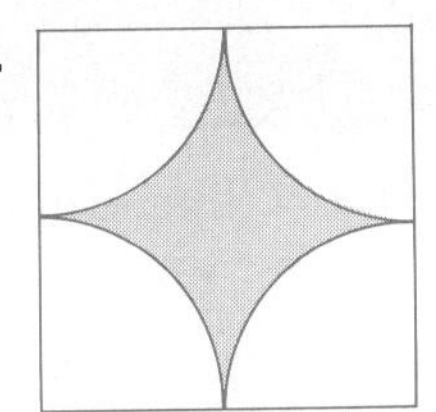

32.

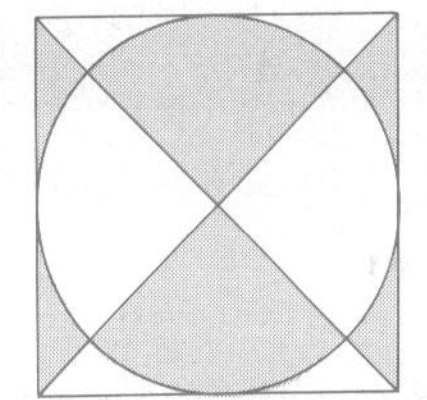

33. Write a formula for the area of an annulus in terms of radii r_1 and r_2, where $r_1 \geq r_2$.

34. Find, to the nearest hundredth, the area of a segment of a circle with 10-cm radius and an arc of degree measure 54. (HINT: Use appropriate trigonometric ratios.)

35. Find, to the nearest tenth, the area of a segment of a circle with a 16-cm radius and an arc of degree measure 137.

36. Write a formula for the area of a segment of a circle.

Mixed Review

1. Find the area of a regular hexagon with sides 12 in. long. *12.6*
2. Find the area of a trapezoid with a 12-cm upper base, a 16-cm lower base, and a 5-cm altitude. *12.5*
3. What is the distance between two points with coordinates -9 and 17? *1.2*
4. The measure of an angle is 50 more than its supplement. Find the measure of the angle. *1.6*

Application: *Estimating Area*

Example 1

The aerial photograph below, which was taken from an altitude of 20,000 ft, shows a piece of lakefront property which is being considered for development. How can the photograph be used to estimate the area of the land in acres?

First, it is necessary to know the scale ratio. In the enlargement below, 1 in. = 1,650 ft or .3125 mi. Therefore, one square inch represents $(.3125)^2$ or .098 mi^2.

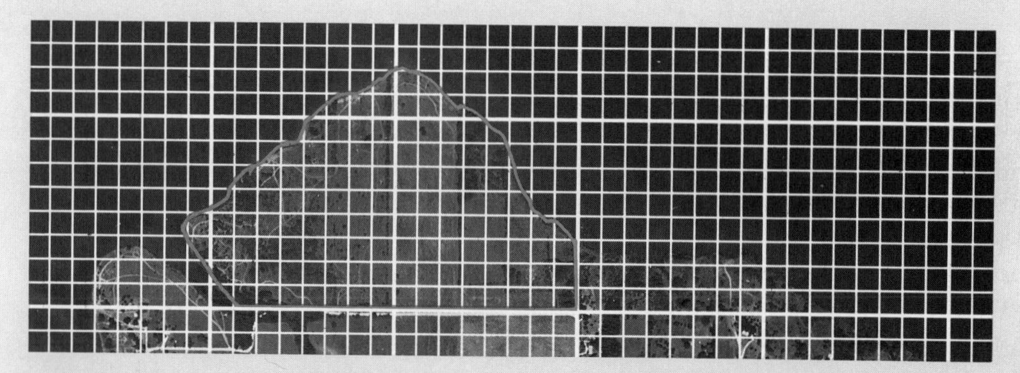

Place a grid over the photograph as shown and count the number of squares that fall inside the boundary of the property. Then count the number of squares that straddle the boundary and divide that number by 2:

$$105 + (24 \div 2) = 117 \text{ squares}$$

The grid shown has 64 squares to the square inch, so the area of each square is $\frac{1}{64}$ in^2. Thus, an estimate of the area of the property as shown in the photograph is $117 \times \frac{1}{64} \approx 1.83$ in^2. Since 1 in^2 represents .098 mi^2, the actual area of the property is about $1.83 \times .098 \approx .179$ mi^2. There are 640 acres in a square mile, so the estimated area in acres of the property is $.179 \times 640 \approx 115$ acres.

Other methods are used to approximate irregular areas. In one laboratory, the weight of a plywood cutout of a shape is compared with the weight of a piece of plywood of known area.

Example 2

Another method used to approximate area is the *trapezoid method.* To find the approximate area of $ABCD$ below, divide $\overline{AB}$ into n congruent segments each d units long. Draw parallel segments across the region at these points. Each region is approximately a trapezoid.

$$\begin{aligned} \text{Area} &= \tfrac{1}{2}(b_1 + b_2)d + \tfrac{1}{2}(b_2 + b_3)d + \cdots + \tfrac{1}{2}(b_{n-1} + b_n) \\ &= d(\tfrac{1}{2}b_1 + b_2 + b_3 + \cdots + b_{n-1} + \tfrac{1}{2}b_n) \end{aligned}$$

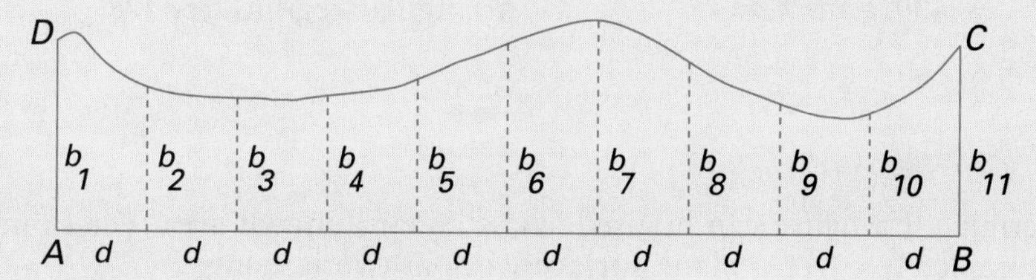

Use the trapezoid method to estimate the area of the region shown at right.

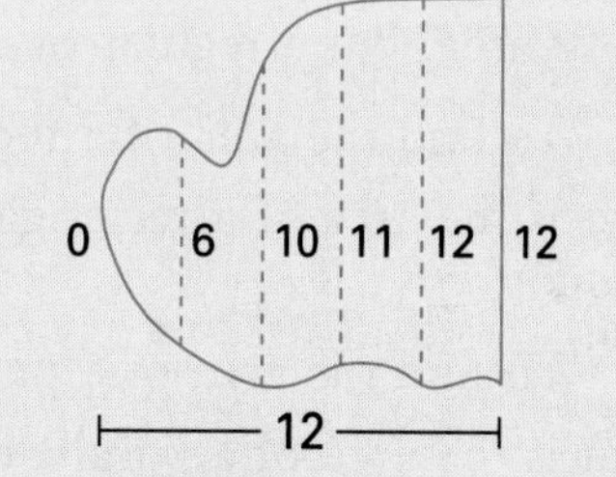

$$\begin{aligned} \text{Area} &= \tfrac{12}{5}[\tfrac{1}{2}(0) + 6 + 10 + 11 + 12 + \tfrac{1}{2}(12)] \\ &= \tfrac{12}{5}(45) \\ &= 108 \text{ sq units} \end{aligned}$$

A plowed field is outlined in the photograph below. If 1 ton of limestone per acre is required to treat the soil for planting hay, what is the weight of limestone required for the field shown? Use two different methods of area approximation and compare the results. (The scale ratio is the same as in Example 1.)

Chapter 12 Review

Key Terms

altitude (p. 483)
annulus (p. 517)
apothem of a regular polygon (p. 499)
area of a circle (p. 512)
area of a polygon (p. 478)
area of a sector (p. 516)
base (p. 483)
center of a regular polygon (p. 498)
circumference of a circle (p. 510)
Heron's Formula (p. 490)
kite (p. 489)
length of an arc (p. 515)
limit (p. 510)
perimeter (p. 499)
pi (π) (p. 511)
polygonal region (p. 478)
radius of a regular polygon (p. 499)
sector of a circle (p. 516)
segment of a circle (p. 517)
standard unit (p. 475)
triangular region (p. 478)

Key Ideas and Review Exercises

12.1 To determine if a figure can be used as a standard unit of area, determine if it can completely cover a plane surface without overlapping.

1. Make a sketch to show that a rectangle can be used as a standard unit of area, but a regular octagon cannot be used.

12.2 To find the area of a rectangle, use the formula: Area $= bh$.

2. What is the area of a rectangle with a 3.5-cm base and a 4.7-cm altitude?

12.3 To find the area of a parallelogram, use the formula: Area $= bh$.

3. What is the area of a parallelogram with a 7.9-mm base and a 2.3-mm altitude?

12.4 To find the area of a triangle, use the formula: Area $= \frac{1}{2}bh$.

4. Find the area of a triangle with a 4.6-in. base and 2.3-in. altitude.
5. Find the area of a right triangle with sides of length 5, 12, and 13.

12.5 To find the area of a trapezoid, use the formula: Area $= \frac{1}{2}h(b_1 + b_2)$.

6. Find the area of a trapezoid with a 9-cm lower base, a 17-cm upper base, and a 6-cm altitude.
7. Find the area of an isosceles trapezoid with bases of length 10 and 20 and a base angle measuring 60.

12.6 To find the perimeter of a regular polygon, add the lengths of the sides. To find the area of a regular polygon, use the formula: Area $= \frac{1}{2}ap$.

8. Find the perimeter of a regular nonagon with sides 4 m long.
9. Find the area of a regular octagon with a 160-in. perimeter and an apothem of 24.14 in.

12.7 To find the ratio of the perimeters of similar polygons, find the ratio of the lengths of corresponding sides.

To find the ratio of the areas of two similar polygons, find the square of the ratio of the lengths of corresponding sides.

10. The lengths of two corresponding sides of two similar octagons are 8 cm and 12 cm. If the perimeter of the first is 70 cm, what is the perimeter of the second?

11. The lengths of two corresponding sides of two similar hexagons are 12 mm and 15 mm. If the area of the first hexagon is 40 mm^2, what is the area of the second?

12.8 To find the circumference of a circle, use the formula $C = \pi d$. To find the area of a circle, use the formula: $A = \pi r^2$.

12. What is the circumference of a circle with a 5-in. radius? Use 3.14 as an approximation of π.

13. What is the area of a circle with a 3-in. radius? Use 3.14 as an approximation of π.

12.9 To find the length of an arc of a circle with measure m, use $l = \frac{m}{360} \cdot 2\pi r$.

arc

sector

To find the area of a sector of a circle with an arc of degree measure m, use $A = \frac{m}{360} \cdot \pi r^2$.

To find the areas of the segments of circles and annuli, subtract appropriate portions of the circle.

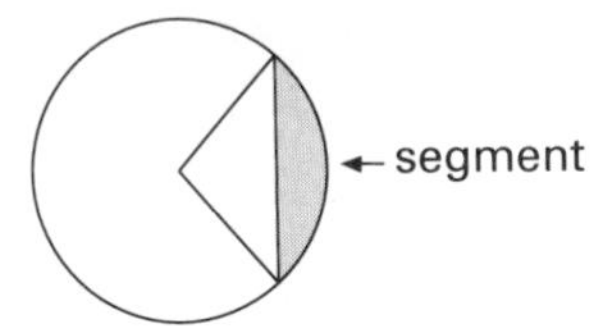

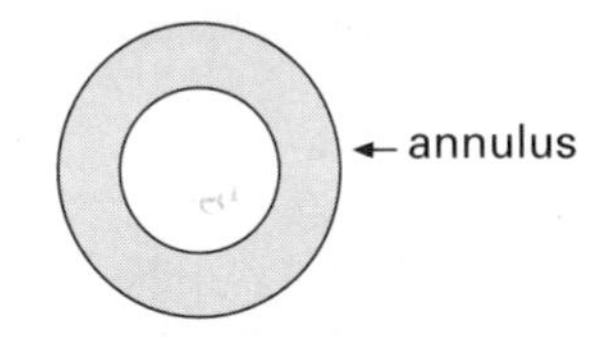

A(segment) = A(sector) − A(triangle) A(annulus) = A(outer ⊙) − A(inner ⊙)

14. What is the length of an arc, with measure 30, of a circle with a 12-cm radius?

15. Find the area of a sector of a circle with a 10-in. radius determined by an arc of degree measure 30.

16. Find the area of a segment of a circle with a 6-cm radius determined by an arc of degree measure 60.

17. Find the area of an annulus bounded by concentric circles with the radii of 10 cm and 8 cm.

Chapter 12 Test

Find the area of the figure. (Items 1–12)

1. rectangle *ABCD*

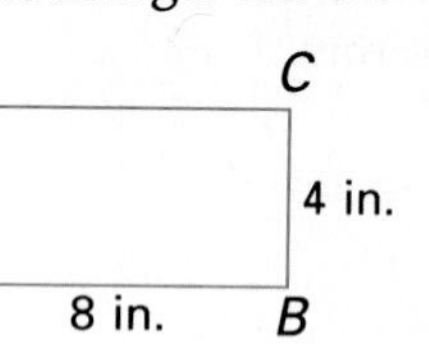

2. parallelogram *EFGH*

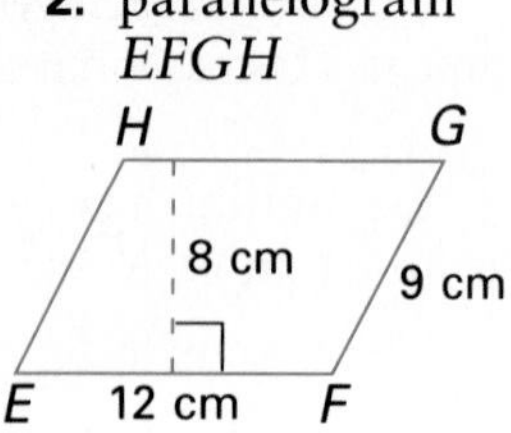

3. triangle *IJK*

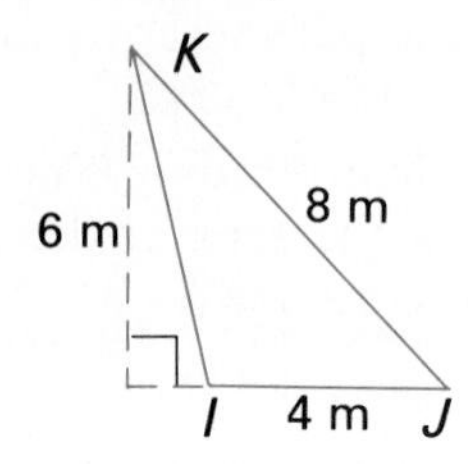

4. triangle *LMN*

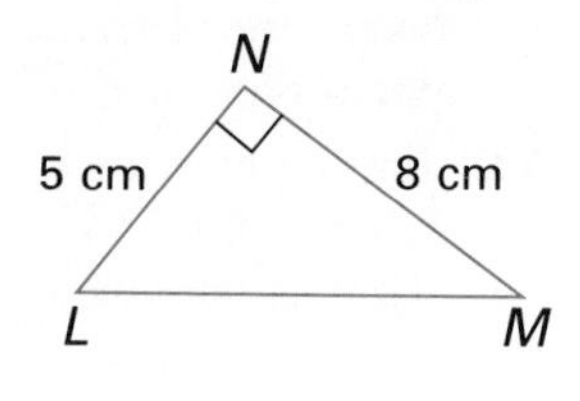

5. rhombus *OPQR* **6.** kite *STUV* **7.** equilateral $\triangle WXY$ **8.** trapezoid *ZABC*

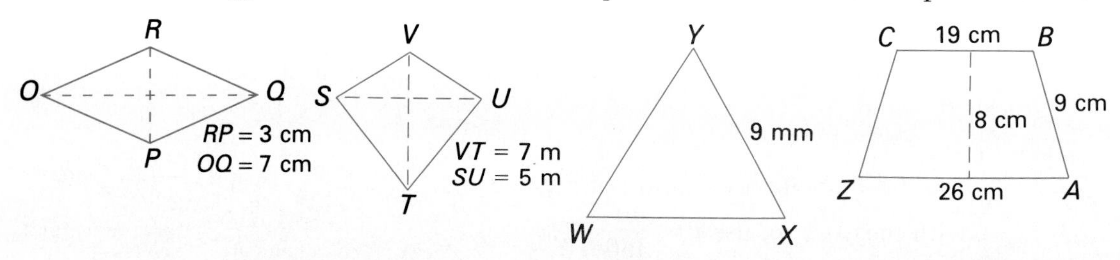

9. trapezoid *DEFG* **10.** circle *H*

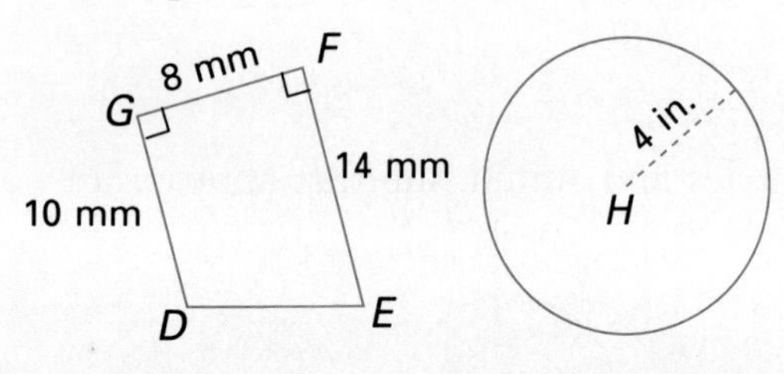

11. regular octagon *IJKLMNOP*

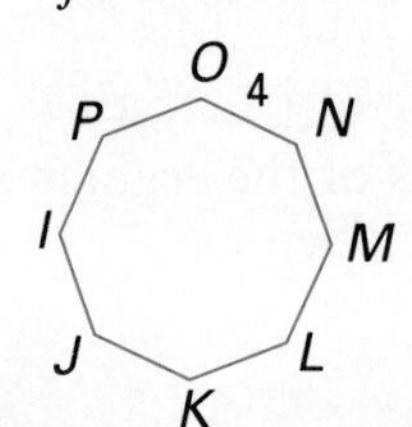

12. sector *QRS*

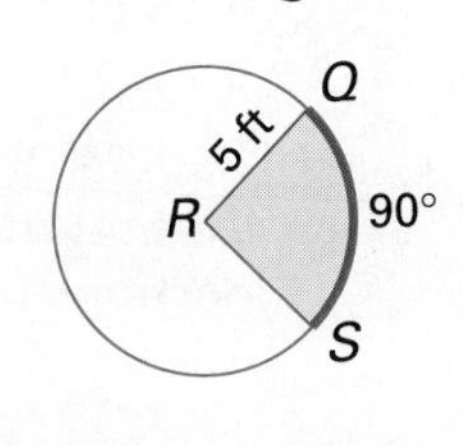

13. Find the area of a square with 9-in. sides.

14. Find the area of a circle with a 12-cm diameter.

15. Find the area of an annulus bounded by circles with 10-cm and 12-cm diameters.

16. Find the area of a regular octagon with an apothem of 10 m and a side length of 8.28 m.

17. Find the area of a segment of a circle with 10-mm radii determined by an arc of degree measure 60.

18. Find the length of an arc measuring 60 in a circle with 12-cm radii.

19. The circumference of a circle is 8π. Find the perimeter of an inscribed square.

College Prep Test

1. The area of square *DEFH* is 64, and the area of square AHGK is 25. Find the area of square *ABCD*.
(A) 9
(B) 89
(C) 144
(D) 169
(E) 225

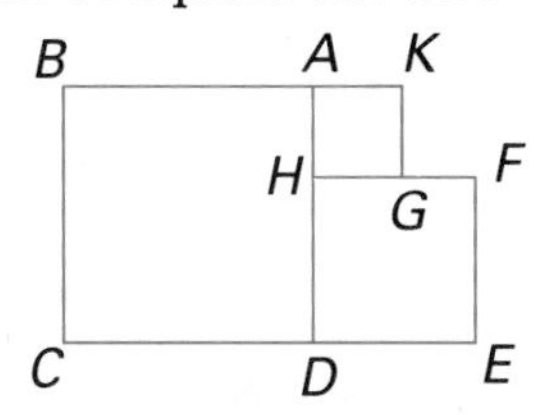

2. In $\triangle ABC$, $BR = RP$, $AC = 12$, $RP = 4$. The area of $ACED = 40$. Find the area of $\triangle DEB$.
(A) 8
(B) 10
(C) 12
(D) 14
(E) 16

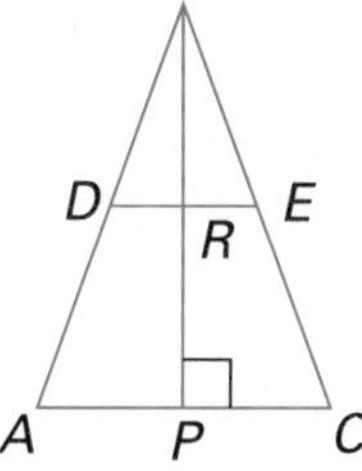

3. In $\odot O$, $OA = 6$ and $\overline{AO} \perp \overline{OB}$ Find the area of the shaded portion.
(A) 2π
(B) $\pi - 2$
(C) $6\pi - 9\sqrt{3}$
(D) $9\pi - 18$
(E) $36\pi - 9\sqrt{3}$

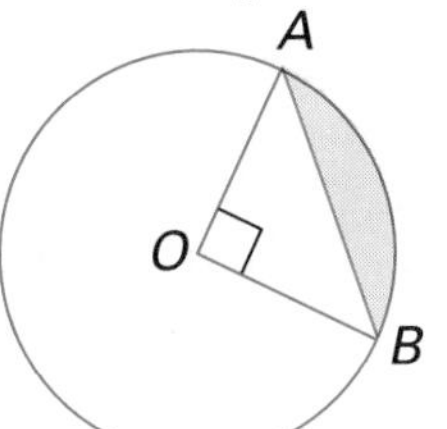

4. The larger circle has a diameter of 4 ft. The smaller circles are congruent and touch at the center of the larger circle. Find the area of the shaded portion.
(A) π (B) 2π (C) 4π
(D) 6π (E) 8π

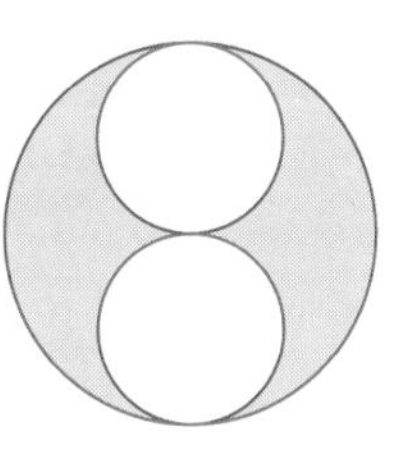

5. A picture is 6 ft wide by 8 ft long. If the frame has a width of 6 in., what is the ratio of the area of the frame to the area of the picture?
(A) $\frac{5}{16}$ (B) $\frac{5}{4}$ (C) $\frac{4}{5}$
(D) $\frac{5}{12}$ (E) $\frac{16}{5}$

6. On each side of a square, an isosceles right triangle is drawn. If the perimeter of the square is 16, what is the area of the shaded portion?
(A) 4
(B) 8
(C) 12
(D) 16
(E) 32

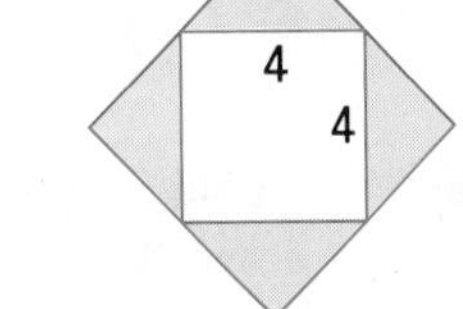

7. Given: $\overline{AB} \parallel \overline{CD}$, $DC = 4$, $AB = 12$, $m\angle A = m\angle B = 45$. Find the area of trapezoid *ABCD*.
(A) 16
(B) 32
(C) $16\sqrt{2}$
(D) $32\sqrt{2}$
(E) 64

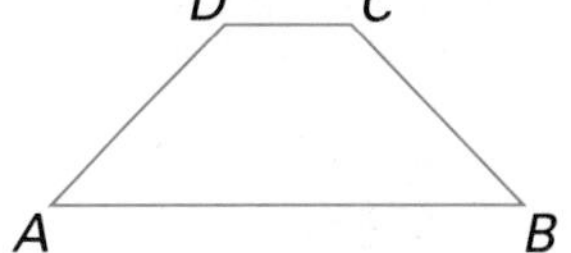

8. A rectangular field is half as wide as it is long. It is enclosed by x yards of fencing. What is the area of the field?
(A) $\frac{x^2}{2}$ (B) $2x^2$ (C) $\frac{x^2}{18}$
(D) $3\frac{1}{2}$ (E) $\frac{x^2}{72}$

9. What is the area of *PQRS*?
(A) 5
(B) 8
(C) 20
(D) 16
(E) 10

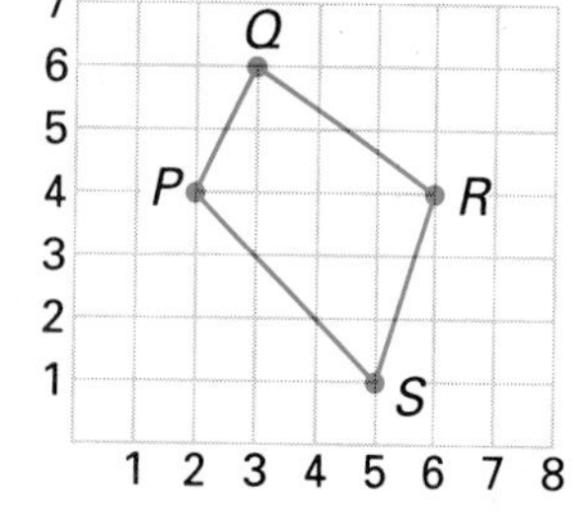

Cumulative Review (Chapters 1–12)

1. State whether true or false. *1.2*
(A) $AB = BC$
(B) $AB + BC = CD$
(C) $BD = -7$

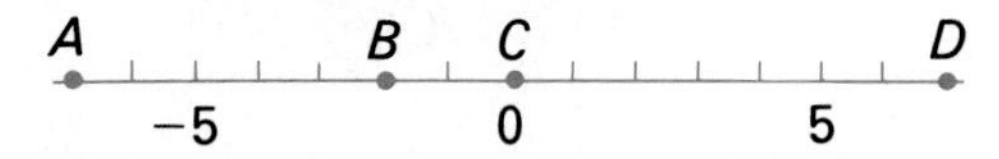

2. Based on the diagram, *1.4, 1.6*
(A) m $\angle AOC$ = ______.
(B) m $\angle DOC$ = ______.
(C) supp of $\angle DOX$ is $\angle$______.

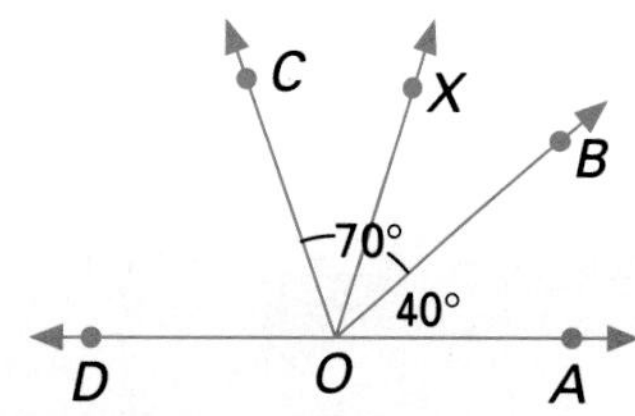

3. Based on the diagram, find: *1.8*
(A) m $\angle COD$.
(B) m $\angle COB$.
(C) m $\angle XOD$.

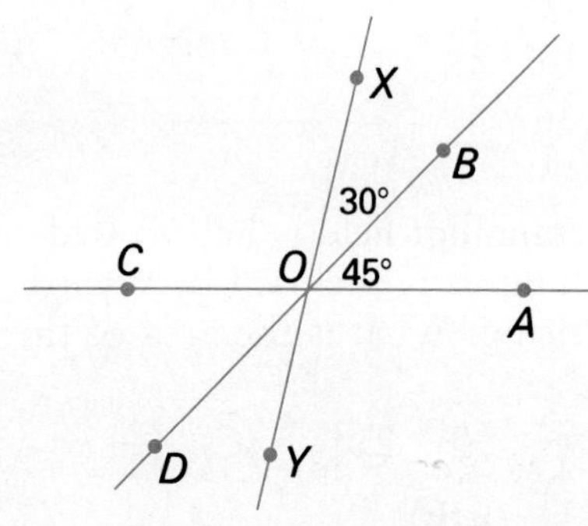

4. *m* and *n* are parallel, find: *3.2*
(A) m $\angle 5$. 70 (B) m $\angle 2$.
(C) m $\angle 4$. 110 (D) m $\angle 3$.

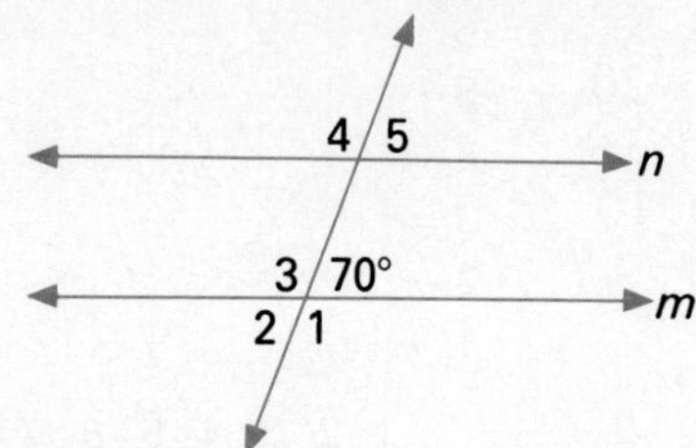

5. Prove that in any triangle the measure of an exterior angle is equal to the sum of the measures of the two remote interior angles. *3.7*

6. Given: $\overline{DB}$ and $\overline{AE}$ bisect each other. *4.3*
Prove: $\triangle ACB \cong \triangle ECD$

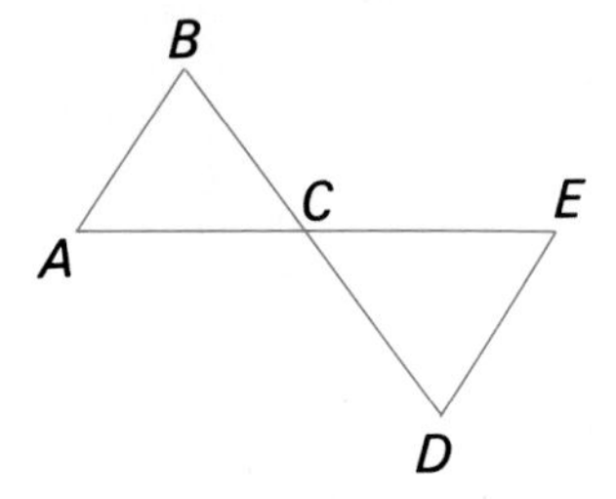

7. Given: $AD = CE$, $AB = CB$, $DB = EB$ *5.3*
Prove: $\triangle ACD \cong \triangle CAE$

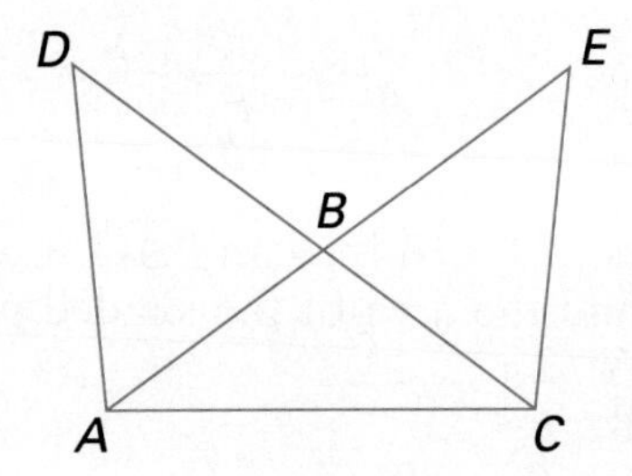

8. If $\triangle ABC$ is equilateral and the length of each side is 10, find the length of the altitude. *5.5*

9. For each of the following, state whether a triangle can be formed. If so, is it right, obtuse, or acute? Sides are: *5.8*
(A) 2, 3, 2.
(B) 7, 8, 17.
(C) 6, 8, 10.
(D) 4, 4, 4.

10. What is the sum of the measures of the interior angles of a pentagon? *6.2*

11. State whether true or false: *7.3*
(A) Every rhombus is a square.
(B) Every square is a rhombus.
(C) Every trapezoid can be divided into two congruent triangles.

12. Given: $\overline{AB} \parallel \overline{CD}$ *8.3*
Prove: $\triangle AXB \sim \triangle CXD$

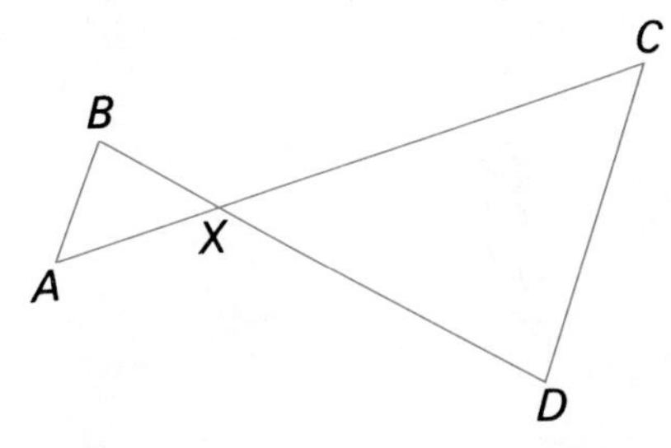

13. Given the triangles in Exercise 12, complete the proportion: $\frac{AB}{CD} = \frac{BX}{?}$. *8.4*

14. For the triangle below, find: *9.5, 9.6*
(A) $\sin \angle A$. (B) $\cos \angle A$.
(C) $\tan \angle B$.

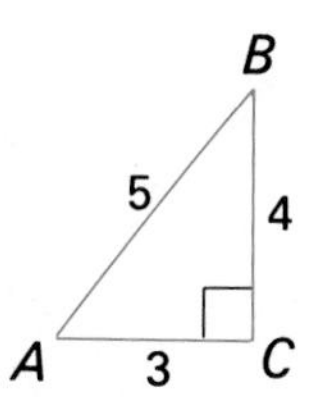

15. Find the midpoint of the segment joining the points $A(2,3)$ and $B(-4,-7)$. *10.2*

16. What is the slope of the line whose equation is $2y + 2x = 5$? *10.3*

17. Find $m\widehat{AB}$, for ⊙O. *11.4*

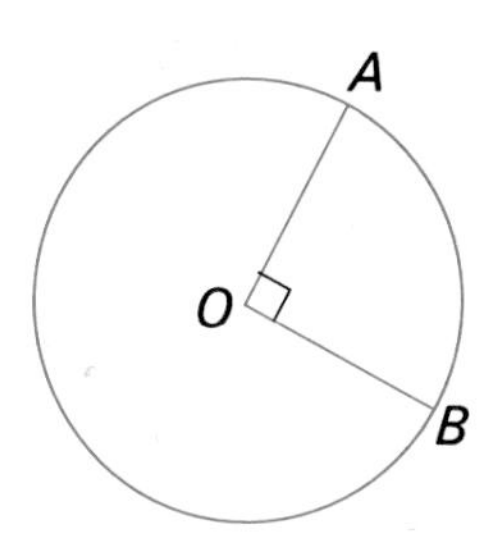

18. Find $m\widehat{AB}$. *11.6*

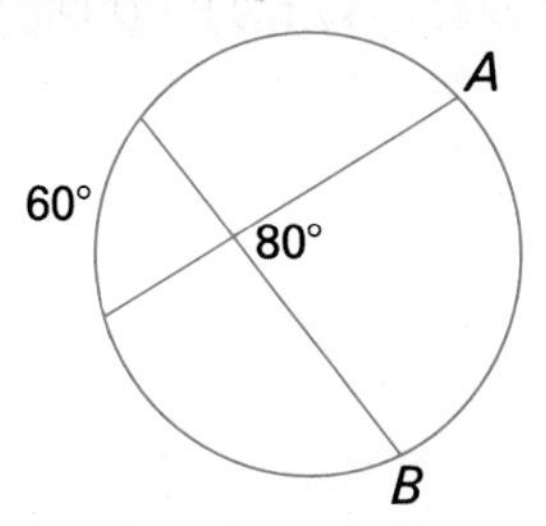

19. $\overline{OA}$ is a tangent and $\overline{OB}$ is a secant. Find m $\angle AOB$. *11.8*

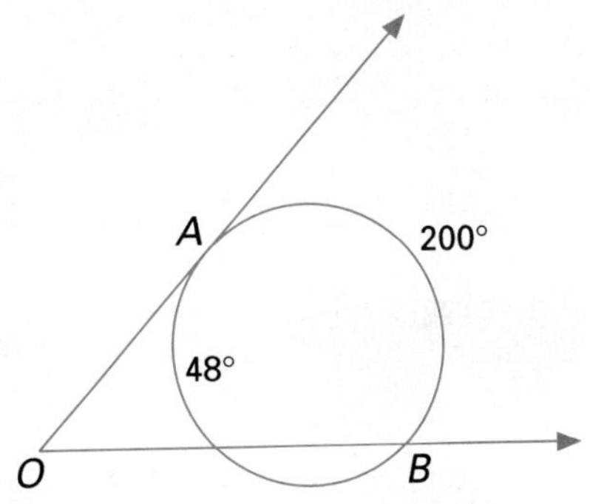

20. Find the area of a triangle with base 12 ft and altitude 6 ft. *12.4*

21. Find the area of the trapezoid. *12.5*

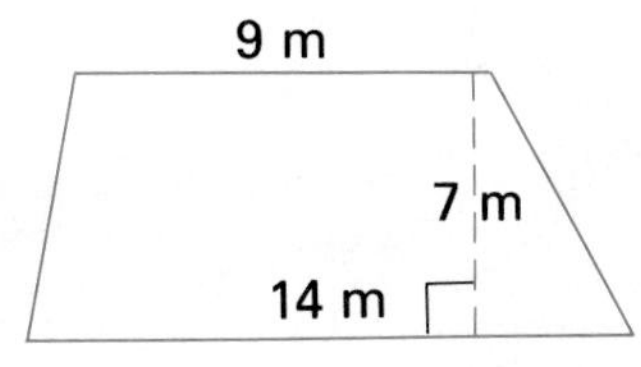

22. Find the area of the shaded portion if the measure of the central angle is 120. *12.9*

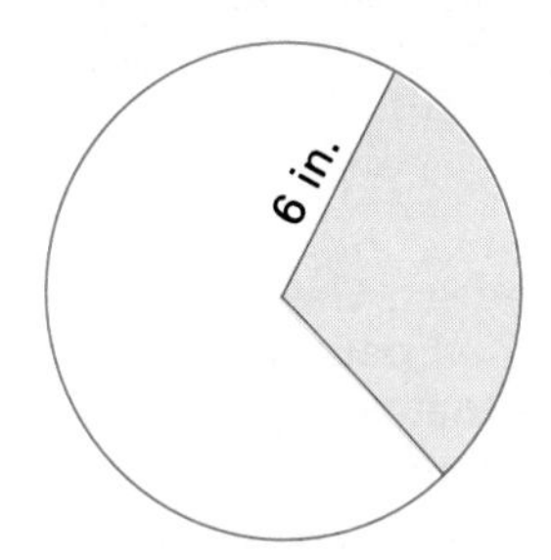

13 LOCI

Karl Friedrich Gauss (1777–1855)—Gauss, the son of uneducated parents, is widely regarded as one of the three or four greatest mathematicians of all time. At age three, he corrected an arithmetic mistake in his father's payroll.

13.1 Finding Locations

Objective To sketch and describe loci

In baseball, a pitch is considered a strike if it passes over any part of home plate at a height between the knee and the shoulder of the batter. Since home plate is 17 in. wide, the strike zone is a rectangle 17 in. wide with a height depending on the batter's height.

Definition A **locus** is the set of all points that satisfy a given set of conditions. (The plural of *locus* is *loci*.)

EXAMPLE 1 Find the locus of points in a plane exactly 2 cm from a given point P. Give a reason for your answer.

Solution (1) Draw point P.
(2) Locate several points 2 cm from P.
(3) Draw a smooth curve through these points.

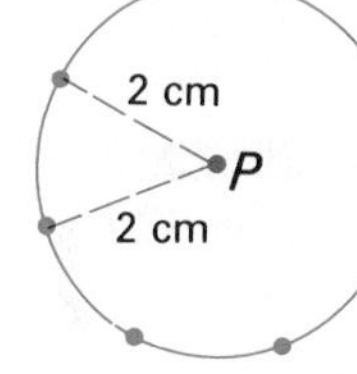

The locus is a circle with P as its center and a 2-cm radius.

EXAMPLE 2 Find the locus of points in a plane less than 3 cm from a given point.

Solution The locus is the interior of a circle with point Q as the center and with a 3-cm radius. (A dashed circle is drawn to show that the points of the circle are not included in the locus.)

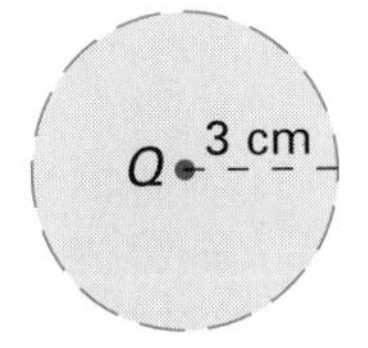

To determine a locus:
(1) Start with the given figure.
(2) Locate several points that satisfy the specified conditions.
(3) Draw a smooth curve through the points.
(4) Describe the figure and its location.

EXAMPLE 3 Sketch and describe the locus of points in a plane that are equidistant from two given parallel lines m and n.

Solution

(1) Draw two parallel lines m and n.

(2) Locate several points the same distance from m as from n. (Distance is measured along the perpendicular from a point to the line.)

(3) Draw a line through these points.

The locus is a line parallel to m and n and halfway between them.

Drawing a picture does not prove that a locus is correct. To establish a locus, it is necessary to prove that the locus contains all points, and only those points, that satisfy the given conditions. The statement of a locus theorem is a biconditional. Its proof involves two parts.

Theorem 13.1 The **locus of points** in a plane equidistant from two given points is the perpendicular bisector of the segment having the two points as endpoints.

Part I: If a point lies on the perpendicular bisector of a segment, then it is equidistant from the endpoints of the segment.

Given: P is on the perpendicular bisector of $\overline{AB}$.

Prove: $PA = PB$

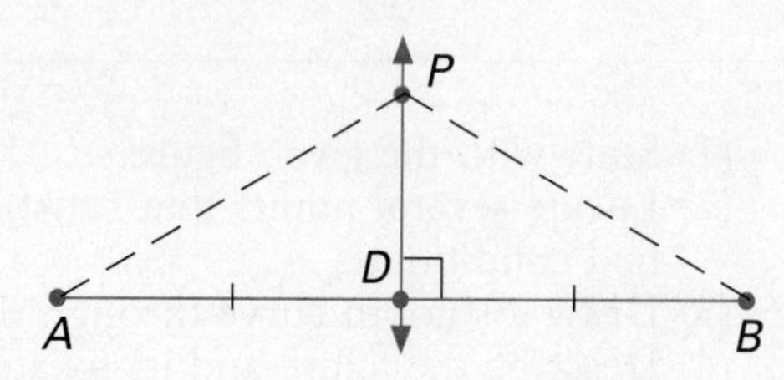

Part II: If a point is equidistant from the endpoints of a segment, then it lies on the perpendicular bisector of the segment.

Given: $PA = PB$

Prove: P is on the perpendicular bisector of $\overline{AB}$.

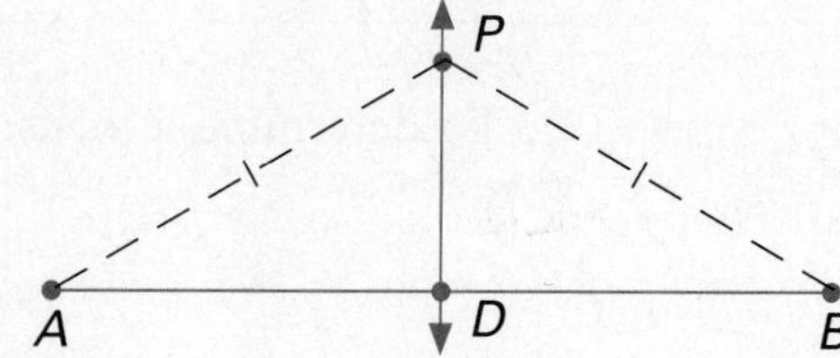

Theorem 13.2 In a plane, the locus of points equidistant from the sides of an angle is the bisector of the angle.

Part I: If a point is on the bisector of an angle, then it is equidistant from the sides of the angle.

Given: $\overrightarrow{OC}$ bisects $\angle AOB$;
P is on $\overrightarrow{OC}$; $\overline{PD} \perp \overrightarrow{OA}$,
$\overline{PE} \perp \overrightarrow{OB}$.

Prove: $PD = PE$

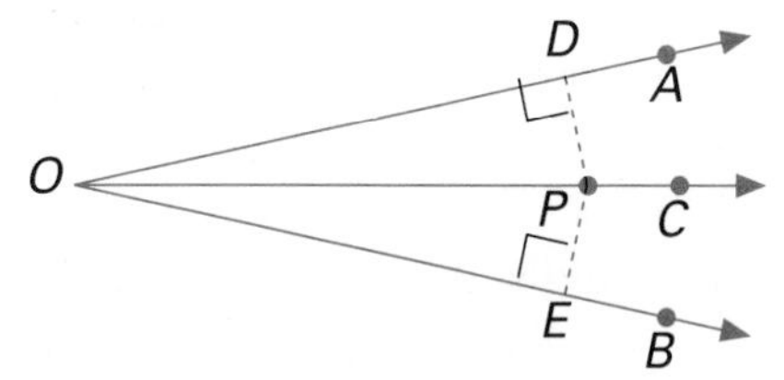

Plan The distance from a point to a line is measured along the perpendicular to the line. Show that $\overline{PD}$ and $\overline{PE}$ are congruent corresponding sides of congruent triangles ODP and OEP.

Part II: If a point is equidistant from the sides of an angle, then it is on the bisector of the angle.

Given: P is in the interior of $\angle AOB$; $\overline{PD} \perp \overrightarrow{OA}$,
$\overline{PE} \perp \overrightarrow{OB}$, and $PD = PE$.

Prove: $\overrightarrow{OP}$ bisects $\angle AOB$.

Plan To show that $\overrightarrow{OP}$ is the bisector, show that $\angle DOP$ and $\angle EOP$ are congruent corresponding angles of congruent triangles ODP and OEP.

Classroom Exercises

Describe the locus of points in a plane. (Exercises 1–4)

1. equidistant from two parallel lines that are 10 cm apart
2. 7 cm from a given point
3. equidistant from two points 8 cm apart
4. less than 5 mm from a given point
5. Explain the meaning of the word *locus*.

Written Exercises

Sketch and describe the given locus of points in a plane. (Exercises 1–14)

1. 5 cm from a given point
2. less than 4 cm from a given point
3. greater than 3 cm from a given point
4. equidistant from two points that are 4 cm apart

5. equidistant from the sides of a right angle
6. equidistant from the sides of an obtuse angle
7. 3 cm from a given line
8. less than 4 cm from a given line
9. more than 7 cm from a given line
10. closer to point A than to point B, if A and B are 8 cm apart
11. closer to side $\overrightarrow{OB}$ of $\angle AOB$ than to side $\overrightarrow{OA}$
12. of the center of a circle rolling along a line
13. of the center of a circle rolling around another circle
14. equidistant from the four vertices of a square

15. A power line runs in a straight line for several miles. According to the description of the right-of-way, the power company has a right-of-way that extends 10 ft on either side of the line. Describe in mathematical terms the boundary of this right-of-way.
16. An experimental steering system for automobiles uses a magnetic wire buried in the center of a lane of the highway. Such a wire is placed in a perfectly straight stretch of highway with a lane 24 ft wide. Describe in mathematical terms the location of this wire.
17. A goat is tied to a stake with a 10-ft rope. While tied, the goat eats all the grass that it can reach. If there are no obstacles in the way, describe the patch of ground from which the goat could eat.
18. Suppose that the same goat in Exercise 17 was tied to a corner of a barn that was 20 ft by 40 ft. Describe the patch of ground from which the goat could eat the grass.
19. Prove Theorem 13.1.
20. Prove Theorem 13.2, (Part I).
21. Prove Theorem 13.2, (Part II).
22. Prove the locus of the midpoints of all congruent chords of a given circle is a circle concentric to the original circle.
23. Prove the locus of vertices of all right triangles sharing a given segment as hypotenuse is a circle with the hypotenuse as a diameter. The endpoints of the diameter are not part of the locus.

Mixed Review

Construct the following.

1. the perpendicular bisector of a given segment ***5.6***
2. the bisector of a given obtuse angle ***1.5***
3. a perpendicular to a line from a point not on the line ***5.6***

13.2 Multiple Conditions for Loci

Objective To sketch and describe loci satisfying two or more conditions

Descriptions of most loci involve distances. The distance between two points is the length of the segment that has the two points as endpoints. The distance from a point to a line is measured along the perpendicular from the point to the line. The distance between two parallel lines is the distance from any point on one of the lines to the other line along the perpendicular. In general, the distance between two geometric figures is the length of the shortest segment between the figures.

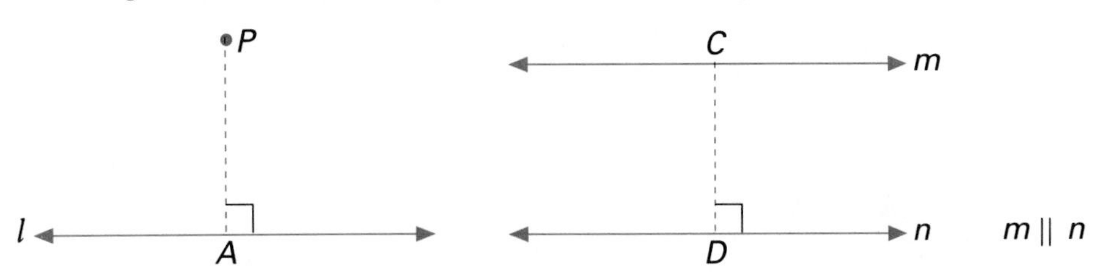

Definition

The distance from a point P to a circle O having radii of length r is $|OP - r|$.

Case 1: P is in the exterior of $\odot$O.
$PA = |OP - r|$

Case 2: P is the interior of $\odot$O.
$PA = |r - OP| = |OP - r|$

Case 3: If P is on $\odot$O, then $PA = |OP - r| = 0$.

EXAMPLE 1 Find the locus of points in a plane that are 6 cm from a circle having a 9-cm radius.

Solution

$$6 = |OP - r|$$
$$6 = |OP - 9|$$
$$OP + 9 = 6 \text{ or } OP - 9 = -6$$
$$OP = 15 \qquad OP = 3$$

The locus is two circles concentric to the given circle, one with a 3-cm radius and one with a 15-cm radius.

EXAMPLE 2 Sketch and describe the locus of points in a plane that are a distance d from a circle having a radius of length r.

Plan Because the lengths r and d are variables, three cases must be investigated: $d > r$, $d = r$, $d < r$.

Solution

Case 1: $d > r$
The locus is a circle concentric to the given circle, having radius of length $r + d$.

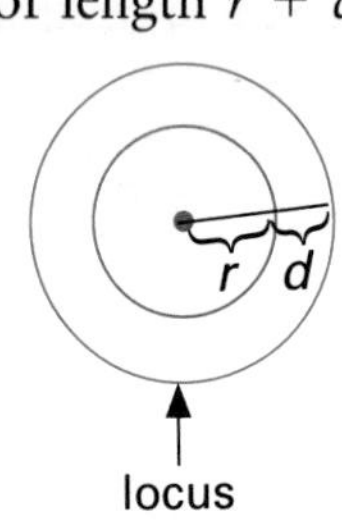

Case 2: $d = r$
The locus is the center of the given circle and a circle concentric to the given circle, having radius of length $r + d$, or $2d$.

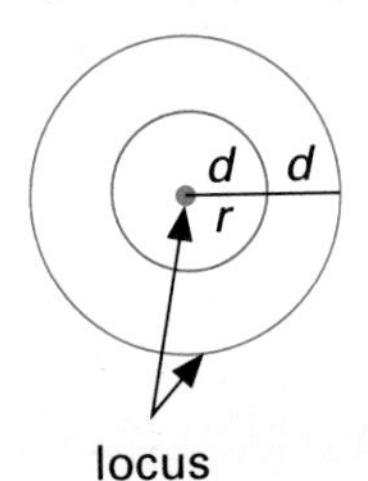

Case 3: $d < r$
The locus is two circles concentric to the given circle. The two circles have radii of lengths $r + d$ and $r - d$.

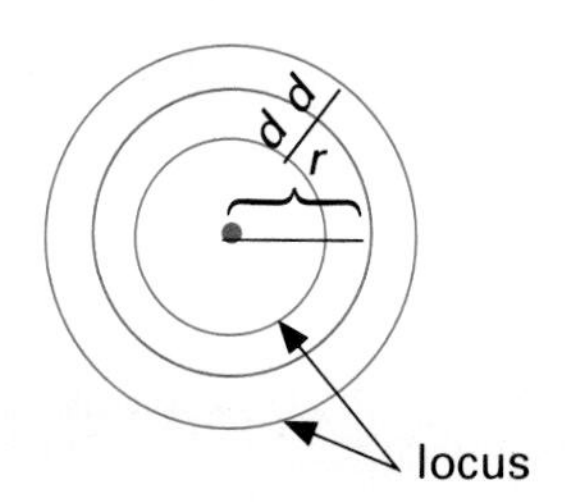

A locus may involve more than one condition.

EXAMPLE 3 Find the locus of points in a plane that are 3 cm from line m *and* 4 cm from point P on line m.

Plan
(1) The locus of points 3 cm from line m is two lines, each parallel to m and 3 cm from m.
(2) The locus of points 4 cm from point P is a circle with P as center and radius of length 4 cm.
(3) The intersection of these two loci is the required locus.

Solution

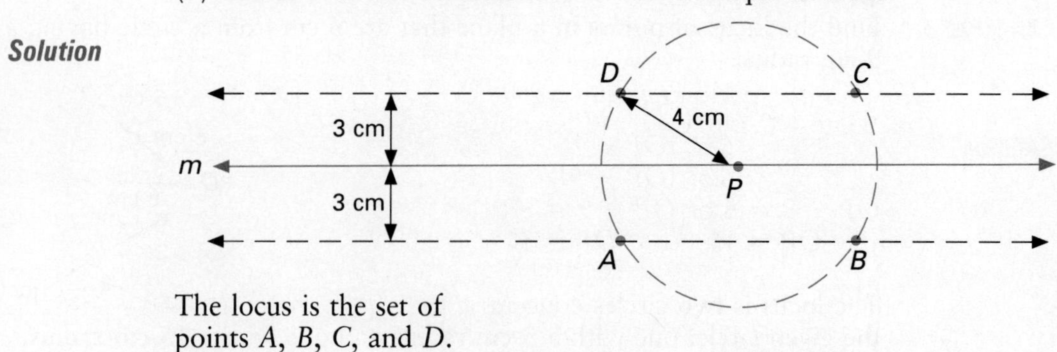

The locus is the set of points A, B, C, and D.

EXAMPLE 4 Find the locus of points in a plane that are 3 cm from point P *and* 4 cm from point Q, if $PQ = 6$ cm.

Solution $\odot P$ is the locus of points 3 cm from P.
$\odot Q$ is the locus of points 4 cm from Q.

The locus is the set of points A and B, the intersections of $\odot P$ and $\odot Q$.

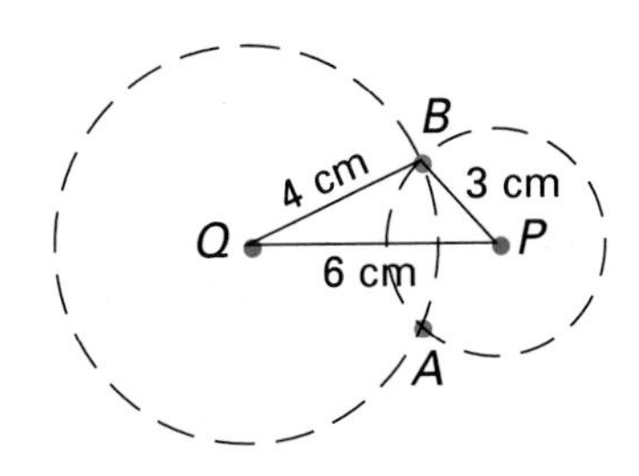

There may be several possibilities for the intersection of two or more loci, depending on the distance involved. If they do not intersect at all, then there are no points that satisfy all of the given conditions.

EXAMPLE 5 Find the locus of points in a plane that are a distance r from a point P and a distance s from point Q, if P and Q are a distance d apart ($d \neq 0$).

Solution **Case 1:** $r + s < d$
There are no points in the locus.

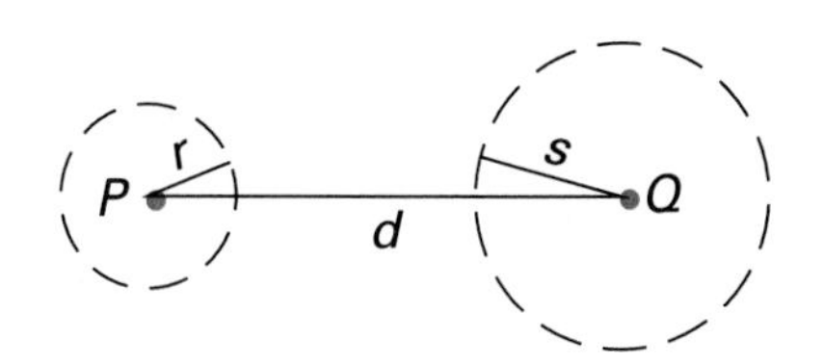

Case 2: $r + s = d$
The locus is one point, the point of tangency of the two circles with centers P and Q having radii of lengths r and s, respectively.

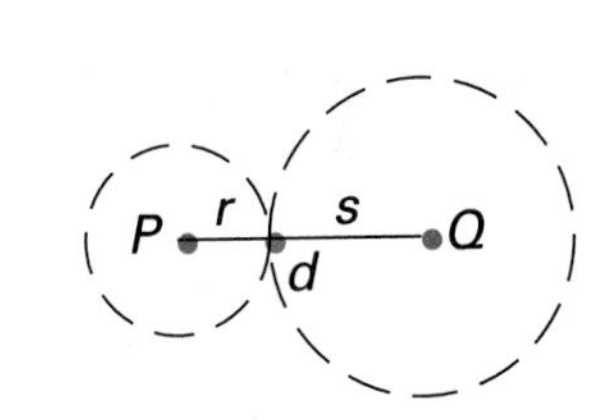

Case 3: $r + s > d$
The locus is the set of two points, which are the intersections of the circles with centers P and Q having radii of lengths r and s, respectively.

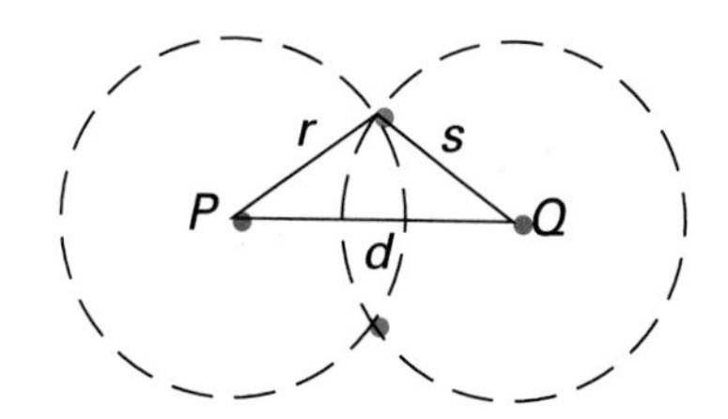

There are two other possibilities for the locus. If $s > r + d$ *or* $r > s + d$, then one circle lies in the interior of the other, with no intersection. If $s = r + d$ *or* $r = s + d$, then one circle is internally tangent to the other with one point of intersection.

Classroom Exercises

Describe the locus of points in a plane.

1. 10 cm from a given point
2. equidistant from two parallel lines
3. 5 cm from a circle with a 10-cm radius
4. 4 cm from P and 6 cm from Q, where P and Q are 12 cm apart
5. greater than 6 cm from a circle having a 3-cm radius

Written Exercises

Sketch and describe the locus of points in a plane.

1. 2 cm from $\odot O$
2. 5 cm from $\odot O$
3. 8 cm from $\odot O$
4. less than 3 cm from $\odot O$
5. in the exterior of $\odot O$ *and* greater than 2 cm from the circle

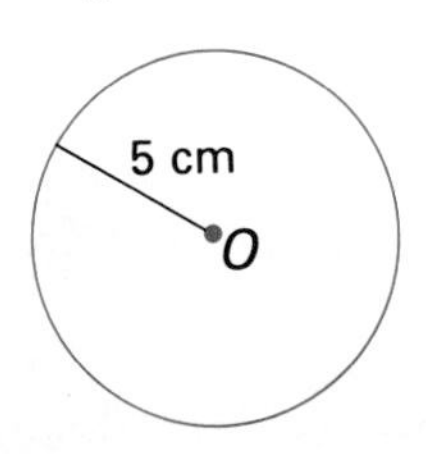

6. 2 cm from P *and* 5 cm from Q
7. 6 cm from P *and* 6 cm from Q
8. 5 cm from P *and* 5 cm from Q
9. 9 cm from P *and* 6 cm from Q
10. equidistant from P and Q
11. more than 8 cm from P *and* more than 7 cm from Q
12. less than 7 cm from P *and* less than 6 cm from Q

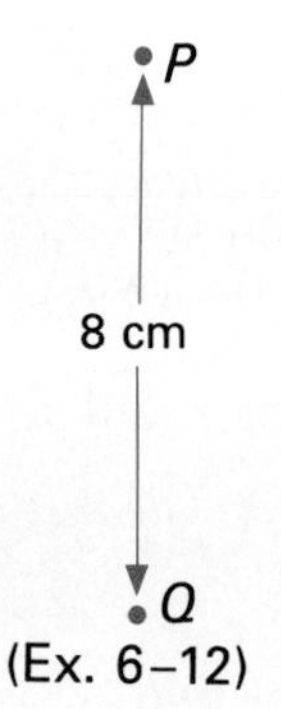

(Ex. 6–12)

13. on l *and* 5 cm from P
14. on m *and* 8 cm from P
15. 5 cm from l *and* 5 cm from m
16. 4 cm from P *and* 4 cm from l
17. 2 cm from m *and* 9 cm from P
18. less than 8 cm from P *and* less than 3 cm from m

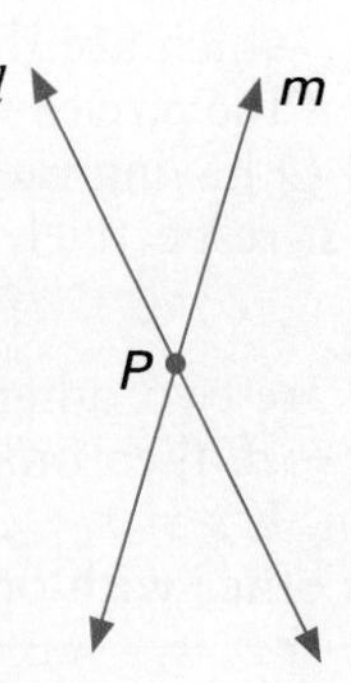

Sketch and describe the locus of points in a plane. Give all possible cases.

19. 3 cm from point P *and* 5 cm from point Q, given that P and Q are a distance d apart

20. 8 cm from P *and* t cm from point Q, given that P *and* Q are 8 cm apart

21. in the interior of an angle *and* equidistant from the sides and a distance d from the vertex of the angle

22. equidistant from two intersecting lines

23. equidistant from all points of $\odot O$

24. equidistant from two concentric circles

25. a distance d from a given line *and* a distance r from a point on the line

26. a distance d from each of two points that are a distance t from each other

27. less than a distance d from each of two points that are a distance t from each other

Sketch and describe the locus in a plane. (Exercises 28–29)

28. determined by a point on a circle as it rolls along a line

29. determined by a point on a circle as it rolls around outside another circle

30. A Spirograph® is a device for drawing designs using a pen mounted in a moving wheel. The designs are actually loci. The locus of points illustrated below is determined by an interior point that is a distance d from a circle as the circle rolls along a line. How does the design differ as the distance d changes?

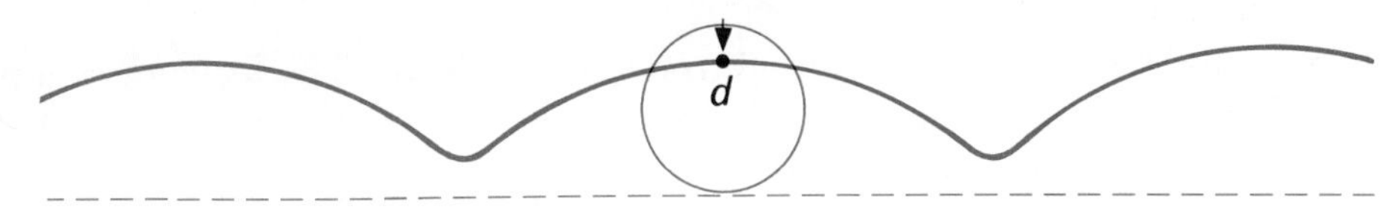

Mixed Review

Is the statement true or false?

1. Two triangles are similar if two angles of one are congruent to two angles of the other. *8.3*

2. Two triangles are similar if two sides of one are congruent to two sides of the other. *8.4*

3. Two triangles are congruent if two angles of one are congruent to two angles of the other. *4.6*

13.3 Loci in Space

Objective To sketch and describe loci in space

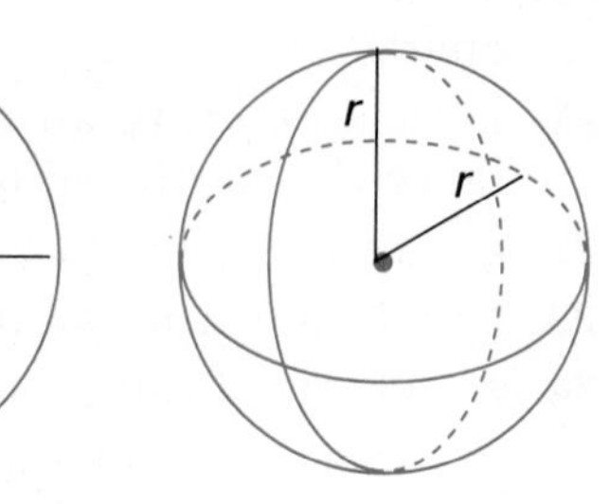

Loci (plural of locus) can be restricted to one plane, or they can be generalized to space. The locus of points a given distance r from a fixed point was described as a circle. This is true only if the locus is restricted to a plane. If this restriction is removed, the locus is a sphere with the fixed point as its center.

EXAMPLE 1 Sketch and describe the locus of points in space a given distance d from a fixed line l. How is this different from the locus when restricted to one plane?

Solution

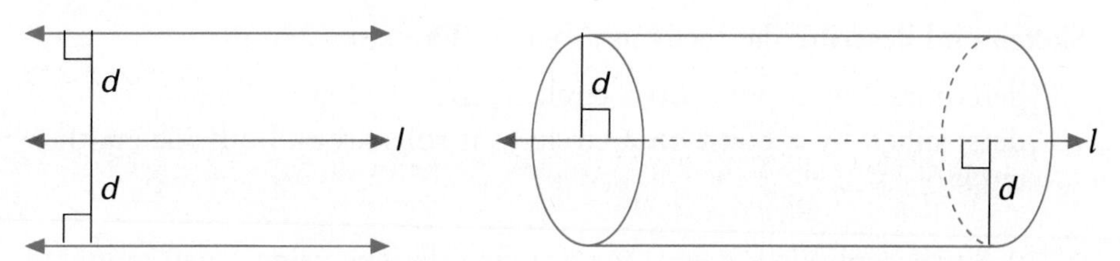

The locus in one plane is a pair of parallel lines with l halfway between them. In space the locus is a cylindrical surface with l as its axis and d as the length of a radius.

Sometimes an inequality is used to describe a locus. The locus of points in space a given distance from a fixed point is a sphere. A sphere is a surface. The locus of points in space less than a given distance from a fixed point is the *interior* of a sphere, a solid. The locus of points in a space greater than a given distance from a fixed point is the *exterior* of a sphere.

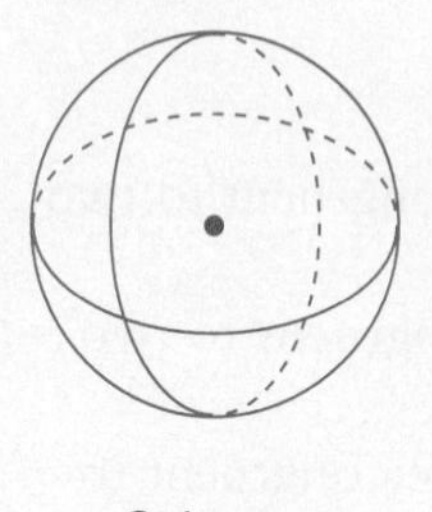

Sphere

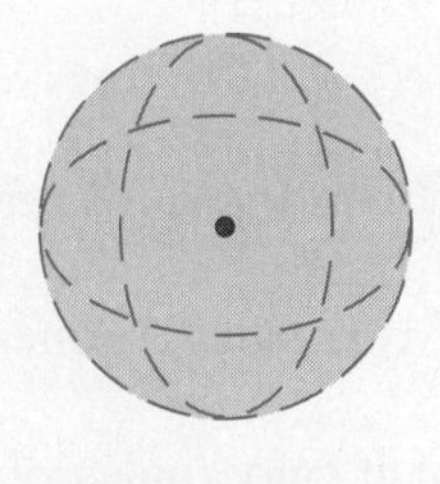

Interior of sphere

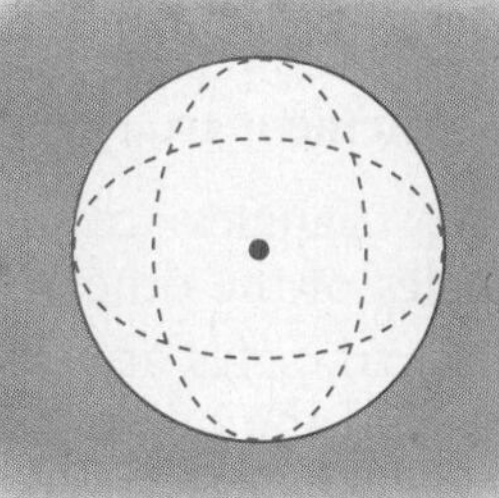

Exterior of sphere

EXAMPLE 2 Points A and B are 10 cm apart. Point P is the midpoint of $\overline{AB}$. Sketch and describe the locus of points in space equidistant from A and B *and* more than 4 cm from point P.

Solution

(1) In a plane, the locus of points equidistant from A and B is the perpendicular bisector of $\overline{AB}$. In space, the locus of points equidistant from A and B is a plane $\mathcal{Q}$ that is the perpendicular bisector of $\overline{AB}$. Note that P is in $\mathcal{Q}$, since P is the midpoint of $\overline{AB}$.

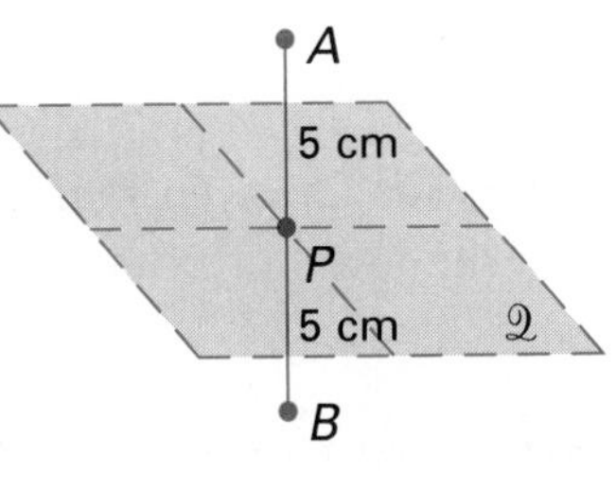

(2) The locus of points in $\mathcal{Q}$ that are more than 4 cm from point P is the exterior of the circle in $\mathcal{Q}$ with a radius of length 4 cm and P as its center.

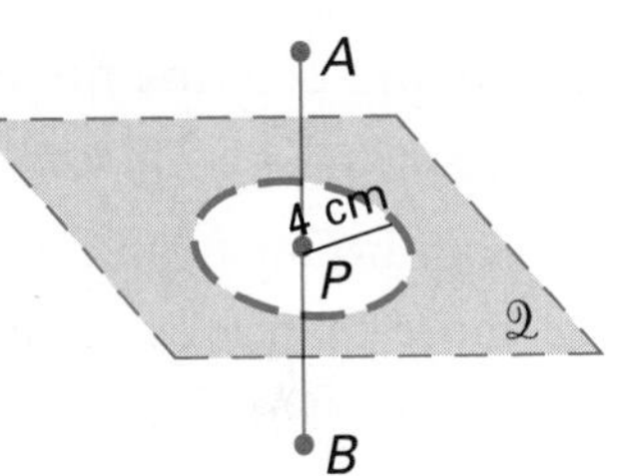

Thus, the required locus is the set of points in the perpendicular bisector $\mathcal{Q}$ of $\overline{AB}$ and in the exterior of $\odot P$, where $\odot P$ has a 4-cm radius.

Classroom Exercises

Describe the locus of points in space.

1. 5 cm from a given point
2. 5 cm from a given line
3. 5 cm from a given plane
4. equidistant from two given points

Written Exercises

Sketch and describe the locus of points in space.

1. 6 m from a given point
2. 10 m from a given line
3. less than 5 cm from a given point
4. greater than 5 cm from a given point
5. less than 5 cm from a given line
6. greater than 5 cm from a given line
7. equidistant from two points that are 10 cm apart
8. equidistant from two parallel lines
9. equidistant from two parallel planes
10. less than 5 cm from a given plane
11. equidistant from the sides of given angle

Sketch and describe the locus of points in space.

12. 5 cm from each of two points that are 6 cm apart
13. 5 cm from each of two parallel lines that are 6 cm apart
14. 10 cm from a given point *and* 10 cm from a given plane that is 5 cm from the given point
15. the locus of points a distance d from point P *and* a distance t from point Q, given that the distance between P and Q is s
16. the locus of points 5 cm from a given segment that is 4 cm long

Midchapter Review

Sketch and describe the given locus of points in a plane.

1. 3 cm from a given point *13.1*
2. equidistant from two rays that form a 60-degree angle *13.1*
3. more than 5 cm from the circumference of a circle with radius 4 cm *13.2*

Sketch and describe the given locus of points in space. *13.3*

4. equidistant from two points that are 8 cm apart
5. 5 in. from line m *and* 10 in. from point P on m

Application: *Site for a House*

A landscape architect is developing a site plan for a new house. The local building code specifies that no part of the house can be closer than 30 ft to the street right-of-way line, nor 10 ft to any of the other property lines. A tree is to remain in the front yard. The house must be at least 10 ft away from the trunk of the tree.

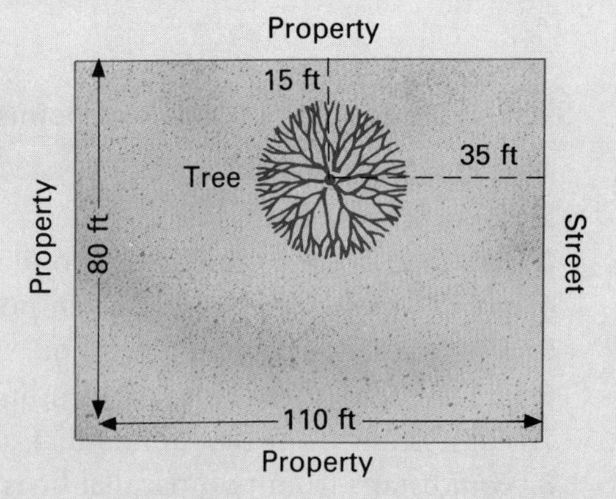

1. Suppose that the house is rectangular in shape, with a front 50 ft long and parallel to the street and a depth of 40 ft. Copy the map and show where this house could be built.
2. Suppose that the house could be turned in any direction. Show on the map where it now could be located.

13.4 Concurrent Bisectors in Triangles

Objectives

To use constructions to find loci

To apply perpendicular bisector and angle bisector concurrency theorems

To locate and use incenters and circumcenters

A piece of machinery has a broken gear. To replace the gear, it is necessary to know the diameter of the gear and the total number of teeth on the gear. A piece of the broken gear was salvaged. How can the diameter and the number of teeth be determined from this piece?

In Chapter 11, one approach to this problem was suggested. In this lesson you will investigate geometric constructions and loci that will reveal an alternate method.

EXAMPLE 1 Use constructions to find the locus of points equidistant from the three vertices of $\triangle ABC$.

Plan In a plane, the locus of points equidistant from two points is the perpendicular bisector of the segment having the points as endpoints. Construct the perpendicular bisectors of $\overline{AB}$ and $\overline{AC}$.

Solution Notice that $DB = AD$ and $AD = DC$. So, by the Transitive Property, $DB = DC$. This means that D will also be on the perpendicular bisector of $\overline{BC}$. Therefore, the locus of points equidistant from the three vertices of $\triangle ABC$ is a single point D.

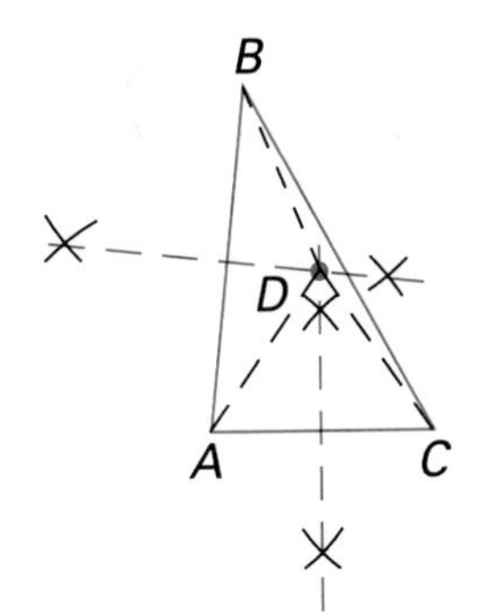

Three or more lines that intersect or converge at a common point are **concurrent lines.**

Theorem 13.3 The perpendicular bisectors of the sides of a triangle are concurrent at a point equidistant from the vertices of the triangle.

Given: $\triangle ABC$; l, m, and n are the perpendicular bisectors of $\overline{AB}$, $\overline{BC}$, and $\overline{CA}$, respectively.

Prove: l, m, and n are concurrent at a point equidistant from A, B, and C.

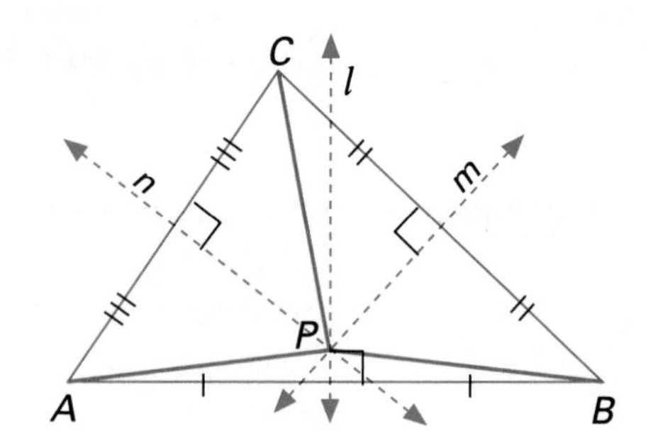

Proof

Statement	Reason
1. l, m, and n are the $\perp$ bis of $\overline{AB}$, $\overline{BC}$, and $\overline{CA}$, respectively.	1. Given
2. $\overline{AB} \nparallel \overline{CA}$	2. Sides of $\triangle$ intersect.
3. $l \nparallel n$	3. Coplanar lines $\perp$ to nonparallel lines are $\nparallel$.
4. l and n intersect at point P.	4. $\nparallel$ lines intersect.
5. $PB = PA$, $PA = PC$	5. A point on $\perp$ bis of seg is equidist from its endpts.
6. $PB = PC$	6. Trans Prop
7. m contains P.	7. A point equidist from endpts of seg lies on its $\perp$ bis.
8. l, m, and n are concurrent at a point equidistant from A, B, and C.	8. Def of concurrent lines

Definition

The **circumcenter** of a triangle is the intersection of the perpendicular bisectors of the sides of the triangle.

EXAMPLE 2 Construct the locus of points equidistant from the sides of $\triangle EFG$.

Plan The locus of points equidistant from the sides of an angle is the bisector of the angle. Construct the bisector of $\angle E$ and $\angle F$.

Solution Point H is equidistant from sides $\overline{EG}$ and $\overline{EF}$. It is equidistant from sides $\overline{EF}$ and $\overline{FG}$. Therefore, it is equidistant from sides $\overline{EG}$ and $\overline{FG}$.

The locus is point H.

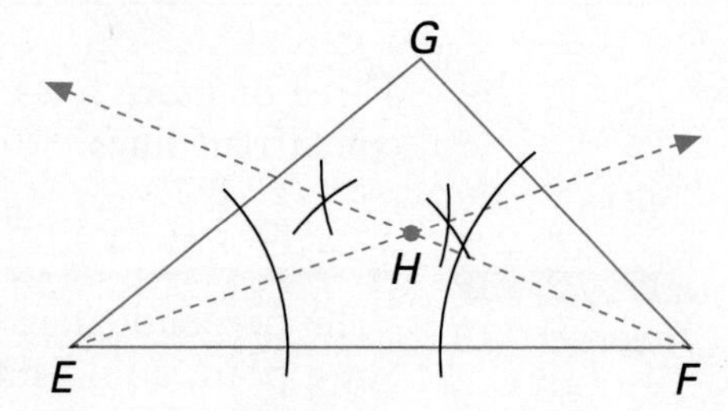

The locus of points equidistant from the sides of the triangle is the intersection of the three angle bisectors of the triangle.

Theorem 13.4 The bisectors of the angles of a triangle are concurrent at a point equidistant from the sides of the triangle.

Given: $\triangle ABC$, $\overrightarrow{AD}$, $\overrightarrow{BE}$, and $\overrightarrow{CF}$ are the bisectors of $\angle A$, $\angle B$, and $\angle C$, respectively.
Prove: $\overrightarrow{AD}$, $\overrightarrow{BE}$, and $\overrightarrow{CF}$ are concurrent at a point equidistant from $\overline{AB}$, $\overline{BC}$, and $\overline{CA}$.

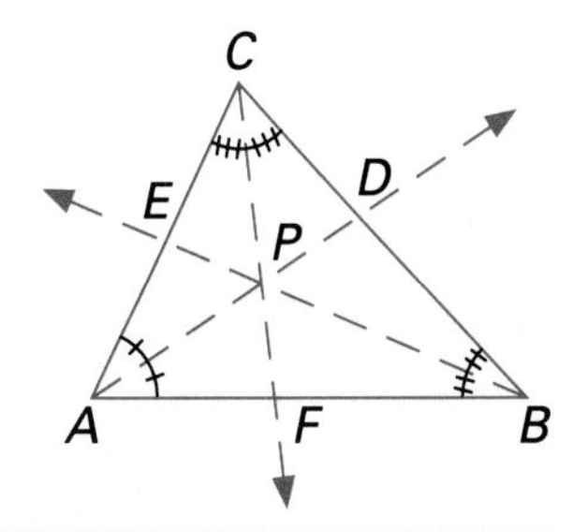

Definition The **incenter** of a triangle is the intersection of the bisectors of the angles of the triangle.

The two previous theorems prove that an incenter and a circumcenter exist in every triangle. This fact suggests some useful constructions.

EXAMPLE 3 Draw a triangle. Construct a circle that contains the three vertices of the triangle.

Construction (1) Draw $\triangle ABC$. Construct perpendicular bisectors of $\overline{AB}$ and $\overline{BC}$, intersecting at point P.

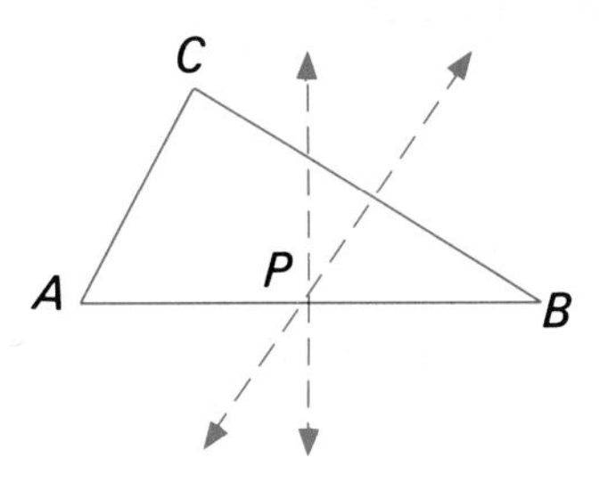

(2) Using PA as the length of a radius, draw $\odot P$.

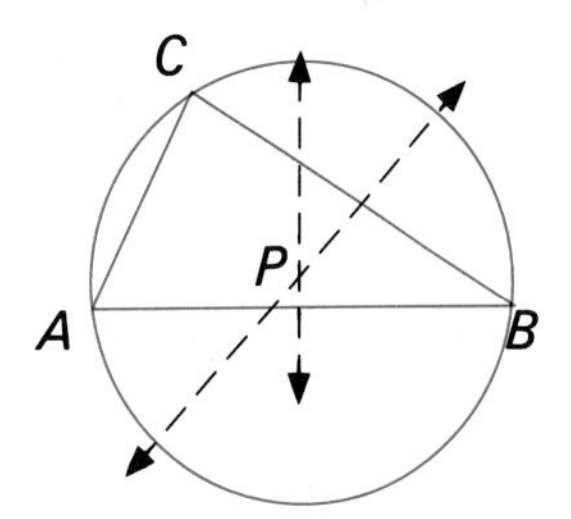

Result: $\odot P$ is the desired circle.

The circle constructed is the **circumcircle**, which **circumscribes** the triangle. P is the circumcenter of the triangle.

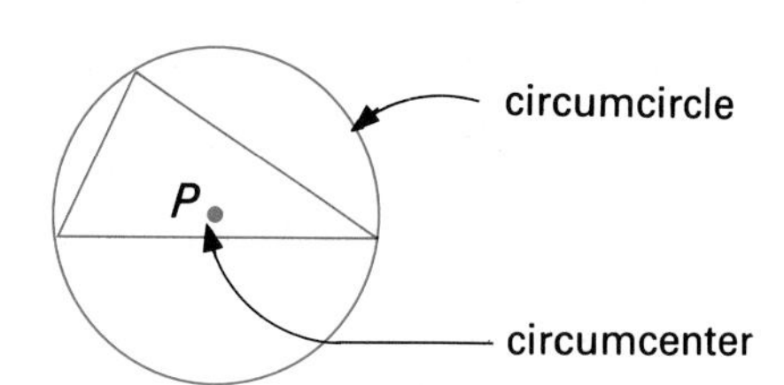

EXAMPLE 4 Draw a triangle. Inscribe a circle in the triangle.

Construction (1) Draw $\triangle ABC$. Construct the bisectors of $\angle A$ and $\angle B$, intersecting at point P.

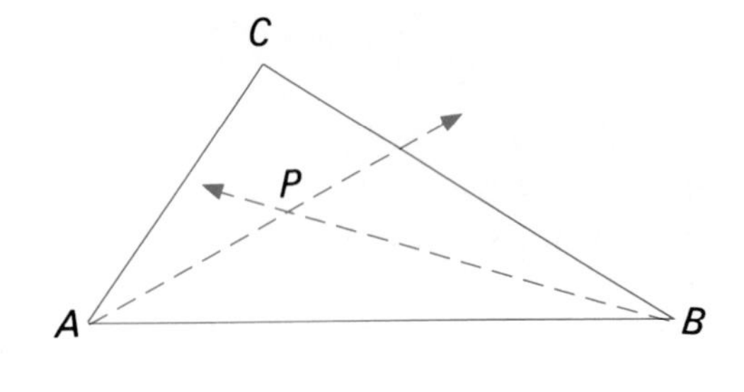

(2) Construct $\overline{PD} \perp \overline{AB}$. Using PD as the length of a radius, draw $\odot P$.

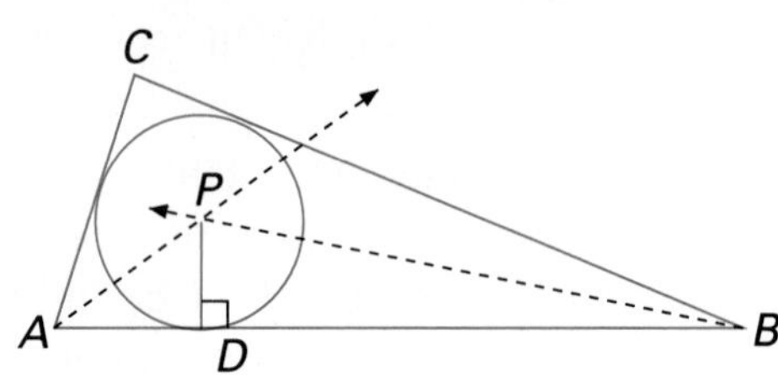

Result: $\odot P$ is inscribed in $\triangle ABC$.

The circle constructed is called the **incircle** of the triangle. It is *inscribed* in the triangle. P is the **incenter** of the triangle.

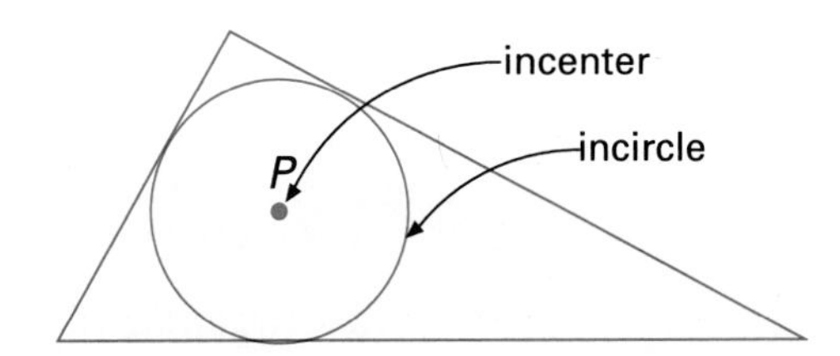

Focus on Reading

The given word has both a prefix and a root word. In a dictionary that has derivations, find the meanings of the prefix and root word.

1. circumscribed
2. circumcircle
3. circumcenter
4. inscribed
5. incircle
6. incenter

Classroom Exercises

Is the statement true or false?

1. The point of intersection of the bisectors of the sides of a triangle is a circumcenter.

2. The bisectors of the angles of a triangle are concurrent at a point equidistant from the vertices.

3. The locus of points equidistant from the three vertices of a triangle is the intersection of the perpendicular bisectors of the sides.

4. The perpendicular bisectors of the sides of an obtuse triangle are not concurrent.

Written Exercises

1. Draw an acute triangle. Construct the incenter of the triangle.
2. Draw an obtuse triangle. Construct the circumcenter of the triangle.
3. Draw a right triangle. Construct the incenter of the triangle.
4. Draw a right triangle. Construct the circumcenter of the triangle.
5. Draw an obtuse triangle. Construct the inscribed circle.
6. Draw a right triangle. Construct the circumscribed circle.
7. Construct an isosceles triangle. Construct both the incenter and the circumcenter of the triangle. Where do the two centers lie?
8. Construct an equilateral triangle. Construct both the incenter and the circumcenter of the triangle. Where do the two centers lie?
9. Trace a circle. Use a construction to locate the center of the circle.
10. Using the diagram in the lesson introduction, draw the circle containing the gear. Find the number of gear teeth.

True or false? Draw a diagram to justify your answer.

11. No triangle has its incenter on one side of the triangle.
12. No triangle has its circumcenter on one side of the triangle.
13. The incenter of a triangle can never be the circumcenter.
14. The circumcenter of a triangle can never be in the exterior of the triangle.
15. Prove Theorem 13.4.

The equilateral triangle shown has sides of length 10. (Exercises 16–20)

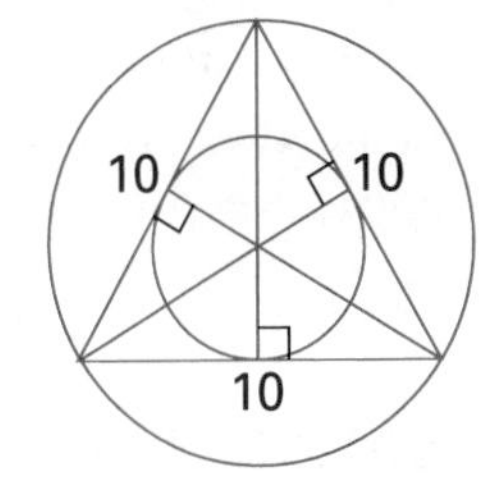

16. Find the radius of the circumcircle of the triangle.
17. Find the radius of the incircle of the triangle.
18. Find the difference between the areas of the triangle and its incircle.
19. Find the difference between the areas of the triangle and its circumcircle.
20. Find the difference between the areas of the two circles.

Mixed Review

1. Draw a line and a point approximately 5 cm away from the line. Construct the line through the point parallel to the given line. *3.3*

Find the value of x that makes a true proportion. *8.1*

2. $\frac{2}{3} = \frac{x}{12}$

3. $\frac{8}{12} = \frac{12}{x}$

Computer Investigation

Concurrence in Triangles

Use a computer software program that draws random triangles by classification (acute, obtuse, or right), constructs medians, labels points, and measures line segments.

In the triangle shown at the right, three *medians* meet at one point, G. In the computer investigation of Chapter 5, you discovered that the medians of any triangle always meet, or *concur*. This will be formally stated in the next lesson. However, there is an even more interesting relationship between the three medians of any triangle and their point of intersection. The activities below will lead you to discover this important numerical relationship.

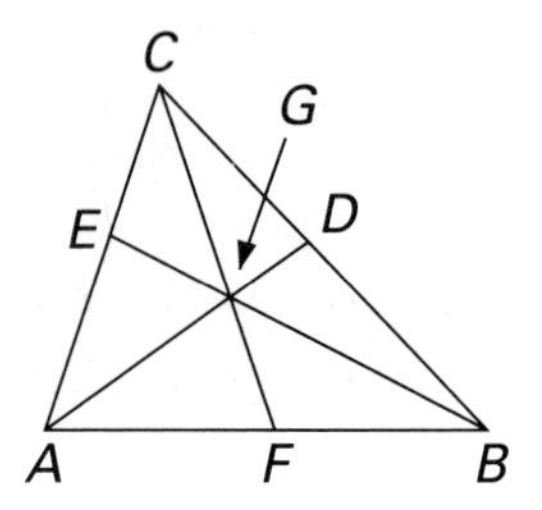

Activity 1

Draw an acute triangle and the indicated medians.

1. median $\overline{AD}$ from A to $\overline{BC}$
2. median $\overline{BE}$ from B to $\overline{AC}$
3. median $\overline{CF}$ from C to $\overline{AB}$
4. Label the point of intersection of the three medians G.

Activity 2

Find the indicated measures for the segments of the medians constructed in Activity 1. Then compute the indicated ratio to the nearest tenth. Keep a record of the ratios.

5. DG, GA; $\frac{DG}{GA}$
6. FG, GC; $\frac{FG}{GC}$
7. EG, GB; $\frac{EG}{GB}$

Activity 3

Repeat the steps of Exercises 1–7 for the following new triangles.

8. obtuse triangle
9. right triangle
10. isosceles triangle

Summary

The three medians of any triangle *concur* at one point. The point of *concurrence* of the three medians of any triangle divides each median into two segments with lengths in the ratio ___.

13.5 Concurrent Altitudes and Medians

Objectives

To apply altitude and median concurrency theorems
To construct orthocenters and centroids of triangles

Three lines usually do not intersect in a point. However, in the last lesson, two cases of concurrency were developed. The altitudes and medians of a triangle are also concurrent.

Theorem 13.5 The lines containing the altitudes of a triangle are concurrent.

Given: $\triangle ABC$, $\overline{AD}$, $\overline{BE}$, and $\overline{CF}$ are altitudes.
Prove: $\overline{AD}$, $\overline{BE}$, and $\overline{CF}$ are concurrent.

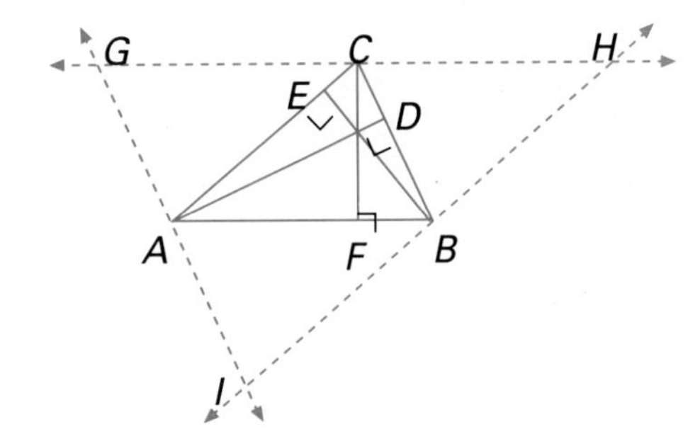

Plan

Draw $\overleftrightarrow{GH} \parallel \overline{AB}$ through C, $\overleftrightarrow{HI} \parallel \overline{AC}$ through B, and $\overleftrightarrow{IG} \parallel \overline{BC}$ through A, to form $\triangle GHI$. By definition, $ABHC$ and $ABCG$ are parallelograms, and $CH = AB = GC$. Since a line perpendicular to one of two parallel lines is perpendicular to the other, $\overline{CF} \perp \overline{GH}$. Therefore, $\overline{CF}$ is the perpendicular bisector of $\overline{GH}$.

Similarly, it can be shown that $\overline{AD}$ and $\overline{BE}$ are perpendicular bisectors of $\overline{GI}$ and $\overline{HI}$, respectively. So, $\overline{AD}$, $\overline{BE}$, and $\overline{CF}$ are concurrent, being the perpendicular bisectors of the sides of $\triangle GHI$.

Definition The **orthocenter** of a triangle is the intersection of the lines containing the altitudes of the triangle.

The orthocenter of a triangle can be either in the interior of the triangle, on the triangle itself, or in the exterior of the triangle, depending on whether the triangle is acute, right, or obtuse, respectively.

All triangles have three perpendicular bisectors of sides, three angle bisectors, and three altitudes. For each of these, the three intersect in a common point. All triangles also have three medians. These also intersect in a common point.

Theorem 13.6 Two medians of a triangle intersect at a point two-thirds of the distance from each vertex to the midpoint of the opposite side.

Given: Medians $\overline{AD}$ and $\overline{BE}$ of $\triangle ABC$ intersect at O.
Prove: $AO = \frac{2}{3}(AD)$, $BO = \frac{2}{3}(BE)$

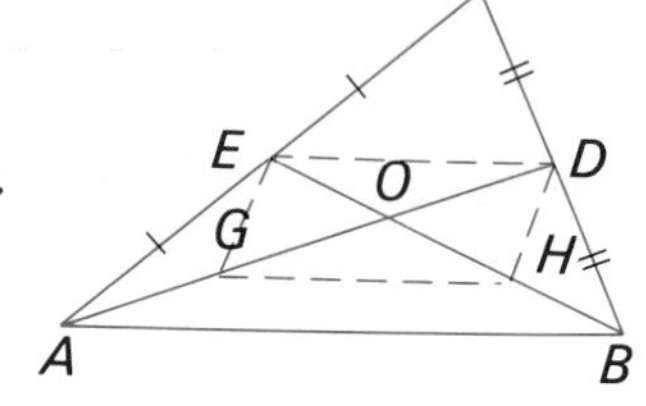

Plan Diagonals of a parallelogram bisect each other. Construct a parallelogram with two of its vertices at the points where the medians intersect the sides.

Flow diagram

Median $\overline{AD}$ → $BD = DC$
Median $\overline{BE}$ → $AE = EC$
→ $ED = \frac{1}{2}AB$
→ $\overline{ED} \parallel \overline{AB}$

Draw G, midpt of $\overline{AO}$. → $AG = GO$
Draw H, midpt of $\overline{BO}$. → $BH = HO$
→ $\overline{GH} \parallel \overline{AB}$
→ $GH = \frac{1}{2}AB$

→ $\overline{ED} \parallel \overline{GH}$
→ $ED = GH$
→ $EDHG$ is a ▱

→ $GO = OD$ → $AG = GO = OD$ → $AO = \frac{2}{3}AD$
→ $HO = OE$ → $BH = HO = OE$ → $BO = \frac{2}{3}BE$

Theorem 13.7 The medians of a triangle are concurrent at a point that is two-thirds the distance from each vertex to the midpoint of the opposite side.

Definition The **centroid** of a triangle is the point of intersection of the medians of the triangle.

EXAMPLE 1 Construct the centroid of $\triangle XYZ$.

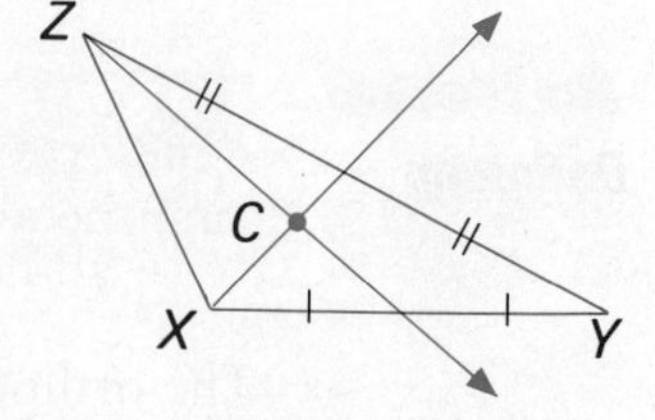

Solution A median is a segment from a vertex of the triangle to the midpoint of the opposite side. Bisect $\overline{XY}$ and $\overline{YZ}$. Draw the two medians.

The centroid is the intersection, point C.

EXAMPLE 2 Medians $\overline{IM}$ and $\overline{JL}$ of $\triangle IJK$ intersect at N; $JN = 10x + 4$, $NL = 7x + 1$. Find JN.

Plan By Theorem 13.6, $JN = \frac{2}{3}(JL)$ and $NL = \frac{1}{3}(JL)$. So, $JN = 2(NL)$.

Solution

$$\begin{aligned} JN &= 2(NL) \\ 10x + 4 &= 2(7x + 1) \\ 10x + 4 &= 14x + 2 \\ x &= \frac{1}{2} \end{aligned}$$

Therefore, $JN = 10x + 4 = 10 \cdot \frac{1}{2} + 4$, or 9.

Classroom Exercises

$\overline{AD}$, $\overline{BE}$, and $\overline{CF}$ are medians of $\triangle ABC$.

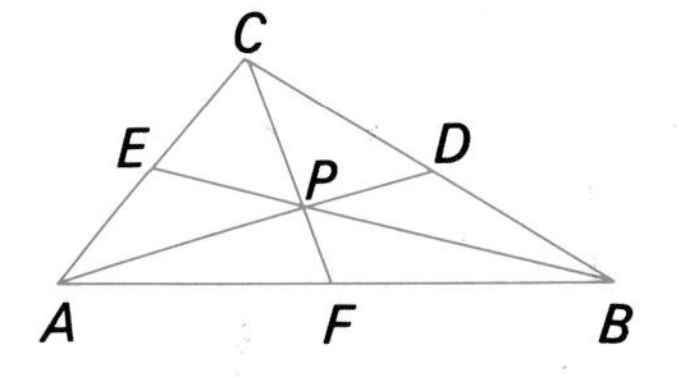

1. If $AP = 16$, find PD.
2. If $BP = 9$, find PE.
3. If $PD = 7$, find AP.
4. If $PF = 6$, find CP.
5. If $PE = 3$, find BE.
6. If $PF = 8$, find CF.
7. If $BE = 9$, find PE.
8. If $AD = 24$, find PD.

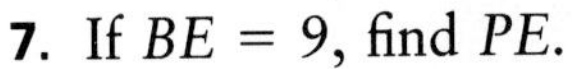

Written Exercises

1. Draw an obtuse triangle. Construct the centroid of the triangle.
2. Draw a right triangle. Construct the centroid of the triangle.
3. Draw an acute triangle. Construct the orthocenter of the triangle.
4. Draw a right triangle. Construct the orthocenter of the triangle.
5. Construct an equilateral triangle. Construct both the orthocenter and the centroid of the triangle. Where do these two points lie?

$\triangle ABC$ has medians $\overline{AD}$, $\overline{BE}$, and $\overline{CF}$ intersecting at P.

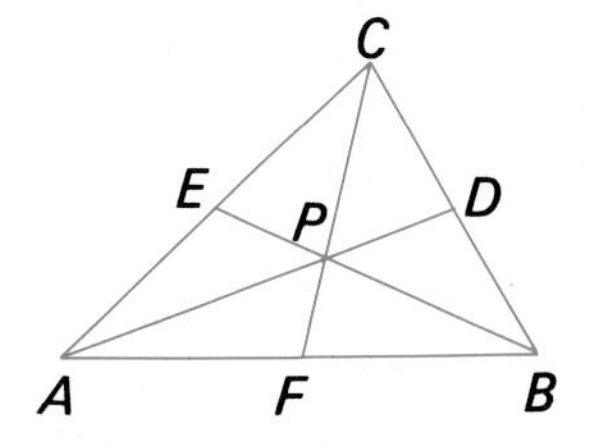

6. If $CP = 4x - 6$ and $PF = x + 1$, find PF.
7. If $DP = 2x + 3$ and $PA = 6x - 14$, find PA.
8. If $PF = x - 3$ and $CF = 2x + 1$, find PC.
9. If $BE = 2x + 9$ and $EP = x - 2$, find BP.
10. If $BP = 4x$ and $BE = 5x + 5$, find EP.
11. If $AP = 2x - 6$ and $AD = 2x + 9$, find DP.
12. Draw an acute, a right, and an obtuse triangle. Show that the orthocenter may lie within, on, or outside such triangles, respectively. Then show that the centroid always lies in the interior of the triangle.

$\triangle GHI$ has altitude $\overline{IJ}$, median $\overline{IK}$, and centroid C.

13. If $JK = 9$ and $IJ = 12$, find IC.
14. If $JK = 15$ and $IJ = 36$, find CK.
15. If $CK = 2$ and $JK = 3$, find IJ.
16. If $IC = 6$ and $IJ = 8$, find JK.

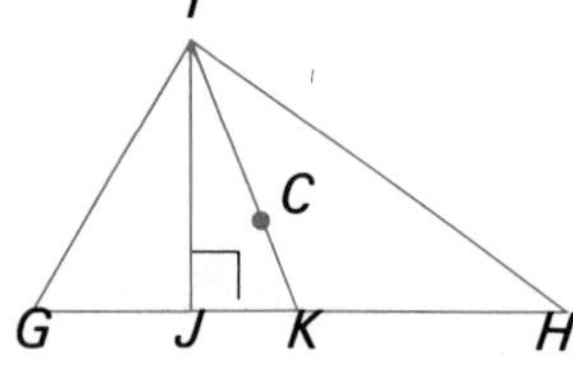

17. An isosceles triangle has sides measuring 10 cm, 10 cm, and 12 cm. How far above the base is the centroid?
18. An equilateral triangle has sides measuring 12 cm. How far above the base is the centroid?
19. Based on Theorem 13.7, present an argument explaining why a centroid could never lie on or outside a triangle.
20. Prove Theorem 13.5.
21. Prove Theorem 13.7.
22. Prove that the three medians of an equilateral triangle form, with the sides, six congruent triangles.
23. Prove that in an equilateral triangle the incenter, the circumcenter, the orthocenter, and the centroid are the same point.

Mixed Review

1. Draw two parallel lines 2 in. apart. Sketch and describe the locus of points in a plane equidistant from these two lines. *13.1*
2. Draw two points 2 in. apart. Sketch and describe the locus of points in a plane equidistant from these two points. *13.1*
3. If the lengths of two legs of a right triangle are 5 cm and 12 cm, what is the length of the hypotenuse? *9.2*
4. If the lengths of the hypotenuse of a right angle is 12 cm and the length of one leg is 4 cm, what is the length of the other leg? *9.2*

Brainteaser

Shadows are a special type of locus. Consider shadows formed by an ordinary light such as a street light. In the illustration, the shadow shown is the intersection of the pavement with a cone, where the light source is considered to be a point. If the shadow had been formed by sunlight, it would be (for all practical purposes) the intersection of the pavement with a cylinder. Why would there be this difference for the two light sources? What shapes of shadows would be formed on a flat surface by holding a circular disk in different positions? A square piece of cardboard? A right triangular piece of cardboard?

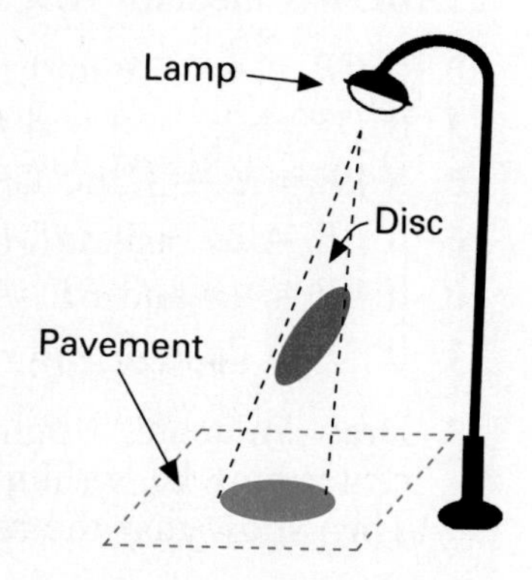

Chapter 13 Review

Key Terms

centroid (p. 548)
circumcenter (p. 542)
circumcircle (p. 543)
circumscribed (p. 543)
concurrent lines (p. 541)
incenter (p. 543)
incircle (p. 544)
inscribed (p. 544)
locus, loci (p. 529)
orthocenter (p. 547)

Key Ideas and Review Exercises

13.1 To determine a locus, locate those points and only those points that satisfy the condition of the locus.

1. Sketch and describe the locus of points in a plane that is 9 cm from a given point P.
2. Sketch and describe the locus of points in a plane that is less than 2 cm from a given line.

13.2 To determine a locus that satisfies two or more given conditions, locate the set of points that satisfies each condition. The intersection of these sets of points is the desired locus.

3. Sketch and describe the locus of points in a plane that are 5 cm from point P *and* 12 cm from point Q if P and Q are 13 cm apart.

13.3 To determine a locus in space, locate those points in all planes that satisfy the given conditions.

4. Sketch and describe the locus of points in space that are 2 cm from a given line.
5. Sketch and describe the locus of points in space that are less than 2 cm from a given point.

13.4 To locate the *circumcenter* of a triangle, construct the perpendicular bisectors of two sides. To locate the *incenter* of a triangle, construct the bisectors of two angles.

6. Draw an obtuse triangle. Construct the circumcenter of the triangle.
7. Draw a right triangle. Construct the incenter of the triangle.

13.5 To construct the *orthocenter* of a triangle, construct two altitudes. The intersection of the lines containing the altitudes is the orthocenter.

To construct the *centroid* of a triangle, construct two medians. The point of intersection is the centroid.

8. Draw an obtuse triangle. Construct the orthocenter of the triangle.
9. Draw an acute triangle. Construct the centroid of the triangle.

Chapter 13 Test

Sketch and describe the locus of points in a plane (Ex. 1–6).

1. equidistant from two points 6 cm apart
2. equidistant from two parallel lines 4 cm apart
3. less than 5 cm from a given point
4. 2 cm from a circle with a 5-cm radius
5. 3 cm from point P *and* 2 cm from line l

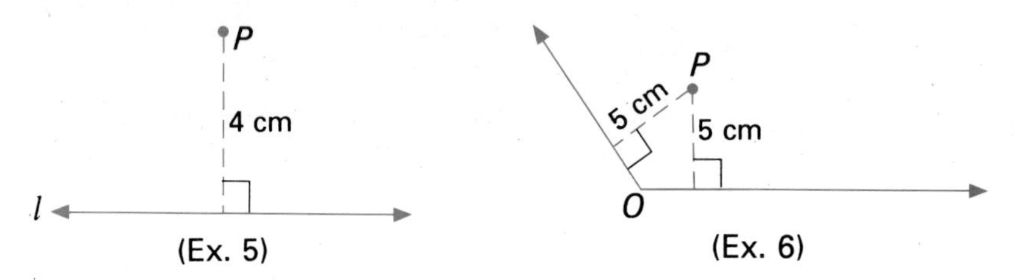

6. less than 2 cm from point P *and* equidistant from the sides of $\angle O$

7. Describe the locus of points in a plane equidistant from the sides *and* in the interior of an obtuse angle *and* less than a given distance r from the vertex of the angle.

8. Describe the locus of points in space 3 cm from a given point P.

9. Describe the locus of points in a plane equidistant from two given concentric circles *and* at a given distance d from a given line l.

10. Copy $\triangle ABC$. Construct the incenter of the triangle.

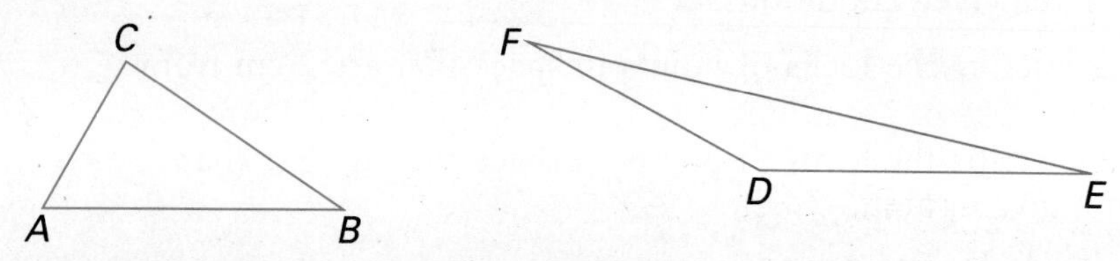

11. Copy $\triangle DEF$. Construct the centroid of the triangle.

$\overline{GJ}$, $\overline{HK}$, and $\overline{IL}$ are medians of $\triangle GHI$. (Ex. 12–14)

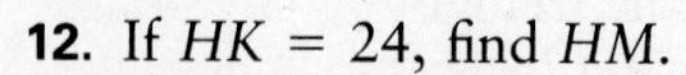

12. If $HK = 24$, find HM.
13. If $IM = 12$, find IL.
14. If $GM = 8x - 12$ and $MJ = x + 3$, find GM.

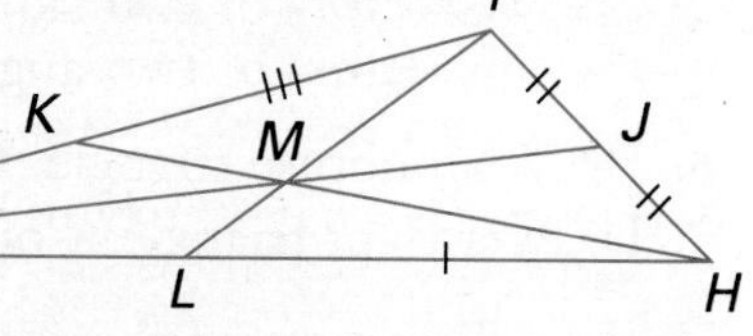

* **15.** Copy the arc of the circle shown. Using appropriate constructions, find its center.

College Prep Test

Choose the one best answer to each question or problem.

1. The locus of points 3 in. from a given line and 5 in. from a point on that line is exactly
(A) 1 point. (B) 2 points.
(C) 3 points. (D) 4 points.
(E) 5 points.

2. The locus of points in a plane at a given distance d from a given point in the plane represents which of the following?
(A) One line (B) Two lines
(C) A circle (D) Two circles
(E) A point

3. Which of the following is an equation of the locus of points a distance of 6 from the origin?
(A) $x = 6$ (B) $y = 6$
(C) $x^2 + y^2 = 6$ (D) $x^2 + y^2 = 36$
(E) $x^2 + y^2 = 0$

4. The locus of points in a plane of a given triangle equidistant from the 3 vertices of the triangle depicts which of the following?
(A) One point (B) One circle
(C) One line (D) Two lines
(E) Two circles

5. How many points in a plane are equidistant from 2 given points A and B *and* 3 in. from the line $\overleftrightarrow{AB}$?
(A) 0 (B) 1 (C) 2
(D) 3 (E) 4

6. The midpoint of the hypotenuse of a right triangle is
(A) equidistant from all 3 vertices.
(B) the intersection of the 3 angle bisectors.
(C) the intersection of the 3 medians.
(D) the center of the incircle.
(E) the incenter.

7. Given: $\triangle ABC$, with medians $\overline{AY}$, $\overline{BX}$, and $\overline{CZ}$ concurrent at P, $XP = 4$. Find BX.
(A) 4
(B) 8
(C) 12
(D) 6
(E) 10

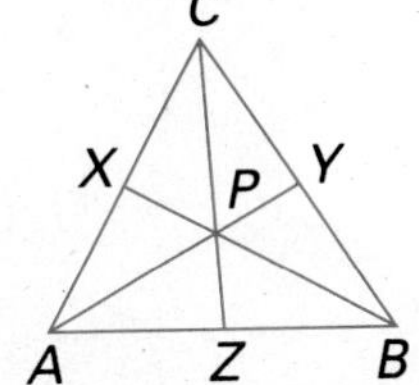

8. What is the equation of the locus of points equidistant from A and B?
(A) $y = x + 1$
(B) $y = x - 1$
(C) $y = 1$
(D) $y = -x$
(E) $y = x$

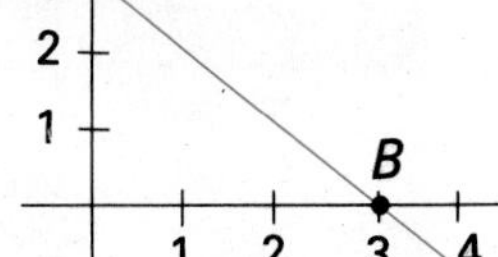

9. Given: Point P on line l
How many points are 2 in. from P *and* 4 in. from l?
(A) 0
(B) 1
(C) 2
(D) 3
(E) 4

10. Two concentric circles have radii of 2 cm and 6 cm. Line m is tangent to the smaller circle. How many points are equidistant from the circles and 2 in. from m?
(A) 0
(B) 1
(C) 2
(D) 3
(E) 4

14 FIGURES IN SPACE

Grace Chisholm Young (1868–1944)—Grace Young wrote a geometry book that, like all the texts of that time, dealt only with two-dimensional figures in a plane. However, she included patterns for folding paper to make three-dimensional models.

14.1 Polyhedra and Rigidity

Objectives

To determine whether a given figure is rigid
To determine the number of faces, edges, and vertices of a polyhedron

A newly built garage looked like the picture on the left below when it was first built. Several months later, after a heavy snow, it looked like the one on the right.

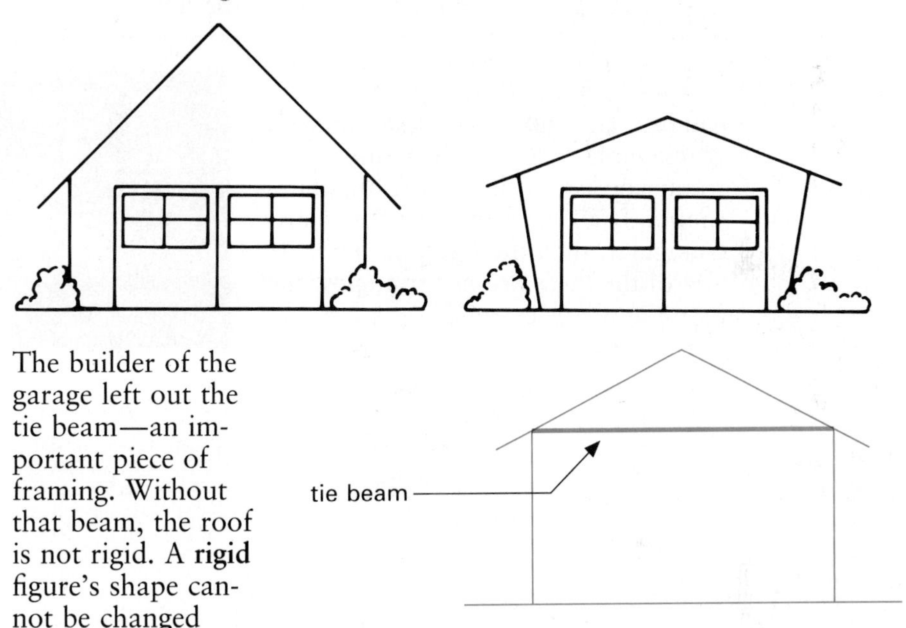

The builder of the garage left out the tie beam—an important piece of framing. Without that beam, the roof is not rigid. A **rigid** figure's shape cannot be changed without changing the lengths of the sides.

A triangle is a rigid figure. To change the shape of a triangle, the lengths of the sides must be changed. A quadrilateral is not a rigid figure. The shape of a quadrilateral can change without changing the lengths of the sides. This is to be expected since triangles with congruent corresponding sides are congruent, while quadrilaterals with congruent corresponding sides are not necessarily congruent.

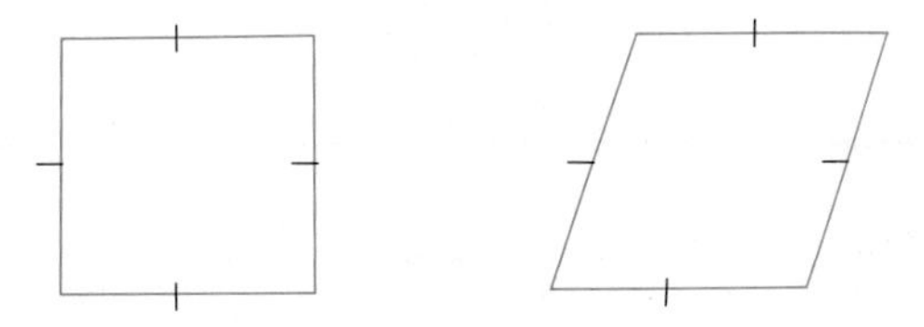

You can demonstrate that a plane figure is not rigid if there is a noncongruent figure with congruent corresponding sides.

EXAMPLE 1 Use sketches to show that a trapezoid is not a rigid figure.

Solution

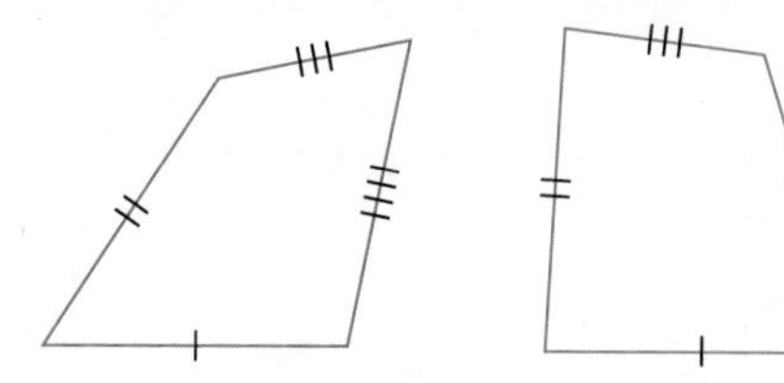

The counterexample at the right has congruent corresponding sides but not the same shape as the trapezoid at the left.

Three-dimensional or *space* figures can also be rigid. A geodesic dome is an example of a rigid structure. Its frame consists entirely of triangular shapes. If all the faces of a figure are triangular, then the figure is rigid. If any of the faces are not triangles, the figure may not be rigid.

EXAMPLE 2 Is the solid rigid? Explain why or why not.

Solution

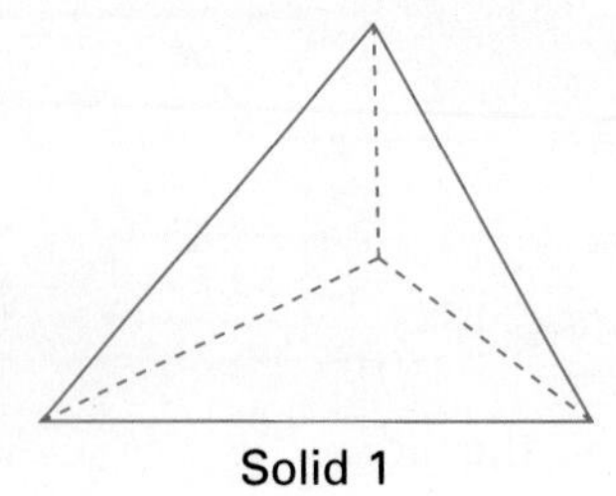

Solid 1

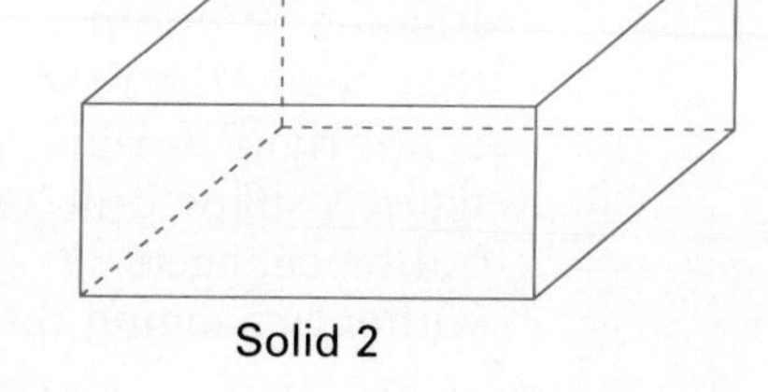

Solid 2

All faces of Solid 1 are triangles.
Triangles are rigid.
The solid is rigid.

The faces of Solid 2 are not rigid.
There exists a solid with a different shape and congruent corresponding sides of faces.
Solid 2 is not rigid.

Definition

A **polyhedron** is a three-dimensional figure with polygonal regions as its faces. The plural of polyhedron is *polyhedra*.

To simplify discussion, the faces of polyhedra will be referred to as polygons, rather than polygonal regions.

EXAMPLE 3 State whether the space figure is a polyhedron.

Solution

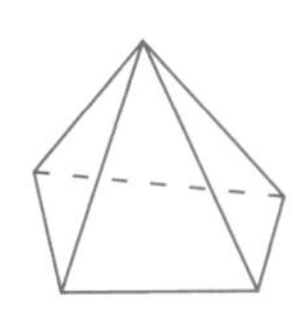

A polyhedron

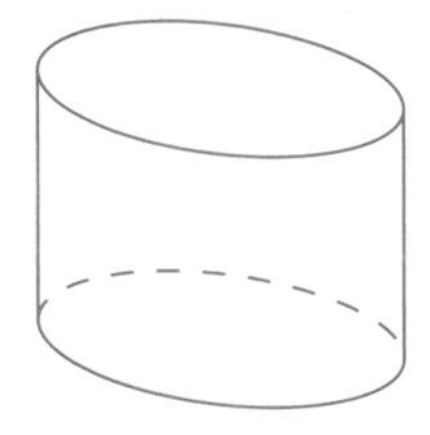

Not a polyhedron

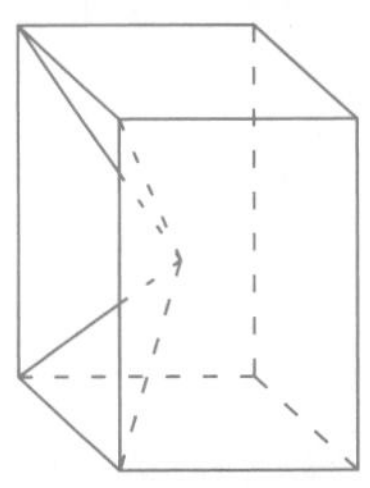

A polyhedron

Polyhedra are named by the number of faces they contain. If all faces of a polyhedron are congruent regular polygons, the polyhedron is called a **regular polyhedron**. There are exactly five regular polyhedra, as illustrated below.

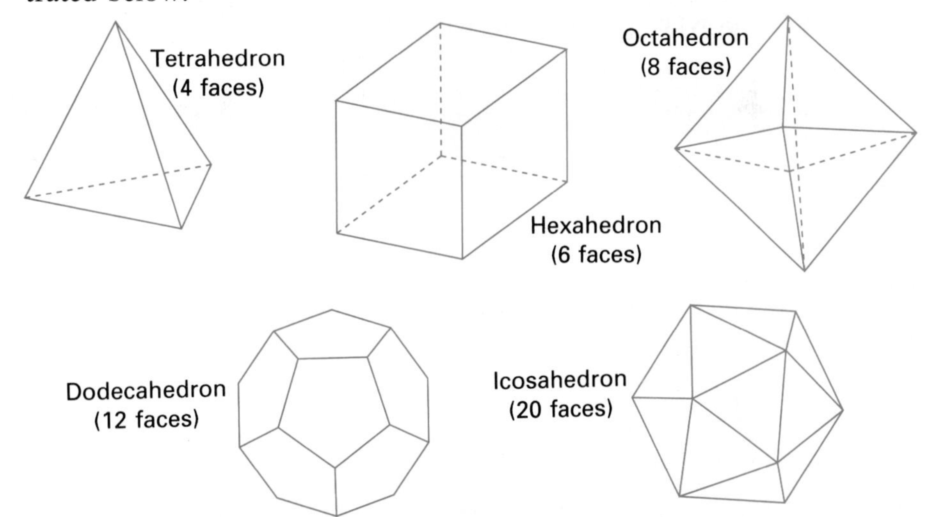

Dodecahedron
(12 faces)

Icosahedron
(20 faces)

The faces of a polyhedron intersect in segments called **edges**. The intersections of the edges are called the **vertices**.

EXAMPLE 4 What shape are the faces of a regular dodecahedron? How many edges and how many vertices does it have?

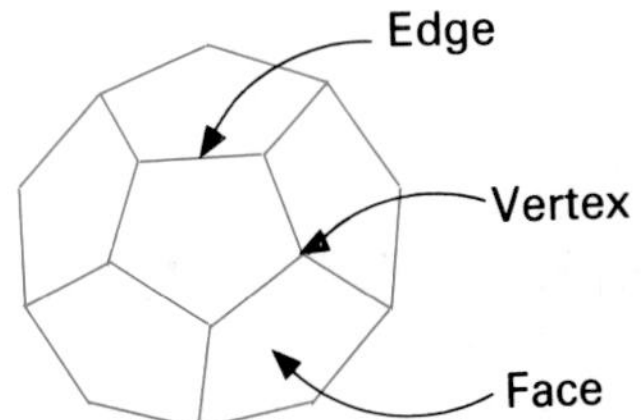

Solution The faces of a regular dodecahedron are regular pentagons. A dodecahedron has 12 faces, each of which has 5 sides. Since each edge is shared by two faces, there are $(12 \cdot 5) \div 2 = 30$ edges. Each pentagon has 5 vertices. Since each vertex is shared by three faces, there are $(12 \cdot 5) \div 3 = 20$ vertices.

Focus on Reading

Look up the word *polygon* in a dictionary. Underline the prefix. Using this prefix, write the name for the polyhedron with the same number of faces. From your reading of this lesson, tell whether there is a regular polyhedron of that name. (Exercises 1–6)

1. pentagon
2. heptagon
3. nonagon
4. decagon
5. undecagon
6. dodecagon
7. Find several other words, from outside the area of geometry, that use the same prefixes as in Exercises 1–6.

Classroom Exercises

Is the figure a polyhedron? Explain why or why not.

1.

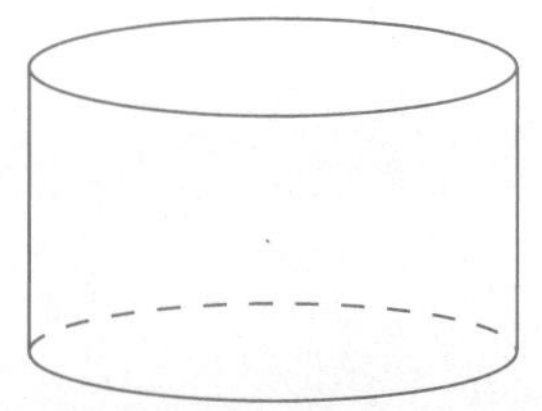

2.

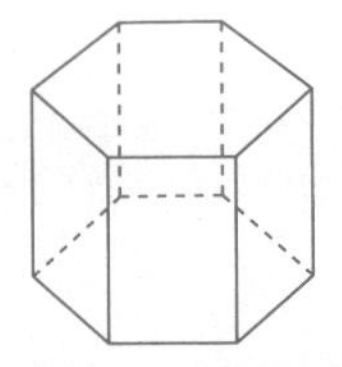

3.

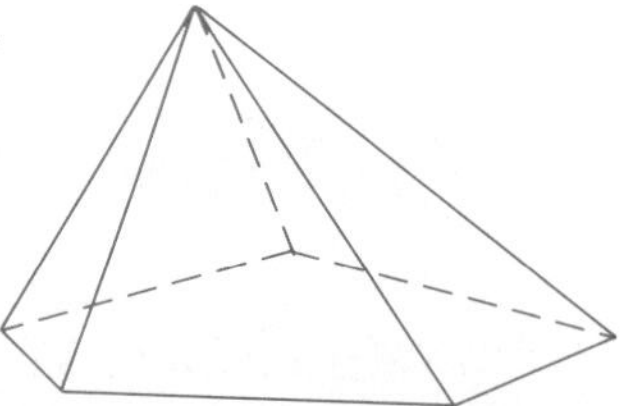

4. Name the five regular polyhedrons. Describe each polyhedron.

Written Exercises

Make sketches to show that the polygon is not rigid.

1. a square
2. a rectangle
3. a pentagon
4. a hexagon

Tell whether the figure is rigid. Explain why or why not.

5.

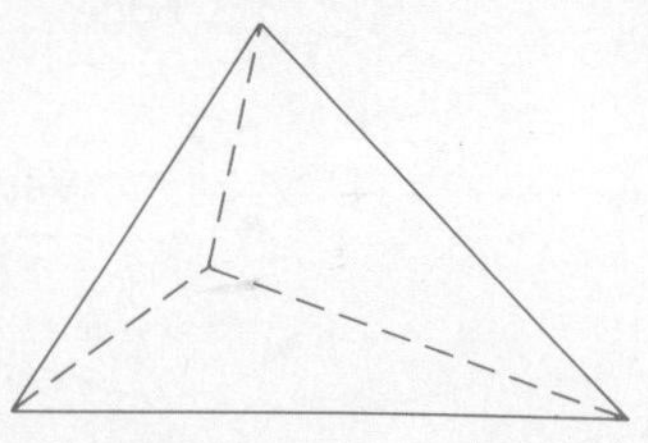

6.

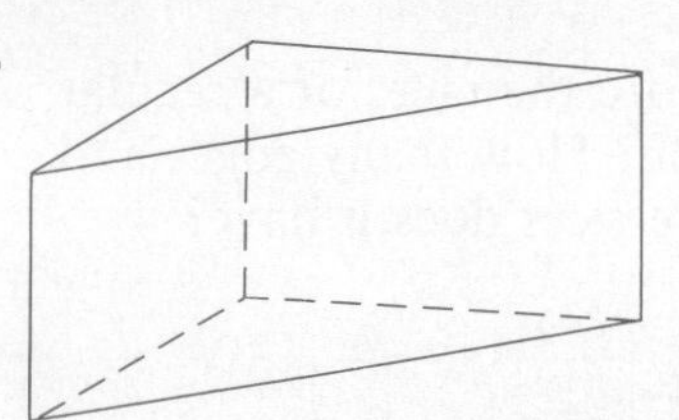

7.

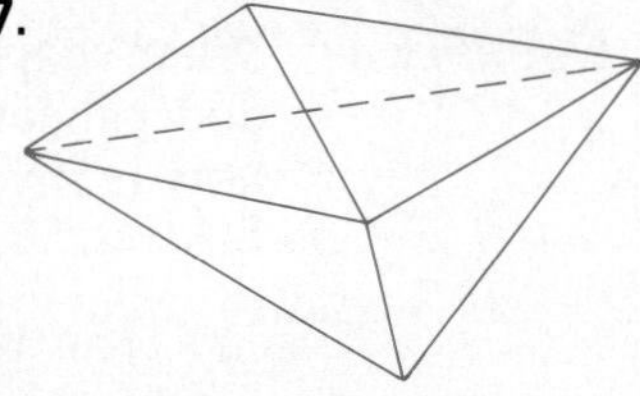

Do the edges form a rigid figure for the given regular polyhedron? Why or why not?

8. a regular tetrahedron
9. a regular hexahedron
10. a regular octahedron
11. a regular icosahedron

For the regular polyhedron, find the number of faces (F), the number of edges (E), and the number of vertices (V).

12. a regular tetrahedron

13. a regular octahedron

14. a regular dodecahedron

15. a regular icosahedron

16. Using your results from the Exercises 12–15, find a formula relating F, E, and V for regular polyhedra. Does this formula work for polyhedra that are not regular?

17. Explain how SSS congruency guarantees that any given triangle is rigid.

18. A regular icosahedron can be assembled using the pattern shown. Copy the pattern on a larger scale, cut it out, and build your own icosahedron.

19. Design a pattern for building a regular hexahedron. Verify the pattern by assembling it.

20. The sum of the angle measures around a common vertex in a plane is 360. What is the sum of the measures of the angles of each regular polyhedron around a common vertex? Does this sum appear to be proportional to the number of faces of the polyhedron?

Mixed Review

Find the area of the polygon. *12.3, 12.5, 12.6*

1. parallelogram

6 cm
7 cm
8 cm

2. trapezoid

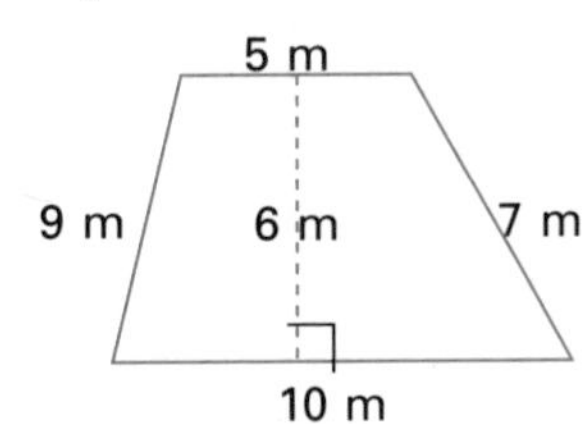

3. regular hexagon

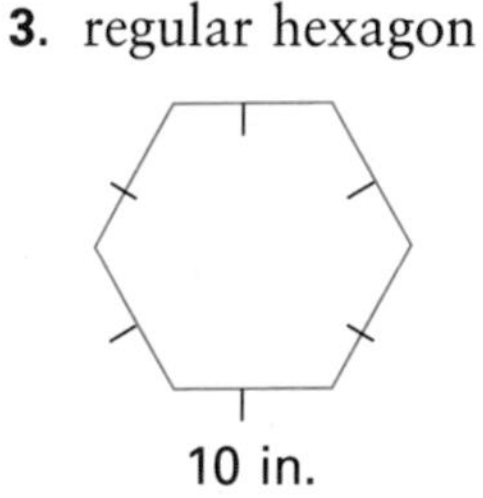

4. Write the truth table for $(p \vee q) \wedge \sim p$. *1.7, 3.8*

Brainteaser

In the figure below, can you reposition two matches to make three triangles?

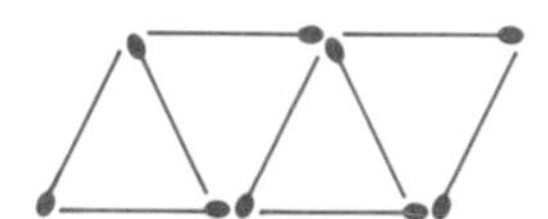

In the figure at the right, can you reposition just three matches to make three squares?

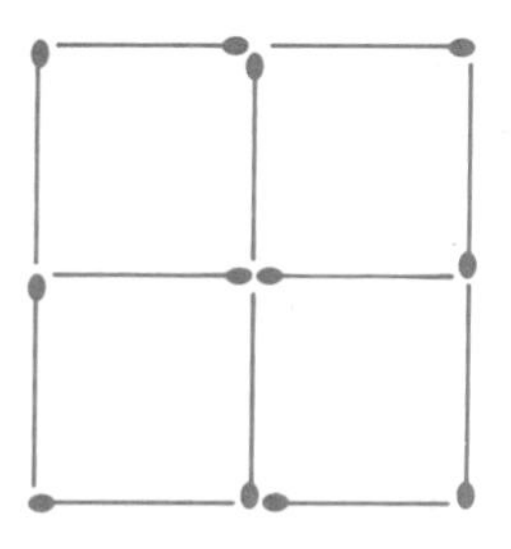

14.2 Prisms and Cylinders

Objectives To classify and draw prisms and cylinders
To identify parts of prisms and cylinders

Geometric figures contained in a plane, such as polygons or circles, are *two-dimensional* figures. *Three-dimensional* figures, or solids, cannot be contained in a plane. Prisms, which have polygonal faces, and cylinders, which have curved surfaces, are special three-dimensional figures.

Definitions

A **prism** is a polyhedron with two congruent polygonal faces, called the **bases,** in parallel planes. The remaining faces, called the **lateral faces,** are parallelograms. Each lateral face shares an edge with each of the bases.

Prisms are often classified by their bases. For example, a triangular prism has triangular bases.

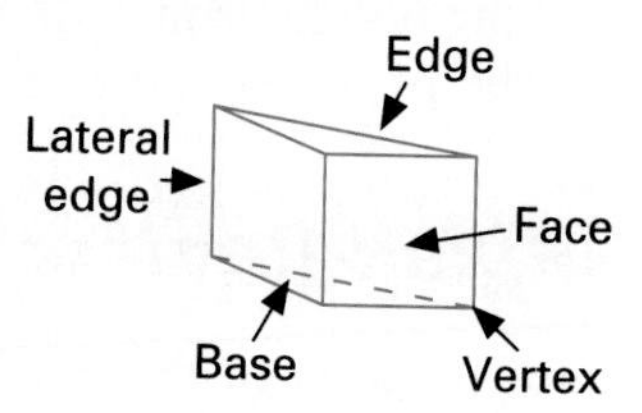

EXAMPLE 1 How many faces, edges, and vertices does a quadrilateral prism have?

Solution Sketch such a prism and count.
The prism has 6 faces, 12 edges, and 8 vertices.

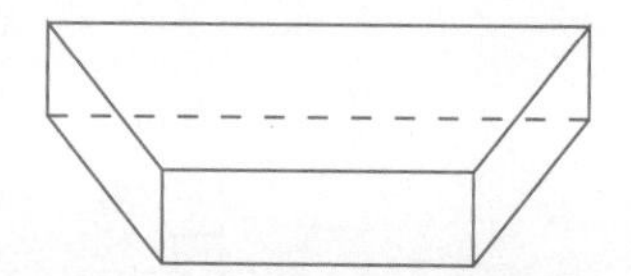

In a **right prism,** the lateral edges are *perpendicular* to the bases. The lateral faces are rectangular. A prism that is not a right prism is called **oblique prism.** A **rectangular solid** is a right prism in which the bases and lateral faces are rectangles. A **parallelepiped** is a prism in which the bases and the lateral faces are parallelograms.

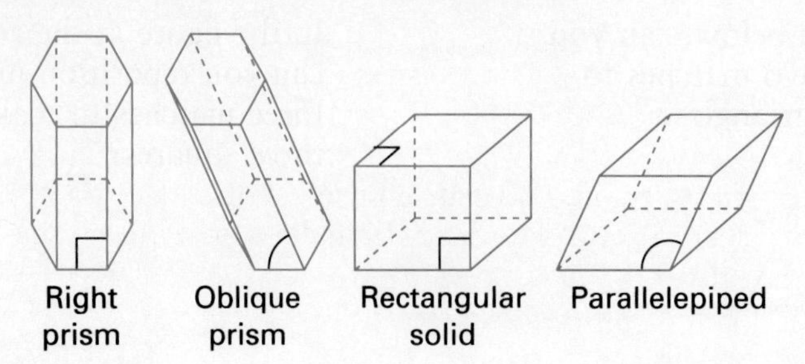

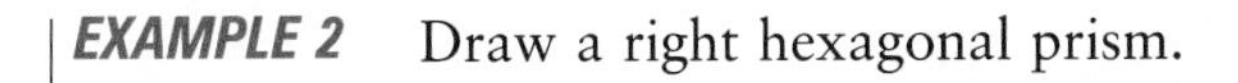

EXAMPLE 2 Draw a right hexagonal prism.

Solution

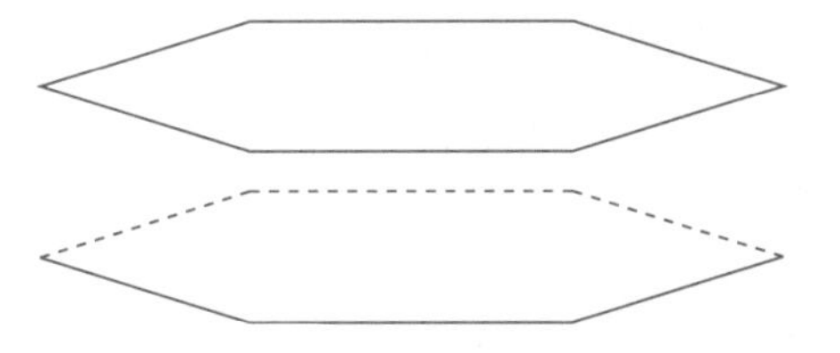

Draw two congruent hexagons. The corresponding sides should be parallel and vertically aligned.

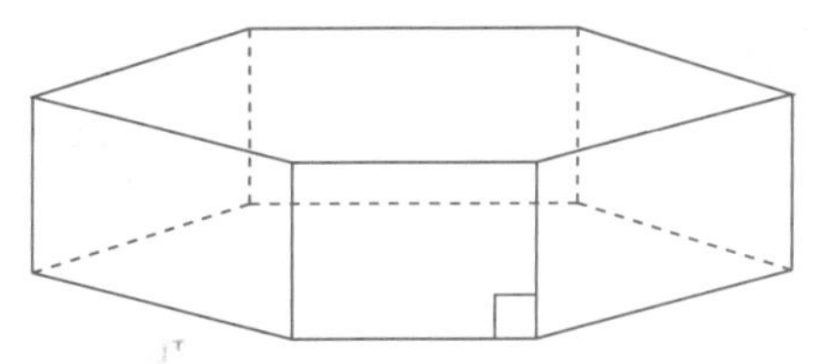

Connect the corresponding vertices. Use dashed lines to indicate edges hidden from view.

Like a prism, a **cylinder** consists of two congruent, parallel bases. However, the bases of a cylinder are circles instead of polygons. The *lateral surface* of the cylinder is the curved surface between the bases. It can be thought of as consisting of an infinite number of parallel segments with their endpoints on the bases.

If a segment connecting a pair of corresponding points of the bases is perpendicular to the bases, then the cylinder is a *right cylinder*. A cylinder that is not a right cylinder is an *oblique cylinder*.

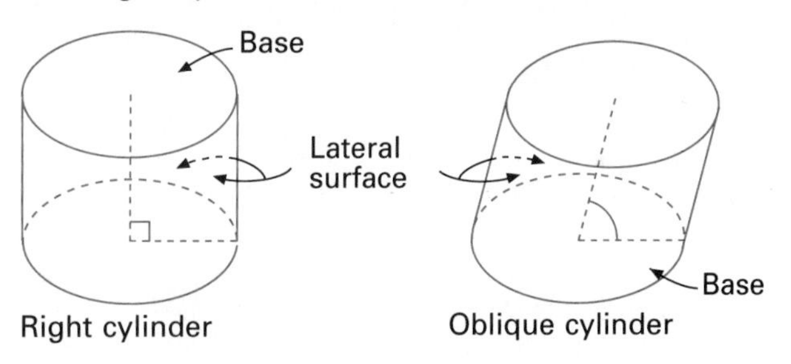

EXAMPLE 3 Draw a right cylinder.

Solution

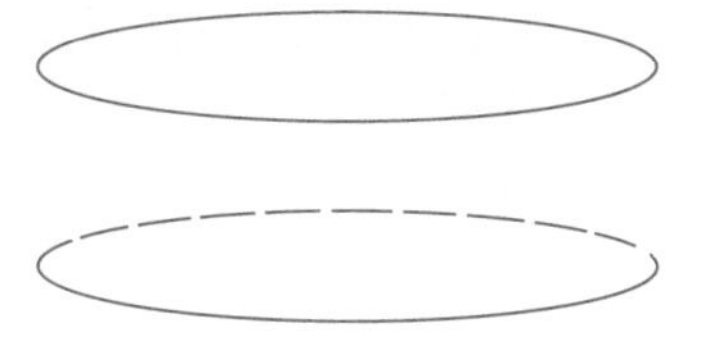

Draw two congruent ellipses, with corresponding points aligned vertically.

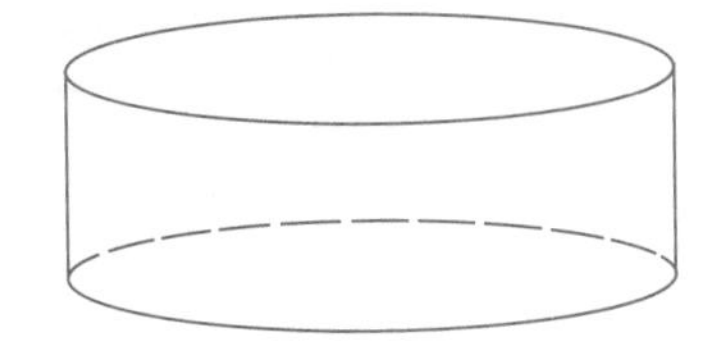

Connect two pairs of corresponding points.

Classroom Exercises

Does the figure appear to be a prism, a cylinder, or neither?

1. 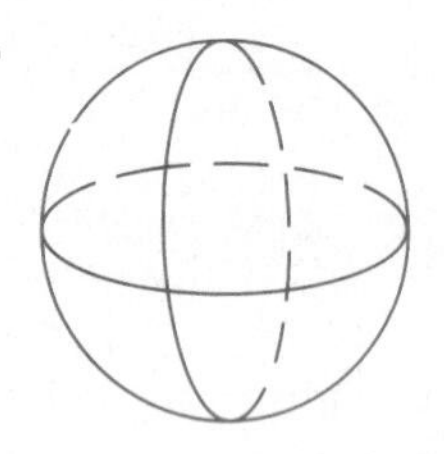

2. 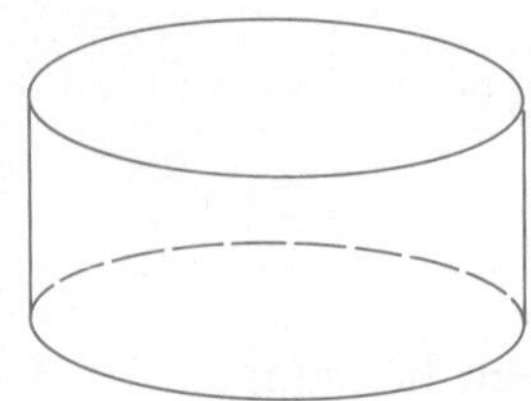

3. 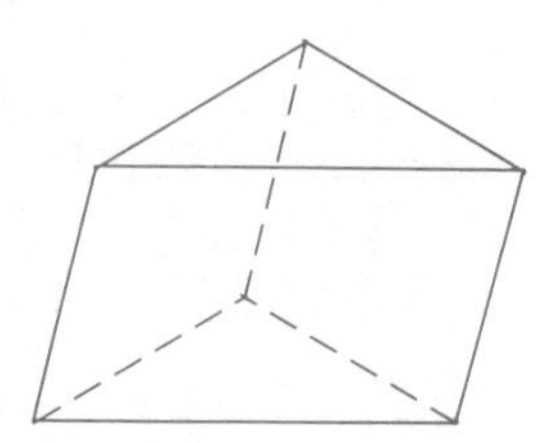

4. 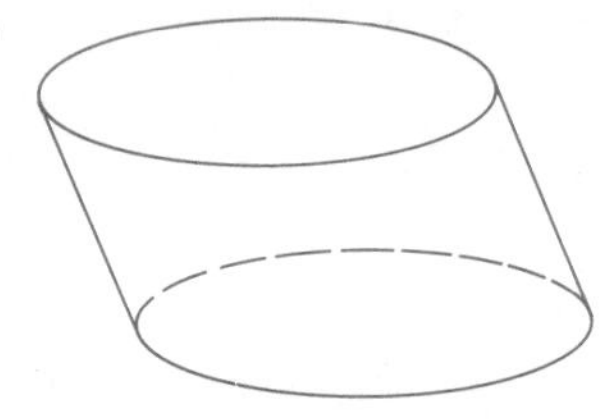

5. 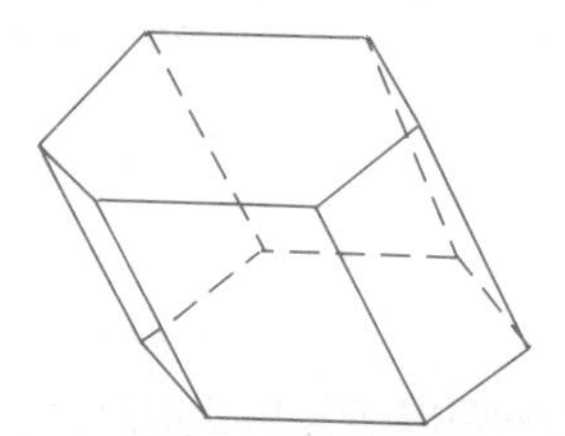

6. 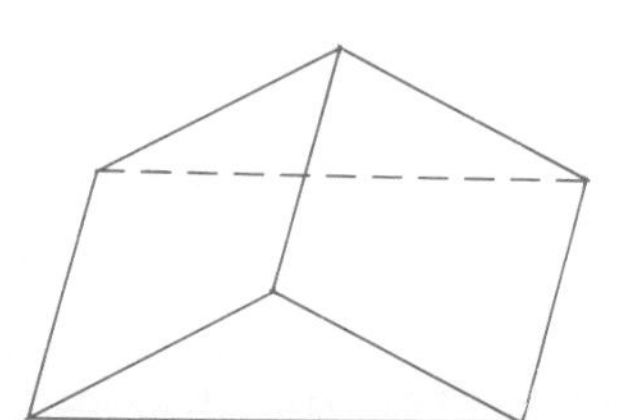

Classify the prism by its base.

7. 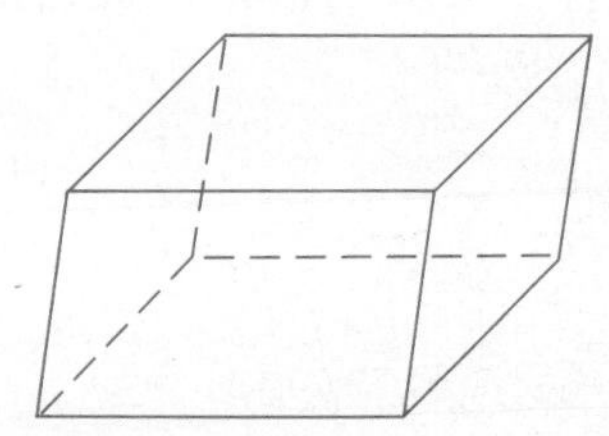

8. 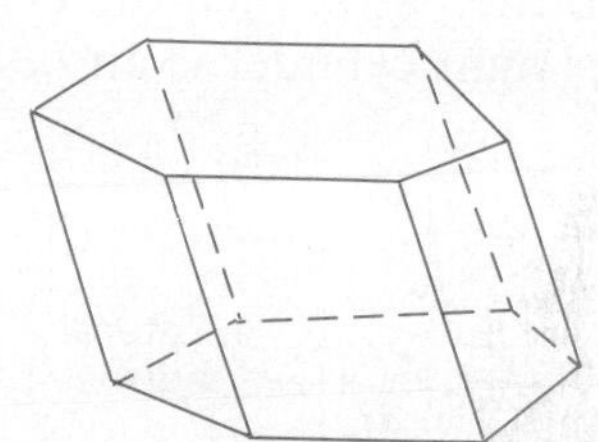

9. 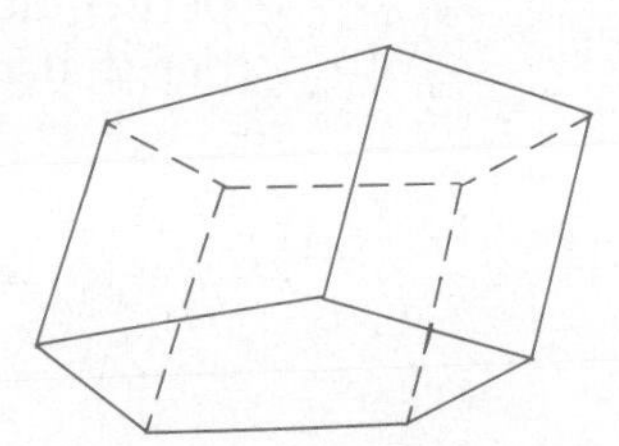

Written Exercises

How many faces, edges, and vertices does the prism have?

1. 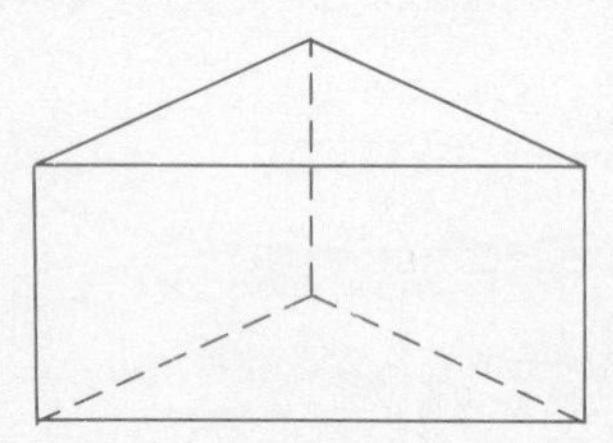

2. 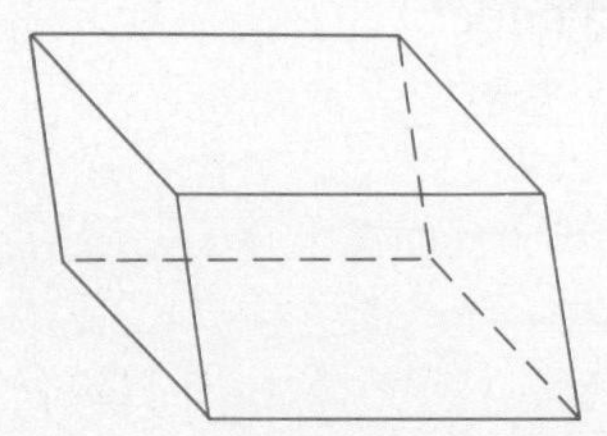

3. 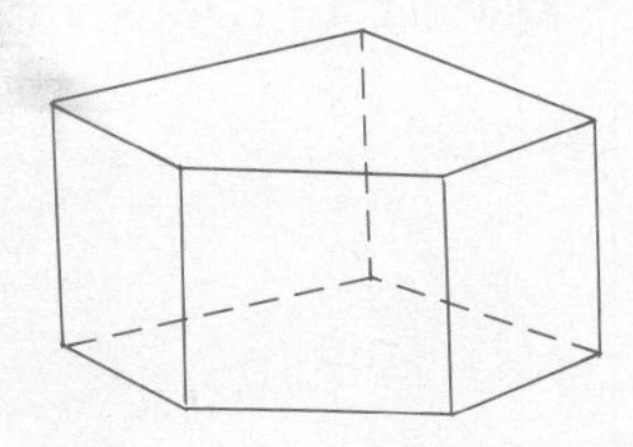

How many lateral faces and lateral edges does the prism have?

4. a triangular prism

5. a pentagonal prism

6. an octagonal prism

7. a dodecagonal prism

Draw an example of the described figure.

8. a right triangular prism

9. an oblique rectangular prism

10. an oblique pentagonal prism

11. a right heptagonal prism

12. a right cylinder

13. an oblique cylinder

True or false? (Exercises 14–17)

14. A prism has the same number of lateral faces as lateral edges.

15. A prism has the same number of faces as vertices.

16. An octagonal prism has twice the number of faces as a quadrilateral prism.

17. An octagonal prism has twice the number of vertices as a quadrilateral prism.

18. Prove that any two lateral edges of a prism are parallel and congruent.

19. Prove that any two lateral faces of a right prism with square bases are congruent.

20. Prove that two opposite lateral faces of a rectangular solid are congruent.

21. Find a formula that relates the number of vertices in a prism to the number of sides in each base.

22. Find a formula that relates the number of edges in a prism to the number of sides in each base.

23. A diagonal of a prism is a segment with two vertices in different faces as endpoints. Prove that any two diagonals of a rectangular solid are congruent.

24. A right triangular prism can be assembled using the pattern shown. Design a similar pattern for a rectangular solid, a parallelepiped, a right hexagonal prism, an oblique hexagonal prism, and a right cylinder.

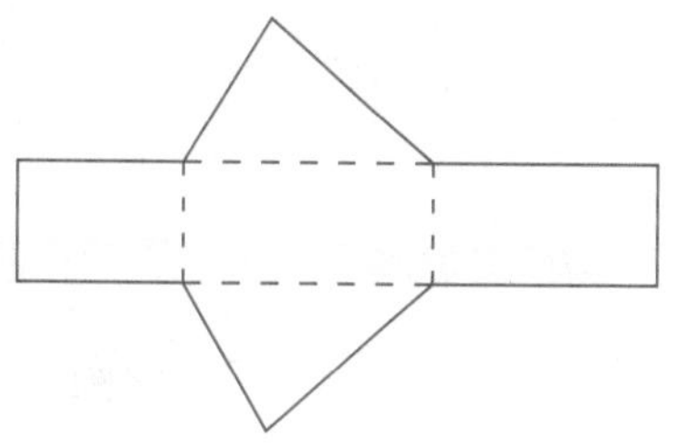

Mixed Review

1. The measure of an angle is 3 less than twice the measure of its complement. What is the measure of the angle? *1.6*

2. The measure of an angle is 12 more than twice the measure of its supplement. What is the measure of the angle? *1.6*

3. Write the conjunction of the following statements: Lines are long. Segments are short. *1.7*

4. Write an indirect proof for this statement: A triangle cannot contain two obtuse angles. *3.6*

5. Prove that the measure of an exterior angle of a triangle is equal to the sum of the measures of the two remote interior angles. *3.7*

14.3 Areas of Prisms and Cylinders

Objectives To find the areas of right prisms and cylinders
To compare areas of similar solids

The *total area of a prism* is the sum of the areas of its faces. This is called the *surface area*. The sum of the areas of the lateral faces only is called the *lateral area* of the prism.

EXAMPLE 1 Find the total area of the rectangular solid.

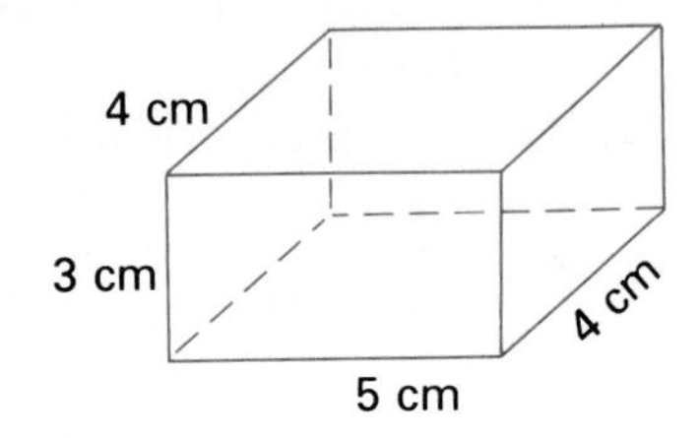

Plan All faces are rectangles. Find the sum of the areas of the six faces.

Solution $A = 2 \cdot (5 \cdot 4) + 2 \cdot (4 \cdot 3) + 2 \cdot (5 \cdot 3)$
$A = 40 + 24 + 30 = 94 \text{ cm}^2$

In finding areas of prisms it is helpful to visualize the surface as being flattened out. The figure below shows how a right hexagonal prism would appear when unfolded. The resulting surface consists of one large rectangle connected to two hexagons.

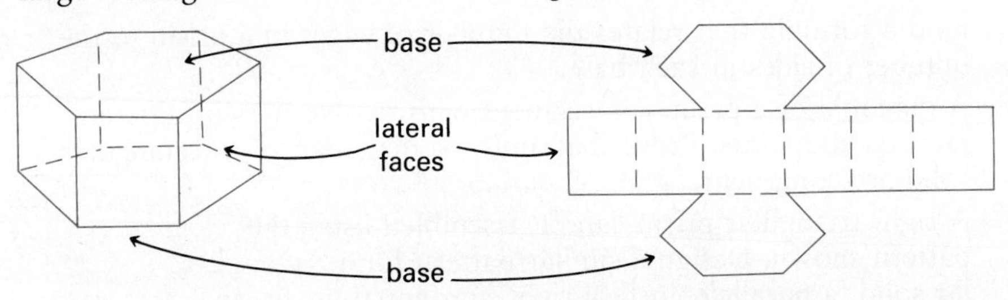

Definition An **altitude** of a prism or cylinder is any segment perpendicular to the planes containing the two bases with endpoints in these planes.

Theorem 14.1 The lateral area of a right prism is the product of the perimeter of a base and the length of an altitude ($L = ph$).

EXAMPLE 2 Find the lateral area of a right square prism with 6-cm base edges and a 10-cm altitude.

Solution The perimeter of the base is $4 \cdot 6 = 24$ cm.
$L = ph$
$L = (4 \cdot 6) \cdot 10 = 240 \text{ cm}^2$

If a right cylinder is cut open and unwrapped, the resulting surface consists of a rectangle and two circles. The length of one side of the rectangle equals the circumference of the circle.

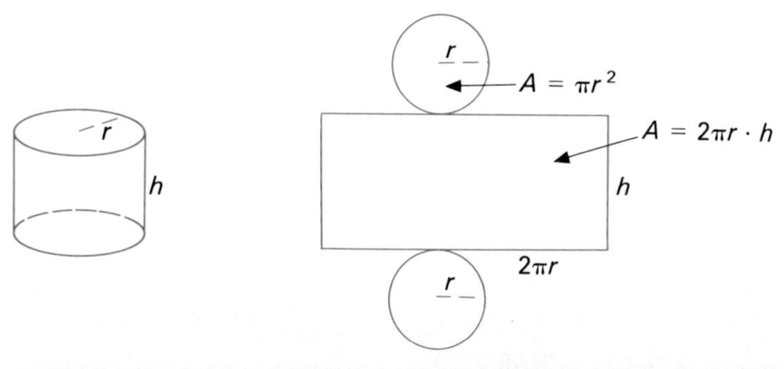

Theorem 14.2 The lateral area of a right cylinder is the product of the circumference of a base and the length of an altitude. The lateral area of a right cylinder with radius r and altitude of length h is $2\pi rh$.

Theorem 14.3 The total area of a right cylinder with radius of length r and altitude of length h is $2\pi r^2 + 2\pi rh$, or $2\pi r(r + h)$.

EXAMPLE 3 A diameter of a base of a right cylinder is 16 cm long. An altitude is 5 cm long. Find the lateral area and the total area.

Solution

Since $d = 16$ cm, $r = 8$ cm

$$\text{lateral area} = 2\pi rh = 2\pi \cdot 8 \cdot 5 = 80\pi \text{ cm}^2$$

$$\text{total area} = 2\pi r(r + h) = 2\pi \cdot 8 \cdot (8 + 5) = 208\pi \text{ cm}^2$$

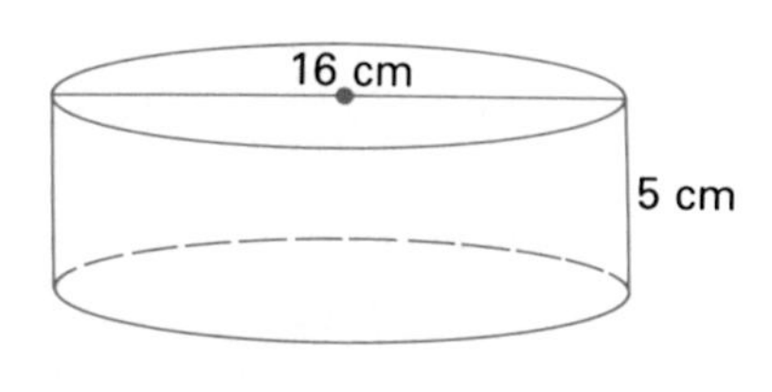

Similar solids have the same shape but may have different sizes.

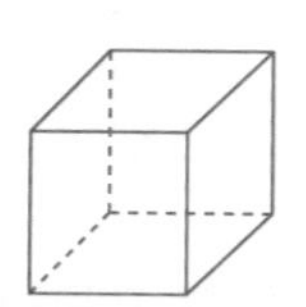

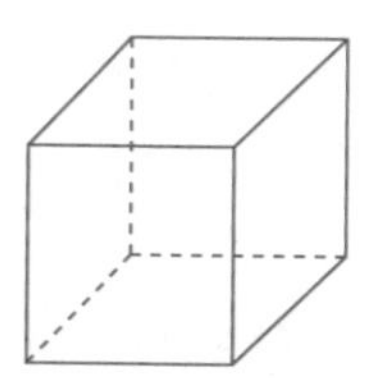

Definition Two polyhedra are **similar** if all corresponding angles between faces and between edges are congruent, and the lengths of corresponding edges are proportional.

EXAMPLE 4 The lengths of edges of two rectangular solids have the ratio 3 to 2. What is the ratio of their areas?

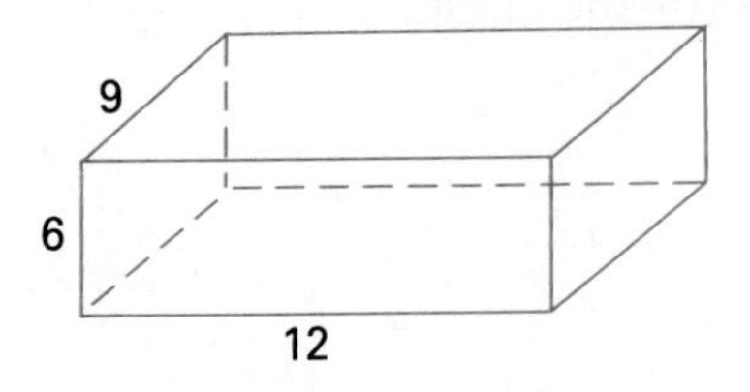

Prism 1

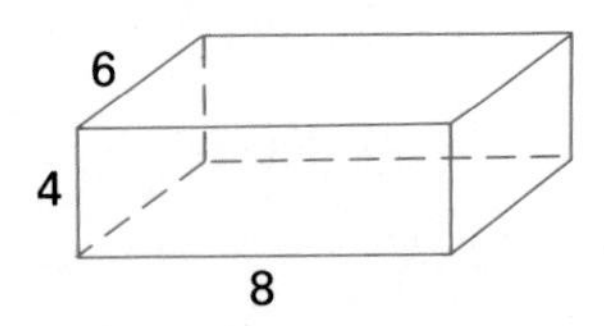

Prism 2

Solution

$$\text{Area}_1 = 2 \cdot (6 \cdot 12) + 2 \cdot (9 \cdot 12) + 2 \cdot (6 \cdot 9) = 468$$
$$\text{Area}_2 = 2 \cdot (4 \cdot 8) + 2 \cdot (6 \cdot 8) + 2 \cdot (4 \cdot 6) = 208$$
$$\frac{\text{Area}_1}{\text{Area}_2} = \frac{468}{208} = \frac{9}{4}$$

The ratio of the areas in Example 4 is the square of the ratio of the lengths of the edges of the two similar polyhedra: $(\frac{3}{2})^2 = \frac{9}{4}$. This example suggests a more general relationship of linear and area measures.

Theorem 14.4 If the ratio of the lengths of corresponding edges of two similar polyhedra is $\frac{a}{b}$, then the ratio of the lateral areas and total areas is $(\frac{a}{b})^2$.

Classroom Exercises

Find the lateral area of the right prism. (The bases are shaded.)

1.

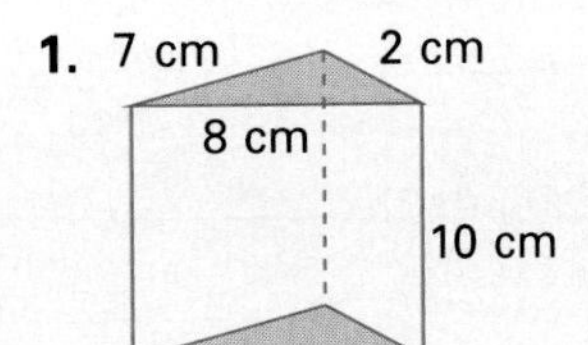

2.

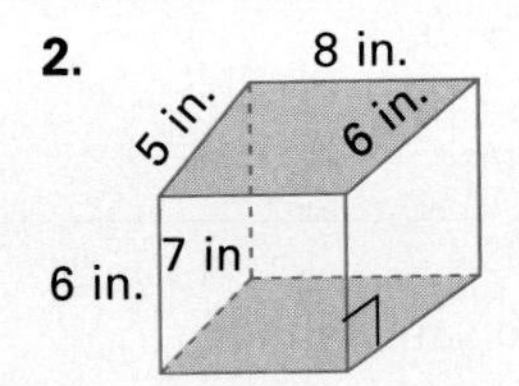

3.

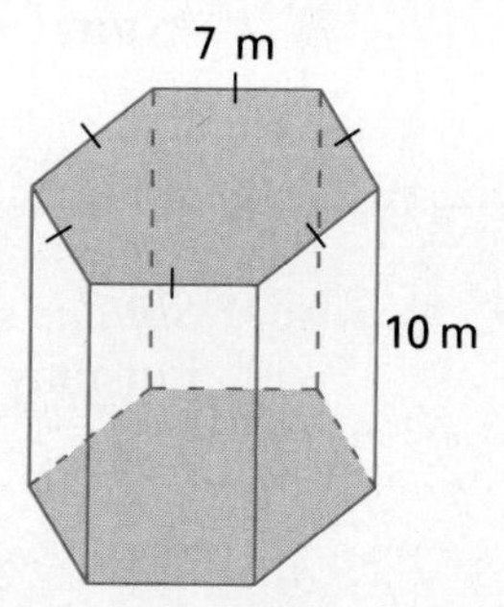

For Exercises 4–6, use the right cylinder.

4. Find the lateral area.

5. Find the area of the bases.

6. Find the total area.

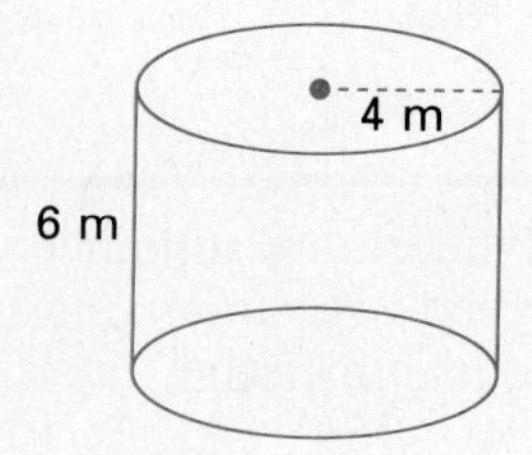

Written Exercises

Find the lateral area and the total area of the rectangular solid, the bases of which are shaded.

1.

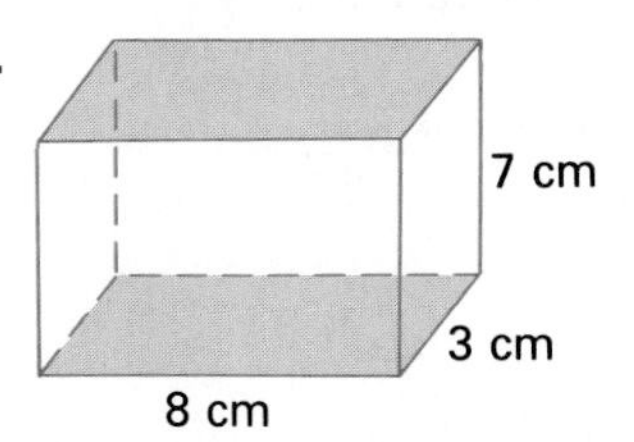

2.

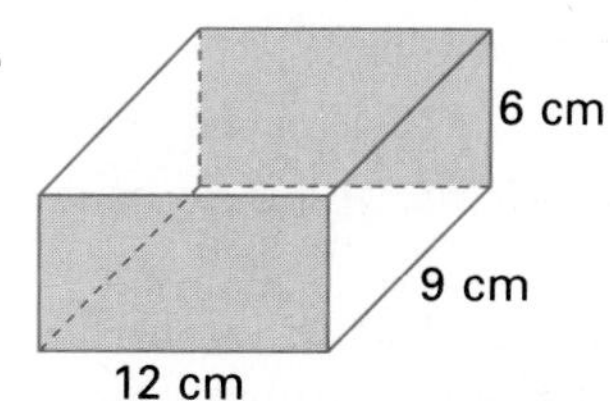

3.

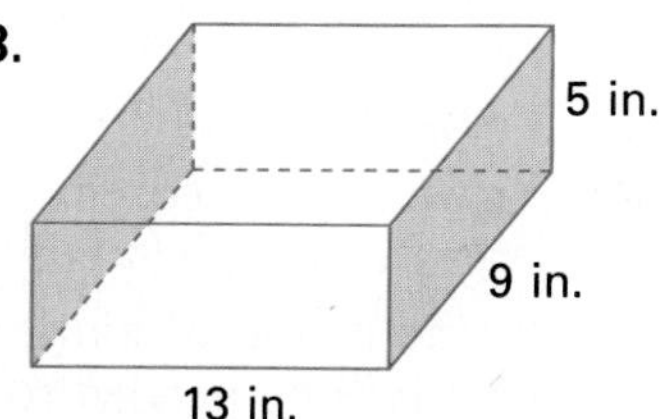

Draw a picture to illustrate how the surface of the right prism would appear when unfolded.

4.

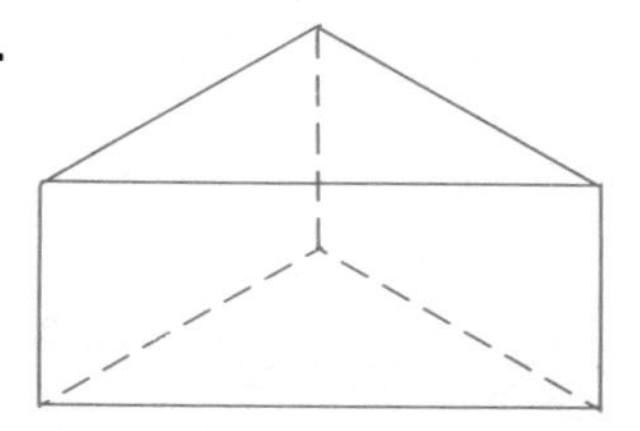

5.

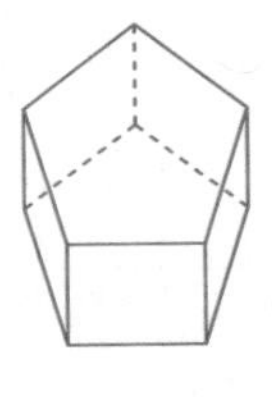

6.

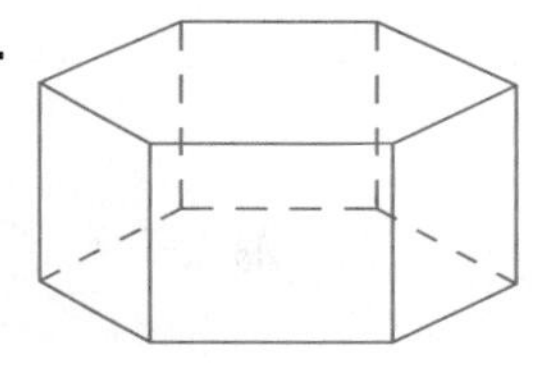

Find the lateral area and total area of the right cylinder given the radius and height.

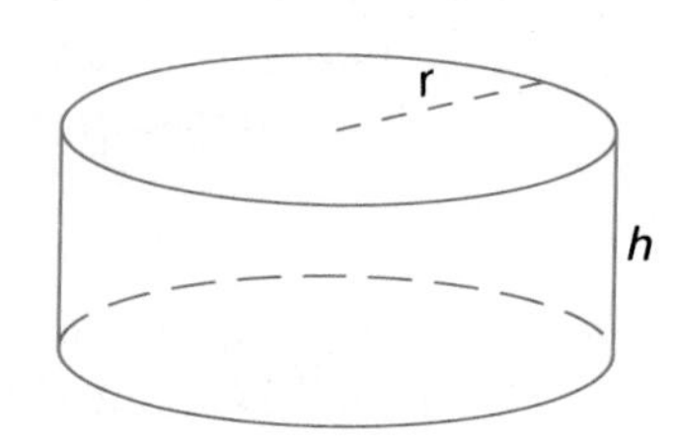

7. $r = 10$ cm, $h = 6$ cm
8. $r = 5$ in., $h = 8$ in.
9. $r = 7$ m, $h = 9$ m
10. $r = 9$ cm, $h = 7$ cm

Find the ratio of the areas of two similar prisms for the given ratio of lengths of corresponding edges.

11. 3 to 5 **12.** 9 to 5 **13.** 1 to 4 **14.** 15 to 21

15. The walls and ceiling of a warehouse are to be painted. How many square meters must be covered if the warehouse is 120 m by 96 m with a ceiling 3 m high?

C **Make a sketch of the right prism described. Find the lateral area and the total area. (Exercises 16–19)**

16. The length of an altitude is 12 cm. Each base is a right triangle with legs measuring 6 cm and 8 cm.

17. The length of an altitude is 15 cm. Each base is a right triangle with legs measuring 8 cm and 9 cm. (Round to the nearest tenth.)

18. The length of an altitude is 10 cm. Each base is a regular hexagon with an edge measuring 4 cm.

19. The length of an altitude is 5 m. Each base is a regular pentagon with an edge measuring 2 m and an apothem measuring approximately 1.4 m.

20. A concrete roller is 3 m long. Its diameter is 2 m long. How much area will it cover in 400 revolutions?

21. Prove Theorem 14.1.

22. Prove Theorem 14.3.

23. Prove that Theorem 14.4 holds for the case of two cubes with edges of lengths a and b.

24. If one gallon of paint will cover 400 ft^2 of surface, how many gallons will be needed to paint the swimming pool shown below?

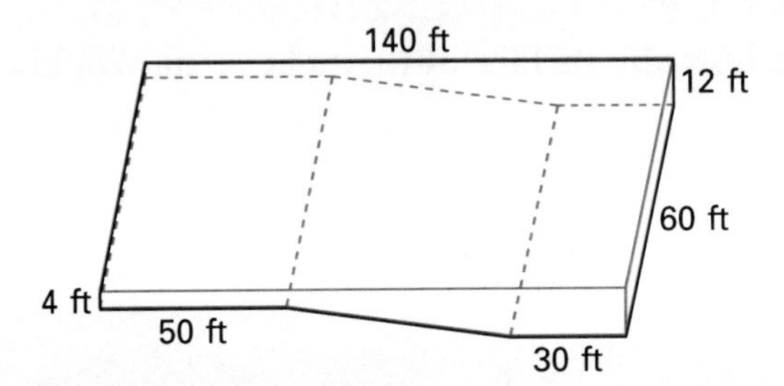

25. A right square prism is inscribed in a right circular cylinder of the same height. A radius and an altitude of the cylinder are each 10 cm long. What is the ratio of their total areas?

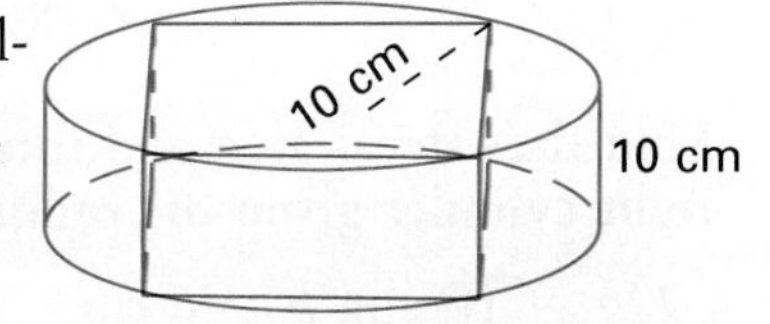

26. A right cylinder is inscribed in a right square prism of the same height. A radius and an altitude of the cylinder are each 10 cm long. What is the ratio of their total areas?

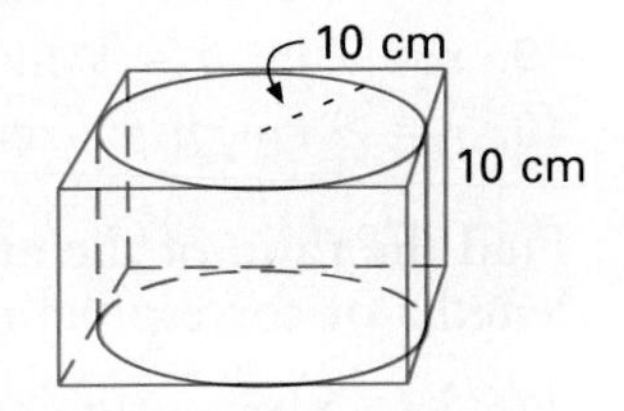

27. If the lateral surface of an oblique cylinder were "unrolled," what would its shape be? Is there a simple formula for its lateral area?

Mixed Review

If a and b are the lengths of the two legs of a right triangle and c is the length of the hypotenuse, find the missing length. *9.2*

1. $a = 3$, $b = 4$, $c =$ ______

2. $a =$ ______, $b = 12$, $c = 13$

3. $a = 9$, $b =$ ______, $c = 41$

4. $a = 2$, $b = 3$, $c =$ ______

14.4 Volumes of Prisms and Cylinders

Objectives

To find the volume of a prism
To find the volume of a cylinder

Direct comparison can be used to compare the sizes of some solids. If one solid can fit inside a second solid, then the second is larger than the first. Many solids cannot be compared in this way, however. Their *volumes* can be used to compare their sizes.

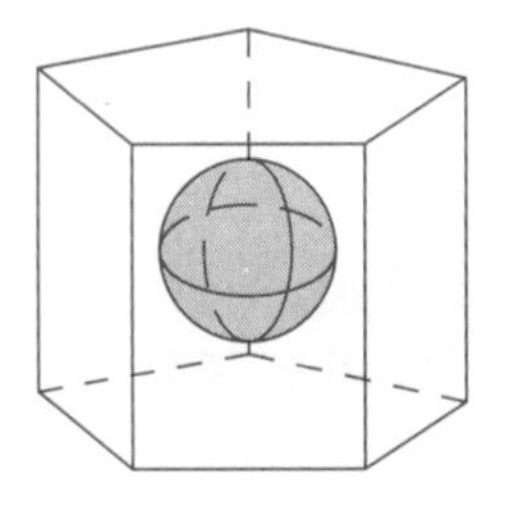

The figure below consists of a number of cubes of the same size. If one of these cubes is considered to be a *unit* of measure, then the total number of these cubes can be considered to be a *measure* of the solid. This measure is the **volume** of the solid.

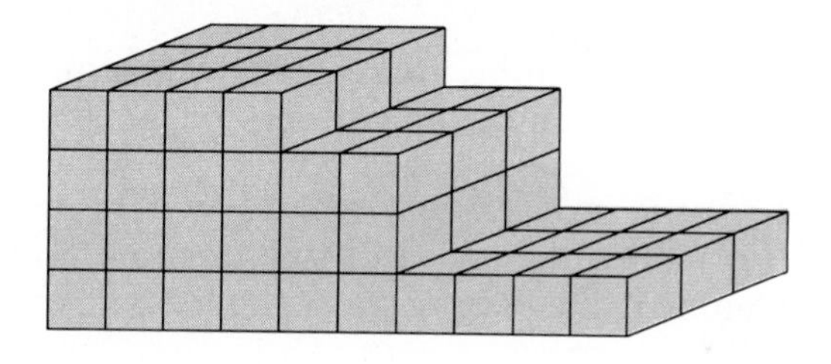

To find any measure, an appropriate unit must be used. *Square units* are used to measure *area. Cubic units* are used to measure volume. One cubic unit is the volume of a cube in which each edge measures one unit. A cube with each edge one centimeter (1 cm) long has a volume of one *cubic centimeter* (cm^3).

The rectangular solid below contains four layers of unit cubes. Each layer contains 5 rows of 6, or 30 unit cubes. (Thirty also is the area of a five-by-six unit rectangle.) Four layers of 30 cubic units makes a total of 120 cubic units. This suggests a formula for the volume V of a rectangular solid.

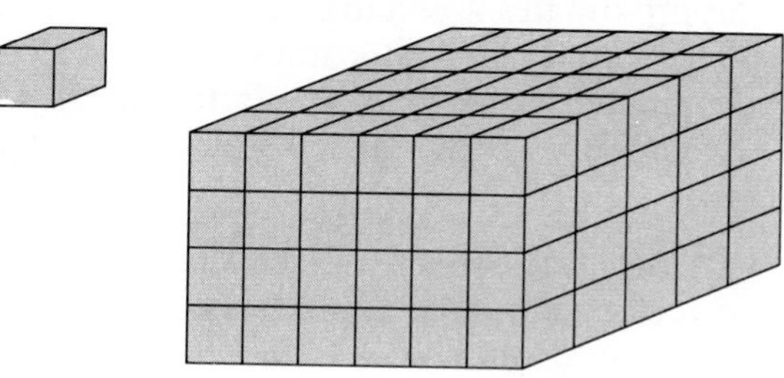

Postulate 23 For any rectangular solid, the volume $V = lwh$, where l, w, and h are the lengths of three edges with a common vertex.

EXAMPLE 1 Find the volume of a rectangular solid measuring 6 cm by 8 cm by 10 cm.

Solution Volume $= 6 \cdot 8 \cdot 10 = 480 \text{ cm}^3$

A cube is a special rectangular solid in which all edges are congruent.

Theorem 14.5 The volume of a cube with edges of length s is s^3.

EXAMPLE 2 If the volume of a cube is 8 m^3, what is the length of an edge?

Solution

$$\text{Volume} = s^3 = 8$$
$$s = \sqrt[3]{8}$$
$$s = 2 \text{ m}$$

Some solids can be broken down into smaller solids. The volume of the solid is the sum of the volumes of the smaller solids.

Postulate 24 If a solid is the union of two or more nonoverlapping solids, then its volume is the sum of the volumes of these nonoverlapping parts.

EXAMPLE 3 Find the volume of the solid. Each angle is a right angle.

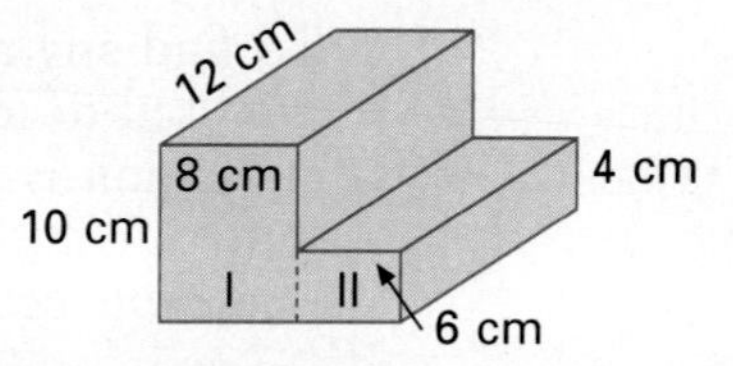

Solution Volume (Solid I) $= 8 \cdot 10 \cdot 12 = 960 \text{ cm}^3$
Volume (Solid II) $= 6 \cdot 4 \cdot 12 = 288 \text{ cm}^3$
By Postulate 24, total volume $= 1{,}248 \text{ cm}^3$.

A stack of index cards closely approximates a rectangular solid. The volume of such a stack can be computed using the formula for the volume of a rectangular solid.

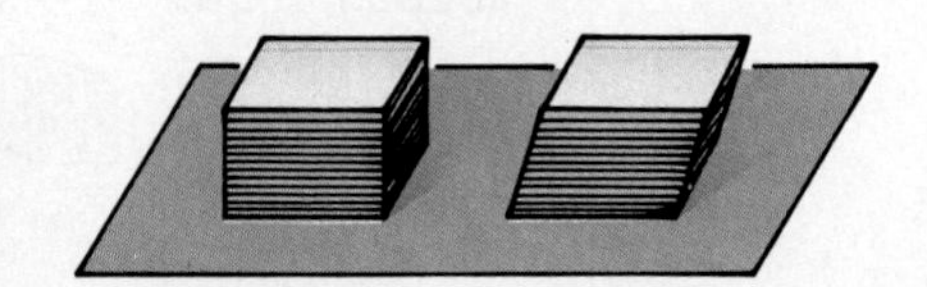

Notice that the stack can be slanted so that it no longer has the shape of a rectangular solid. This can be done in many ways.

Even though the shape of the stack of cards has changed, it should be clear that the volume has not. The area of every layer (the *cross-sectional area* at a particular level) has stayed the same. Based on the example of the cards, a generalization can be made.

Postulate 25

Cavalieri's Principle: If two solids have equal heights, and if the cross sections formed by every plane parallel to the bases of both solids have equal areas, then the volumes of the solids are equal.

The solids in the illustration will have equal volumes if all their cross-sectional areas are equal.

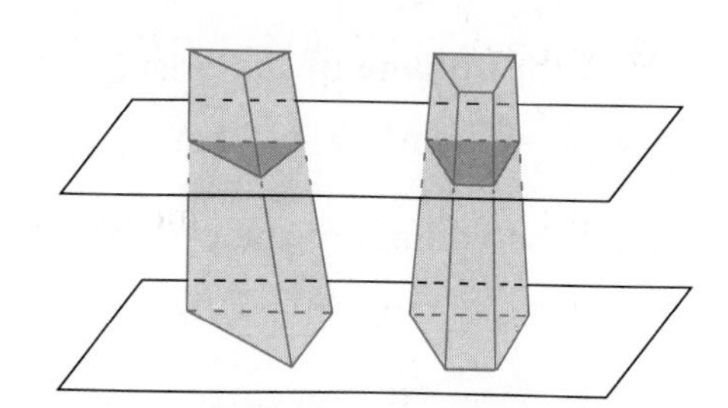

Using Cavalieri's Principle, a formula for the volume of any prism or cylinder can be obtained. It is stated here without proof.

Theorem 14.6

For any prism or cylinder, the volume is the product of the area of a base and the length of an altitude ($V = Bh$, where B = area of base and h = altitude).

EXAMPLE 4 Find the volume of the prism if the area of its base is 30 cm^2 and its altitude is 4 cm.

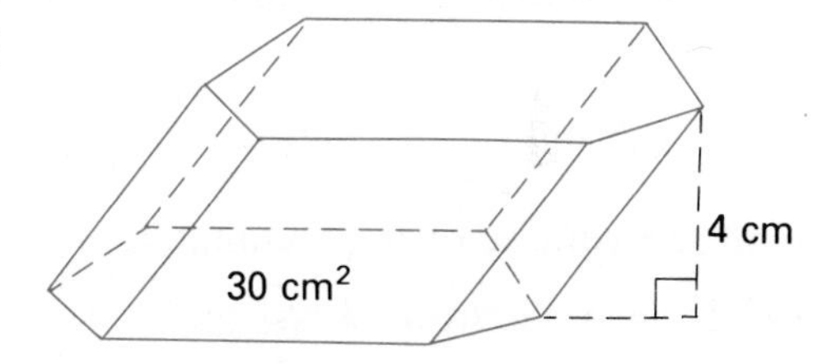

Solution

$$V = Bh$$
$$= 30 \cdot 4$$
$$= 120 \text{ cm}^3$$

Since the area of a circle is πr^2, the following is easily proved.

Corollary

The volume of a cylinder is Bh, or $\pi r^2 h$.

EXAMPLE 5 A radius of a base of a right cylinder is 5 cm long. An altitude is 6 cm long. Find the volume.

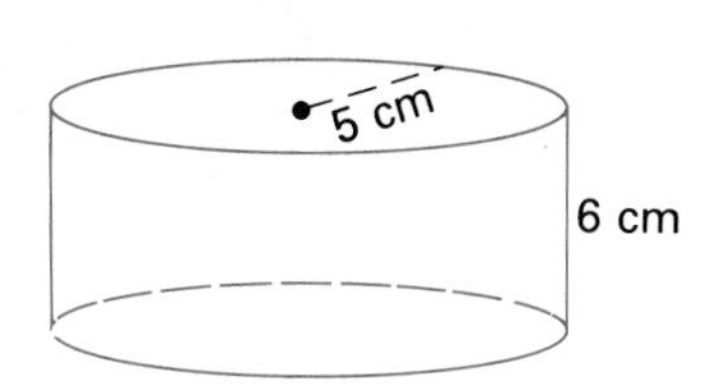

Solution

$$V = Bh = \pi r^2 \cdot h$$
$$V = \pi \cdot 5^2 \cdot 6$$
$$V = 150\pi \text{ cm}^3$$

Classroom Exercises

Find the volume of the rectangular solid.

1.

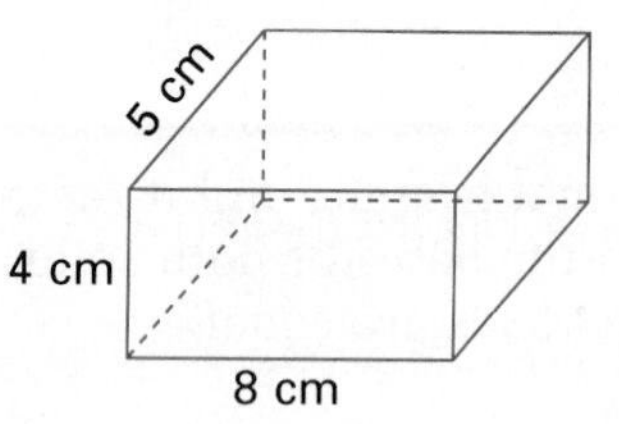

2.

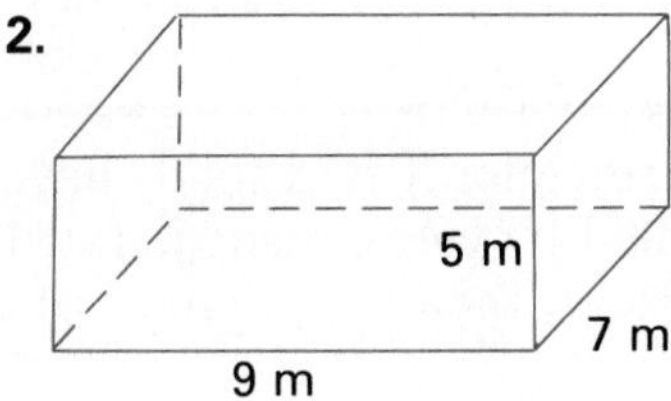

3.

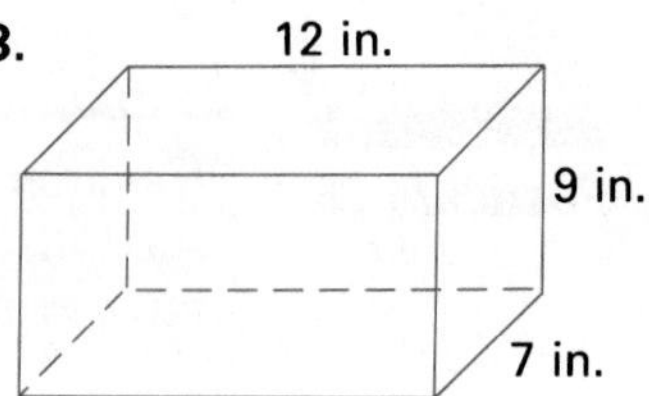

Find the volume of the prism.

4. $B = 7\text{ cm}^2$, $h = 5$ cm

5. $B = 9\text{ m}^2$, $h = 6$ m

Find the volume of the cylinder.

6. $r = 10$ mm, $h = 5$ mm

7. $r = 2$ cm, $h = 3$ cm

Written Exercises

Find the volume of the prism.

1. $B = 17\text{ cm}^2$, $h = 23$ cm

2. $B = 32\text{ ft}^2$, $h = 17$ ft

Find the volume of the cylinder.

3. $r = 12$ in., $h = 7$ in.

4. $r = 15$ ft, $h = 8$ ft

5. $d = 8$ mm, $h = 3$ mm

6. $d = 16$ cm, $h = 9$ cm

Find the volume of the rectangular solid.

7. 5 cm by 8 cm by 7 cm

8. 9 mm by 4 mm by 6 mm

9. 13 in. by 12 in. by 17 in.

10. 27 yd by 13 yd by 7 yd

Find the volume of the solid. Each angle is a right angle.

11.

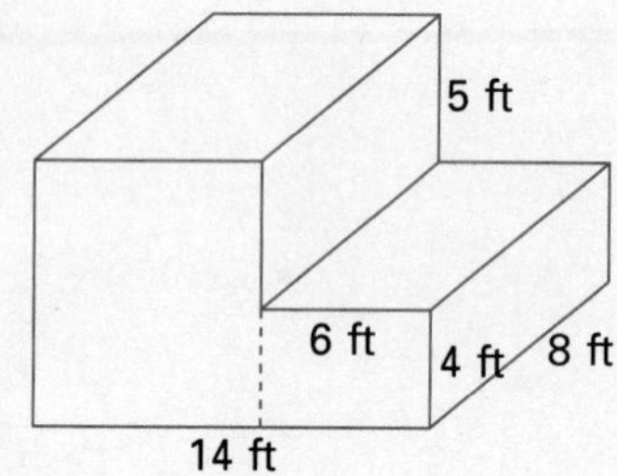

Find the volume.

12. a cube with 6-cm edges

13. a cube with 12-mm edges

14. a prism with an isosceles right triangular base with 10-cm legs and a 17-cm altitude

15. a prism with an equilateral triangle as its base with 10-cm sides and a 17-cm altitude

16. a prism with a regular hexagonal base with 10-cm sides and a 17-cm altitude

17. Prove Theorem 14.5.

18. Prove the corollary to Theorem 14.6.

A volume of 231 in^3 has a capacity of 1 gallon. (Exercises 19–20)

19. A large bucket in the shape of a right cylinder has a diameter of 16 in. and a height of 20 in. How many gallons of liquid can the bucket hold?

20. A can of juice has a diameter of 4 in. and a height of 7 in. How many pints of juice can the can hold?

21. Find the ratio of the area of a cube to its volume.

22. A right square prism is inscribed in a right circular cylinder of the same height. A radius and an altitude of the cylinder are each 10 cm long. What is the ratio of their volumes?

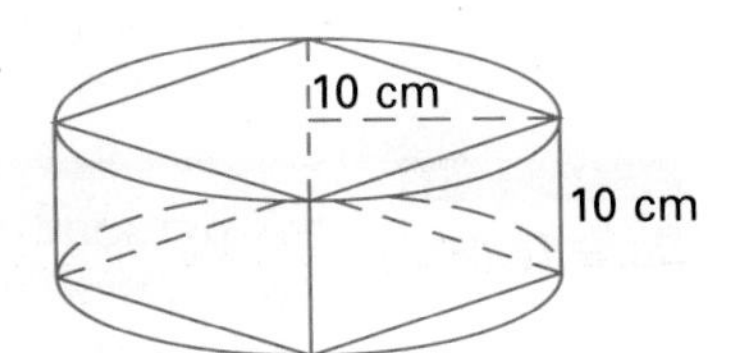

23. A cylinder of revolution is formed when a rectangle with sides of length m and n is revolved about one of its sides. Find the ratio of the volumes of the cylinders formed by the revolution in the diagram at the right.

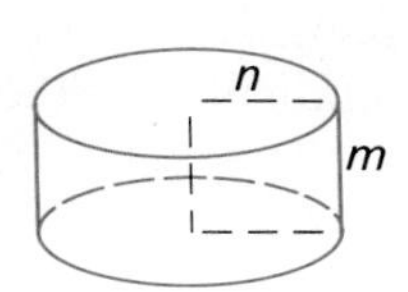

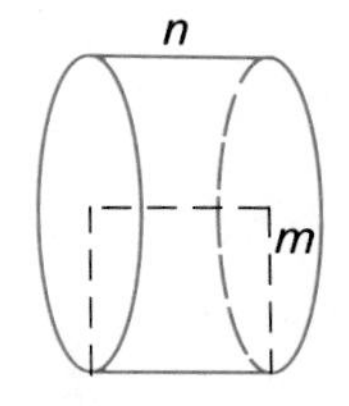

Midchapter Review

1–2. Find the number of faces (F), the number of edges (E), and the number of vertices (V) for each of the polyhedra shown. ***14.1***

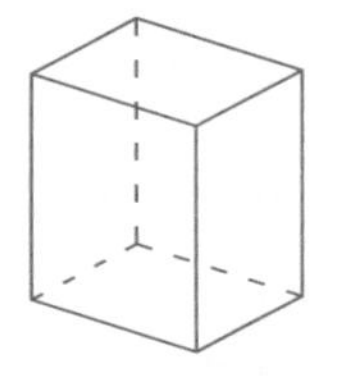

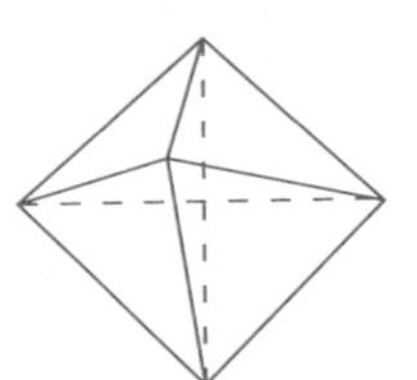

3. Draw a triangular prism. How many lateral faces and lateral edges does it have? ***14.2***

4. Find the lateral area of the prism. The bases are shaded. ***14.3***

5. Find the lateral area of the cylinder. ***14.3***

6. Find the volume of the prism. ***14.4***

7. Find the volume of the cylinder. ***14.4***

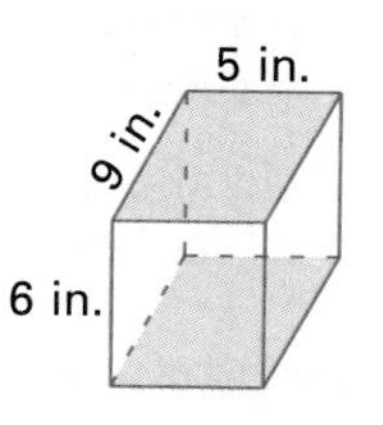

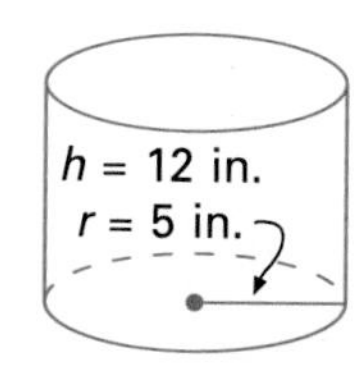

14.5 Areas of Pyramids and Cones

Objectives To find the lateral and total areas of a regular pyramid
To find the lateral and total areas of a right cone

The pyramids of Egypt have fascinated people for centuries. Because of the rigidity of the triangular faces of pyramids, such shapes are sometimes used in modern building designs.

Definitions

A **pyramid** is a polyhedron composed of a polygonal region, called the *base*, and triangular regions, called the *lateral faces*. The lateral faces intersect in a common point called the *vertex*. The intersections of each pair of lateral faces are the **lateral edges**. The **altitude** of the pyramid is the segment from the vertex to the plane containing and perpendicular to the base.

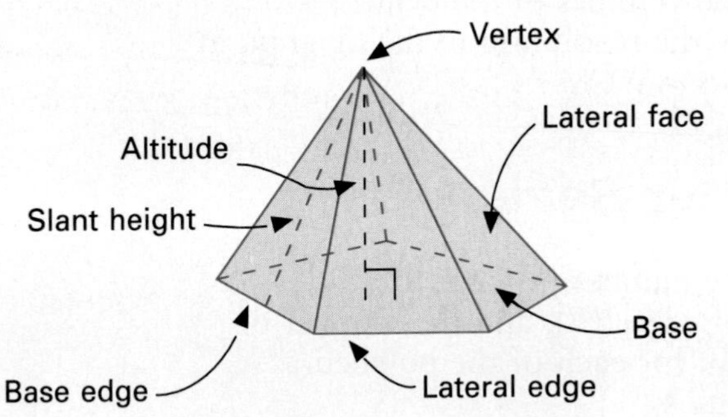

In a *right pyramid*, the altitude contains and is perpendicular to the center of the base. A **regular pyramid** is a right pyramid where the base is a regular polygon and the lateral faces are congruent isosceles triangles.

The **lateral area** of a pyramid is the sum of the areas of its lateral faces. The length of the altitude of each lateral face is called the **slant height** of the regular pyramid.

Regular
hexagonal pyramid

In finding areas of pyramids, it may be helpful to think of the lateral faces of the solid as flattened out into a plane with the base.

EXAMPLE 1 Find the lateral area of a regular square pyramid with 4-cm base edges and a 5-cm slant height.

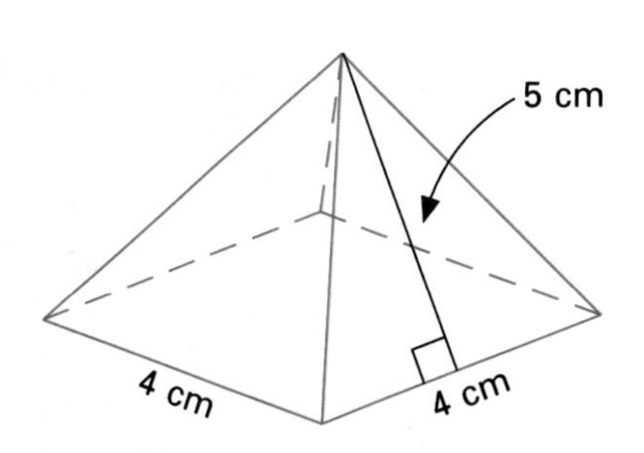

Plan Multiply the area of a triangular face by 4.

Solution Lateral area $= 4 \cdot (\frac{1}{2} \cdot 4 \cdot 5)$
$= 4 \cdot 10 = 40 \text{ cm}^2$

As suggested by Example 1, the lateral area of a regular pyramid is the product of the number of lateral faces and the area of a lateral face. The lateral area $L = n \cdot (\frac{1}{2}bl)$, or $\frac{1}{2}l(bn)$, where b is the length of a side of the base, l is the slant height, and n is the number of sides of the base. But bn is the perimeter p of the base. Therefore, $L = \frac{1}{2}pl$. This reasoning constitutes a proof of the theorem that follows.

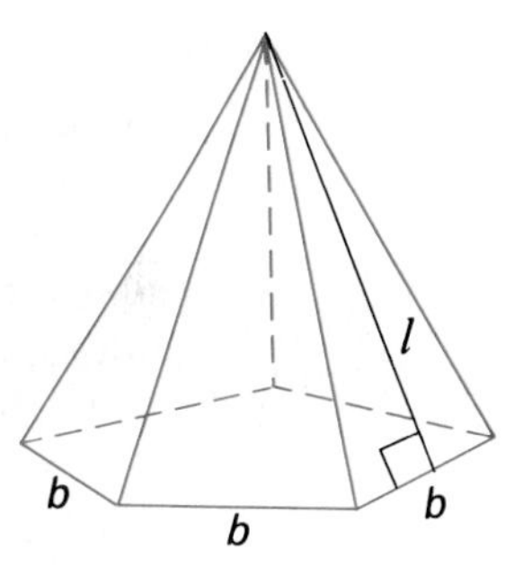

Theorem 14.7 The lateral area of a regular pyramid is one-half the product of the perimeter of the base and the slant height ($L = \frac{1}{2}pl$).

The *total area* of any pyramid is the sum of the lateral area and the area of the base. The base of a regular pyramid is a regular polygon, the area of which is one-half the product of its perimeter and its apothem.

EXAMPLE 2 Find the total area of a regular pentagonal pyramid if the length of a side of the base is 2 m, the length of an apothem a of the base is 1.4 m, and the slant height is 4 m.

Solution Total area = Lateral area + Base area
$= \frac{1}{2}pl + \frac{1}{2}pa$
$= \frac{1}{2}(5 \cdot 2) \cdot 4 + \frac{1}{2}(5 \cdot 2) \cdot 1.4$
$= 20 + 7 = 27 \text{ m}^2$

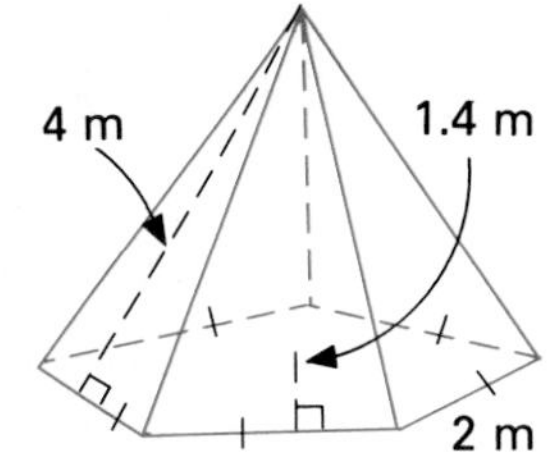

EXAMPLE 3 Find the total area of a regular square pyramid whose lateral sides are equilateral triangles with 4-cm edges.

Plan To find the area of one lateral face, use the formula for the area of an equilateral triangle.

Solution

$$\begin{aligned}\text{Total area} &= \text{Lateral area} + \text{Base area}\\ &= 4 \cdot \left(\frac{s^2}{4}\sqrt{3}\right) + s^2\\ &= 4 \cdot \left(\frac{4^2}{4}\sqrt{3}\right) + 4^2\\ &= 4 \cdot (4\sqrt{3}) + 16 = (16\sqrt{3} + 16)\ \text{cm}^2\end{aligned}$$

As a cylinder resembles a prism, so a *cone* resembles a pyramid. The *base* of a cone is a circle. The *lateral surface* of a cone is made up of an infinite number of segments that connect a single point, called the *vertex*, and all points on the edge of the base. The *altitude* of the cone is the segment from the vertex to the base and perpendicular to the base. If an endpoint of the altitude is the center of the circle, then it is a *right cone*. If it is not a right cone, then it is an *oblique cone*.

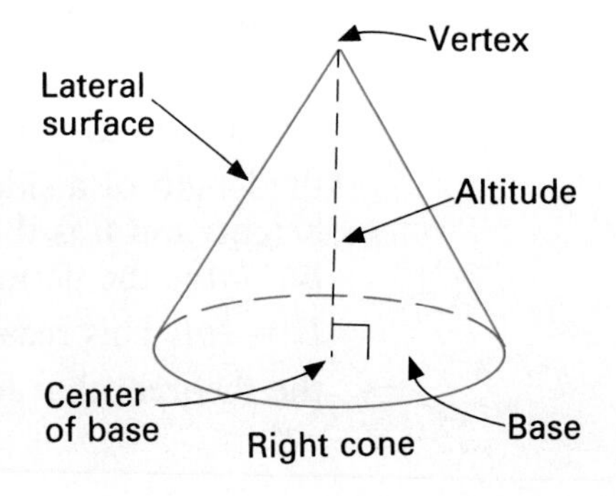

Right cone

The approach used to find the lateral area of a regular pyramid can be generalized to find the lateral area of a right cone. This formula will be presented without proof.

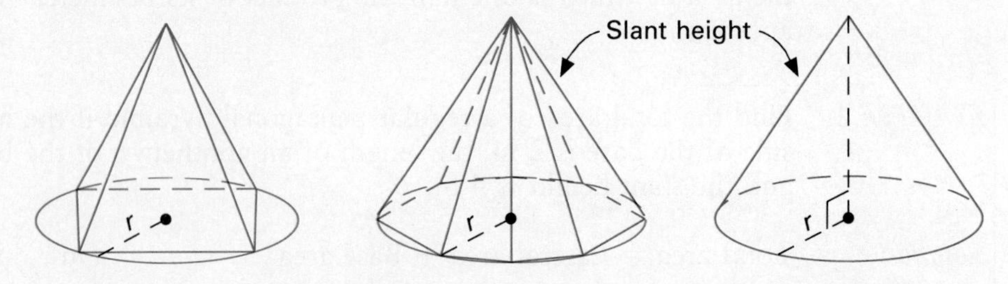

A regular pyramid has lateral area of $\frac{1}{2}pl$, where p is the perimeter of the base and l is the slant height. In a right cone, $p = 2\pi r$, where r is the radius of the base because the base is a circle. Therefore, a cone has a lateral area of $\frac{1}{2}(2\pi r)l$, or πrl.

Theorem 14.8 The lateral area (L) of a right cone is πrl.
The total area (A) is $\pi rl + \pi r^2 = \pi r(l + r)$.

EXAMPLE 4 Find the total area of a right cone with a 7-cm radius and a 12-cm slant height.

Solution Total area $= \pi r(l + r)$
$= \pi \cdot 7(12 + 7) = 133\pi \text{ cm}^2$

Classroom Exercises

Use sketches to illustrate how the surface of the right pyramid would appear when unfolded and flattened.

1. a square pyramid
2. an equilateral triangular pyramid
3. a regular hexagonal pyramid
4. a regular pentagonal pyramid
5. Find the lateral area of a hexagonal pyramid with 10-cm base edges and an 8-cm slant height.

Written Exercises

Find the lateral area of the regular pyramid.

1.
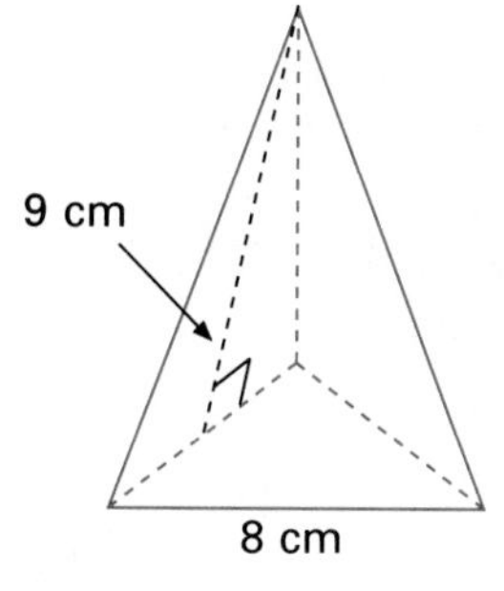

2.
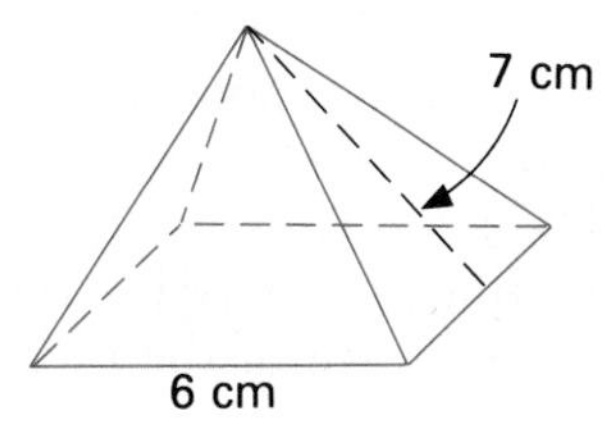

3.
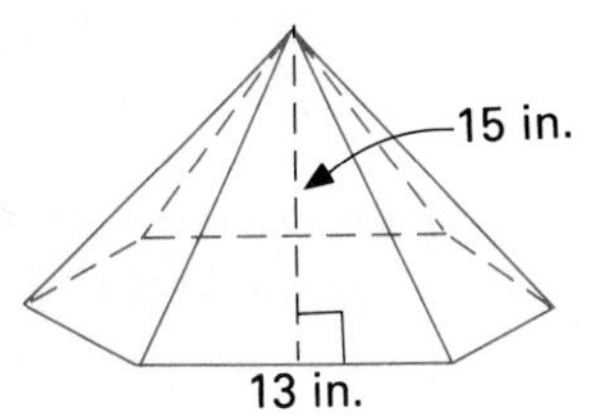

Find the total area of the regular pyramid.

4.
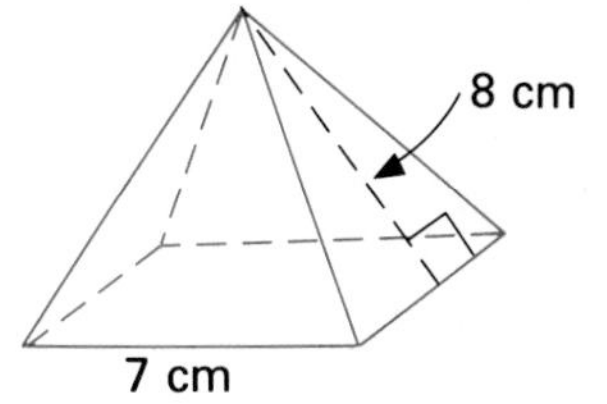

5.
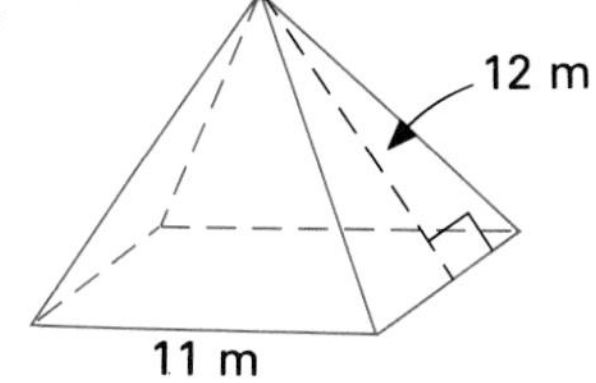

6.
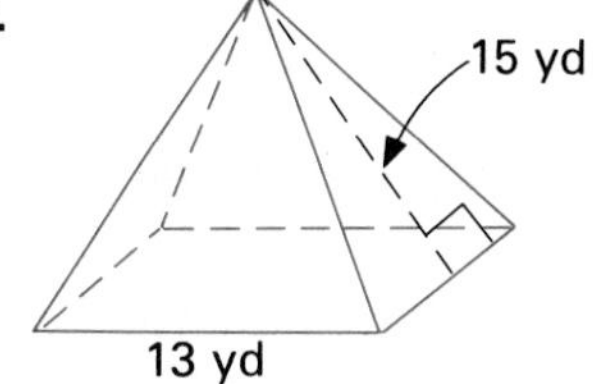

Find the lateral area and the total area of the right cone.

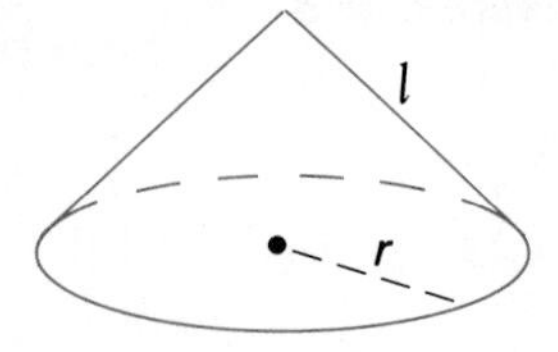

7. $r = 10$ cm, $l = 6$ cm
8. $r = 8$ ft, $l = 4$ ft
9. $r = 12$ m, $l = 11$ m
10. $r = 17$ in., $l = 13$ in.

Find the lateral area and the total area of the pyramid. (Exercises 11–14)

11. a regular triangular pyramid with 5-cm base edges and 8-cm slant height
12. a regular triangular pyramid with 5-cm base edges and 8-cm lateral edges
13. a regular hexagonal pyramid with 5-cm base edges and 8-cm slant heights
14. a regular hexagonal pyramid with 5-cm base edges and 8-cm lateral edges

15. The altitude of a regular square pyramid is 8 cm long, and the length of a side if the base is 20 cm. What is the lateral area of the pyramid?
16. The altitude of a right cone is 12 cm long, and the length of a radius of the base is 16 cm. What is the lateral area of the cone?
17. The lateral area of a regular pyramid is 144 cm^2, and the slant height is 16 cm. What is the perimeter of the base?
18. The lateral area of a regular pyramid is 196 mm^2, and the perimeter of the base is 14 mm. What is the slant height of the pyramid?
19. Prove: If the base of a pyramid is a regular polygon and the foot of the altitude is at the center of the base, then all the lateral faces are congruent isosceles triangles.
20. Derive a formula for the total area of a right cone in terms of the length of its altitude h and the length of the radius r of the base.
21. A right cone with radius of length r and altitude of length h is inscribed in a right cylinder of the same radius and altitude lengths. What is the ratio of their lateral areas? What is the ratio of their total areas?

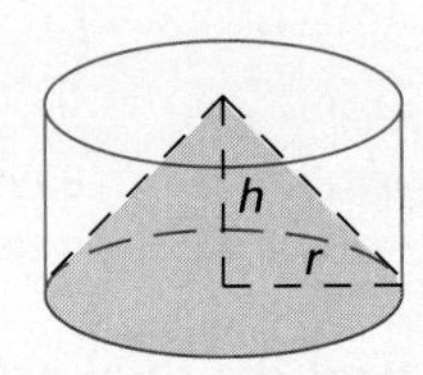

Mixed Review

1. What is the lateral area of a right cylinder with a 25-cm base radius and a 12-cm altitude? ***14.3***
2. What is the total area of a right regular triangular prism with 5-cm base edges and a 10-cm altitude? ***14.3***

14.6 Volumes of Pyramids and Cones

Objectives To find the volume of a pyramid and a cone
To find the ratios of volumes of solids

Taking models of a cone and a cylinder with the same base and height, fill up the cone with sand or water and pour the contents of that into the cylinder. Repeat this process until the cylinder is full. How do the volumes compare? Try this for other such pairs of cones and cylinders. Does the same relation hold? Try this experiment for a pyramid and a prism with the same base and height.

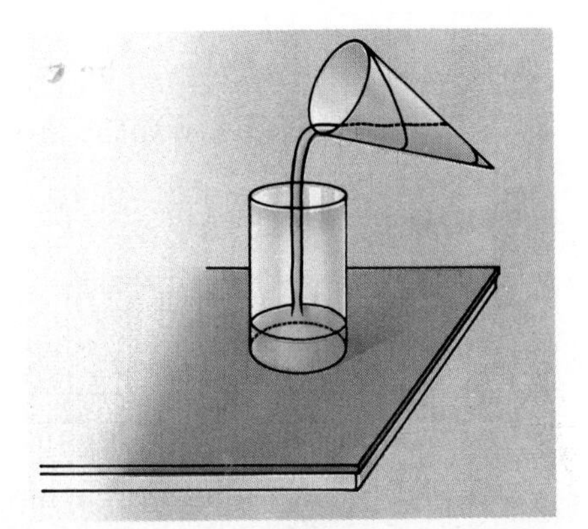

Theorem 14.9 The volume of a pyramid is one-third the volume of a prism with the same base and altitude as the pyramid. The volume of a cone is one-third the volume of a cylinder with the same base and altitude as the cone ($V = \frac{1}{3}Bh$).

Corollary 1 The volume of a pyramid or cone is one-third the product of the area of its base and the length of its altitude ($V = \frac{1}{3}Bh$).

EXAMPLE 1 Find the volume of the triangular pyramid.

Solution

$V = \frac{1}{3}Bh$

$V = \frac{1}{3}(\frac{1}{2} \cdot 5 \cdot 6) \cdot 8$

$V = \frac{1}{3} \cdot 15 \cdot 8$

$= 40 \text{ cm}^3$

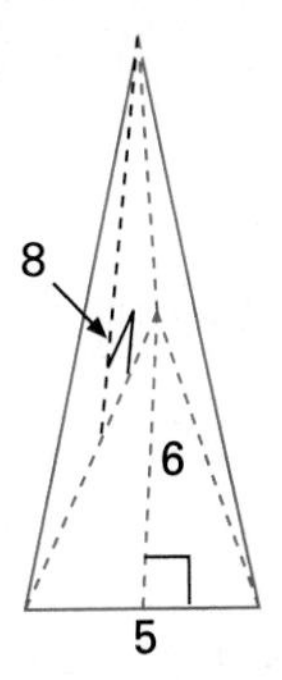

Corollary 2 The volume of a cone with a base radius of length r and an altitude of length h is $\frac{1}{3}\pi r^2 h$.

EXAMPLE 2 Find the volume of a cone with a 5-cm radius and a 6-cm altitude.

Solution

$$V = \frac{1}{3}\pi r^2 h$$
$$= \frac{1}{3}\pi(5^2) \cdot 6$$
$$= \frac{1}{3}\pi 25 \cdot 6 = 50\pi \text{ cm}^3$$

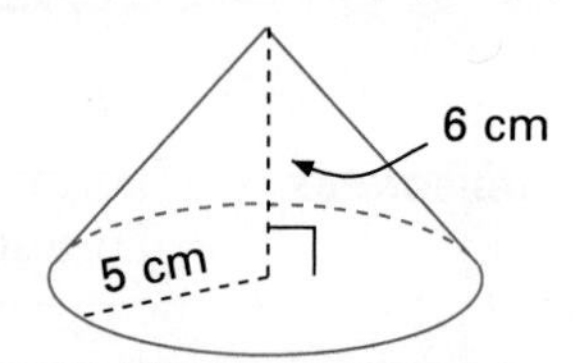

EXAMPLE 3 If the volume of a pyramid is 20 cm³, and the area of the base is 5 cm², what is the length of the altitude?

Solution

$$V = \frac{1}{3}Bh$$
$$20 = \frac{1}{3}(5)h$$
$$60 = 5h$$
$$h = \frac{60}{5} = 12 \text{ cm}$$

EXAMPLE 4 A right triangle has legs 3 cm and 4 cm long. Find the volumes of the cones formed by revolving the triangle around each leg.

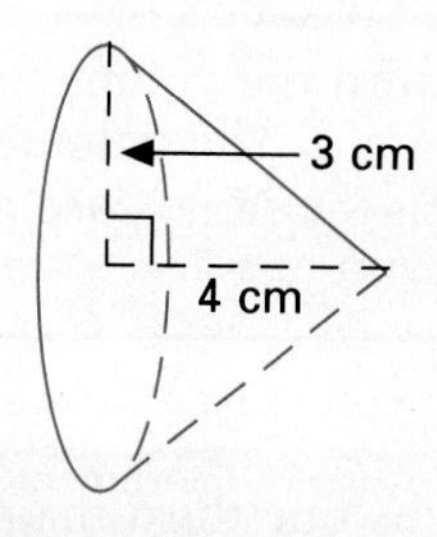

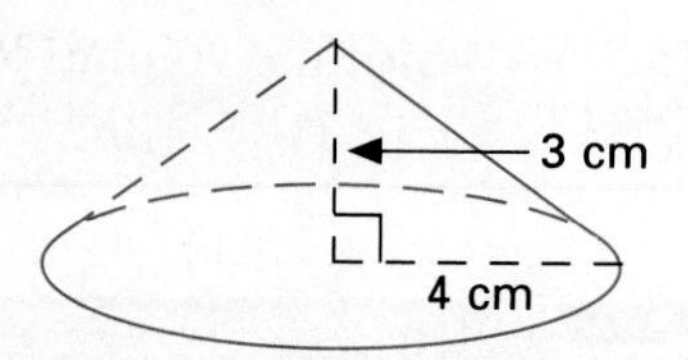

Solution

Revolution around 4-cm leg

$$V = \frac{1}{3}\pi r^2 h$$
$$V = \frac{1}{3}\pi(3^2)4$$
$$V = 12\pi \text{ cm}^3$$

Revolution around 3-cm leg

$$V = \frac{1}{3}\pi r^2 h$$
$$V = \frac{1}{3}\pi(4^2)3$$
$$V = 16\pi \text{ cm}^3$$

Classroom Exercises

Find the volume of the pyramid.

1. $B = 20 \text{ cm}^2, h = 6 \text{ cm}$

2. $B = 15 \text{ mm}^2, h = 8 \text{ mm}$

Find the volume of the cone.

3. $B = 15\pi \text{ in}^2, h = 2 \text{ in.}$

4. $r = 2 \text{ mm}, h = 6 \text{ mm}$

Find the altitude of the pyramid.

5. $V = 100 \text{ cm}^3, B = 60 \text{ cm}^2$

6. $V = 144 \text{ in}^3, B = 3 \text{ in}^2$

Written Exercises

Find the volume of the rectangular pyramid.

1.

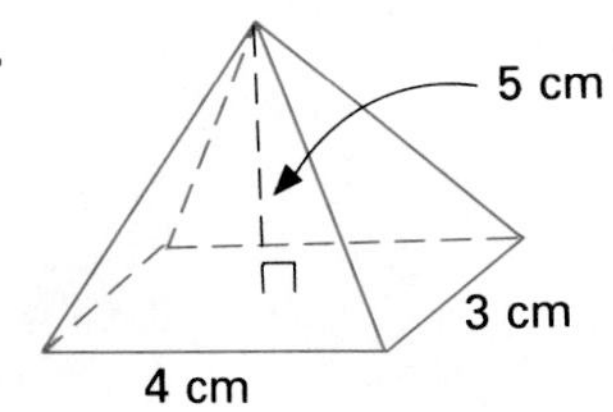

2.

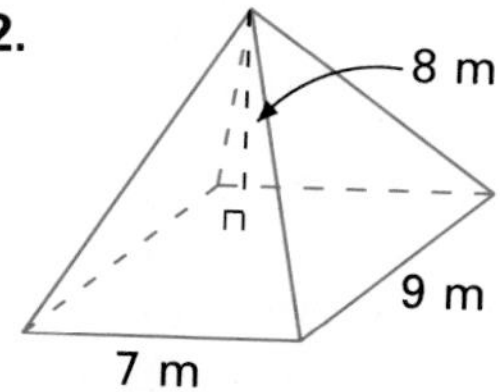

3.

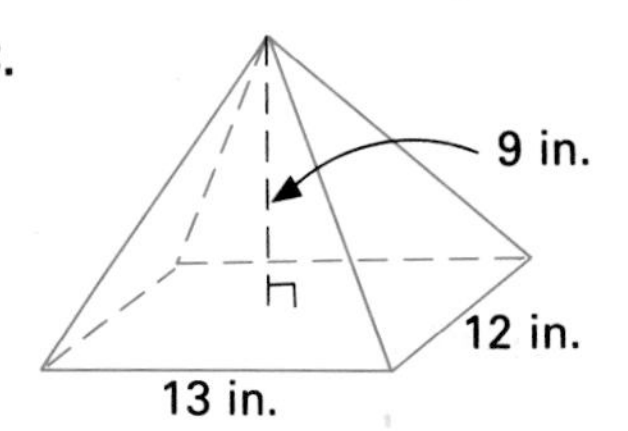

Find the volume of the pyramid.

4. $B = 9\text{ cm}^2$, $h = 7\text{ cm}$,

5. $B = 7\text{ mm}^2$, $h = 12\text{ mm}$

6. $B = 5\text{ in}^2$, $h = 8\text{ in.}$

7. $B = 10\text{ ft}^2$, $h = 7\text{ ft}$

Using a calculator, find the volume of the cone, correct to the nearest tenth. For π, use 3.14. (Exercises 8–11)

8. $r = 9\text{ mm}$, $h = 10\text{ mm}$

9. $r = 5\text{ cm}$, $h = 12\text{ cm}$

10. $r = 8.5\text{ m}$, $h = 5.2\text{ m}$

11. $r = 7.7\text{ cm}$, $h = 4.6\text{ cm}$

12. If the volume of a pyramid is 12 cm^3, and the area of the base is 4 cm^2, what is the length of the altitude?

13. If the volume of a square pyramid is 18 mm^3, and the length of the altitude is 6 mm, what is the length of an edge of the square base?

14. If the volume of a right cone is $32\pi\text{ m}^3$, and the length of an altitude is 6 m, what is the length of the radius of the base?

15. If the volume of a right cone is $75\pi\text{ ft}^3$, and the area of the base is $25\pi\text{ ft}^2$, what is the length of the altitude?

Find the volume.

16. In a pyramid with a rectangular base, adjacent sides of the base are 7 cm and 9 cm long. The length of the altitude of the pyramid is 8 cm.

17. In a pyramid with a square base, a side of the square is 5 mm long. The length of the altitude of the pyramid is 15 mm.

18. A pyramid has an isosceles right-triangular base. The legs of the triangle are 12 cm long. The length of the altitude of the pyramid is 7 cm.

19. A pyramid has a right-triangular base. The two legs are 6 cm and 8 cm long. The length of the altitude of the pyramid is 10 cm.

20. A pyramid has an equilateral triangular base. The sides of the triangle are 10 cm long. The length of the altitude of the pyramid is 18 cm.

21. A pyramid has a regular hexagonal base. The sides of the hexagon are 6 mm long. The length of the altitude is 10 mm.

Find the ratio of the volumes of the two solids.

22. a square pyramid with 2-cm base edges and a 2-cm altitude and a square pyramid with 3-cm base edges and a 3-cm altitude

23. a right cone with a 2-mm base radius and a 2-mm altitude and a right cone with a 3-mm base radius and a 3-mm altitude

24. The owner of an ice cream parlor is choosing cups for her new store. Of the two cups shown, which has the greater volume? Express the difference in the volumes in terms of π. How do the amounts of ice cream heaped above the tops of the cups compare?

25. Find the ratio of the total area of a regular tetrahedron with edge length s to its volume.

26. A right square pyramid is inscribed in a right cone of the same height. A radius of the base and the altitude of the cone are each 10 cm long. What is the ratio of the volume of the pyramid to that of the cone?

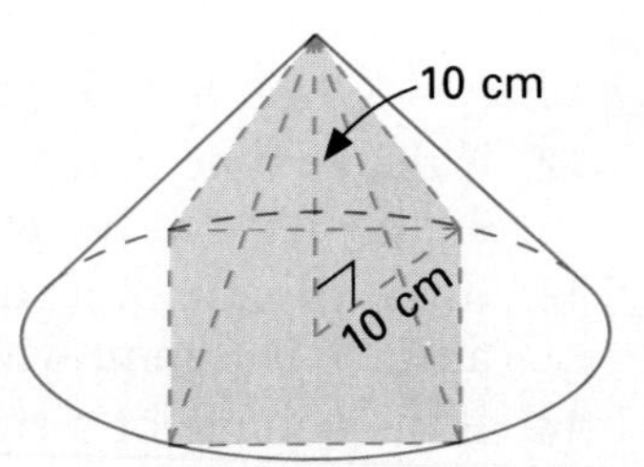

27. A right cone is inscribed in a right square pyramid of the same height. A radius and the altitude of the cone are each 10 cm long. What is the ratio of the volumes of the two solids?

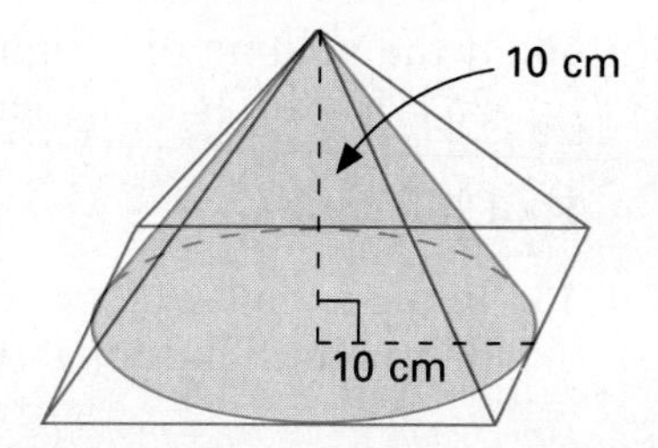

28. A right cone is inscribed in a cube with edge length s. What is the ratio of the volumes?

Mixed Review

1. Draw a segment 2 in. long. Construct the bisector of the segment. *1.3*

2. Draw an obtuse angle. Construct the angle bisector. *1.5*

3. Draw a line and a point not on the line. Construct a line through the given point parallel to the given line. *3.3*

Application: *Manufacturing Paper Cups*

Cone-shaped paper cups can be manufactured from patterns in the shape of sectors of circles. The figures below show two patterns. The first is cut with a straight angle. The second is cut with an angle measuring 120.

1. If each sector has a radius of 6 cm, what will be the area of each sector? (HINT: What part of the circle is used by the sector?)

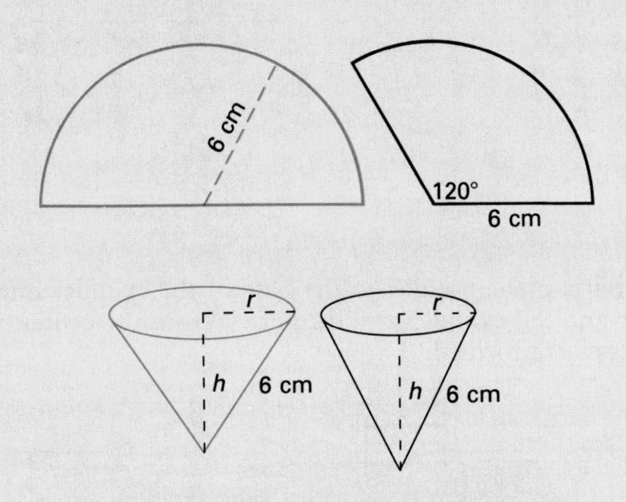

To find the volume of a cone-shaped cup formed from the first sector, we must first consider that the length of the arc of the sector becomes the circumference of the base of the cone.

$$\text{Arc length} = \frac{1}{2} \cdot 2\pi \cdot 6 = 6\pi = \text{Circumference of cone base}$$

Using the circumference formula again, we find the radius of the base.

$$\text{If } C = 2\pi r = 6\pi, \text{ then } r = 3$$

Using the Pythagorean Theorem, we can find the height of the cone.

$$h = \sqrt{36 - 9} = \sqrt{27} = 3\sqrt{3}$$

We can now find the volume of the cone.

$$V = \frac{1}{3}\pi r^2 h = \frac{1}{3}\pi \cdot 3^2 \cdot 3\sqrt{3}$$
$$= \pi \cdot 9 \cdot \sqrt{3}$$
$$= 48.97 \text{ cm}^3$$

2. Find the volume of the cup formed by the sector measuring 120.

3. As the area increases by 50% from the smaller to the larger sector, by how much does the volume of the cup increase?

14.7 Areas and Volumes of Spheres

Objectives

To find areas and volumes of spheres
To find the ratios of areas and volumes of spheres

A **sphere** is the set of all points in space at a given distance from a given point called the *center*. A formula for the volume of a sphere can be found by comparing a sphere to a cylinder with a double cone removed. Let the sphere and the cylinder have equal radii r and equal heights $2r$.

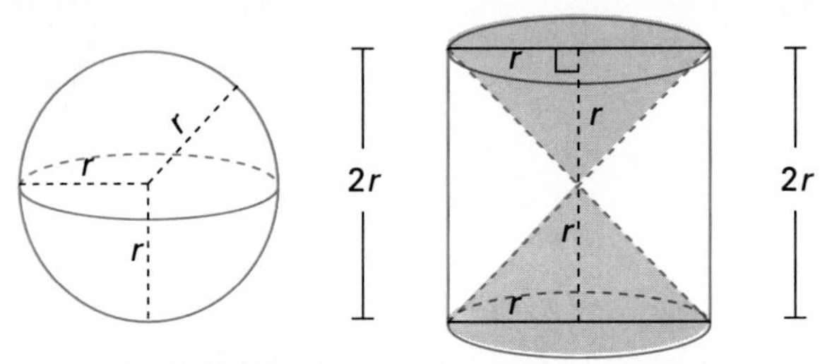

Suppose that a plane parallel to the base of the cylinder intersects both the sphere and the cylinder at a distance y from the center of each. Two cross sections are formed.

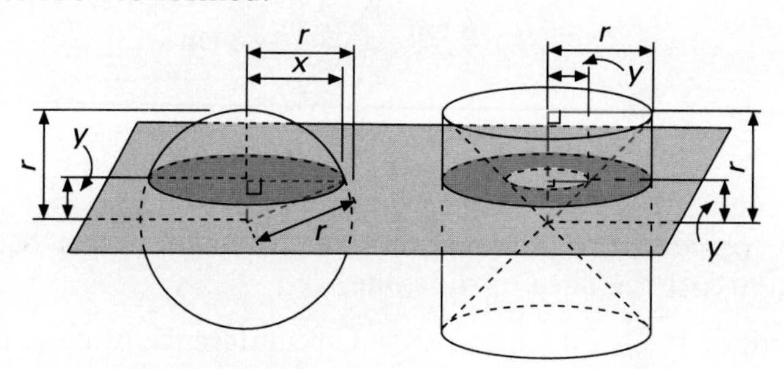

The plane cuts a circle in the sphere, and a ring-shaped figure called an *annulus* in the cylinder with the double cone removed. Let the radius of the circle be x. To find the dimensions of the annulus, notice that the altitude r of one of the cones forms an isosceles right triangle with a radius and the side of the cone. A similar triangle is formed by a radius of the inner circle of the annulus with the side and altitude of the cone. By similarity the triangle is isosceles, and so the radius of the inner circle is equal to the distance y.

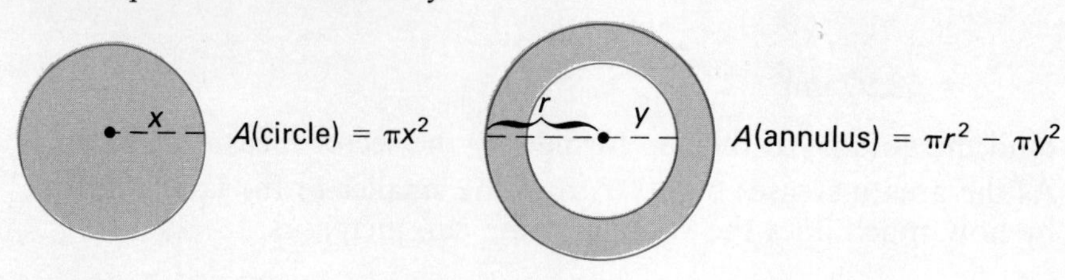

In the sphere, a right triangle relationship exists between the distances x, y, and the radius r. By the Pythagorean Theorem, $r^2 = x^2 + y^2$, or $x^2 = r^2 - y^2$. Thus the area πx^2 of the circle may be expressed as $\pi(r^2 - y^2)$ or $\pi r^2 - \pi y^2$. But this is the same as the area of the annulus. Therefore, *all* corresponding cross sections of the sphere and the solid formed by the cylinder minus the double cone have equal areas. By Cavalieri's Principle, the sphere and this solid have equal volumes.

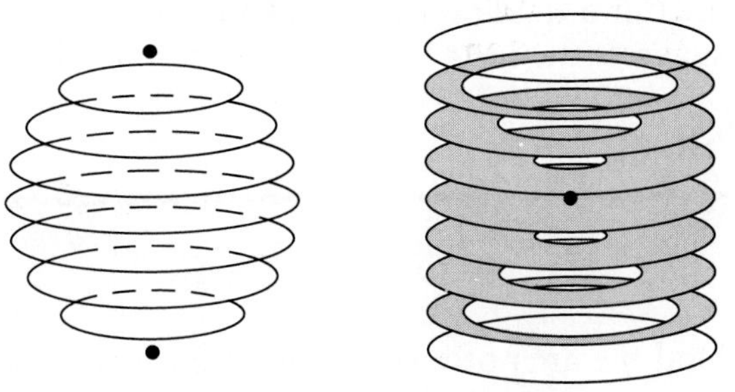

Corresponding cross sections.

The volume of this solid is the difference between the volumes of the cylinder and the double cone cut out of it.

$$\begin{aligned}
\text{Volume(sphere)} &= \text{Volume(cylinder)} - \text{Volume(double cone)} \\
&= \pi r^2 \cdot (2r) - 2 \cdot (\tfrac{1}{3}\pi r^2 \cdot r) \\
&= 2\pi r^3 - \tfrac{2}{3}\pi r^3 \\
&= \tfrac{4}{3}\pi r^3
\end{aligned}$$

This essentially proves the theorem that follows.

Theorem 14.10 The volume of a sphere with radius of length r is $\frac{4}{3}\pi r^3$.

EXAMPLE 1 Find the volume of a sphere with a 3-cm radius.

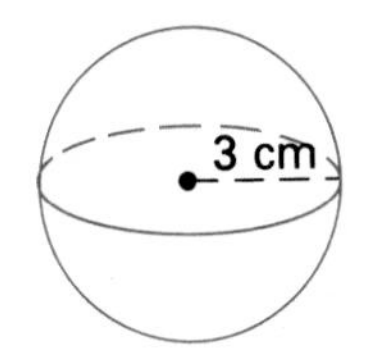

$$\begin{aligned}
V &= \tfrac{4}{3}\pi r^3 \\
&= \tfrac{4}{3}\pi \cdot 3^3 \\
&= 36\pi \text{ cm}^3
\end{aligned}$$

A *great circle of a sphere* is the intersection of the sphere and any plane that contains the center of the sphere. Thus, a radius of a great circle is also a radius of the sphere.

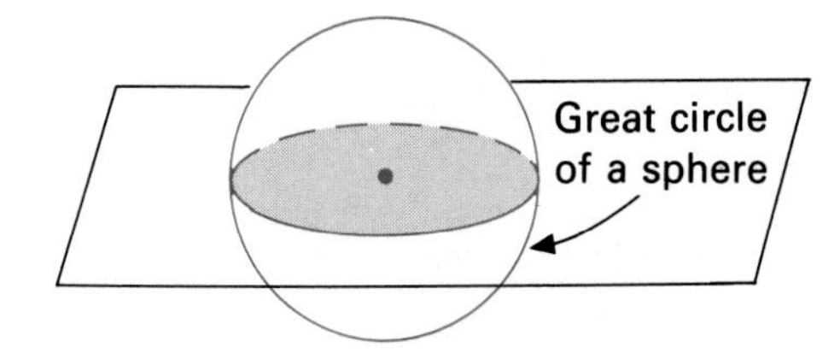

Plane intersecting sphere and passing through its center

Two intersecting great circles formed by perpendicular planes divide a sphere into four *quadrants*. It is proven in advanced geometry that the surface area of a quadrant is equal to the area of a great circle. Thus, the area of a quadrant is πr^2. This suggests the next theorem.

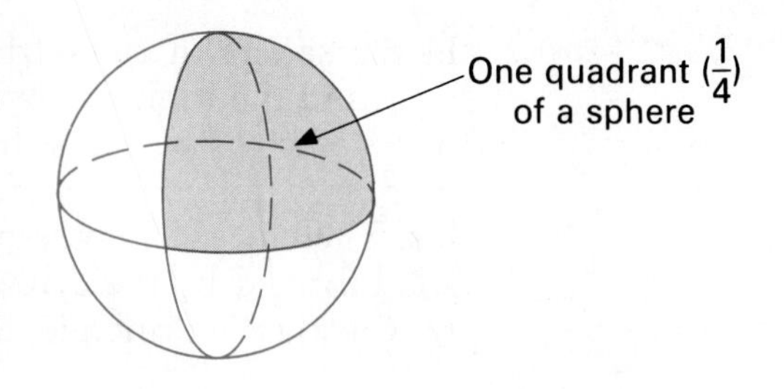

Two intersecting great circles formed by perpendicular planes

Theorem 14.11 The area of a sphere is $4\pi r^2$.

EXAMPLE 2 Find the area of a sphere with a 6-cm radius.

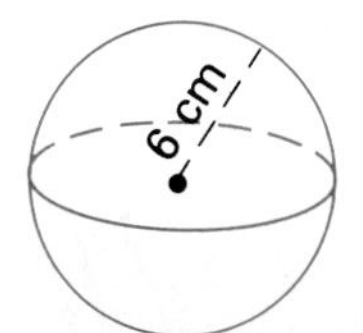

Solution

$$\begin{aligned} \text{Area} &= 4\pi r^2 \\ &= 4\pi(6^2) = 4\pi(36) \\ &= 144\pi \text{ cm}^2 \end{aligned}$$

EXAMPLE 3 Find the length of a radius of a sphere with an area of 64π cm^2.

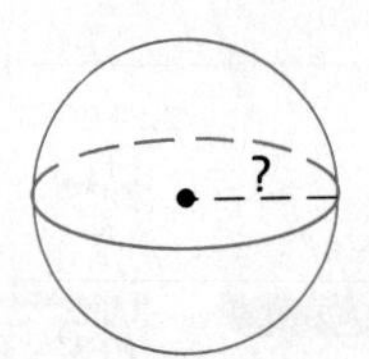

$$\begin{aligned} \text{Area} &= 4\pi r^2 \\ 64\pi \text{ cm}^2 &= 4\pi r^2 \\ r^2 &= \frac{64\pi}{4\pi} = 16 \\ r &= \sqrt{16} = 4 \text{ cm} \end{aligned}$$

Classroom Exercises

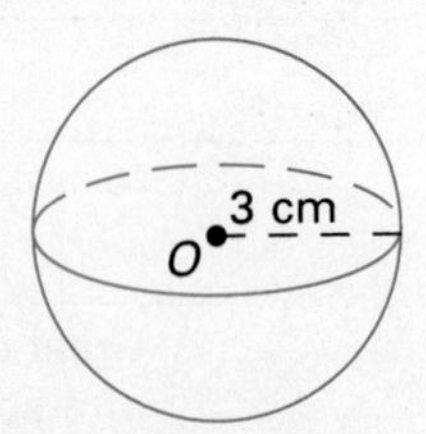

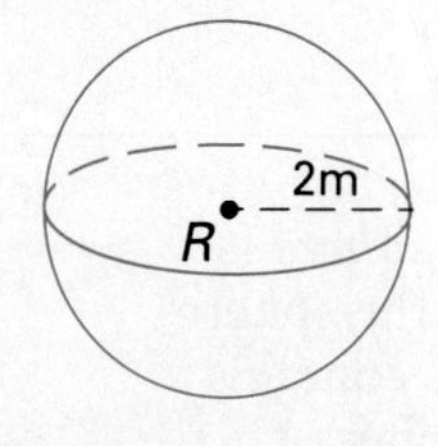

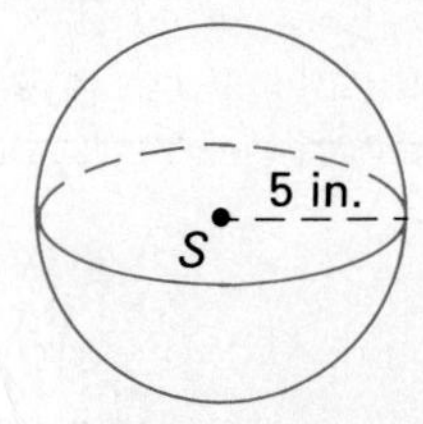

1. Find the area of Sphere *O*.
2. Find the volume of Sphere *O*.
3. Find the area of Sphere *R*.
4. Find the volume of Sphere *R*.
5. Find the area of Sphere *S*.
6. Find the volume of Sphere *S*.

Written Exercises

Find the area and volume of the sphere.

1.

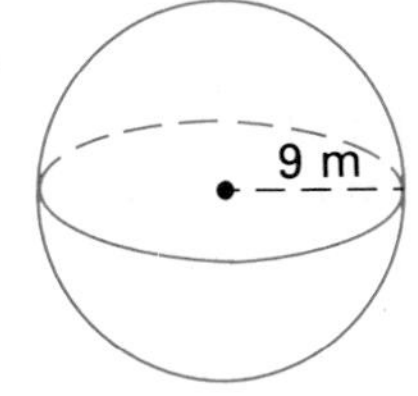

2.

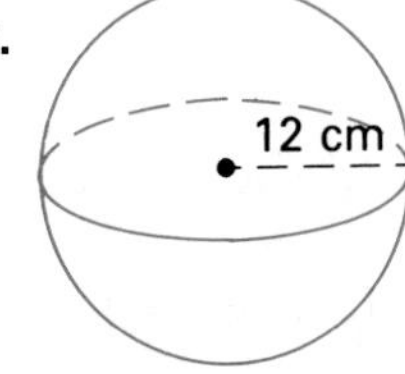

3.

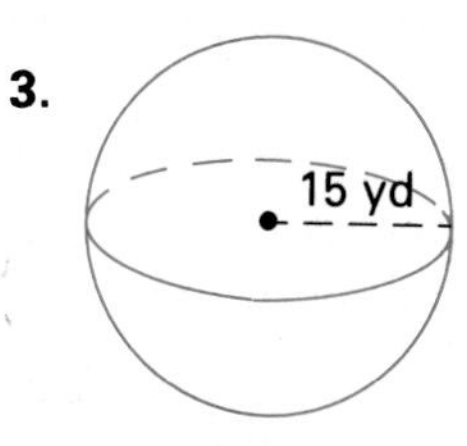

Find the area and volume of a sphere with the given radius.

4. $r = 7$ cm
5. $r = 4$ mm
6. $r = 13$ in.
7. $r = 2.3$ ft
8. $r = 1.5$ m
9. $r = 4.3$ yd

Find the area and volume of a sphere with the given diameter.

10. $d = 4$ m
11. $d = 12$ mm
12. $d = 16$ ft

Find the length of a radius of the sphere.

13. volume $= 36\pi$ cm^3
14. volume $= 324\pi$ m^3
15. area $= 100\pi$ cm^2
16. area $= 10$ mm^2
17. volume $= 64$ ft^3
18. area $= 100$ in^2

19. The radius of a sphere is 10 cm. The radius of a second sphere is 20 cm. What is the ratio of the volumes of the two spheres?

20. The radius of a sphere is 5 cm. The radius of a second sphere is 15 cm. What is the ratio of their volumes?

21. The radius of a sphere is 4 cm. The radius of a second sphere is 8 cm. What is the ratio of their areas?

22. The radius of a sphere is 2 cm. The radius of a second sphere is 6 cm. What is the ratio of their areas?

23. What is the volume of a hollow spherical shell with a 5-cm inner radius and a 6-cm outer radius?

24. What is the volume of a hollow spherical shell with a 6-cm inner radius and a 7-cm outer radius?

25. If the area of the great circle is 64π cm^2, what is the area of the sphere? What is the volume of the sphere?

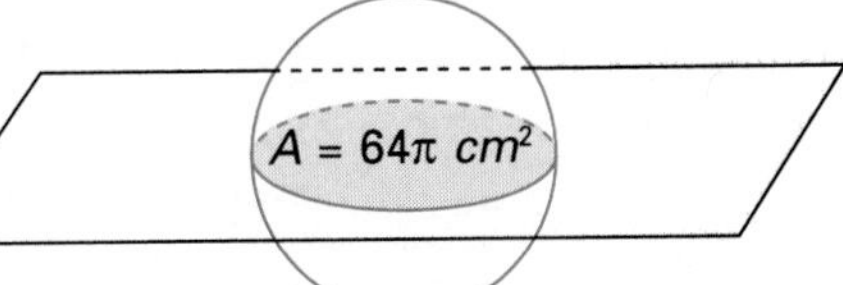

26. Find the length of a radius of a sphere whose volume in cubic units is equal to its area in square units.

27. The earth is almost spherical in shape. If the circumference of a great circle of the earth is about 40,000 km, what is the surface area of the earth?

28. The atmosphere of the earth has an altitude of about 550 km. Use Exercise 27 to find the volume of the earth and its atmosphere.

29. A steel gas tank is in the shape of a sphere. A radius to the inner surface of the tank is 2 ft long. The tank itself is made of $\frac{1}{4}$-in. thick steel. Find the difference between the outside area of the tank and the inside area.
30. If a gallon of paint will cover 400 ft^2 of surface, how many gallons of paint will be needed to paint both the inside and the outside of the tank described in Exercise 29?
31. A sphere with a radius of length r is inscribed in a cube. What is the ratio of the volumes?
32. A sphere with 13-cm radius is intersected by a plane. How far from the plane must the center of the sphere be so that the intersection is a circle with an area of 25π cm^2?

Mixed Review

Using the number line below, find the distance between the pair of points. *1.2*

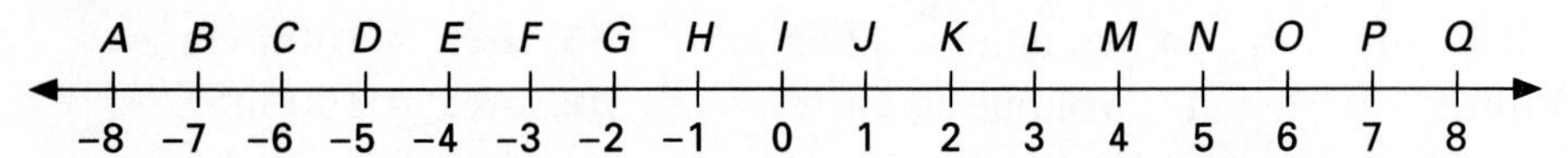

1. J and M
2. I and Q
3. A and F
4. C and H
5. State the symmetric property of congruence. *2.3*

Algebra Review

To solve inequalities of the form $ax + b < cx + d$:
(1) Combine like terms.
(2) Solve for x.

Example:

$$3x - 5 < 5x + 7$$
$$-2x < 12$$
$$x > -6$$

← Direction of inequality changes when multiplying or dividing by a negative number.

Solve the inequality.

1. $3x + 5 > 2x + 1$
2. $7x - 1 < 6x + 3$
3. $-2x - 3 > -3x - 6$
4. $5x + 4 < 2x + 1$
5. $9x + 2 < 4x - 8$
6. $-4x + 1 > -7x - 8$
7. $3x + 5 > 5x - 1$
8. $4x - 4 > 7x - 10$
9. $x - 9 < 5x - 1$
10. $3 + 2x < 4 - 3x$
11. $4 - x > x - 4$
12. $3 - 2x < 4x + 7$

14.8 Coordinates in Space

Objectives To locate points in space
To find the distance between two points in space

Two coordinates are needed to locate a point in a plane. The x-coordinate, or *abscissa*, shows the distance and direction from the y-axis to the point. The y-coordinate, or *ordinate*, shows the distance and direction from the x-axis to the point.

Coordinates can also be used to describe figures in space. To locate points in space, a third coordinate is necessary. In space, three lines can be perpendicular to each other at a given point. Three such perpendicular lines, an x-axis, a y-axis, and a z-axis, are used to set up the coordinate system.

An **ordered triple** of numbers (x,y,z) locates a point with reference to the x-, y-, and z-axes. The x-coordinate gives the distance of the point from the plane containing the y- and z-axes (usually called the yz-plane). The y-coordinate gives the distance from the xz-plane, and the z-coordinate gives the distance from the xy-plane. Positive and negative directions are indicated on the axes.

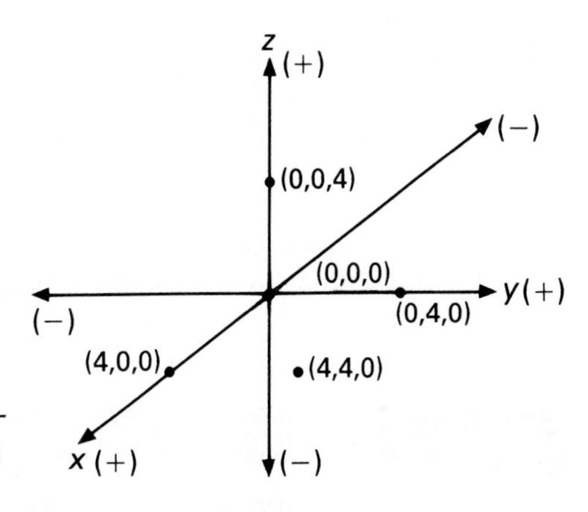

EXAMPLE 1 Graph $P(3,-2,4)$.

x-coord. forward 3   *y*-coord. left 2   *z*-coord. up 4

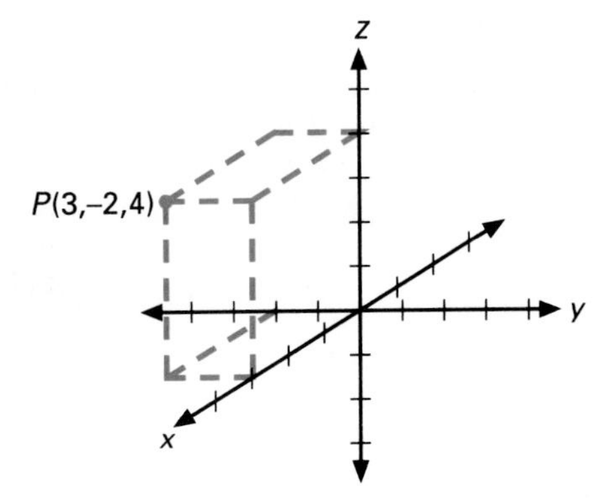

The Pythagorean Theorem gives a formula for the length of the hypotenuse of a right triangle in terms of the lengths of the two legs. The length of a diagonal of a rectangular solid can be found by applying the Pythagorean Theorem twice. Consider the diagonal of a rectangular solid with edges of lengths a, b, and c.

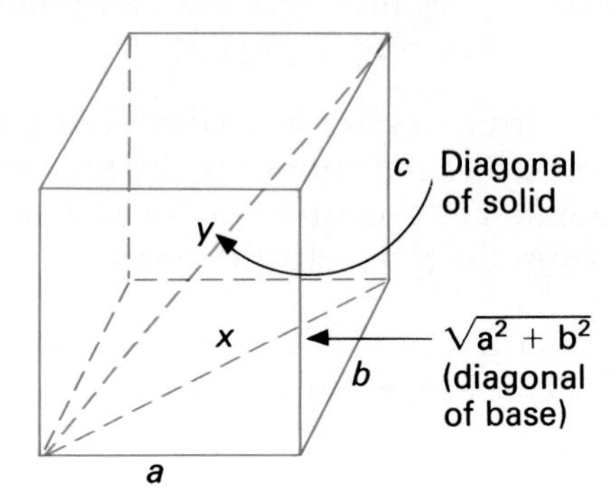

$$x^2 = a^2 + b^2$$
$$y^2 = x^2 + c^2$$
$$y^2 = a^2 + b^2 + c^2$$
$$y = \sqrt{a^2 + b^2 + c^2}$$

This extension of the Pythagorean Theorem can be used to prove a formula for the distance between points in space.

Theorem 14.12 The distance between points $P(x_1,y_1,z_1)$ and $Q(x_2,y_2,z_2)$ is given by the formula $d = \sqrt{(x_2 - x_1)^2 + (y_2 - y_1)^2 + (z_2 - z_1)^2}$.

EXAMPLE 2 Find the distance between $P(3,-2,4)$ and $Q(-3,3,1)$.

$P(3,-2,4)$ $\quad Q(-3,3,1)$

$x_1,y_1,z_1 \quad x_2,y_2,z_2$

$$d = \sqrt{(x_2 - x_1)^2 + (y_2 - y_1)^2 + (z_2 - z_1)^2}$$
$$PQ = \sqrt{(-3 - 3)^2 + (3 - (-2))^2 + (1 - 4)^2}$$
$$= \sqrt{(-6)^2 + (5)^2 + (-3)^2}$$
$$= \sqrt{36 + 25 + 9} = \sqrt{70}$$

So, $PQ = \sqrt{70}$.

Classroom Exercises

On which axis does the point lie?

1. $A(0,0,2)$
2. $B(-5,0,0)$
3. $C(0,-1,0)$

Which coordinate plane contains the point?

4. $(0,7,-3)$
5. $(1,0,-4)$
6. $(2,-5,0)$

Written Exercises

Graph the ordered triple.

1. $(0,0,0)$
2. $(3,1,4)$
3. $(0,3,4)$
4. $(-5,2,3)$
5. $(-2,-3,-4)$
6. $(0,-5,3)$

Find the distance between the pair of points.

7. $J(3,-1,2)$ and $K(0,2,-1)$
8. $L(0,3,-2)$ and $M(2,0,-1)$
9. $P(8,-2,6)$ and $Q(2,5,3)$
10. $R(2,8,6)$ and $S(4,-2,-3)$
11. $T(-2,1,-1)$ and $U(-9,6,2)$
12. $V(7,3,0)$ and $W(-4,1,6)$

Find the coordinates of the midpoint of the segment $\overline{PQ}$.

13. $P(0,0,0)$, $Q(-2,6,4)$
14. $P(-2,4,11)$, $Q(8,0,3)$
15. $P(-2,1,3)$, $Q(1,-5,4)$
16. $P(0,5,-9)$, $Q(-7,2,1)$
17. A certain sphere is the set of all points in space a given distance r from a point (h,j,k). Use the distance formula for space to write an equation of a sphere.
18. Write an equation of a sphere with center $(3,-1,2)$ and a radius of length 2.

Mixed Review

1. What is the area of a rectangle with sides 3 cm and 7 cm long? *12.2*
2. What is the area of a triangle with base 7 in. long and an altitude of 5 in.? *12.4*
3. What is the area of a trapezoid with bases 7 m and 13 m long and an altitude of 3 m? *12.5*
4. What is the distance between the points $P(2,-7)$ and $Q(-4,7)$? *10.1*
5. What is the midpoint of the segment with endpoints $A(2,-5)$ and $B(-8,7)$? *10.2*

Chapter 14 Review

Key Terms

altitude (p. 564)
area of a prism (p. 564)
base (p. 560)
cone (p. 579)
cone of revolution (p. 580)
cylinder (p. 561)
distance in space (p. 589)
edge (p. 557)
face (p. 560)
lateral face (p. 560)
lateral surface (p. 561)
oblique (p. 560)
ordered triple (p. 589)
parallelepiped (p. 560)
polyhedron (p. 556)
prism (p. 560)
pyramid (p. 574)
rectangular solid (p. 560)
regular polyhedron (p. 557)
right prism (p. 560)
rigid (p. 555)
similar polyhedra (p. 565)
slant height (p. 574)
sphere (p. 584)
three-dimensional (p. 560)
two-dimensional (p. 560)
vertices (p. 557)
volume (p. 569)

Key Ideas and Review Exercises

14.1 To show that a polyhedron is rigid, show that all faces are triangular.

To show that a polyhedron is not rigid, show a counterexample.

1. Use sketches to show that a square is not rigid.

14.2 A prism has lateral faces that are parallelograms.

To draw a figure that is a prism or a cylinder, draw congruent parallel bases. For a rigid prism or cylinder, the corresponding sides should be parallel and vertically aligned.

2. Draw a right square prism.

3. Draw an oblique cylinder.

How many faces, edges, and vertices does the prism have?

4. a triangular prism

5. a pentagonal prism

14.3 In the following formulas, p is the perimeter of the base h is the length of the altitude, and r is the radius of the base.

To find the lateral area of a right prism, use the formula $L = ph$.

To find the lateral area of a right cylinder, use the formula $L = 2\pi rh$.

To find the total area of a right prism or cylinder, add the lateral area to the sum of the areas of the bases.

6. Find the lateral area of a right-square prism with a 12-cm altitude and 5-cm base edges.

7. Find the total area of a right cylinder with a 4-cm altitude and a 5-cm base radius.

14.4 To find the volume of a prism or a cylinder, use the formula $V = Bh$.

8. Find the volume of a prism with a base area of 8 cm^2 and a 5-cm altitude.

9. Find the volume of a right cylinder with a 3-m base radius and a 8-m altitude.

10. Find the height of a right cylinder with a 4-m radius and a volume of 48π m^3.

14.5 In the following formulas, p is the perimeter of the base, l is the slant height, and r is the radius of the base.

To find the lateral area of a regular pyramid, use the formula $L = \frac{1}{2}pl$.

To find the lateral area of a right cone, use the formula $L = \pi rl$.

To find the total area of a pyramid or cone, add the lateral area and the base area.

11. Find the lateral area of a regular pyramid with a 20-cm base perimeter and a 3-cm slant height.

12. Find the lateral area of a right cone with a 6-cm base radius and a 9-cm slant height.

13. Find the total area of a regular pyramid with a 3-mm base edge and an 8-mm slant height.

14. Find the total area of a right cone with a 5-mm base radius and a 7-mm slant height.

14.6 To find the volume of a pyramid or cone, use the formula $V = \frac{1}{3}Bh$.

15. Find the volume of a pyramid with a 25-cm^2 base area and a 9-cm altitude.

16. Find the volume of a cone with a 3-cm base radius and a 5-cm altitude.

17. Find the radius of a cone with height of 10 in. and a volume of 30π in^3.

14.7 To find the volume of a sphere, use the formula $V = \frac{4}{3}\pi r^3$.

To find the area of a sphere, use the formula $A = 4\pi r^2$.

18. Find the volume and area of a sphere with a 6-mm radius.

19. Find the radius of a sphere with a volume of 972π cm^3.

14.8 To locate a point in space, three coordinates (x,y,z) are needed.

To find the distance between two points in space, use the formula $d = \sqrt{(x_2 - x_1)^2 + (y_2 - y_1)^2 + (z_2 - z_1)^2}$.

20. Draw a set of x-, y-, and z-axes. Draw the point $(3,-1,4)$.

21. Find the distance between $A(3,-5,2)$ and $B(-5,3,4)$.

Chapter 14 Test

Sketch the solid. Tell how many faces, edges, and vertices it has.

1. a cube

2. a regular tetrahedron

3. a regular quadrilateral pyramid

Find the total area of the solid. State the units of measure.

4. rectangular solid

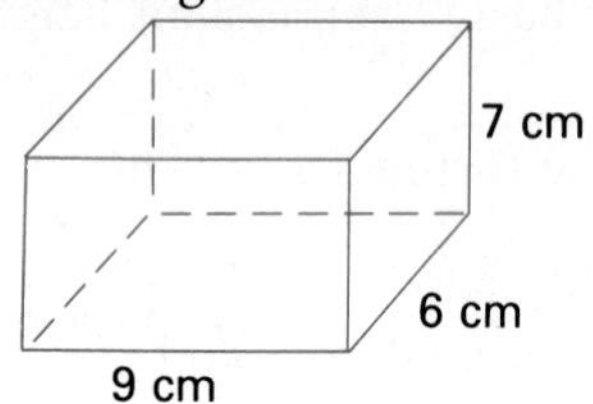

5. sphere

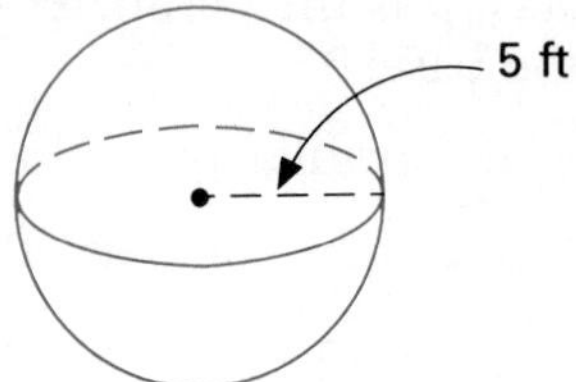

6. regular square pyramid

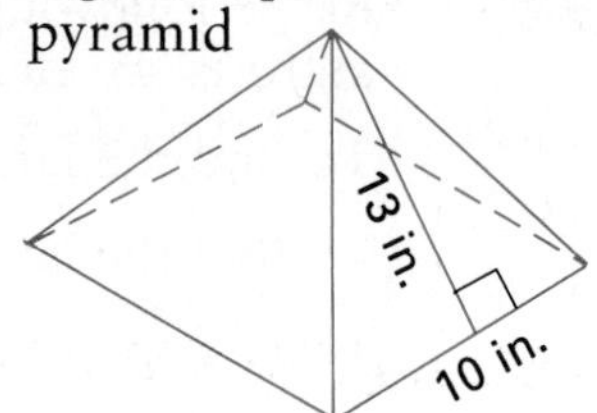

Find the lateral area of the solid. State the units of measure.

7. right triangular prism

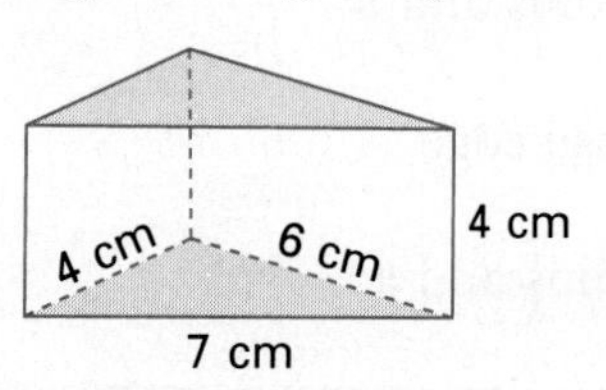

8. right cone

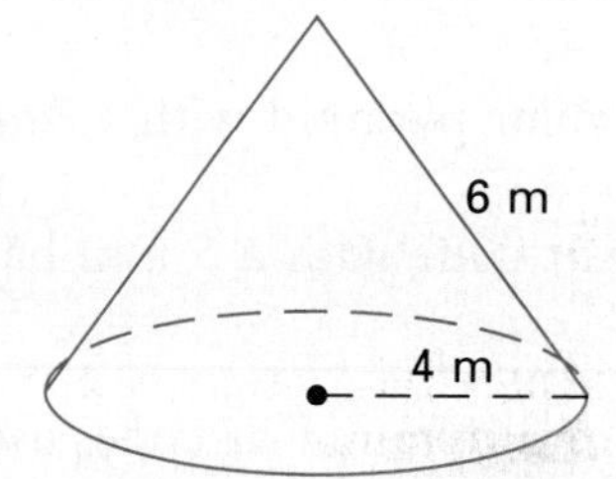

9. right cylinder

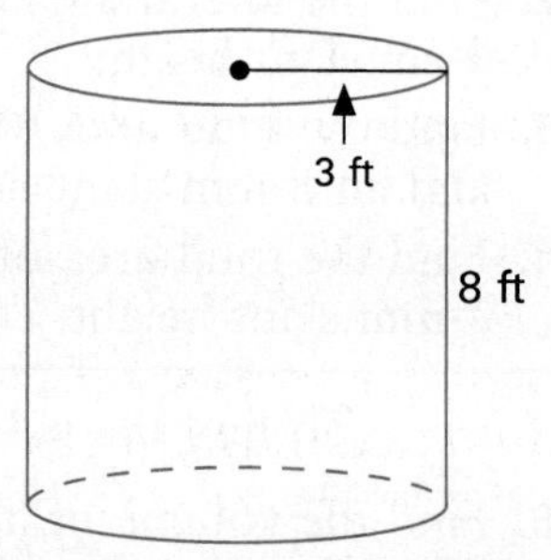

Find the volume of the solid.

10. rectangular solid

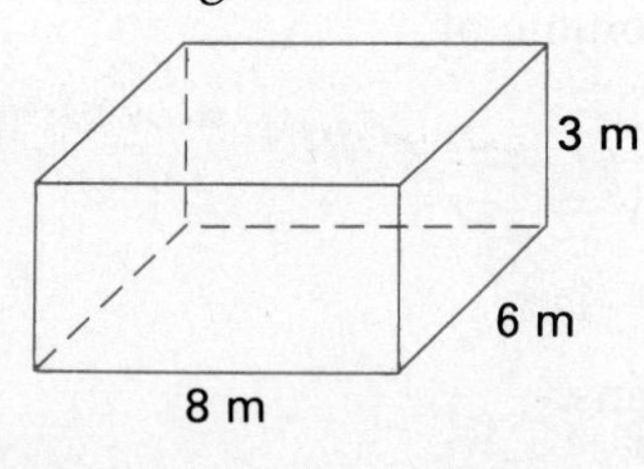

11. cone

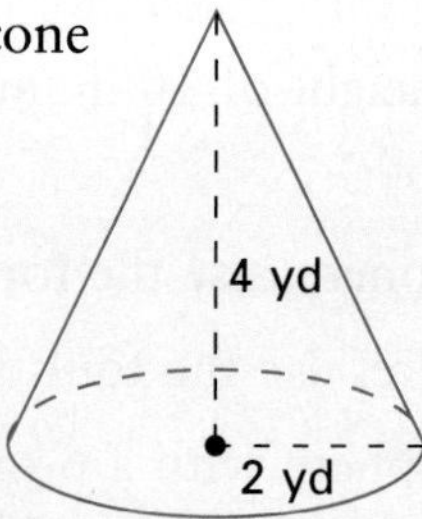

12. sphere

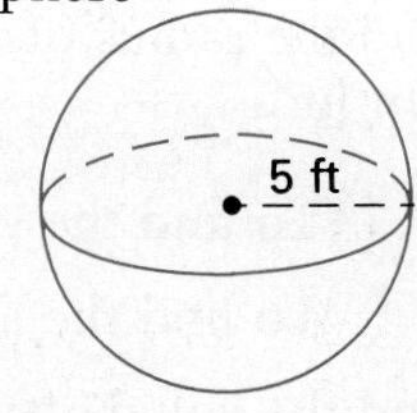

***13.** Use a sketch to illustrate Cavalieri's Principle. In your own words, write a brief, clear explanation of the principle.

***14.** The volume of a cube is 64π cm^3. Find its area.

15. If the lengths of the altitude and the radius of a right cylinder are multiplied by 5, what is the ratio of the new volume to the original?

16. Find the distance between $P(-1,3,5)$ and $Q(4,2,-1)$.

College Prep Test

In each item compare a quantity in Column 1 with a quantity in Column 2. Write the letter of the correct answer from the following choices.

A—The quantity in Column 1 is greater than the quantity in Column 2.
B—The quantity in Column 2 is greater than the quantity in Column 1.
C—The quantity in Column 1 is equal to the quantity in Column 2.
D—The relationship cannot be determined from the given information.

	Column 1	Column 2
1.	Volume of Cylinder A	Volume of Cylinder B
2.	Total area of A	Total area of B
3.	Number of edges in a triangular prism	Number of vertices in a hexagonal prism
4.	Volume of above prism	Volume of above prism
5.	Volume of cylinder	Volume of sphere
6.	Lateral area of cylinder	Area of sphere
7. $BC = 25$, $DC = 10$	Area of $ABCD$	Area of $\odot O$
8.	Area of $ABCD$	Area of $\odot O$

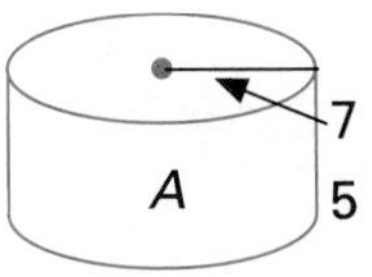

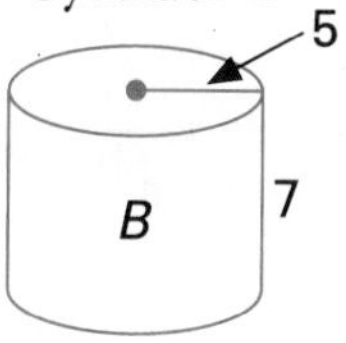

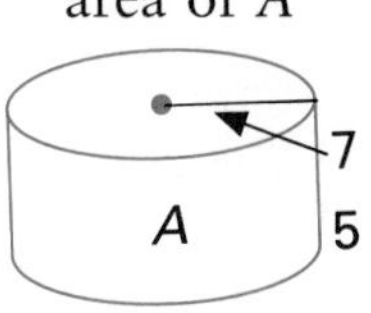

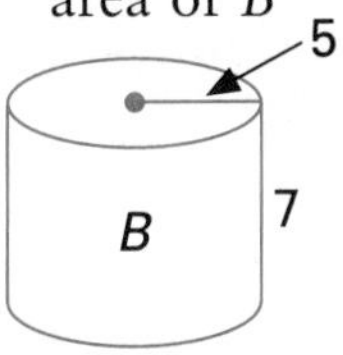

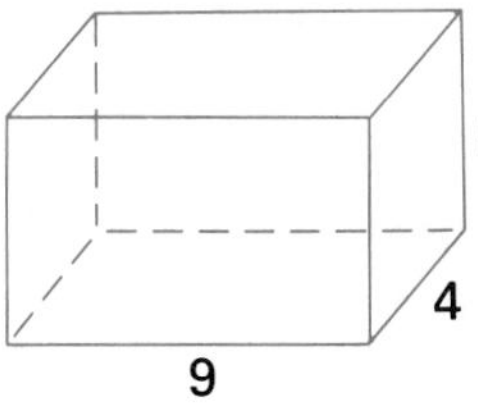

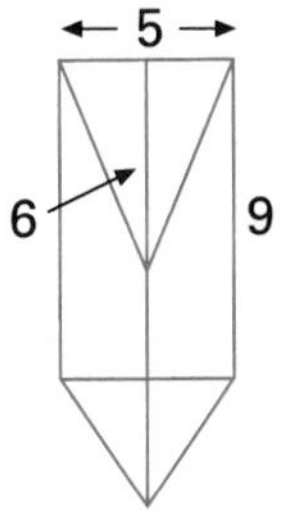

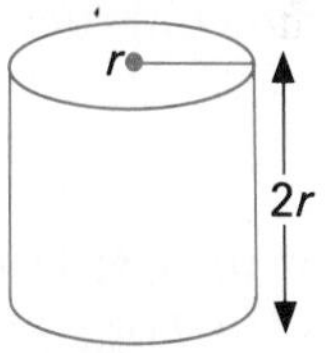

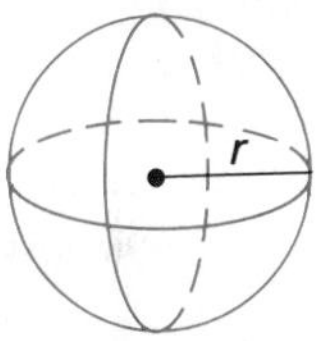

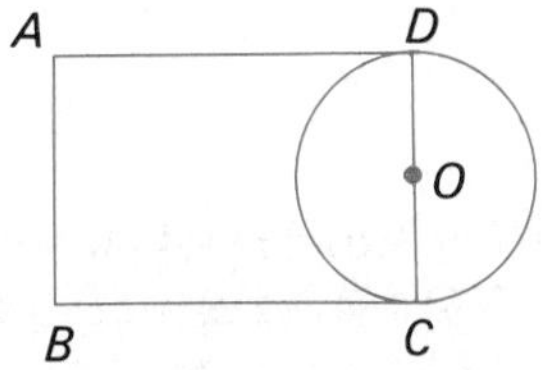

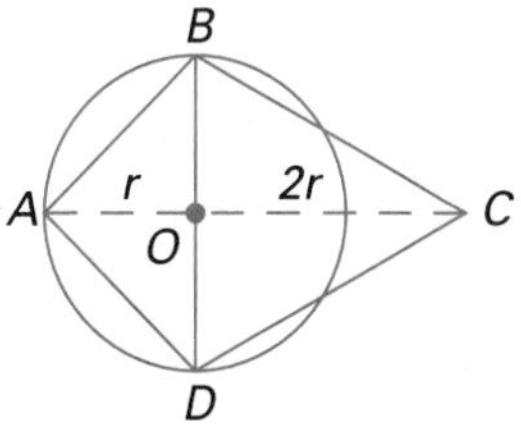

Cumulative Review (Chapters 1–14)

1. Find the measure of a supplement of an angle with a measure of 116. **1.6**

2. The measure of an angle is $3x + 40$. The measure of its vertical angle is $2x + 50$. What is the measure of the angle? **2.7**

3. Write the inverse of the statement: If two angles are congruent, then their supplements are congruent. **3.8**

4. Given: $\triangle ABC \cong \triangle DEF$, G is the midpoint of $\overline{AB}$; H is the midpoint of $\overline{DE}$. **5.6**
   Prove: $\overline{CG} \cong \overline{FH}$

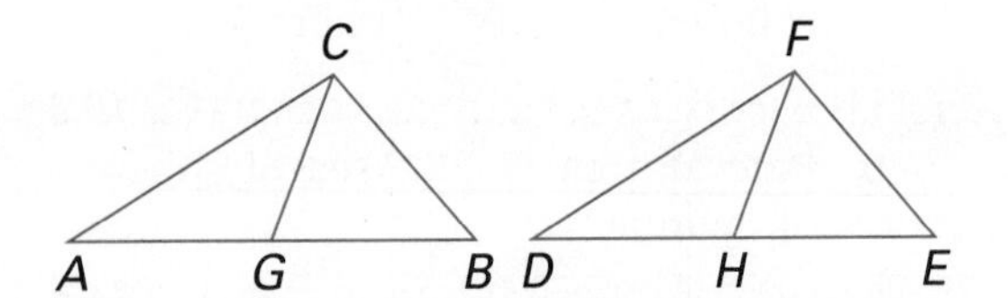

5. Construct a triangle with sides congruent to the three segments below. **5.8**

6. True or false? If $\angle A$ of $\triangle ABC$ is an obtuse angle, then $BC > AC$. **5.8**

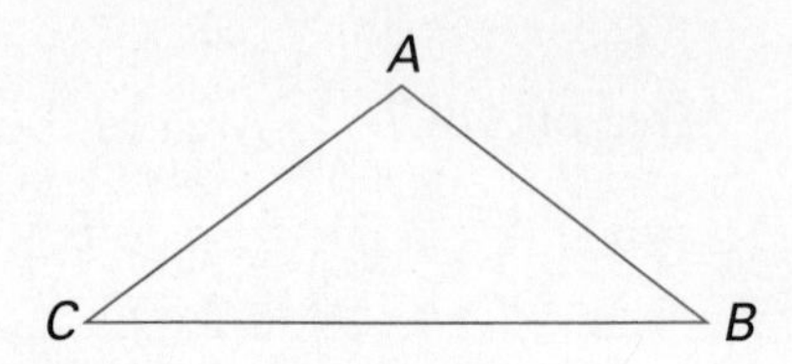

7. True or false? The property that *consecutive angles of a quadrilateral are supplementary* is a sufficient condition for the quadrilateral to be a parallelogram. **6.5**

8. Given: Trap $ABCD$ with median $\overline{EF}$, $AB = 26$, $DC = 13$ **7.4**
   Find EF.

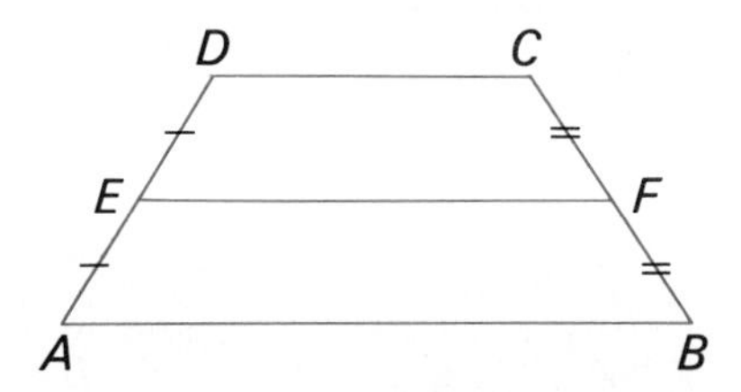

9. Given: Trap $ABCD$, $\overline{AB} \parallel \overline{DC}$, $\overline{AD} \cong \overline{BC}$, altitudes $\overline{DE}$ and $\overline{CF}$ **7.5**
   Prove: $\overline{DE} \cong \overline{CF}$

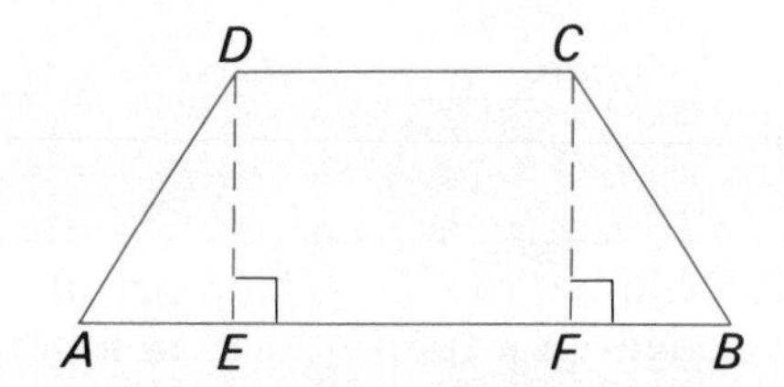

10. Copy the segment $\overline{AB}$ below. Construct the point P such that $\frac{AP}{PB} = \frac{3}{5}$. **8.5**

A B

11. If the lengths of the sides of a triangle are 9, 40, and 41, is the triangle a right triangle? **9.2**

12. Find the length of an altitude of an equilateral triangle if the length of a side is 4. **9.4**

13. Given: Right triangle ABC, $m\angle A = 38$ and $AB = 10$. Find AC. *9.5*

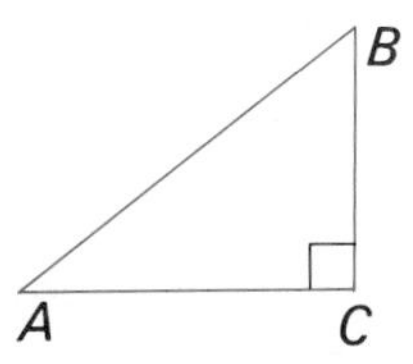

14. Prove using coordinate geometry that the diagonals of a rectangle bisect each other. *10.6*

15. A chord of a circle is 2 cm long and is 2 cm from the center of the circle. What is the length of a radius of the circle? *11.2*

16. Find the measure of an inscribed angle of a circle if the angle intercepts an arc of measure 42. *11.6*

17. What is the area of triangle ABC below? *12.4*

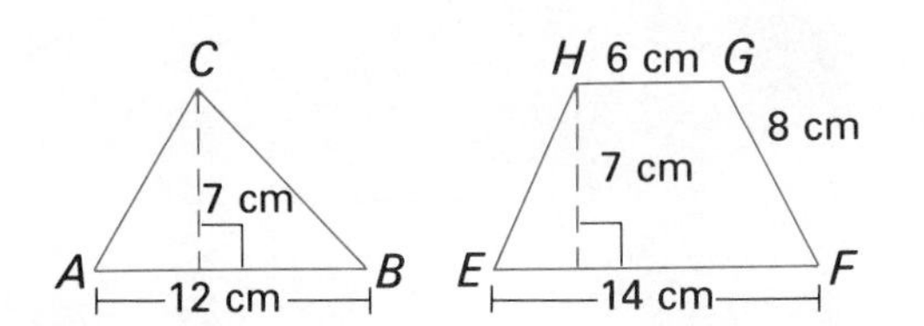

18. Given: Trapezoid $EFGH$ with $\overline{EF} \parallel \overline{HG}$. Find the area of $EFGH$. *12.5*

19. Sketch and describe the locus of points in a plane 3 cm from a circle with a 7-cm radius. *13.2*

20. Find the total area of the cylinder. *14.3*

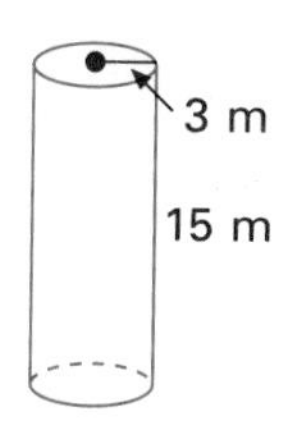

21. Find an equation in standard form of a line that passes through $(4,-2)$ and is perpendicular to the line $3x + 2y = 6$. *10.5*

22. Find the area of $\square ABCD$ if $m\angle A = 45$, $AB = 12$, and $AD = 8$. *12.3*

23. Tangents $\overline{PA}$ and $\overline{PB}$ to a circle from external point P form an angle measuring 70. What is the measure of the minor arc $\widehat{AB}$? *11.7*

24. If secant $PC = 9$ cm and chord $CB = 5$ cm, how long is tangent $\overline{PA}$? *11.9*

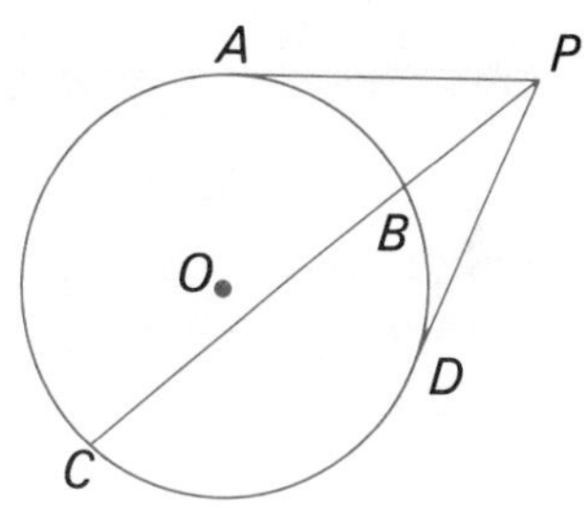

25. The apothem of a regular hexagon is $6\sqrt{3}$ cm. What is its area? *12.6*

26. A regular square pyramid has a base 6 cm on a side and an altitude of 4 cm. What is its total area? *14.5*

27. Find the diagonal of a room that is 16 ft $\times$ 12 ft $\times$ 8 ft. *14.7*

28. Two circles with radii 8 cm and 3 cm have their centers 13 cm apart. How long is their common external tangent? *11.3*

29. Find the area of a segment that intercepts an arc of measure 60 in a circle with a radius of 12 in. *12.9*

30. A balloon is inflated to a radius of 12 in. If more air is added to increase the radius by 3 in., what will be the change in volume? *14.7*

31. Describe the triangle formed by $A(-6,5)$, $B(-2,-3)$, $C(6,1)$. *10.5*

15 TRANSFORMATIONS

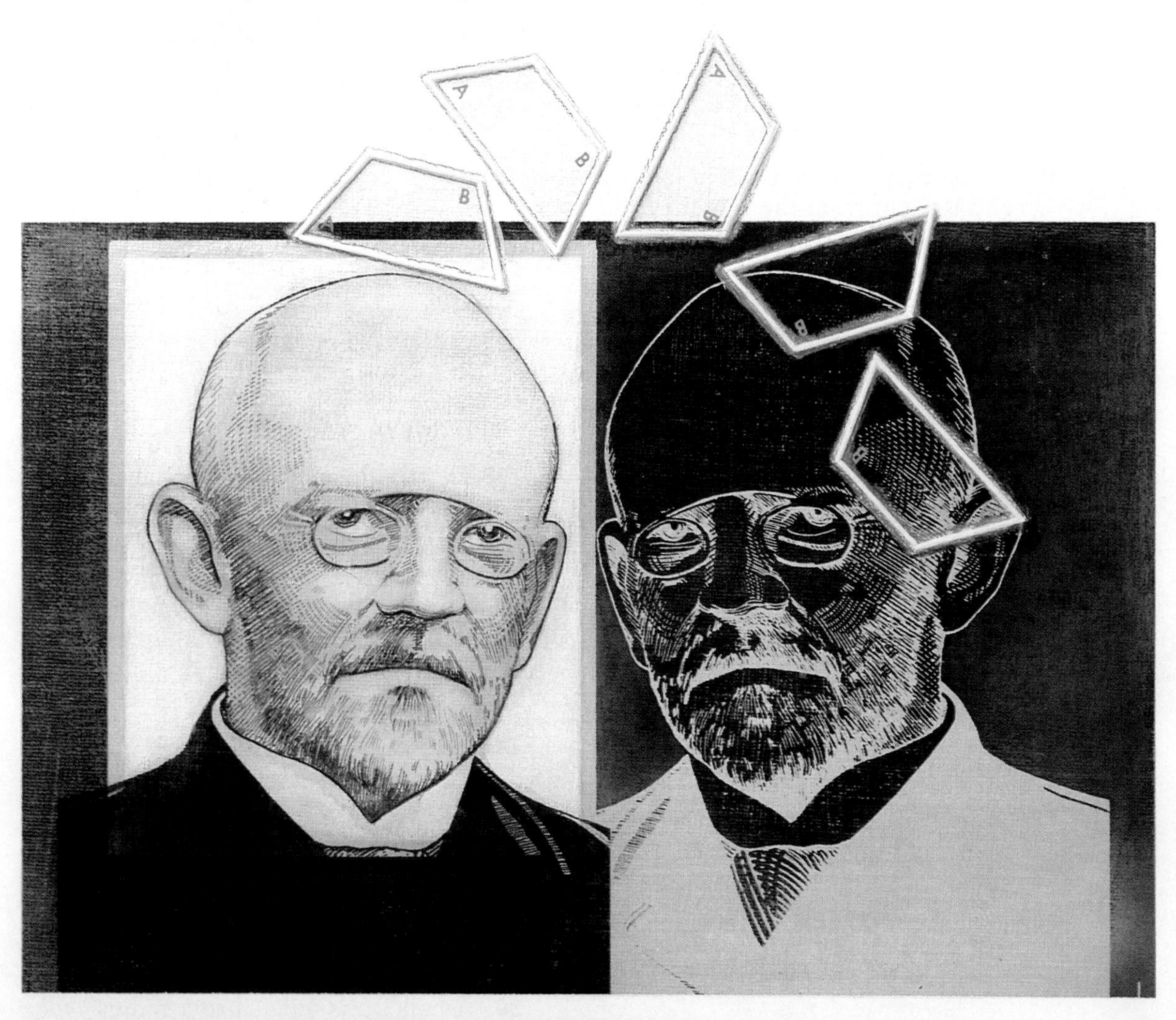

David Hilbert (1862–1943)—Hilbert studied mathematical properties called invariants that stay the same when a geometric figure is rotated, stretched, and reflected. When a segment is stretched, its length varies, but the "betweenness" of its points stays the same.

15.1 Reflections

Objectives

To find reflections of figures
To locate and construct lines of symmetry
To determine whether or not figures are reflections of each other

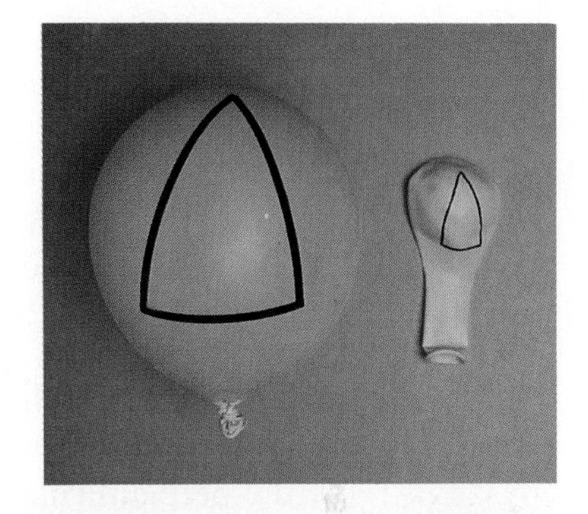

If you were to draw a triangle on a balloon, blowing up the balloon would stretch the figure into a new shape. However, each point of the new figure, or *image*, would still correspond to exactly one point of the original triangle. Such a matching of points is a **geometric transformation.**

One kind of transformation is a *reflection*.

Definition

In a plane, a **reflection** about line l is a transformation which maps each point P into a point P' such that (1) if P is on l, then $P' = P$; and (2) if P is not on l, then l is the perpendicular bisector of $\overline{PP'}$.

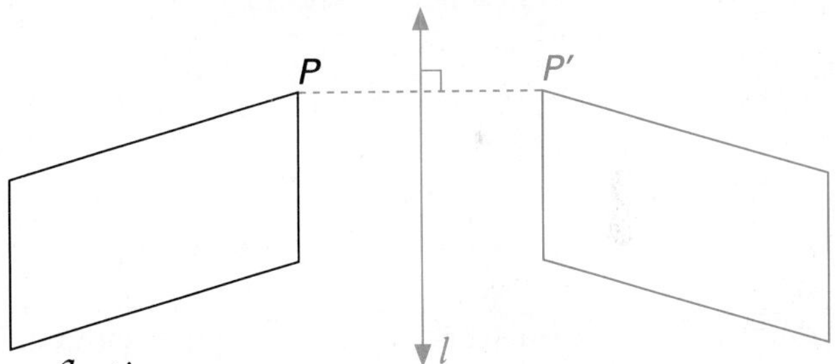

In this book, the term *reflection* will also be used to refer to the reflection image itself.

When you see a reflection of yourself in a mirror, your image seems to be the same size as your body. One way to show that the size of the image is the same size as the original figure is to use a transparent mirror. Using this, you can see both the original figure and its mirror image at the same time. It is possible to reach behind the mirror and draw the image that you see. The image drawn will be the same size as the original figure.

A transformation is distance-preserving if and only if the distance between two points is the same as the distance between the corresponding images of these points. Such a transformation is called an **isometry.**

Theorem 15.1 A reflection is an isometry.

Given: Points A and B with reflections A' and B', respectively, about l
Prove: $AB = A'B'$

Plan Show that $\overline{AB}$ and $\overline{A'B'}$ are corresponding sides of congruent quadrilaterals.

Proof Since A' is a reflection of A about l and B' is a reflection of B about l, then l is the perpendicular bisector of $\overline{AA'}$ and $\overline{BB'}$. This means that $AP = A'P$, $BQ = B'Q$, and that $\angle 1$, $\angle 2$, $\angle 3$, and $\angle 4$ are congruent right angles. $PQ = PQ$. Therefore, quad $APQB \cong$ quad $A'PQB'$ by SASAS, and $AB = A'B'$ for any two points A and B of the figure.

If a property is preserved, it is called an **invariant condition.** Therefore, congruence of segments, or distance, is an invariant condition under a reflection.

Using the definition of a reflection, you can construct a figure which is the reflection of a given figure about a given line.

EXAMPLE 1 Construct a figure which is a reflection about l of $\triangle ABC$.

Plan Construct a perpendicular segment from each vertex to l such that l bisects the segment.

Solution

(1) Swing an arc with center A, intersecting l at X and Y.
(2) Swing arcs, with the same radius as the first arc, with centers X and Y, intersecting at A' on the opposite side of l.
(3) Repeat this for points B and C.
(4) Draw $\overline{A'B'}$, $\overline{B'C'}$, and $\overline{C'A'}$.
(5) $\triangle A'B'C'$ is the reflection of $\triangle ABC$ about the line l.

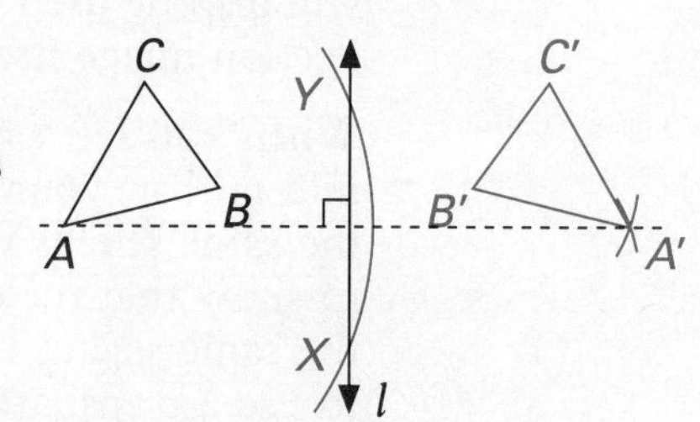

The reflection or mirror line is called the **line of symmetry.**

EXAMPLE 2 Find the line of symmetry of the figure $ABCD$ and its reflection $A'B'C'D'$.

Plan Construct the perpendicular bisector to the segment joining a point and its image.

Solution

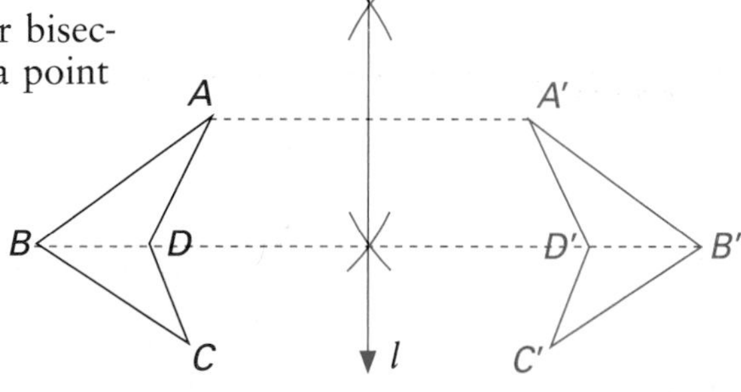

(1) Draw $\overline{AA'}$. Using compass and straightedge, construct its perpendicular bisector l. l is the line of symmetry.

(2) Verify by drawing $\overline{BB'}$ and constructing its perpendicular bisector. The two perpendicular bisectors should be the same line.

EXAMPLE 3 Determine whether or not the given pair of figures are reflections. If they are, then find the line of symmetry.

a.

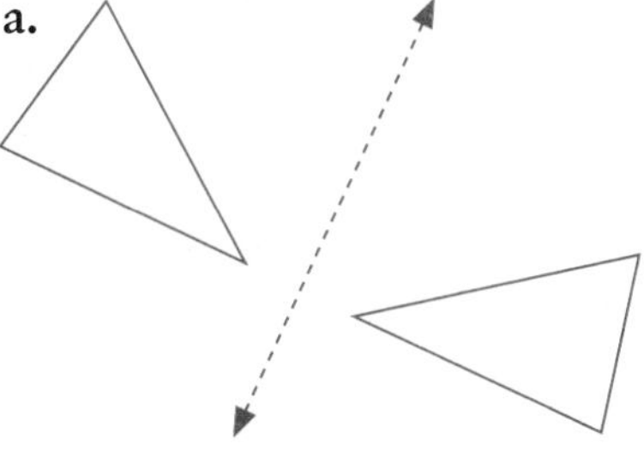

b.

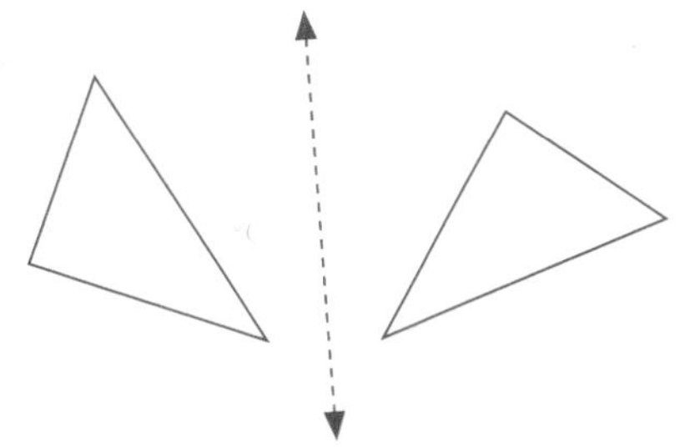

Plan

(1) Trace the pair of figures.

(2) Fold the paper so that the two figures come together.

(3) If the figures coincide, then the crease in the paper is the line of symmetry of the two figures.

Solutions

a. The figures coincide. They are reflections of each other, and the crease is the line of symmetry.

b. The figures do *not* coincide. They are not reflections of each other.

Some figures are their own mirror images about some line. The pentagon is its own image about l. A hexagon is its own mirror image about any diagonal and about each line connecting the midpoints of opposite sides. Such figures are called **symmetric figures.** A symmetric polygon has *line symmetry.* A symmetric polyhedron has *plane symmetry.*

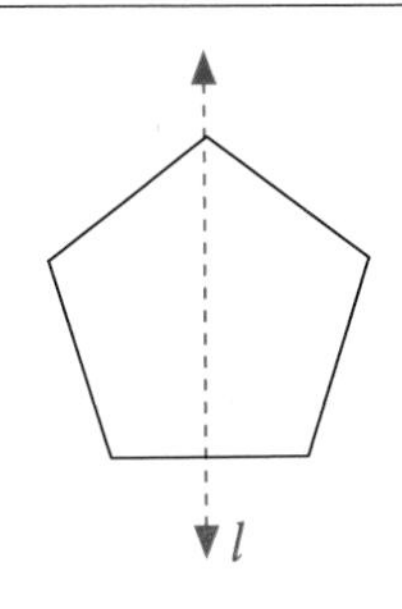

EXAMPLE 4 How many lines or planes of symmetry does each figure have?

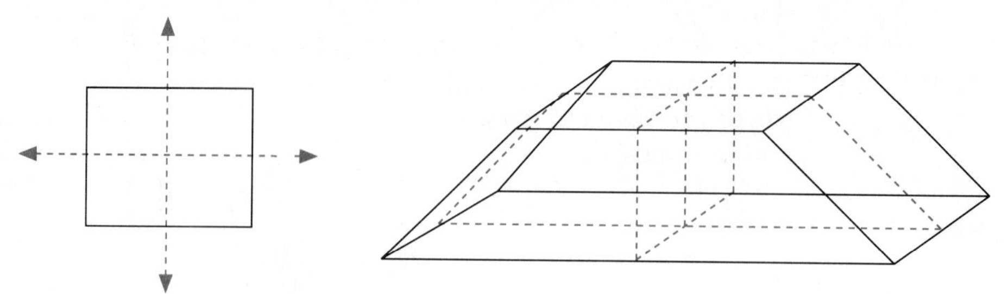

The rectangle has two lines of symmetry.

This polyhedron has two planes of symmetry.

Classroom Exercises

Do the pairs of figures below appear to be reflections about the given line?

1.

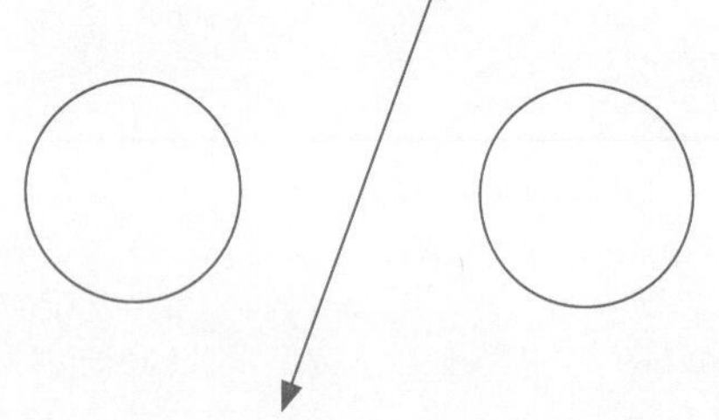

2.

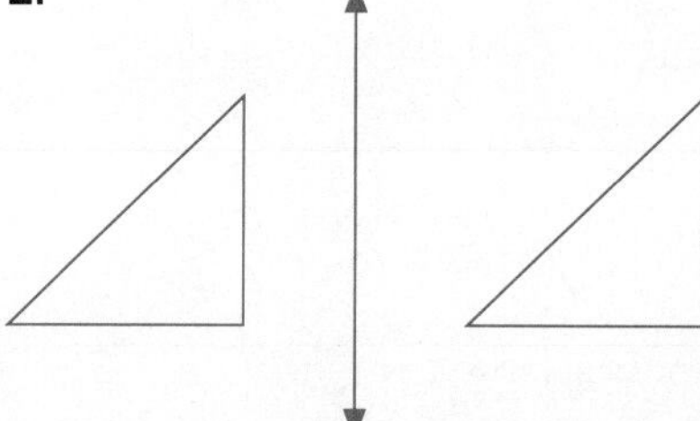

Which of the figures below have at least one line of symmetry?

3.

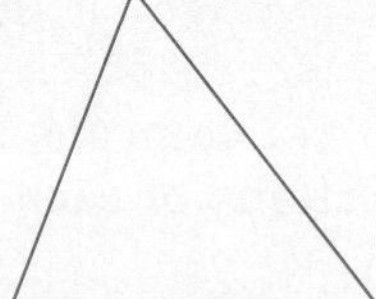

4.

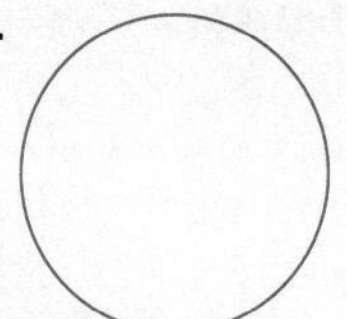

5.

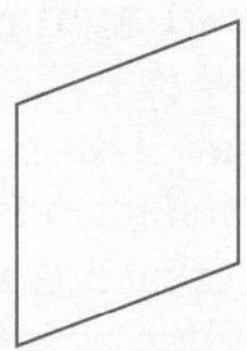

Which of the polyhedra below appear to have planes of symmetry?

6. tetrahedron **7.** rectangular solid **8.** volleyball

Tell whether the indicated property is an invariant condition under a reflection.

9. length of a side **10.** angle measure **11.** parallel sides **12.** position

Written Exercises

Copy each given figure and line l. Construct its reflection using the method shown in Example 1.

1.

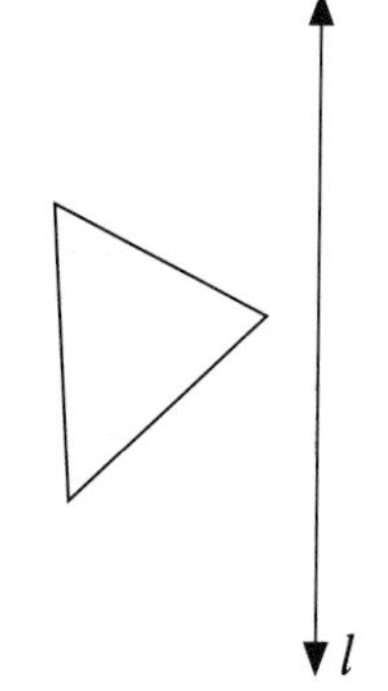

2.

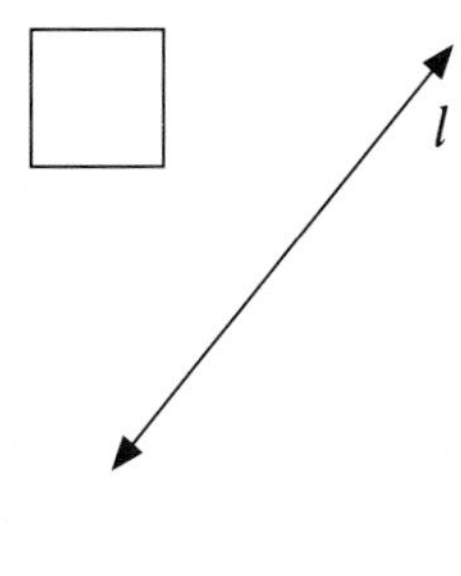

3.

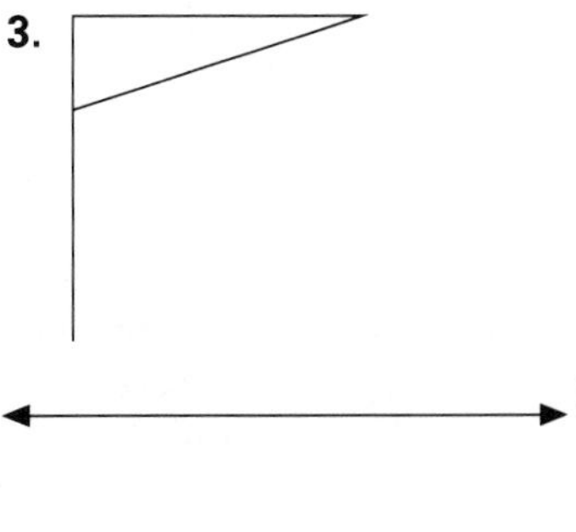

Copy each figure and its reflection. Find the line of symmetry using geometric constructions.

4.

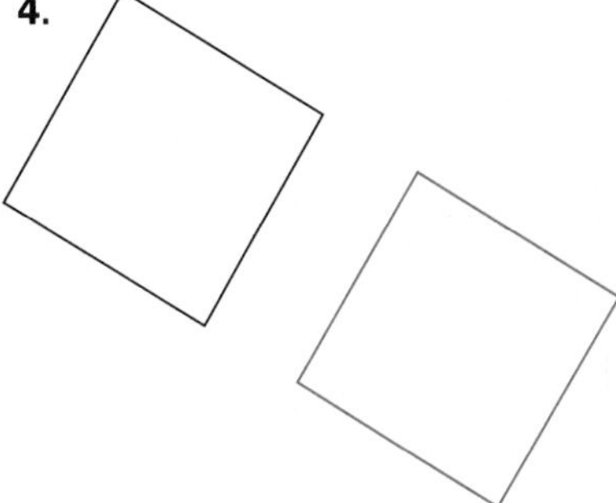

5.

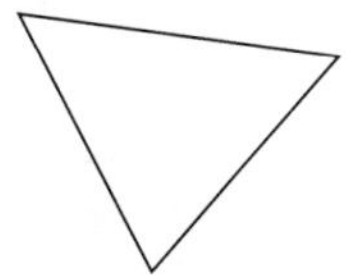

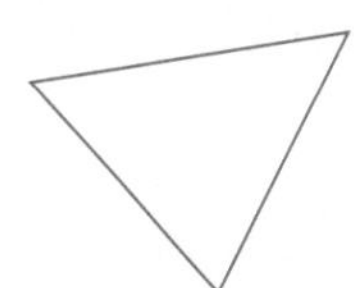

Copy each pair of figures. Determine which figures are reflections of each other.

6.

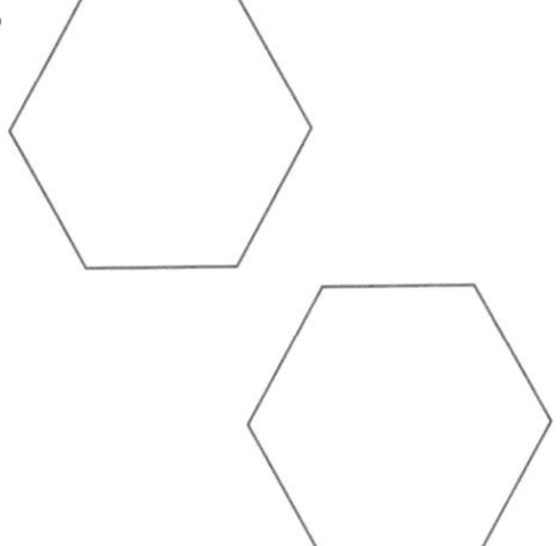

7.

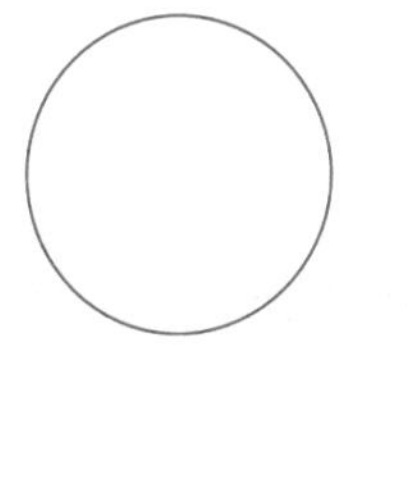

Copy each figure. Construct all lines of symmetry for each figure. How many lines of symmetry are there?

8.

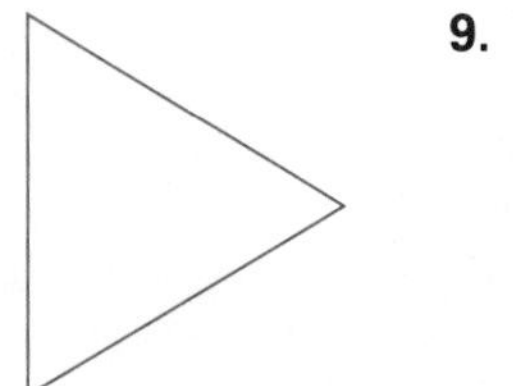

9.

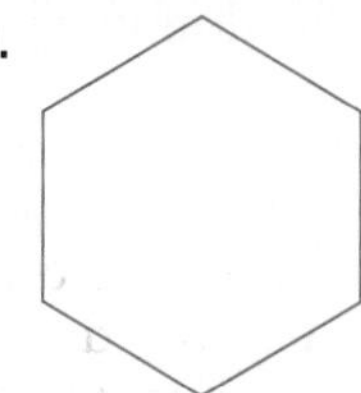

10.

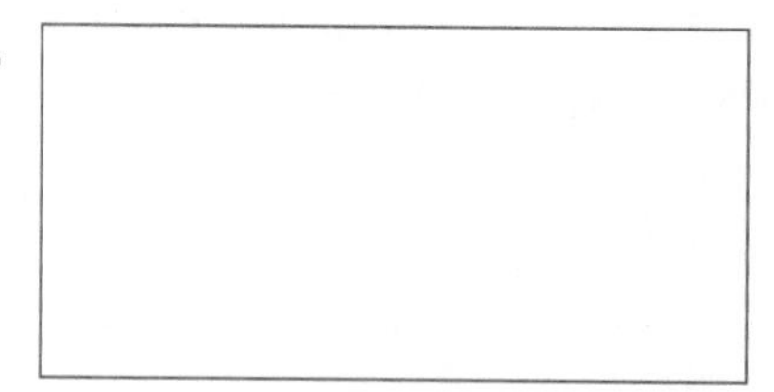

Copy each figure. Construct all planes of symmetry.

11.

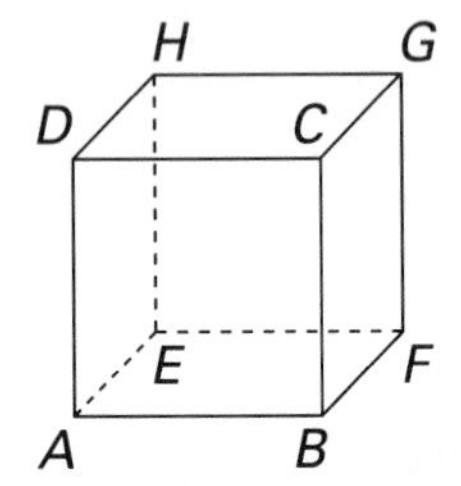

12.

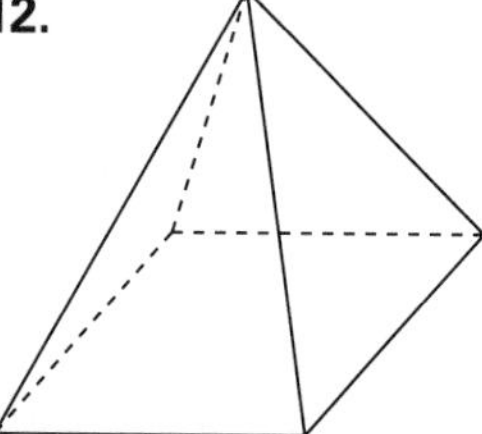

13. Prove Theorem 15.1 for the case where A lies on line l and B does not lie on line l.

14. Prove Theorem 15.1 for the case where A and B both lie on line l.

15. Prove that a triangle and its reflection are congruent.

16. Explain why area is an invariant condition under reflections.

17. Justify the construction used in Example 1 by proving that l is the perpendicular bisector of $\overline{AA'}$.

18. Will two congruent circles always be reflections of each other? Why or why not?

19. Will two congruent squares always be reflections of each other? Why or why not?

20. Name some man-made products that do not have line or plane symmetry.

Mixed Review

Given: $\overline{AB} \cong \overline{DC}$, $\overline{HG} \cong \overline{EF}$, $\overline{HG} \perp \overline{GB}$, $\overline{EF} \perp \overline{FC}$, $\overline{AB} \perp \overline{GB}$, $\overline{DC} \perp \overline{FC}$, $\overline{AG} \cong \overline{DF}$

Prove each of the following. *4.3, 5.4*

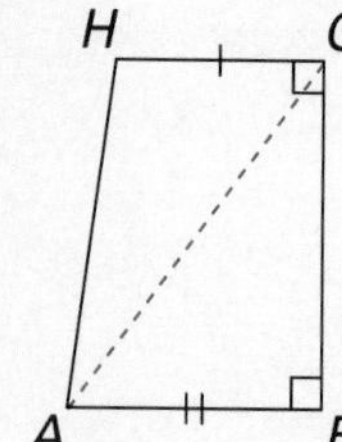

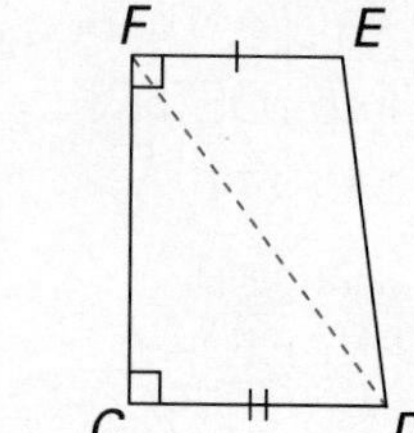

1. $\triangle ABG \cong \triangle DCF$

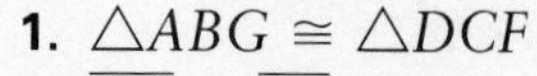

2. $\overline{BG} \cong \overline{CF}$
3. $\angle AGH \cong \angle DFC$
4. $\triangle AGH \cong \triangle DFE$
5. $\overline{AH} \cong \overline{DE}$

Computer Investigation

Reflections

Use a computer software program that draws random triangles and quadrilaterals by classification (acute, obtuse, etc.); moves and labels points; constructs line segments between two given points; draws reflections about a given line segment; and measures line segments and angles.

The following activities make use of a computer software program to draw reflected polygons. Then the program's measuring tool can be used to verify congruence to the nearest tenth of a unit.

Activity 1

Draw and label an obtuse triangle *ABC*. Draw and label any segment $\overline{ED}$ that is in the exterior of the triangle. $\overline{ED}$ will be the axis of symmetry for drawing the reflection of $\triangle ABC$.

1. Draw the reflection of $\overline{AC}$ about $\overline{ED}$.
2. Draw the reflection of $\overline{AB}$ about $\overline{ED}$.
3. Draw the reflection of $\overline{BC}$ about $\overline{ED}$.

The new triangle is the reflection of $\triangle ABC$ about $\overline{ED}$.

4. Verify that all corresponding points are symmetric with respect to $\overline{ED}$. (Use the measuring tool to show that $\overline{ED}$ is the perpendicular bisector of a segment formed by two corresponding points of the triangles.)
5. Use the measuring tool of the software program to verify that all corresponding parts of the two triangles are indeed congruent.

Activity 2

Repeat the steps of Exercises 1–5 of Activity 1 above for each of the following new triangles.

6. acute triangle
7. right triangle
8. isosceles triangle

Activity 3

Draw and label a random quadrilateral *ABCD*. Draw and label any segment $\overline{EF}$ that is in the exterior of *ABCD*. Draw the reflection of this quadrilateral about $\overline{EF}$.

9. Verify that a pair of corresponding points are symmetric with respect to $\overline{EF}$.
10. Verify that the two quadrilaterals are congruent.

15.2 Translations

Objectives

To draw translation images of figures
To find the resultant image of two reflections

If you were to push a box along a straight path, the box in its final position would be the *slide image* of the box in its original position. Any such motion of an object along a line that does not result in the object's turning in any way is a slide, or *translation*. A translation is another kind of transformation.

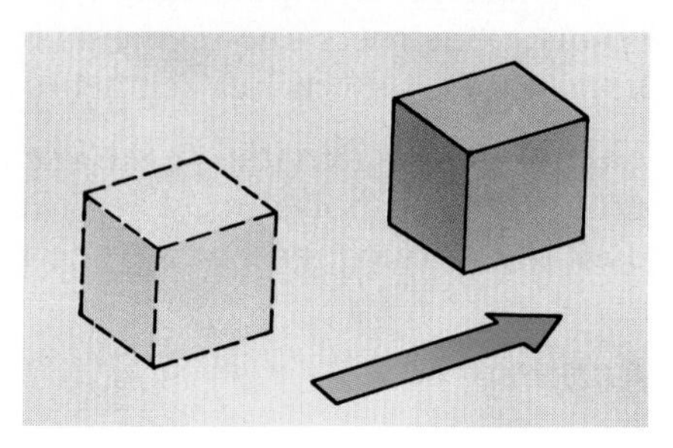

To describe a translation, you need to know which way and how far to *slide* the given figure. An informal way to show this is to use a **slide arrow** to indicate direction and distance.

EXAMPLE 1 Draw the slide image of the given figure using the given slide arrow. Do lengths of sides appear to be invariant?

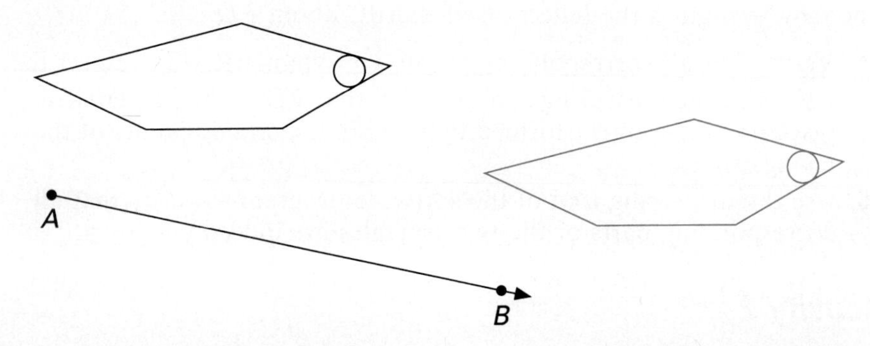

Solution

(1) Trace the given figure (black) on a sheet of paper. Draw a dot on your tracing at point A of the slide arrow.
(2) Without turning your paper, slide your drawing so that the dot moves along the arrow to point B.
(3) Notice that your traced image is now positioned directly over the slide image (red).
(4) Make a drawing of your own, with a given image and slide arrow of your choice. Repeat steps (1) and (2).
(5) Using carbon paper, or other blackened material, transfer your traced image in its new position onto your original drawing. The resulting image is the slide image of the original image with respect to the slide arrow.

To construct the image of $ABCD$ using the slide arrow $\overrightarrow{OP}$, construct a line through each vertex of $ABCD$ parallel to $\overrightarrow{OP}$. Construct $\overline{AA'} \cong \overline{OP}$, $\overline{BB'} \cong \overline{OP}$, $\overline{CC'} \cong \overline{OP}$, and $\overline{DD'} \cong \overline{OP}$. Draw $\overline{A'B'}$, $\overline{B'C'}$, $\overline{C'D'}$, and $\overline{D'A'}$. $A'B'C'D'$ is the image of $ABCD$ with respect to the given slide arrow.

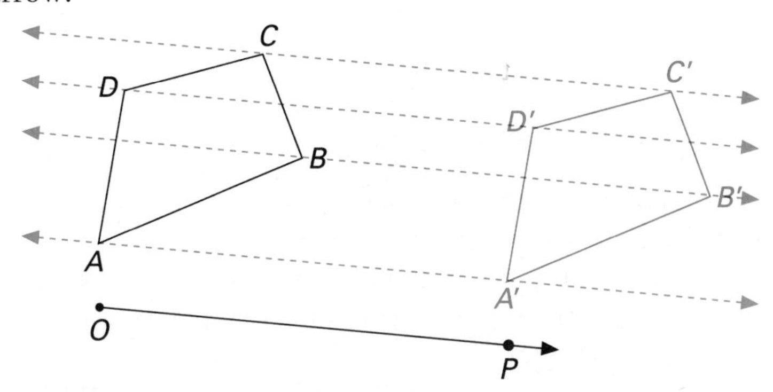

Definition

A ***translation*** in a plane from A to A' is a transformation which maps any point P into a point P' such that $\overline{PP'} \cong \overline{AA'}$ and $\overline{PP'} \parallel \overline{AA'}$.

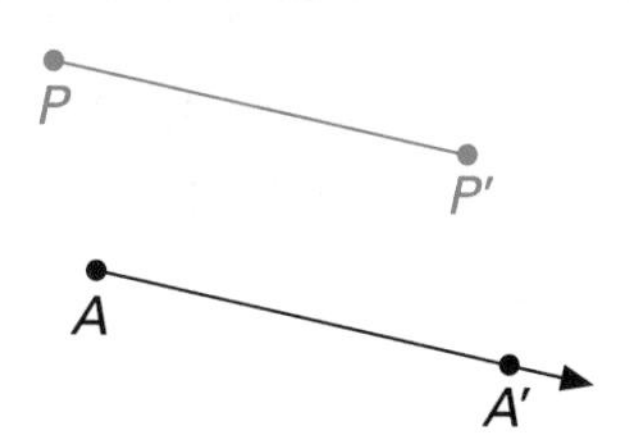

Theorem 15.2

A translation is an isometry.

Given: Points A and B, with translation images A' and B' from O to P
Prove: $AB = A'B'$

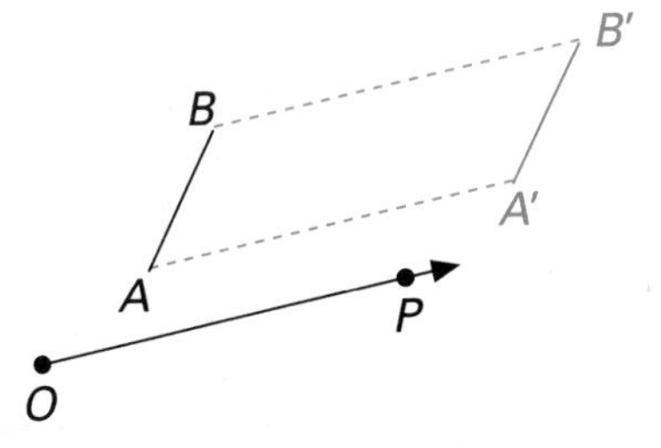

Plan

Show that $\overline{AB}$ and $\overline{A'B'}$ are opposite sides of a parallelogram.

Proof (Paragraph Form)

Since A' is the image of A, and B' is the image of B under the translation from O to P, then $\overline{AA'} \cong \overline{OP}$, $\overline{BB'} \cong \overline{OP}$, $\overline{AA'} \parallel \overline{OP}$, and $\overline{BB'} \parallel \overline{OP}$. Then by transitivity, $\overline{AA'} \cong \overline{BB'}$ and $\overline{AA'} \parallel \overline{BB'}$. These are congruent, parallel sides of $AA'B'B$. Therefore, $AA'B'B$ is a parallelogram, and opposite sides $\overline{AB}$ and $\overline{A'B'}$ are congruent. Thus, $AB = A'B'$.

Since a translation is an isometry, lengths of sides and angle measures of polygons are invariant. Different slide arrows may lead to the same slide image. Slide arrows m, n, and t below all result in the same image. All three slide arrows are congruent, parallel, and point in the same direction.

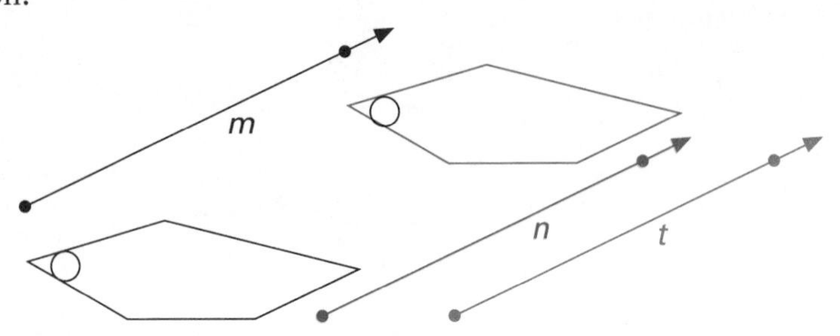

Translations can be determined by a pair of reflections about parallel lines.

EXAMPLE 2 Find the translation image determined by two reflections, first about line l and then about line m, where l and m are parallel.

Plan Construct the reflection $\triangle A'B'C'$ of $\triangle ABC$ about line l.

Construct the reflection $\triangle A''B''C''$ of $\triangle A'B'C'$ about line m.

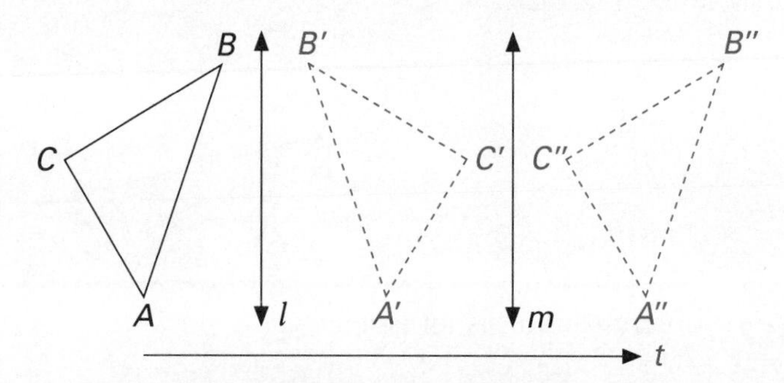

Solution $\triangle A''B''C''$ is the translation image of $\triangle ABC$ using slide arrow t, which is parallel and congruent to $\overline{AA''}$.

Example 2 suggests the following theorem.

Theorem 15.3 The resultant image determined by two successive reflections about parallel lines is a translation.

Successive reflections are sometimes referred to as the *product* of reflections.

Classroom Exercises

Which pairs of figures below appear to be translations?

1.

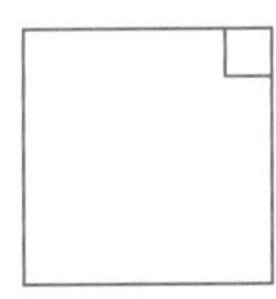

2.

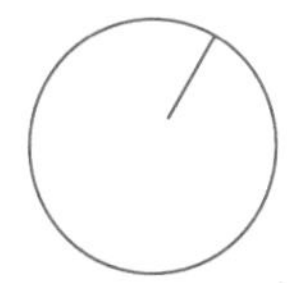

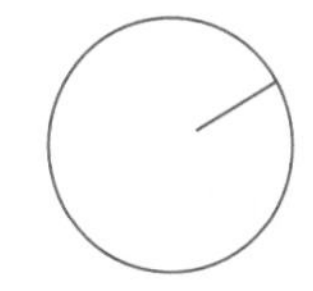

Using slide arrow *t*, find the translation image of each point.

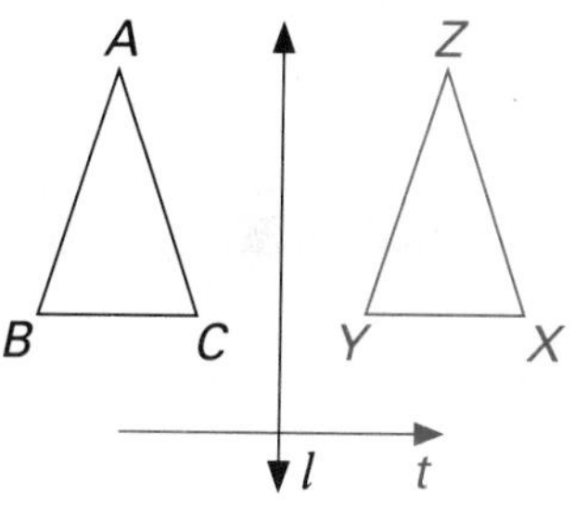

3. A **4.** B **5.** C

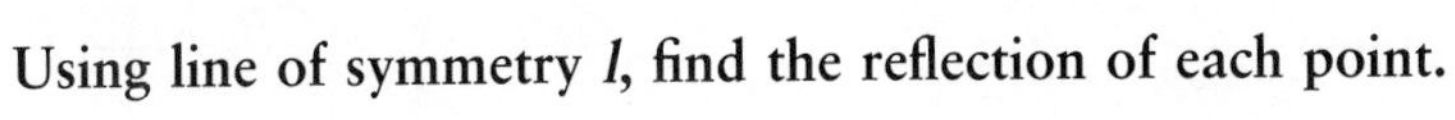

Using line of symmetry *l*, find the reflection of each point.

6. A **7.** B **8.** C

Written Exercises

Copy each given figure and the slide arrow. Draw a slide image for each.

1.

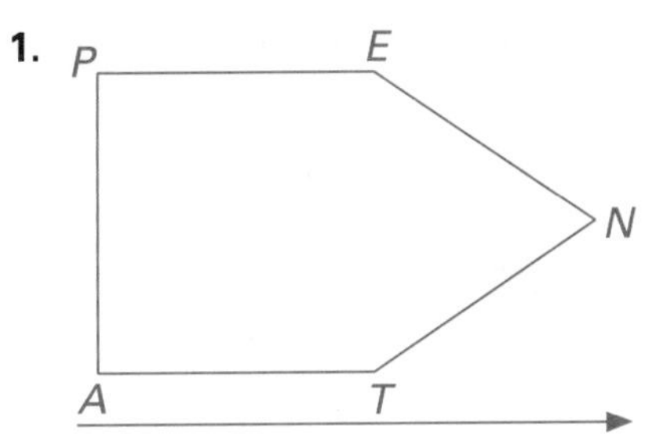

2.

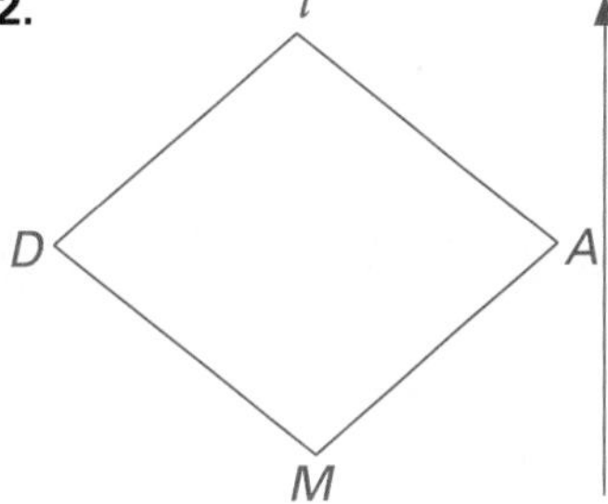

3.

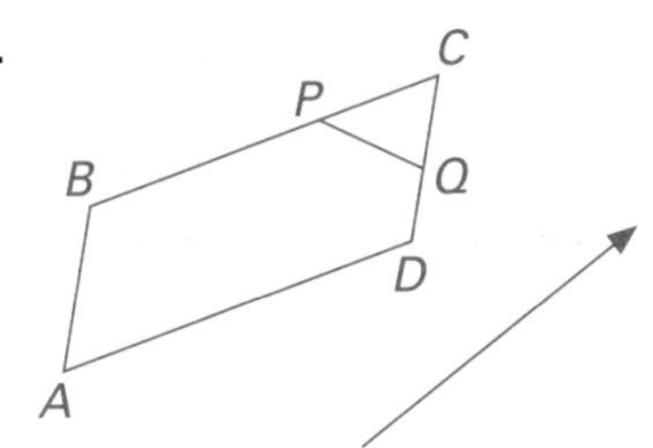

4.

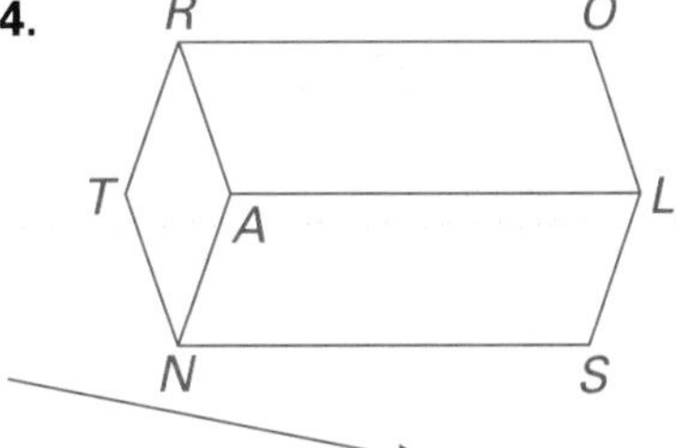

Find the slide image determined by two reflections, first about line *m*, and then about line *n*.

5.

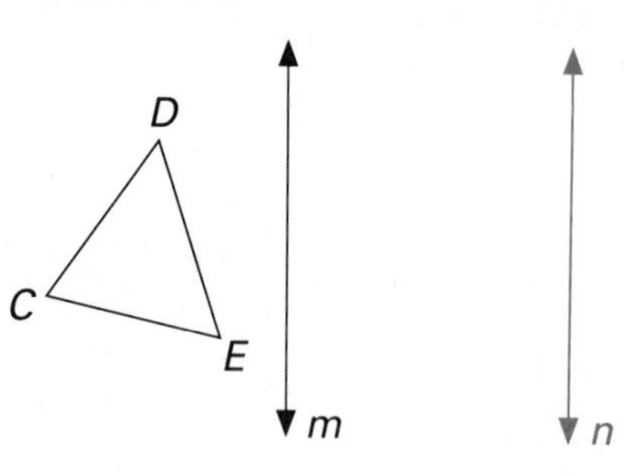

6.

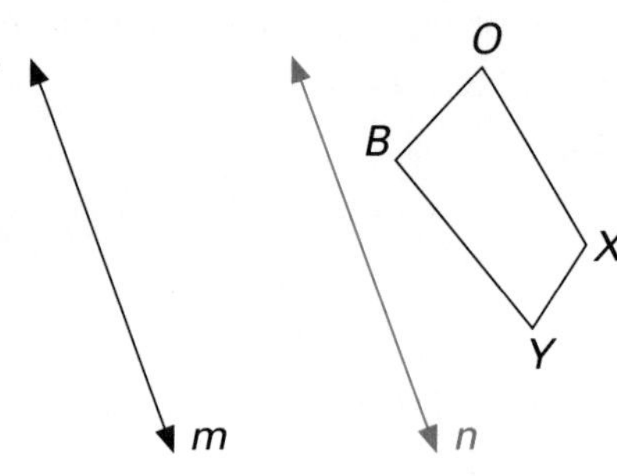

7. Find the image of $\triangle ABC$ determined by two reflections, first about line l, and then about line m.

8. Find the image of $\triangle ABC$ determined by two reflections, first about line m, and then about line l.

9. Based on the results of Exercises 7 and 8, does it appear that order is important when performing successive reflections? Explain.

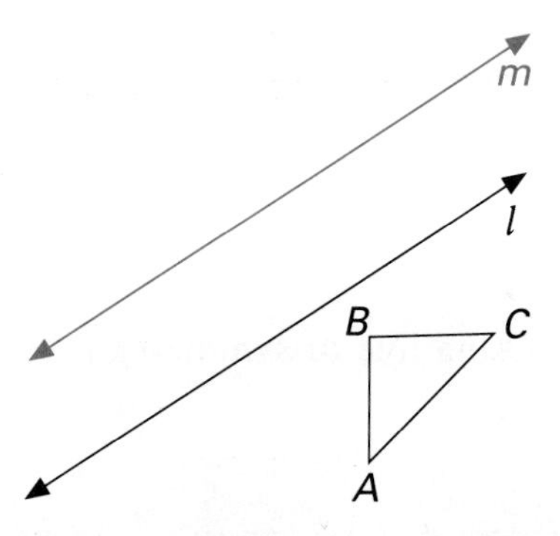

10. Find the translation image of $\triangle XYZ$ using t. Then find the translation image of the result using v.

11. Find the translation image of $\triangle XYZ$ using v. Then find the translation image of the result using t.

12. Based on the results of Exercises 10 and 11, does it appear that order is important when performing successive translations? Explain.

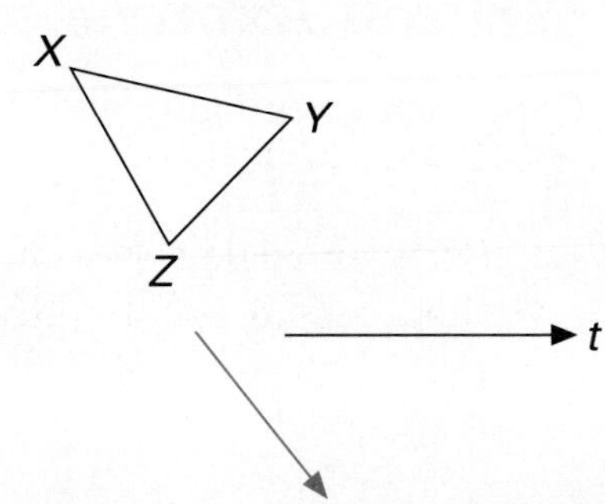

Find the reflection of the figure about *m*. Find the reflection of the image about *n*. Draw a slide arrow that results in the same final image.

13.

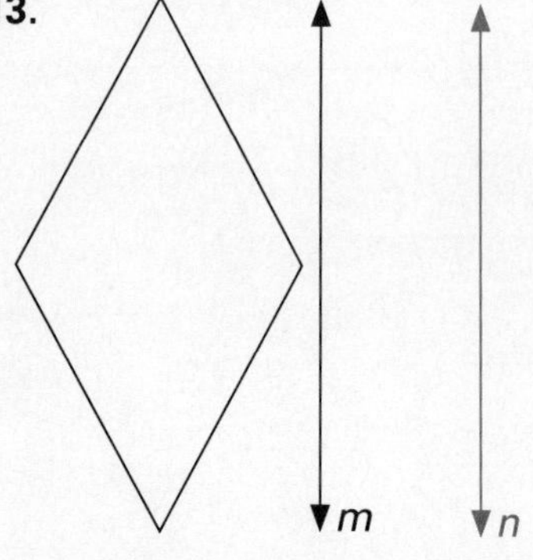

14.

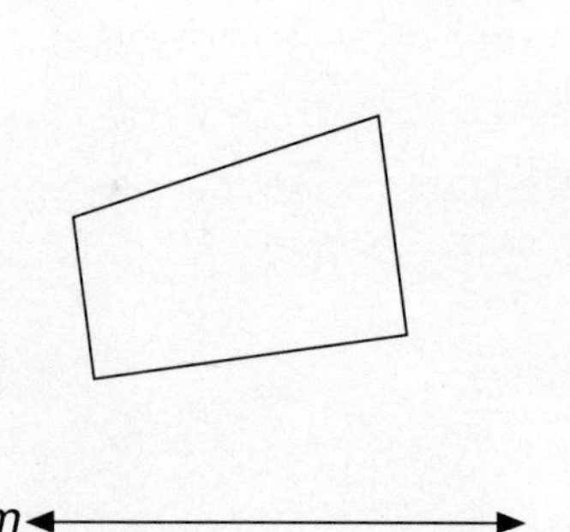

15. Copy the given figures and lines. Find the image of $ABCD$ determined by two reflections, first about line l and then about line m. Is the resultant image a reflection? Is the resultant image a translation?

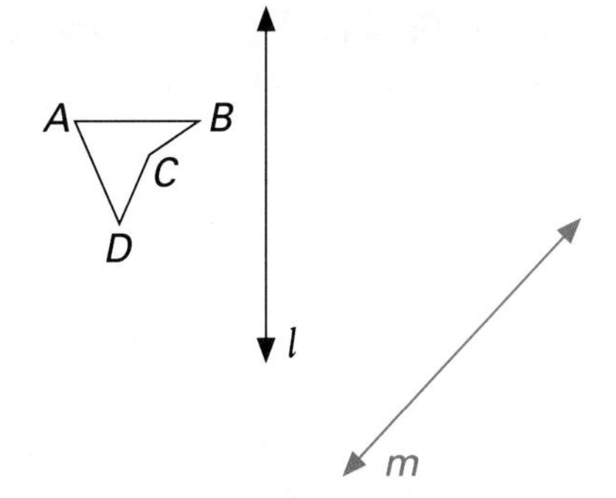

16. What properties of figures can be proved invariant under translations?

17. Prove that the translation image of a segment is parallel to that segment.

18. Prove Theorem 15.3 for $\overline{AB}$ about parallel lines m and n.

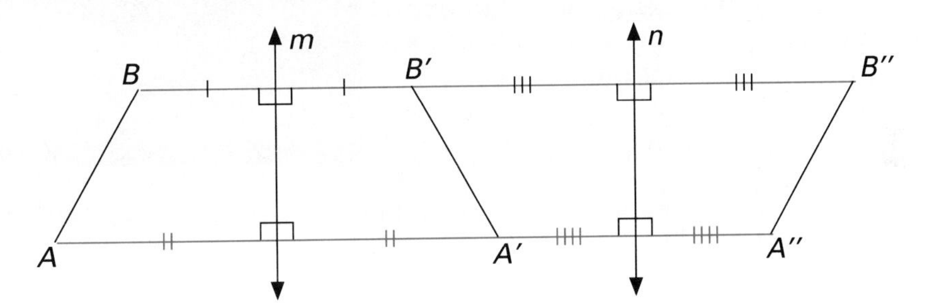

Mixed Review

1. Draw an angle and then construct an angle congruent to it. **1.4**
2. Construct an angle with degree measure 90. **5.4**
3. Draw a line segment $\overline{AB}$ and construct the perpendicular bisector of the segment. **5.6**

Application: *Designs*

Some designs are the result of translating a simple figure several times. For each of the following basic designs, draw the design that is found by sliding each given basic design 1 in. to the right a number of times.

15.3 Rotations

Objectives To draw rotation images of figures
To find the centers of rotations
To determine measures of angles of rotations

All wheels share a common property. Points on a wheel rotate in a circular path as the wheel turns on its axis. A pendulum has a similar property. The weight on the end of the pendulum moves in a path which is an arc of a circle. This motion is a *rotation*. The weight *rotates* around the pivot of the pendulum, which is the **center of rotation.** The center of rotation will be a point for coplanar figures or a line for figures in space.

To describe a rotation, you need to know the center of rotation and the magnitude of the rotation. To show this, use a turn arrow.

EXAMPLE 1 Draw a figure that is the rotation image of the given figure about P using the given turn arrow, which is an arc.

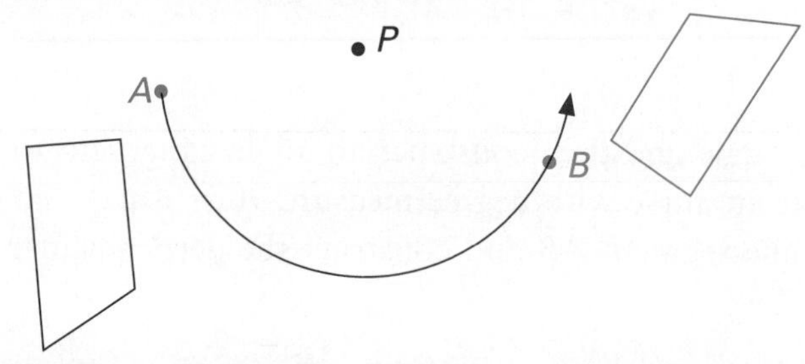

Solution

(1) Trace the given figure (black) on a sheet of paper. Draw a dot on your tracing at point A of the rotation arrow.
(2) Place a pen or pencil on your tracing at point P and hold it firmly in place. Turn your copy around point P until the dot is over the point B.
(3) Notice that your traced image is now positioned over the rotation image (red).
(4) Make a drawing of your own, with a given image and rotation arrow of your choice. Repeat steps (1) and (2).
(5) Using carbon paper, or other blackened material, transfer your traced image in its rotated position onto your original drawing. The resulting image is the rotation image of the original image, with respect to the rotation arrow and its pivot point.

Some figures are their own rotation images. If the square below is rotated 90 degrees or 180 degrees about P, the point of intersection of the two diagonals, the rotation image coincides with the original square. If the figure at the right below is rotated 120 degrees about O, the rotation image coincides with the original figure. Each figure has **rotational symmetry.**

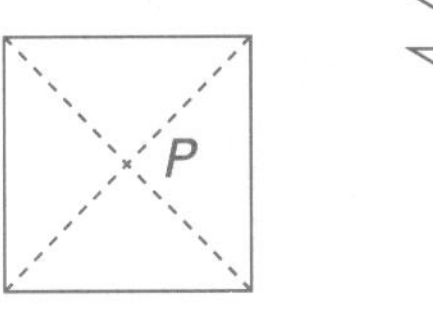

EXAMPLE 4 A regular pentagonal prism has rotational symmetry about a line l. What is the measure of the angle of rotation necessary for the figure to coincide with itself?

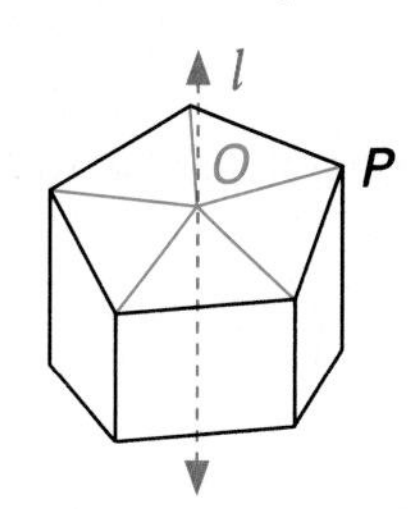

Solution Since each of the angles shown at O has a measure of 72, then the necessary angle of rotation measures 72, or any multiple of 72.

The figure at the right has *point symmetry.* The bisector of the segment determined by a point and its image is P for all points of the figure. Each point is rotated about P into its image. Point symmetry is a special case of rotational symmetry in which the angle of rotation is 180.

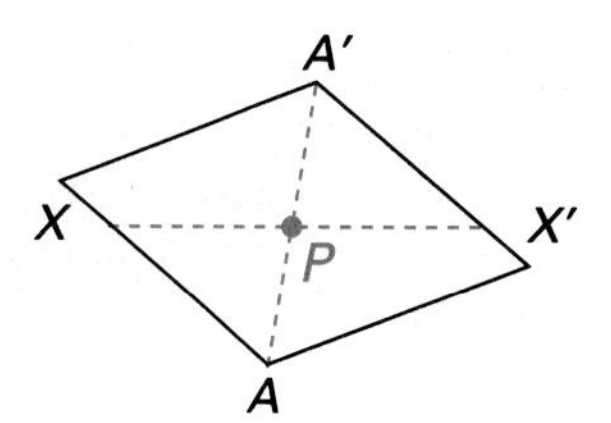

Classroom Exercises

In the diagram, ABC and $JKLM$ have been reflected about lines l and m, successively.

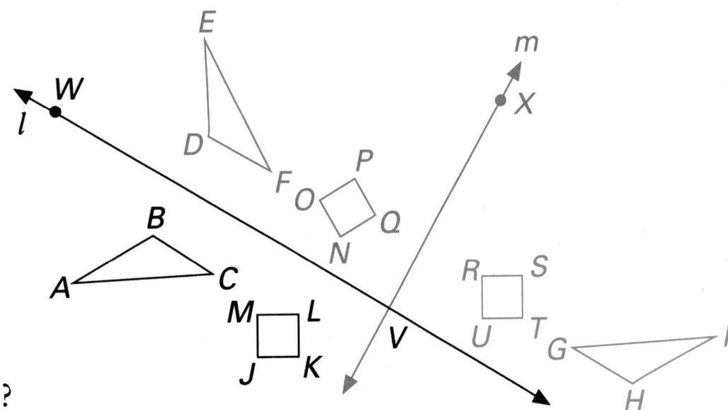

1. What is the center of rotation?

Find the rotational image of each point.

2. A 3. B 4. C 5. J 6. K 7. L 8. M

Written Exercises

Find the rotation image of each given figure about P by finding successive reflections about m and n. Measure the angle of rotation.

1. 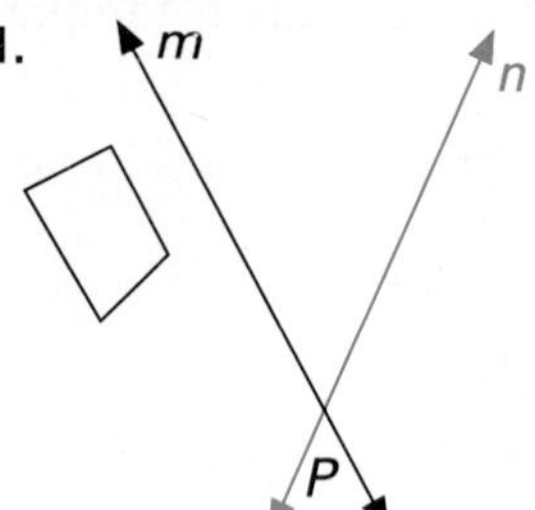

2. 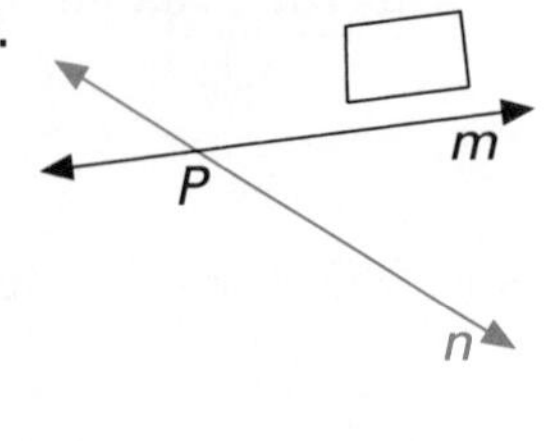

3. 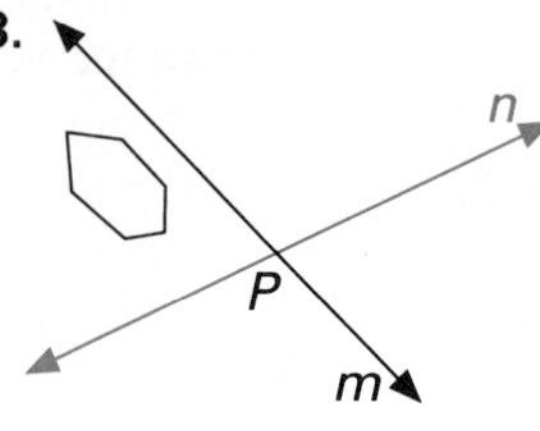

Copy each pair of figures. Find the center of rotation.

4. 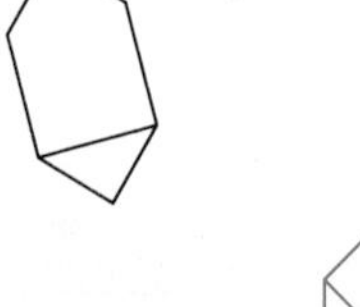

5. 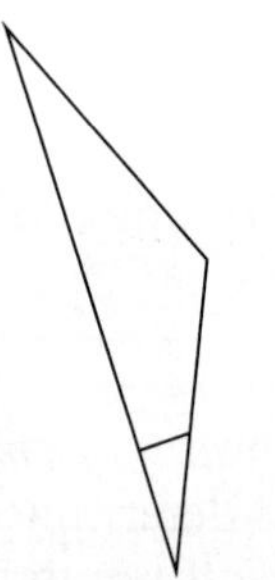

6. 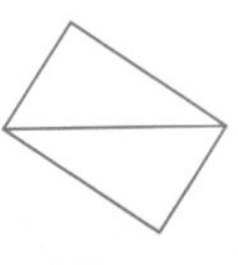
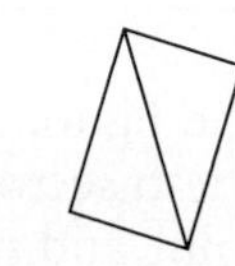

7. Draw the rotational image obtained by reflecting $\triangle ABC$ successively about l and m.
8. Draw the rotational image obtained by reflecting $\triangle ABC$ successively about m and l.
9. Based on the results of Exercises 7 and 8, is order of reflection important when performing a rotation?

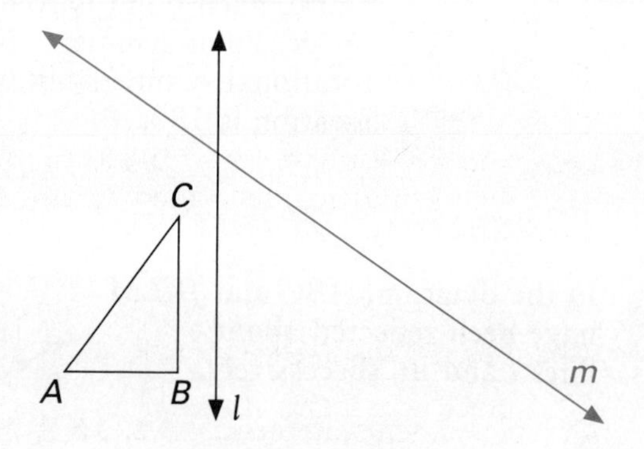

Each figure has rotational symmetry. What is the measure of the angle of rotation necessary for the figure to coincide with itself? (Exercises 10–15)

10. 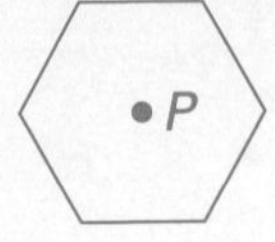

11. 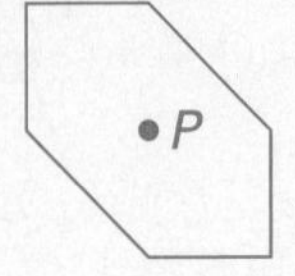

12. 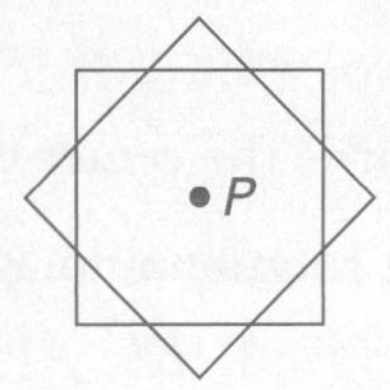

13.

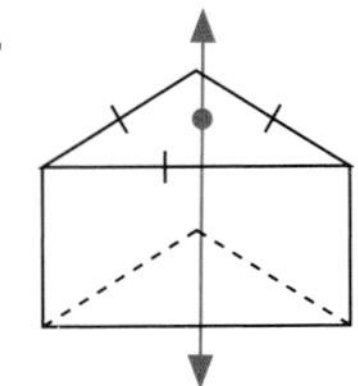

14.

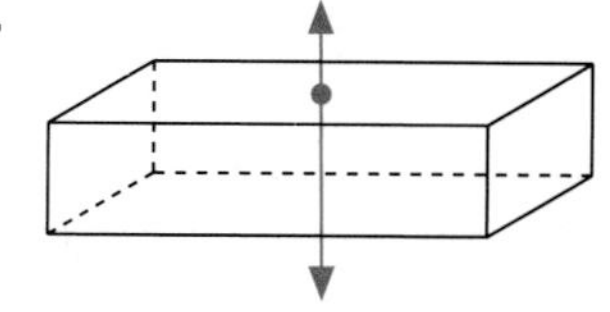

15.

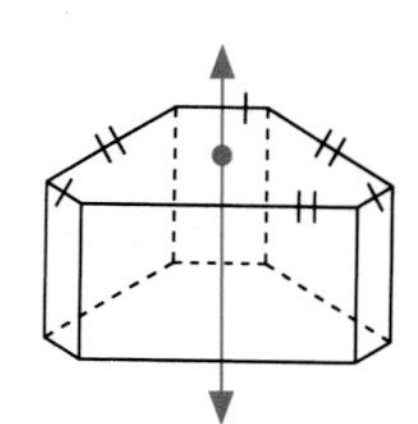

What product of two transformations will result in the image for each figure?

16.

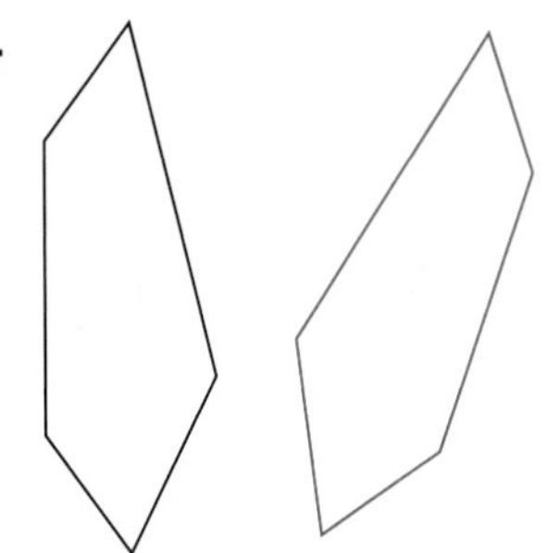

17.

18.

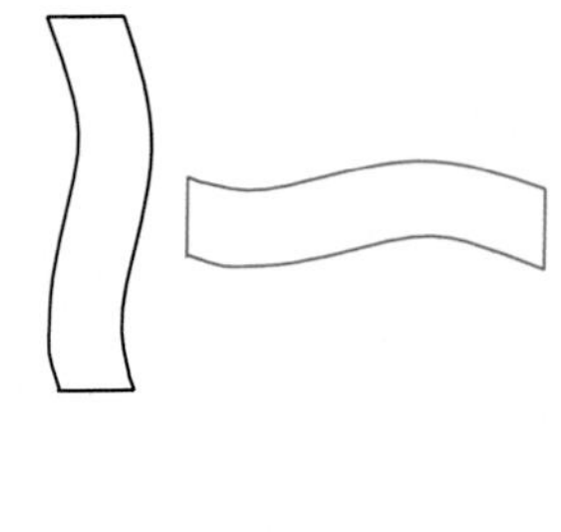

19. A segment is rotated 180 degrees about a point not on the line containing the segment. Prove that the segment and its rotation image are parallel.
20. Prove that an equilateral triangle has rotational symmetry.
21. Prove that a parallelogram has point symmetry.
22. Prove Theorem 15.4.
23. Prove Theorem 15.5.
24. Prove or disprove: If a polygon has both line symmetry and rotational symmetry, then the polygon is regular.

Midchapter Review

Complete each statement. ***15.1, 15.2, 15.3***

1. A transformation that preserves distances is called an __________.
2. For a reflection the perpendicular bisectors of segments joining each point and its image form the line of __________.
3. The distance between any two points and their image will be equal for a __________.
4. An equilateral triangle has __________ symmetry.
5. Two successive reflections about lines that are not parallel result in a __________.

15.4 Transformations and Coordinates

Objectives To describe transformations using coordinates
To draw transformation images using coordinates

Transformations can be described using coordinate geometry. When describing a transformation, or *mapping*, the notation $(x, y) \rightarrow (x + 6, y + 3)$ means that a point in the original figure with coordinates x and y maps onto the point with coordinates $x + 6$ and $y + 3$, respectively.

EXAMPLE 1 Draw the image of $ABCD$ under the mapping $(x, y) \rightarrow (x + 6, y + 3)$.

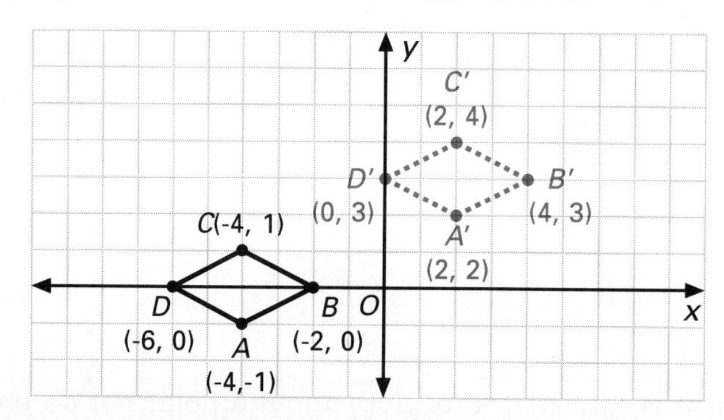

(1) Draw $ABCD$ on graph paper.
(2) For each vertex, add 6 to the abscissa and 3 to the ordinate.
(3) Mark the points corresponding to the new coordinates and label them A', B', C', and D'.
(4) Draw $A'B'C'D'$, the image of $ABCD$.

EXAMPLE 2 Using Example 1, show that the distance between any point of the figure and its image is constant. Show that the slope of the segment, determined by any point of this figure and its image, is also constant.

Plan Choose two points $P_1(x_1, y_1)$ and $P_2(x_2, y_2)$ and their images $P_1'(x_1 + 6, y_1 + 3)$ and $P_2'(x_2 + 6, y_2 + 3)$. Use the distance formula and the slope formula.

Proof

$P_1P_1' = \sqrt{(x_1 + 6 - x_1)^2 + (y_1 + 3 - y_1)^2}$, or $\sqrt{6^2 + 3^2}$
$P_2P_2' = \sqrt{(x_2 + 6 - x_2)^2 + (y_2 + 3 - y_2)^2}$, or $\sqrt{6^2 + 3^2}$

Therefore, $P_1P' = P_2P_2'$.

$m_1 = \frac{y_1 + 3 - y_1}{x_1 + 6 - x_1}$, or $\frac{3}{6}$ $\quad m_2 = \frac{y_2 + 3 - y_2}{x_2 + 6 - x_2}$, or $\frac{3}{6}$

Therefore, $m_1 = m_2$.

Theorem 15.6 The transformation defined by adding a constant to the coordinates of each point is a translation.

Plan Show that the distance between each pair of points is equal to the distance between their images. Show that the segments connecting all points and their images are parallel.

EXAMPLE 3 Draw the reflection of $\triangle ABC$ about the y-axis. Find the coordinates of the vertices and describe the mapping using coordinate notation.

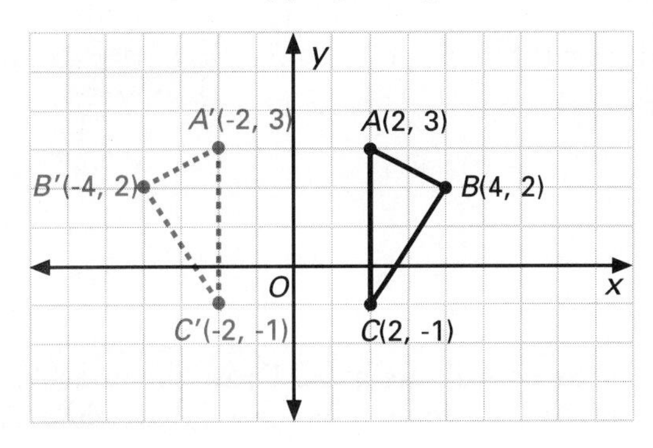

Solution The coordinates of $\triangle A'B'C'$ are $(-2, 3)$, $(-4, 2)$, and $(-2, -1)$. This reflection is described by $(x, y) \rightarrow (-x, y)$.

EXAMPLE 4 Classify the transformation shown on the graph below. Describe the mapping using coordinate notation.

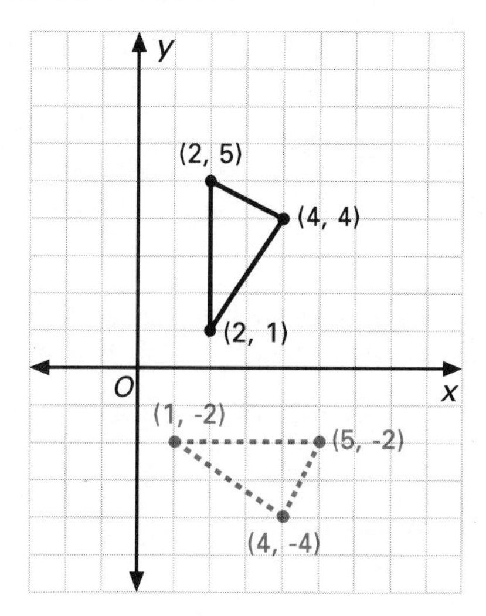

Solution The transformation is a 90-degree clockwise rotation about the origin. This rotation is described by $(x, y) \rightarrow (y, -x)$.

Classroom Exercises

Classify the transformation shown on the graph. Describe the mapping using coordinate notation.

1.

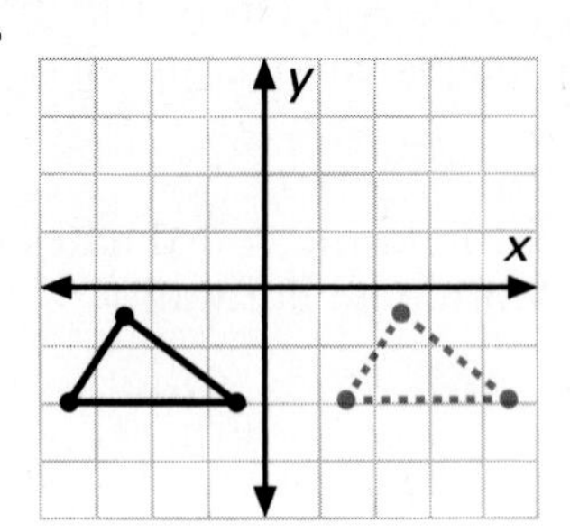

2.

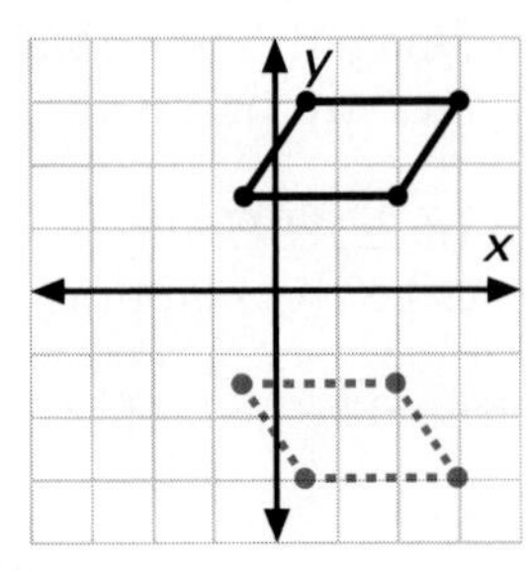

3.

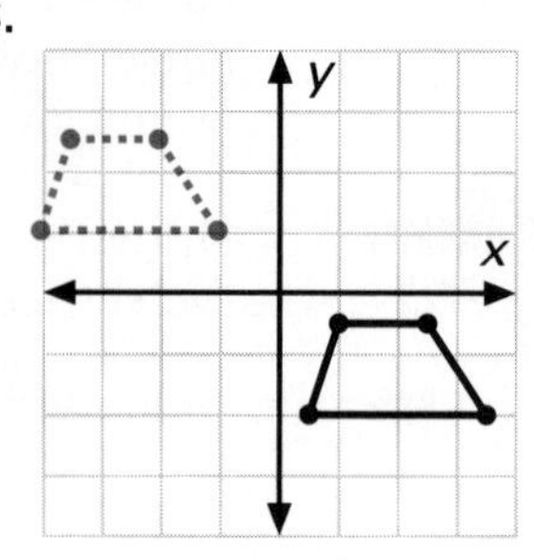

Written Exercises

Copy $\triangle ABC$ and the coordinate axes. Draw the image of $\triangle ABC$ according to the given coordinate change. Classify the transformation.

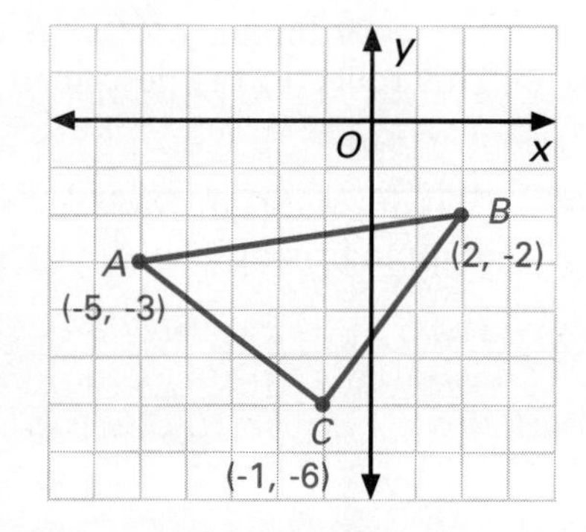

1. $(x, y) \rightarrow (x + 5, y)$
2. $(x, y) \rightarrow (x, y - 3)$
3. $(x, y) \rightarrow (x + 2, y + 1)$
4. $(x, y) \rightarrow (x, -y)$
5. $(x, y) \rightarrow (y, x)$
6. $(x, y) \rightarrow (y, -x)$
7. $(x, y) \rightarrow (-y, -x)$
8. $(x, y) \rightarrow (x, -y + 4)$
9. $(x, y) \rightarrow (-x - 3, -y - 1)$
10. $(x, y) \rightarrow (-y, -x + 2)$

Show that the distances between two points and the distances between the corresponding images are equal for each transformation.

11. $(x, y) \rightarrow (x + 1, y + 2)$
12. $(x, y) \rightarrow (x - 4, y + 5)$
13. $(x, y) \rightarrow (x + 6, y - 7)$
14. $(x, y) \rightarrow (x - 9, y - 7)$
15. $(x, y) \rightarrow (y, x)$
16. $(x, y) \rightarrow (-y, x)$

Classify the resulting transformation.

17. adding a constant to each x-coordinate
18. replacing each x-coordinate by its opposite
19. replacing both the x- and the y-coordinates by their opposites
20. interchanging the x- and y-coordinates of each point

Describe the mapping using coordinate notation.

21. a rotation of 180 degrees about the origin
22. a reflection about the x-axis
23. a reflection about the line $y = x$
24. a reflection about the line $y = -x$
25. a reflection about the line $x = 5$
26. a reflection about the line $y = -3$
27. a reflection about the line $x = c$
28. a rotation of 180 degrees about the point (3, 4)

Mixed Review

Write the truth table for each of the following. *1.7*

1. $(p \vee q) \rightarrow p$
2. $(p \wedge q) \rightarrow q$
3. $[(p \rightarrow q) \wedge p] \rightarrow q$
4. Write an equation of a circle with a radius of 3 having its center at the origin. *11.11*
5. Write an equation in standard form of a circle with a radius of 5 having its center at the point (−3,2). *11.11*

Application: *Stadium Lights*

A new lighting system is being installed at the Superior High School football stadium. Four large banks of lights mounted on 60-ft-tall poles are being placed around the field, two on each side. The poles, with the lights already installed on them, have been delivered to the stadium and are lying on the ground as shown. The holes for the foundations have been dug and the concrete footings poured. A large crane has arrived to position the poles in place. What combination of rotations, translations, and/or reflections will it take to position the poles properly? What instructions would you give the crane operator?

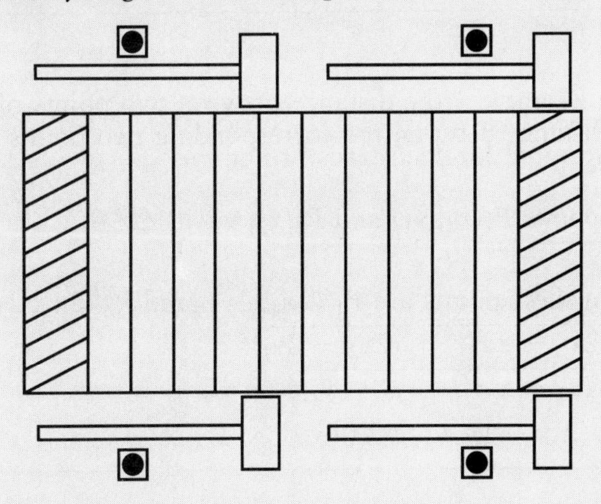

15.5 Dilations and Other Transformations

Objectives To find a dilation of a figure
To find a shear image of a figure

The map shown to the right represents the state of Iowa. If we disregard the curvature of the earth, this map is similar to the actual state. That is, distances are proportional and the measures of angles are equal. Such a map is a dilation.

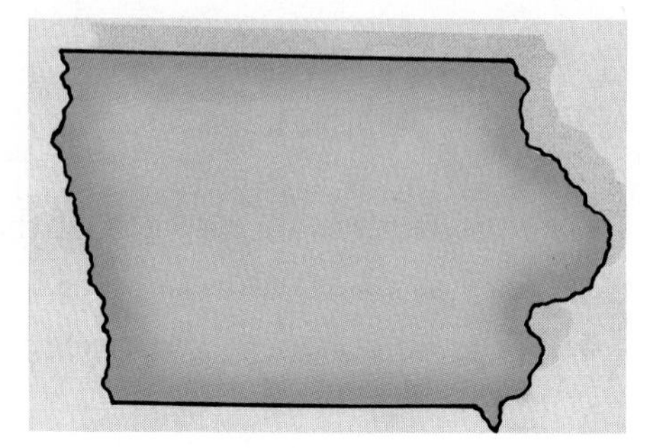

Definition A **dilation** is a transformation which produces an image similar to the original figure.

EXAMPLE 1 Draw the image of $\triangle ABC$ under the mapping $(x, y) \rightarrow (2x, 2y)$.

(1) Draw the figure on graph paper.
(2) Replace each x-coordinate with $2x$ and each y-coordinate with $2y$.
(3) Draw the points corresponding to the new coordinates.
(4) Connect these points.

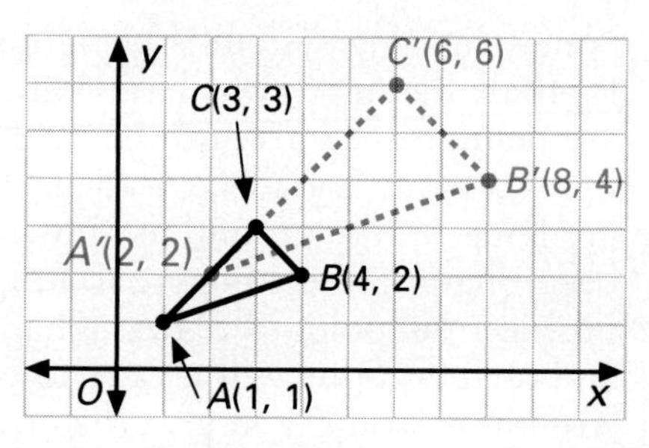

EXAMPLE 2 Show that in Example 1 the distance between two points of the image is twice the distance between the corresponding two points of the original figure.

Solution Choose two points $P_1(x_1, y_1)$ and $P_2(x_2, y_2)$ in $\triangle ABC$.

$P_1P_2 = \sqrt{(x_2 - x_1)^2 + (y_2 - y_1)^2}$

The images of these points are $P_1'(2x_1, 2y_1)$ and $P_2'(2x_2, 2y_2)$.

$$P_1'P_2' = \sqrt{(2x_2 - 2x_1)^2 + (2y_2 - 2y_1)^2}$$
$$= 2\sqrt{(x_2 - x_1)^2 - (y_2 - y_1)^2}$$

Therefore, $P_1'P_2' = 2P_1P_2$.

Theorem 15.7 The transformation defined by multiplying each coordinate of each point by a constant is a dilation.

EXAMPLE 3 What changes in coordinates will determine a dilation image with lengths of sides that are half the original lengths?

(1) Draw $\triangle ABC$ on graph paper.
(2) Take half of each coordinate for each vertex.
(3) Draw the points corresponding to the new coordinates.
(4) Connect these points.
(5) The new triangle is similar to the original but with sides half as long.

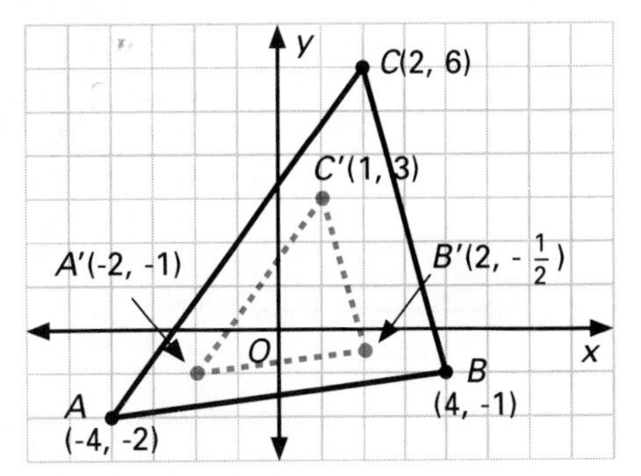

Some transformations do not preserve either size or shape. The **shear** transformation preserves horizontal distances or vertical distances, but not both. Also, the images of angles formed by perpendiculars to horizontal segments will be congruent to one another.

EXAMPLE 4 Draw the image of quad $ABCD$ under the mapping $(x, y) \rightarrow (x + y, y)$.

(1) Draw the figure on graph paper. Mark the axes.
(2) Replace each x-coordinate with $x + y$.
(3) Draw the points corresponding to the new coordinates.
(4) Connect these points.

Quadrilateral $A'B'C'D'$ is the shear image of quadrilateral $ABCD$.

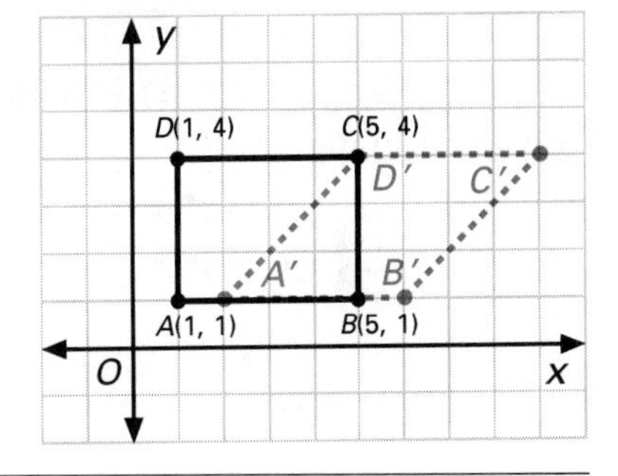

Focus on Reading

For each of the following, indicate whether the statement is always, sometimes, or never true.

1. The product of two reflections is a translation.
2. A square will remain a square after a rotation.
3. A trapezoid has point symmetry.
4. The product of two reflections is a rotation.

Classroom Exercises

1. What changes in coordinates will determine the given dilation image of $\triangle ABC$ if sides of the image are three times as long as the sides of the original figure.

What changes in coordinates determine the given dilation from the solid line to the dashed line figure?

2.

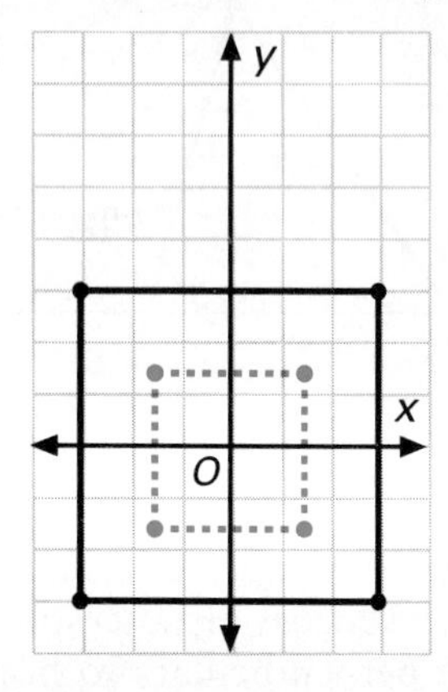

3.

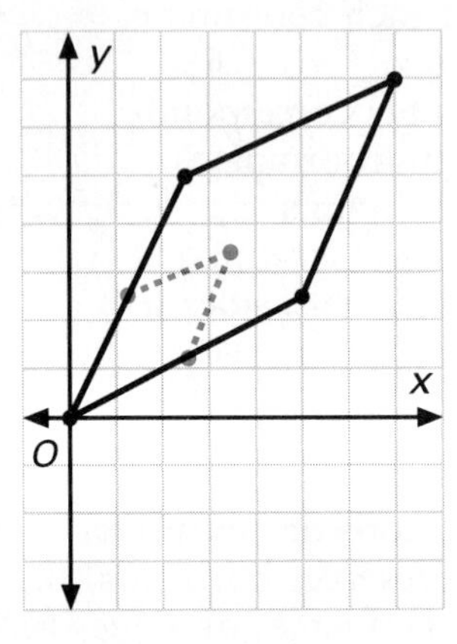

4.

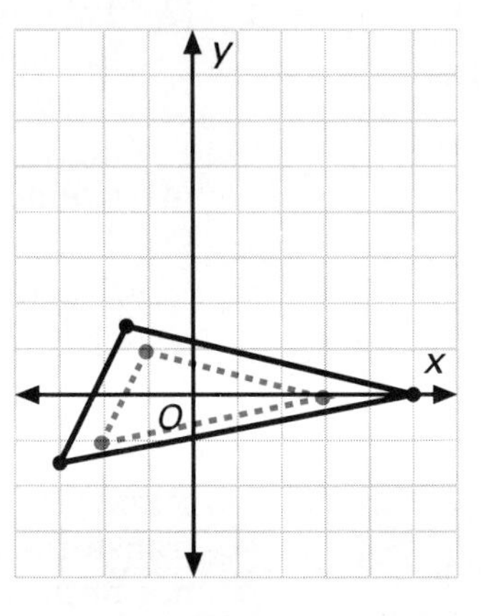

Written Exercises

Draw the image of $\triangle ABC$ under the given mapping.

1. $(x, y) \rightarrow (2x, 2y)$
2. $(x, y) \rightarrow (3x, 3y)$
3. $(x, y) \rightarrow (1\frac{1}{2}x, 1\frac{1}{2}y)$
4. $(x, y) \rightarrow (\frac{3}{4}x, \frac{3}{4}y)$

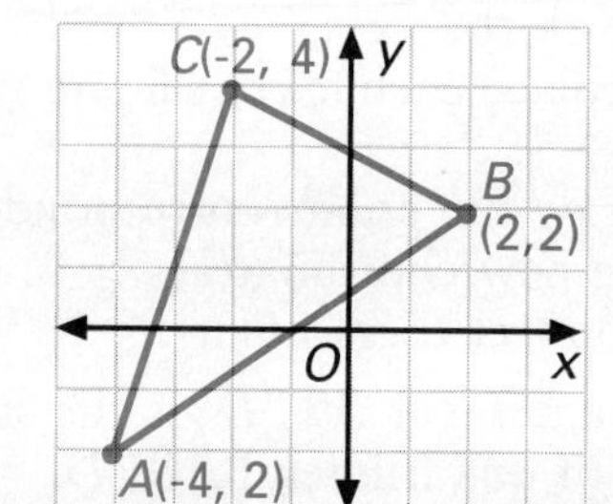

Draw the image of quad $ABCD$, replacing x with $x + y$.

5.

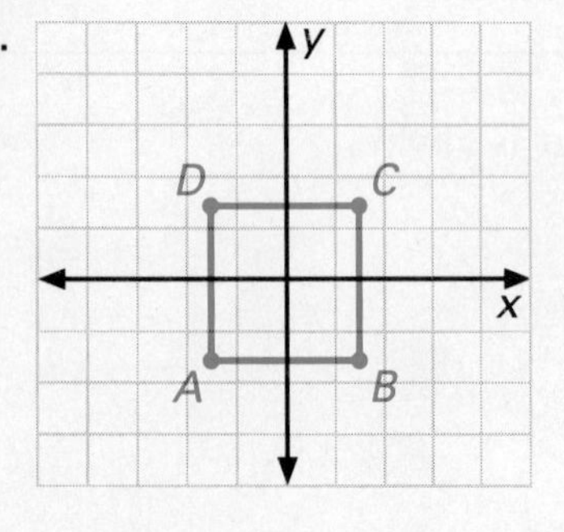

6.

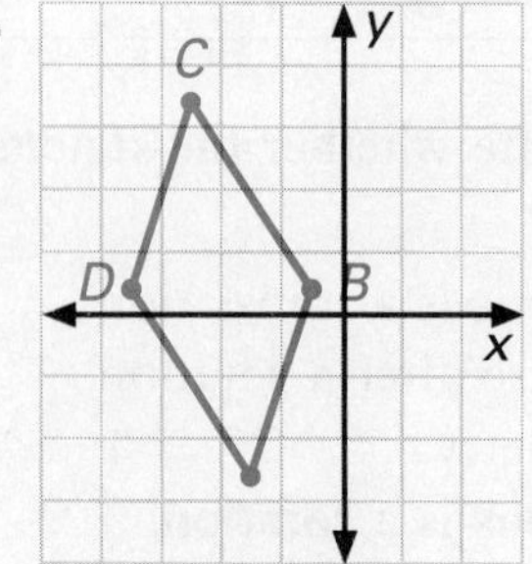

7.

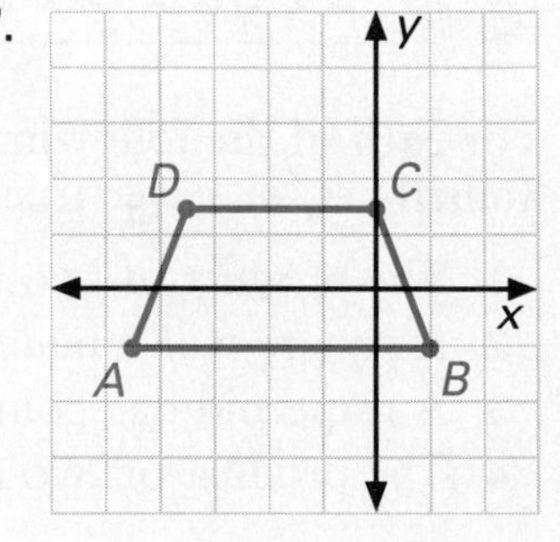

8. Show that if $(x, y) \to (3x, 3y)$, the distance between two points of the image is three times the distance between the corresponding two points of the original figure.
9. Prove Theorem 15.7.

Draw the image $\triangle A'B'C'$ of $\triangle ABC$ under the given transformation. Is the transformation a dilation, a shear, or neither of these?

10. $(x, y) \to (3x, y)$
11. $(x, y) \to (x, 3y)$
12. $(x, y) \to (2x, 2y)$
13. $(x, y) \to (3x, 2y)$
14. $(x, y) \to (\frac{1}{2}x, \frac{1}{2}y)$
15. $(x, y) \to (0, 2y)$

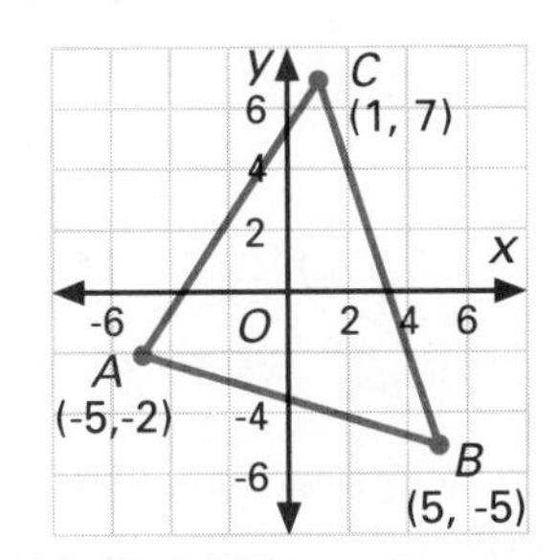

Draw the image of $A'B'C'D'$ of $ABCD$ under the given transformation. Is the transformation a dilation, a shear, or neither of these?

16. $(x, y) \to (x + y, y)$
17. $(x, y) \to (x, x + y)$
18. $(x, y) \to (x + y, x + y)$
19. $(x, y) \to (x - y, y)$
20. $(x, y) \to (x, y - x)$
21. $(x, y) \to (x + y, y - x)$
22. $(x, y) \to (x^2, y)$
23. $(x, y) \to (x^2, 2y)$

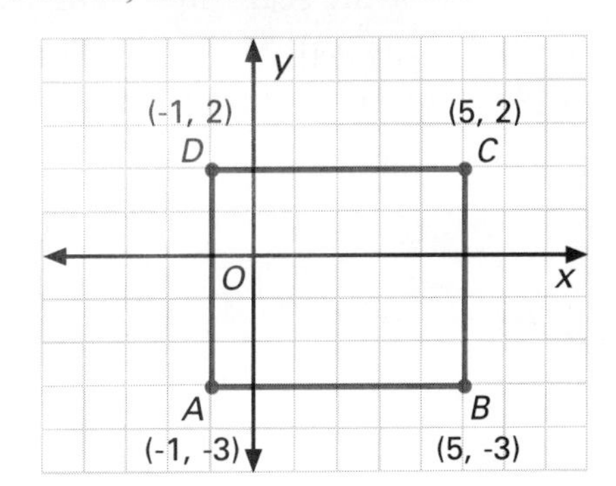

Mixed Review

1. Draw an obtuse angle. Construct the bisector of the angle. ***1.5***
2. Draw a segment approximately 4 in. long. Construct a segment two-thirds as long. ***8.6***

In $\odot O$, m$\widehat{AD} = 74$ and m$\widehat{BC} = 28$.

3. Find m $\angle DBA$.
4. Find m $\angle DFA$.
5. Find m $\angle DEA$.

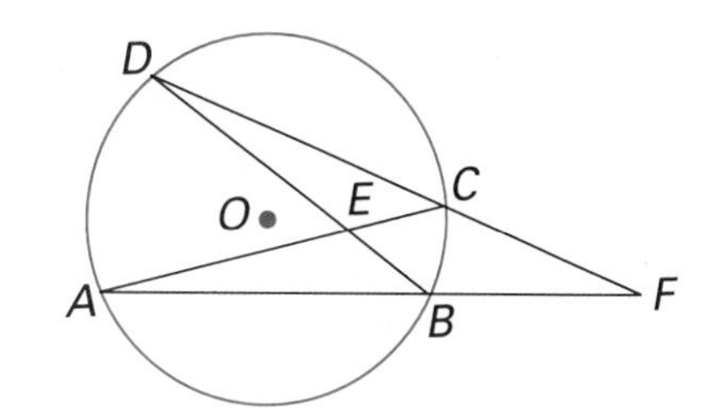

Chapter 15 Review

Key Terms

center of rotation (p. 612)
dilation (p. 622)
geometric transformation (p. 599)
image (p. 599)
invariant condition (p. 600)
isometry (p. 600)
line of symmetry (p. 600)
mapping (p. 618)
reflection (p. 599)
rotation (p. 613)
rotational symmetry (p. 615)
shear (p. 623)
slide arrow (p. 606)
slide image (p. 606)
symmetric figure (p. 601)
translation (p. 607)

Key Ideas and Review Exercises

15.1 To find the reflection of a figure, construct the image so that the reflection line is the perpendicular bisector of the segment joining corresponding points.
To locate the line of symmetry of a figure and its image, construct the perpendicular bisector of the segment joining corresponding points.

1. Construct a figure that is the reflection image of $\triangle PQR$ about line m.

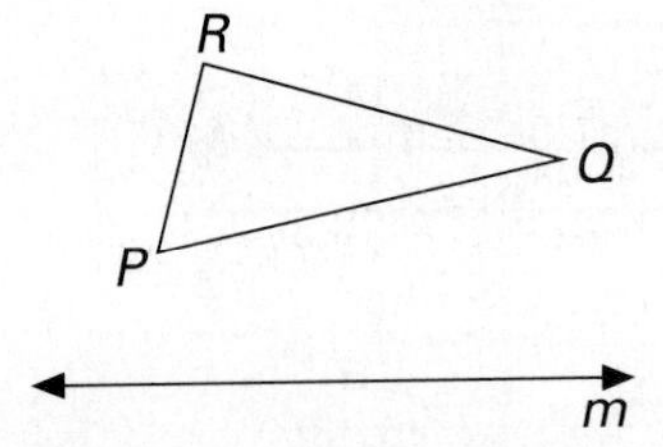

2. Find the line of symmetry of the given figure and its image.

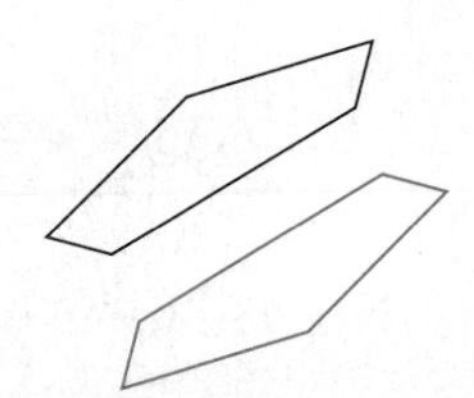

15.2 To find the translation image of a figure, construct the image so that segments joining corresponding points are congruent and parallel.
To find the product of two reflections, find the image of the first reflection; then find the image of that image by the second reflection.

3. Find the translation image of the given figure using the given slide arrow.

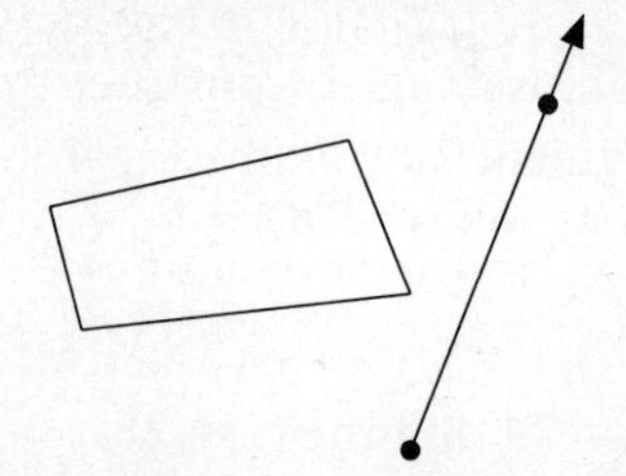

15.3 To find the rotation image of a figure, construct the image so that the distances from the center of rotation to corresponding points are equal.
To find the center of rotation, trace the figures on paper. Fold the paper so that a point and its image coincide. Fold again.

4. Find the center of rotation of the given figure and its image.

15.4 To find the product of two transformations, find the image of the original figure by the first transformation. Then find the image of that image by the second transformation.

5. Find the image of the given figure through a 60 degree clockwise rotation about P. Find the translation of the rotation image using the given slide arrow.

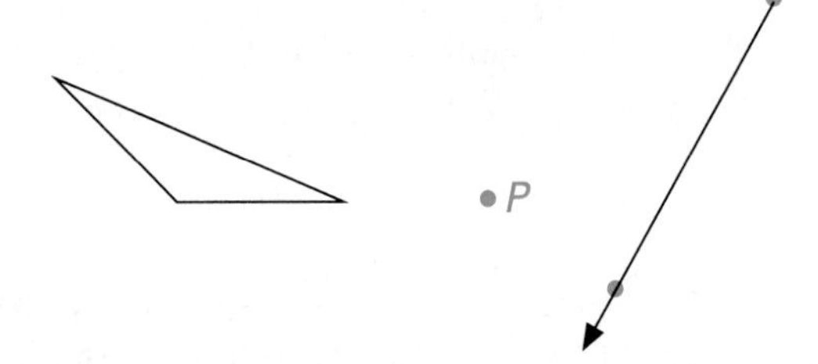

15.5 To map a transformation using coordinates, determine the necessary change in coordinates needed to describe the image.

To draw a transformation image using coordinates, draw the points corresponding to the new coordinates.

To describe a dilation, multiply each coordinate of the figure by a constant.

To describe a horizontal shear image, replace each x-coordinate with a new coordinate that is the sum of multiples of the x- and y-coordinates.

6. Draw the image of quad $ABCD$ so that each x-coordinate is increased by 2 and each y-coordinate is decreased by 3.
7. What changes in coordinates will give a dilation image with sides 4 times as long as those of the original polygon?
8. Draw the image of quad $ABCD$ replacing x with $x + y$.

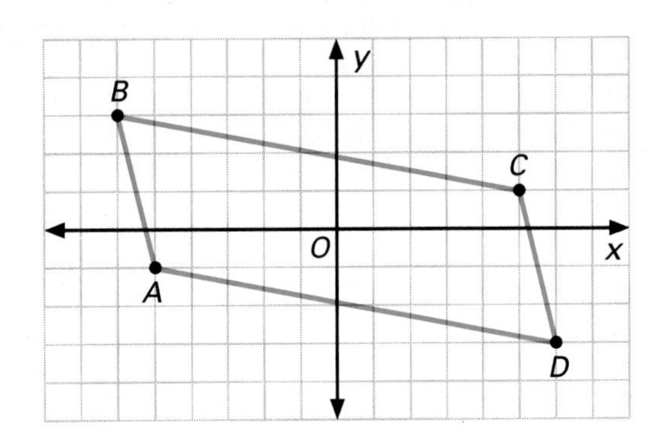

Chapter 15 Test

For each figure below, tell whether the figure on the right appears to be a reflection, a translation image, a rotation image, or neither of the figure on the left.

1.

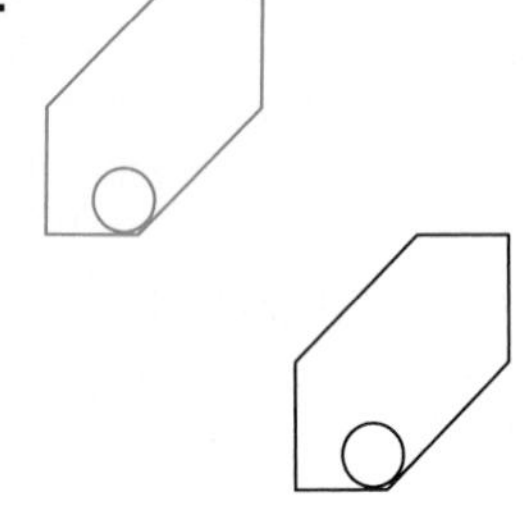

2.

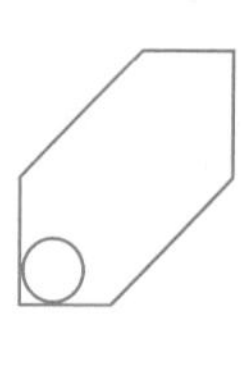

3.

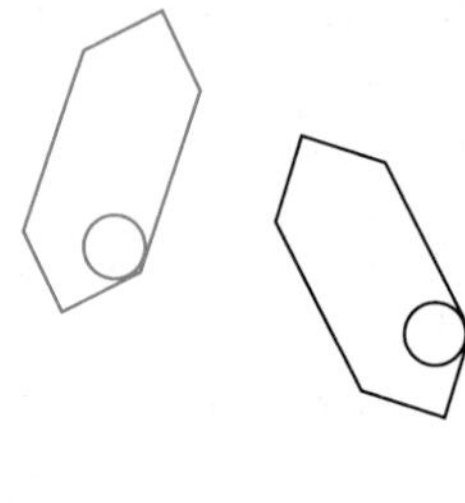

4. Find the line of symmetry or center of rotation of the figures in Exercise 1–3.

5. Find the center of rotation of the figure and its image.

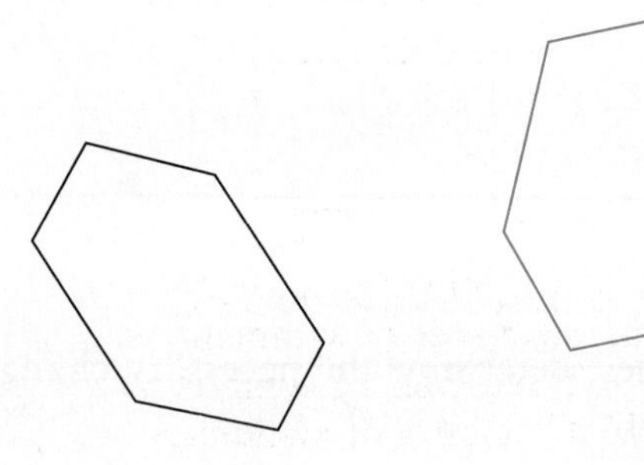

6. Each triangle is the image of the other by a combination of two transformations. Tell what these most likely are.

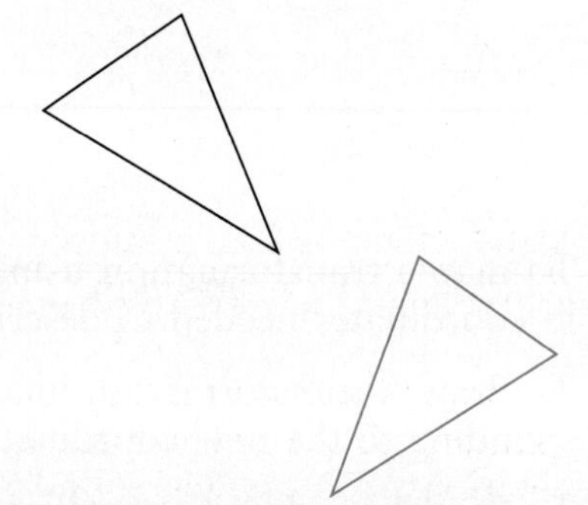

Replace the vertices of the given triangle with each set of vertices. Is the new figure a reflection, a translation, a rotation, a shear image, a dilation, or neither of the original triangle?

7. $(x, y) \rightarrow (x + 2, y)$

8. $(x, y) \rightarrow (2x, 2y)$

9. $(x, y) \rightarrow (x, x + y)$

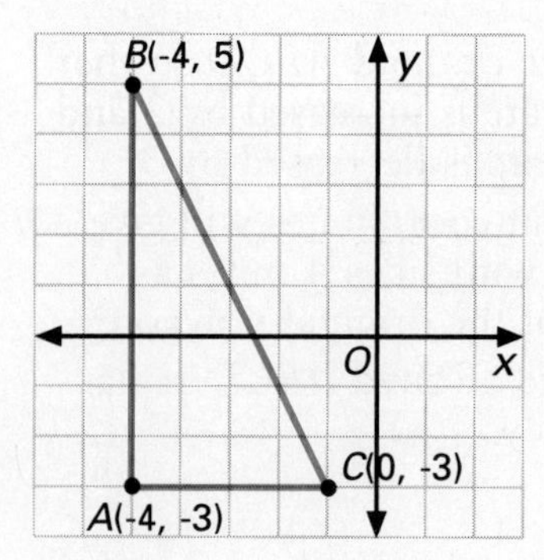

***10.** Prove that a rotation is an isometry.

College Prep Test

1. Which of the following is not true for a translation?
   (A) Figure and image are congruent.
   (B) Figure slides to become an image.
   (C) Figure flips over to become an image.
   (D) It is an isometry.

2. To turn a figure upside down and move it 60 degrees down and left requires ______.
   (A) a translation and a rotation
   (B) a reflection and a translation
   (C) a reflection and a rotation
   (D) a rotation and a dilation

3. 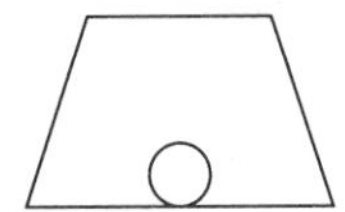 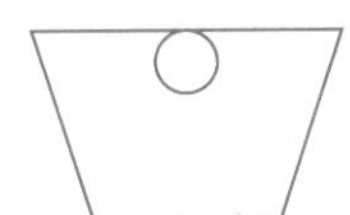

   The image is obtained through ______.
   (A) a translation (B) a rotation
   (C) a dilation (D) a reflection

4. $(x, y) \rightarrow (3x, 3y)$ will give ______.
   (A) a dilation (B) a reflection
   (C) a rotation (D) a translation

5. Which of the figures below have point symmetry?

   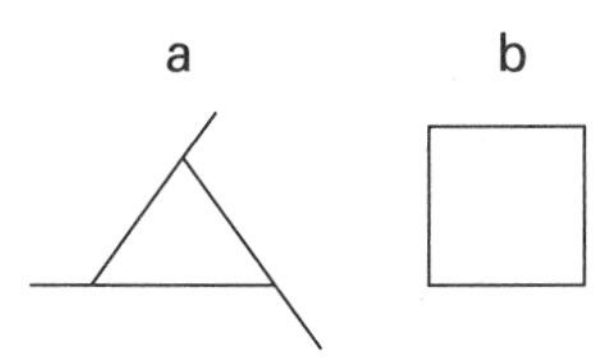

   (A) Neither a nor b
   (B) Both a and b
   (C) a only
   (D) b only

6. 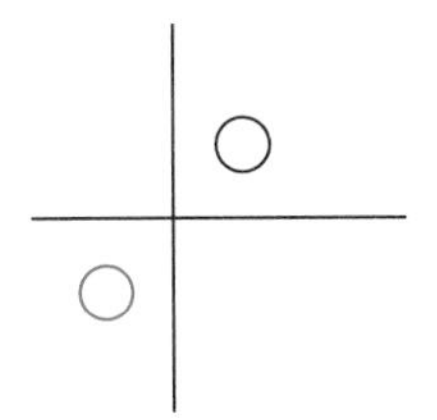

   The mapping for the figure and its image is ______.
   (A) $(x, y) \rightarrow (-x, y)$
   (B) $(x, y) \rightarrow (x, -y)$
   (C) $(x, y) \rightarrow (-x, -y)$
   (D) $(x, y) \rightarrow (2x, y)$

7. The figure below has ______.

   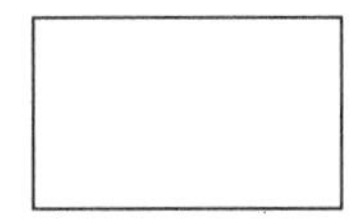

   (A) one line of symmetry
   (B) two lines of symmetry
   (C) three lines of symmetry
   (D) four lines of symmetry

8. The following is not an isometry.
   (A) a reflection (B) a translation
   (C) a rotation (D) a dilation

9. 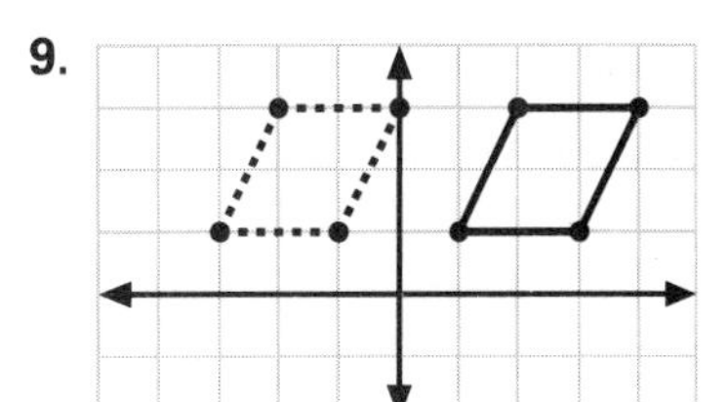

   The mapping above is ______.
   (A) $(x - 4, y)$ (B) $(-x, y)$
   (C) $(x - 2, y)$ (D) $(x, -y)$

Postulates, Theorems, and Corollaries

Postulate 1 *Ruler Postulate:*
1. Any two distinct points on a line can be assigned coordinates 0 and 1.
2. There is a one-to-one correspondence between the real numbers and all points on the line.
3. To every pair of points, there corresponds exactly one positive number called the distance between the two points.

Postulate 2 *Segment Addition Postulate:* If C is between A and B, then $AC + CB = AB$.

Postulate 3 Any segment has exactly one midpoint.

Postulate 4 *Protractor Postulate:* In a given plane, select any line $\overleftrightarrow{AB}$ and any point C between A and B. Also select any two points R and S on the same side of $\overleftrightarrow{AB}$ such that S is not on $\overrightarrow{CR}$. Then there is a pairing of rays to real numbers from 0 to 180 as follows.
1. $\overrightarrow{CA}$ is paired with 0 and $\overrightarrow{CB}$ is paired with 180.
2. If $\overrightarrow{CR}$ is paired with x, then $0 < x < 180$.
3. If $\overrightarrow{CR}$ is paired with x and $\overrightarrow{CS}$ is paired with y, then $m\angle RCS = |x - y|$.

Postulate 5 *Angle Addition Postulate:* If D is in the interior of $\angle ABC$, then $m\angle ABC = m\angle ABD + m\angle DBC$.

Postulate 6 Every angle, except a straight angle, has exactly one bisector.

Postulate 7 If the outer rays of two adjacent angles form a straight angle, then the sum of the measures of the angles is 180.

Theorem 1.1 If the outer rays of two acute adjacent angles are perpendicular, then the sum of the measures of the angles is 90.

Theorem 2.1 If two angles are supplements of congruent angles, then they are congruent. (Supplements of congruent angles are congruent.)

Corollary If two angles are supplements of the same angle, then they are congruent. (Supplements of the same angle are congruent.)

Theorem 2.2 If two angles are complements of congruent angles, then they are congruent. (Complements of congruent angles are congruent.)

Corollary If two angles are complements of the same angle, then they are congruent. (Complements of congruent angles are congruent.)

Theorem 2.3 If two angles are right angles, then they are congruent.

Theorem 2.4 *Vertical Angles Theorem:* Vertical angles are congruent.

Corollary If two lines are perpendicular, then four right angles are formed.

Postulate 8 A line contains at least two points. A plane contains at least three noncollinear points. Space contains at least four noncoplaner points.

Postulate 9 For any two points, there is exactly one line containing them.

Theorem 2.5 Two lines intersect at exactly one point.

Postulate 10 If two points of a line are in a given plane, then the line itself is in the plane.

Theorem 2.6 If a line intersects a plane, but is not contained in the plane, then the intersection is exactly one point.

Postulate 11 If two planes intersect, then they intersect in exactly one line.

Postulate 12 Three noncollinear points are contained in exactly one plane.

Theorem 2.7 A line and a point not on the line are contained in exactly one plane.

Theorem 2.8 Two intersecting lines are contained in exactly one plane.

Postulate 13 *Alternate Interior Angles Postulate:* If a transversal intersects two lines such that alternate interior angles are congruent (equal in measure), then the lines are parallel.

Theorem 3.1 If a transversal intersects two lines such that corresponding angles are congruent, then the lines are parallel.

Theorem 3.2 If two lines are intersected by a transversal such that interior angles on the same side of the transversal are supplementary, then the lines are parallel.

Theorem 3.3 In a plane, if two lines are perpendicular to the same line, then they are parallel.

Postulate 14 *Parallel Postulate:* Through a point not on a line, there is exactly one line parallel to the given line.

Theorem 3.4 If two parallel lines are intersected by a transversal, then alternate interior angles are congruent.

Theorem 3.5 If two parallel lines are intersected by a transversal, then corresponding angles are congruent.

Theorem 3.6 If two parallel lines are intersected by a transversal, then interior angles on the same side of the transversal are supplementary.

Theorem 3.7 If a transversal is perpendicular to one of two parallel lines, then it is perpendicular to the other.

Theorem 3.8 In a plane, if two lines are parallel to the line, then they are parallel to each other.

Theorem 3.9 The sum of the measures of the angles of a triangle is 180.

Theorem 3.10 *Exterior Angle Theorem:* The measure of an exterior angle of a triangle is equal to the sum of the measures of its two remote interior angles.

Theorem 3.11 If two parallel planes are intersected by a third plane, then the lines of intersection are parallel.

Theorem 4.1 In a right triangle, the two angles other than the right angle are complementary and acute.

Postulate 15 *SAS Postulate for Congruence of Triangles:* If two sides and the included angle of one triangle are congruent to the corresponding two sides and included angle of a second triangle, then the triangles are congruent.

Postulate 16 *SSS Postulate for Congruence of Triangles:* If the three sides of one triangle are congruent to the corresponding three sides of a second triangle, then the triangles are congruent.

Postulate 17 *ASA Postulate for Congruence of Triangles:* If two angles and the included side of one triangle are congruent to the corresponding two angles and included side of a second triangle, then the triangles are congruent.

Theorem 4.2 *Third Angle Theorem:* If two angles of one triangle are congruent to two angles of a second triangle, then the third angles of the triangles are congruent.

Theorem 4.3 *AAS Theorem:* If two angles and a nonincluded side of one triangle are congruent to the corresponding two angles and side of a second triangle, then the triangles are congruent.

Theorem 5.1 If two sides of a triangle are congruent, then angles opposite these sides are congruent. (The base angles of an isosceles triangle are congruent.)

Corollary If a triangle is equilateral, then it is also equiangular, and the measure of each angle is 60.

Theorem 5.2 If two angles of a triangle are congruent, then the sides opposite these angles are congruent.

Corollary If a triangle is equiangular, then it is also equilateral.

Theorem 5.3 *Hypotenuse-Leg (HL) Theorem:* Two right triangles are congruent if the hypotenuse and a leg of one are congruent, respectively, to the hypotenuse and corresponding leg of the other.

Theorem 5.4 The altitude from the vertex angle to the base of an isosceles triangle is a median. (The altitude bisects the base.)

Theorem 5.5 Corresponding medians of congruent triangles are congruent.

Theorem 5.6 Corresponding altitudes of congruent triangles are congruent.

Theorem 5.7 The bisector of the vertex angle of an isosceles triangle is the perpendicular bisector of the base.

Corollary The bisector of the vertex angle of an isosceles triangle is also a median and an altitude of the triangle.

Theorem 5.8 A line containing two points, each equidistant from the endpoints of a given segment, is the perpendicular bisector of the segment.

Theorem 5.9 Any point on the perpendicular bisector of a segment is equidistant from the endpoints of the segment.

Theorem 5.10 *Exterior Angle Inequality Theorem:* The measure of an exterior angle of a triangle is greater than the measure of either of its remote interior angles.

Theorem 5.11 If one side of a triangle is longer than another side, then the measure of the angle opposite the longer side is greater than the measure of the angle opposite the shorter side.

Theorem 5.12 If one angle of a triangle has a greater measure than a second angle, then the side opposite the greater angle is longer than the side opposite the smaller angle.

Theorem 5.13 In a scalene triangle, the longest side is opposite the largest angle and the largest angle is opposite the longest side.

Theorem 5.14 The perpendicular segment from a point to a line is the shortest segment from the point to the line.

Corollary The longest side of a right triangle is the hypotenuse.

Theorem 5.15 *Triangle Inequality Theorem:* The sum of the lengths of any two sides of a triangle is greater than the length of the third side.

Theorem 5.16 *SAS Inequality Theorem:* If two sides of one triangle are congruent, respectively, to two sides of a second triangle, and the included angle of the first triangle has a greater measure than the included angle of the second triangle, then the third side of the first triangle is longer than the third side of the second triangle.

Theorem 5.17 *SSS Inequality Theorem:* If two sides of one triangle are congruent, respectively, to two sides of a second triangle, and the length of the third side of the first triangle is greater than the length of the third side of the second triangle, then the angle opposite the third side of the first triangle has a greater measure than the angle opposite the third side of the second triangle.

Theorem 6.1 The sum of the measures of the interior angles of a convex polygon with n sides is $(n - 2)180$.

Corollary 1 The sum of the measures of the interior angles of a convex quadrilateral is 360.

Corollary 2 The measure of an angle of a regular polygon with n sides is $\frac{(n - 2)180}{n}$.

Theorem 6.2 The sum of the measures of the exterior angles, one at each vertex, of any convex polygon is 360.

Corollary The measure of an exterior angle of a regular polygon with n sides is $\frac{360}{n}$.

Theorem 6.3 A diagonal of a parallelogram forms two congruent triangles.
Corollary 1 Opposite sides of a parallelogram are congruent.
Corollary 2 Opposite angles of a parallelogram are congruent.

Theorem 6.4 Consecutive angles of a parallelogram are supplementary.

Theorem 6.5 The diagonals of a parallelogram bisect each other.

Theorem 6.6 If both pairs of opposite sides of a quadrilateral are congruent, then the quadrilateral is a parallelogram.

Theorem 6.7 If the diagonals of a quadrilateral bisect each other, then the quadrilateral is a parallelogram.

Theorem 6.8 If two sides of a quadrilateral are parallel and congruent, then the quadrilateral is a parallelogram.

Theorem 6.9 If both pairs of opposite angles of a quadrilateral are congruent, then the quadrilateral is a parallelogram.

Theorem 6.10 *Midsegment Theorem:* The segment joining the midpoints of two sides of a triangle is parallel to the third side, and its length is half the length of the third side.

Theorem 6.11 If two lines are parallel, then all points of each line are equidistant from the other line.

Theorem 6.12 If three parallel lines cut off congruent segments on one transversal, then they cut off congruent segments on every transversal.
Corollary If any number of parallel lines cut off congruent segments on one transversal, then they cut off congruent segments on every transversal.

Theorem 6.13 If a segment is parallel to one side of a triangle and contains the midpoint of a second side, then this segment bisects the third side.

Theorem 7.1 *SASAS for Congruent Quadrilaterals:* Two quadrilaterals are congruent if any three sides and the included angles of one are congruent, respectively, to three sides and the included angles of the other.

Theorem 7.2 *ASASA for Congruent Quadrilaterals:* Two quadrilaterals are congruent if any three angles and the included sides of one are congruent, respectively, to three angles and the included sides of the other.

Theorem 7.3 The diagonals of a rhombus are perpendicular.

Theorem 7.4 The diagonals of a rectangle are congruent.

Theorem 7.5 Each diagonal of a rhombus bisects two angles of the rhombus.

Theorem 7.6 A parallelogram with one right angle is a rectangle.

Theorem 7.7 A parallelogram with two adjacent, congruent sides is a rhombus.

Theorem 7.8 A parallelogram with perpendicular diagonals is a rhombus.

Theorem 7.9 A parallelogram with congruent diagonals is a rectangle.

Theorem 7.10 A parallelogram with a diagonal that bisects opposite angles is a rhombus.

Theorem 7.11 A quadrilateral with four congruent sides is a rhombus.

Theorem 7.12 All altitudes of a trapezoid are congruent.

Theorem 7.13 The median of a trapezoid is parallel to its bases. Its length is one-half the sum of the lengths of the two bases.

Theorem 7.14 The base angles of an isosceles trapezoid are congruent.

Theorem 7.15 If the base angles of a trapezoid are congruent, then the trapezoid is isosceles.

Theorem 7.16 The diagonals of an isosceles trapezoid are congruent.

Theorem 7.17 If the diagonals of a trapezoid are congruent, then the trapezoid is isosceles.

Theorem 8.1 In a proportion, the product of the extremes equals the product of the means.

Corollary If the product of the extremes equals the product of the means, then a proportion exists.

Theorem 8.2 If $\frac{a}{b} = \frac{c}{d}$ (assume no denominator equals zero), then:

1. $\frac{b}{d} = \frac{d}{c}$
2. $\frac{a}{c} = \frac{b}{d}$
3. $\frac{a+b}{b} = \frac{c+d}{d}$
4. $\frac{a-b}{b} = \frac{c-d}{d}$
5. $\frac{a}{b} = \frac{a+c}{b+d}$

Theorem 8.3 Congruent triangles are similar.

Theorem 8.4 *Transitive Property of Triangle Similarity:* If $\triangle ABC \sim \triangle DEF$ and $\triangle DEF \sim \triangle GHI$, then $\triangle ABC \sim \triangle GHI$.

Postulate 18 *AA Similarity Postulate:* If two angles of a triangle are congruent to two angles of another triangle, then the two triangles are similar.

Theorem 8.5 *Triangle Proportionality Theorem:* If a line is parallel to one side of a triangle and intersects the other two sides, then it divides the two sides proportionally.

Theorem 8.6 If a line divides two sides of a triangle proportionally, then the line is parallel to the third side of the triangle.

Theorem 8.7 *SAS Similarity Theorem:* If an angle of one triangle is congruent to an angle of another triangle and the corresponding sides that include these angles are proportional, then the triangles are similar.

Theorem 8.8 *SSS Similarity Theorem:* If all three pairs of corresponding sides of two triangles are proportional, then the two triangles are similar.

Theorem 8.9 Corresponding altitudes of similar triangles are proportional to corresponding sides.

Theorem 8.10 Corresponding medians of similar triangles are proportional to corresponding sides.

Theorem 8.11 The bisector of an angle of a triangle divides the opposite side of the triangle into segments proportional to the other two sides.

Theorem 9.1 In a right triangle, the altitude to the hypotenuse forms two similar right triangles, each of which is also similar to the original triangle.

Corollary 1 In a right triangle, the square of the length of the altitude to the hypotenuse equals the product of the lengths of the segments formed on the hypotenuse.

Corollary 2 If the altitude is drawn to the hypotenuse of a right triangle, then the square of the length of either leg equals the product of the lengths of the hypotenuse and the segment of the hypotenuse adjacent to that leg.

Theorem 9.2 *Pythagorean Theorem:* In any right triangle, the sum of the squares of the lengths of the legs is equal to the square of the length of the hypotenuse.

Theorem 9.3 *Converse of the Pythagorean Theorem:* If the sum of the squares of the lengths of two sides of a triangle is equal to the square of the length of the third side, then the triangle is a right triangle.

Theorem 9.4 If the square of the longest side of a triangle is greater (less) than the sum of the squares of the lengths of the other two sides, then the triangle is obtuse (acute).

Theorem 9.5 In a 45-45-90 triangle, the hypotenuse is $\sqrt{2}$ times as long as a leg.

Theorem 9.6 In a 30-60-90 triangle, the hypotenuse is twice as long as the leg opposite the 30 angle. The leg opposite the 60 angle is $\sqrt{3}$ times as long as the leg opposite the 30 angle.

Theorem 10.1 *Distance Formula:* The distance d between $P_1(x_1,y_1)$ and $P_2(x_2,y_2)$ is given by the formula $d = \sqrt{(x_2 - x_1)^2 + (y_2 - y_1)^2}$.

Theorem 10.2 *Midpoint Formula:* Given $P(x_1,y_1)$ and $Q(x_2,y_2)$, the coordinates (x_m,y_m) of M, the midpoint of $\overline{PQ}$, are $\left(\frac{x_1 + x_2}{2}, \frac{y_1 + y_2}{2}\right)$.

Theorem 10.3 All segments of a non-vertical line have equal slopes.

Theorem 10.4 An equation of a line with slope m containing the point $P_1(x_1,y_1)$ is $y - y_1 = m(x - x_1)$.

Theorem 10.5 If a line has slope m and y-intercept b, then an equation of the line is $y = mx + b$.

Theorem 10.6 If two non-vertical lines have the same slope, then they are parallel.

Theorem 10.7 If two non-vertical lines are parallel, then they have equal slopes.

Theorem 10.8 If the product of the slopes of two non-vertical lines is -1, then the lines are perpendicular.

Theorem 10.9 The product of the slopes of two non-vertical perpendicular lines is -1.

Theorem 11.1 If a line or segment contains the center of a circle and is perpendicular to a chord, then it *bisects* the chord.

Theorem 11.2 In the same circle or in congruent circles, congruent chords are equidistant from the center(s).

Theorem 11.3 In the same circle or in congruent circles, chords that are equidistant from the center(s) are congruent.

Theorem 11.4 In the same circle or congruent circles, if two chords are unequally distant from the center(s), then the chord nearer its corresponding center is the longer chord.

Theorem 11.5 In the same circle or congruent circles, if two chords are unequal in length, then the longer chord is nearer the center of its circle.

Theorem 11.6 If a line is perpendicular to a radius at its endpoint on the circle, then the line is tangent to the circle.

Theorem 11.7 If a line is tangent to a circle, then the line is perpendicular to the radius drawn to the point of tangency.

Theorem 11.8 Two segments drawn tangent to a circle from an exterior point are congruent.

Corollary The angle between two tangents to a circle from an exterior point is bisected by the segment joining its vertex and the center of the circle.

Postulate 19 If P is a point on $\widehat{APB}$, then $m\widehat{AP} + m\widehat{PB} = m\widehat{APB}$.

Theorem 11.9 In the same circle or in congruent circles:

1. If chords are congruent, then their corresponding arcs and central angles are congruent;
2. If arcs are congruent, then their corresponding chords and central angles are congruent;
3. If central angles are congruent, then their corresponding arcs and chords are congruent.

Theorem 11.10 *Inscribed Angle Theorem:* The measure of an inscribed angle is one-half of the degree measure of its intercepted arc.

Corollary 1 If two inscribed angles intercept the same arc or congruent arcs, then the angles are congruent.

Corollary 2 An angle inscribed in a *semicircle* is a *right* angle.

Corollary 3 If two arcs of a circle are included between parallel chords or secants, then the arcs are congruent.

Corollary 4 The opposite angles of an inscribed quadrilateral are supplementary.

Theorem 11.11 The measure of an angle formed by two secants or chords intersecting in the interior of a circle is one-half the sum of the measures of the arcs intercepted by the angle and its vertical angle.

Theorem 11.12 The measure of an angle formed by two secants intersecting in the exterior of the circle is one-half the difference of the measures of the intercepted arcs.

Theorem 11.13 If a tangent and a secant (or a chord) intersect at the point of tangency on a circle, then the measure of the angle formed is one-half the measure of its intercepted arc.

Theorem 11.14 The measure of an angle formed either by (1) a tangent and secant intersecting at a point exterior to a circle, or (2) two tangents intersecting at a point exterior to a circle equals one-half the difference of the measures of the intercepted arcs.

Theorem 11.15 If two chords of a circle intersect, then the product of the lengths of the segments of one chord equals the product of the lengths of the segments of the other chord.

Theorem 11.16 If a tangent and a secant intersect in the exterior of a circle, then the square of the length of the tangent segment equals the product of the lengths of the secant segment and the external secant segment.

Corollary If two secants intersect in the exterior of a circle, then the product of the lengths of one secant segment and its external segment equals the product of the lengths of the other secant segment and its external segment.

Theorem 11.17 The equation of a circle with the coordinates of the center (h,k) and a radius of length r is $(x - h)^2 + (y - k)^2 = r^2$.

Postulate 20 Congruent polygons have equal areas.

Postulate 21 The area of a rectangle is the product of the lengths of a base and a corresponding altitude (Area of rectangle $= bh$).

Theorem 12.1 The area of a square is the square of the length s of a side ($A = s^2$).

Postulate 22 *Area Addition Postulate:* If a region is the union of two or more nonoverlapping regions, then its area is the sum of the areas of these nonoverlapping regions.

Theorem 12.2 The area of a parallelogram is the product of the lengths of a base and a corresponding altitude (Area of parallelogram $= bh$).

Theorem 12.3 The area of a triangle is one-half the product of the lengths of a base and a corresponding altitude (Area of triangle $= \frac{1}{2}bh$).

Theorem 12.4 The area of a kite is one-half the product of the lengths of the diagonals (Area of kite $= \frac{1}{2}d_1d_2$).

Corollary The area of a rhombus is one-half the product of the lengths of the two diagonals.

Theorem 12.5 If s is the length of a side of an equilateral triangle, then the area is $\frac{s^2}{4}\sqrt{3}$.

Theorem 12.6 *Heron's Formula:* If a, b, and c are the lengths of the sides of a triangle and s is the semiperimeter, such that $s = \frac{1}{2}(a + b + c)$, then Area(triangle) $= \sqrt{s(s - a)(s - b)(s - c)}$.

Theorem 12.7 The area of a trapezoid is one-half the product of the sum of the lengths of the upper and lower bases and the length of an altitude.

Theorem 12.8 A circle can be circumscribed about any regular polygon.

Theorem 12.9 The area of a regular polygon is one-half the product of the apothem and the perimeter [Area (n-gon) $= \frac{1}{2}ap$)].

Theorem 12.10 The area of a regular polygon is $n[\sin(\frac{180}{n})]\,[\cos(\frac{180}{n})]r^2$, or $\frac{ns^2}{4\tan\left(\frac{180}{n}\right)}$, where n is the number of sides, s is the length of a side, and r is the length of a radius.

Theorem 12.11 The ratio of the perimeters of two similar polygons is the same as the ratio of the lengths of any two corresponding sides.

Theorem 12.12 The ratio of the areas of two similar triangles is the square of the ratio of the lengths of any two corresponding sides.

Theorem 12.13 The ratio of the areas of two similar polygons is the square of the ratio of the lengths of any two corresponding sides.

Theorem 12.14 The ratio of the circumference to the length of a diameter is the same for all circles.

Corollary The circumference of a circle with radius of length r is $2\pi r$.

Theorem 12.15 The area of a circle with radius of length r is πr^2.

Theorem 12.16 The area of a sector of a circle is one-half the product of the length s of the arc and the length r of its radius ($A = \frac{1}{2}rs$).

Theorem 13.1 The locus of points in a plane equidistant from two given points is the perpendicular bisector of the segment having the two points as endpoints.

Theorem 13.2 In a plane, the locus of points equidistant from the sides of an angle is the bisector of the angle.

Theorem 13.3 The perpendicular bisectors of the sides of a triangle are concurrent at a point equidistant from the vertices of the triangle.

Theorem 13.4 The bisectors of the angles of a triangle are concurrent at a point equidistant from the sides of the triangle.

Theorem 13.5 The lines containing the altitudes of a triangle are concurrent.

Theorem 13.6 Two medians of a triangle intersect at a point two-thirds of the distance from each vertex to the midpoint of the opposite side.

Theorem 13.7 The medians of a triangle are concurrent at a point that is two-thirds the distance from each vertex to the midpoint of the opposite side.

Theorem 14.1 The lateral area of a right prism is the product of the perimeter of a base and the length of an altitude ($L = ph$).

Theorem 14.2 The lateral area of a right cylinder is the product of the circumference of a base and the length of an altitude. The lateral area of a right cylinder with radius r and altitude of length h is $2\pi rh$.

Theorem 14.3 The total area of a right cylinder with radius of length r and altitude of length h is $2\pi r^2 + 2\pi rh$, or $2\pi r(r + h)$.

Theorem 14.4 If the ratio of the lengths of corresponding edges of two similar polyhedra is $\frac{a}{b}$, then the ratio of the lateral areas and of the total areas is $(\frac{a}{b})^2$.

Postulate 23 For any rectangular solid, the volume $V = lwh$, where l, w, and h are the lengths of three edges with a common vertex.

Theorem 14.5 The volume of a cube with edges of length s is s^3.

Postulate 24 If a solid is the union of two or more nonoverlapping solids, then its volume is the sum of the volumes of these nonoverlapping parts.

Postulate 25 *Cavalieri's Principle:* If two solids have equal heights, and if the cross sections formed by any plane parallel to the bases of both solids have equal areas, then the volumes of the solids are equal.

Theorem 14.6 For any prism or cylinder, the volume is the product of the area of a base and the length of an altitude ($V = Bh$, where B = area of base and h = altitude).

Corollary The volume of a cylinder is Bh, or $\pi r^2 h$.

Theorem 14.7 The lateral area of a regular pyramid is one-half the product of the perimeter of the base and the slant height ($L = \frac{1}{2}pl$).

Theorem 14.8 The lateral area L of a right cone is πrl. The total area (A) is $\pi rl + \pi r^2 = \pi r(l + r)$.

Theorem 14.9 The volume of a pyramid is one-third the volume of a prism with the same base and altitude as the pyramid. The volume of a cone is one-third the volume of a cylinder with the same base and altitude as the cone ($V = \frac{1}{3}Bh$).

Corollary 1 The volume of a pyramid or cone is one-third the product of the area of its base and the length of its altitude ($V = \frac{1}{3}Bh$).

Corollary 2 The volume of a cone with a base radius of length r and an altitude of length h is $\frac{1}{3}\pi r^2 h$.

Theorem 14.10 The volume of a sphere with radius of length r is $\frac{4}{3}\pi r^3$.

Theorem 14.11 The area of a sphere is $4\pi r^2$.

Theorem 14.12 The distance between points $P(x_1,y_1,z_1)$ and $Q(x_2,y_2,z_2)$ is given by the formula $d = \sqrt{(x_2 - x_1)^2 + (y_2 - y_1)^2 + (z_2 - z_1)^2}$.

Theorem 15.1 A reflection is an isometry.

Theorem 15.2 A translation is an isometry.

Theorem 15.3 The resultant image determined by two successive reflections about parallel lines is a translation.

Theorem 15.4 A rotation is an isometry.

Theorem 15.5 The measure of the angle of rotation formed by two successive reflections (or by the product of two reflections) is twice the measure of the non-obtuse angle between the two lines of symmetry.

Theorem 15.6 The transformation defined by adding a constant to the coordinates of each point is a translation.

Theorem 15.7 The transformation defined by multiplying each coordinate of each point by a constant is a dilation.

Table of Roots and Powers

No.	Sq.	Sq. Root	Cube	Cu. Root	No.	Sq.	Sq. Root	Cube	Cu. Root
1	1	1.000	1	1.000	51	2,601	7.141	132,651	3.708
2	4	1.414	8	1.260	52	2,704	7.211	140,608	3.733
3	9	1.732	27	1.442	53	2,809	7.280	148,877	3.756
4	16	2.000	64	1.587	54	2,916	7.348	157,564	3.780
5	25	2.236	125	1.710	55	3,025	7.416	166,375	3.803
6	36	2.449	216	1.817	56	3,136	7.483	175,616	3.826
7	49	2.646	343	1.913	57	3,249	7.550	185,193	3.849
8	64	2.828	512	2.000	58	3,364	7.616	195,112	3.871
9	81	3.000	729	2.080	59	3,481	7.681	205,379	3.893
10	100	3.162	1,000	2.154	60	3,600	7.746	216,000	3.915
11	121	3.317	1,331	2.224	61	3,721	7.810	226,981	3.936
12	144	3.464	1,728	2.289	62	3,844	7.874	238,328	3.958
13	169	3.606	2,197	2.351	63	3,969	7.937	250,047	3.979
14	196	3.742	2,744	2.410	64	4,096	8.000	262,144	4.000
15	225	3.875	3,375	2.466	65	4,225	8.062	274,625	4.021
16	256	4.000	4,096	2.520	66	4,356	8.124	287,496	4.041
17	289	4.123	4,913	2.571	67	4,489	8.185	300,763	4.062
18	324	4.243	5,832	2.621	68	4,624	8.246	314,432	4.082
19	361	4.359	6,859	2.668	69	4,761	8.307	328,509	4.102
20	400	4.472	8,000	2.714	70	4,900	8.367	343,000	4.121
21	441	4.583	9,261	2.759	71	5,041	8.426	357,911	4.141
22	484	4.690	10,648	2.802	72	5,184	8.485	373,248	4.160
23	529	4.796	12,167	2.844	73	5,329	8.544	389,017	4.179
24	576	4.899	13,824	2.884	74	5,476	8.602	405,224	4.198
25	625	5.000	15,625	2.924	75	5,625	8.660	421,875	4.217
26	676	5.099	17,576	2.962	76	5,776	8.718	438,976	4.236
27	729	5.196	19,683	3.000	77	5,929	8.775	456,533	4.254
28	784	5.292	21,952	3.037	78	6,084	8.832	474,552	4.273
29	841	5.385	24,389	3.072	79	6,241	8.888	493,039	4.291
30	900	5.477	27,000	3.107	80	6,400	8.944	512,000	4.309
31	961	5.568	29,791	3.141	81	6,561	9.000	531,441	4.327
32	1,024	5.657	32,768	3.175	82	6,724	9.055	551,368	4.344
33	1,089	5.745	35,937	3.208	83	6,889	9.110	571,787	4.362
34	1,156	5.831	39,304	3.240	84	7,056	9.165	592,704	4.380
35	1,225	5.916	42,875	3.271	85	7,225	9.220	614,125	4.397
36	1,296	6.000	46,656	3.302	86	7,396	9.274	636,056	4.414
37	1,369	6.083	50,653	3.332	87	7,569	9.327	658,503	4.431
38	1,444	6.164	54,872	3.362	88	7,744	9.381	681,472	4.448
39	1,521	6.245	59,319	3.391	89	7,921	9.434	704,969	4.465
40	1,600	6.325	64,000	3.420	90	8,100	9.487	729,000	4.481
41	1,681	6.403	68,921	3.448	91	8,281	9.539	753,571	4.498
42	1,764	6.481	74,088	3.476	92	8,464	9.592	778,688	4.514
43	1,849	6.557	79,507	3.503	93	8,649	9.644	804,357	4.531
44	1,936	6.633	85,184	3.530	94	8,836	9.695	830,584	4.547
45	2,025	6.708	91,125	3.557	95	9,025	9.747	857,375	4.563
46	2,116	6.782	97,336	3.583	96	9,216	9.798	884,736	4.579
47	2,209	6.856	103,823	3.609	97	9,409	9.849	912,673	4.595
48	2,304	6.928	110,592	3.634	98	9,604	9.899	941,192	4.610
49	2,401	7.000	117,649	3.659	99	9,801	9.950	970,299	4.626
50	2,500	7.071	125,000	3.684	100	10,000	10.000	1,000,000	4.642

Trigonometric Ratios

Angle Measure	*Sin*	*Cos*	*Tan*	*Angle Measure*	*Sin*	*Cos*	*Tan*
0°	0.000	1.000	0.000	46°	.7193	.6947	1.036
1°	.0175	.9998	.0175	47°	.7314	.6820	1.072
2°	.0349	.9994	.0349	48°	.7431	.6691	1.111
3°	.0523	.9986	.0524	49°	.7547	.6561	1.150
4°	.0698	.9976	.0699	50°	.7660	.6428	1.192
5°	.0872	.9962	.0875	51°	.7771	.6293	1.235
6°	.1045	.9945	.1051	52°	.7880	.6157	1.280
7°	.1219	.9925	.1228	53°	.7986	.6018	1.327
8°	.1392	.9903	.1405	54°	.8090	.5878	1.376
9°	.1564	.9877	.1584	55°	.8192	.5736	1.428
10°	.1736	.9848	.1763	56°	.8290	.5592	1.483
11°	.1908	.9816	.1944	57°	.8387	.5446	1.540
12°	.2079	.9781	.2126	58°	.8480	.5299	1.600
13°	.2250	.9744	.2309	59°	.8572	.5150	1.664
14°	.2419	.9703	.2493	60°	.8660	.5000	1.732
15°	.2588	.9659	.2679	61°	.8746	.4848	1.804
16°	.2756	.9613	.2867	62°	.8829	.4695	1.881
17°	.2924	.9563	.3057	63°	.8910	.4540	1.963
18°	.3090	.9511	.3249	64°	.8988	.4384	2.050
19°	.3256	.9455	.3443	65°	.9063	.4226	2.145
20°	.3420	.9397	.3640	66°	.9135	.4067	2.246
21°	.3584	.9336	.3839	67°	.9205	.3907	2.356
22°	.3746	.9272	.4040	68°	.9272	.3746	2.475
23°	.3907	.9205	.4245	69°	.9336	.3584	2.605
24°	.4067	.9135	.4452	70°	.9397	.3420	2.747
25°	.4226	.9063	.4663	71°	.9455	.3256	2.904
26°	.4384	.8988	.4877	72°	.9511	.3090	3.077
27°	.4540	.8910	.5095	73°	.9563	.2924	3.270
28°	.4695	.8829	.5317	74°	.9613	.2756	3.487
29°	.4848	.8746	.5543	75°	.9659	.2588	3.732
30°	.5000	.8660	.5774	76°	.9703	.2419	4.010
31°	.5150	.8572	.6009	77°	.9744	.2250	4.331
32°	.5299	.8480	.6249	78°	.9781	.2079	4.704
33°	.5446	.8387	.6494	79°	.9816	.1908	5.145
34°	.5592	.8290	.6745	80°	.9848	.1736	5.671
35°	.5736	.8192	.7002	81°	.9877	.1564	6.314
36°	.5878	.8090	.7265	82°	.9903	.1392	7.115
37°	.6018	.7986	.7536	83°	.9925	.1219	8.144
38°	.6157	.7880	.7813	84°	.9945	.1045	9.514
39°	.6293	.7771	.8098	85°	.9962	.0872	11.43
40°	.6428	.7660	.8391	86°	.9976	.0698	14.30
41°	.6561	.7547	.8693	87°	.9986	.0523	19.08
42°	.6691	.7431	.9004	88°	.9994	.0349	28.64
43°	.6820	.7314	.9325	89°	.9998	.0175	57.29
44°	.6947	.7193	.9657	90°	1.000	0.000	
45°	.7071	.7071	1.000				

Glossary

acute angle: An angle with a measure less than 90. (p. 16)

acute triangle: A triangle with three acute angles. (p. 139)

adjacent angles: Two coplanar angles with one common side and a common vertex, but no common interior points. (p. 21)

adjacent sides of a polygon: Sides which intersect at a vertex. (p. 225)

alternate interior/exterior angles: Interior/Exterior, nonadjacent angles which lie on opposite sides of a transversal. (pp. 97, 98)

altitude of a prism: Any segment perpendicular to the planes containing the two bases with endpoints in these planes. (p. 564)

altitude of a triangle: A segment from a vertex of the triangle perpendicular to the opposite side. (p. 194)

altitude of a trapezoid: A perpendicular segment from any point on one base to the other base. (p. 286)

angle: Two rays with a common endpoint. The rays are the *sides*, the endpoint is the *vertex*. (p. 15)

angle bisector: A segment, line, ray, or plane which divides an angle into two congruent angles. (p. 23)

apothem of a regular polygon: The length of a perpendicular segment from the center of the polygon to a side. (p. 499)

arc: An unbroken part of a circle measured by referring to an angle whose vertex is the center of the circle. (p. 432)

area of a circle: The limit of the areas of inscribed regular polygons as the number of sides increases indefinitely. (p. 512)

area of a polygon: The number of square units in the region bounded. (p. 478)

base angles of an isosceles triangle: The angles of an isosceles triangle opposite the legs or congruent sides. (p. 177)

base of an isosceles triangle: The side opposite the vertex angle in an isosceles triangle. (p. 177)

bases of a prism: Two congruent polygonal faces that lie in parallel planes. (p. 560)

bisector of a segment: A line, ray, segment, or plane that intersects a segment at its midpoint. (p. 10)

center of a regular polygon: The common center of its inscribed and circumscribed circles. (p. 498)

central angle: An angle whose vertex is the center of the circle. (p. 432)

chord: A segment within a circle with endpoints on the circle. (p. 416)

circle: The set of all points in a plane that are a given distance from a fixed point in that plane. (p. 415)

circumcenter: The intersection of perpendicular bisectors of sides of a triangle. (p. 423)

circumscribed circle: A circle in which a polygon is inscribed. (p. 543)

circumscribed polygon: A polygon whose sides are tangent to a circle. (p. 427)

collinear: Points contained within the same line. (p. 2)

complementary angles: Two angles with measures whose sum is 90. (p. 28)

concentric circles: Coplanar circles of different radii with a common center. (p. 438)

concurrent lines: Lines that converge, or intersect, in one common point. (p. 541)

conditional: A statement that can be written in the form "If p, then q." (p. 63)

cone: A pyramid-like object whose base is a circle and whose lateral surface is made up of an infinite number of segments between the circle and a vertex. (p. 576)

congruent angles: Angles that have the same measure. (p. 17)

congruent arcs: Arcs that have the same degree measure. (p. 438)

congruent circles: Circles that have congruent radii. (p. 415)

congruent quadrilaterals: Quadrilaterals with congruent corresponding sides. (p. 269)

congruent segments: Segments that have the same length. (p. 10)

congruent triangles: Triangle with congruent, corresponding sides and angles. (p. 143)

conjunction: If p and q are both statements, the statement "*p and q*" is their conjunction. (p. 32)

converse: The statement formed by interchanging the hypothesis and conclusion of a conditional. (p. 109)

convex: A polygon is convex if a segment joining any two interior points of the polygon is in the interior of the polygon. (p. 227)

coordinates: A corresponding ordered pair (x,y) for every point in a plane. (p. 379)

coplanar: Points within the same plane. (p. 2)

corollary: A theorem whose proof follows from another theorem in a few steps. (p. 71)

corresponding angles: Angles which lie on the same side of a transversal. (p. 98)

cosine: In right $\triangle ABC$ with acute $\angle A$, cosine $\angle A = \dfrac{\textit{length of adjacent leg}}{\textit{length of hypotenuse}}$ (p. 366)

cylinder: A cylinder consists of two congruent and parallel circular bases whose lateral surface is the infinite number of segments connecting the circles. (p. 561)

deductive reasoning: Inferring from general principles to prove a statement. (p. 67)

diagonal: A segment which joins two nonconsecutive vertices. (p. 226)

diameter: A chord that contains the center of a circle. (p. 416)

dilation: A transformation which produces an image similar to the original. (p. 622)

disjunction: The logical union "*p or q*," where p and q are statements. (p. 33)

equiangular: Has congruent angles. (p. 139)

equidistant: Equally distant from. (p. 11)

equilateral: With congruent sides. (p. 139)

exterior angle: An angle that is adjacent and supplementary to one of the angles of a triangle. (p. 121)

geometric mean: The geometric mean of two positive numbers, a and b, is the positive number m, such that $\frac{a}{m} = \frac{m}{b}$. (p. 304)

geometric transformation: Matching the points of a figure with the points of a second figure called the *image*. (p. 599)

hypotenuse: The side opposite the right angle in a right triangle. (p. 188)

hypothesis: The hypothesis is the "if" part of a conditional statement. (p. 63)

indirect proof: A proof which assumes that the desired conclusion is not true and shows that this assumption leads to a contradiction. (p. 106)

inductive reasoning: Generalizations based on repeated observations. (p. 68)

inscribed angle: An angle whose vertex lies on a circle and whose sides contain chords of the circle. (p. 442)

inscribed circle: A circle about which a polygon is circumscribed. (p. 427)

inscribed polygon: A polygon whose sides are chords of a circle. (p. 423)

intersection: A set of points contained in two or more intersecting figures. (p. 3)

intercepted arc: The arc $\overparen{AB}$ is called the intercepted arc of inscribed angle ACB. (p. 442)

inverse: The statement formed by negating the hypothesis and the conclusion. (p. 128)

isometry: A transformation that is distance-preserving. The distance between two points is the same as the distance between the corresponding images of these points. (p. 600)

isosceles trapezoid: A trapezoid with congruent legs. (p. 290)

isosceles triangle: A triangle with at least two congruent sides. (p. 139)

kite: A quadrilateral in which one diagonal is the perpendicular bisector of the other. (p. 489)

lateral faces of a prism: The faces of a prism that are formed by parallelograms. (p. 560)

lateral surface: The curved surface between the bases of a geometric figure. (p. 561)

legs of an isosceles triangle: The two congruent sides of an isosceles triangle. (p. 177)

legs of a right triangle: The sides forming the right angle of a right triangle. (p. 188)

linear equation: An equation whose graph is a line in which the set of points has the coordinates that satisfy the given equation. (p. 393)

linear pair: Two adjacent angles whose outer rays are opposite rays. (p. 26)

line of symmetry: The reflection or mirror line of a reflection transformation. (p. 600)

locus: The set of all points that satisfy a given set of conditions. (p. 529)

major arc: The major arc $\overparen{ACB}$ consists of points A and B and all points of the circle in the *exterior* of central angle AOB. (p. 433)

means: b and c are the means of the proportion $\frac{a}{b} = \frac{c}{d}$ $(b \neq 0, d \neq 0)$. (p. 303)

median of a trapezoid: A segment that joins the midpoints of the legs. (p. 286)

midpoint of an arc: A point M is the midpoint of $\overparen{AMB}$ if $\overparen{AM} \cong \overparen{MB}$. (p. 439)

midpoint of a segment: The point at which a segment is divided into two congruent segments. (p. 10)

minor arc: The minor arc, $\overparen{AB}$, of central angle AOB consists of points A, B, and all points on the circle that lie in the interior of the central angle. (p. 432)

oblique cylinder: A cylinder that is not a right cylinder. (p. 561)

oblique prism: A prism that is not a right prism. (p. 560)

obtuse angle: An angle with a measure greater than 90. (p. 16)

obtuse triangle: A triangle with one obtuse angle. (p. 139)

opposite rays: If point O is between points A and B on $\overleftrightarrow{AB}$, then $\overrightarrow{OA}$ and $\overrightarrow{OB}$ are called opposite rays. (p. 26)

opposite sides of a quadrilateral: The nonadjacent sides of a quadrilateral. (p. 241)

parallel lines: Coplanar lines that do not intersect. (p. 93)

parallelogram: A quadrilateral with both pairs of opposite sides parallel. (p. 241)

parallel planes: Planes that do not intersect. (p. 130)

perimeter of a triangle: The sum of the lengths of its three sides. (p. 178)

perpendicular: Two lines which intersect to form a right angle. (p. 27)

perpendicular bisector: A perpendicular bisector of a side is perpendicular to that side at its midpoint. (p. 199)

polygon: The union of three or more coplanar segments such that each endpoint is shared by exactly two segments; segments intersect only at their endpoints; and intersecting segments are noncollinear. (p. 225)

postulate: A statement that is accepted without proof. (p. 5)

prism: A polyhedron with two congruent polygonal faces, called the bases, in parallel planes. (p. 560)

proportion: An equation of the form $\frac{a}{b} = \frac{c}{d}$ $(b \neq 0, d \neq 0)$. (p. 303)

pyramid: A polyhedron composed of a polygonal region, called the base, and triangular regions, called the lateral faces which intersect at a vertex. (p. 574)

quadrilateral: A quadrilateral is a polygon with four sides. (p. 241)

radius: In a circle, a radius is a segment from the center to any point on the circle. (plural: radii) (p. 415)

radius of a regular polygon: A radius of the regular polygon is a segment from the center to a vertex of the polygon. (p. 499)

ratio: A ratio is a comparison of two numbers by division. The ratio of a to b $(b \neq 0)$ may be written a to b, $\frac{a}{b}$, or $a{:}b$. (p. 303)

ray: $\overrightarrow{XY}$ consists of $\overline{XY}$ and all points P such that Y is between X and P. (p. 7)

rectangle: A parallelogram with four right angles. (p. 273)

reflection: In a plane, a reflection about line l is a transformation which maps each point P into a point P' such that (1) if P is on l, then $P' = P$, and (2) if P is not on l, then l is the perpendicular bisector of $\overline{PP'}$. (p. 599)

regular polygon: A convex polygon that is both equilateral and equiangular. (p. 227)

regular polyhedron: A polyhedron in which all the faces are congruent regular polygons. (p. 557)

remote interior angle: An interior angle that is not adjacent to the given exterior angle. (p. 121)

rhombus: A parallelogram with four congruent sides. (p. 273)

right angle: An angle with a measure of 90. (p. 16)

right cylinder: A cylinder in which a segment connecting a pair of corresponding points of the bases is perpendicular to the bases. (p. 561)

right prism: A prism in which the lateral edges are perpendicular to the bases. (p. 560)

right triangle: A triangle with one right angle. (p. 139)

rotation: A rotation $R_{P,m}$ of point A about point P through an angle of measure m is a transformation which *maps* A into its *rotation image* A' such that $PA = PA'$ and m $\angle APA' = m$. (p. 613)

rotational symmetry: A figure has rotational symmetry if there is a rotation in which the figure and its image coincide under the rotation. (p. 615)

scalene triangle: A triangle with no congruent sides. (p. 139)

secant: A line that intersects the circle in two points. (p. 417)

sector of a circle: The region bounded by an arc of a circle and two radii whose endpoints are endpoints of that arc. (p. 516)

segment: Segment $\overline{AB}$ is the set of points consisting of points A, B, and all points between A and B. (p. 6)

similar polygons: Polygons in which corresponding angles are congruent and the ratios of the lengths of corresponding sides are equal. (p. 308)

similar triangles: Triangles in which corresponding angles are congruent and corresponding sides are proportional. (p. 313)

sine: The sine of an acute angle of a right triangle is the ratio of the leg opposite the angle to the length of the hypotenuse. (p. 360)

skew lines: Noncoplanar lines. (p. 93)

slope: The slope of a segment $\overline{P_1P_2}$, with endpoints $P_1(x_1,y_1)$ and $P_2(x_2,y_2)$, is the ratio $\frac{y_2 - y_1}{x_2 - x_1}$ $(x_2 \neq x_1)$. (p. 388)

space: The set of all points. (p. 2)

sphere: The set of all points in space at a given distance from a given point called the center. (p. 584)

square: A rectangle with four congruent sides. (p. 273)

standard form: The standard form of an equation of a line is an equation of the form $ax + by = c$, where a and b are not both zero, and where $a > 0$. (p. 393)

straight angle: An angle with a measure of 180. (p. 16)

supplementary angles: Two angles with measures whose sum is 180. Each angle is called a *supplement* of the other. (p. 27)

symmetric figures: Figures that are their own mirror images about a line or about a plane. (p. 601)

tangent: The tangent of an angle is the ratio of the length of the opposite leg to the length of the adjacent leg. (p. 366)

tangent circles: Two coplanar circles which are tangent to the same line at the same point. (p. 428)

tangent segment: A segment that contains a point of tangency and another point of a tangent line to a circle. (p. 425)

tangent to a circle: A line that is coplanar with the circle and intersects the circle in exactly one point. (p. 417)

theorem: A statement that has been proved true. (p. 27)

translation: A translation in a plane from A to A' is a transformation which maps any point P into a point P' such that $\overline{PP'} \cong \overline{AA'}$ and $\overline{PP'} \parallel \overline{AA'}$. (p. 607)

transversal: A line, ray, or segment that intersects two or more coplanar lines, rays, or segments, each at a different point. (p. 97)

trapezoid: A quadrilateral with exactly one pair of parallel sides. (p. 285)

triangle: A figure formed by three segments joining three noncollinear points. (p. 115)

vertex: The common endpoint of the sides of an angle. (p. 15)

vertex angle of an isosceles triangle: The angle formed by the legs of an isosceles triangle. (p. 177)

vertical angles: Two nonadjacent angles formed by intersecting lines. (p. 77)

volume of a prism: The product of the area of a base and the length of an altitude. ($V = Bh$, where B = area of base and h = altitude.) (p. 569)

x-axis: The horizontal line in a plane. (p. 379)

y-axis: The vertical line in a plane. (p. 379)

y-intercept: The point at which the line intersects the y-axis. (p. 394)

Selected Answers to Written Exercises

Chapter 1

Written Exercises, page 4

1. $\overleftrightarrow{PQ}, \overleftrightarrow{QP}$ **3.** $\mathcal{N}$ **5.** True **7.** False. E lies below M. **9.** B **11.** $\overleftrightarrow{SQ}$ **13.** S

15.

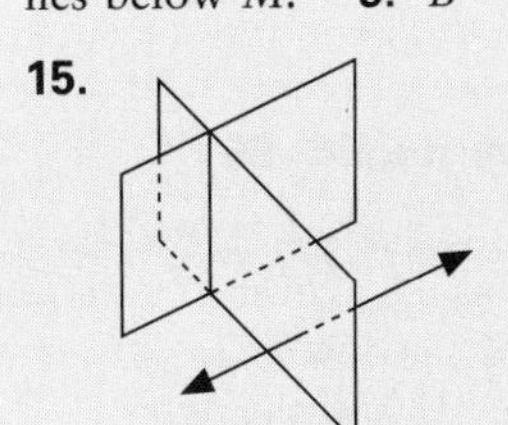

17.

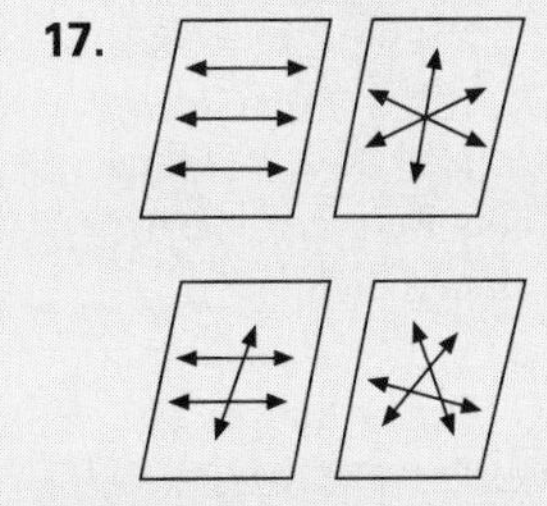

Written Exercises, pages 8–9

1. 6 **3.** 7 **5.** 14 **7.** 17 **9.** 12 **11.** 3.7

13. A R V

$AR + RV = AV$

15. $\overrightarrow{WR}, \overrightarrow{WY}, \overrightarrow{WP}$ **17.** $\overline{AB}$, or $\overline{BA}$ **19.** HW, or WH **21.** 12 **23.** 1, −13 **25.** 14.44 **27.** 12 **29.** 8 *or* −16

Written Exercises, pages 13–14

1. $\overline{TL} \cong \overline{LU}$, $TL = LU$

3. G H, P Q **5.** M, P Q

7. 2 **9.** $8\frac{1}{2}$ **11.** −2.3 **13.** $2\frac{1}{5}$ **15.** A: 0, B: 16 **17.** A: 3, B: 17 **19.** M: 1, B: 6 **21.** G: 4 **23.** No, G can be negative. **25.** 12

Written Exercises, pages 18–19

1. $\angle 3, \angle K, \angle JKL, \angle LKJ$ **3.** $\angle 1, \angle MFA, \angle AFM$; $\angle 2$ $\angle WFM, \angle MFW, \angle AFW$; $\angle WFA$ **5–7.** See Construction on page 17. **9.** 60, acute **11.** 110, obtuse **13.** 20, acute **15.** 180, st **17.** $\angle FAB$ **19.** 50 **21.** $\angle AFE, \angle DGE, \angle CFB, \angle FGC$ **23.** No. m $\angle TUV = 40$

Written Exercises, pages 24–25

1. No; no common side **3.** No; no common side **5.** 35 **7.** 72 **9.** 45 **11.** 70 **13.** See Construction on page 23. **15.** 14 **17.** 52 **19.** 76 **21.** 25

Written Exercises, pages 29–30

1. Comp; the outer rays of 2 acute adj $\angle$s are $\perp$. **3.** Neither; the sum of the meas of the $\angle$s is neither 90 nor 180. **5.** Comp = 16, supp = 106 **7.** No comp, supp = 81 **9.** No comp, no supp **11.** Comp = 58.3, supp = 148.3 **13.** If $3 < m < 93$, comp $= 93 - m$; if $3 < m < 183$, supp $= 183 - m$. **15.** If $5 < x < 50$, comp $= 100 - 2x$; if $5 < x < 95$, supp $= 190 - 2x$. **17.** No. The supp of an $\angle$ greater than 90 must be less than 90. **19.** Supp **21.** 125, 55 **23.** 90, 90 **25.** 140, 40 **27.** 15, 75 **29.** 42, 48, 138 **31.** Assume $\angle ABC$ and $\angle CBD$ are adj and comp. By the Angle Add Post, we know that m $\angle ABC$ + m $\angle CBD$ = m $\angle ABD$. Since $\angle ABC$ and $\angle CBD$ are comp, we know that m $\angle ABC$ + m $\angle CBD$ = 90. So, m $\angle ABD$ = 90. Thus, by def $\perp$, $\overrightarrow{BA} \perp \overrightarrow{BD}$.

Written Exercises, page 34

1. Conj: A potato is a vegetable *and* West Berlin is a country. (False) Disj: A potato is a vegetable *or* West Berlin is a country. (True) **3.** Conj: $3 + 7 < 10$ *and* a plane is a flat surface. (True) Disj: $3 + 7 < 10$ *or* a plane is a flat surface. (True) **5.** T **7.** T **9.** When both p and q are false **11.** Conj: Plane $\mathcal{M}$ intersects plane $\mathcal{N}$ in $\overleftrightarrow{BE}$ *and* pts F, A, and B are coll. (False) Disj: Plane $\mathcal{M}$ intersects plane $\mathcal{N}$ in $\overleftrightarrow{BE}$ *or* pts F, A and B are coll. (True) **13.** Conj: $\overline{DE}$ lies in plane $\mathcal{P}$ *and* $\overleftrightarrow{BE}$ contains only two pts, E and B. (False) Disj: $\overline{DE}$ lies in plane $\mathcal{P}$ *or* $\overleftrightarrow{BE}$ contains only two pts, E and B. (False)

15.

p	q	$p \lor q$	r	$(p \lor q) \lor r$
T	T	T	T	T
T	T	T	F	T
T	F	T	T	T
T	F	T	F	T
F	T	T	T	T
F	T	T	F	T
F	F	F	T	T
F	F	F	F	F

$(p \vee q) \vee r$ is false only when p, q, and r are all false. **17.** 16

Written Exercises, page 37

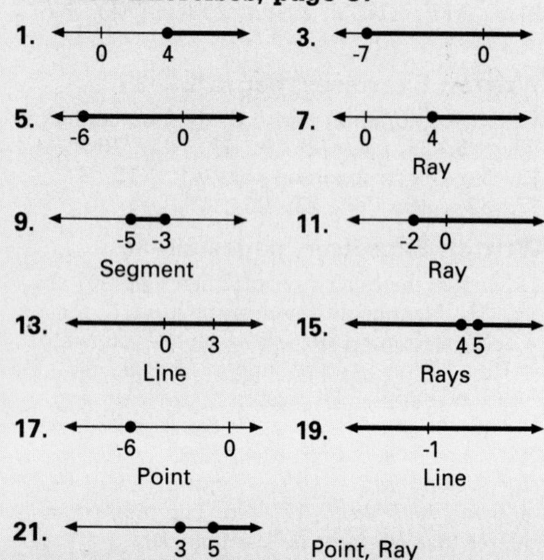

Chapter 2

Written Exercises, pages 45–47

1. $HK + KL = HL$ (Seg Add Post) **3.** $AB = BC$, $\overline{AB} \cong \overline{BC}$ (Def of seg bis) **5.** m $\angle 1$ + m $\angle 2$ = m $\angle XYZ$ (Angle Add Post) **7.** m $\angle 1$ + m $\angle 2$ = 90 (If the outer rays of 2 acute adj $\angle$s are $\perp$, then the sum of the meas of the $\angle$s is 90.) **9.** m $\angle 1$ + m $\angle 2$ = 180 (If the outer rays of two adj $\angle$s form a st $\angle$, the sum of their measures is 180.) **11.** m $\angle BEC$ = m $\angle CED$, $\angle BEC \cong \angle CED$ (Def of $\angle$ bis) **13.** m $\angle 3$ + m $\angle 4$ = 90 (If the outer rays of two acute adj $\angle$s are $\perp$, then the sum of the meas of the $\angle$s is 90.)

15.

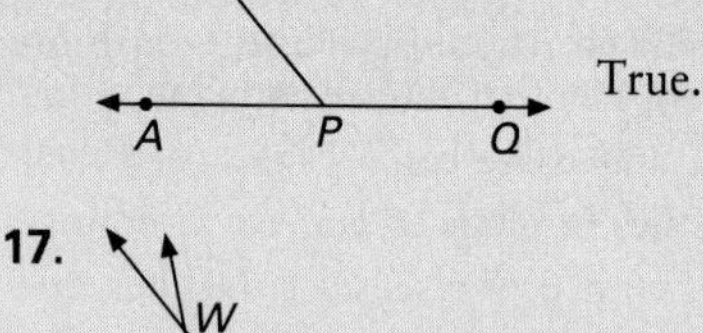

True.

17.

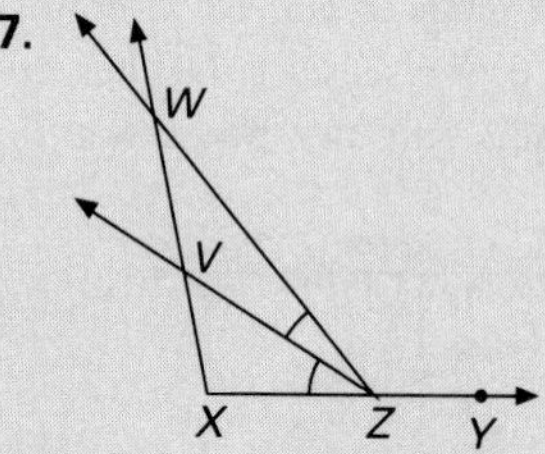

True. By Angle Add Post, m $\angle VZY$ = m $\angle WZY$ + m $\angle VZW$; $\overrightarrow{ZV}$ bis $\angle XZW$, so m $\angle VZW$ = m $\angle XZV$; $\therefore$ by sub, m $\angle VZY$ = m $\angle WZY$ + m $\angle XZV$.

Written Exercises, pages 51–52

1. Reflex Prop of Eq **3.** Trans Prop of Congr **5.** Equations may be added. **7.** $7x = 28$ (Given); $x = 4$ (Div Prop of Eq) **9.** (2) Subt Prop of Eq **11.** (2) If the outer rays of 2 adj $\angle$s form a st $\angle$, then the sum of their meas is 180. (3) Sub **13.** No; not reflexive and not symmetric **15.** No; not reflexive and not transitive

Written Exercises, pages 54–56

1. $\overrightarrow{OB}$ bis $\angle TOY$ (Given); $\therefore$ m $\angle 1$ = m $\angle 2$ (Def of $\angle$ bis) **3.** $\overline{UR}$ bis $\overline{YK}$ (Given); $\therefore \overline{YU} \cong \overline{KU}$ (Def of seg bis) **5.** m $\angle 3$ = m $\angle 4$ (Given); m $\angle 3$ + m $\angle 2$ = 180 (If the outer rays of 2 adj $\angle$s form a st $\angle$, then the sum of their meas is 180.); $\therefore$ m $\angle 2$ + m $\angle 4$ = 180 (Sub) **7.** $\overrightarrow{AB}$ bis $\angle PAR$, $\angle 3 \cong \angle 2$ (Given); $\angle 1 \cong \angle 3$ (Def of $\angle$ bis); $\therefore \angle 1 \cong \angle 2$ (Trans Prop of Congr) **9.** $\overline{TL}$ bis $\overline{BG}$; $\overline{TA} \cong \overline{BA}$ (Given); $\overline{BA} \cong \overline{AG}$ (Def of seg bis); $\therefore \overline{TA} \cong \overline{AG}$ (Trans Prop of Congr) **11.** B is the midpt of $\overline{AC}$ (Given); $AB = BC$ (Def of midpt); $BC + CD = BD$ (Seg Add Post); $\therefore AB + CD = BD$ (Sub) **13.** $\overline{BA} \perp \overline{BD}$; $\overline{BD}$ bis $\angle EBC$ (Given); m $\angle 2$ = m $\angle 3$ (Def of $\angle$bis); m $\angle 1$ + m $\angle 2$ = 90 (If the outer rays of 2 adj $\angle$s are $\perp$, then the sum of their meas is 90.); $\therefore$ m $\angle 1$ + m $\angle 3$ = 90 (Sub) **15.** $\overline{AC} \perp \overline{AB}$; m $\angle ADB$ = m $\angle CAD$ (Given); m $\angle CAD$ + m $\angle DAB$ = 90 (If the outer rays of 2 acute adj $\angle$s are $\perp$, then the sum of their meas is 90.); $\therefore$ m $\angle ADB$ + m $\angle DAB$ = 90 (Sub)

Written Exercises, pages 59–61

1. (Given); (Equations may be subtracted.); MP, NQ; (Seg Add Post); (Sub) **3.** Subt Prop of Eq **5.** m $\angle 5$ = m $\angle 8$, m $\angle 6$ = m $\angle 7$ (Given); m $\angle 5$ + m $\angle 6$ = m $\angle ABC$, m $\angle 7$ + m $\angle 8$ = m $\angle EFG$ (Angle Add Post); m $\angle 5$ + m $\angle 6$ = m $\angle 7$ + m $\angle 8$ (Equations may be added.); $\therefore$ m $\angle ABC$ = m $\angle EFG$ (Sub) **7.** m $\angle 3$ = m $\angle 5$ (Given); m $\angle 3$ + m $\angle 4$ = 180 (If the outer rays of 2 adj $\angle$s from a st $\angle$, then the sum of their meas is 180.) $\therefore$ m $\angle 4$ + m $\angle 5$ = 180 (Sub) **9.** $AB = RS$ (Given); $AB + BR = AR$, $RS + BR = BS$ (Seg Add Post); $AB + BR = RS + BR$ (Add Prop of Eq); $\therefore AR = BS$ (Sub) **11.** $\overline{WX} \perp \overline{WY}$, m $\angle 5$ = m $\angle 7$ (Given); m $\angle 5$ + m $\angle 6$ = 90 (If the outer rays of 2 acute adj $\angle$s are $\perp$, then the sum of their meas is 90.); $\therefore$ m $\angle 7$ + m $\angle 6$ = 90 (Sub)
13. $\overline{PQ} \perp \overline{PS}$, $\overline{PR} \perp \overline{PT}$ (Given); m $\angle 1$ + m

$\angle 2 = 90$, m $\angle 1$ + m $\angle 3 = 90$ (If the outer rays of 2 acute adj $\angle$s are $\perp$, then the sum of their meas is 90.); m $\angle 1$ + m $\angle 2$ = m $\angle 1$ + m $\angle 3$ (Sub); $\therefore$ m $\angle 2$ = m $\angle 3$ (Subt Prop of Eq) **15.** $\overrightarrow{BE}$ bis $\angle ABD$ (Given); m $\angle 7$ = m $\angle 8$ (Def of $\angle$ bis); m $\angle 6$ + m $\angle 7$ = m $\angle EBC$ (Angle Add Post); $\therefore$ m $\angle 6$ + m $\angle 8$ = m $\angle EBC$ (Sub) **17.** $AC = AD$ (Given); $AE + ED = AD$ (Seg Add Post); $BE = AE$ (Given); $\therefore BE + ED = AC$ (Sub)

Written Exercises, pages 65–66

1. If $\angle A$ is obtuse, then $90 <$ m $\angle A < 180$. **3.** If $x^3 = -27$, then $x = -3$. (True) **5.** If $x < 4$, then $x < 5$. (True) **7.** If the sum of the measures of two angles is 180, then the angles are supplementary. *or*: If two angles are supplementary, then the sum of their measures is 180. **9.** An angle is acute if its measure is less than 90. **11.** If $\overrightarrow{AB} \perp \overrightarrow{AC}$, then m $\angle 1$ + m $\angle 2 = 90$.

13.

p	q	$p \rightarrow q$	$p \wedge (p \rightarrow q)$
T	T	T	T
T	F	F	F
F	T	T	F
F	F	T	F

Written Exercises, pages 69–70

1. Mary will not develop cavities. **3.** $\angle A \cong \angle B$ **5.** If you have a cold, then you should not exercise strenuously. **7.** $AD = BD$ in first three, but not in the last triangle. Inductive reasoning does not always lead to true conclusions. **9.** Deductively **11.** Deductively **13.** 15 **15.** 190

Written Exercises, pages 73–75

1. m $\angle EBA$ = m $\angle ECD$. Supps of $\cong$ $\angle$s are $\cong$. **3.** m $\angle DCA$ = m $\angle CAB$, Comps of $\cong$ $\angle$s are $\cong$. **5.** m $\angle 1 = 60$, m $\angle 2 = 60$ (Given); m $\angle 1$ = m $\angle 2$ ($\angle 1 \cong \angle 2$) (Sub); $\angle EBA$ is supp $\angle 1$, $\angle ECD$ is supp $\angle 2$ (If the outer rays of adj $\angle$s form a st $\angle$, the $\angle$s are supp.); $\angle EBA \cong \angle ECD$ (m $\angle EBA$ = m $\angle ECD$) (Supps of $\cong$ $\angle$s are $\cong$.) **7.** $\overline{AB} \perp \overline{AD}$, $\overline{BC} \perp \overline{CD}$, $\angle DAC \cong \angle ACB$ (Given); $\angle DAC$ is comp $\angle CAB$ (If the outer rays of 2 acute adj $\angle$s are $\perp$, the $\angle$s are comp.); $\therefore$ $\angle DCA \cong \angle CAB$ (m $\angle DCA$ = m $\angle CAB$) (Comps of $\cong$ $\angle$s are $\cong$.) **9.** m $\angle 1 = 48$. Proof: $\overline{AB} \perp \overline{CD}$, $\angle ABE \cong \angle ABF$ (Given); m $\angle 1$ + m $\angle 2 = 90$ (If the outer rays of 2 acute adj $\angle$s are $\perp$, the sum of their meas is 90.); m $\angle 2 = 90$ (Def of $\cong$ $\angle$s); $\therefore$ m$\angle 1 = 48$ (Eq may be added.) **11.** $\overrightarrow{QP} \perp \overline{QR}$, $\angle 3$ and $\angle 2$ are comp (Given); $\angle 1$ and $\angle 2$ are comp (If the outer rays of 2 acute adj $\angle$s are $\perp$, then the $\angle$s are comp.); $\therefore$ $\angle 1 \cong \angle 3$ (Comps of the same $\angle$ are $\cong$.) **13.** m $\angle 2$ = m $\angle 5$, m $\angle 3$ = m $\angle 5$ (Given); m $\angle 2$ = m $\angle 3$ (Sub); $\angle 1$ and $\angle 2$ are supp; $\angle 4$ and $\angle 3$ are supp (If the outer rays of 2 adj $\angle$s form st $\angle$, then the $\angle$s are supp.); $\therefore$ $\angle 1 \cong \angle 4$ (Supps of $\cong$ $\angle$s are $\cong$.) **15.** Using the diagram in the middle of page 72: $\angle A$ and $\angle B$ are comp; $\angle C$ and $\angle D$ are comp; m $\angle B$ = m $\angle C$ (Given); m $\angle A$ + m $\angle B = 90$, m $\angle C$ + m $\angle D = 90$ (Def of comp $\angle$s); m $\angle A$ + m $\angle B$ = m $\angle C$ + m $\angle D$ (Sub); m $\angle A$ + m $\angle B$ = m $\angle B$ + m $\angle D$ (Sub); $\therefore$ m $\angle A$ = m $\angle D$ (Subt Prop of Eq) **17.** $\angle 1$ and $\angle 2$ are rt $\angle$s (Given); m $\angle 1 = 90$, m $\angle 2 = 90$ (Def of rt $\angle$s); m $\angle 1$ = m $\angle 2$ (Sub) **19.** $\angle 1$ is supp of $\angle 4$ (Given); $\angle 3$ is supp of $\angle 4$ (If the outer rays of 2 adj $\angle$s form a st line, the $\angle$s are supp.); $\angle 1 \cong \angle 3$ (Supps of same $\angle$ are $\cong$.); $l \parallel m$ (If corr $\angle$s are $\cong$, lines are $\parallel$.); $\therefore$ $\angle 2$ is supp of $\angle 3$ (If lines are $\parallel$, then int $\angle$s on the same side of transv are supp.)

Written Exercises, pages 80–81

1. 65 **3.** 60 **5.** 70 **7.** 140 **9.** 140 **11.** 23 **13.** 113 **15.** m $\angle 3$ = m $\angle 5$ (Vert $\angle$s are $\cong$.); m $\angle 3$ + m $\angle 4$ = m $\angle DBF$ (Angle Add Post); m $\angle 5$ + m $\angle 4$ = m $\angle DBF$ (Sub) **17.** 30 **19.** Vert $\angle$s are two nonadj $\angle$s with a common vertex. The sides of both $\angle$s must be contained in 2 lines or segs which intersect at the common vertex of the $\angle$s.

Written Exercises, page 85

1. No, Thm 2.6 **3.** Infinitely many **5.** No, Post 12 **7.** Sometimes **9.** Sometimes **11.** Lines l and m intersect (Given); l and m intersect at a pt, call it A (Thm 2.4); l contains a pt B other than A, m has a pt C other than A (Post 8); A, B, and C are noncoll (Def of noncoll pts); Exactly one plane $\mathcal{N}$ contains A, B and C (Post 12); $\mathcal{N}$ contains l, $\mathcal{N}$ contains m (Post 10); $\therefore$ Exactly one plane contains l and m (Steps 6 & 7).

Chapter 3

Written Exercises, page 96

1. $\overline{ST}$ and $\overline{PQ}$ **3.** $\overline{PQ}$ and $\overline{QR}$
5.

7.

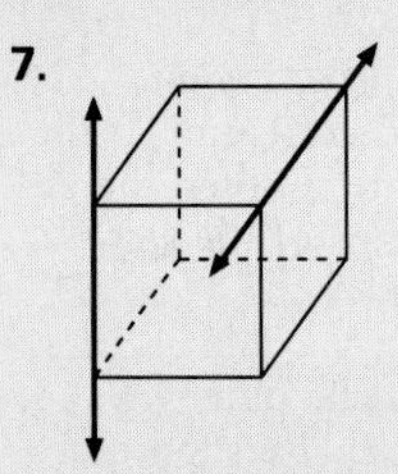

9. Always (by definition) **11.** Never (Two parallel planes share no points in common.)
13.

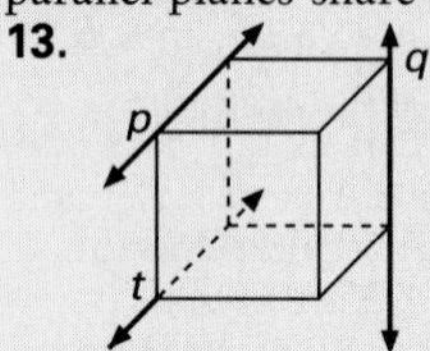

Written Exercises, page 99

1. Corr **3.** Alt ext **5.** Corr **7.** Corr **9.** Alt int **11.** Corr **13.** Alt int **15.** Alt int **17.** None **19.** Corr

Written Exercises, pages 104–105

1. $\overline{PQ} \parallel \overline{RS}$; corr $\angle$s are $\cong$. **3.** $\overline{RS} \parallel \overline{PQ}$; int $\angle$s on the same side of the transv are supp. **5.** $l \parallel m$; alt int $\angle$s are $\cong$. **7.** $\overline{PQ} \parallel \overline{RS}$; corr $\angle$s are $\cong$. **9.** $l \parallel m$; corr $\angle$s are $\cong$. **11.** $\overline{PT}$ is a transv to $\overline{PQ}$ and $\overline{RS}$. **13.** $\overline{PT}$ is a transv to the vertical studs. $\angle 1 \cong \angle 2$ (corr $\angle$s). **15.** 30 **17.** 80, 100 **19.** 85 **21.** m $\angle 3$ = m $\angle 2$, $\overline{BC}$ bis $\angle DBE$ (Given); m $\angle 1$ = m $\angle 2$ (Def of $\angle$ bis); m $\angle 3$ = m $\angle 1$ (Sub); $\therefore \overline{AD} \parallel \overline{BC}$ (If alt int $\angle$s are $\cong$, then lines are $\parallel$.)
23.

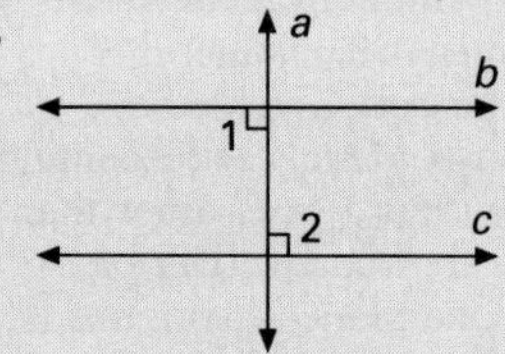

$b \perp a$, $c \perp a$ (Given); m $\angle 1$ = 90, m $\angle 2$ = 90 (Def of $\perp$); m $\angle 1$ = m $\angle 2$ (Sub); $b \parallel c$ (If alt int $\angle$s are $\cong$, then lines are $\parallel$.) **25.** $x^2 + 12x + 72 = 180$; $x^2 + 12x - 108 = 0$; $(x - 6)(x + 18) = 0$; $x = 6$

Written Exercises, pages 107–108

1. Given: P is not the midpt of $\overline{AB}$. Assume: $AP = PB$. Then $AP \cong PB$, by def of $\cong$ segs, and so P is a midpt of $\overline{AB}$. But P is not a midpt of $\overline{AB}$ and P is a midpt of $\overline{AB}$ is a contradiction. So the assumption that $AP = PB$ must be false. Therefore, $AP \neq PB$. **3.** Given: $l \not\perp m$. Assume: m $\angle 1$ = m $\angle 3$. Then m $\angle 1$ + m $\angle 3$ = 180 (if the outer rays of 2 adj $\angle$s form a st $\angle$, the sum of the meas of the $\angle$s is 180), and so m $\angle 1$ + m $\angle 1$ = 180 (subs). Thus 2(m $\angle 1$) = 180, and m $\angle 1$ = 90 (div prop of eq). Hence, $l \perp m$ (def of $\perp$ lines). But $l \perp m$ and $l \not\perp m$ is a contradiction. So the assumption that m $\angle 1$ = m $\angle 3$ must be false. Therefore, m $\angle 1$ $\neq$ m $\angle 3$. **5.** Given: $l \perp m$, $m \nparallel n$. Assume: $l \perp n$. Then l is a transv $\perp$ to both m and n. So $m \parallel n$. But $m \perp n$ and $m \parallel n$ is a contradiction. So the assumption that $l \perp n$ is false. Therefore, $l \not\perp n$. **7.** Given: $m \nparallel n$. Assume: m $\angle 1$ + m $\angle 2$ = 180. Then, since $\angle 1$ and $\angle 3$ form a st $\angle$, m $\angle 1$ + m $\angle 3$ = 180. Since $\angle 2$ and $\angle 3$ are both supp $\angle$s of $\angle 1$, they are $\cong$. Since they are $\cong$ corr $\angle$s, $m \parallel n$. But $m \parallel n$ and $m \nparallel n$ is a contradiction. So the assumption that m $\angle 1$ + m $\angle 2$ = 180 is false. Therefore, m $\angle 1$ + m $\angle 2$ $\neq$ 180. **9.** Given: m$\angle 1$ $\neq$ m $\angle 2$. Assume: $\angle 1$ and $\angle 2$ are vert $\angle$s. Then $\angle 1 \cong \angle 2$ (m $\angle 1$ = m$\angle 2$). But m $\angle 1$ $\neq$ m $\angle 2$ and m $\angle 1$ = m $\angle 2$ is a contradiction. So the assumption that $\angle 1$ and $\angle 2$ are vert $\angle$s must be false. Therefore, $\angle 1$ and $\angle 2$ are not vertical $\angle$s.
11. Given: $\overleftrightarrow{AB}$ and $\overleftrightarrow{CD}$ are skew. Assume: $\overleftrightarrow{AC}$ and $\overleftrightarrow{BD}$ are not skew. Since $\overleftrightarrow{AC}$ and $\overleftrightarrow{BD}$ are not skew, they either intersect or are $\parallel$. For $\overleftrightarrow{AC}$ and $\overleftrightarrow{BD}$ to either intersect or be $\parallel$, they must lie in the same plane, and therefore pts A, B, C, and D must also lie in the same plane. Since $\overleftrightarrow{AB}$ and $\overleftrightarrow{CD}$ are skew, A, B, C, and D cannot all lie in the same plane. But A, B, C and D lie in the same plane and A, B, C and D do not all lie in the same plane is a contradiction. So, the assumption that $\overleftrightarrow{AC}$ and $\overleftrightarrow{BD}$ are skew is false. Therefore, $\overleftrightarrow{AC}$ and $\overleftrightarrow{BD}$ are skew.

Written Exercises, pages 112–113

1. 37 **3.** 100 **5.** 105 **7.** 100 **9.** If the sum of the meas of 2 $\angle$s is 90, then the $\angle$s are comp. (True) **11.** If 2 lines are not $\parallel$, then they are skew. (False) **13.** m $\angle 1$ = 25, not enough info to find m $\angle 2$. **15.** m $\angle 3$ = 56, m $\angle 5$ = 75 **17.** $r \parallel s$ (Given); m $\angle 3$ = m $\angle 6$ (Vert $\angle$s are $\cong$.); m $\angle 6$ = m $\angle 5$ (If lines are $\parallel$, then corr $\angle$s are $\cong$.); m $\angle 3$ = m $\angle 5$ (Sub); m $\angle 1$ + m $\angle 5$ = 180 (If the outer rays of 2 adj $\angle$s form a st $\angle$, then the $\angle$s are supp.); m $\angle 1$ + m $\angle 3$ = 180 (Sub); $\therefore \angle 1$ and $\angle 3$ are supp (Def of supp $\angle$s) **19.** $l \parallel m$, m $\angle 4$ = m $\angle 3$ (Given); m $\angle 4$ = m $\angle 5$ (If lines are $\parallel$, then alt int $\angle$s are $\cong$.); m $\angle 6$ = m $\angle 3$ (Vert $\angle$s are

≅.); m ∠6 = m ∠5 (Sub); $r \parallel s$ (If corr ∠s are ≅, then lines are ∥.) **21.** $\overline{BC}$ bis ∠ABD, $\overline{AB} \parallel \overline{CD}$ (Given); m ∠1 = m ∠2 (Def of ∠ bis); m ∠1 = m ∠3 (If lines are ∥, then alt int ∠s are ≅.); ∴ m ∠2 = m ∠3 (Trans Prop of Eq) **23.** $\overline{AB} \parallel \overline{CD}$, $\overline{BD} \perp \overline{CD}$ (Given); $\overline{BD} \perp \overline{AB}$ (If a transv is ⊥ to one of 2 ∥ lines, then it is ⊥ to the other.); m ∠1 + m ∠2 = 90 (If the outer rays of 2 adj acute ∠s are ⊥, then the sum of their meas is 90.); m ∠1 = m ∠3 (If lines are ∥, then alt int ∠s are ≅.); m ∠2 + m ∠3 = 90 (Sub); ∴ ∠2 and ∠3 are comp (Def of comp ∠s)

25.

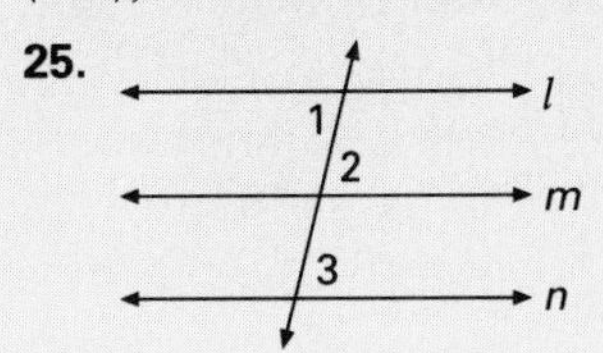

$l \parallel m$, $m \parallel n$ (Given); m ∠1 = m ∠2 (Alt int ∠s are ≅.); m ∠2 = m ∠3 (If lines are ∥, then corr ∠s are ≅.); m ∠1 = m ∠3 (Sub); $l \parallel n$ (If alt int ∠s are ≅, then lines are ∥.) **27.** $\overleftrightarrow{AD} \parallel \overleftrightarrow{EG}$. l bis ∠ABF, m bis ∠BFG (Given); m ∠ABF = m ∠BFG (If lines are ∥, alt int ∠s are ≅.); m ∠ABF = m ∠1 + m ∠2, m ∠BFG = m ∠3 + m ∠4 (Angle Add Post); m ∠1 + m ∠2 = m ∠3 + m ∠4 (Sub); m ∠1 = m ∠2, m ∠3 = m ∠4 (Def of ∠ bis); m ∠2 + m ∠2 = m ∠3 + m ∠3 (2 × m ∠2 = 2 × m ∠3)(Sub); m ∠2 = m ∠3 (Div Prop of Eq); ∴ $l \parallel m$ (If alt int ∠s are ≅, then lines are ∥.)

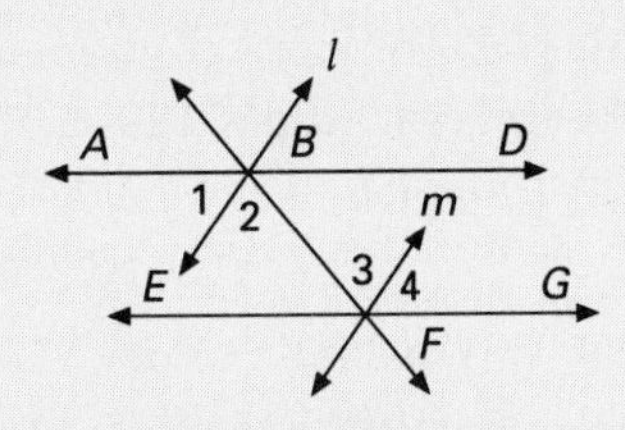

Written Exercises, pages 117–119

1. 60, 60, 60 **3.** 36, 36, 108 **5.** 40, 60, 80 **7.** 27, 63, 90 **9.** 55 **11.** 92 **13.** 140 **15.** $\overline{CD} \parallel \overline{AB}$ (Given); m ∠5 + m ∠ACD = 180 (If lines are ∥, then int ∠s on same side of the transv are supp.); m ∠ACD = m ∠2 + m ∠3 (Angle Add Post); m ∠5 + m ∠2 + m ∠3 = 180 (Sub); m ∠2 = m ∠4 (If lines are ∥, then alt int ∠s are ≅.); ∴ m ∠5 + m ∠3 + m ∠4 = 180 (Sub) **17.** m ∠1 = 51, m ∠2 = 51, m ∠3 = 97, m ∠4 = 83 **19.** 125 **21.** 20 **23.** 90 **25.** are ⊥

Written Exercises, pages 123–124

1. 80 **3.** 102 **5.** 55 **7.** 42 **9.** 100 **11.** 12 **13.** 50 **15.** 120 **17.** $\overleftrightarrow{PQ} \parallel \overleftrightarrow{ST}$ (Given); m ∠3 = m ∠4 (If 2 lines are ∥, then alt int ∠s are ≅.); m ∠1 = m ∠2 + m ∠3 (Meas of ext ∠ of a △ = sum of meas of remote int ∠s.); m ∠1 = m ∠2 + m ∠4 (Sub). **19.** 143 **21.** 80 **23.** 15

Written Exercises, page 129

1. ∥ lines do not intersect. (True) **3.** The sides of a rt ∠ are not ⊥. (False) **5.** Inv: If the outer rays of 2 adj ∠s do not form a st ∠, then the ∠s are not supp. (True) Contr: If 2 adj ∠s are not supp, then their outer rays do not form a line. (True) **7.** Inv: If an ∠ is not a rt ∠, then its meas is 20. (False) Contr: If the meas of an ∠ is 20, then it is not a rt ∠. (True) **9.** Conv: If lines are ∥, then corr ∠s are ≅. Bicond: Corr ∠s are ≅ if and only if lines are ∥. (True)

11.

p	$\sim p$	$\sim(\sim p)$
T	F	T
F	T	F

Written Exercises, page 132

1. True

3.

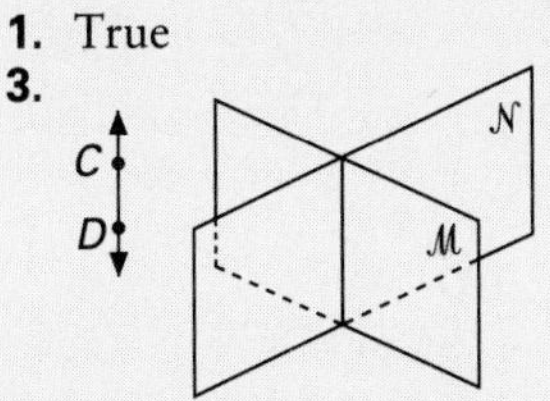

False. ℳ and 𝒩 are both ∥ $\overleftrightarrow{CD}$, but ℳ ∦ 𝒩. **5.** False. $\overleftrightarrow{CD} \parallel \overleftrightarrow{AB}$, but ℳ ∦ 𝒩. **7.** Given: ℳ and 𝒩 intersect in $\overleftrightarrow{AB}$; $\overleftrightarrow{CD} \parallel$ 𝒩, $\overleftrightarrow{CD} \parallel$ ℳ, $\overleftrightarrow{CD}$ is not skew to $\overleftrightarrow{AB}$. Assume: $\overleftrightarrow{CD} \nparallel \overleftrightarrow{AB}$. Since $\overleftrightarrow{CD}$ is not skew or ∥ to $\overleftrightarrow{AB}$, $\overleftrightarrow{CD}$ intersects $\overleftrightarrow{AB}$. Thus, $\overleftrightarrow{CD}$ shares at least one pt with both 𝒩 and ℳ, since $\overleftrightarrow{AB}$ lies in each. But $\overleftrightarrow{CD}$ shares no pts with 𝒩 or ℳ, since $\overleftrightarrow{CD}$ ∥ to both. But $\overleftrightarrow{CD}$ shares a pt with 𝒩 and ℳ and $\overleftrightarrow{CD}$ shares no pts with 𝒩 or ℳ is a contradiction. So, the assumption that $\overleftrightarrow{CD} \nparallel \overleftrightarrow{AB}$ is false, so $\overleftrightarrow{CD} \parallel \overleftrightarrow{AB}$. **9.** Given: 𝒫 ∥ 𝒬, and line m intersects 𝒫 at (just the one) point G of line m. Assume: m does not intersect 𝒬, i.e. m ∥ 𝒬. There is a ℛ which contains m and is ∥ 𝒬. ℛ contains m, so it contains the point G. But there is just one plane that contains the point G and is ∥ to 𝒬, so planes ℛ and 𝒫 must be the same plane. Thus 𝒫 contains m and hence all points of m (infintely many), which contradicts the assertion that m intersects 𝒫 (at just one point). So the assumption that m does not intersect $\mathcal{Q}$ must be false. Therefore, m intersects $\mathcal{Q}$.

Chapter 4

Written Exercises, page 141

1. Equilateral **3.** Scalene **5.** Right **7.** 28 **9.** 40 **11.** 24 **13.** $\triangle ABC$, m $\angle C = 90$ (Given); m $\angle A$ + m $\angle B$ + m $\angle C = 180$ (Sum of meas of a $\triangle$ is 180.); m $\angle A$ + m $\angle B = 90$ (Subt prop of eq.); ∴ $\angle A$ and $\angle B$ are comp (Def of comp $\angle$s). Also, m $\angle A > 0$ and m $\angle B > 0$, so m $\angle A < 90$ and m $\angle B < 90$ (Subt prop of eq), i.e. $\angle A$ and $\angle B$ are acute. **15.** Given: an obtuse $\triangle$. Assume: the $\triangle$ has a rt $\angle$. Then, since it is an obtuse $\triangle$, it has an angle with meas > 90. Now the sum of the meas of an obtuse $\angle$ and a rt $\angle$ is > 180, so the sum of the meas of the $\angle$s of the given $\triangle$ must be > 180. But the sum of the meas of the $\angle$s of the $\triangle$ must be 180. So the assumption that the obtuse $\triangle$ has a rt $\angle$ is false. Therefore, an obtuse $\triangle$ cannot have a rt $\angle$.

Written Exercises, page 145

1. $\triangle HRP$ **3.** $\triangle OMP$ **5.** $\angle Q$ **7.** $\overline{YW}$ **9.** $\overline{KT}$

11.

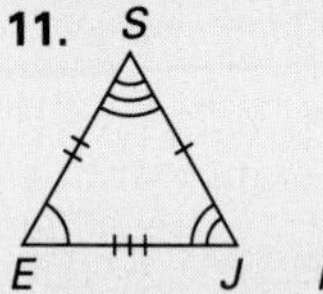

(Answers will vary.)

13. 6 **15.** 11

Written Exercises, pages 150–152

1. $\angle W \cong \angle U$ **3.** $\angle C \cong \angle G$ **5.** $\overline{QP} \perp \overline{QR}$, $\overline{TS} \perp \overline{TU}$, $\overline{QR} \cong \overline{TU}$, $\overline{QP} \cong \overline{TS}$ (Given); m $\angle Q = 90$, m $\angle T = 90$ (Def of $\perp$); m $\angle Q$ = m $\angle T$ (Sub); ∴ $\triangle PQR \cong \triangle STU$ (SAS). **7.** $\angle 1 \cong \angle 2$, $\overline{EC} \cong \overline{ED}$, E is the midpt of $\overline{AB}$ (Given); $\overline{AE} \cong \overline{BE}$ (Def of midpt); ∴ $\triangle ACE \cong \triangle BDE$ (SAS) **9.** $\overline{AB} \cong \overline{EF}$, $\overline{AC} \cong \overline{ED}$, $\overline{AB} \parallel \overline{FE}$ (Given); $\angle A \cong \angle E$ (If lines are ∥, then alt int $\angle$s are ≅.); ∴ $\triangle ABC \cong \triangle EFD$ (SAS) **11.** $\overline{DB} \perp \overline{AC}$, $\overline{DB}$ bis $\overline{AC}$ (Given); m $\angle DBC = 90$, m $\angle DBA = 90$ (Def of $\perp$); m $\angle DBC$ = m $\angle DBA$ (Sub); $\overline{DB} \cong \overline{DB}$ (Reflex Prop of Congr); $\overline{BC} \cong \overline{BA}$ (Def of seg bis); ∴ $\triangle ABD \cong \triangle CBD$ (SAS) **13.** $\overline{BA} \cong \overline{CA}$, $\overline{AD}$ bis $\angle A$ (Given); $\angle BAD \cong \angle CAD$ (Def of $\angle$ bis); $\overline{AD} \cong \overline{AD}$ (Reflex Prop of Congr); ∴ $\triangle BAD \cong \triangle CAD$ (SAS) **15.** If the $\angle$ included by the ≅ sides of an isosc $\triangle$ is bis, then the resulting $\triangle$s are ≅. **17.** $\overline{CF} \parallel \overline{BE}$, $\angle 3 \cong \angle 4$, $\overline{BC} \cong \overline{BE}$ (Given); $\angle 4 \cong \angle 5$ (If lines are ∥, then alt int $\angle$s are ≅.); $\angle 3 \cong \angle 5$ (Trans Prop of Congr); $\overline{BF} \cong \overline{BF}$ (Reflex Prop of Congr); ∴ $\triangle CBF \cong \triangle EBF$ (SAS)

Written Exercises, pages 155–157

1. $\angle P$, $\angle M$, $\overline{AH} \cong \overline{PM}$ **3.** $\overline{TU} \cong \overline{QP}$; (Given); (Given); $\angle T \cong \angle Q$; (If lines are ∥, then alt int $\angle$s are ≅.); (SAS) **5.** $\overline{CD}$ bis $\angle C$, $\angle 1 \cong \angle 2$ (Given); $\angle ACD \cong \angle BCD$ (Def of $\angle$ bis); $\overline{CD} \cong \overline{CD}$ (Reflex Prop of Congr); ∴ $\triangle CDB \cong \triangle CDA$ (ASA) **7.** $\overline{PG} \cong \overline{SG}$, $\overline{TP} \cong \overline{TS}$ (Given); $\overline{TG} \cong \overline{TG}$ (Reflex Prop of Eq); ∴ $\triangle TPG \cong \triangle TSG$ (SSS)

9.

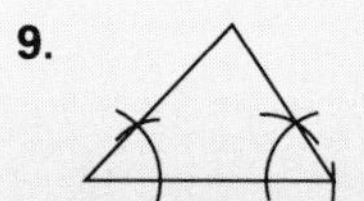

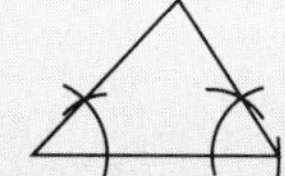

11. $\overline{AD} \parallel \overline{BC}$, $\overline{DC} \parallel \overline{BA}$ (Given); $\angle ADB \cong \angle CBD$, $\angle CDB \cong \angle ABD$ (If lines are ∥, then alt int $\angle$s are ≅.); $\overline{DB} \cong \overline{DB}$ (Reflex Prop of Congr); ∴ $\triangle ADB \cong \triangle CBD$ (ASA) **13.** $\overline{RS}$ bis $\angle PRQ$, $\overline{RS} \perp \overline{PQ}$ (Given); $\angle PRS \cong \angle QRS$ (Def $\angle$ bis); $\overline{RS} \cong \overline{RS}$ (Reflex Prop of Congr); m $\angle PSR = 90$, m $\angle QSR = 90$ (Def of $\perp$); m $\angle PSR$ = m $\angle QSR$ (Sub); ∴ $\triangle PRS \cong \triangle QRS$ (ASA) **15.** $GL = LH$, $\overline{GM} \perp \overline{GL}$, $\overline{HN} \perp \overline{HL}$ (Given); m $\angle G = 90$, m $\angle H = 90$ (Def of $\perp$); m $\angle G$ = m $\angle H$ (Sub); $\angle MLG \cong \angle KLH$ (Vert $\angle$s are ≅.); ∴ $\triangle GLM \cong \triangle HLK$ (ASA). They are corr parts of ≅ $\triangle$.

Written Exercises, pages 159–160

1. $\overline{AC} \cong \overline{CE}$, $\overline{AB} \cong \overline{DE}$, $\angle 1 \cong \angle 2$, $\overline{BG} \cong \overline{DF}$ (Given); $AC - AB = CE - DE$ (Equations may be subtracted.); $AC - AB = BC$, $CE - DE = CD$ (Seg Add Post); $BC = CD$ (Sub); $\triangle BGC \cong \triangle DFC$ (SAS) **3.** $\overline{AB} \cong \overline{CD}$ ($AB = CD$), $\angle 3 \cong \angle 2$, $\overline{EC} \cong \overline{FB}$ (Given); $AB + BC = CD + BC$ (Add Prop of Eq); $AB + BC = AC$, $CD + BC = BD$ (Seg Add Post); $AC = DB$ ($\overline{AC} \cong \overline{DB}$) (Sub); ∴ $\triangle AEC \cong \triangle DFB$ (SAS) **5.** $\overline{AG} \cong \overline{DG}$, $\overline{FG} \cong \overline{EG}$, $\overline{AB} \cong \overline{DC}$, $\angle A \cong \angle D$ (Given); $AG - FG = DG - EG$ (Equations may be subt.); $AG - FG = AF$, $DG - EG = DE$ (Seg Add Post); $AF = DE$ ($\overline{AF} \cong \overline{DE}$) (Sub); ∴ $\triangle FAB \cong \triangle EDC$ (SAS) **7.** $\overline{RP} \cong \overline{US}$, $\overline{RQ} \cong \overline{UT}$, $\overline{QW} \cong \overline{TV}$, $\overline{RP} \parallel \overline{US}$ (Given); $RP - RQ = US - UT$ (Equations may be subtracted.); $RP - RQ = PQ$, $US - UT = ST$

(Seg Add Post); $PQ = ST$ (Sub); $\angle Q \cong \angle T$ (If lines are $\parallel$, then alt int $\angle$s are $\cong$.); $\therefore \triangle PQW \cong \triangle STV$ (SAS) **9.** $\overline{AD} \cong \overline{CB}$, $\overline{CE} \parallel \overline{BF}$, $\angle 1$ and $\angle 2$ are supp (Given); $\angle 1$ and $\angle 3$ are supp (If the outer rays of 2 adj $\angle$s form a st $\angle$, then the $\angle$s are supp.); $\angle 2 \cong \angle 3$ (Supps of the same $\angle$ are $\cong$.); $AD - BD = CB - BD$ (Subt Prop of Eq); $AD - BD = AB$, $CB - BD = CD$ (Seg Add Post); $AB = DC$ ($\overline{AB} \cong \overline{DC}$) (Sub); $\angle DCE \cong \angle ABF$ (If lines are $\parallel$, then corr $\angle$s are $\cong$.); $\therefore \triangle CED \cong \triangle BFA$ (ASA) **11.** $\overline{PQ} \cong \overline{RS}$, $\overline{WP} \perp \overline{PS}$, $\overline{SX} \perp \overline{PS}$, $\angle 3 \cong \angle 4$, $\angle 1 \cong \angle 2$ (Given); $PQ + QR = RS + QR$ (Add Prop of Eq); $PQ + QR = PR$, $RS + QR = QS$ (Seg Add Post); $PR = SQ$ ($\overline{PR} \cong \overline{SQ}$) (Sub); $\angle 3$ and $\angle UPR$ are comp, $\angle 4$ and $\angle TSQ$ are comp (If the outer rays of 2 adj acute $\angle$s are $\perp$, then the $\angle$s are comp.); $\angle UPR \cong \angle TSQ$ (Comps of $\cong$ $\angle$s are $\cong$.); $\angle 1$ and $\angle TQR$ are supp, $\angle 2$ and $\angle URQ$ are supp (If the outer rays of 2 adj $\angle$s form a st $\angle$, then the $\angle$s are supp.); $\angle TQR \cong \angle URQ$ (Supps of $\cong$ $\angle$s are $\cong$.); $\triangle PUR \cong \triangle STQ$ (ASA) **13.** $\overline{AB} \cong \overline{DC}$, $\overline{AE} \cong \overline{DF}$, $\overline{CE} \cong \overline{BF}$ (Given); $AB + BC = DC + BC$ (Add Prop of Eq); $AB + BC = AC$, $DC + BC = BD$ (Seg Add Post); $AC = DB$ ($\overline{AC} \cong \overline{DB}$) (Sub); $\therefore \triangle ACE \cong \triangle DBF$ (SSS)

Written Exercises, pages 164–166

1. Yes, SAS **3.** No, AAA not sufficient **5.** 40 **7.** 40 **9.** (Def of seg bis); (Def of $\angle$ bis); $\triangle PQR \cong \triangle PQS$ (AAS) **11.** $\overline{PQ} \perp \overline{RS}$, $\angle R \cong \angle S$ (Given); $m\angle RQP = 90$, $m\angle SQP = 90$ (Def of $\perp$); $m\angle RQP = m\angle SQP$ (Sub); $\overline{PQ} \cong \overline{PQ}$ (Reflex Prop of Congr); $\therefore \triangle PQR \cong \triangle PQS$ (AAS) **13.** $\angle P \cong \angle S$, $\overline{PU} \cong \overline{SR}$, $\overline{PR} \parallel \overline{US}$, $\overline{QU} \parallel \overline{TR}$ (Given); $\angle 2 \cong \angle 1$ (If lines are $\parallel$, then alt int $\angle$s are $\cong$.); $\angle 2 \cong \angle 3$ (If lines are $\parallel$, then corr $\angle$s are $\cong$.); $\angle 1 \cong \angle 3$ (Trans Prop of Congr); $\therefore \triangle PUQ \cong \triangle SRT$ (AAS) **15.** $\overline{AB} \perp \overline{BE}$, $\overline{EF} \perp \overline{BE}$, $\angle 1 \cong \angle 2$, $\angle 3 \cong \angle 4$, $\overline{BD} \cong \overline{EC}$ (Given); $\angle GBC$ and $\angle 1$ are comp, $\angle HED$ and $\angle 2$ are comp (If the outer rays of 2 adj acute $\angle$s are $\perp$, then the $\angle$s are comp.); $\angle GBC \cong \angle HED$ (Comp $\angle$s of $\cong$ $\angle$s are $\cong$.); $\angle GCB$ and $\angle 3$ are supp, $\angle HDE$ and $\angle 4$ are supp (If the outer rays of 2 adj $\angle$s form a st $\angle$, then the $\angle$s are supp.); $\angle GCB \cong \angle HDE$ (Supps of $\cong$ $\angle$s are $\cong$.); $BD - CD = EC - CD$ (Subt Prop of Eq); $BD - CD = BC$, $EC - CD = ED$ (Seg Add Post); $BC = ED$ ($\overline{BC} \cong \overline{ED}$) (Sub); $\therefore \triangle BCG \cong \triangle EDH$ (ASA) **17.** Using the $\cong$ $\triangle$s ABC and PQR on page 162: $\angle P \cong \angle A$, $\angle Q \cong \angle B$, $\overline{PR} \cong \overline{AC}$ (Given); $\angle R \cong \angle C$ (If 2 $\angle$s of $\cong$ $\triangle$s are $\cong$, then the third $\angle$s of the $\triangle$s are $\cong$.); $\therefore \triangle PQR \cong \triangle ABC$ (ASA)

Written Exercises, pages 168–169

1. $\overline{PS} \cong \overline{RQ}$, $\overline{SR} \cong \overline{QP}$ (Given); $\overline{SQ} \cong \overline{SQ}$ (Reflex Prop of Congr); $\triangle SQP \cong \triangle QSR$ (SSS); $\therefore \angle 1 \cong \angle 2$ (CPCTC) **3.** $\overline{AC}$ and $\overline{DE}$ bis each other (Given); $\overline{AB} \cong \overline{CB}$, $\overline{DB} \cong \overline{EB}$ (Def of seg bis); $\angle DBA \cong \angle EBC$ (Vert $\angle$s are $\cong$.); $\triangle BAD \cong \triangle BCE$ (SAS), $\angle A \cong \angle C$ (CPCTC); $\therefore \overline{AD} \parallel \overline{CE}$ (If alt int $\angle$s are $\cong$, then lines are $\parallel$.) **5.** $\overline{FS} \cong \overline{UQ}$, $\overline{FP} \parallel \overline{UR}$, $\overline{FP} \cong \overline{UR}$ (Given); $FS + SQ = UQ + SQ$ (Add Prop of Eq); $FS + SQ = FQ$, $UQ + SQ = US$ (Seg Add Post); $FQ = US$ ($\overline{FQ} \cong \overline{US}$) (Sub); $\angle F \cong \angle U$ (If lines are $\parallel$, then alt int $\angle$s are $\cong$.); $\triangle PFO \cong \triangle RUS$ (SAS); $\angle PQF \cong \angle RSU$ (CPCTC); $\therefore \overline{PQ} \parallel \overline{RS}$ (If alt int $\angle$s are $\cong$, then lines are $\parallel$.) **7.** $\overline{TQ} \perp \overline{PR}$, $\angle 1 \cong \angle 2$, $\overline{TQ} \perp \overline{US}$ (Given); $\angle TQU$ and $\angle 1$ are comp, $\angle TQS$ and $\angle 2$ are comp (If the outer rays of 2 adj acute $\angle$s form a st $\angle$, then the $\angle$s are comp.); $\angle TQU \cong \angle TQS$ (Comps of $\cong$ $\angle$s are $\cong$.); $\overline{TQ} \cong \overline{TQ}$ (Reflex Prop of Congr); $m\angle UTQ = 90$, $m\angle STQ = 90$ (Def of $\perp$); $m\angle UTQ = m\angle STQ$ (Sub); $\triangle UTQ \cong \triangle STQ$ (ASA); $\therefore \overline{UQ} \cong \overline{SQ}$ (CPCTC) **9.** $\overline{CB} \perp \overline{AB}$, $\overline{CB} \perp \overline{BH}$, $\overline{DB} \cong \overline{IB}$, $\angle 1 \cong \angle 2$ (Given); $\angle CBD$ and $\angle 1$ are comp, $\angle CBI$ and $\angle 2$ are comp (If the outer rays of 2 adj acute $\angle$s are $\perp$, then the $\angle$s are comp.); $\angle CBD \cong \angle CBI$ (Comps of $\cong$ $\angle$s are $\cong$.); $\overline{CB} \cong \overline{CB}$ (Reflex Prop); $\triangle DBC \cong \triangle IBC$ (SAS); $\therefore \angle 5 \cong \angle 6$ (CPCTC)

Chapter 5

Written Exercises, pages 180–181

1. $m\angle W = 40$, $m\angle A = 100$ **3.** $m\angle U = 60$, $m\angle R = 60$ **5.** $m\angle C = 40$, $m\angle A = 70$, $m\angle B = 70$ **7.** 80 **9.** 45 **11.** 35 **13.** 14 **15.** $\angle 4 \cong \angle 2$ (Given); $\angle 1 \cong \angle 2$ (Vert $\angle$s are $\cong$.); $\angle 4 \cong \angle 1$ (Trans Prop of Congr); $\therefore \overline{AB} \cong \overline{AC}$ (Sides opp $\cong$ $\angle$s are $\cong$.) **17.** 81 cm **19.** $\overline{QV} \cong \overline{RW}$, $\overline{PQ} \cong \overline{PR}$, $\overline{PS} \cong \overline{PT}$ (Given); $\angle Q \cong \angle R$ ($\angle$s opp $\cong$ sides are $\cong$.); $PQ - PS = PR - PT$ (Equations may be subtracted.); $PQ - PS = SQ$, $PR - PT = TR$ (Seg Add Post); $SQ = TR$ (Sub); $\triangle SQV \cong \triangle TRW$ (SAS); $\therefore \angle 1 \cong \angle 2$ (CPCTC)

21. $\triangle ABC$ is equilateral (Given); $\overline{AB} \cong \overline{AC}$, $\overline{CA} \cong \overline{CB}$ (Def of equil $\triangle$); $\angle C \cong \angle B$, $\angle B \cong \angle A$ ($\angle$s opp $\cong$ sides are $\cong$.); $\angle C \cong \angle A$ (Trans Prop of Congr); $m\angle C + m\angle B + m\angle A =$

180 (Sum of meas of $\angle$s of $\triangle = 180$.); m $\angle B$ + m $\angle B$ + m $\angle B$ = 180, or 3 × m $\angle B$ = 180 (Sub); m $\angle B$ = 60 (Div Prop of Eq); ∴ m $\angle A$ = 60, m $\angle C$ = 60 (Sub) **23.** m $\angle GKL$ = 50, m $\angle GLK$ = 90 **25.** $\overline{BC} \cong \overline{AC}$, $\overrightarrow{CE}$ bis $\angle DCA$ (Given);

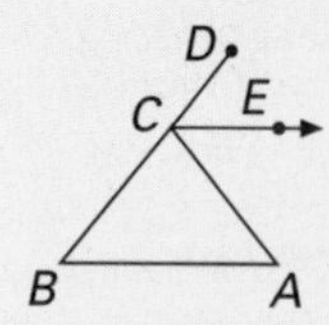

m $\angle A$ = m $\angle B$ ($\angle$s opp ≅ sides are ≅.); m $\angle A$ + m $\angle B$ = m $\angle DCA$ (Meas of ext $\angle$ = sum of meas of 2 remote int $\angle$s.); m $\angle B$ + m $\angle B$ = m $\angle DCA$, or 2 × m $\angle B$ = m $\angle DCA$ (Sub); m $\angle B = \frac{1}{2}$(m $\angle DCA$) (Div Prop Eq); m $\angle DCE = \frac{1}{2}$(m $\angle DCA$) (Def of $\angle$ bis); m $\angle B$ = m $\angle DCE$ (Sub); ∴ $\overrightarrow{CE} \parallel \overline{AB}$ (If corr $\angle$s are ≅, then lines are ∥.) **27.** $\overline{AE} \perp \overline{CD}$, E is midpt of $\overline{CD}$, $\overline{BA} \perp$ all lines in $\mathcal{M}$ through A (Given); m $\angle AED$ = 90, m $\angle AEC$ = 90 (Def of ⊥); m $\angle AED$ = m $\angle AEC$ (Sub); $\overline{CE} \cong \overline{ED}$ (Def of midpt); $\overline{AE} \cong \overline{AE}$ (Reflex Prop of Congr); $\triangle CEA \cong \triangle DEA$ (SAS); $\overline{AC} \cong \overline{AD}$ (CPCTC); m $\angle BAD$ = 90, m $\angle BAC$ = 90 (Def of ⊥); m $\angle BAD$ = m $\angle BAC$ (Sub); $\overline{BA} \cong \overline{BA}$ (Reflex Prop of Congr); $\triangle CAB \cong \triangle DAB$ (SAS); ∴ $\overline{BC} \cong \overline{BD}$ (CPCTC)

Written Exercises, page 183

1. $\triangle DAE, \triangle CBE; \triangle DBA, \triangle CAB$ **3.** $\triangle DAE, \triangle CBE$ **5.** $\triangle DAE, \triangle CBE$ **7.** $\triangle PUV, \triangle RSV$ **9.** $\triangle TUR, \triangle TSP; \triangle TUV, \triangle TSV$ **11.** $\triangle PVQ, \triangle RVQ; \triangle PVT, \triangle RVT; \triangle PVU, \triangle RVS$ **13.** $\triangle TPR, \triangle TRP; \triangle UPR, \triangle SRP; \triangle TPQ, \triangle TRQ$ **15.** $\triangle IDE, \triangle JDC; \triangle IDB, \triangle JDA$ **17.** $\triangle EBA, \triangle CAB; \triangle EBC, \triangle CAE; \triangle EBD, \triangle CAD; \triangle EBG, \triangle FAC$ **19.** $\triangle EHA, \triangle CHB; \triangle EHC, \triangle CHE$ **21.** $\triangle IAH, \triangle JBH; \triangle DAJ, \triangle DBI; \triangle AFC, \triangle BGE; \triangle DAC, \triangle DBE$

Written Exercises, pages 186–187

1. $\angle 3 \cong \angle 4$, $\overline{AE} \cong \overline{CE}$ (Given); $\overline{DE} \cong \overline{DE}$ (Reflex Prop of Congr); $\triangle ADE \cong \triangle CDE$ (SAS); ∴ $\overline{AD} \cong \overline{CD}$ (CPCTC) **3.** $\overline{PS}$ bis $\angle RPQ$, $\overline{PR} \cong \overline{PQ}$ (Given); $\angle RPS \cong \angle QPS$ (Def of $\angle$ bis); $\overline{PS} \cong \overline{PS}$ (Reflex Prop of Congr); $\triangle RPS \cong \triangle QPS$ (SAS); ∴ $\overline{RS} \cong \overline{QS}$ (CPCTC) **5.** $\overline{AD} \parallel \overline{CD}$, $\overline{HA} \cong \overline{GC}$ (Given); $\angle HAB \cong \angle GCB$ (If lines are ∥, then alt int $\angle$s are ≅.); $\angle 4 \cong \angle 3$ (Vert $\angle$s are ≅.); $\triangle HAB \cong \triangle GCB$ (AAS); ∴ $\overline{HB} \cong \overline{GB}$ (CPCTC) **7.** B is the midpt of $\overline{AC}$, $\overline{AD} \perp \overline{AC}$, $\overline{EC} \perp \overline{AC}$ (Given); $\overline{AB} \cong \overline{CB}$ (Def of midpt); $\angle DBA \cong \angle EBC$ (Vert $\angle$s are ≅.); m $\angle DAB$ = 90, m $\angle ECB$ = 90 (Def of ⊥); m $\angle DAB$ = m $\angle ECB$ (Sub); $\triangle DAB \cong \triangle ECB$ (ASA); $\overline{DB} \cong \overline{EB}$ (CPCTC); ∴ B is midpt of $\overline{DE}$ (Def of midpt) **9.** $\angle A \cong \angle D$, $\overline{AE} \cong \overline{DE}$ (Given); $\angle 5 \cong \angle 6$ (Vert $\angle$s are ≅.); $\triangle ABE \cong \triangle DCE$ (ASA); ∴ $\overline{AB} \cong \overline{DC}$ (CPCTC) **11.** $\overline{PR}$ bis $\angle SPT$, $\angle 3 \cong \angle 4$ (Given); $\angle 9 \cong \angle 10$ (Def of $\angle$ bis); $\overline{PR} \cong \overline{PR}$ (Reflex Prop of Congr); $\triangle PSR \cong \triangle PTR$ (ASA); $\overline{PS} \cong \overline{PT}$ (CPCTC); $\overline{PQ} \cong \overline{PQ}$ (Reflex Prop of Congr); $\triangle PSQ \cong \triangle PTQ$ (SAS); ∴ $\overline{QS} \cong \overline{QT}$ (CPCTC) **13.** $\overline{PS} \cong \overline{PT}$, $\overline{SR} \cong \overline{TR}$ (Given); $\overline{PR} \cong \overline{PR}$ (Reflex Prop of Congr); $\triangle PSR \cong \triangle PTR$ (SSS); $\angle 3 \cong \angle 4$ (CPCTC); $\overline{RQ} \cong \overline{RQ}$ (Reflex Prop of Congr); $\triangle RSQ \cong \triangle RTQ$ (SAS); ∴ $\angle 2 \cong \angle 5$ (CPCTC) **15.** $\overline{AD}$ bis $\overline{BE}$, $\angle 5 \cong \angle 6$ (Given); $\overline{BG} \cong \overline{EG}$ (Def of seg bis); $\angle 1 \cong \angle 3$ (Vert $\angle$s are ≅.); $\triangle ABG \cong \triangle DEG$ (ASA); $\overline{AG} \cong \overline{DG}$, $\angle CAG \cong \angle FDG$ (CPCTC); $\angle AGC \cong \angle DGF$ (Vert $\angle$s are ≅.); $\triangle CAG \cong \triangle FDG$ (ASA); ∴ $\overline{AC} \cong \overline{DF}$ (CPCTC) **17.** $\overline{AC} \parallel \overline{FD}$, $\overline{BC} \cong \overline{EF}$ (Given); $\angle F \cong \angle C$ (If lines are ∥, then alt int $\angle$s are ≅.); $\angle 2 \cong \angle 4$ (Vert $\angle$s are ≅.); $\triangle BCG \cong \triangle EFG$ (AAS); $\overline{CG} \cong \overline{FG}$ (CPCTC); $\angle AGC \cong \angle DGF$ (Vert $\angle$s are ≅.); $\triangle ACG \cong \triangle DFG$ (ASA); ∴ $\overline{AC} \cong \overline{DF}$ (CPCTC) **19.** $AX = BX$, $CX = EX$ (Given); m $\angle BAX$ = m $\angle ABX$, m $\angle CEX$ = m $\angle ECX$ ($\angle$s opp ≅ sides are ≅.); m $\angle AXB$ = m $\angle CXE$ (Vert $\angle$s are ≅.); m $\angle BAX$ + m $\angle ABX$ + m $\angle AXB$ = 180, m $\angle CEX$ + m $\angle ECX$ + m $\angle CXE$ = 180 (Sum of meas of all $\angle$s of $\triangle$ = 180.); m $\angle BAX$ + m $\angle ABX$ + m $\angle AXB$ = m $\angle CEX$ + m $\angle ECX$ + m $\angle CXE$ (Sub); 2 × m $\angle BAX$ + m $\angle AXB$ = 2 × m $\angle ECX$ + m $\angle CXE$ (Sub); 2 × m $\angle BAX$ + m $\angle AXB$ = 2 × m $\angle CEX$ + m $\angle AXB$ (Sub); 2 × m $\angle BAX$ = 2 × m $\angle CEX$ (Subt Prop of Eq); m $\angle BAX$ = m $\angle ECX$ (Div Prop of Eq); ∴ $\overline{AB} \parallel \overline{EC}$ (If alt int $\angle$s are ≅, then lines are ∥.)

Written Exercises, pages 190–192

1. $\overline{QR} \perp \overline{QP}$, $\overline{HF} \perp \overline{HG}$, $\overline{QR} \cong \overline{HF}$, $\overline{QP} \cong \overline{HG}$ (Given); m $\angle Q$ = 90, m $\angle H$ = 90 (Def of ⊥); m $\angle Q$ = m $\angle H$ (Sub); ∴ $\triangle RQP \cong \triangle FHG$ (SAS) **3.** $\overline{QR} \perp \overline{QP}$, $\overline{HF} \perp \overline{HG}$, $\angle R \cong \angle F$,

$\overline{PR} \cong \overline{GF}$ (Given); m $\angle Q = 90$, m $\angle H = 90$ (Def of $\perp$); m $\angle Q =$ m $\angle H$ (Sub); $\triangle PQR \cong \triangle GHF$ (AAS); $\therefore \overline{RQ} \cong \overline{FH}$ (CPCTC) **5.** $\overline{AF} \perp \overline{AD}$, $\overline{DE} \perp \overline{AD}$, $\overline{FA} \cong \overline{ED}$, $\overline{FB} \parallel \overline{EC}$ (Given); m $\angle FAB = 90$, m $\angle EDC = 90$ (Def of $\perp$); m $\angle FAB =$ m $\angle EDC$ (Sub); $\angle FBA \cong \angle ECD$ (If lines are $\parallel$, then alt ext $\angle$s are $\cong$.); $\therefore \triangle ABF \cong \triangle DCE$ (SAS) **7.** $\overline{AP} \cong \overline{BP}$, $\overline{PO} \perp \overline{AO}$, $\overline{PO} \perp \overline{BO}$ (Given); $\overline{PO} \cong \overline{PO}$ (Reflex Prop of Congr); $\triangle POA \cong \triangle POB$ (HL); $\therefore \overline{AO} \cong \overline{BO}$ (CPCTC) **9.** $\overline{PQ} \perp \overline{PS}$, $\overline{SR} \perp \overline{SP}$, $\overline{PT} \cong \overline{ST}$ (Given); m $\angle QPS = 90$, m $\angle RSP = 90$ (Def of $\perp$); m $\angle QPS =$ m $\angle RSP$ (Sub); $\angle 1 \cong \angle 2$ ($\angle$s opp $\cong$ sides are $\cong$.); $\overline{PS} \cong \overline{PS}$ (Reflex Prop of Congr); $\triangle PSR \cong \triangle SPQ$ (ASA); $\therefore \angle Q \cong \angle R$ (CPCTC) **11.** $\overline{PQ} \perp \overline{PS}$, $\overline{SR} \perp \overline{SP}$, $\angle 3 \cong \angle 4$ (Given); m $\angle QPS = 90$, m $\angle RSP = 90$ (Def of $\perp$); m $\angle QPS =$ m $\angle RSP$ (Sub); $\angle 1$ and $\angle 3$ are comp, $\angle 2$ and $\angle 4$ are comp (If the outer rays of 2 adj acute $\angle$s are $\perp$, then $\angle$s are comp.); $\angle 1 \cong \angle 2$ (Comp $\angle$s of $\cong$ $\angle$s are $\cong$.); $\overline{PS} \cong \overline{PS}$ (Reflex Prop of Congr); $\triangle QPS \cong \triangle RSP$ (ASA); $\therefore \angle Q \cong \angle R$ (CPCTC) **13.** $\overline{AB} \perp \overline{AE}$, $\overline{CB} \perp \overline{CD}$, $\overline{AB} \cong \overline{CB}$, $\angle ABD \cong \angle EBC$ (Given); m $\angle BAE = 90$, m $\angle BCD = 90$ (Def of $\perp$); m $\angle BAE =$ m $\angle BCD$ (Sub); m $\angle ABD -$ m $\angle 4 =$ m $\angle EBC -$ m $\angle 4$ (Subt Prop of Eq); m $\angle ABD -$ m $\angle 4 =$ m $\angle 3$, m $\angle EBC -$ m $\angle 4 =$ m $\angle 5$ (Angle Add Post); m $\angle 3 =$ m $\angle 5$ (Sub); $\triangle ABE \cong \triangle CBD$ (ASA); $\overline{EB} \cong \overline{BD}$ (CPCTC); $\therefore \triangle EBD$ is isos (Def of isos $\triangle$) **15.** In rt $\triangle$s ABC and DEF: $\overline{AC} \cong \overline{DF}$, $\overline{BC} \cong \overline{EF}$, $\angle C$ is a rt $\angle$, $\angle F$ is a rt $\angle$ (Given); $\angle C \cong \angle F$ (All $\angle$s are $\cong$.); $\therefore \triangle ABC \cong \triangle DEF$ (SAS) **17.** Answers may vary. **19.** Answers may vary. **21.** $\overline{PR} \perp \overline{SQ}$, $\overline{QT} \perp \overline{SP}$, $\overline{UT} \cong \overline{UR}$ (Given); m $\angle UTP = 90$, m $\angle URQ = 90$ (Def of $\perp$); m $\angle UTP =$ m $\angle URQ$ (Sub); $\angle TUP \cong \angle RUQ$ (Vert $\angle$s are $\cong$.); $\triangle TUP \cong \triangle RUQ$ (ASA); $\overline{UP} \cong \overline{UQ}$ (CPCTC); $UP + UR = UQ + UT$ (Add Prop of Eq); $UP + UR = PR$, $UQ + UT = QT$ (Seg Add Post); $PR = QT$ (Sub); $\angle S \cong \angle S$ (Reflex Prop of Congr); m $\angle SRP = 90$, m $\angle STQ = 90$ (Def of $\perp$); m $\angle SRP =$ m $\angle STQ$ (Sub); $\triangle SRP \cong \triangle STQ$ (AAS); $\therefore \overline{SR} \cong \overline{ST}$ (CPCTC)

Written Exercises, pages 197–198

1. $\triangle ADG$ **3.** $\overline{AB}$, $\overline{GB}$ **5.** $\overline{GC}$ **7.** $\overline{RS}$ is a median, $\overline{PR} \cong \overline{QR}$ (Given); $\overline{PS} \cong \overline{QS}$ (Def of median); $\overline{RS} \cong \overline{RS}$ (Reflex Prop of Congr); $\therefore \triangle PSR \cong \triangle QSR$ (SSS) **9.** 17 **11.** $\overline{PR}$ is an alt of $\triangle PQS$ (Given); m $\angle 1 +$ m $\angle 3 +$ m $\angle PRQ = 180$, m $\angle 2 +$ m $\angle 4 +$ m $\angle PRS = 180$ (Sum of meas of $\angle$s of $\triangle = 180$); m $\angle 2 +$ m $\angle 4 +$ m $\angle PRS =$ m $\angle 1 +$ m $\angle 3 +$ m $\angle PQR$ (Sub); $\angle PRS$ and $\angle PRQ$ are rt $\angle$s (Def of alt); m $\angle PRS =$ m $\angle PRQ$ (All rt $\angle$s are $\cong$.); m $\angle 2 +$ m $\angle 4 +$ m $\angle PRS =$ m $\angle 1 +$ m $\angle 3 +$ m $\angle PRS$ (Sub); $\therefore$ m $\angle 2 +$ m $\angle 4 =$ m $\angle 1 +$ m $\angle 3$ (Subt Prop of Eq) **13.** $\overline{PR}$ is a median, $\angle 3 \cong \angle 4$ (Given); $QR = SR$ (Def of median); $\overline{PQ} \cong \overline{PS}$ (Sides opp $\cong$ $\angle$s are $\cong$.); $\triangle PQR \cong \triangle PSR$ (SAS); $\angle 1 \cong \angle 2$ (CPCTC); $\therefore \overline{PR}$ bis $\angle QPS$ (Def of $\angle$ bis)
15. $\overline{DB}$ bis $\angle ADC$, $\overline{BE}$ and $\overline{BF}$ are altitudes of $\triangle$s ABD and CBD, respectively (Given); $\angle EDB \cong \angle FDB$ (Def of $\angle$ bis); $\angle BED$ and $\angle BFD$ are rt $\angle$s (Def of alt); $\angle BED \cong \angle BFD$ (All rt $\angle$s are $\cong$.); $\overline{BD} \cong \overline{BD}$ (Reflex Prop of Congr); $\triangle BED \cong \triangle BFD$ (AAS); $\therefore \overline{BE} \cong \overline{BF}$ (CPCTC) **17.** Answers may vary.
19. True. In $\triangle ABC$ with pt D on $\overline{BC}$, $\overline{AD}$ is alt, $\overline{AD}$ bis $\angle BAC$ (Given); $\angle BAD \cong \angle CAD$ (Def $\angle$ bis); $\angle ADB$ and $\angle ADC$ are rt $\angle$s (Def alt); $\angle ADB \cong \angle ADC$ (All rt $\angle$s are $\cong$.); $\overline{AD} \cong \overline{AD}$ (Reflex Prop of Congr); $\triangle ABC \cong \triangle ADC$ (ASA); $\overline{AB} \cong \overline{AC}$ (CPCTC); $\therefore \triangle ABC$ is isos (Def isos $\triangle$) **21.** False, except in the special case of an equil $\triangle$. Isos $\triangle ABC$ with $AB = AC$, $AB \neq BC$, and $\overline{BM}$ bis $\angle ABC$ (Given). Assume that $\overline{BM}$ is also an alt. Then $\triangle BMA \cong \triangle BMC$ by ASA. So, $\angle A \cong \angle C$ by CPCTC; hence, $\triangle ABC$ is equil with $AB = BC$. But this contradicts the given information, so the assumption that $\overline{BM}$ is an alt is false.

Written Exercises, page 203

1.

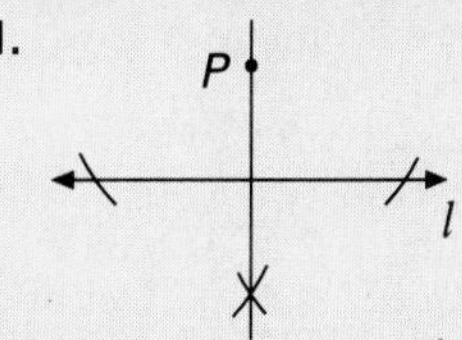

3. $\overline{TQ}$ bis $\angle PTR$, $\overline{TP} \cong \overline{TR}$ (Given); $\triangle PTR$ is isos (Def of isos$\triangle$); $\therefore \overline{TQ}$ is the $\perp$ bis of $\overline{PR}$ (Bis of vertex $\angle$ of isos $\triangle$ is $\perp$ bis of base.)
5. $\angle 3 \cong \angle 4$, $\overline{TQ}$ bis $\angle PTR$ (Given); $\angle 5$ and $\angle 3$ are supp, $\angle 6$ and $\angle 4$ are supp (If the outer rays of 2 adj $\angle$s form a st $\angle$, then the $\angle$s are supp.); $\angle 5 \cong \angle 6$ (Supp $\angle$s of $\cong$ $\angle$s are $\cong$.); $\overline{TP} \cong \overline{TR}$ (Sides opp $\cong$ $\angle$s are $\cong$); $\overline{TQ}$ is $\perp$ bis

of $\overline{PR}$ (Bis of vertex ∠ of isos △ is ⊥ bis of base.); ∴ $\overline{PQ} \cong \overline{RQ}$ (Def of seg bis) **7.** $\overline{CD}$ is ⊥ bis of $\overline{AB}$ (Given); $\overline{AD} \cong \overline{BD}$ (Any pt on ⊥ bis of a seg is equidist from endpts of seg.); ∴ $\angle 1 \cong \angle 2$ (∠s opp ≅ sides are ≅.) **9.** Assume there are two distinct line segments from a point P perpendicular to a line at points A and B. Then $\triangle APB$ is formed, with m $\angle PAB$ = m $\angle PBA = 90$. But m $\angle APB > 0$, since $\overline{PA}$ and $\overline{PB}$ are distinct. Thus the sum of the measures of the angles of the △ will be greater than 180, which contradicts the theorem that the sum of the measures of the angles is 180. Therefore, the assumption that there can be two distinct perpendiculars from a point to a line must be false; we conclude that there can be only one.

Written Exercises, pages 208–209

1. $\overline{CB}$ **3.** $\overline{BK}$ **5.** m $\angle 2 >$ m $\angle P$ (Given); m $\angle 2 =$ m $\angle 1$ (Vert ∠s are ≅.); m $\angle 1 >$ m $\angle P$ (Sub Prop of Ineq); ∴ $PT > QT$ (Side opp larger ∠ is longer.) **7.** $\overrightarrow{SQ}$ bis $\angle PSR$ (Given); $\angle 1 \cong \angle 2$ (Def ∠ bis); m $\angle 3 >$ m $\angle 2$ (Ext ∠ Ineq Thm); m $\angle 3 >$ m $\angle 1$ (Sub Prop of Ineq); ∴ $RS > RQ$ (Side opp larger ∠ is longer.) **9.** Scalene $\triangle ABC$, m $\angle B >$ m $\angle C$, m $\angle C >$ m $\angle A$ (Given); $AC > AB$, $AB > BC$ (Side opp larger ∠ is longer.); $AC > BC$ (Trans Prop of Ineq); ∴ $\overline{AC}$ is longest side and is opp the largest ∠ ($\overline{AC}$ is opp $\angle B$) **11.** m $\angle B$, m $\angle C$ **13.** $AC < AB$ **15.** $SQ > SP$, $SR = SQ$ (Given); m $\angle P >$ m $\angle SQP$ (∠ opp longer side is larger.); m $\angle R =$ m $\angle SQR$ (∠ opp ≅ sides are ≅.); m $\angle P$ + m $\angle R >$ m $\angle SQP$ + m $\angle SQR$ (Add Prop of Ineq); m $\angle SQP$ + m $\angle SQR$ = m $\angle PQR$ (Angle Add Post); ∴ m $\angle P$ + m $\angle R >$ m $\angle PQR$ (Sub Prop of Ineq) **17.** $AB = BC$ (Given); draw $\overline{BD} \perp \overline{AC}$ (Only one seg from pt not on line is ⊥ to the line.); $\triangle ABD$ is rt △(Def of ⊥); m $\angle ABD < 90$ (non–rt ∠s of a rt △ are acute.); m $\angle ADB = 90$ (Def of ⊥); m $\angle ABD <$ m $\angle ADB$ (Sub); $AB > AD$ (Side opp larger ∠ is larger); $\triangle CBD$ is rt △ (Def of ⊥); m $\angle CBD < 90$ (Non–rt ∠s of a rt △ are acute.); m $\angle CDB = 90$ (Def of ⊥); m $\angle CBD <$ m $\angle CDB$ (Sub); $BC > DC$ (Side opp larger angle is longer.); $AB + BC >$ $AD + DC$ (Inequalities may be added.); $AD + DC = AC$ (Seg Add Post); ∴ $AB + BC > AC$ (Sub)

Written Exercises, pages 213–214

1. Yes **3.** Yes **5.** ⊥ seg **7.**

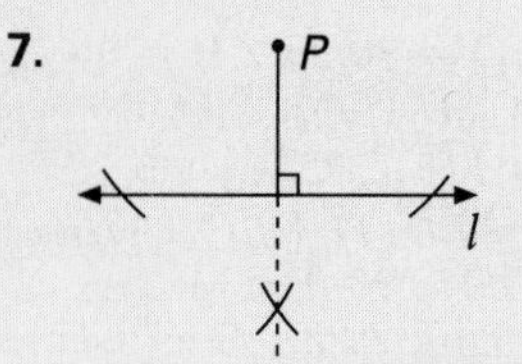

9. $4 < x < 14$ **11.** $0 < x < 12$ **13.** $7 < x < 23$ **15.** $661 < CD < 2{,}087$ **17.** $379 < CM < 1{,}805$ **19.** Rt $\triangle ABC$, hyp is $\overline{AB}$ (Given). $\angle C$ is rt ∠, so $\angle A$ and $\angle B$ are acute ∠s. m $\angle A <$ m $\angle C$ and m $\angle B <$ m $\angle C$. Since m $\angle A <$ m $\angle C$ then $BC < AB$, since for two sides of a △, the longer side is opp the larger ∠. Also, since m $\angle B <$ m $\angle C$, $AC < AB$. Therefore, the hyp $\overline{AB}$ is the longest side. **21.** $37.48 < AC < 92.7$ **23. Part One:** Obtuse $\triangle ABC$, $\angle B$ is obtuse, $\angle A$ and $\angle C$ are acute (Given); Construct $\overline{BD} \perp \overline{AC}$ (From a pt, a ⊥ can be constructed to a line.); $AB > AD$, $BC > DC$ (Longest side of rt △ is hyp.); $AB + BC > AD + DC$ (Add Prop of Ineq); $AD + DC = AC$ (Seg Add Post); ∴ $AB + BC > AC$ (Sub Prop of Ineq) **Part Two:** $AC > AB$ (Side opp larger ∠ is longer.); $BC > 0$ (Length is pos.); ∴ $AC + BC > AB$ (Add Prop of Ineq) **Part Three:** $AC > BC$ (Side opp larger ∠ is longer.); $AB > 0$ (Length is pos.); ∴ $AC + AB > BC$ (Add Prop of Ineq) **25.** $RQ + RS > SQ$, $SP + ST > PT$ (△ Ineq Thm); $RQ + RS + SP + ST > SQ + PT$ (Add Prop of Ineq); $RS + SP = RP$, $ST + TQ = SQ$ (Seg Add Post); $RQ + RP + ST > ST + TQ + PT$ (Sub Prop of Ineq); ∴ $RQ + RP > PT + TQ$ (Subt Prop of Ineq)

Written Exercises, pages 218–219

1. $<$ **3.** $>$ **5.** m $\angle P >$ m $\angle R$, $PQ = QR$ (Given); $SR > SP$ (Side opp larger ∠ is longer.); $\overline{SQ} \cong \overline{SQ}$ (Reflex Prop of Congr); ∴ m $\angle 2 >$ m $\angle 1$ (SSS Ineq Thm) **7.** $QR = PS$, $QS < PR$ (Given); $\overline{PQ} \cong \overline{PQ}$ (Reflex Prop of Congr); ∴ m $\angle PQR >$ m $\angle SPQ$ (SSS Ineq Thm) **9.** $TP > TQ$, $SQ = PR$ (Given); $\overline{PQ} \cong \overline{PQ}$ (Reflex Prop of Congr); m $\angle 1 >$ m $\angle 4$ (∠ opp longer side is larger.);∴ $PS > QR$ (SAS Ineq Thm) **11.** As the hinge opens wider (∠ measure increases), the third side becomes longer. The two ≅ sides are hinged together. SAS refers to the 2 ≅ sides and the included ∠. **13.** $\overline{ED} \cong \overline{DF}$, m $\angle 1 >$ m $\angle 2$, D is midpt of $\overline{CB}$, $\overline{AE} \cong \overline{AF}$ (Given); $\overline{DB} \cong \overline{DC}$ (Def of midpt); $EC > FB$ (SAS Ineq Thm); $EC + AE > FB + AE$ (Add Prop of Ineq); $EC + AE > FB + AF$ (Sub); $EC + AE = AC$, $FB + AF =$

AB (Seg Add Post); $\therefore AC > AB$ (Sub Prop of Ineq) **15.** $AE = AB = ED = DC, EB < EC$ (Given); $m\angle A + m\angle B + m\angle AEB = 180$, $m\angle C + m\angle D = m\angle CED = 180$ (Sum of meas of $\triangle = 180$); $m\angle A + m\angle B + m\angle AEB = m\angle C + m\angle D + m\angle CED$ (Sub); $m\angle B = m\angle AEB$, $m\angle C = m\angle CED$ ($\angle$s opp $\cong$ sides of $\triangle$ are $\cong$); $m\angle A + 2(m\angle AEB) = m\angle D + 2(m\angle C)$ (Sub); $m\angle A < m\angle D$ (SAS Ineq Thm); $-1(m\angle A) > -1(m\angle D)$ (Mult Prop of Ineq); $2(m\angle AEB) > 2(m\angle DCE)$; $\therefore m\angle AEB > m\angle DCE$ (Mult Prop of Ineq)

Chapter 6

Written Exercises, pages 228–229

1. Pentagon $ABCDE$; convex **3.** Quad $LMNO$; convex **5.** Equiangular **7.–15.** See chart on page 226 and definitions on page 227. **17.** 0; 0 **19.** 2; 5 **21.** 4; 14 **23.** 47: 1.175
25.

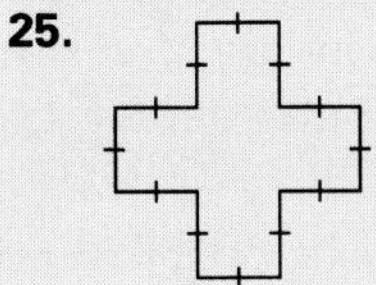

Written Exercises, pages 232–234

1. 540 **3.** 1,440 **5.** 10,800 **7.** 140 **9.** Not possible **11.** Not possible **13.** 7 **15.** 13 **17.** 27 **19.** 90 **21.** 144 **23.** 168 **25.** 3 **27.** 5 **29.** 77, 95, 82, 106 **31.** 120 **33.** 90, 45, 135, 135, 135 **35.** Yes **37.** Yes, 4 **39.** Yes, 6 **41.** Yes, 18 **43.** Quad $ABCD$ is convex (Given); $m\angle DAB = m\angle DAC + m\angle CAB$, $m\angle DCB = m\angle DCA + m\angle BCA$ (Angle Add Post); $m\angle D + m\angle DAC + m\angle DCA = 180$, $m\angle B + m\angle BCA + m\angle BAC = 180$ (Sum of the meas of the $\angle$s of a $\triangle = 180$.); $m\angle D + m\angle DAC + m\angle DCA + m\angle B + m\angle BCA + m\angle BAC = 360$ (Equations may be added.); $\therefore m\angle D + m\angle DAB + m\angle B + m\angle DCB = 360$ (Sub) **45.** Quad $ABCD$ is convex (Given); $m\angle EAB + m\angle AEB + m\angle EBA = 180$, $m\angle BEC + m\angle ECB + m\angle CBE = 180$, $m\angle ECD + m\angle CED + m\angle EDC = 180$, $m\angle DEA + m\angle DAE + m\angle ADE = 180$ (Sum of the meas of the $\angle$s of a $\triangle = 180$.); $m\angle EAB + m\angle EBA + m\angle AEB + m\angle BEC + m\angle ECB + m\angle CBE + m\angle ECD + m\angle CED + m\angle EDC + m\angle DEA + m\angle DAE + m\angle ADE = 720$ (Equations may be added.); $m\angle DEC + m\angle CEB + m\angle BEA + m\angle AED = 360$ (Sum of the meas of the $\angle$s around a pt $= 360$.); $m\angle EAB + m\angle EBA + m\angle ECB + m\angle EBC + m\angle ECD + m\angle EDC + m\angle DAE + m\angle ADE = 360$ (Equations may be subtracted.); $m\angle DAB = m\angle DAE + m\angle BAE$, $m\angle ABC = m\angle ABE + m\angle CBE$, $m\angle BCD = m\angle BCE + m\angle DCE$, $m\angle CDA = m\angle CDE + m\angle ADE$ (Angle Add Post); $\therefore m\angle ABC + m\angle BCD + m\angle CDA + m\angle DAB = 360$ (Sub) **47.** Reg polygon with n sides (Given); polygon has n $\angle$s (Number of sides of polygon = number of $\angle$s); measure of each $\angle$ is equal, call it x (Def of reg polygon); $nx = (n - 2)180$ (Sum of meas of int $\angle$s of convex polygon $= (n - 2)180$); $\therefore x = \dfrac{(n-2) \cdot 180}{n}$ (Div Prop of Eq) **49.** No

Written Exercises, pages 238–239

1. 360, yes **3.** 36, 72, 108, 144 **5.** 120 **7.** 60 **9.** 18 **11.** 6 **13.** 12 **15.** Meas of each ext $\angle = 180 -$ (meas int $\angle$). Since int $\angle$s are $\cong$, $180 -$ (meas int $\angle$) is same for each ext $\angle$. **17.** Octagon **19.** 3
21. Yes, a square

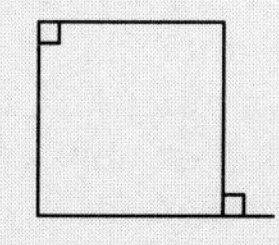

Written Exercises, pages 244–246

1. No **3.** Yes, diag forms 2 $\cong$ $\triangle$s. **5.** 17 **7.** 61 **9.** 72 **11.** 16 **13.** 122 **15.** 120 **17.** ▱$ABCD$, $\angle A \cong \angle DBA$ (Given); $DA = DB$ (Sides opp $\cong$ $\angle$s are $\cong$.); $DA = BC$ (Opp sides of ▱ are $\cong$.); $DB = BC$ (Sub) **19.** Opp $\angle$s are not $\cong$. **21.** Opp sides are not $\cong$.
23. ▱$ABCD$ with diags $\overline{AC}$ AND $\overline{BD}$ (Given); $\triangle ADC \cong \triangle CBA$ (Diag of ▱ forms 2 $\cong$ $\triangle$s.); $\angle CBA \cong \angle CDA$ (CPCTC); $\triangle ADB \cong \triangle CBD$ (Diag of ▱ forms 2 $\cong$ $\triangle$s.); $\therefore \angle BAD \cong \angle DCB$ (CPCTC) **25.** ▱$ACDF$, ▱$BCEF$ (Given); $FD = AC$, $AF = DC$, $BF = CE$, $BC = FE$ (Opp sides of ▱ are $\cong$.); $FD - FE = AC - BC$ (Equations may be subtracted.); $FD - FE = ED$, $AC - BC = AB$ (Seg Add Post); $ED = AB$ (Sub); $\therefore \triangle ABF \cong \triangle DEC$ (SSS) **27.** ▱$JLMP$, K is midpt of $\overline{JL}$, N is midpt of $\overline{PM}$ (Given); $NM = \frac{1}{2}PM$, $JK = \frac{1}{2}JL$ (Def of midpt); $PM = JL$ (Opp sides ▱ are $\cong$.); $\frac{1}{2}PM = \frac{1}{2}JL$ (Mult Prop of Eq); $NM = JK$ (Sub); $\overline{PM} \parallel \overline{JL}$ (Def of ▱); $\angle NMO \cong \angle KJO$ (If lines

are ∥, then alt int ∠s are ≅.); ∠*JOK* ≅ ∠*MON* (Vert ∠s are ≅.); △*MON* ≅ △*JOK* (AAS); $\overline{JO} \cong \overline{MO}$ (CPCTC); *O* is midpt of $\overline{JM}$ (Def of midpt)

29.

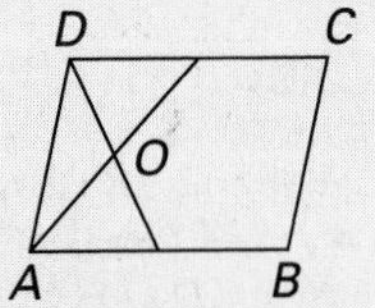

▱*ABCD*, $\overline{DO}$ bis ∠*ADC*, $\overline{AO}$ bis ∠*DAB* (Given); m ∠*DAB* + m ∠*ADC* = 180 (Consec ∠s of ▱ are supp.); m ∠*OAD* = $\frac{1}{2}$ × m ∠*DAB*, m ∠*ODA* = $\frac{1}{2}$ × m ∠*ADC* (Def of ∠ bis); m ∠*OAD* + m ∠*ODA* = $\frac{1}{2}$(m ∠*ADC* + m ∠*DAB*) (Equations may be added.); m ∠*OAD* + m ∠*ODA* = 90 (Sub); m ∠*OAD* + m ∠*ODA* + m ∠*DOA* = 180 (Sum of meas of △ = 180); m ∠*DOA* = 90 (Equations may be subtracted.); ∴ $\overline{DO} \perp \overline{AO}$ (Def of ⊥)

Written Exercises, pages 250–251

1. Yes; both pairs opp sides are ∥. **3.** Yes; both pairs opp sides are ≅. **5.** No; same pair opp sides must be ∥ and ≅. **7.** No; both pair opp ∠s must be ≅. **9.** No; only adj sides are ≅. **11.** Yes; both pair opp ∠s are ≅. **13.** No; same sides that are ≅ must be ∥. **15.** Quad *ABCD*, m ∠*A* = m ∠*C* (∠*A* ≅ ∠*C*), m ∠*B* = m ∠*D* (∠*B* ≅ ∠*D*) (Given); m ∠*A* + m ∠*B* + m ∠*C* + m ∠*D* = 360 (Sum of meas of int ∠s of a convex quad = 360.); m ∠*A* + m ∠*B* + m ∠*A* + m ∠*B* = 360 (2 × m ∠*A* + 2 × m∠*B* = 360) (Sub); m ∠*A* + m ∠*B* = 180 (Div Prop of Eq); m ∠*C* + m ∠*B* = 180 (Sub); $\overline{AB} \parallel \overline{CD}$, $\overline{BC} \parallel \overline{AD}$ (If int ∠s on same side of a transv are supp, then lines are ∥.); ∴ *ABCD* is a ▱ (Def of ▱) **17.** ▱*ACDF*, $\overline{FE} \cong \overline{CB}$ (*FE* = *CB*) (Given); $\overline{FD} \parallel \overline{AC}$ (Def of ▱); *FD* = *AC* (Opp sides of ▱ are ≅.); *FD* − *FE* = *AC* − *BC* (Equations may be subtracted.); *FD* − *FE* = *ED*, *AC* − *BC* = *AB* (Seg Add Post); *ED* = *AB* (Sub); ∴ *ABDE* is ▱ (If one pair opp sides are ∥ and ≅, then quad is ▱.) **19.** $\overline{AF} \parallel \overline{CD}$, △*AFE* ≅ △*DCB* (Given); $\overline{AF} \cong \overline{CD}$, $\overline{AE} \cong \overline{DB}$, $\overline{FE} \cong \overline{CB}$ (CPCTC); *ACDF* is ▱ (If 1 pair opp sides are both ∥ and ≅, then quad is ▱.); *FD* = *AC* (Def of ▱); *FD* − *FE* = *AC* − *CB* (Equations may be subtracted.); *FD* − *FE* = *ED*, *AC* − *CB* = *AB* (Seg Add Post); *ED* = *AB* ($\overline{ED} \cong \overline{AB}$) (Sub); ∴ *ABDE* is ▱ (If both pairs of opp sides are ≅, then quad is ▱.) **21.** True, quad *ABCD*, *A* and *D* are supp, ∠*D* and ∠*C* are supp (Given); $\overline{AB} \parallel \overline{CD}$, $\overline{BC} \parallel \overline{AD}$ (If int ∠s on same side of transv are supp, then lines are ∥.); ∴ *ABCD* is a ▱ (Def of ▱) **23.** False

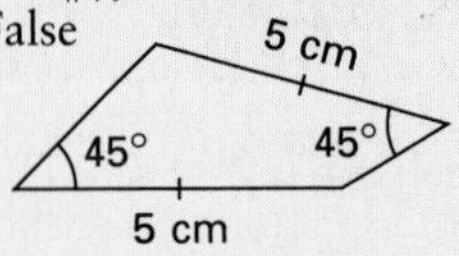

25. True; quad *ABCD*, m ∠*A* + m ∠*B* = *K*, m ∠*B* + m ∠*C* = *K*, m ∠*C* + m ∠*D* = *K* (Given); m ∠*A* + m ∠*B* = m ∠*B* + m ∠*C*, m ∠*B* + m ∠*C* = m ∠*C* + m ∠*D* (Sub); m ∠*A* = m ∠*C*, m ∠*B* = m ∠*D* (Subt Prop of Eq); ∴ *ABCD* is ▱ (If both pair opp ∠s are ≅, then quad is ▱.)

Written Exercises, pages 254–255

1. $\overline{FG}$ **3.** 6 **5.** 6 **7.** 14 **9.** $\overline{AD}$ **11.** 10 **13.** 3 **15.** *O* is midpt of $\overline{LM}$, *P* is midpt of $\overline{MN}$, *Q* is midpt of $\overline{NL}$ (Given); $\overline{OP} \parallel \overline{NL}$, $\overline{OQ} \parallel \overline{MN}$ (Midseg Thm); $\overline{OP} \parallel \overline{NQ}$, $\overline{OQ} \parallel \overline{NP}$ (If seg is ∥ to line, it is also ∥ to any seg of the line.); ∴ *OPNQ* is ▱ (Def of ▱) **17.** *O* is midpt of $\overline{LM}$, *P* is midpt of $\overline{MN}$, *Q* is midpt of $\overline{NL}$ (Given); $\overline{QL} \parallel \overline{PO}$, $\overline{PQ} \parallel \overline{LO}$ (Midseg Thm); *OPQL* is ▱ (Def of ▱); ∴ △*LOQ* ≅ △*PQO* (Diag of ▱ forms 2 ≅ △s.) **19.** *U* is midpt of $\overline{RS}$, *V* is midpt of $\overline{ST}$, $\overline{RT} \cong \overline{ST}$ (Given); *VU* = $\frac{1}{2}$*RT* (Midseg Thm); *VS* = $\frac{1}{2}$*ST* (Def of midpt); *VS* = $\frac{1}{2}$*RT* (Sub); *VS* = *VU* (Sub); ∴ △*USV* is isos (Def of isos △) **21.** Quad *ABCD*, *E*, *F*, *G*, *H* are midpts of $\overline{AB}$, $\overline{BC}$, $\overline{DC}$, $\overline{DA}$, respectively (Given); $\overline{GF} \parallel \overline{DB}$, *GF* = $\frac{1}{2}$*DB*, $\overline{HE} \parallel \overline{DB}$, *HE* = $\frac{1}{2}$*DB* (Midseg Thm); $\overline{HE} \parallel \overline{GF}$ (2 lines ∥ to the same line are ∥.); *HE* = *GF* (Sub); *EFGH* is ▱ (If same pair opp sides are ∥ and ≅, then quad is ▱.); $\overline{HF}$ and $\overline{GE}$ bis each other (Diags of ▱ bis each other.) **23.** *H*, *I*, *J*, *K*, and *L* are the midpts of $\overline{DF}$, $\overline{EF}$, $\overline{GF}$, $\overline{DG}$, and $\overline{DE}$, respectively. *JH* = $\frac{1}{2}$*DG*, *HI* = $\frac{1}{2}$*DE*, *JI* = $\frac{1}{2}$*GE*, *KL* = $\frac{1}{2}$*GE* (Midseg Thm); *DK* = $\frac{1}{2}$*DG*, *DL* = $\frac{1}{2}$*DE* (Def of midpt); *JH* = *DK*, *HI* = *DL*, *JI* = *KL* (Sub); ∴ △*JHI* ≅ △*KDL* (SSS)

Written Exercises, pages 260–261

1. 21 **3.** 31 **5.** 29 **7.–8.** See Construction on page 260. **9.** ∠1 ≅ ∠5,

$\angle 5 \cong \angle 9$, $\overline{GH} \cong \overline{HI}$ (Given); $\overline{LG} \parallel \overline{KH}$, $\overline{KH} \parallel \overline{JI}$ (If corr $\angle$s are $\cong$, then lines are $\parallel$.); $\overline{LG} \parallel \overline{JI}$ (If 2 lines are $\parallel$ to same line, then they are $\parallel$.); $\therefore$ $\overline{LK} \cong \overline{KJ}$ (If 3 $\parallel$ lines make $\cong$ segs on 1 transv, then they make $\cong$ segs on every transv.) **11.** $\angle 3 \cong \angle 10$, $\overline{KH} \parallel \overline{JI}$, $\overline{GH} \cong \overline{HI}$ (Given); $\overline{JI} \parallel \overline{LG}$ (If alt int $\angle$s are $\cong$, then lines are $\parallel$.); $\overline{KH} \parallel \overline{LG}$ (If 2 lines are $\parallel$ to the same line, then they are $\parallel$.); $\therefore$ $\overline{LK} \cong \overline{KJ}$ (If 3 $\parallel$ lines make $\cong$ segs on 1 transv, then they make $\cong$ segs on every transv.)

13.

Construct $\perp$ from 1 line to the other.

15. Answers may vary.

17.

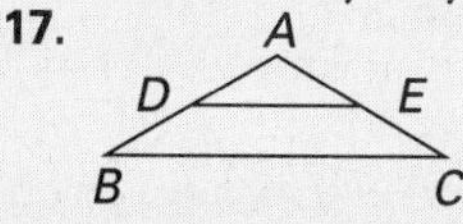

$\overline{AD} \cong \overline{BD}$, $\overline{DE} \parallel \overline{BC}$ (Given); Draw $l \parallel \overline{DE}$ through pt A (Through a given pt, there exists exactly 1 line $\parallel$ to given line.); $l \parallel \overline{BC}$ (If 2 lines are $\parallel$ to the same line, then they are $\parallel$.); $\therefore$ $\overline{AE} \cong \overline{EC}$ (If 3 $\parallel$ lines make $\cong$ segs on 1 transv, then they make $\cong$ segs on every transv.)

19. $DE = \frac{1}{2}BC$; $\overline{DE} \nparallel \overline{BC}$.

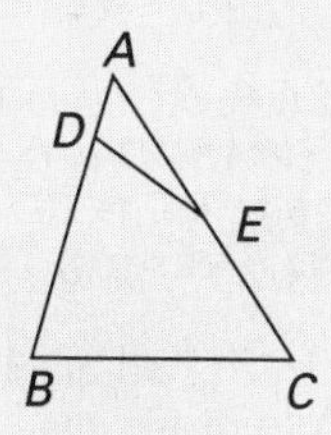

Chapter 7

Written Exercises, pages 271–272

1. $\overline{AB} \cong \overline{WX}$, $\overline{BC} \cong \overline{XY}$, $\overline{CD} \cong \overline{YZ}$, $\overline{DA} \cong \overline{ZW}$; $\angle A \cong \angle W$, $\angle B \cong \angle X$, $\angle C \cong \angle Y$, $\angle D \cong \angle Z$ **3.** $\overline{EB}$ is $\perp$ bis of $\overline{AC}$, $\overline{AF} \cong \overline{CD}$, $\angle A \cong \angle C$ (Given); $\angle ABE$, $\angle CBE$ are rt $\angle$s (Def of $\perp$); $\angle ABE \cong \angle CBE$ (Rt $\angle$s are $\cong$.); $\overline{AB} \cong \overline{BC}$ (Def bis); $\overline{BE} \cong \overline{BE}$ (Reflex Prop of Congr); $\therefore$ quad $ABEF \cong$ quad $CBED$ (SASAS)
5. $\overline{AF} \cong \overline{BD}$, $\overline{AG} \cong \overline{BC}$, $\angle A \cong \angle B$ (Given); $\overline{AB} \cong \overline{AB}$ (Reflex Prop of Congr); $\therefore$ quad $ABDG \cong$ quad $BAFC$ (SASAS) **7.** $\overline{TX} \cong \overline{WU}$, $\triangle TZY \cong \triangle WZV$ (Given); $\overline{YZ} \cong \overline{VZ}$, $\overline{TZ} \cong \overline{WZ}$, $\overline{TY} \cong \overline{WV}$, $\angle TYZ \cong \angle WVZ$ (CPCTC); $TX - TY = WU - WV$ (Equations may be subtracted.); $TX - TY = YX$, $WU - WV = VU$ (Seg Add Post); $YX = VU$ (Sub); $\angle XYZ$ and $\angle TYZ$ are supp, $\angle UVZ$ and $\angle WVZ$ are supp (If outer rays of 2 adj $\angle$s form st $\angle$, then $\angle$s are supp.); $\angle XYZ \cong \angle UVZ$ (Supp of $\cong$ $\angle$s are $\cong$.); $\angle YZW \cong \angle VZT$ (Vert $\angle$s are $\cong$.); $\therefore$ quad $TUVZ \cong$ quad $WXYZ$ (SASAS) **9.** $\square TUWX$, $\overline{YV}$ and $\overline{TW}$ bis each other at Z (Given); $\overline{TU} \cong \overline{XW}$ (Opp sides of $\square$ are $\cong$.); $\overline{TU} \parallel \overline{XW}$ (Def of $\square$); $\angle XWZ \cong \angle UTZ$ (If lines are $\parallel$, then alt int $\angle$s are $\cong$.); $\overline{WZ} \cong \overline{TZ}$, $\overline{YZ} \cong \overline{VZ}$ (Def of seg bis); $\angle WZY \cong \angle TZV$ (Vert $\angle$s are $\cong$.); $\therefore$ quad $TUVZ \cong$ quad $WXYZ$ (SASAS) **11.** $\angle F \cong \angle J$, $\angle G \cong \angle K$, $\angle H \cong \angle L$, $\overline{FG} \cong \overline{JK}$, $\overline{GH} \cong \overline{KL}$ (Given); Draw $\overline{HF}$, $\overline{LJ}$ (2 pts determine a line.); $\triangle HFG \cong \triangle LJK$ (SAS); m $\angle 1$ = m $\angle 5$, m $\angle 2$ = m $\angle 6$, $\overline{HF} \cong \overline{LJ}$ (CPCTC); m $\angle 1$ + m $\angle 4$ = m $\angle H$, m $\angle 5$ + m $\angle 8$ = m $\angle L$, m $\angle 2$ + m $\angle 3$ = m $\angle F$, m $\angle 6$ + m $\angle 7$ = m $\angle J$ (Angle Add Post); m $\angle 1$ + m $\angle 4$ = m $\angle 5$ + m $\angle 8$, m $\angle 2$ + m $\angle 3$ = m $\angle 6$ + m $\angle 7$ (Sub); m $\angle 4$ = m $\angle 8$, m $\angle 3$ = m $\angle 7$ (Equations may be subtracted.); $\triangle HEF \cong \triangle LIJ$ (ASA); $\overline{HE} \cong \overline{LI}$ (CPCTC); $\therefore$ quad $EFGH \cong$ quad $IJKL$ (SASAS)

Written Exercises, pages 276–277

1. True **3.** True **5.** True **7.** True
9. False

11. Yes **13.** Yes **15.** 23 **17.** 90
19. False

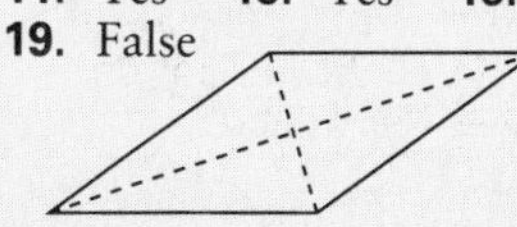

21. True. A square is a rhomb. The diags of a rhomb are $\perp$. **23.** 14 **25.** 45 **27.** Rect $EFGH$, $\overline{HF}$ and $\overline{GE}$ are diags (Given); $\angle HEF$ and $\angle GFE$ are rt $\angle$s (Def of rect); $\angle HEF \cong \angle GFE$ (Rt $\angle$s are $\cong$.); $\overline{HE} \cong \overline{GF}$ (Opp sides of $\square$ are $\cong$.); $\overline{EF} \cong \overline{EF}$ (Reflex Prop of Congr); $\triangle HEF \cong \triangle GFE$ (SAS); $\therefore$ $\overline{HF} \cong \overline{GE}$ (CPCTC)

29. 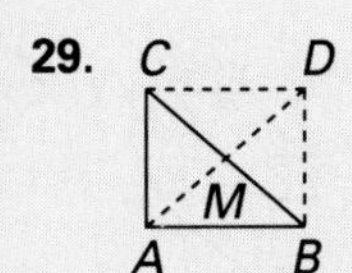

$\triangle ABC$, m $\angle CAB = 90$, M is midpt of $\overline{BC}$ (Given); draw $\overline{BD} \perp \overline{AB}$, and draw $\overline{CD} \perp \overline{AC}$ (At given pt on line, $\perp$ may be drawn.); $\overline{CA} \parallel \overline{BD}$, $\overline{CD} \parallel \overline{BA}$ (In plane, 2 lines $\perp$ to same line are $\parallel$.); $ABCD$ is ▱ (Def ▱); $\angle CDB$ is rt $\angle$ (Opp $\angle$s in ▱ are $\cong$.); $ABCD$ is rect (Def rect); Draw $\overline{AD}$ (2 pts determine line.); $\overline{AD}$ contains M (Diags of ▱ bis each other; a line has 1 midpt.); $\overline{AD} \cong \overline{CB}$ (Diags of rect are $\cong$.); $AM = \frac{1}{2}AD$, $BM = \frac{1}{2}CB$ (Def bis); $\therefore \overline{AM} \cong \overline{CM} \cong \overline{BM}$ (Sub) **31.** Plan for proof: First, prove $\triangle OKP \cong \triangle PLQ \cong \triangle QMR \cong \triangle RNO$ by SAS. Then state that $\overline{OP} \cong \overline{PQ} \cong \overline{QR} \cong \overline{RO}$ by CPCTC. Then state that quad $OPQR$ is a ▱. Proceed by proving that meas of $\angle ORQ = 90$, and by def of supp $\angle$s, show that $\angle OPQ$, $\angle PQR$, $\angle QRP$, and $\angle ROP$ are rt $\angle$s. Thus, it can be proved that quad $OPQR$ is a square by def of square.

Written Exercises, pages 281–282

1. False **3.** True **5.** True

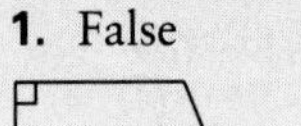

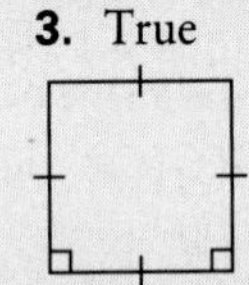

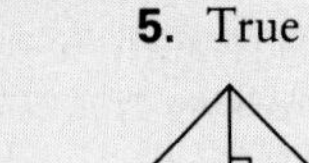

7. Not sufficient **9.** Not sufficient

11. Not sufficient

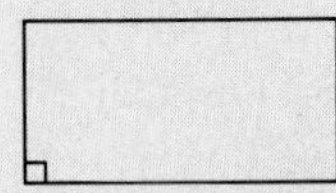

13. $\overline{AB} \cong \overline{DC}$, $\overline{AB} \parallel \overline{DC}$, $\overline{AC} \cong \overline{BD}$ (Given); $ABCD$ is ▱ (If 2 sides of quad are $\parallel$ and $\cong$, it is ▱.); $\therefore ABCD$ is rect (▱ with $\cong$ diags is rect.)

15. 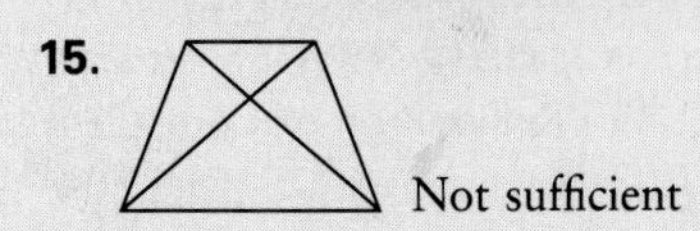 Not sufficient

17. 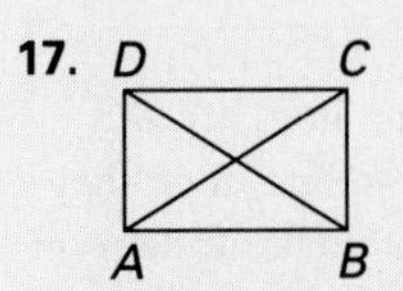

Sufficient. Proof: Quad $ABCD$ with $\overline{BD} \cong \overline{AC}$, $\overline{BD}$ and $\overline{AC}$ bis each other (Given); $ABCD$ is ▱ (If diags bis each other, quad is ▱.); $\therefore ABCD$ is rect (If diags of ▱ are $\cong$, ▱ is rect.)

19. 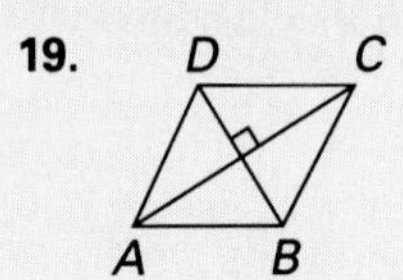

Sufficient. Proof: Quad $ABCD$, $\overline{BD}$ and $\overline{AC}$ bis each other, $\overline{BD} \perp \overline{AC}$ (Given); $ABCD$ is ▱ (If diags bis each other, quad is ▱.); $\therefore ABCD$ is rhom (▱ with $\perp$ diags is rhom.)

21. 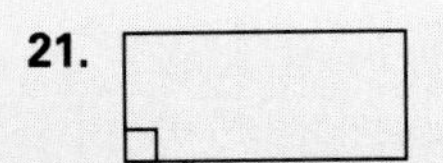

Not sufficient

23. 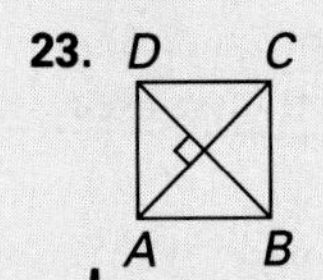

Sufficient. Proof: Rect $ABCD$ with $\overline{AC} \perp \overline{DB}$ (Given); $ABCD$ is ▱ (Def rect); $ABCD$ is rhom (▱ with $\perp$ diags is rhom.); $AB \cong BC \cong CD \cong DA$ (Def rhom); $\therefore ABCD$ is sq (Def sq)

25. A *necessary* condition is true of all quads that are rects, but not enough to prove quad is a rectangle. A *sufficient* condition is enough for quad to be rectangle.

27. 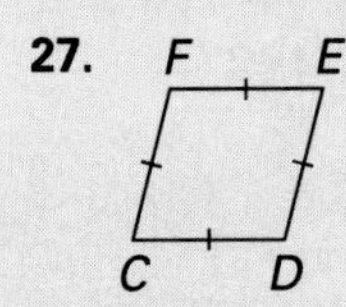

▱ $CDEF$, $\overline{CD} \cong \overline{CF}$ (Given); $\overline{CD} \cong \overline{FE}$, $\overline{CF} \cong \overline{DE}$ (Opp sides of ▱ are $\cong$.); $\overline{CD} \cong \overline{DE} \cong \overline{FE} \cong \overline{CF}$ (Trans Prop of Congr); $\therefore$ ▱$CDEF$ is rhom (Def of rhom)

29. D C 1 2 3 4 A B

$\square$ $ABCD$, $\overline{DB}$ bis opp $\angle$s (Given); $\angle 1 \cong \angle 2$, $\angle 3 \cong \angle 4$ (Def $\angle$ bis); $\overline{BD} \cong \overline{BD}$ (Reflex Prop of Congr); $\triangle ABD \cong \triangle CBD$ (ASA); $\overline{AB} \cong \overline{CB}$ (CPCTC); $\therefore$ $\square$ $ABCD$ is rhom ($\square$ with adj $\cong$ sides is rhom.)

31.

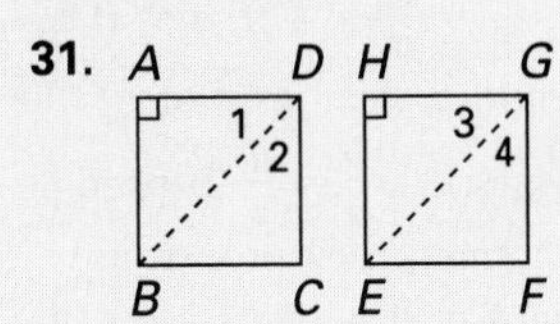

Sufficient. Sq $ABCD$ and sq $EFGH$, $\overline{BD} \cong \overline{EG}$ (Given); m $\angle A = 90$, m $\angle D = 90$, m $\angle H = 90$, m $\angle G = 90$, m $\angle B = 90$, m $\angle E = 90$ (Def sq); m $\angle 1$ = m $\angle 2$, m $\angle 3$ = m $\angle 4$ (Diags of rhom bis opp $\angle$s.); m $\angle 1$ + m $\angle 2$ = m $\angle D$, m $\angle 3$ + m $\angle 4$ = m $\angle G$ (Angle Add Post); 2(m $\angle 1$) = 90, 2(m $\angle 3$) = 90 (Sub); m $\angle 1$ = 45, m $\angle 3$ = 45 (Div Prop of Eq); $\triangle ABD \cong \triangle HEG$ (AAS); $\overline{AD} \cong \overline{HG}$, $\overline{AB} \cong \overline{HE}$ (CPCTC); $\therefore$ $ABCD \cong EFGH$ (ASASA)

33.

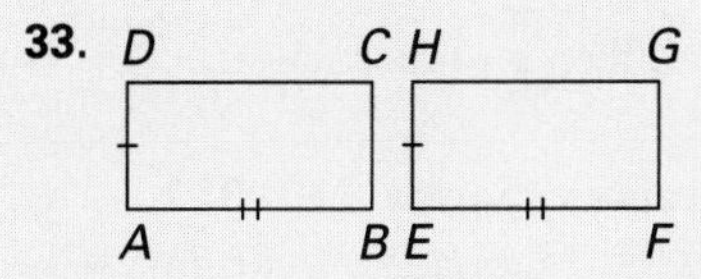

Sufficient. Rects $ABCD$ and $EFGH$, $\overline{DA} \cong \overline{HE}$, $\overline{AB} \cong \overline{EF}$ (Given); m $\angle D = 90$, m $\angle A = 90$, m $\angle B = 90$, m $\angle H = 90$, m $\angle E = 90$, m $\angle F = 90$ (Def rect); m $\angle D$ = m $\angle H$, m $\angle A$ = m $\angle E$, m $\angle B$ = m $\angle F$ (Sub); $\therefore$ $ABCD \cong EFGH$ (ASASA)

Written Exercises, pages 288–289

1. 11 **3.** 10 **5.** 16 **7.** 19

9. Possible

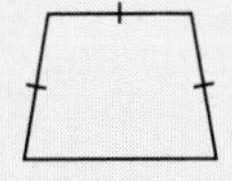

11. Possible

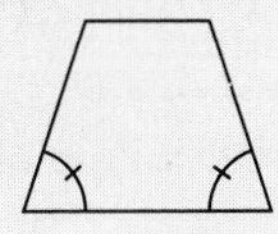

13. Not possible; figure would be rect.

15. Possible

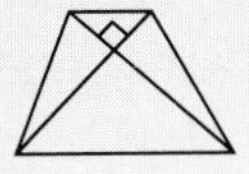

17. GH varies with value of x.

19.

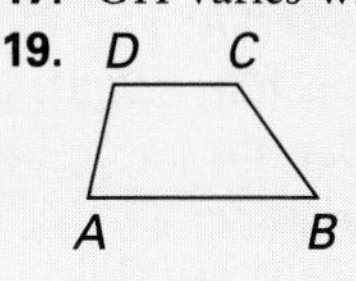

$\angle A \not\cong \angle B$.

21.

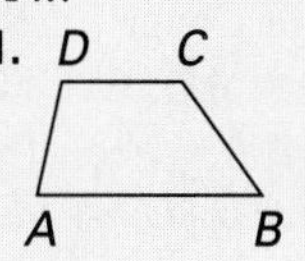

$\overline{AD} \not\cong \overline{BC}$.

23. Trap $ABCD$ with median $\overline{GF}$, $\overline{AC}$ intersects $\overline{GF}$ at H, $\overline{DE} \perp \overline{AC}$ at H (Given); G is midpt $\overline{AD}$ (Def median); $\overline{AG} \cong \overline{GD}$ (Def of midpt); $\overline{DC} \parallel \overline{GF} \parallel \overline{AB}$ (Median of trap is $\parallel$ to bases.); $\overline{HD} \cong \overline{HE}$ (If 3 $\parallel$ lines cut $\cong$ segs on 1 transv, they cut $\cong$ segs on every transv.); $\angle DHA$ and $\angle EHA$ are rt $\angle$s (Def of $\perp$); $\angle DHA \cong \angle EHA$ (Rt $\angle$s are $\cong$.); $\overline{AH} \cong \overline{AH}$ (Reflex Prop of Congr); $\triangle AHD \cong \triangle AHE$ (SAS); $\overline{AE} \cong \overline{AD}$ (CPCTC); $\therefore$ $\triangle AED$ is isos (Def of isos $\triangle$)

25. Counterexample

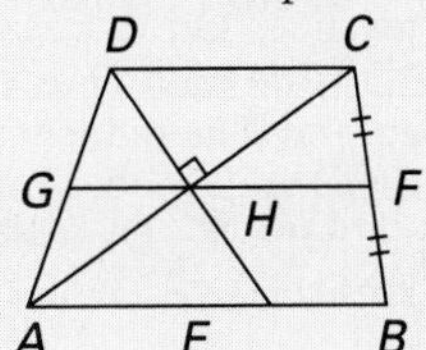

Written Exercises, pages 293–294

1. 83 **3.** 58 **5.** 39 **7.** 36 **9.** 30 **11.** 87 **13.** $\overline{AC} \cong \overline{BC}$, $\overline{DE} \parallel \overline{AB}$ (Given); $ABED$ is a trap (Def trap); $\angle A \cong \angle B$ ($\angle$s opp $\cong$ sides of $\triangle$ are $\cong$.); $\therefore$ $ABED$ is isos trap (If base $\angle$s of trap $\cong$, trap is isos.) **15.** 110 **17.** 54 **19.** 22 **21.** Not necessary, not sufficient **23.** Necessary and sufficient

25.

Necessary, not sufficient

27.

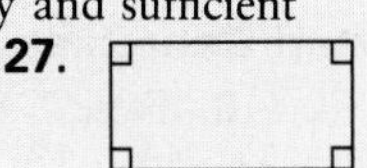

Necessary, not sufficient

29. Trap $ABCD$, $\angle A \cong \angle B$ (Given); Draw $\overline{DE} \perp \overline{AB}$, and draw $\overline{CF} \perp \overline{AB}$ (There is exactly 1 $\perp$ from pt to line.); $\angle AED$ and $\angle BFC$ are rt $\angle$s (Def of $\perp$); $\angle AED \cong \angle BFC$ (Rt $\angle$s are $\cong$); $\triangle AED \cong \triangle BFC$ (AAS); $\overline{AD} \cong \overline{BC}$ (CPCTC); $\therefore$ $ABCD$ is isos trap (Def isos trap) **31.** Trap $ABCD$, $\overline{DC} \parallel \overline{AB}$, $\overline{AC} \cong \overline{BD}$ (Given); draw $\overline{CF} \perp \overline{AB}$ and $\overline{DE} \perp \overline{AB}$ (There is 1 $\perp$ from pt to line.); $\angle AFC$ and $\angle BED$ are rt $\angle$s (Def rt $\angle$); $\triangle AFC$ and $\triangle BED$ are rt $\triangle$s (Def rt $\triangle$); $\overline{CF} \cong \overline{DE}$ (Alts of trap are $\cong$.); $\triangle AFC \cong \triangle BED$ (HL);

$\angle BAC \cong \angle ABD$ (CPCTC); $\overline{BA} \cong \overline{AB}$ (Reflex Prop of Congr); $\triangle BAC \cong \triangle ABD$ (SAS); $\therefore \overline{BC} \cong \overline{AD}$ (CPCTC) **33.** Isos trap $ABCD$, $\overline{DC} \parallel \overline{AB}$ (Given); $\overline{DA} \cong \overline{CB}$ (Def isos trap); $\overline{AC} \cong \overline{BD}$ (Diags of isos trap are $\cong$.); $\overline{DC} \cong \overline{DC}$ (Reflex Prop of Congr); $\triangle ACD \cong \triangle BDC$ (SSS); $\angle DAC \cong \angle CBD$ (CPCTC); $\angle DEA \cong \angle CEB$ (Vert $\angle$s are $\cong$.); $\therefore \triangle AED \cong \triangle BEC$ (AAS) **35.** M is midpt of $\overline{AF}$, N is midpt of $\overline{FC}$, $\triangle AFE \cong \triangle CFG$ (Given); $\overline{AF} \cong \overline{CF}$ (CPCTC); $\frac{1}{2}AF = \frac{1}{2}CF$ (Mult Prop of Eq); $AM = \frac{1}{2}AF$, $NC = \frac{1}{2}FC$ (Def midpt); $AM = NC$ (Sub); $\overline{MN} \parallel \overline{AC}$ (Midseg Thm); $\therefore ACNM$ is isos trap (Def isos trap)

Written Exercises, pages 296–297

1. Follow steps of Example 1 on p. 295. **3.** Construct $\angle A \cong$ to given $\angle$. On $\angle$ sides construct $AB = b$ and $AD = a$. With D as ctr and b as rad, and B as ctr and a as rad, swing arcs intersecting at C. $ABCD$ is $\square$. **5.** Follow steps of Example 2 on p. 296. **7.** On a line construct $A'B' = AB$. Construct $\angle A' \cong \angle A$, $\angle B' \cong \angle B$. On side of $\angle A'$, construct $A'D' = AD$ and on side of $\angle B'$, construct $B'C' = BC$. Draw $\overline{C'D'}$. $A'B'C'D' \cong ABCD$. **9.** Not possible **11.** Construct $AB = s$. With A as ctr and s as rad, and B as ctr and d as rad, draw two arcs intersecting at D. With B as ctr and s as rad, and A as ctr and d as rad, draw two arcs intersecting at C on the same side of $\overleftrightarrow{AB}$ as D. $ABCD$ is required trap.

Chapter 8

Written Exercises, pages 306–307

1. True **3.** False **5.** True **7.** True **9.** 18 **11.** 60 **13.** ± 2 **15.** ± 16 **17.** $\frac{4}{1} = \frac{8}{2}$; $\frac{1}{2} = \frac{4}{8}$; $\frac{1+4}{4} = \frac{2+8}{8}$; $\frac{1-4}{4} = \frac{2-8}{8}$; $\frac{1}{4} = \frac{1+2}{4+8}$ **19.** 4 **21.** 3 **23.** 6 **25.** $3\sqrt{5}$ **27.** pq **29.** True **31.** True **33.** 40, 50 **35.** 45, 54, 81 **37.** $b = 12.5$, $a = 17.5$ **39.** \$4.83 **41.** 12 **43.** False. $\frac{4}{5} = \frac{8}{10}$, but $\frac{4+2}{5+2} \neq \frac{8+2}{10+2}$ **45.** True. Proof: $\frac{x}{y} = \frac{r}{s} = \frac{m}{n}$ (Given); $xs = ry$ (If proport, then prod of means = prod of extremes.); $xy + xs = xy + ry$ (Add Prop of Eq); $x(y + s) = y(x + r)$ (Distr Prop); $\frac{x}{y} = \frac{x + r}{y + s}$ (If prod of means = prod of extremes, then a proport exists.); $\frac{x + r}{y + s} = \frac{m}{n}$ (Sub); $n(x + r) = m(y + s)$ (If proport, then prod of means = prod of extremes.); $n(x + r) + (x + r)(y + s) = m(y + s) + (x + r)(y + s)$ (Add Prop of Eq); $(x + r)[n + (y + s)] = (y + s)[m + (x + r)]$ (Distr Prop); $(x + r)(y + s + n) = (y + s)(x + r + m)$ (Comm Prop, Assoc Prop); $\frac{x + r}{y + s} = \frac{x + r + m}{y + s + n}$ (If prod of means = prod of extremes, then a proport exists.); $\therefore \frac{x}{y} = \frac{x + r + m}{y + s + n}$ (Sub) **47.** $\frac{a}{b} = \frac{c}{d}$, $b \neq 0$, $d \neq 0$ (Given); $bc = ad$ (If proport, then prod of means = prod of extremes.); $\therefore \frac{b}{a} = \frac{d}{c}$ (If prod of means = prod of extremes, then proportion exists.) **49.** $\frac{a}{b} = \frac{c}{d}$, $b \neq 0$, $d \neq 0$ (Given); $ad = bc$ (If proport, then prod of means = prod of extremes.); $ad + bd = bc + bd$ (Add Prop of Eq); $(a + b)d = (c + d)b$ (Distr Prop); $\therefore \frac{a + b}{b} = \frac{c + d}{d}$ (If prod of means = prod of extremes, then a proport exists.) **51.** $\frac{a}{b} = \frac{c}{d}$, $b \neq 0$, $d \neq 0$ (Given); $ad = bc$ (If proport, then prod of means = prod of extremes.); $ab + ad = ab + bc$ (Add Prop of Eq); $a(b + d) = b(a + c)$ (Distr Prop); $\therefore \frac{a}{b} = \frac{a + c}{b + d}$ (If prod of means = prod of extremes, then a proport exists.)

Written Exercises, pages 310–311

1. $HE = 12$, $FG = 15$, $GH = 9$ **3.** $UV = 15$, XU and $VW = 8\frac{1}{3}$
5.

D C H G
55°
A 7 B E 12 F

$FG = 12$, m $\angle F = 125$
7. $ST = 4$, $TU = 4\frac{1}{2}$ **9.** Sq $ABCD$, sq $XYZW$ (Given); $ABCD$, $XYZW$ are rects with $AB = BC = CD = DA$ and $XY = YZ = ZW = WX$ (Def sq); $ABCD$, $XYZW$ each have 4 rt $\angle$s (Def of rect); $\angle A \cong \angle X$, $\angle B \cong \angle Y$, $\angle C \cong \angle Z$, $\angle D \cong \angle W$ (All rt $\angle$s are $\cong$.); $\frac{AB}{XY} = \frac{AB}{XY}$ (Reflex Prop of Eq); $\frac{AB}{XY} = \frac{BC}{YZ} = \frac{CD}{ZW} = \frac{DA}{WX}$ (Sub); $\therefore ABCD \sim XYZW$ (Def of $\sim$ polygons)
11. Yes. $\cong$ polygons have $\cong$ $\angle$s and corr sides in the const ratio $\frac{1}{1}$. **13.** 13 ft **15.** 4 ft

17.

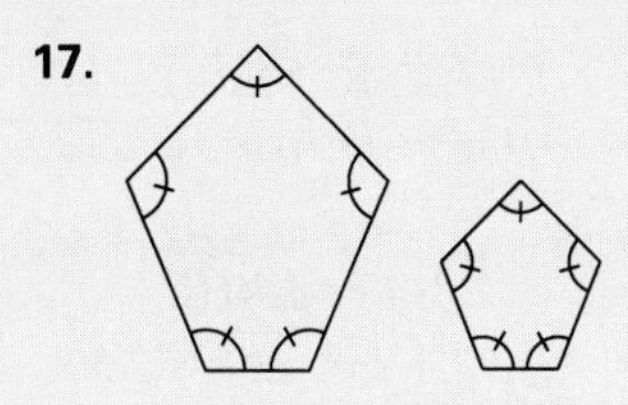

Written Exercises, pages 315–316

1. 15 **3.** 80, 40 **5.** 40, 80 **7.** $\overline{CD} \parallel \overline{AB}$ (Given); $\angle D \cong \angle A$, $\angle C \cong \angle B$ (If 2 lines are $\parallel$, then alt int $\angle$s are $\cong$.); $\therefore \triangle ABE \sim \triangle DCE$ (AA~) **9.** $\overline{CD} \parallel \overline{AB}$ (Given); $\angle D \cong \angle A$, $\angle C \cong \angle B$, (If 2 lines are $\parallel$, then alt int $\angle$s are $\cong$.); $\triangle ABE \sim \triangle DCE$ (AA~); $\therefore \frac{AB}{DC} = \frac{AE}{DE}$ (Def of ~ $\triangle$s) **11.** Yes **13.** $\overline{IJ} \perp \overline{FG}$, $\overline{HG} \perp \overline{FG}$ (Given); $\angle IFJ$ and $\angle G$ are rt $\angle$s. (Def of $\perp$); m $\angle IJF$ = m $\angle G$ (Sub); m $\angle F$ = m $\angle F$ (Reflex Prop of Eq); $\therefore \triangle FJI \sim \triangle FGH$ (AA~) **15.** $\overline{EB} \perp \overline{AC}$, $\overline{CF} \perp \overline{AE}$ (Given); $\angle ABE$ and $\angle AFC$ are rt $\angle$s. (Def of $\perp$); m $\angle ABE$ = m $\angle AFC$ (Sub); $\angle A \cong \angle A$ (Reflex Prop of Congr); $\therefore \triangle ABE \sim \triangle AFC$ (AA~) **17.** $GHIJ$ is a trap (Given); $\overline{JI} \parallel \overline{GH}$ (Def of trap); $\angle JIG \cong \angle HGI$, $\angle IJH \cong \angle GHJ$ (If 2 lines are $\parallel$, then alt int $\angle$s are $\cong$.); $\triangle JIK \sim \triangle HGK$ (AA~); $\frac{GH}{IJ} = \frac{HK}{JK}$ (Def of ~ $\triangle$s); $\therefore GH \cdot JK = IJ \cdot HK$ (If proport, then prod of means = prod of extremes.) **19.** $\triangle LMN \sim \triangle PQR$, alts $\overline{NO}$ and $\overline{RS}$, respectively (Given); $\angle L \cong \angle P$ (Def of ~ $\triangle$s); $\overline{NO} \perp \overline{LM}$, $\overline{RS} \perp \overline{PQ}$ (Def of alt); $\angle NOL$ and $\angle RSP$ are rt $\angle$s. (Def of $\perp$); m $\angle NOL$ = m $\angle RSP$ (Sub); $\triangle LON \sim \triangle PSR$ (AA~); $\therefore \frac{NL}{RP} = \frac{NO}{RS}$ (Def of ~ $\triangle$s) **21.** Rt $\triangle ABC$, $\overline{ED} \perp \overline{AB}$, $\overline{BE}$ bis $\angle ABC$ (Given); m $\angle EDB$ = 90 (Def of $\perp$); m $\angle ABC$ = 90 (Def rt $\triangle$); m $\angle CBE$ + m $\angle DBE$ = m $\angle ABC$ (Angle Add Post); m $\angle CBE$ + m $\angle DBE$ = 90 (Sub); m $\angle CBE$ = m $\angle DBE$ (Def of $\angle$ bis); 2 × m $\angle DBE$ = 90 (Sub); m $\angle DBE$ = 45 (Div Prop of Eq);
m $\angle DBE$ + m $\angle EDB$ + m $\angle DEB$ = 180 (Sum of meas of $\angle$s of $\triangle$ = 180.);
45 + 90 + m $\angle DEB$ = 180 (Sub); m $\angle DEB$ = 45 (Subt Prop of Eq); m $\angle DBE$ = m $\angle DEB$ (Sub); $ED = BD$ (Sides opp $\cong$ $\angle$s of $\triangle$ are $\cong$.); $\angle A \cong \angle A$ (Reflex Prop of Congr); m $\angle EDA$ = 90 (Def of $\perp$); m $\angle EDA$ = m $\angle ABC$ (Sub); $\triangle ADE \sim \triangle ABC$ (AA~); $\frac{AE}{AC} = \frac{ED}{BC}$ (Def of ~ $\triangle$s); $\therefore \frac{AE}{AC} = \frac{BD}{BC}$ (Sub) **23.** $\triangle ABC \cong \triangle DEF$ (Given); $\angle A \cong \angle D$, $\angle B \cong \angle E$, $\angle C \cong \angle F$, $AB = DE$, $BC = EF$, $CA = FD$ (Def of $\cong$ $\triangle$s); $\frac{AB}{AB} = \frac{BC}{BC} = \frac{CA}{CA}$ (A nonzero number div by itself = 1.); $\frac{AB}{DE} = \frac{BC}{EF} = \frac{CA}{FD}$ (Sub); $\therefore \triangle ABC \sim \triangle DEF$ (Def of ~ $\triangle$s)
25. ▱$XYZW$ (Given); $\overline{WZ} \parallel \overline{XY}$ (Opp sides of ▱ are $\parallel$.); $\angle 1 \cong \angle 3$, $\angle 2 \cong \angle 4$, (If lines are $\parallel$, then alt int $\angle$s are $\cong$.); $\triangle XRQ \sim \triangle ZPQ$ (AA~); $\frac{XQ}{ZQ} = \frac{RQ}{PQ}$ (Def of ~ $\triangle$s); $\therefore XQ \cdot PQ = ZQ \cdot RQ$ (If proport, then prod of extremes = prod of means.)

Written Exercises, pages 320–321

1. 8 **3.** 6 **5.** 4 **7.** Yes **9.** Yes **11.** Yes **13.** Vert: yes; horiz: no **15.** 16 **17.** 18 **19.** $\angle A \cong \angle C$, $\angle BDE \cong \angle BED$ (Given); m $\angle A$ + m $\angle C$ + m $\angle B$ = 180, m $\angle BDE$ + m $\angle BED$ + m $\angle B$ = 180 (Sum of meas of $\angle$s of $\triangle$ = 180.);
m $\angle A$ + m $\angle C$ + m $\angle B$ =
m $\angle BDE$ + m $\angle BED$ + m $\angle B$ (Sub); m $\angle B$ = m $\angle B$ (Reflex Prop of Eq); m $\angle A$ + m $\angle C$ = m $\angle BDE$ + m $\angle BED$ (Subt Prop of Eq);
2 × m $\angle A$ = 2 × m $\angle BDE$ (Sub); m $\angle A$ = m $\angle BDE$ (Div Prop of Eq); $\overline{DE} \parallel \overline{AC}$ (If 2 lines are inters by a transv so that corr $\angle$s are $\cong$, then lines are $\parallel$.); $\therefore \frac{AD}{DB} = \frac{CE}{EB}$ (Triangle Prop Thm) **21.** $\triangle ABC$ with D between A and C, E between B and C, $\frac{AD}{DC} = \frac{BE}{EC}$ (Given); through A draw $\overleftrightarrow{AF} \parallel \overline{DE}$ with F on $\overleftrightarrow{CB}$ (through a pt not on a line, a line can be drawn $\parallel$ to that line); $\frac{AD}{DC} = \frac{FE}{EC}$ ($\triangle$ Prop Thm); $\frac{BE}{EC} = \frac{FE}{EC}$ (Sub); $BE = FE$ (Mult Prop of Eq); B and F are the same pt (Ruler Post); $\overleftrightarrow{AB}$ and $\overleftrightarrow{AF}$ are the same line (For any 2 pts there is exactly one line containing them); $\overleftrightarrow{AB} \parallel \overline{DE}$ (Renaming); $\therefore \overline{AB} \parallel \overline{DE}$ (Segments of $\parallel$ lines are $\parallel$.) **25.** $\overleftrightarrow{AB} \parallel \overleftrightarrow{CD} \parallel \overleftrightarrow{EF}$, with transv l and m (Given); There are 2 possibilities.

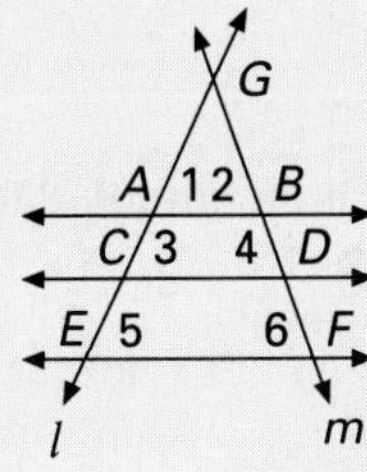

First: l and m intersect at G (2 lines inters in at

most 1 pt.); m $\angle 1$ = m $\angle 3$, m $\angle 2$ = m $\angle 4$, m $\angle 3$ = m $\angle 5$, m $\angle 4$ = m $\angle 6$ (If 2 $\parallel$ lines are inters by a trans, then corr $\angle$s are $\cong$.); $\triangle GAB \sim \triangle GCD$, $\triangle GCD \sim \triangle GEF$ (AA$\sim$); $\triangle GAB \sim \triangle GCD \sim \triangle GEF$ (Trans Prop of $\sim$); $\frac{EG}{CG} = \frac{FG}{DG}$ (Def of $\sim \triangle$s); $\frac{EG - CG}{CG} = \frac{FG - DG}{DG}$ (If $\frac{a}{b} = \frac{c}{d}$, then $\frac{a-b}{b} = \frac{c-d}{d}$); $EG - CG = EC$, $FG - DG = FD$ (Angle Add Post); $\frac{EC}{CG} = \frac{FD}{DG}$ (Sub); $\frac{EC}{FD} = \frac{CG}{DG}$ (If $\frac{a}{b} = \frac{c}{d}$, then $\frac{a}{c} = \frac{b}{d}$); $\frac{CG}{AG} = \frac{DG}{BG}$ (Def of $\sim \triangle$s); $\frac{CG - AG}{AG} = \frac{DG - BG}{BG}$ (If $\frac{a}{b} = \frac{c}{d}$, then $\frac{a-b}{b} = \frac{c-d}{d}$); $CG - AG = CA$, $DG - BG = BD$ (Angle Add Post); $\frac{CA}{AG} = \frac{BD}{BG}$ (Sub); $\frac{AG}{CA} = \frac{BG}{BD}$ (If $\frac{a}{b} = \frac{c}{d}$, then $\frac{b}{a} = \frac{d}{c}$); $\frac{AG + CA}{CA} = \frac{BG + BD}{BD}$ (If $\frac{a}{b} = \frac{c}{d}$, then $\frac{a+b}{b} = \frac{c+d}{d}$); $AG + CA = CG$, $BG + BD = GD$ (Angle Add Post); $\frac{CG}{CA} = \frac{GD}{DB}$ (Sub); $\frac{CG}{DG} = \frac{CA}{DB}$ (If $\frac{a}{b} = \frac{c}{d}$, then $\frac{a}{c} = \frac{b}{d}$); $\frac{EC}{FD} = \frac{CA}{DB}$ (Sub); $\therefore \frac{EC}{CA} = \frac{FD}{DB}$ (If $\frac{a}{b} = \frac{c}{d}$, then $\frac{a}{c} = \frac{b}{d}$)

Second: $l \parallel m$ (Parallel Post); $ABDC$, $CDFE$ are $\square$s (Def of $\square$); $AC = BD$, $CE = DF$ (Opp sides of $\square$ are $\cong$.); $\frac{AC}{AC} = \frac{CE}{CE} = 1$ (Any nonzero number div by itself = 1.); $\frac{AC}{BD} = \frac{CE}{DF}$ (Sub); $\therefore \frac{AC}{CE} = \frac{BD}{DF}$ (If $\frac{a}{b} = \frac{c}{d}$, then $\frac{a}{c} = \frac{b}{d}$)

27. $\overline{FD} \parallel \overline{CA}$, $\overline{DE} \parallel \overline{AB}$ (Given); $\frac{GF}{FC} = \frac{GD}{DA}$, $\frac{GD}{DA} = \frac{GE}{EB}$ (Triangle Proport Thm); $\therefore \frac{GF}{FC} = \frac{GE}{EB}$ (Sub)

Written Exercises, pages 326–327

1. Yes, AA$\sim$ **3.** No **5.** Yes, AA$\sim$ **7.** No **9.** Equiangular $\triangle$s ABC and DEF (Given); m $\angle A$ = m $\angle B$ = m $\angle D$ = m $\angle E$ = 60 (Meas of $\angle$ of reg polygon with n sides = $\frac{(n-2)180}{n}$; $\therefore \triangle ABC \sim \triangle DEF$ (AA$\sim$)

11. $\square GKLJ$ (Given); $\overline{JL} \parallel \overline{GK}$, $\overline{JG} \parallel \overline{LK}$ (Def of $\square$); $\angle JLI \cong \angle KHL$, $\angle IJL \cong \angle JGK$, $\angle JGK \cong \angle LKH$ (If 2 $\parallel$ lines are inters by transv, then corr $\angle$s are $\cong$.); $\angle IJL \cong \angle LKH$ (Trans Prop of Congr); $\therefore \triangle JLI \sim \triangle KHL$ (AA$\sim$) **13.** $\frac{NQ}{NO} = \frac{NP}{NM}$ (Given); $\angle N \cong \angle N$ (Reflex Prop of Congr); $\triangle NPQ \sim \triangle NMO$ (SAS$\sim$); $\therefore \frac{NQ}{NO} = \frac{QP}{OM}$ (Def of $\sim \triangle$s)

15.

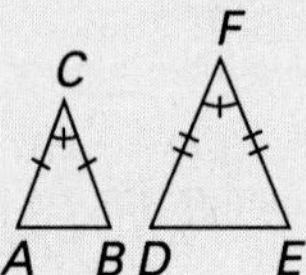

Isos $\triangle$s ABC and DEF with $\cong$ vert $\angle$s C and F (Given); $AC = BC$, $DF = EF$ (Def isos $\triangle$); $\frac{AC}{AC} = \frac{DF}{DF} = 1$ (A number div by itself = 1.); $\frac{AC}{BC} = \frac{DF}{EF}$ (Sub); $\frac{AC}{DF} = \frac{BC}{EF}$ (If $\frac{a}{b} = \frac{c}{d}$, then $\frac{a}{c} = \frac{b}{d}$); $\therefore \triangle ABC \sim \triangle DEF$ (SAS$\sim$) **17.** $\triangle$s ABC and DEF, $\frac{CA}{FD} = \frac{AB}{DE} = \frac{CB}{FE}$ (Given); locate D' on $\overline{CA}$ and E' on $\overline{CB}$ such that $\overline{CD'} \cong \overline{FD}$ and $\overline{CE'} \cong \overline{FE}$ (Ruler Post); draw $\overline{DE}$ (2 pts determine a line.); $\frac{CA}{CD'} = \frac{CB}{CE'}$ (Sub); $\angle C \cong \angle C$ (Reflex Prop of Congr); $\triangle ABC \sim \triangle D'E'C$ (SAS$\sim$); $\frac{AB}{D'E'} = \frac{CA}{CD'}$ (Def of $\sim \triangle$s); $\frac{AB}{D'E'} = \frac{CA}{FD}$ (Sub); $\frac{AB}{D'E'} = \frac{AB}{DE}$ (Sub); $\frac{D'E'}{AB} = \frac{DE}{AB}$ (If $\frac{a}{b} = \frac{c}{d}$, then $\frac{b}{a} = \frac{d}{c}$); $D'E' = DE$ (Mult Prop of Eq); $\triangle D'E'C \cong \triangle DEF$ (SSS); $\triangle D'E'C \sim \triangle DEF$ ($\cong \triangle$s are $\sim$.); $\therefore \triangle ABC \sim \triangle DEF$ (Trans Prop of $\sim$) **19.** $\triangle ABC$, $\triangle EDF$ such that $\overleftrightarrow{ED} \perp \overline{AB}$, $\overleftrightarrow{DF} \perp \overline{BC}$, $\overleftrightarrow{FE} \perp \overline{CA}$ (Given); $\angle DJH \cong \angle BJK$ (Vert $\angle$s are $\cong$.); $\angle DHJ \cong \angle BKJ$ (Rt $\angle$s are $\cong$.); $\triangle DHJ \sim \triangle BKJ$ (AA$\sim$); $\angle JDH \cong \angle B$ (Def $\sim \triangle$s); $\angle EKL \cong \angle AGL$ (Rt $\angle$s are $\cong$.); $\angle KLE \cong \angle GLA$ (Vert $\angle$s are $\cong$.); $\triangle EKL \sim \triangle AGL$ (AA$\sim$); $\angle KEL \cong \angle A$ (Def of $\sim \triangle$s); $\therefore \triangle ABC \sim \triangle EDF$ (AA$\sim$)

21.

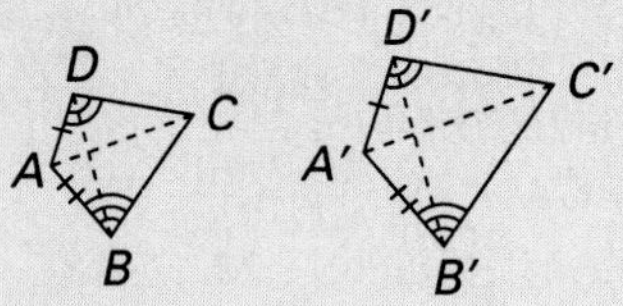

Quads $ABCD$ and $A'B'C'D'$ with m $\angle A$ = m $\angle A'$, m $\angle D$ = m $\angle D'$, m $\angle B$ = m $\angle B'$, $\frac{AD}{A'D'} = \frac{AB}{A'B'}$ (Given); $\triangle ADB \sim \triangle A'D'B'$ (SAS~); m $\angle ADB$ = m $\angle A'D'B'$, m $\angle ABD$ = m $\angle A'B'D'$, $\frac{DB}{D'B'} = \frac{AD}{A'D'}$ (Def ~ $\triangle$s); m $\angle D$ − m $\angle ADB$ = m $\angle D'$ − m $\angle A'D'B'$, m $\angle B$ − m $\angle ABD$ = m $\angle B'$ − m $\angle A'B'D'$ (Equations may be subtracted.); m $\angle D$ − m $\angle ADB$ = m $\angle D$ − m $\angle ADB$ = m $\angle CDB$, m $\angle D'$ − m $\angle A'D'B'$ = m $\angle C'D'B'$ (Angle Add Post); m $\angle CDB$ = m $\angle C'D'B'$ (Sub); $\triangle CDB \sim \triangle C'D'B'$, $\triangle CBD \sim \triangle C'B'D'$ (AA~); $\frac{DC}{D'C'} = \frac{CB}{C'B'} = \frac{DB}{D'B'}$, m $\angle C$ = m $\angle C'$ (Def ~ $\triangle$s); $\frac{AD}{A'D'} = \frac{AB}{A'B'} = \frac{BC}{B'C'} = \frac{DC}{D'C'}$ (Sub); $\therefore$ quad $ABCD \sim$ quad $A'B'C'D'$ (All corr $\angle$s are $\cong$ and all corr sides are proport.)

Written Exercises, pages 330–331

1. 6 **3.** 8 **5.** 24 **7.** 16 **9.** 3 **11.** 6 **13.** 21 **15.** 8 **17.** 1.4 **19.** 1.5

21. $\overline{AE}$ bis $\angle DAB$, $\overline{CE}$ bis $\angle DCB$ (Given); $\frac{AD}{AB} = \frac{DE}{BE}$, $\frac{DE}{BE} = \frac{DC}{CB}$ (Bis of $\angle$ div opp side of $\triangle$ proport to other 2 sides); $\therefore \frac{AD}{AB} = \frac{DC}{CB}$ (Sub)

23.

C F A x B D y E

$\triangle ABC \sim \triangle DEF$ with medians $\overline{CX}$ and $\overline{FY}$, respectively (Given); $AX = BX$, $DY = EY$ (Def median); $AX + BX = AB$, $DY + EY = DE$ (Seg Add Post); $2AX = AB$, $2DY = DE$ (Sub); $\frac{CA}{FD} = \frac{AB}{DE}$, $\angle A \cong \angle D$ (Def of ~ $\triangle$s); $\frac{CA}{FD} = \frac{2AX}{2DY}$ (Sub); $\frac{CA}{FD} = \frac{AX}{DY}$ (Id Prop for Mult); $\triangle CAX \sim \triangle FDY$ (SAS~); $\therefore \frac{CA}{FD} = \frac{CX}{FY}$ (Def of ~ $\triangle$s) **25.** $\triangle ABC \sim \triangle DEF$ with alts $\overline{CW}$ and $\overline{FY}$ and medians $\overline{CX}$ and $\overline{FZ}$, respectively (Given); $\frac{CW}{FY} = \frac{CB}{FE}$ (Corr alts of ~ $\triangle$s are proport to corr sides.); $\frac{CB}{FE} = \frac{CX}{FZ}$ (Corr medians of ~ $\triangle$s are proport to corr sides.); $\therefore \frac{CW}{FY} = \frac{CX}{FZ}$ (Sub) **27.** See illustration for Thm 8.9. $\frac{CG}{FH} = \frac{AB}{DE}$, $\frac{CG}{FH} = \frac{BC}{EF}$, $\frac{CG}{FH} = \frac{CA}{FD}$ (Given); $\frac{AB}{DE} = \frac{BC}{EF} = \frac{CA}{FD}$ (Sub); $\therefore \triangle ABC \sim \triangle DEF$ (SSS~)

Chapter 9

Written Exercises, pages 342–343

1. 8 **3.** $2\sqrt{6}$ **5.** 4 **7.** 4 **9.** 18 **11.** 4 **13.** $2\sqrt{7}$ **15.** 4 **17.** 8 or 2 **19.** $2\sqrt{2}$ **21.** $2\sqrt{10}, 2\sqrt{6}, 2\sqrt{15}$ **23.** $3\sqrt{2}$

Rt $\triangle RST$ with rt $\angle RTS$ and alt $\overline{TU}$ (Given); $\triangle RST \sim \triangle RTU$ (Alt to hyp in rt $\triangle$ forms 2 ~ $\triangle$s, each ~ to orig $\triangle$.); $\frac{s}{t} = \frac{m}{s}$ (Def ~ $\triangle$s); $\therefore s^2 = mt$ (If proport, then prod of means = prod of extremes.). Similarly, use $\triangle STR \sim \triangle SUT$ to prove $r^2 = nt$. **27.** Let alt = h. Then $\frac{a}{c} = \frac{h}{b}$; $h = \frac{ab}{c}$ **29.** Rt $\triangle ACE$ with rt $\angle EAC$, $\overline{DB} \perp \overline{AC}$ (Given); $\angle DBC$ is a rt $\angle$ (Def of $\perp$); $\angle EAC \cong \angle DBC$ (All rt $\angle$s are $\cong$.); $\angle C \cong \angle C$ (Reflex Prop of Congr); $\triangle EAC \sim \triangle DBC$ (AA~); $\frac{x}{y+u} = \frac{w}{u}$ (Def ~ $\triangle$s); $\therefore$ $w(y + u) = xu$ (If proport, then prod of means = prod of extremes.)

31. $x = \frac{9}{2}\sqrt{5}$

Written Exercises, pages 347–348

1. $c = 13$ **3.** $c = 6\sqrt{2}$ **5.** $b = 2\sqrt{13}$ **7.** $a = \sqrt{11}$ **9.** $2\sqrt{5}$ **11.** $2\sqrt{7}$ **13.** $10\sqrt{2}$ **15.** 18.4 **17.** 26.5 **19.** 18.0 **21.** 16 in **23.** $4\sqrt{15}$ ft **25.** $2\sqrt{5}$ in **27.** $7\frac{1}{17}$ cm **29.** $2\sqrt{26}$ **31.** 64.8 ft **33.** $\sqrt{l^2 + w^2 + h^2}$ **35.** $x + 6$

Written Exercises, pages 352–353

1. Yes, $\angle C$ **3.** No **5.** Obtuse **7.** Acute **9.** Right **11.** Acute **13.** Right **15.** Right **17.** Yes **19.** Yes **21.** Yes **23.** If sq of longest side = sum of sqs of other 2 sides, then $\triangle$ is rt. If sq of longest side > (or <) sum of sqs of other 2 sides, then $\triangle$ is obtuse (or acute). **25.** Alt $\overline{AD}$ to $\overline{BC}$, $(AC)^2 = BC \cdot DC$ (Given); $\frac{AC}{BC} = \frac{DC}{AC}$ (If $ad = bc$, then $\frac{a}{b} = \frac{c}{d}$); $\angle C \cong \angle C$ (Reflex Prop of Congr); $\triangle ABC \sim \triangle ADC$ (SAS~); m $\angle BAC$ = m $\angle ADC$ (Def ~ $\triangle$s); $\overline{AD} \perp \overline{BC}$ (Def of alt); $\angle ADC$ is a rt $\angle$. (Def of rt $\angle$); m $\angle BAC$ = 90 (Sub) **27.** 6, 8, 10

Written Exercises, pages 357–358

1. $y = 6\sqrt{3}, z = 12$ **3.** $x = 7, y = 7\sqrt{3}$ **5.** $q = 6, r = 6\sqrt{2}$ **7.** $p = 4\sqrt{2}, r = 8$ **9.** $g = 6.8, h = 3.4\sqrt{3}$ **11.** $h = \frac{17}{2}\sqrt{3}$, $k = \frac{17}{2}$ **13.** $9\sqrt{3}$ **15.** $3\sqrt{2}$ **17.** $3\sqrt{2}$ **19.** $2\sqrt{2}$ **21.** 6 **23.** $4\sqrt{2}$ **25.** Rt

$\triangle PQR$ with rt $\angle PRQ$, m $\angle Q = 60$, m $\angle QPR = 30$ (Given); Extend $\overrightarrow{QR}$ to T so that $RT = RQ$ (2 pts determine a line.); $\overline{PR} \cong \overline{PR}$ (Reflex Prop of Congr); $\angle PRQ \cong \angle PRT$ (Rt $\angle$s are $\cong$.); $\triangle PQR \cong \triangle PTR$ (SAS); m $\angle T$ = m $\angle Q = 60$, m $\angle TPR$ = m $\angle QPR = 30$ (CPCTC); m $\angle TPR$ + m $\angle QPR$ = m $\angle QPT$ (Angle Add Post); m $\angle QPT = 60$ (Sub); $\triangle QPT$ is equiang (Def equiang $\triangle$); $\triangle QPT$ is equil (If $\triangle$ is equiang, then $\triangle$ is equil); let $RQ = RT = s$ (Notation); $QT = 2s$ (Seg Add Post); $QP = QT$ (Def equil $\triangle$); $QP = 2s$ (Trans Prop of Eq or Sub); $s^2 + (PR)^2 = (2s)^2$ (Pythag Thm); $(PR)^2 = 3s^2$ (Subt Prop of Eq); $\therefore PR = s\sqrt{3}$ ($\sqrt{a^2} = \sqrt{a} \cdot \sqrt{a} = a$ for $a > 0$) **27.** $64\sqrt{2} - 64$

Written Exercises, pages 363–364

1. 0.7547 **3.** 0.3256 **5.** 0.8829 **7.** 39 **9.** 33 **11.** 73 **13.** 12.0 **15.** 56 **17.** 23.7 **19.** 71 **21.** 36.6 m **23.** 7.8 **25.** 15.4 **27.** 2 min 12 s **29.** 8.5 **31.** 23

Written Exercises, pages 368–369

1. 7.7 **3.** 6.1 **5.** 42 **7.** 34 **9.** 12.7 **11.** 16.5 **13.** 54.4 **15.** 49 **17.** 61 **19.** 117.0 **21.** 66 **23.** 15 **25.** 32 **27.** 29 **29.** 23, 8 **31.** 5

Written Exercises, pages 371–373

1. 12,388 ft **3.** 37 ft **5.** 98 ft **7.** 87 ft **9.** 225 ft **11.** 22,835 ft **13.** 1,248 m **15.** $AB = h\left(\frac{1}{\tan \angle 1} - \frac{1}{\tan \angle 2}\right)$ **17.** 489 ft

Chapter 10

Written Exercises, pages 382–383

1. (5,3) I **3.** (−2,1) II **5.** (−4,4) II
7.–9.

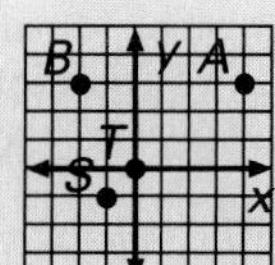

11. 2, 4, $2\sqrt{5}$ **13.** 5, 5, $5\sqrt{2}$ **15.** 5 **17.** 13 **19.** $4\sqrt{2}$ **21.** 13 **23.** $\sqrt{41}$ **25.** Yes **27.** Yes **29.** No, scalene **31.** Yes **33.** Yes. $AB = \sqrt{26}$, $BC = \sqrt{20}$, $CD = \sqrt{26}$, $AD = \sqrt{20}$ **35.** Yes. $AB = \sqrt{26}$, $BC = \sqrt{10}$, $CD = \sqrt{26}$, $AD = \sqrt{10}$ **37.** Yes, diag $AC = \sqrt{145}$, diag $BD = \sqrt{145}$ **39.** 0, −2 **41.** (−2,0) **43.** (3,4), (4,3), (−3,4), (−4,3), (−3,−4), (−4,−3), (3,−4), (4,−3) (5,0) (−5,0) (0,5) (0,−5)

Written Exercises, pages 386–387

1. (2,2) **3.** (1,−1) **5.** (−3,4) **7.** $\left(3\frac{1}{2}, 2\frac{1}{2}\right)$ **9.** $\left(2\frac{1}{2}, \frac{1}{2}\right)$ **11.** $\left(-3\frac{1}{2}, -3\right)$ **13.** (6,6) **15.** (−2,−6) **17.** (7,−11) **19.** Midpt of both: (3,3) **21.** Midpt of both: $\left(1, 1\frac{1}{2}\right)$ **23.** $\sqrt{10}$ **25.** $M(-3,3)$; $N(-1,5)$; $MN = \frac{1}{2}BC$ **27.** $M(6.55, 1.15)$ **29.** (1,5), (5,2) **31.** $\left(\frac{2x_2 + x_1}{3}, \frac{2y_2 + y_1}{3}\right)$ **33.** (6,5)

Written Exercises, pages 390–391

1. $\frac{4}{3}$ **3.** $\frac{5}{3}$ **5.** −2 **7.** $-\frac{3}{4}$ **9.** $-\frac{3}{5}$ **11.** $\frac{1}{3}$ **13.** 0 **15.** $\frac{1}{2}$
17. (−2, 4), (0, 5) **19.** (1, 4), (6, 2)

21. Yes; $m(\overline{DE}) = \frac{2}{3}$, $m(\overline{EF}) = \frac{2}{3}$, $m(\overline{DF}) = \frac{2}{3}$ **23.** No; $m(\overline{JK}) = -\frac{3}{2}$, $m(\overline{KL}) = 2$, $m(\overline{LJ}) = -\frac{1}{3}$ **25.** 27 **27.** 6 **29.** 4 **31.** $\frac{1}{5}$ **33.** (2,6), (8,−4), (0,0)

Written Exercises, pages 395–396

1. $y = 4$ **3.** $y = -2$ **5.** $x = -4$ **7.** $y = -3x + 2$ **9.** $y = \frac{1}{3}x - \frac{4}{3}$ **11.** $y = -2x + 0$ **13.** $y - 2 = -3(x - 1)$ **15.** −3 **17.** $y - 4 = 3(x + 1)$ **19.** $y + 1 = -1(x + 3)$ **21.** $y - 0 = -\frac{4}{3}(x - 4)$ **23.** $x + 4y = 13$ **25.** $5x + 2y = -10$ **27.** $x - 10y = 14$
29. **31.** **33.**

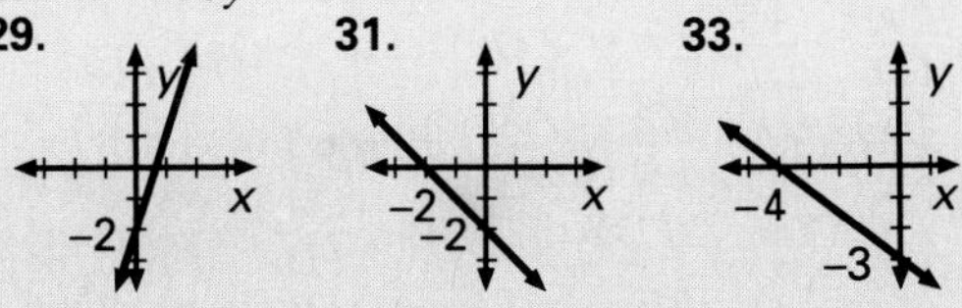

35. $y = -2x - 2$ **37.** $y = \frac{2}{3}x + 2$ **39.** $y = -\frac{1}{2}x - 1$ **41.** $m = \pm 1$ **43.** $y - 1 = \frac{1}{\sqrt{3}}(x - 2)$, or $y - 1 = \frac{\sqrt{3}}{3}(x - 2)$

Written Exercises, pages 400–401

1. Neither **3.** ∥ **5.** $y = 2x$ **7.** $y = 5x - 19$ **9.** $y = \frac{3}{4}x - 5$ **11.** $y = -\frac{1}{2}x$ **13.** $y = -x - 1$ **15.** $y = -\frac{4}{3}x + \frac{5}{3}$ **17.** ⊥

19. $\parallel$ **21.** Yes **23.** Yes **25.** No
27. 11 **29.** 1 **31.** Kite
33. Parallelogram, Rectangle, Square
35. $9x - 8y = 11$ **37.** $3x + 4y = 2$

Written Exercises, pages 406–407

1. $(-3,0)$ **3.** $(2p,2q)$

5. **7.**

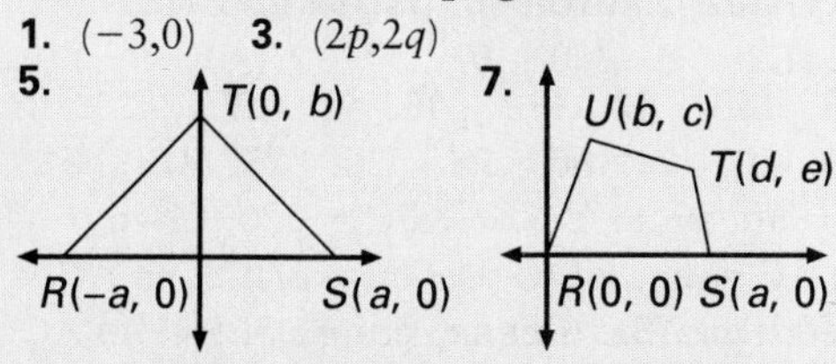

9. Two of the 4 slopes are equal (2 $\parallel$ sides).

11.

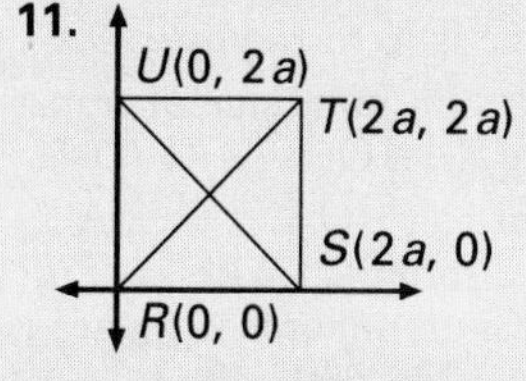

Midpt($\overline{US}$) = midpt($\overline{TR}$) = (a,a)

13. **15.**

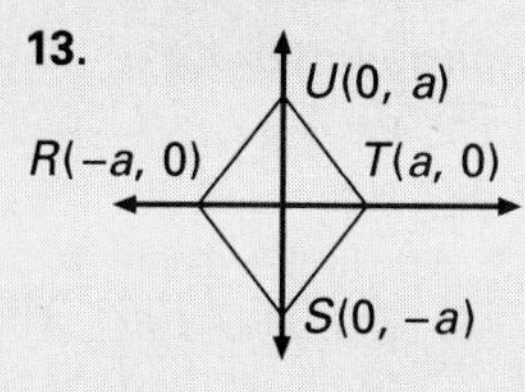

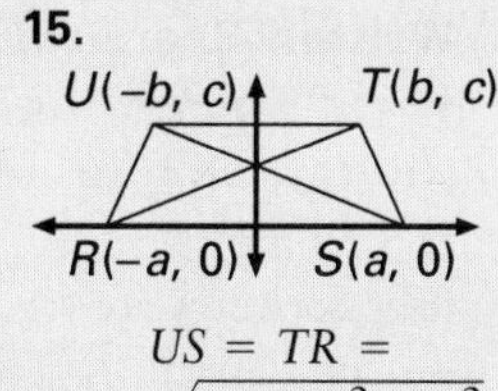

$US = TR = \sqrt{(b + a)^2 + c^2}$

17.

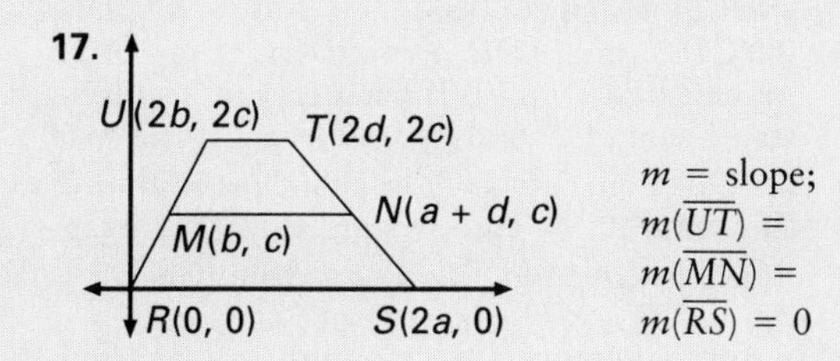

m = slope; $m(\overline{UT}) = m(\overline{MN}) = m(\overline{RS}) = 0$

19.

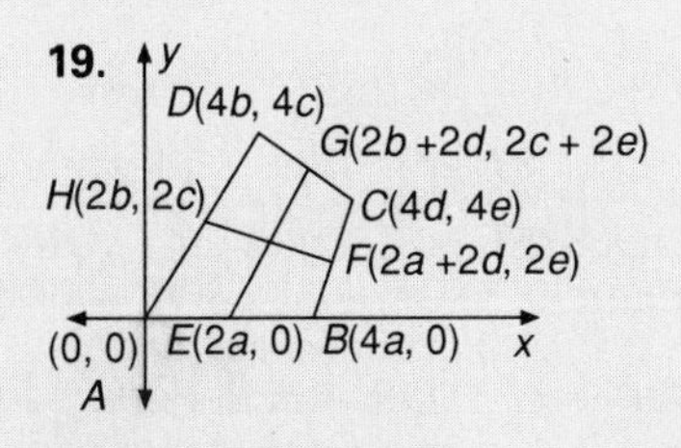

Midpt of $\overline{EG} = (a + b + d, c + e)$
Midpt of $\overline{FH} = (a + b + d, c + e)$
Since the midpts are the same pt, the segments bisect each other.

21.

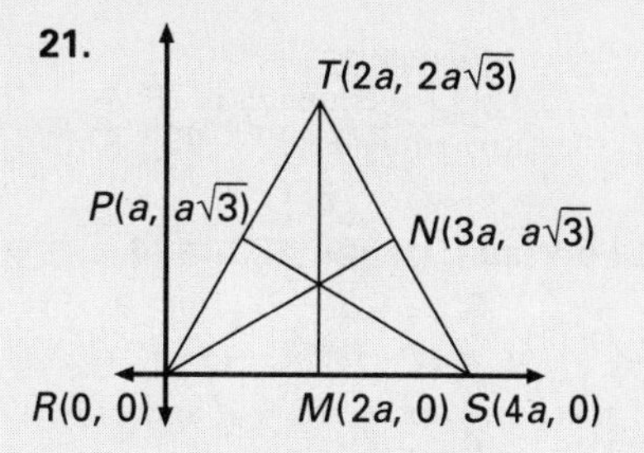

$RN = PS = TM = 2a\sqrt{3}$

23.

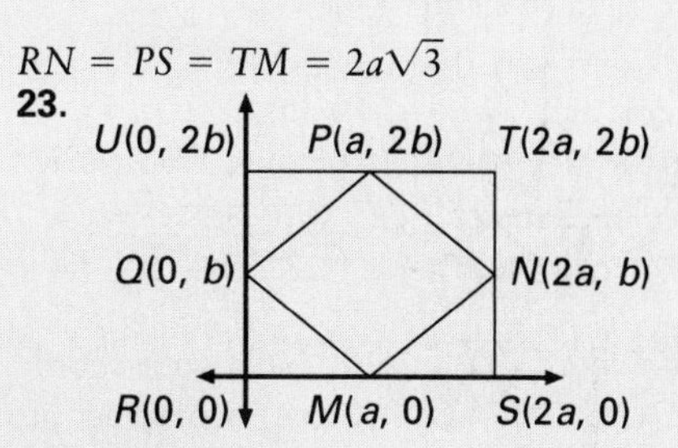

$PQ = QM = MN = NP = \sqrt{a^2 + b^2}$; m = slope; $m(\overline{PQ}) = m(\overline{MN}) = \frac{b}{a}$, $m(\overline{QM}) = m(\overline{PN}) = -\frac{b}{a}$

25.

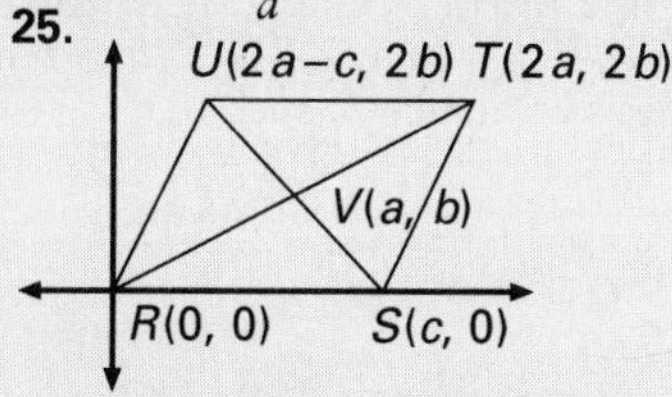

$m(\overline{UR}) = m(\overline{TS}) = \frac{2b}{2a - c}$, so $\overline{UR} \parallel \overline{TS}$. $m(\overline{UT}) = m(\overline{RS}) = \frac{0}{c} = 0$, so $\overline{UT} \parallel \overline{RS}$. $\therefore$ Quad $RSTU$ is a $\square$.

Chapter 11

Written Exercises, pages 418–419

1. $x > 6$ **3.** $x > 5$ **5.** $\overline{FG}$; $\overline{AB}$ **7.** $\overleftrightarrow{DE}$; $\overleftrightarrow{CD}$ **9.** 7.4 **11.** $6x$ **13.** $12x - 8$
15. $5\sqrt{3}$ **17.** $2a - 6$ **19.** Chord for $\odot R$ and $\odot D$ **21.** Sec for $\odot R$, tan to $\odot S$
23. Chord for $\odot R$, tan to $\odot S$ **25.** 8
27. m $\angle POQ = 60$ (Given); m $\angle POQ$ + m $\angle OPQ$ + m $\angle PQO = 180$ (Sums of meas of $\angle$s of $\triangle = 180$.); m $\angle OPQ$ + m $\angle PQO = 120$ (Equations may be subtracted.); $OP = OQ$ (Def $\odot$); m $\angle OPQ$ = m $\angle PQO$ (If 2 sides of $\triangle \cong$, then meas of $\angle$s opp sides are =.); m $\angle OPQ$ + m $\angle OPQ = 120$ ($2 \times$ m $\angle OPQ = 120$) (Sub); m $\angle OPQ = 60$ (Div Prop of Eq); m $\angle PQO = 60$ (Trans Prop

of Eq); $\triangle OPQ$ is equiangular (Def equiangular); $\therefore \triangle OPQ$ is equilateral. (If $\triangle$ equiangular, then it is equilateral.) **29.** $8\sqrt{2}$ **31.** 5

Written Exercises, pages 423–424

1. 13 **3.** $8\sqrt{3}$ **5.** $\frac{10}{3}\sqrt{3}$ **7.** 16 **9.** 55 **11.** $\overline{TS}$ **13.** 14.4 **15.** $2\sqrt{7}$ **17.** In the same ⊙, or in ≅ ⊙s, chords are ≅ if and only if they are equidistant from the ctr.
19.

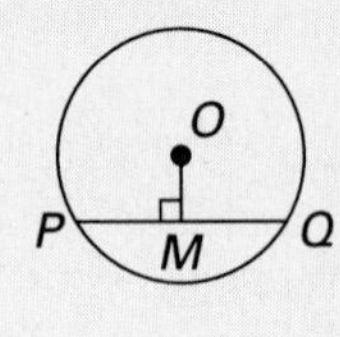

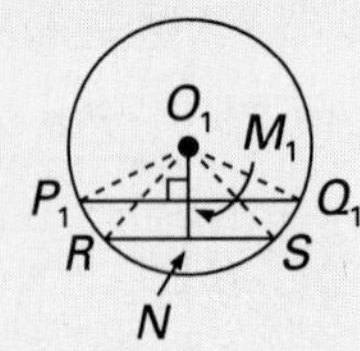

⊙O with chord $\overline{PQ} \perp \overline{OM}$, ⊙$O_1$ with chord $\overline{RS} \perp \overline{O_1N}$, $OM < ON$ (Given); Construct pt M_1 on $\overline{O_1N}$ such that $OM_1 = OM$ (Ruler Post); construct chord $\overline{P_1Q_1}$ through pt $M_1 \perp \overline{O_1N}$ (Parallel Post); $PQ = P_1Q_1$ (In ≅ ⊙s, equidistant chords are ≅.); $m\angle P_1O_1Q_1 = m\angle P_1O_1R + m\angle RO_1S + m\angle SO_1Q_1$ (Angle Add Post); $m\angle P_1O_1Q_1 > m\angle RO_1S$ ($a > b$ if there exists a positive number c such that $b + c = a$.); $O_1P_1 = O_1R = O_1S = O_1Q_1$ (Radii of ⊙ are ≅.); $P_1Q_1 > RS$ (SAS Ineq); $\therefore PQ > RS$ (Sub)
21. 320.5 cm

Written Exercises, pages 429–430

1. $4\sqrt{2}$ **3.** $2\sqrt{3}$ **5.** 3 **7.** $8\sqrt{6}$ **9.** 17 **11.** 36 **13.** Use the figure above Thm 11.8. Use in Ex. 12 to justify $\triangle OAT \cong \triangle OBT$ (Ex. 12); then $\angle 1 \cong \angle 2$ (CPCTC); $\therefore \overrightarrow{OT}$ bis $\angle ATB$ (Def ∠ bis) **15.** ⊙O and ⊙P are tan to $\overline{AC}$ and $\overline{AD}$ (Given); $AC = AD$, $AB = AE$ (2 segs tan to a circle from ext pt are ≅.); $AC = AB + CB$, $AD = AE + DE$ (Seg Add Post); $AB + CB = AE + DE$ (Sub); $\therefore CB = DE$ (Equations may be subtracted.) **17.** 15
19.

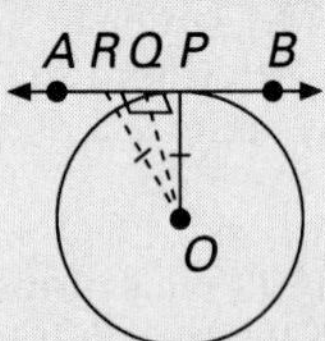

Indirect Proof: $\overline{OP}$ is *not* ⊥ to $\overleftrightarrow{AB}$ (Assume conclusion is false.); Draw $\overline{OQ} \perp$ to $\overleftrightarrow{AB}$ (From pt to line there is unique ⊥.); Locate pt R on $\overleftrightarrow{AB}$ such that $RQ = PQ$ (Ruler Post); $\overline{QO} \cong \overline{QO}$ (Reflex Prop of Congr); $\triangle RQO \cong \triangle PQO$ (SAS); $OR = OP$ (CPCTC); pt R is on ⊙O (Def ⊙); $m\angle RQO = m\angle PQO = 90$ (Rt ∠s have = meas.); 2 pts at tan $\overleftrightarrow{AB}$ lie on ⊙O. (This is a contradiction of def of tan.); $\therefore \overline{OP} \perp \overleftrightarrow{AB}$ (Conclusion must be true.) **21.** $d \approx 21$ m

Written Exercises, pages 436–437

1. 120 **3.** 280 **5.** 230 **7.** 40 **9.** 120 **11.** 180 **13.** 180 **15.** 52 **17.** 282 **19.** 40 **21.** 60 **23.** 320 **25.** $\overline{AD}$ and $\overline{CD}$ are tan to ⊙O, $m\widehat{ABC} = 270$ (Given); $m\widehat{AC} = m\angle AOC$ (Def meas of arc); $m\widehat{AC} = 360 - m\widehat{ABC}$ (Def meas of major arc); $m\widehat{AC} = 360 - 270 = 90$ (Sub, Subt Prop of Eq); $m\angle AOC = 90$ (Sub); $m\angle OAD = m\angle OCD = 90$ (If line tan to ⊙, line is ⊥ to radius.); $\overline{OA} \parallel \overline{CD}$, $\overline{OC} \parallel \overline{AD}$ (If int ∠s on same side of transv are supp, then lines are ∥.); $AOCD$ is ▱ (Def ▱); $AOCD$ is rect (Def rect); $\overline{OA} \cong \overline{OC}$ (All radii are ≅.); $AOCD$ is rhom (▱ with 2 ≅ adj sides is rhom.); $\therefore AOCD$ is square (A square is a rect that is a rhom.) **27.** 300

Written Exercises, pages 440–441

1. $\widehat{UT}$ **3.** $\widehat{UW}$ **5.** $\widehat{VT}$, $\widehat{VP}$, $\widehat{PR}$ **7.** $\angle B \cong \angle C$ (Given); $\overline{AC} \cong \overline{AB}$ (If 2 sides of △ ≅, sides opp ∠s are ≅.); $\therefore \widehat{AC} \cong \widehat{AB}$ (If chords are ≅, their corr arcs are ≅.) **9.** ⊙O, $\overline{DC} \cong \overline{BC}$, $\overline{AC}$ is diam (Given); $\overline{DO} \cong \overline{BO}$ (Radii of ⊙ are ≅.); $\overline{OC} \cong \overline{OC}$ (Reflex Prop of Congr); $\triangle DCO \cong \triangle BCO$ (SSS); $m\angle BOC = m\angle DOC$ (CPCTC); $m\angle AOD + m\angle DOC = m\angle AOB + m\angle BOC = 180$ (If outer rays of ∠s form a str ∠, sum of ∠ meas = 180.); $m\angle AOD = m\angle AOB$ (Equations may be subtr.); $\therefore \widehat{AD} \cong \widehat{AB}$ (If central ∠s ≅, corr arcs are ≅.) **11.** Trap $PQRS$ inscribed in ⊙O, $\overline{PQ}$ is diam (Given); $\overline{PO} \cong \overline{SO} \cong \overline{RO} \cong \overline{QO}$ (Radii of ⊙ are ≅.); $\angle RSO \cong \angle SRO$ (If 2 sides of △ ≅, ∠s opp those sides are ≅.); $\overline{SR} \parallel \overline{PQ}$ (Def trap); $\angle SOP \cong \angle RSO$; $\angle ROQ \cong \angle SRO$ (If transv intersects ∥ lines, alt int ∠s are ≅); $\angle SOP \cong \angle ROQ$ (Trans Prop of Congr); $\triangle SOP \cong \triangle ROQ$ (SAS); $m\angle SOR = m\angle SOR$ (Reflex Prop of Eq); $m\angle SOP + m\angle SOR = m\angle ROQ + m\angle SOR$ (Equations may be added.); $m\angle SOP + m\angle SOR = m\angle POR$, $m\angle ROQ + m\angle SOR = m\angle QOS$ (Angle Add Post); $m\angle POR = m\angle QOS$ (Sub); $\therefore \overline{PR} \cong \overline{SQ}$ (If central ∠s ≅, corr chords are ≅.)

Written Exercises, pages 445–446

1. 18 **3.** 130 **5.** $\frac{x}{2}$ **7.** 55 **9.** 30

11. 90 **13.** 90 **15.** 110 **17.** 45
19. 30 **21.** 60 **23.** 100 **25.** 100

27. $\odot O$ with inscribed $\angle ACB$, pt O in int of $\angle C$ (Given); Draw diam $\overline{CP}$ (2 pts determine line.); $\angle ACP$ and $\angle BCP$ contain ctr O (Def diam); m $\angle ACP = \frac{1}{2}$ m $\widehat{AP}$; m $\angle BCP = \frac{1}{2}$ m $\widehat{BP}$ (Case 1 of Thm 11.10); m $\angle ACP$ + m $\angle BCP = \frac{1}{2}$ m $\widehat{AP} + \frac{1}{2}$ m $\widehat{BP}$ (Equations may be added.); m $\angle ACB = \frac{1}{2}$ m $\widehat{AP} + \frac{1}{2}$ m $\widehat{BP}$ (Angle Add Post); m $\angle ACB = \frac{1}{2}$(m $\widehat{AP}$ + m $\widehat{BP}$) (Distr Prop); $\therefore$ m $\angle ACB = \frac{1}{2}$ m $\widehat{AB}$ (Arc Add Post)

29. $\odot O$ with inscr $\angle$s BAC, DEF; $\widehat{BC} \cong \widehat{DF}$ (Given); m $\widehat{BC}$ = m $\widehat{DF}$ (Def $\cong$ arcs); $\frac{1}{2}$ m $\widehat{BC} = \frac{1}{2}$ m $\widehat{DF}$ (Mult Prop of Eq); m $\angle A = \frac{1}{2}$ m $\widehat{BC}$, m $\angle E = \frac{1}{2}$ m $\widehat{DF}$ (Meas inscr $\angle = \frac{1}{2}$ meas of intercepted arc.); $\therefore$ m $\angle A$ = m $\angle E$ (Sub)

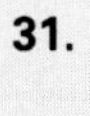

31.

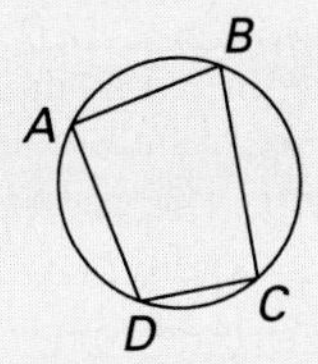

Inscr quad $ABCD$ (Given); m $\widehat{ABC}$ + m $\widehat{ADC}$ = 360 (Arc meas $\bigcirc$ = 360.); $\frac{1}{2}$ m $\widehat{ABC} + \frac{1}{2}$ m $\widehat{ADC}$ = 180 (Mult Prop of Eq); m $\angle D = \frac{1}{2}$ m $\widehat{ABC}$, m $\angle B = \frac{1}{2}$ m $\widehat{ADC}$ (Meas inscr $\angle = \frac{1}{2}$ meas of intercepted arc.); m $\angle D$ + m $\angle B$ = 180 (Sub) $\angle D$ and $\angle B$ are supp (Def of supp $\angle$s) **33.** m $\angle 1 = \frac{1}{2}$ m $\widehat{CD}$, diam $\overline{BD}$ (Given); $\frac{1}{2}$ m $\widehat{AD}$ = m $\angle 1$ (Meas inscr $\angle = \frac{1}{2}$ meas intercepted arc.); $\frac{1}{2}$ m $\widehat{CD} = \frac{1}{2}$ m $\widehat{AD}$ (Sub); m $\widehat{CD}$ = m $\widehat{AD}$ (Sub); m $\widehat{BCD}$ = m $\widehat{BAD}$ (Semicircles have meas of 180.); m $\widehat{BCD}$ − m $\widehat{CD}$ = m $\widehat{BAD}$ − m $\widehat{AD}$ (Equations may be subtr.); $\therefore$ m $\widehat{BC}$ = m $\widehat{BA}$ (Arc Add Post)

35.

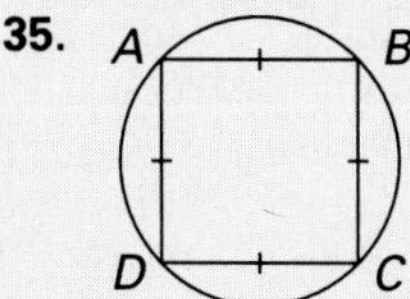

Inscr rhombus $ABCD$ (Given); Prove rhombus $ABCD$ is rect (See Ex 34); $\overline{AB} \cong \overline{BC} \cong \overline{CD} \cong \overline{DA}$ (Def of rhombus); $\therefore$ $ABCD$ is square (Def square)

37.

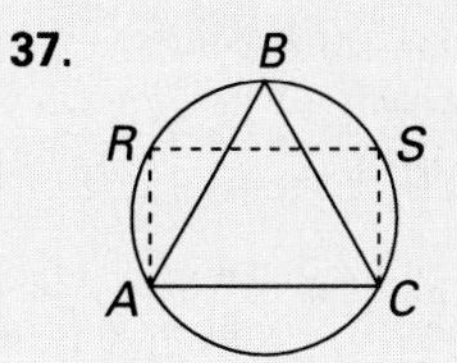

Inscr equilateral $\triangle ABC$, R and S are midpts of $\widehat{AB}$ and $\widehat{BC}$ resp (Given); m $\angle A$ = m $\angle B$ = m $\angle C$ = 60 (Def equilat $\triangle$); m $\widehat{AB}$ = m $\widehat{BC}$ = m $\widehat{CA}$ = 120 (Meas inscr $\angle = \frac{1}{2}$ meas intercepted arc.); m $\widehat{CAR}$ = m $\widehat{ACS}$ = 180 (Arc Add Post); m $\angle ARS$ = m $\angle RSC$ = 90 (Meas inscr $\angle = \frac{1}{2}$ meas intercepted arc.); m $\angle RAC$ = m $\angle SCA$ = 90 (In inscr quad, opp $\angle$s are supp.); $\therefore$ $ARSC$ is rect (Def rect)

Written Exercises, pages 449–451

1. 107 **3.** 60 **5.** 90 **7.** 62 **9.** 5
11. 20 **13.** 60

15.

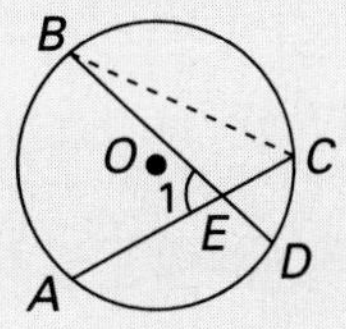

$\odot O$ with secs $\overleftrightarrow{AC}$ and $\overleftrightarrow{BD}$ intersecting at int pt E (Given); m $\angle 1$ = m $\angle CBE$ + m $\angle BCE$ (Meas ext $\angle$ = sum of meas of 2 remote int $\angle$s.); m $\angle CBE = \frac{1}{2}$ m $\widehat{CD}$, m $\angle BCE = \frac{1}{2}$ m $\widehat{AB}$ (Meas inscr $\angle = \frac{1}{2}$ meas interc arc; m $\angle 1 = \frac{1}{2}$ m $\widehat{CD} + \frac{1}{2}$ m $\widehat{AB}$ (Sub); $\therefore$ m $\angle 1 = \frac{1}{2}$ (m $\widehat{CD}$ + m $\widehat{AB}$) (Distr Prop)

17. 110 **19.** 10 **21.** 60 **23.** 60
25. 54 **27.** 54 **29.** No. If m $\widehat{AF}$ = 60 and m $\angle 4$ = 30, then 30 = $\frac{1}{2}$(60 = m $\widehat{EB}$), or m $\widehat{EB}$ = 0. This is a contradiction.

31.

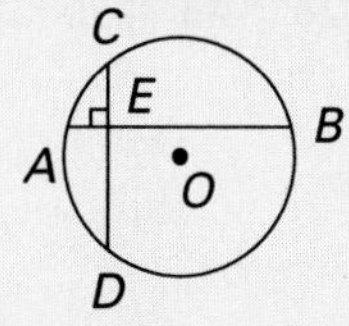

Chords $\overline{AB}$ and $\overline{CD}$ of $\odot O$ are $\perp$ at pt E (Given); m $\angle AEC$ = 90 (Def $\perp$ lines); m $\angle AEC = \frac{1}{2}$(m $\widehat{AC}$ + m $\widehat{BD}$) (Meas of $\angle$ formed by 2 chords is $\frac{1}{2}$ sum of meas of arcs intercepted); 90 = $\frac{1}{2}$(m $\widehat{AC}$ + m $\widehat{BD}$) (Sub); $\therefore$ 180 = m $\widehat{AC}$ + m $\widehat{BD}$ (Mult Prop of Eq)

33. P is in ext of $\odot O$, $\overline{PB}$ and $\overline{PA}$ are secs, C is on ⊙ (Given); m $\angle C = \frac{1}{2}$m $\widehat{AB}$ (Meas inscr $\angle = \frac{1}{2}$ meas intercepted arc.); m $\angle P = \frac{1}{2}$(m $\widehat{AB}$ − m $\widehat{CD}$) (Meas of $\angle$ formed by 2 secs is $\frac{1}{2}$ diff of meas of intercepted arcs.); $\frac{1}{2}$(m $\widehat{AB}$ − m $\widehat{CD}$) < $\frac{1}{2}$ m $\widehat{AB}$ ($a - b < a$ for all positive numbers b.); m $\angle P <$ m $\angle C$ (Sub) **35.** m $\widehat{DE}$ = 50, m $\widehat{CB}$ = 90

37.

D 3
2 E
C W Y

$\overline{CD}$ and $\overline{CE}$ are ≅ chords in ⊙; W is on $\widehat{CE}$; $\overleftrightarrow{CW}$ and $\overleftrightarrow{DE}$ meet at Y (Given); m $\widehat{CD}$ = m $\widehat{CE}$ (If chords ≅, then corr arcs are ≅.); m $\angle 2$ + m $\angle 3 = \frac{1}{2}$ m $\widehat{CE}$ (Meas inscr $\angle = \frac{1}{2}$ meas intercepted arc.); m $\angle 2$ + m $\angle 3 = \frac{1}{2}$ m $\widehat{CD}$ (Sub); m $\angle 3 = \frac{1}{2}$ m $\widehat{WE}$ (Meas inscr $\angle = \frac{1}{2}$ meas intercepted arc); m $\angle 1 = \frac{1}{2}$(m $\widehat{CD}$ − m $\widehat{WE}$) (Meas of $\angle$ formed by 2 sec in ext = $\frac{1}{2}$ diff of meas of intercepted arcs.); m $\angle 1 = \frac{1}{2}$ m $\widehat{CD} - \frac{1}{2}$ m $\widehat{WE}$ (Distr Prop); ∴ m $\angle 1$ = m $\angle 2$ + m $\angle 3$ − m $\angle 3$ = m $\angle 2$, or m $\angle CYD$ = m $\angle CDW$ (Sub)

Written Exercises, pages 455–456

1. 60 **3.** 60 **5.** 10 **7.** 70 **9.** 10 **11.** 60 **13.** 110 **15.** 10 **17.** 65 **19.** 65 **21.** Tan $\overleftrightarrow{QR}$ and $\overleftrightarrow{QS}$ intersecting at Q (Given); m $\angle 1$ + m $\angle 2$ = m $\angle 3$ (Meas ext $\angle$ = sum of meas of 2 remote int $\angle$s.); m $\angle 1$ = m $\angle 3$ − m $\angle 2$ (Subtr Prop of Eq); m $\angle 3 = \frac{1}{2}$ m $\widehat{RTS}$, m $\angle 2 = \frac{1}{2}$ m $\widehat{RS}$ (If tan and sec intersect at pt of tan, then $\angle$ formed is $\frac{1}{2}$ meas of intercepted arc.); m $\angle 1 = \frac{1}{2}$ m $\widehat{RTS} - \frac{1}{2}$ m $\widehat{RS}$ (Sub); ∴ m $\angle 1 = \frac{1}{2}$(m $\widehat{RTS}$ − m $\widehat{RS}$) (Distr Prop) **23.** 20 **25.** 120 **27.** 180

29.

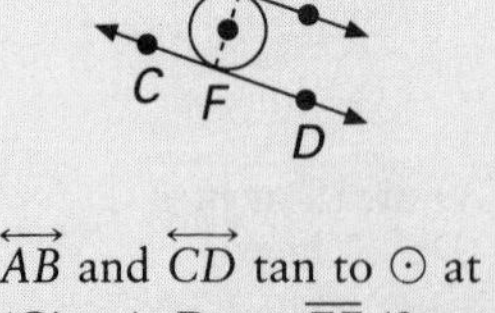

$\overleftrightarrow{AB}$ and $\overleftrightarrow{CD}$ tan to ⊙ at E and F, resp. $\overleftrightarrow{AB} \parallel \overleftrightarrow{CD}$ (Given): Draw $\overline{EF}$ (2 pts determine line.); m $\angle BEF$ = m $\angle CFE$ (If 2 ∥ lines are cut by transv, alt int $\angle$s are ≅.); m $\angle BEF = \frac{1}{2}$ m $\widehat{EF}$, m $\angle CFE = \frac{1}{2}$ m $\widehat{FE}$ (If tan and sec intersect at pt of tan, $\angle$ formed = $\frac{1}{2}$ meas of intercepted arc.); $\frac{1}{2}$ m $\widehat{EF} = \frac{1}{2}$ m $\widehat{FE}$ (Sub); ∴ m $\widehat{EF}$ = m $\widehat{FE}$ (Mult Prop of Eq)

31.

A
O P
B

From pt P in ext ⊙, two tan intersect ⊙O at A and B (Given); Draw $\overline{OA}$, $\overline{OB}$ (2 pts determine line.); $\overline{AO} \perp \overline{AP}$, $\overline{OB} \perp \overline{BP}$(Line tan to ⊙ is ⊥ to radius at pt of tan.); m $\angle OAP$ = m $\angle OBP$ = 90 (Def ⊥); m $\angle OAP$ + m $\angle OBP$ + m $\angle AOB$ + m $\angle P$ = 360 (Sum of $\angle$ meas of polygon = $(n - 2)180$.); m $\angle AOB$ + m $\angle P$ = 180 (Equations may be subtracted.); m $\widehat{AB}$ = m $\angle AOB$ (Arc meas = meas of central $\angle$.); ∴ m $\widehat{AB}$ + m $\angle$P = 180 (Sub)

Written Exercises, pages 460–461

1. 16 **3.** $2\sqrt{10}$ **5.** 21 **7.** $2\sqrt{10}$ **9.** $2\sqrt{2}$ **11.** 1.7 **13.** 4 **15.** Chords $\overline{AC}$ and $\overline{BD}$ intersecting at E (Given); m $\angle AEB$ = m $\angle DEC$ (Vertical $\angle$s are ≅.); m $\angle A$ = m $\angle D$ (If 2 inscr $\angle$s intercept same arc, $\angle$s are = in meas.); $\triangle AEB \sim \triangle DEC$ (AA for ~△s); $\frac{BE}{AE} = \frac{EC}{ED}$ (Def ~ △s); ∴ $BE \cdot ED = AE \cdot EC$ (In true proportion, prod of extremes = prod of means.) **17.** ⊙O with sec $\overleftrightarrow{BC}$ and $\overleftrightarrow{DE}$ intersecting at A (Given); Draw $\overleftrightarrow{AF}$ tan at F (2 pts determine line.); $AC \cdot AB = (AF)^2$, $(AF)^2 = AE \cdot AD$ (If tan and sec intersect, then square of tan seg = prod of sec seg and external sec seg.); ∴ $AC \cdot AB = AE \cdot AD$ (Sub) **19.** 10 in. **21.** $\overleftrightarrow{PR}$ tan to ⊙O at Q, $\overline{VU} \parallel \overline{PR}$ (Given); m $\angle S = \frac{1}{2}$ m $\widehat{QT}$ (Meas inscr $\angle = \frac{1}{2}$ meas intercepted arc.); m $\angle RQT = \frac{1}{2}$ m $\widehat{QT}$ (Angle formed by tan and sec at pt of tan $\frac{1}{2}$ meas intercepted arc.); m $\angle S$ = m $\angle RQT$ (Sub); m $\angle RQT$ = m $\angle$ VUQ (If lines ∥, alt int $\angle$s are ≅.); m $\angle S$ = m $\angle VUQ$ (Trans Prop of Eq); m $\angle Q$ = m $\angle Q$ (Reflex); $\triangle QVU \sim \triangle QTS$ (AA △ ~); $\frac{VQ}{QU} = \frac{TQ}{QS}$ (Def~); ∴ $VQ \cdot QS = QU \cdot TQ$ (In true proportion, prod of extremes = prod of means.)

23.

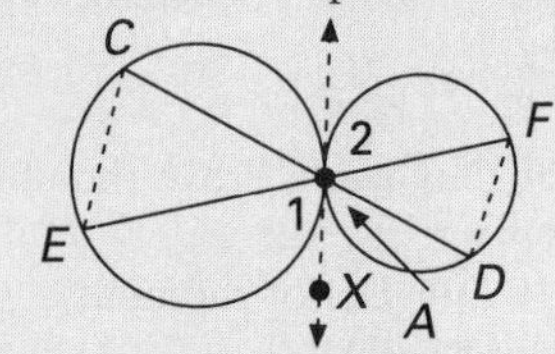

Two non $\cong$ $\odot$s externally tan at A with $\overleftrightarrow{CAD}$ and $\overleftrightarrow{EAF}$ (Given); Draw $\overleftrightarrow{AX}$ tan to both $\odot$s (There is 1 line tan at given pt.); m $\angle 1 = \frac{1}{2}$ m $\widehat{AE}$, m $\angle 2 = \frac{1}{2}$ m $\widehat{AF}$ (Angle formed by sec and tan at pt of tan $= \frac{1}{2}$ meas intercepted arc.); m $\angle C = \frac{1}{2}$ m $\widehat{AE}$, m $\angle D = \frac{1}{2}$ m $\widehat{AF}$ (Meas inscr $\angle = \frac{1}{2}$ meas intercepted arc.); m $\angle C =$ m $\angle 1$, m $\angle D =$ m $\angle 2$ (Sub); m $\angle CAE =$ m $\angle FAD$, m $\angle 1 =$ m $\angle 2$ (Vert $\angle$s are $\cong$.); m $\angle C =$ m $\angle D$ (Trans Prop of Congr); $\triangle ACE \sim \triangle ADF$ (AA $\triangle \sim$); $\frac{CA}{EA} = \frac{AD}{AF}$ (Def $\sim$ $\triangle$s); $\therefore CA \cdot AF = EA \cdot AD$ (In true proportion, prod of extremes = prod of means.)

Written Exercises, pages 464–465

1. Construct midpt M of $\overline{AB}$. Using $\overline{MB}$ as radius, draw a $\odot$ with M as ctr. **3.** Construct midpt M of $\overline{PQ}$. Using $\overline{MQ}$ as radius, draw a $\odot$ with M as ctr. Then construct a line through Q $\perp$ $\overline{MQ}$. **5.** Follow procedure for inscribing a regular hexagon in a $\odot$ on page 462. **7.** See procedure on page 463. **9.** Tan seg should measure 4 in.

11.

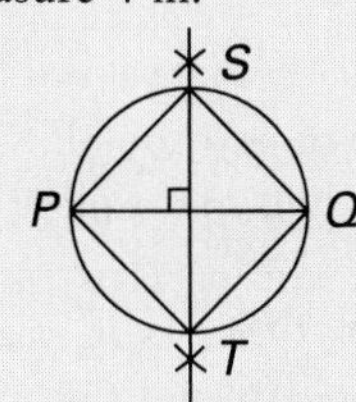

The diameters $\overline{PQ}$ and $\overline{ST}$, together with $\overline{SP}$, $\overline{PT}$, $\overline{TQ}$, and $\overline{QS}$, form four rt $\triangle$s, since $\overline{ST} \perp \overline{PQ}$. The legs are all radii of the $\odot$, and thus congr. So the 4 $\triangle$s are $\cong$ by SAS. By CPCTC the hyp (sides of the quad $PTQS$) are $\cong$. The 2 acute $\angle$s of each $\triangle$ meas 45, since the legs are equal in length. Thus, each angle of $PTQS$ is $45 + 45 = 90$. As a result $PTQS$ is a rect and a square.

13.

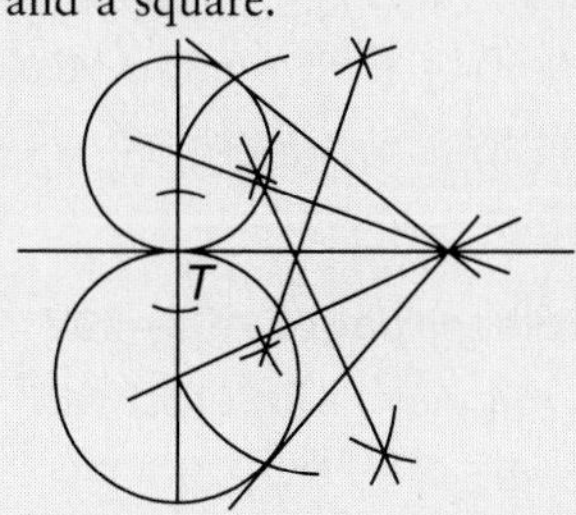

Construct a line containing ctr O of original $\odot$ and pt T. Construct a $\perp$ to $\overline{OT}$ at T, including a pt A. Mark off any pt Q on $\overrightarrow{OT}$ on the other side of T from O. Draw $\odot O$ with radius QT. From pt A on $\overleftrightarrow{TA}$ construct a tan to $\odot O$ and a tan to $\odot Q$, using the procedure on page 462.

15. (a) $\overleftrightarrow{QT}$ is a diam of $\odot M$ because it has endpts on the $\odot$ and contains ctr M. **(b)** $\angle OAT$ is rt because an angle inscribed in a semicircle is a rt $\angle$. **(c)** $\overline{TA}$ is $\perp$ to $\odot O$ at pt A because $\angle OAT$ is rt.

17.

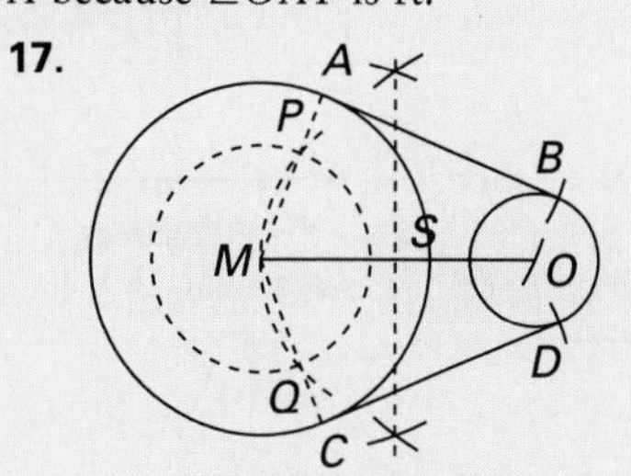

Construct a seg $\overline{OM}$ 13 units in length. Construct C, the midpt of $\overline{OM}$. From O draw a $\odot$ of radius 3 and from M draw a $\odot$ of radius 8. From M draw an auxiliary circle with a radius of $8 - 3$ or 5. Use CO as the length to swing 2 arcs that intersect the auxiliary circle at pts P and Q. Draw $\overrightarrow{MP}$ and $\overrightarrow{MQ}$ intersecting $\odot M$ at A. From pts A and C swing arcs of length PO intersecting circle O at pts B and D, respectively. Draw tans $\overline{AB}$ and $\overline{CD}$, which should measure 12 units in length by Pythagorean Thm.

Written Exercises, pages 468–469

1. $x^2 + y^2 = 25$

3. $(x - 1)^2 + (y - 1)^2 = 4$

5. $(x - 1)^2 + (y + 3)^2 = 16$

7. $(x + 2)^2 + (y + 4)^2 = 1$ **9.** $x^2 + y^2 - 4 = 0$

11. $x^2 + y^2 + 6x - 8y = 0$

13. Ctr$(-2,1)$, $r = 2$ **15.** Ctr$(2,0)$, $r = 1$

17. Ctr$(0,0)$, $r = 2$ **19.** Ctr$(3,0)$, $r = 2$

21. Ctr$(1,4)$, $r = 2$ **23.** Ctr$(-9,10)$, $r = 2$

25. $(x - 3)^2 + (y - 4)^2 = 25$

27. $(x + 4)^2 + (y + 3)^2 = 25$ **29.** Yes

31. No **33.** $(x + 1)^2 + (y + 1)^2 = 13$

35. No; $r_1 = 3$, $r_2 = 4$

37. $4x - 3y + 42 = 0$ **39.** No

Chapter 12

Written Exercises, page 477

1.

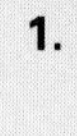
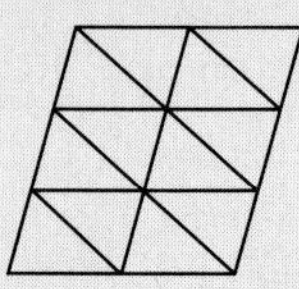

3.

5. Yes

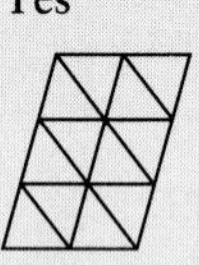

7. Yes

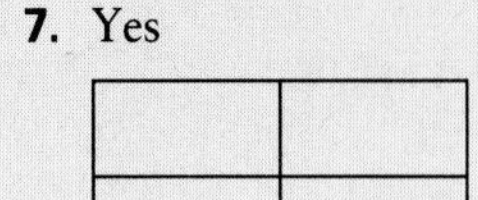

9. Yes

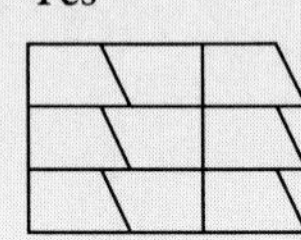

11. Yes (Answers may vary.)

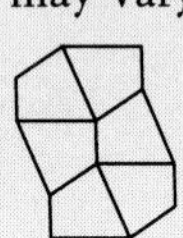

13. No **15.** No **17.** No **19.** No

Written Exercises, pages 480–482

1. 17.5 sq units **3.** 14.4 m^2 **5.** 12.42 in^2 **7.** 6.25 mm^2 **9.** 75.69 yd^2 **11.** 8 in. **13.** 4 yd **15.** $x^2 - 9$ **17.** $y^2 - z^2$ **19.** $b = 3$ cm **21.** $h = \sqrt{6}$ m **23.** $h = 2x - y$ **25.** 12 cm^2 **27.** 20 in^2 **29.** 12 **31.** $\sqrt{6}$ **33.** $A = 128$ ft^2 **35.** 2.97 **37.** 18 mm, 12 mm **39.** True. Area 1 = bh, Area 2 = $(2b)h$ (Area of rect = bh); $\frac{\text{Area 1}}{\text{Area 2}} = \frac{bh}{2bh} = \frac{1}{2}$; Area 2 = 2 · Area 1 **41.** False. $A_1 = (cb)(ch) = c^2bh$, $A_2 = c(bh) = cbh$; $c^2bh \neq cbh$.

Written Exercises, pages 485–486

1. 56 cm^2 **3.** 17.68 in^2 **5.** 6 mm **7.** 9.18 ft^2 **9.** $\sqrt{5}$ m **11.** 3.4 cm **13.** 9 **15.** 20 **17.** $2\sqrt{2}$ **19.** 36 sq units **21.** $40\sqrt{2}$ sq units **23.** $\triangle FAB \sim \triangle EAD$; $\frac{AB}{AD} = \frac{BF}{DE}$; $\frac{15}{12} = \frac{BF}{10}$; $BF = 12.5$. **25.** 12 sq units **27.** 7.7 sq units **29.** 15.7 sq units **31.** ▱$ABCD$ with base $\overline{AB}$ and alt $\overline{CE}$, $AB = b$, $CE = h$ (Given); construct $\overline{DF} \perp$ to $\overleftrightarrow{AB}$ with F on $\overline{AB}$ (from pt to line there is unique ⊥); $\overline{AB} \parallel \overline{CD}$ (Def of ▱); $DF = CE$ (Distance between ∥ lines is constant); $DA = CB$ (Opp sides of ▱ have eq length); ∠DFA and ∠CEB are rt ∠s (Def of ⊥); $\triangle DFA \cong \triangle CEB$ (HL); $FA = EB$ (CPCTC); Area $\triangle DFA$ = Area $\triangle CEB$ (≅ polygons have eq area); Area ▱$ABCD$ = Area $AECD$ + Area $\triangle CEB$ (Area Add Post); Area ▱$ABCD$ = Area $AECD$ + Area $\triangle DFA$ (Sub); Area $AECD$ + Area $\triangle DFA$ = Area rect $FECD$ (Area Add Post); Area ▱$ABCD$ = Area rect $FECD$ (Sub); Area rect $FECD = FE \cdot CE$ (Area rect = bh); $FE = AB$ (Seg Add Post); Area rect $FECD = AB \cdot CE = b \cdot h$ (Sub); Area ▱$ABCD = bh$ (Sub). **33.** 3, 16, 8. Area = $\frac{1}{2}D_s + D_i - 1$ where D_s is number of dots on sides of ▱ and D_i is number of dots in interior.

Written Exercises, pages 491–492

1. 10 m^2 **3.** 30 in^2 **5.** 24 m^2 **7.** 20 cm **9.** 10 mm **11.** 63 mm^2 **13.** $25\sqrt{3}$ in^2 **15.** $CA = 27$ **17.** $BC = 6.75$ **19.** 30 sq units **21.** $\frac{27}{4}\sqrt{15}$ sq units **23.** alt = $4\sqrt{3}$, $A = 16\sqrt{3}$ sq units **25.** $d_2 = 6.25$ cm **27.** 12 sq units **29.** 20 sq units **31.** Square $ABCD$ with side of length s and diag d (Given); Area $(ABCD) = s^2$ ($A = bh$); $d = s\sqrt{2}$ (In 45–45–90 △, hyp = $s\sqrt{2}$); $s = \frac{d}{\sqrt{2}} = \frac{d\sqrt{2}}{2}$ (Rationalization of denominator); Area $(ABCD) = \left(\frac{d\sqrt{2}}{2}\right)^2$ (Sub); Area $(ABCD) = \frac{2d^2}{4} = \frac{d^2}{2}$. Alternate Proof using rhombus: Since a square is a rhombus, $A = \frac{1}{4} \cdot 2d^2 = \frac{1}{2} \cdot d^2$ (diagonals of a square are equal); $A = \frac{1}{2}d^2$ (Sub)

33. $\triangle MNO$, $MN = NO = OM = s$ (Given); Draw $\overline{OP} \perp \overline{MN}$, with P on $\overline{MN}$ and $OP = h$ (From pt to line there is unique ⊥); $\overline{OP} \cong \overline{OP}$ (Reflex); ∠OPM ≅ ∠OPN (Rt ∠s are ≅); $\triangle OPM \cong \triangle OPN$ (HL); m ∠NOM = 60 (Meas of ∠ of equil △ is 60); m ∠NOM = m ∠NOP + m ∠MOP (Angle Add Post); m ∠NOP = m ∠MOP (CPCTC); m ∠NOM = 2 · m ∠NOP (Sub); 2 · m ∠NOP = 60 (Sub); m ∠NOP = 30 (Div Prop of Eq); $OP = \frac{1}{2}s\sqrt{3}$ (Prop of 30–60–90 △); Area $(\triangle MNO) = \frac{1}{2} MN \cdot OP$ (Area △ = $\frac{1}{2}bh$); Area $(\triangle MNO) = \frac{1}{2}s\left(\frac{1}{2}s\sqrt{3}\right)$ (Sub); Area $(\triangle MNO) = \frac{s^2}{4}\sqrt{3}$ (Comm Assoc Prop).

Written Exercises, pages 495–497

1. 104 m^2 **3.** 54 mm^2 **5.** 112.5 mm^2 **7.** 18 ft **9.** 7 cm **11.** 54.25 sq units **13.** $43.75\sqrt{3}$ sq units **15.** $14.25\sqrt{2}$ sq units **17.** 420 sq units **19.** $27\sqrt{3}$ sq units **21.** $b = 30$, $h = 9$ **23.** $b = 15$, $h = 9$

25. 15 sq units **27.** 37.31 cm^2 **29.** 170.43m^2 **31.** 10 cm, 32 cm **33.** Trap with bases b_1 and b_2, s the length of a side adj to $\angle A$ (Given); Let h be the length of alt of trap (Distance between $\|$ lines is constant); $\sin A = \frac{h}{s}$(Def of sine); $h = s(\sin A)$ (Mult Prop of Eq); Area (trap) $= \frac{1}{2}(b_1 + b_2)h$ (Formula for area of trap); Area (trap) $= \frac{1}{2}h(b_1 + b_2)$ (Comm Prop); Area (Trap) $= \frac{1}{2}s(\sin A)(b_1 + b_2)$ (Sub).

Written Exercises, pages 501–502

1. $\sqrt{3}$ cm **3.** 3 mm **5.** 65 cm **7.** 1,820 in **9.** 16 cm^2 **11.** 3.6 in^2 **13.** 4.33 cm **15.** 32 m, 4 m, 64 m^2 **17.** $\frac{2}{3}\sqrt{30}$ ft, $4\sqrt{30}$ ft, $20\sqrt{3}$ ft^2 **19.** 172.0 cm^2 **21.** 695.3 yd^2 **23.** 7,135.8 mm^2 **25.** A regular polygon with ctr O and side $\overline{AB}$ (Given); Draw $\overline{OD} \perp$ from O to D on $\overline{AB}$(From pt to line there is unique $\perp$); $\odot O$ circumscribes the given polygon ($\odot$ can be circumscribed around any reg polygon); $OA = OB$ (Radii are $\cong$); $\triangle ABO$ is isos (Def of isos $\triangle$); $\overline{OD}$ is alt of $\triangle ABO$ (Def of alt); $\overline{OD}$ bis $\overline{AB}$ (Alt from vertex $\angle$ to base of isos $\triangle$ bis the base). **27.** Reg polygon with ctr O and sides $\overline{AB}$, $\overline{BC}$, $\overline{CD}$ · · · (Given); $\odot O$ circumscribes the reg polygon ($\odot$ can be circumscribed about any reg polygon); $AB = BC = CD = \cdots$ (Def of reg polygon); $OA = OB = OC = \cdots$ (Radii of $\odot$ are $\cong$); $\triangle AOB \cong \triangle BOC \cong \triangle COD \cong \cdots$ (SSS). **29.** Reg polygon with ctr O, n sides, radius r, side of length s, $\overline{OM}$ a $\perp$ bis of side $\overline{AB}$ at pt M (Given); Area (reg polygon) $= n\left[\sin\left(\frac{180}{n}\right)\cos\left(\frac{180}{n}\right)\right]r^2$ (Proved in Ex 28); Area (reg polygon) $= n\left[r \cdot \sin\left(\frac{180}{n}\right)\right] \cdot \left[r \cdot \cos\left(\frac{180}{n}\right)\right]$ (Comm Prop); $r \cdot \sin\left(\frac{180}{n}\right) = r \cdot \frac{MB}{r} = MB$ (Def of sin); $MB = \frac{1}{2}s$ (Def of seg bis); Area (reg polygon) $= n\left(\frac{1}{2}s\right)\left[r \cdot \cos\left(\frac{180}{n}\right)\right]$ (Sub); $\tan\left(\frac{180}{n}\right) = \sin\left(\frac{180}{n}\right) \div \cos\left(\frac{180}{n}\right)$ (Trigonometric Identity); $\cos\left(\frac{180}{n}\right) = \sin\left(\frac{180}{n}\right) \div \tan\left(\frac{180}{n}\right)$ (Mult Prop of Eq); Area (reg polygon) $= n\left(\frac{1}{2}s\right) \cdot \left[r \cdot \sin\left(\frac{180}{n}\right)\right] \div \tan\left(\frac{180}{n}\right)$ (Sub); Area (reg polygon) $= n\left(\frac{1}{2}s\right)\left(\frac{1}{2}s\right) \div \tan\left(\frac{180}{n}\right)$ (Sub); Area (reg polygon) $= \dfrac{ns^2}{4\tan\left(\frac{180}{n}\right)}$ **31.** Let rad = 1. Then Area (inscribed square) = 2, Area (circumscribed square) = 4. Ratio = 1:2 **33.** 3,696 tiles

Written Exercises, 506–508

1. 9:7; 81:49 **3.** 13:11; 169:121 **5.** $\frac{5}{3}$; $\frac{25}{9}$ **7.** 14 cm **9.** 78 ft **11.** $\frac{7}{8}$ **13.** $\frac{17}{19}$ **15.** $\frac{2}{1}$; $\frac{4}{1}$ **17.** 9 cm, 27 cm **19.** $\frac{5}{2}\sqrt{6}$ in., $6\sqrt{6}$ in. **21.** 4 cm^2 **23.** Similar rects A and B, with corr adj sides of lengths a and b, a_1 and b_1, respectively (Given); $\frac{a}{a_1} = \frac{b}{b_1}$ (Def ~ poly); A(rect A) $= ab$, A(rect B) $= a_1b_1$ (A of rect $= bh$); $\frac{\text{A(rect } A)}{\text{A(rect } B)} = \frac{ab}{a_1b_1} = \frac{a}{a_1} \cdot \frac{b}{b_1} = \frac{a}{a_1} \cdot \frac{a}{a_1}$ (Sub); $\frac{\text{A(rect } A)}{\text{A(rect } B)} = \left(\frac{a}{a_1}\right)^2$ (Notation). **25.** p($\triangle ABC$) = 12 cm, p($\triangle DEF$) = 15 cm **27.** Length of side = $\sqrt{2}$ cm **29.** $\triangle ABC \sim \triangle DEF$, with corr sides of lengths a, b, and c, and d, e, and f, respectively, and semiperimeters s_1 and s_2, respectively (Given); $\frac{\text{A}(\triangle ABC)}{\text{A}(\triangle DEF)} = \frac{\sqrt{s_1(s_1 - a)(s_1 - b)(s_1 - c)}}{\sqrt{s_2(s_2 - d)(s_2 - e)(s_2 - f)}}$ (Heron's formula; Div Prop of Eq); $\frac{\text{p}(\triangle ABC)}{\text{p}(\triangle DEF)} = \frac{a}{d}$ (Ratio of perim of ~ poly = ratio of corr sides); $\dfrac{\frac{1}{2}\text{p}(\triangle ABC)}{\frac{1}{2}\text{p}(\triangle DEF)} = \frac{a}{d}$ (Any nonzero real div by self = 1); $\frac{s_1}{s_2} = \frac{a}{d}$ (Def of semiperimeter, sub); $\frac{s_1}{a} = \frac{s_2}{d}$ (If $\frac{a}{b} = \frac{c}{d}$, then $\frac{a}{c} = \frac{b}{d}$); $\frac{s_1 - a}{a} = \frac{s_2 - d}{d}$ (If $\frac{a}{b} = \frac{c}{d}$, then $\frac{a - b}{b} = \frac{c - d}{d}$); $\frac{s_1 - a}{s_2 - d} = \frac{a}{d}$ (If $\frac{a}{b} = \frac{c}{d}$, then $\frac{a}{c} = \frac{b}{d}$); similarly, show that $\frac{s_1 - b}{s_2 - e} = \frac{b}{e}$ and $\frac{s_1 - c}{s_2 - f} = \frac{c}{f}$; then by substitution:

$\frac{s_1}{s_2} \cdot \frac{s_1 - a}{s_2 - d} \cdot \frac{s_1 - b}{s_2 - e} \cdot \frac{s_1 - c}{s_2 - f} = \frac{a}{d} \cdot \frac{a}{d} \cdot \frac{a}{c} \cdot \frac{b}{c} \cdot \frac{c}{f}$; $\frac{a}{d} = \frac{b}{c} = \frac{c}{f}$ (Def of ~ △s); $\frac{s_1(s_1 - a)(s_1 - b)(s_1 - c)}{s_2(s_2 - d)(s_2 - e)(s_2 - f)} = \frac{a}{d} \cdot \frac{a}{d} \cdot \frac{a}{d} \cdot \frac{a}{d}$ (Sub); $\frac{\sqrt{s_1(s_1 - a)(s_1 - b)(s_1 - c)}}{\sqrt{s_2(s_2 - d)(s_2 - e)(s_2 - f)}} = \sqrt{\left(\frac{a}{d}\right)^4} = \left(\frac{a}{d}\right)^2$ (Prop of roots); $\frac{A(\triangle ABC)}{A(\triangle DEF)} = \left(\frac{a}{d}\right)^2$ (Sub).

Written Exercises, pages 513–514

1. 62.8 cm **3.** 314.0 yd **5.** 12.6 mm^2 **7.** 314.0 in^2 **9.** 25π cm^2 **11.** 8 in, 8π in, 16π in^2 **13.** 5 m, 10π m, 25π m^2 **15.** 5 in, 10 in, 25π in^2 **17.** 10 cm, 20 cm, 20π cm **19.** 5 m, 5π m, 6.25 m^2 **21.** $\frac{50}{\pi}$ in, $\frac{100}{\pi}$ in, $\frac{2{,}500}{\pi}$ in^2 **23.** $100(\pi - 2) \approx 114$ cm^2 **25.** $(300\sqrt{3} - 100\pi)cm^2 \approx 205.6$ cm^2 **27.** $3\frac{10}{71}$ **29.** $100 - 25\pi$ **31.** $\frac{1}{4}$ **33.** 40,192 km; 19 km; 19 km **35.** False. Counterexample: ⊙O with $r = 1$, ⊙P with $r = 2$. $\frac{C(\odot O)}{C(\odot P)} = \frac{2\pi}{4\pi} = \frac{1}{2}$; $\frac{A(\odot O)}{A(\odot P)} = \frac{\pi}{4\pi} = \frac{1}{4}$; $\frac{1}{2} \neq \frac{1}{4}$. **37.** True. ⊙Q with r_1 and circumscr sq Q, ⊙R with r_2 and circumscr sq R (Given); $A(\odot Q) = \pi r_1^2$, $A(\odot R) = \pi r_2^2$ ($A(\odot) = \pi r^2$); sides of sq R are tan to ⊙R (Def of inscr ⊙); diam of ⊙R including pts of tan to sq R is ⊥ to sides of sq R (Radii are ⊥ to tan at pt of tan); such diam of ⊙R is ∥ to side of sq R (lines ⊥ to same line are ∥); length of diam of ⊙R = length of side of sq R (Opp sides of rect are ≅); length of diam of ⊙R = $2 \cdot r_2$ (Def of diam); $2r_2$ = length of s of sq R (Sub); $A(\text{sq } R) = 4r_2^2$ ($A(\text{sq}) = s^2$); $A(\text{sq } Q) = 4r_1^2$ (Proceed similarly); $\frac{A(\odot Q)}{A(\odot R)} = \frac{\pi r_1^2}{\pi r_2^2} = \frac{r_1^2}{r_2^2}$; $\frac{A(\text{sq } Q)}{A(\text{sq } R)} = \frac{4r_1^2}{4r_2^2} = \frac{r_1^2}{r_2^2}$; $\frac{A(\odot Q)}{A(\odot R)} = \frac{A(\text{sq } Q)}{A(\text{sq } R)}$ (Sub)

Written Exercises, pages 518–519

1. Yes; arcs of eq meas in same ⊙ have eq length. **3.** Yes; arcs of eq meas in ≅ ⊙s have eq length. **5.** No; arcs of eq meas in non≅ ⊙s have uneq length. **7.** 4 **9.** 6π **11.** $\frac{10}{3}\pi$ m, $\frac{50}{3}\pi$ m^2 **13.** $\frac{16}{3}\pi$ ft, $\frac{64}{3}\pi$ ft^2 **15.** $\frac{7}{2}\pi$ mm, $\frac{21}{4}\pi$ mm^2 **17.** 75π cm^2 **19.** 72π mm^2 **21.** 6 ft^2 **23.** $\frac{1}{4}$ in **25.** 60, 6π sq units, $6\pi - 9\sqrt{3}$ sq units **27.** 90, 90, $25\pi - 50\sqrt{3}$ sq units **29.** ⊙P with radius r and arc of meas m with arc length s (Given); $s = \frac{m}{360} \cdot 2\pi r$ (Def of arc length); $\frac{s}{2\pi r} = \frac{m}{360}$ (Div Prop of Eq); Let sector_1 of ⊙P be bounded by arc s and radii with endpts on s (Def of sector of a ⊙); $A(\text{sector}_1) = \frac{m}{360} \cdot \pi r^2$ (Def of area of sector); $A(\text{sector}_1) = \frac{s}{2\pi r} \cdot \pi r^2$ (Sub); $A(\text{sector}_1) = \frac{s \cdot r \cdot \pi r}{2 \cdot \pi r} = \frac{1}{2}rs$. **31.** $(100 - 25\pi)$ sq units **33.** $\pi(r_1^2 - r_2^2)$ **35.** 218.6 cm^2

Chapter 13

Written Exercises, pages 531–532

1. ⊙ with radius of 5 cm. **3.** All pts in the ext of a ⊙ with radius of 3 cm. **5.** The bis of the ∠ through the vertex of the rt ∠. **7.** Two ∥ lines 3 cm from the given line. **9.** The region outside two ∥ lines that are 14 cm apart, excluding the lines. **11.** The region that is closer to side $\overrightarrow{OB}$ on one side of the ∠ bis of ∠AOB, excluding the line which makes up the bis.

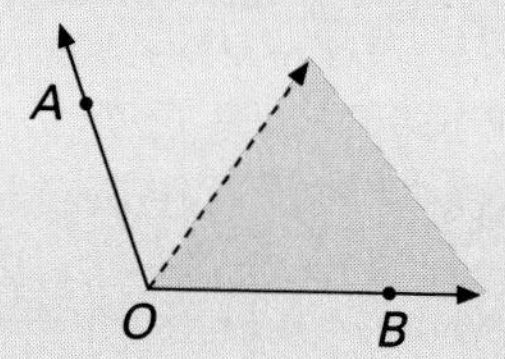

13. A ⊙ concentric to the int ⊙ and having a radius that is the sum of the radii of the two given circles.

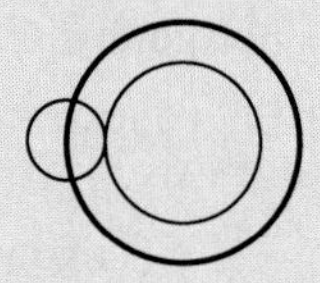

15. The boundary is a pair of ∥ lines that are 20 ft apart, each 10 ft from the power line.

17.

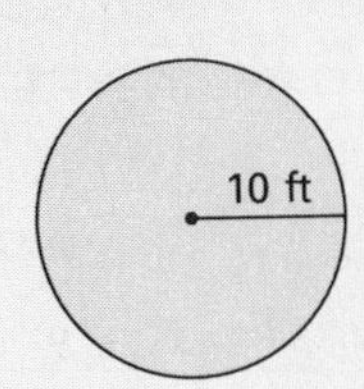

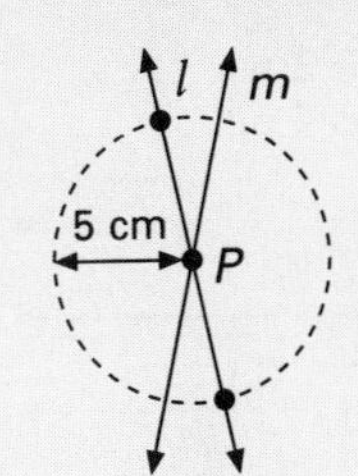

19. Part I. P lies on the $\perp$ bis of $\overline{AB}$ (Given); $\overline{AD} \cong \overline{BD}$ (Def of bis); $\angle PDA \cong \angle PDE$ ($\perp$s form $\cong \angle$s); $\overline{PD} \cong \overline{PD}$ (Reflex); $\triangle PDA \cong \triangle PDB$ (SAS); $PA = PB$ (CPCTC). Part II. $PA = PB$ (Given); Draw $\overrightarrow{PD} \perp \overline{AB}$ (There is exactly one $\perp$ from a pt to a line); $\overline{PD} \cong \overline{PD}$ (Reflex); $\overrightarrow{PDA} \cong \overrightarrow{PDB}$ (HL); $AD = BD$ (CPCTC); $\therefore$ $\overline{PD}$ is $\perp$ bis of AB (Def of $\perp$ bis). **21.** (Ref to illus in text) P is in the int of $\angle AOB$ equidistant from P to OA, PE is distance from P to OB (Def of distance); $PD = PE$ (Given); $\overline{OP} \cong \overline{OP}$ (Reflex); $\triangle DOP$ and $\triangle EOP$ are rt $\triangle$s (Def of $\perp$ and def of rt $\triangle$); $\triangle POD \cong \triangle POE$ (HL); $\angle POD \cong \angle POE$ (CPCTC); $\therefore$ $\overline{OP}$ is the bis of $\angle AOB$ (Def of $\angle$ bis.) **23.** Part I. $\odot P$, V lies on $\odot P$, $\overline{AB}$ diam of $\odot P$ (Given); Draw $\overline{VA}$, $\overline{VB}$ (2 pts determine a line); $\triangle AVB$ is a rt $\triangle$ (An $\angle$ inscr in a semicircle is a rt $\triangle$). Part II. $\odot$P, $\overline{AB}$ a diam of $\odot P$, $\overline{AB}$ a hyp of $\triangle AVB$ (Given); Draw $\overline{CV}$, $\overline{AV}$ so that X lies on $\overline{BV}$ (|| Post); $\angle PXB$ is a rt $\angle$ (Corr formed by || lines are $\cong$); $BP = AP$ (Radii are $\cong$); $BX = VX$ (line || base of a $\triangle$ cuts off seg proportional to the sides); $PX = PX$ (Reflex); $\triangle BXP \cong \triangle VXP$ (Rt $\angle$s are $\cong$); $\triangle BXP \cong \triangle VXP$ (SAS); $PV = PB$ (CPCTC); PV is a radius of $\odot P$ (Radii are $\cong$); V is on $\odot P$ (Def of $\odot$).

Written Exercises, pages 536–537

1. $\odot$s with ctr 0, radius of 3 and 7. **3.** $\odot$ with 0 as ctr, radius of 13 cm. **5.** Ext of $\odot$ with ctr 0, radius of 7. **7.** 2 pts; inters of 2 $\odot$s, ctrs of P and Q and each radius of 6. **9.** 2 pts; inters of $\odot$s with ctr P, radius of 9 cm and with ctr Q, radius of 6 cm. **11.** Inters of ext of $\odot$s with P as ctr, radius of 8 cm and with Q as ctr, radius of 7 cm. **13.** 2 pts; inters of l and the $\odot$ with P as ctr and radius of 5 cm.

15. 4 pts; inters of 2 pairs of || lines, 1 on either side of l, || to and 5 cm from l, the other pair the same with respect to m. **17.** 4 pts; inters of 2 lines || m, 1 on each side 2 cm from m and $\odot$ with P as ctr, radius of 9 cm. **19.** Inters of $\odot$ with P as ctr, radius of 3 cm and $\odot$ with Q as ctr, radius of 5 cm.
If $d > 8$ cm, locus contains no pts.
If $d = 8$ cm, locus is pt of tangency.
If 2 cm $< d <$ 8 cm, locus is 2 pts.
If $d = 2$ cm, locus is pt of tangency.
If $d < 2$ cm, locus contains no pts.

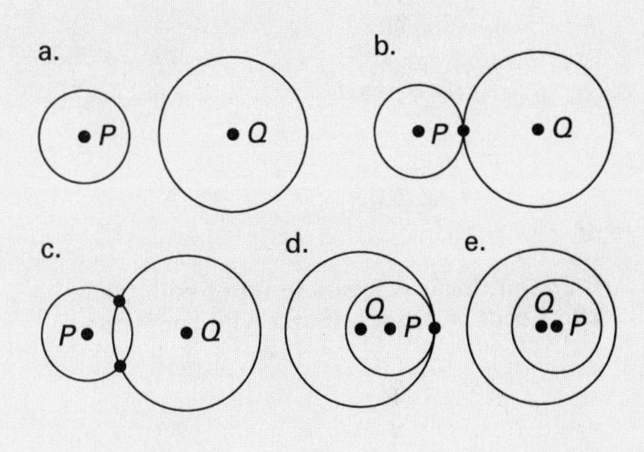

21. If $d = 0$, then locus has no pts.
If $d > 0$, then locus is pt; inters of $\angle$ bis and $\odot$ with vert as ctr, r of d.

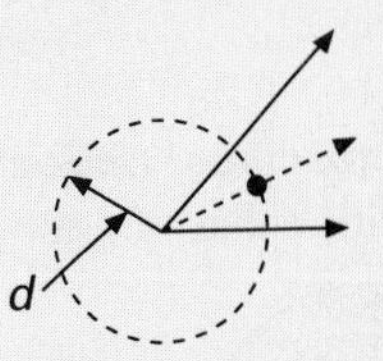

23. Center of $\odot$ 0.
25. If $d = 0$ and $r = 0$ locus is a pt.
If $d = 0$ and $r > 0$ then locus is 2 pts.
If $d > r$ locus contains no pts.

If $d = r$ locus is 2 pts.
If $d < r$ locus is 4 pts.

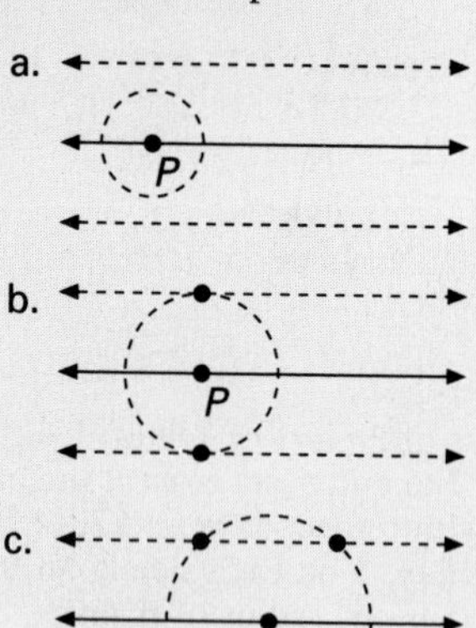

27. If $2d < t$ or $2d = t$, locus contains no pts.

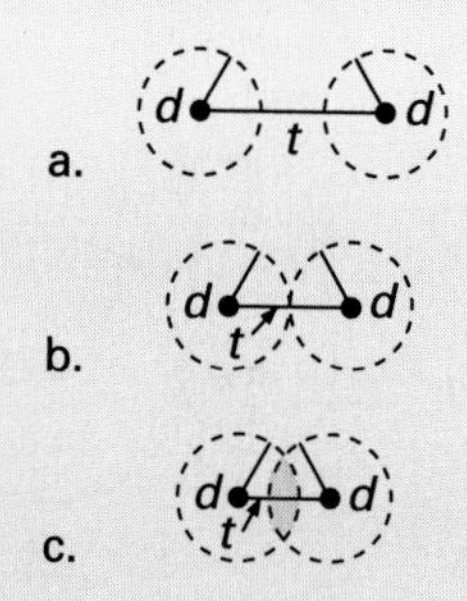

If $2d > t$, locus is region in int of both ⊙s.

29. Locus is a series of arcs with endpts on ⊙.

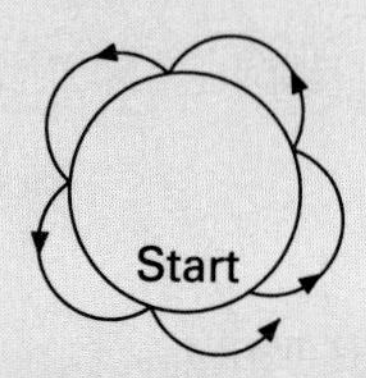

Written Exercises, pages 539–40

1. Sphere with radius of 6 m.

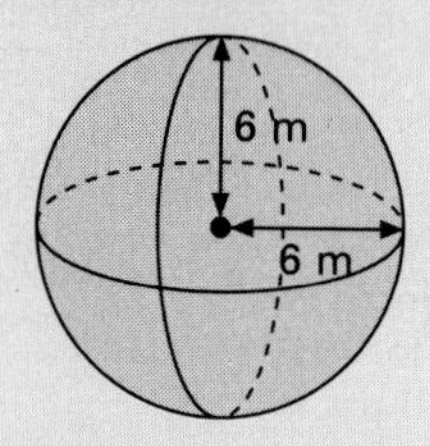

3. Int of sphere with pt as ctr, radius of 5 cm.

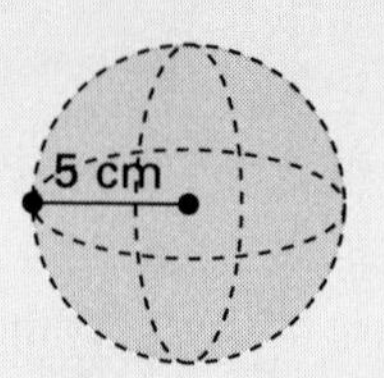

5. Int of cyl surface with given line axis, radius of 5 cm.

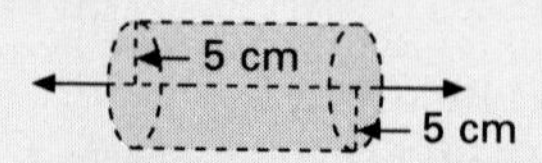

7. Plane that is the ⊥ bis of $\overline{AB}$.

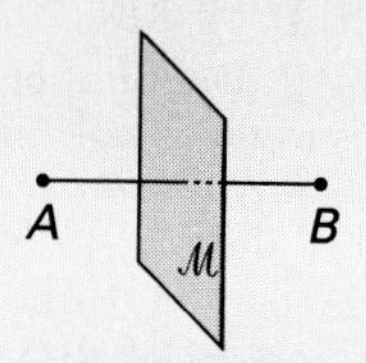

9. A third plane that is ∥ to the other two and halfway between them.

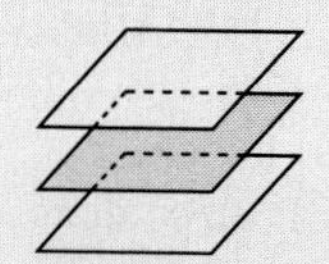

11. Plane that bis the angle.

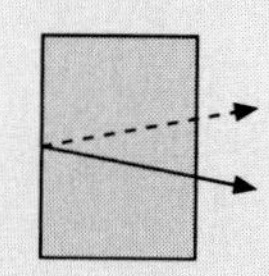

13. Two lines formed by the inters of cyl surfaces around the given lines.

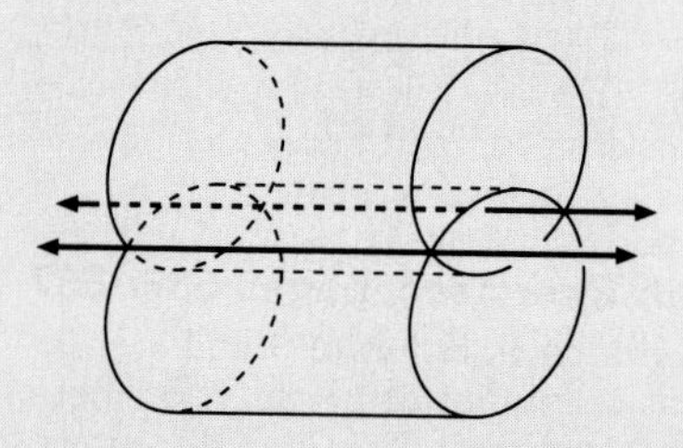

15. Locus is inters of sphere with P as ctr and sphere with Q as ctr. Depending on the radii of the spheres and the distance apart of P and Q, there will be no inters, a point of inters, a circle of inters, or a sphere of inters. One sphere may be inside the other.

Written Exercises, page 545

1. Incenter

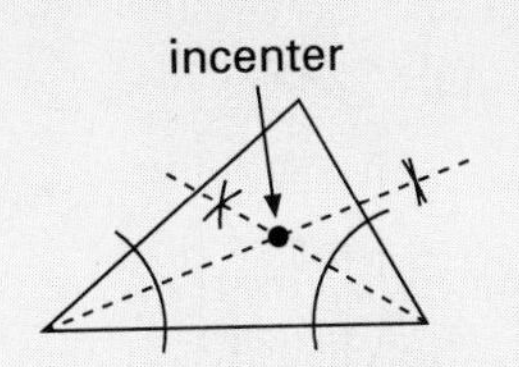

3. Incenter

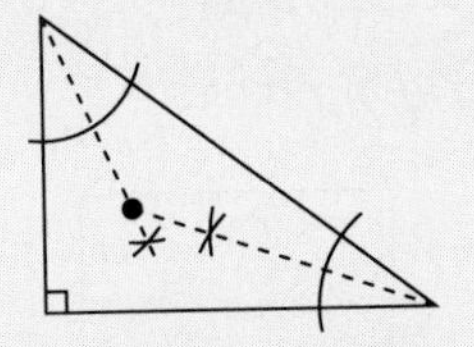

5. Inscribed circle

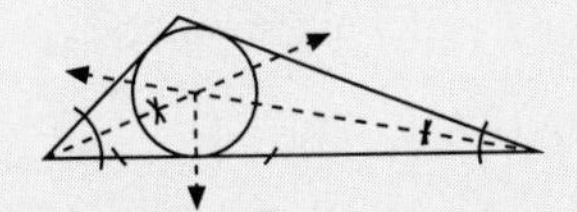

7. Both lie on ⊥ bis of base.

9.

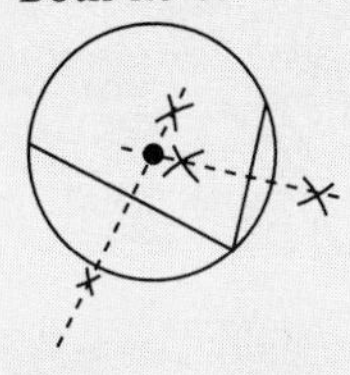

11. True, inctr is on ∠ bis, so in int of △ **13.** False, equil △ **15.** △*ABC*; $\overrightarrow{AD}$, $\overrightarrow{BE}$, and $\overrightarrow{CF}$ are the bisectors of ∠*A*, ∠*B*, and ∠*C*, respectively (Given): $\overrightarrow{AD}$ intersects $\overrightarrow{BE}$ at pt *P*. (The bis of the ∠s of the triangles intersect at the incenter.); dist from *P* to *AC* = dist from *P* to *AB*, *P* to *CB* = *P* to *AB* (If a pt is on the bis of an ∠, then it is equidistant from the sides of the ∠); dist from *P* to *AC* = dist from *P* to *CB* (Trans Prop); ∴ $\overrightarrow{CF}$ intersects pt *P* (If a pt is equidistant from the sides of an angle, then it is on the bis of the ∠). **17.** $\frac{5\sqrt{3}}{3}$ **19.** $\frac{25}{3}(4\pi - 3\sqrt{3})$

Written Exercises, pages 549–550

1. Construct ⊥ bis to 2 sides. Draw median to each of the 2 sides. Pt of inters will be centroid. **3.** Constr alt to 2 sides; pt of inters is orthocenter. **5.** In equil △, alt is also median; centroid, orthocenter same pt. **7.** 46 **9.** 26 **11.** 15 **13.** 10 **15.** $3\sqrt{3}$ **17.** $2\frac{2}{3}$ cm **19.** Except for its endpts, a median contains only pts in int of △. Intersection of medians must be int pt. **21.** Given: △*ABC* with meds $\overline{AD}$, $\overline{BE}$, $\overline{CF}$. Prove: $\overline{AD}$, $\overline{BE}$, $\overline{CF}$ are concrnt at *P*, $AP = \frac{2}{3}AD$, $BP = \frac{2}{3}BE$, $CP = \frac{2}{3}CF$. $\overline{AD}$, $\overline{BE}$ inters at *P*, $AP = \frac{2}{3}AD$, $BP = \frac{2}{3}BE$ (Thm 13.6); $\overline{AD}$, $\overline{CF}$ inters at *Q*, $AQ = \frac{2}{3}AD$, $CQ = \frac{2}{3}CF$ (Thm 13.6); *AP* = *AQ* (Subst); *P* = *Q* (Ruler Post); $\overline{AD}$, $\overline{BE}$, $\overline{CF}$ are concrnt (Def of concrnt); ∴ $CP = \frac{2}{3}CF$ (Sub). **23.** Use fig for Ex 22. Given: Equil △*ABC*, $\overline{AE}$, $\overline{BF}$. $\overline{CD}$ are alt inters at *P*. Prove: *P* is incenter, circumctr, orthoctr, centroid. *P* is orthoctr (Def orthoctr); $\overline{AE}$, $\overline{BF}$, $\overline{CD}$ are med (Alt of equil △ are med); *P* is centroid (Def of centroid); $\overline{AE}$, $\overline{BF}$, $\overline{CD}$ are ∠ bis (Med of equil △ is bis vert ∠); *P* is inctr (Def of inctr); $\overline{AE}$, $\overline{BF}$, $\overline{CD}$ are ⊥ bis of sides (Alt of equil △ are ⊥ bis of sides); ∴ *P* is circumctr.

Chapter 14

Written Exercises, pages 558–559

1.

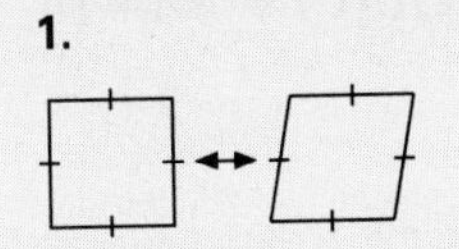

3.

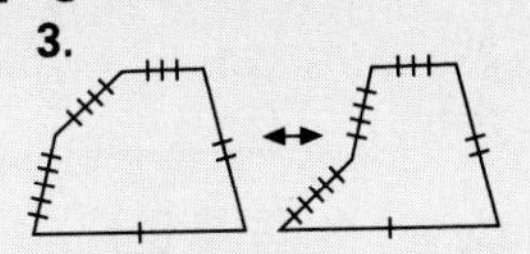

5. Yes, faces are △s. **7.** Yes, faces are △s. **11.** Yes, faces are △s. **13.** *F* = 8, *E* = 12, *V* = 6 **15.** *F* = 20, *E* = 30, *V* = 12 **17.** No noncongruent counterexample of a triangle with congruent sides can be provided. **19.** Check students' patterns and models.

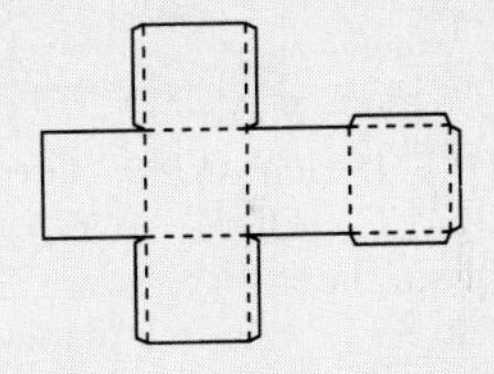

Written Exercises, pages 562–563

1. *F* = 5, *E* = 9, *V* = 6 **3.** *F* = 7, *E* = 15, *V* = 10 **5.** 5 lateral faces, 5 lateral edges **7.** 12 lateral faces, 12 lateral edges

9.

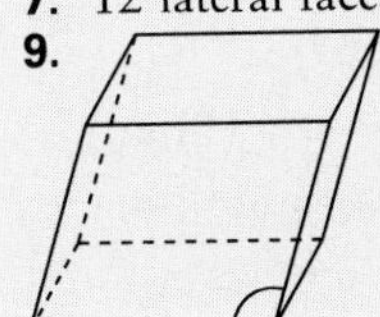

11.

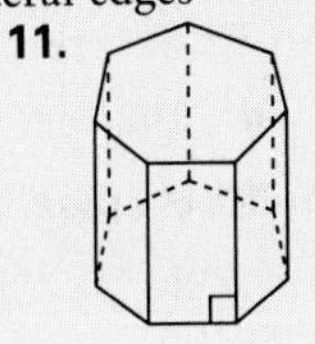

13.

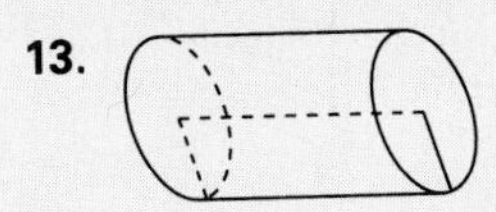

15. F **17.** T

19.

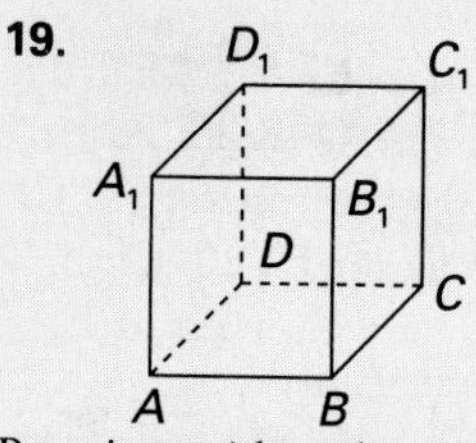

Rt prism with sq bases $ABCD \cong A_1B_1C_1D_1$ (Given); ABB_1A_1, BCC_1B_1, CDD_1C_1, DAA_1D_1 are ▱s (Def prism); $\overline{AA_1} \perp \overline{AB}$, $\overline{BB_1} \perp \overline{AB}$, $\overline{BB_1} \perp \overline{BC}$, $\overline{CC_1} \perp \overline{BC}$, $\overline{CC_1} \perp \overline{CD}$, $\overline{DD_1} \perp \overline{CD}$, $\overline{DD_1} \perp \overline{DA}$, $\overline{AA_1} \perp \overline{DA}$ (In rt prism, lateral edges are ⊥ to bases); $\angle A_1\,AB$, $\angle ABB_1$, $\angle B_1\,BC$, $\angle BCC_1$, $\angle C_1CD$, $\angle D_1DA$, $\angle DAA_1$, are rt ∠s. (Def of ⊥); $\overline{AA_1} \cong \overline{BB_1} \cong \overline{CC_1} \cong \overline{DD_1}$ (Opp sides of ▱ are ≅.); $\overline{AB} \cong \overline{BC} \cong \overline{CD} \cong \overline{DA}$ (Def square); $\therefore ABB_1A_1 \cong BCC_1B_1 \cong CDD_1C_1 \cong DAA_1D_1$ (SASAS) **21.** $V = 2 \cdot S$

23.

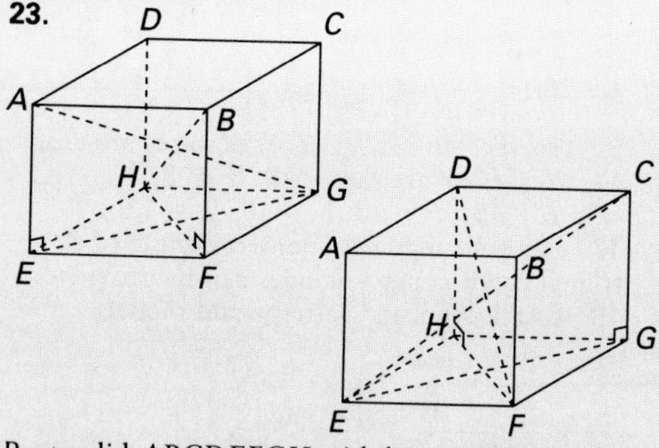

Rect solid *ABCDEFGH* with bases *ABCD* and *EFGH* (Given); ∠*AEG*, ∠*BFH*, ∠*CGE*, ∠*DHF* are rt ∠s (In rt prism, lateral edges are ⊥ to bases.); *ABCD* and *EFGH* lie in ∥ planes (Def prism); $AE = BF = CG = DH$ (Distance between ∥ planes is constant); $EG = FH$ (Diags of rect are ≅.); $\triangle AEG \cong \triangle BFH \cong \triangle CGE \cong \triangle DHF$ (SAS); $\therefore \overline{AG} \cong \overline{BH} \cong \overline{CE} \cong \overline{DF}$ (CPCTC).

Written Exercises, page 567

1. 154 cm^2, 202 cm^2 **3.** 364 in^2, 454 in^2

5.

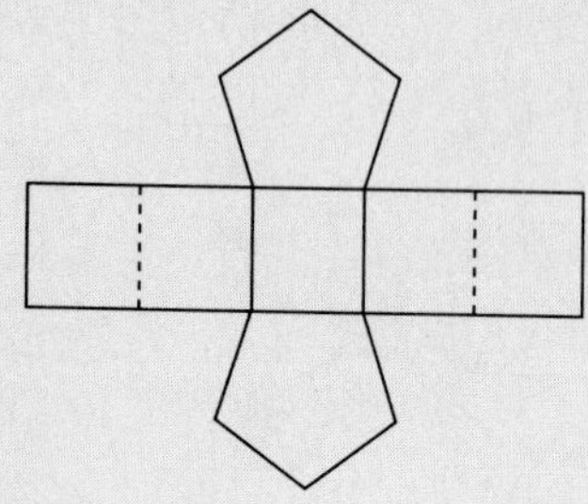

7. 120π cm^2, 320π cm^2 **9.** 126π m^2, 224π m^2 **11.** 9 to 25 **13.** 1 to 16

15. 12.816 m^2

17.

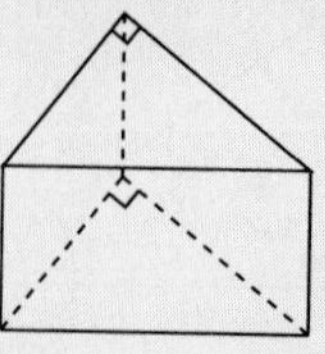

435.6 cm^2, 507.6 cm^2;

19.

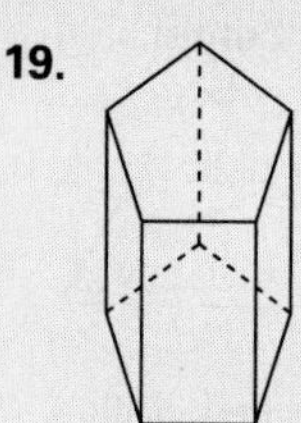

50 m^2, 64 m^2;

21.

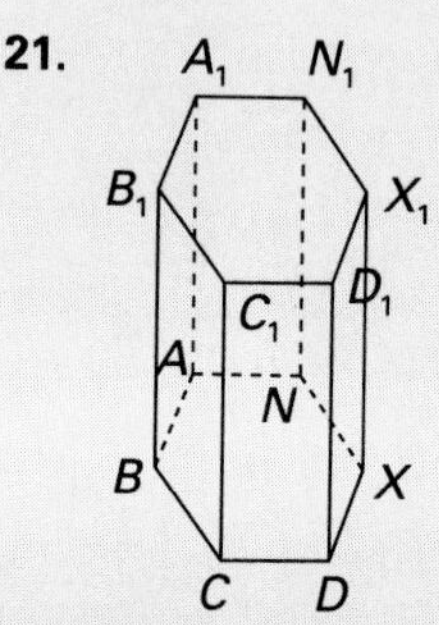

Rt prism with alt *h* and bases *ABCDXN* and $A_1B_1C_1D_1X_1N_1$ (Given); Lateral area L = Area (ABB_1A_1 + BCC_1B_1 + *CDD*, *C*, + *DXX*, *D*, (XNN_1X_1) + *ANN*, *A*, (Def lateral area of prism); $\overline{AA_1}$, $\overline{BB_1}$, $\overline{CC_1}$, $\overline{DD_1}$, $\overline{XX_1}$, $\overline{NN_1}$ are ⊥ to bases (Def rt prism); $h = AA_1 = BB_1 = CC_1 = DD_1, = XX_1 = NN_1$ (Def alt of prism); $\overline{AB}$, $\overline{BC}$, $\overline{CD}$, $\overline{DX}$, $\overline{XN}$, and $\overline{NA}$ are bases of ▱s (Def prism); Area $(ABB_1A_1) = AB \cdot h$, Area $(BCC_1B_1) = BC \cdot h$, Area $(CDD_1\,C_1) = CD \cdot h$, Area $(DXX_1D_1) = DX \cdot h$, Area $(XNN_1X_1) = XN \cdot h$, Area $(NAA_1\,N_1) = NA \cdot h$ (Area ▱ = *bh*); $L = AB \cdot h + BC \cdot h + CD \cdot h + DX \cdot h + XN \cdot h + AN \cdot h$ (Sub); $L = (AB + BC + CD + DX + XN + AN)h$ (Distr Prop); $AB + BC + CD + DX + XN + AN$ = perim of base of prism (Def perim); $L = ph$ (Sub). **23.** Cubes 1 and 2 with edges of lengths *a* and *b*, respectively (Given); *L* (Cube 1) $= 4a \cdot a = 4a^2$, *L* (Cube 2) $= 4b \cdot b = 4b^2$ $(L = ph)$; $\frac{L\ (\text{Cube } 1)}{L\ (\text{Cube } 2)} = \frac{4a^2}{4b^2} = \frac{a^2}{b^2}$ (Equations may be divided.); $\frac{L\ (\text{Cube } 1)}{L\ (\text{Cube } 2)} = \left(\frac{a}{b}\right)^2$ (Algebra); *A* (Cube 1) $= 4a^2 + 2a^2$, *A* (Cube 2) $= 4b^2 + 2b^2$ (Def total area of prism); *A* (Cube 1) $= 6a^2$, *A* (Cube 2) $= 6b^2$ (Distr

Prop); $\frac{A \text{ (Cube 1)}}{A \text{ (Cube 2)}} = \frac{6a^2}{6b^2} = \frac{a^2}{b^2}$ (Equations may be divided.); $\therefore \frac{A \text{ (Cube 1)}}{A \text{ (Cube 2)}} = \left(\frac{a}{b}\right)^2$ (Algebra) **25.** $(1 + \sqrt{2})$ to π
27. Parallelogram, $L = 2\pi rh$

Written Exercises, pages 572–573

1. 391 cm^3 **3.** 1,008π in^3 **5.** 48π mm^3
7. 280 cm^3 **9.** 2,652 in^3 **11.** 768 ft^3
13. 1,728 mm^3 **15.** $425\sqrt{3}$ cm^3
17.

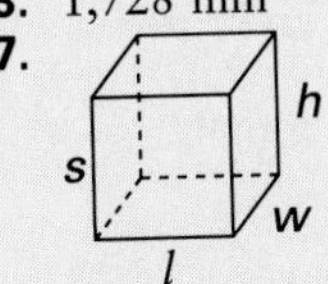

Cube with edges of length s (Given); Each face of cube is rect (A sq is a rect.); Cube is rect solid (Def rect solid); Vol of cube = lwh (For rect solid, $V = lwh$); $s = l = w = h$ (All edges of cube are $\cong$.); Vol of cube = $s \cdot s \cdot s = s^3$ (Sub, alg). **19.** 17.4 gal **21.** 6 to s **23.** n to m

Written Exercises, page 577–578

1. 108 cm^2 **3.** 585 in^2 **5.** 385 m^2 **7.** 60π cm^2, 160π cm^2 **9.** 132π m^2, 276π m^2 **11.** 60 cm^2, $(60 + \frac{25}{4}\sqrt{3})$ cm^2 **13.** 120 cm^2, $(120 + \frac{75}{2}\sqrt{3})$ cm^2 **15.** $80\sqrt{41}$ cm^2 **17.** 18 cm **19.** Pyramid P with reg poly base $ABC \cdots N$, alt $\overline{VX}$, X at ctr of $ABC \cdots N$ (Given); $\overline{VX}$ is $\perp$ to $\overline{AX}$, $\overline{BX}$, $\overline{CX}$, $\cdots$ $\overline{NX}$ (Alt of pyramid $\perp$ to base.); $\angle VXA$, $\angle VXB$, $\angle VXC$, . . . $\angle VXN$ are rt $\angle$s. (Def of $\perp$); $\angle VXA$, $\angle VXB$, $\angle VXC \cdots \angle VXN$ are $\cong$ (Rt $\angle$s are $\cong$.); $\overline{VX} \cong \overline{VX}$ (Reflex Prop of Congr); $\overline{AX}$, $\overline{BX}$, $\overline{CX} \cdots \overline{NX}$ are $\cong$ (Radii reg poly are $\cong$.); $\triangle VXA$, $\triangle VXB$, $\triangle VXC \cdots$ $\triangle VXN$ are $\cong$ (SAS); $\overline{VA}$, $\overline{VB}$, $\overline{VC} \cdots \overline{VN}$ are $\cong$ (CPCTC); $AB = BC = CD = \cdots$ (Def reg poly); $\triangle AVB \cong \triangle BVC \cong \triangle CVD \cong \cdots$ (SSS); $\triangle AVB$, $\triangle BVC$, $\cdots$ are isos $\triangle$s (Def isos $\triangle$) **21.** L(cyl): L(cone) is $2h$: $\sqrt{r^2 + h^2}$ A(cyl): A(cone) is $2(h + r)$: $\sqrt{r^2 + h^2} + r$

Written Exercises, page 581–582

1. 20 cm^3 **3.** 468 in^3 **5.** 28 mm^3
7. $\frac{70}{3}$ ft^3 **9.** 314.0 cm^3 **11.** 285.5 cm^3
13. 3 mm **15.** 9 ft **17.** 125 mm^3
19. 80 cm^3 **21.** $180\sqrt{3}$ mm^3 **23.** 8 to 27
25. $6\sqrt{6}$ to s **27.** π to 4

Written Exercises, page 587–588

1. $A = 324\pi$ m^2, $V = 972\pi$ m^3 **3.** $A = 900\pi$ yd^2, $V = 4{,}500\pi$ yd^3 **5.** $A = 64\pi$ mm^2, $V = 85\frac{1}{3}\pi$ mm^3 **7.** $A = 21.16\pi$ ft^2, $V = 16.22\pi$ ft^3 **9.** $A = 73.96$ yd^2, $V = 106.0$ yd^3 **11.** $A = 144\pi$ mm^2, $V = 288\pi$ mm^3 **13.** 3 cm **15.** 5 cm **17.** $\frac{2\sqrt[3]{6\pi^2}}{\pi}$ ft
19. 1 to 8 **21.** 1 to 4 **23.** $121\frac{1}{3}\pi$ cm^3
25. $A = 256\pi$ cm^2, $V = 682\frac{2}{3}\pi$ cm^3
27. $\frac{1.6}{\pi} \times 10^9$ km^2 **29.** 151 in^2
31. $\frac{Vs}{Vc} = \frac{\pi}{6}$

Written Exercises, page 591

1–5.

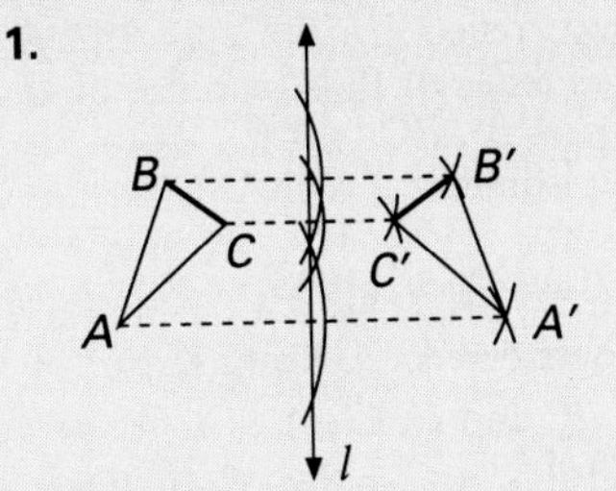

7. $3\sqrt{3}$ **9.** $\sqrt{94}$ **11.** $\sqrt{83}$
13. $(-1, 3, 2)$ **15.** $(-\frac{1}{2}, -2, \frac{7}{2})$
17. $r = \sqrt{(x - h)^2 + (y - j)^2 + (z - k)^2}$

Chapter 15

Written Exercises, pages 603–604

1.

3.

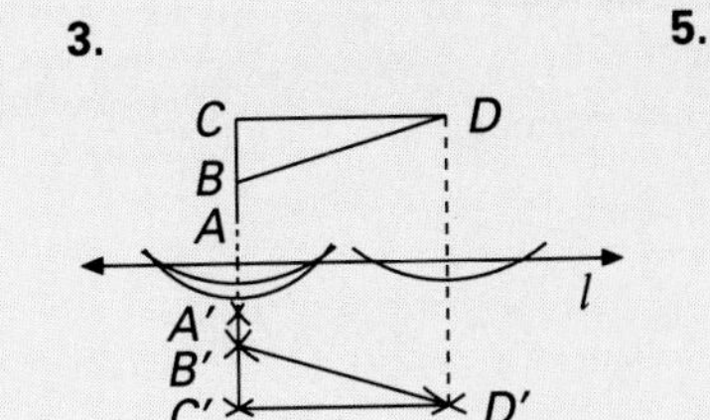

5.

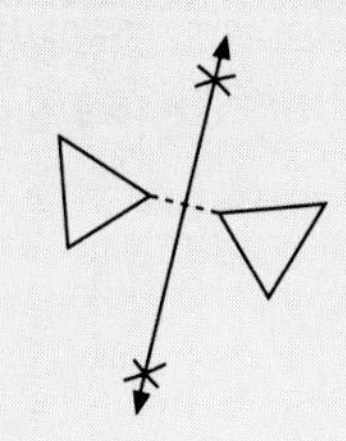

Selected Answers

7. Yes

9.

11.

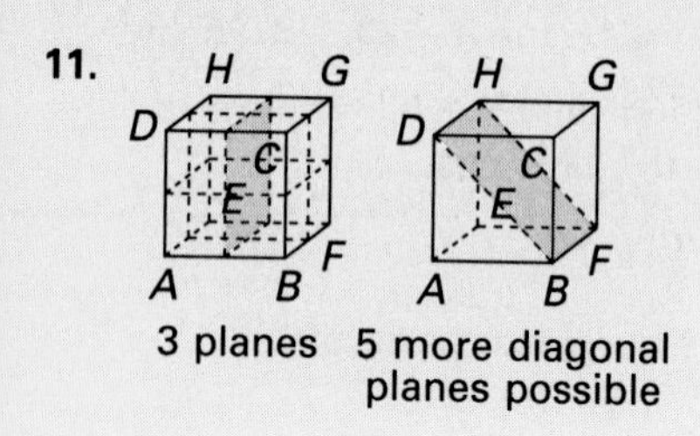

13.

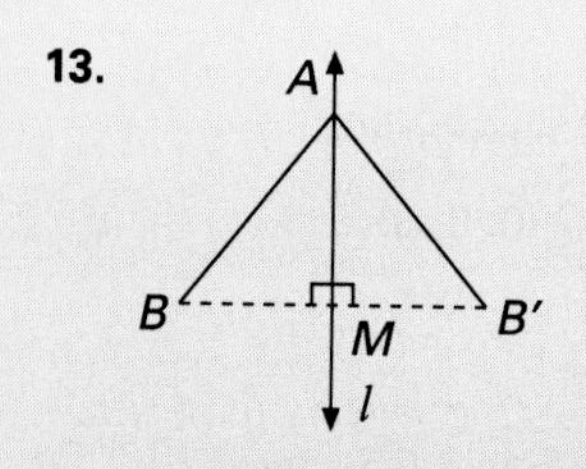

Pts A & B with reflec $A'B'$ about l with A on l and B not on l (Given); Since A is on l, $A = A'$, l is $\perp$ bis of BB' (Def reflec); $BM = B'M$, $BB' \perp l$ (Def $\perp$ bis); $\angle AMB'$ and $\angle AMB$ are rt $\angle$s (Def $\perp$); $\angle AMB' \cong \angle AMB$ (Rt $\angle$s are $\cong$.); $\overline{AM} \cong \overline{AM}$ (Reflex Prop of Congr); $\triangle AMB \cong \triangle AMB'$ (SAS); $\therefore A'B = AB'$ (CPCTC)

15. $\triangle ABC$ with its reflec $\triangle A'B'C'$ (Given); A' is reflec of A, B' is reflec of B, C' is reflec of C (Def reflec); $AB = A'B'$, $BC = B'C'$, $AC = A'C'$ (Reflec is an isometry.); $\therefore \triangle ABC \cong \triangle A'B'C'$ (SSS) **17.** $\overline{AY} \cong \overline{A'Y}$, $\overline{AX} \cong \overline{A'X}$ (Constr with $\cong$ arcs); l is $\perp$ bis $\overline{AA'}$ (2 pts each equidist from endpts determ $\perp$ bis of seg.); Repeat for pts B, B' and C, C'; $\overline{A'B'}$ reflec of $\overline{AB}$, $\overline{B'C'}$ reflec of $\overline{BC}$, $\overline{A'C'}$ reflec of $\overline{AC}$ (Def reflec); $\therefore \triangle A'B'C'$ is reflec of $\triangle ABC$ (Def reflec)

19. No, see counterexample.

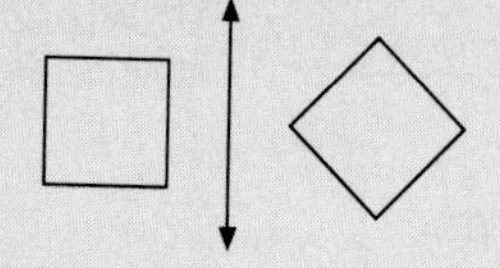

Written Exercises, page 609–611

1–3. Check students' drawings.

5.

7.

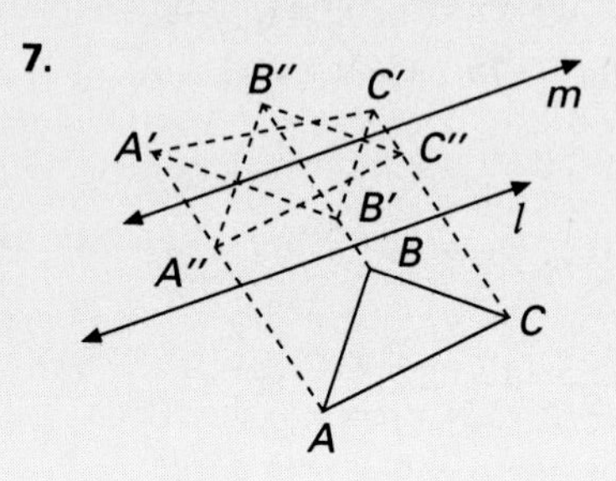

9. Yes, order is important. Direction of translation is changed. Reflec is not comm.

11. **13.**

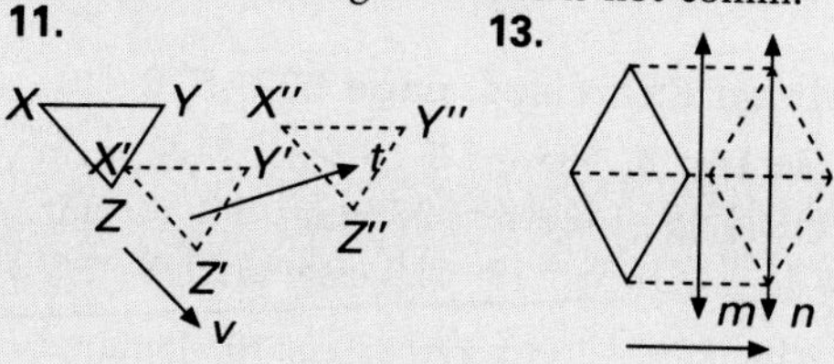

15. Neither

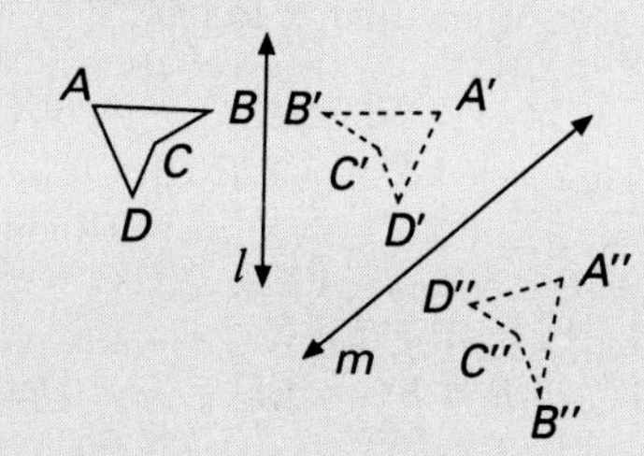

17.

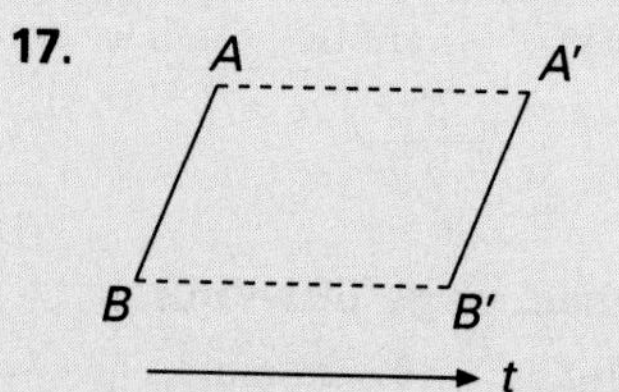

$\overline{AA'} \parallel \overline{BB'}$ and $\overline{AA'} \cong \overline{BB'}$ (Def transl); $ABB'A'$ is $\square$ (Quad with 1 pair opp sides $\cong$ and $\parallel$ is $\square$.) $\therefore \overline{AB} \parallel \overline{A'B'}$ (Opp sides $\square$ are $\parallel$.)

Written Exercises, pages 616–617

1.

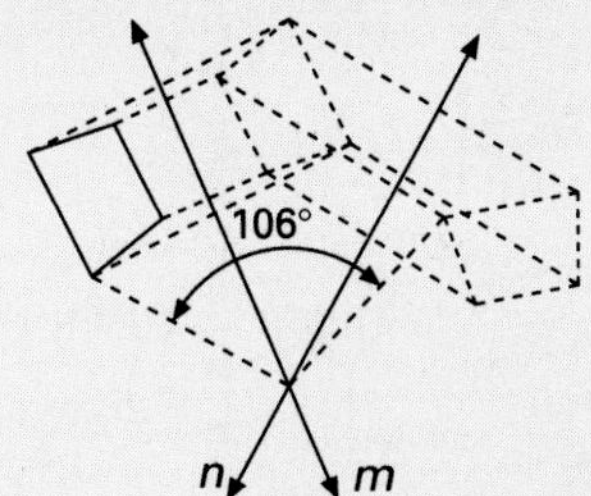

3.

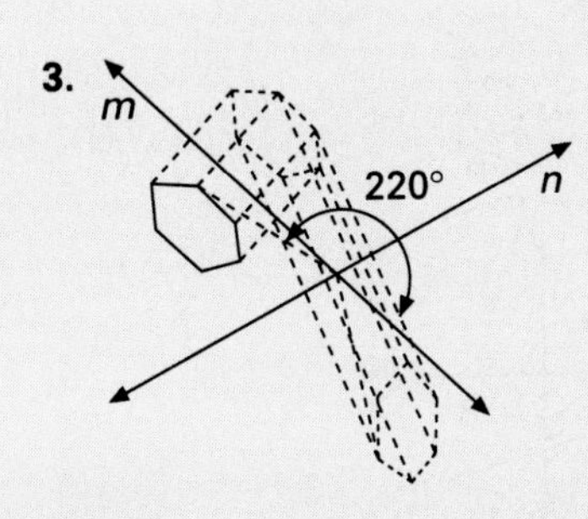

5.

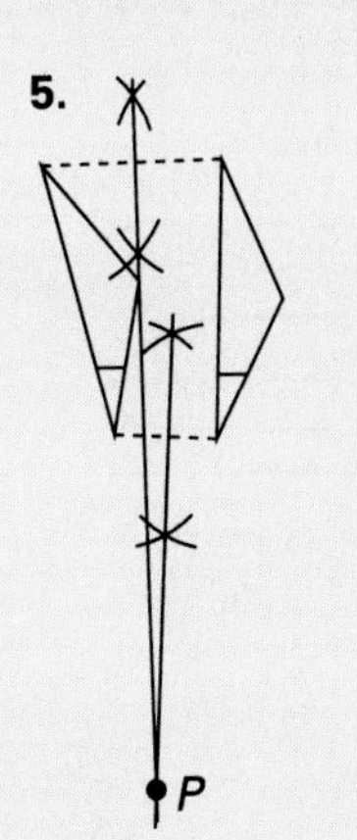

7.

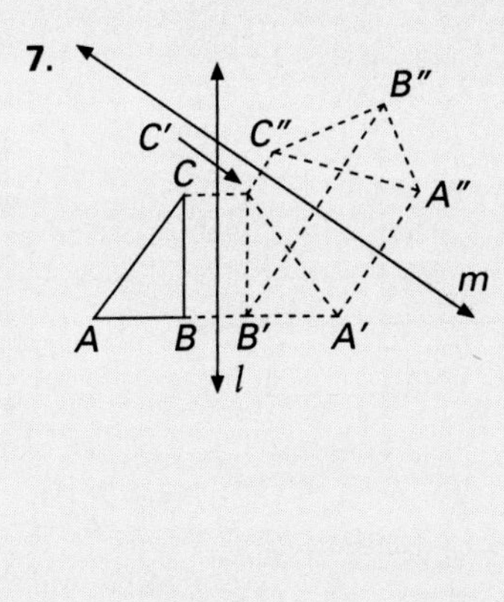

9. Yes, order is important. **11.** 180
13. 120 **15.** 120 **17.** Rotation, translation
19.

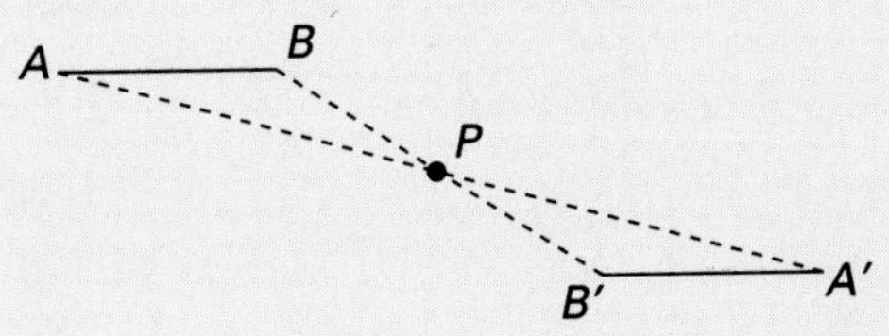

$\overline{AB} \cong \overline{A'B'}$ (Rot is an isometry.); $\overline{AP} \cong \overline{A'P'}$, $\overline{BP} \cong \overline{B'P}$ (Def of rot); $\triangle ABP \cong \triangle A'B'P$ (SSS); $\angle PBA \cong \angle PB'A'$ (CPCTC) $\therefore \overline{AB} \parallel \overline{A'B'}$ (If 2 alt int $\angle$s are $\cong$, lines are $\parallel$.)

21.

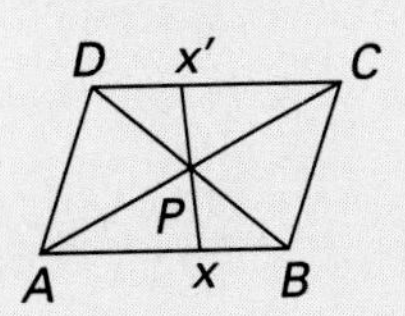

$\square ABCD$ with diags $\overline{AC}$ and $\overline{BD}$ which intersect at P. From any pt X of $\square ABCD$, draw $\overleftrightarrow{XP}$. Label the other pt of intersection of $\square ABCD$ and $\overleftrightarrow{XP}$ as X'. $\overline{AB} \parallel \overline{DC}$ (Def of $\square$) $\angle PAX \cong \angle PCX'$ (If lines $\parallel$, alt int $\angle$s are $\cong$.); $\angle APX \cong \angle CPX'$ (Vert $\angle$s are $\cong$.); $AP = CP$ (Diags of $\square$ bis each other.); $\triangle APX \cong \triangle CPX'$ (ASA); $\overline{PX} \cong \overline{PX'}$ (CPCTC); $\therefore \square ABCD$ has pt symmetry. (Def pt symmetry)

23.

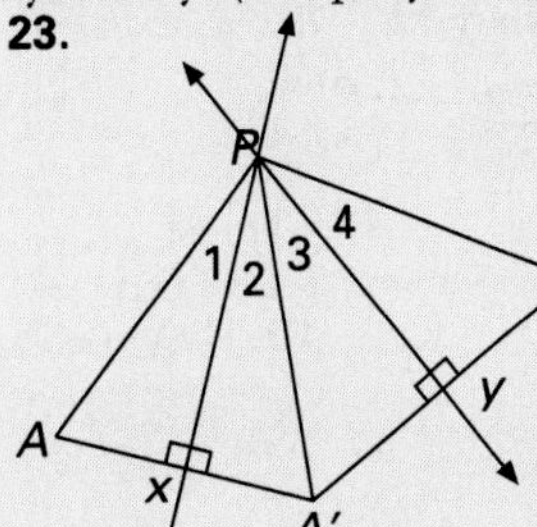

$\overline{AX} \cong \overline{A'X}$, $\overline{A'Y} \cong \overline{A''Y}$ (Def reflec); $\overline{PX} \cong \overline{PX}$, $\overline{PY} \cong \overline{PY}$ (Reflex Prop of Congr); $\angle AXP \cong \angle A'XP$, $\angle A'YP \cong \angle A''YP$ (Def reflec, rt $\angle$s are $\cong$.); $\triangle AXP \cong \triangle A'XP$, $\triangle P'AY \cong \triangle A''YP$ (SAS); m $\angle 1$ = m $\angle 2$, m $\angle 3$ = m $\angle 4$ (CPCTC); m $\angle APA''$ = 2 × m $\angle 2$ + 2 × m $\angle 3$ (Angle Add Post, Sub); $\therefore$ m $\angle APA''$ = 2 × m $\angle XPY$ (Distr Prop, Angle Add Post)

Written Exercises, pages 620–621

1. Transl

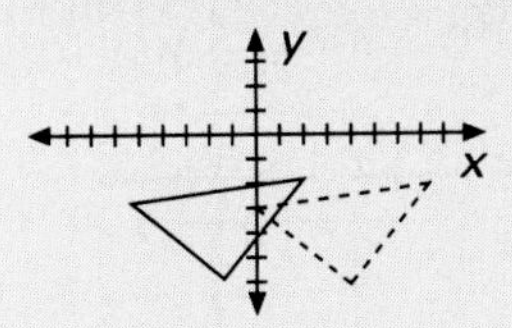

3. Transl

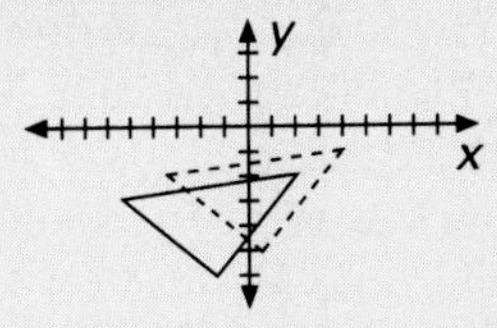

5. Reflec

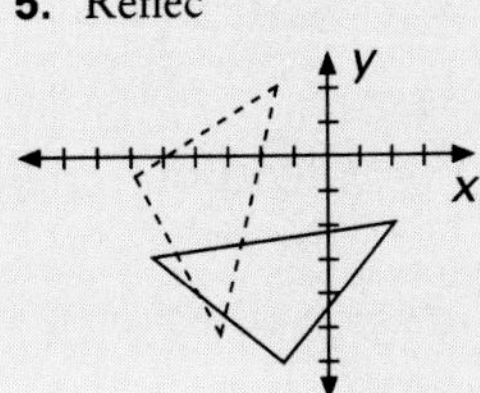

7. Reflec

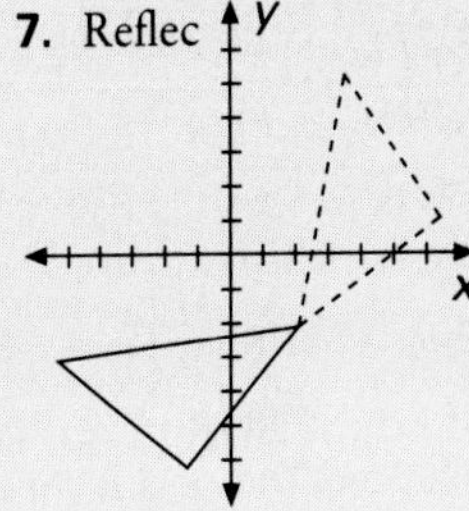

9. Rot, Reflec

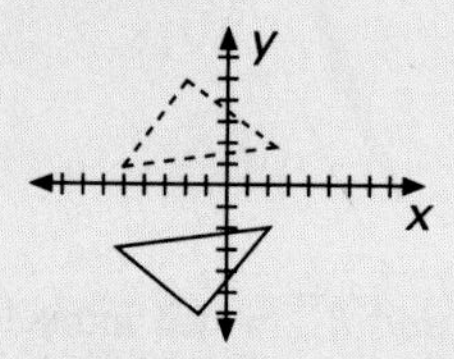

11. $d = \sqrt{(x_1 - x_2)^2 + (y_1 - y_2)^2}$; $d' = \sqrt{[(x_1 + 1) - (x_2 + 1)]^2 + [(y_1 + 2) - (y_2 + 2)]^2} = \sqrt{(x_1 - x_2)^2 + (y_1 - y_2)^2}$ **13.** $d = \sqrt{(x_1 - x_2)^2 + (y_1 - y_2)^2}$; $d' = \sqrt{[(x_1 + 6) - (x_2 + 6)]^2 + [(y_1 - 7) - (y_2 - 7)]^2} = \sqrt{(x_1 - x_2)^2 + (y_1 - y_2)^2}$ **15.** $d = \sqrt{(x_1 - x_2)^2 + (y_1 - y_2)^2}$; $d' = \sqrt{(y_1 - y_2)^2 + (x_1 - x_2)^2} = \sqrt{(x_1 - x_2)^2 + (y_1 - y_2)^2}$ **17.** Transl **19.** Rot **21.** $(x,y) \rightarrow (-x,-y)$ **23.** $(x,y) \rightarrow (y,x)$ **25.** $(x,y) \rightarrow (10 - x,y)$ **27.** $(x,y) \rightarrow (2c - x,y)$

Written Exercises, page 624

1.

3.

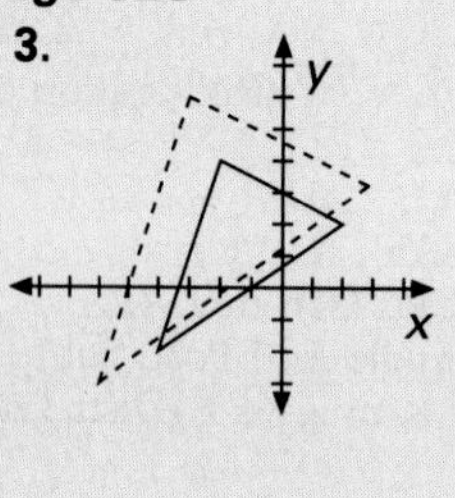

5.

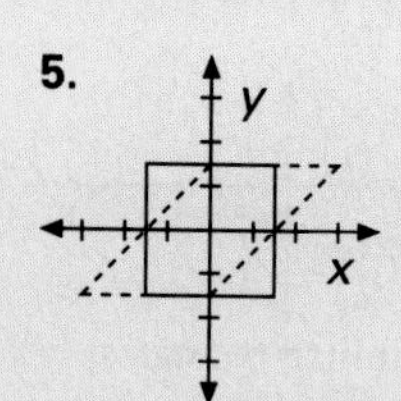

7.

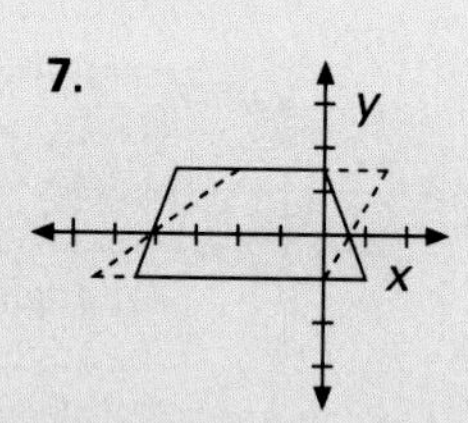

9. $PQ = \sqrt{(x_1 - x_2)^2 + (y_1 - y_2)^2}$; $P'Q' = \sqrt{(kx_1 - kx_2)^2 + (ky_1 - ky_2)^2} = k\sqrt{(x_1 - x_2)^2 + (y_1 - y_2)^2} = kPQ$ (Distr form); In the same way, prove $Q'R' = kQR$, $P'R' = kPR$; $\frac{PQ}{P'Q'} = \frac{QR}{Q'R'} = \frac{PR}{P'R'}$ (Div Prop of Eq); $\triangle PQR \sim \triangle P'Q'R'$ (SSS ~); Transl is a dilation (Def of dilation)

11. Neither,

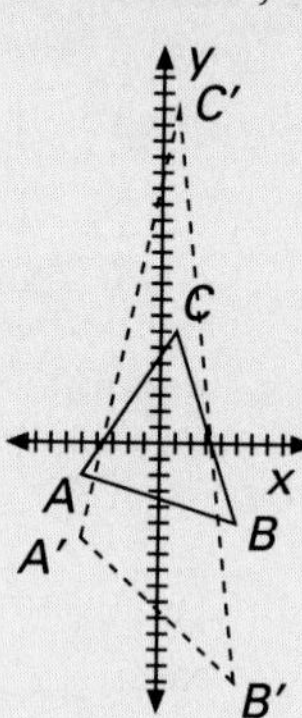

13. Neither,

15. Neither,

17. Shear,

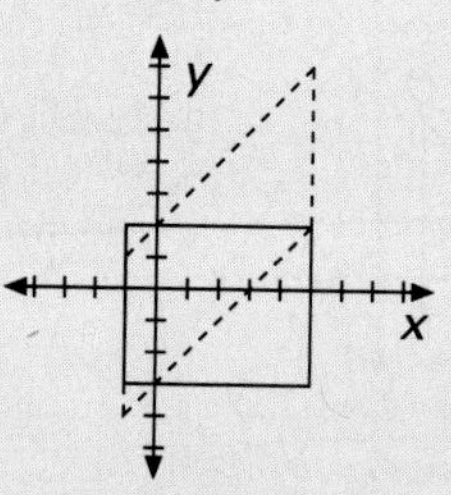

19. Shear,

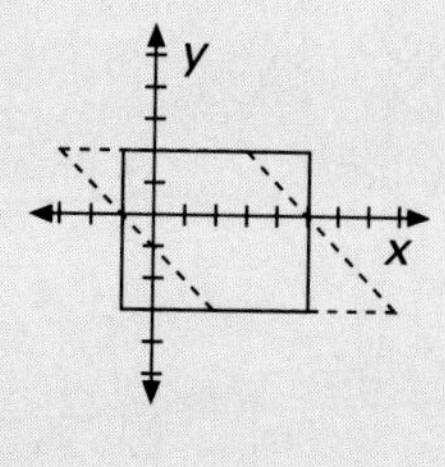

21. Neither,

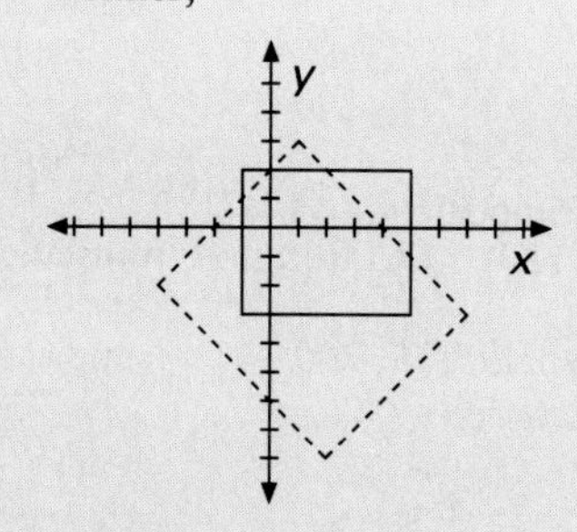

23. Neither,

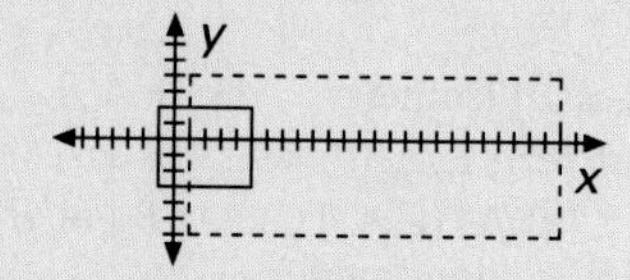

Index

Boldfaced numerals indicate the pages that contain definitions.